SOLUTIONS MANUAL

Volume I Chapters 1-7

TO ACCOMPANY

FUNDAMENTALS OF ENGINEERING THERMODYNAMICS

4Th EDITION

MICHAEL J. MORAN
The Ohio State University

HOWARD N. SHAPIRO
Iowa State University of Sciences and Technology

John Wiley & Sons, Inc.
New York/Chichester/Weinheim/Brisbane/Singapore/Toronto

To order books or for customer service call 1-800-CALL-WILEY (225-5945).

ISBN 0-471-35411-2

Printed in the United States of America

10 9 8 7 6 5 4 3 2 1

Printed and bound by Victor Graphics, Inc.

PREFACE

This Instructor's Manual accompanies the fourth edition of *Fundamentals of Engineering Thermodynamics* by M.J. Moran and H.N. Shaprio. The manual is in two volumes, and provides the following instructional aids:

- Volume I
  - Sample syllabi for two-course sequences and one-term survey courses in engineering thermodynamics on both semester and quarter bases.
  - Solutions to end-of-chapter problems for Chaps. 1 through 7 using the same format as in the solved text examples.
- Volume II
  - Comments on the design and open-ended problems, Chaps. 1 through 14, to *assist* instructors in guiding students as they develop *individual* solutions.
  - Solutions to end-of-chapter problems for Chaps. 8 through 14 using the same format as in the solved text examples.

The Instructor's Resource CD-ROM that accompanies the fourth edition also provides solutions for selected end-of-chapter problems using our software *Interactive Thermodynamics: IT.*

We have exercised considerable care that the solutions provided are both error free and accurate in methodology. Still, mistakes are inevitable and we would like to know about any that come to light. Updates and corrections are available through *www.wiley.com/college/moranshapiro.*

We hope that the solutions and other features of this manual will be useful to you, and we welcome your comments, criticisms, or suggestions.

Michael J. Moran
Howard N. Shapiro

September, 1999

SAMPLE SYLLABI

In this part of the Instructor's Manual, sample syllabi are presented as follows:

1. Two-course sequence on a semester basis.

2. Two-course sequence on a semester basis - no coverage of exergy.

3. Two-course sequence on a quarter basis.

4. Two-course sequence on a quarter basis - no coverage of exergy.

5. One-course surveys

   A. semester basis

   B. quarter basis

In courses where computer use is not an objective, discussions involving computer software can be omitted.

The pace of presentation is typical of what is followed at the authors' home institutions: At Iowa State there is a two-course sequence on a *semester* basis and at Ohio State there is a two course sequence on a *quarter* basis.

Although the sample syllabi exhibit several possible variations, the topic sequencings presented are not exhaustive, and other arrangements are possible. The authors would appreciate knowing about alternative arrangements that work well.

1. *Engineering Thermodynamics:* Two-course sequence on a semester basis

Each course has 45 meetings (3 semester credits each). The course plans shown below account for 40 meetings. The additional meetings allow for midterm exams, review sessions, etc.

Meeting	First Course	Second Course
1	1.1, 1.2, 1.3, 1.4	9.1, 9.2
2	1.5, 1.6, 1.7, 1.8, 1.9	9.3, 9.4
3	2.1, 2.2.1	9.5, 9.6
4	2.2.2, 2.2.3, 2.2.4	9.7
5.	2.3, 2.4	9.8
6	2.5	9.9
7	2.6, 2.7	9.10, 9.11, 9.15 (options B, C)
8	3.1, 3.2	10.1, 10.2.1
9	3.3.1, 3.3.2, 3.3.3	10.2.2, 10.3
10	3.3.4, 3.3.5, 3.3.6	10.4, 10.5
11	3.4, 3.5	10.6, 10.7, 10.8
12	3.6, 3.7	12.1, 12.2
13	3.8, 3.9	12.3, 12.4.1
14	4.1	12.4.2, 12.5.1
15	4.2	12.5.2, 12.5.3
16	4.3	12.5.4, 12.6
17	4.3 (continued)	12.7, 12.8
18	4.3 (continued)	12.9.1, 12.9.2, 12.9.3
19	4.4, 4.5	12.9.4, 12.9.5, 12.10
20	5.1, 5.2	13.1
21	5.3	13.2.1, 13.2.2
22	5.4	13.2.3, 13.3
23	5.5, 5.6, 5.7, 5.8	13.4, 13.5
24	6.1, 6.2	13.6
25	6.3.1, 6.3.2	13.7, 13.8
26	6.3.3, 6.4	13.9, 13.10
27	6.5	14.1
28	6.6	14.2, 14.3.1
29	6.7, 6.8	14.3.2, 14.3.3
30	6.9, 6.10	14.7 (option C only)
31	7.1, 7.2.1, 7.2.2	
32	7.2.3, 7.2.4, 7.2.5	
33	7.3	
34	7.4, 7.5	
35	7.6, 7.7, 7.8	see next page for
36	8.1, 8.2.1, 8.2.2	meetings 31-40
37	8.2.3, 8.2.4	
38	8.3, 8.4.1	
39	8.4.2, 8.4.3, 8.5	
40	8.6, 8.7	

Options for meeting 31 through 39 of the second course[a]

	A	B	C
31	14.4 (selected topics)	14.4 (selected topics)	11.1
32	14.4 (selected topics)	14.4 (selected topics)	11.2, 11.3.1, 11.3.2
33	14.5, 14.6, 14.7	14.5, 14.6, 14.7	11.3.3, 11.4.1
34	9.12.1, 9.12.2	11.1	11.4.2, 11.5.1
35	9.12.3, 9.13.1	11.2, 11.3.1, 11.3.2	11.5.2, 11.5.3
36	9.13.2	11.3.3, 11.4.1	11.6, 11.7
37	9.13.3	11.4.2, 11.5.1	11.8
38	9.14	11.5.2, 11.5.3	11.9
39	9.15	11.6, 11.7	11.10
40	---	11.10 (omit 11.8, 9)	---

a. Alternatively, the compressible flow coverage of Option A can be considered with the other Chap. 9 topics, and the thermodynamic relations coverage of Options B and C can be treated in sequence, following Chap. 10 and prior to Chap. 12.

2. *Engineering Thermodynamics:* Two-course sequence on a semester basis with no coverage of exergy principles.

Each course has 45 meetings (3 semester credits each). The course plans shown below account for 39 meetings. Six additional meetings allow for midterm exams, review sessions, etc.

Meeting	**First Course**	**Second Course**
1	1.1, 1.2, 1.3, 1.4	9.1, 9.2
2	1.5, 1.6, 1.7, 1.8, 1.9	9.3, 9.4
3	2.1, 2.2.1	9.5, 9.6
4	2.2.2, 2.2.3, 2.2.4	9.7
5	2.3, 2.4	9.8
6	2.5	9.9
7	2.6, 2.7	9.10, 9.11
8	3.1, 3.2	10.7
9	3.3.1, 3.3.2, 3.3.3	9.12.1, 9.12.2
10	3.3.4, 3.3.5, 3.3.6	9.12.3, 9.13.1
11	3.4, 3.5	9.13.2
12	3.6, 3.7	9.13.3
13	3.8, 3.9	9.14
14	4.1	9.15
15	4.2	11.1
16	4.3	11.2, 11.3.1, 11.3.2
17	4.3 (continued)	11.3.3, 11.4.1
18	4.3 (continued)	11.4.2, 11.5.1
19	4.4, 4.5	11.5.2, 11.5.3
20	5.1, 5.2	11.6, 11.7, 11.10 (omit 11.8, 9)
21	5.3	12.1, 12.2
22	5.4	12.3, 12.4.1
23	5.5, 5.6, 5.7, 5.8	12.4.2, 12.5.1
24	6.1, 6.2	12.5.2, 12.5.3
25	6.3.1, 6.3.2	12.5.4, 12.6
26	6.3.3, 6.4	12.7, 12.8
27	6.5	12.9.1, 12.9.2, 12.9.3
28	6.6	12.9.4, 12.9.5, 12.10
29	6.7, 6.8	13.1
30	6.9, 6.10	13.2.1, 13.2.2
31	8.1, 8.2.1, 8.2.2	13.2.3, 13.3
32	8.2.3, 8.2.4	13.4, 13.5
33	8.3, 8.4.1	13.10, 14.1
34	8.4.2, 8.4.3	14.2, 14.3.1
35	8.5, 8.7	14.3.2, 14.3.3
36	10.1, 10.2.1	14.4 (selected topics)
37	10.2.2, 10.3	14.4 (selected topics)
38	10.4, 10.5	14.5, 14.6
39	10.6, 10.8	14.7

3. *Engineering Thermodynamics:* Two-course sequence on a quarter basis

Each course has 40 meetings (4 quarter credits each). The course plans shown below account for 36 meetings. Four additional meetings allow for midterm exams, review sessions, etc.

Meeting	First Course	Second Course
1	1.1, 1.2, 1.3, 1.4	8.1, 8.2.1, 8.2.2
2	1.5, 1.6, 1.7, 1.8, 1.9	8.2.3, 8.2.4
3	2.1, 2.2.1	8.3, 8.4.1
4	2.2.2, 2.2.3, 2.2.4	8.4.2, 8.4.3, 8.5
5	2.3, 2.4	8.6, 8.7
6	2.5	9.1, 9.2
7	2.6, 2.7	9.3, 9.4
8	3.1, 3.2	9.5, 9.6
9	3.3.1, 3.3.2, 3.3.3	9.7
10	3.3.4, 3.3.5, 3.3.6	9.8, 9.9
11	3.4, 3.5	9.10, 9.11, 9.15 (options B, C)
12	3.6, 3.7	10.1, 10.2.1
13	3.8, 3.9	10.2.2, 10.3
14	4.1	10.4, 10.5
15	4.2	10.6, 10.7, 10.8
16	4.3	12.1, 12.2
17	4.3 (continued)	12.3, 12.4.1
18	4.3 (continued)	12.4.2, 12.5.1
19	4.4, 4.5	12.5.2, 12.5.3
20	5.1, 5.2	12.5.4, 12.6
21	5.3	12.7, 12.8
22	5.4	12.9.1, 12.9.2, 12.9.3
23	5.5, 5.6, 5.7, 5.8	12.9.4, 12.9.5, 12.10
24	6.1, 6.2	13.1
25	6.3.1, 6.3.2	13.2.1, 13.2.2
26	6.3.3, 6.4	13.2.3, 13.3
27	6.5	13.4, 13.5
28	6.6	13.6
29	6.7, 6.8	13.7, 13.8
30	6.9, 6.10	13.9, 13.10
31	7.1, 7.2.1, 7.2.2	
32	7.2.3, 7.2.4, 7.2.5	
33	7.3	see next page for
34	7.4, 7.5	meetings 31-36
35	7.6, 7.7	
36	7.8	

Options for meetings 31 through 36 of the second course[a]

	A Compressible Flow	B Thermodynamic Relations	C Chemical and Phase Equilibrium
31	9.12.1, 9.12.2	11.1	14.1
32	9.12.3, 9.13.1	11.2, 11.3.1, 11.3.2	14.2, 14.3.1
33	9.13.2	11.3.3, 11.4.1	14.3.2, 14.3.3
34	9.13.3	11.4.2, 11.5.1	14.4 (Selected topics)
35	9.14	11.5.2, 11.5.3	14.5, 14.6
36	9.15	11.6, 11.7, 11.10 (omit 11.8, 9)	14.7

[a] Alternatively, the compressible flow coverage of Option A can be considered with the other Chap. 9 topics, and the thermodynamic relations coverage of Options B and C can be treated in sequence, following Chap. 10 and prior to Chap. 12.

4. *Engineering Thermodynamics:* Two-course sequence on a quarter basis with no coverage of exergy principles.

Each course has 40 meetings (4 quarter credits each). The course plans shown below account for 36 meetings. Four additional meetings allow for midterm exams, review sessions, etc.

Meeting	**First Course**	**Second Course**
1	1.1, 1.2, 1.3, 1.4	9.1, 9.2
2	1.5, 1.6, 1.7, 1.8, 1.9	9.3, 9.4
3	2.1, 2.2.1	9.5, 9.6
4	2.2.2, 2.2.3, 2.2.4	9.7
5	2.3, 2.4	9.8
6	2.5	9.9, 9.11, 9.15 (options B, C)
7	2.6, 2.7	10.1, 10.2.1
8	3.1, 3.2	10.2.2, 10.3
9	3.3.1, 3.3.2, 3.3.3	10.4, 10.5
10	3.3.4, 3.3.5, 3.3.6	10.6, 10.7, 10.8
11	3.4, 3.5	12.1, 12.2
12	3.6, 3.7	12.3, 12.4.1
13	3.8, 3.9	12.4.2, 12.5.1
14	4.1	12.5.2, 12.5.3
15	4.2	12.5.4, 12.6
16	4.3	12.7, 12.8
17	4.3 (continued)	12.9.1, 12.9.2, 12.9.3
18	4.3 (continued)	12.9.4, 12.9.5, 12.10
19	4.4, 4.5	13.1
20	5.1, 5.2	13.2.1, 13.2.2
21	5.3	13.2.3, 13.3
22	5.4	13.4, 13.5, 13.10
23	5.5, 5.6, 5.7, 5.8	14.1
24	6.1, 6.2	14.2, 14.3.1
25	6.3.1, 6.3.2	14.3.2, 14.3.3
26	6.3.3, 6.4	14.7 (option C only)
27	6.5	---
28	6.6	
29	6.7, 6.8	See below for
30	6.9, 6.10	meetings 28-36
31	8.1, 8.2.1, 8.2.2	
32	8.2.3, 8.2.4	
33	8.3, 8.4.1	
34	8.4.2, 8.4.3	
35	8.5, 8.7	
36	---	

Options for meetings 28 through 36 of the second course[a]

	A	B	C
28	14.4 (selected topics)	14.4 (selected topics)	11.1
29	14.4 (selected topics)	14.4 (selected topics)	11.2, 11.3.1, 11.3.2
30	14.5, 14.6, 14.7	14.5, 14.6, 14.7	11.3.3, 11.4.1
31	9.12.1, 9.12.2	11.1	11.4.2, 11.5.1
32	9.12.3, 9.13.1	11.2, 11.3.1, 11.3.2	11.5.2, 11.5.3
33	9.13.2	11.3.3, 11.4.1	11.6, 11.7
34	9.13.3	11.4.2, 11.5.1	11.8
35	9.14	11.5.2, 11.5.3	11.9
36	9.15	11.6, 11.7, 11.10 (omit 11.8, 9)	11.10

[a] Alternatively, the compressible flow coverage of Option A can be considered with the other Chap. 9 topics, and the thermodynamic relations coverage of Options B and C can be treated in sequence, following Chap. 10 and prior to Chap. 12.

5. Surveys of *Engineering Thermodynamics*

Meeting	(A) Semester Basis: 45 meetings (3 sem. credits)[a]	(B) Quarter Basis: 30 meetings (3 quar. credits)[b]
1	1.1, 1.2, 1.3, 1.4	1.1, 1.2, 1.3, 1.4
2	1.5, 1.6, 1.7, 1.8, 1.9	1.5, 1.6, 1.7, 1.8, 1.9
3	2.1, 2.2	2.1, 2.2
4	2.3, 2.4	2.3,2.4
5	2.5, 2.6, 2.7	2.5, 2.6, 2.7
6	3.1, 3.2	3.1, 3.2
7	3.3	3.3
8	3.4, 3.5	3.4, 3.5
9	3.6, 3.7, 3.8, 3.9	3.6, 3.7, 3.8, 3.9
10	4.1	4.1
11	4.2	4.2
12	4.3	4.3
13	4.3 (continued), 4.5 Omit 4.4	4.3 (continued), 4.5 Omit 4.4
14	5.1, 5.2, 5.3	5.1,5.2, 5.3
15	5.4	5.4
16	5.5, 5.6, 5.7, 5.8	5.5, 5.6, 5.7, 5.8
17	6.1, 6.2	6.1, 6.2
18	6.3.1, 6.3.2	6.3.1, 6.3.2
19	6.3.3, 6.4	6.3.3, 6.4
20	6.5	6.5
21	6.6, 6.7	6.6, 6.7, 6.10
22	6.8, 6.9, 6.10	
23	8.1, 8.2.1, 8.2.2	
24	8.2.3, 8.2.4, 8.3	see next page
25	8.4, 8.5, 8.7 (Omit 8.6)	for meetings 22-27
26	9.1, 9.2	
27	9.3, 9.4	
28	9.5, 9.6	
29	9.7, 9.8, 9.15	
30	10.1, 10.2	
31	10.3, 10.5, 10.8	
32	12.1, 12.2	
33	12.3, 12.4	
34		
35	see next page	
36	for meetings	
37	34-39	
38		
39		

[a] Six additional class meetings allow for midterm exams, review sessions, etc.
[b] Three additional class meetings allow for midterm exams, review sessions, etc.

Semester options

Meeting	Property Relations	Psychrometrics	Combustion
34	11.1	12.5	13.1
35	11.2, 11.3.1, 11.3.2	12.6, 12.7	13.2.1, 13.2.2
36	11.3.3, 11.4.1	12.8, 12.9.1	13.2.3, 13.3
37	11.4.2, 11.5.1	12.9.2, 12.9.3	13.4.1, 13.4.2
38	11.5.2, 11.5.3	12.9.4, 12.9.5	13.4.3, 13.5
39	11.6, 11.7, 11.10 (omit 11.8, 9)	12.10	13.10

Quarter Options

Meeting	Power	Combustion
22	6.8, 6.9	12.1, 12.2
23	8.1, 8.2.1, 8.2.2	12.3, 13.1
24	8.2.3, 8.2.4, 8.3	13.2.1, 13.2.2
25	8.4, 8.5, 8.7	13.2.3, 13.3
26	9.5, 9.6	13.4.1, 13.4.2
27	9.7, 9.8, 9.15	13.4.3, 13.5, 13.10

SOLUTIONS TO END-OF CHAPTER PROBLEMS

Each of the more than 1400 end-of-chapter problems have been solved using the same methodology as employed in the 141 solved examples of the text. This solution methodology, described in Sec. 1.7.3, is intended to help students think systematically about engineering systems, develop judgmental skills, and develop sound engineering practices. Solutions begin by listing assumptions and proceed step-by-step using fundamentals. Unit conversions are explicitly included when numerical evaluations are made. The solutions are suitable for photocopying and/or preparing transparencies for classroom use.

In this fourth edition, problems have been classified under headings. By referring to the headings, instructors can tailor problem assignments to achieve a desired coverage of selected topics.

We typically assign three or four problems per class meeting; one or more are collected for grading either on a daily or weekly basis. Copies of the solutions to assigned problems are posted or placed on reserve in the library. Students use these solutions extensively for study. We recommend that instructors consult the solutions before assigning problems. This ensures that students gain experience in the intended principles at an appropriate level of difficulty and also allows instructors to anticipate what hints may be appropriate for some of the more challenging problems.

CHAPTER ONE GETTING STARTED: INTRODUCTORY CONCEPTS AND DEFINITIONS

Chapter 1 - Third and Fourth edition problem correspondence.

3rd	4th	3rd	4th
1.1	---	1.31	1.32*
1.2	---	1.32	---
1.3	---	1.33	---
1.6	1.1*	1.34	---
1.4	1.2	---	1.33
1.5	---	---	1.34
---	1.3	---	1.35
---	1.4	---	1.36
---	1.5	1.35	1.37
1.7	1.6*	1.36	1.38
1.8	1.7	1.37	---
1.9	1.8	1.38	---
1.10	1.9	1.39	1.39
1.11	1.10	1.40	1.40
1.12	1.11	1.41	1.41
1.13	1.12*	1.42	1.42
1.14	1.13*	1.43	---
---	1.14	1.44	---
1.15	---	---	1.43
1.16	1.15*	---	1.44
1.17	1.16	1.45	1.45
1.18	1.17	1.46	1.46
1.19	---	1.47	---
---	1.18	1.48	---
1.20	---	---	1.47
---	1.19	---	1.48
1.21	1.20	---	1.49
1.22	1.21	1.49	1.50*
1.23	1.22	1.50	---
1.24	1.23	1.51	---
1.25	---	1.52	1.51*
---	1.24	1.53	1.52*
---	1.25	1.54	1.53
---	1.26	1.55	1.54
1.26	1.27	1.56	1.55
1.27	---	1.57	1.56*
---	1.28	1.58	1.57
1.28	1.29*	---	1.58
1.29	1.30*	1.59	---
1.30	1.31		

*Revised

PROBLEM 1.1

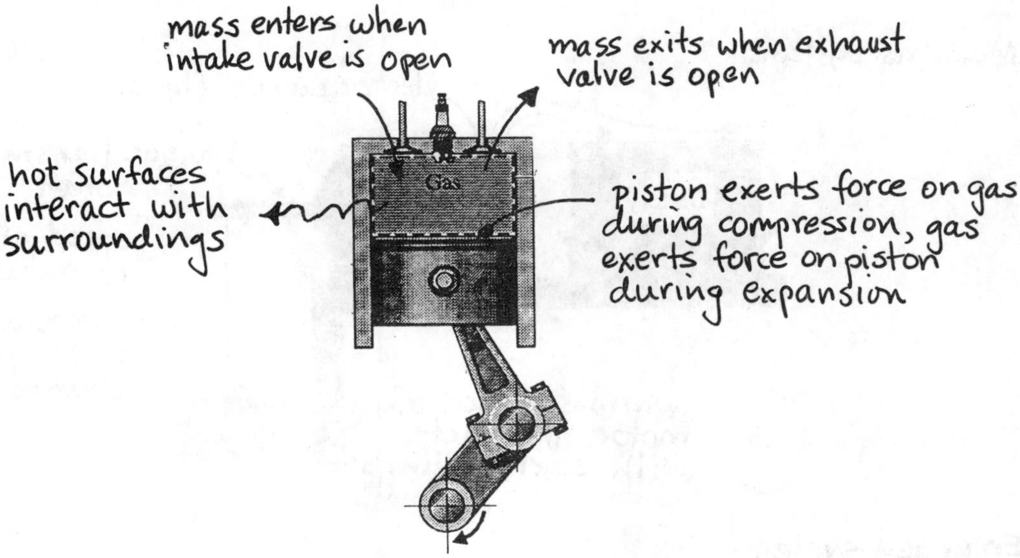
mass enters when intake valve is open
mass exits when exhaust valve is open
hot surfaces interact with surroundings
Gas
piston exerts force on gas during compression, gas exerts force on piston during expansion

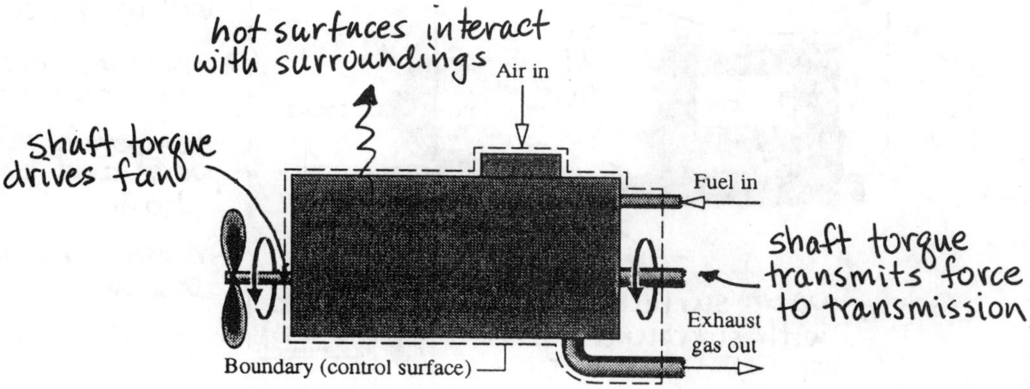
hot surfaces interact with surroundings
Air in
shaft torque drives fan
Fuel in
shaft torque transmits force to transmission
Exhaust gas out
Boundary (control surface)

PROBLEM 1.2

Motor as system:

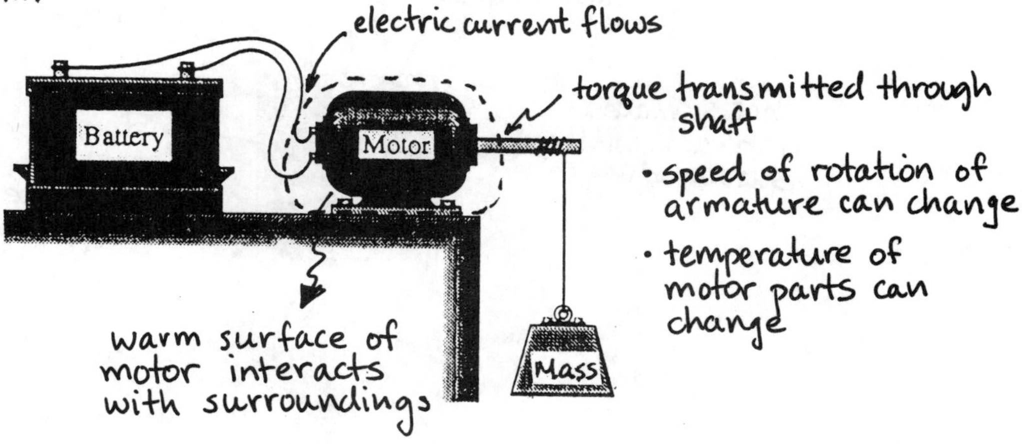
electric current flows
torque transmitted through shaft
Battery
Motor
Mass
• speed of rotation of armature can change
• temperature of motor parts can change
warm surface of motor interacts with surroundings

Enlarged system:

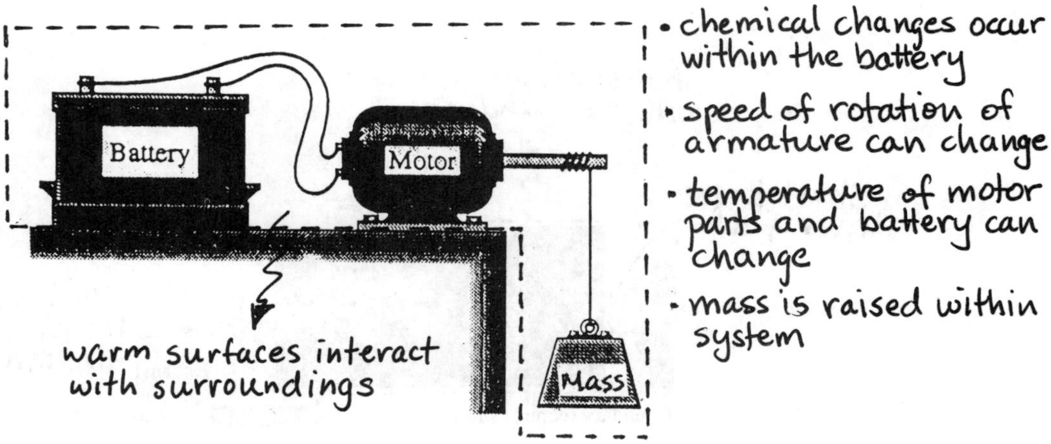
Battery
Motor
Mass
• chemical changes occur within the battery
• speed of rotation of armature can change
• temperature of motor parts and battery can change
• mass is raised within system
warm surfaces interact with surroundings

COMMENT: The shaft torque and current flow interactions become internal to the enlarged system.

PROBLEM 1.3

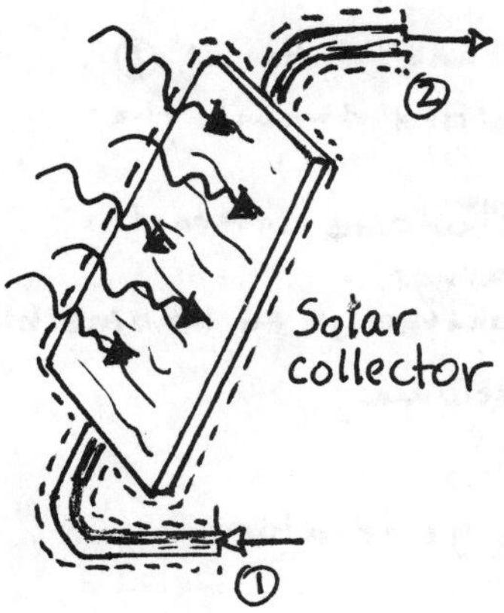

A control volume encloses the solar collector.

- Cool water enters the collector at ①, and hot water exits at ②.
- solar radiation impinges on the front of the collector.
- Warm surfaces of the collector interact with the surroundings.
- some of the incoming radiation is reflected away, and some is absorbed in the collector surface.

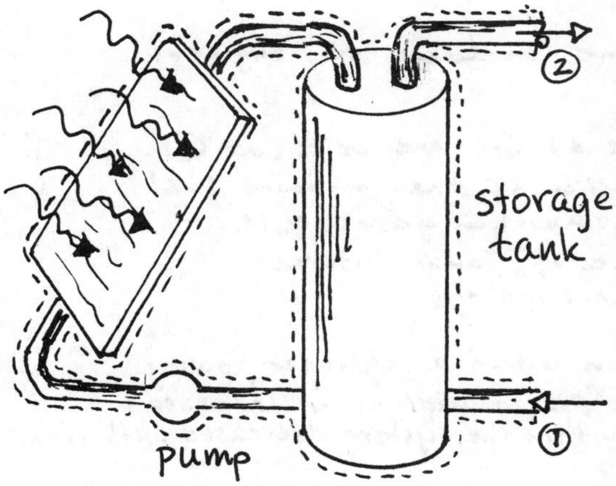

A control volume encloses the solar collector, the tank, and the interconnected piping.

- Cold water enters the tank at ①, and hot water exits at ②.
- Warm surfaces of the collector, storage tank, and interconnected piping interact with the surroundings.
- Solar radiation impinges on the front of the collector; some is reflected and some is absorbed.
- The temperature of the water in the storage tank changes with time.

PROBLEM 1.4

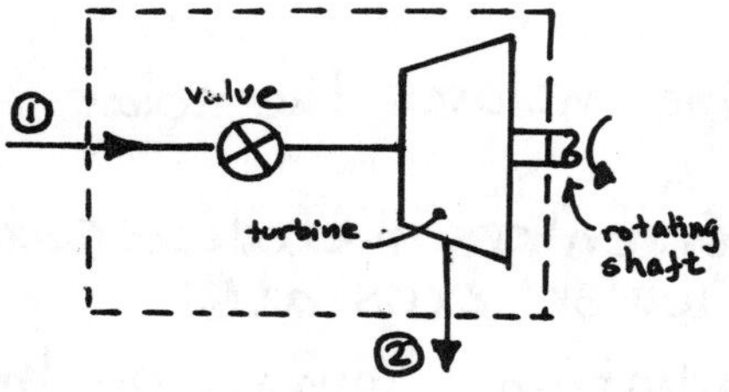

A control volume encloses the valve and turbine.

- Steam enters at ① and exits at ②.
- A torque is transmitted through the rotating shaft.
- Warm surfaces of the turbine interact with the surroundings.

Within the control volume, steam flows across the valve and through the turbine blades.

When the generator is included in the control volume,

- Steam enters at ① and exits at ②.
- Warm surfaces of the turbine and the generator interact with the surroundings.
- Electric current flows from the generator.

Note that the transmitted torque does not cross the boundary of the enlarged control volume.

PROBLEM 1.5

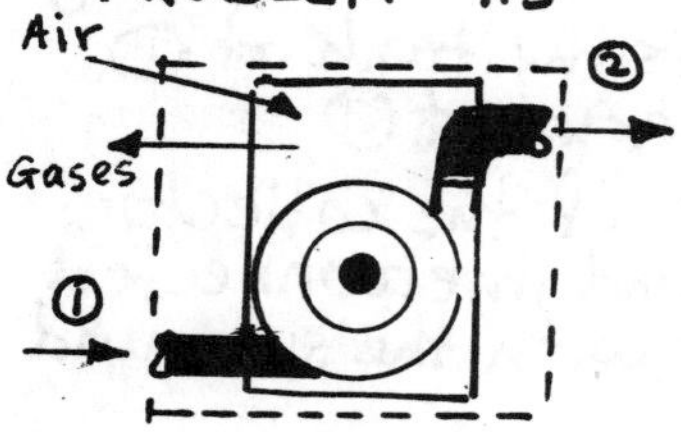

A control volume encloses the engine-driven pump.

- Water enters at ① and exits at ②
- Air for combustion of the on-board fuel enters, and combustion gases exit.
- Warm surfaces of the pump interact with the surroundings.

Within the pump, a piston is kept in motion within a cylinder owing to combustion of the on-board fuel. The piston motion is harnessed to pump the liquid. The amount of fuel within the system decreases with time.

When the hose and nozzle are included, a high-speed water jet exits the extended control volume at the nozzle exit.

PROBLEM 1.6

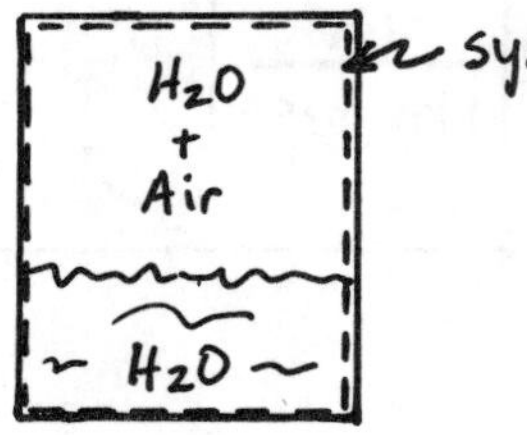

- two phases are present (liquid and gas).
- not a pure substance because composition is different in each phase.

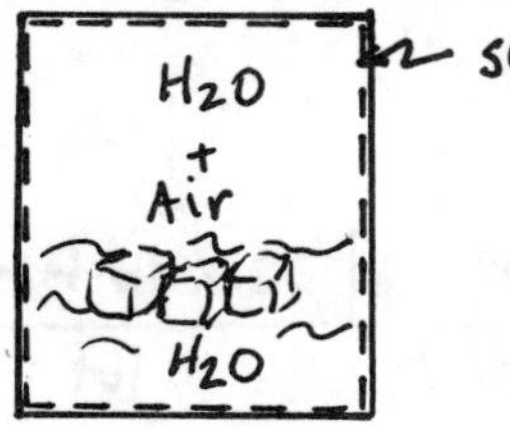

- three phases are present (solid, liquid, and gas).
- not a pure subtance because composition of gas phases is different than that of the solid and liquid phases.

PROBLEM 1.7

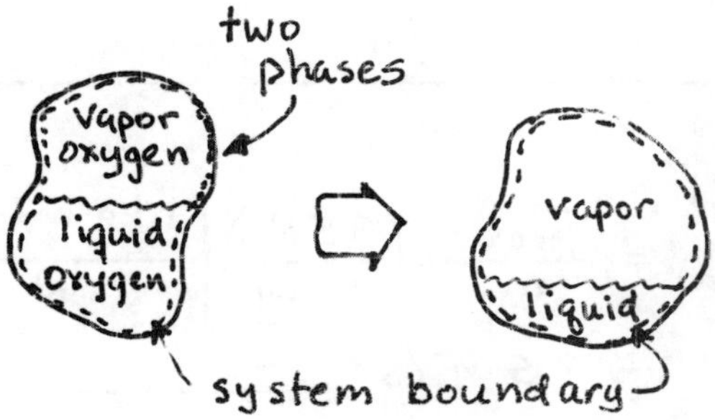

The system is a pure substance. Although the liquid is vaporized, the system remains fixed in chemical composition and is chemically homogeneous.

PROBLEM 1.8

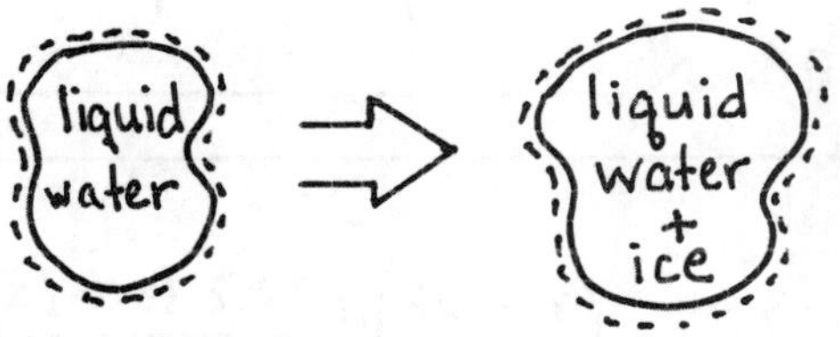

The system is a pure substance. Although the phases change, the system remains of fixed chemical composition and is chemically homogeneous.

PROBLEM 1.9

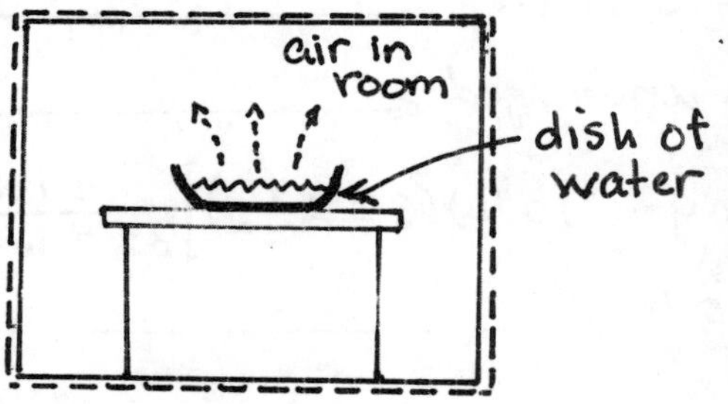

The system is not a pure substance during the process since the composition of the gas phase changes as water evaporates into the air. Once all of the water evaporates, the gas phase comes to equilibrium and the composition becomes homogeneous. At this point, the gas phase can be treated as a pure substance.

PROBLEM 1.10

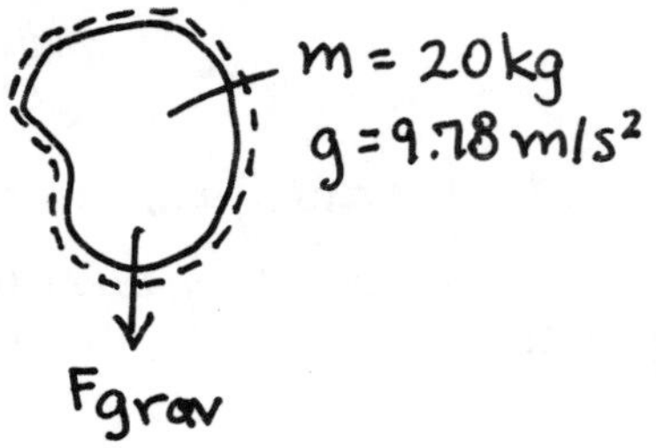

$$F_{grav} = mg = (20\,\text{kg})(9.78\,\tfrac{\text{m}}{\text{s}^2})\left|\frac{1\,\text{N}}{1\,\text{kg}\cdot\text{m/s}^2}\right|$$

$$= 195.6\ \text{N} \longleftarrow F_{grav}$$

PROBLEM 1.11

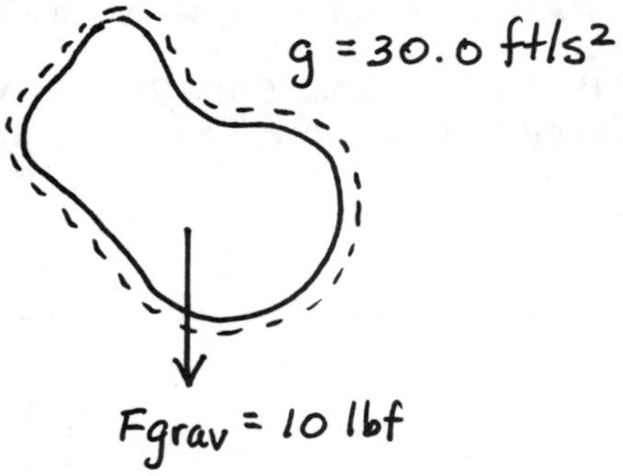

$$m = \frac{F_{grav}}{g} = \left(\frac{10\ \text{lbf}}{30.0\ \text{ft/s}^2}\right)\left|\frac{32.2\ \text{lb}\cdot\text{ft/s}^2}{1\ \text{lbf}}\right|$$

$$= 10.73\ \text{lb} \longleftarrow m$$

PROBLEM 1.12

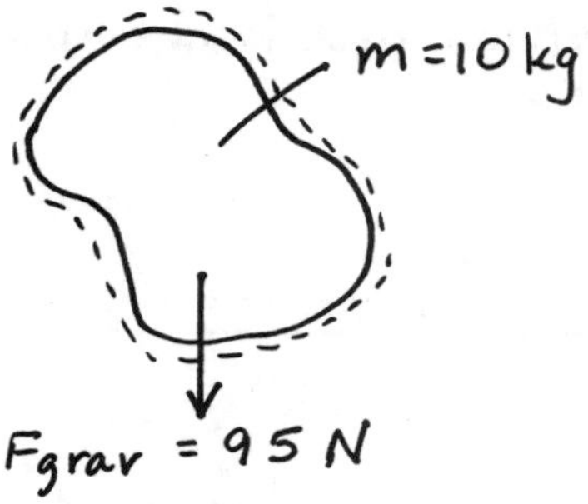

(a) $$g_{local} = \frac{F_{grav}}{m} = \left(\frac{95\ \text{N}}{10\ \text{kg}}\right)\left|\frac{1\ \text{kg}\cdot\text{m/s}^2}{1\ \text{N}}\right|$$

$$= 9.5\ \text{m/s}^2 \longleftarrow g_{local}$$

(b) mass is unchanged. ← m

$$F_{grav} = mg = (10\ \text{kg})(9.81\ \text{m/s}^2)\left|\frac{1\ \text{N}}{1\ \text{kg}\cdot\text{m/s}^2}\right|$$

$$= 98.1\ \text{N} \longleftarrow F_{grav}$$

PROBLEM 1.13

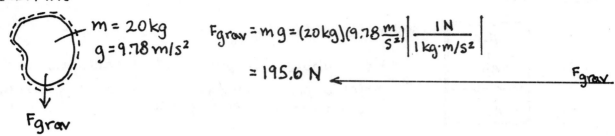

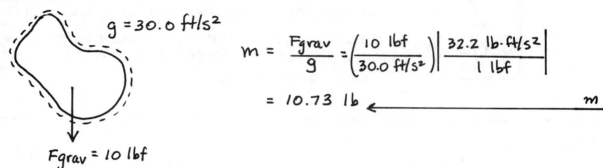

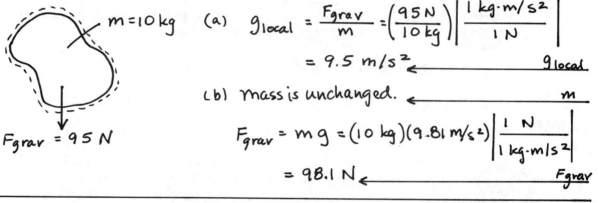

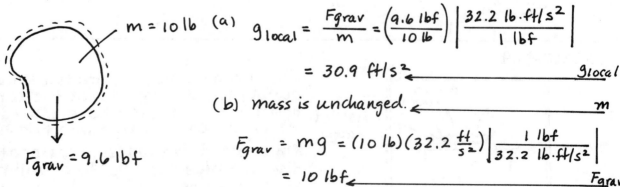

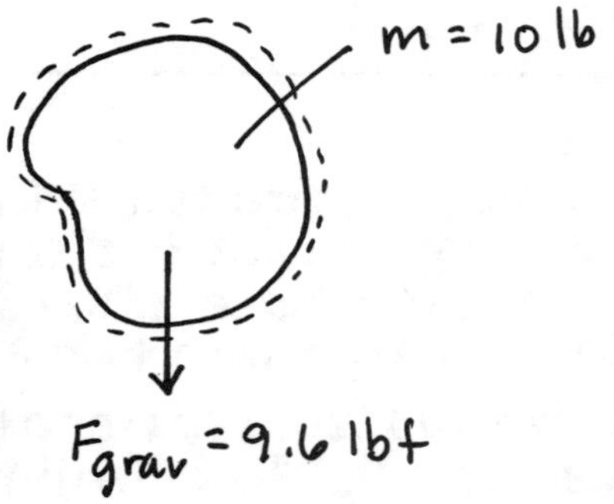

(a) $$g_{local} = \frac{F_{grav}}{m} = \left(\frac{9.6\ \text{lbf}}{10\ \text{lb}}\right)\left|\frac{32.2\ \text{lb}\cdot\text{ft/s}^2}{1\ \text{lbf}}\right|$$

$$= 30.9\ \text{ft/s}^2 \longleftarrow g_{local}$$

(b) mass is unchanged. ← m

$$F_{grav} = mg = (10\ \text{lb})(32.2\ \tfrac{\text{ft}}{\text{s}^2})\left|\frac{1\ \text{lbf}}{32.2\ \text{lb}\cdot\text{ft/s}^2}\right|$$

$$= 10\ \text{lbf} \longleftarrow F_{grav}$$

PROBLEM 1.14

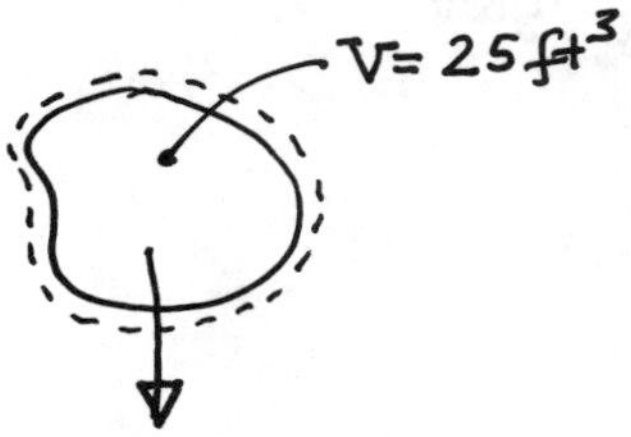

$F_{grav} = 3.5\,lbf$

$g_{moon} = 5.47\,ft/s^2$

$g_{mars} = 12.86\,ft/s^2$

In general, $F_{grav} = m\,g$. So,

$$m = \frac{F_{grav}}{g} \quad (*)$$

Since the mass is the same on mars as on the moon,

$$\left(\frac{F_{grav}}{g}\right)_{mars} = \left(\frac{F_{grav}}{g}\right)_{moon}$$

Accordingly

$$(F_{grav})_{mars} = \left(\frac{g_{mars}}{g_{moon}}\right)(F_{grav})_{moon}$$

$$= \left(\frac{12.86\,ft/s^2}{5.47\,ft/s^2}\right)(3.5\,lbf) = 8.23\,lbf \longleftarrow$$

The density is $\rho = m/V$. Applying Eq. (*) with data on mars

$$m = \left(\frac{8.23\,lbf}{12.86\,ft/s^2}\right)\left|\frac{32.2\,lb\cdot ft/s^2}{1\,lbf}\right| = 20.61\,lb$$

Then

$$\rho = \frac{20.61\,lb}{25\,ft^3} = 0.824\,\frac{lb}{ft^3} \longleftarrow$$

PROBLEM 1.15

Eq. 1.10 is used on both parts: $n = m/M$, where M is from Tables A-1.

(a) $m = Mn$, $n = 20$ kmol:

Air: $m = (28.97\,kg/kmol)(20\,kmol) = 579.4\,kg$

C: $m = (12.01\,kg/kmol)(20\,kmol) = 240.2\,kg$

H_2O: $m = (18.02\,kg/kmol)(20\,kmol) = 360.4\,kg$

CO_2: $m = (44.01\,kg/kmol)(20\,kmol) = 880.2\,kg$

(b) $n = m/M$, $m = 50\,lb$:

H_2: $n = (50\,lb)/(2.016\,lb/lbmol) = 24.802\,lbmol$

N_2: $n = (50\,lb)/(28.01\,lb/lbmol) = 1.785\,lbmol$

NH_3: $n = (50\,lb)/(17.03\,lb/lbmol) = 2.936\,lbmol$

C_3H_8: $n = (50\,lb)/(44.09\,lb/lbmol) = 1.134\,lbmol$

PROBLEM 1.16

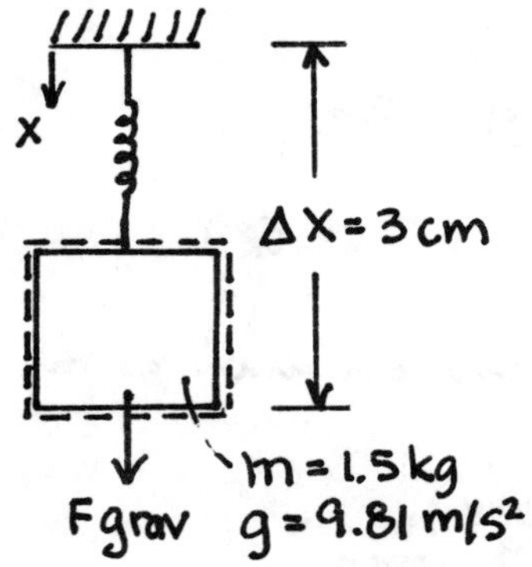

$F_{spring} = K(\Delta x)$ and $F_{spring} = F_{grav} = mg$

Thus

$K(\Delta x) = mg$

and

$$K = \frac{mg}{\Delta x}$$

$$= \frac{(1.5\ kg)(9.81\ m/s^2)}{(3\ cm)} \left| \frac{1\ N}{kg \cdot m/s^2} \right|$$

$$= 4.905\ N/cm \longleftarrow \text{proportionality constant}$$

PROBLEM 1.17

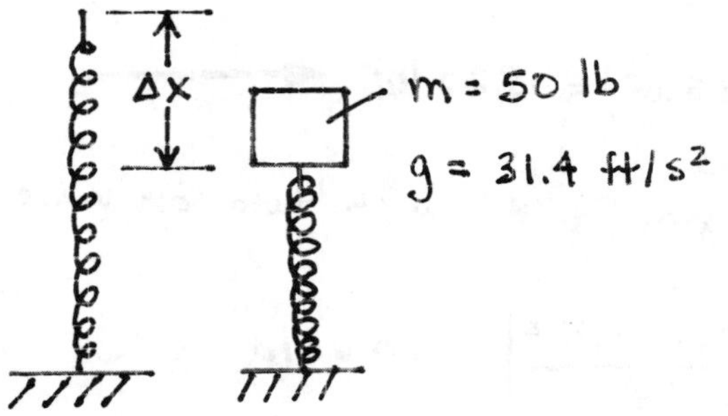

The spring is known to deflect 0.04 in for every 1 lbf of applied force. Thus, we begin by determining the weight of the object

$$F_{grav} = mg$$

$$= (50\ lb)(31.4\ ft/s^2) \left| \frac{1\ lbf}{32.2\ ft \cdot lb/s^2} \right|$$

$$= 48.76\ lbf$$

The deflection is

$$\Delta x = (0.04\ in/lbf)(48.76\ lbf) = 1.95\ in \longleftarrow \Delta x$$

PROBLEM 1.18

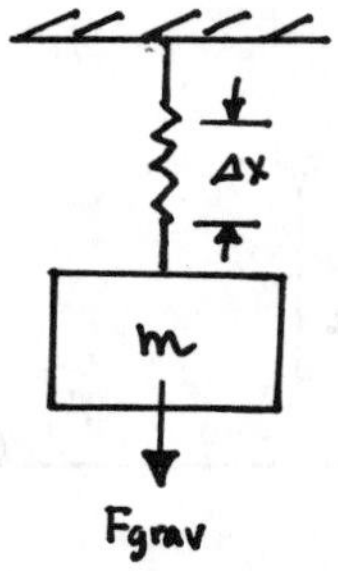

For a linear spring, $F_{spring} = K(\Delta x)$, where Δx is the spring extension. Since $F_{spring} = F_{grav} = mg$, we have $K(\Delta x) = mg$. Since m and K are independent of location, the local acceleration of gravity is proportional to the deflection. Thus

mars:

$$\frac{g_{mars}}{g_{earth}} = \frac{(\Delta x)_{mars}}{(\Delta x)_{earth}} = \frac{0.116\text{in}}{0.291\text{in}} \Rightarrow g_{mars} = \left(\frac{0.116}{0.291}\right)\left(32.174\frac{ft}{s^2}\right)$$

$$= 12.825\frac{ft}{s^2} \longleftarrow$$

moon:

$$\frac{g_{moon}}{g_{earth}} = \frac{(\Delta x)_{moon}}{(\Delta x)_{earth}} \Rightarrow (\Delta x)_{moon} = \left(\frac{g_{moon}}{g_{earth}}\right)(\Delta x)_{earth}$$

$$= \left(\frac{5.471}{32.174}\right)(0.291\text{in}) = 0.049\text{in} \longleftarrow$$

PROBLEM 1.19

Deceleration occurs from 5 mi/h to rest in 0.1 s. The average acceleration magnitude is

$$|a|_{avg} = \frac{\Delta V}{\Delta t} = \left(\frac{5\text{ mi/h} - 0}{0.1\text{ s}}\right)\left|\frac{5280\text{ ft}}{1\text{ mi}}\right|\left|\frac{1\text{ h}}{3600\text{ s}}\right|$$

$$= 73.33\text{ ft/s}^2$$

or, in g's

$$|a|_{avg} = \left(\frac{73.33\text{ ft/s}^2}{32.2\text{ ft/s}^2}\right) = 2.28\text{ g's} \longleftarrow \text{deceleration (in g's)}$$

Thus, the magnitude of the average force applied is

$$|F|_{avg} = m|a|_{avg} = (50\text{ lb})\left(73.33\frac{ft}{s^2}\right)\left|\frac{1\text{ lbf}}{32.2\text{ lb}\cdot\text{ft/s}^2}\right|$$

$$= 113.9\text{ lbf} \longleftarrow |F|_{avg}$$

PROBLEM 1.20

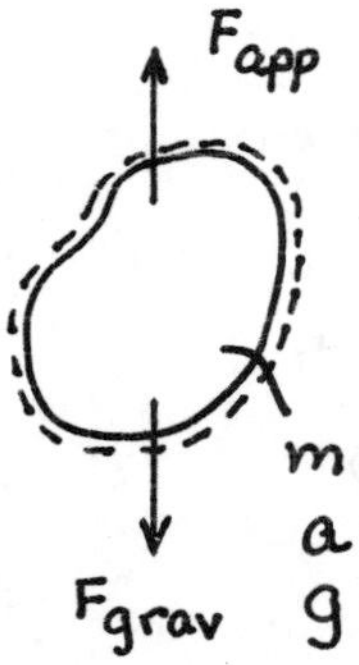

$m = 3\,kg$
$a = 7\ m/s^2$ (up)
$g = 9.81\ m/s^2$

$$F_{app} - F_{grav} = ma$$

$$F_{app} = ma + mg = m(a+g)$$

$$= (3\ kg)(7 + 9.81)\frac{m}{s^2}\left|\frac{1\ N}{1\ kg\cdot m/s^2}\right|$$

$$= 50.43\ N \leftarrow F_{app}$$

PROBLEM 1.21

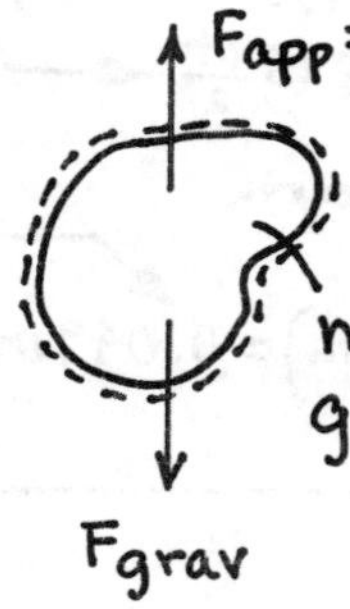

$m = 7\ lb$
$g = 32.2\ ft/s^2$

$$F_{app} - F_{grav} = ma$$

$$a = \frac{F_{app} - F_{grav}}{m} = \frac{F_{app} - mg}{m}$$

$$= \frac{F_{app}}{m} - g$$

$$= \left(\frac{20\ lbf}{7\ lb}\right)\left|\frac{32.2\ lb\cdot ft/s^2}{1\ lbf}\right| - 32.2\ ft/s^2$$

$$= 59.8\ ft/s^2\ \text{(upwards)} \leftarrow a$$

PROBLEM 1.22

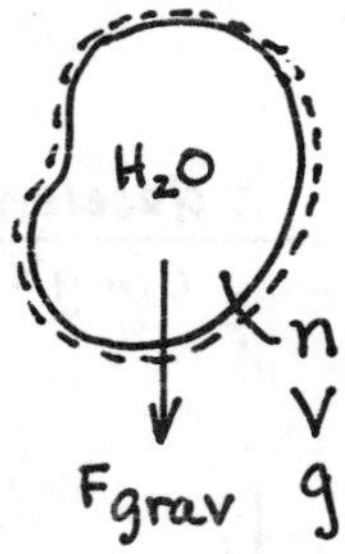

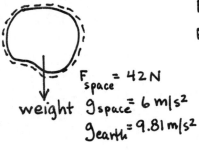

$n = 0.5\ lbmol$
$V = 0.145\ ft^3$
$g = 30.5\ ft/s^2$

From Table A-1E : $M = 18.02\ lb/lbmol$

$$F_{grav} = mg = nMg$$

$$= (0.5\ lbmol)\left(18.02\ \frac{lb}{lbmol}\right)\left(30.5\ \frac{ft}{s^2}\right)\left|\frac{1\ lbf}{32.2\ lb\cdot ft/s^2}\right|$$

$$= 8.534\ lbf \leftarrow F_{grav}$$

$$\rho_{ave} = \frac{m}{V} = \frac{(0.5)(18.02)}{(0.145)} = 62.14\ lb/ft^3 \leftarrow \rho_{ave}$$

PROBLEM 1.23

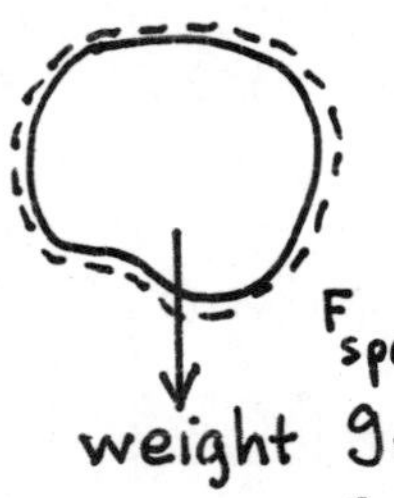

$F_{space} = 42\,N$
$g_{space} = 6\ m/s^2$
$g_{earth} = 9.81\ m/s^2$

$$\left.\begin{aligned} F_{space} &= m\, g_{space} \\ F_{earth} &= m\, g_{earth} \end{aligned}\right\}\ \frac{F_{space}}{g_{space}} = \frac{F_{earth}}{g_{earth}}$$

$$F_{earth} = F_{space}\left(\frac{g_{earth}}{g_{space}}\right) = (42\ N)\left(\frac{9.81}{6}\right)$$

$$= 68.67\ N \leftarrow F_{earth}$$

PROBLEM 1.24

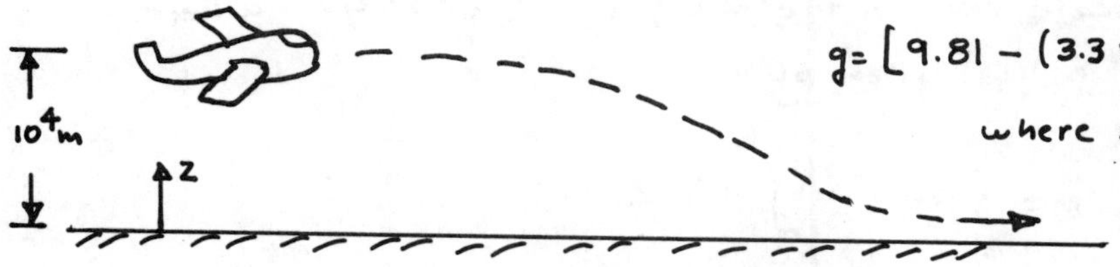

$g = [9.81 - (3.3 \times 10^{-6}/s^2)z]\ m/s^2$,

where z is in m.

Since weight is $\mathcal{W} = mg$, the percent change in weight on landing is

$$\%\ \text{change} = \left[\frac{\mathcal{W}|_{z=0} - \mathcal{W}}{\mathcal{W}}\right](100)$$

$$= \left[\frac{(9.81\ m/s^2) - [(9.81\ m/s^2) - (3.3\times10^{-6}/s^2)(10^4\ m)]}{[9.81 - (3.3\times10^{-6})(10^4)]}\right](100) = \left[\frac{3.3\times10^{-2}}{9.777}\right](100)$$

$$= 0.34\% \longleftarrow$$

PROBLEM 1.25

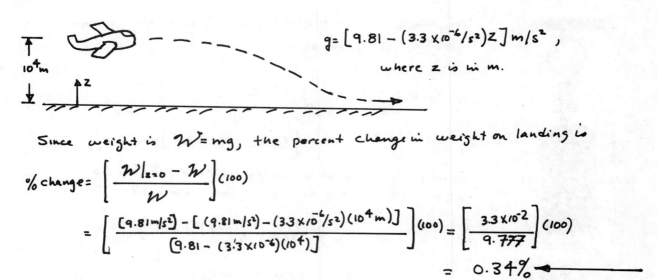

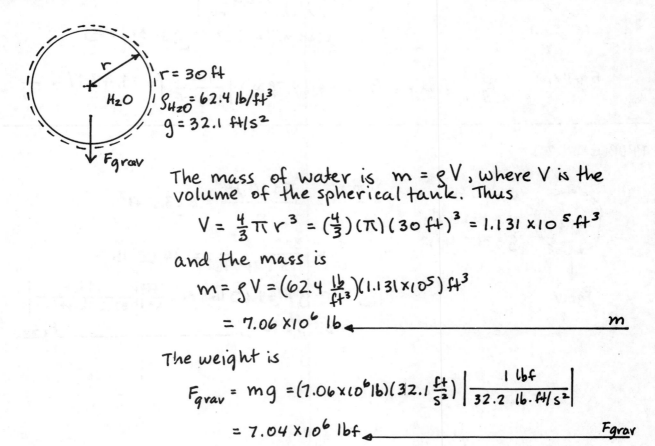

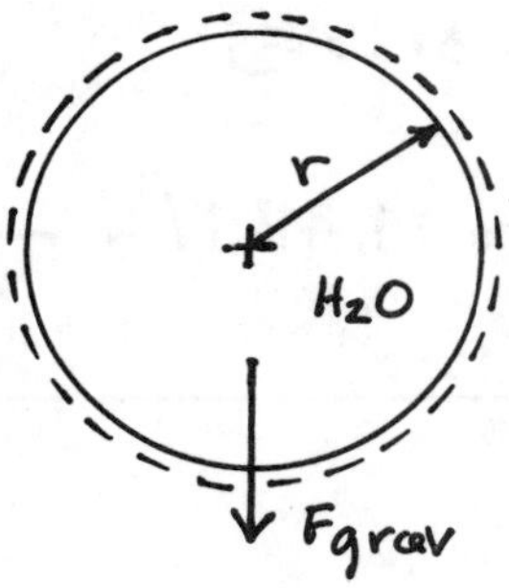

$r = 30$ ft

$\rho_{H_2O} = 62.4\ lb/ft^3$

$g = 32.1\ ft/s^2$

The mass of water is $m = \rho V$, where V is the volume of the spherical tank. Thus

$$V = \frac{4}{3}\pi r^3 = \left(\frac{4}{3}\right)(\pi)(30\ ft)^3 = 1.131\times10^5\ ft^3$$

and the mass is

$$m = \rho V = \left(62.4\ \frac{lb}{ft^3}\right)(1.131\times10^5)\ ft^3$$

$$= 7.06\times10^6\ lb \longleftarrow \quad m$$

The weight is

$$F_{grav} = mg = (7.06\times10^6\ lb)\left(32.1\ \frac{ft}{s^2}\right)\left|\frac{1\ lbf}{32.2\ lb\cdot ft/s^2}\right|$$

$$= 7.04\times10^6\ lbf \longleftarrow \quad F_{grav}$$

PROBLEM 1.26

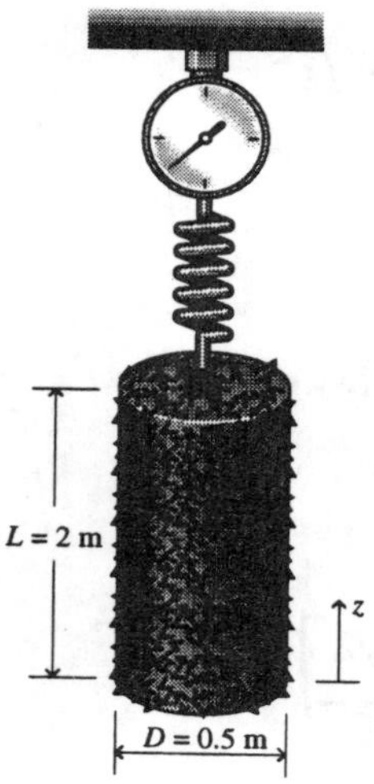

$g = 9.78\ m/s^2$

$\rho = 7800 - 360(z/L)^2\ kg/m^3$, where z is in m.

The scale records the weight, $F_{grav} = mg$, where mass m is given by

$$m = \int_{Vol} \rho\, dV = \int_0^L \rho A\, dz$$

$$= A \int_0^L [7800 - 360(z/L)^2]\, dz$$

For simplicity, introduce a new variable: $\hat{z} = z/L$, so $dz = L\, d\hat{z}$ and the expression for mass reads

$$m = AL \int_0^1 [7800 - 360(\hat{z})^2]\, d\hat{z}$$

$$= AL \left[7800\hat{z} - \frac{360(\hat{z})^3}{3} \right]_0^1$$

$$= \left(\frac{\pi D^2}{4}\right) L \left[7680 \frac{kg}{m^3} \right]$$

$$= \frac{\pi}{4}(0.5\ m)^2 (2m) \left[7680 \frac{kg}{m^3}\right] = 3016\ kg$$

Finally,

$$F_{grav} = mg = (3016\ kg)\left(9.78 \frac{m}{s^2}\right) \left| \frac{1\ N}{1\ kg \cdot m/s^2} \right| = 29{,}496\ N \leftarrow$$

PROBLEM 1.27

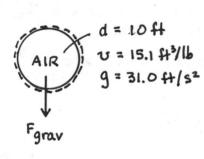

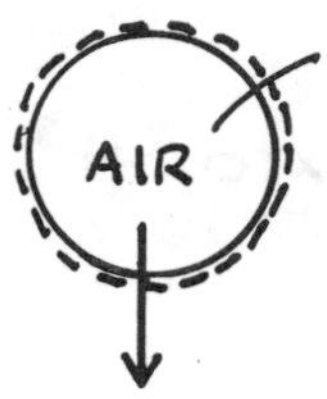

$d = 10\ ft$

$v = 15.1\ ft^3/lb$

$g = 31.0\ ft/s^2$

$$V = \frac{\pi d^3}{6} = \frac{\pi (10^3) ft^3}{6} = 523.6\ ft^3$$

$$m = \frac{V}{v} = \frac{523.6\ ft^3}{15.1\ ft^3/lb} = 34.68\ lb$$

$$F_{grav} = mg = (34.68\ lb)\left(31.0 \frac{ft}{s^2}\right) \left| \frac{1\ lbf}{32.2\ lb \cdot ft/s^2} \right|$$

$$= 33.39\ lbf \leftarrow F_{grav}$$

PROBLEM 1.28

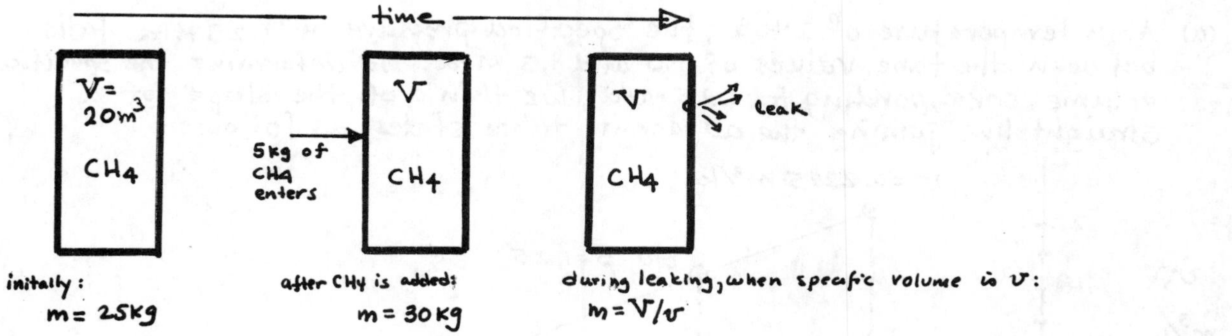

(a) Using $v = V/m$

initial specific volume: $v = \frac{20\ m^3}{25\ kg} = 0.8\ \frac{m^3}{kg}$ ←

after an additional 5 kg enters: $v = \frac{20\ m^3}{30\ kg} = 0.67\ \frac{m^3}{kg}$ ←

(b) Beginning with 30 kg of CH_4 in the cylinder, the amount of mass that has leaked is

$$\begin{bmatrix}\text{Amount} \\ \text{leaked}\end{bmatrix} = [30\ kg] - \begin{bmatrix}\text{mass in cylinder} \\ \text{when the specific} \\ \text{volume in the cylinder} \\ \text{is } v\end{bmatrix}$$

$$= [30\ kg] - \left[\frac{20 m^3}{v}\right]$$

In writing this, we assume that the specific volume does not vary with location within the cylinder: the gas leaks slowly, so the specific volume within the cylinder varies only with time.

Since the specific volume v varies from 0.67 m³/kg, when 30 kg is in the cylinder, to the specified upper limit specified: 1.0 m³/kg, the plot is

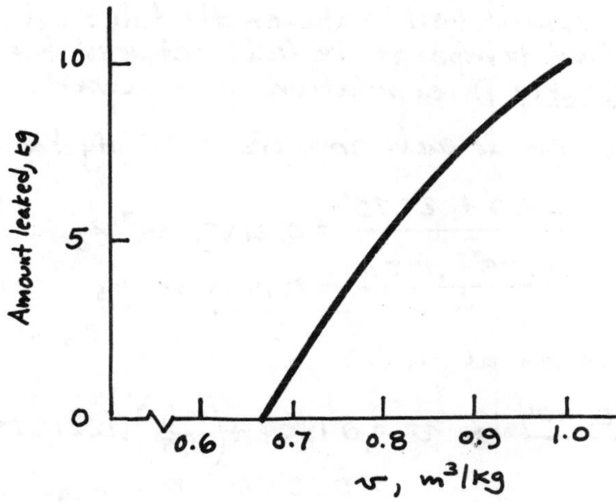

PROBLEM 1.29

(a) At a temperature of 240°C, the specified pressure of 1.25 MPa falls between the tabe values of 1.0 and 1.5 MPa. To determine the specific volume corresponding to 1.25 MPa, we think of the slope of a straight line joining the adjacent table states, as follows:

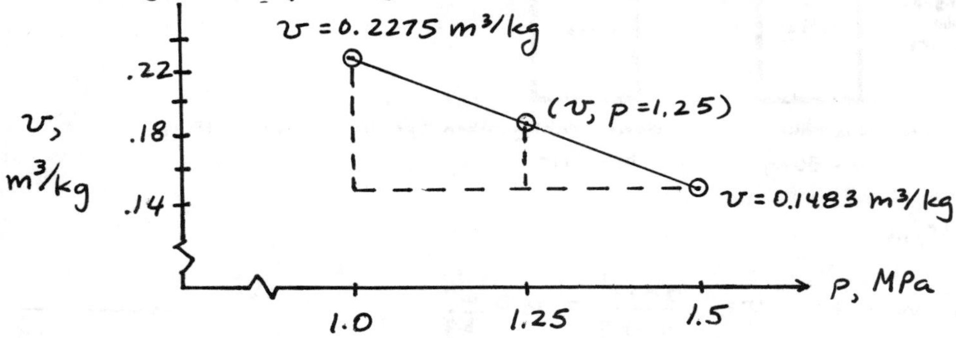

similar triangles:

$$|\text{slope}| = \frac{v - 0.1483}{1.5 - 1.25} = \frac{0.2275 - 0.1483}{1.5 - 1.0} \Rightarrow v = 0.1483 + \left(\frac{0.25}{0.50}\right)(0.2275 - 0.1483)$$

$$= 0.1879 \text{ m}^3/\text{kg} \quad \text{(a)}$$

(b) At a pressure of 1.5 MPa, the given specific volume of 0.1555 m³/kg falls between the table values of 240 and 280°C. To determine the temperature corresponding to the given specific volume, we think of the slope of a straight line joining the adjacent table states, as follows:

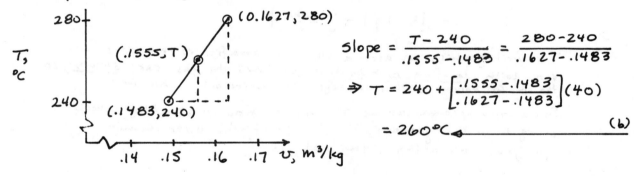

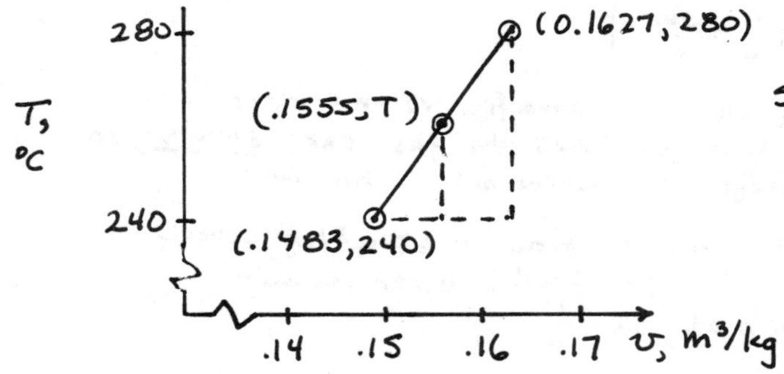

$$\text{Slope} = \frac{T - 240}{.1555 - .1483} = \frac{280 - 240}{.1627 - .1483}$$

$$\Rightarrow T = 240 + \left[\frac{.1555 - .1483}{.1627 - .1483}\right](40)$$

$$= 260°\text{C} \quad \text{(b)}$$

(c) In this case, the specified pressure falls between the table values of 1.0 and 1.5 MPa <u>and</u> the specified temperature falls between the table values of 200 and 240°C. Thus, <u>double</u> interpolation is required.

- At 220°C, the specific volume at each pressure is simply the average over the interval:

at 1.0 MPa, 220°C; $v = \frac{.2060 + .2275}{2} = 0.21675$ m³/kg

at 1.5 MPa, 220°C $v = \frac{.1325 + .1483}{2} = 0.1404$ m³/kg

- Thus, with the same approach as in (a)

$$\frac{v - 0.1404}{1.5 - 1.4} = \frac{0.21675 - 0.1404}{1.5 - 1.0} \Rightarrow v = 0.1404 + \left(\frac{0.1}{0.5}\right)(0.21675 - 0.1404)$$

$$= 0.15567 \text{ m}^3/\text{kg} \quad \text{(c)}$$

PROBLEM 1.30

(a) At a temperature of 120°C, the specified pressure of 54 lbf/in.² falls between the table values of 50 and 60 lbf/in.² To determine the specific volume corresponding to 54 lbf/in.², we think of the slope of a straight line joining the adjacent table states, as follows:

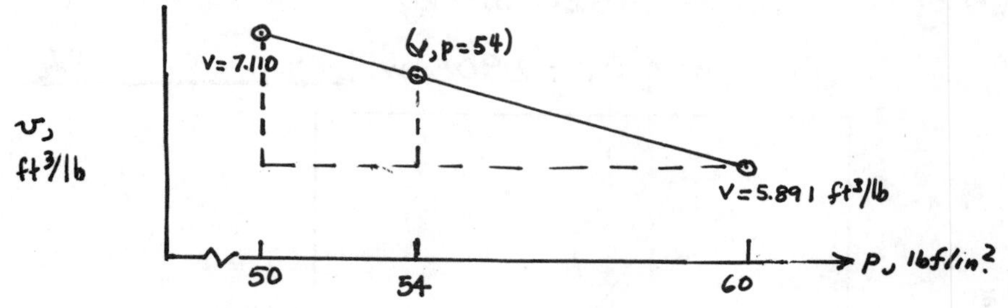

similar triangles:

$$|slope| = \frac{v - 5.891}{60-54} = \frac{7.110-5.891}{60-50} \Rightarrow v = 5.891 + \frac{6}{10}(7.110 - 5.891) = 6.622 \frac{ft^3}{lb}$$ ← (a)

(b) At a pressure of 60 lbf/in.², the given specific volume of 5.982 ft³/lb falls between the table values of 120 and 140°F. To determine the temperature corresponding to the given specific volume, we think of the slope of a straight line joining the adjacent table states, as follows:

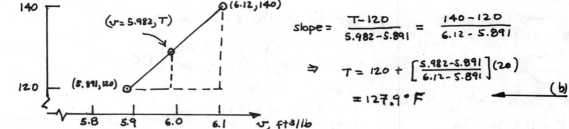

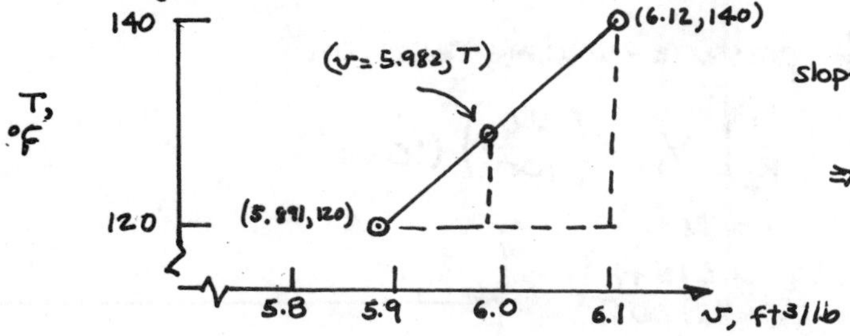

$$slope = \frac{T-120}{5.982-5.891} = \frac{140-120}{6.12-5.891}$$

$$\Rightarrow T = 120 + \left[\frac{5.982-5.891}{6.12-5.891}\right](20)$$

$$= 127.9°F$$ ← (b)

(c) In this case, the specified pressure falls between the table values of 50 and 60 lbf/in² and the specified temperature falls between the table values of 100 and 120°F. Thus, double interpolation is required.

- At 110°F, the specific volume at each pressure is simply the average over the interval:

  at 50 $\frac{lbf}{in^2}$, 110°F; $v = \frac{7.110 + 6.836}{2} = 6.973$ ft³/lb

  at 60 $\frac{lbf}{in^2}$, 110°F; $v = \frac{5.891 + 5.659}{2} = 5.775$ ft³/lb

- Then, with the same approach as in (a)

$$\frac{v - 5.775}{60-58} = \frac{6.973 - 5.775}{60-50} \Rightarrow v = 5.775 + \frac{2}{10}[6.973 - 5.775]$$

$$= 6.015 \; ft^3/lb$$ ← (c)

PROBLEM 1.31

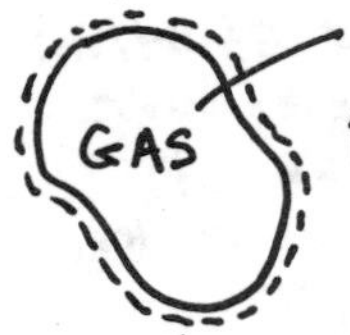

$m = 2\,kg$

$pv^{1.3} = const.$

$p_1 = 1\,bar$, $v_1 = 0.5\,m^3/kg$

$p_2 = 0.25\,bar$

From the pressure-specific volume relation

$$v_2 = \left(\frac{p_1}{p_2}\right)^{\frac{1}{1.3}} v_1 = \left(\frac{1}{0.25}\right)^{\frac{1}{1.3}} (0.5\,m^3/kg)$$

$$= 1.4524\,m^3/kg$$

$$V_2 = v_2 m = 2.905\,m^3 \longleftarrow V_2$$

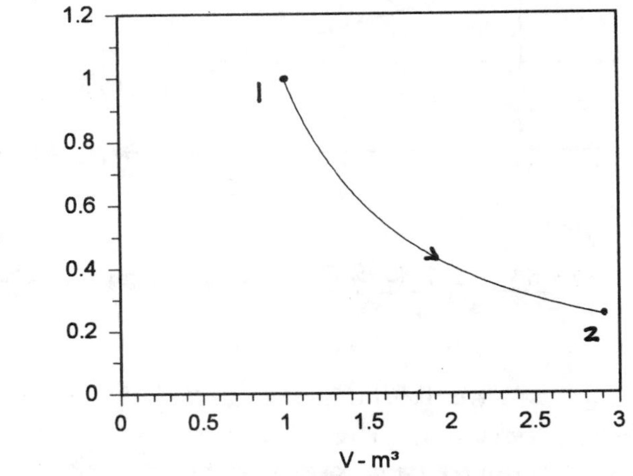

PROBLEM 1.32

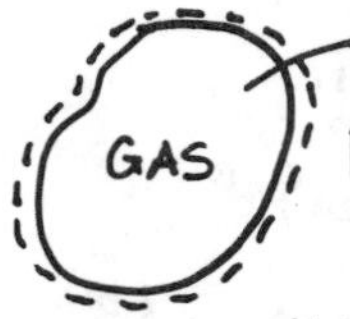

$m = 1\,lb$

$pV^n = const.$

$p_1 = 20\,lbf/in^2$, $V_1 = 10\,ft^3$

$p_2 = 100\,lbf/in^2$

From the pressure-volume relation

$$V_2 = \left(\frac{p_1}{p_2}\right)^{\frac{1}{n}} V_1 = \left(\frac{20}{100}\right)^{\frac{1}{n}} (10\,ft^3)$$

$n = 1$; $V_2 = 2\,ft^3$

$n = 1.2$; $V_2 = 2.615\,ft^3$

$n = 1.3$; $V_2 = 2.900\,ft^3$

$n = 1.4$; $V_2 = 3.168\,ft^3$

$\longleftarrow V_2$

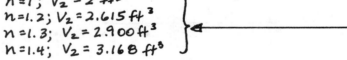

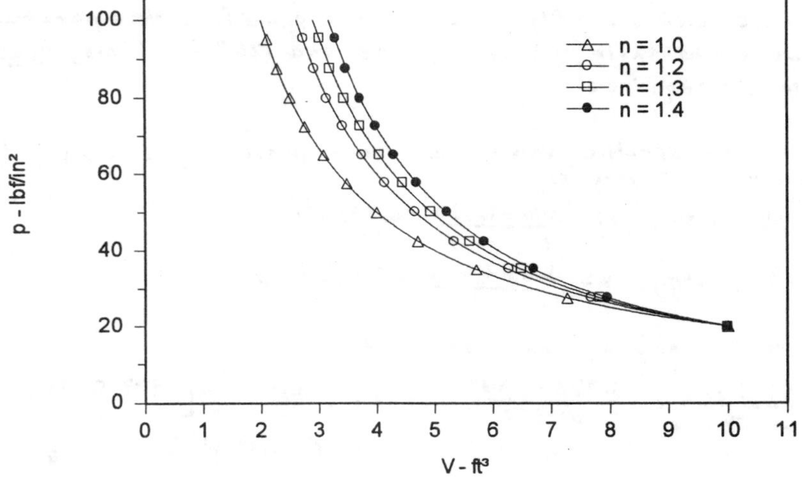

PROBLEM 1.33

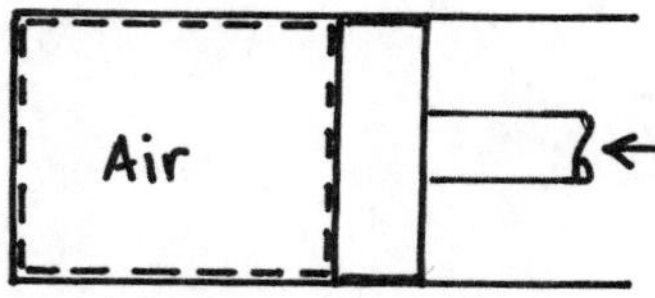

$p_1 = 20$ lbf/in²
$V_1 = 1.5$ ft³
$p_2 = 100$ lbf/in²
$V_2 = 0.5$ ft³

The pressure-volume relation is linear during the process. Thus

$$p = p_1 + \left(\frac{p_2 - p_1}{V_2 - V_1}\right)(V - V_1)$$

Or, using given data

$$p = 20\ \frac{\text{lbf}}{\text{in}^2} + \frac{(100-20)\ \text{lbf/in}^2}{(0.5-1.5)\ \text{ft}^3}(V - 1.5)\ \text{ft}^3$$

$$= 140 - 80V$$

When $V = 1.2$ ft³

$$p = 140 - 80(1.2) = 44\ \text{lbf/in}^2 \longleftarrow p$$

On p-V coordinates

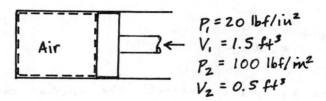

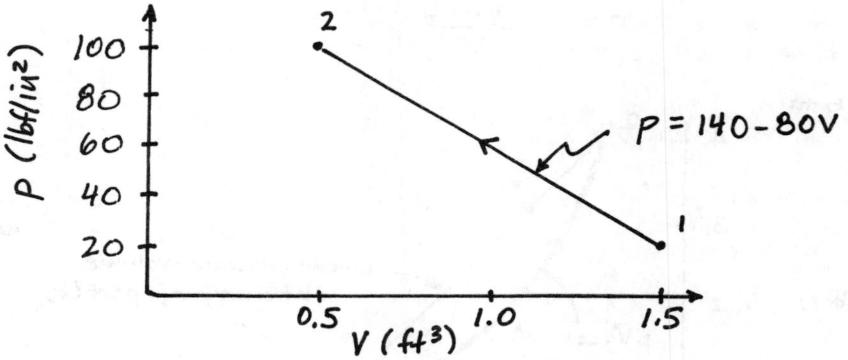

PROBLEM 1.34

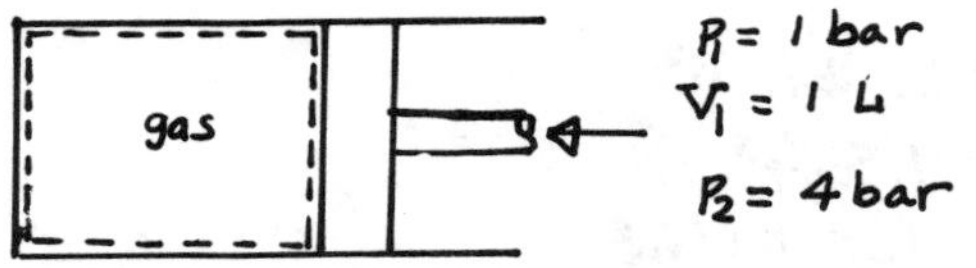

(a) The process is described by pV = constant. The constant can be evaluated using data at state 1:

$$pV = \text{constant}$$
$$= p_1 V_1$$
$$= (1\,\text{bar})(1\,\text{L}) = 1\ \text{bar}\cdot\text{L}$$

So, for every state during the process, we have the relation

$$pV = 1\ \text{bar}\cdot\text{L}$$

When $p = 3$ bar,

$$V = \frac{1\ \text{bar}\cdot\text{L}}{3\ \text{bar}} = 0.33\ \text{L}$$ ◄

when $p = 4$ bar, $$V_2 = \frac{1\ \text{bar}\cdot\text{L}}{4\ \text{bar}} = 0.25\ \text{L}$$

Plotting the relation on pressure-volume coordinates we use

$$p = \frac{1\ \text{bar}\cdot\text{L}}{V}$$

to obtain

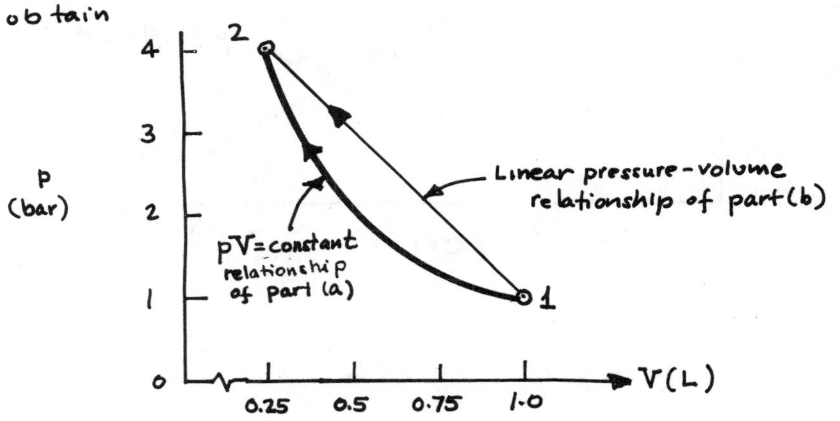

(b) For comparison, the linear pressure-volume relationship is shown on the plot above. The volume corresponding to $p = 3$ bar can be obtained simply using the slope of the straight line between 1 and 2:

$$|\text{slope}| = \frac{(4-1)\,\text{bar}}{(1.0-0.25)\,\text{L}} = \frac{(3-1)}{(1.0-V)} \rightarrow V = 0.5\ \text{L}$$ ◄

This value also can be read from the plot.

PROBLEM 1.35

Thermodynamic cycle:

1-2: pV = constant

$p_1 = 1\,\text{bar}$, $V_1 = 1\,m^3$, $V_2 = 0.2\,m^3$

2-3 p = constant, $V > V_2$ (expansion), $V_3 = 1.0\,m^3$

3-1 V = constant

For process 1-2, pV = constant. The constant can be evaluated using data at state 1:

$$pV = \text{constant}$$
$$= p_1 V_1$$
$$= (1\,\text{bar})(1\,m^3) = 1\,\text{bar}\cdot m^3$$

Accordingly, on a pressure-volume plot process 1-2 is described by

$$p = \frac{1\,\text{bar}\cdot m^3}{V}$$

In particular, when $V_2 = 0.2\,m^3$, $p = 5\,\text{bar}$.

The thermodynamic cycle takes the form

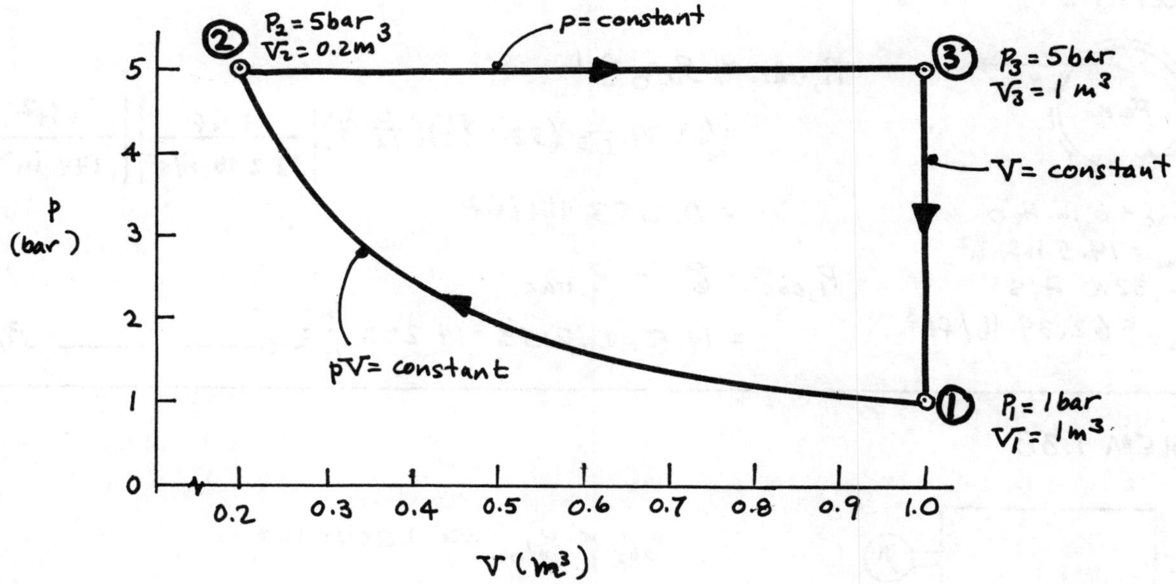

PROBLEM 1.36

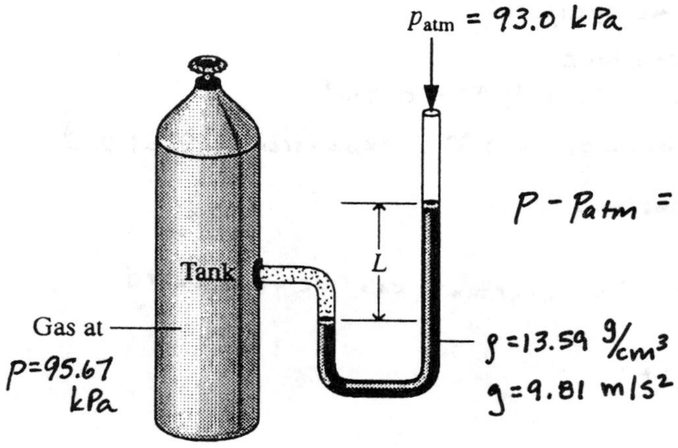

$$p - p_{atm} = \rho g L$$

(a)

$$L = \frac{p - p_{atm}}{\rho g}$$

$$= \frac{(95.67-93.0)\ kPa}{(13.59\ g/cm^3)(9.81\ m/s^2)}\left|\frac{10^3 N/m^2}{1\ kPa}\right|\left|\frac{10^3 g}{1\ kg}\right|\left|\frac{1\ m^3}{10^6\ cm^3}\right|\left|\frac{1\ kg\cdot m/s^2}{1\ N}\right|$$

$$= 0.02\ m = 2\ cm \longleftarrow L$$

(b) $p_{gage} = p - p_{atm} = 95.67 - 93.0 = 2.67\ kPa \longleftarrow p_{gage}$

PROBLEM 1.37

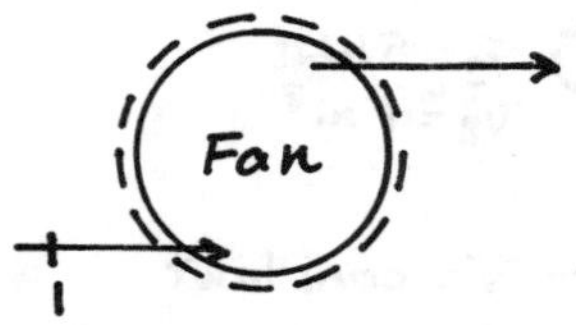

$p_{1,vac}$ = 6 in. H_2O
p_{atm} = 14.5 lbf/in²
g = 32.0 ft/s²
ρ_{H_2O} = 62.39 lb/ft³

$$p_{1,vac} = \rho_{H_2O}\, g\, L$$

$$= \left(62.39\ \frac{lb}{ft^3}\right)\left(32.0\ \frac{ft}{s^2}\right)\left(\frac{6}{12}\ ft\right)\left|\frac{1\ lbf}{32.2\ lb\cdot ft/s^2}\right|\left|\frac{1\ ft^2}{144\ in^2}\right|$$

$$= 0.2153\ lbf/in^2$$

$$p_{1,abs} = p_{atm} - p_{1,vac}$$

$$= 14.5 - 0.2153 = 14.285\ \frac{lbf}{in^2} \longleftarrow p_{1,abs}$$

PROBLEM 1.38

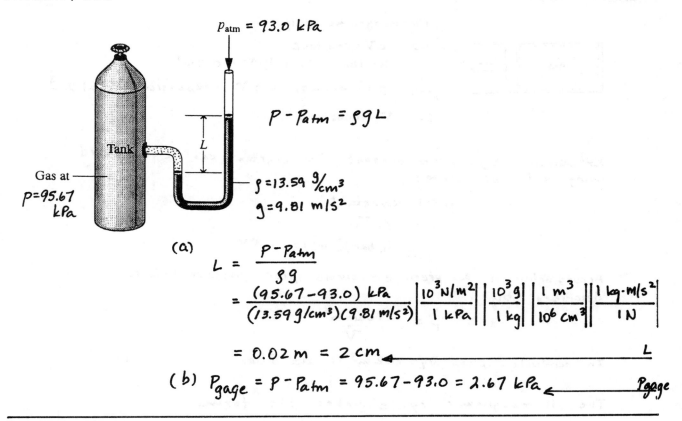

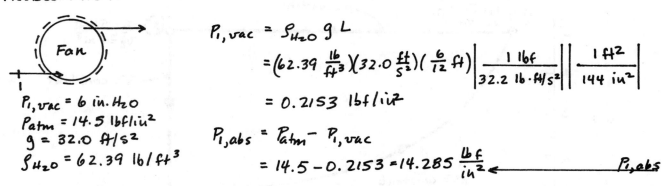

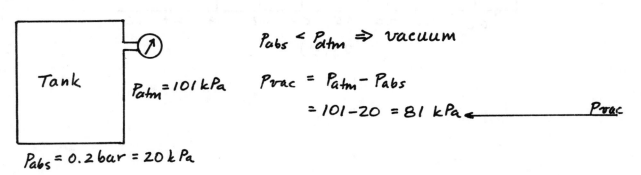

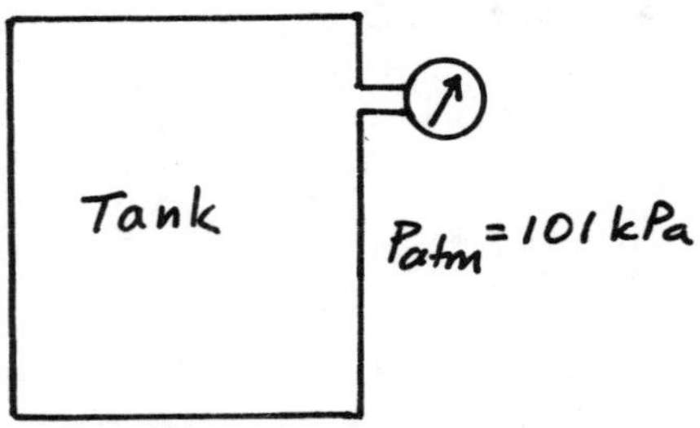

$p_{abs} < p_{atm} \Rightarrow$ vacuum

$$p_{vac} = p_{atm} - p_{abs}$$

$$= 101 - 20 = 81\ kPa \longleftarrow p_{vac}$$

PROBLEM 1.39

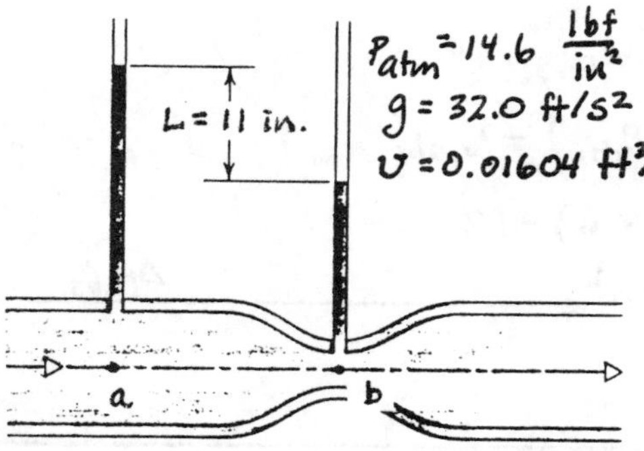

$p_{atm} = 14.6 \ \frac{lbf}{in^2}$

$g = 32.0 \ ft/s^2$

$v = 0.01604 \ ft^3/lb$

$$|\Delta p_{gage}| = \rho_{H_2O}\, g\, L$$

$$\rho_{H_2O} = 1/v_{H_2O} = 62.344 \ lb/ft^3$$

$$|\Delta p_{gage}| = \left(62.344 \frac{lb}{ft^3}\right)\left(32.0 \frac{ft}{s^2}\right)\left(\frac{11}{12} ft\right)\left|\frac{1 \ lbf}{32.2 \ lb \cdot ft/s^2}\right|\left|\frac{1 \ ft^2}{144 \ in^2}\right|$$

$$= 0.3944 \ lbf/in^2 \longleftarrow \Delta p_{gage}$$

(decreases)

PROBLEM 1.40

$p_{surr} = 101 \ kPa$

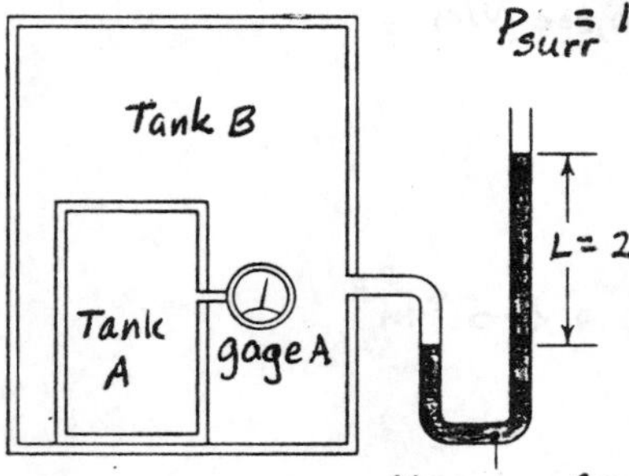

$p_{gage,A} = 1.4$ bars

Mercury ($\rho = 13.59 \ g/cm^3$)

$g = 9.81 \ m/s^2$

$$L = 20 \ cm = 0.2 \ m$$

$$p_{gage,B} = \rho_{Hg}\, g\, L$$

$$= \left(13.59 \frac{g}{cm^3}\right)\left(9.81 \frac{m}{s^2}\right)(0.2 \ m)\left|\frac{1 \ kg}{10^3 g}\right|\left|\frac{10^6 \ cm^3}{1 \ m^3}\right|\left|\frac{1 \ N}{1 \ kg \cdot m/s^2}\right|$$

$$= 26{,}664 \ N/m^2 = 0.2666 \ bar$$

$$p_{abs,B} = p_{surr} + p_{gage,B}$$

$$= 1.01 + .2666 = 1.2766 \ bar \longleftarrow$$

$$p_{abs,A} = p_{abs,B} + p_{gage,A}$$

$$= 1.2766 + 1.4 = 2.6766 \ bar \longleftarrow$$

PROBLEM 1.41

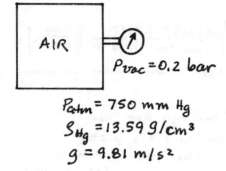

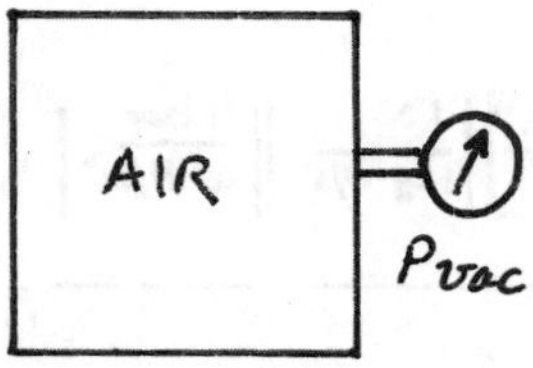

$p_{vac} = 0.2$ bar

$p_{atm} = 750$ mm Hg

$\rho_{Hg} = 13.59 \ g/cm^3$

$g = 9.81 \ m/s^2$

$$p_{atm} = \rho_{Hg}\, g\, L_{Hg}$$

$$= \left(13.59 \frac{g}{cm^3}\right)(9.81 \ m/s^2)\left(\frac{750}{1000}\right) m \left|\frac{1 \ kg}{10^3 g}\right|\left|\frac{10^6 \ cm^3}{1 \ m^3}\right|\left|\frac{1 \ N}{1 \ kg \cdot m/s^2}\right|\left|\frac{1 \ bar}{10^5 \ N/m^2}\right|$$

$$= 1 \ bar$$

$$p_{abs} = p_{atm} - p_{vac}$$

$$= 1 \ bar - 0.2 \ bar = 0.8 \ bar \longleftarrow p_{abs}$$

PROBLEM 1.42

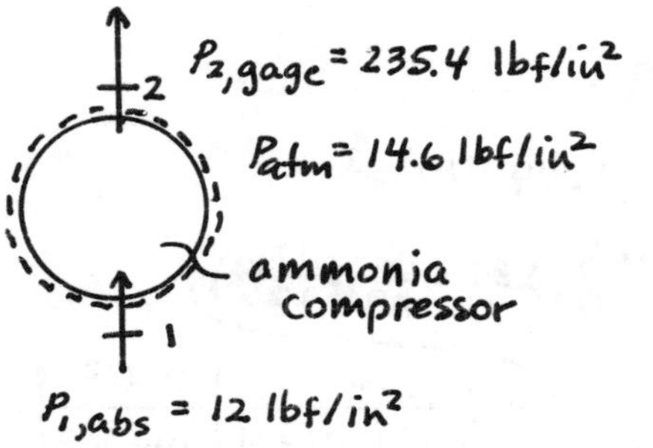

$$\Delta p_{abs} = p_{2,abs} - p_{1,abs}$$
$$= (p_{2,gage} + p_{atm}) - p_{1,abs}$$
$$= (235.4 + 14.6) - 12$$
$$= 238 \text{ lbf/in}^2 \longleftarrow \Delta p_{abs}$$

PROBLEM 1.43

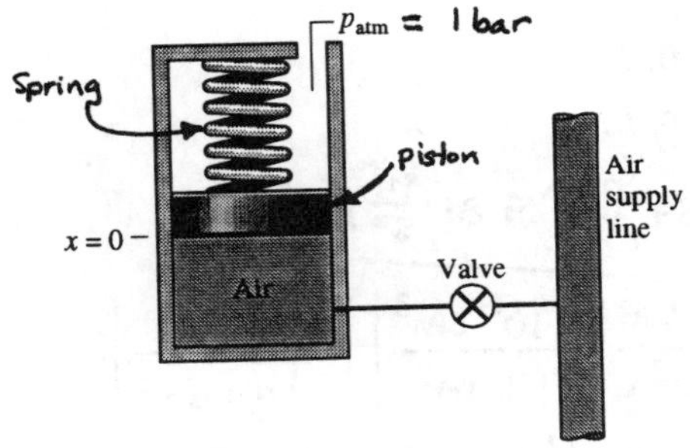

$g = 9.81 \text{ m/s}^2$

$F_{spring} = kx$, where $k = 10{,}000$ N/m

For the piston:

$m = 10$ kg

$A = 7.8 \times 10^{-3} \text{ m}^2$

For the air: $\Delta V = 3.9 \times 10^{-4} \text{ m}^3$

Initially, x = 0 and there is no spring force acting on the piston. Also, friction between the piston and the cylinder wall can be ignored. Accordingly, the force exerted by the air within the cylinder on the bottom of piston is equal to the weight of the piston plus the force exerted by the atmosphere on the top of the piston:

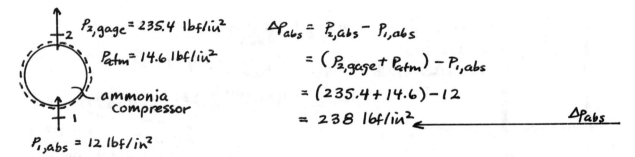

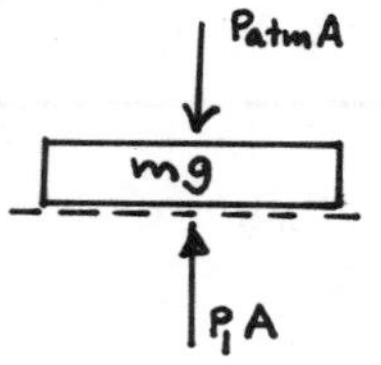

$\Sigma F_x = 0$:

$$p_1 A = p_{atm} A + mg$$

$$p_1 = p_{atm} + \frac{mg}{A}$$

$$p_1 = 1 \text{ bar} + \left[\frac{(10 \text{ kg})(9.81 \text{ m/s}^2)}{7.8 \times 10^{-3} \text{ m}^2}\right] \left|\frac{1 \text{ N}}{1 \text{ kg}\cdot\text{m/s}^2}\right| \left|\frac{1 \text{ bar}}{10^5 \text{ N/m}^2}\right|$$

$$p_1 = 1.126 \text{ bar} \longleftarrow p_1$$

Finally, the force exerted by the air within the cylinder on the bottom of the piston is equal to the weight of the piston plus the force exerted by the atmosphere on the top of the piston plus the force exerted by the spring on the top of the piston:

PROBLEM 1.43 (Cont'd)

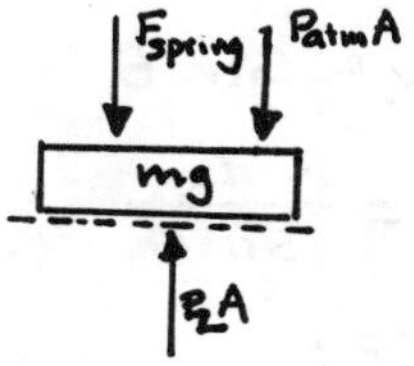

$\Sigma F_x = 0$:

$$P_2 A = P_{atm} A + mg + F_{spring}$$

$$P_2 = P_{atm} + \frac{mg}{A} + \frac{F_{spring}}{A}$$

For the spring, $F_{spring} = kx$, where x is found using the increase in volume of the air:

$$x = \frac{\Delta V}{A} = \frac{3.9 \times 10^{-4}\,m^3}{7.8 \times 10^{-3}\,m^2} = 0.05\ m$$

Collecting results

$$P_2 = P_{atm} + \frac{mg}{A} + \left[\frac{(10{,}000\,N/m)(0.05\,m)}{(7.8 \times 10^{-3}\,m^2)}\left|\frac{1\,bar}{10^5\,N/m^2}\right|\right]$$

$$= 1.126\ bar + 0.641\ bar$$

$$= 1.767\ bar \longleftarrow P_2$$

PROBLEM 1.44

$P_{atm} = 0.98\ bar$

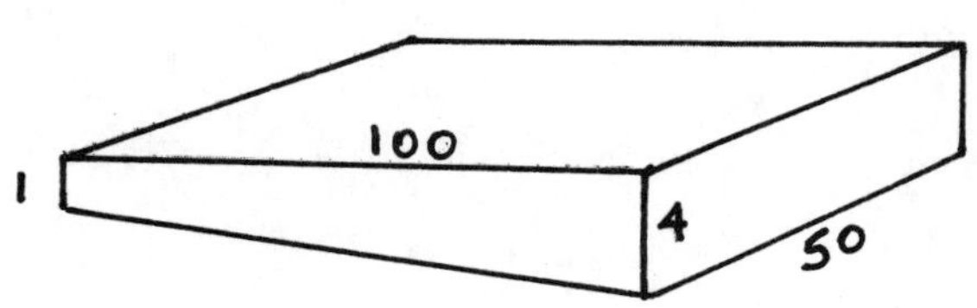

$\rho_{H_2O} = 998.2\ kg/m^3$
$g = 9.8\ m/s^2$

The total force on the bottom of the pool is the sum of the weight of the water and the downward force of atmospheric pressure on the surface of the water;

$$F_{tot} = F_{grav} + F_{atm} \qquad (a)$$

To get the weight of the water, first find the mass, as follows

$$m = \rho V = \left(998.2\ \frac{kg}{m^3}\right)\left[(1)(100)(50) + \tfrac{1}{2}(100)(3)(50)\right] m^3$$

$$= (998.2)(12500) = 1.25 \times 10^7\ kg$$

Thus

$$F_{grav} = mg = (1.25 \times 10^7\,kg)(9.81\,m/s^2)\left|\frac{1\,N}{1\,kg \cdot m/s^2}\right|\left|\frac{1\,kN}{10^3\,N}\right| = 1.226 \times 10^5\ kN$$

$$F_{atm} = P_{atm} A_{surface} = (0.98\,bar)(100 \times 50)\,m^2\left|\frac{10^5\,N/m^2}{1\,bar}\right|\left|\frac{1\,kN}{10^3\,N}\right| = 4.9 \times 10^5\ kN$$

Or

$$F_{tot} = 1.226 \times 10^5 + 4.9 \times 10^5 = 6.126 \times 10^5\ kN \longleftarrow F_{tot}$$

The depth at the center of the pool is $h = 2.5\ m$. Thus

$$P = P_{atm} + \rho g h$$

$$= (0.98\,bar)\left|\frac{100\,kPa}{1\,bar}\right| + \left(998.2\,\frac{kg}{m^3}\right)\left(9.8\,\frac{m}{s^2}\right)(2.5\,m)\left|\frac{1\,N}{1\,kg \cdot m/s^2}\right|\left|\frac{1\,kPa}{10^3\,N/m^2}\right| = 122.5\ kPa \longleftarrow P$$

PROBLEM 1.45

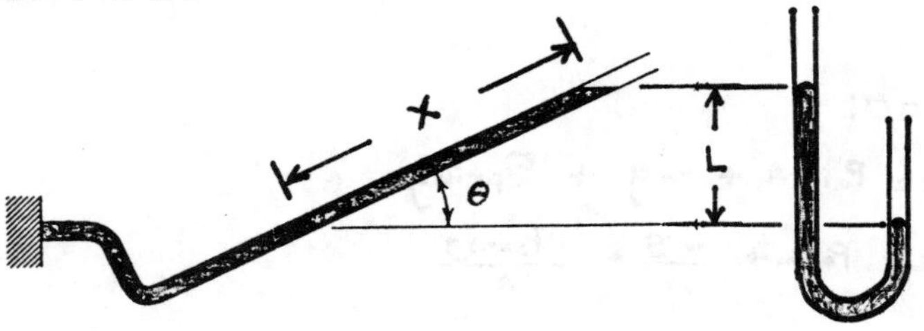

$L/x = \sin\theta$

$$x = \frac{L}{\sin\theta}$$

$\therefore x > L$

inclined manometer gives greater resolution

PROBLEM 1.46

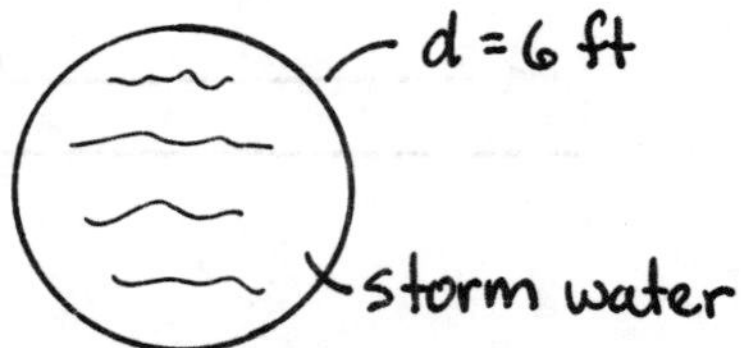

$\rho = 62.2 \text{ lb/ft}^3$

$g = 32.2 \text{ ft/s}^2$

$$\Delta p = \rho g d$$

$$= \left(62.2 \frac{\text{lb}}{\text{ft}^3}\right)\left(32.2 \frac{\text{ft}}{\text{s}^2}\right)(6 \text{ ft})\left|\frac{1 \text{ lbf}}{32.2 \text{ lb}\cdot\text{ft/s}^2}\right|\left|\frac{1 \text{ ft}^2}{144 \text{ in}^2}\right|$$

$$= 2.59 \text{ lbf/in}^2 \longleftarrow \Delta p$$

PROBLEM 1.47

Consider the variation of pressure owing to the effect of gravity. For an element in a motionless gas or liquid, the forces acting are the forces of pressure on the upper and lower surfaces and the weight of the system:

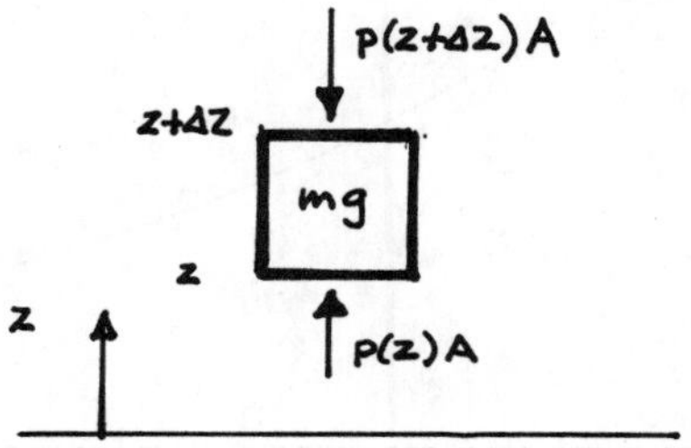

$\Sigma F_z = 0$: $\quad 0 = p(z+\Delta z)A + mg - p(z)A$

where
$$m = \rho V = \rho[A\Delta z] = \frac{[A\Delta z]}{v}$$

Rearranging
$$A\left[p(z+\Delta z) - p(z)\right] = -\left(\frac{A\Delta z}{v}\right)g$$

or
$$\frac{p(z+\Delta z) - p(z)}{\Delta z} = -\frac{g}{v}$$

In the limit as $\Delta z \to 0$. This gives
$$\frac{dp}{dz} = -\frac{g}{v} \qquad (*)$$

(a) Atmosphere. If $v = C/p$, where $C = 72.435(m^3/kg)(kPa) = 72{,}435\ \frac{N\cdot m}{kg}$, is inserted in Eq. (*), we have

$$\frac{dp}{dz} = -\frac{gp}{C} \Rightarrow \frac{d\ln p}{dz} = -g/C \Rightarrow \ln p = -\frac{gz}{C} + \ln K$$

or
$$p = K\exp\left(-\frac{gz}{C}\right)$$

When $z=0$, $p = p_0$ (1atm), giving $p = p_0 \exp(-gz/C)$.

Inserting known values gives p in atm when z is in km

$$p = p_0 \exp\left(\frac{-(9.81\,m/s^2)(z\ km)}{72{,}435\ \frac{N\cdot m}{kg}}\left|\frac{10^3 m}{1 km}\right|\left|\frac{1N}{1 kg\cdot m/s^2}\right|\right) = p_0 \exp(-0.135 z)$$

This relationship is shown in the accompanying plot.

PROBLEM 1.47 (Cont'd)

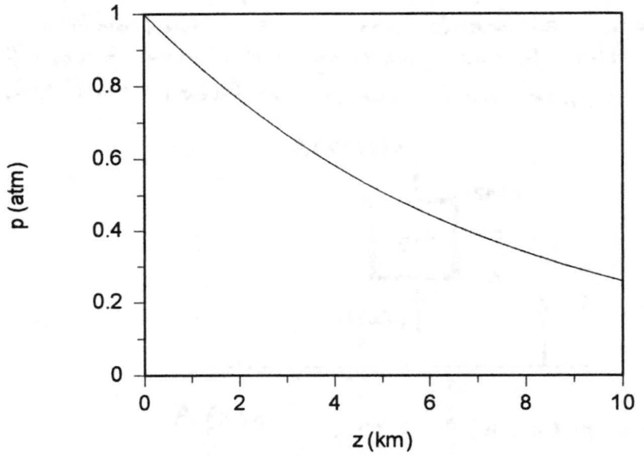

(b) <u>Ocean</u>. Letting $\hat{z}\,(=-z)$ denote depth, Eq. (*) reads

$$\frac{dp}{d\hat{z}} = \frac{g}{v}$$

Then, assuming v is constant

$$p = \left(\frac{g}{v}\right)\hat{z} + K$$

When $\hat{z} = 0$, $p = p_0$ (1atm), giving $p = \left(\frac{g}{v}\right)\hat{z} + p_0$.

Inserting known values gives p in atm when $\hat{z}$ is in km

$$p = p_0 + \left[\frac{(9.81\ \text{m/s}^2)\,\hat{z}(\text{km})}{0.956\times10^{-3}\ \text{m}^3/\text{kg}}\left|\frac{10^3\,\text{m}}{1\,\text{km}}\right|\left|\frac{1\,\text{atm}}{1.01325\times10^5\,\text{N/m}^2}\right|\left|\frac{1\,\text{N}}{1\,\text{kg}\cdot\text{m/s}^2}\right|\right]$$

$$= p_0 + 101.3\,\hat{z}$$

This relationship is shown in the accompanying plot.

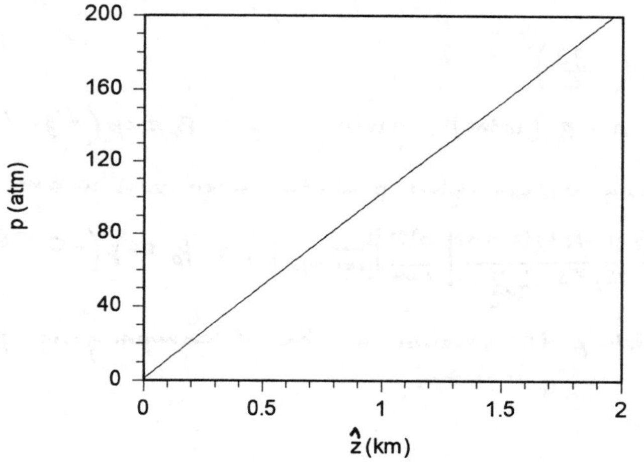

PROBLEM 1.48

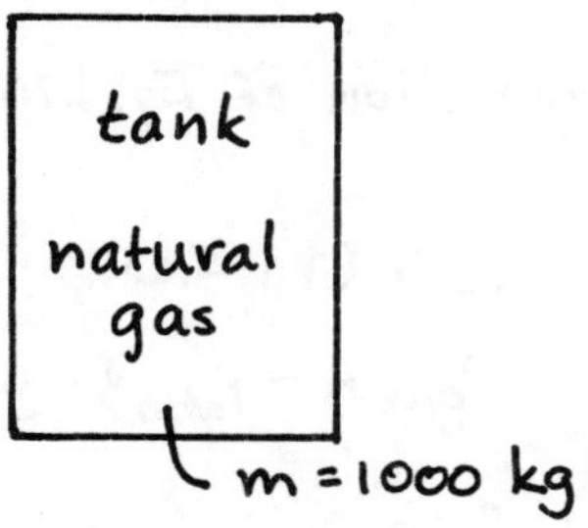

For the gas in the tank, the p-v-T relation is

$$p = [(5.18\times10^{-3})\,T/(v-0.002668)] - (8.91\times10^{-3})/v^2$$

Solving iteratively for v at $p = 100$ bar, $T = 255$ K we get

$$v = 0.00884\ \text{m}^3/\text{kg}$$

and

$$V = mv = 8.84\ \text{m}^3 \quad \longleftarrow V$$

The following plots can be constructed for T = 250, 500, and 1000 K:

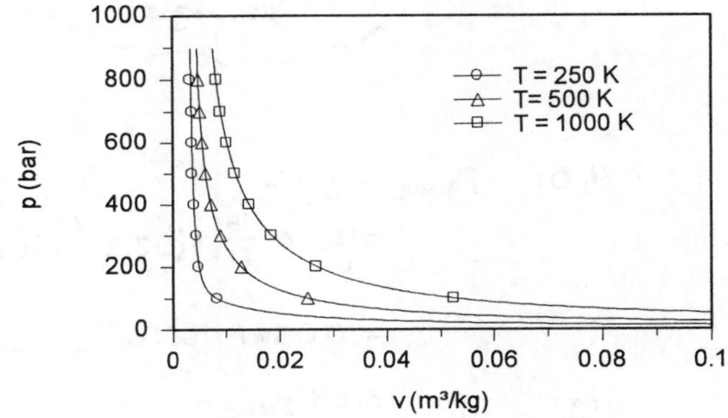

PROBLEM 1.49

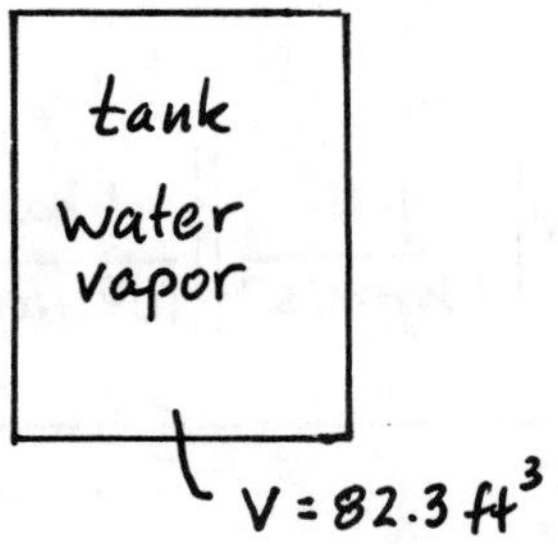

For the water vapor in the tank, the p-v-T relation is

$$p = [(0.5954)\,T/(v-0.2708)] - 63.36/v^2$$

where v is in ft³/lb, T is in °R, and p is in lbf/in²

Solving iteratively for v at $p = 1500$ lbf/in², $T = 1140$°R we get

$$v = 0.686\ \text{ft}^3/\text{lb}$$

Thus, with the given value for V

$$m = V/v = 82.3/0.686 = 120\ \text{lb} \quad \longleftarrow m$$

The following plots can be constructed for T = 1200, 1400, 1600 °R:

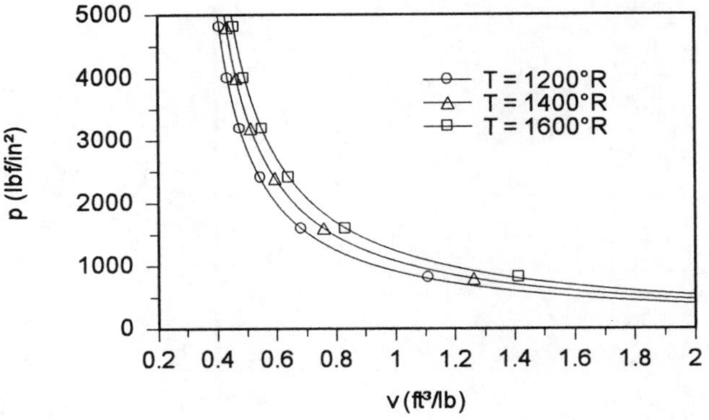

PROBLEM 1.50

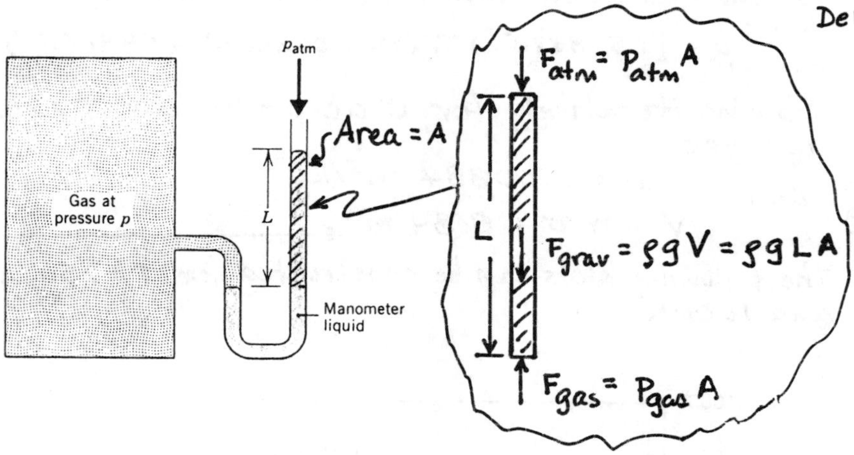

Derivation of Eq. 1.15:

$$\Sigma F(\uparrow) = 0$$

$$P_{gas} A - P_{atm} A - \rho g L A = 0$$

$$\therefore P_{gas} - P_{atm} = \rho g L$$

(a)

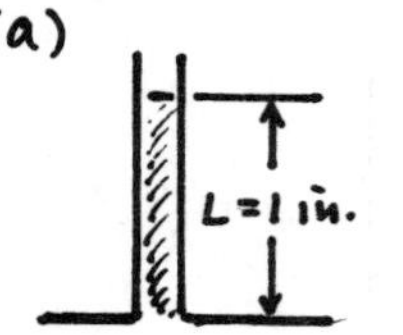

H_2O: $P_{gage} = \rho g L$

$$= \left(6.24 \frac{lb}{ft^3}\right)\left(32.2 \frac{ft}{s^2}\right)\left(\frac{1}{12} ft\right)\left|\frac{1\ lbf}{32.2\ lb \cdot ft/s^2}\right|\left|\frac{1\ ft^2}{144\ in^2}\right|$$

$$= 0.0361\ lbf/in^2 \longleftarrow P_{gage}$$

Hg: $\rho = 13.59\ \rho_{H_2O}$

$$\Rightarrow P_{gage} = 13.59\ (0.0361) = 0.491\ lbf/in^2 \longleftarrow P_{gage}$$

(b)

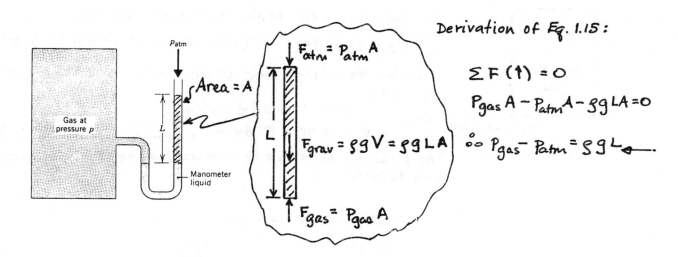

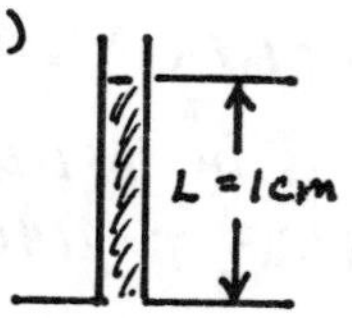

H_2O: $P_{gage} = \rho g L$

$$= \left(1000 \frac{kg}{m^3}\right)\left(9.81 \frac{m}{s^2}\right)(0.01\ m)\left|\frac{1\ N}{1\ kg \cdot m/s^2}\right|\left|\frac{1\ bar}{10^5\ N/m^2}\right|$$

$$= 9.81 \times 10^{-4}\ bars \longleftarrow P_{gage}$$

Hg: $\rho_{Hg} = 13.59\ \rho_{H_2O}$

$$\Rightarrow P_{gage} = 13.59\ (9.81 \times 10^{-4}) = 0.01333\ bars \longleftarrow P_{gage}$$

PROBLEM 1.51

Using Eq. 1.22

$$T(°F) = 1.8\, T(°C) + 32$$

(a) $\underline{\underline{T(°C) = 21}}$

$$T(°F) = (1.8)(21) + 32 = 69.8$$

$$T(°R) = T(°F) + 459.67 = 529.47$$

(b) $\underline{\underline{T(°C) = -17.78}}$

$$T(°F) = (1.8)(-17.78) + 32 = 0$$

$$T(°R) = 0 + 459.67 = 459.67$$

(c) $\underline{\underline{T(°C) = -50}}$

$$T(°F) = (1.8)(-50) + 32 = -58$$

$$T(°R) = -58 + 459.67 = 401.67$$

(d) $\underline{\underline{T(°C) = 300}}$

$$T(°F) = (1.8)(300) + 32 = 572$$

$$T(°R) = 572 + 459.67 = 1031.67$$

(e) $\underline{\underline{T(°C) = 100}}$

$$T(°F) = (1.8)(100) + 32 = 212$$

$$T(°R) = 212 + 459.67 = 671.67$$

(f) $\underline{\underline{T(°C) = -273.15}}$

$$T(°F) = (1.8)(-273.15) + 32 = -459.67$$

$$T(°R) = -459.67 + 459.67 = 0$$

PROBLEM 1.52

Using Eq. 1.22

$$T(°C) = \frac{T(°F)}{1.8} - \frac{32}{1.8}$$

$$= \frac{T(°F)}{1.8} - 17.78$$

(a) $T(°F) = 212$

$$T(°C) = \frac{212}{1.8} - 17.78 = 100$$

$$T(K) = 100 + 273.15 = 373.15$$

(b) $T(°F) = 68$

$$T(°C) = \frac{68}{1.8} - 17.78 = 20$$

$$T(K) = 20 + 273.15 = 293.15$$

(c) $T(°F) = 32$

$$T(°C) = \frac{32}{1.8} - 17.78 = 0$$

$$T(K) = 0 + 273.15 = 273.15$$

(d) $T(°F) = 0$

$$T(°C) = \frac{0}{1.8} - 17.78 = -17.78$$

$$T(K) = -17.78 + 273.15 = 255.37$$

(e) $T(°F) = -40$

$$T(°C) = -\frac{40}{1.8} - 17.78 = -40$$

$$T(K) = -40 + 273.15 = 233.15$$

(f) $T(°F) = -459.67$

$$T(°C) = -\frac{459.67}{1.8} - 17.78 = -273.15$$

$$T(K) = 0$$

PROBLEM 1.53

$$T_2(°C) - T_1(°C) = [T_2(°C) + 273.15] - [T_1(°C) + 273.15] = T_2(K) - T_1(K)$$

$$T_2(°F) - T_1(°F) = [T_2(°F) + 459.67] - [T_1(°F) + 459.67] = T_2(°R) - T_1(°R)$$

PROBLEM 1.54

The expression for resistance is

$$R = R_o \exp\left[\beta\left(\frac{1}{T} - \frac{1}{T_o}\right)\right]$$

where $R_o = 2.2\ \Omega$ and $T_o = 310$ K

Since $R = 0.31\ \Omega$ at $T = 422$ K

$$.31 = (2.2)\exp\left[\beta\left(\frac{1}{422} - \frac{1}{310}\right)\right]$$

Solving for β

$$\beta = \frac{\ln(.31/2.2)}{\left(\frac{1}{422} - \frac{1}{310}\right)} = 2288.9\ \text{K}$$

Thus

$$R = 2.2 \exp\left[2288.9\left(\frac{1}{T} - \frac{1}{310}\right)\right]$$

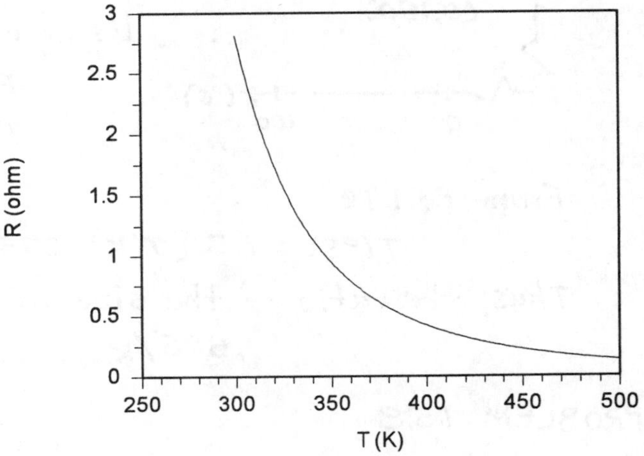

PROBLEM 1.55

GIVEN DATA:

	T (°C)	R (Ω)
Test 1	0	51.39
Test 2	91	51.72

$$R = R_o[1 + \alpha(T - T_o)]$$

Let $T_o = 0°C$

From the data at $T = 0°C$

$$51.39 = R_o[1 + \alpha(0 - 0)] \Rightarrow R_o = 51.39\ \Omega$$

At $T = 91°C$, $R = 51.72$. Thus

$$51.72 = 51.39[1 + \alpha(91 - 0)] \Rightarrow \alpha = 7.0565 \times 10^{-5}\ (°C)^{-1}$$

Finally

$$R = 51.39[1 + (7.0565 \times 10^{-5})T]$$

$$R(T = 50°C) = 51.57\ \Omega \longleftarrow R$$

PROBLEM 1.56

Using Eq. 1.22, $T(°F) = 1.8\,T(°C) + 32$. For $T(°F) = T(°C) = \hat{T}$

$$\hat{T} = 1.8\hat{T} + 32$$

Solving $\hat{T} = -40$ °F or °C $\longleftarrow \hat{T}$

In K: $T(K) = T(°C) + 213.15 = -40 + 273.15 = 233.15 \longleftarrow \hat{T}(K)$

In °R: $T(°R) = T(°F) + 459.67 = -40 + 459.67 = 419.67 \longleftarrow \hat{T}(°R)$

PROBLEM 1.57

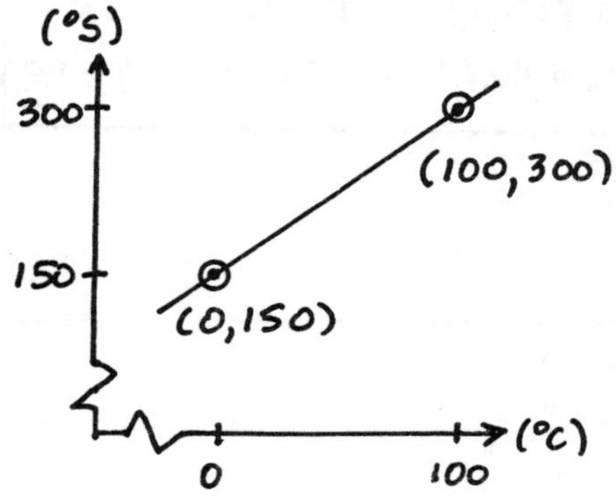

From the data, the relation can be expressed as

$$T(°S) = \left(\frac{300-150}{100-0}\right) T(°C) + 150°C$$
$$= 1.5\ T(°C) + 150$$

or

$$T(°C) = \frac{T(°S) - 150}{1.5}$$

Using this relation

$T = 100°S \rightarrow -33.33\ °C$ ◄———— $T(°C)$

$T = 400°S \rightarrow 166.67\ °C$ ◄———— $T(°C)$

From Eq. 1.19

$$T(°S) = 1.5\,[T(K) - 273.15] + 150$$

Thus, the ratio of the size of the °S to the kelvin is

$1.5\ °S/K$ ◄———— $°S/K$

PROBLEM 1.58

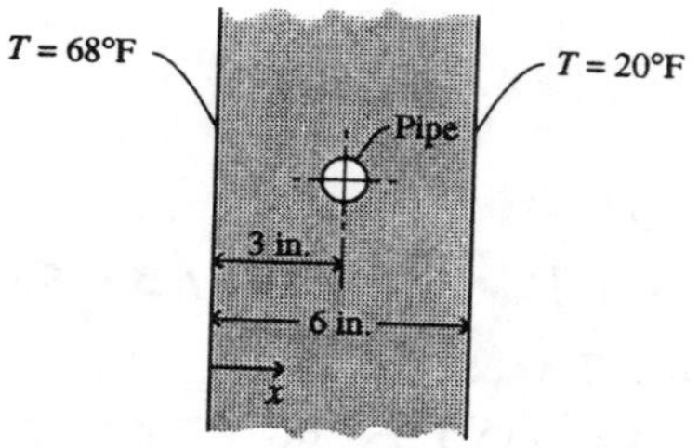

Since the temperature variation is linear

$$T = mx + b$$

where m is the slope and b is the value when $x = 0$.

Slope:

$$m = \left[\frac{20°F - 68°F}{6\ in.}\right] = -8\ °F/in.$$

When $x = 0$, $T = 68°F$

Accordingly, the temperature variation through the wall is

$$T = -(8\ °F/in.)\,x + 68°F$$

Checking the temperature at the pipe: $x = 3\ in.$

$$T = -(8°F/in.)(3\ in.) + 68°F$$
$$= 44°F$$

There is no danger of freezing. ◄————

CHAPTER TWO
ENERGY AND THE FIRST LAW OF THERMODYNAMICS

Chapter 2 - Third and fourth edition problem correspondence.

3rd	4th	3rd	4th	3rd	4th	3rd	4th
2.1	2.1*	---	2.25	2.45	2.48	2.69	---
2.2	2.2	2.20	---	2.46	2.49	---	2.74
2.3	2.3*	2.21	2.26	2.47	---	2.70	2.75
2.4	2.4	2.22	2.27	---	2.50	2.71	2.76
2.5	2.5*	2.23	2.28*	2.48	2.51	2.72	2.77
2.6	---	2.24	2.29	2.49	2.52	2.73	2.78*
---	2.6	2.25	2.30	2.50	2.53	---	2.79
2.7	2.7	2.26	---	2.51	2.54	---	2.80
2.8	---	---	2.31	2.52	2.55*	2.74	---
---	2.8	2.27	2.32	2.53	2.56*	2.75	---
2.9	2.9	2.28	2.33	2.54	---	2.76	---
2.10	2.10	2.29	---	---	2.57	---	2.81
2.11	---	2.30	---	2.55	2.58	---	2.82
---	2.11	---	2.34	2.56	---	2.77	2.83
2.12	2.12	2.31	2.35*	---	2.59	2.78	---
2.13	2.13	2.32	2.36	2.57	2.60*	2.79	---
2.14	2.14	2.33	2.37*	2.58	2.61*	2.80	2.84
2.15	2.15	2.34	---	2.59	2.62*	2.81	2.85
2.16	---	---	2.38	2.60	---	2.82	2.86
---	2.16	2.35	2.39	---	2.63	---	2.87
2.17	2.17*	2.36	---	2.61	2.64*	2.83	---
2.18	2.18	2.37	2.40	2.62	---	2.84	2.88*
2.19	---	2.38	2.41*	---	2.65	---	2.89
---	2.19	2.39	2.42	2.63	2.66		
---	2.20	2.40	2.43	2.64	2.67		
---	2.21	2.41	2.44	2.67	2.68		
---	2.22	2.42	2.45	2.66	2.69		
---	2.23	2.43	---	2.65	2.70*		
---	2.24	---	2.46	---	2.71		
		2.44	2.47*	---	2.72		
				2.68	2.73*		

* Revised

PROBLEM 2.1

KNOWN: An automobile of known mass accelerates from a given velocity to another.

FIND: Determine the initial kinetic energy and the change in kinetic energy.

SCHEMATIC & GIVEN DATA:

$m = 1200$ kg
$V_1 = 50$ km/h
$V_2 = 100$ km/h

ASSUMPTIONS: (1) The automobile is the closed system. (2) The velocities and kinetic energies are relative to the road.

ANALYSIS: The initial kinetic energy (assumption 2) is

$$KE_1 = \frac{1}{2} m V_1^2$$

$$= \frac{1}{2}(1200\,\text{kg})\left(50\ \frac{\text{km}}{\text{h}}\right)^2 \left|\frac{10^3\,\text{m/km}}{3600\,\text{s/h}}\right|^2 \left|\frac{1\,\text{kJ}}{10^3\,\text{N}\cdot\text{m}}\right| \left|\frac{1\,\text{N}}{1\,\text{kg}\cdot\text{m/s}^2}\right|$$

$$= 115.74\ \text{kJ} \qquad \longleftarrow KE_1$$

The change in kinetic energy is

$$KE_2 - KE_1 = \frac{1}{2} m \left[V_2^2 - V_1^2\right]$$

$$= \frac{1}{2}(1200)\left[(100)^2 - (50)^2\right] \left|\frac{10^3}{3600}\right|^2 \left|\frac{1}{10^3}\right|$$

$$= 347\ \text{kJ} \qquad \longleftarrow KE_2 - KE_1$$

PROBLEM 2.2

KNOWN: An object of known weight is located at a specified elevation relative to the surface of the earth.

FIND: Determine gravitational potential energy of the object.

SCHEMATIC & GIVEN DATA:

ASSUMPTIONS: (1) The object is a closed system. (2) The acceleration of gravity is constant.

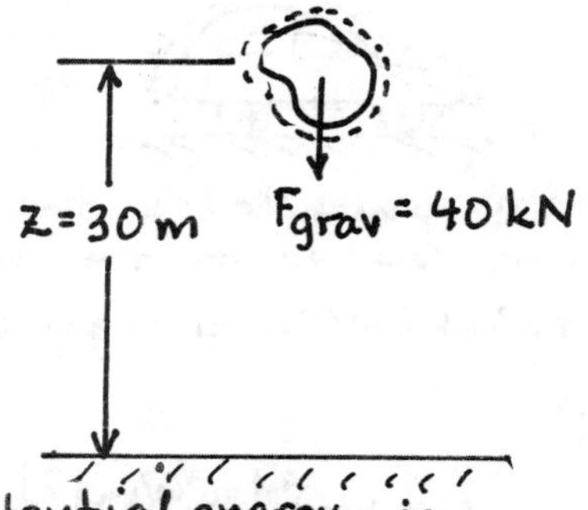

$g = 9.78\ m/s^2$

ANALYSIS: The gravitational potential energy is

$$PE = mgz$$

with $F_{grav} = mg$ we get

$$PE = F_{grav} \cdot z$$

$$= (40\ kN)(30\ m)\left|\frac{1\ kJ}{1\ kN \cdot m}\right| = 1200\ kJ \longleftarrow \quad PE$$

PROBLEM 2.3

KNOWN: An object of known mass undergoes a specified change in its kinetic energy while its potential energy increases.

FIND: Determine the final velocity.

SCHEMATIC & GIVEN DATA:

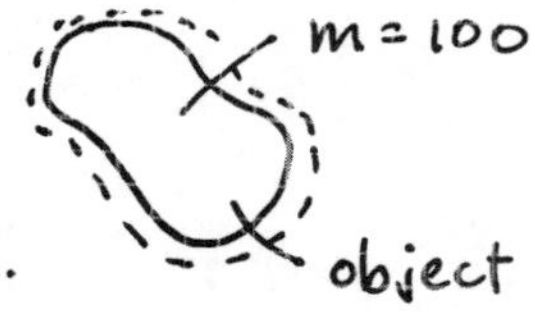

$V_1 = 50\ ft/s$

$\Delta KE = -1000\ ft \cdot lbf$

ASSUMPTION: The object is a closed system.

ANALYSIS: The change in kinetic energy is

① $$\Delta KE = \tfrac{1}{2} m (V_2^2 - V_1^2)$$

Thus, solving for V_2 and inserting values and a unit conversion factor

$$V_2 = \sqrt{\frac{2\,\Delta KE}{m} + V_1^2}$$

$$= \sqrt{\frac{2(-500\ ft \cdot lbf)}{(100\ lb)}\left|\frac{32.2\ lb \cdot ft/s^2}{1\ lbf}\right| + (50\ ft/s)^2}$$

$$= 46.67\ ft/s \longleftarrow \quad V_2$$

---

1. The analysis makes no use of the information related to potential energy.

PROBLEM 2.4

KNOWN: A body of known volume and density experiences a given decrease in gravitational potential energy.

FIND: Determine the change in elevation.

SCHEMATIC & GIVEN DATA:

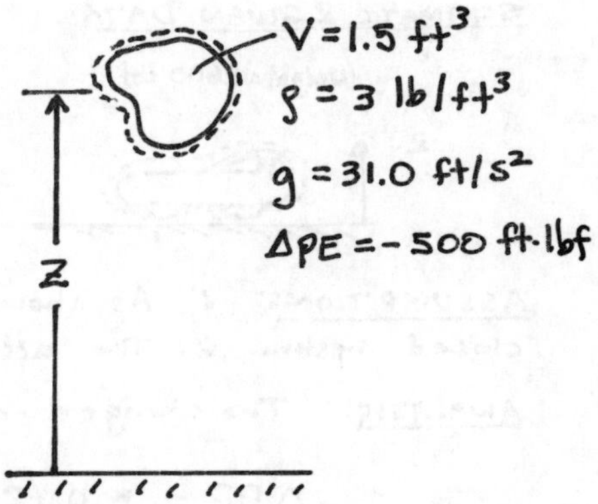

ASSUMPTIONS: (1) The body is a closed system. (2) The acceleration of gravity is constant. (3) The density of the body is uniform throughout.

ANALYSIS: Based on assumption (3)

$$m = \rho V = (3\ \text{lb/ft}^3)(1.5\ \text{ft}^3) = 4.5\ \text{lb}$$

The change in potential energy and the elevation are related by

$$\Delta PE = mg\Delta z$$

Thus, solving for Δz

$$\Delta z = \frac{\Delta PE}{mg} = \frac{(-500\ \text{ft}\cdot\text{lbf})}{(4.5\ \text{lb})(31.0\ \text{ft/s}^2)}\left|\frac{32.2\ \text{lb}\cdot\text{ft/s}^2}{1\ \text{lbf}}\right|$$

① $$= -115.4\ \text{ft} \longleftarrow \Delta z$$

1. The negative sign denotes a decrease in elevation.

PROBLEM 2.5

KNOWN: An automobile of known weight travels from sea level to a known elevation.

FIND: Determine the change in potential energy.

SCHEMATIC & GIVEN DATA:

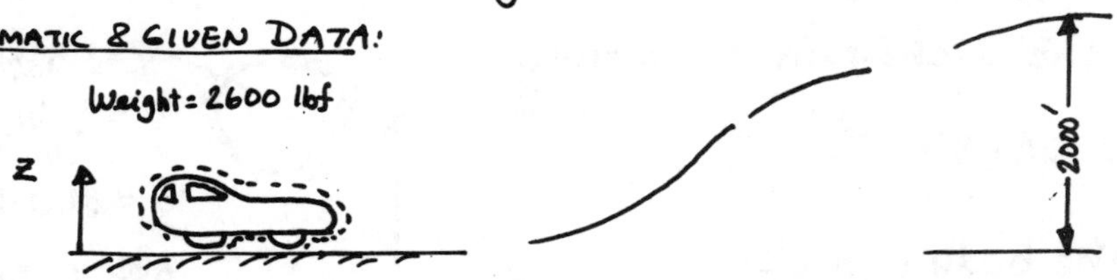

ASSUMPTIONS: 1. As shown in the schematic, the automobile is the closed system. 2. The acceleration of gravity is constant.

ANALYSIS: The change in potential energy is

$$\Delta PE = mg\,(z_2 - z_1)$$

The quantity mg is recognized as the vehicle weight. Thus, inserting known values

$$\Delta PE = (2600\ \text{lbf})(2000\ \text{ft}) = 5.2\times10^6\ \text{ft}\cdot\text{lbf} \longleftarrow \Delta PE$$

PROBLEM 2.6

KNOWN: An object of known mass decelerates from a given initial velocity to a known final velocity.

FIND: Determine the change in kinetic energy of the object.

SCHEMATIC & GIVEN DATA:

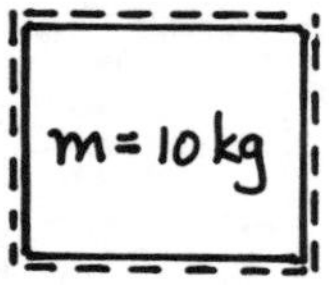

$V_1 = 500$ m/s

$V_2 = 100$ m/s

ASSUMPTION: The object is a closed system.

ANALYSIS: The change in kinetic energy is

$$\Delta KE = \tfrac{1}{2}m\left[V_2^2 - V_1^2\right]$$

Inserting known values and converting units

$$\Delta KE = \tfrac{1}{2}(10\ \text{kg})\left[100^2 - 500^2\right]\frac{\text{m}^2}{\text{s}^2}\left|\frac{1\ \text{N}}{1\ \text{kg}\cdot\text{m/s}^2}\right|\left|\frac{1\ \text{kJ}}{10^3\ \text{N}\cdot\text{m}}\right|$$

① $$= -1200\ \text{kJ} \longleftarrow \Delta KE$$

---

1. The negative sign denotes a <u>decrease</u> in kinetic energy, as expected.

PROBLEM 2.7

KNOWN: An airplane of known mass flies with a given velocity at a given altitude, both measured relative to the surface of the earth.

FIND: Calculate (a) the kinetic and potential energies of the airplane, and (b) the final velocity for a given change in kinetic energy.

SCHEMATIC & GIVEN DATA:

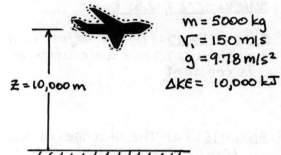

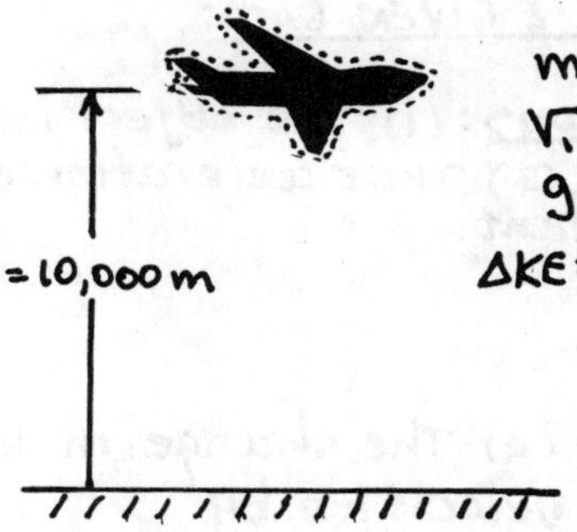

ASSUMPTIONS: (1) The airplane is a closed system. (2) The acceleration of gravity is constant. (3) In part (b), there is no change in elevation.

ANALYSIS: (a) The kinetic energy is

$$KE_1 = \frac{1}{2} m V_1^2 = \frac{1}{2}(5000\ \text{kg})(150^2)\frac{\text{m}^2}{\text{s}^2}\left|\frac{1\ \text{N}}{1\ \text{kg}\cdot\text{m/s}^2}\right|\left|\frac{1\ \text{kJ}}{10^3\ \text{N}\cdot\text{m}}\right|$$

$$= 56{,}520\ \text{kJ} \quad \longleftarrow KE_1$$

The gravitational potential energy is

$$PE = mgz = (5000\ \text{kg})(9.78\ \text{m/s}^2)(10{,}000\ \text{m})\left|\frac{1\ \text{N}}{1\ \text{kg}\cdot\text{m/s}^2}\right|\left|\frac{1\ \text{kJ}}{10^3\ \text{N}\cdot\text{m}}\right|$$

$$= 489{,}000\ \text{kJ} \quad \longleftarrow PE$$

(b) The change in kinetic energy is related to the initial and final velocities by

$$\Delta KE = \frac{1}{2} m (V_2^2 - V_1^2)$$

Thus, solving for V_2^2 and converting units

$$V_2^2 = \frac{2\,\Delta KE}{m} + V_1^2$$

$$= \frac{2(10{,}000\ \text{kJ})}{(5000\ \text{kg})}\left|\frac{1\ \text{kg}\cdot\text{m/s}^2}{1\ \text{N}}\right|\left|\frac{10^3\ \text{N}\cdot\text{m}}{1\ \text{kJ}}\right| + (150^2)\frac{\text{m}^2}{\text{s}^2}$$

$$= 26{,}500\ \text{m}^2/\text{s}^2$$

or

① $$V_2 = 162.8\ \text{m/s} \quad \longleftarrow V_2$$

1. In part (b), the velocity increases, as expected.

PROBLEM 2.8

KNOWN: An object of known mass moves with a given velocity.

FIND: Determine (a) the final velocity for a given change in kinetic energy, and (b) the change in elevation for a given change in potential energy.

SCHEMATIC & GIVEN DATA:

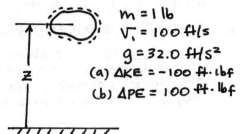

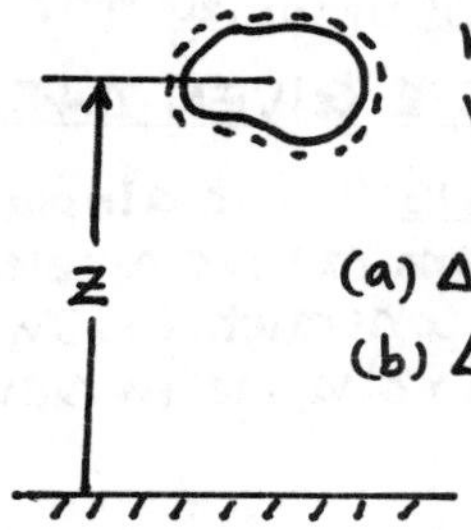

$m = 1$ lb
$V_1 = 100$ ft/s
$g = 32.0$ ft/s²
(a) $\Delta KE = -100$ ft·lbf
(b) $\Delta PE = 100$ ft·lbf

ASSUMPTIONS: (1) The object is a closed system. (2) The acceleration of gravity is constant.

ANALYSIS: (a) The change in kinetic energy is related to the initial and final velocities by

$$\Delta KE = \tfrac{1}{2} m \left[V_2^2 - V_1^2\right]$$

Thus, solving for the final velocity V_2

$$V_2 = \sqrt{\frac{2\,\Delta KE}{m} + V_1^2}$$

Inserting values and converting units accordingly

$$V_2 = \sqrt{\frac{(2)(-100\ \text{ft}\cdot\text{lbf})}{(1\ \text{lb})}\left|\frac{32.2\ \text{lb}\cdot\text{ft/s}^2}{1\ \text{lbf}}\right| + 100^2\ \frac{\text{ft}^2}{\text{s}^2}}$$

① $= 59.67$ ft/s ← V_2

(b) The change in potential energy is related to the change in elevation by

$$\Delta PE = mg\,\Delta z$$

Thus, the change in elevation is

$$\Delta z = \frac{\Delta PE}{mg}$$

$$= \frac{(100\ \text{ft}\cdot\text{lbf})}{(1\ \text{lb})(32.0\ \text{ft/s}^2)}\left|\frac{32.2\ \text{lb}\cdot\text{ft/s}^2}{1\ \text{lbf}}\right|$$

② $= 100.6$ ft ← Δz

1. The velocity decreases, as expected.
2. The elevation increases, as expected.

PROBLEM 2.9

KNOWN: An object of known mass accelerates from a given initial velocity to a given final velocity due to the action of a resultant force.

FIND: Determine the work done by the resultant force.

SCHEMATIC & GIVEN DATA:

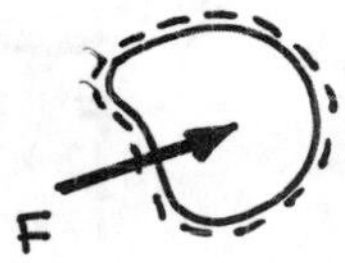

$m = 50\,kg$

$V_1 = 20\,m/s$

$V_2 = 50\,m/s$

ASSUMPTIONS: (1) The object is a closed system. (2) The resultant force is the only interaction between the object an its surroundings.

ANALYSIS: By assumption(2), the work of the resultant force must equal the change in kinetic energy. Thus, using Eq. 2.6

① $$\text{work} = \frac{1}{2} m (V_2^2 - V_1^2)$$

$$= \frac{1}{2}(50\,kg)(50^2 - 20^2)\frac{m^2}{s^2}\left|\frac{1\,N}{1\,kg\cdot m/s^2}\right|\left|\frac{1\,kJ}{10^3\,N\cdot m}\right|$$

$$= 52.5\,kJ \quad \leftarrow \text{work}$$

1. The increase in kinetic energy of the object is the result of energy transferred to it by the work of the resultant force.

PROBLEM 2.10

KNOWN: An object of known mass undergoes a change of kinetic energy due to the action of a resultant force. The final velocity and the work done by the force are given.

FIND: Determine the initial velocity.

SCHEMATIC & GIVEN DATA:

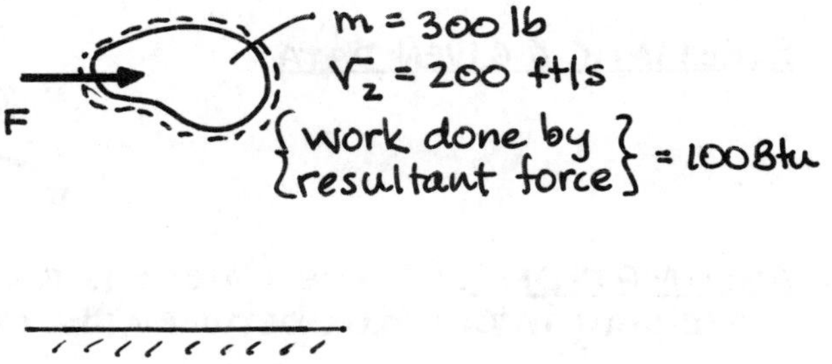

ASSUMPTIONS: (1) The object is a closed system. (2) There is no change in elevation. (3) The resultant force is the only interaction between the object and its surroundings.

ANALYSIS: By assumption (3), the work of the resultant force must equal the change in kinetic energy. Thus, using Eq. 2.6

$$\text{work} = \frac{1}{2} m (V_2^2 - V_1^2)$$

Solving for V_1^2, inserting values, and converting units

$$V_1^2 = -\frac{2W}{m} + V_2^2$$

$$= -\frac{2(100\ \text{Btu})}{(300\ \text{lb})}\left|\frac{778\ \text{ft}\cdot\text{lbf}}{1\ \text{Btu}}\right|\left|\frac{32.2\ \text{lb}\cdot\text{ft/s}^2}{1\ \text{lbf}}\right| + 200^2\ \frac{\text{ft}^2}{\text{s}^2}$$

$$= 23{,}300\ \text{ft}^2/\text{s}^2$$

or

① $V_1 = 152.6$ ft/s ← V_1

1. The increase in velocity reflects the increase in kinetic energy of the object as a result of energy transferred to it by the work of the resultant force.

PROBLEM 2.11

KNOWN: Data are provided for a disk-shaped flywheel.

FIND: (a) Obtain appropriate expressions for the moment of inertia and the kinetic energy. (b) Using given data, determine the kinetic energy and mass for a steel flywheel. (c) Using results from part (b), determine the radius and mass of an aluminum flywheel.

SCHEMATIC & GIVEN DATA:

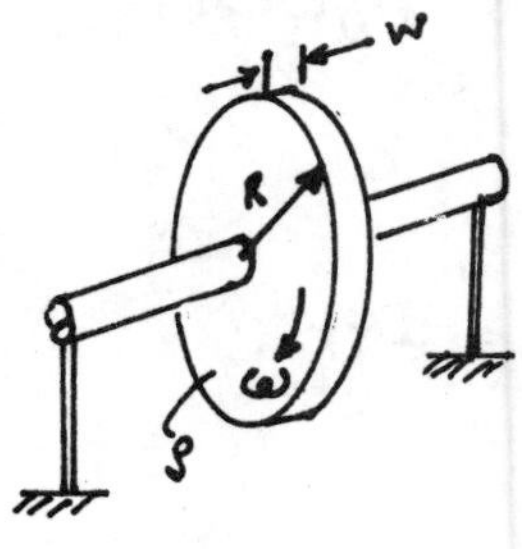

Steel flywheel:

$\omega = 3000$ RPM

$R = 0.38$ m

$w = 0.025$ m

Aluminum flywheel:

$\omega = 3000$ RPM

$w = 0.025$ m

ASSUMPTIONS: 1. The flywheel is the closed system. 2. Motion is relative to the flywheel support structure.

ANALYSIS:

(a) Evaluating the moment of inertia

$$I = \int_{vol} \rho r^2 dV$$

For the disk, $dV = (2\pi r dr)w$. Thus, since ρ is constant

$$I = \rho(2\pi)w\int_0^R r^3 dr = \rho\pi w R^4/2 \longleftarrow I$$

The kinetic energy is

$$KE = \int_{vol} \left(\tfrac{1}{2}\rho V^2\right) dV$$

and $V = r\omega$, so

$$KE = \int_0^R \left(\tfrac{1}{2}\rho r^2\omega^2\right)(2\pi r dr)w$$

$$= \tfrac{1}{2}\rho\omega^2(2\pi)w\int_0^R r^3 dr$$

$$= \tfrac{1}{2}\underbrace{\left(\rho\pi\frac{R^4}{2}w\right)}_{I}\omega^2$$

$$= \tfrac{1}{2}I\omega^2 \longleftarrow KE$$

(b) From Table A-19, the density of steel is $\rho = 8060\ \text{kg/m}^3$. Thus, the mass is

$$m = \rho V = \rho[w\cdot\pi R^2]$$

$$= \left(8060\ \frac{\text{kg}}{\text{m}^3}\right)\left[(0.025\text{m})\cdot\pi\cdot(0.38\text{m})^2\right] = 91.41\ \text{kg} \longleftarrow m$$

Using the result of part (a), $KE = \frac{1}{2}I\omega^2$, where

$$I = \pi\rho w\frac{R^4}{2} = \frac{\pi}{2}\left(8060\ \frac{\text{kg}}{\text{m}^3}\right)(0.025\text{m})(0.38\text{m})^4 = 6.6\ \text{kg}\cdot\text{m}^2$$

PROBLEM 2.11 (Contd.)

Accordingly,

$$KE = \frac{1}{2} I \omega^2 = \frac{1}{2}(6.6\ kg \cdot m^2)\left(3000\ \frac{Rev}{min}\left|\frac{2\pi\ rad}{REV}\right|\left|\frac{1\ min}{60\ s}\right|\right)^2 \left|\frac{1\ N}{1\ kg \cdot m/s^2}\right|$$

$$= 32.57 \times 10^4\ N \cdot m$$ ← KE

(c) If ω, W, and KE are the same for the aluminum flywheel as for the steel flywheel

$$(KE)_{AL} = (KE)_{ST}$$

$$\left(\frac{1}{2} I \omega^2\right)_{AL} = \left(\frac{1}{2} I \omega^2\right)_{ST} \Rightarrow I_{AL} = I_{ST}$$

or

$$\left(\pi \rho W \frac{R^4}{2}\right)_{AL} = \left(\pi \rho W \frac{R^4}{2}\right)_{ST}$$

$$\Rightarrow \quad (\rho R^4)_{AL} = (\rho R^4)_{ST}$$

$$R_{AL} = \left(\frac{\rho_{ST}}{\rho_{AL}}\right)^{1/4} R_{ST}$$

with ρ_{AL} from Table A-19, $\rho_{AL} = 2700\ kg/m^3$

$$R_{AL} = \left(\frac{8060}{2700}\right)^{\frac{1}{4}}(0.38\ m)$$

$$= 0.5\ m$$ ← R_{AL}

Then, the mass of the aluminum flywheel is

$$m = \rho V = \rho\left[W \pi R^2\right]$$

$$= \left(2700\ \frac{kg}{m^3}\right)(0.025\ m)(\pi)(0.5\ m)^2$$

$$= 53.01\ kg$$ ← m

PROBLEM 2.12

KNOWN: An object of known mass moves along a straight line with a known velocity.

FIND: Determine the rotational speed of a flywheel whose rotational kinetic energy is equal in magnitude to the object's linear kinetic energy.

SCHEMATIC & GIVEN DATA:

$KE_{obj} = KE_{fw}$

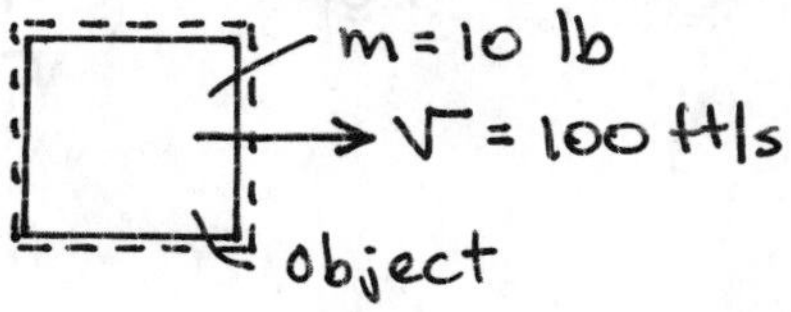

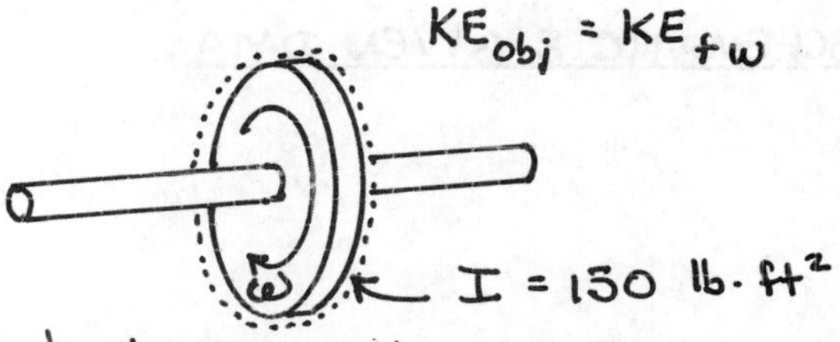

ASSUMPTION: (1) The object and the flywheel are both closed systems.

ANALYSIS: The kinetic energy of the object is

$$KE_{obj} = \frac{1}{2} m V^2$$

$$= \frac{1}{2}(10\ \text{lb})(100^2)\ \frac{\text{ft}^2}{\text{s}^2} \left| \frac{1\ \text{lbf}}{32.2\ \text{lb}\cdot\text{ft/s}^2} \right| = 1553\ \text{ft}\cdot\text{lbf}$$

For the flywheel (See the solution to Problem 2.11)

$$KE_{fw} = \frac{1}{2} I \omega^2$$

or

$$\omega = \sqrt{\frac{2\, KE_{fw}}{I}} = \sqrt{\frac{2(1553\ \text{ft}\cdot\text{lbf})}{(150\ \text{lb}\cdot\text{ft}^2)} \left| \frac{32.2\ \text{lb}\cdot\text{ft/s}^2}{1\ \text{lbf}} \right|}$$

$$= 25.82\ \frac{1}{\text{s}}$$

In terms of RPM

$$\omega = \left(25.82\ \frac{1}{\text{s}}\right) \left| \frac{\text{rev}}{2\pi} \right| \left| \frac{60\ \text{s}}{1\ \text{min}} \right| = 246.6\ \text{rev/min} \longleftarrow \omega$$

PROBLEM 2.13

KNOWN: Two objects fall freely under the infuence of gravity from rest and the same initial elevation.

FIND: Show that the magnitudes of the velocities are equal at the moment just before they stike the earth.

SCHEMATIC & GIVEN DATA:

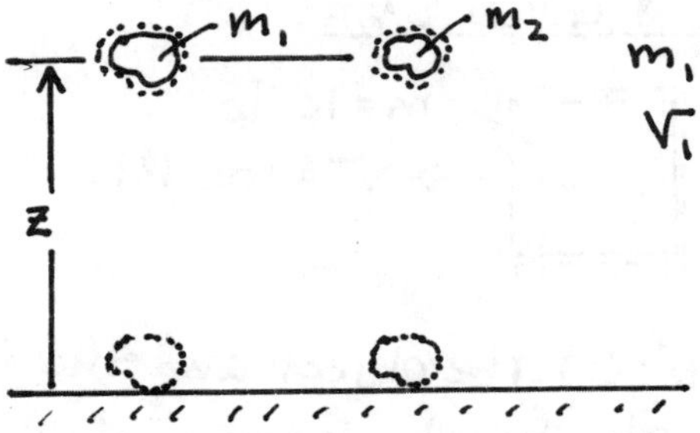

$m_1 \neq m_2$

$V_1 = 0$

ASSUMPTIONS: (1) An object in free fall is a closed system. (2) The acceleration of gravity is constant. (3) There is no affect of air resistance. (4) The only force acting is that due to gravity.

ANALYSIS: For an object falling freely under the infuence of gravity, Eq 2.11 applies

$$\frac{1}{2} m (V_2^2 - V_1^2) + mg(z_2 - z_1) = 0$$

For $V_1 = 0$ and $z_2 = 0$

$$\frac{1}{2} m V_2^2 = mgz_1$$

Thus

$$V_2 = \sqrt{2gz_1}$$

Since the final velocity doesn't depend on mass, both objects will have identical velocities at the moment just before they strike the earth.

PROBLEM 2.14

KNOWN: An object of known mass is projected upward from the surface of the earth with a known initial velocity. The only force acting on the object is the force of gravity.

FIND: Plot the velocity of the object versus elevation and determine the elevation when its velocity reaches zero.

SCHEMATIC & GIVEN DATA:

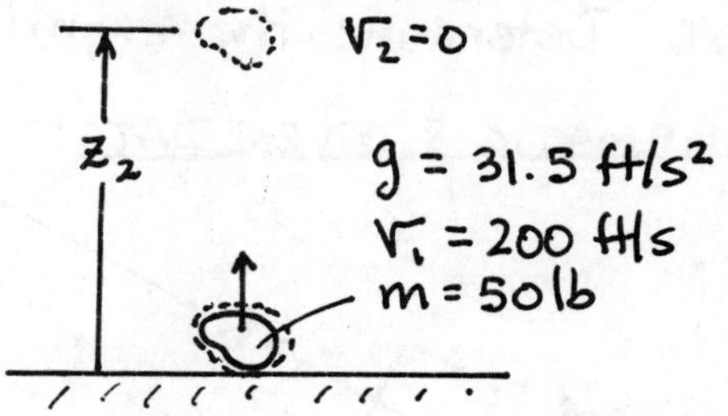

ASSUMPTIONS: (1) The object is a closed system. (2) The acceleration of gravity is constant. (3) The only force acting on the object is the force of gravity.

ANALYSIS: Since the only force acting on the body is the force of gravity, Eq. 2.10 applies. Thus, the velocity and elevation are related to the initial condition by

$$\frac{1}{2} m V^2 + m g z = \frac{1}{2} m V_1^2 + m g z_1^{\,\nearrow 0}$$

Thus, solving for V

$$V = \sqrt{V_1^2 - 2gz} \quad \longleftarrow \text{Expression for } V \text{ in terms of } z$$

When $V_2 = 0$, z_2 is

$$z_2 = \frac{V_1^2}{2g} = \frac{200^2 \text{ ft}^2/\text{s}^2}{2(31.5 \text{ ft/s}^2)} = 634.92 \text{ ft} \quad \longleftarrow z_2$$

Plotting the above relationship

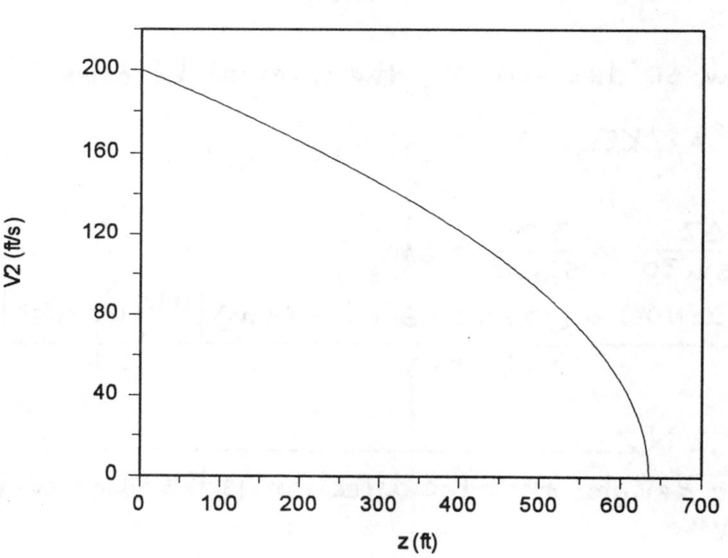

PROBLEM 2.15

KNOWN: A block of known mass moves along an inclined surface. The change in elevation and the change in kinetic energy of the block are given. The block is acted upon by a force parallel to the incline and the force of gravity.

FIND: Determine the magnitude and direction of applied force.

SCHEMATIC & GIVEN DATA:

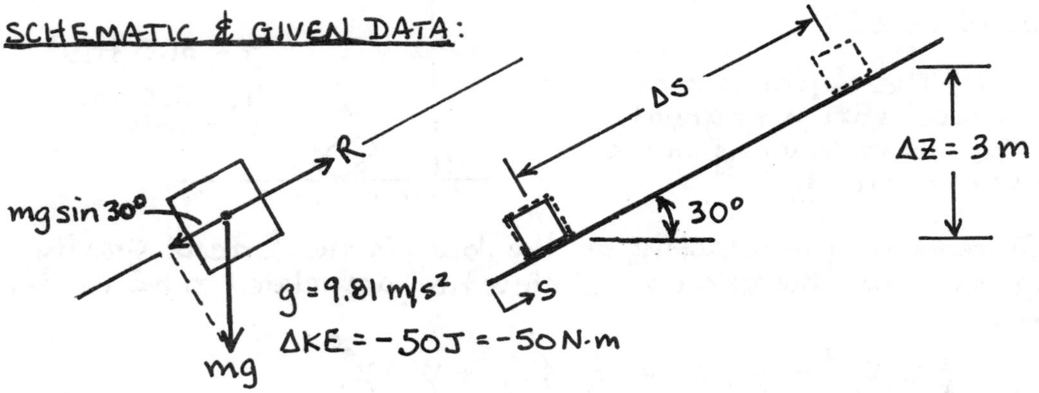

ASSUMPTIONS: (1) The block is a closed system. (2) The acceleration of gravity is constant. (3) The force R is constant. (4) There is no frictional force between the block and the incline.

ANALYSIS: Beginning with Eq. 2.6

$$\int \underline{F} \cdot d\underline{s} = \Delta KE$$

From the free body diagram, the dot product $\underline{F} \cdot d\underline{s}$ can be expressed as

$$\underline{F} \cdot d\underline{s} = (R - mg \sin 30°)\,ds$$

so

$$\int_{s_1}^{s_2} (R - mg \sin 30°)\,ds = \Delta KE$$

Noting that $mg \sin 30° ds = mg \Delta z$, the integral becomes

$$R\,\Delta s = \Delta KE + mg\Delta z$$

Evaluating Δs

$$\Delta s = \frac{\Delta z}{\sin 30°} = \frac{3\text{ m}}{\sin 30°} = 6\text{ m}$$

Thus

$$R = \frac{(-50\text{ N·m}) + (10\text{ kg})(9.81\text{ m/s}^2)(3\text{ m})\left|\frac{1\text{ N}}{1\text{ kg·m/s}^2}\right|}{(6\text{ m})}$$

$$= 40.72\text{ N} \quad \leftarrow R$$

The positive value denotes that the direction is the same as indicated on the above figure.

PROBLEM 2.16

KNOWN: Beginning from rest, and object of known mass slides down an inclined plane. The length of the ramp is given.

FIND: Determine the velocity of the object at the bottom of the ramp.

SCHEMATIC & GIVEN DATA:

$m = 20 kg$

$g = 9.81\ m/s^2$

$V_1 = 0$

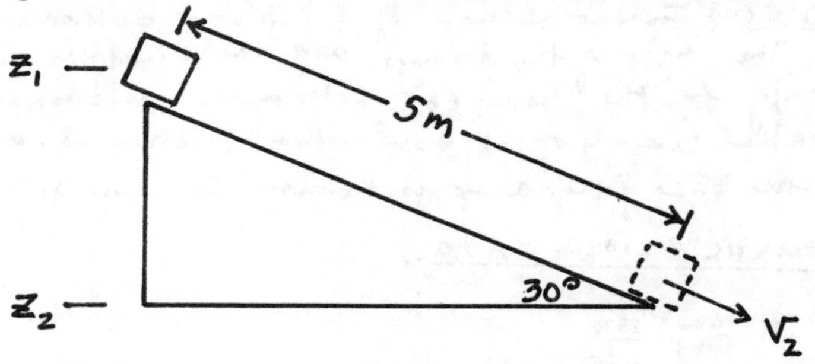

ASSUMPTIONS: (1) The mass is a closed system. (2) here is no friction between the mass and the ramp, and air resistance is negligible. (3) The acceleration of gravity is constant.

ANALYSIS: By assumption (2), the only force acting on the system is the force of gravity. Thus, Eq. 2.11 applies

① $$\frac{1}{2} m (V_2^2 - V_1^{2\,0}) + m g (z_2 - z_1) = 0$$

Solving for V_2

$$V_2 = \sqrt{2g(z_1 - z_2)}$$

From trigonometric relationships

$$z_1 - z_2 = (5\ m) \sin 30°$$

Thus

$$V_2 = \sqrt{2(9.81\ m/s^2)(5\ m) \sin 30°}$$

$$= 7.00\ m/s$$ ◄ V_2

---

1. Even though the object travels along an inclined path, the vertical distance appears in this expression.

PROBLEM 2.17

KNOWN: A box slides down a ramp. The mass of the box and its velocity at the top of the ramp are known. The ramp geometry is also specified.

FIND: (a) In the absence of friction, determine the velocity of the box at the base of the ramp and the changes in kinetic and potential energy for the box. (b) Determine the changes in kinetic and potential energy of the box when friction is acting and the velocity at the base of the ramp is known. Compare with the results of part (a).

SCHEMATIC & GIVEN DATA:

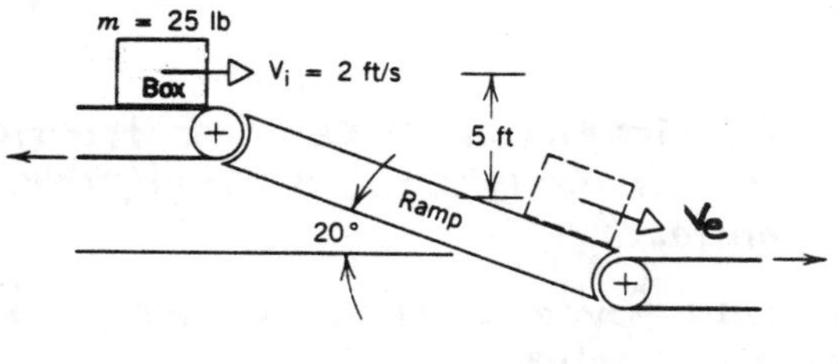

ASSUMPTIONS: 1. The box is the closed system. 2. In part (a), friction is negligible. 3. The acceleration of gravity is 32.0 ft/s².

ANALYSIS: (a) In the absence of any resultant force acting on the system, including friction, Eq. 2.9 reduces to

$$\frac{1}{2} m (V_e^2 - V_i^2) + mg(z_e - z_i) = 0 \qquad (*)$$

Solving

$$V_e = \sqrt{V_i^2 + 2g(z_i - z_e)}$$

$$= \sqrt{\left(2\frac{ft}{s}\right)^2 + 2\left(32\frac{ft}{s^2}\right)(5 ft)} = 18 \text{ ft/s} \longleftarrow V_e$$

As shown by Eq (*), the kinetic and potential energy changes have the same magnitude but are opposite in sign. Thus

$$\Delta PE = mg(z_e - z_i) = (25 lb)\left(32.0 \frac{ft}{s^2}\right)(-5 ft)\left|\frac{1 lbf}{32.2 lb \cdot ft/s^2}\right|$$

$$= -124.2 \text{ ft} \cdot \text{lbf} \longleftarrow \Delta PE$$

and $\Delta KE = +124.2 \text{ ft} \cdot \text{lbf}$. ⟵ ΔKE

(b) Whether there is friction or not, the potential energy change in this case is the same as determined in part (a): −124.2 ft·lbf. If V_e = 9 ft/s, the change in kinetic energy is smaller in magnitude:

$$\Delta KE = \frac{1}{2} m [V_e^2 - V_i^2] = \frac{1}{2}(25 lb)\left[(9 ft/s)^2 - (2 ft/s)^2\right]\left|\frac{1 lbf}{32.2 lb \cdot ft/s^2}\right| = 29.9 \text{ ft} \cdot \text{lbf} \longleftarrow \Delta KE$$

Referring again to Eq. 2.9, the decrease in potential energy in this case can be accounted for in terms of the increase in kinetic energy of the box and work to overcome friction.

PROBLEM 2.18

KNOWN: A system of known mass and a given initial velocity experiences a constant deceleration due to the action of a resultant force and comes to rest.

FIND: Determine the length of time the force is applied and the work.

SCHEMATIC & GIVEN DATA:

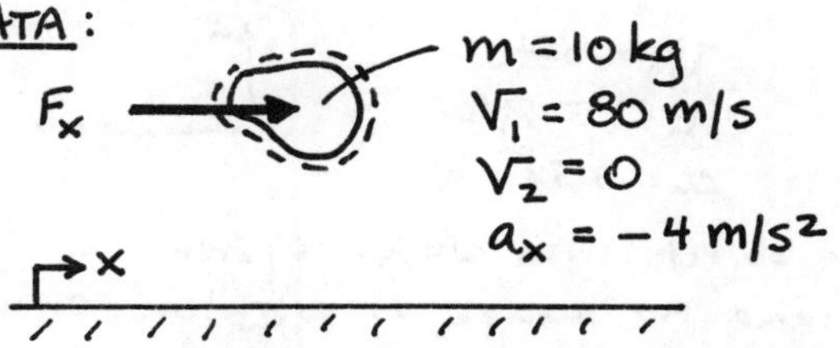

ASSUMPTIONS: (1) The system is closed. (2) The horizontal deceleration is constant.

ANALYSIS: To find the time, use the fact that the acceleration is constant, as follows

$$a_x = \frac{dV}{dt} \Rightarrow dV = a_x\, dt$$

$$\int_{V_1}^{V_2} dV = \int_{t_1}^{t_2} a_x\, dt$$

or

$$\cancel{V_2}^{\,0} - V_1 = a_x\, \Delta t$$

Thus

$$\Delta t = \frac{-V_1}{a_x} = \frac{(-80\ m/s)}{(-4\ m/s^2)} = 20\ s \longleftarrow \quad \Delta t$$

The work of the force F_x is found from Eq. 2.6

$$\text{Work} = \frac{1}{2} m (\cancel{V_2^2}^{\,0} - V_1^2)$$

$$= -\frac{1}{2}(10\ kg)(80^2)\frac{m^2}{s^2}\left|\frac{1\ N}{1\ kg\cdot m/s^2}\right|\left|\frac{1\ kJ}{10^3\ N\cdot m}\right|$$

① $$= -32\ kJ \longleftarrow \quad \text{work}$$

1. The negative denotes energy transfer out of the system.

PROBLEM 2.19

KNOWN: An object of known mass, acted on by a resultant force for a specified time, experiences a constant acceleration from a known initial velocity.

FIND: Determine the work of the resultant force.

SCHEMATIC & GIVEN DATA:

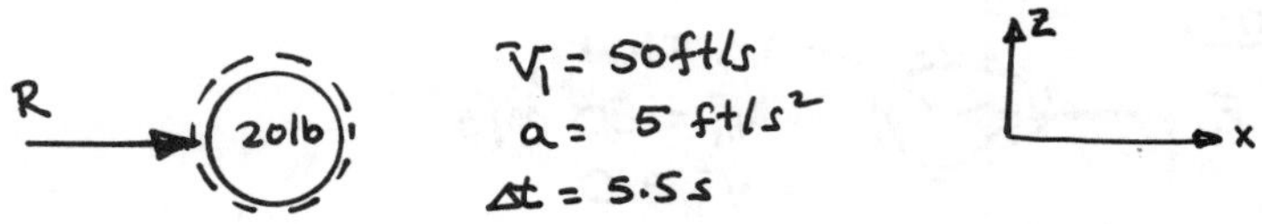

$V_1 = 50 \text{ ft/s}$

$a = 5 \text{ ft/s}^2$

$\Delta t = 5.5 \text{ s}$

ASSUMPTIONS: 1. The object is the closed system. 2. Motion is horizontal, so the system experiences no change in potential energy. 3. The horizontal acceleration is constant.

ANALYSIS: Applying Eq. 2.9 with assumption 2, the work of the force R is given by

$$\text{Work} = \frac{1}{2} m (V_2^2 - V_1^2)$$

The final velocity V_2 can be determined using assumption 3:

$$\frac{dV}{dt} = a \quad \text{(constant)}$$

Thus

$$dV = a\,dt \Rightarrow V_2 - V_1 = a(t_2 - t_1)$$

or

$$V_2 = V_1 + a(t_2 - t_1)$$

$$= 50\frac{\text{ft}}{\text{s}} + 5\frac{\text{ft}}{\text{s}^2}(5.5\text{ s}) = 77.5 \text{ ft/s}$$

Then

$$W = \frac{1}{2}(20 \text{ lb})\left[\left(77.5\frac{\text{ft}}{\text{s}}\right)^2 - \left(50\frac{\text{ft}}{\text{s}}\right)^2\right]\left|\frac{1 \text{ lbf}}{32.2 \text{ lb}\cdot\text{ft/s}^2}\right|$$

$$= 1089 \text{ ft}\cdot\text{lbf}$$

W ←

PROBLEM 2.20

KNOWN: The drag force on a truck is known as a function of velocity and other parameters.

FIND: Determine the power required by the truck to overcome drag.

SCHEMATIC & GIVEN DATA:

$C_d = 0.65$
$A = 10\ m^2$

$\rho = 1.1\ kg/m^3$

ASSUMPTION: The truck is the system.

ANALYSIS: Applying Eq. 2.13, the power required to overcome drag is

$$\dot{W}_d = F_d \cdot V = \left(\tfrac{1}{2} C_d A \rho V^2\right) V$$

$$= \tfrac{1}{2} C_d A \rho V^3$$

$$= \tfrac{1}{2}(0.65)(10\ m^2)\left(1.1\ \frac{kg}{m^3}\right)\left[\left(110\ \frac{km}{h}\right)\left|\frac{10^3\ m}{km}\right|\left|\frac{1h}{3600s}\right|\right]^3 \left|\frac{1\ N}{1\ kg \cdot m/s^2}\right|\left|\frac{1\ kJ}{10^3 N \cdot m}\right|$$

$$= 102\ \frac{kJ}{s}\left|\frac{1\ kW}{1\ kJ/s}\right|$$

$$= 102\ kW \longleftarrow \dot{W}_d$$

PROBLEM 2.21

KNOWN: The force associated with the rolling resistance of the tires of a truck is known as a function of the truck weight.

FIND: Determine the power required by the truck to overcome rolling resistance.

SCHEMATIC & GIVEN DATA:

$f = 0.0069$

ASSUMPTION: The truck is the system.

ANALYSIS: Applying Eq. 2.13, the power required to overcome rolling resistance is

$$\dot{W}_r = F_r \cdot V = (f\mathcal{W})V$$

$$= (0.0069)\left(322.5\ kN\left|\frac{10^3 N}{1 kN}\right|\right)\left(110\ \frac{km}{h}\left|\frac{10^3 m}{1 km}\right|\left|\frac{1\ h}{3600 s}\right|\right)\left|\frac{1\ kJ}{10^3 N \cdot m}\right|\left|\frac{1\ kW}{1\ kJ/s}\right|$$

$$= 68\ kW \longleftarrow \dot{W}_r$$

PROBLEM 2.22

KNOWN: The drag force and the force associated with rolling resistance are known as functions of variables associated with a vehicle in motion.

FIND: (a) Determine the power required to overcome drag and rolling resistance when the vehicle is moving at 55 mi/h. (b) Plot the quantities of part (a) and their sum versus vehicle velocity ranging from 0 to 75 mi/h. Discuss the implication for vehicle fuel economy.

$\mathcal{W} = 3550\ \text{lbf}$
$A = 23.3\ \text{ft}^2$
$C_d = 0.34$
$f = 0.02$

$\rho = 0.08\ \text{lb/ft}^3$

ASSUMPTIONS: The vehicle is the system.

ANALYSIS: Applying Eq. 2.13, the power required to overcome drag is

$$\dot{W}_d = F_d \cdot V = \left(\tfrac{1}{2} C_d A \rho V^2\right) V = \tfrac{1}{2} C_d A \rho V^3$$

$$= \tfrac{1}{2}(0.34)(23.3\ \text{ft}^2)\left(0.08\ \tfrac{\text{lb}}{\text{ft}^3}\right)\left[V\left(\tfrac{\text{mi}}{\text{h}}\right)\left|\tfrac{5280\ \text{ft}}{\text{mi}}\right|\left|\tfrac{\text{h}}{3600\ \text{s}}\right|\right]^3\left|\tfrac{1\ \text{lbf}}{32.2\ \text{lb}\cdot\text{ft/s}^2}\right|\left|\tfrac{\text{hp}}{550\ \text{ft}\cdot\text{lbf/s}}\right|$$

$$= 5.65\times10^{-5}\,[V(\text{mi/h})]^3\ \text{hp} \qquad (*)$$

The power required to overcome rolling resistance is

$$\dot{W}_r = F_r \cdot V = (f\,\mathcal{W})V = (0.02)(3550\ \text{lbf})\left[V\left(\tfrac{\text{mi}}{\text{h}}\right)\left|\tfrac{5280\ \text{ft}}{\text{mi}}\right|\left|\tfrac{\text{h}}{3600\ \text{s}}\right|\right]\left|\tfrac{\text{hp}}{550\ \text{ft}\cdot\text{lbf/s}}\right|$$

$$= 0.189\ V\left(\tfrac{\text{mi}}{\text{h}}\right)\ \text{hp} \qquad (**)$$

(a) When $V = 55$ mi/h, we have

$$\dot{W}_d = 5.65\times10^{-5}[55]^3 = 9.4\ \text{hp}$$
$$\dot{W}_r = 0.189[55] = 10.4\ \text{hp}$$

← $\dot{W}_d, \dot{W}_r$

PROBLEM 2.22 (Cont'd)

(b) Letting V range from 0 to 75 mi/h, the accompanying plots can be developed.

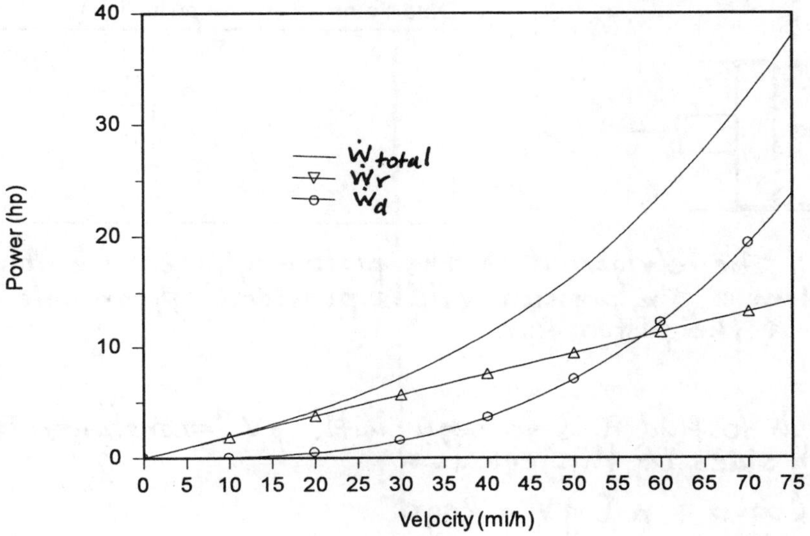

We see from the plots that up to about 58 mi/h, the power required to overcome rolling resistance is more significant than the power to overcome drag. At higher speeds, the drag effect becomes dominant because of the V^3 term in the expression for $\dot{W}_d$. Since the power to overcome these effects is developed by the engine from the fuel stored on board the vehicle, high-speed driving has an especially significant effect on fuel consumption.

PROBLEM 2.23

KNOWN: Measured data for pressure versus volume during the compression of a refrigerant are given.

FIND: (a) Determine n for a fit of the data by pV^n = constant. (b) Use the result of part (a) to evaluate the work done during the compression. (c) Evaluate the work using graphical or numerical integration of the data, (d) compare and discuss parts (b) and (c).

SCHEMATIC & GIVEN DATA:

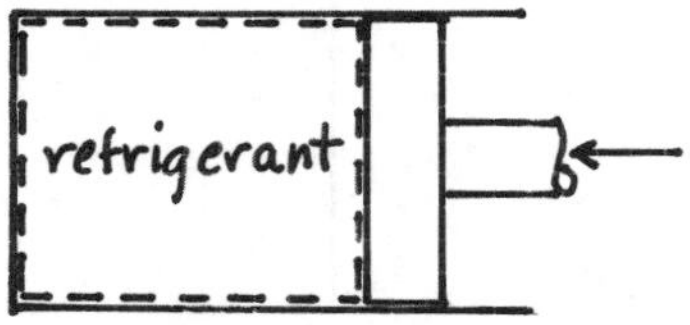

Data Point	p (lbf/in.²)	V (in.³)
1	112	13.0
2	131	11.0
3	157	9.0
4	197	7.0
5	270	5.0
6	424	3.0

ASSUMPTIONS: 1. The refrigerant in the piston-cylinder assembly form a closed system. 2. The pressure values provided approximate the pressure at the piston face.

ANALYSIS:

(a) One approach to find n is to begin with pV^n = constant. Taking the log of both sides of this equation

$$\log p + n \log V = \log c$$

or

$$\log p = (-n)\log V + \log c$$

Thus, $(-n)$ corresponds to the slope of a plot of log p vs. log V. Using
① a spreadsheet program to obtain the plot and the least squares best fit curve:

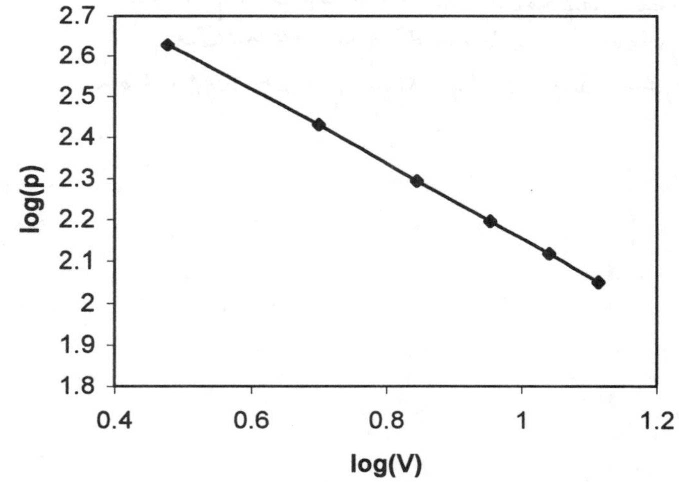

From the curve fit

$(-n) = -0.90887$

or

$n = 0.90887$ ← n

Thus

$pV^{0.90887}$ = constant

PROBLEM 2.23 (Cont'd)

(b) Using the results of part (a) and the procedure of Example 2.1, the work is

$$W = \int_{V_1}^{V_2} p\,dV = \frac{p_2V_2 - p_1V_1}{(1-n)}$$

$$= \frac{(424\ \text{lbf/in}^2)(3.0\ \text{in}^3) - (112)(13.0)}{(1-0.90887)}\left|\frac{1\ \text{ft}}{12\ \text{in}}\right|\left|\frac{1\ \text{Btu}}{778\ \text{ft}\cdot\text{lbf}}\right|$$

$$= -0.2163\ \text{Btu} \leftarrow W$$

② (c) A graphical evaluation of the work involves a plot of the tabulated data and a smooth curve drawn through the data points:

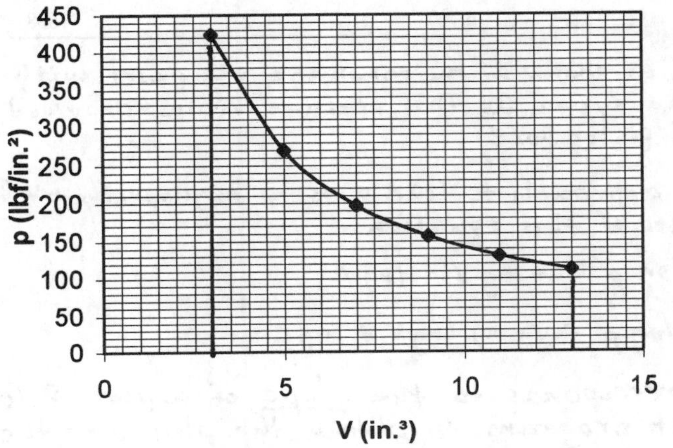

Each elemental rectangle in the plot contributes the following to the area under the curve:

$$(10\ \frac{\text{lbf}}{\text{in}^2})(0.5\ \text{in}^3)\left|\frac{1\ \text{ft}}{12\ \text{in}}\right|\left|\frac{1\ \text{Btu}}{778\ \text{ft}\cdot\text{lbf}}\right| = 5.356\times10^{-4}\ \text{Btu}$$

The number of retangles is approximately 401.1, thus

$$W \approx (401.1)(5.356\times10^{-4}) = -0.2148\ \text{Btu} \leftarrow W$$

(d) The results obtained in parts (b) and (c) are in good agreement. Each should be considered a plausible estimate for the reasons presented on page 42 in the discussion of actual expansion and compression processes.

---

1. The software IT could be used to obtain the least squares curve fit by programming the equations for curve fitting. It is easier to use a spreadsheet program in this instance, however.
2. The only measured data are the tabulated data points, shown as filled circles. The smooth curve does not necessarily represent the actual pressure at the piston face for the corresponding volume.

PROBLEM 2.24

KNOWN: Measured pressure-volume data for an expansion of gases within the cylinder of an internal combustion engine are given.

FIND: (a) Determine n for a fit of the data by pV^n = constant. (b) Use the result of part (a) to evaluate the work done in the expansion. (c) Evaluate the work done using graphical or numerical integration of the data. (d) Compare and discuss parts (c), (d).

SCHEMATIC & GIVEN DATA:

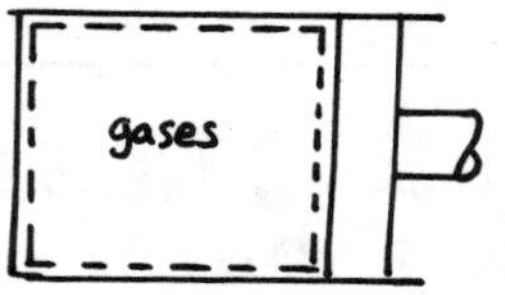

Data Point	p (bar)	V (cm³)
1	15	300
2	12	361
3	9	459
4	6	644
5	4	903
6	2	1608

ASSUMPTIONS: 1. As shown in the schematic, the gases within the piston-cylinder form the closed system. 2. The pressure values provided approximate the pressure at the piston face.

ANALYSIS: (a) One approach to find n is to begin with pV^n= constant. Taking the log of both sides of this equation

$$\log p + n \log V = \log c$$

or

$$\log p = (-n) \log V + \log c$$

Thus, (−n) corresponds to the slope of a plot of log p vs. log V. Using

① a spreadsheet program to obtain the plot and the least squares best fit curve:

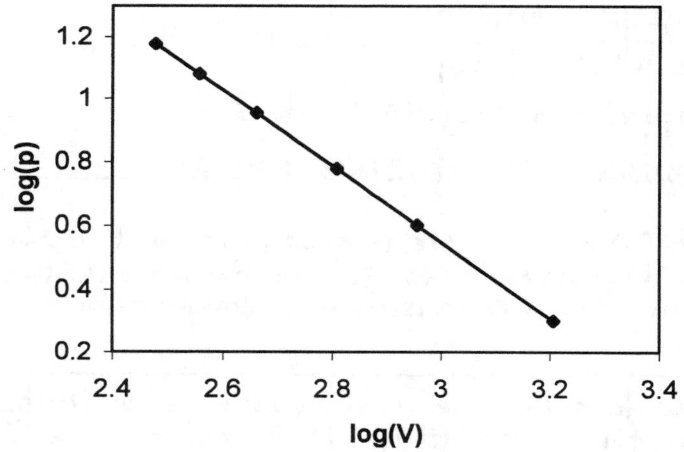

From the curve fit

$(-n) = -1.1996$

or

$n = 1.1996$ ← n

Thus

$pV^{1.1996}$ = constant

PROBLEM 2.24 (Cont'd)

(b) Using the results of part (a) and the procedure of Example 2.1, the work is

$$W = \int_{V_1}^{V_2} p\,dV = \frac{p_2 V_2 - p_1 V_1}{(1-n)}$$

$$= \frac{(2\text{ bar})(1608\text{ cm}^3) - (15)(300)}{(1-1.1996)}\left|\frac{10^5\text{ N/m}^2}{1\text{ bar}}\right|\left|\frac{1\text{ m}^3}{10^6\text{ cm}^3}\right|\left|\frac{1\text{ kJ}}{10^3\text{ N}\cdot\text{m}}\right|$$

$= 0.643$ kJ ◄ W

② (c) A graphical evaluation of the work involves a plot of the tabulated data and a smooth curve drawn through the data points:

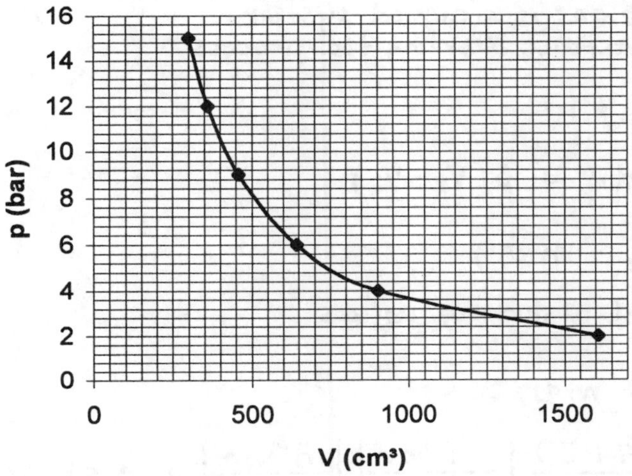

Each elemental rectangle in the plot contributes the following to the area under the curve:

$$(.4\text{ bar})(50\text{ cm}^3)\left|\frac{10^5\text{ N/m}^2}{1\text{ bar}}\right|\left|\frac{1\text{ m}^3}{10^6\text{ cm}^3}\right|\left|\frac{1\text{ kJ}}{10^3\text{ N}\cdot\text{m}}\right| = 0.002\text{ kJ}$$

The number of rectangles is approximately 324, thus

$W \approx (324)(0.002) = 0.648$ kJ ◄ W

(d) The results obtained in parts (b) and (c) are in good agreement. Each should be considered a plausible estimate for the reasons presented on page 42 in the discussion of actual expansion and compression processes.

---

1. The software IT could be used to obtain the least squares curve fit by programming the equations for curve fitting. It is easier to use a spreadsheet program in this instance, however.

2. The only measured data are the tabulated data points, shown as filled circles. The smooth curve does not necessarily represent the actual pressure at the piston face for the corresponding volume.

PROBLEM 2.25

KNOWN: A known amount of gas undergoes a constant-pressure process in a piston-cylinder assembly beginning at a specified specific volume. The work is known.

FIND: Determine the final volume.

SCHEMATIC & GIVEN DATA:

① 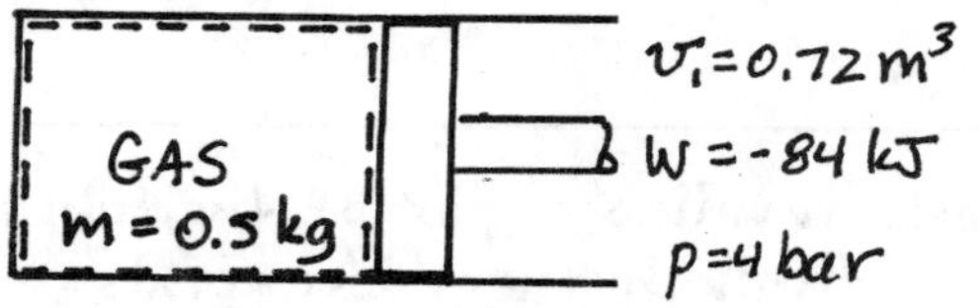

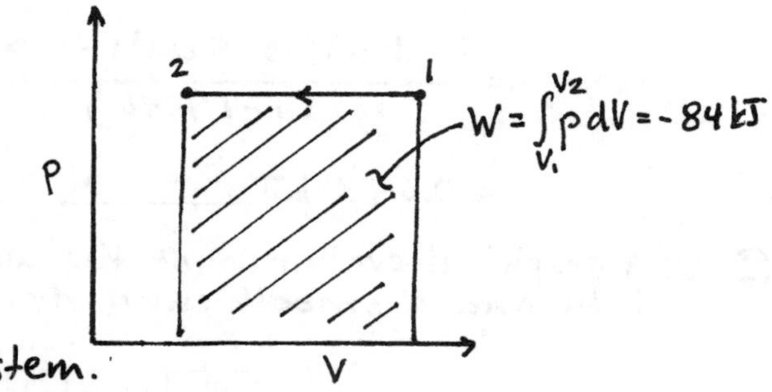

ASSUMPTIONS: (1) The gas is a closed system.
(2) Pressure is constant during the process.

ANALYSIS: Using Eq. 2.17

$$W = \int_{V_1}^{V_2} p\,dV = p(V_2 - V_1)$$

$$= p(V_2 - m v_1)$$

Solving for V_2 and inserting values

$$V_2 = \frac{W}{p} + m v_1$$

$$= \frac{(-84\text{ kJ})}{(4\text{ bar})}\left|\frac{1\text{ bar}}{10^5\text{ N/m}^2}\right|\left|\frac{10^3\text{ N}\cdot\text{m}}{1\text{ kJ}}\right| + (0.5\text{ kg})(0.72\text{ m}^3/\text{kg})$$

$$= 0.15\text{ m}^3 \leftarrow V_2$$

---

1. In the first printing of the fourth edition, the pressure of 4 bar was omitted inadvertently.

PROBLEM 2.26

KNOWN: Air in a piston-cylinder assembly undergoes a process for which $pV^{1.4}$ = constant. The work is known.

FIND: Determine the final volume and pressure.

SCHEMATIC & GIVEN DATA:

$p_1 = 30\ \frac{lbf}{in^2}$

$V_1 = 25\ ft^3$

(1)

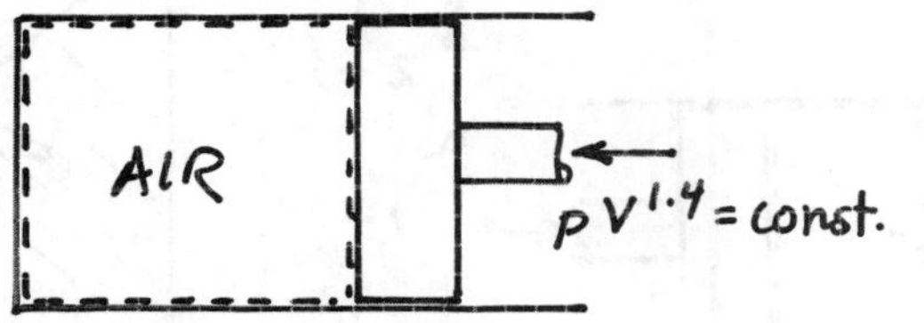

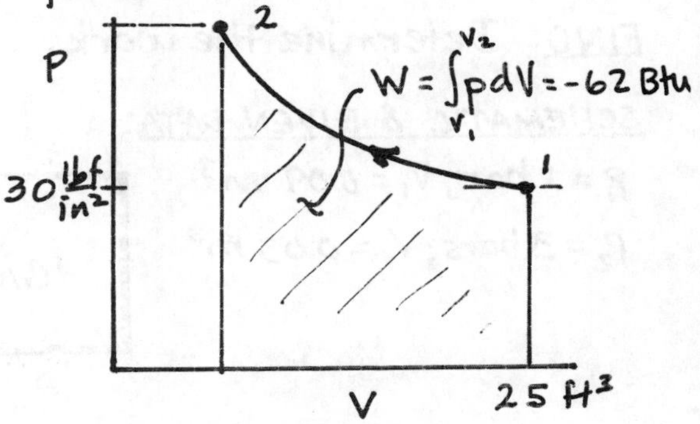

ASSUMPTIONS: (1) The air is a closed system. (2) The process is polytropic.

ANALYSIS: To determine V_2, substitute the p-V relation into Eq. 2.17 and integrate

$$W = \int_{V_1}^{V_2} p\,dV = \int_{V_1}^{V_2} \left(\frac{\text{const.}}{V^{1.4}}\right) dV = \text{const.}\ \frac{V_2^{-.4} - V_1^{-.4}}{(-.4)}$$

Now, const. $= p_1 V_1^{1.4}$. With this expression, and solving for $V_2^{1.4}$

$$V_2^{-.4} = \frac{(-.4)\,W}{p_1 V_1^{1.4}} + V_1^{-.4}$$

$$= \frac{(-.4)(-62\ \text{Btu})}{(30\ \text{lbf/in}^2)(25\ \text{ft}^3)^{1.4}} \left|\frac{778\ \text{ft}\cdot\text{lbf}}{1\ \text{Btu}}\right| \left|\frac{1\ \text{ft}^2}{144\ \text{in}^2}\right| + (25\ \text{ft}^3)^{-.4}$$

$$= 0.32524$$

$\Rightarrow V_2 = 16.58\ ft^3$ ← V_2

Now, we can use the p-V relation to get p_2

$$p_2 = p_1\left(\frac{V_1}{V_2}\right)^{1.4} = \left(30\ \frac{\text{lbf}}{\text{in}^2}\right)\left(\frac{25}{16.58}\right)^{1.4} = 53.31\ \text{lbf/in}^2$$ ← p_2

---

1. The shaded area on the p-V diagram represents the work for the polytropic process.

PROBLEM 2.27

KNOWN: A gas undergoes a compression process. Pressure and volume are given at the initial and final states. Pressure and volume are related linearly during the process.

FIND: Determine the work.

SCHEMATIC & GIVEN DATA:

$p_1 = 1$ bar, $V_1 = 0.09\ m^3$

$p_2 = 3$ bars, $V_2 = 0.03\ m^3$

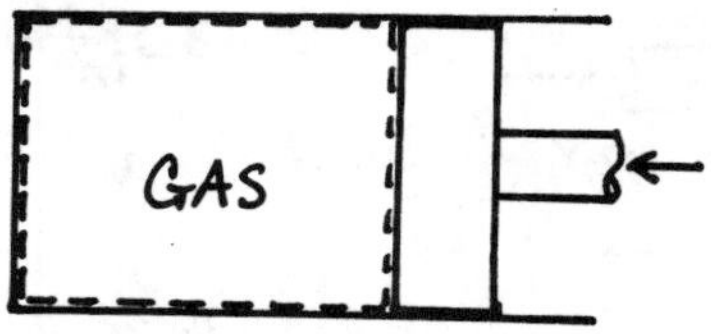

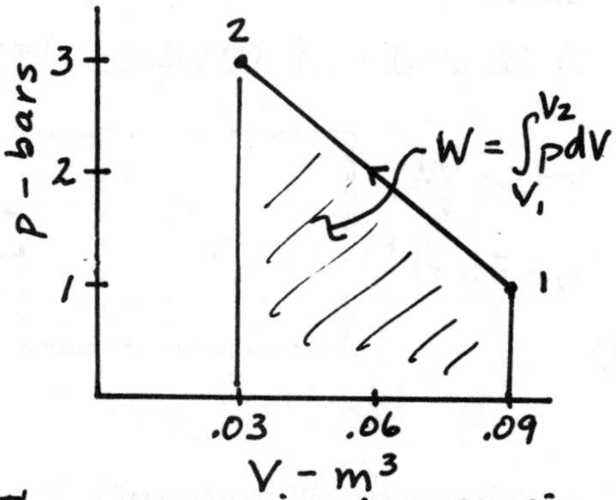

ASSUMPTIONS: (1) The gas is a closed system. (2) The compression is a quasi-equilibrium process. with a linear relation between pressure and volume.

ANALYSIS: Based on the given data, the p-V relation can be expressed as

$$p = 4 - \left(\frac{2}{0.06}\right) V$$

where p is in bars and V is in m³. The work is determined using Eq. 2.17

$$W = \int_{V_1}^{V_2} p\,dV$$

Inserting the p-V relation and integrating

$$W = \int_{V_1 = .09\,m^3}^{V_2 = .03\,m^3} \left[4 - \left(\frac{2}{.06}\right)V\right] \left|\frac{10^5\,N/m^2}{1\,bar}\right| \left|\frac{1\,kJ}{10^3\,N\cdot m}\right| dV$$

$$= \left[4V - \left(\frac{1}{.06}\right)V^2\right]\Bigg|_{V_1 = 0.09}^{V_2 = 0.03} |100|$$

$$= \left[4(0.03 - 0.09) - \left(\frac{0.03^2 - 0.09^2}{0.06}\right)\right] |100|$$

① $= -12$ kJ ← W

1. The negative sign for work denotes energy transfer to the system.

PROBLEM 2.28

KNOWN: Carbon dioxide expands from a known initial state to a known final pressure. The pressure-volume relation is specified.

FIND: For the gas, determine the work.

SCHEMATIC & GIVEN DATA:

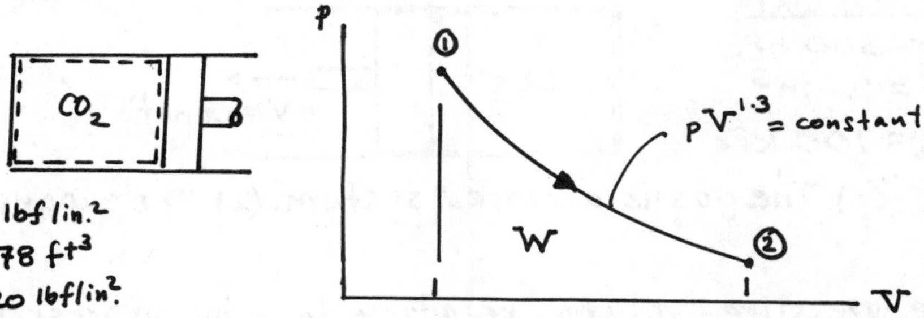

$p_1 = 60$ lbf/in.²
$V_1 = 1.78$ ft³
$p_2 = 20$ lbf/in.²

ASSUMPTIONS: 1. The gas is a closed system. 2. The moving boundary is the only work mode. 3. The expansion is polytropic.

ANALYSIS: Using Eq. 2.17 and the procedure discussed in part (a) of Example 2.1

$$W = \int_{V_1}^{V_2} p\,dV = \int_{V_1}^{V_2} \frac{C}{V^{1.3}}\,dV = \frac{C V_2^{-0.3} - C V_1^{-0.3}}{(-0.3)}$$

The constant C can be evaluated at either end state: $C = p_1 V_1^{1.3} = p_2 V_2^{1.3}$, giving

$$W = \frac{(p_2 V_2^{1.3}) V_2^{-0.3} - (p_1 V_1^{1.3})(V_1^{-0.3})}{(-0.3)} = \frac{p_2 V_2 - p_1 V_1}{(-0.3)}$$

To complete the calculation, the final volume is required. That is

$$p_1 V_1^{1.3} = p_2 V_2^{1.3} \Rightarrow V_2 = \left(\frac{p_1}{p_2}\right)^{1/1.3} V_1 = \left(\frac{60}{20}\right)^{1/1.3} (1.78\text{ ft}^3) = 4.14\text{ ft}^3$$

Inserting values, the work is

$$W = \frac{(20\text{ lbf/in.}^2)\left|144\text{ in}^2/1\text{ ft}^2\right|(4.14\text{ ft}^3) - (60)|144|(1.78)}{(-0.3)}$$

① $= 11{,}520$ ft·lbf ← W (ft·lbf)

Converting to Btu

$$W = (11{,}520\text{ ft}\cdot\text{lbf})\left|\frac{1\text{ Btu}}{778\text{ ft}\cdot\text{lbf}}\right|$$

$= 14.81$ Btu ← W(Btu)

---

1. In this case, the work is the area under the process line from state 1 to state 2 on the p-V diagram. Also, note that the work is positive, denoting energy transfer from the gas.

PROBLEM 2.29

KNOWN: A gas expands from a known initial state to a known final pressure. The pressure-volume relation for the process is specified.

FIND: Sketch the process on a p-V diagram and determine the work.

SCHEMATIC & GIVEN DATA:

$p_1 = 500$ kPa
$V_1 = 0.1\ m^3$
$p_2 = 100$ kPa

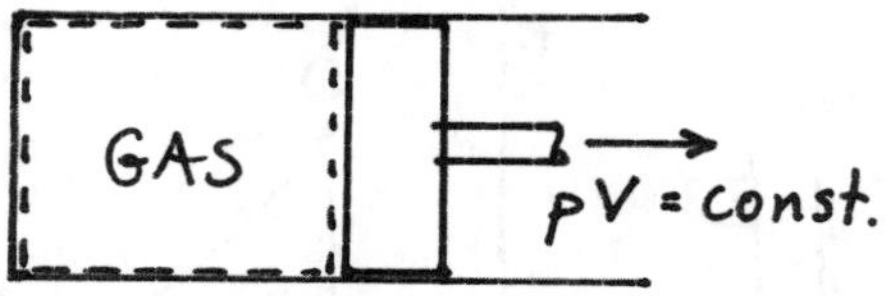

ASSUMPTIONS: (1) The gas is a closed system. (2) The expansion is polytropic.

ANALYSIS: The pressure-volume relation for the process is

$$pV = \text{const.} \Rightarrow p = \frac{\text{const.}}{V}$$

Thus, the p-V diagram is:

$$V_2 = \left(\frac{p_1}{p_2}\right) V_1 = 0.5\ m^3$$

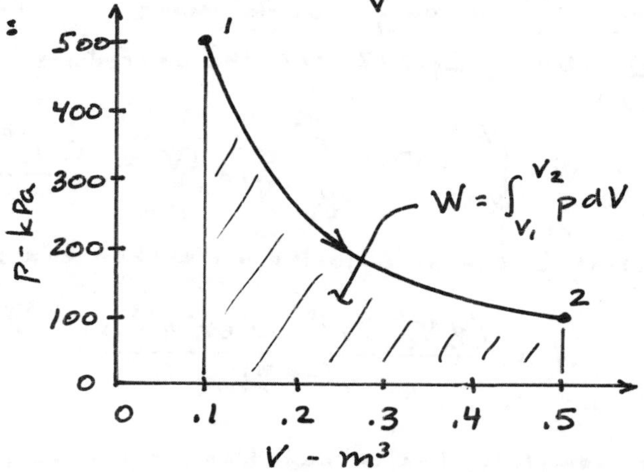

Next, to determine the work, substitute the p-V relation into Eq. 2.17 and integrate.

$$W = \int_{V_1}^{V_2} p\,dV = \int_{V_1}^{V_2} \left(\frac{\text{const.}}{V}\right) dV = p_1 V_1 \ln \frac{V_2}{V_1}$$

$$= (500\ \text{kPa})(0.1\ m^3) \ln\left(\frac{0.5}{0.1}\right) \left|\frac{10^3\ N/m^2}{1\ \text{kPa}}\right| \left|\frac{1\ \text{kJ}}{10^3\ N\cdot m}\right|$$

① $= +80.47$ kJ ← W

1. The work is positive for the expansion, as expected.

PROBLEM 2.30

KNOWN: Air undergoes a polytropic process between two specified states.

FIND: Determine the work.

SCHEMATIC & GIVEN DATA:

$n = 0.5$ lbmol

$p_1 = 20$ lbf/in.^{2}; $v_1 = 9.26$ ft^3/lb

$p_2 = 60$ lbf/in.^{2}; $v_2 = 3.98$ ft^3/lb

$pV^n =$ constant

ASSUMPTIONS: (1) The air is a closed system. (2) The system undergoes a polytropic process.

ANALYSIS: From the pressure-volume relation for a polytropic process

$$p_1 V_1^n = p_2 V_2^n \Rightarrow p_1 v_1^n = p_2 v_2^n$$

Solving for n

$$n = \frac{\log(p_1/p_2)}{\log(v_2/v_1)} = \frac{\log(20/60)}{\log(3.98/9.26)} = 1.301$$

Now, using Eq. 2.17 to determine work and with the molecular weight of air from Table A-1E

$$W = \int_{V_1}^{V_2} p\,dV = m\int_{v_1}^{v_2} p\,dv = m(\text{const.})\int_{v_1}^{v_2} \frac{dv}{v^n}$$

$$= m\left[\frac{(p_2 v_2^n)v_2^{1-n} - (p_1 v_1^n)v_1^{1-n}}{1-n}\right] = m\left(\frac{p_2 v_2 - p_1 v_1}{1-n}\right)$$

Now $m = (0.5 \text{ lbmol})(28.97 \text{ lb/lbmol}) = 14.485 \text{ lb}$

and

$$W = (14.485 \text{ lb})\left[\frac{(60 \text{ lbf/in}^2)(3.98 \text{ ft}^3/\text{lb}) - (20)(9.26)}{(1-1.301)}\right]\left|\frac{144 \text{ in}^2}{1 \text{ ft}^2}\right|\left|\frac{1 \text{ Btu}}{778 \text{ ft}\cdot\text{lbf}}\right|$$

① $= -477.4$ Btu ← W

1. The negative sign for work denotes energy transfer into the system.

PROBLEM 2.31

KNOWN: Warm air cools slowly in a piston-cylinder assembly from a known initial volume to a known final volume. During the process, a spring exerts a force on the piston that varies linearly from a known initial value to a final value of zero.

FIND: Determine the initial and final pressures of the air, and the work.

SCHEMATIC & GIVEN DATA:

$V_1 = 0.003\ m^3$

$V_2 = 0.002\ m^3$

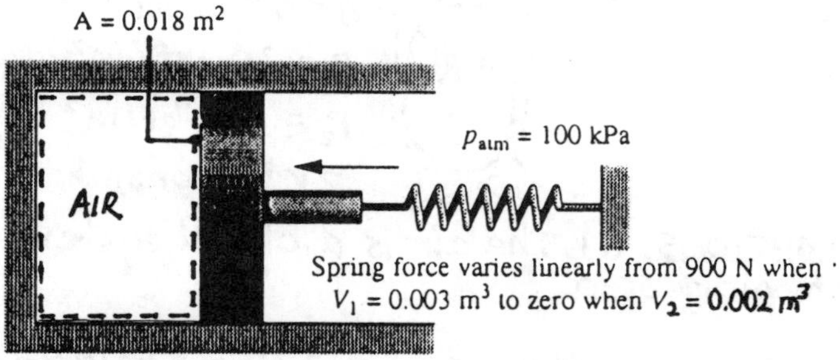

ASSUMPTIONS: (1) The air is a closed system. (2) The process occurs slowly, so there is no acceleration of the piston. (3) There is no friction between the piston and the cylinder wall. (4) The spring force varies linearly with volume.

ANALYSIS: The initial and final pressures of the air are determined from a free-body diagram of the piston, as follows. That is, $\Sigma F = 0$, so

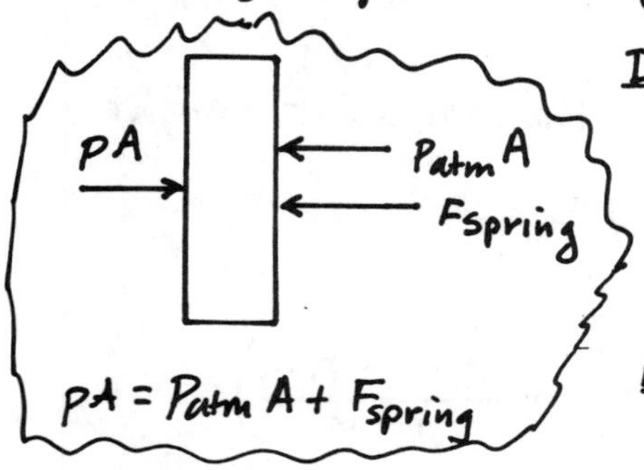

Initially: $F_{spring} = 900\ N$

$$p_1 = p_{atm} + \frac{F_{spring,1}}{A}$$

$$= 100\ kPa + \frac{(900\ N)}{(0.018\ m^2)}\left|\frac{1\ kPa}{10^3\ N/m^2}\right| = 150\ kPa \quad \leftarrow p_1$$

Finally: $F_{spring} = 0 \Rightarrow p_2 = 100\ kPa \quad \leftarrow p_2$

Now, the work is determined using Eq. 2.17, $W = \int_{V_1}^{V_2} p\,dV$, but from above

$$p = p_{atm} + \frac{F_{spring}}{A}$$

Since the spring force varies linearly from 900 N to zero as volume goes from $V_1 = 0.003\ m^3$ to $V_2 = 0.002\ m^3$

$$F_{spring} = \left(\frac{900\ N}{0.001}\right)(V - 0.002)$$

and

$$W = \int_{V_1}^{V_2}\left(p_{atm} + \frac{F_{spring}}{A}\right)dV = \int_{V_1}^{V_2}\left[100 + \left(\frac{900}{0.001}\right)\frac{(V-0.002)}{(0.018)|10^3|}\right]dV$$

$$= \int_{V_1}^{V_2}\left[\not{1}\not{0}\not{0} + 50000V - \not{1}\not{0}\not{0}\right]dV = \left(\frac{50000}{2}\right)V^2\Big|_{V_1 = 0.003\ m^3}^{V_2 = 0.002\ m^3}$$

① $$= -0.125\ kPa\cdot m^3\left|\frac{10^3\ N/m^2}{1\ kPa}\right|\left|\frac{1\ kJ}{10^3\ N\cdot m}\right| = -0.125\ kJ \quad \leftarrow W$$

---

1. The negative sign denotes that the piston does work on the air as the air cools. Also, the atmosphere and the spring do work on the piston.

PROBLEM 2.32

KNOWN: Air undergoes two processes in series.

FIND: Sketch the processes on a p-v diagram and determine the work per unit mass of air.

SCHEMATIC AND GIVEN DATA:

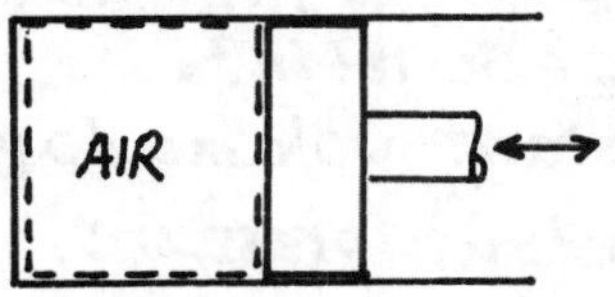

PROCESS 1-2: $pv^{1.3}$ = const.
$p_1 = 100$ kPa, $v_1 = 0.04$ m³/kg
$v_2 = 0.02$ m³/kg

PROCESS 2-3: p = const., $v_3 = v_1$

ASSUMPTIONS: (1) The air is a closed system. (2) Both processes are quasi-equilibrium processes.

ANALYSIS: For process 1-2, $p_2 = p_1\left(\frac{v_1}{v_2}\right)^{1.3} = 246.23$ kPa. Thus, the p-v diagram is

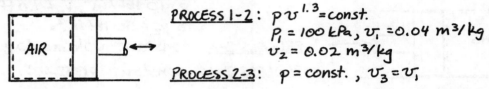

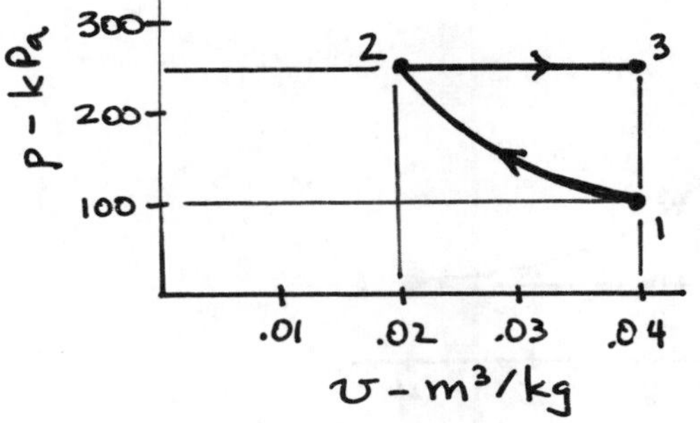

The work for each process is determined using Eq. 2.17

$$W = \int p\,dV = m\int p\,dv \Rightarrow \frac{W}{m} = \int p\,dv$$

Thus

$$\frac{W_{12}}{m} = \int_{v_1}^{v_2} \frac{\text{const.}}{v^{1.3}}\,dv = \left(\frac{p_2 v_2 - p_1 v_1}{1-1.3}\right)$$

$$= \left[\frac{(246.23\text{ kPa})(0.02\text{ m}^3/\text{kg}) - (100)(0.04)}{(1-1.3)}\right]\left|\frac{10^3\text{ N/m}^2}{1\text{ kPa}}\right|\left|\frac{1\text{ kJ}}{10^3\text{ N}\cdot\text{m}}\right|$$

$$= -3.082\text{ kJ/kg}$$

$$\frac{W_{23}}{m} = \int_{v_2}^{v_3} p\,dv = p_2(v_3 - v_2)$$

$$= (246.23\text{ kPa})(0.04-0.02)\frac{\text{m}^3}{\text{kg}}\left|\frac{10^3}{10^3}\right|$$

$$= +4.9246\text{ kJ/kg}$$

Finally

① $$\frac{W_{13}}{m} = \frac{W_{12}}{m} + \frac{W_{23}}{m} = +1.8426\ \frac{\text{kJ}}{\text{kg}} \longleftarrow \frac{W_{13}}{m}$$

---

1. The result is positive, denoting that the net energy transfer by work is from the system to the surroundings.

PROBLEM 2.33

KNOWN: A gas undergoes three processes that complete a cycle.

FIND: Sketch the processes on a p-V diagram and determine the net work.

SCHEMATIC & GIVEN DATA:

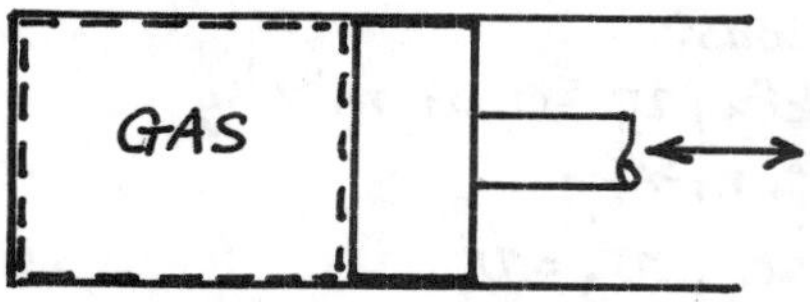

PROCESS 1-2: compression with $pV = \text{const.}$ $p_1 = 10\ \text{lbf/in}^2$, $V_1 = 4.0\ \text{ft}^3$ to $p_2 = 50\ \text{lbf/in}^2$

PROCESS 2-3: constant volume to $p_3 = p_1$

PROCESS 3-1: constant pressure

ASSUMPTIONS: (1) The gas is a closed system. (2) Processes 1-2 and 3-1 are quasi-equilibrium processes.

ANALYSIS: For process 1-2, $V_2 = (p_1/p_2)V_1 = 0.8\ \text{ft}^3$. Thus, the p-V diagram is

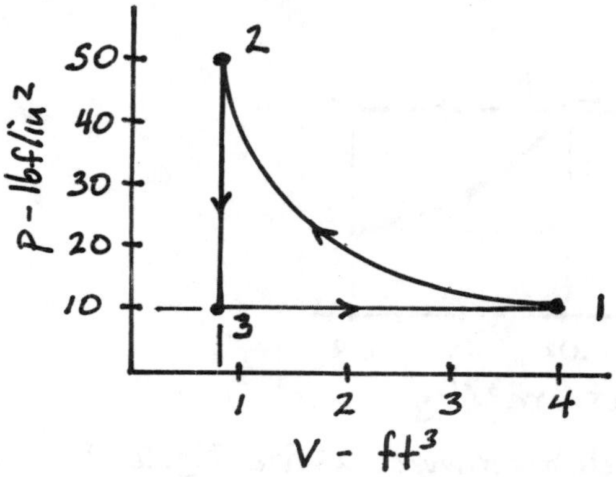

The work for process 1-2 is determined using Eq. 2.17

$$W_{12} = \int_{V_1}^{V_2} p\,dV = \int_{V_1}^{V_2} \frac{\text{const.}}{V}\,dV = (p_1V_1)\ln\left(\frac{V_2}{V_1}\right)$$

$$= (10\ \text{lbf/in}^2)(4.0\ \text{ft}^3)\ln\left(\frac{0.8}{4.0}\right)\left|\frac{144\ \text{in}^2}{1\ \text{ft}^2}\right|\left|\frac{1\ \text{Btu}}{778\ \text{ft}\cdot\text{lbf}}\right|$$

$$= -11.92\ \text{Btu}$$

For process 2-3, $W_{23} = 0$. Finally, for process 3-1

$$W_{31} = \int_{V_3}^{V_1} p\,dV = p_3(V_1 - V_3) = p_1(V_1 - V_2)$$

$$= (10\ \text{lbf/in}^2)(4 - 0.8)\ \text{ft}^3\left|\frac{144}{778}\right| = 5.923\ \text{Btu}$$

Thus

① $W_{net} = W_{12} + W_{23} + W_{31} = -11.92 + 0 + 5.923 = -5.997\ \text{Btu} \leftarrow W_{net}$

---

1. The negative sign denotes that the net energy transfer by work for the cycle is into the system.

PROBLEM 2.34

KNOWN: A gas contained within a piston-cylinder assemby undergoes a thermodynamic cycle consisting of three processes.

FIND: Determine the work for each process and the net work for the cycle.

SCHEMATIC & GIVEN DATA:

From Problem 1.35:

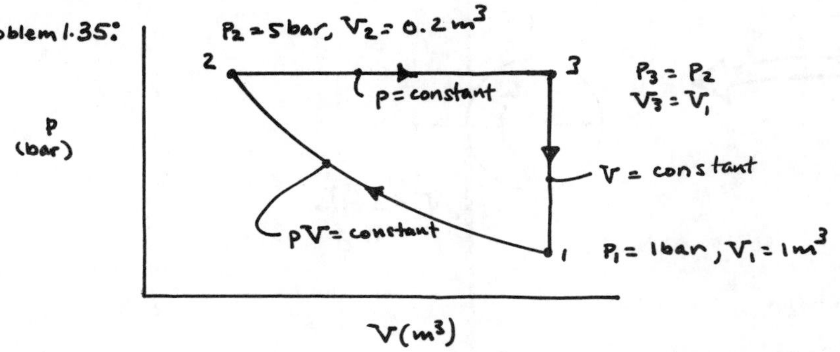

ASSUMPTIONS: 1. The gas is the closed system. 2. Volume change is the only work mode.

ANALYSIS: Using Eq. 2.17

Process 1-2: $$W_{12} = \int_{V_1}^{V_2} p\,dV = \int_{V_1}^{V_2} \frac{c}{V}\,dV = c \ln \frac{V_2}{V_1} = P_1 V_1 \ln \frac{V_2}{V_1}$$

$$= \left[(1 \text{ bar}) \left| \frac{10^5 \text{ N/m}^2}{1 \text{ bar}} \right| (1 \text{ m}^3)\right] \ln \left[\frac{0.2}{1.0}\right] \left| \frac{1 \text{ kJ}}{10^3 \text{ N·m}} \right|$$

(1) $$= -160.9 \text{ kJ}$$ ← W_{12}

Process 2-3: $$W_{23} = \int_{V_2}^{V_3} p\,dV = p_2[V_3 - V_2]$$

$$= (5 \text{ bar}) \left| \frac{10^5 \text{ N/m}^2}{1 \text{ bar}} \right| \left[1 \text{ m}^3 - 0.2 \text{ m}^3\right] \left| \frac{1 \text{ kJ}}{10^3 \text{ N·m}} \right| = 400 \text{ kJ}$$ ← W_{23}

Process 3-1: $$W_{31} = \int_{V_3}^{V_1} p\,dV = 0$$ since $V_1 = V_3$ ← W_{31}

The net work is

$$W_{net} = W_{12} + W_{23} + W_{31}$$

(2) $$= -160.9 + 400 + 0 = 239.1 \text{ kJ}$$ ← W_{net}

---

1. The minus denotes energy transfer to the gas as it is compressed.

2. The value of work depends on the details of the process between two states and is not determined only by the end states. In this case, the net work for the overall cycle is positive. By contrast, there is no net change in the properties p and V for the overall cycle. Property values are determined only by the end states. Work is not a property, and Σ W's ≠ 0.

PROBLEM 2.35

KNOWN: An object of known mass is attached to a rope wound around a pulley and falls at constant speed.

FIND: Determine the power transmitted to the pulley and the rotational speed of the pulley.

SCHEMATIC & GIVEN DATA:

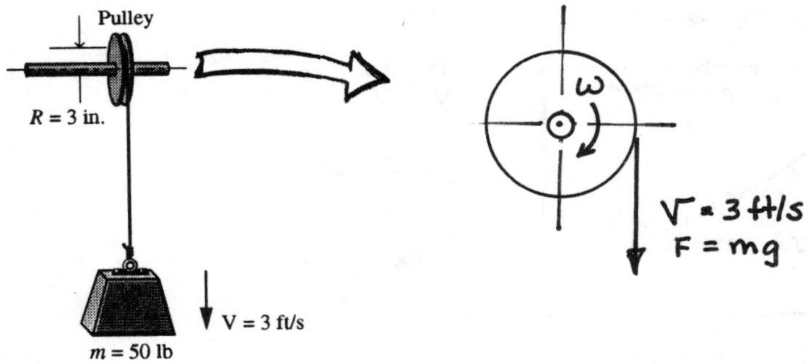

ASSUMPTIONS: 1. The object falls at constant speed. 2. The acceleration of gravity is constant, $g = 32.0\ ft/s^2$.

ANALYSIS: The power transmitted to the pulley can be determined using Eq. 2.13

$$\dot{W} = F \cdot V = mgV$$

$$= (50\ lb)\left(32.0\ \frac{ft}{s^2}\right)\left(3\ \frac{ft}{s}\right)\left|\frac{1\ lbf}{32.2\ lb \cdot ft/s^2}\right|$$

$$= \left(149.1\ \frac{ft \cdot lbf}{s}\right)\left|\frac{1\ hp}{550\ ft \cdot lbf/s}\right|$$

$$= 0.27\ hp \longleftarrow \dot{W}$$

Since $V = R\omega$, we have

$$\omega = \frac{V}{R} = \frac{3\ ft/s}{(3/12)\ ft}\left|\frac{1\ Rev}{2\pi\ rad}\right|\left|\frac{60\ s}{1\ min}\right|$$

$$= 115\ RPM \longleftarrow \omega$$

PROBLEM 2.36

KNOWN: The rotational speed and diameter of a drive shaft pulley are known. The net force applied by the belt on the pulley is also given.

FIND: Determine the applied torque and the power transmitted.

SCHEMATIC & GIVEN DATA:

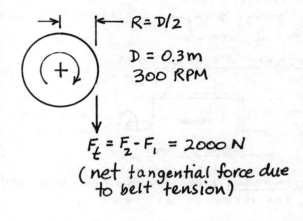

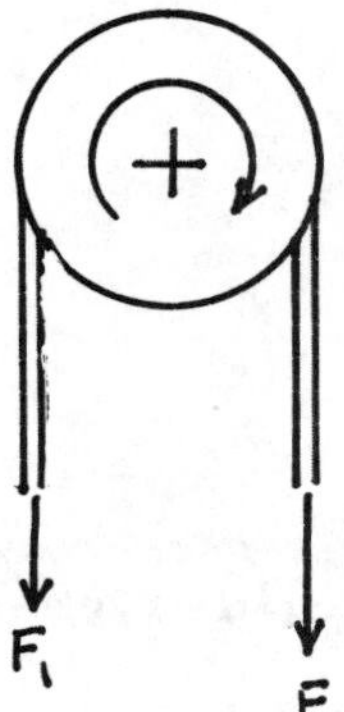

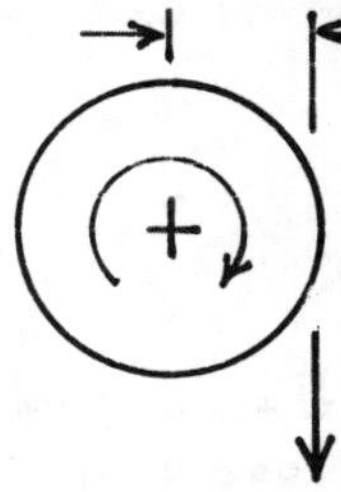

ANALYSIS: The torque is calculated using the tangential force and the radius at which it acts

$$\mathcal{T} = F_t \cdot R$$

$$= (2000 \text{ N})\left(\frac{0.3}{2} \text{ m}\right) = 300 \text{ N}\cdot\text{m} \longleftarrow \mathcal{T}$$

Thus, with Eq. 2.20 the power transmitted is

$$\dot{W}_{shaft} = \mathcal{T} \cdot \omega$$

$$= (300 \text{ N}\cdot\text{m})\left(300 \frac{\text{rev}}{\text{min}}\right)\left(2\pi \frac{1}{\text{rev}}\right)\left|\frac{1 \text{ min}}{60 \text{ s}}\right|\left|\frac{1 \text{ J}}{1 \text{ N}\cdot\text{m}}\right|\left|\frac{1 \text{ kW}}{10^3 \text{ J/s}}\right|$$

$$= 9.42 \text{ kW} \longleftarrow \dot{W}$$

PROBLEM 2.37

KNOWN: Operating data are provided for an electric motor at steady state.

FIND: Determine the electric power required by the motor and the power developed by the output shaft. Determine the net power input to the motor. Also, determine the amounts of energy transfer by electrical work and by the shaft during 2h of operation

SCHEMATIC & GIVEN DATA:

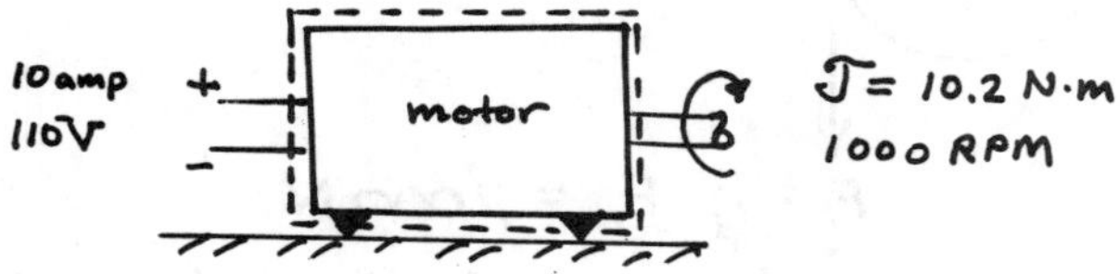

ASSUMPTIONS 1. As shown in the schematic, the motor is the closed system.
2. The system is at steady state.

ANALYSIS: (a) Using Eq. 2.21

$$\dot{W}_{electric} = -(\text{Voltage})(\text{current})$$

$$= -(110 \text{ volts})(10 \text{ amp}) \left| \frac{1 \text{ watt/amp}}{1 \text{ volt}} \right| = -1100 \text{ watt} \left| \frac{1 \text{kW}}{10^3 \text{W}} \right|$$

① $$= -1.1 \text{ kW} \quad \longleftarrow \dot{W}_{electric}$$

Using Eq. 2.20

$$\dot{W}_{shaft} = (\text{Torque})(\text{angular velocity})$$

$$= (10.2 \text{ N·m})\left(1000 \frac{\text{Rev}}{\text{min}} \left| \frac{2\pi \text{ rad}}{\text{Rev}} \right| \left| \frac{1 \text{ min}}{60 \text{ s}} \right| \right) \left| \frac{1 \text{ kW}}{10^3 \text{ N·m/s}} \right|$$

$$= 1.07 \text{ kW} \quad \longleftarrow \dot{W}_{shaft}$$

(b) $$\dot{W}_{net} = \dot{W}_{electric} + \dot{W}_{shaft}$$

② $$= (-1.1 \text{ kW}) + (1.07 \text{ kW}) = -0.03 \text{ kW} \quad \longleftarrow \dot{W}_{net}$$

(c) Integrating to find energy transfer amounts

$$W_{electric} = \int_0^{2h} \dot{W}_{elec} dt = (-1.1 \text{ kW})(2h) = -2.2 \text{ kW·h} \quad \longleftarrow W_{electric}$$

$$W_{shaft} = \int_0^{2h} \dot{W}_{shaft} dt = (1.07 \text{ kW})(2h) = 2.14 \text{ kW·h} \quad \longleftarrow W_{shaft}$$

---

1. The minus sign is needed because energy is transferred to the motor electrically.

2. This value represents the portion of the electric power input that is not obtained as a shaft power output because of effects within the motor such as electrical resistance and friction.

PROBLEM 2.38

KNOWN: An electric storage battery is charged with a constant current for a known length of time.

FIND: Determine the cost of electricity to charge the battery.

SCHEMATIC & GIVEN DATA:

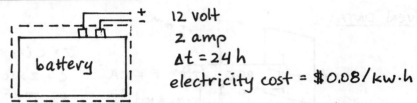

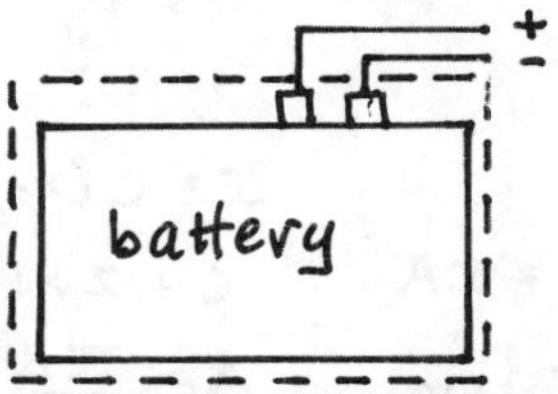

12 volt
2 amp
$\Delta t = 24$ h
electricity cost = \$0.08/kw·h

ASSUMPTION: The voltage and current are constant.

ANALYSIS: The electric power is obtained using Eq. 2.21

$$|\dot{W}| = \mathcal{E} i$$

Since the voltage and current are constant, the electricity used in 24 h is

$$|W| = \int_{t_1}^{t_2} \dot{W}\,dt = |\dot{W}|\Delta t = \mathcal{E} i \Delta t$$

$$= (2\,\text{amp})(12\,\text{volt})\left|\frac{1\ \text{Watt/amp}}{1\ \text{volt}}\right|(24\,\text{h})\left|\frac{1\,\text{kW}}{10^3\,\text{Watt}}\right|$$

$$= 0.576\ \text{kW·h}$$

The cost is

$$\text{cost} = (0.576\ \text{kW·h})(\$0.08/\text{kW·h}) \approx \$.05 \leftarrow \text{electric cost}$$

PROBLEM 2.39

The students will need to estimate the typical power for each of the devices. They also will need to estimate how long they each device in the course of a month. The details are left to the reader.

PROBLEM 2.40

KNOWN: A solid cylindrical bar is slowly stretched. The initial and final length and the stress acting at the end of the bar are known.

FIND: Determine the work done on the bar.

SCHEMATIC & GIVEN DATA:

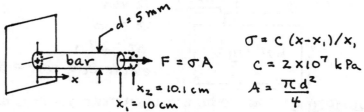

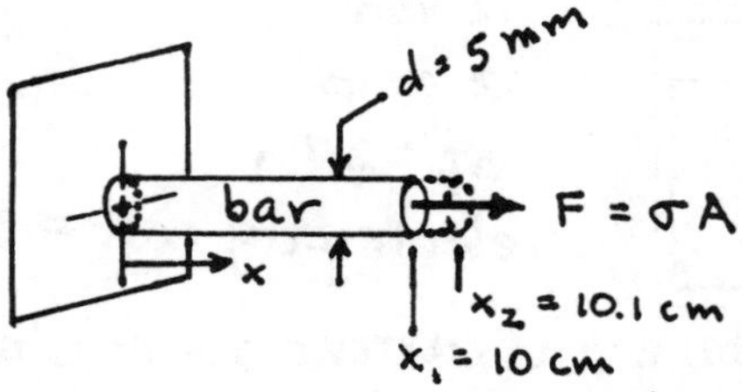

$\sigma = C(x - x_1)/x_1$

$C = 2\times10^7$ kPa

$A = \frac{\pi d^2}{4}$

ASSUMPTIONS: (1) The bar is a closed system. (2) The moving boundary is the only work mode. (3) The known normal stress acts at the end of the bar. (4) The change in area A is neglected.

ANALYSIS: The work done is given by Eq. 2.18

$$W = -\int_{x_1}^{x_2} \sigma A\, dx = -\frac{CA}{x_1}\int_{x_1}^{x_2} (x - x_1)\, dx$$

$$= -\frac{CA}{x_1}\left(\frac{x^2}{2} - x_1 x\right)\Big|_{x_1}^{x_2} dx$$

$$= -\frac{C\pi d^2}{4x_1}\left[\left(\frac{x_2^2 - x_1^2}{2}\right) - x_1(x_2 - x_1)\right]$$

Now, with $C = 2\times10^7$ kPa, $d = 5$ mm $= 0.005$ m, $x_1 = 10$ cm $= 0.1$ m, and $x_2 = 0.101$ m

$$W = -\frac{(2\times10^7\ \text{kPa})\pi(5\times10^{-3}\ \text{m})^2}{4(0.1\ \text{m})\left|\frac{1\ \text{kPa}}{10^3\ \text{N/m}^2}\right|}\left[\left(\frac{0.101^2 - 0.1^2}{2}\right)\text{m}^2 - (0.1)(.101 - 0.1)\text{m}^2\right]$$

① $= -1.963$ N·m $= -1.963$ J ← W

---

1. The negative sign denotes energy tranfer by work to the bar.

PROBLEM 2.41

KNOWN: A wire suspended vertically is stretched by an applied force.

FIND: Determine (a) the work done, and (b) the Young's modulus.

SCHEMATIC & GIVEN DATA:

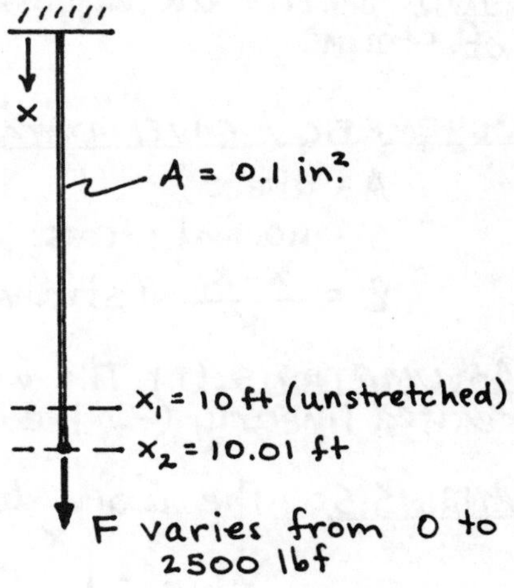

ASSUMPTIONS: (1) The wire is a closed system. (2) The moving boundary is the only work mode. (3) The applied force varies linearly with x. (4) The change in area A is neglected.

ANALYSIS: (a) The applied force varies with x according to

$$F = \left(\frac{2500 \text{ lbf}}{0.01 \text{ ft}}\right)(x - 10 \text{ ft}) = 2.5\times10^5\, x - 2.5\times10^6$$

where x is in ft and F is in lbf.

Thus, applying Eq. 2.12 to evaluate the work of the force F

$$W = -\int_{x_1}^{x_2} F\,dx$$

$$= -\int_{10\text{ ft}}^{10.01\text{ ft}} (2.5\times10^5 x - 2.5\times10^6)\,dx$$

$$= -\left[\frac{2.5\times10^5}{2}(10.01^2 - 10^2) - 2.5\times10^6(10.01 - 10)\right]$$

$$= -12.5 \text{ ft}\cdot\text{lbf} \qquad \longleftarrow W$$

(b) From Problem 2.40, the Young's modulus c can be expressed as

$$c = \sigma\left(\frac{x_0}{x - x_0}\right)$$

where $\sigma = F/A$ is the normal stress, and x_0 is the unstretched length. Thus

$$c = \frac{F}{A}\left(\frac{x_0}{x - x_0}\right)$$

$$= \frac{(2500 \text{ lbf})}{(0.1 \text{ in}^2)}\frac{(10 \text{ ft})}{(10.01 - 10)\text{ ft}}$$

$$= 2.5\times10^7 \text{ lbf/in}^2 \qquad \longleftarrow \text{Young's modulus}$$

PROBLEM 2.42

KNOWN: A wire of constant cross-sectional area and a given initial length is stretched. The stress-strain relation is known.

FIND: Derive an expression for the work done on the wire as a function of strain.

SCHEMATIC & GIVEN DATA:

A - area

σ - normal stress

$\varepsilon = \dfrac{x - x_o}{x_o}$ (strain)

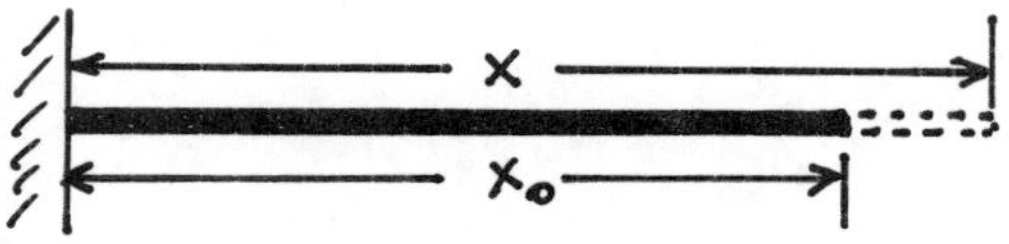

ASSUMPTIONS: (1) The wire is a closed system. (2) Stress and strain are related linearly. (3) The cross-sectional area remains constant.

ANALYSIS: The work done on the wire is given by Eq. 2.18

$$W = -\int_{x_o}^{x} \sigma A \, dx$$

From the given stress-strain relation

$$\sigma = C\varepsilon = C\left(\frac{x - x_o}{x_o}\right)$$

where C is a constant (Young's modulus). From this expression

$$d\varepsilon = \frac{dx}{x_o} \Rightarrow dx = x_o \, d\varepsilon$$

Substituting into the work expression

$$W = -\int_0^{\varepsilon} (C\varepsilon) A (x_o \, d\varepsilon) = -CAx_o \int_0^{\varepsilon} \varepsilon \, d\varepsilon$$

Finally

$$W = -\frac{CAx_o\varepsilon^2}{2}$$ ← work expression

PROBLEM 2.43

KNOWN: A soap film on a wire frame is stretched.

FIND: Determine the work done.

SCHEMATIC & GIVEN DATA:

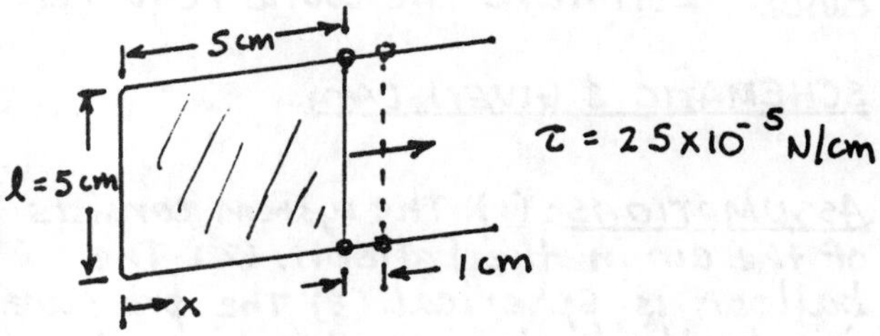

ASSUMPTIONS: (1) The film is a closed system. (2) The moving boundary is the only work mode. (3) The surface tension is constant, acting on both sides of the film.

ANALYSIS: The work is determined using Eq. 2.19

$$W = -\int_{A_1}^{A_2} \tau\, dA = -\int_{x_1}^{x_2} \tau\, 2\ell\, dx$$

For constant surface tension

$$W = -\tau\, 2\ell\, \Delta x$$

$$= -(25\times10^{-5}\ \tfrac{N}{m})\, 2\,(5\ cm)(1\ cm)\left|\frac{1\ m}{10^2\ cm}\right|\left|\frac{1\ J}{1\ N\cdot m}\right|$$

① $$= -2.5\times10^{-5}\ J \quad \longleftarrow W$$

1. The negative sign denotes work done on the film. Note the small magnitude of the work required to stretch the film.

PROBLEM 2.44

KNOWN: A liquid film on a wire frame is stretched.

FIND: Determine the work done.

SCHEMATIC & GIVEN DATA:

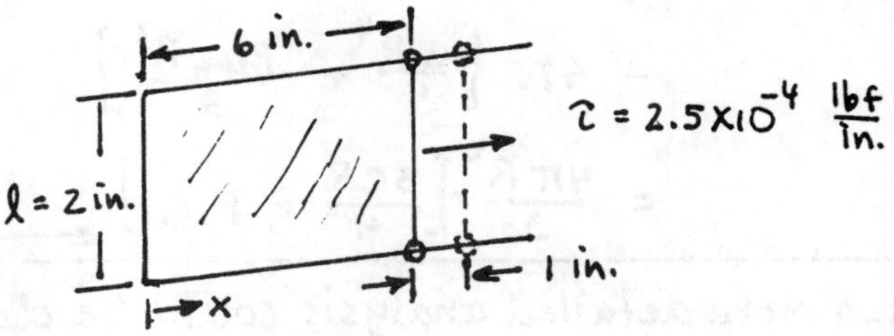

ASSUMPTIONS: (1) The film is a closed system. (2) The moving boundary is the only work mode. (3) The surface tension is constant, acting on both sides of the film.

ANALYSIS: The work is determined using Eq. 2.19

$$W = -\int_{A_1}^{A_2} \tau\, dA = -\int_{x_1}^{x_2} \tau\, 2\ell\, dx$$

For constant surface tension

$$W = -\tau\, 2\ell\, \Delta x$$

$$= -(2.5\times10^{-4}\ \tfrac{lbf}{in.})\, 2\,(2\ in.)(1\ in.)\left|\frac{1\ ft}{12\ in.}\right|$$

① $$= -8.33\times10^{-5}\ ft\cdot lbf \quad \longleftarrow W$$

1. The negative sign denotes work done on the film. Note the small magnitude of the work required to stretch the film.

PROBLEM 2.45

KNOWN: A common baloon is being inflated.

FIND: Estimate the work required.

SCHEMATIC & GIVEN DATA:

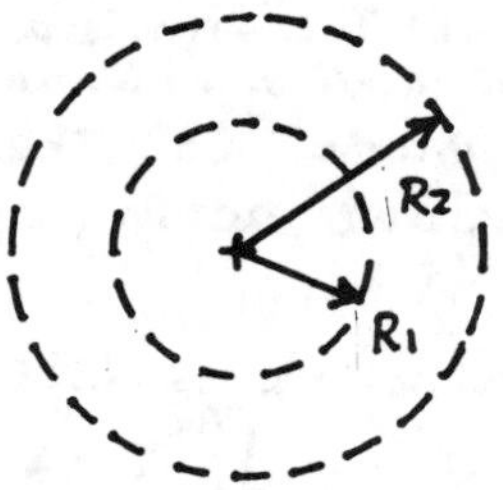

ASSUMPTIONS: (1) The system consists of the air in the balloon. (2) The balloon is spherical. (3) The pressure inside the balloon varies linearly with the balloon's radius.

ANALYSIS: With the above assumptions, the work can be estimated using Eq. 2.17; $W = \int p\,dV$.

If the pressure is assumed vary linearly with radius, then the pressure when $R=0$ is atmospheric, and

$$p = CR + p_{atm}$$

where c is a constant. Further, for a sphere

$$V = \frac{4}{3}\pi R^3$$

and

$$dV = 4\pi R^2 dR$$

Thus

$$W = \int_{R=0}^{R} (CR + p_{atm})\, 4\pi R^2 dR \qquad (CR + p_{atm})$$

$$= 4\pi \left[\frac{cR^4}{4} + \frac{p_{atm} R^3}{3} \right]$$

① $$= \frac{4\pi R^3}{3} \left[\frac{3CR}{4} + p_{atm} \right] \longleftarrow \text{work expression}$$

1. A more detailed analysis could be done to incorporate a more accurate stress-strain relation for the balloon material and to account more accurately for geometry.

PROBLEM 2.46

KNOWN: Data are provided for a brick wall at steady state.

FIND: Determine the temperature difference across the wall.

SCHEMATIC & GIVEN DATA:

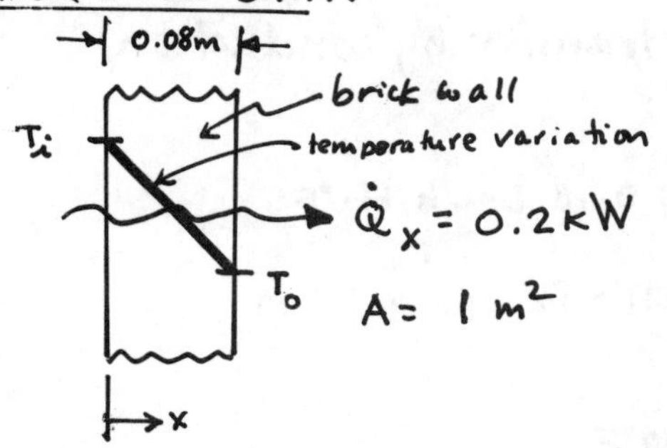

ASSUMPTIONS: 1. The wall is at steady state. 2. The temperature varies linearly through the wall. 3. Energy transfer is by conduction.

ANALYSIS: Using Eq. 2.32 together with assumption 2

$$\dot{Q}_x = -kA\frac{dT}{dx}, \quad \frac{dT}{dx} = \frac{T_o - T_i}{L}$$

where $L = 0.08\,m$ and $(T_o - T_i)$ is the temperature difference across the wall.

Thus

$$\dot{Q}_x = -kA\left[\frac{T_o - T_i}{L}\right]$$

$$\Rightarrow \quad (T_o - T_i) = -\left[\frac{\dot{Q}_x L}{kA}\right]$$

With $k = 0.72\ W/m\cdot K$ from Table A-19

$$(T_o - T_i) = -\left[\frac{(200\,W)(0.08\,m)}{\left(0.72\frac{W}{m\cdot K}\right)(1\,m^2)}\right]$$

$$= -22.2\,K \qquad \longleftarrow \quad T_o - T_i$$

where the minus sign is required because $T_o < T_i$.

PROBLEM 2.47

KNOWN: Energy transfer by conduction occurs at steady state through a plane wall. The dimensions, thermal conductivity, and surface temperatures are specified.

FIND: Determine the rate of energy transfer by conduction.

SCHEMATIC & GIVEN DATA:

① 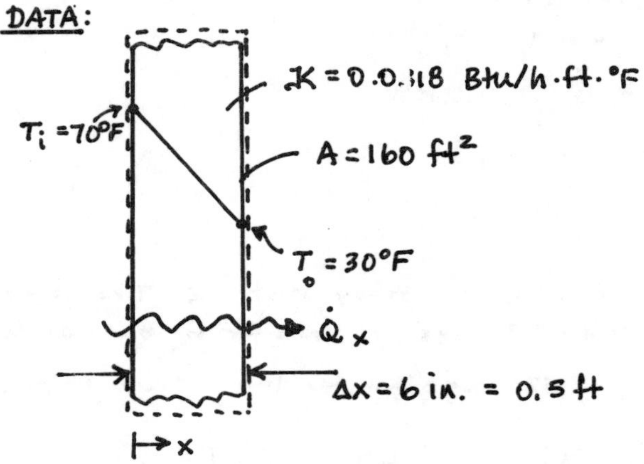

ASSUMPTIONS: (1) The wall is at steady state. (2) The temperature varies linearly through the wall. (3) The thermal conductivity is uniform and heat transfer is by conduction.

ANALYSIS: Using Eq. 2.32 together with assumption 2

$$\dot{Q}_x = -\kappa A \frac{dT}{dx} \quad , \quad \frac{dT}{dx} = \frac{T_o - T_i}{\Delta x}$$

Thus

$$\dot{Q}_x = -\kappa A \left(\frac{T_o - T_i}{\Delta x}\right)$$

Inserting values

$$\dot{Q}_x = -\left(0.0318 \frac{\text{Btu}}{\text{h}\cdot\text{ft}\cdot{}^\circ\text{F}}\right)(160\ \text{ft}^2)\left(\frac{30-70}{0.5}\right)\frac{{}^\circ\text{F}}{\text{ft}}$$

$$= 407\ \text{Btu/h} \longleftarrow \dot{Q}_x$$

---

1. The values given are for a typical insulated frame wall.

PROBLEM 2.48

KNOWN: A surface of given diameter emits thermal radiation at a known rate at a specified temperature.

FIND: Determine the emissivity of the surface and plot the rate of radiant emission vs. surface temperature.

SCHEMATIC & GIVEN DATA:

ASSUMPTIONS: (1) The Stefan – Boltzmann law applies. (2) The emissivity is constant.

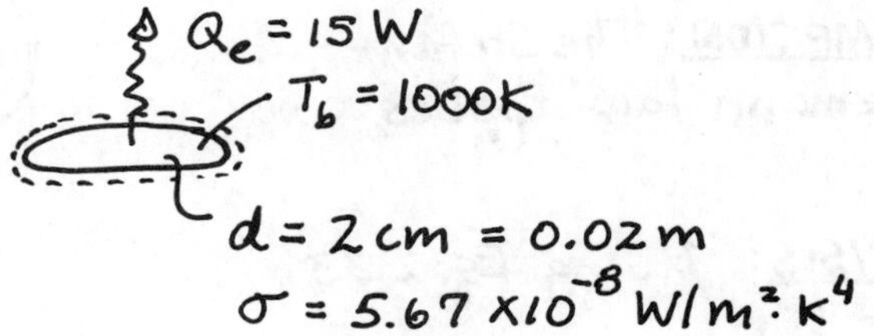

ANALYSIS: Using Eq. 2.33

$$\dot{Q}_e = \varepsilon \sigma A T_b^4 \qquad (*)$$

Solving for ε

$$\varepsilon = \frac{\dot{Q}_e}{\sigma A T^4} = \frac{\dot{Q}_e}{\sigma\left(\frac{\pi d^2}{4}\right)T^4}$$

$$= \frac{(15\ W)}{\left(5.67\times10^{-8}\ \frac{W}{m^2\cdot K^4}\right)\left(\frac{\pi(0.02m)^2}{4}\right)(1000\ K)^4}$$

$= 0.842 \longleftarrow$ ε

Using the emissivity just calculated, and varying T_b from 0 to 2000K, the following plot can be constructed based on Eq. (*) above

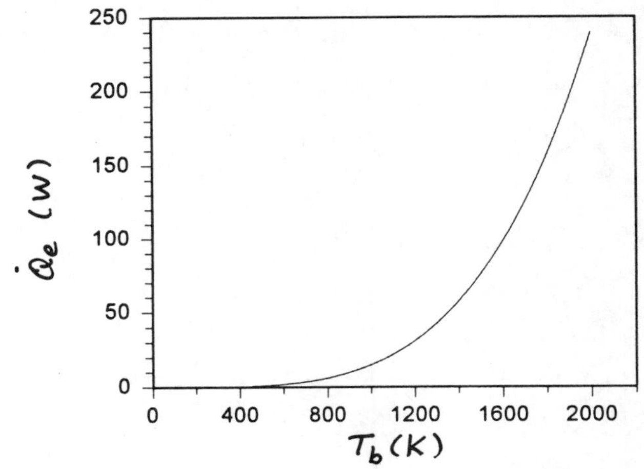

Notice the rapid growth in $\dot{Q}_e$ with T_b.

PROBLEM 2.49

KNOWN: A sphere of known surface area, temperature, and emissivity emits thermal radiation.

FIND: Determine the rate of thermal emission.

SCHEMATIC & GIVEN DATA:

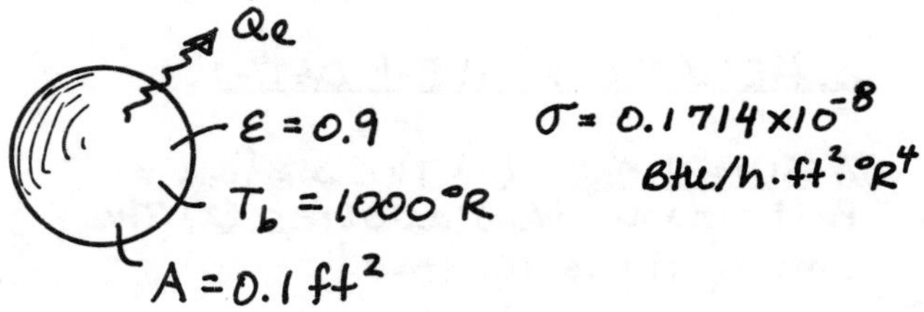

ASSUMPTION: The Stefan-Boltzmann law applies.

ANALYSIS: Using Eq. 2.33

$$\dot{Q}_e = \varepsilon \sigma A T_b^4$$

$$= (0.9)\left(0.1714 \times 10^{-8} \frac{\text{Btu}}{\text{hr} \cdot \text{ft}^2 \cdot °\text{R}^4}\right)(0.1\ \text{ft}^2)(1000\ °\text{R})^4$$

$$= 154.3\ \text{Btu/h} \longleftarrow \dot{Q}_e$$

PROBLEM 2.50

KNOWN: Data are provided for a flat surface cooled convectively by a gas.

FIND: Determine the range of the heat transfer rate for cooling by free convection. By forced convection.

SCHEMATIC & GIVEN DATA:

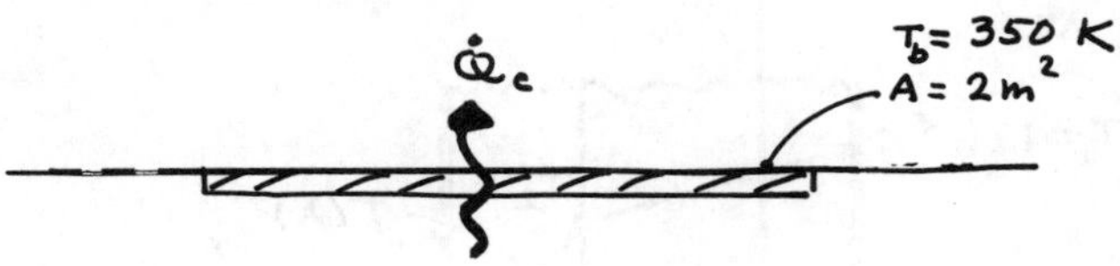

ASSUMPTION: Heat transfer between the surface and the gas is by convection.

ANALYSIS: Using Eq. 2.34

$$\dot{Q}_c = hA(T_b - T_f)$$
$$= h(2\,m^2)(350-300)\,K$$
$$= h(100\ m^2\cdot K) \qquad (*)$$

With data from Table 2.1

	h (W/m²·K)	$\dot{Q}_c$ from Eq(*), in kW
free convection:	2 − 25	0.2 − 2.5 ←
forced convection:	25 − 250	2.5 − 25 ←

PROBLEM 2.51

KNOWN: Energy transfer occurs by conduction through a composite plane wall consisting of two layers.

FIND: Determine the steady-state heat flux and the temperature at the interface between the layers.

SCHEMATIC & GIVEN DATA:

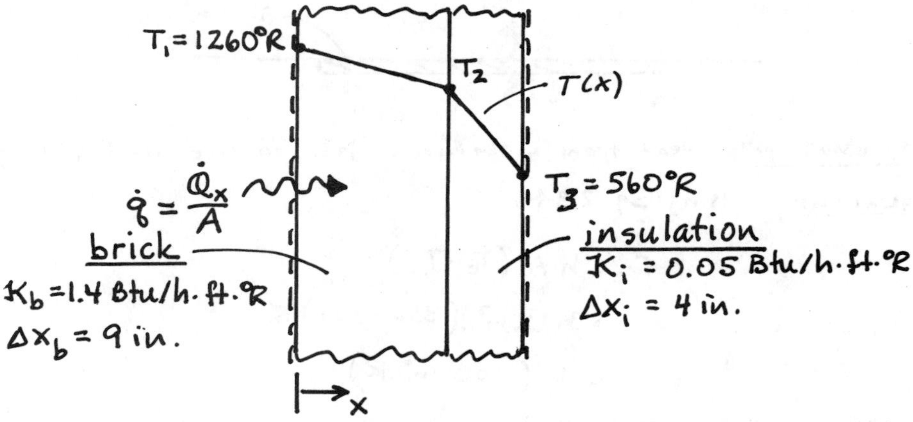

ASSUMPTIONS: (1) The wall is a closed system at steady state. (2) The temperature distributions are linear in both layers. (3) The two layers are in perfect thermal contact. (4) The thermal conductivities of both layers are uniform.

ANALYSIS: From Eq. 2.32

$$\dot{q} = -K_b \left.\frac{dT}{dx}\right)_{brick} = -K_i \left.\frac{dT}{dx}\right)_{insul.}$$

Since the temperature distributions are linear

$$\dot{q} = K_b \frac{T_1 - T_2}{\Delta x_b} = K_i \frac{T_2 - T_3}{\Delta x_i}$$

or

$$\dot{q} = \frac{T_1 - T_2}{R_b} = \frac{T_2 - T_3}{R_i} \qquad (*)$$

where $R_b = \Delta x_b / K_b$ and $R_i = \Delta x_i / K_i$. From (*)

① $$T_2 = T_1 - \dot{q} R_b \qquad (**)$$

Combining (*) and (**)

$$\dot{q} = \frac{T_1 - T_3}{R_b + R_i} \qquad (***)$$

Inserting values

$$R_b = \frac{(9/12)\ \text{ft}}{(1.4\ \text{Btu/h}\cdot\text{ft}\cdot{}^\circ\text{R})} = 0.5357\ \frac{\text{h}\cdot{}^\circ\text{R}}{\text{Btu}}\ ;\ R_i = \frac{4/12}{0.05} = 6.667\ \frac{\text{h}\cdot{}^\circ\text{R}}{\text{Btu}}$$

PROBLEM 2.51 (Cont'd)

Thus

$$\dot{q} = \frac{1260 - 560}{(0.5357 + 6.667)} = 97.19 \text{ Btu/h} \longleftarrow \quad \dot{q}$$

Now, inserting values in (**)

$$T_2 = 1260 - (97.19)(0.5357)$$

$$= 1208\,^\circ R \longleftarrow \quad T_2$$

---

1. The form of (***) illustrates the analogy between heat conduction through a composite wall and electric current flow through a series of resistances. The temperature difference in the numerator is analogous to a voltage difference, and the value of R_b and R_i are "thermal resistances" analogous to the electrical resistances.

PROBLEM 2.52

KNOWN: Energy transfer occurs from the inside air to the outside air through an insulated frame wall.

FIND: Determine the steady-state heat transfer rate through the wall.

SCHEMATIC & GIVEN DATA:

inside air

$T_i = 70°F$

$h_i = 1.5\ Btu/h\cdot ft^2\cdot °R$

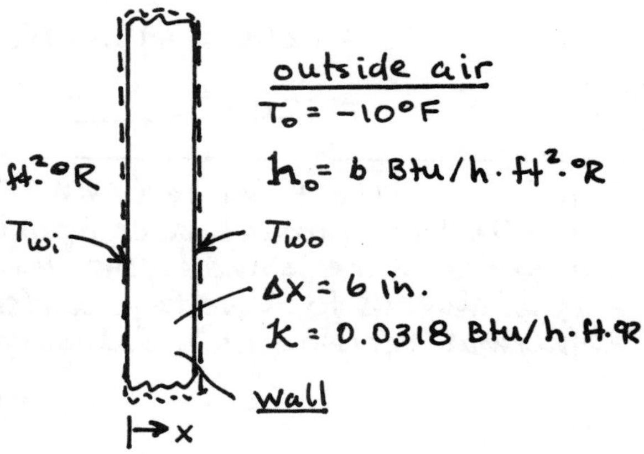

ASSUMPTIONS: (1) The system is at steady state. (2) Newton's law of cooling applies for heat from the air to the wall. (3) The temperature distribution is linear through the wall. (4) The thermal conductivity of the wall is uniform.

ANALYSIS: For energy transfer between the inside air and the wall

$$\dot{Q}_i = h_i A (T_i - T_{wi}) \qquad (1)$$

For conduction through the wall

$$\dot{Q}_x = -k A \frac{dT}{dx} = kA\left(\frac{T_{wi} - T_{wo}}{\Delta x}\right) \qquad (2)$$

Finally, for energy transfer between the wall and the outside air

$$\dot{Q}_o = h A (T_{wo} - T_o) \qquad (3)$$

Now, from (1)

$$T_{wi} = T_i - \dot{Q}_i\left(\frac{1}{h_i A}\right) \qquad (4)$$

And, from (3)

$$T_{wo} = T_o + \dot{Q}_o\left(\frac{1}{h_o A}\right) \qquad (5)$$

Combing (2), (4), and (5), and noting that $\dot{Q}_i = \dot{Q}_x = \dot{Q}_o \equiv \dot{Q}$

① $$\dot{Q} = \frac{A\,(T_o - T_i)}{\left(\frac{1}{h_i} + \frac{\Delta x}{k} + \frac{1}{h_o}\right)}$$

Inserting values

$$\dot{Q} = \frac{(120\ ft^2)(530 - 450)°R}{\left(\frac{1}{1.5\ \frac{Btu}{h\cdot ft^2\cdot °R}} + \frac{.5\ ft}{.0318\ \frac{Btu}{h\cdot ft\cdot °R}} + \frac{1}{6\ \frac{Btu}{h\cdot ft^2\cdot °R}}\right)}$$

$$= \frac{(120)(80)}{(16.56)} = 579.7\ Btu/h \longleftarrow \dot{Q}$$

1. The quantity in the denominator is called the overall resistance, or "R-value."

PROBLEM 2.53

KNOWN: A hot surface is covered with insulation. Energy transfer occurs from outer surface of the insulation and the surrounding air.

FIND: Determine the minimum thickness of insulation to maintain the outer surface temperature below a desired value at steady state.

SCHEMATIC & GIVEN DATA:

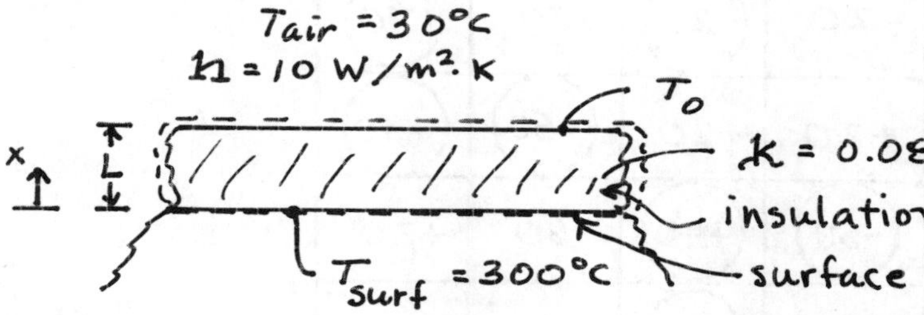

ASSUMPTIONS: (1) The system is at steady state. (2) Newton's law of cooling applies for heat transfer from the insulation to the air. (3) The temperature distribution through the insulation is linear. (4) The thermal conductivity of the insulation is uniform.

ANALYSIS: For energy transfer by conduction through the insulation

$$\dot{Q}_x = -kA\frac{dT}{dx} = kA\frac{(T_{surf} - T_o)}{L}$$

Further, for energy transfer from the insulation to the air

$$\dot{Q}_o = hA(T_o - T_{air})$$

At steady state, $\dot{Q}_x = \dot{Q}_o$, so

$$\frac{kA}{L}(T_{surf} - T_o) = hA(T_o - T_{air})$$

Solving for L

$$L = \frac{k(T_{surf} - T_o)}{h(T_o - T_{air})}$$

With $k = 0.08$ W/m·K, $h = 10$ W/m²·K, $T_{surf} = 300$°C, and $T_{air} = 30$°C the plot below can be constructed

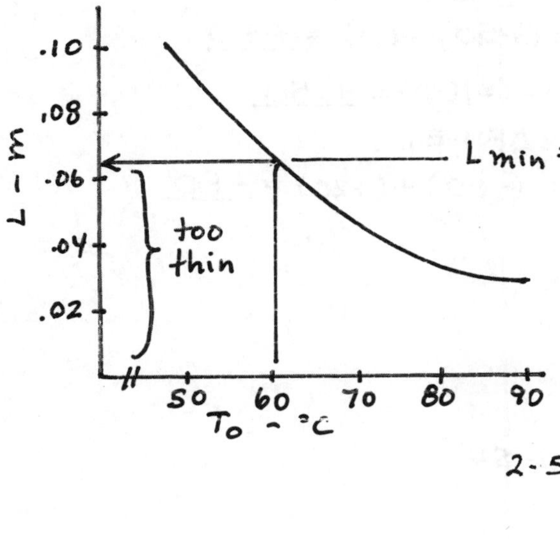

Lmin

PROBLEM 2.54

Process	Q	W	E_1	E_2	ΔE
a	+50	−20	(−20)	+50	(+70)
b	+50	+20	+20	(+50)	(+30)
c	−40	(−60)	(+40)	+60	+20
d	(−90)	−90	(+50)	+50	0
e	+50	(+150)	+20	(−80)	−100

Process a: $\Delta E = Q - W = (+50) - (-20) = \underline{+70}$

$E_2 - E_1 = \Delta E \Rightarrow E_1 = E_2 - \Delta E$

$= (+50) - (+70) = \underline{-20}$

Process b: $\Delta E = Q - W = (+50) - (+20) = \underline{+30}$

$E_2 - E_1 = \Delta E \Rightarrow E_2 = \Delta E + E_1$

$= (+30) + (+20) = \underline{+50}$

Process c: $W = Q - \Delta E = (-40) - (20) = \underline{-60}$

$E_2 - E_1 = \Delta E \Rightarrow E_1 = E_2 - \Delta E$

$= (+60) - (+20) = \underline{+40}$

Process d: $Q = \Delta E + W = (0) + (-90) = \underline{-90}$

$E_2 - E_1 = \Delta E \Rightarrow E_1 = E_2 - \Delta E$

$= (+50) - (0) = \underline{+50}$

Process e: $W = Q - \Delta E = (+50) - (-100) = \underline{+150}$

$E_2 - E_1 = \Delta E \Rightarrow E_2 = \Delta E + E_1$

$= (-100) + (+20) = \underline{-80}$

PROBLEM 2.55

KNOWN: A system of known mass undergoes a process for which the heat transfer is specified. In addition, the changes in specific internal energy, elevation, and kinetic energy are given.

FIND: Determine the work.

SCHEMATIC & GIVEN DATA:

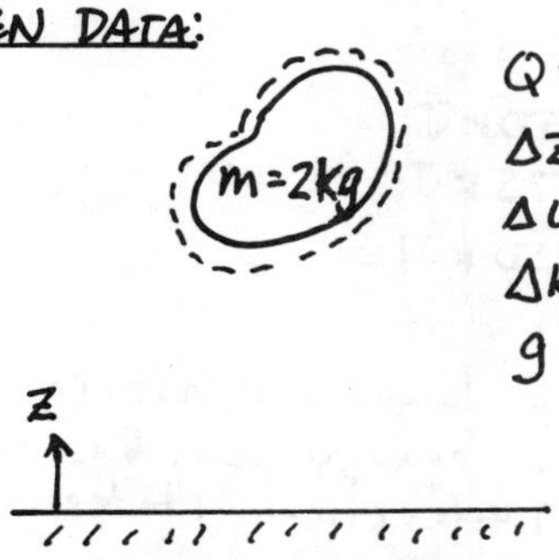

$Q = -25\text{ kJ}$
$\Delta z = 700\text{ m}$
$\Delta u = -15\text{ kJ/kg}$
$\Delta KE = 0$
$g = 9.6\text{ m/s}^2$

ASSUMPTIONS: (1) The system is closed. (2) There is no change in kinetic energy. (3) The acceleration of gravity is constant.

ANALYSIS: The work is determined by applying the energy balance and assumption 2

$$\cancelto{0}{\Delta KE} + \Delta PE + \Delta U = Q - W$$

The potential energy change can be expressed $\Delta PE = mg\Delta z$, and the internal energy change is $\Delta U = m\Delta u$. Thus

$$mg\Delta z + m\Delta u = Q - W$$

Solving for W and inserting values

$$W = Q - mg\Delta z - m\Delta u$$

$$= (-25\text{ kJ}) - (2\text{ kg})(9.6\tfrac{\text{m}}{\text{s}^2})(700\text{ m})\left|\frac{1\text{ N}}{1\text{ kg}\cdot\text{m/s}^2}\right|\left|\frac{1\text{ kJ}}{10^3\text{ N}}\right| - (2\text{ kg})(-15\text{ kJ/kg})$$

① $= -8.44\text{ kJ}$ ← W

1. The negative sign denotes energy transfer to the system by work.

PROBLEM 2.56

KNOWN: A system of known mass undergoes a process for which the heat transfer and work are known. The initial specific internal energy is also known.

FIND: Determine the final specific internal energy.

SCHEMATIC & GIVEN DATA:

(1)

$Q = -150\,kJ$
$W = -75\,kJ$
$u_1 = 450\,kJ/kg$

ASSUMPTIONS: 1. A closed system of known mass is under consideration. 2. Changes in kinetic and potential energy can be neglected. 3. The initial and final states are equilibrium states.

ANALYSIS: The change in internal energy can be determined from the energy balance using assumption 2

$$\cancel{\Delta KE} + \cancel{\Delta PE} + \Delta U = Q - W$$

$$\Rightarrow \quad \Delta U = Q - W$$

Then, with assumption 3, $\Delta U = m(u_2 - u_1)$, so

$$m(u_2 - u_1) = Q - W$$

or

$$u_2 = u_1 + \left[\frac{Q - W}{m}\right]$$

$$= 450\,\frac{kJ}{kg} + \left[\frac{(-150) - (-75)}{3\,kg}\right] kJ$$

$$= 425\,\frac{kJ}{kg} \longleftarrow u_2$$

---

1. According to the sign conventions for Q and W, Q is negative when there is a net heat transfer of energy from the system and W is negative when there is a net work transfer of energy to the system.

PROBLEM 2.57

KNOWN: Five kg of steam undergo an expansion in a piston-cylinder assembly from state 1 to state 2. During the process there is a known heat transfer to the steam and a known work transfer of energy to the steam by a paddle wheel. The change in specific internal energy of the steam is also known.

FIND: Determine the amount of energy transfer by work from the steam to the piston during the process.

SCHEMATIC & GIVEN DATA:

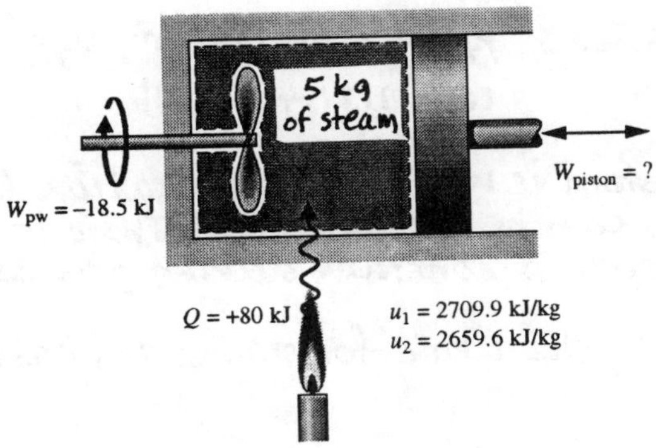

ASSUMPTIONS: 1. The steam is the closed system. 2. There is no change in the kinetic or potential energy from state 1 to state 2.

ANALYSIS: The net work can be determined from an energy balance. That is, with assumption 2

$$\Delta KE^{0} + \Delta PE^{0} + \Delta U = Q - W$$

or

$$W = Q - \Delta U$$

The net work is the sum of the work associated with the paddlewheel W_{pw} and the work done on the piston W_{piston}:

$$W = W_{pw} + W_{piston}$$

From the given information $W_{pw} = -18.5$ kJ, where the minus sign is required because the paddle wheel transfers energy to the system. Collecting results

$$W_{pw} + W_{piston} = Q - \Delta U$$

or

$$W_{piston} = Q - \Delta U - W_{pw}$$

$$= Q - m(u_2 - u_1) - W_{pw}$$

$$= 80\,\text{kJ} - 5\,\text{kg}\,(2659.6 - 2709.9)\frac{\text{kJ}}{\text{kg}} - (-18.5\,\text{kJ})$$

$$= +350\,\text{kJ} \longleftarrow W_{piston}$$

where the positive sign indicates that energy is transferred from the steam to the piston as the steam expands during the process.

PROBLEM 2.58

KNOWN: A system of known mass undergoes two processes in series.

FIND: Determine the work and heat transfer for the second process.

SCHEMATIC & GIVEN DATA:

Process 1-2: $v_1 = v_2 = 4.434\ \text{ft}^3/\text{lb}$, $p_1 = 100\ \text{lbf/in}^2$
$u_1 = 1105.8\ \text{Btu/lb}$, $Q_{12} = -581.36\ \text{Btu}$

Process 2-3: $p_2 = p_3 = 60\ \text{lbf/in}^2$, $v_3 = 7.82\ \text{ft}^3/\text{lb}$
$u_3 = 1121.4\ \text{Btu/lb}$

ASSUMPTIONS: (1) A closed system is under consideration. (2) Kinetic and potential energy effects can be neglected. (3) There is no work for process 1-2. (4) The pressure is constant during process 2-3.

ANALYSIS: By assumption (4), the work for process 2-3 can be found using Eq. 2.17

$$W_{23} = \int_{V_2}^{V_3} p\,dV = p_2(V_3 - V_2) = p_2 m (v_3 - v_2)$$

$$= \left(60\ \frac{\text{lbf}}{\text{in}^2}\right)(2\ \text{lb})(7.82 - 4.434)\frac{\text{ft}^3}{\text{lb}}\left|\frac{144\ \text{in}^2}{1\ \text{ft}^2}\right|\left|\frac{1\ \text{Btu}}{778\ \text{ft}\cdot\text{lbf}}\right|$$

$$= 75.21\ \text{Btu} \longleftarrow \quad W_{23}$$

Use the energy balance to find Q_{23}

$$\Delta U_{23} = Q_{23} - W_{23}$$

With $\Delta U = m\Delta u$

$$Q_{23} = m(u_3 - u_2) + W_{23} \qquad (*)$$

To get u_2, use the energy balance for process 1-2

$$m(u_2 - u_1) = Q_{12} - \cancelto{0}{W_{12}}$$

$$u_2 = \frac{Q_{12}}{m} + u_1 = \frac{(-581.36\ \text{Btu})}{(2\ \text{lb})} + 1105.8\ \text{Btu/lb}$$

$$= 815.12\ \text{Btu/lb}$$

Thus, from (*)

$$Q_{23} = (2)(1121.4 - 815.12) + (75.21)$$

$$= 687.8\ \text{Btu} \longleftarrow \quad Q_{23}$$

PROBLEM 2.59

KNOWN: A storage battery is charged by the electric power output from a windmill. The work and heat transfer rates are known.

FIND: Determine for 8 h of operation (a) the total amount of energy stored and (b) the value of the stored energy.

SHEMATIC & GIVEN DATA:

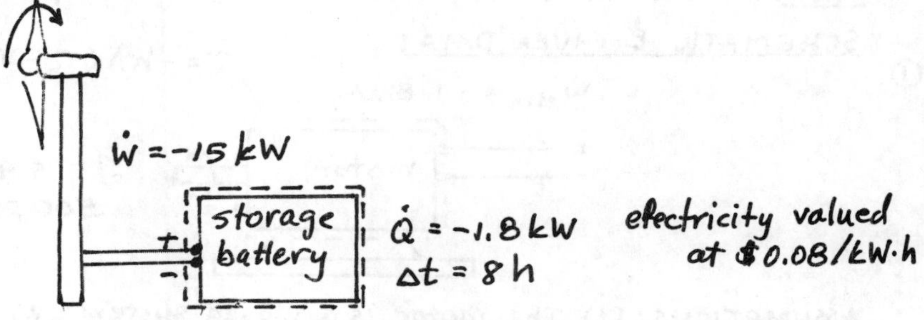

ASSUMPTIONS: (1) The battery is a closed system. (2) The work and heat transfer rates are constant.

ANALYSIS: (a) The amount of energy stored is found from an energy balance

$$\Delta E = Q - W \qquad (*)$$

To evaluate Q and W, respectively

$$Q = \int_{t_1}^{t_2} \dot{Q}\,dt = \dot{Q}\,\Delta t = (-1.8 \text{ kW})(8 \text{ h}) = -14.4 \text{ kW}\cdot\text{h}$$

$$= (-14.4 \text{ kW}\cdot\text{h})\left|\frac{1 \text{ kJ/s}}{1 \text{ kW}}\right|\left|\frac{3600 \text{ s}}{1 \text{ h}}\right|$$

$$= -51{,}840 \text{ kJ}$$

$$W = \int_{t_1}^{t_2} \dot{W}\,dt = \dot{W}\,\Delta t = (-15 \text{ kW})(8 \text{ h}) = -120 \text{ kW}\cdot\text{h}$$

$$= (-120)\left|\frac{1}{1}\right|\left|\frac{3600}{1}\right| = -4.32 \times 10^5 \text{ kJ}$$

Inserting these results in (*)

$$\Delta E = (-51{,}840) - (-4.32 \times 10^5) = 3.8 \times 10^5 \text{ kJ} \longleftarrow \Delta E$$

(b) If electricity is valued at \$0.08/kW·h

$$\text{value of stored energy} = (3.8 \times 10^5 \text{ kJ})\left|\frac{1 \text{ kW}}{1 \text{ kJ/s}}\right|\left|\frac{1 \text{ h}}{3600 \text{ s}}\right|(\$0.08/\text{kW}\cdot\text{h})$$

$$= \$8.45 \longleftarrow \text{value}$$

PROBLEM 2.60

KNOWN: An electric motor operates at steady state. The electric power input is given and the shaft power output can be determined. An expression is given for the heat transfer rate.

FIND: Determine the average surface temperature of the motor.

SCHEMATIC & GIVEN DATA:

1

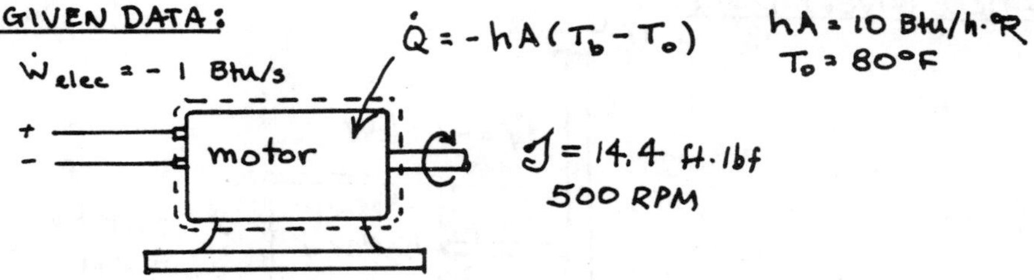

ASSUMPTIONS: (1) The motor is a closed system. (2) The system is at steady state; all operating data are constant with time. (3) The normal sign conventions for heat and work are used.

ANALYSIS: The rate form of the energy balance reduces as follows

$$\cancel{\frac{dKE}{dt}}^{0} + \cancel{\frac{dPE}{dt}}^{0} + \cancel{\frac{dU}{dt}}^{0} = \dot{Q} - (\dot{W}_{elec} + \dot{W}_{shaft})$$

$$\dot{Q} = \dot{W}_{elec} + \dot{W}_{shaft}$$

or, using the given expression for heat transfer rate

$$-hA(T_b - T_o) = \dot{W}_{elec} + \dot{W}_{shaft}$$

Solving for T_b

$$T_b = \frac{\dot{W}_{elec} + \dot{W}_{shaft}}{(-hA)} + T_o$$

Evaluating the work terms

$$\dot{W}_{elec} = -1 \text{ Btu/s} = -3600 \text{ Btu/h}$$

$$\dot{W}_{shaft} = \mathcal{T}\omega = (14.4 \text{ ft·lbf})\left(500 \frac{\text{rev}}{\text{min}}\right)\left(2\pi \frac{\text{rad}}{\text{rev}}\right)\left(\frac{60 \text{ min}}{1 \text{ h}}\right)\left(\frac{1 \text{ Btu}}{778 \text{ ft·lbf}}\right)$$

$$= 3489 \text{ Btu/h}$$

Finally

$$T_b = \frac{(-3600 + 3489)\text{ Btu/h}}{(-10 \text{ Btu/h·°R})} + 540°R = 551°R$$

$$= 91°F \longleftarrow T_b$$

---

1. Note that when $T_b > T_o$, $\dot{Q}$ is negative, denoting energy transfer by heat **from** the motor **to** the surroundings.

PROBLEM 2.61

KNOWN: A closed system undergoes a process with a known heat transfer rate, and the power varies as a specified function of time.

FIND: Determine (a) the rate of change of system energy at $t = 0.6\,h$ and (b) the change in system energy after 2 h.

SCHEMATIC & GIVEN DATA:

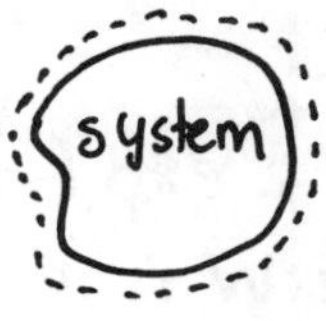

$$\dot{Q} = -10\ \text{kW}$$

$$\dot{W} = \begin{cases} -8t & 0 < t \le 1\,h \\ -8 & t > 1\,h \end{cases}$$

where t is in h and $\dot{W}$ is in kW

ASSUMPTION: The system is closed.

ANALYSIS: (a) The time rate of change of energy at any time t is given by

$$\frac{dE}{dt} = \dot{Q} - \dot{W} \qquad (*)$$

At $t = 0.6\,h$

$$\left.\frac{dE}{dt}\right|_{t=0.6h} = \left[\dot{Q} - (-8t)\right]_{t=0.6h} = (-10\ \text{kW}) - (-8 \cdot 0.6)\ \text{kW}$$

① $$= -5.2\ \text{kW} \longleftarrow dE/dt$$

(b) The change in system energy is obtained by integrating (*) over the time period of 2 h. That is

$$\Delta E = \int_{t=0}^{t=2h} (\dot{Q} - \dot{W})\,dt$$

$$= \dot{Q}\Delta t - \left[\int_{t=0}^{t=1h} (-8t)\,dt + \int_{t=1h}^{t=2h} (-8)\,dt\right]$$

$$= (-10)(2) - \left[\frac{(-8)}{2}t^2\right]_0^1 - \left[(-8)t\right]_1^2$$

$$= -20 - [(-4)] - [(-8)(2-1)]$$

$$= -8\ \text{kW}\cdot\text{h}$$

Thus

② $$\Delta E = (-8\ \text{kW}\cdot\text{h})\left|\frac{1\ \text{kJ/s}}{1\ \text{kW}}\right|\left|\frac{3600\,s}{1\,h}\right| = -28{,}800\ \text{kJ} \longleftarrow \Delta E$$

---

1. At $t = 0.6\,h$, the energy of the system is decreasing at a rate of 5.2 kW because the rate of energy transfer out by heat exceeds the rate of energy transfer in by work.

2. The negative sign denotes a net decrease of energy over the time period.

PROBLEM 2.62

KNOWN: A storage battery develops a known power output as a function of time.

FIND: Plot the power output and the change in energy of the battery, each as a function of time.

SCHEMATIC & GIVEN DATA:

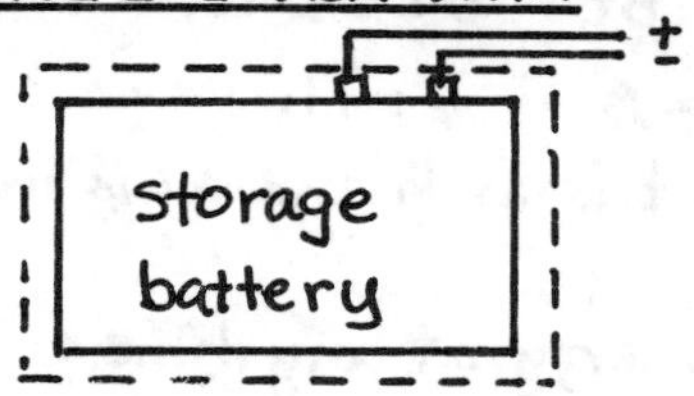

$\dot{W} = 1.2 \exp(-t/60)$

(where $\dot{W}$ is in kW and t is time in s)

$\dot{Q} = 0$

ASSUMPTIONS: (1) A closed system is under consideration. (2) There is no heat transfer.

ANALYSIS: The time rate of change of system energy is

$$\frac{dE}{dt} = \dot{Q}^{\,0} - \dot{W} \Rightarrow \frac{dE}{dt} = -\dot{W}$$

or

$$\frac{dE}{dt} = -1.2 \exp(-t/60)$$

Integrating from t=0 to any time t

$$\Delta E = -\int_0^t 1.2 \exp(-t/60)\, dt = \frac{-1.2}{(-1/60)}\left[\exp(-t/60) - 1\right]$$

$$= 72\left[\exp(-t/60) - 1\right] \text{ kJ}$$

(a) Using software to plot $\dot{W}$ and ΔE vs. t ranging from 0 to 250 s:

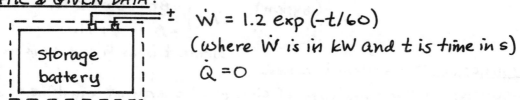

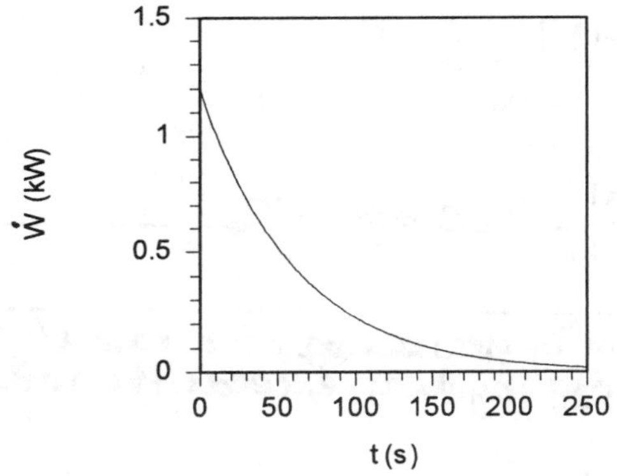

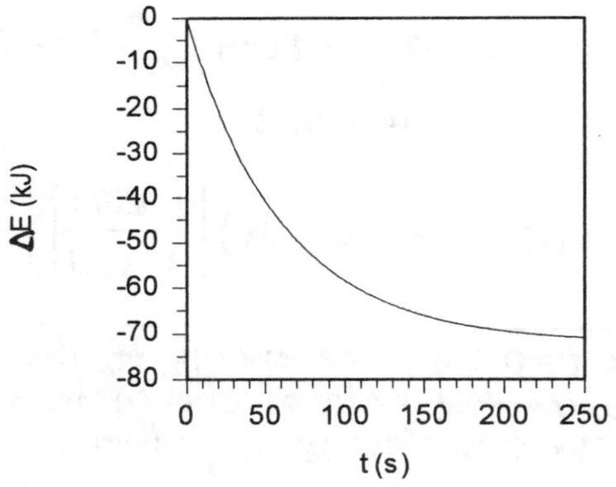

(b) In the limit as $t \to \infty$, we have $\dot{W} \to 0$ and $\Delta E \to -72$ kJ. The plots show that these limits are approached closely after about 225 s. For times greater than this, no more electric power is developed by the battery and the energy remains constant.

PROBLEM 2.63

KNOWN: A gas of known mass expands in a piston-cylinder from a specified initial state to a known final pressure. The pressure-volume for the process is given. Also, the specific internal energy change is known.

FIND: Determine the heat transfer for the process.

SCHEMATIC & GIVEN DATA:

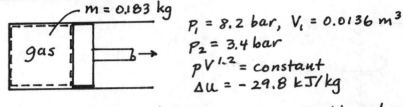

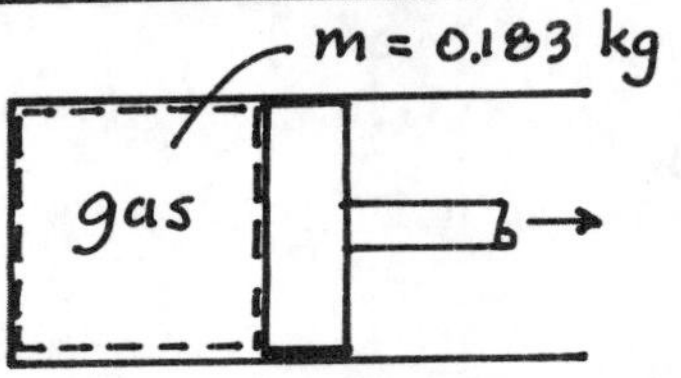

$p_1 = 8.2$ bar, $V_1 = 0.0136\ m^3$
$p_2 = 3.4$ bar
$pV^{1.2} = \text{constant}$
$\Delta u = -29.8$ kJ/kg

ASSUMPTION: (1) The gas is a closed system. (2) Kinetic and potential energy effects are negligible. (3) The process is polytropic with $pV^{1.2} = \text{constant}$.

ANALYSIS: The heat transfer can be found using an energy balance. First, find the work using Eq. 2.17, as follows:

$$W = \int_{V_1}^{V_2} p\,dV = \int_{V_1}^{V_2} \frac{\text{const.}}{V^{1.2}}\,dV$$

Integrating and simplifying

$$W = \frac{p_2V_2 - p_1V_1}{(1-1.2)}$$

The volume at the final state is $V_2 = (p_1/p_2)^{1/1.2} V_1 = 0.02832\ m^3$. Thus

$$W = \frac{(3.4\text{ bar})(0.02832\ m^3) - (8.2)(0.0136)}{(1-1.2)}\left|\frac{10^5\ N/m^2}{1\text{ bar}}\right|\left|\frac{1\text{ kJ}}{10^3\ N\cdot m}\right|$$

$$= 7.616\text{ kJ}$$

Now, to get the heat transfer, begin with the energy balance

$$\Delta KE^{\,0} + \Delta PE^{\,0} + \Delta U = Q - W$$

Solving for Q and noting that $\Delta U = m\,\Delta u$

$$Q = m\,\Delta u + W$$

$$= (0.183\text{ kg})(-29.8\text{ kJ/kg}) + (7.616\text{ kJ})$$

① $= 2.163\text{ kJ}$ ← Q

1. The heat transfer is positive, denoting energy transfer by heat to the gas as it expands.

PROBLEM 2.64

KNOWN: Air contained in a rigid well-insulated tank receives energy at a specified rate from a paddle wheel.

FIND: Determine the specific volume at the final state, the energy transfer by work, and the change in specific internal energy of the air.

SCHEMATIC & GIVEN DATA:

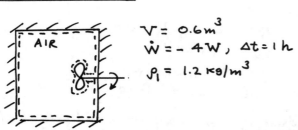

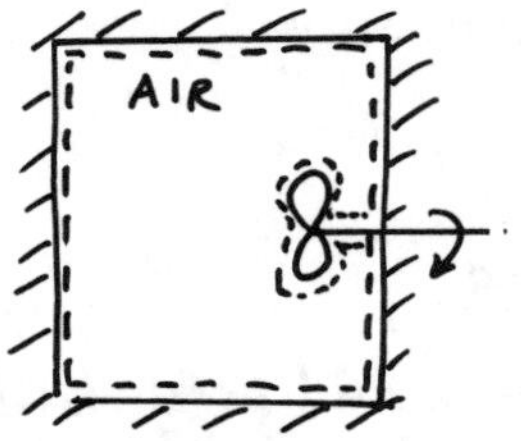

$V = 0.6\,m^3$

$\dot{W} = -4\,W$, $\Delta t = 1\,h$

$\rho_1 = 1.2\,kg/m^3$

ASSUMPTIONS: 1. The air is the closed system. 2. For the system, $Q = 0$ and there are no changes in kinetic energy and potential energy. 3. The initial and final states are equilibrium states.

ANALYSIS: (a) Since the mass and volume are each constant, the specific volume at states 1 and 2 are the same: $v_2 = v_1$. Thus, with $v_1 = 1/\rho_1$, we have

$$v_2 = \frac{1}{(1.2\,kg/m^3)} = 0.83\,m^3/kg \qquad \longleftarrow v_2$$

(b) To evaluate W, integrate

$$W = \int_0^{1h} \dot{W}\,dt = \int_0^{1h} (-4W)\,dt = (-4W)(1h)\left|\frac{1\,J/s}{1\,W}\right|\left|\frac{3600\,s}{1\,h}\right|\left|\frac{kJ}{10^3\,J}\right|$$

$$= -14.4\,kJ \qquad \longleftarrow W$$

(c) The change in specific internal energy can be found from an energy balance:

$$\cancel{\Delta KE} + \cancel{\Delta PE} + \Delta U = \cancel{Q} - W \quad \Rightarrow \quad m\,\Delta u = -W$$

or

$$u_2 - u_1 = \frac{(-W)}{m}$$

To find m

$$m = \rho_1 V = \left(1.2\,\frac{kg}{m^3}\right)(0.6\,m^3) = 0.72\,kg$$

So

$$u_2 - u_1 = \frac{-(-14.4\,kJ)}{0.72\,kg} = 20\,\frac{kJ}{kg} \qquad \longleftarrow u_2 - u_1$$

PROBLEM 2.65

KNOWN: An electric resistor transfers energy at a constant rate to a gas contained in a rigid tank. Heat transfer between the gas and the surroundings occurs as a specified function of time.

FIND: (a) Plot the time rate of change of energy of the gas for $0 \le t \le 20$ min, (b) Determine the net change in energy of the gas after 20 min., (c) Determine the cost of the electrical input to the resistor.

SCHEMATIC & GIVEN DATA:

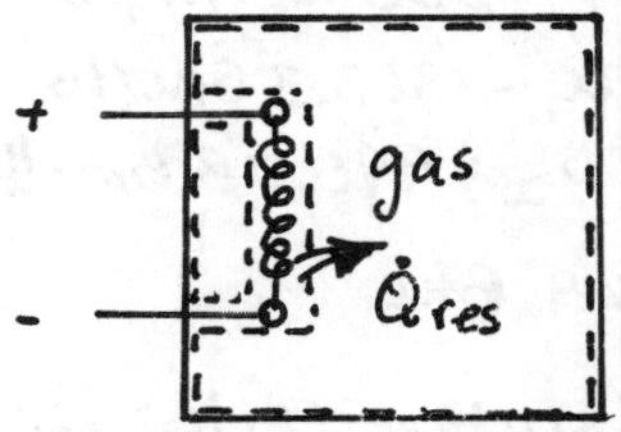

$\dot{Q}_{surr} = -50t$
where $\dot{Q}_{surr}$ is in Watts and t is in min.

$\dot{Q}_{res} = 1000$ W

cost of electricity = \$0.08/kW·h

ASSUMPTIONS: (1) The gas is a closed system. (2) The energy transfer from the resistor to the gas occurs at a constant rate. (3) There is no work.

ANALYSIS: (a) The time rate of change of energy is

$$\frac{dE}{dt} = \dot{Q} - \cancel{\dot{W}}^{\,0} \Rightarrow \frac{dE}{dt} = \dot{Q}_{surr} + \dot{Q}_{res}$$

or

$$\frac{dE}{dt} = -50t + 1000 \quad (t \text{ in min., } dE/dt \text{ in W}) \qquad (*)$$

Using software to plot dE/dt vs. t ranging from 0 to 20 min:

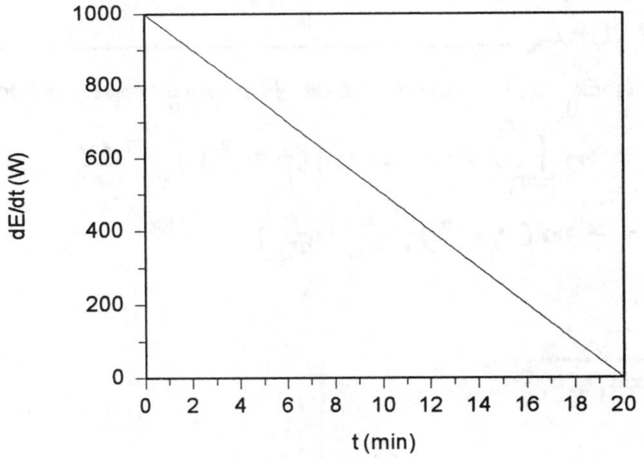

(b) Integrating (*) from t = 0 to 20 min

$$\Delta E = \int_0^{20} [-50t + 1000] = \left[-\left(\frac{50}{2}\right)t^2 + 1000t\right]_0^{20} = 10000 \text{ W·min}$$

$$= (10000 \text{ W·min}) \left|\frac{1 \text{ kJ/s}}{10^3 \text{ W}}\right| \left|\frac{60 \text{ s}}{1 \text{ min}}\right| = 600 \text{ kJ} \longleftarrow \Delta E$$

(c) Cost of electric input $= (1 \text{ kW})(20 \text{ min}) \left|\frac{1 \text{ h}}{60 \text{ min}}\right| (\$0.08/\text{kW·h})$

$= \$0.027 \longleftarrow$ cost of electric input

PROBLEM 2.66

KNOWN: A known mass of steam undergoes a polytropic process in a piston-cylinder assembly. The initial and final states and the heat transfer are specified.

FIND: Determine the work and the final specific volume.

SCHEMATIC & GIVEN DATA:

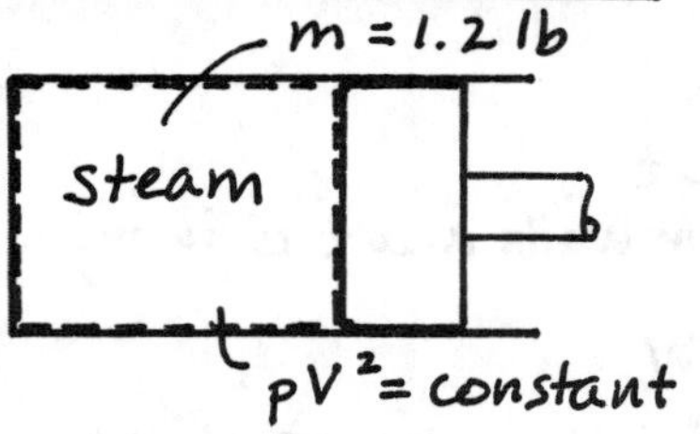

State 1: $p_1 = 500$ lbf/in², $v_1 = 1.701$ ft³/lb
$u_1 = 1363.3$ Btu/lb

State 2: $u_2 = 990.58$ Btu/lb

$Q = -342.9$ Btu

ASSUMPTIONS: (1) The steam is a closed system. (2) Kinetic and potential energy effects are negligible. (3) The process is polytropic with n=2.

ANALYSIS: To determine the work, begin with the energy balance

$$\Delta KE^{\,0} + \Delta PE^{\,0} + \Delta U = Q - W$$

With $\Delta U = m(u_2 - u_1)$, and solving for W

$$W = Q - m(u_2 - u_1)$$

$$= (-342.9 \text{ Btu}) - (1.2 \text{ lb})(990.58 - 1363.3) \text{ Btu/lb}$$

$$= 104.4 \text{ Btu} \longleftarrow \qquad W$$

To get v_2, begin with Eq. 2.17 and use the polytropic process expression

$$W = \int_{V_1}^{V_2} p\,dV = m\int_{v_1}^{v_2} p\,dv = m(p_1 v_1^2)\int_{v_1}^{v_2} \frac{dv}{v^2}$$

$$= -m(p_1 v_1^2)\left(\frac{1}{v_2} - \frac{1}{v_1}\right)$$

Solving for $1/v_2$

$$\frac{1}{v_2} = \frac{1}{v_1} - \frac{W}{m p_1 v_1^2}$$

$$= \left(\frac{1}{1.701 \text{ ft}^3/\text{lb}}\right) - \frac{(104.4 \text{ Btu})}{(1.2 \text{ lb})(500 \text{ lbf/in}^2)(1.701 \text{ ft}^3/\text{lb})^2}\left|\frac{778 \text{ ft·lbf}}{1 \text{ Btu}}\right|\left|\frac{1 \text{ ft}^2}{144 \text{ in}^2}\right|$$

$$= 0.2630$$

Thus

① $$v_2 = \frac{1}{0.2630} = 3.802 \text{ ft}^3/\text{lb} \longleftarrow \qquad v_2$$

1. The increase in volume is consistent with the positive sign for work, denoting energy transfer out of the system by work.

PROBLEM 2.67

KNOWN: A known mass of gas undergoes a polytropic process between two specified states. The relationship between pressure, volume, and internal energy is known for the gas.

FIND: Determine the heat transfer.

SCHEMATIC & GIVEN DATA:

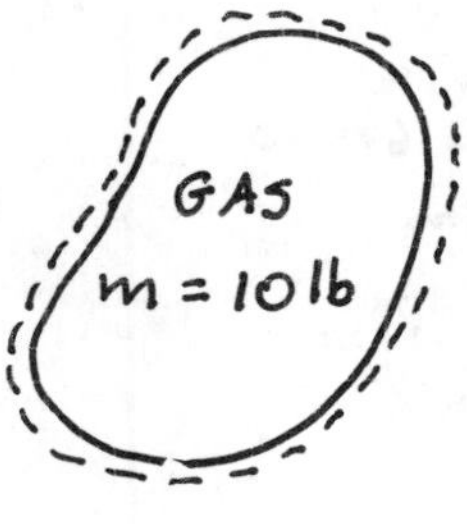

$p_1 = 60$ lbf/in², $v_1 = 6.0$ ft³/lb

$p_2 = 20$ lbf/in²

$pv^{1.3} =$ constant

For the gas

$u = (0.2651)\,pv - 95.436$

(where v is in ft³/lb, p is in lbf/in², and u is in Btu/lb)

ASSUMPTIONS: (1) The gas is a closed system. (2) Kinetic and potential energy effects are negligible. (3) The process is polytropic with n = 1.3.

ANALYSIS: The heat transfer is determined from an energy balance. First, determine the work beginning with Eq. 2.17

$$W = \int_{V_1}^{V_2} p\,dV = \int_{V_1}^{V_2} \frac{\text{const.}}{V^{1.3}}\,dV = \left(\frac{p_2V_2 - p_1V_1}{1-1.3}\right) = m\left(\frac{p_2v_2 - p_1v_1}{1-1.3}\right)$$

Evaluating v_2

$$v_2 = \left(\frac{p_1}{p_2}\right)^{\frac{1}{1.3}} v_1 = \left(\frac{60}{20}\right)^{\frac{1}{1.3}}(6.0) = 13.97 \text{ ft}^3/\text{lb}$$

Thus

$$W = (10\text{ lb})\left[\frac{(20\text{ lbf/in}^2)(13.97\text{ ft}^3/\text{lb}) - (60)(6.0)}{(1-1.3)}\right]\left|\frac{144\text{ in}^2}{1\text{ ft}^2}\right|\left|\frac{1\text{ Btu}}{778\text{ ft}\cdot\text{lbf}}\right|$$

$$= 497.3 \text{ Btu}$$

Now, writing the energy balance

$$\Delta KE + \Delta PE + \Delta U = Q - W$$

Noting that $\Delta U = m(u_2 - u_1)$ and solving for Q

$$Q = m(u_2 - u_1) + W$$

Evaluating $(u_2 - u_1)$

$$(u_2 - u_1) = [(0.2651)(20)(13.97) - 95.436] - [(0.2651)(60)(6.0) - 95.436]$$

$$= -21.37 \text{ Btu/lb}$$

Finally

$$Q = (10\text{ lb})(-21.37\text{ Btu/lb}) + (497.3\text{ Btu})$$

① $= 283.6$ Btu ← Q

1. The signs for work and heat denote energy from the system by work to the system by heat, respectively.

PROBLEM 2.68

KNOWN: An electrical resistor transfers energy to a gas contained in a vertical piston-cylinder assembly.

FIND: Determine the change in elevation of the piston.

SCHEMATIC & GIVEN DATA:

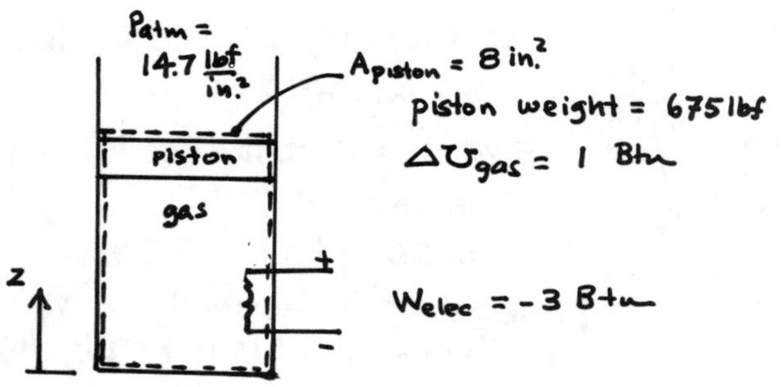

ASSUMPTIONS: 1. The system is the piston plus the gas. 2. For the piston, $\Delta U = 0$, $\Delta KE = 0$. For the gas, $\Delta PE = 0$, $\Delta KE = 0$. 3. For the system, $Q = 0$. 4. Friction between piston and cylinder can be ignored. (5) The mass of the electrical resistor is negligible. So $\Delta U = 0$ for the resistor.

ANALYSIS: An energy balance for the system reads

$$\cancel{\Delta KE}^{0} + [(\Delta PE)_{piston} + \cancel{(\Delta PE)_{gas}}^{0}] + [\cancel{(\Delta U)_{piston}}^{0} + (\Delta U)_{gas} + \cancel{(\Delta U)_{resistor}}^{0}] = \cancel{Q}^{0} - W$$

or

$$(\Delta PE)_{piston} + (\Delta U)_{gas} = -W$$

For the piston

$$(\Delta PE)_{piston} = mg(z_2 - z_1)$$

For the system consisting of freely-moving piston (assumption 4) and gas, the net work is the sum of the electrical work and the work done at the top of the piston in displacing the surrounding atmosphere:

$$W = W_{elect} + \int_{z_1}^{z_2} (P_{atm} A_{piston})\,dz \Rightarrow W = W_{elect} + (P_{atm} A_{piston})(z_2 - z_1)$$

Collecting results

$$mg(z_2 - z_1) + (\Delta U)_{gas} = -[W_{elect} + (P_{atm} A_{piston})(z_2 - z_1)]$$

Solving for the change in elevation of the piston

$$(z_2 - z_1) = \frac{-(\Delta U)_{gas} - W_{elect}}{[(mg) + (P_{atm} A_{piston})]}$$

$$= \frac{[-(1\ \text{Btu}) - (-3\ \text{Btu})]\left|\frac{778\ \text{ft}\cdot\text{lbf}}{1\ \text{Btu}}\right|}{[675\ \text{lbf} + (14.7\ \frac{\text{lbf}}{\text{in.}^2})(8\ \text{in.}^2)]} = 1.96\ \text{ft} \longleftarrow z_2 - z_1$$

PROBLEM 2.69

KNOWN: A known quantity of air undergoes a process in a vertical piston-cylinder assembly. The initial and final volumes are given, and the heat transfer is specified.

FIND: Determine the change in specific internal energy.

SCHEMATIC & GIVEN DATA:

$p_{atm} = 100$ kPa

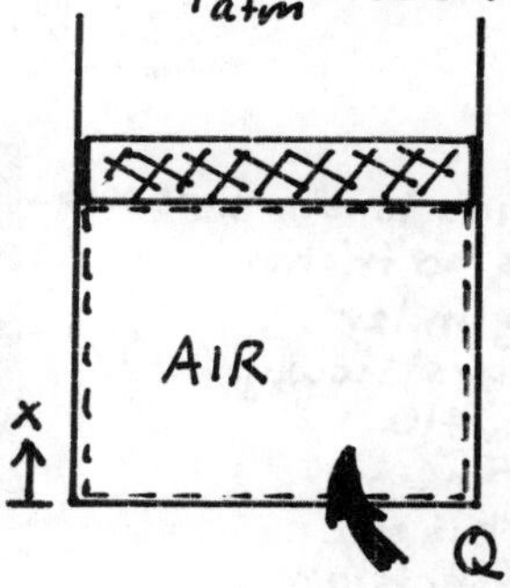

$m_{pist} = 50$ kg

$A_{pist} = 0.01\ m^2$

$m_{air} = 4$ g $= 0.004$ kg

$V_1 = 5$ L $= 0.005\ m^3$

$V_2 = 0.0025\ m^3$

$Q = -1.41$ kJ

ASSUMPTIONS: (1) The air is a closed system. (2) Kinetic and potential energy effects are negligible for the air. (3) There is no friction between the piston and the cylinder wall. (4) The process occurs slowly with no acceleration of the piston. (5) The acceleration of gravity is constant, $g = 9.81\ m/s^2$.

ANALYSIS: For the piston, $\Sigma F_x = 0$. Thus, if p is the pressure exerted by the air

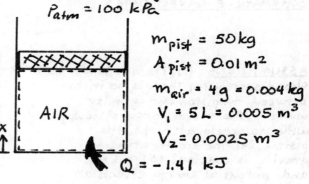

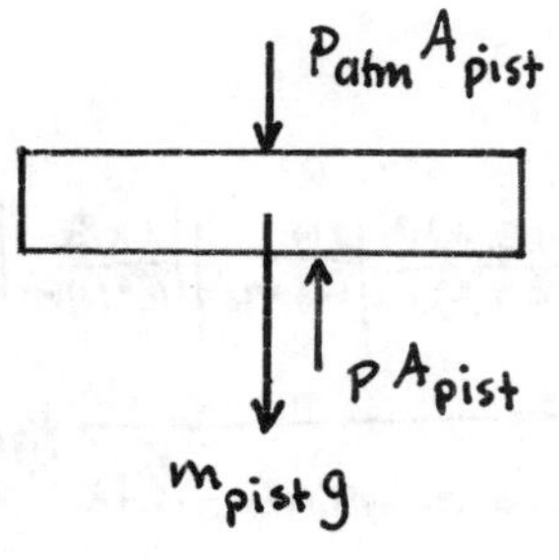

$$pA_{pist} = p_{atm} A_{pist} + m_{pist}\, g$$

$$p = p_{atm} + \frac{m_{pist}\, g}{A_{pist}}$$

$$= 100\ kPa + \frac{(50\ kg)(9.81\ m/s^2)}{(0.01\ m^2)} \left|\frac{1\ N}{1\ kg \cdot m/s}\right| \left|\frac{1\ kPa}{10^3\ N/m^2}\right|$$

$$= 149.05\ kPa$$

To find the work for the process, use Eq. 2.17. Noting that the pressure is constant

$$W = \int_{V_1}^{V_2} p\,dV = p(V_2 - V_1)$$

$$= (149.05\ kPa)(0.0025 - 0.005)\ m^3 \left|\frac{10^3\ N/m^2}{1\ kPa}\right| \left|\frac{1\ kJ}{10^3\ N \cdot m}\right|$$

$$= -0.3726\ kJ$$

Now, the energy balance reduces to $\cancel{\Delta KE}^{0} + \cancel{\Delta PE}^{0} + \Delta U = Q - W$. Thus, with $\Delta U = m_{air}\, \Delta u$

$$\Delta u = \frac{Q - W}{m_{air}} = \frac{(-1.41\ kJ) - (-0.3726\ kJ)}{(0.004\ kg)}$$

① $$= -259.3\ kJ/kg \longleftarrow \Delta u$$

1. The net amount of energy transfer is from the gas to the surroundings resulting in a decrease of the internal energy of the system.

PROBLEM 2.70

KNOWN: A gas undergoes a process in a piston-cylinder assembly. The piston is constrained by a spring with a linear force-displacement relation.

FIND: Determine (a) the initial pressure of the gas, (b) the work done by the gas on the piston, and (c) the heat transfer.

SCHEMATIC & GIVEN DATA:

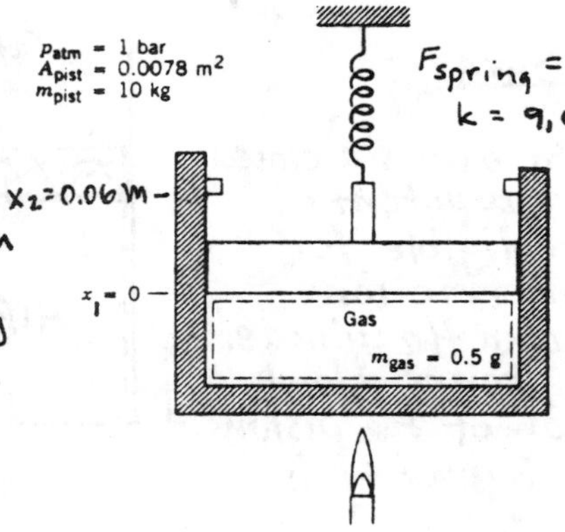

$F_{spring} = kx$

$k = 9{,}000$ N/m

$g = 9.81$ m/s²

for the gas
$u_1 = 210$ kJ/kg
$u_2 = 335$ kJ/kg

ASSUMPTIONS: (1) The gas is a closed system. (2) There is no friction between the piston and cylinder wall. (3) The process occurs slowly with no acceleration of the piston. (4) The acceleration of gravity is constant. (5) Kinetic and potential energy effects are negligible.

ANALYSIS: (a) Initially, the spring exerts no force on the piston, which is at rest. Thus, with assumption 2 the force exerted by the gas on the bottom of the piston equals the piston weight plus the force of the atmosphere acting on the top of the piston. That is

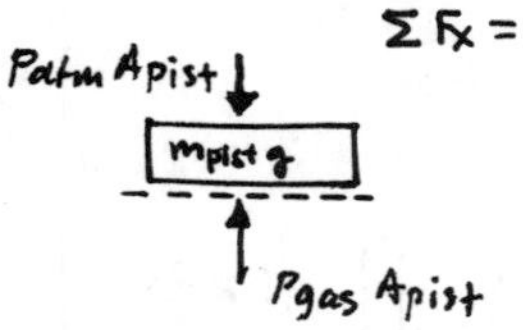

$$\Sigma F_x = 0 \Rightarrow \quad p_{gas} A_{pist} = (m_{pist}\, g) + p_{atm} A_{pist}$$

$$p_{gas} = p_{atm} + \frac{m_{pist} g}{A_{pist}} = 1\text{ bar}\left|\frac{100\text{ kPa}}{1\text{ bar}}\right| + \frac{(10\text{ kg})(9.81\text{ m/s}^2)}{(7.8\times10^{-3}\text{ m}^2)}\left|\frac{1\text{ N}}{1\text{ kg}\cdot\text{m/s}^2}\right|\left|\frac{1\text{ kPa}}{10^3\text{ N/m}^2}\right|$$

$$= 112.6\text{ kPa} \longleftarrow p_{gas}$$

(b) As the piston moves from $x = 0$ to $x = 0.06$ m, the spring force acts. Then, with assumptions 2 and 3, $\Sigma F_x = 0$ reads

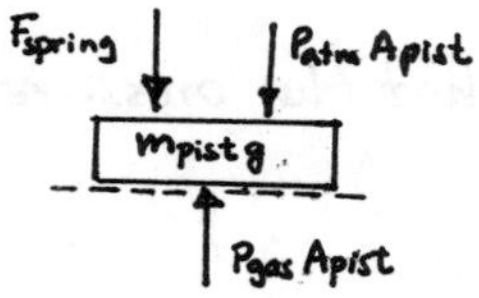

$$p_{gas} A_{pist} = (m_{pist}\, g) + p_{atm} A_{pist} + F_{spring}$$
$$= (m_{pist}\, g) + p_{atm} A_{pist} + kx$$

The work done by the gas on the piston is given by

$$W = \int_{x_1}^{x_2} (p_{gas} A_{pist})\, dx = \int_{x_1}^{x_2} [p_{atm} A_{pist} + m_{pist}\, g + kx]\, dx$$

$$= \left[[p_{atm} A_{pist} + m_{pist}\, g]\, x + \frac{kx^2}{2}\right]_0^{0.06}$$

$$= \left[\left(10^5 \frac{\text{N}}{\text{m}^2}\right)(7.8\times10^{-3}\text{ m}^2) + (10\text{ kg})\left(9.81 \frac{\text{m}}{\text{s}^2}\right)\left|\frac{1\text{ N}}{1\text{ kg}\cdot\text{m/s}^2}\right|\right](0.06\text{ m}) + \left(9\times10^3 \frac{\text{N}}{\text{m}}\right)\left(\frac{(0.06\text{ m})^2}{2}\right)$$

$$= [780\text{ N} + 98.1\text{ N}](0.06\text{ m}) + 16.2\text{ N}\cdot\text{m} = 68.89\text{ N}\cdot\text{m}$$

$$= 68.89\text{ N}\cdot\text{m}\left|\frac{1\text{ J}}{1\text{ N}\cdot\text{m}}\right| = 68.89\text{ J} \longleftarrow W$$

(c) The energy balance for the system consisting of the gas reduces to $\Delta U = Q - W$. Then, with $\Delta U = m_{gas}(u_2 - u_1)$

$$Q = m_{gas}(u_2 - u_1) + W \Rightarrow Q = (0.5\text{ g})(335 - 210)\frac{\text{J}}{\text{g}} + 68.89\text{ J} = 131.39\text{ J} \longleftarrow Q$$

PROBLEM 2.71

KNOWN: A system undergoes a cycle consisting of four pocesses in series.

FIND: (a) Complete the table of energy values for the cycle, and (b) detcermine whether the cycle is a power cycle or a refrigeration cycle.

SCHEMATIC & GIVEN DATA:

Process	ΔU	Q	W
1–2	a	c	−610
2–3	670	d	230
3–4	b	0	920
4–1	−360	e	0

ASSUMPTIONS: (a) The system is a closed system. (b) Kinetic and potential energy effects can be neglected.

ANALYSIS: (a) Beginning with Process 3-4

b. $\Delta U = Q - W$: $\Delta U = 0 - 920 = -920 \Rightarrow b = -920$

Now, for the cycle, $\Sigma(\Delta U) = 0$. Thus

a. $a + 670 + (-920) + (-360) = 0 \Rightarrow a = 610$

$\Delta U = Q - W$

c. $610 = c - (-610) \Rightarrow c = 0$

d. $670 = d - 230 \Rightarrow d = 900$

e. $-360 = e - 0 \Rightarrow e = -360$

(b) For the cycle

$$W_{cycle} = W_{12} + W_{23} + W_{34} + W_{41}$$

$$= (-610) + (230) + (920) + 0 = 540$$

① Since $W_{cycle} > 0$, the cycle is a <u>power cycle</u>.

---

1. For the cycle, note that

$$Q_{cycle} = Q_{12} + Q_{23} + Q_{34} + Q_{41} = (0) + (900) + (0) + (-360) = 540$$

Thus $Q_{cycle} = W_{cycle}$, as expected.

PROBLEM 2.72

KNOWN: A system undergoes a thermodynamic cycle consisting of four processes in series.

FIND: Complete the table of energy values provided for the cycle and determine whether the cycle is a power cycle or a refrigeration cycle.

SCHEMATIC & GIVEN DATA:

Process	ΔU	ΔKE	ΔPE	ΔE	Q	W
1	950	50	0	c	1000	g
2	a	0	50	−450	e	450
3	−650	b	0	−600	f	0
4	200	−100	−50	d	0	h

ANALYSIS: For a cycle, the overall change in every property value is zero.

a. $\Sigma(\Delta U) = 0$: $950 + a + (-650) + 200 = 0 \Rightarrow a = -500$

b. $\Sigma(\Delta KE) = 0$: $50 + 0 + b + (-100) = 0 \Rightarrow b = 50$

$\Delta E = \Delta U + \Delta KE + \Delta PE$:

c. $c = 950 + 50 + 0 \Rightarrow c = 1000$

d. $d = 200 + (-100) + (-50) \Rightarrow d = 50$

$\Delta E = Q - W$:

e. $(-450) = e - (450) \Rightarrow e = 0$

f. $(-600) = f - 0 \Rightarrow f = -600$

g. $1000 = 1000 - g \Rightarrow g = 0$

h. $50 = 0 - h \Rightarrow h = -50$

For the overall cycle, $W_{cycle} = Q_{cycle}$

$W_{cycle} = 0 + 450 + 0 + (-50) = 400$

$Q_{cycle} = 1000 + 0 + (-600) + 0 = 400$

} Checks the calculations above. Also, note that $\Sigma(\Delta E) = 0$.

Since $W_{cycle} > 0$, the cycle is a power cycle

PROBLEM 2.73

KNOWN: A gas undergoes a thermodynamic cycle consisting of three processes.

FIND: Determine the heat transfer and work for process 2-3, and whether the cycle is a power cycle or a refrigeration cycle.

SCHEMATIC & GIVEN DATA: The following data are given for each process:

Process 1-2: compression with $pV = \text{const.}$
$p_1 = 1$ bar, $V_1 = 1.6\ m^3$ to $V_2 = 0.2 m^3$
$U_2 - U_1 = 0$

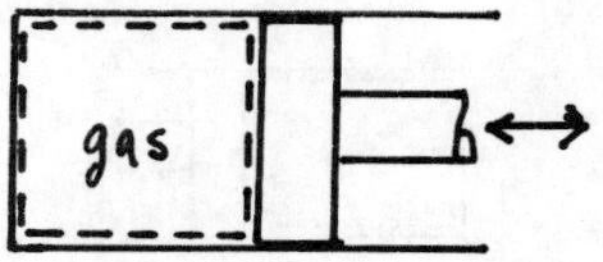

Process 2-3: Constant pressure to $V_3 = V_1$

Process 3-1: Constant volume, $U_1 - U_3 = -3549$ kJ

ASSUMPTIONS: (1) The gas is a closed system. (2) Neglect kinetic and potential energy changes. (3) The compression from state 1 to 2 is a polytropic process.

ANALYSIS: To find the work for Process 2-3, use Eq. 2.17, with constant pressure

$$W_{23} = \int_{V_2}^{V_3} p\,dV = p_2(V_3 - V_2)$$

Using the p-V relation for process 1-2; $p_2 = \left(\frac{V_1}{V_2}\right)p_1 = 8$ bars. Thus, with $V_3 = V_1$

$$W_{23} = (8\text{ bars})(1.6 - 0.2)m^3 \left|\frac{10^5\ N/m^2}{1\text{ bar}}\right|\left|\frac{1\text{ kJ}}{10^3\ N\cdot m}\right| = 1120\text{ kJ} \leftarrow W_{23}$$

The energy balance for process 2-3 reduces to

$$Q_{23} = (U_3 - U_2) + W_{23}$$

To get $U_3 - U_2$, note that for the cycle, $\Delta U\Big]_{cycle} = 0$. Thus

$$(U_2 - U_1)^{\nearrow 0} + (U_3 - U_2) + (U_1 - U_3) = 0 \Rightarrow (U_3 - U_2) = -(U_1 - U_3) = 3549\text{ kJ}$$

Finally, for process 2-3

$$Q_{23} = (3549\text{ kJ}) + (1120\text{ kJ}) = 4669\text{ kJ} \leftarrow Q_{23}$$

Next, for process 1-2, $\Delta U = 0$, so $Q_{12} = W_{12}$. Using Eq. 2.17

$$Q_{12} = W_{12} = \int_{V_1}^{V_2} p\,dV = p_1 V_1 \ln\frac{V_2}{V_1} = (1)(1.6)\ln\left(\frac{0.2}{1.6}\right)\left|\frac{10^5}{10^3}\right| = -332.7\text{ kJ}$$

And, for process 3-1: $W_{31} = 0$, and $Q_{31} = U_1 - U_3 = -3549$ kJ. Collecting various results

$$W_{cycle} = W_{12} + W_{23} + W_{31}^{\nearrow 0} = (-332.7) + (1120) = +787.3\text{ kJ}$$

Since $W_{cycle} > 0$, the cycle is a power cycle.

PROBLEM 2.74

KNOWN: A gas undergoes a thermodynamic cycle consisting of three processes.

FIND: Sketch the cycle on a p-V diagram. Calculate W_{cycle}, Q_{23}, Q_{31}, and determine whether the cycle is a power or refrigeration cycle.

SCHEMATIC & GIVEN DATA:

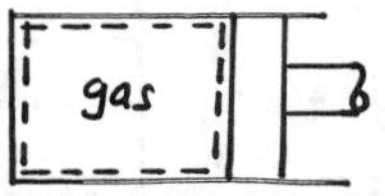

1-2: $V = 0.028\ m^3$, $U_2 - U_1 = 26.4\ kJ$

2-3: $pV = \text{constant}$, $U_3 = U_2$

3-1: $p = 1.4$ bar, $W_{31} = -10.5\ kJ$

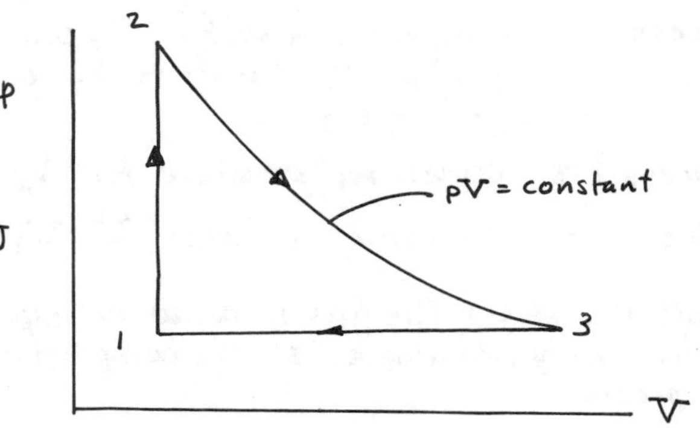

ASSUMPTIONS: 1. The gas is the closed system. 2. For the system, $\Delta KE = \Delta PE = 0$. 3. Volume change is the only work mode.

ANALYSIS: (b) $W_{cycle} = W_{12} + W_{23} + W_{31}$. Since volume is constant and volume change is the only work mode, $W_{12} = 0$. To find W_{23}

$$W_{23} = \int_{V_2}^{V_3} p\,dV = \int_{V_2}^{V_3} \frac{c}{V}\,dV = c \ln\frac{V_3}{V_2} = p_3 V_3 \ln\frac{V_3}{V_2} \quad , V_2 = V_1$$

To evaluate V_3

$$W_{31} = \int_{V_3}^{V_1} p\,dV = p(V_1 - V_3) \Rightarrow V_3 = V_1 - W_{31}/p$$

or

$$V_3 = 0.028\,m^3 - \frac{(-10.5\,kJ)}{(1.4\text{ bar})}\left|\frac{1\text{ bar}}{10^5\,N/m^2}\right|\left|\frac{10^3\,N\cdot m}{1\,kJ}\right| = 0.103\ m^3$$

Then

$$W_{23} = (1.4\,\text{bar})\left|\frac{10^5\,N/m^2}{1\text{ bar}}\right|(0.103\,m^3)\ln\left(\frac{0.103}{0.028}\right)\left|\frac{1\,kJ}{10^3\,N\cdot m}\right| = 18.78\ kJ$$

Finally,

$$W_{cycle} = 0 + 18.78 + (-10.5) = 8.28\ kJ.$$ The cycle is a power cycle. ← W_{cycle}

(c) An energy balance for 2-3 gives Q_{23}: $\cancel{\Delta U}^{0} + \cancel{\Delta KE}^{0} + \cancel{\Delta PE}^{0} = Q_{23} - W_{23} \Rightarrow Q_{23} = W_{23}$ ← Q_{23}

(d) To evaluate Q_{31} begin with an energy balance together with assumption 2

$$U_1 - U_3 = Q_{31} - W_{31} \Rightarrow Q_{31} = (U_1 - U_3) + W_{31}.$$

Since there is no overall change in internal energy for the cycle, $\Sigma(\Delta U) = 0$

$$(U_2 - U_1) + (U_3 - U_2) + (U_1 - U_3) = 0 \Rightarrow$$

$$(U_1 - U_3) = -26.4 - (0) = -26.4\ kJ$$

Then

① $$Q_{31} = (-26.4) + (-10.5) = -36.9\ kJ$$ ← Q_{31}

1. As a check, note that $Q_{cycle} = W_{cycle}$, $Q_{cycle} = Q_{12} + Q_{23} + Q_{31}$. An energy balance for 1-2 gives $Q_{12} = 26.4\ kJ$. So $Q_{cycle} = 26.4 + 18.78 + (-36.9) = 8.28\ kJ$, which checks the result of part (b).

PROBLEM 2.75

KNOWN: A closed system undergoes a cycle consisting of three processes.

FIND: Sketch the cycle on a p-V diagram and calculate the net work for the cycle and the heat transfer for process 2-3.

SCHEMATIC & GIVEN DATA: The following data are given for each process:

Process 1-2: adiabatic compression with $pV^{1.4} = \text{const.}$ from $p_1 = 50\ \text{lbf/in}^2$, $V_1 = 3\ \text{ft}^3$ to $V_2 = 1\ \text{ft}^3$

Process 2-3: constant volume

Process 3-1: Constant pressure, $U_1 - U_3 = 46.7$ Btu

SYSTEM

ASSUMPTIONS: (1) The system is closed. (2) Kinetic and potential energy effects are negligible. (3) Process 1-2 is polytropic.

ANALYSIS: (a) Since process 1-2 is a polytropic compression, the p-V diagram for the cycle is

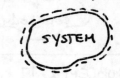

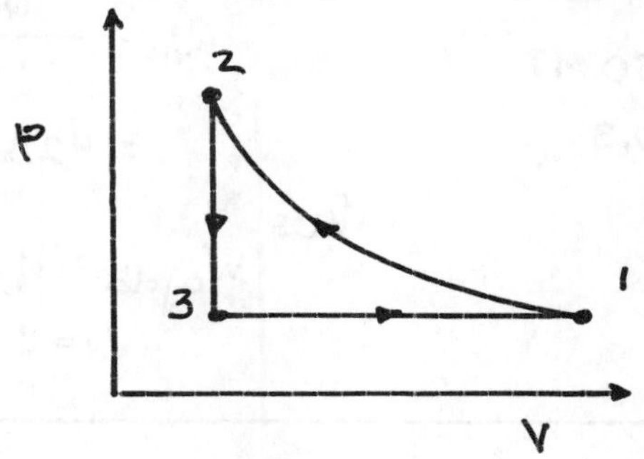

(b) Use Eq. 2.17 to evaluate the work for process 1-2

$$W_{12} = \int_{V_1}^{V_2} p\,dV = \text{const.} \int_{V_1}^{V_2} \frac{dV}{V^{1.4}} = p_1 V_1^{1.4} \frac{(V_2^{-.4} - V_1^{-.4})}{(-.4)}$$

$$= \left(50\ \frac{\text{lbf}}{\text{in}^2}\right)(3\ \text{ft}^3)^{1.4} \left[\frac{(1\ \text{ft}^3)^{-.4} - (3\ \text{ft}^3)^{-.4}}{(-.4)}\right] \left|\frac{144\ \text{in}^2}{1\ \text{ft}^2}\right| \left|\frac{1\ \text{Btu}}{778\ \text{ft-lbf}}\right|$$

$$= -38.3\ \text{Btu}$$

For process 2-3 : $W_{23} = 0$

Finally, for process 3-1 use Eq. 2.17 : $W_{31} = \int_{V_3}^{V_1} p\,dV = p_1 (V_1 - V_3)$

$$W_{31} = \left(50\ \frac{\text{lbf}}{\text{in}^2}\right)(3-1)\ \text{ft}^3 \left|\frac{144\ \text{in}^2}{1\ \text{ft}^2}\right| \left|\frac{1\ \text{Btu}}{778\ \text{ft-lbf}}\right| = +18.51\ \text{Btu}$$

Thus

$$W_{cycle} = W_{12} + W_{23} + W_{31} = -19.79\ \text{Btu} \longleftarrow \underline{W_{cycle}}$$

(c) For the overall cycle $Q_{cycle} = W_{cycle}$

$$\cancelto{0}{Q_{12}} + Q_{23} + Q_{31} = W_{cycle}$$

$$Q_{23} = W_{cycle} - Q_{31}$$

For process 3-1 : $\Delta KE + \Delta PE + (U_1 - U_3) = Q_{31} - W_{31} \Rightarrow Q_{31} = U_1 - U_3 + W_{31}$

$$Q_{31} = 46.7 + 18.51 = +65.21\ \text{Btu}$$

Finally $\quad Q_{23} = W_{cycle} - Q_{31} = -19.79 - 65.21 = -85\ \text{Btu} \longleftarrow \underline{Q_{23}}$

PROBLEM 2.76

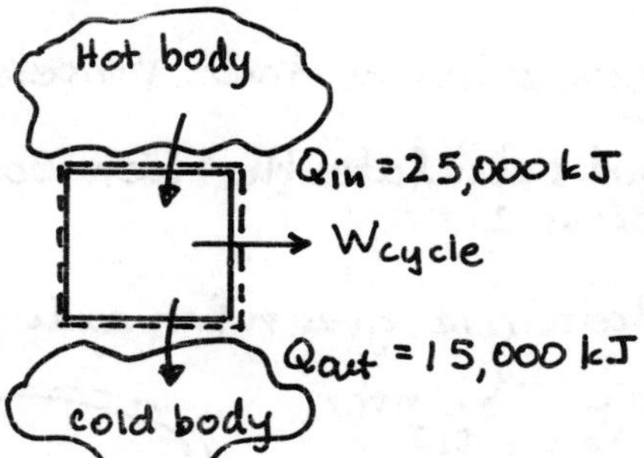

$$W_{cycle} = Q_{in} - Q_{out}$$
$$= 25{,}000 - 15{,}000 = 10{,}000 \text{ kJ} \leftarrow W_{cycle}$$
$$\eta = \frac{W_{cycle}}{Q_{in}} = \frac{10{,}000}{25{,}000} = 0.4\ (40\%) \leftarrow \eta$$

PROBLEM 2.77

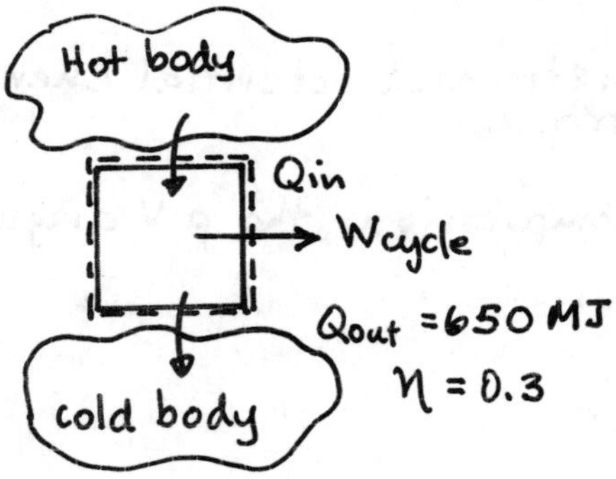

$$\eta = \frac{W_{cycle}}{Q_{in}} = \frac{Q_{in} - Q_{out}}{Q_{in}} = 1 - \frac{Q_{out}}{Q_{in}}$$

Solving for Q_{in}

$$Q_{in} = \frac{Q_{out}}{1-\eta} = \frac{650}{(1-.3)}$$
$$= 928.6 \text{ MJ} \leftarrow Q_{in}$$

Thus

$$W_{cycle} = \eta Q_{in} = (.3)(928.6)$$
$$= 278.6 \text{ MJ} \leftarrow W_{cycle}$$

PROBLEM 2.78

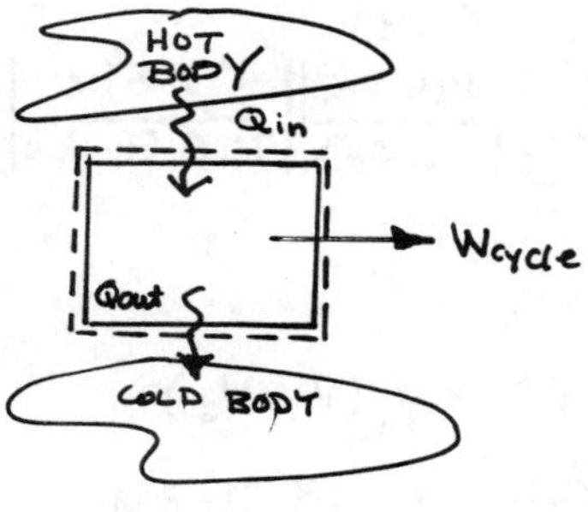

$W_{cycle} = 8 \times 10^6$ Btu

$Q_{out} = 12 \times 10^6$ Btu

Energy Balance: $W_{cycle} = Q_{cycle}$
$$= Q_{in} - Q_{out}$$

$$\Rightarrow \quad Q_{in} = W_{cycle} + Q_{out}$$
$$= 8 \times 10^6 + 12 \times 10^6$$
$$= 20 \times 10^6 \text{ Btu}$$

$$\eta = \frac{W_{cycle}}{Q_{in}} = \frac{8 \times 10^6 \text{ Btu}}{20 \times 10^6 \text{ Btu}} = 0.4\ (40\%) \leftarrow \eta$$

PROBLEM 2.79

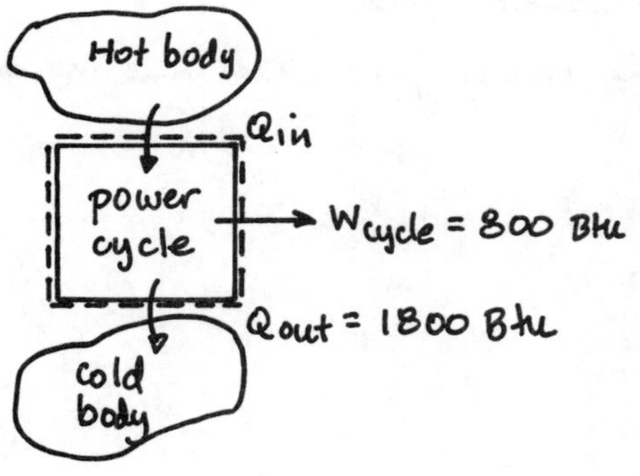

Energy balance: $W_{cycle} = Q_{in} - Q_{out}$

$$Q_{in} = W_{cycle} + Q_{out}$$
$$= 800 + 1800 = 2600 \text{ Btu}$$

$$\eta = \frac{W_{cycle}}{Q_{in}} = \frac{800}{2600}$$

$$= 0.308 \; (30.8\%) \longleftarrow \eta$$

PROBLEM 2.80

KNOWN: Operating data are provided for a power cycle.

FIND: Determine the net rate power is developed, the net work output annually, and the value of the net work, in \$/year.

SCHEMATIC & GIVEN DATA:

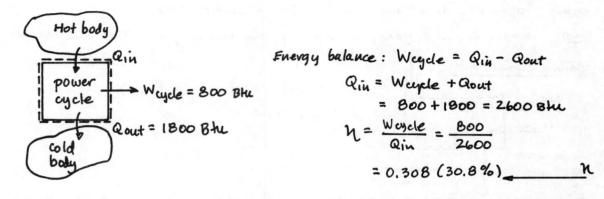

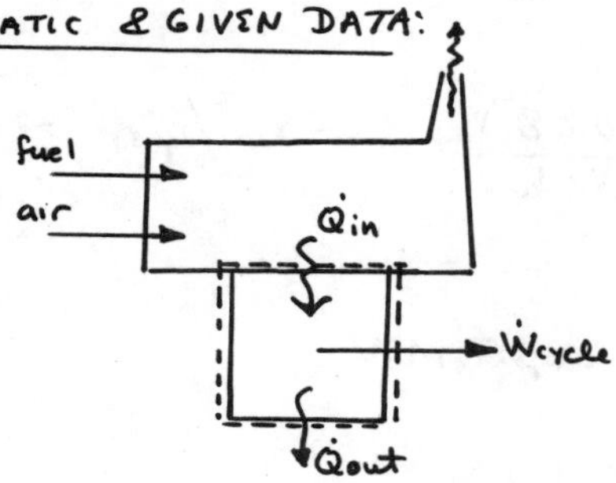

$\dot{Q}_{in} = 300\text{MW}$

$\eta = 33.3\%$

ASSUMPTIONS: 1. The system undergoes a power cycle. 2. The cycle operates steadily for 8000 h annually. 3. The value of the net work is \$0.08 /kW·h

ANALYSIS: (a) To determine $\dot{W}_{cycle}$

$$\eta = \frac{\dot{W}_{cycle}}{\dot{Q}_{in}} \Rightarrow \dot{W}_{cycle} = \eta \dot{Q}_{in} = (0.333)(300\text{MW}) = 99.9 \text{ MW} \longleftarrow \eta$$

(b) With assumption 2

$$W_{cycle} = (99.9 \text{ MW})\left|\frac{10^3 \text{ kW}}{1 \text{ MW}}\right|\left(\frac{8000 \text{ h}}{\text{year}}\right) = 799.2 \times 10^6 \frac{\text{kW·h}}{\text{year}} \longleftarrow W_{cycle}$$

(c) With assumption 3

$$\dot{\$} = \left(799.2 \times 10^6 \frac{\text{kW·h}}{\text{year}}\right)\left(\frac{\$0.08}{\text{kW·h}}\right) = \$63.94 \times 10^6 \text{/year} \longleftarrow \$$$

PROBLEM 2.81

KNOWN: Operating data are provided for a power cycle.

FIND: Determine, in \$/year, the value of the power generated and the cost of $\dot{Q}_{in}$.

SCHEMATIC & GIVEN DATA:

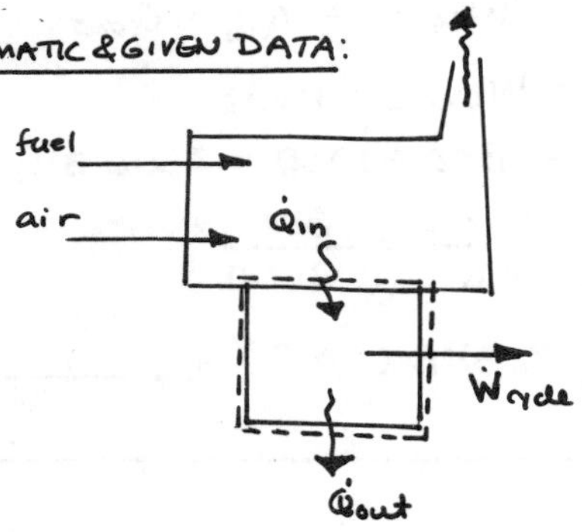

$\eta = 35\%$

$\dot{W}_{cycle} = 100 MW$

ASSUMPTIONS: 1. The system undergoes a thermodynamic cycle. 2. The power cycle operates steadily for 8000 h per year. 3. Electricity is valued at \$0.08/kW·h and $\dot{Q}_{in}$ at \$4.50/GJ.

ANALYSIS: (a) Using assumptions 2,3

$$\dot{\$}_{electricity} = (100\ MW)\left|\frac{10^3\ kW}{1\ MW}\right|\left(\frac{8000\ h}{year}\right)\left(\frac{\$0.08}{kW\cdot h}\right) = \$64\times10^6/year \leftarrow \dot{\$}_{electricity}$$

(b) With

$$\eta = \frac{\dot{W}_{cycle}}{\dot{Q}_{in}} \Rightarrow \dot{Q}_{in} = \frac{\dot{W}_{cycle}}{\eta} = \frac{100\ MW}{0.35} = 285.7\ MW$$

① $$\dot{\$}_{Q_{in}} = (285.7\ MW)\left|\frac{1\ GJ/s}{10^3\ MW}\right|\left|\frac{3600\ s}{h}\right|\left(\frac{8000\ h}{year}\right)\left(\frac{\$4.50}{GJ}\right) = \$37\times10^6/year \leftarrow \dot{\$}_{Q_{in}}$$

1. This is an operating cost. Additional operating costs include the cost for the plant, a capital cost.

PROBLEM 2.82

(a) Window air conditioner.

cold body : environment inside the building served
hot body : outside environment

(b) Nuclear power plant.

hot body : coolant circulated through the reactor core
cold body : sea water

(c) Ground source heat pump.

hot body : environment inside the building served
cold body : ground

PROBLEM 2.83

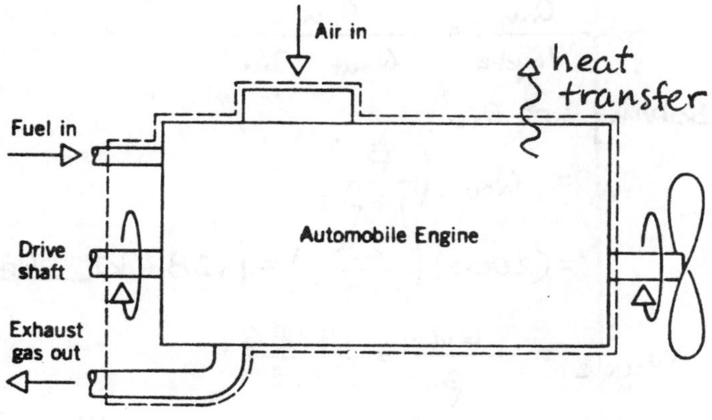

ASSUMPTION: Engine operates at steady state.

An engine operating at steady state has continuous flows of fuel and air in as well as exhaust gas out. It also develops a steady power output. Although the operation of an engine does not conform to the strict definition of a thermodynamic cycle, there are similarities between an engine and a power cycle as illustrated schematically in Fig. 2.15(a). Specifically, we can think of the hot gases formed during combustion as playing the role of the hot body, and the surroundings, which interact with the engine through heat transfer and the discharge of hot exhaust gas, can be viewed as the cold body.

PROBLEM 2.84

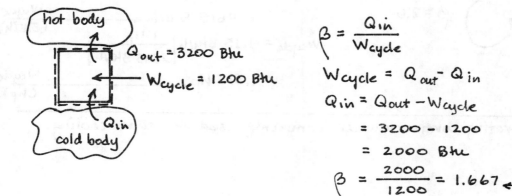

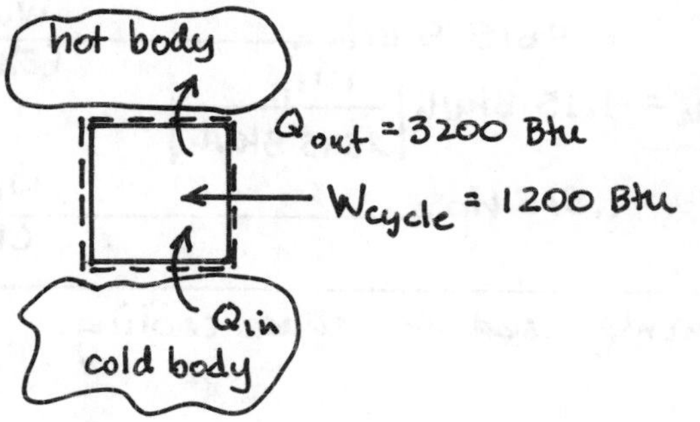

$$\beta = \frac{Q_{in}}{W_{cycle}}$$

$$W_{cycle} = Q_{out} - Q_{in}$$

$$Q_{in} = Q_{out} - W_{cycle}$$

$$= 3200 - 1200$$

$$= 2000 \text{ Btu}$$

$$\beta = \frac{2000}{1200} = 1.667 \longleftarrow \beta$$

PROBLEM 2.85

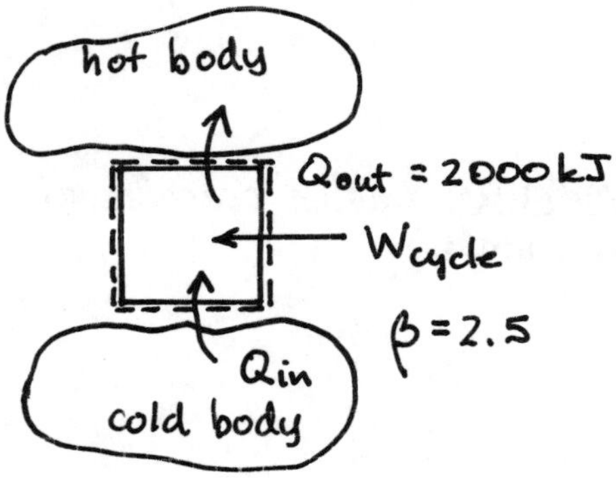

$$\beta = \frac{Q_{in}}{W_{cycle}} = \frac{Q_{in}}{Q_{out} - Q_{in}}$$

Solving for Q_{in}

$$Q_{in} = Q_{out}\left(\frac{\beta}{1+\beta}\right)$$

$$= (2000)\left(\frac{2.5}{1+2.5}\right) = 1428.6 \text{ kJ} \leftarrow Q_{in}$$

$$W_{cycle} = \frac{Q_{in}}{\beta} = \frac{1428.6}{2.5}$$

$$= 571.4 \text{ kJ} \leftarrow W_{cycle}$$

PROBLEM 2.86

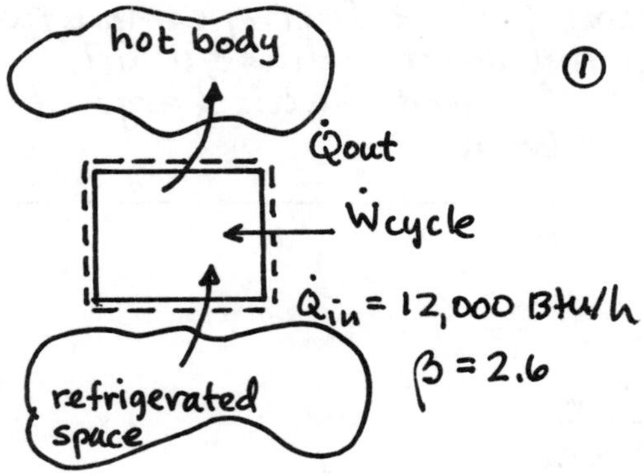

① Since the cycle operates continuously

$$\beta = \frac{\dot{Q}_{in}}{\dot{W}_{cycle}}$$

$$\dot{W}_{cycle} = \dot{Q}_{in}/\beta$$

$$= 12{,}000/2.6$$

$$= 4615 \text{ Btu/h} \leftarrow \dot{W}_{cycle} \text{ (Btu/h)}$$

$$\dot{W}_{cycle} = 4615 \text{ Btu/h} \left| \frac{1 \text{ hp}}{2545 \text{ Btu/h}} \right|$$

$$= 1.813 \text{ hp} \leftarrow \dot{W}_{cycle} \text{ (hp)}$$

---

1. Cycles operating continuously are commonly used for space cooling.

PROBLEM 2.87

KNOWN: Operating data are provided for a heat pump.

FIND: Determine the net power required to operate the heat pump and its monthly cost.

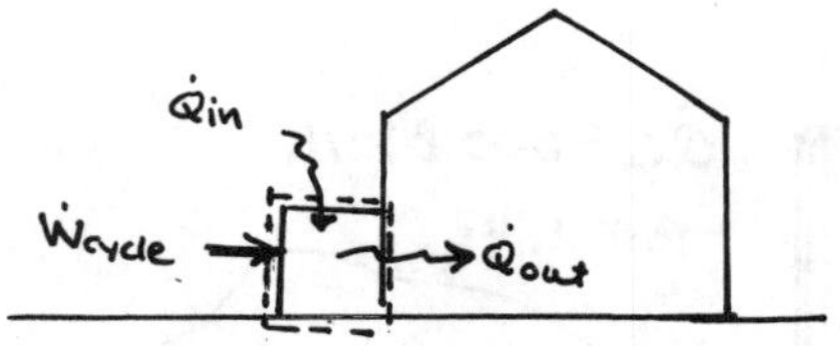

ASSUMPTIONS: 1 The system undergoes a heat pump cycle. 2. The cycle operates steadily for 200 h monthly. 3. Electricity is valued at \$0.08/kW·h.

ANALYSIS: (a) The coefficient of performance for the heat pump is

$$\gamma = \frac{\dot{Q}_{out}}{\dot{W}_{cycle}} \Rightarrow \dot{W}_{cycle} = \frac{\dot{Q}_{out}}{\gamma} = \frac{20\,kW}{2.5} = 8\,kW \longleftarrow \gamma$$

(b) With assumptions 2,3

$$\dot{\$} = (8\,kW)\left(\frac{200\,h}{month}\right)\left(\frac{\$0.08}{kW \cdot h}\right) = \$128/month \longleftarrow \dot{\$}$$

PROBLEM 2.88

KNOWN: Operating data are provided for a heat pump.

FIND: Determine the coefficient of performance and the monthly cost to operate the heat pump.

SCHEMATIC & GIVEN DATA:

$\dot{W}_{cycle} = 7.8\,hp$

$\dot{Q}_{out} = 60{,}000\,Btu/h$

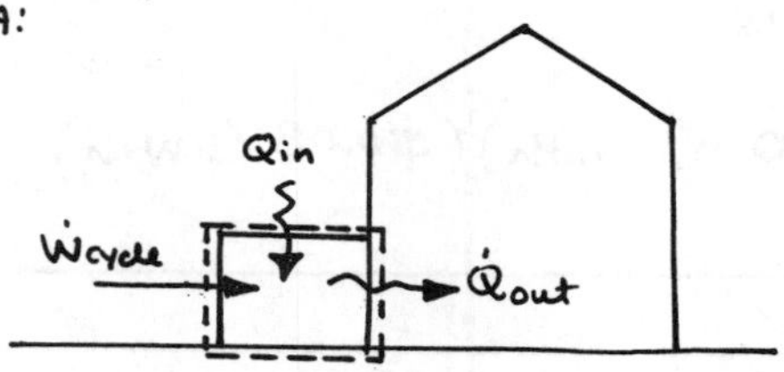

ASSUMPTIONS: 1. The system undergoes a heat pump cycle. 2. The cycle operates steadily for 200 h monthly. 3. Electricity is valued at \$0.08/kW·h

ANALYSIS: (a) The coefficient of performance for a heat pump is

$$\gamma = \frac{\dot{Q}_{out}}{\dot{W}_{cycle}} = \frac{60{,}000\,Btu/h}{(7.8\,hp)\left|\frac{2545\,Btu/h}{hp}\right|} = 3.02 \longleftarrow \gamma$$

(b) Using assumptions 2, 3

$$\dot{\$} = (7.8\,hp)\left|\frac{1\,kW}{1.341\,hp}\right|\left(\frac{200\,h}{month}\right)\left(\frac{\$0.08}{kW \cdot h}\right) = \$93/month \longleftarrow \dot{\$}$$

PROBLEM 2.89

KNOWN: Operating data are given for a household refrigerator.

Find: Determine the cost of operating the refrigerator for 360 h per month.

SCHEMATIC & GIVEN DATA:

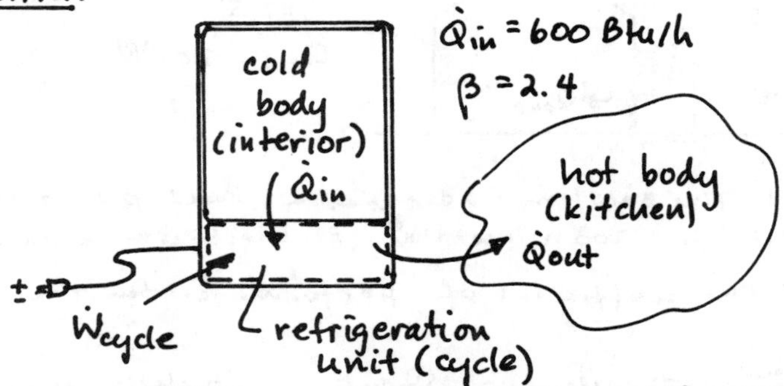

ASSUMPTIONS: (1) The refrigeration unit operates as a refrigeration cycle. (2) The cycle operates steadily for 360 h monthly. (3) Electricity is valued at \$0.08/kWh.

ANALYSIS: To determine the electric power input, begin with the expression for coefficient of performance: $\beta = \dot{Q}_{in}/\dot{W}_{cycle}$. Solving for the power

$$\dot{W}_{cycle} = \dot{Q}_{in}/\beta$$

$$= \left(\frac{600 \text{ Btu/h}}{2.4}\right)\left|\frac{1 \text{ kW}}{3413 \text{ Btu/h}}\right| = 0.0732 \text{ kW}$$

The monthly cost rate, $\dot{\$}$, is

$$\dot{\$} = (0.0732 \text{ kW})(360 \text{ h/month})(\$0.08/\text{kW}\cdot\text{h})$$

① $= \$2.11/\text{month}$ ← $\dot{\$}$

---

1. The actual electric cost to operate a refrigerator is likely to deviate significantly from this value due to such factors as the electric rate, door openings, inserting room-temperature foods, etc.

CHAPTER THREE
EVALUATING PROPERTIES

Chapter 3. Third and fourth editon problem correspondence.

3rd	4th	3rd	4th	3rd	4th
3.1	---	3.38	---	3.76	3.76
3.3	3.1	---	3.38	3.77	3.77
3.2	3.2	3.39	---	3.78	3.78*
3.5	3.3*	---	3.39	3.79	---
3.4	3.4	3.40	3.40*	---	3.79
3.7	3.5	3.41	3.41*	3.80	---
3.6	3.6*	3.42	3.42*	---	3.80
3.9	3.7*	3.43	3.43	3.81	3.81
3.8	3.8	3.44	3.44	3.82	3.82
3.11	3.9*	3.45	3.45*	3.83	3.83
3.10	---	3.46	3.46*	3.84	3.84
---	3.10	3.47	3.47*	3.85	3.85
3.13	3.11	3.48	3.48	3.86	3.86*
3.12	3.12	3.49	---	3.87	3.87*
3.15	3.13	---	3.49	3.88	3.88*
3.14	3.14	3.50	3.50*	3.89	3.89
3.17	3.15	3.51	3.51*	3.90	3.90
3.16	---	3.52	3.52*	3.91	3.91
---	3.16	3.53	3.53*	3.92	3.92
3.19	3.17	3.54	3.54	3.93	---
3.18	3.18	3.55	3.55	---	3.93
3.21	3.19	3.56	3.56*	3.94	3.94
3.20	3.20	3.57	---	3.95	3.95
3.23	3.21	---	3.57	3.98	3.96
3.22	---	3.58	3.58*	3.97	3.97
---	3.22	3.59	3.59	3.96	3.98*
---	3.23	3.60	3.60	3.99	---
3.24	3.24	3.61	3.61*	---	3.99
3.25	---	3.62	3.62	3.100	3.100
3.27	3.25	3.63	3.63	3.101	3.101
3.26	3.26	3.64	3.64	3.102	3.102*
3.33	3.27*	3.65	---	3.103	3.103*
3.34	3.28*	---	3.65	3.104	3.104*
3.28	---	3.66	3.66	3.105	3.105
3.29	---	3.67	---	3.106	3.106
---	3.29	---	3.67	3.107	---
3.30	---	3.68	3.68	---	3.107
---	3.30	3.69	3.69	3.108	---
---	3.31	3.70	---	---	3.108
3.32	3.32*	---	3.70	3.109	3.109
3.31	3.33*	3.71	3.71	3.110	---
---	3.34	3.72	3.72	---	3.110
3.35	3.35	3.73	---	3.111	---
---	3.36	---	3.73	---	3.111
3.36	---	3.74	3.74		
3.37	3.37	3.75	3.75		

* Revised

PROBLEM 3.1

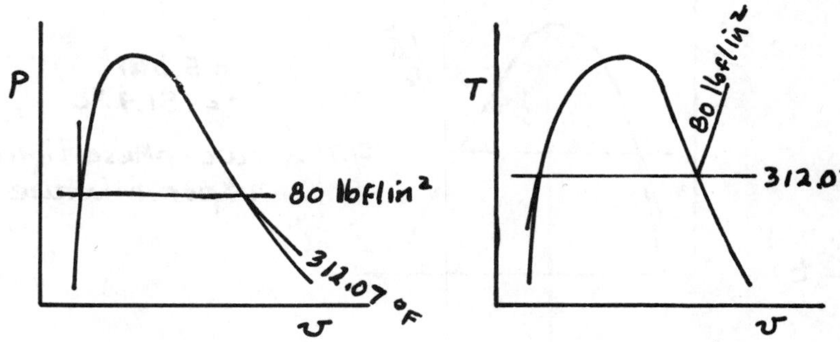

$p = 80$ lbf/in²
$T = 312.07$°F
two-phase liquid-vapor mixture

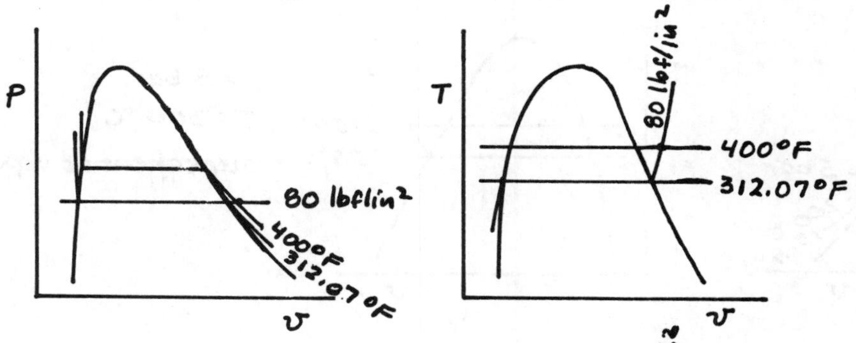

$p = 80$ lbf/in²
$T = 400$°F
superheated vapor

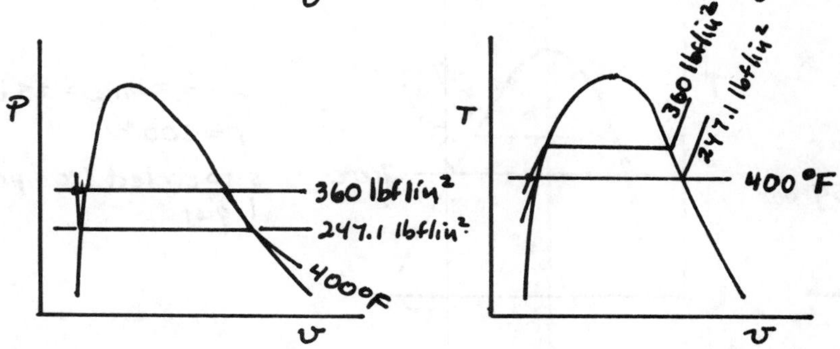

$T = 400$°F
$p = 360$ lbf/in²
subcooled (compressed) liquid

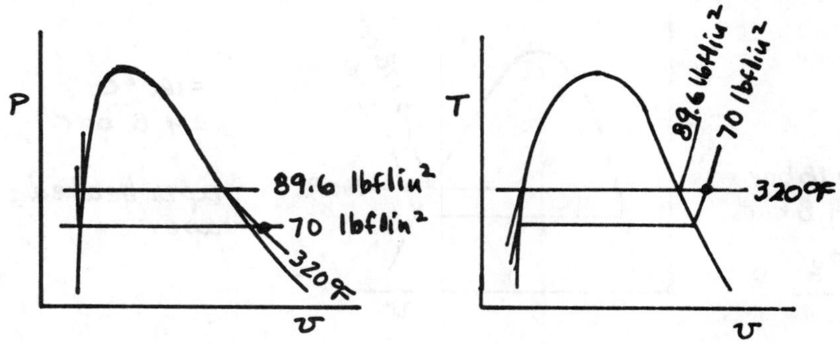

$T = 320$°F
$p = 70$ lbf/in²
superheated vapor

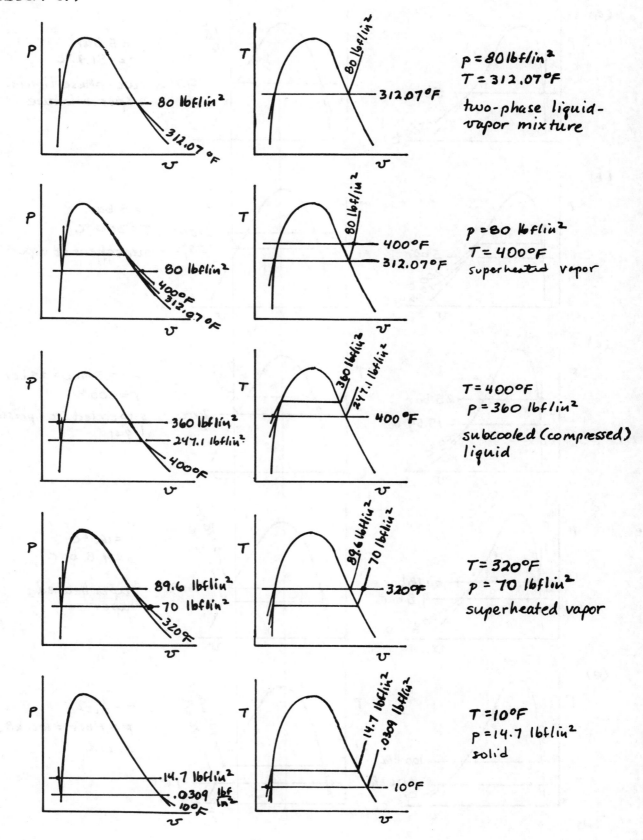

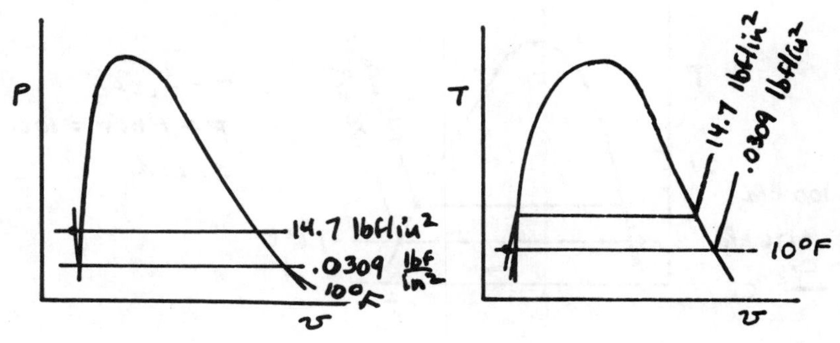

$T = 10$°F
$p = 14.7$ lbf/in²
solid

PROBLEM 3.2

(a)

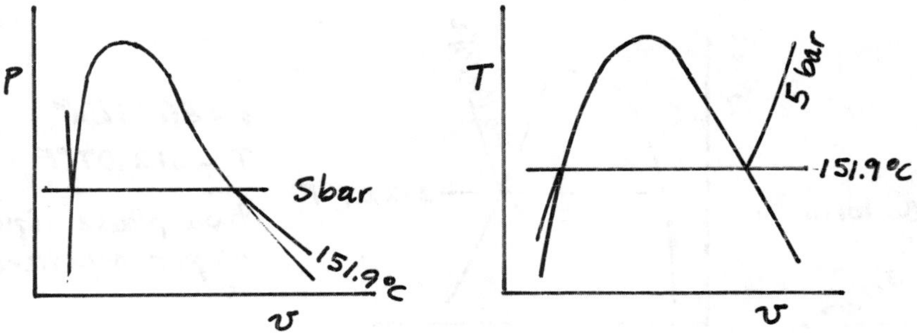

p = 5 bar
T = 151.9 °C
two-phase liquid-vapor mixture

(b)

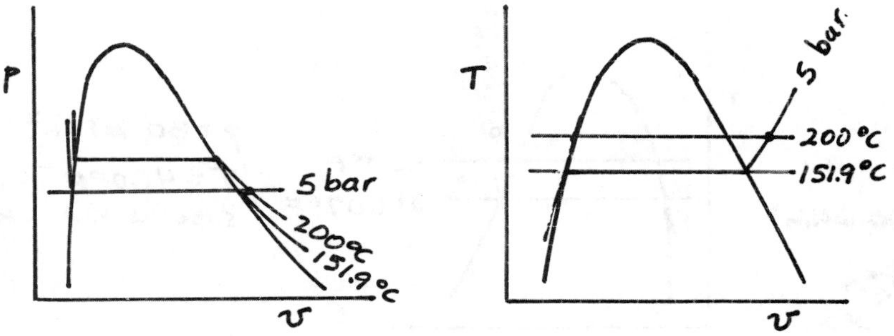

p = 5 bar
T = 200 °C
superheated vapor

(c)

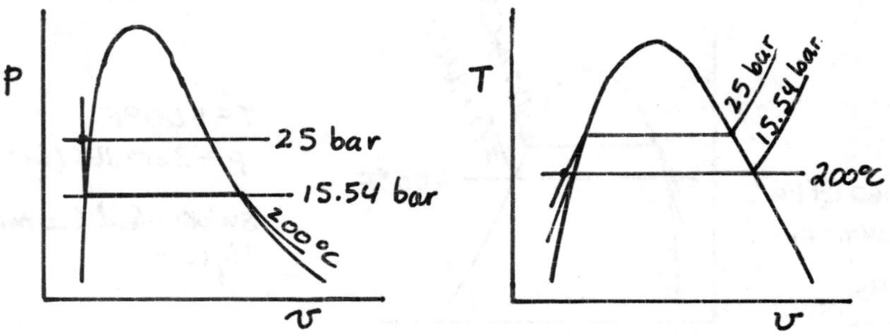

p = 2.5 MPa = 25 bar
T = 200 °C
subcooled (compressed) liquid

(d)

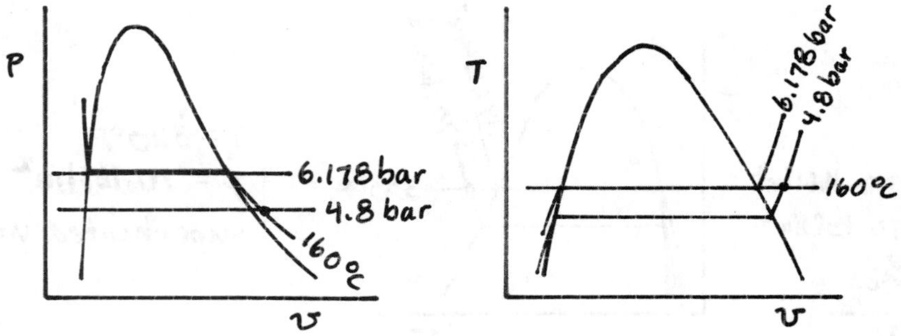

T = 160 °C
p = 4.8 bar
superheated vapor

(e)

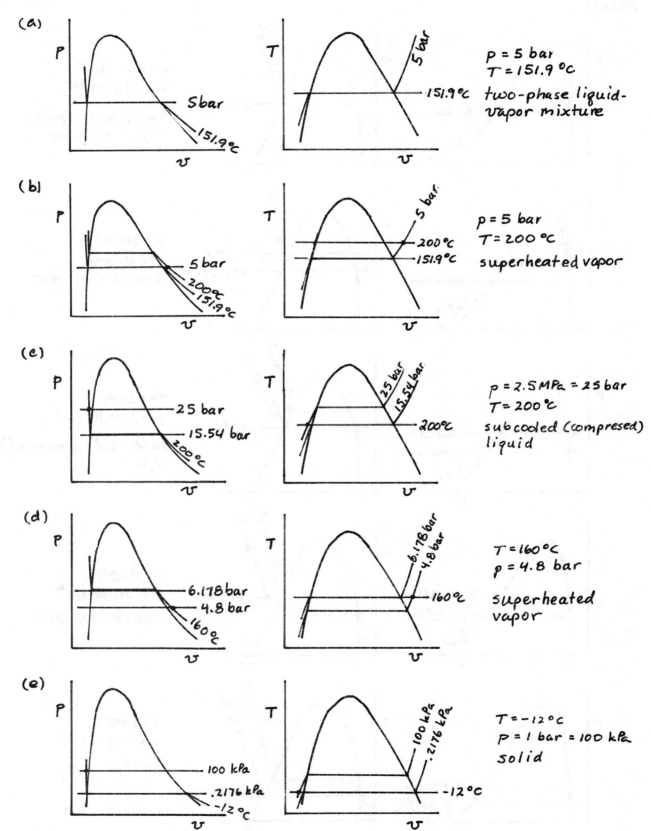

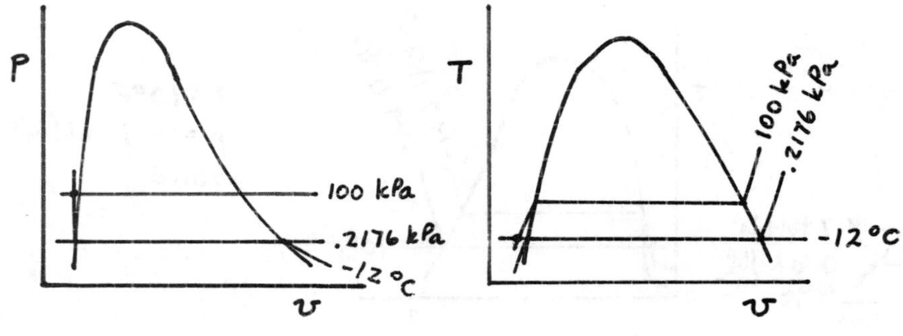

T = -12 °C
p = 1 bar = 100 kPa
solid

PROBLEM 3.3

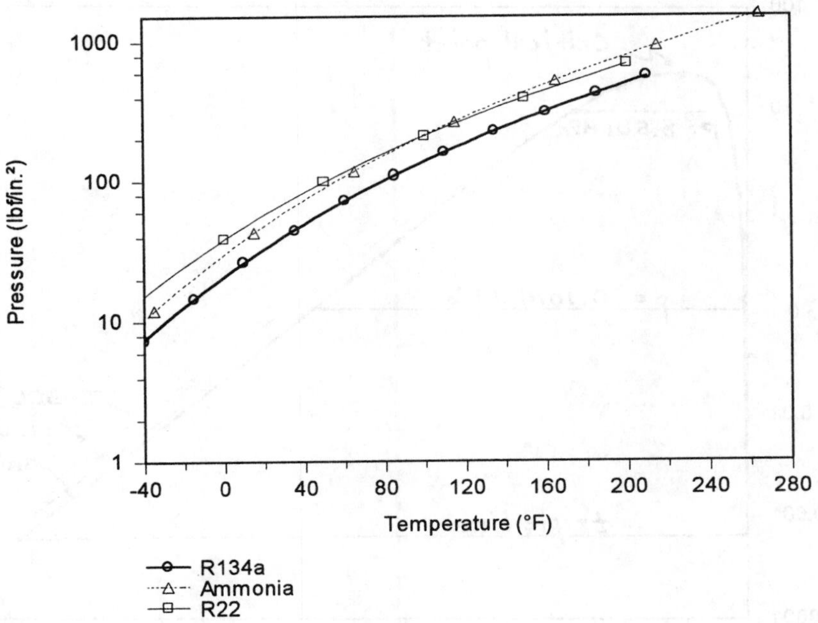

PROBLEM 3.4

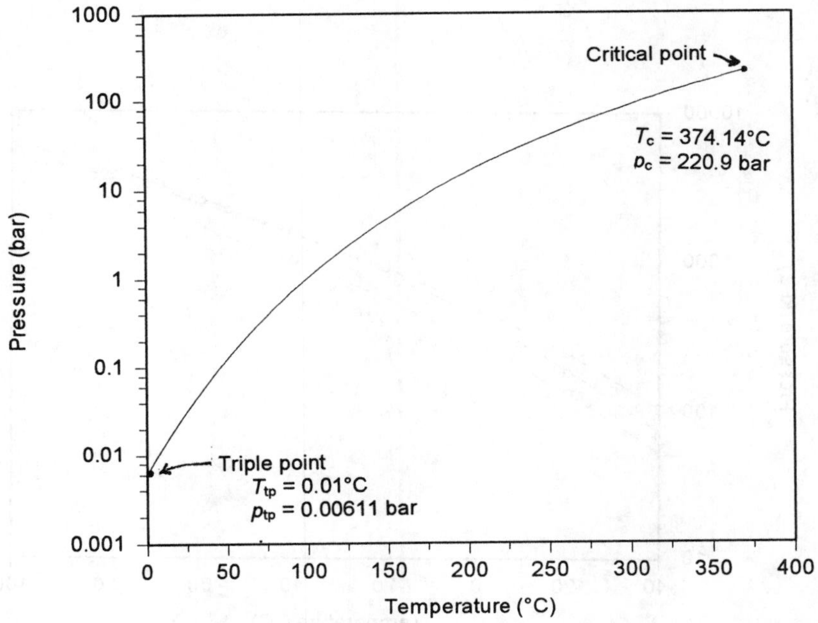

PROBLEM 3.5

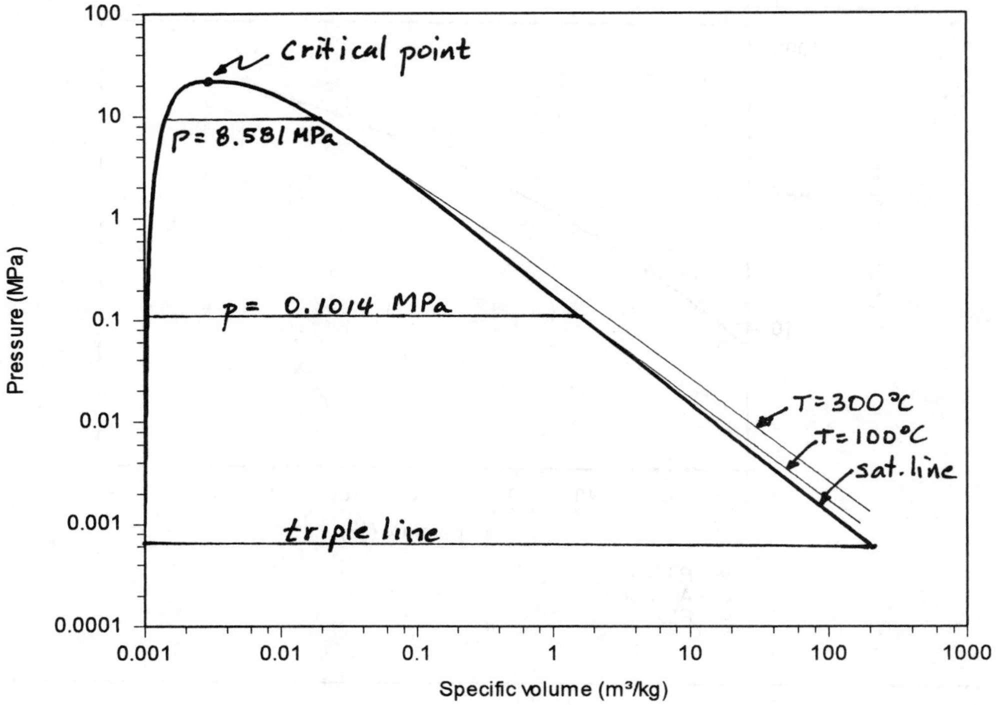

PROBLEM 3.6

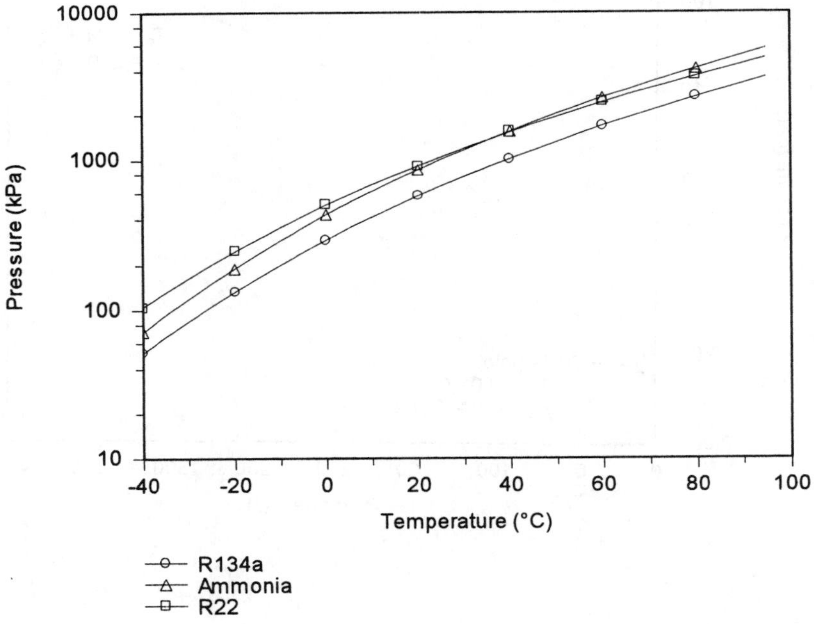

PROBLEM 3.7

(a) H_2O; $T = 100°C$, $v = 0.8\ m^3/kg$

Table A-2: $v_f = 1.0435 \times 10^{-3}\ m^3/kg$, $v_g = 1.673\ m^3/kg$

$$x = \frac{v - v_f}{v_g - v_f} = \frac{0.8 - 1.0435 \times 10^{-3}}{1.673 - 1.0435 \times 10^{-3}} = 0.478\ (47.8\%) \longleftarrow x$$

(b) Refrigerant 134a; $T = 0°C$, $v = 0.7721\ cm^3/g$

$$v = \left(0.7721\ \frac{cm^3}{g}\right)\left|\frac{10^3\ g}{1\ kg}\right|\left|\frac{1\ m^3}{10^6\ cm^3}\right| = 0.7721 \times 10^{-3}\ m^3/kg$$

Table A-10: $v_f = 0.7721 \times 10^{-3}\ m^3/kg \Rightarrow x = 0 \longleftarrow x$

(c) Ammonia; $T = -40°C$, $v = 1\ kg/m^3$

Table A-13: $v_f = 1.4493 \times 10^{-3}\ m^3/kg$, $v_g = 1.5524\ m^3/kg$

$$x = \frac{1 - 1.4493 \times 10^{-3}}{1.5524 - 1.4493 \times 10^{-3}} = 0.644\ (64.4\%) \longleftarrow x$$

(d) Refrigerant 22; $p = 1\ MPa = 10\ bar$, $v = 0.0054\ m^3/kg$

Table A-7: $v_f = 0.8352 \times 10^{-3}\ m^3/kg$, $v_g = 0.0236\ m^3/kg$

$$x = \frac{v - v_f}{v_g - v_f} = \frac{0.0054 - 0.8352 \times 10^{-3}}{0.0236 - 0.8352 \times 10^{-3}} = 0.2\ (20\%) \longleftarrow x$$

PROBLEM 3.8

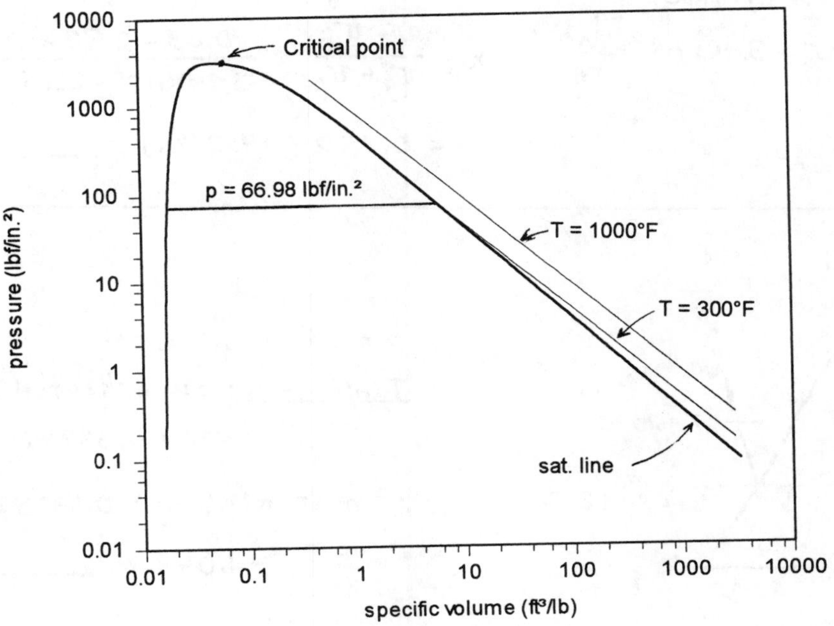

PROBLEM 3.9

(a) H_2O; $p = 100\ \text{lbf/in.}^2$, $v = 3.0\ \text{ft}^3/\text{lb}$

Table A-3E: $v_f = 0.01774\ \text{ft}^3/\text{lb}$, $v_g = 4.434\ \text{ft}^3/\text{lb}$

$$x = \frac{v - v_f}{v_g - v_f} = \frac{3.0 - 0.01774}{4.434 - 0.01774} = 0.675 (67.5\%) \longleftarrow x$$

(b) Refrigerant 134a; $T = -40\,°F$, $v = 5.7173\ \text{ft}^3/\text{lb}$

Table A-10E: $v = v_g = 5.7173\ \text{ft}^3/\text{lb}$

saturated vapor $\Rightarrow x = 1$ (100%) $\longleftarrow x$

(c) Ammonia; $p = 200\ \text{lbf/in}^2$, $v = 1.0\ \text{ft}^3/\text{lb}$

Table A-14E: $v_f = 0.02732\ \text{ft}^3/\text{lb}$, $v_g = 1.5010\ \text{ft}^3/\text{lb}$

$$x = \frac{v - v_f}{v_g - v_f} = \frac{1.0 - 0.02732}{1.5010 - 0.02732} = 0.66 (66\%) \longleftarrow x$$

(d) Refrigerant 22; $T = 30°F$, $v = 0.1\ \text{ft}^3/\text{lb}$

Table A-7E: $v_f = 0.01246\ \text{ft}^3/\text{lb}$, $v_g = 0.7804\ \text{ft}^3/\text{lb}$

$$x = \frac{v - v_f}{v_g - v_f} = \frac{0.1 - 0.01246}{0.7804 - 0.01246} = 0.114 (11.4\%) \longleftarrow x$$

PROBLEM 3.10

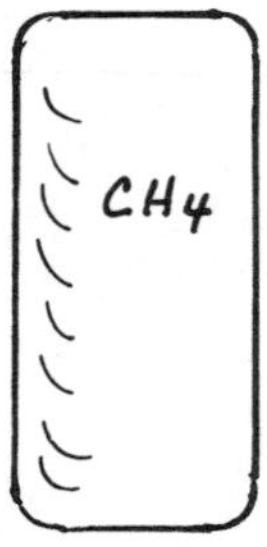

$T = 160\,K$
$m = 10\,kg$
$V = 0.3\,m^3$
$v_f = 2.97\times10^{-3}\ \frac{m^3}{kg}$
$v_g = 3.94\times10^{-2}\ \frac{m^3}{kg}$

First, find the specific volume

$$v = \frac{V}{m} = \frac{0.3\ m^3}{10\ kg} = 0.03\ m^3/kg$$

Now, the quality is

$$x = \frac{v - v_f}{v_g - v_f} = \frac{0.03 - 2.97\times10^{-3}}{3.94\times10^{-2} - 2.97\times10^{-3}}$$

$$= 0.742 (74.2\%) \longleftarrow x$$

PROBLEM 3.11

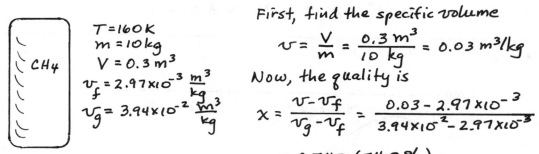

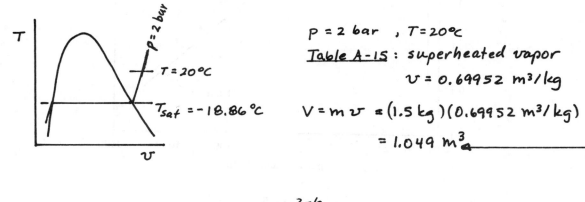

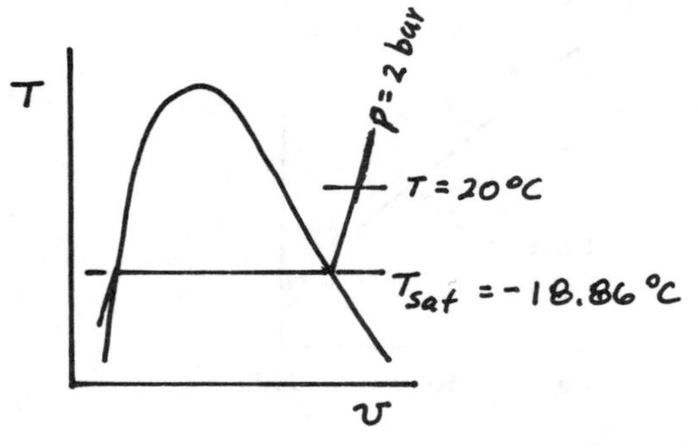

$p = 2\ \text{bar}$, $T = 20°C$

Table A-15: superheated vapor

$v = 0.69952\ m^3/kg$

$$V = m v = (1.5\ kg)(0.69952\ m^3/kg)$$

$$= 1.049\ m^3 \longleftarrow V$$

PROBLEM 3.12

$v_1 = v_2 = 1.5 \text{ ft}^3/\text{lb}$

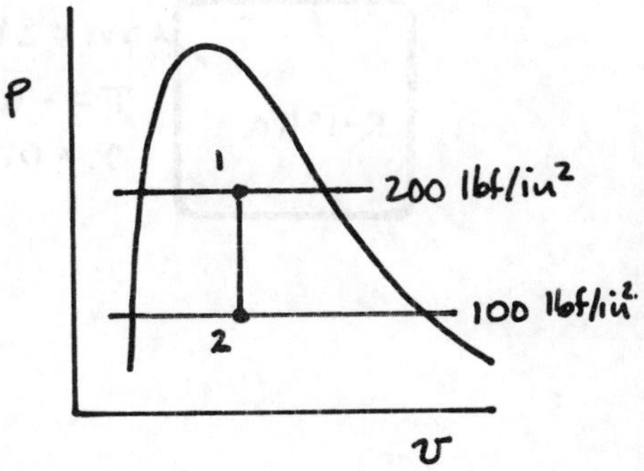

Table A-3E: $v_{f_2} = 0.01774 \text{ ft}^3/\text{lb}, v_{g_2} = 4.434 \text{ ft}^3/\text{lb}$

$v_2 = 1.5 \text{ ft}^3/\text{lb} \Rightarrow$

$$x_2 = \frac{v_2 - v_{f_2}}{v_{g_2} - v_{f_2}}$$

$$= \frac{1.5 - 0.01774}{4.434 - 0.01774}$$

$$= 0.3356 \; (33.56\%) \quad \longleftarrow x_2$$

PROBLEM 3.13

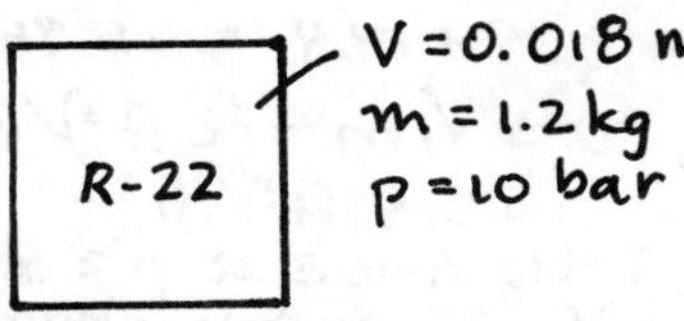

R-22, $V = 0.018 \text{ m}^3$, $m = 1.2 \text{ kg}$, $p = 10 \text{ bar}$

$$v = \frac{V}{m} = \frac{0.018 \text{ m}^3}{1.2 \text{ kg}}$$

$$= 0.015 \text{ m}^3/\text{kg}$$

Table A-7: $v_f < v < v_g \Rightarrow$ two-phase

at 10 bar; $T_{sat} = 23.40°C \quad \longleftarrow T$

PROBLEM 3.14

H_2O, $m = 2 \text{ lb}$, $V = ?$, $p = 1000 \text{ lbf/in}^2$

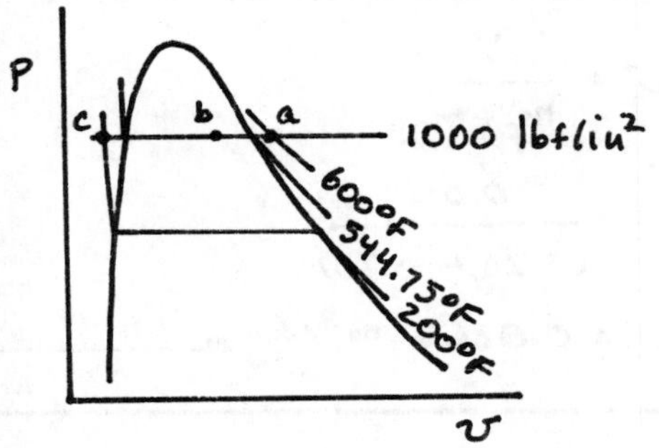

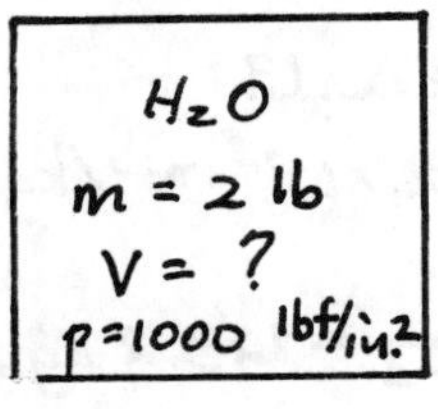

(a) $T = 600°F \Rightarrow$ superheated vapor

Table A-4E: $v = 0.514 \text{ ft}^3/\text{lb}$

$V = mv = (2 \text{ lb})(0.514 \text{ ft}^3/\text{lb})$

$= 1.028 \text{ ft}^3 \quad \longleftarrow V$

(b) $x = 0.8 \Rightarrow$ two-phase liquid-vapor mixture

Table A-3E: $v_f = 0.02159 \text{ ft}^3/\text{lb}$

$v_g = 0.446 \text{ ft}^3/\text{lb}$

$v = v_f + x(v_g - v_f) = 0.3612 \text{ ft}^3/\text{lb}$

$V = (2 \text{ lb})(0.3612 \text{ ft}^3/\text{lb})$

$= 0.7224 \text{ ft}^3 \quad \longleftarrow V$

(c) $T = 200°F \Rightarrow$ subcooled (compressed) liquid

Table A-5E: $v = 0.016580 \text{ ft}^3/\text{lb}$

$V = (2)(0.016580) = 0.0332 \text{ ft}^3 \quad \longleftarrow V$

PROBLEM 3.15

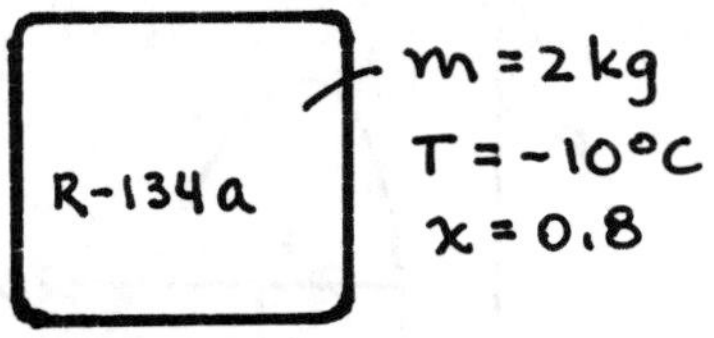

Using data from Table A-10:

$v = v_f + x(v_g - v_f)$

$= 0.75335\times10^{-3} + (.8)(0.09935 - 0.75335\times10^{-3})$

$= 0.07963\ m^3/kg$

$V = mv = (2\ kg)(0.07963\ m^3/kg)$

$= 0.1593\ m^3$ ← V

PROBLEM 3.16

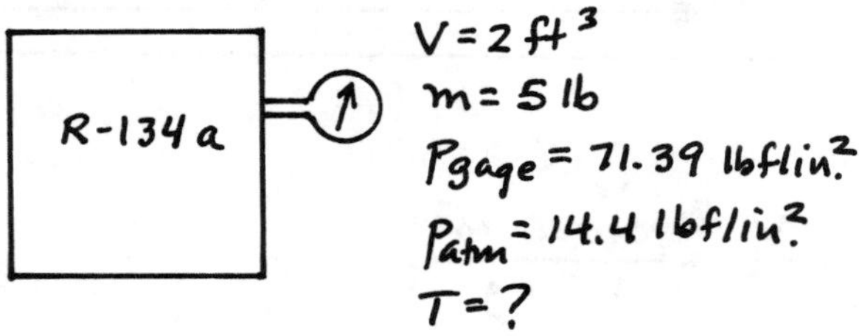

First determine the absolute pressure p

$p = p_{gage} + p_{atm}$

$= 71.39 + 14.4 = 85.79\ lbf/in.^2$

Now, $v = V/m = (2\ ft^3)/(5\ lb)$

$= 0.4\ ft^3/lb$

From Table A-10E at $p \approx 85.79$

$v_f < v < v_g \Rightarrow$ two-phase liquid-vapor mixture

Thus, $T = T_{sat} \approx 70°F$ ← T

PROBLEM 3.17

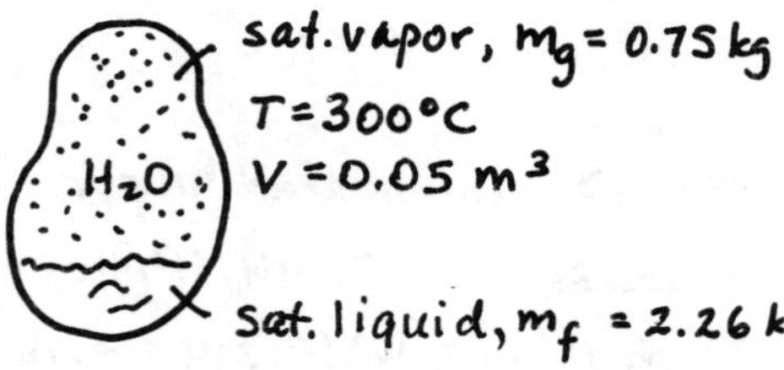

For the two-phase liquid-vapor mixture, the specific volume is

$$v = \frac{V}{m_f + m_g}$$

$$= \frac{0.05}{(2.26 + 0.75)}$$

$= 0.0166\ m^3/kg$ ← v

PROBLEM 3.18

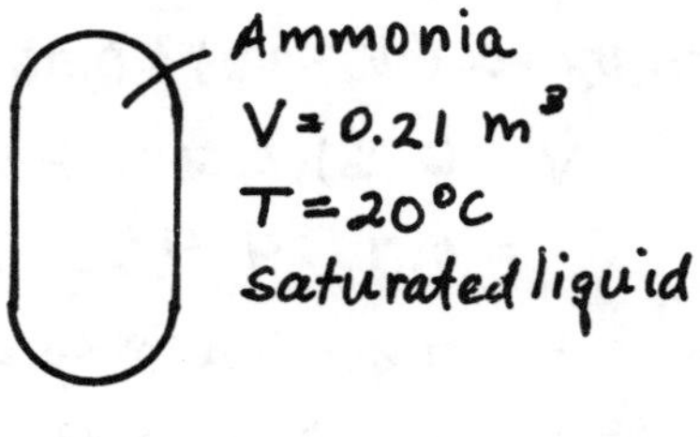

Using data from Table A-13:

$v = v_{f@20°C} = 1.6386\times10^{-3}\ m^3/kg$

$$m = \frac{V}{v} = \frac{0.21}{1.6386\times10^{-3}} = 128.2\ kg$$ ← m

at $T = 20°C$; $p_{sat} = 8.5762\ bar\left(\frac{100\ kPa}{1\ bar}\right)$

$= 857.62\ kPa$ ← P

PROBLEM 3.19

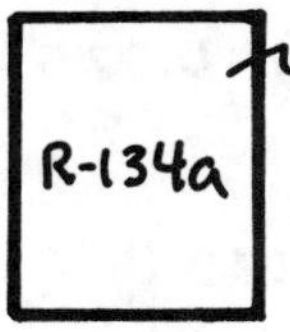

$V = 0.006 \text{ m}^3$
$p = 180 \text{ kPa}$
$= 1.8 \text{ bar}$
$0 \le x \le 1$

First, using data from Table A-11 at $p = 1.8$ bar: the specific volume varies with x as follows:

$$v = v_f + x(v_g - v_f)$$
$$= 0.7485 \times 10^{-3} + x\,(0.1098 - 0.7485 \times 10^{-3})$$

Thus

$$m = \frac{V}{v} = \frac{0.006}{0.7485 \times 10^{-3} + x\,(0.10905)} \longleftarrow m$$

Now $m_g = x\,m$; $m_f = m - m_g$

$$\frac{V_g}{V} = \frac{v_g\,m_g}{V} \; ; \; \frac{V_f}{V} = 1 - \frac{V_g}{V} \longleftarrow \text{Volume fractions}$$

Sample calculation: $x = 0.9$; $m = 0.0607$ kg, $V_g/V = 0.9993$, $V_f/V = 0.0007$

The following IT code is used to develop the plots below:

```
p = 180 // kPa
V=0.006 //m³
v = vsat_Px("R134A", p, x)
vg = vsat_Px("R134A", p, 1)
x=0.9
m=V/v
mg=x*m
mf=m*(1-x)
vfrac_g=(vg*mg)/V
vfrac_f=1-vfrac_g
mm=mf+mg
```

Results for sample calculation

x	0.9
m	0.06065 kg
vfrac_f	0.0007566
vfrac_g	0.9992

These results compare well with the results of the hand calculation above.

To develop data for the following plots, sweep x from 0 to 1 in steps of 0.01.

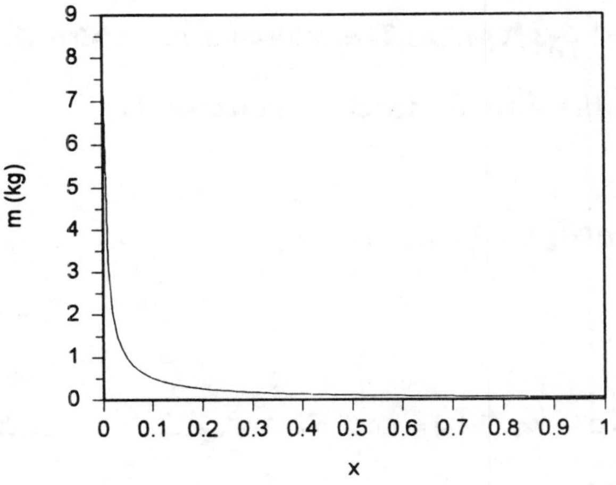

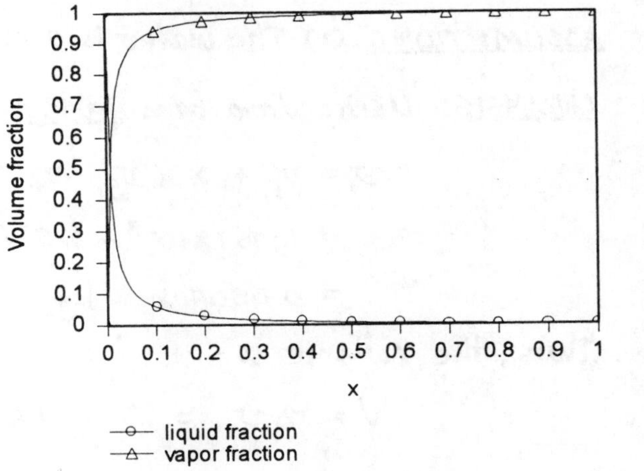

The mass contained in the tank decreases rapidly with increasing quality. The corresponding volume fractions of liquid and vapor are strong functions of quality as well.

PROBLEM 3.20

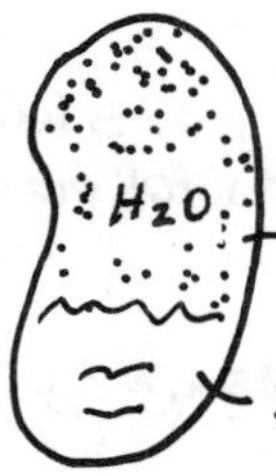

From the definition of quality

$$m_g = \left(\frac{x}{1-x}\right) m_f = \left(\frac{0.2}{1-0.2}\right)(0.1\text{ lb}) = 0.025\text{ lb} \leftarrow m_g$$

The total mass m is

$$m = m_f + m_g = 0.1 + 0.025 = 0.125\text{ lb}$$

Using data from Table A-2E at 400°F:

$$v = v_f + x(v_g - v_f) = 0.01864 + (.2)(1.866 - 0.01864)$$
$$= 0.3881\text{ ft}^3/\text{lb}$$
$$V = mv = (0.125\text{ lb})(0.3881\text{ ft}^3/\text{lb})$$
$$= 0.0485\text{ ft}^3 \leftarrow V$$

PROBLEM 3.21

KNOWN: A two-phase liquid-vapor mixture of water, initially at a given pressure and quality, is heated in a closed rigid tank until only saturated water vapor remains. The mass is known.

FIND: Determine the volume of the tank and the final pressure.

SCHEMATIC & GIVEN DATA:

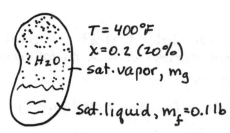

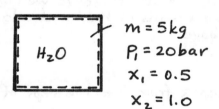

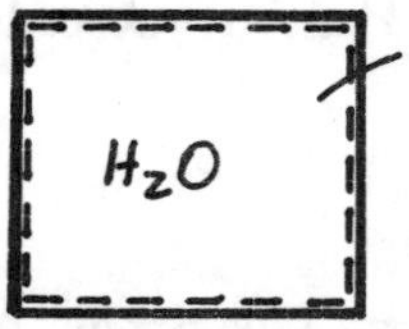

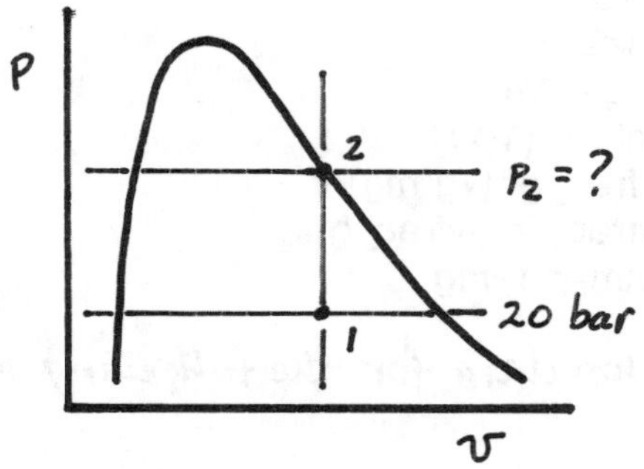

ASSUMPTIONS: (1) The water is a closed system. (2) The volume is constant.

ANALYSIS: Using data from Table A-3, the initial specific volume is

$$v_1 = v_{f1} + x_1(v_{g1} - v_{f1})$$
$$= 1.1767\times10^{-3} + (.5)(0.09963 - 1.1767\times10^{-3})$$
$$= 0.0504\text{ m}^3/\text{kg}$$

Thus, the volume is

$$V = mv_1 = (5\text{ kg})(0.0504\text{ m}^3/\text{kg}) = 0.252\text{ m}^3 \leftarrow V$$

By assumptions (1) and (2), $v_2 = v_1$. Thus

$$P_2 \approx P_{sat\,@\,v_2} = 39.6\text{ bar} \leftarrow P_2$$

PROBLEM 3.22

KNOWN: A specified amount of water is heated in a closed, rigid tank from a known initial state to a specified final temperature.

FIND: Determine the mass of vapor initially present and the final pressure.

SCHEMATIC & GIVEN DATA:

State 1: $T_1 = 260°F$, $x_1 = 0.6$
State 2: $T_2 = 320°F$

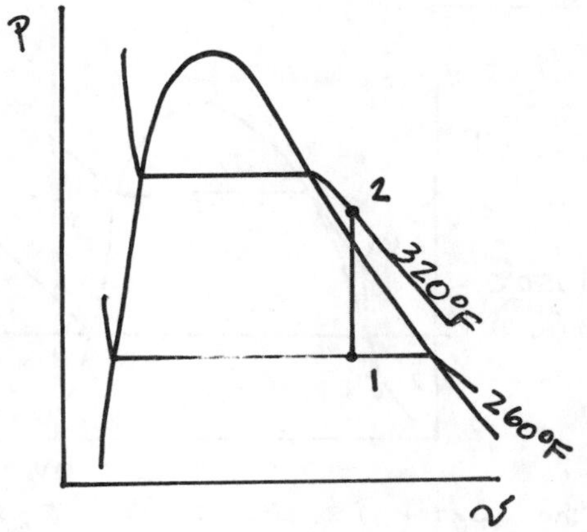

ASSUMPTIONS: (1) The water is the closed system. (2) the volume is constant.

ANALYSIS: The initial mass of vapor in the tank is found using the known total mass and the quality x_1 at the initial state

$$m_{g1} = x_1 m = (0.6)(5\text{ lb}) = 3\text{ lb} \longleftarrow m_{g1}$$

Now, since the volume and total mass are constant during the process, the specific volumes at states 1 and 2 are equal: $v_2 = v_1$. From Table A-2E at 260°F, $v_{f1} = 0.01708$ ft³/lb and $v_{g1} = 11.77$ ft³/lb. Thus

$$v_2 = v_1 = v_{f1} + x_1(v_{g1} - v_{f1})$$

$$= 0.01708 + (0.6)(11.77 - 0.01708) = 7.0688\text{ ft}^3/\text{lb}$$

Now, from Table A-2E at 320°F, $v_g = 4.919$ ft³/lb. Since $v_2 > v_{g@320°F}$ state 2 is in the superheated vapor region. Thus, interpolating in Table A-4 at 320°F, 7.0688 ft³/lb

$$P_2 \approx 64.27\text{ lbf/in.}^2 \longleftarrow P_2$$

PROBLEM 3.23

KNOWN: A specified amount of water is heated in a closed rigid tank from a known initial state to a final pressure.

FIND: Sketch the process on T-v and p-v diagrams. Determine the volume of the tank and the temperature at the final state.

SCHEMATIC & GIVEN DATA:

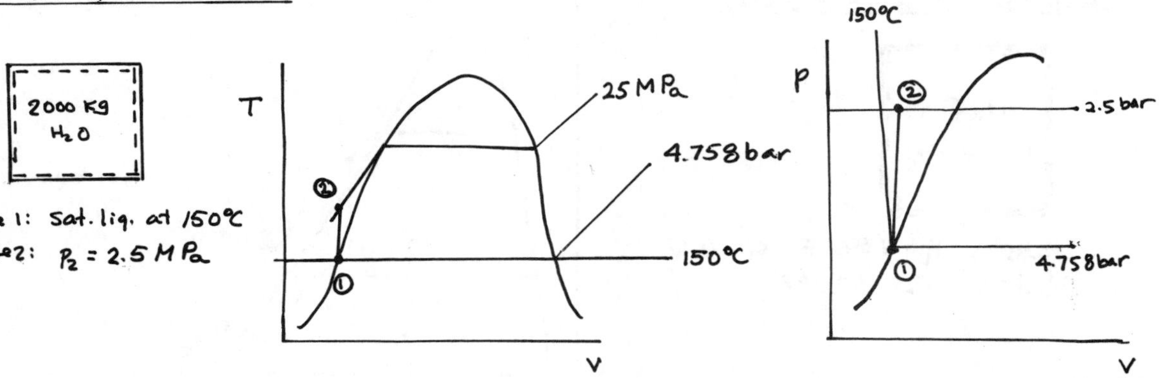

State 1: Sat. liq. at 150°C

State 2: $p_2 = 2.5$ MPa

ASSUMPTIONS: 1. The water is the closed system. 2. Volume remains constant.

ANALYSIS: Since volume and total mass remain constant during the process, the specific volume at states 1, 2 are equal: $v_2 = v_1$. From Table A-2 at 150°C

$$v_1 = v_f(150°C) = 1.0905 \times 10^{-3} \text{ m}^3/\text{kg}$$

As shown by the T-v and p-v diagrams, state 2 is in the liquid region. Interpolation in Table A-5 gives $T_2 = 150.15°C$ ◄

The total volume is

$$V = mv = (2000 \text{ kg})\left(\frac{1.0905}{10^3} \frac{\text{m}^3}{\text{kg}}\right) = 2.181 \text{ m}^3$$ ◄

PROBLEM 3.24

KNOWN: Steam is cooled in a closed rigid container from a known initial state to a known final temperature.

FIND: Determine the pressure at which condensation first occurs, the fraction of the total mass that condenses, and percentage of the volume occupied by saturated liquid at the final state.

SCHEMATIC & GIVEN DATA:

$p_1 = 15$ bar
$T_1 = 240°C$

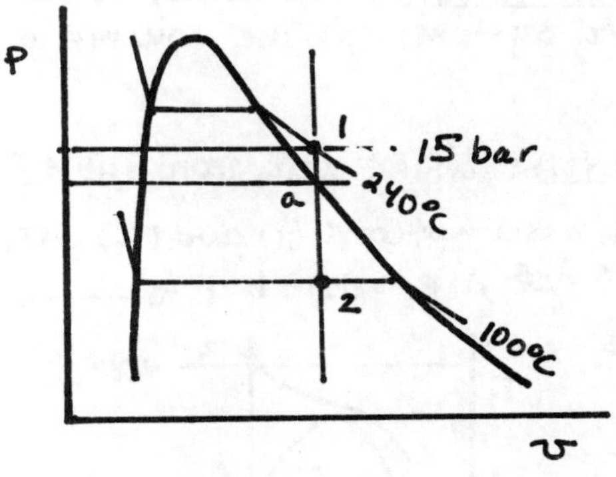

ASSUMPTIONS: (1) The steam is a closed system. (2) The volume is constant.

ANALYSIS: By assumptions (1) and (2), the specific volume is constant. Thus, using data from Table A-2

$$v = 0.1483 \text{ m}^3/\text{kg}$$

$$p_a = p_{sat@v} \approx 13.68 \text{ bar} \longleftarrow$$

Next, the fraction of the total mass that condenses is

$$\text{fraction condensed} = \frac{m_{f2}}{m} = 1 - x_2$$

$$x_2 = \frac{v_2 - v_{f2}}{v_{g2} - v_{f2}} = \frac{0.1483 - 1.0435\times10^{-3}}{1.673 - 1.0435\times10^{-3}} = 0.0881$$

Thus

$$\text{fraction condensed} = 1 - 0.0881 = 0.9119 \longleftarrow$$

The fraction of the volume occupied by saturated liquid at the final state is

$$\frac{V_{f2}}{V} = \frac{(m_{f2} v_{f2})}{(m v)} = (1 - x_2)\frac{v_{f2}}{v}$$

$$= (0.9119)\frac{1.0435\times10^{-3}}{0.1483} = 0.00642\ (0.642\%) \longleftarrow$$

PROBLEM 3.25

KNOWN: Water vapor is heated in a closed, rigid tank from satured vapor at a known temperature to a known final temperature.

FIND: Determine the initial and final pressures and sketch the process on T-v and p-v diagrams.

ASSUMPTIONS: (1) The water vapor is a closed system. (2) The volume is constant.

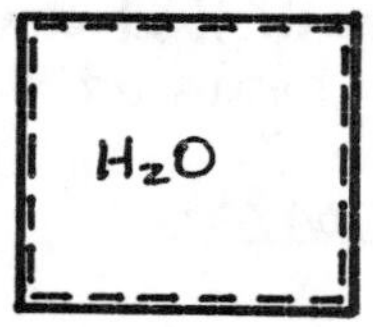

State 1: $T_1 = 160°C$, sat. vapor

State 2: $T_2 = 400°C$

ANALYSIS: Using data from Table A-2, $p_1 = 6.178$ bar ← p_1

With assumptions (1) and (2), $v_2 = v_1 = 0.3071$ m³/kg. Interpolating in Table A-3, $p_2 \approx 9.99$ bar ← p_2

Thus

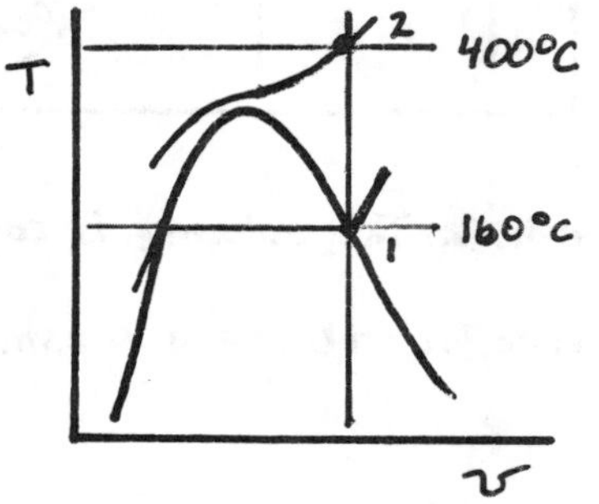

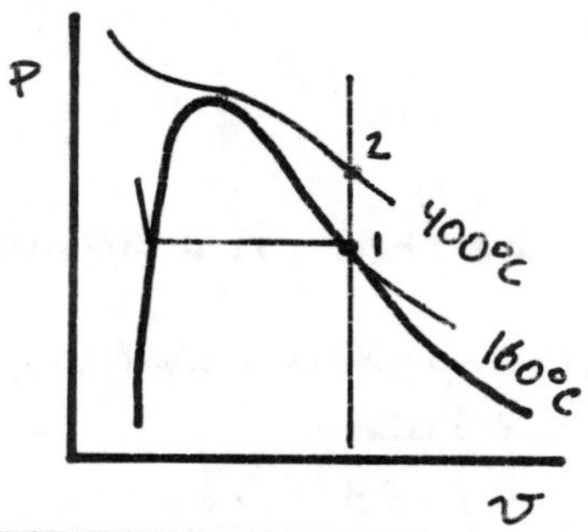

PROBLEM 3.26

KNOWN: Ammonia undergoes an isothermal process from a known initial state to the saturated vapor state.

FIND: Determine the initial and final pressures, and sketch the process on T-v and p-v diagrams.

ASSUMPTIONS: (1) The ammonia is a closed system. (2) The process is isothermal.

State 1: $T_1 = 80°F$, $v_1 = 10$ ft³/lb

State 2: $T_2 = T_1$, Sat. vapor

ANALYSIS: Interpolating in Table A-15E, $p_1 \approx 33.86$ lbf/in² ← p_1

Using data from Table A-13E, $p_2 = 153.13$ lbf/in² ← p_2

Thus

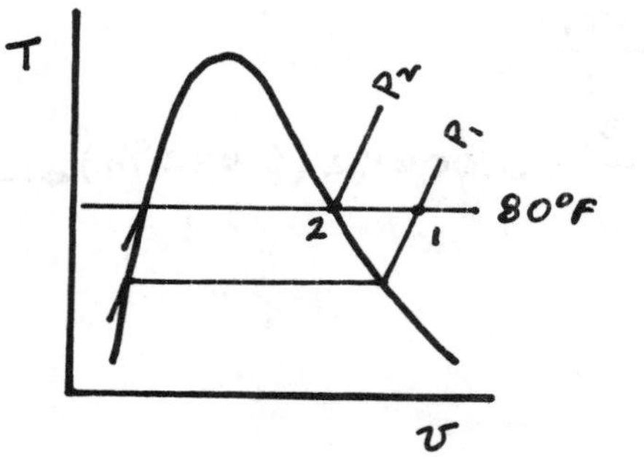

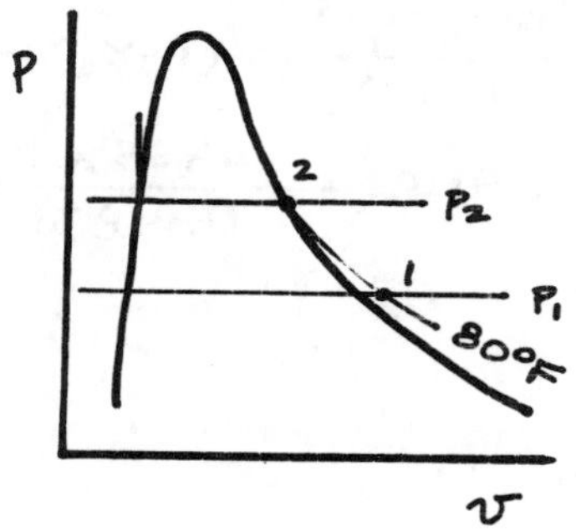

PROBLEM 3.27

KNOWN: A two-phase liquid-vapor mixture is heated at fixed volume from an initial pressure to the critical point.

FIND: Determine the quality at the initial state.

SCHEMATIC & GIVEN DATA:

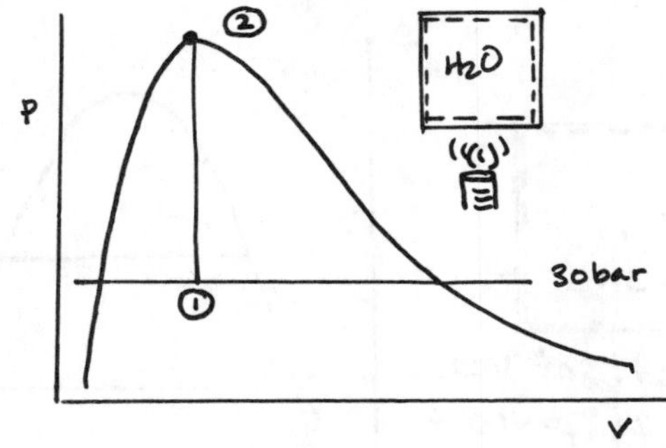

ASSUMPTIONS: 1. The quantity of water under consideration is the closed system. 2. Volume remains constant.

ANALYSIS: Since volume and mass remain constant, $v_2 = v_1$. From Table A-3 at the critical point, state 2, $v_2 = 3.155 \times 10^{-3}\ m^3/kg$. Thus, with $v_2 = v_1$, we have with data from Table A-3 at 30 bar

$$v_1 = v_f + x_1(v_g - v_f) \Rightarrow$$

$$x_1 = \frac{v_1 - v_f}{v_g - v_f} = \frac{(3.155 - 1.2165)10^{-3}}{(66.68 - 1.2165)10^{-3}} = 0.0296 \quad (2.96\%) \leftarrow x_1$$

PROBLEM 3.28

KNOWN: A two-phase liquid-vapor mixture is heated at fixed volume from an initial pressure to the critical point.

FIND: Determine the quality at the initial state.

SCHEMATIC & GIVEN DATA:

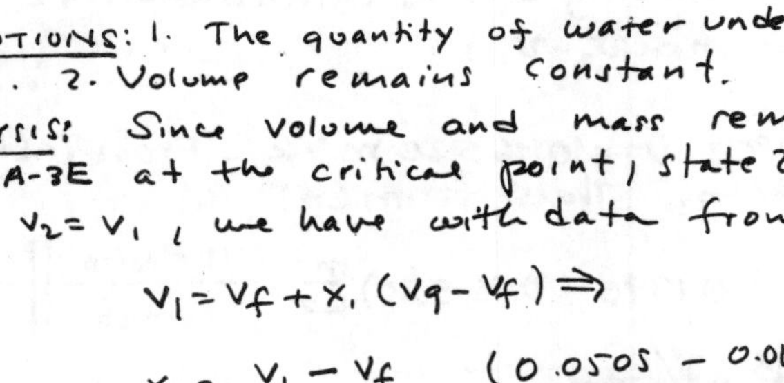

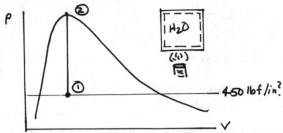

ASSUMPTIONS: 1. The quantity of water under consideration is the closed system. 2. Volume remains constant.

ANALYSIS: Since volume and mass remain constant, $v_2 = v_1$. From Table A-3E at the critical point, state 2, $v_2 = 0.0505\ ft^3/lb$. Thus, with $v_2 = v_1$, we have with data from Table A-3E at 450 lbf/in.²

$$v_1 = v_f + x_1(v_g - v_f) \Rightarrow$$

$$x_1 = \frac{v_1 - v_f}{v_g - v_f} = \frac{(0.0505 - 0.01955)}{1.033 - 0.01955} = 0.0305 \quad (3.05\%) \leftarrow x_1$$

PROBLEM 3.29

KNOWN: A specified amount of water is heated in a closed rigid tank. The volume is known, and data are given for the initial and final states.

FIND: Determine the pressures at the initial and final states.

SCHEMATIC & GIVEN DATA:

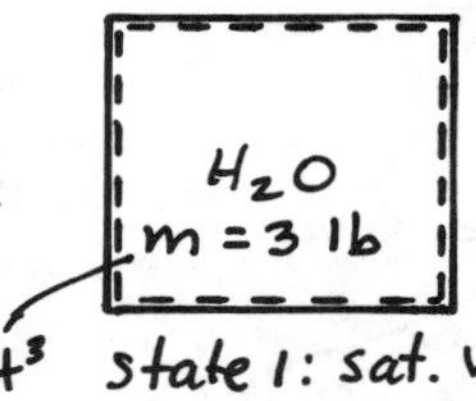

state 1: sat. vap.
State 2: $T_2 = 400°F$

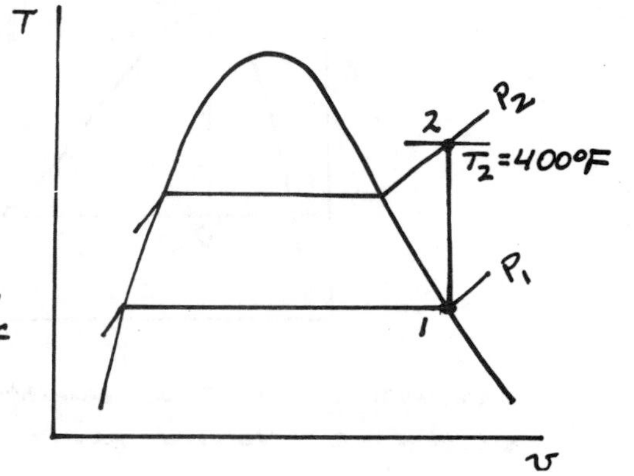

ASSUMPTIONS: (1) The water is the closed system. (2) The volume is constant.

ANALYSIS: The T-v diagram is shown to the right. To determine the pressures, first fix state 2. Since the volume and mass are constant, $v_1 = v_2$. Thus

$$v_2 = V/m = (13.3\ ft^3)/(3\ lb) = 4.4333\ ft^3/lb$$

Now, interpolating in Table A-4E at 400°F, 4.4333 ft³/lb; $p_2 \approx 111.7\ lbf/in.^2$ ← p_2

Then, from Table A-3E, at $v_g = 4.4333\ ft^3/lb$; $p_1 = p_{sat} \approx 100\ lbf/in.^2$ ← p_1

PROBLEM 3.30

KNOWN: Refrigerant 134a undergoes a constant pressure process. The initial and final states are fixed.

FIND: Determine the work for the process, per unit mass of refrigerant.

SCHEMATIC & GIVEN DATA:

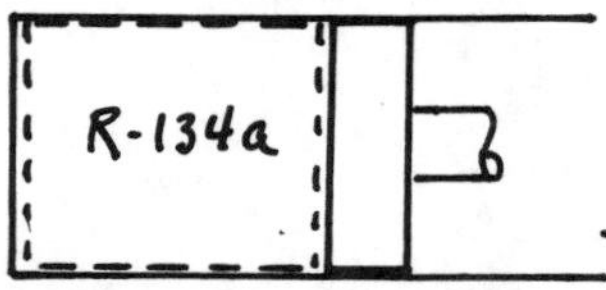

p = 1.4 bar
$T_1 = 20°C$
State 2: sat. vapor

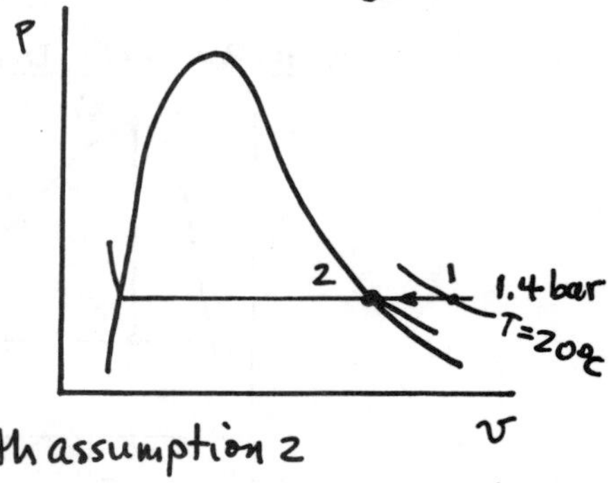

ASSUMPTIONS: (1) The refrigerant is the closed system. (2) The pressure remains constant.

ANALYSIS: The work is determined using Eq. 2.17 with assumption 2

$$W = \int p\,dV = m\int_{v_1}^{v_2} p\,dv = mp(v_2 - v_1) \qquad (*)$$

From Table A-12 at 1.4 bar, 20°C; $v_1 = 0.16520\ m^3/kg$. From Table A-11, $v_2 = v_{g@1.4\,bar} = 0.1395\ m^3/kg$. Thus, from (*)

$$\frac{W}{m} = p(v_2 - v_1) = (1.4\ bar)(0.1395 - 0.16520)\frac{m^3}{kg}\left|\frac{10^5\ N/m^2}{1\ bar}\right|\left|\frac{1\ kJ}{10^3\ N\cdot m}\right|$$

$$= -3.598\ kJ/kg$$ ← W/m

PROBLEM 3.31

KNOWN: Water is compressed at a constant pressure between two specified states.

FIND: Determine the temperatures at the initial and final states, and the work for the process.

SCHEMATIC & GIVEN DATA:

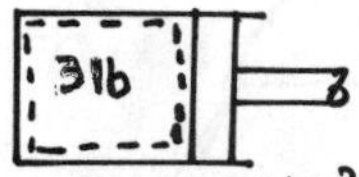

$p = 100$ lbf/in²

$V_1 = 14.8$ ft³

$V_2 = 13.3$ ft³

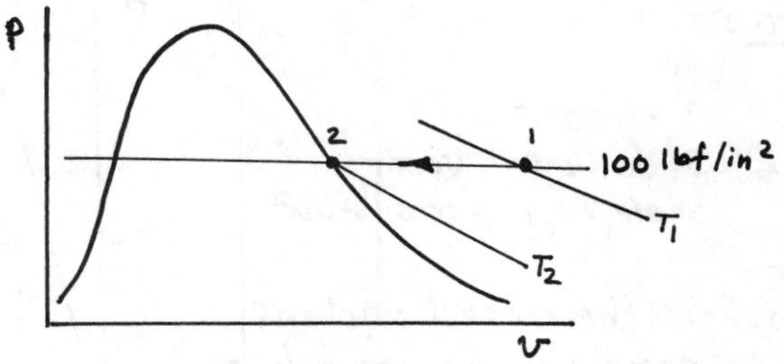

ASSUMPTIONS: 1. The given quantity of water is the closed system. 2. Volume change is the only work mode. 3. The process occurs at constant pressure.

ANALYSIS: The initial and final states are fixed by the pressure, 100 lbf/in², and the initial and final specific volumes:

$$v_1 = \frac{V_1}{m} = \frac{14.8 \text{ ft}^3}{3 \text{ lb}} = 4.933 \text{ ft}^3/\text{lb}$$

$$v_2 = \frac{V_2}{m} = \frac{13.3 \text{ ft}^3}{3 \text{ lb}} = 4.433 \text{ ft}^3/\text{lb}$$

Checking Table A-3E at 100 lbf/in², $v_g = 4.434$ ft³ lb. Thus, state 1 is in the superheat region and state 2 is saturated vapor, as shown in the p-v diagram. Accordingly, the temperature at state 2 is the saturation temperature corresponding to 100 lbf/in²: $T_2 = 327.86°F$. Referring to Table A-4E with v_1 at 100 lbf/in², $T_1 = 400°F$.

To evaluate the work

$$W = \int_1^2 p\,dV = p(V_2 - V_1)$$

$$= \left(100 \frac{\text{lbf}}{\text{in}^2}\right)\left|\frac{144 \text{ in}^2}{1 \text{ ft}^2}\right| [13.3 - 14.8] \text{ ft}^3 \left|\frac{1 \text{ Btu}}{778 \text{ ft}\cdot\text{lbf}}\right|$$

$$= -27.76 \text{ Btu}$$

The water is compressed, thus the work is negative.

PROBLEM 3.32

KNOWN: Water is heated in a piston-cylinder assembly at constant temperature from saturated vapor to a given final pressure.

FIND: Determine the work per unit mass by (a) numerical integration using steam table data and (b) using IT.

SCHEMATIC & GIVEN DATA:

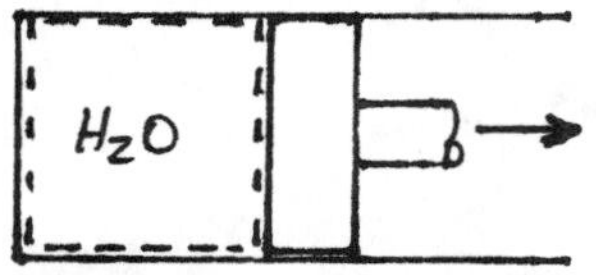

$T = 400°F$
State 1: sat. vapor
State 2: $p_2 = 100$ lbf/in²

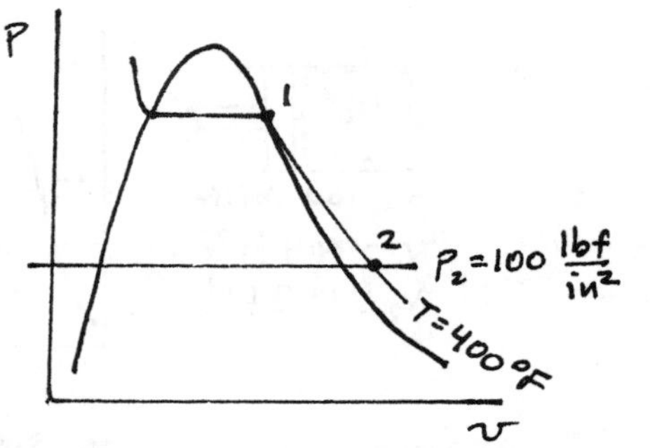

ASSUMPTIONS: (1) The water is the closed system. (2) The process occurs at constant temperature.

ANALYSIS: (a) To determine the work, use Eq. 2.17

$$\frac{W}{m} = \int_{v_1}^{v_2} p\,dv \qquad (*)$$

A numerical scheme is employed with data from Table A-4E at $T = 400°F$.

p (lbf/in²)	v (ft³/lb)
247.1	1.866 (sat. vapor, Table A-2E)
200	2.361
180	2.648
160	3.007
140	3.466
120	4.079
100	4.934

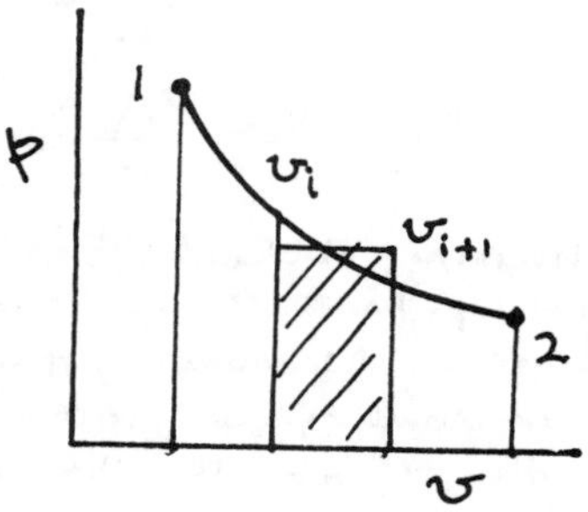

$$W/m \approx \sum_{i=1}^{6} \left(\frac{p_i + p_{i+1}}{2}\right)(v_{i+1} - v_i)$$

$$= \left[\left(\frac{247.1+200}{2}\right)\frac{lbf}{in^2}(2.361-1.866)\frac{ft^3}{lb} + \left(\frac{200+180}{2}\right)(2.648-2.361)\right.$$

$$+ \left(\frac{180+160}{2}\right)(3.007-2.648) + \left(\frac{160+140}{2}\right)(3.466-3.007) + \left(\frac{140+120}{2}\right)(4.079-3.466)$$

$$\left. + \left(\frac{120+100}{2}\right)(4.934-4.079)\right]\left|\frac{144\ in^2}{1\ ft^2}\right|\left|\frac{1\ Btu}{778\ ft\cdot lbf}\right| = 86.77\ Btu \leftarrow \frac{W/m}{\text{(part a)}}$$

(b) The following code uses the integration capability of IT to integrate (*) above:

```
T=400  // °F
v=v_PT("Water",p,T)  // ft³/lb
W=conv*Integral(p,v)
conv = 144/778
/*
Using the Explore button, sweep v from 1.864 (sat. vapor)
to 4.934 in steps of 0.01
Result:  W/m = 86.48  Btu/lb ←
*/
```

W/m (part b)

Note that the results of parts (a) and (b) are in agreement.

PROBLEM 3.33

KNOWN: Refrigerant 134a undergoes a constant pressure process from a specified initial state to a given final quality.

FIND: Determine the work.

SCHEMATIC & GIVEN DATA:

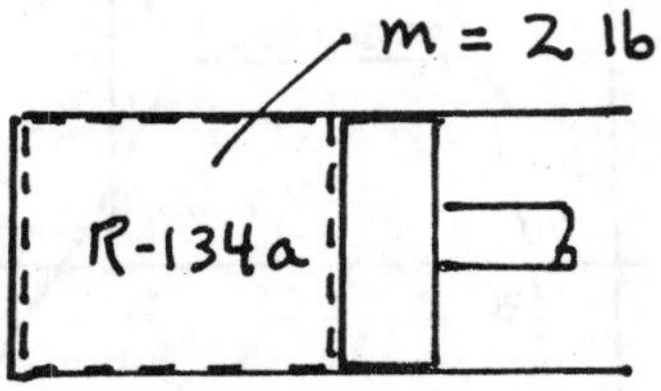

$p_1 = p_2 = 180$ lbf/in²

$T_1 = 120°F$

$x_2 = 0.765$

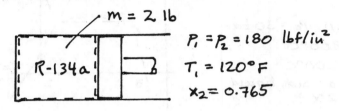

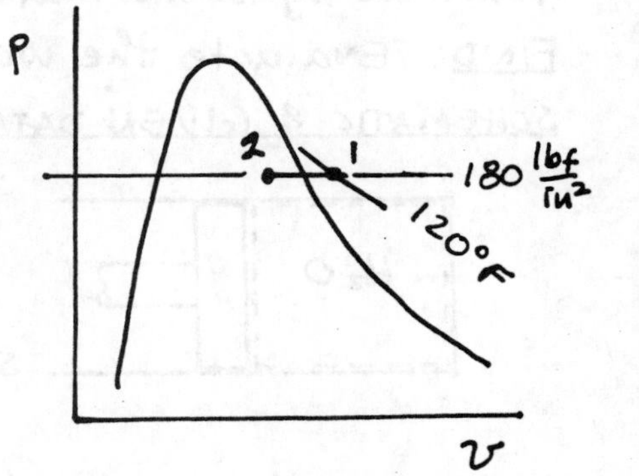

ASSUMPTIONS: (1) The refrigerant is a closed system. (2) The process occurs at constant pressure. (3) Volume change is the only work mode.

ANALYSIS: State 1 is a superheated vapor. Referring to Table A-12E

$$v_1 = 0.2595 \text{ ft}^3/\text{lb}$$

Using the given quality at state 2 and data from Table A-11E

$$v_2 = v_f + x(v_g - v_f) = 0.01438 + (0.765)[0.2569 - 0.01438]$$
$$= 0.2 \text{ ft}^3$$

Using Eq. 2.17 to determine the work

$$W = \int_{V_1}^{V_2} p\,dV = p(V_2 - V_1) = mp(v_2 - v_1)$$

Inserting values

$$W = (2 \text{ lb})(180 \text{ lbf/in}^2)(0.2 - 0.2595)\frac{\text{ft}^3}{\text{lb}}\left|\frac{144 \text{ in}^2}{1 \text{ ft}^2}\right|\left|\frac{1 \text{ Btu}}{778 \text{ ft}\cdot\text{lbf}}\right|$$

① $= -3.96$ Btu ◄ W

1. The refrigerant is compressed, thus the work is negative.

PROBLEM 3.34

KNOWN: Water vapor undergoes a constant volume cooling process followed by isothermal condensation to saturated liquid.

FIND: Evaluate the work per unit mass.

SCHEMATIC & GIVEN DATA:

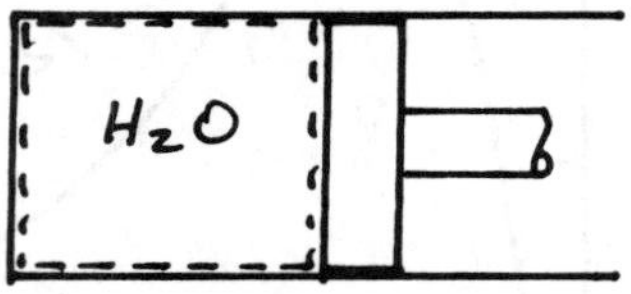

$p_1 = 3.0$ MPa = 30 bar
$T_1 = 300°C$
$T_2 = 200°C$
State 3: Sat. liquid at 200°C

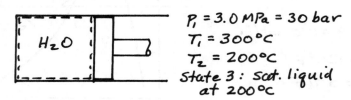

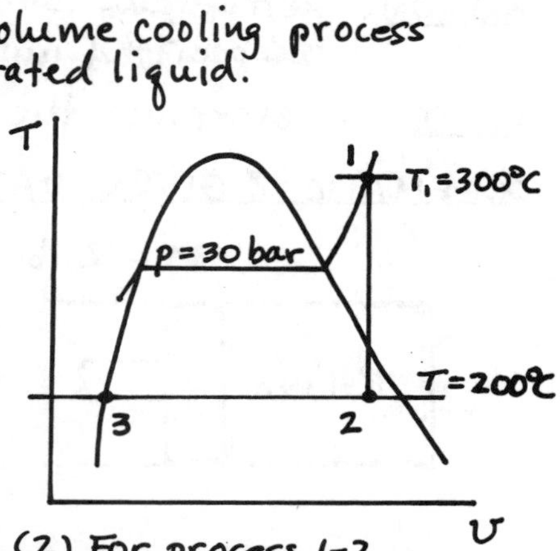

ASSUMPTIONS: (1) The water is the closed system. (2) For process 1-2, the volume is constant. (3) For process 2-3, the temperature is constant. (4) Volume change is the only work mode.

ANALYSIS: By assumptions (2) and (4), only process 2-3 involves work. To evaluate the work, begin with Eq. 2.17

$$W = \int_{V_1}^{V_2} p\,dV \Rightarrow \frac{W}{m} = \int_{v_1}^{v_2} p\,dv$$

Next, we fix each state and obtain relevant data. Since the mass and volume are constant to process 1-2, $v_2 = v_1$. Interpolating in Table A-4; $v_1 = 0.08105$ m³/kg.

With $v_2 = v_1 = 0.08105$, we see in Table A-2 at 200°C that $v_f < v_2 < v_g$. Thus, state 2 is in the two-phase, liquid-vapor region. Since the temperature is constant for process 2-3, so is the pressure. That is

$$p_2 = p_3 = p_{sat@200°C} = 15.54 \text{ bar}$$

Finally, from Table A-2, $v_3 = v_{f@200°C} = 1.1565 \times 10^{-3}$ m³/kg. Thus

$$\frac{W}{m} = \int_{v_1}^{v_2} p\,dv = p_2(v_3 - v_2)$$

$$= (15.54 \text{ bar})(1.1565\times10^{-3} - 0.08105)\frac{m^3}{kg}\left|\frac{10^5 N/m^2}{1 \text{ bar}}\right|\left|\frac{1 kJ}{10^3 N\cdot m}\right|$$

① $= -124.2$ kJ/kg ← W/m

1. The negative sign denotes energy transfer <u>to</u> the system by work during process 2-3.

PROBLEM 3.35

KNOWN: Refrigerant 22 undergoes a process with a known pressure - volume relation from a given initial state to a given final pressure.

FIND: Calculate the work for the process.

SCHEMATIC & GIVEN DATA:

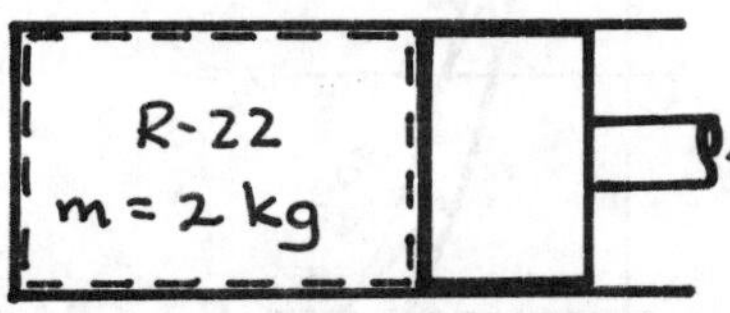

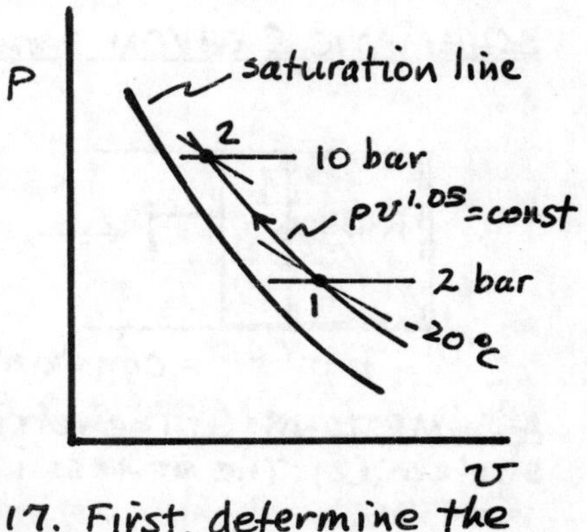

ASSUMPTIONS: (1) The refrigerant is a closed system. (2) The process is polytropic, with $n = 1.05$

ANALYSIS: The work is obtained by using Eq. 2.17. First, determine the specific volumes. From Table A-9, $v_1 = 0.11520\ m^3/kg$, and

$$v_2 = \left(\frac{p_1}{p_2}\right)^{\frac{1}{1.05}} v_1 = \left(\frac{2}{10}\right)^{\frac{1}{1.05}} (0.11520) = 0.024875\ m^3/kg$$

The work is

$$W = \int_{V_1}^{V_2} p\,dV = m\int_{v_1}^{v_2} p\,dv = m\int_{v_1}^{v_2} \frac{\text{const.}}{v^{1.05}}\,dv$$

$$= \frac{m\,(p_2 v_2 - p_1 v_1)}{(1-1.05)}$$

$$= \frac{(2\ kg)\left[(10\ bar)(0.024875\ m^3/kg) - (2)(0.11520)\right]}{(1-1.05)} \left|\frac{10^5\ N/m^2}{1\ bar}\right| \left|\frac{1\ kJ}{10^3\ N\cdot m}\right|$$

① $= -73.4\ kJ$ ◀ W

1. The negative value for work indicates that the process is a compression process

PROBLEM 3.36

KNOWN: Refrigerant 134a undergoes a polytropic process from a specified initial state to a given final temperature.

FIND: Determine the final pressure and the work per unit mass.

SCHEMATIC & GIVEN DATA:

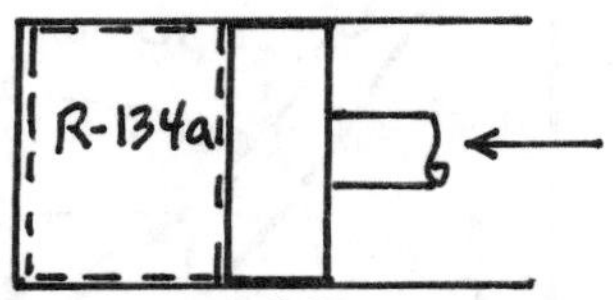

$p_1 = 200\text{ kPa} = 2\text{ bar}$
$T_1 = -10°C$
$T_2 = 50°C$

$pv^{1.05} = \text{constant}.$

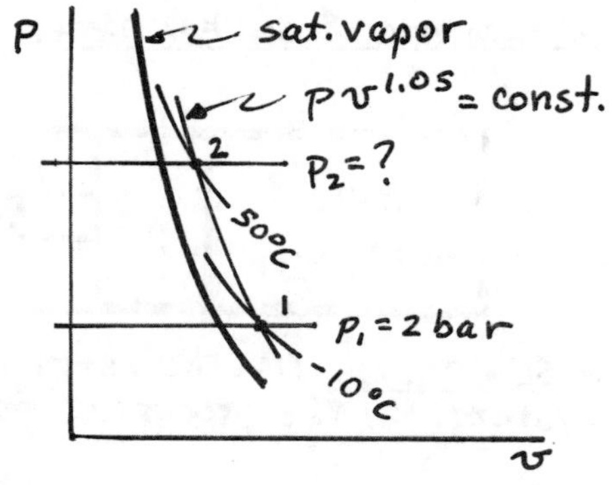

ASSUMPTIONS: (1) The refrigerant is the closed system. (2) The process is polytropic, with $n = 1.05$.

ANALYSIS: State 1 is fixed by $T_1 = -10°C$, $p_1 = 2$ bar. From Table A-12; $v_1 = 0.09938$ ft³/lb. Thus, with the polytropic process expression

$$v_2 = (p_1/p_2)^{1/1.05}\, v_1 \qquad (*)$$

In Table A-12 at $T_2 = 50°C$, there is a unique v associated with each pressure, that is

$$v_{table} = v(p, T_2 = 50°C) \qquad (**)$$

The following table of data and plot are obtained using (*) and (**). Where the lines intersect, both expressions are satisfied simultaneously.

P(bar)	(*)	(**)
8	0.02654	0.02846
9	0.02372	0.02472
10	0.02146	0.02171
11	0.01960	---
12	0.01804	0.01712

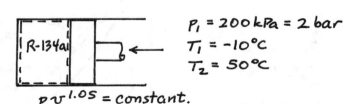

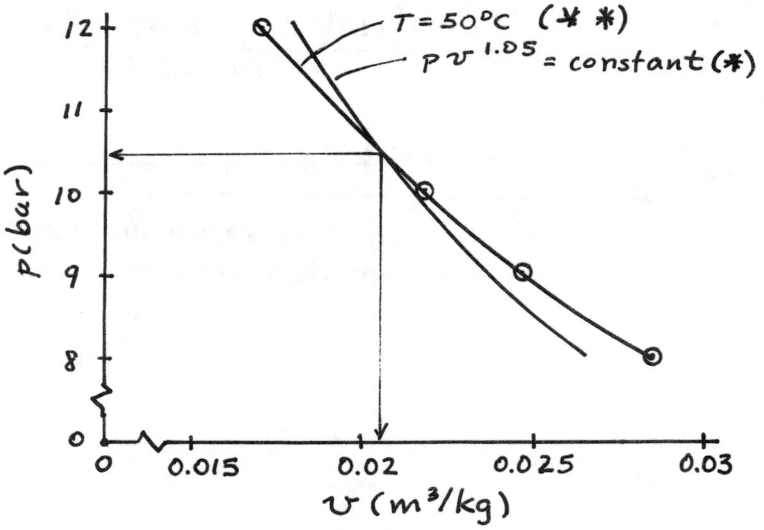

From the plot:

$p_2 \approx 10.4$ bar ← p_2

$v_2 \approx 0.0205$ m³/kg

The expressions (*) and (**) can be solved much more easily using the following IT code:

```
p1 = 2  // bar
T1 = -10  // °C
T2 = 50  // °C

v1 = v_PT("R134A", p1, T1)  // m³/kg
p2 * v2^n = p1 * v1^n                     (*)
n = 1.05
v2 = v_PT("R134A", p2, T2)               (**)
```

The results from the computer solution are $p_2 = 10.38$ bar, $v_2 = 0.0207$ m³/kg, which agree well with the graphical results.

Now, the work is determined using Eq. 2.17 and the procedure of Example 2.1 for the polytropic process to get

$$\frac{W}{m} = \int_{v_1}^{v_2} p\,dv = \left(\frac{p_2 v_2 - p_1 v_1}{1-n}\right)$$

Using the values from the computer solution

$$\frac{W}{m} = \frac{(10.38\text{ bar})(0.0207\text{ m}^3/\text{kg}) - (2)(0.09938)}{(1-1.05)}\left|\frac{10^5\text{ N/m}^2}{1\text{ bar}}\right|\left|\frac{1\text{ kJ}}{10^3\text{ N}\cdot\text{m}}\right|$$

$$= -32.21\text{ kJ} \longleftarrow \frac{W}{m}$$

PROBLEM 3.37

KNOWN: Refrigerant 134a is compressed in a piston-cylinder assembly from a known initial state to a known final pressure. The pressure-specific volume relationship for the process is given.

FIND: Determine the work.

SCHEMATIC & GIVEN DATA:

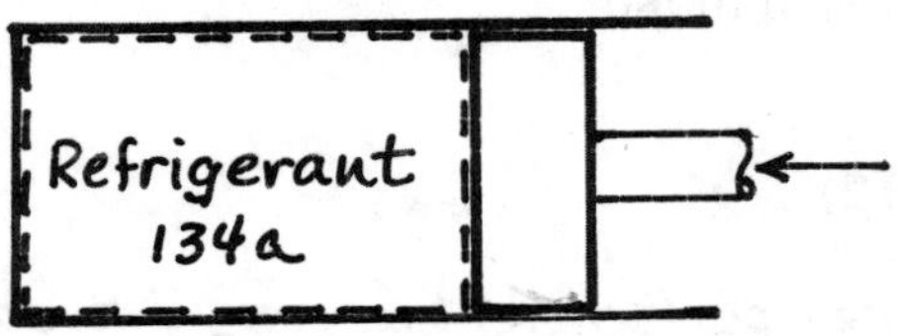

$m = 0.04$ lb
$p_1 = 10$ lbf/in^2
$T_1 = 20$°F
$p_2 = 160$ lbf/in^2
pv = const.

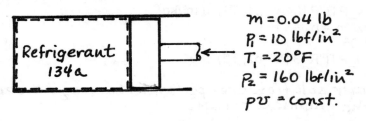

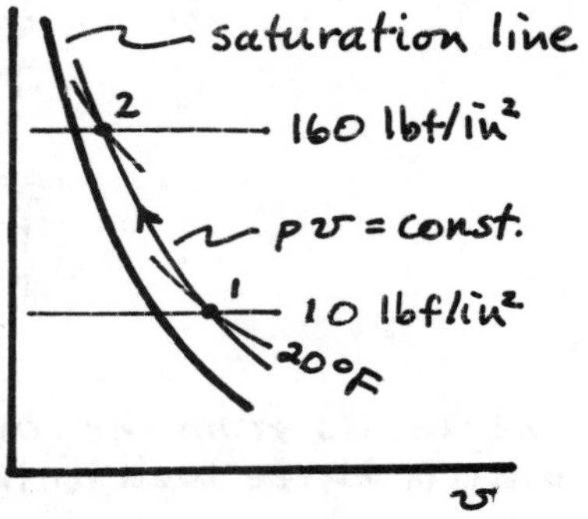

ASSUMPTIONS: (1) The refrigerant is a closed system. (2) The process is polytropic with pv = const.

ANALYSIS: The work is obtained by using Eq. 2.17. First, determine the specific volumes. From Table A-12E, $v_1 = 4.9297$ ft^3/lb, and

$$v_2 = \left(\frac{p_1}{p_2}\right) v_1 = \left(\frac{10}{160}\right)(4.9297) = 0.3081 \text{ ft}^3/\text{lb}$$

The work is

$$W = \int_{V_1}^{V_2} p\,dV = m\int_{v_1}^{v_2} p\,dv = m\int_{v_1}^{v_2} \frac{\text{const.}}{v}\,dv$$

$$= m(p_1 v_1)\ln\left(\frac{v_2}{v_1}\right) = m(p_1 v_1)\ln\left(\frac{p_1}{p_2}\right)$$

$$= (0.04 \text{ lb})(10 \text{ lbf/in}^2)(4.9297 \tfrac{\text{ft}^3}{\text{lb}})\ln\left(\frac{10}{160}\right)\left|\frac{144 \text{ in}^2}{1 \text{ ft}^2}\right|\left|\frac{1 \text{ Btu}}{778 \text{ ft}\cdot\text{lbf}}\right|$$

(1) $= -1.012$ Btu ◄ W

1. The work is negative for the compression, as expected.

PROBLEM 3.38 Water is the substance

(a) $p = 3$ bar, $T = 240°C$

Table A-4

$v = 0.781\ m^3/kg$

$u = 2713.1$ kJ/kg

IT Results

v = 0.7805 m³/kg
u = 2713 kJ/kg

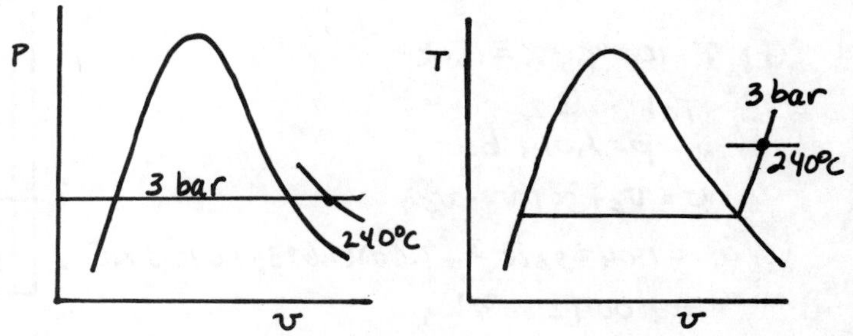

(b) $p = 3$ bar, $v = 0.5\ m^3/kg$

Table A-3; $v_g < v < v_f$

$\Rightarrow T = 133.6°C$

$$x = \frac{v - v_f}{v_g - v_f}$$

$$= \frac{0.5 - 1.0732 \times 10^{-3}}{0.6058 - 1.0732 \times 10^{-3}}$$

$= 0.825$

$\therefore u = u_f + x(u_g - u_f)$

$= 561.15 + (.825)(2543.6 - 561.5)$

$= 2196.7$ kJ/kg

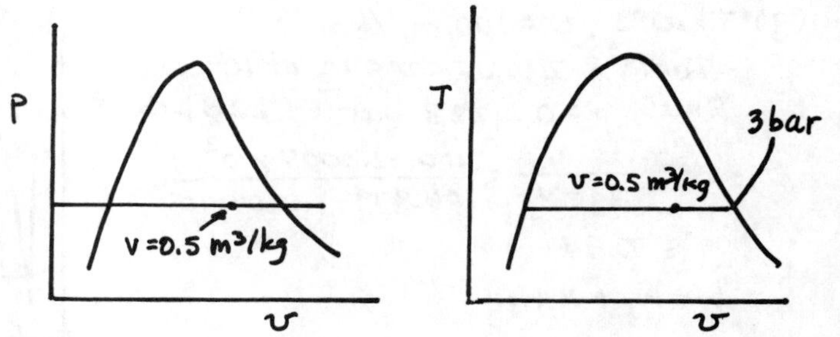

IT Results

x = 0.825
2196 kJ/kg

(c) $T = 400°C$, $p = 10$ bar

Table A-4

$v = 0.3066\ m^3/kg$

$h = 3263.9$ kJ/kg

IT Results

v = 0.3066 m³/kg
h = 3263 kJ/kg

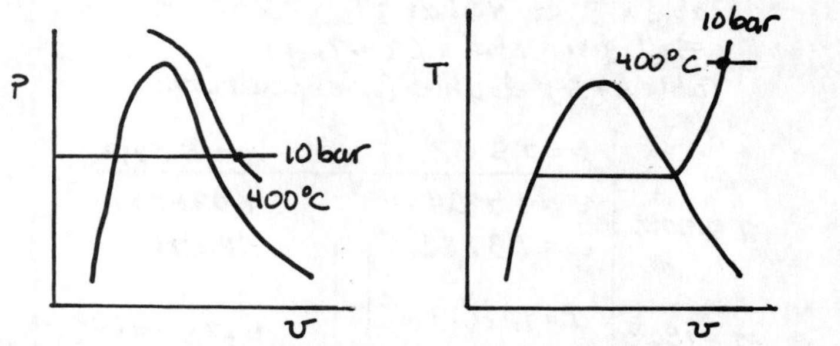

(d) $T = 320°C$, $v = 0.03\ m^3/kg$

Table A-2; $v > v_g$ at 320°C

$\Rightarrow$ Table A-4; At 320°C the state falls between 60 and 80 bar. Interpolating,

$p = 74.67$ bar $= 7.467$ MPa

$h = 2678.0$ kJ/kg

IT Results; p = 7.356 MPa, u = 2682 kJ/kg

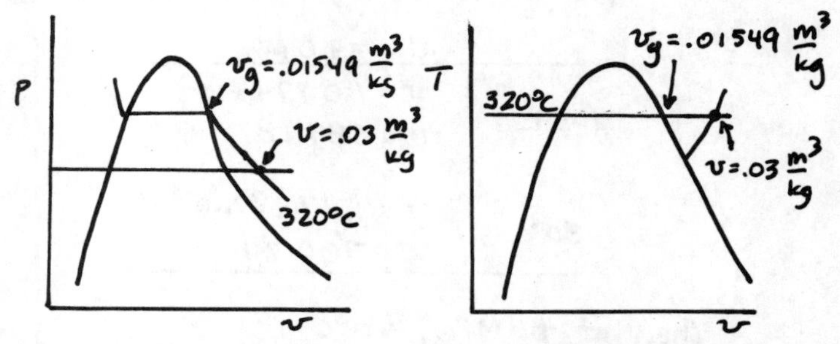

(e) $p = 28\ MPa = 280$ bar, $T = 520°C$

Table A-2

$v = 0.01020\ m^3/kg$

$h = 3192.3$ kJ/kg

IT Results

v = 0.0102 m³/kg
h = 3192 kJ/kg

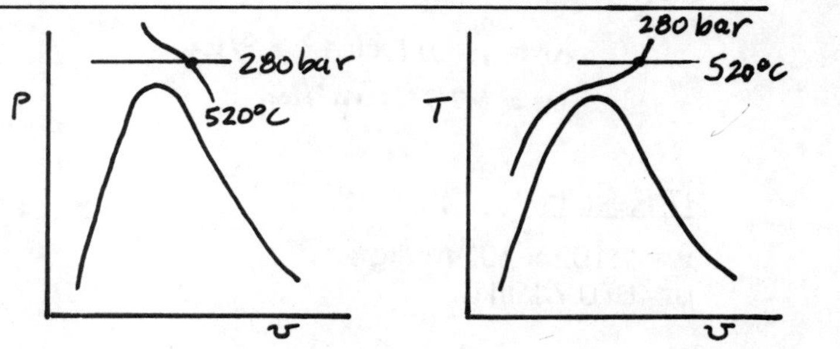

PROBLEM 3.38 (Contd.)

(f) $T = 100°C$, $x = 0.60$

Table A-2

$p = 1.014$ bar

$v = v_f + x(v_g - v_f)$

$= 1.0435 \times 10^{-3} + (.60)(1.673 - 1.0435 \times 10^{-3})$

$= 1.0042\ m^3/kg$

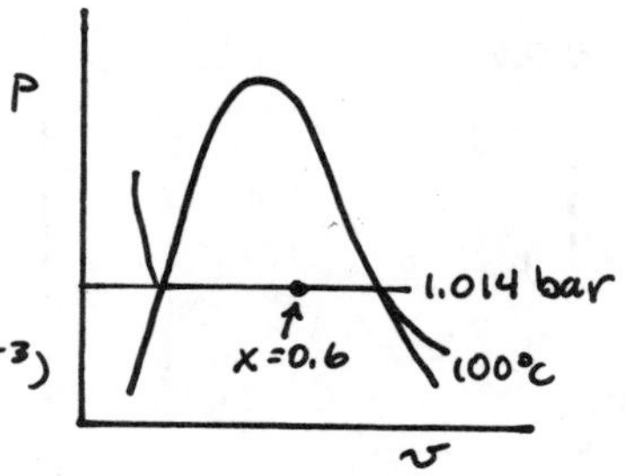

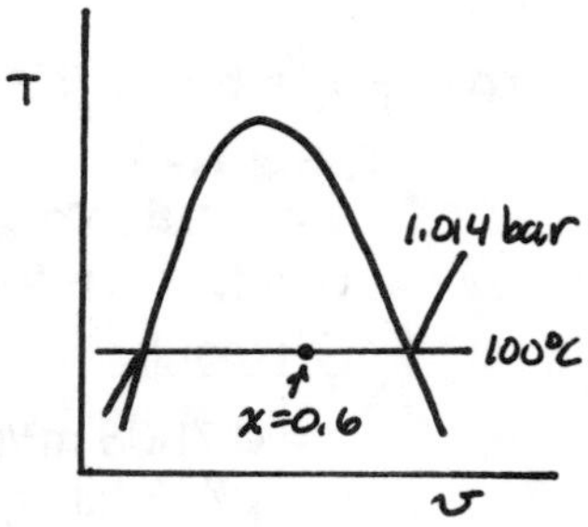

IT Results; p = 1.013 bar, v = 1.004 m³/kg

(g) $T = 10°C$, $v = 100\ m^3/kg$

Table A-2; $v_f < v < v_g$ at 10°C.

Thus, $p = 0.01228$ bar $= 1.228$ kPa

$$x = \frac{v - v_f}{v_g - v_f} = \frac{100 - 1.0004 \times 10^{-3}}{106.379 - 1.0004 \times 10^{-3}}$$

$= 0.94$

$h = h_f + x\, h_{fg}$

$= 42.01 + (.94)(2477.7) = 2371$ kJ/kg

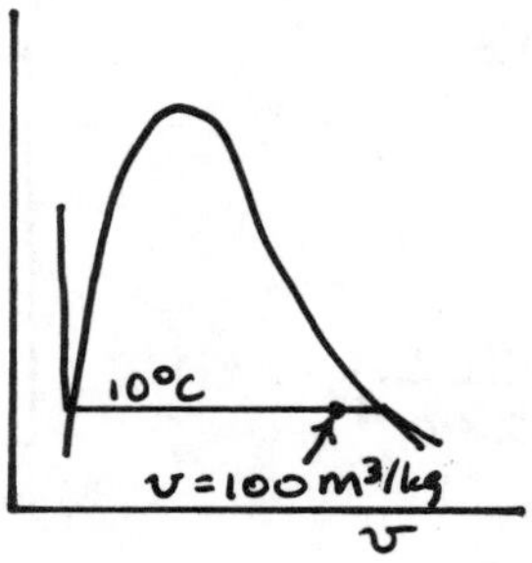

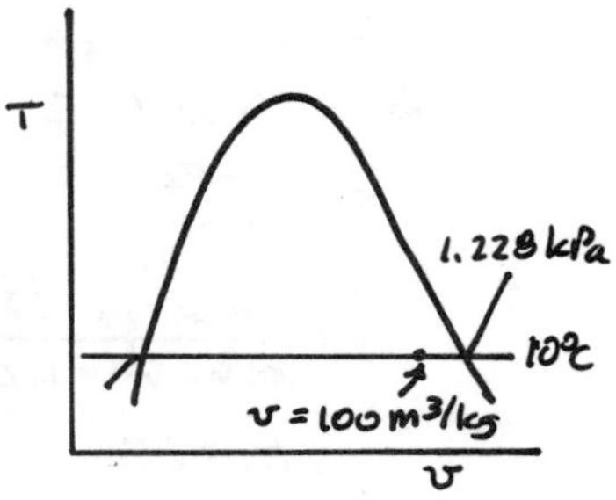

IT Results; x = 0.94, h = 2371 kJ/kg

(h) $p = 4$ MPa $= 40$ bar, $T = 160°C$

Table A-3 at 40 bar; $T_{sat} = 250.4°C$

$\Rightarrow$ liquid state ($T < T_{sat}$)

Table A-5; double interpolation

	p = 2.5 MPa	p = 5.0 MPa
T=140°C	$v = 1.0784 \times 10^{-3}$	1.0768×10^{-3}
	$u = 587.82$	586.76
T=180°C	$v = 1.1261 \times 10^{-3}$	1.1240×10^{-3}
	$u = 761.16$	759.63

	at p=4.0 MPa
140°C	$v = 1.0774 \times 10^{-3}$
	$u = 587.18$
180°C	$v = 1.1248 \times 10^{-3}$
	$u = 760.24$

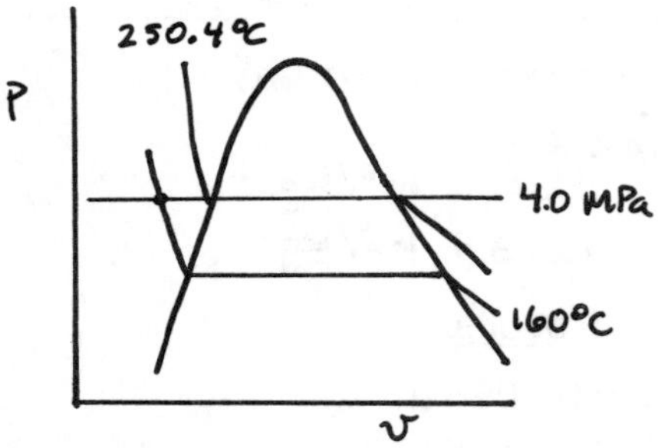

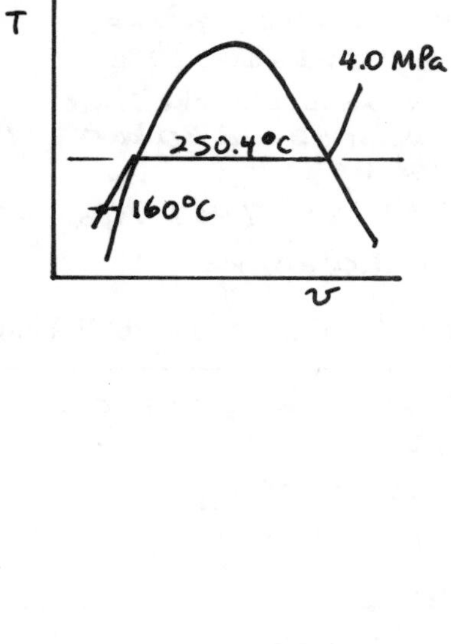

Then, at 4.0 MPa, 160°C

$v = 1.1011 \times 10^{-3}\ m^3/kg$

$u = 673.71\ m^3/kg$

IT Results

v = 1.102 x 10^{-3} m³/kg

u = 670.7 kJ/kg

PROBLEM 3.39 Water is the substance.

(a) $p = 20$ lbf/in.², $T = 400°F$

Table A-4E

$v = 25.43$ ft³/lb

$u = 1145.1$ Btu/lb

IT Results

$v = 25.43$ ft³/lb

$u = 1145$ Btu/lb

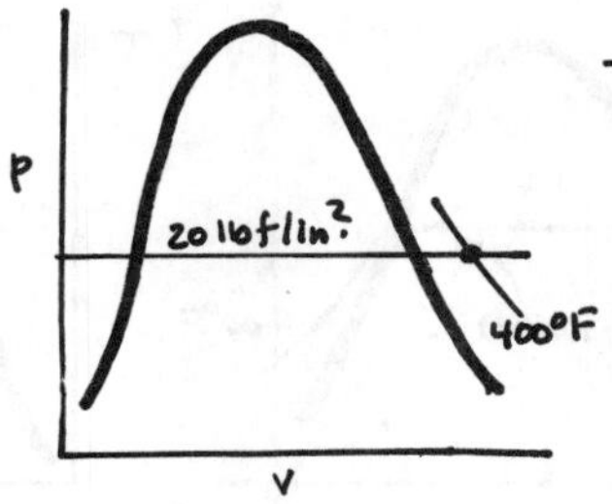

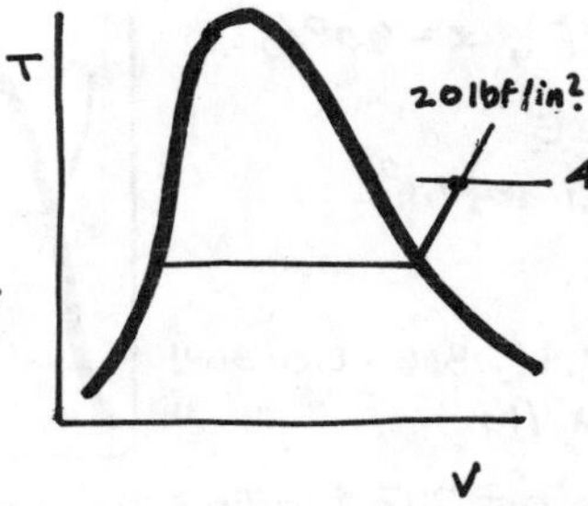

(b) $p = 20$ lbf/in.², $v = 16$ ft³/lb

Table A-3E $v_f < v < v_g$

$\Rightarrow T = 227.96°F$

$$x = \frac{v - v_f}{v_g - v_f} = \frac{16 - 0.01683}{20.09 - 0.01683} = 0.796$$

$$\therefore u = u_f + x(u_g - u_f) = 196.19 + 0.796(1082 - 196.19) = 901.29 \text{ Btu/lb}$$

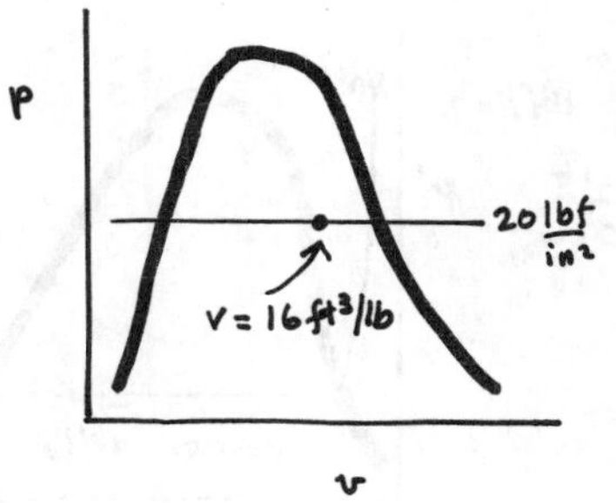

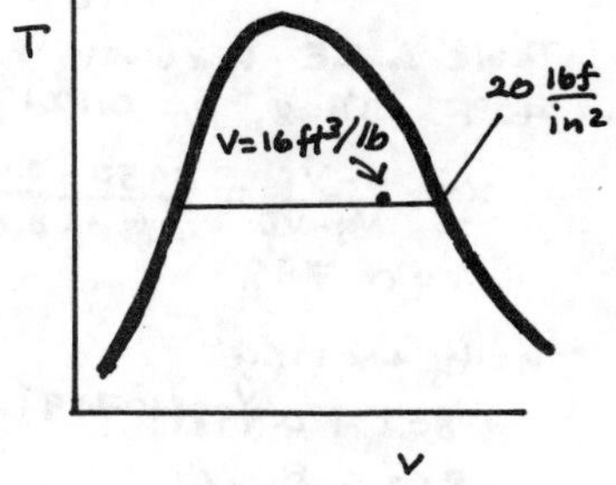

IT Results; T = 228 °F, x = 0.7962, u = 901.3 Btu/lb

(c) $T = 900°F$, $p = 170$ lbf/in.²

Table A-4E, interpolate at 900°F

$v = 4.734$ ft³/lb

$h = 1478.05$ Btu/lb

IT Results

$v = 4.718$ ft³/lb

$h = 1478$ Btu/lb

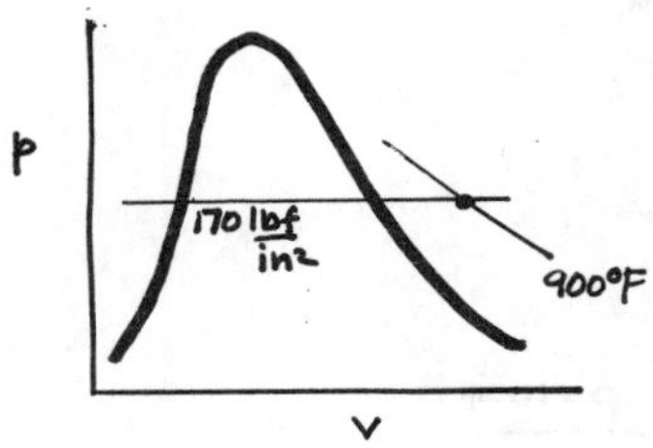

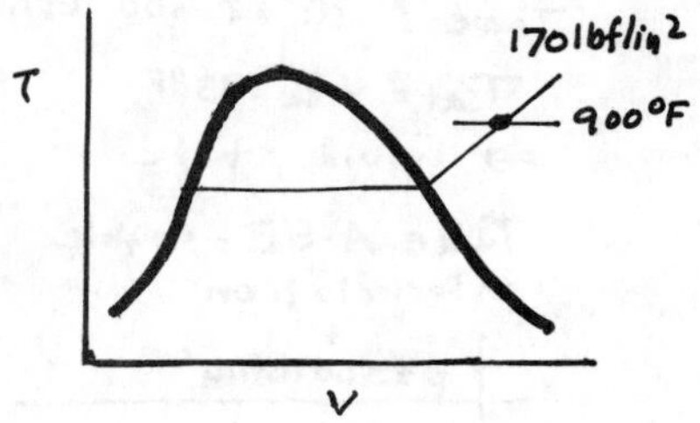

(d) $T = 600°F$, $v = 0.6$ ft³/lb

Table A-2E, $v > v_g$ at 600°F.

$\Rightarrow$ Table A-4E. At 600°F the state falls between 800 and 900 lbf/in².

Interpolating,

$p = 885.6$ lbf/in²

$u = 1163.34$ Btu/lb

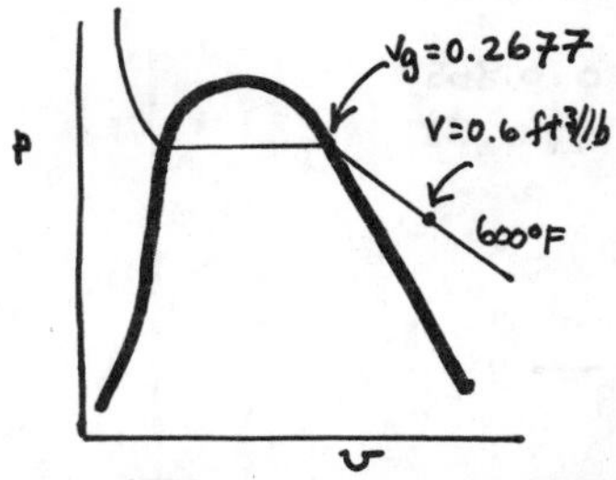

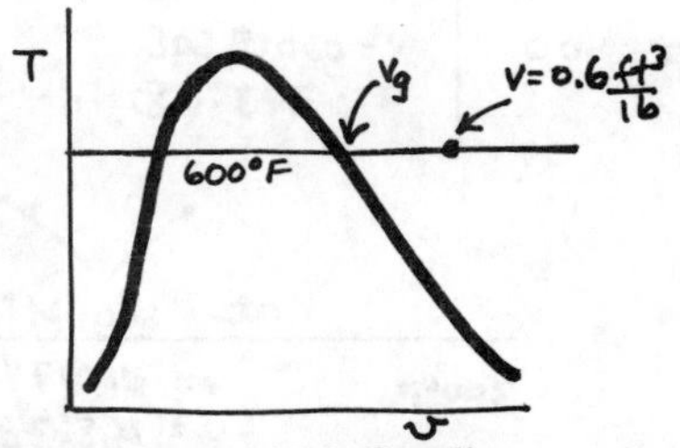

IT Results; p = 884.3 lbf/in.², u = 1163 Btu/lb

(e) $p = 700$ lbf/in.², $T = 650°F$

Table A-4E, interpolation at 700 lbf/in.²

$v = 0.85$ ft³/lb

$h = 1312.3$ Btu/lb

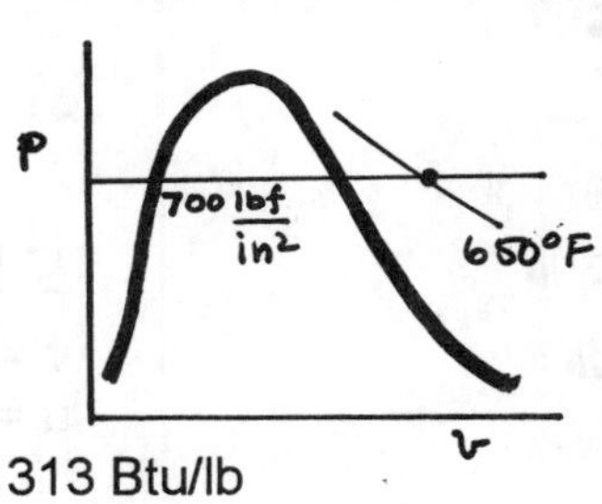

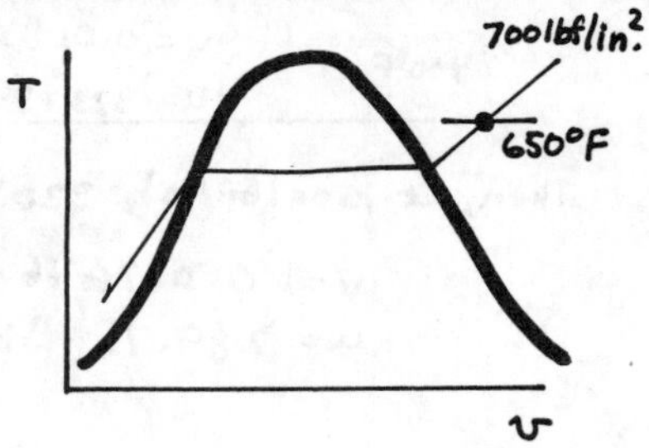

IT Results; v = 0.852 ft³/lb, h = 1313 Btu/lb

PROBLEM 13.39 (Contd.)

(f) $T = 400°F$, $x = 90\%$

Table A-2E

$p = 247.1$ lbf/in.²

$v = v_f + x\, v_{fg}$

$= 0.01864 + .9(1.866 - 0.01864)$

$= 1.6813$ ft³/lb

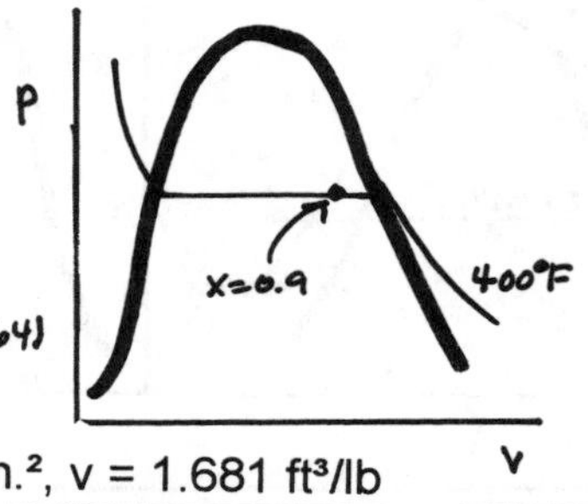

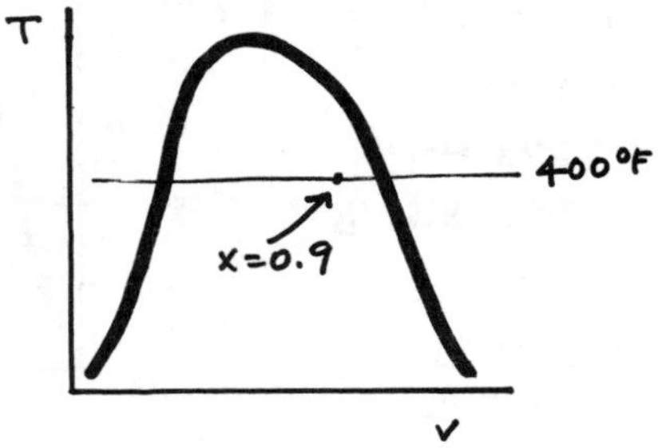

IT Results; p = 247.1 lbf/in.², v = 1.681 ft³/lb

(g) $T = 40°F$, $v = 1950 \frac{ft^3}{lb}$

Table A-2E $v_f < v < v_g$ at 40°F. Thus, $p = 0.1217$ lbf/in.²

$x = \frac{v - v_f}{v_g - v_f} = \frac{1950 - 0.016}{2445 - 0.016}$

$= 0.798$

$h = h_f + x\, h_{fg}$

$= 8.02 + 0.798(1070.9)$

$= 862.6$ Btu/lb

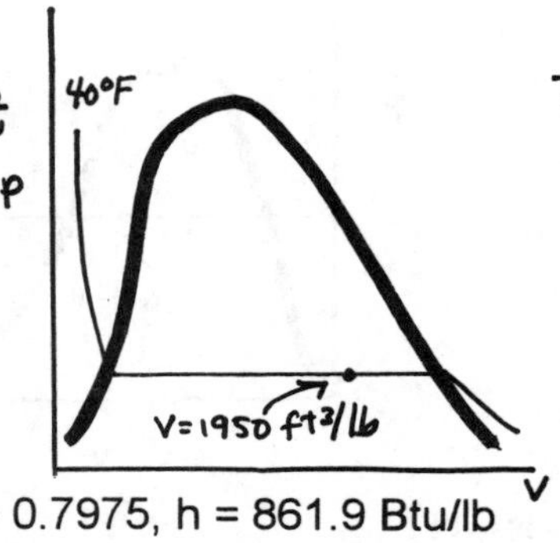

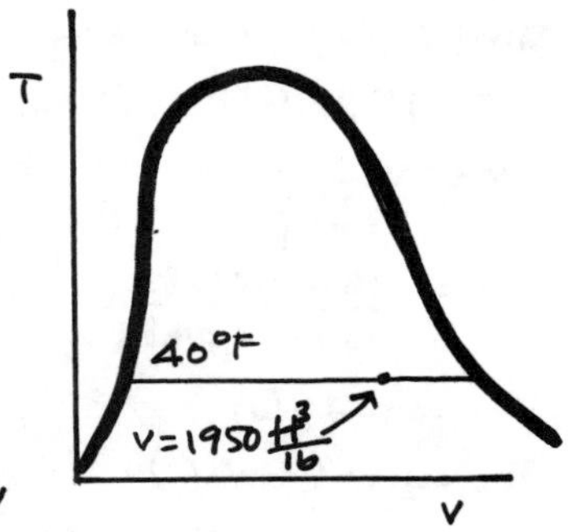

IT Results; p = 0.1217 lbf/in.², x = 0.7975, h = 861.9 Btu/lb

---

(h) $p = 600$ lbf/in.²; $T = 320°F$

Table A-3E at 600 lbf/in²

$T_{sat} = 486.33°F$

⇒ liquid state

Table A-5E – double interpolation

	p=500 lbf/in²	p=1000
T= 300°F	v= 0.017416 u= 268.92	0.017379 268.24
T= 400	v= 0.018608 u= 373.68	0.018550 372.55

	at 600 lbf/in²
300°F	v= 0.017409 u= 268.78
400°F	v= 0.018596 u= 373.45

Then, at 600 lbf/in², 320°F

$v = 0.017646$ ft³/lb

$u = 289.71$ Btu/lb

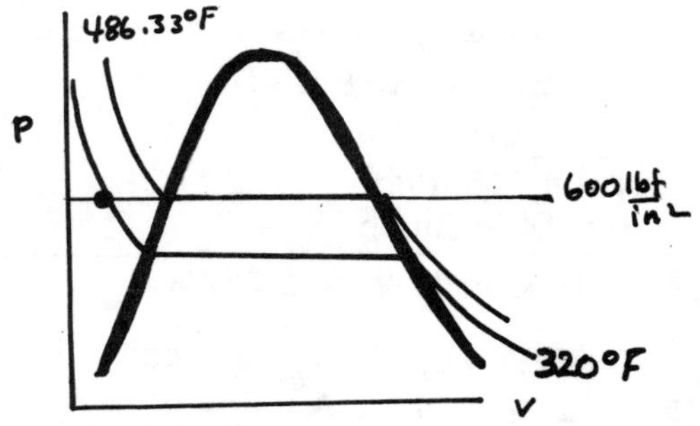

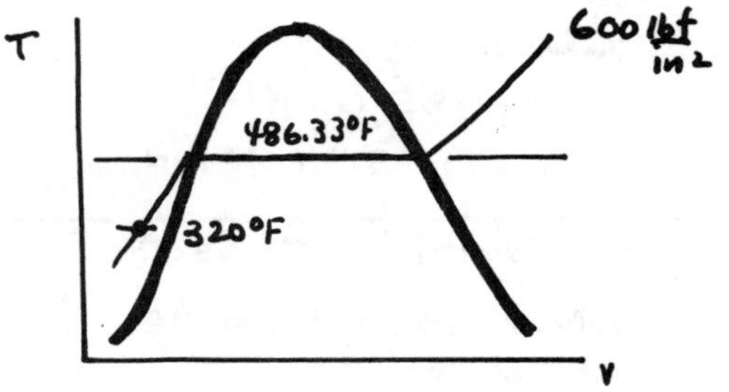

IT Results

v = 0.01766 ft³/lb

u = 288.3 Btu/lb

PROBLEM 3.40

(a) Refrigerant 134a; $T = 60°C$, $v = 0.072\ m^3/kg$
from Table A-10 at 60°C; $v_g = 0.0114\ m^3/kg$
$v > v_g \Rightarrow$ superheated vapor
Interpolating in Table A-12; $p = 3.63$ bar $= \underline{363\ kPa}$

$h = \underline{302.1\ kJ/kg}$

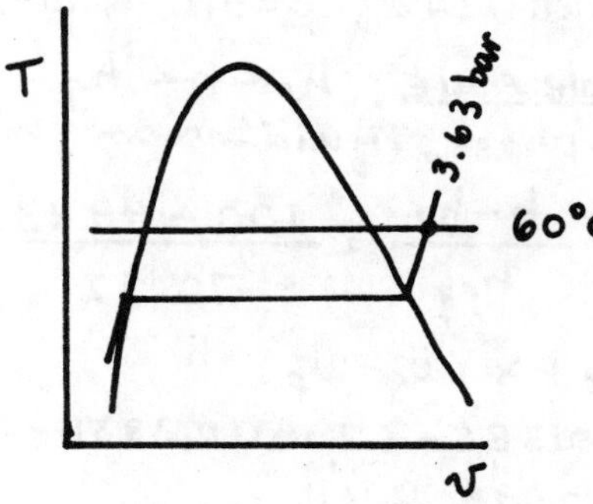

(b) Ammonia; $p = 8$ bar, $v = 0.005\ m^3/kg$
from Table A-14 at 8 bar; $v_g = 0.1596\ m^3/kg$
$v_f < v < v_g \Rightarrow$ two-phase, liquid-vapor mixture

$T = T_{sat} = \underline{17.84°C}$

$$x = \frac{v - v_f}{v_g - v_f} = \frac{0.005 - 1.6302\times10^{-3}}{0.1596 - 1.6302\times10^{-3}} = 0.02133$$

$$u = u_f + x(u_g - u_f)$$
$$= 262.64 + (0.02133)(1330.64 - 262.64)$$
$$= \underline{285.4\ kJ/kg}$$

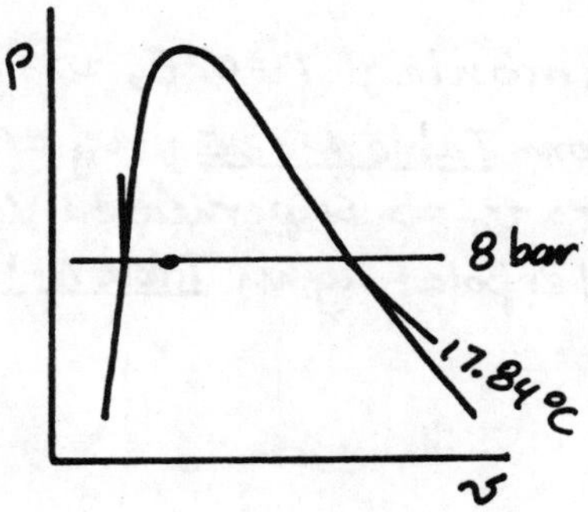

(c) Refrigerant 22; $T = -10°C$, $u = 200$ kJ/kg
From Table A-7 at $-10°C$, $u_f < u < u_g$
$\Rightarrow$ 2-phase liquid-vapor mixture

$p = p_{sat} = 3.8485$ bar

$$x = \frac{u - u_f}{u_g - u_f} = \frac{200 - 33.27}{223.02 - 33.27}$$
$$= 0.879$$

$$\therefore\ v = v_f + x(v_g - v_f)$$
$$= \frac{0.7606}{10^3} + 0.879\left(0.0652 - \frac{0.7606}{10^3}\right)$$
$$= 0.0574\ m^3/kg$$

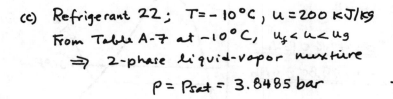

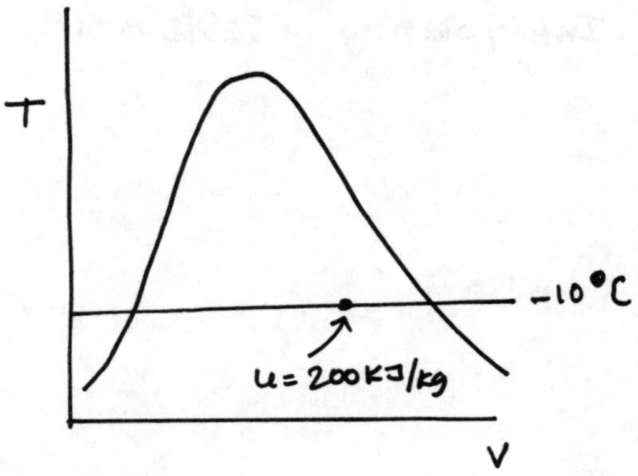

PROBLEM 3.41

(a) Refrigerant 134a; $p = 140$ lbf/in², $h = 100$ Btu/lb

from Table A-11E; $h_f < h < h_g$

⇒ two-phase, liquid-vapor mixture

$$x = \frac{h - h_f}{h_{fg}} = \frac{100 - 44.43}{70.52} = 0.788$$

$$v = v_f + x(v_g - v_f)$$

$$= 0.01386 + (.788)(0.3358 - 0.01386)$$

$$= \underline{0.2675 \text{ ft}^3/\text{lb}}$$

$$T = T_{sat} = \underline{100.56\,°F}$$

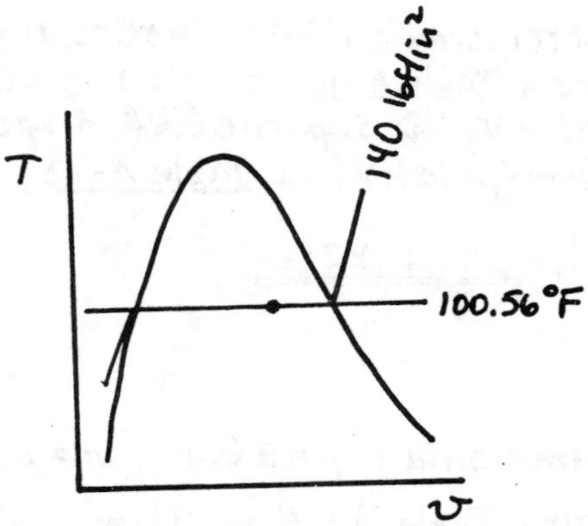

(b) Ammonia; $T = 0°F$, $v = 15$ ft³/lb

from Table A-13E; $v_g = 9.1100$ ft³/lb

$v > v_g$ ⇒ superheated vapor

Interpolating in Table A-15E: $\underline{p = 18.85 \text{ lbf/in}^2}$

$\underline{h = 615.2 \text{ Btu/lb}}$

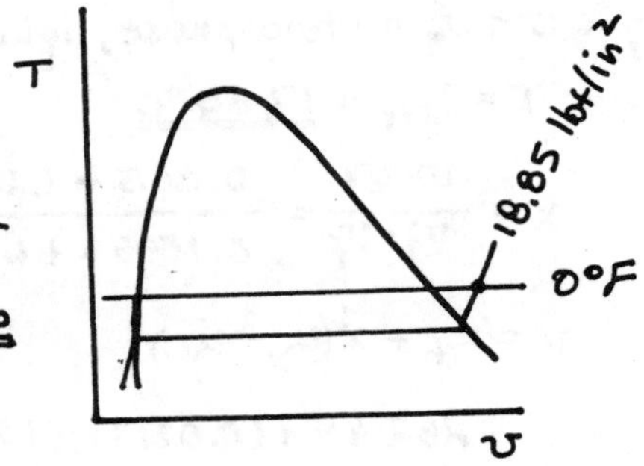

(c) Refrigerant 22; $T = 30°F$, $v = 1.2$ ft³/lb

from Table A-7E, $v_g = 0.7804$ ft³/lb

$v > v_g$ ⇒ superheated vapor

Interpolating in Table A-9E, $\underline{p = 47.60 \text{ lbf/in}^2}$

$\underline{h = 108.80 \text{ Btu/lb}}$

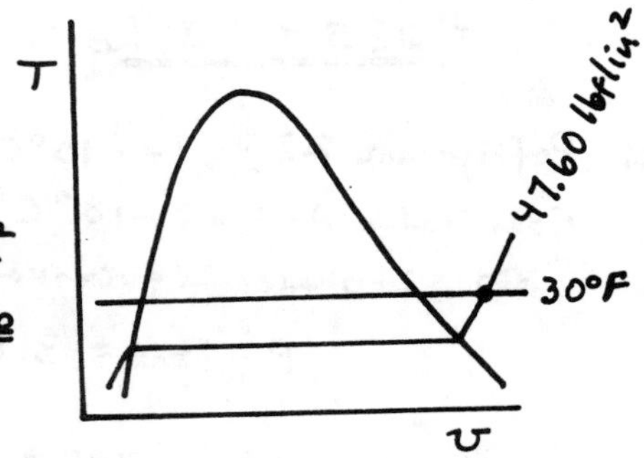

PROBLEM 3.42

(a) H_2O at $p = 15\ MPa = 150\ bar$, $T = 100°C$

from Table A-5; $v = 1.0361 \times 10^{-3}\ m^3/kg$, $h = 430.28\ kJ/kg$

(b) Using saturated liquid data from Table A-2;

$$v \approx v_{f@100°C} = 1.0435 \times 10^{-3}\ m^3/kg$$

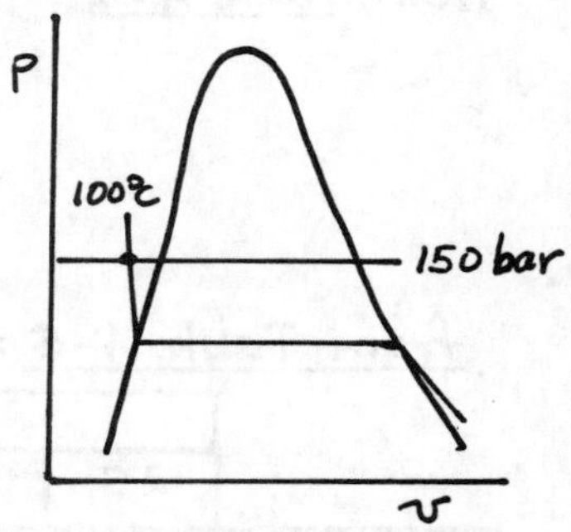

Now, using Eq. 3.13 to estimate specific enthalpy

$$h \approx h_{f@100°C} + v_{f@100°C}\,(p - p_{sat@100°C})$$

$$= 419.04\ \frac{kJ}{kg} + 1.0435 \times 10^{-3}\ \frac{m^3}{kg}(150 - 1.014)\ bar \left|\frac{10^5\ N/m^2}{1\ bar}\right|\left|\frac{1\ kJ}{10^3\ N \cdot m}\right|$$

(1) $= 419.04 + 15.55 = 434.59\ kJ/kg$

1. Note that $h_{f@100°C}$ is a plausible estimate for h in some applications (within about 2.6% of the value from Table A-5).

PROBLEM 3.43

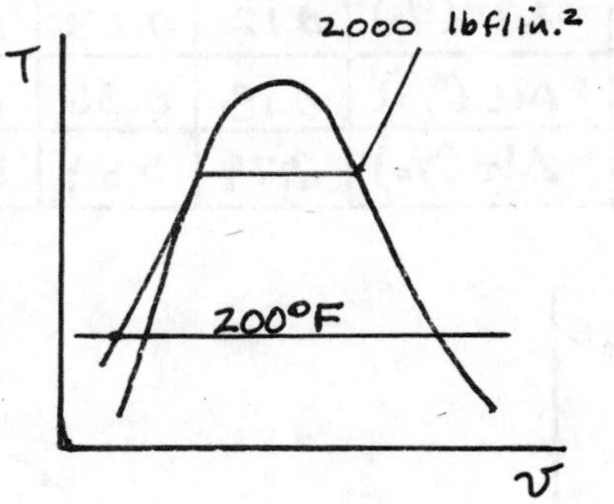

H_2O at $T = 200°F$, $p = 2000\ lbf/in^2$

from Table A-5E: $v = 0.016527\ ft^3/lb$

$h = 172.60\ Btu/lb$

The specific volume and enthalpy values can also be estimated using saturation data from Table A-2E and Eqs. 3.11 and 3.14, respectively.

$$v \approx v_f(200°F) = 0.01663\ ft^3/lb$$

and

$$h \approx h_f(200°F) + v_f(200°F)\left[p - p_{sat@200°F}\right]$$

$$= 168.07\ \frac{Btu}{lb} + 0.01663\ \frac{ft^3}{lb}\left[2000\ \frac{lbf}{in^2} - 11.529\ \frac{lbf}{in^2}\right]\left|\frac{144\ in^2}{1\ ft^2}\right|\left|\frac{1\ Btu}{778\ ft \cdot lbf}\right|$$

$$= 174.19\ Btu/lb$$

PROBLEM 3.44

H_2O at $T = 20°C$, $p_{sat} \le p \le 300$ bar

from Table A-2: $p_{sat} = 0.02339$ bar
$v_f = 1.0018 \times 10^{-3}$ m³/kg
$u_f = 83.95$ kJ/kg
$h_f = 83.96$ kJ/kg

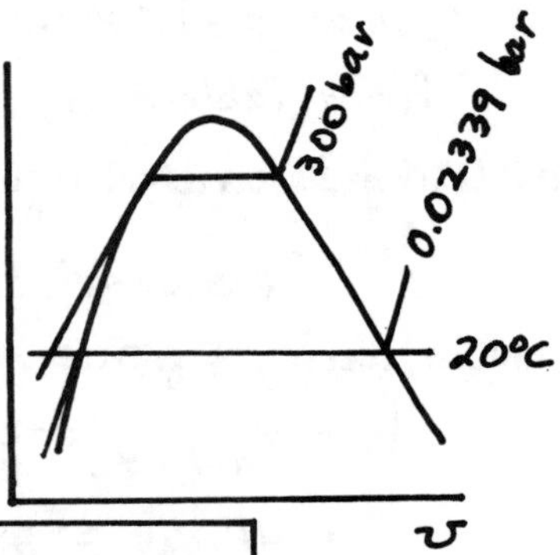

from Table A-5:

	P - bar (T=20°C).							
	25	50	75	100	150	200	250	300
$v \times 10^3$	1.0006	.9995	.9984	.9972	.9950	.9928	.9907	.9886
u	83.80	83.65	83.50	83.36	83.06	82.77	82.47	82.17
h	86.30	88.65	90.99	93.33	97.99	102.62	107.24	111.84

	25	50	75	100	150	200	250	300
Δv (%)	0.12	0.23	0.34	0.46	0.68	0.90	1.11	1.32
Δu (%)	0.18	0.36	0.54	0.70	1.06	1.41	1.76	2.12
Δh (%)	2.79	5.59	8.37	11.12	16.69	22.22	27.73	33.21

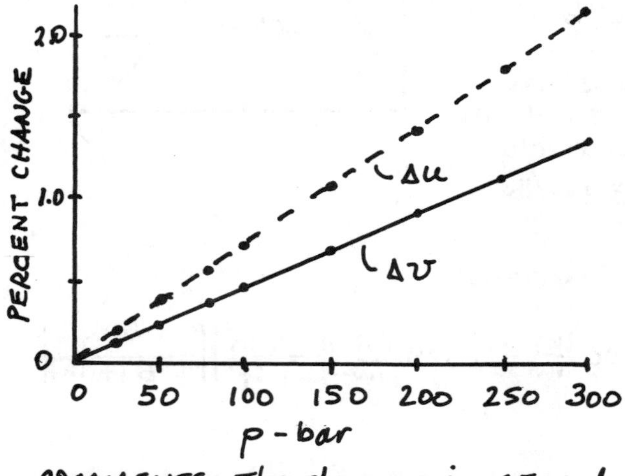

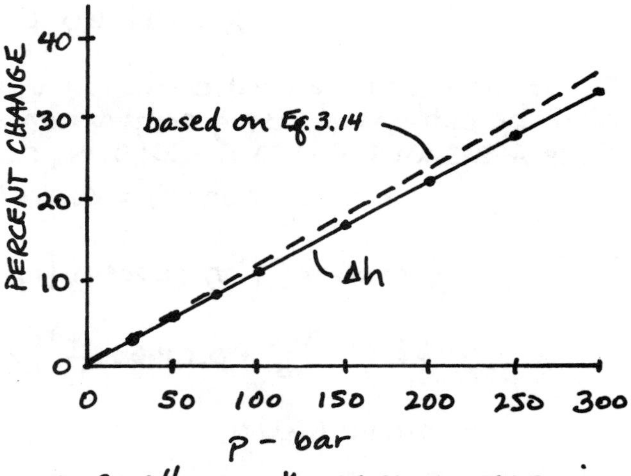

COMMENTS: The changes in v and u are small over the pressure range in this problem. Thus, v_f and u_f are reasonable approximations, respectively. However, h deviates by over 30% from h_f, and the approximation $h \approx h_f$ is not reasonable. Note that Eq. 3.14 provides a plausible estimate of h for compressed liquids when tabular data are not available.

PROBLEM 3.45

Refrigerant 134a at 95°F, 150 lbf/in.²

Table A-10E at 95°F; p_{sat} = 128.62 lbf/in.²

⇒ liquid state when p = 150 lbf/in.²

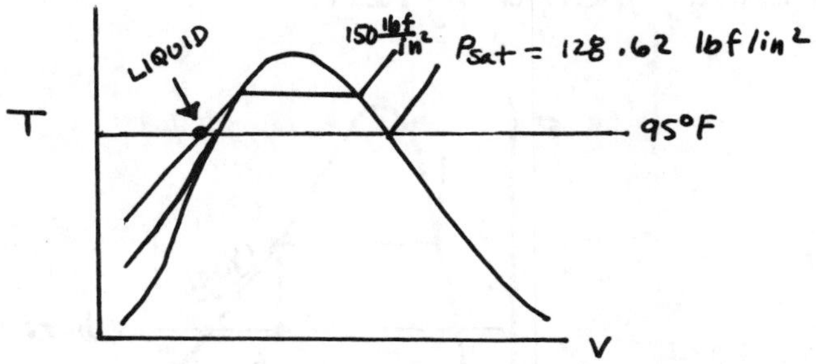

Using Eq. 3.11, $v \approx v_f(95°F)$ = 0.01371 ft³/lb ← v

Using Eq. 3.14, $h \approx h_f(95°F)$ = 42.47 Btu/lb ← h

PROBLEM 3.46

Refrigerant 22 at 30°C, 2000 kPa = 20 bar

Table A-7 at 30°C; p_{sat} = 11.9345 bar

⇒ liquid state when p = 20 bar

Using Eq. 3.11; $v \approx v_f(30°C) = 0.85395 \times 10^{-3}\ m^3/kg$

Using Eq. 3.14; $h \approx h_f(30°C)$ = 81.595 kJ/kg

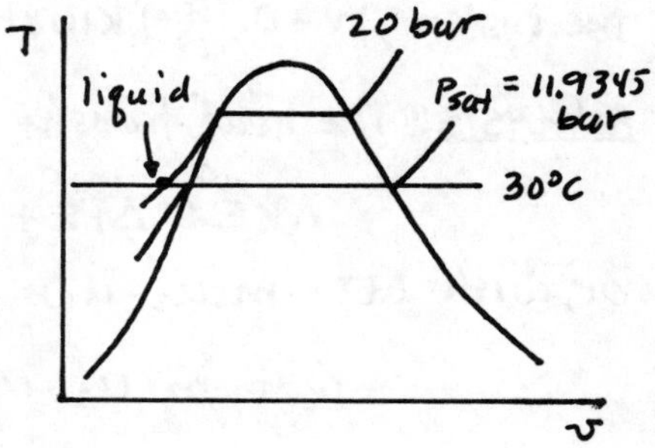

PROBLEM 3.47

Refrigerant 134a at 41°C, 1.4 MPa

Table A-11 at 14 bar, T_{sat} = 52.43°C.

⇒ liquid state

With Eq. 3.11, $v \approx v_f(41°C)$
= 0.8747 m³/kg

With Eq. 3.14, $h \approx h_f(41°C)$
= 107.69 kJ/kg

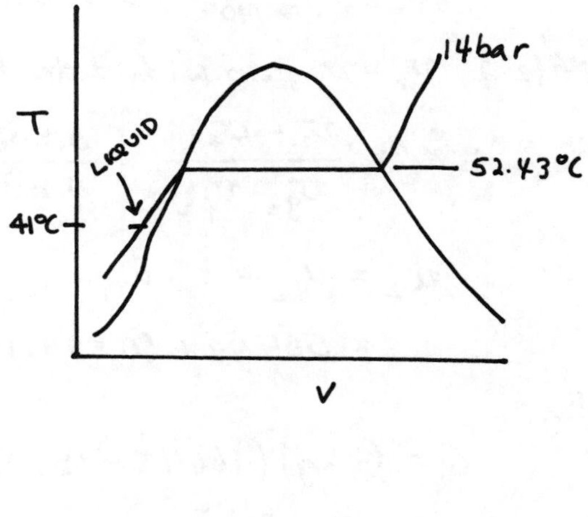

PROBLEM 3.48

KNOWN: Saturated water vapor is contained in a rigid tank at a known initial temperature. The pressure drops as a result of heat transfer to a known final value.

FIND: Determine the amount of energy transfer by heat.

SCHEMATIC & GIVEN DATA:

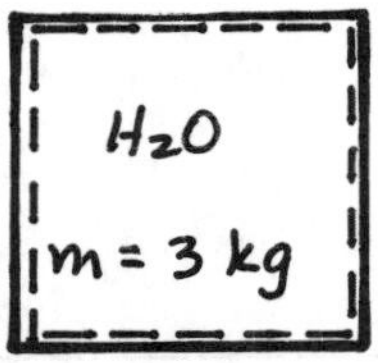

$T_1 = 140°C$
Sat. vapor

$P_2 = 200$ kPa = 2 bars

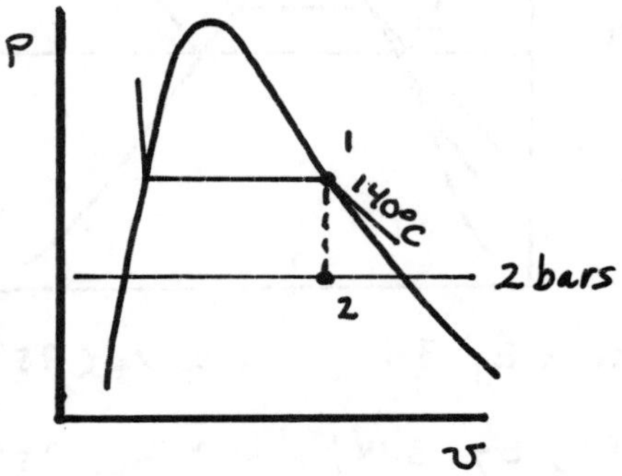

ASSUMPTIONS: (1) The H_2O is a closed system. (2) The volume is constant. (3) For the system, W = 0. (4) Kinetic and potential energy effects are negligible.

ANALYSIS: The heat transfer is determined using the energy balance

$$\Delta KE^{\,0} + \Delta PE^{\,0} + \Delta U = Q - W^{\,0}$$

or, with $\Delta U = m(u_2 - u_1)$

$$Q = m(u_2 - u_1)$$

Using data from Table A-2

$$u_1 = u_{g@140°C} = 2550.0 \text{ kJ/kg}$$

$$v_1 = v_{g@140°C} = 0.5089 \text{ m}^3/\text{kg}$$

For state 2, $v_2 = v_1$, so with data from Table A-3

$$x_2 = \frac{v_2 - v_{f2}}{v_{g2} - v_{f2}} = \frac{0.5089 - 1.0605\times10^{-3}}{0.8857 - 1.0605\times10^{-3}} = 0.5741$$

and

$$u_2 = u_{f2} + x_2(u_{g2} - u_{f2})$$
$$= 504.49 + (0.5741)(2529.5 - 504.49) = 1667.0 \text{ kJ/kg}$$

Finally

$$Q = (3 \text{ kg})(1667.0 - 2550.0) \text{ kJ/kg}$$

① $= -2649$ kJ ← Q

1. The negative sign for Q denotes energy transfer out of the system.

PROBLEM 3.49

KNOWN: Water contained in a rigid, insulated tank is heated from a specified initial state.

FIND: Determine the final temperature of the water.

SCHEMATIC & GIVEN DATA:

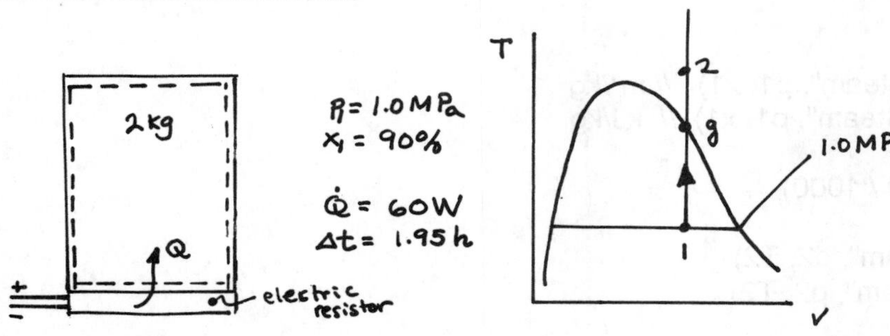

ASSUMPTIONS: 1. The water is the closed system. 2. For the system, W = 0, and ΔKE = ΔPE = 0. 3. Volume remains constant. 4. The initial and final states are equilibrium states.

ANALYSIS: The values of two independent intensive properties are required to fix the final state. Since the total volume and mass do not change, one of these is the final specific volume: $v_2 = v_1$. Using data from Table A-3,

$$v_1 = v_f + x(v_g - v_f) = \frac{1.1273}{10^3} + 0.9\left[0.1944 - \frac{1.1273}{10^3}\right]$$

$$= 0.1751 \text{ m}^3/\text{kg}$$

The other property is u_2, found from an energy balance. With assumption 2,

$$\Delta U + \cancel{\Delta KE} + \cancel{\Delta PE} = Q - \cancel{W}$$

$$\Rightarrow \quad m(u_2 - u_1) = Q \Rightarrow \quad u_2 = u_1 + Q/m \qquad (*)$$

With data from Table A-3

$$u_1 = u_f + x(u_g - u_f) = 761.68 + .9\,(2583.6 - 761.68)$$

$$= 2401.4 \text{ kJ/kg}$$

Thus, since $Q = \int \dot{Q}\,dt = \dot{Q}\Delta t$,

$$u_2 = 2401.4 + \frac{(60\text{ W})(1.95\text{h})}{2\text{kg}}\left|\frac{1\text{ J/s}}{1\text{ W}}\right|\left|\frac{3600\text{s}}{1\text{h}}\right|\left|\frac{\text{kJ}}{10^3\text{ J}}\right|$$

$$= 2612 \frac{\text{kJ}}{\text{kg}}$$

Using state g shown on the T-v diagram for reference; $v_g = v_1$. Interpolating in Table A-3 gives, $u_g = 2587$ kJ/kg. Since $u_2 > u_g$, state 2 is in the superheated vapor region, as shown on the figure.

Thus, state 2 is fixed by $u_2 = 2612$ kJ/kg, $v_2 = 0.1751$ m³/kg. Since interpolating in Table A-4 with u_2, v_2 is inconvenient, we resort to using Interactive Thermodynamics: IT to determine T_2, as follows:

PROBLEM 3.49 (Contd.)

The IT code is

```
p1 = 10  // bar
x1 = 0.9
Qdot = 60  // W
delt = 1.95  // h
m = 2  // kg
v1 = vsat_Px("Water/Steam", p1, x1)   // m³/kg
u1 = usat_Px("Water/Steam", p1, x1)  //  kJ/kg
m * (u2 - u1) = Q
Q = Qdot * delt * (3600 / 1000)
v2 = v1
v2 = v_PT("Water/Steam", p2, T2)
u2 = u_PT("Water/Steam", p2, T2)
```

<u>*IT* Results</u>

T2	198.7 °C ← T_2
p2	11.59 bar

PROBLEM 3.50

KNOWN: Refrigerant 134a is compressed from a given initial state to a given final pressure. There is no heat transfer, and the work is known.

FIND: Determine the final temperature.

SCHEMATIC & GIVEN DATA:

$P_1 = 30$ lbf/in²
$T_1 = 20$°F
$P_2 = 160$ lbf/in²
$W = -0.56$ Btu

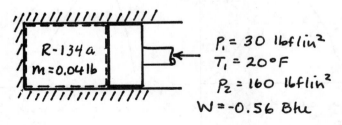

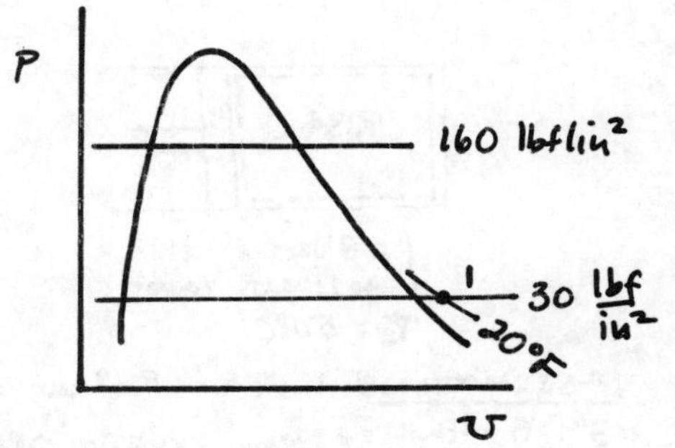

ASSUMPTIONS: (1) The R-134a is a closed system. (2) There is no heat transfer. (3) Kinetic and potential energy effects are negligible.

ANALYSIS: The final state is fixed by determining u_2 using the energy balance.

$$\Delta KE^{0} + \Delta PE^{0} + \Delta U = Q^{0} - W$$

or, with $\Delta U = m(u_2 - u_1)$

$$u_2 = -\frac{W}{m} + u_1$$

Referring to Table A-12E; $u_1 = 96.26$ Btu/lb. Thus

$$u_2 = -\frac{(-0.56 \text{ Btu})}{(0.04 \text{ lb})} + 96.26 \text{ Btu/lb}$$

$$= 110.26 \text{ Btu/lb}$$

Referring again to Table A-12E; $T_2 \approx 121.6$°F ← T_2

PROBLEM 3.51

KNOWN: Refrigerant 134a in a piston-cylinder assembly undergoes a process at constant pressure between two known states.

FIND: Determine the work and heat transfer, per unit mass

SCHEMATIC & GIVEN DATA:

$p = 8\,\text{bar}$
state 1: sat. vapor
$T_2 = 50°C$

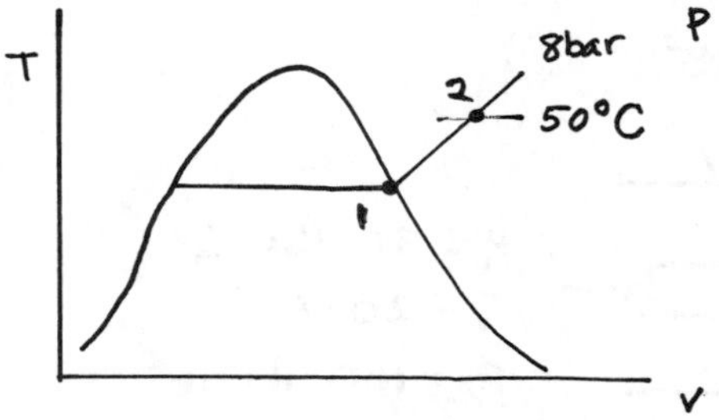

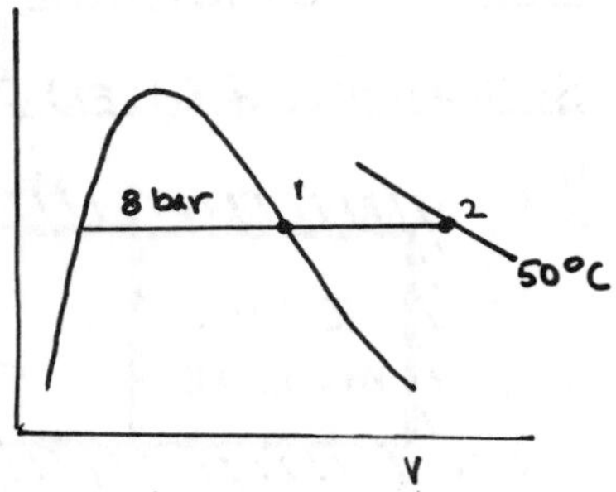

ASSUMPTIONS: 1. The R134a is a closed system. 2. Pressure is constant. 3. For the system $\Delta KE = \Delta PE = 0$.

ANALYSIS: With assumption 2 and data from Table A-12

$$W = \int_1^2 p\,dV = p(V_2 - V_1) = m\,p(v_2 - v_1)$$

$$\Rightarrow W/m = p(v_2 - v_1) = 8\,\text{bar}\left|\frac{10^5\,N/m^2}{1\,\text{bar}}\right|\left(0.02846 - 0.02547\right)\frac{m^3}{kg}\left|\frac{1\,kJ}{10^3\,N\cdot m}\right|$$

$$= 2.392\ kJ/kg$$

Applying an energy balance and data from Table A-12

$$\Delta U + \Delta KE + \Delta PE = Q - W$$

$$\Rightarrow \quad Q = \Delta U + W$$

$$= m(u_2 - u_1) + W$$

(1) $$\Rightarrow \quad \frac{Q}{m} = (u_2 - u_1) + \frac{W}{m}$$

$$= (261.62 - 243.78) + 2.392$$

$$= 20.23\ kJ/kg$$

---

1. Using $\frac{W}{m} = p(v_2 - v_1)$, this also reads

$$\frac{Q}{m} = (u_2 - u_1) + p(v_2 - v_1)$$

$$= h_2 - h_1 \qquad (\text{since } p_2 = p_1 = p)$$

with data from Table A-12

$$\frac{Q}{m} = 284.39 - 264.15 = 20.24\ kJ/kg$$

PROBLEM 3.52

KNOWN: Saturated liquid water is cooled in a closed, rigid tank to a final state where saturated liquid and vapor are present.

FIND: Determine the heat transfer for the process.

SCHEMATIC & GIVEN DATA:

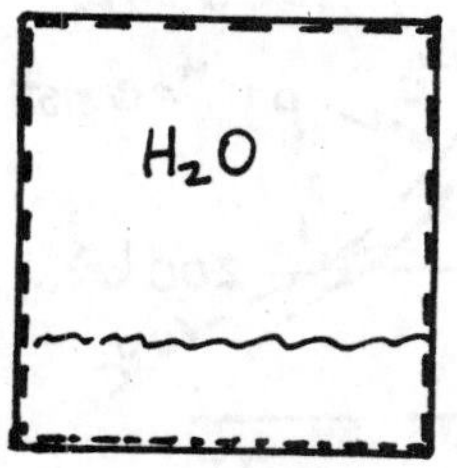

$T_2 = 50°C$

$m_{f2} = 1999.97$ kg

$m_{g2} = 0.03$ kg

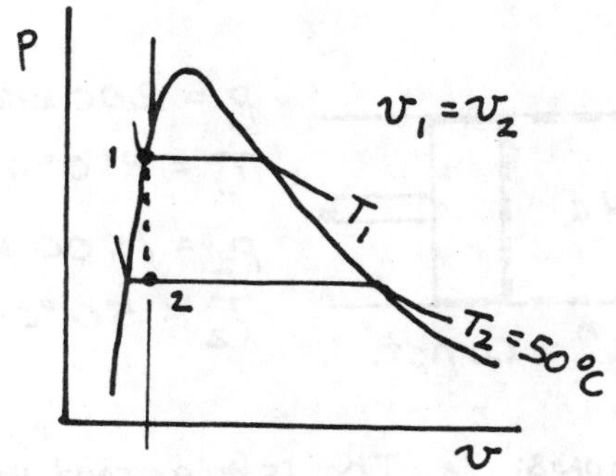

ASSUMPTIONS: 1. The water is the closed system. 2. The volume is constant. 3. Kinetic and potential energy changes are absent. 4. The initial and final states are equilibrium states.

ANALYSIS: The heat transfer is found using an energy balance:

$$\cancel{\Delta KE} + \cancel{\Delta PE} + \Delta U = Q - \cancel{W}$$

$$\Rightarrow \quad Q = \Delta U = U_2 - U_1$$

where

$$\begin{cases} U_1 = m u_1 = m\, u_f(T_1) \\ U_2 = m_{f2} u_{f2} + m_{g2} u_{g2} \end{cases}$$

State 1 is fixed by saturated liquid and $v_1 = v_2$, where with data from Table A-2

$$v_2 = \frac{m_{f2} v_{f2} + m_{g2} v_{g2}}{m_{f2} + m_{g2}}$$

$$= \frac{(1999.97\ \text{kg})\left(\frac{1.0121}{10^3}\frac{\text{m}^3}{\text{kg}}\right) + (0.03\ \text{kg})\left(12.032\ \frac{\text{m}^3}{\text{kg}}\right)}{2000\ \text{kg}}$$

$$= 1.193 \times 10^{-3}\ \frac{\text{m}^3}{\text{kg}}$$

Using $v_1 = v_2$, and noting that state 1 is saturated liquid, Table A-2 gives $T_2 = 222°C$ and $u_1 = u_{f1} = 948.2$ kJ/kg

Thus,

$$U_1 = (2000\ \text{kg})(948.2\ \text{kJ/kg}) = 1.896 \times 10^6\ \text{kJ}$$

$$U_2 = (1999.97)(209.32) + (0.03)(2443.5) = 0.419 \times 10^6\ \text{kJ}$$

$$\Rightarrow \quad Q = (0.419 \times 10^6 - 1.896 \times 10^6)\ \text{kJ}$$

$$= -1.477 \times 10^6\ \text{kJ} \quad \longleftarrow$$

PROBLEM 3.53

KNOWN: Refrigerant 134a undergoes a process for which pv^n = constant between two specified states.

FIND: Determine the work and heat transfer for the process, each per unit of mass.

SCHEMATIC & GIVEN DATA:

pv^n = const.

$p_1 = 200$ kPa
$T_1 = -10°C$
$p_2 = 1000$ kPa
$T_2 = 50°C$

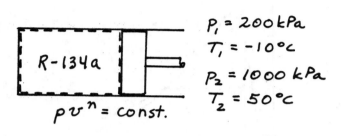

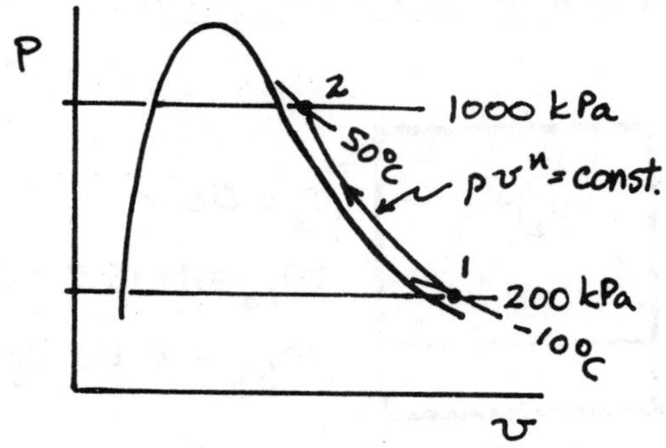

ASSUMPTIONS: 1. The refrigerant is a closed system. 2. The process is described by pv^n = constant. 3. Kinetic and potential energy effects are negligible.

ANALYSIS: Using Eq. 2.17

$$\frac{W}{m} = \int_{v_1}^{v_2} p\,dv = \int_{v_1}^{v_2} \frac{\text{const.}}{v^n}\,dv$$

$$= \frac{p_2 v_2 - p_1 v_1}{1-n} \qquad (*)$$

To find the polytropic exponent n, we first must determine the initial and final specific volumes. Using Table A-12

$p_1 = 200$ kPa, $T_1 = -10°C \Rightarrow v_1 = 0.09938$ m³/kg

$p_2 = 1000$ kPa, $T_2 = 50°C \Rightarrow v_2 = 0.02171$ m³/kg

Thus, for the polytropic process

$$p_1 v_1^n = p_2 v_2^n \Rightarrow \log(p_2/p_1) = n \log(v_1/v_2)$$

$$n = \frac{\log(p_2/p_1)}{\log(v_1/v_2)} = \frac{\log(1000/200)}{\log(.09938/.02171)} = 1.058$$

Inserting values in Eq. (*)

$$\frac{W}{m} = \frac{(1000\text{ kPa})(0.02171\text{ m}^3/\text{kg}) - (200)(.09938)}{(1-1.058)} \left|\frac{10^3\text{ N/m}^2}{1\text{ kPa}}\right| \left|\frac{1\text{ kJ}}{10^3\text{ N}\cdot\text{m}}\right|$$

$= -31.62$ kJ/kg ← W/m

The heat transfer is found by using an energy balance

$$\Delta KE^{\,0} + \Delta PE^{\,0} + \Delta U = Q - W$$

with $\Delta U = m(u_2 - u_1)$

$$\frac{Q}{m} = (u_2 - u_1) + \frac{W}{m}$$

From Table A-12; $u_1 = 221.50$ kJ/kg and $u_2 = 258.48$ kJ/kg. With the result from Problem 3.36 for W/m

$$\frac{Q}{m} = (258.48 - 221.50) + (-31.62 \text{ kJ/kg})$$

$$= +5.36 \text{ kJ/kg} \leftarrow \text{Q/m}$$

PROBLEM 3.54

KNOWN: Refrigerant 134a is compressed in a piston-cylinder assembly. The initial and final states are known, and the work is known.

FIND: Determine the heat transfer.

SCHEMATIC & GIVEN DATA:

See solution to Problem 3.37.

ASSUMPTIONS: (1) The refrigerant is a closed system. (2) Kinetic and potential energy effects are negligible.

ANALYSIS: The heat transfer is found by using an energy balance

$$\cancel{\Delta KE}^{0} + \cancel{\Delta PE}^{0} + \Delta U = Q - W$$

with $\Delta U = m(u_2 - u_1)$

$$Q = m(u_2 - u_1) + W$$

From Table A-12E ; $u_1 = 97.67$ Btu/lb and $u_2 = 110.68$ Btu/lb. With the result from Problem 3.37 for W

$$Q = (0.04\text{ lb})(110.68 - 97.67)\text{Btu/lb} + (-1.012\text{ Btu})$$

① $$= -0.4916\text{ Btu} \quad\longleftarrow\quad Q$$

---

1. The heat transfer is out of the system.

PROBLEM 3.55

KNOWN: A two-phase, liquid-vapor mixture of H_2O is stirred in a rigid, well-insulated tank until only saturated vapor remains.

FIND: Determine the amount of energy transfer by work.

SCHEMATIC & GIVEN DATA:

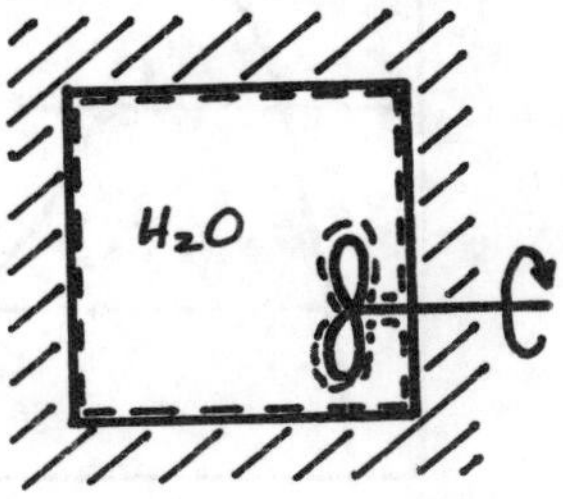

$m_{f_1} = 0.07$ lb

$m_{g_1} = 0.07$ lb

$p_1 = 20$ lbf/in²

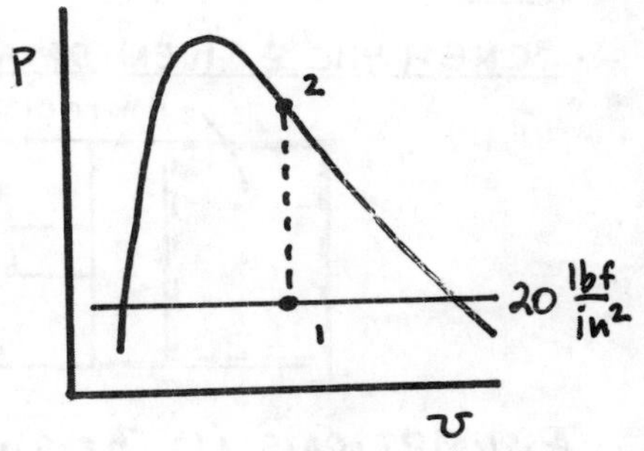

ASSUMPTIONS: (1) The H_2O is a closed system. (2) There is no heat transfer. (3) The volume is constant. (4) Kinetic and potential energy effects are negligible.

ANALYSIS: The work is determined using the energy balance

$$\cancel{\Delta KE}^{0} + \cancel{\Delta PE}^{0} + \Delta U = \cancel{Q}^{0} - W$$

or, with $\Delta U = m(u_2 - u_1)$

$$W = m(u_1 - u_2)$$

From the given data

$$x_1 = \frac{m_{g_1}}{m_{f_1} + m_{g_1}} = \frac{0.07}{0.14} = 0.5$$

Using data from Table A-3E

$$u_1 = u_{f_1} + x_1(u_{g_1} - u_{f_1}) = 196.19 + (.5)(1082.0 - 196.19) = 639.10 \text{ Btu/lb}$$

$$v_2 = v_1 = v_{f_1} + x_1(v_{g_1} - v_{f_1})$$
$$= 0.01683 + (.5)(20.09 - 0.01683) = 10.053 \text{ ft}^3/\text{lb}$$

Interpolating in Table A-3E with $v_g = 10.053$ ft³/lb $\Rightarrow u_2 = 1093.0$ Btu/lb

Finally

$$W = (0.07 + 0.07)(639.10 - 1093.0)$$

① $$= -63.55 \text{ Btu} \leftarrow W$$

1. The negative sign for work denotes energy transfer into the system, as expected.

PROBLEM 3.56

KNOWN: Ammonia expands in a piston-cylinder assembly from a known initial state to a known final pressure. The pressure-specific volume relation is specified.

FIND: Determine the work and heat transfer, each per unit mass.

SCHEMATIC & GIVEN DATA:

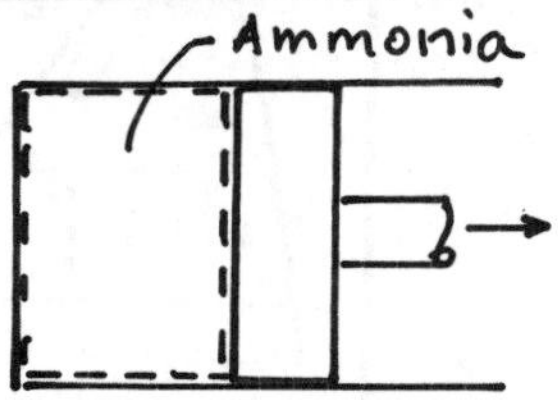

$p_1 = 500 \text{ kP} = 5 \text{ bar}$

$x_1 = 0.98 (98\%)$

$p_2 = 150 \text{ kPa} = 1.5 \text{ bar}$

$pv = \text{constant}$

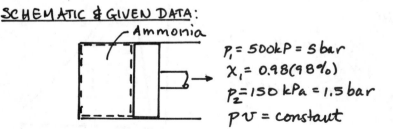

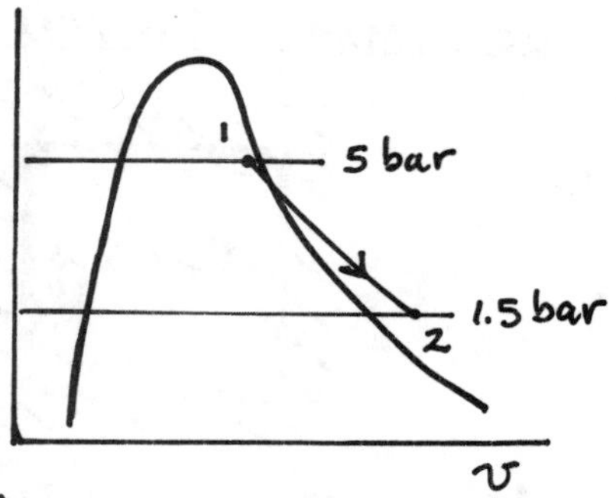

ASSUMPTIONS: (1) The ammonia is the closed system. (2) During the process, $pv = \text{constant}$. (3) Kinetic and potential energy effects are negligible.

ANALYSIS: The work is determined using Eq. 2.17 along with the given p-v relation

$$W = \int_{V_1}^{V_2} p\,dV = m\int_{v_1}^{v_2} p\,dv = m\int_{v_1}^{v_2} \frac{\text{const.}}{v}\,dv$$

With $p \cdot v = \text{const.} = p_1 v_1$

$$\frac{W}{m} = (p_1 v_1) \ln\left(\frac{v_2}{v_1}\right)$$

Using data from Table A-14; $v_1 = v_{f1} + x_1(v_{g1} - v_{f1}) = 0.24533 \text{ m}^3/\text{kg}$. Thus

$$v_2 = \left(\frac{p_1}{p_2}\right) v_1 = \frac{(5)(0.24533)}{(1.5)} = 0.81777 \text{ m}^3/\text{kg}$$

① and

$$\frac{W}{m} = (5 \text{ bar})\left(0.24533 \frac{\text{m}^3}{\text{kg}}\right)\left|\frac{10^5 \text{ N/m}^2}{1 \text{ bar}}\right|\left|\frac{1 \text{ kJ}}{10^3 \text{ N}\cdot\text{m}}\right| \ln\left(\frac{0.81777}{0.24533}\right) = 147.7 \frac{\text{kJ}}{\text{kg}} \longleftarrow \frac{W}{m}$$

The heat transfer is found using the energy balance. With assumption 3

$$\cancel{\Delta KE}^{0} + \cancel{\Delta PE}^{0} + \Delta U = Q - W$$

and $\Delta U = m(u_2 - u_1)$

$$Q/m = (u_2 - u_1) + W$$

From Table A-14 at 5 bar, $x_1 = 0.98$; $u_1 = u_{f1} + x_1(u_{g1} - u_{f1}) = 1298.57 \text{ kJ/kg}$. Also, interpolating in Table A-15 at $p_2 = 1.5 \text{ bar}$, $v_2 = 0.81777 \text{ m}^3/\text{kg}$ gives $u_2 = 1312.72 \text{ kJ/kg}$. Finally

$$Q/m = (1312.72 - 1298.57)\frac{\text{kJ}}{\text{kg}} + (147.7 \text{ kJ/kg})$$

① $$= 161.8 \text{ kJ/kg} \longleftarrow Q/m$$

---

1. The respective positive signs for heat and work denote that for this process there is energy transfer by heat <u>into</u> the system and energy transfer by work <u>out</u> of the system.

PROBLEM 3.57

<u>KNOWN</u>: A known amount of H_2O at a specified initial pressure is confined to one side of a rigid, well insulated container by a partition. The parition is removed and the H_2O expands into the initially evacuated side, reaching a specified pressure when equilibrium is attained.

<u>FIND</u>: Determine the initial quality and the overall volume of the container.

<u>SCHEMATIC & GIVEN DATA</u>:

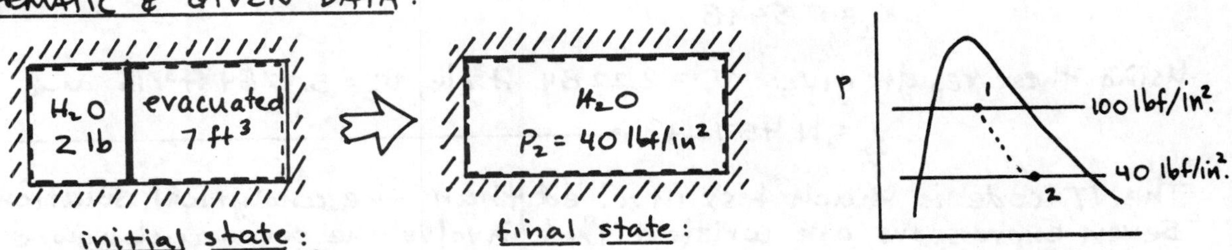

initial state: $P_1 = 100$ lbf/in², $x_1 = ?$

final state: $P_2 = 40$ lbf/in²

<u>ASSUMPTIONS</u>: (1) The contents of the tank are the closed system. (2) The expansion occurs with no work or heat transfer. (3) The volume of the tank is constant. (4) The are no changes in kinetic or potential energy.

<u>ANALYSIS</u>: First, we solve this problem analytically. Then, we use the software <u>Interactive Thermodynamics: IT</u>. Begin with the total volume

$$V = m v_1 + 7 \text{ ft}^3$$

where m = 2 lb. Also, noting that $V = m v_2$, we get

$$v_2 = \frac{V}{m} = v_1 + \frac{7 \text{ ft}^3}{2 \text{ lb}} \Rightarrow v_2 = v_1 + 3.5 \text{ ft}^3/\text{lb} \qquad (1)$$

① Now, v_1 and v_2 can be expressed in terms of the qualities x_1 and x_2, respectively

$$v_1 = v_{f1} + x_1(v_{g1} - v_{f1})$$
$$v_2 = v_{f2} + x_2(v_{g2} - v_{f2})$$

From (1)

$$v_{f2} + x_2(v_{g2} - v_{f2}) = v_{f1} + x_1(v_{g1} - v_{f1}) + 3.5$$

Using data from Table A-3E

$$0.01715 + x_2(10.50 - 0.01715) = 0.01774 + x_1(4.434 - 0.01774) + 3.5$$

or

$$x_2(10.483) = x_1(4.4163) + 3.50059 \qquad (2)$$

Now, an overall energy balance reads $\cancel{\Delta KE} + \cancel{\Delta PE} + \Delta U = \cancel{Q} - \cancel{W} \Rightarrow \Delta U = 0$, or $u_2 = u_1$. In terms of saturated liquid and saturated vapor data

$$u_{f2} + x_2(u_{g2} - u_{f2}) = u_{f1} + x_1(u_{g1} - u_{f1})$$

Inserting values from Table A-3E

PROBLEM 3.57 (Contd.)

$$236.03 + x_2(1092.3 - 236.03) = 298.3 + x_1(1105.8 - 298.3)$$

or

$$x_2(856.27) = x_1(807.5) + 62.27 \qquad (3)$$

Expressions (2) and (3) each involve the unknown qualities x_1 and x_2. Solving simutaneously

$x_1 = 0.5006$ } analytical resuts ← x_1

$x_2 = 0.5448$

Using these results gives $v_1 = 2.2284$ ft³/lb, $v_2 = 5.7284$ ft³/lb, and

$V = 11.457$ ft³ ← V

The IT code is much less involved than the analytical solution. Seven expressions are written that involve the seven unknowns: $V, v_1, v_2, x_1, x_2, u_1, u_2$. The equation solver solves automatically, without the need to further manipulate the expressions algebraically. The code follows

```
p1 = 100  // lbf/in.²
p2 = 40
m = 2  // lb
V = m * v1 + 7 // ft³
v2 = v1 + 3.5
v1 = vsat_Px("Water/Steam", p1,x1)
v2 = vsat_Px("Water/Steam", p2, x2)
u1 = usat_Px("Water/Steam", p1, x1)
u1 = u2
u2 = usat_Px("Water/Steam", p2, x2)
```

② V 11.46 } IT Resuts
x1 0.5007

---

1. This assumes that state 2 is in the two-phase liquid vapor region. If state 2 is actually in the superheated vapor region, x_1 and/or x_2 would turn out to be physically unreasonable, and it would be necessary to modify the solution method.
2. There is excellent agreement between the analytical and computer results.

PROBLEM 3.58

KNOWN: Water is cooled at constant pressure from saturated vapor to saturated liquid.

FIND: Determine the work and heat transfer and show that the heat transfer equals the change in enthalpy.

SCHEMATIC & GIVEN DATA:

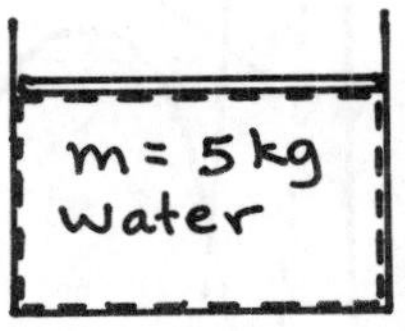

$p = 100$ kPa = 1 bar

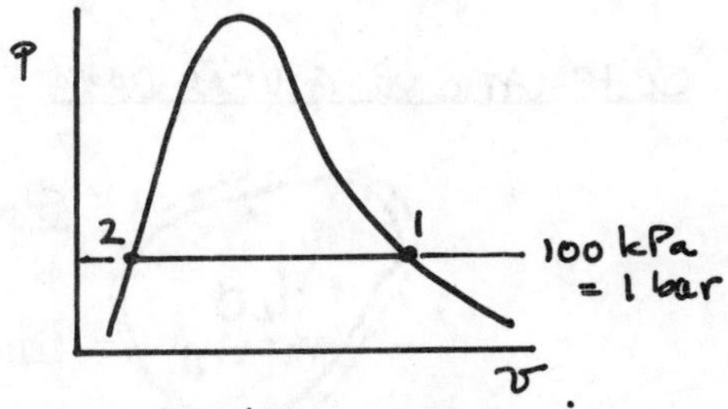

ASSUMPTIONS: (1) The water is the closed system. (2) The pressure is constant. (3) Kinetic and potential energy effects are neglected.

ANALYSIS: To determine the work, use Eq. 2.17 along with assumption 2

$$W = \int_{V_1}^{V_2} p\,dV = p(V_2 - V_1) = mp(v_2 - v_1) \qquad (1)$$

Using data from Table A-3 at $p = 1$ bar; $v_1 = v_g = 1.694\ \text{m}^3/\text{kg}$ and $v_2 = v_f = 1.0432\times10^{-3}\ \text{m}^3/\text{kg}$. Thus

$$W = (5\ \text{kg})(100\ \text{kPa})\left|\frac{10^3\ \text{N/m}^2}{1\ \text{kPa}}\right|(1.0432\times10^{-3} - 1.694)\frac{\text{m}^3}{\text{kg}}\left|\frac{1\ \text{kJ}}{10^3\ \text{N}\cdot\text{m}}\right|$$

$$= -846.5\ \text{kJ} \longleftarrow W$$

The heat transfer is found using the energy balance; $\Delta KE^{\,0} + \Delta PE^{\,0} + \Delta U = Q - W$

$$Q = \Delta U + W$$

or, with $\Delta U = m(u_2 - u_1)$

$$Q = m(u_2 - u_1) + W \qquad (2)$$

With $u_1 = u_g = 2506.1$ kJ/kg and $u_2 = u_f = 417.36$ kJ/kg from Table A-3

$$Q = (5\ \text{kg})(417.36 - 2506.1)\frac{\text{kJ}}{\text{kg}} + (-846.5\ \text{kJ})$$

$$= -11290\ \text{kJ} \longleftarrow Q$$

Combining (1) and (2)

$$Q = m(u_2 - u_1) + mp(v_2 - v_1)$$

$$= m\left[(u_2 + p_2 v_2) - (u_1 + p_1 v_1)\right]$$

With $h = u + pv$ and data from Table A-3

① $$Q = m\left[h_2 - h_1\right] = (5)\left[417.46 - 2675.5\right] = -11290\ \text{kJ}$$

---

1. The result developed here for a closed system is <u>only</u> valid for a constant pressure process.

PROBLEM 3.59

KNOWN: Saturated solid water is heated at constant pressure to saturated liquid.

FIND: Determine the work and heat transfer. Show that the heat transfer equals the change in enthalpy.

SCHEMATIC & GIVEN DATA:

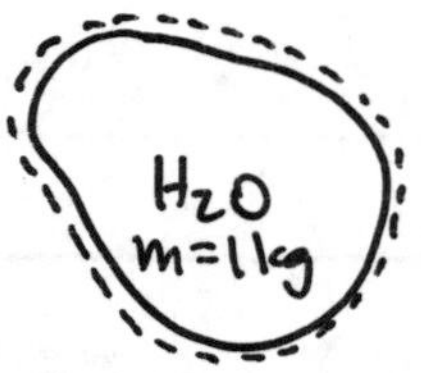

State 1: triple point sat. solid

State 2: sat. liquid

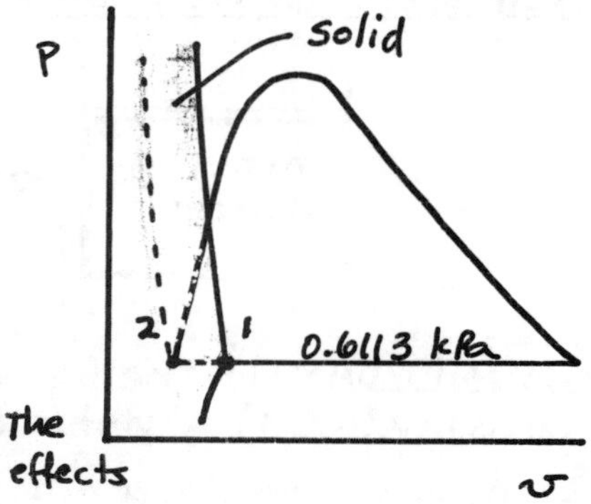

ASSUMPTIONS: (1) The H_2O is a closed system. (2) The pressure is constant. (3) Kinetic and potential energy effects are negligible.

ANALYSIS: The work is determined using Eq. 2.17

$$W = \int_{V_1}^{V_2} p\,dV = mp(v_2 - v_1) \qquad (1)$$

From Table A-6; $v_1 = v_{i@T_{tp}} = 1.0908\times10^{-3}$ m³/kg. Also, from Table A-2; $v_2 = v_{f@T_{tp}} = 1.002\times10^{-3}$ m³/kg. Thus

$$W = (1\text{ kg})(0.6113\text{ kPa})(1.002\times10^{-3} - 1.0908\times10^{-3})\frac{\text{m}^3}{\text{kg}}\left|\frac{1\text{ kN/m}^2}{1\text{ kPa}}\right|\left|\frac{1\text{ kJ}}{1\text{kN}\cdot\text{m}}\right|$$

$$= -5.428\times10^{-5}\text{ kJ} \longleftarrow W$$

The heat transfer is found using an energy balance.

$$\Delta KE + \Delta PE + \Delta U = Q - W$$

or, with $\Delta U = m(u_2 - u_1)$

$$Q = m(u_2 - u_1) + W \qquad (2)$$

From Table A-6; $u_1 = -333.40$ kJ/kg, and from Table A-2; $u_2 = 0.00$. Thus

$$Q = (1\text{ kg})[0 - (-333.40)]\frac{\text{kJ}}{\text{kg}} + (-5.428\times10^{-5}\text{ kJ})$$

$$= 333.4\text{ kJ} \longleftarrow Q$$

Combining Eqs. (1) and (2)

$$Q = m(u_2 - u_1) + mp(v_2 - v_1)$$

$$= m[(u_2 + pv_2) - (u_1 + pv_1)]$$

$$= m(h_2 - h_1)$$

PROBLEM 3.60

KNOWN: A three-phase, solid-liquid-vapor mixture of H_2O at the triple point is heated at constant pressure to saturated vapor.

FIND: Determine the work and heat transfer.

SCHEMATIC & GIVEN DATA:

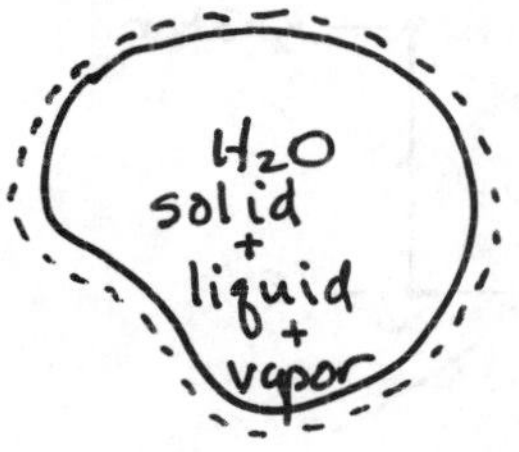

$m_{i_1} = 1$ lb

$m_{f_1} = 1$ lb

$m_{g_1} = 0.2$ lb

$m_{g_2} = 2.2$ lb

triple point

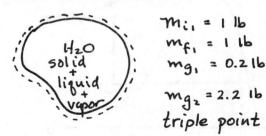

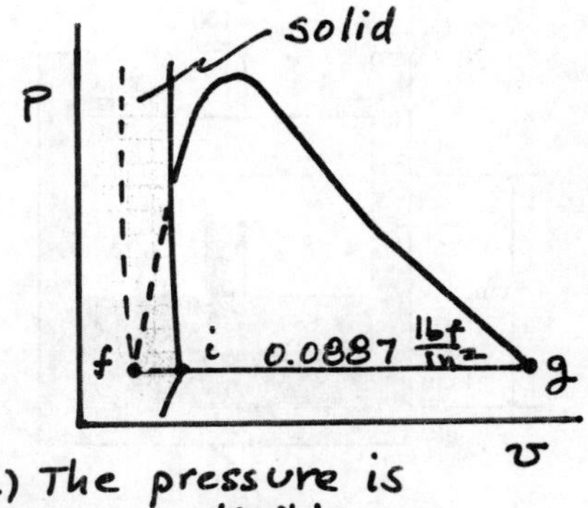

ASSUMPTIONS: (1) The H_2O is a closed system. (2) The pressure is constant. (3) Kinetic and potential energy effects are negligible.

ANALYSIS: The work is determined using Eq. 2.17

$$W = \int_{V_1}^{V_2} p\,dV = p(V_2 - V_1)$$

From Table A-6E; $p = 0.0887$ lbf/in², $T = 32.018$ °F. Also

$$v_i = 0.01747 \text{ ft}^3/\text{lb}, \quad v_g = 3302 \text{ ft}^3/\text{lb}$$

From Table A-2E; $v_f = 0.01602$ ft³/lb. Thus

$$V_1 = m_{i_1} v_i + m_{f_1} v_f + m_{g_1} v_g$$

$$= (1)(0.01747) + (1)(0.01602) + (.2)(3302) = 660.4 \text{ ft}^3$$

$$V_2 = (2.2)(3302) = 7264.4 \text{ ft}^3$$

and

$$W = \left(0.0887 \frac{\text{lbf}}{\text{in}^2}\right)(7264.4 - 660.4)\text{ ft}^3 \left|\frac{144 \text{ in}^2}{1 \text{ ft}^2}\right| \left|\frac{1 \text{ Btu}}{778 \text{ ft}\cdot\text{lbf}}\right|$$

$= 108.4$ Btu ← W

To determine the heat transfer use the energy balance

$$\Delta KE^{\,0} + \Delta PE^{\,0} + \Delta U = Q - W$$

or,

$$Q = \Delta U + W$$

Again, with data from the tables

$$U_1 = m_{i_1} u_i + m_{f_1} u_f + m_{g_1} u_g = (1)(-143.34) + (1)(0) + (.2)(1021.2)$$

$$= 60.9 \text{ Btu}$$

$$U_2 = m_{g_2} u_g = (2.2)(1021.2) = 2246.6 \text{ Btu}$$

Finally

$$Q = (2246.6 - 60.9) + (108.4) = 2294.1 \text{ Btu}$$ ← Q

PROBLEM 3.61

KNOWN: Water contained in a piston-cylinder assemby is heated at constant pressure followed by heating at constant volume.

FIND: Determine the heat transfer.

SCHEMATIC & GIVEN DATA:

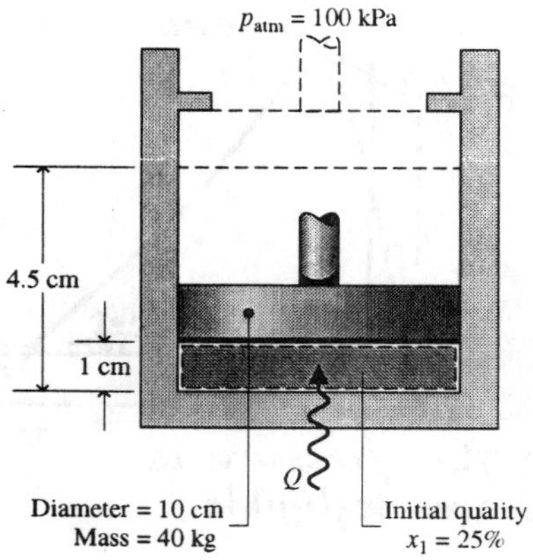

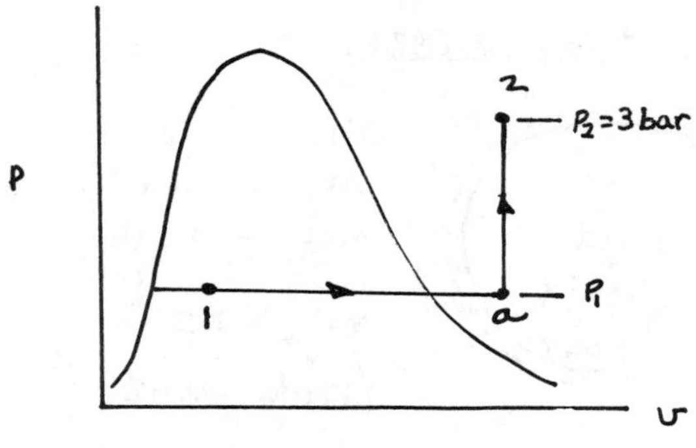

ASSUMPTIONS: 1. The water is a closed system. 2. The pressure is constant until the piston hits the stops. 3. Friction between the piston and cylinder wall can be ignored. 4. For the system, ΔPE, ΔKE can be ignored. 5. g is constant.

ANALYSIS: To begin, fix the three states located by dots on the p-v diagram. State 1 is fixed by $x_1 = 25\%$ and p_1, which is found from a force balance on the piston: The force exerted by the water on the lower face of the piston equals the piston weight plus the force exerted on the top face by the atmosphere:

$$P_1 A = m_{pist}\, g + P_{atm} A \quad , \quad A = \pi D_{pist}^2/4$$

$$\Rightarrow \quad P_1 = \frac{m_{pist}\, g}{A} + P_{atm} = \frac{(40\text{ kg})(9.81\text{ m/s}^2)}{\frac{\pi}{4}(0.1\text{ m})^2}\left|\frac{1\text{ N}}{\text{kg}\cdot\text{m/s}^2}\right|\left|\frac{\text{bar}}{10^5\text{ N/m}^2}\right| + (100\text{ kPa})\left|\frac{1\text{ bar}}{10^2\text{ kPa}}\right|$$

$$= 1.5\text{ bar}$$

State a is fixed by $p_a = p_1 = 1.5$ bar and the specific volume, v_a. From the given geometry, $V_a = 4.5 V_1$ or $v_a = 4.5 v_1$, where

$$v_1 = v_{f1} + x_1(v_{g1} - v_{f1}) = \frac{1.0528}{10^3} + 0.25\left(1.159 - \frac{1.0528}{10^3}\right) = 0.29054\text{ m}^3/\text{kg}$$

$$\rightarrow v_a = 4.5(0.29054) = 1.3074\text{ m}^3/\text{kg}$$

Since $v_a > v_g(1.5\text{ bar})$, state a falls in the superheated vapor region, as shown in the p-v diagram. State 3 is fixed by $v_2 = v_a$ and $p_2 = 3$ bar.

The total mass of water is

$$m = \frac{V}{v_1} = \frac{\pi/4\,(0.1\text{ m})^2(0.01\text{ m})}{0.29054\text{ m}^3/\text{kg}} = 2.703\times10^{-4}\text{ kg}$$

Since there is no work associated with the constant volume portion of the process, the total work is obtained as the water undergoes the constant pressure expansion from 1 to a:

$$W = \int_1^a p\,dV = p[V_a - V_1] = p[V_2 - V_1] = mp(v_2 - v_1)$$

That is, with $v_2 = 4.5 v_1$

$$W = (2.703\times10^{-4}\ kg)(1.5\ bar)\left|\frac{10^5\ N/m^2}{1\ bar}\right|(3.5)\left(0.29054\frac{m^3}{kg}\right)\left|\frac{1\ J}{1\ N\cdot m}\right|$$

$$= 41.23\ J$$

An energy balance reads

$$\Delta U + \cancel{\Delta KE} + \cancel{\Delta PE} = Q - W$$

$$\Rightarrow \quad Q = \Delta U + W$$

$$= m(u_2 - u_1) + W$$

where

$$u_1 = u_{f1} + x_1(u_{g1} - u_{f1})$$

$$= 466.94 + 0.25(2519.7 - 466.94) = 980.13\ kJ/kg$$

To find u_2, interpolate in Table A-4 at 3 bar using v_2: $u_2 = 3263.53$ kJ/kg. Then

$$Q = (2.703\times10^{-4}\ kg)(3263.53 - 980.13)\frac{kJ}{kg}\left|\frac{10^3\ J}{kJ}\right| + 41.23\ J$$

$$= 658.43\ J$$

PROBLEM 3.62

KNOWN: A system consisting of H_2O is compressed isothermally and then is heated at constant volume. Data are known at each of the principal states, and the work is known for process 1-2.

FIND: Determine the heat transfer for each process.

SCHEMATIC & GIVEN DATA:

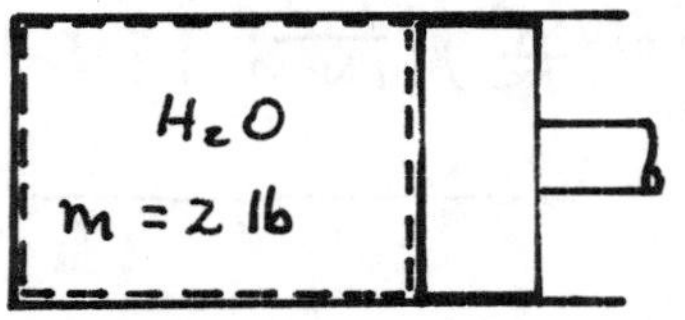

State 1: $T_1 = 300°F$
$V_1 = 20\ ft^3$
$W_{12} = -90.8$ Btu

State 2: $T_2 = T_1$
$V_2 = 9.05\ ft^3$

State 3: $V_3 = V_2$
$P_3 = 120\ lbf/in^2$

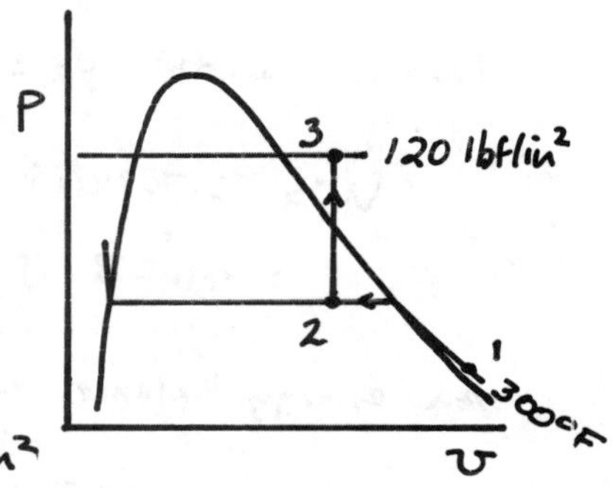

ASSUMPTIONS: (1) The H_2O is a closed system. (2) Process 1-2 occurs at constant temperature and process 2-3 occurs at constant volume. (3) Kinetic and potential energy effects are negligible.

ANALYSIS: First, fix each of the principal states.

State 1: $T_1 = 300°F$, $v_1 = V_1/m = 10\ ft^3/lb$. From Table A-4E; $u_1 = 1104.1$ Btu/lb

State 2: $T_2 = 300°F$, $v_2 = V_2/m = 9.05/2 = 4.525\ ft^3/lb \Rightarrow$ two-phase (Table A-2E)

$$x_2 = \frac{v_2 - v_{f2}}{v_{g2} - v_{f2}} = \frac{4.525 - 0.01745}{6.472 - 0.01745} = 0.6984$$

$$u_2 = u_{f2} + x_2(u_{g2} - u_{f2}) = 269.5 + (.6984)(1100.0 - 269.5) = 849.5 \text{ Btu/lb}$$

State 3: $P_3 = 120\ lbf/in^2$, $v_3 = 4.525$. Interpolating in Table A-4E; $u_3 = 1166.3$ Btu/lb

Now, using energy balances for each process

Process 1-2: $\cancel{\Delta KE}^0 + \cancel{\Delta PE}^0 + \Delta U = Q_{12} - W_{12}$

$$Q_{12} = m(u_2 - u_1) + W_{12}$$
$$= (2\ lb)(849.5 - 1104.1)\ Btu/lb + (-90.8\ Btu)$$
$$= -600 \text{ Btu}$$ ← Q_{12}

Process 2-3: $\cancel{\Delta KE}^0 + \cancel{\Delta PE}^0 + \Delta U = Q_{23} - \cancel{W_{23}}^0$

$$Q_{23} = m(u_3 - u_2)$$
$$= (2)(1166.3 - 849.5)$$
$$= 633.6 \text{ Btu}$$ ← Q_{23}

PROBLEM 3.63

KNOWN: Refrigerant 22 undergoes a constant-pressure process. The pressure and the initial and final volumes are specified.

FIND: Determine the work and heat transfer for the process.

SCHEMATIC & GIVEN DATA:

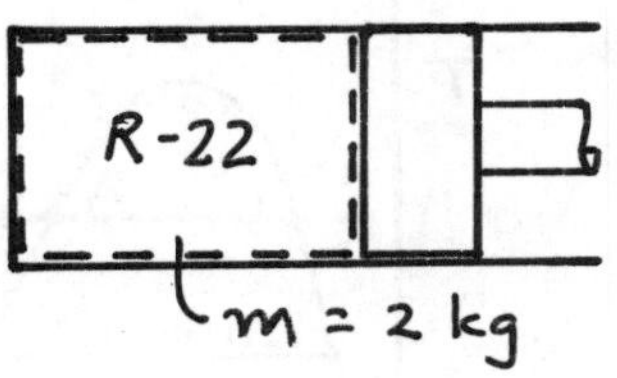

$p_1 = p_2 = 6$ bar

$V_1 = 0.06\ m^3$

$V_2 = 1.5\,V_1$

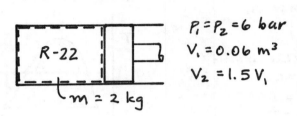

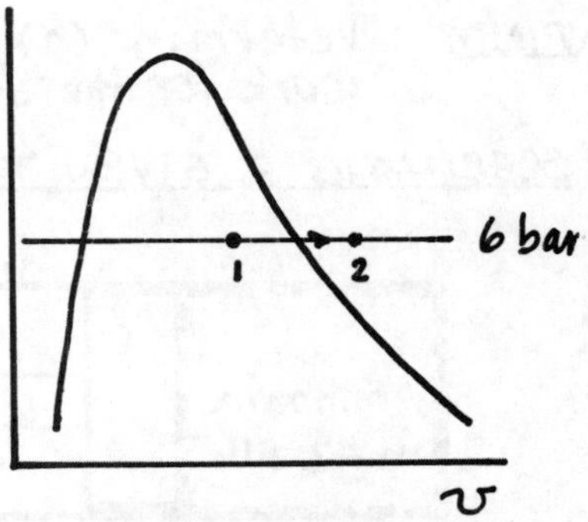

ASSUMPTIONS: (1) The refrigerant is a closed system. (2) The pressure is constant. (3) Kinetic and potential energy effects are constant.

ANALYSIS: The work is determined using Eq. 2.17

$$W = \int_{V_1}^{V_2} p\,dV = p(V_2 - V_1)$$

$$= (6\ \text{bar})(0.045 - 0.03)\,m^3 \left|\frac{10^5\ N/m^2}{1\ \text{bar}}\right|\left|\frac{1\ kJ}{10^3\ N\cdot m}\right|$$

$$= 18\ kJ \quad \longleftarrow W$$

The heat transfer is found using the energy balance

$$\Delta KE^{\,0} + \Delta PE^{\,0} + \Delta U = Q - W$$

or, with $\Delta U = m(u_2 - u_1)$

$$Q = m(u_2 - u_1) + W$$

From Table A-8, with $v_1 = V_1/m = 0.06/2 = 0.03\ m^3/kg$ and $p = 6$ bar

$$x_1 = \frac{v_1 - v_{f1}}{v_{g1} - v_{f1}} = \frac{0.03 - 0.7927\times10^{-3}}{0.0392 - 0.7927\times10^{-3}} = 0.7605$$

$$u_1 = u_{f1} + x_1(u_{g1} - u_{f1}) = 51.53 + (0.7605)(228.44 - 51.53)$$

$$= 186.07\ kJ/kg$$

And, from Table A-9, at $v_2 = 0.045\ m^3/kg$ and $p = 6$ bar

$$u_2 = 245.64\ kJ/kg$$

Thus

$$Q = (2\ kg)(245.64 - 186.07)\ kJ/kg + (18\ kJ)$$

$$= 137.1\ kJ \quad \longleftarrow Q$$

PROBLEM 3.64

KNOWN: Ammonia undergoes two processes in series in a piston-cylinder assembly; the first at constant volume and the second at constant temperature. Data are known at each of three end states and the heat transfer is given for the second process.

FIND: Determine (a) the heat transfer for the first process, and (b) the work for the second process.

SCHEMATIC & GIVEN DATA:

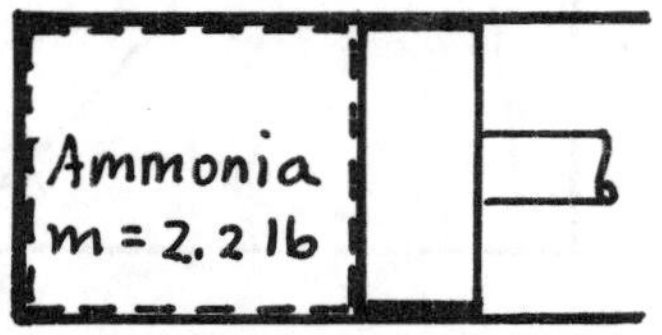

$p_1 = 120$ lbf/in²
$x_1 = 1.0$
$T_2 = 100°F$
$v_2 = v_1$
$T_3 = T_2$
$x_3 = 1.0$
$Q_{23} = -98.9$ Btu

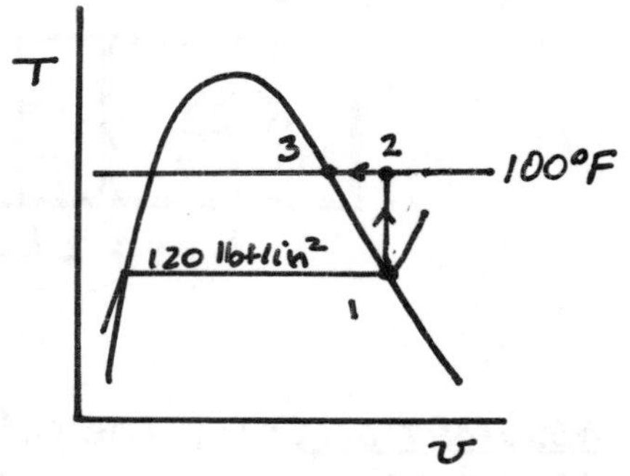

ASSUMPTIONS: (1) The ammonia is closed system. (2) The processes occur at constant volume and constant temperature, respectively. (3) Kinetic and potential energy effects are negligible.

ANALYSIS: The heat transfer Q_{12} and work W_{23} are found using energy balances. First, fix each of the principal states.

State 1: $p_1 = 120$ lbf/in², $x_1 = 1.0 \Rightarrow u_1 = 572.73$ Btu/lb (Table A-14E)
$v_1 = 2.4745$ ft³/lb

State 2: $v_2 = v_1$, $T_2 = 100°F$; Interpolating in Table A-15E
$u_2 \approx 589.17$ Btu/lb

State 3: $T_3 = 100°F$, $x_3 = 1.0 \Rightarrow u_3 = 576.51$ Btu/lb (Table A-13E)

(a) The energy balance for process 1-2 reduces to

$$\Delta KE^{0} + \Delta PE^{0} + m(u_2 - u_1) = Q_{12} - W_{12}$$

and

$$Q_{12} = m(u_2 - u_1)$$
$$= (2.2 \text{ lb})(589.17 - 572.73) \text{ Btu/lb} = 36.17 \text{ Btu}$$ ← Q_{12}

(b) Similarly, for process 2-3

$$\Delta KE^{0} + \Delta PE^{0} + m(u_3 - u_2) = Q_{23} - W_{23}$$

$$W_{23} = Q_{23} - m(u_3 - u_2)$$
$$= (-98.9 \text{ Btu}) - (2.2 \text{ lb})(576.51 - 589.17) \text{ Btu/lb}$$
$$= -71.05 \text{ Btu}$$ ← W_{23}

PROBLEM 3.65

KNOWN: Ammonia vapor is compressed according to pv^n = constant between specified end states.

FIND: Determine the work and heat transfer for the process, each per unit mass.

SCHEMATIC & GIVEN DATA:

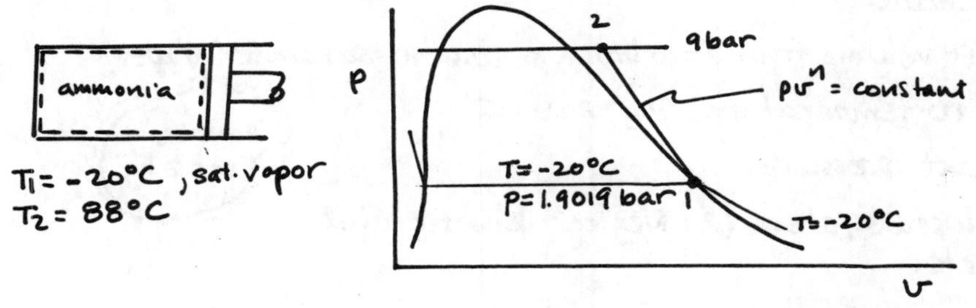

ASSUMPTIONS: 1. The ammonia is the closed system. 2. For the system, kinetic and potential energy effects can be ignored. 3. The process is described by pv^n = constant.

ANALYSIS: The work can be evaluated as

$$W = \int_1^2 p\,dV \Rightarrow \frac{W}{m} = \int_1^2 p\,dv = \int_1^2 \frac{c}{v^n}\,dv = \frac{p_2 v_2 - p_1 v_1}{1-n}$$

This requires n, which can be evaluated using data from Tables A-13, 15:

$$p_2 v_2^n = p_1 v_1^n \Rightarrow \ln p_2 + n \ln v_2 = \ln p_1 + n \ln v_1$$

$$\Rightarrow \quad n \ln v_1/v_2 = \ln p_2/p_1 \quad \text{or} \quad n = \frac{\ln(p_2/p_1)}{\ln(v_1/v_2)}$$

Then

$$n = \frac{\ln(9/1.9019)}{\ln(0.6233/0.1868)} \Rightarrow n = 1.29$$

So,

$$\frac{W}{m} = \frac{[(9\text{ bar})(0.1868\frac{m^3}{kg}) - (1.9019)(0.6233)]}{1-1.29}\left|\frac{10^5 N/m^2}{1 bar}\right|\left|\frac{1 kJ}{10^3 N\cdot m}\right|$$

$$= -170.9 \text{ kJ/kg}$$

An energy balance reduces to $\Delta U = Q - W$, or

$$Q = \Delta U + W \Rightarrow Q/m = (u_2 - u_1) + W/m$$

$$\frac{Q}{m} = (1469.63 - 1299.23) + (-170.9)$$

$$= -0.5 \text{ kJ/kg}$$

PROBLEM 3.66

<u>KNOWN</u>: A system consisting of ammonia undergoes a cycle composed of three processes.

<u>FIND</u>: Sketch p-v and T-v diagrams. Determine the cycle net work and the heat transfer for each process.

<u>SCHEMATIC & GIVEN DATA</u>:

Process 1-2: constant volume from $p_1 = 10$ bar, $x_1 = 0.6$ to saturated vapor

Process 2-3: constant temperature, $Q_{23} = 228$ kJ

Process 3-1: constant pressure

Ammonia $m = 2$ kg

<u>ASSUMPTIONS</u>: (1) Closed system. (2) Neglect kinetic and potential energy effects.

<u>ANALYSIS</u>: First, fix each of the principal states.

State 1: $p_1 = 10$ bar, $x_1 = 0.6$. Thus $u_1 = u_{f1} + x_1(u_{g1} - u_{f1}) = 296.10 + (.6)(1334.66 - 296.10)$
$= 919.24$ kJ/kg (Table A-14)

$v_1 = v_{f1} + x_1(v_{g1} - v_{f1}) = 0.07776$ m³/kg

State 2: Sat. vapor, $v_2 = v_1$. Interpolating in Table A-13, at $v_g = 0.07776$ m³/kg gives $u_2 \approx 1341.26$ kJ/kg and $T_2 \approx 42.52$°C.

State 3: $p_3 = p_1 = 10$ bar, $T_3 = T_2 = 42.52$°C. Interpolating in Table A-15
$u_3 = 1374.95$ kJ/kg and $v_3 \approx 0.14027$ m³/kg

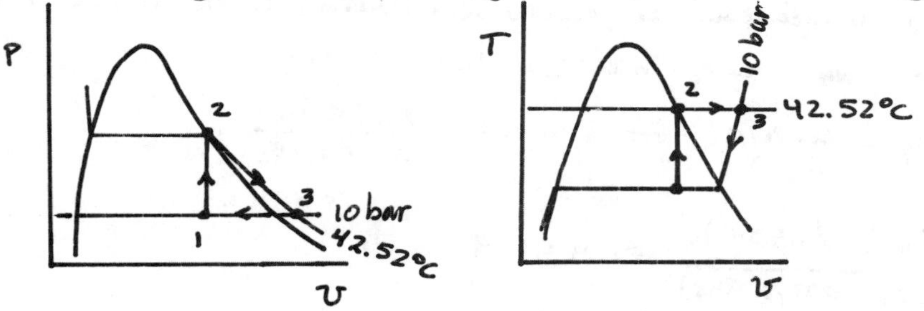

The required energy transfers are determined by using energy balances.

Process 1-2: $\cancel{\Delta KE}^0 + \cancel{\Delta PE}^0 + m(u_2 - u_1) = Q_{12} - \cancel{W_{12}}^0 \Rightarrow Q_{12} = m(u_2 - u_1) = 844.0 \text{ kJ}$ ← Q_{12}

Process 2-3: $\cancel{\Delta KE}^0 + \cancel{\Delta PE}^0 + m(u_3 - u_2) = Q_{23} - W_{23} \Rightarrow W_{23} = Q_{23} - m(u_3 - u_2)$
$= 160.6 \text{ kJ}$

Process 3-1: Using Eq. 2.17; $W_{31} = \int_{V_3}^{V_1} p\,dV = \int_{v_3}^{v_1} m p\,dv = mp(v_1 - v_3)$

$$= (2 \text{ kg})(10 \text{ bar})(0.07776 - 0.14027)\frac{\text{m}^3}{\text{kg}}\left|\frac{10^5 \text{ N/m}^2}{1 \text{ bar}}\right|\left|\frac{1 \text{ kJ}}{10^3 \text{ N-m}}\right|$$

$= -125.0$ kJ

and $\cancel{\Delta KE}^0 + \cancel{\Delta PE}^0 + m(u_1 - u_3) = Q_{31} - W_{31} \Rightarrow Q_{31} = m(u_1 - u_3) + W_{31}$
$= -1036.4 \text{ kJ}$ ← Q_{31}

① ② $W_{cycle} = W_{12} + W_{23} + W_{31} = 0 + 160.6 + (-125.0) = 35.6 \text{ kJ}$ ← W_{cycle}

1. The positive value for W_{cycle} indicates the cycle is a power cycle.
2. A check of the net heat transfer confirms that $Q_{cycle} = W_{cycle}$.

PROBLEM 3.67

KNOWN: One kg of water undergoes a thermodynamic cycle composed of four processes.

FIND: Determine the thermal efficiency

SCHEMATIC & GIVEN DATA:

1-2: constant pressure at 10 bar from sat. vap.

2-3: constant volume to 5 bar, 160°C

3-4 isothermal compression, $Q_{34} = -815.8$ kJ.

4-1 constant volume

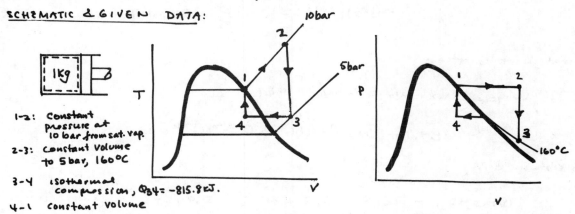

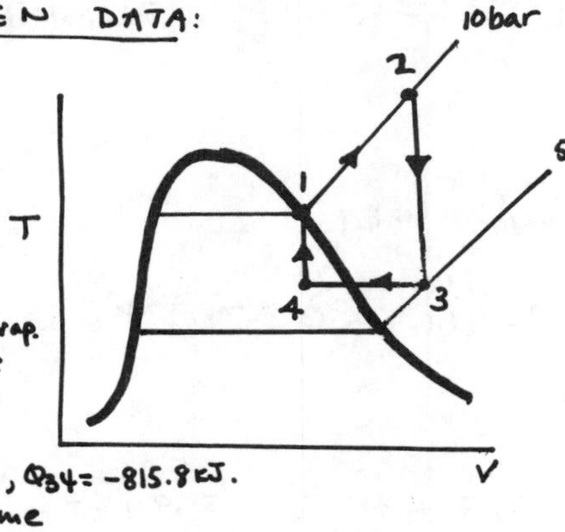

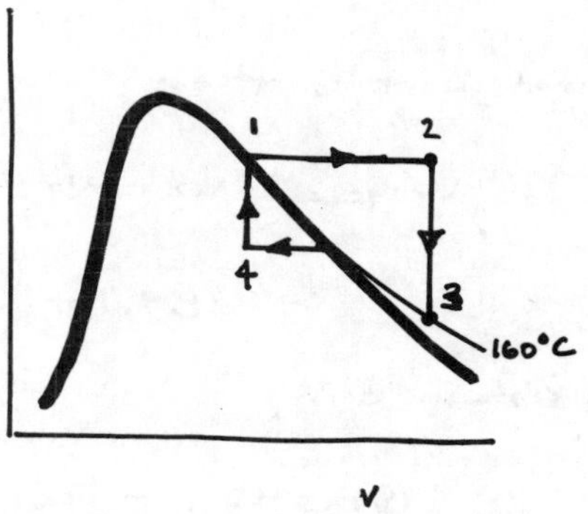

ASSUMPTIONS: 1. The water is the closed system. 2. Volume change is the only work mode. 3. Kinetic and potential energy effects can be ignored.

ANALYSIS: The thermal efficiency of a power cycle is $\eta = W_{cycle}/Q_{in}$ (see discussion of Eqs. 2.42, 43), where $W_{cycle} = W_{12} + W_{23} + W_{34} + W_{41}$.

Process 1-2: State 1 is fixed. State 2 is fixed by $P_2 = 10$ bar, $v_2 = v_3$. From Table A-4

$$v_3 = 0.3835 \frac{m^3}{kg}, \quad u_3 = 2575.2 \text{ kJ/kg}$$

$$\Rightarrow W_{12} = \int_1^2 p\,dV = pm(v_2 - v_1) = (10 \text{ bar})\left|\frac{10^5 N/m^2}{1 \text{ bar}}\right|(1 kg)(0.3835 - 0.1944)\frac{m^3}{kg}\left|\frac{1 kJ}{10^3 N\cdot m}\right|$$

$$= 189.1 \text{ kJ}$$

An energy balance reduces to give $Q = \Delta U + W$, or

$$Q_{12} = m(u_2 - u_1) + W = (1 kg)[3231.8 - 2583.6] + 189.1 \text{ kJ} = 837.3 \text{ kJ}$$

where u_2 is obtained from Table A-3 using $P_2, v_2 (= v_3)$.

Process 2-3: $W_{23} = 0$

$$Q_{23} = m(u_3 - u_2) + W_{23}^{\,0} = (1 kg)[2575.2 - 3231.8] = -656.6 \text{ kJ}$$

Process 3-4: $Q_{34} = -815.8$ kJ (given). Then, $\Delta U = Q_{34} - W_{34}$ gives

$$W_{34} = Q_{34} - \Delta U = Q_{34} - m(u_4 - u_3)$$

State 4 is fixed by $T_4 = 160°C$, $v_4 = v_1$

$$\therefore x_4 = \frac{v_4 - v_f}{v_g - v_f} = \frac{0.1944 - 1.102/10^3}{0.3071 - 1.102/10^3} = 0.6317$$

$$\Rightarrow u_4 = u_f + x_4(u_g - u_f) = 674.86 + (0.6317)(2568.4 - 674.86) = 1871 \text{ kJ/kg}$$

$$\Rightarrow W_{34} = -815.8 - (1)[1871 - 2575.2] = -111.6 \text{ kJ}$$

PROBLEM 3.67 (Contd.)

Process 4–1: $W_{41} = 0$, and

$$Q_{41} = \Delta U + \cancel{W_{41}}^{0} = m(u_1 - u_4)$$

$$= (1\,kg)[2583.6 - 1871]\frac{kJ}{kg} = 712.6\,kJ$$

The net work is then

$$W_{cycle} = W_{12} + W_{23} + W_{34} + W_{41}$$

(1) $$= 189.1 + 0 + (-111.6) + 0 = 77.5\,kJ$$

To obtain Q_{in},

$$Q_{in} = Q_{12} + Q_{41} = 837.3 + 712.6 = 1549.9\,kJ$$

Then

$$\eta = \left(\frac{77.5}{1549.9}\right) = 0.05 \quad (5\%)$$

---

(1) As a check, note that for every cycle $Q_{cycle} = W_{cycle}$.

$$Q_{cycle} = Q_{12} + Q_{23} + Q_{34} + Q_{41}$$

$$= 837.3 + (-656.6) + (-815.8) + 712.6 = 77.5\,kJ$$

which agrees with W_{cycle} calculated using the work quantities.

PROBLEM 3.68

KNOWN: A system consisting of Refrigerant 22 undergoes a cycle composed of three processes.

FIND: Sketch p-v and T-v diagrams. Determine the cycle net work and the heat transfer for each process.

SCHEMATIC & GIVEN DATA:

Process 1-2: Constant pressure from $p_1 = 30$ lbf/in^2, $x_1 = 0.95$ to $T_2 = 40°F$

Process 2-3: isothermal with $W_{23} = -11.82$ Btu to saturated vapor

Process 3-1: Adiabatic expansion; $Q_{31} = 0$

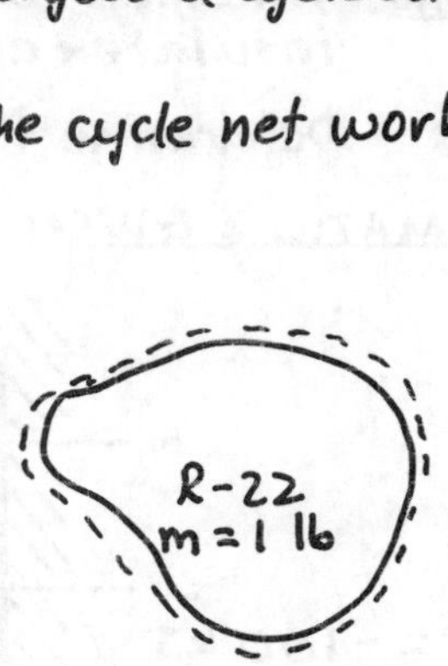

ASSUMPTIONS: (1) The R-22 is a closed system. (2) kinetic and potential energy effects are negligible.

ANALYSIS: First, using data from the refrigerant tables, fix each state.

State 1: Table A-8E at $p_1 = 30$ lbf/in^2 and $x_1 = 0.95$

$$u_1 = u_{f1} + x_1(u_{g1} - u_{f1}) = 7.38 + (.95)(93.67 - 7.38) = 89.36 \text{ Btu/lb}$$

$$v_1 = v_{f1} + x_1(v_{g1} - v_{f1}) = 0.01178 + (.95)(1.7430 - 0.01178) = 1.6564 \text{ ft}^3\text{/lb}$$

State 2: from Table A-9E at $p_2 = p_1$, $T_2 = 40°F$; $u_2 = 100.40$ Btu/lb, $v_2 = 1.9858$ ft^3/lb

State 3: $T_3 = 40°F$, sat. vapor. From Table A-7E; $u_3 = 98.08$ Btu/lb

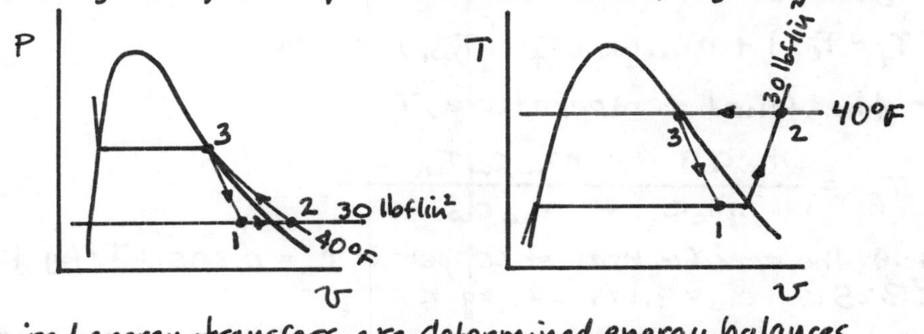

The required energy transfers are determined energy balances.

Process 1-2: Using Eq. 2.17, $W_{12} = \int_{V_1}^{V_2} p\,dV = m\,p_1(v_2 - v_1)$

$$W_{12} = (1\text{ lb})(30\text{ lbf/in}^2)(1.9858 - 1.6564)\text{ ft}^3\text{/lb}\left|\frac{144\text{ in}^2}{1\text{ ft}^2}\right|\left|\frac{1\text{ Btu}}{778\text{ ft}\cdot\text{lbf}}\right| = 1.829 \text{ Btu}$$

$$\Delta KE^{\,0} + \Delta PE^{\,0} + m(u_2 - u_1) = Q_{12} - W_{12} \Rightarrow Q_{12} = m(u_2 - u_1) + W_{12} = 12.87 \text{ Btu} \leftarrow Q_{12}$$

Process 2-3: $W_{23} = -11.82$, $\Delta KE^{\,0} + \Delta PE^{\,0} + m(u_3 - u_2) = Q_{23} - W_{23}$

$$Q_{23} = m(u_3 - u_2) + W_{23} = -14.14 \text{ Btu} \leftarrow Q_{23}$$

Process 3-1: $\Delta KE^{\,0} + \Delta PE^{\,0} + m(u_1 - u_3) = Q_{31}^{\,0} - W_{31}$

$$W_{31} = -m(u_1 - u_3) = 8.72 \text{ Btu}$$

① ② $W_{cycle} = W_{12} + W_{23} + W_{31} = (1.829) + (-11.82) + (8.72) = -1.27$ Btu $\leftarrow W_{cycle}$

1. The negative value for W_{cycle} indicates the cycle is a refrigeration cycle.
2. A check of the net heat transfer confirms that $Q_{cycle} = W_{cycle}$.

PROBLEM 3.69

KNOWN: An electric resistor transfers energy to water in a well-insulated copper tank.

FIND: Determine the final equilibrium temperature.

SCHEMATIC & GIVEN DATA:

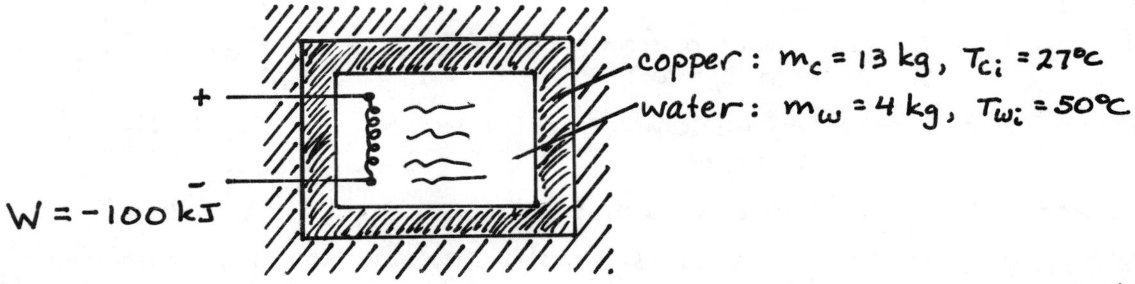

ASSUMPTIONS: (1) The copper tank and water are a closed system. (2) There is no heat transfer. (3) The copper and water behave as incompressible substances with constant specific heats. (4) Kinetic and potential energy effects are negligible. (5) No energy is stored in the resistor; $\Delta U_{res} = 0$

ANALYSIS: The energy balance reduces as follows

$$\cancel{\Delta KE}^{0} + \cancel{\Delta PE}^{0} + \Delta U = \cancel{Q}^{0} - W \Rightarrow \Delta U = -W$$

or

$$\cancel{\Delta U_{res}}^{0} + \Delta U_c + \Delta U_w = -W$$

$$m_c c_c (T_f - T_{ci}) + m_w c_w (T_f - T_{wi}) = -W$$

Solving for the final temperature, T_f

$$T_f = \frac{m_c c_c T_{ci} + m_w c_w T_{wi} - W}{m_c c_c + m_w c_w}$$

Using Table A-19, the specific heat of copper is $c_c = 0.385$ kJ/kg·K. And for water at 325K, $c_w = 4.179$ kJ/kg·K. Thus

① $$T_f = \frac{(13 \text{ kg})(0.385 \text{ kJ/kg·K})(27+273)\text{K} + (4)(4.179)(50+273) - (-100)}{(13 \text{ kg})(0.385 \text{ kJ/kg·K}) + (4)(4.179)}$$

$$= 322.3 \text{ K} = 49.3°\text{C}$$ ← T_f

---

1. Note that temperature in °C could have been used directly in this expression.

PROBLEM 3.70

KNOWN: A hot steel bar is placed in an open tank of cool water. The bar and the water come to equilibrium with no heat transfer with the surroundings.

FIND: Determine the final temperature.

SCHEMATIC & GIVEN DATA:

$m_s = 50$ lb
$T_{si} = 200°F$
$V_w = 5\ ft^3$
$T_{wi} = 70°F$

ASSUMPTIONS: (1) The steel bar and water are the closed system. (2) The steel and water behave as incompressible substances with constant specific heats. (3) Kinetic and potential energy effects are neglected. (4) There is no heat transfer between the tank and its surroundings, and W = 0.

ANALYSIS: The energy balance reduces as follows:

$$\cancel{\Delta KE} + \cancel{\Delta PE} + \Delta U = \cancel{Q} - \cancel{W} \Rightarrow \Delta U = 0$$

or $\Delta U_s + \Delta U_w = 0$. Thus, if T_f is the final equilibrium temperature

① $$m_s c_s (T_f - T_{si}) + m_w c_w (T_f - T_{wi}) = 0$$

Solving for T_f

$$T_f = \frac{m_s c_s T_{si} + m_w c_w T_{wi}}{m_s c_s + m_w c_w} \qquad (*)$$

To evaluate m_w, use v_f @ 70°F from Table A-2E to get

$$m_w = \frac{V_w}{v_f} = \frac{5\ ft^3}{.01605\ ft^3/lb} = 311.5\ lb$$

Now, with $c_s = 0.115$ Btu/lb·°R and $c_w = 0.998$ Btu/lb·°R (at 540°R) from Table A-19E, we can evaluate (*)

$$T_f = \frac{(50\ lb)(0.115\ Btu/lb·°R)(200+460)°R + (311.5)(0.998)(70+460)}{(50\ lb)(0.115\ Btu/lb·°R) + (311.5)(0.998)}$$

$= 532°R = 72°F$ ← T_f

---

1. Note that temperature in °F could be used in this expression because it involves temperature differences.

PROBLEM 3.71

KNOWN: An isolated system is formed from saturated water vapor and a copper slab initially at different temperatures.

FIND: Determine the final equilibrium temperature of the system.

SCHEMATIC & GIVEN DATA:

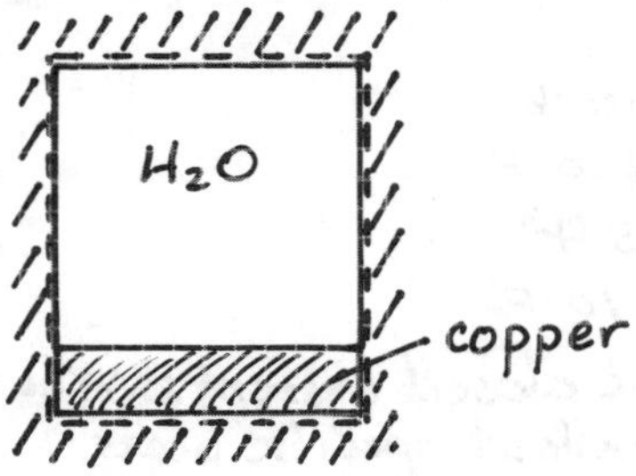

$T_{w1} = 130°C$, $m_w = 0.2 kg$

$T_{c1} = 30°C$, $m_c = 10 kg$

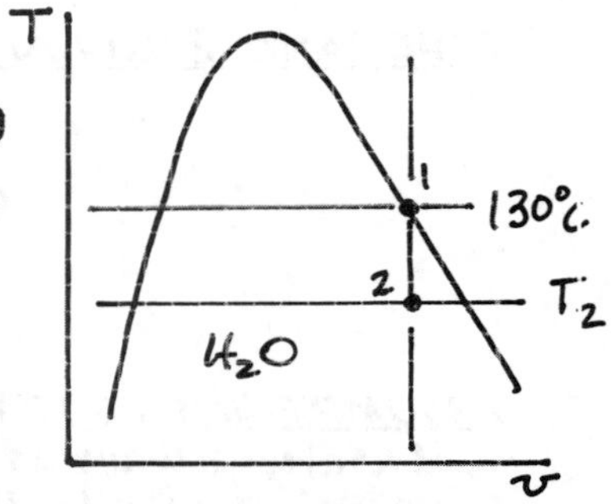

ASSUMPTIONS: (1) The H_2O and copper form an isolated system. (2) The volume of the system is constant. (3) The copper behaves as an incompressible substance with constant specific heats. (4) kinetic and potential energy effects are negligible.

ANALYSIS: The energy balance reduces as follows:

$$\Delta KE^{0} + \Delta PE^{0} + \Delta U = Q^{0} - W^{0} \Rightarrow \Delta U = 0$$

or

$$\Delta U_c + \Delta U_w = 0$$

$$m_c c (T_2 - T_{1c}) + m_w (u_{2w} - u_{1w}) = 0$$

With $u_{1w} = u_{g@130°C} = 2539.9$ kJ/kg and $c = 0.385$ kJ/kg·K from Table A-19

$$(10 \text{ kg})(.385 \text{ kJ/kg·K})(T_2 - 30) + (.2 \text{ kg})(u_{2w} - 2539.9) = 0$$

or, solving for T_2

$$T_2 = 30 - \frac{(.2)(u_{2w} - 2539.9)}{(10)(.385)} \qquad (1)$$

The final specific volume of the water vapor, v_{2w}, is the same as the initial value; $v_1 = v_{g@130°C} = 0.6685$ m³/kg. The final quality of the H_2O is

$$x_2 = \frac{v_{2w} - v_{f2}}{v_{g2} - v_f}$$

and

$$u_{2w} = u_{f2} + x_2 (u_g - u_f)$$

Thus, an iterative solution is required. First, guess a trial value of T_2. Calculate x_2 and u_{2w} based on the trial T_2 and saturation data. Use (1) to calculate a corresponding value of T_2. Repeat until the trial value equals the one predicted by (1).

Result: $T_2 \approx 98°C$ ← T_2

PROBLEM 3.72

KNOWN: A liquid in a rigid tank is stirred by a paddle wheel and experiences energy transfer by heat with its surroundings.

FIND: Obtain a differential equation for temperature T in terms of time t and relevant parameters. Solve for T(t).

SCHEMATIC & GIVEN DATA:

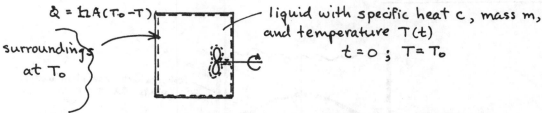

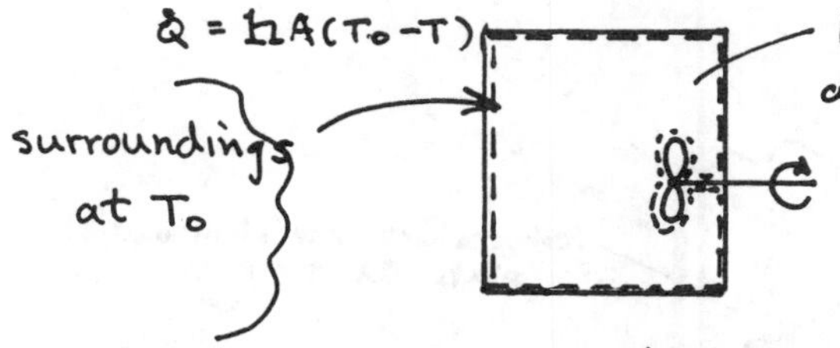

ASSUMPTIONS: (1) The liquid is the closed system. (2) The liquid behaves as an incompressible substance with constant specific heat c. (3) The energy transfer from the paddle wheel to the system occurs at a constant rate. (4) There are no changes in kinetic or potential energy.

ANALYSIS: The energy rate balance is

$$\cancel{\frac{dKE}{dt}}^{0} + \cancel{\frac{dPE}{dt}}^{0} + \frac{dU}{dt} = \dot{Q} - \dot{W}$$

with $dU/dt = mc\, dT/dt$ and $\dot{Q} = hA(T_0 - T)$

$$mc\frac{dT}{dt} = hA(T_0 - T) - \dot{W} \longleftarrow \text{differential equation}$$

To solve, let $\theta = T_0 - T$; $dT = -d\theta$. Thus

$$\frac{d\theta}{dt} + \left(\frac{hA}{mc}\right)\theta - \frac{\dot{W}}{mc} = 0$$

The solution of this differential equation is of the form

$$\theta(t) = C_1 \exp\left[-\left(\frac{hA}{mc}\right)t\right] + \frac{\dot{W}}{hA}$$

at $t = 0$, $T = T_0 \Rightarrow \theta = 0$. Thus $C_1 = -\dot{W}/hA$, so

$$\theta(t) = \frac{\dot{W}}{hA}\left\{1 - \exp\left[-\left(\frac{hA}{mc}\right)t\right]\right\}$$

or

① ② $$T(t) = T_0 - \frac{\dot{W}}{hA}\left\{1 - \exp\left[-\left(\frac{hA}{mc}\right)t\right]\right\} \longleftarrow T(t)$$

---

1. The solution can be verified by direct substitution into the differential equation. 2. The solution satisfies the initial condition that $T(0) = T_0$.

PROBLEM 3.73

KNOWN: A steel plate of known thickness and initial temperature is quenched in a bath whose temperature is also known.

FIND: Derive a partial differential equation for the variation of temperature within the plate as a function of time and position. Evaluate the physical parameters using data from Table A-19.

SCHEMATIC & GIVEN DATA:

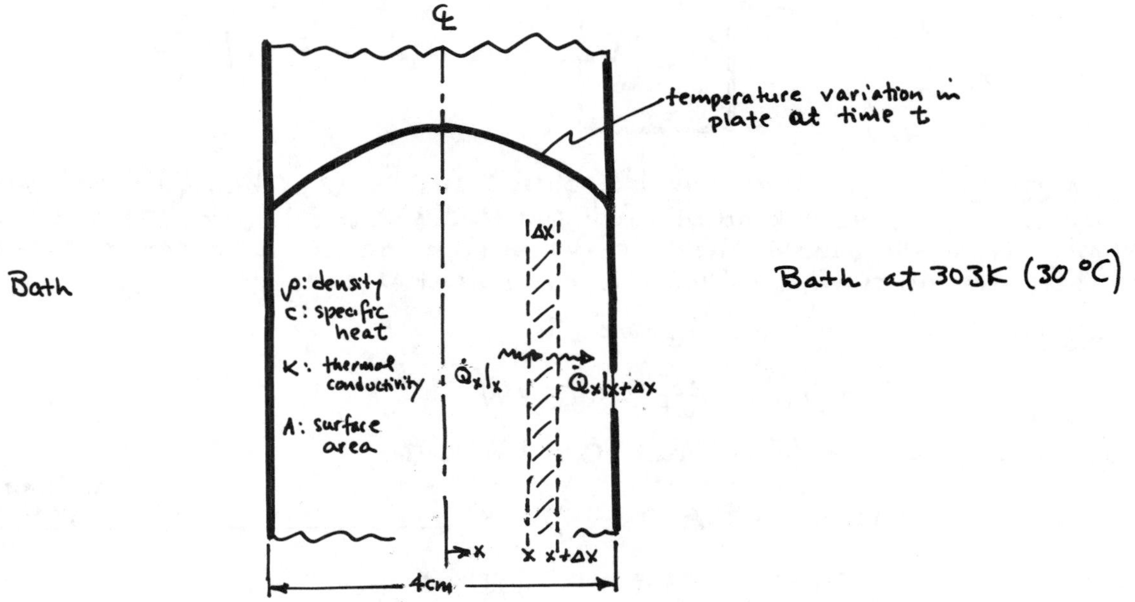

ASSUMPTIONS: 1. The element of thickness Δx is the closed system. 2. The incompressible model applies. 3. Specific heat and thermal conductivity are each constant. 4. There is no work and no kinetic/potential energy effects.

ANALYSIS: The energy rate balance applied to the element of thickness Δx reads

$$\cancel{\frac{dKE}{dt}} + \cancel{\frac{dPE}{dt}} + \frac{dU}{dt} = \left[\dot{Q}_x|_x - \dot{Q}_x|_{x+\Delta x}\right] - \cancel{\dot{W}}$$

The temperature within the plate varies with position and time. Thus, with $U = mu = (\rho A \Delta x)u$, where $u = u(T)$,

$$\frac{dU}{dt} = (\rho A \Delta x)\frac{du}{dT}\frac{\partial T}{\partial t} = (\rho A \Delta x)\, c\,\frac{\partial T}{\partial t}$$

Collecting results

$$\rho A c\,\frac{\partial T}{\partial t} = -\left[\frac{\dot{Q}_x|_{x+\Delta x} - \dot{Q}_x|_x}{\Delta x}\right]$$

In the limit as $\Delta x \to 0$

$$\rho A c\,\frac{\partial T}{\partial t} = -\frac{\partial \dot{Q}_x}{\partial x} \qquad (*)$$

PROBLEM 3.73 (Contd.)

When $T = T(x,t)$, Fourier's conduction model can be expressed as

$$\dot{Q}_x = -\kappa A \frac{\partial T}{\partial x} \qquad (**)$$

Combining Eqs. (*), (**)

$$\rho c \frac{\partial T}{\partial t} = \frac{\partial}{\partial x}\left(\kappa \frac{\partial T}{\partial x}\right)$$

Since thermal conductivity is assumed constant

$$\rho c \frac{\partial T}{\partial t} = \kappa \frac{\partial^2 T}{\partial x^2}$$

With data from Table A-19

$\rho = 8060 \text{ kg/m}^3$

$c = 0.480 \text{ kJ/kg}\cdot\text{K}$

$\kappa = 15.1 \text{ W/m}\cdot\text{K}$

Defining

$$\alpha = \frac{\kappa}{\rho c} = \frac{\left(15.1 \frac{W}{m\cdot K}\right)\left|\frac{1\ J/s}{1\ W}\right|}{\left(8060 \frac{kg}{m^3}\right)\left(0.48 \frac{kJ}{kg\cdot K}\right)\left(\frac{10^3 J}{1\ kJ}\right)} = 3.9\times10^{-6} \frac{m^2}{s}$$

we have

① $$\frac{\partial T}{\partial t} = \alpha \frac{\partial^2 T}{\partial x^2}$$ ← differential equation

---

1. This is a second order partial differential equation giving $T(x,t)$. The solution requires an initial condition:

$$T(x, t=0) = 673\text{ K } (400°C)$$

and appropriate boundary conditions. For further discussion, see <u>Fundamentals of Heat and Mass Transfer</u> by F. P. Incropera and D. P. DeWitt.

PROBLEM 3.74

H_2O at $p = 100$ bars, $T = 400°C$

(a) $p_R = p/p_c = 100 \text{ bar}/220.9 \text{ bar} = 0.45$, $T_R = T/T_c = 673\text{K}/647.3\text{K} = 1.04$ } Fig. A-1: $Z \approx 0.86$ ← Z_{chart}

(b) Table A-4 at $p = 100$ bar, $T = 400°C \Rightarrow v = 0.02641 \text{ m}^3/\text{kg}$

$$Z = \frac{pv}{RT} = \frac{(100 \text{ bar})(0.02641 \text{ m}^3/\text{kg})}{\left(\frac{8.314}{18.02}\right)\frac{\text{kJ}}{\text{kg}\cdot\text{K}}(673\text{ K})}\left|\frac{10^5 \text{N/m}^2}{1 \text{ bar}}\right|\left|\frac{1 \text{ kJ}}{10^3 \text{ N}\cdot\text{m}}\right|$$

$= 0.851$ ← Z_{table}

PROBLEM 3.75

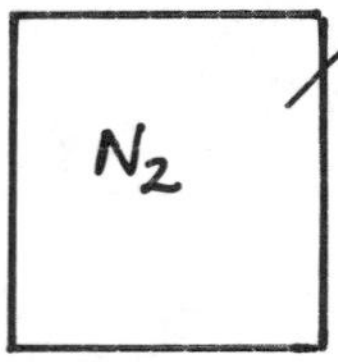

$m = 40$ kg
$p = 17$ MPa
$T = 180$ K
$V = ?$

Determine Z: $p_R = 170/33.9 = 5.02$, $T_R = 180/126.2 = 1.43$ } Fig. A-2: $Z \approx 0.82$

$$V = \frac{ZmRT}{p} = \frac{(.82)(40\text{kg})\left(\frac{8.314}{28.01}\frac{\text{kJ}}{\text{kg}\cdot\text{K}}\right)(180\text{K})}{(17\times10^6 \text{ N/m}^2)}\left|\frac{10^3 \text{ N}\cdot\text{m}}{1 \text{ kJ}}\right|$$

$= 0.103 \text{ m}^3$ ← V

PROBLEM 3.76

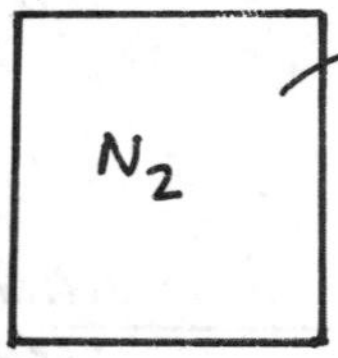

$V = 6 \text{ ft}^3$
$T = 360°R$
$p = 3000 \text{ lbf/in}^2$
$m = ?$

Determine Z: $p_R = \frac{3000/14.7}{33.5} = 6.1$, $T_R = 360/227 = 1.6$ } Fig A-2: $Z \approx 0.93$

$$m = \frac{pV}{ZRT} = \frac{(3000 \text{ lbf/in}^2)(6 \text{ ft}^3)}{(0.93)\left(\frac{1545}{28.01}\frac{\text{kJ}}{\text{kg}\cdot\text{K}}\right)}\left|\frac{144 \text{ in}^2}{1 \text{ ft}^2}\right|$$

$= 140$ lb ← m

PROBLEM 3.77

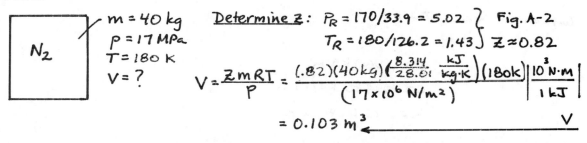

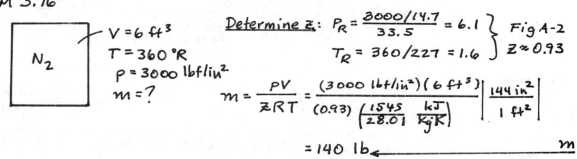

$T = 600°R$
$v = 0.172 \text{ ft}^3/\text{lb}$
$p = ?$

Determine Z: $T_R = 600/548 = 1.09$

$$v_R' = \frac{v p_c}{R T_c}$$

$$= \frac{(0.172 \text{ ft}^3/\text{lb})(72.9 \text{ atm})}{\left(\frac{1545}{44.01}\frac{\text{ft}\cdot\text{lbf}}{\text{lb}\cdot°\text{R}}\right)(548°\text{R})}\left|\frac{14.7 \text{ lbf/in}^2}{1 \text{ atm}}\right|\left|\frac{144 \text{ in}^2}{1 \text{ ft}^2}\right|$$

$= 1.38$ — Fig A-1: $Z \approx 0.82$

$$p = \frac{ZRT}{v} = \frac{(0.82)\left(\frac{1545}{44.01}\frac{\text{ft}\cdot\text{lbf}}{\text{lb}\cdot°\text{R}}\right)(600°\text{R})}{(0.172 \text{ ft}^3/\text{lb})}\left|\frac{1 \text{ ft}^2}{144 \text{ in}^2}\right|$$

$= 697 \text{ lbf/in}^2$ ← p

PROBLEM 3.78

KNOWN: Oxygen (O_2) contained in a closed, rigid tank is cooled to a final state where pressure is specified.

FIND: Determine the tank volume and the final temperature.

SCHEMATIC & GIVEN DATA:

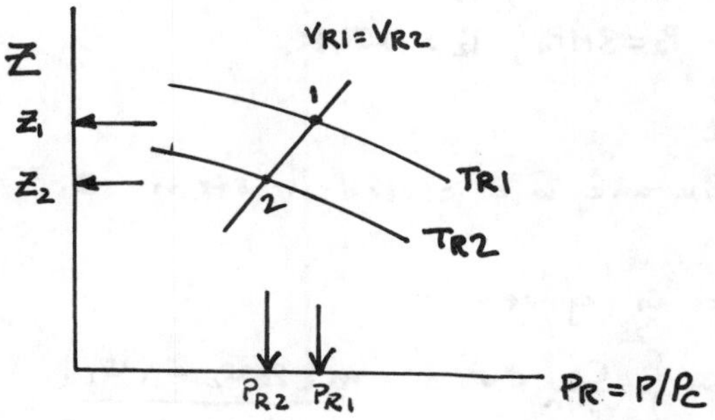

$P_1 = 30$ bar
① $T_1 = 200$ K
$P_2 = 20$ bar

ASSUMPTIONS: 1. The oxygen is a closed system. 2. The volume of the tank is constant.

ANALYSIS: The volume of the tank can be evaluated using the compressibility factor at state 1:

$$Z_1 = \frac{P_1(V/m)}{(\bar{R}/M)\,T_1} \Rightarrow V = \frac{Z_1 m (\bar{R}/M) T_1}{P_1}$$

The value of Z_1 can be obtained by inspection of the compressibility chart using P_{R1} and T_{R1}. With T_c, P_c from Table A-1

$$P_{R1} = \frac{30\text{ bar}}{50.5\text{ bar}} = 0.59 \quad , \quad T_{R1} = \frac{200\text{K}}{154\text{ K}} = 1.3$$

From Fig A-1, $Z_1 = 0.91$. Thus

$$V = 0.91\,\frac{(0.5\text{kg})(8314/32\ \text{N·m/kg·K})\,200\text{K}}{(30\times10^5\ \text{N/m}^2)}$$

$$= 7.88\times10^{-3}\ \text{m}^3 \quad \longleftarrow$$

Since volume and mass are constant, $v_2 = v_1$ and thus $v'_{R1} = v'_{R1}$, as shown in the sketch above. Returning to Fig A-1 with
② $P_R = 20/50.5 = 0.396$ and following a line of constant v'_R from 1 to 2, we have $T_{R2} = 0.96$. Thus, $T_2 = 0.96\,T_c$

$$T_2 = (0.96)(154) = 148\text{ K} \quad \longleftarrow$$

---

1. These values for P_1, T_1, T_2 revise the values given in the first printing of the book.

2. Alternatively, $Z_2 = 0.825$ can be read from the figure, from which T_2 can be evaluated as $T_2 = 147$ K.

PROBLEM 3.79

KNOWN: Five kg of C_4H_{10} contained in a piston-cylinder arrangement undergo a process between known states for which pv = constant.

FIND: Determine the work.

SCHEMATIC & GIVEN DATA:

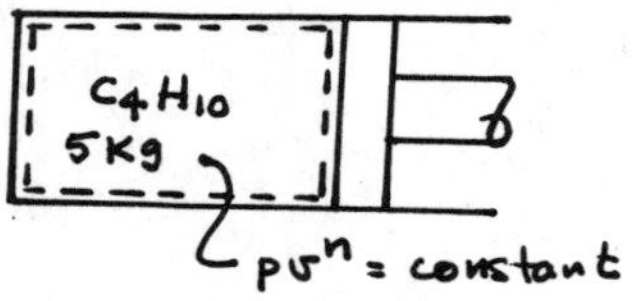

$p_1 = 5\text{ MPa}$, $T_1 = 500\text{K}$
$p_2 = 3\text{ MPa}$, $T_2 = 450\text{K}$

ASSUMPTIONS: 1. The Butane is a closed system. 2. The process is described by pv^n = constant.

ANALYSIS: The work is given by

$$W = \int_1^2 p\,dV = m\int_1^2 \frac{C}{v^{1/n}}\,dv = \frac{m(p_2v_2 - p_1v_1)}{1-n}$$

To evaluate W requires v_1, v_2 and n. The compressibility chart can be used to obtain v_1 and v_2: From Table A-1, $p_c = 38\text{ bar}$, $T_c = 425\text{K}$, $M = 58.12$.

- $p_{R1} = \frac{p_1}{p_c} = \frac{50\text{ bar}}{38\text{ bar}} = 1.32$, $T_{R1} = \frac{T_1}{T_c} = \frac{500\text{K}}{425\text{K}} = 1.18$ } Fig. A-2: $v_R' = 0.6$

$$\Rightarrow v_1 = v_R'\left(\frac{RT_c}{p_c}\right) = 0.6\frac{\left(\frac{8314}{58.12}\frac{\text{N}\cdot\text{m}}{\text{kg}\cdot\text{K}}\right)(425\text{K})}{(38\times10^5\text{ N/m}^2)} = 0.0096\ \frac{\text{m}^3}{\text{kg}}$$

- $p_{R2} = \frac{p_2}{p_c} = \frac{30}{38} = 0.79$, $T_{R2} = \frac{T_2}{T_c} = \frac{450}{425} = 1.06$ } Fig. A-1 $v_{R2}' = 0.99$

$$\Rightarrow v_2 = \frac{(0.99)(8314/58.12)(425)}{(38\times10^5)} = 0.0158\ \frac{\text{m}^3}{\text{kg}}$$

To find n, use $p_2v_2^n = p_1v_1^n \Rightarrow \left(\frac{v_2}{v_1}\right)^n = \frac{p_1}{p_2} \Rightarrow n = \frac{\ln(p_1/p_2)}{\ln(v_2/v_1)}$

$$\therefore\ n = \frac{\ln(5/3)}{\ln(0.0158/.0096)} = 1.025$$

Thus, the work is

$$W = \frac{m(p_2v_2 - p_1v_1)}{1-n} = (5\text{ kg})\frac{[(3\text{ MPa})(0.0158\text{ m}^3/\text{kg}) - (5)(0.0096)]}{1-1.025}\left|\frac{10^6\text{ N/m}^2}{1\text{ MPa}}\right|\left|\frac{1\text{ kJ}}{10^3\text{ N}\cdot\text{m}}\right|$$

$$= 120\text{ kJ} \longleftarrow W$$

PROBLEM 3.80

KNOWN: Two lbmol of C_2H_4 undergo a constant-pressure compression process in a piston-cylinder assembly. The work, pressure, and initial gas temperature are known.

FIND: Evaluate the final temperature.

SCHEMATIC & GIVEN DATA:

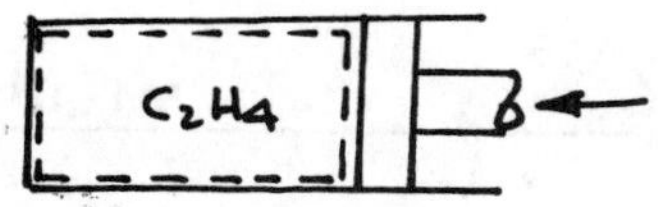

$p = 213$ lbf/in^2

$T_1 = 612°R$

$W = -800$ Btu

$n = 2$ lbmol

ASSUMPTIONS: 1. The C_2H_4 is a closed system. 2. Pressure remains constant during the compression process.

ANALYSIS: The final state of the gas is fixed by pressure, 213 lbf/in^2, and the specific volume $\bar{v}_2$, which can be evaluated from the work.

$$W = \int_1^2 p\,dV = p(V_2 - V_1) = n\,p(\bar{v}_2 - \bar{v}_1) \Rightarrow \bar{v}_2 = \bar{v}_1 + \frac{W}{np} \quad (*)$$

$\bar{v}_1$ can be found from the compressibility chart. From Table A-1E, $p_c = 50.5$ atm, $T_c = 510°R$. Then

$$p_R = \frac{(213/14.7)\,\text{atm}}{50.5\,\text{atm}} = 0.287, \quad T_{R1} = \frac{612°R}{510°R} = 1.2.$$

Using these, Fig. A-1 gives $v'_{R1} = 4.0$. Then

$$\bar{v}_1 = v'_{R1}\left(\frac{\bar{R}T_c}{p_c}\right) = 4.0\left(\frac{\left(1545\,\frac{\text{ft}\cdot\text{lbf}}{\text{lbmol}\cdot°R}\right)(510°R)}{(50.5 \times 14.7)\,\text{lbf/in}^2}\right)\left|\frac{1\,\text{ft}^2}{144\,\text{in}^2}\right| = 29.48\ \text{ft}^3/\text{lbmol}$$

Returning to Eq. (*)

$$\bar{v}_2 = \bar{v}_1 + \frac{W}{np} = 29.48\,\frac{\text{ft}^3}{\text{lbmol}} + \frac{(-800\,\text{Btu})\left|\frac{778\,\text{ft}\cdot\text{lbf}}{1\,\text{Btu}}\right|}{(2\,\text{lbmol})(213 \times 144\,\text{lbf/ft}^2)}$$

$$= 19.33\ \text{ft}^3/\text{lbmol}$$

So, at state 2, $p_{R2} = p_{R1} = 0.287$ and $v'_{R2} = \bar{v}_2 p_c / \bar{R}T_c$

$$v'_{R2} = \frac{(19.33\,\text{ft}^3/\text{lbmol})((50.5)(14.7)(144)\,\text{lbf/ft}^2)}{\left(1545\,\frac{\text{ft}\cdot\text{lbf}}{\text{lbmol}\cdot°R}\right)(510°R)} = 2.62$$

Returning to Fig. A-1, $T_{R2} = 0.9$. Thus

$$T_2 = T_{R2}T_c = (0.9)(510°R) = 459°R \quad \longleftarrow$$

PROBLEM 3.81

(a) Using the ideal gas equation of state

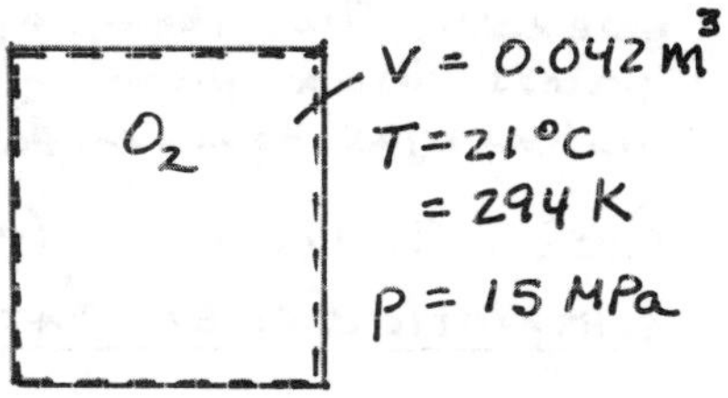

$$m = \frac{pV}{RT}$$

$$= \frac{(15\ MPa)(0.042\ m^3)}{\left(\frac{8.314}{32.00}\ \frac{kJ}{kg \cdot K}\right)(294\ K)} \left|\frac{10^6\ N/m^2}{1\ MPa}\right| \left|\frac{1\ kJ}{10^3\ N \cdot m}\right|$$

$= 8.248\ kg$ ← m(ideal gas)

(b) Using data from Table A-1

$$p_R = p/p_c = \frac{150\ bar}{50.5\ bar} = 2.97$$

$$T_R = T/T_c = \frac{294\ K}{154\ K} = 1.91$$

Fig. A-2 $v_R' \approx 0.60$

Thus

$$v = v_R' \left(\frac{RT_c}{p_c}\right) = (0.6)\left[\frac{\left(\frac{8.314}{32.00}\right)(154)}{(50.5)}\right]\left(\frac{1}{10^2}\right) = 4.754 \times 10^{-3}\ m^3/kg$$

and

(1) $$m = \frac{V}{v} = \frac{0.042\ m^3}{4.754 \times 10^{-3}\ m^3/kg} = 8.835\ kg$$ ← m (chart)

1. Assuming that the chart value is correct, the ideal gas model under-predicts the mass by about 6.6%.

PROBLEM 3.82

For all states in the shaded region on the T-v diagram $p < 0.1994$ bar. Since $Z \to 1$ as $p_R \to 0$, the ideal gas model can be checked at state 1. If it is valid there it will be even better at other states in the shaded region. From Table A-2

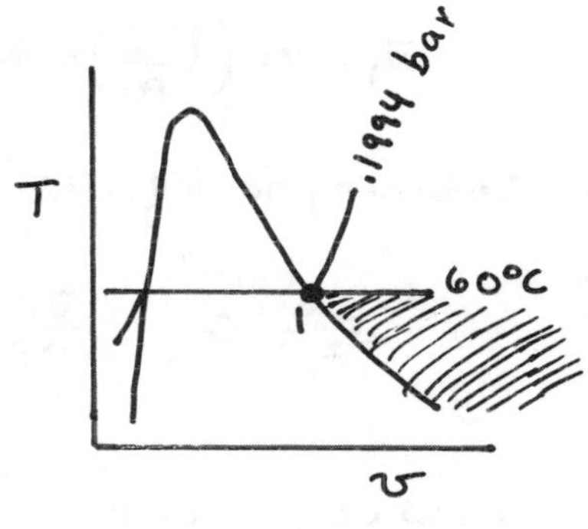

$$v_1 = v_{g@60°C} = 7.671\ m^3/kg$$

$$Z_1 = \frac{p_1 v_1}{RT_1}$$

$$= \frac{(0.1994\ bar)(7.671\ m^3/kg)}{\left(\frac{8.314}{18.02}\ \frac{kJ}{kg \cdot K}\right)(60+273)\ K}\left(\frac{10^5\ N/m^2}{1\ bar}\right)\left(\frac{1\ kJ}{10^3\ N \cdot m}\right)$$

$= 0.996 \Rightarrow$ Ideal gas model is accurate in shaded region.

PROBLEM 3.83

The ideal gas equation is accurate for ranges of pressure and temperature for which $Z \approx 1$. We might arbitrarily consider the ideal gas model to be satisfactory if $0.95 \le Z \le 1.05$ (5% deviation). These limits are illustrated on the compressibility chart below.

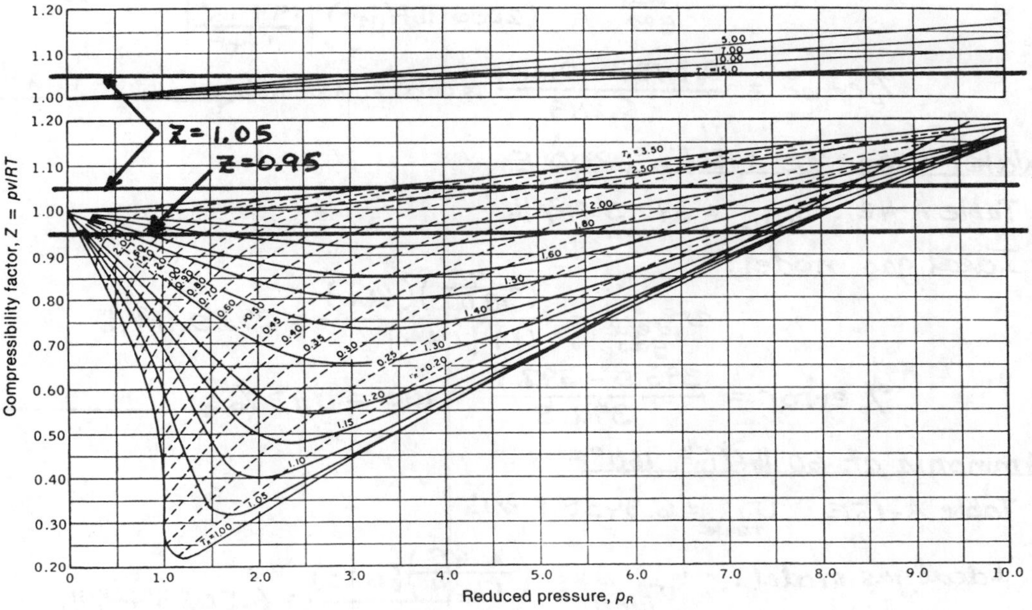

Figure A-2 Generalized compressibility chart, $p_R \le 10.0$. *Source:* E. F. Obert, *Concepts of Thermodynamics*, McGraw-Hill, New York, 1960.

For air, Table A-1 gives; $T_c = 133\,K$, $p_c = 37.7$ bar. The following observations can be made from the chart. The ideal gas model is reasonable when

- $p_R < 0.1$, for all temperatures ($p < 3.77$ bar)
- for $T_R = 2.0$ ($T = 266$ K); $p_R < 7.2$ ($p < 271$ bar)
- for $T_R = 2.5$ ($T = 333$K); $p_R < 5.0$ ($p < 189$ bar)
- for $T_R = 3.5$ ($T = 466$K); $p_R < 3.5$ ($p < 132$ bar)
- for $T_R > 5.0$ ($T = 665$K); $p_R < 3.2$ ($p < 121$ bar)
- for $T_R > 15$ ($T = 1995$K); $p_R < 7.0$ ($p < 264$ bar)
- for $T_R > 2.0$ ($T = 266$K); $p_R < 3.2$ ($p < 121$ bar)

Thus, for many common applications, air can be reasonably modeled as an ideal gas.

PROBLEM 3.84

(a) Water vapor at 2000 lbf/in², 700°F

From Table A-4E : $v_{table} = 0.249\ ft^3/lb$

Ideal gas model:

$$v_{\substack{ideal\\gas}} = \frac{\left(\frac{1545}{18.02}\ \frac{ft \cdot lbf}{lb \cdot {}^\circ R}\right)(1160^\circ R)}{(2000\ lbf/in^2)\left|\frac{144\ in^2}{1\ ft^2}\right|} = 0.3453\ ft^3/lb$$

$$\%\ error = \frac{0.3453 - 0.249}{0.249} \times 100 = 38.7\% \longleftarrow \quad (a)$$

(b) Water vapor at 1 lbf/in², 200°F

Table A-4E : $v_{table} = 392.5\ ft^3/lb$

Ideal gas model:

$$v_{\substack{ideal\\gas}} = \frac{\left(\frac{1545}{18.02}\right)(660)}{(1)|144|} = 393.0\ ft^3/lb$$

$$\%\ error = \frac{393.0 - 392.5}{392.5} \times 100 = 0.13\% \longleftarrow \quad (b)$$

(c) Ammonia at 60 lbf/in², 160°F

Table A-15E : $v_{table} = 6.3458\ ft^3/lb$

Ideal gas model:

$$v_{\substack{ideal\\gas}} = \frac{\left(\frac{1545}{17.04}\right)(620)}{(60)(144)} = 6.5063\ ft^3/lb$$

$$\%\ error = \frac{6.5063 - 6.3458}{6.3458} \times 100 = 2.53\% \longleftarrow \quad (c)$$

(d) Air at 1 atm, 2000°R

$$\left.\begin{aligned} P_R &= P/P_c = 1\ atm/37.2\ atm = 0.027 \\ T_R &= T/T_c = 2000^\circ R/239^\circ R = 8.37 \end{aligned}\right\} \begin{aligned} &\text{Fig. A-2} \\ &Z \approx 1.0 \end{aligned}$$

Ideal gas $\longleftarrow$ (d)

(e) R-22 at 300 lbf/in², 140°F

Table A-9E : $v_{table} = 0.1849\ ft^3/lb$

Ideal gas model:

$$v_{\substack{ideal\\gas}} = \frac{\left(\frac{1545}{86.48}\right)(600)}{(300)|144|} = 0.2481\ ft^3/lb$$

$$\%\ error = \frac{0.2481 - 0.1849}{0.1849} \times 100 = 34.2\% \longleftarrow \quad (e)$$

PROBLEM 3.85

(a) Refrigerant 134a at 80°C, 1.6 MPa

From Table A-12; $v_{table} = 0.01435\ m^3/kg$

Ideal gas model;

$$v_{\substack{ideal\\gas}} = \frac{RT}{P} = \frac{\left(\frac{8.314}{102.03}\frac{kJ}{kg\cdot K}\right)(353\ K)}{(1.6\ MPa)}\left|\frac{1\ MPa}{10^6\ N/m^2}\right|\left|\frac{10^3\ N\cdot m}{1\ kJ}\right|$$

$$= 0.01798\ m^3/kg$$

$$\%\ deviation = \frac{0.01798 - 0.01435}{0.01435} \times 100 = 25.3\%$$

(b) Refrigerant 134a at 80°C, 0.10 MPa

From Table A-12; $v_{table} = 0.28464\ m^3/kg$

Ideal gas model;

$$v_{\substack{ideal\\gas}} = \frac{\left(\frac{8.314}{102.03}\right)(353)}{(0.10)}\left|\frac{1}{10^3}\right| = 0.28765\ m^3/kg$$

$$\%\ deviation = \frac{0.28765 - 0.28464}{0.28464} \times 100 = 1.1\%$$

(Ideal gas model is applicable)

PROBLEM 3.86

KNOWN: Nitrogen is at a state fixed by pressure and specific volume.

FIND: Determine the temperature using generalized compressibility data and compare with the value calculated from the ideal gas model.

ANALYSIS: $p = 100$ bar, $v = 4.5\times10^{-3}\ m^3/kg$. With $T_c = 126$ K and $p_c = 33.9$ bar from Table A-1

$$p_R = \frac{100}{33.9} = 2.95,\quad v_R' = \frac{v\,p_c}{RT_c} = \frac{(4.5\times10^{-3} m^3/kg)(33.9\times10^5 N/m^2)}{\left(\frac{8314}{28.01}\frac{N\cdot m}{kg\cdot K}\right)(126 K)} = 0.408$$

From Fig. A-2, $T_R = 1.51$. Thus $T = T_c T_R = (126K)(1.51) = 190K$ ← T

Using the ideal gas equation of state

$$T = \frac{pv}{R} = \frac{(100\times10^5 N/m^2)(4.5\times10^{-3} m^3/kg)}{\left(\frac{8314}{28.01}\frac{N\cdot m}{kg\cdot K}\right)} = 152 K$$

This value is 20% less than the result obtained with compressibility data.

PROBLEM 3.87

KNOWN: Air is at a state fixed by pressure, volume and total mass.

FIND: Determine the temperature. Check the validity of the ideal gas model.

ANALYSIS: $m = 5$ kg, $p = 0.3$ MPa, $V = 2.2\ m^3$ From Table A-1, $M = 28.97$, $p_c = 37.7$ bar, $T_c = 133$ K. Then

$$p_R = \frac{p}{p_c} = \frac{3\text{ bar}}{37.7\text{ bar}} = 0.08,\quad v_R' = \frac{v\,p_c}{RT_c} = \frac{(2.2m^3/5kg)(37.7\times10^5 N/m^2)}{\left(\frac{8314}{28.97}\frac{N\cdot m}{kg\cdot K}\right)(133 K)} = 43.46$$

Referring to Fig A-1, $Z \sim 1$. Thus, use of the ideal gas equation of state is justified:

$$T = \frac{p(V/m)}{R} = \frac{(3\times10^5 N/m^2)(2.2m^3/5kg)}{(8314/28.97\ N\cdot m/kg\cdot K)} = 460 K$$ ← T

PROBLEM 3.88

KNOWN: A tank of known volume contains air at a specified temperature and pressure.

FIND: Determine the mass of the air. Verify that the ideal gas model applies.

ANALYSIS: $V = 40 ft^3$. $T = 560°R$, $p = 50\ lbf/in.^2$ With T_c, p_c from Table A-1E, $T_c = 239°R$, $p_c = 37.2$ atm

$$T_R = \frac{T}{T_c} = \frac{560°R}{239°R} = 2.34,\quad p_R = \frac{p}{p_c} = \frac{(50/14.7)}{37.2} = 0.09$$

Referring to Fig. A-1, $Z \sim 1$. Thus, use of the ideal gas equation of state is justified: $pV = mRT$, or

$$m = \frac{pV}{RT} = \frac{(50\times144\ lbf/ft^2)(40ft^3)}{\left(\frac{1545}{28.97}\frac{ft\cdot lbf}{lb\cdot °R}\right)(560°R)} = 9.64\ lb$$ ← m

PROBLEM 3.89

For an ideal gas: $\rho = p/\left(\frac{\bar{R}}{M}\right)T = pM/\bar{R}T$

For helium and air, each at the same pressure and temperature

$$\rho_{He} = pM_{He}/\bar{R}T$$

$$\rho_{air} = pM_{air}/\bar{R}T$$

Thus

$$\frac{\rho_{He}}{\rho_{air}} = \frac{M_{He}}{M_{air}} = \frac{4.003}{28.97} = 0.138$$

PROBLEM 3.90

① Using the ideal gas model to determine the volume

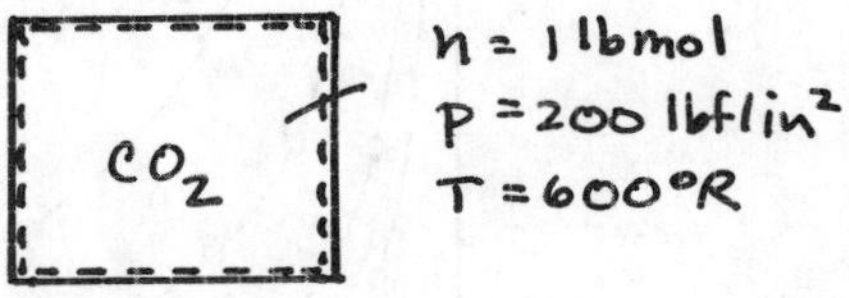

$$V = \frac{n\bar{R}T}{p}$$

$$= \frac{(1 \text{ lbmol})\left(1545 \frac{\text{ft}\cdot\text{lbf}}{\text{lbmol}\cdot{}^\circ\text{R}}\right)(600^\circ\text{R})}{(200 \text{ lbf/in}^2)} \left|\frac{1 \text{ ft}^2}{144 \text{ in}^2}\right|$$

$$= 32.19 \text{ ft}^3 \longleftarrow V$$

---

1. The applicability of the ideal gas model can be verified by referring to the compressibility chart.

PROBLEM 3.91

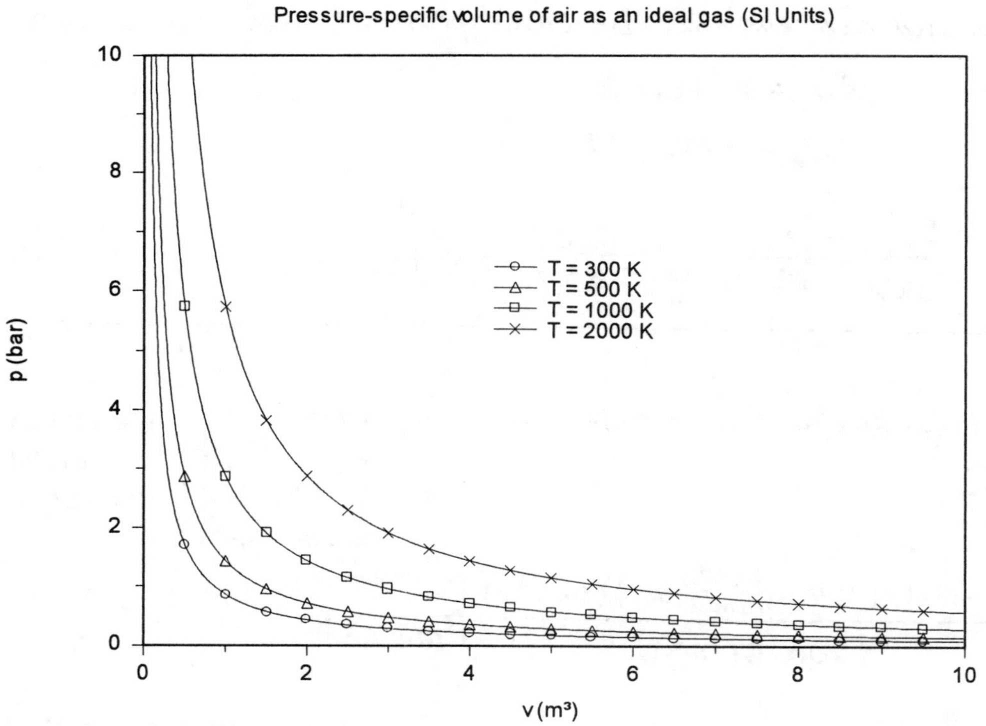

PROBLEM 3.92

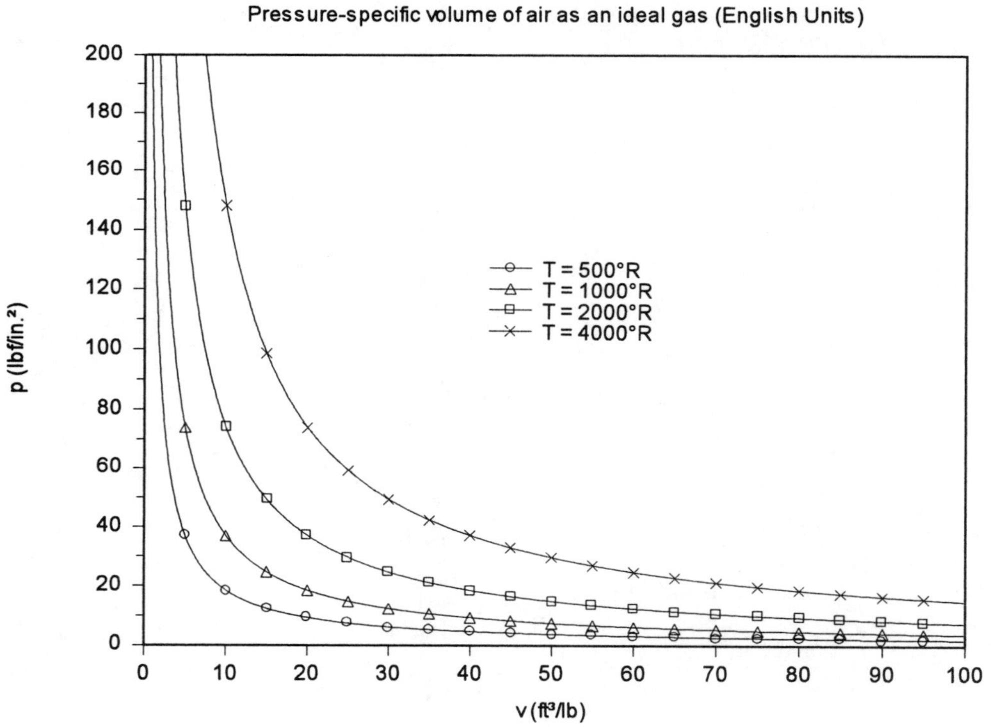

PROBLEM 3.93

KNOWN: Methane undergoes a process between two states, each fixed by known values of pressure and temperature.

FIND: Determine Δh by integrating $\bar{c}_p(T)$ from Table A-1 and check the result using IT.

ANALYSIS: $T_1 = 320\ K$, $p_1 = 2\ bar$; $T_2 = 800\ K$, $p_2 = 10\ bar$. From Table A-1, $T_c = 191\ K$, $p_c = 46.4\ bar$. Thus

$$T_{R1} = \frac{320}{191} = 1.68 \qquad T_{R2} = \frac{800}{191} = 4.19$$

$$p_{R1} = \frac{2}{46.4} = 0.04 \qquad p_{R2} = \frac{10}{46.4} = 0.22$$

Referring to Fig. A-1 at p_{R1}, T_{R1} and p_{R2}, T_{R2}, the ideal gas model is applicable. Thus, $\Delta h = \int_{T_1}^{T_2} c_p\, dT$, where $c_p(T)$ is obtained from Table A-21. That is

$$c_p(T) = R(\alpha + \beta T + \gamma T^2 + \delta T^3 + \epsilon T^4)$$

$$\Rightarrow h(T_2) - h(T_1) = \int_{T_1}^{T_2} c_p(T)\, dT = R\int_{T_1}^{T_2}(\alpha + \beta T + \gamma T^2 + \delta T^3 + \epsilon T^4)\, dT$$

$$= R\left[\alpha(T_2 - T_1) + \frac{\beta}{2}(T_2^2 - T_1^2) + \frac{\gamma}{3}(T_2^3 - T_1^3) + \frac{\delta}{4}(T_2^4 - T_1^4) + \frac{\epsilon}{5}(T_2^5 - T_1^5)\right]$$

Inserting values

$$h(T_2) - h(T_1) = \frac{\left(8.314\ \frac{kJ}{kmol\cdot K}\right)}{\left(16.04\ \frac{kg}{kmol}\right)}\left[3.826(800-320) + \left(\frac{-3.979\times10^{-3}}{2}\right)(800^2 - 320^2)\right.$$

$$+ \left(\frac{24.558\times10^{-6}}{3}\right)(800^3 - 320^3) + \left(\frac{-22.733\times10^{-9}}{4}\right)(800^4 - 320^4)$$

$$\left. + \left(\frac{6.963\times10^{-12}}{5}\right)(800^5 - 320^5)\right]$$

$$= 1489.3\ kJ/kg \longleftarrow \Delta h$$

IT Code

```
T1 = 320 // K
T2 = 800 // K
h1 = h_T("CH4", T1) // kJ/kg
h2 = h_T("CH4", T2) // kJ/kg
delh = h2 - h1
```

IT Result

Δh = 1489 kJ/kg

Alternative solution:

```
cp = cp_T("CH4", T)
dh = Integral(cp,T)
```

Using Explore button, sweep T from 320 K to 800 K in steps of 1.

The analytical and IT results compare very favorably.

PROBLEM 3.94

From Table A-21: $\bar{c}_p/\bar{R} = 2.5 = 5/2$

Now $\bar{c}_v = \bar{c}_p - \bar{R} \Rightarrow \frac{\bar{c}_v}{\bar{R}} = \frac{\bar{c}_p}{\bar{R}} - 1 = \frac{3}{2}$

$$\Rightarrow k = \frac{\bar{c}_p}{\bar{c}_v} = \frac{5/2}{3/2} = 5/3 \longleftarrow k$$

PROBLEM 3.95

KNOWN: A paddle wheel transfers energy to ammonia contained in a closed, rigid, well-insulated tank.

FIND: Determine the final temperature of the ammonia.

SCHEMATIC & GIVEN DATA:

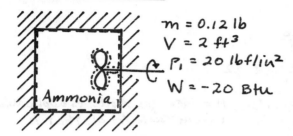

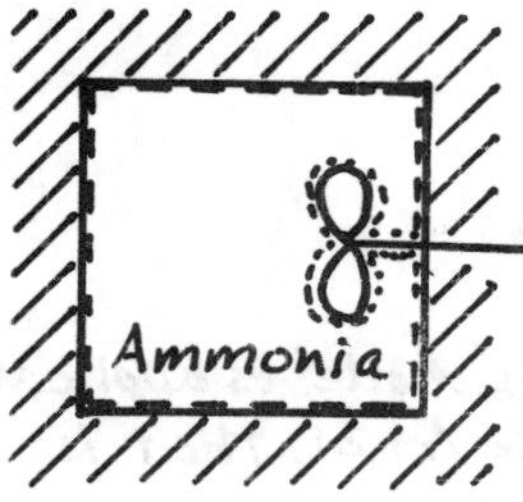

$m = 0.12$ lb
$V = 2$ ft^3
$p_1 = 20$ lbf/in^2
$W = -20$ Btu

ASSUMPTIONS: (1) The ammonia is a closed system. (2) The ammonia behaves
① as an ideal gas. (3) There is no heat transfer. (4) Kinetic and potential energy effects are negligible.

ANALYSIS: To determine the final temperature, begin with the energy balance

$$\Delta KE^{\,0} + \Delta PE^{\,0} + \Delta U = Q^{\,0} - W \qquad (*)$$

Evaluating ΔU

$$\Delta U = m\,\Delta u = m\int_{T_1}^{T_2} c_v(T)\,dT$$

From Table A-21E: $c_p = R(3.591 + 0.274\times10^{-3}T + 2.576\times10^{-6}T^2 - 1.437\times10^{-9}T^3 + 0.2601\times10^{-12}T^4)$
with $c_v = c_p - R$

$$\Delta U = mR\int_{T_1}^{T_2}(2.591 + 0.274\times10^{-3}T + 2.576\times10^{-6}T^3 - 1.437\times10^{-9}T^3 + 0.2601\times10^{-12}T^4)\,dT$$

$$= (0.12\text{ lb})\left(\frac{1545}{17.04}\frac{\text{ft}\cdot\text{lbf}}{\text{lb}\cdot{}^\circ\text{R}}\right)\left|\frac{1\text{ Btu}}{778\text{ ft}\cdot\text{lbf}}\right|\Big[2.591(T_2-T_1) + \frac{0.274\times10^{-3}}{2}(T_2^2-T_1^2)$$
$$+\frac{2.576\times10^{-6}}{3}(T_2^3-T_1^3) - \frac{1.437\times10^{-9}}{4}(T_2^4-T_1^4) + \frac{0.2601\times10^{-12}}{5}(T_2^5-T_1^5)\Big] \qquad (**)$$

The initial temperature is

$$T_1 = \frac{p_1V}{mR} = \frac{(20\text{ lbf/in}^2)(2\text{ ft}^3)}{(.12\text{ lb})\left(\frac{1545}{17.04}\frac{\text{ft}\cdot\text{lbf}}{\text{lb}\cdot{}^\circ\text{R}}\right)}\left|\frac{144\text{ in}^2}{1\text{ ft}^2}\right| = 529.4\,^\circ\text{R}$$

Combining the starred relations and inserting values, T_2 becomes the only unknown. Using an equation solver such as IT,

$$T_2 = 920.1\,^\circ\text{R} \longleftarrow \qquad T_2$$

---

1. For state 1, $p_{R1} = (20)/(111.3\cdot14.7) = 0.00122$, $T_{R1} = (529.4)/(730) = 0.72$. From Fig. A-1; $Z_1 \approx 1$. Thus, assumption 2 is reasonable for state 1. By similar considerations, it can be concluded that the ideal gas model is reasonable for state 2 as well.

PROBLEM 3.96

KNOWN: Air and carbon dioxide are confined to opposite sides of a rigid, well-insulated container. The partition moves and allows conduction from one gas to the other until equilibrium is achieved.

FIND: Determine the final temperature and pressure.

SCHEMATIC & GIVEN DATA: **

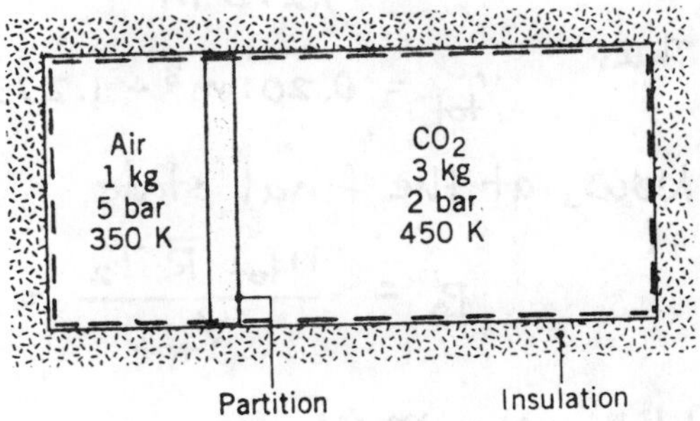

ASSUMPTIONS: (1) The contents of the container form a closed system. (2) The air and CO_2 behaves as ideal gases with constant specific heats. (3) The system is isolated, so Q=0 and W=0. (4) There is no energy stored in the partition. (5) Kinetic and potential energy effects are negligible.

ANALYSIS: To determine the final temperature, begin with the energy balance

$$\Delta KE^{0} + \Delta PE^{0} + \Delta U = Q^{0} - W^{0}$$

or, with $\Delta U = m_{air}\,\Delta u_{air} + m_{CO_2}\,\Delta u_{CO_2}$ and using Eq. 3.50

① $$m_{air}\, c_{v,air}\,(T_2 - T_{1,air}) + m_{CO_2}\, c_{v,CO_2}\,(T_2 - T_{1,CO_2}) = 0$$

Solving for T_2

$$T_2 = \frac{m_{air}\, c_{v,air}\, T_{1,air} + m_{CO_2} c_{v,CO_2}\, T_{1,CO_2}}{m_{air}\, c_{v,air} + m_{CO_2} c_{v,CO_2}}$$

The specific heats are evaluated using data from Table A-20 at a mean temperature of 400K; $c_{v,air} = 0.726$ kJ/kg·K and $c_{v,CO_2} = 0.750$ kJ/kg·K. Thus, the final temperature is

$$T_2 = \frac{(1\text{ kg})(0.726\text{ kJ/kg}\cdot\text{K})(350\text{ K}) + (3)(.750)(450)}{(1\text{ kg})(0.726\text{ kJ/kg}\cdot\text{K}) + (3)(.750)}$$

$$= 425.6\text{ K} \longleftarrow T_2$$

Next, to find the final pressure, the total volume is needed. The initial volume of the air is

$$V_{1,air} = \frac{m_{air}\, R_{air}\, T_{1,air}}{P_{1,air}}$$

$$= \frac{(1\text{ kg})\left(\frac{8.314}{28.97}\frac{\text{kJ}}{\text{kg}\cdot\text{K}}\right)(350\text{ K})}{(5\text{ bar})}\left|\frac{1\text{ bar}}{10^5\text{ N/m}^2}\right|\left|\frac{10^3\text{ N}\cdot\text{m}}{1\text{ kJ}}\right| = 0.201\text{ m}^3$$

PROBLEM 3.96

similarly for the carbon dioxide

$$V_{1,CO_2} = \frac{m_{CO_2} R_{CO_2} T_{1,CO_2}}{P_{1,CO_2}} = \frac{(3)\left(\frac{8.314}{44.01}\right)(450)}{(2)} \left|\frac{10^3}{10^5}\right|$$

$$= 1.275 \text{ m}^3$$

Thus

$$V_{tot} = 0.201 \text{ m}^3 + 1.275 \text{ m}^3 = 1.476 \text{ m}^3$$

Now, at the final state

$$P_2 = \frac{n_{tot} \bar{R} T_2}{V_{tot}} = \frac{(n_{air} + n_{CO_2}) \bar{R} T_2}{V_{tot}}$$

With $n = m/M$

$$P_2 = \frac{(m_{air}/M_{air} + m_{CO_2}/M_{CO_2}) \bar{R} T_2}{V_{tot}}$$

$$= \frac{\left(\frac{1 \text{ kg}}{28.97 \text{ kg/kmol}} + \frac{3}{44.01}\right)\left(8.314 \frac{\text{kJ}}{\text{kmol}\cdot\text{K}}\right)(425.6\text{K})}{(1.476 \text{ m}^3)} \left|\frac{10^3 \text{ N}\cdot\text{m}}{1 \text{ kJ}}\right| \left|\frac{1 \text{ bar}}{10^5 \text{ N/m}^2}\right|$$

$$= 2.462 \text{ bar} \longleftarrow P_2$$

---

1. The assumption of constant specific heats facilitates the determination of T_2. The assumption is reasonable for the relatively small temperature range in this problem.

** Note: In the first printing of this edition, the problem statement contained erroneous data. The statement has been corrected in subsequent printings. **

PROBLEM 3.97

KNOWN: Heat transfer occurs to air in a rigid tank. The initial state is specified.

FIND: Determine the final temperature and pressure using (a) a constant specific value, (b) a specific heat function, (c) the air tables.

SCHEMATIC & GIVEN DATA:

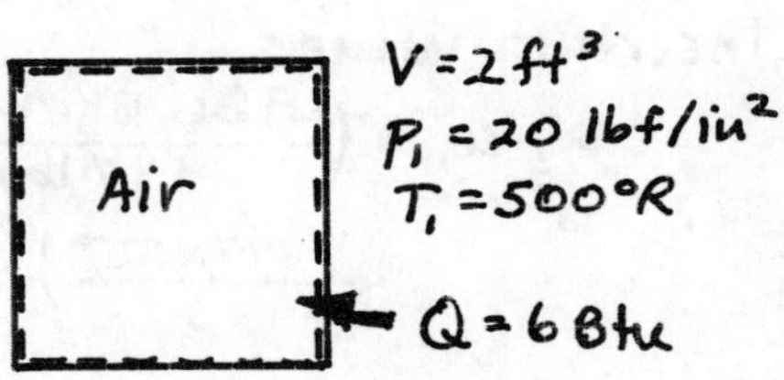

ASSUMPTIONS: (1) The air is a closed system. (2) The air behaves as an ideal gas. (3) There is no work. (4) Kinetic and potential energy effects are negligible.

ANALYSIS: To determine the final temperature, begin with the energy balance

$$\Delta KE^{\,0} + \Delta PE^{\,0} + \Delta U = Q - W^{\,0}$$

or

$$m(u_2 - u_1) = Q \Rightarrow u_2 - u_1 = Q/m \qquad (*)$$

Evaluating the mass

$$m = \frac{pV}{RT} = \frac{(20\ lbf/in^2)(2\ ft^3)}{\left(\frac{1545}{28.97}\frac{ft \cdot lbf}{lb \cdot °R}\right)(500°R)}\left|\frac{144\ in^2}{1\ ft^2}\right| = 0.216\ lb$$

(a) For a constant specific heat c_v

$$u_2 - u_1 = c_v(T_2 - T_1)$$

Thus, with (*)

$$T_2 = \frac{Q}{mc_v} + T_1$$

From Table A-20E, at $T = 100°F$ (600°R), $c_v = 0.172\ Btu/lb \cdot °R$. Thus

$$T_2 = \frac{(6\ Btu)}{(.216\ lb)(.172\ Btu/lb \cdot °R)} + 500°R$$

$$= 661.5°R \longleftarrow \text{(a) } T_2$$

Accordingly, for this temperature interval the value used for c_v is appropriate. The pressure P_2 is

$$P_2 = \frac{mRT_2}{V} = \frac{(.216)\left(\frac{1545}{28.97}\right)(661.5)}{(2)(144)} = 26.46\ lbf/in^2 \longleftarrow \text{(a) } P_2$$

① (b) From Table A-21E

$$c_p = R(\alpha + \beta T + \gamma T^2 + \delta T^3 + \epsilon T^4)$$

with $c_p = c_v + R$

$$c_v = R(\alpha - 1 + \beta T + \gamma T^2 + \delta T^3 + \epsilon T^4)$$

PROBLEM 3.97 (Cont'd)

Thus

$$u_2 - u_1 = \int_{T_1}^{T_2} c_v\, dT$$

$$= R\left[(\alpha-1)(T_2-T_1) + \frac{\beta}{2}(T_2^2-T_1^2) + \frac{\gamma}{3}(T_2^3-T_1^3) + \frac{\delta}{4}(T_2^4-T_1^4) + \frac{\varepsilon}{5}(T_2^5-T_1^5)\right]$$

Inserting values

$$u_2 - u_1 = \left(\frac{1.986\ \text{Btu/lbmol}\cdot{}^\circ\text{R}}{28.97\ \text{lb/lbmol}}\right)\left[2.653\,(T_2-500) - \left(\frac{0.742\times10^{-3}}{2}\right)(T_2^2 - 500^2)\right.$$

$$\left. + \left(\frac{1.017\times10^{-6}}{3}\right)(T_2^3-500^3) - \left(\frac{0.328\times10^{-9}}{4}\right)(T_2^4-500^4) + \left(\frac{0.02632\times10^{-12}}{5}\right)(T_2^5-500^5)\right] \quad (**)$$

Combining the starred relations and inserting values, T_2 becomes the only unknown. Solving iteratively

$T_2 = 661.8\ {}^\circ R$ ← (b) T_2

and $p_2 = 26.47\ \text{lbf/in}^2$ ← (b) p_2

(c) From Table A-22E, $u_1 = 85.20$ Btu/lb. From (*)

$$u_2 = Q/m + u_1$$

$$= \left(\frac{6\ \text{Btu}}{0.216\ \text{lb}}\right) + 85.20 = 112.98\ \text{Btu/lb}$$

Interpolating in Table A-22E

$T_2 = 661.8\ {}^\circ R$ ← (c) T_2

and $p_2 = 26.47\ \text{lbf/in}^2$ ← (c) p_2

---

① *IT* solution (based on part b)

```
T1 = 500  // °R
p1 = 20  // lbf/in.²
V = 2  // ft³
Q = 6  // Btu
delu = Q/m
m = V / v1
v1 = v_TP("Air",T1,p1)
delu = R * ((alpha-1)*(T2-T1)+(beta/2)*(T2^2- T1^2)
      +(gamma/3)*(T2^3-T1^3)+(delta/4)*(T2^4- T1^4)
      +(epsilon/5)*(T2^5-T1^5))
R = 1.986/28.97  // Btu/lb·°R
alpha = 3.653
beta = -0.742E-3
gamma = 1.017E-6
delta = -0.328E-9
epsilon = 0.02632E-12
v2 = v1
v2 = v_TP("Air",T2,p2)
```

IT Results

$T_2 = 661.8$°R

$p_2 = 26.47$ bar

PROBLEM 3.98

KNOWN: A gas mixture expands with a known pressure-volume relation. The initial state is fixed and the heat transfer for the process is known.

FIND: Determine (a) the final temperature, (b) the final pressure, (c) the final volume, and (d) the work.

SCHEMATIC & GIVEN DATA:

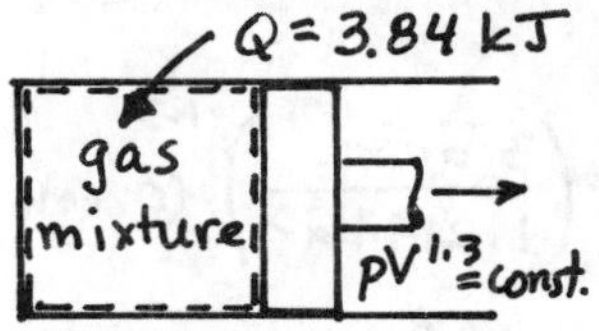

$M = 33$ kg/kmol
$p_1 = 3$ bar
$T_1 = 300$K
$V_1 = 0.1\ m^3$

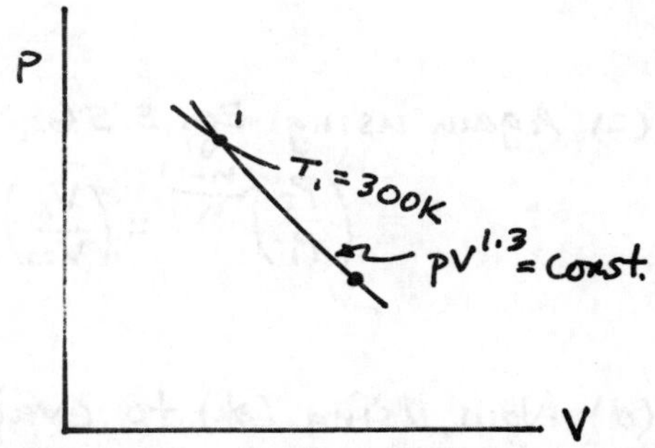

For the mixture: $c_v = 0.6 + (2.5\times10^{-4})T$ (T in K, c_v in kJ/kg·K)

ASSUMPTIONS: (1) The gas mixture is the closed system. (2) The gas mixture behaves as an ideal gas. (3) The process is polytropic with $n = 1.3$. (4) Kinetic and potential energy effects are neglected.

ANALYSIS: (a) To find the final temperature, start with the energy balance and express the internal energy change and work in terms of temperature change, as follows. First

$$\cancel{\Delta KE} + \cancel{\Delta PE} + \Delta U = Q - W$$

From Eq. 3.57 for the polytropic process

$$W = \int_{V_1}^{V_2} p\,dV = \frac{mR(T_2 - T_1)}{1-n} \qquad (*)$$

Also, from the given c_v relation

$$\Delta U = m\int_{T_1}^{T_2} c_v(T)\,dT = m\left[(0.6)(T_2 - T_1) + \left(\frac{2.5\times10^{-4}}{2}\right)(T_2^2 - T_1^2)\right] \qquad (**)$$

Incorporating (*) and (**) into the energy balance

$$m\left[(0.6)(T_2 - T_1) + \left(\frac{2.5\times10^{-4}}{2}\right)(T_2^2 - T_1^2)\right] = Q - \frac{mR(T_2 - T_1)}{1-n} \qquad (***)$$

The mass is found using the ideal gas equation at state 1

$$m = \frac{p_1 V_1}{RT_1} = \frac{(3\text{ bar})(0.1\,m^3)}{\left(\frac{8.314}{33}\right)\frac{kJ}{kg\cdot K}(300K)}\left|\frac{10^5\,N/m^2}{1\text{ bar}}\right|\left|\frac{1\,kJ}{10^3\,N\cdot m}\right| = 0.397\text{ kg}$$

Using IT with $m = 0.397$ kg, $T_1 = 300$K, $Q = 3.84$ kJ, and $R = 8.314/33$ kJ/kg·K to solve for T_2:

```
delU = Q - W
W = (m * R * (T2 - T1)) / (1 - 1.3)
delU = m * (0.6 * (T2 - T1) + ((2.5E-4) / 2) * (T2^2 - T1^2))
```

IT Result
$T_2 = 243.7$ K ← T_2

(b) To find p_2, use Eq. 3.56 for the polytropic process

$$\frac{T_2}{T_1} = \left(\frac{p_2}{p_1}\right)^{n-1/n} \Rightarrow p_2 = \left(\frac{T_2}{T_1}\right)^{\frac{n}{n-1}} p_1 = \left(\frac{243.7\ K}{300\ K}\right)^{\frac{1.3}{.3}} (3\ bar)$$

$$= 1.219\ bar \longleftarrow p_2$$

(c) Again using Eq. 3.56

$$\left(\frac{p_2}{p_1}\right)^{\frac{(n-1)}{n}} = \left(\frac{V_1}{V_2}\right)^{n-1} \Rightarrow V_2 = \left(\frac{p_1}{p_2}\right)^{\frac{1}{n}} V_1 = \left(\frac{3\ bar}{1.219\ bar}\right)^{\frac{1}{1.3}} (0.1\ m^3)$$

$$= 0.1999\ m^3 \longleftarrow V_2$$

(d) Now, using (*) to evaluate the work

$$W = \frac{mR(T_2 - T_1)}{(1-n)}$$

$$= \frac{(0.397\ kg)\left(\frac{8.314}{33}\right)\frac{kJ}{kg \cdot K}(243.7 - 300)}{(1 - 1.3)}$$

$$= 18.77\ kJ \longleftarrow W$$

PROBLEM 3.99

KNOWN: Argon gas undergoes a process for which pv^k = constant from a given initial state to a specified final pressure.

FIND: Determine the work and heat transfer, each per unit of mass.

SCHEMATIC & GIVEN DATA:

$pv^k = \text{const.}$

$P_1 = 1$ bar; $T_1 = 100$K

$P_2 = 15.59$ bar

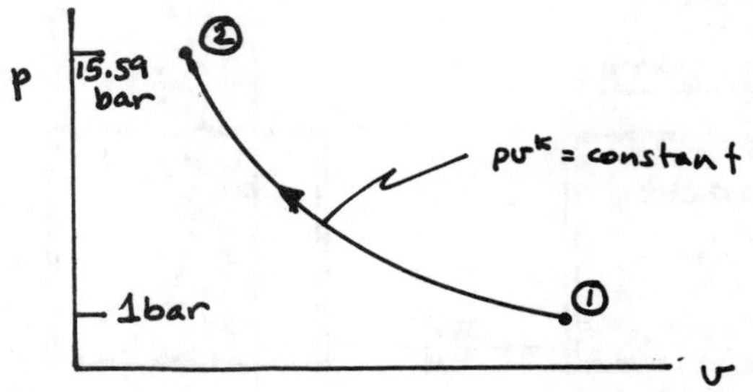

ASSUMPTIONS: 1. The argon is a closed system. 2. The process is described by pv^k = constant. 3. The argon behaves as an ideal gas. 4. Kinetic and potential energy changes are ignored.

ANALYSIS: From Table A-21, $\bar{c}_p = 2.5\bar{R}$. Using Eq. 3.45, $\bar{c}_v = \bar{c}_p - R = 1.5\bar{R}$. Thus $k = \bar{c}_p/\bar{c}_v = 2.5/1.5 = 1.667$.

With assumptions 2 and 3,

$$W = \int_1^2 p\,dV = m\int_1^2 \frac{c}{v^k}\,dv \Rightarrow \frac{W}{m} = \frac{P_2v_2 - P_1v_1}{1-k} = \frac{R(T_2-T_1)}{1-k} \qquad \text{(Eq. 3.57)}$$

To find T_2, use Eq. 3.56 with $n = k$

$$T_2 = T_1\left(\frac{P_2}{P_1}\right)^{(k-1)/k} = 100\text{K}\left(\frac{15.59}{1}\right)^{0.667/1.667} = 300\text{K}$$

Inserting values,

$$\frac{W}{m} = \frac{R(T_2-T_1)}{(1-k)}$$

$$= \frac{\left(\frac{8.314}{39.94}\right)\frac{\text{kJ}}{\text{kg}\cdot\text{K}}(300-100)\text{K}}{(1-1.667)} = -62.45\text{ kJ} \longleftarrow \qquad W/m$$

Next, the heat transfer is

$$\cancel{\Delta KE} + \cancel{\Delta PE} + \Delta U = Q - W$$

$$Q = m(u_2 - u_1) + W$$

or

$$\frac{Q}{m} = (u_2 - u_1) + \frac{W}{m}$$

For a monatomic gas, c_v is <u>constant</u>. Thus

$$\frac{Q}{m} = c_v(T_2 - T_1) + \frac{R(T_2-T_1)}{1-k}$$

With $c_v = R/(k-1)$ (Eq. 3.47b)

$$\frac{Q}{m} = \left(\frac{R}{k-1}\right)(T_2-T_1) + \left(\frac{R}{1-k}\right)(T_2-T_1) = 0 \longleftarrow \qquad Q/m$$

PROBLEM 3.100

KNOWN: The contents of two uninsulated tanks are allowed to mix, and equilibrium is attained at the temperature of the surroundings.

FIND: Determine the amount of energy transfer by heat and the final pressure.

SCHEMATIC & GIVEN DATA:

ASSUMPTIONS: (1) The contents of the two tanks are a closed system. (2) There is no work. (3) The air behaves as an ideal gas. (4) Kinetic and potential energy effects can be neglected.

ANALYSIS: To find the heat transfer, begin with the energy balance.

$$\Delta KE^{\,0} + \Delta PE^{\,0} + \Delta U = Q - W^{\,0}$$

Since the final state is an equilibrium state

$$\Delta U = (m_{1A}+m_{1B})u_2 - (m_{1A}u_{1A} + m_{1B}u_{1B})$$

Thus

$$Q = (m_{1A}+m_{1B})u_2 - (m_{1A}u_{1A} + m_{1B}u_{1B})$$

Using data from Table A-22E

$$Q = (3\text{ lb})(88.62\text{ Btu/lb}) - [(1)(254.66) + (2)(154.57)]$$

$$= -297.9\text{ Btu} \quad \longleftarrow Q$$

To find the final pressure, first determine the volume of each tank

$$V_A = \frac{m_{1A}RT_{1A}}{P_{1A}} = \frac{(1\text{ lb})\left(\frac{1545}{28.97}\frac{\text{ft}\cdot\text{lbf}}{\text{lb}\cdot°R}\right)(1440°R)}{(50\text{ lbf/in}^2)\left|144\text{ in}^2/1\text{ ft}^2\right|} = 10.66\text{ ft}^3$$

$$V_B = \frac{(2)\left(\frac{1545}{28.97}\right)(900)}{(50)\left|144/1\right|} = 13.33\text{ ft}^3$$

Thus, at the final state

$$P_2 = \frac{mRT_2}{(V_A+V_B)} = \frac{(3)\left(\frac{1545}{28.97}\right)(520)}{(10.66+13.33)\left|144/1\right|} = 24.07\text{ lbf/in}^2 \quad \longleftarrow P_2$$

PROBLEM 3.101

KNOWN: A gas is contained in a closed rigid tank fitted with an electric resistor. The constant current and voltage and the constant heat transfer rate are known.

FIND: Determine an average value of specific heat c_p.

SCHEMATIC & GIVEN DATA:

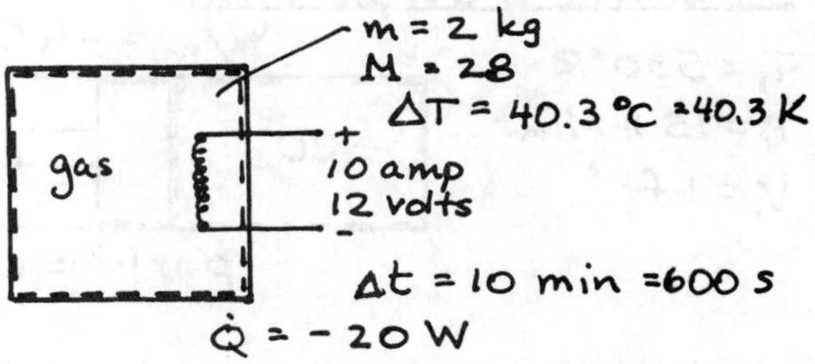

ASSUMPTIONS: (1) The gas is a closed system. (2) The gas behaves as an ideal gas. (3) Kinetic and potential energy effects are negligible.

ANALYSIS: To determine the specific heat, begin with the energy balance

$$\Delta KE^{\nearrow 0} + \Delta PE^{\nearrow 0} + \Delta U = Q - W$$

with $\Delta U = m c_v \Delta T = m(c_p - R)\Delta T$

$$c_p = \frac{Q-W}{m\Delta T} + R$$

Next, evaluate the heat transfer and work, respectively

$$Q = \int_{t_1}^{t_2} \dot{Q}\,dt = \dot{Q}\,\Delta t = (-20\,W)(600\,s)\left|\frac{1\ kJ/s}{10^3\ W}\right|$$

$$= -12\ kJ$$

$$W = \int_{t_1}^{t_2} \dot{W}\,dt = \dot{W}\,\Delta t$$

$$= -(10\ amp)(12\ volts)\left|\frac{1\ W/amp}{1\ volt}\right|(600\ s)\left|\frac{1\ kJ/s}{10^3\ W}\right|$$

$$= -72\ kJ$$

Thus

$$c_p = \frac{(-12\ kJ)-(-72\ kJ)}{(2\ kg)(40.3\ K)} + \left(\frac{8.314}{28}\ \frac{kJ}{kg\cdot K}\right)$$

$$= 1.041\ kJ/kg\cdot K \longleftarrow c_p$$

PROBLEM 3.102

KNOWN: CO_2 gas is compressed from a specified initial state in a process for which $pv^{1.2}$ constant. The work per unit of mass is also known.

FIND: Determine the temperature at the final state and the heat transfer per unit of mass.

SCHEMATIC & GIVEN DATA:

$T_1 = 530°R$

$p_1 = 15 \text{ lbf/in}^2$

$V_1 = 1 \text{ ft}^3$

$W/m = -45 \text{ Btu/lb}$

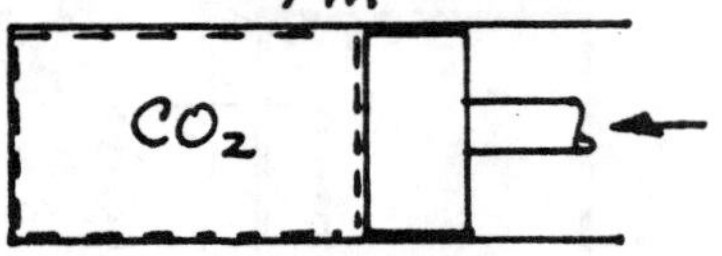

$pv^{1.2} = \text{constant}$

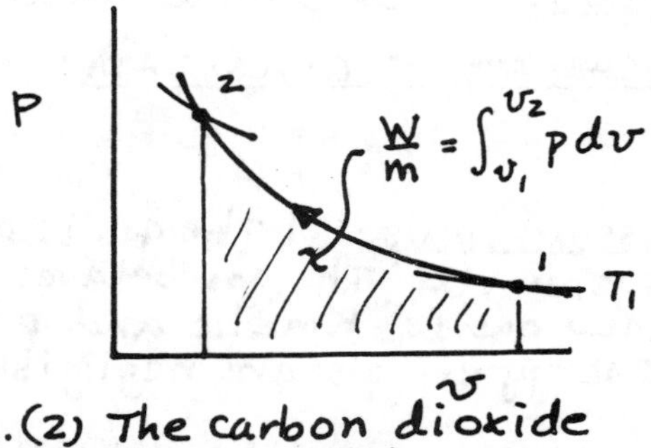

ASSUMPTIONS: (1) The gas is a closed system. (2) The carbon dioxide behaves as an ideal gas. (3) The process is $pv^{1.2}$ = constant. (4) Kinetic and potential energy effects are negligible.

ANALYSIS: The heat transfer is determined using an energy balance. First, the final temperature is found using $W = \int_1^2 p\,dV$ in the form of Eq. 3.57:

$$\frac{W}{m} = \frac{R(T_2 - T_1)}{1-n}$$

Solving for T_2

$$T_2 = \left(\frac{1-n}{R}\right)\frac{W}{m} + T_1$$

$$= \frac{(1-1.2)}{\left(\frac{1545}{44.01}\ \frac{\text{ft}\cdot\text{lbf}}{\text{lb}\cdot°R}\right)}\left(-45\ \frac{\text{Btu}}{\text{lb}}\right)\left|\frac{778\ \text{ft}\cdot\text{lbf}}{1\ \text{Btu}}\right| + 530°R$$

$$= 729.5°R \longleftarrow T_2$$

Now, the energy balance is

$$\Delta KE^{\,0} + \Delta PE^{\,0} + \Delta U = Q - W$$

or

$$\frac{Q}{m} = (\bar{u}_2 - \bar{u}_1)\frac{1}{M} + \frac{W}{m}$$

From Table A-23E, $\bar{u}_1 = 2916.1$ Btu/lbmol and $\bar{u}_2 = 4394.0$ Btu/lbmol.
Inserting values

$$\frac{Q}{m} = \frac{4394.0 - 2916.1}{44.01} + (-45)$$

$$= -11.42 \text{ Btu/lb} \longleftarrow Q/m$$

PROBLEM 3.103

KNOWN: A gas is confined to one side of a rigid, insulated container divided by a partition. The other side is initially evacuated. The partition is removed and the gas expands to fill the entire container.

FIND: Determine the final volume.

SCHEMATIC & GIVEN DATA:

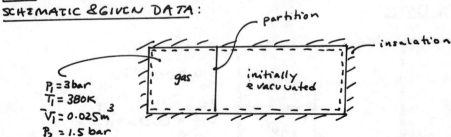

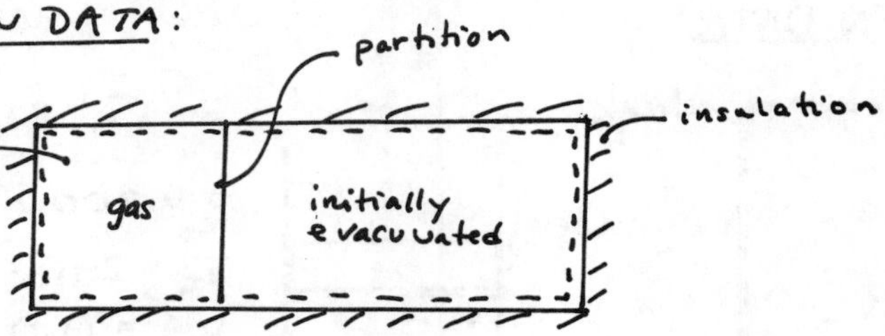

ASSUMPTIONS: 1. The closed system is shown by a dashed line on the above schematic. 2. The gas behaves as an ideal gas. 3. For the process, Q = W = 0, and kinetic and potential energy changes are negligible.

ANALYSIS: Applying the ideal gas equation of state

$$\left.\begin{array}{l} p_1 V_1 = mRT_1 \\ p_2 V_2 = mRT_2 \end{array}\right\rangle \quad \frac{p_2 V_2}{p_1 V_1} = \frac{T_2}{T_1} \Rightarrow V_2 = V_1\left[\frac{p_1}{p_2}\right]\left[\frac{T_2}{T_1}\right]$$

The final temperature is required. Applying an energy balance

$$\cancel{\Delta KE} + \cancel{\Delta PE} + \Delta U = \cancel{Q} - \cancel{W} \Rightarrow \Delta U = 0$$

Since internal energy of an ideal gas depends on temperature alone, $T_2 = T_1$. Thus

$$V_2 = (0.025\,\text{m}^3)\left[\frac{3\,\text{bar}}{1.5\,\text{bar}}\right][1] = 0.05\,\text{m}^3 \quad \longleftarrow$$

PROBLEM 3.104

KNOWN: Air flows through a valve from a rigid tank into a vertical piston-cylinder assembly.

FIND: Determine the total amount of energy transfer by work and heat.

SCHEMATIC & GIVEN DATA:

$m_{1t} = 3$ kg
$P_{1t} = 500$ kPa
$T_t = 290$K

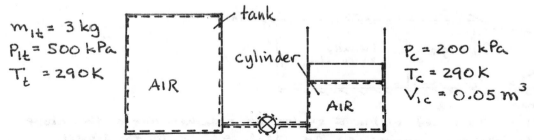

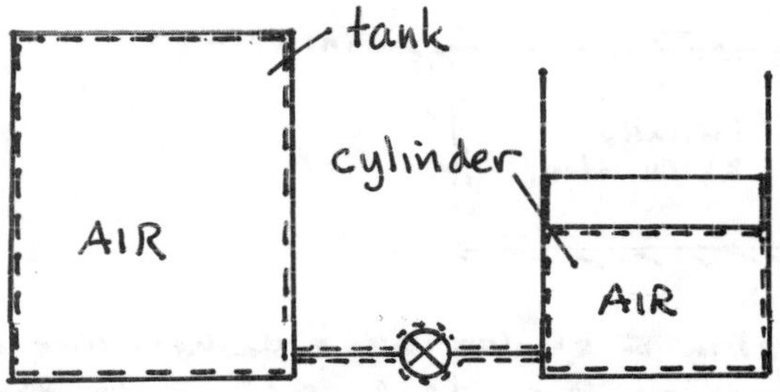

$P_c = 200$ kPa
$T_c = 290$K
$V_{1c} = 0.05\ m^3$

ASSUMPTIONS: (1) The air in the tank and cylinder is a closed system. (2) The air behaves as an ideal gas. (3) The pressure in the cylinder remains constant. (4) Kinetic and potential energy effects are negligible. (5) The temperature of the air is constant.

ANALYSIS: Taking the air in both the tank and the cylinder as the system, the energy balance is

$$\cancel{\Delta KE}^{0} + \cancel{\Delta PE}^{0} + \cancel{\Delta U}^{0} = Q - W$$

where the change in internal energy is zero because the temperature is constant. Thus

$$Q = W$$ ← Q=W

Using $W = \int P_c dV$, with constant pressure P_c

$$Q = P_c (V_2 - V_1)$$

① Now, with the ideal gas model

$$m_{1c} = \frac{P_c V_{1c}}{R T_c} = \frac{(200\ kPa)(0.05\ m^3)}{\left(\frac{8.314}{28.97}\ \frac{kJ}{kg \cdot K}\right)(290K)} \left|\frac{10^3\ N/m^2}{1\ kPa}\right| \left|\frac{1\ kJ}{10^3\ N \cdot m}\right| = 0.12\ kg$$

$$V_1 = V_{1c} + V_t = V_{1c} + \frac{m_{1t} R T_t}{P_{1t}} = 0.05 + \frac{(3)\left(\frac{8.314}{28.97}\right)(290)}{(500)} = 0.549\ m^3$$

$$V_2 = \frac{(m_{tot}) R T_2}{P_2} = \frac{(3 + 0.12)\left(\frac{8.314}{28.97}\right)(290)}{(200)} = 1.298\ m^3$$

Finally

$$Q = (200\ kPa)(1.298 - 0.549)\ m^3 \left|\frac{10^3\ N/m}{1\ kPa}\right| \left|\frac{1\ kJ}{10^3\ N \cdot m}\right| = 149.8\ kJ$$ ← Q

1. The applicability of the ideal gas model can be verified by referring to the compressibility chart.

PROBLEM 3.105

KNOWN: Nitrogen gas undergoes a polytropic expansion in a piston-cylinder assembly.

FIND: Determine the heat transfer using (a) constant specific heat evaluated at 300K, (b) constant specific heat evaluated at 600K, and (c) nitrogen tables.

SCHEMATIC & GIVEN DATA:

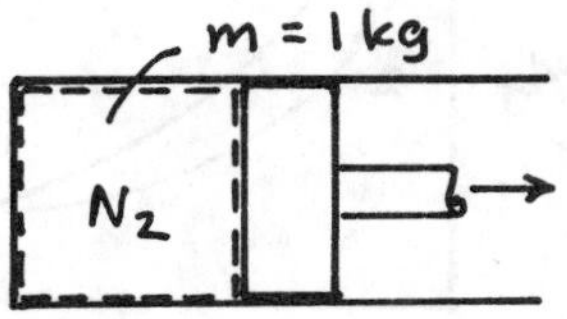

$T_1 = 700$ K
$p_1 = 5$ bars
$p_2 = 2$ bars

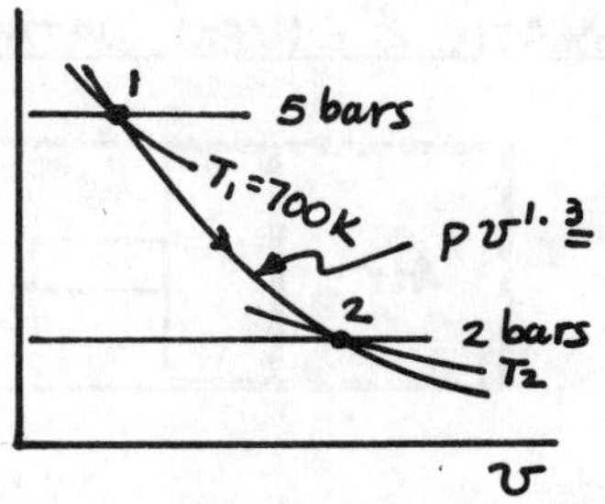

ASSUMPTIONS: (1) The nitrogen is a closed system. (2) The nitrogen behaves as an ideal gas. (3) The process is polytropic, with n = 1.3. (4) Kinetic and potential energy effects are negligible.

ANALYSIS: First, the work is evaluated for the polytropic process. With assumption (2), Eqs. 3.56 and 3.57 apply

$$T_2 = \left(\frac{p_2}{p_1}\right)^{\left(\frac{n-1}{n}\right)} T_1 = \left(\frac{2}{5}\right)^{(.3/1.3)} (700\text{K}) = 566.6\text{ K}$$

and

$$W = \frac{mR(T_2 - T_1)}{1-n} = \frac{(1\text{ kg})\left(\frac{8.314}{28.01}\frac{\text{kJ}}{\text{kg}\cdot\text{K}}\right)(566.6-700)\text{K}}{(1-1.3)} = 132\text{ kJ}$$

To determine the heat transfer, use an energy balance

$$\Delta KE^{\,0} + \Delta PE^{\,0} + \Delta U = Q - W$$

or

$$m(u_2 - u_1) = Q - W$$

$$Q = m(u_2 - u_1) + W$$

(a) From Table A-20 at 300K, $c_v = 0.743$ kJ/kg·K. Thus

$$Q = mc_v(T_2 - T_1) + W$$
$$= (1\text{ kg})(0.743\text{ kJ/kg·K})(566.6-700)\text{K} + (132\text{ kJ})$$
$$= 32.88\text{ kJ} \quad \longleftarrow \quad Q\ (a)$$

(b) From Table A-20 at 600K, $c_v = 0.778$ kJ/kg·K. Thus

$$Q = (1)(0.778)(566.6-700) + (132) = 28.21\text{ kJ} \quad \longleftarrow \quad Q\ (b)$$

(c) Using values for $\bar{u}$ from Table A-23

$$Q = m\left(\frac{\bar{u}_2 - \bar{u}_1}{M}\right) + W$$
$$= (1\text{ kg})\frac{(11{,}850 - 14{,}784)\text{ kJ/kmol}}{(28.01\text{ kg/kmol})} + 132\text{ kJ}$$
$$= 27.25\text{ kJ} \quad \longleftarrow \quad Q\ (c)$$

PROBLEM 3.106

KNOWN: Air undergoes an adiabatic compression followed by constant volume cooling.

FIND: Determine the work for the first process and the heat transfer for the second one, using (a) air table data, (b) constant specific heats.

SCHEMATIC & GIVEN DATA:

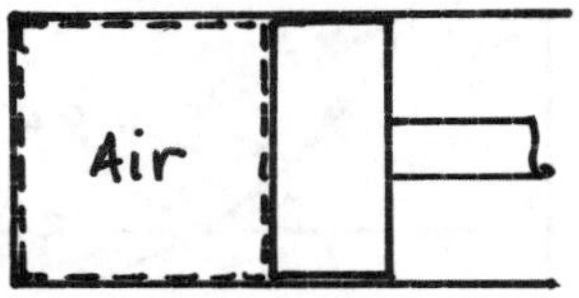

Process 1-2: adiabatic
$p_1 = 1$ bar, $T_1 = 300$K
$p_2 = 15$ bar, $v_2 = .1227$ m³/kg
Process 2-3: Constant volume
$T_3 = 300$K

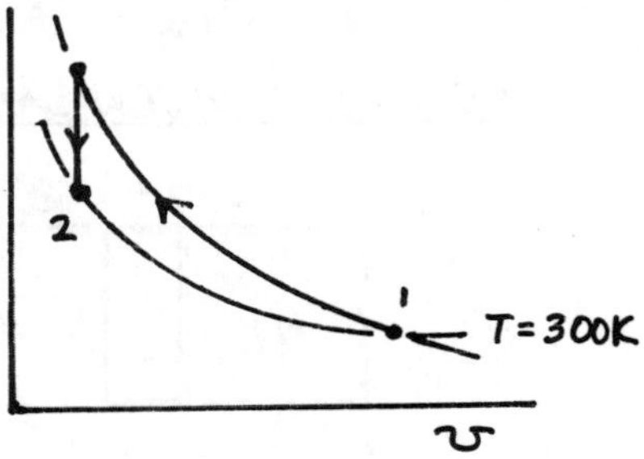

ASSUMPTIONS: (1) The air is a closed system. (2) There is no heat transfer for the first process and no work for the second one. (3) The air behaves as an ideal gas. (4) Kinetic and potential energy effects are negligible.

ANALYSIS: First, using the ideal gas equation of state to determine T_2

$$T_2 = \frac{p_2 v_2}{R} = \frac{(15\text{ bar})(.1227\text{ m}^3/\text{kg})}{\left(\frac{8.314}{28.97}\frac{\text{kJ}}{\text{kg}\cdot\text{K}}\right)}\left|\frac{10^5\text{ N/m}^2}{1\text{ bar}}\right|\left|\frac{1\text{ kJ}}{10^3\text{ N}\cdot\text{m}}\right|$$

$$= 641.3\text{K}$$

(a) Using the energy balance for each process

Process 1-2: $\Delta KE^{\,0} + \Delta PE^{\,0} + m(u_2 - u_1) = Q_{12}^{\,0} - W_{12}$

$$\frac{W_{12}}{m} = u_1 - u_2$$

From Table A-22, $u_1 = 214.07$ kJ/kg and, interpolating, $u_2 = 466.5$ kJ/kg. Thus

$$\frac{W_{12}}{m} = 214.07 - 466.5 = -252.4\text{ kJ/kg} \longleftarrow \frac{W_{12}}{m}$$

Process 2-3: $\Delta KE^{\,0} + \Delta PE^{\,0} + m(u_3 - u_2) = Q_{23} - W_{23}^{\,0}$

$$\frac{Q_{23}}{m} = u_3 - u_2$$

Since $T_3 = T_1$, $u_3 = u_1$ and $Q_{23}/m = W_{12}/m = -252.4$ kJ/kg $\longleftarrow$ Q_{23}/m

(b) Assuming a constant value for c_v

$$\frac{W_{12}}{m} = c_v(T_1 - T_2)$$

From Table A-20, at 300K, $c_v = 0.718$ kJ/kg·K. Thus

$$\frac{W_{12}}{m} = (0.718\text{ kJ/kg}\cdot\text{K})(300 - 641.3)\text{K}$$

$$= -245\text{ kJ/kg} \longleftarrow \frac{W_{12}}{m}$$

and $$\frac{Q_{23}}{m} = \frac{W_{12}}{m} = -245\text{ kJ/kg} \longleftarrow \frac{Q_{23}}{m}$$

PROBLEM 3.107

KNOWN: A system consisting of 2 kg of CO_2 undergoes a power cycle consisting of three processes.

FIND: Skech the cycle on a p-v diagram, and plot the thermal efficiency versus P_2/P_1 ranging from 1.05 to 4.

SCHEMATIC & GIVEN DATA:

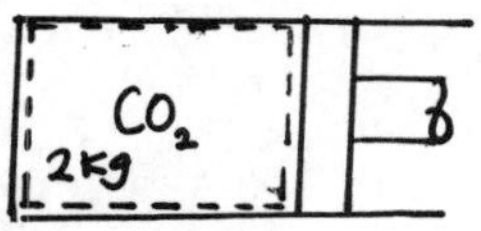

P_1 = 1 bar
T_1 = 300K

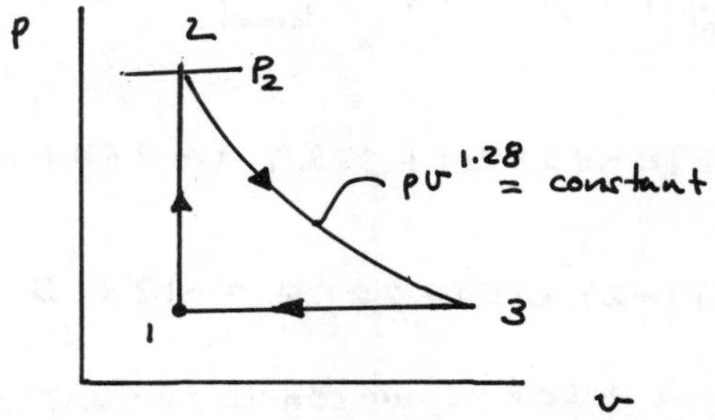

ASSUMPTIONS: 1. The CO_2 is a closed system. 2. The CO_2 behaves as an ideal gas. 3. Process 2-3 is described by $pv^{1.28}$ = constant. 4. Kinetic and potential energy effects are negligible.

ANALYSIS: Begin by developing the key expressions describing the cycle. The thermal efficiency is given by $\eta = W_{net}/Q_{in}$, where $W_{net} = W_{12} + W_{23} + W_{31}$. Since volume is constant $W_{12} = 0$. For process 2-3 and process 3-1

$$W_{23} = m\int_2^3 p\,dv = \frac{m(p_3v_3 - p_2v_2)}{1-n} = \frac{mR(T_3-T_2)}{1-n} \quad (1)$$

$$W_{31} = m\int_3^1 p\,dv = m(P_1 - P_3) = mR(T_1 - T_3) \quad (2)$$

Applying an energy balance:

Process 1-2: $m(u_2-u_1) = Q_{12} - \cancelto{0}{W_{12}} \Rightarrow Q_{12} = m(u_2-u_1)$ (3)

Process 2-3: $m(u_3-u_2) = Q_{23} - W_{23} \Rightarrow Q_{23} = m(u_3-u_2) + W_{23}$ (4)

Process 3-1: $m(u_1-u_3) = Q_{31} - W_{31} \Rightarrow Q_{31} = m(u_1-u_3) + W_{31}$ (5)

Additionally, the ideal gas equation of state gives

$$\left.\begin{array}{l} P_1V = mRT_1 \\ P_2V = mRT_2 \end{array}\right\} \rightarrow \frac{P_2}{P_1} = \frac{T_2}{T_1} \Rightarrow T_2 = \left(\frac{P_2}{P_1}\right)T_1 \quad (6)$$

and with Eq. 3.56 for process 2-3 ($P_3 = P_1$)

$$\frac{T_3}{T_2} = \left(\frac{P_3}{P_2}\right)^{(n-1)/n} \Rightarrow T_3 = T_2\left(\frac{P_1}{P_2}\right)^{(n-1)/n} \quad (7)$$

SAMPLE CALCULATION: $P_2/P_1 = 4$

If $P_2/P_1 = 4$, Eq. (6) gives $T_2 = 4(300) = 1200\text{K}$, and Eq. (7) gives

$$T_3 = (1200\text{K})\left(\frac{1}{4}\right)^{.28/1.28} = 886\text{ K}$$

Eq. (1) then gives

$$W_{23} = \frac{(2\text{ kg})\left(\frac{8.314}{44.01}\frac{\text{kJ}}{\text{kg}\cdot\text{K}}\right)(886-1200)\text{K}}{1-1.28} = 423.7\text{ kJ}$$

PROBLEM 3.107 (Contd.)

Eq.(2) gives

$$W_{31} = (2\,kg)\left(\frac{8.314}{44.01}\frac{kJ}{kg\cdot K}\right)(300-886)\,K = -221.4\,kJ$$

With data from Table A-23, Eq.(3) gives

$$Q_{12} = \left(\frac{2\,kg}{44.01\,kJ/kmol}\right)(43{,}871-6939)\frac{kJ}{kmol} = 1678.3\,kJ$$

Eq.(4) gives

$$Q_{23} = \left(\frac{2}{44.01}\right)(29{,}298-43{,}871) + 423.7 = -238.6\,kJ$$

Eq.(5) gives

$$Q_{31} = \left(\frac{2}{44.01}\right)(6939-29{,}298) - 221.4 = -1237.5\,kJ$$

Since Q_{23} and Q_{31} are both negative, we conclude that $Q_{in} = Q_{12}$. Thus, for the case $p_2/p_1 = 4$

$$\eta = \frac{423.7+(-221.4)}{1678.3} = 0.121\,(12.1\,\%) \longleftarrow \eta\ (p_2/p_1 = 4)$$

Turning now to IT:

```
T1 = 300  // K
p1 = 1  // bar
m = 2  // kg
M = 44.01  // kg/kmol
R = 8.314 / M

W23 = m*R*(T3 - T2) / (1-1.28)
W31 = m*R*(T1 - T3)
m*(u2 - u1) = Q12
Wcycle = W23 + W31
eta = Wcycle / Q12
p2 / p1 = T2 / T1
p2 = r * p1
r = 4
p3 = p1
T3 / T2 = (p3 / p2)^((1.28-1) / 1.28)
u1 = u_T("CO2", T1)
u2 = u_T("CO2", T2)
u3 = u_T("CO2", T3)

/* Using the Explore button, sweep r from
   1.05 to 4 in steps of 0.01.   */
```

IT Results ($r = 4$)

Q_{in} = 1678 kJ
T_2 = 1200 K
T_3 = 886.1 K
W_{23} = 423.6 kJ
W_{31} = -221.4 kJ
η = 0.1205

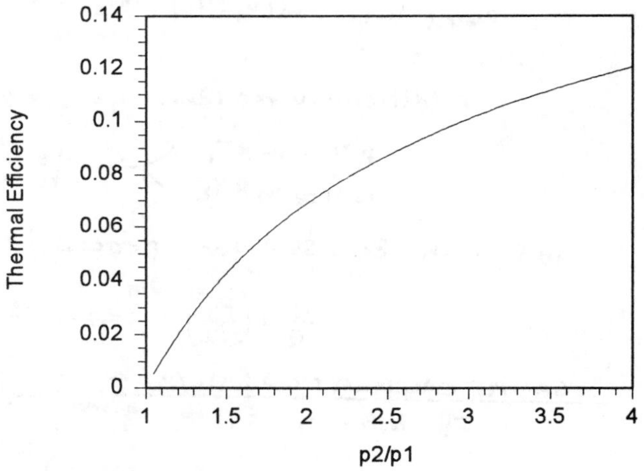

PROBLEM 3.108

KNOWN: Air undergoes a power cycle consisting of three processes.

FIND: Sketch the cycle on a p-v diagram and determine (a) the pressure at state 2, (b) the temperature at state 3, and (c) the thermal efficiency.

SCHEMATIC & GIVEN DATA: The following data are given for each process:

Process 1-2: Constant volume from $p_1 = 20\ \text{lbf/in}^2$, $T_1 = 500°R$ to $T_2 = 820°R$

Process 2-3: Adiabatic expansion to $v_3 = 1.4 v_2$

Process 3-1: Constant-pressure compression

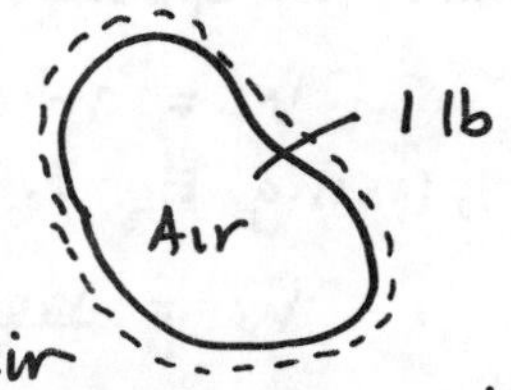

ASSUMPTIONS: (1) The air is the closed system. (2) The air behaves as an ideal gas. (3) Kinetic and potential energy effects are neglected.

ANALYSIS: (a) For process 1-2, the volume is constant. Thus, using the ideal gas equation of state

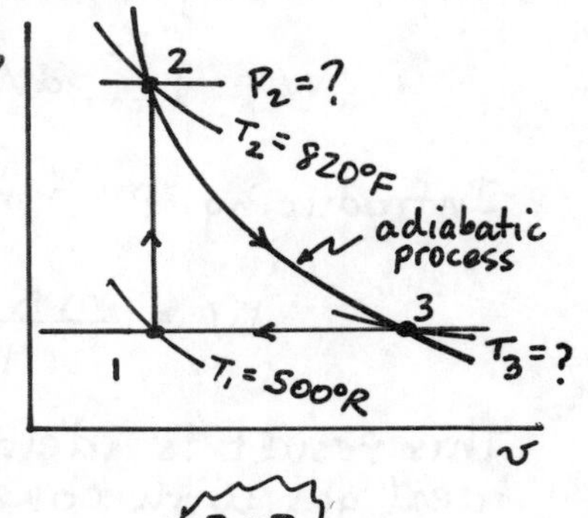

$$\left.\begin{aligned} p_1 V &= mRT_1 \\ p_2 V &= mRT_2 \end{aligned}\right\} \rightarrow \frac{p_1}{p_2} = \frac{T_1}{T_2} \Rightarrow p_2 = \frac{T_2}{T_1} \cdot p_1$$

$$= \frac{820}{500}\left(20\ \frac{\text{lbf}}{\text{in}^2}\right)$$

$$= 32.8\ \frac{\text{lbf}}{\text{in.}^2} \leftarrow p_2$$

(b) Again, using the ideal equation of state

$$\frac{v_3}{v_2} = 1.4 = \frac{mRT_3/p_3}{mRT_2/p_2} = \frac{T_3}{T_2} \cdot \frac{p_2}{p_3} \Rightarrow T_3 = (1.4)\frac{p_3}{p_2} \cdot T_2 \overset{p_3 = p_1}{=} (1.4)\frac{p_1}{p_2} \cdot T_2$$

Thus $T_3 = (1.4)\left(\frac{20}{32.8}\right)(820) = 700°R \leftarrow T_3$

(c) The thermal efficiency is $\eta = W_{cycle}/Q_{in}$. Analyze each process in turn.

Process 1-2: const. volume $\Rightarrow W_{12} = 0$.

Energy Balance: $\cancel{\Delta KE} + \cancel{\Delta PE} + \Delta U = Q_{12} - \cancel{W_{12}} \Rightarrow Q_{12} = m(u_2 - u_1)$

with data from Table A-22

$$Q_{12} = (1\ \text{lb})(140.47 - 85.20)\ \text{Btu/lb} = 55.27\ \text{Btu}$$

Process 2-3: Adiabatic: $m(u_3 - u_2) = \cancel{Q_{23}} - W_{23}$

$$W_{23} = -m(u_3 - u_2) = -(1\ \text{lb})(119.58 - 140.47)\ \text{Btu/lb} = 20.89\ \text{Btu}$$

Process 3-1: p = constant. Using Eq. 2.17 $W_{31} = \int_{V_3}^{V_1} p\,dv = m\,p(v_1 - v_3) = mR(T_1 - T_3)$

$$W_{31} = (1\ \text{lb})\left(\frac{1545}{28.97}\right)\frac{\text{ft}\cdot\text{lbf}}{\text{lb}\cdot°R}\left|\frac{1\ \text{Btu}}{778\ \text{ft}\cdot\text{lbf}}\right|(500 - 700)°R = -13.71\ \text{Btu}$$

Energy Balance: $m(u_1 - u_3) = Q_{31} - W_{31} \Rightarrow Q_{31} = m(u_1 - u_3) + W_{31}$

$$Q_{31} = (1\ \text{lb})(85.20 - 119.58)\ \text{Btu/lb} + (-13.71\ \text{Btu}) = -48.09\ \text{Btu}$$

Thus $W_{cycle} = \cancelto{0}{W_{12}} + W_{23} + W_{31} = 20.87 + (-13.71) = 7.18\ \text{Btu}$

$Q_{in} = Q_{12} = 55.27\ \text{Btu}$

Finally $\eta = W_{cycle}/Q_{in} = 7.18/55.27 = 0.1299\ (12.99\%) \leftarrow \eta$

PROBLEM 3.109

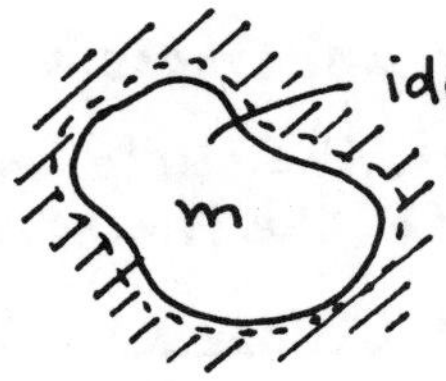

(a) For an adiabatic process with negligible effects of kinetic and potential energy

$$\Delta KE^{0} + \Delta PE^{0} + \Delta U = Q^{0} - W$$

Since the specific heats are constant

$$W = m c_v (T_1 - T_2)$$

or, using Eq. 3.47b, $c_v = R/(k-1)$, so

$$W = \frac{mR(T_2 - T_1)}{1-k}$$ ← (a)

(b) For a polytropic process

$$W = \int_{V_1}^{V_2} p\,dV = \int_{V_1}^{V_2} \frac{\text{const}}{V^n}\,dV = \frac{p_2V_2 - p_1V_1}{1-n}$$

Introducing $PV = mRT$ and letting n be the specific heat ratio k

$$W = \frac{mR(T_2 - T_1)}{1-k}$$

This result is identical to the result of part (a). Thus, for an ideal gas with constant specific heats, the polytropic process pV^k = constant corresponds to an adiabatic process. ← (b)

PROBLEM 3.110

KNOWN: Air undergoes a polytropic process from a known initial state to a given final pressure.

FIND: Plot the heat transfer and work for $1.0 \leq n \leq 1.6$. Investigate the error in heat transfer introduced by assuming constant c_v evaluated at T_1.

SCHEMATIC & GIVEN DAT :

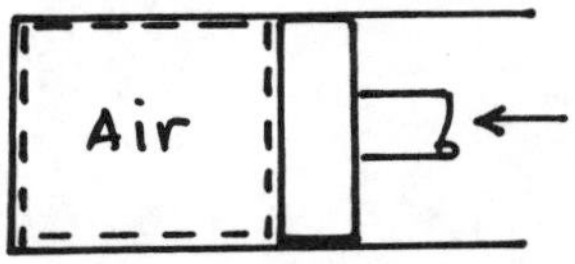

$p_1 = 14.7$ lbf/in.²
$T_1 = 70°F$
$p_2 = 100$ lbf/in.²
$pv^n = \text{constant}$

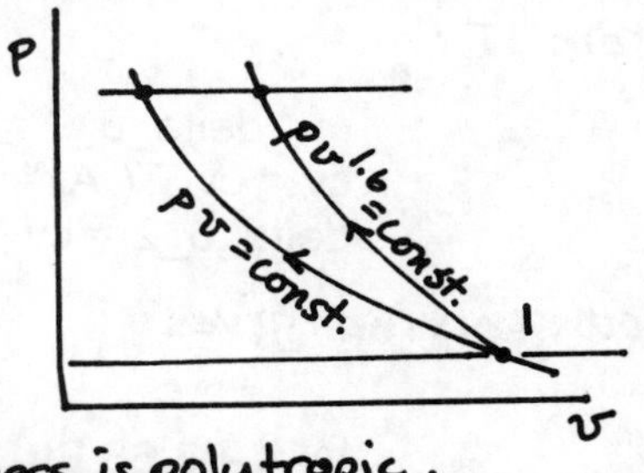

ASSUMPTIONS: (1) The air is a closed system. (2) The process is polytropic. (3) The air behaves as an ideal gas. (4) Neglect kinetic and potential energy effects.

Begin by fixing state 2. From the polytropic process relation

$$p_1 v_1^{\,n} = p_2 v_2^{\,n} \Rightarrow v_2 = (p_1/p_2)^{1/n} v_1$$

Thus, with $v_1 = RT_1/p_1$, and $T_2 = p_2 v_2/R$, state 2 is determined for any value of n. These expressions are evaluated using IT as follows:

```
p1 = 14.7  // lbf/in.²
T1 = 70  // °F
p2 = 100  // lbf/in.²
m = 1  // lb
n = 1.6

p1 * v1^n = p2 * v2^n
v1 = v_TP("Air",T1,p1)
v2 = v_TP("Air",T2,p2)
```

IT Results ($n = 1.6$)

T2 = 627.4 °F
v1 = 13.34 ft³/lb
v2 = 4.026 ft³/lb

The work is evaluated using Eq. 2.17 and Eq. 3.54

$$W = \int_{V_1}^{V_2} p\,dV = m\int_{v_1}^{v_2} p\,dv = \frac{m(p_2 v_2 - p_1 v_1)}{(1-n)} \quad (n \neq 1)$$

Or, as expressed using IT

```
W/m =( (p2 * v2 - p1 * v1) / (1 - n)) * (144 / 778)  // Btu/lb
```

To find Q, use the energy balance: $\cancel{\Delta KE} + \cancel{\Delta PE} + \Delta U = Q - W$. or

$$m\,\Delta u = Q - W$$

<u>Method 1</u>

To evaluate Δu, use data from Table A-22

$$\Delta u = u(T_2) - u(T_1)$$

From IT:

```
m * delta_u_1 = Q_1 - W
u1 = u_T("Air", T1)
u2 = u_T("Air", T2)
delta_u_1 = u2 - u1
```

PROBLEM 3.110 (Contd.)

Method 2

An alternative method to evaluate Δu is to use a constant c_v value from Table A-20 determined at $T_1 = 70°F$. That is

$$\Delta u = c_v (T_2 - T_1)$$

From IT:

```
m * delta_u_2 = Q_2 - W
cv = cv_T("Air", T1)
delta_u_2 = cv * (T2 - T1)
```

Now, solving gives

W = -63.68 Btu

u1 = 90.38 Btu/lb
u2 = 188.3 Btu/lb
delta_u_1 = 97.88 Btu
Q_1 = 34.19 Btu

cv = 0.1711 Btu/lb·°R
delta_u_2 = 95.38 Btu
Q_2 = 31.7 Btu

① Using the Explore button, sweep n from 1.001 to 1.6 in steps of 0.01. The data obtained give the following graphs:

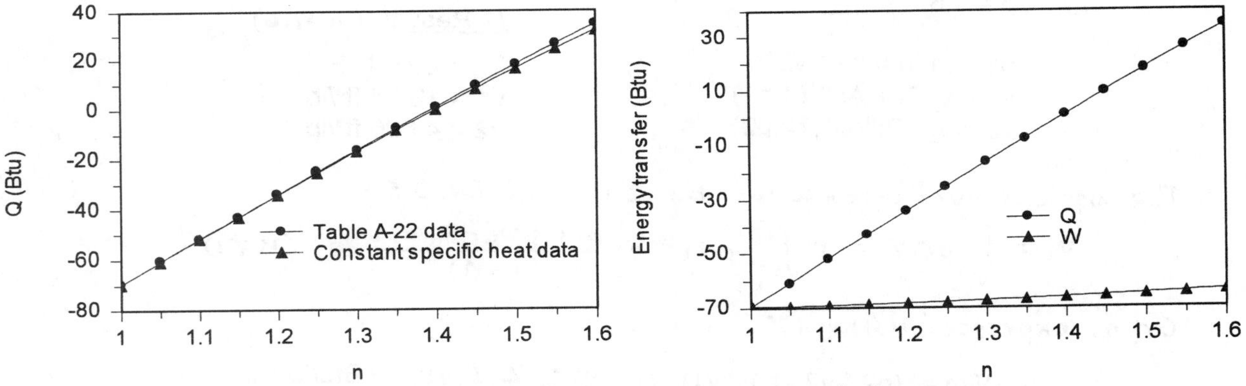

From the graph of Q and W vs. n, we can see several effects:

- Q and W become equal as $n \to 1$, since for $n = 1$ the temperature is constant, and hence $\Delta u = 0$.
- Q changes sign, going from negative (heat out) to positive (heat in) at about $n = 1.4$. W is negative (work in) for all cases.
- W increases only slightly, whereas Q changes significantly.

From the graph of Q vs. n we see that the solutions become identical as $n \to 1$ (constant temperature) and diverge slightly as $n \to 1.6$. For $n = 1.6$, the percentage deviation in Q is about 7.3%.

---

1. By using $n = 1.001$, we avoid the case of $n = 1$ for which Eq. 3.54 cannot be used. However, the slight deviation from $n = 1$ cannot be seen on the plots.

PROBLEM 3.111

KNOWN: Steam undergoes a polytropic process from a known initial state to a given final pressure.

FIND: Plot the heat transfer per unit mass of steam for $1.0 \leq n \leq 1.6$. Investigate the error in heat transfer introduced by assuming the ideal gas model.

SCHEMATIC & GIVEN DATA:

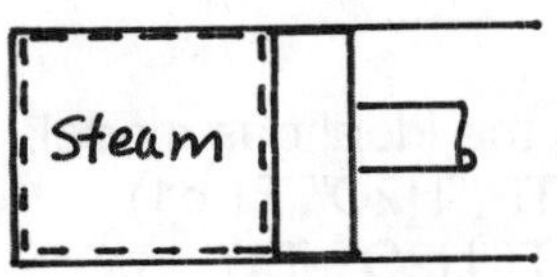

$p_1 = 5\text{MPa} = 50 \text{ bar}$
$T_1 = 280°C$
$p_2 = 20 \text{ MPa} = 200 \text{ bar}$
$pv^n = \text{constant}$

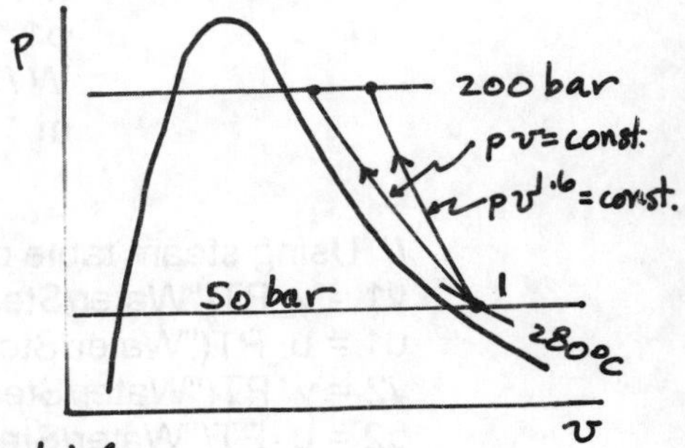

ASSUMPTIONS: (1) The steam is a closed system. (2) The process is polytropic. (3) Kinetic and potential energy effects are neglected.

ANALYSIS: The following relations are used to determine the heat transfer, Q/m:

Process relation: $p_1 v_1^{\,n} = p_2 v_2^{\,n}$ (1)

Work expression (Eqs. 2.17 and 3.54): $\frac{W}{m} = \frac{(p_2 v_2 - p_1 v_1)}{1-n}$ (2)

Energy balance: $\Delta KE + \Delta PE + \Delta U = Q - W$ (KE and PE struck out)

$\Rightarrow \quad (u_2 - u_1) = \frac{Q}{m} - \frac{W}{m}$ (3)

When using steam table data (IT functions or Table A-4):

- $p_1 = 50 \text{ bar}, T_1 = 280°C \Rightarrow v_1, u_1$
- with $p_2 = 200$ bar and (2), v_2 can be determined.
- $p_2, v_2 \Rightarrow u_2, T_2$

When using the ideal gas model (IT "H_2O" functions or Table A-23):

- $p_1 = 50 \text{ bar}, T_1 = 280°C \Rightarrow v_1 = \frac{RT_1}{p_1}$
- $T_1 = 280°C \Rightarrow u(T_1)$
- with $p_2 = 200$ bar and (2), v_2 can be determined.
- $p_2, v_2 \Rightarrow T_2 = \frac{p_2 v_2}{R}$
- $T_2 \Rightarrow u(T_2)$

With either of these schemes, it is then possible to use (2) and (3) to solve for W/m and Q/m.

The corresponding IT code is given on the next page.

PROBLEM 3.111 (Contd.)

```
p1 = 50  // bar
T1 = 280  // °C
p2 = 200  // bar
m = 1 // kg
n = 1.6
p1 * v1^n = p2 * v2^n
W / m = ((p2 * v2 - p1 * v1) / (1 - n)) * 100
m * (u2 - u1) = Q - W
```

```
// Using steam table data:
v1 = v_PT("Water/Steam", p1, T1)
u1 = u_PT("Water/Steam", p1, T1)
v2 = v_PT("Water/Steam", p2, T2)
u2 = u_PT("Water/Steam", p2, T2)
```

```
// Using the ideal gas model:
v1 = v_TP("H2O",T1,p1)
u1 = u_T("H2O", T1)
v2 = v_TP("H2O",T2,p2)
u2 = u_T("H2O", T2)
```

IT Results for n = 1.6

quantity	steam table data	Ideal gas model
Q/m (kJ/kg)	251.9	320.8
T_2 (°C)	583.7	652.1
W/m (kJ/kg)	-240.8	-291
u_1 (kJ/kg)	2645	-1.319E4
u_2 (kJ/kg)	3138	-1.257E4
v_1 (m³/kg)	0.04224	0.05104
v_2 (m³/kg)	0.01766	0.02135

① Using the Explore button, sweep n from 1.001 to 1.6 in steps of 0.01. The data obtained give the following plot:

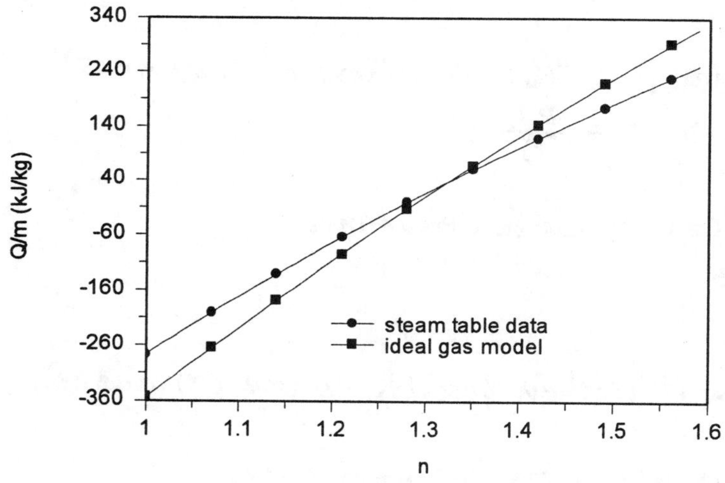

Discussion:

- Q/m varies significantly with n.
- Q/m changes sign, going from negative (heat out) to positive (heat in).
- The steam table and ideal gas solutions exhibit the same general trends. However, the values for Q/m differ significantly in some ranges of n.

---

1. By using n = 1.001, we avoid the case of n = 1 for which (2) above cannot be used. However, the slight deviation from n = 1 cannot be seen on the plot.

CHAPTER FOUR
CONTROL VOLUME ENERGY ANALYSIS

Chapter 4 - Third and fourth edition problem correspondence

3rd	4th	3rd	4th	3rd	4th
4.1	4.1*	4.36	4.37	---	4.73
4.2	4.2*	4.37	4.38*	4.66	4.74*
4.3	4.3	4.38	4.39*	4.67	4.75
4.4	4.4	4.39	4.40	4.68	---
4.5	4.5*	4.40	4.41*	4.69	---
4.6	4.6	4.41	4.42*	---	4.76
4.7	4.7*	4.42	---	4.70	4.77
4.8	---	---	4.43	4.71	---
4.9	---	4.43	4.44	---	4.78
---	4.8	4.44	4.45	---	4.79
---	4.9	4.45	4.46*	4.72	4.80
4.10	4.10	4.46	4.47	4.73	---
4.11	4.11*	4.47	4.48*	---	4.81
4.12	---	4.48	4.49*	4.74	4.82
---	4.12	4.49	4.50	4.75	---
4.13	4.13	4.50	4.51*	---	4.83
4.14	4.14	4.51	---	4.76	4.84*
4.15	4.15*	---	4.52	4.77	4.85*
4.16	---	---	4.53	4.78	4.86*
---	4.16	4.52	4.54	4.79	---
4.17	4.17*	4.53	---	4.80	4.87a
4.18	4.18	---	4.55	4.81	4.87b
4.19	4.19*	4.54	---	4.82	4.88
4.20	---	---	4.56	---	4.89
---	4.20	4.55	---	---	4.90
---	4.21	---	4.57	---	4.91
4.21	4.22	4.56	4.58	4.83	---
4.22	4.23*	4.57	4.59*	---	4.92
4.23	4.24	4.58	4.60	4.84	4.93
4.24	4.25	---	4.61	4.85	---
4.25	---	4.59	4.62	---	4.94
---	4.26	4.60	4.63	4.86	4.95
4.26	4.27	4.61	---	4.87	---
4.27	4.28	---	4.64	---	4.96
4.28	4.29	4.62	4.65*	4.88	4.97
4.29	---	4.63	---	4.89	4.98*
---	4.30	---	4.66	4.90	4.99
4.30	4.31	4.64	4.67	4.91	4.100*
4.31	4.32	4.65	4.68	4.92	4.101
4.32	4.33	---	4.69	4.93	4.102
4.33	4.34	---	4.70	4.94	4.103*
4.34	4.35*	---	4.71		
4.35	4.36	---	4.72		

----

* Revised

PROBLEM 4.1

KNOWN: The mass flow rates into and out of a control volume are known, and the initial mass is given.

FIND: (a) Plot the inlet and exit mass flow rates, the rate of change of mass, and the amount of mass in the control volume, each versus time (b) Estimate the time, in h, when the tank is nearly empty.

SCHEMATIC & GIVEN DATA:

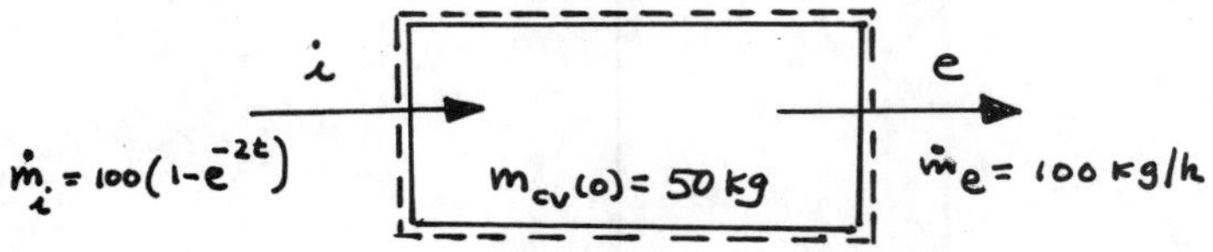

ASSUMPTIONS: (a) As shown in the schematic, a control volume having one inlet and one exit is under consideration.

ANALYSIS: The inlet and exit mass flow rates are given

$$\dot{m}_i = 100(1-e^{-2t}) \text{ kg/h} \quad , \dot{m}_e = 100 \text{ kg/h} \qquad (1)$$

The instantaneous time rate of change of mass is obtained using Eq. 4.6, the mass rate balance

$$\frac{dm_{cv}}{dt} = \dot{m}_i - \dot{m}_e = 100(1-e^{-2t}) - 100$$
$$= -100\, e^{-2t} \qquad (2)$$

The mass contained in the control volume at time t is obtained by integrating Eq. (2)

$$m_{cv}(t) - m_{cv}(0) = -100\int_0^t e^{-2t}dt = 50e^{-2t}\Big]_0^t$$

$$\Rightarrow \quad m_{cv}(t) = 50[e^{-2t} - 1] + 50 = 50e^{-2t} \qquad (3)$$

Plotting Eqs. (1) using IT gives the inlet and exit mass flow rates vs. time:

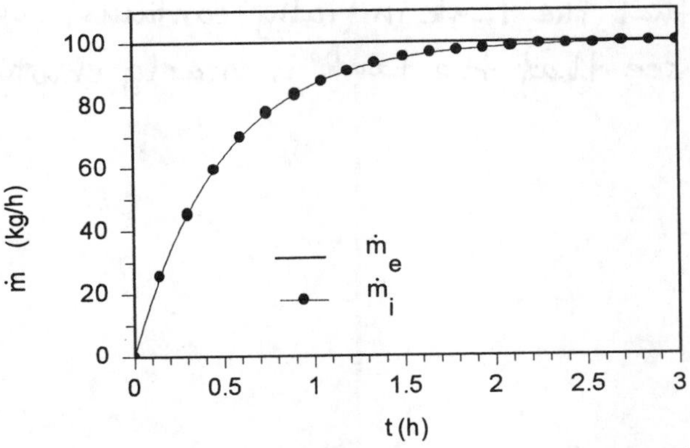

From this plot we see that initially mass flows out at a high rate, and no mass flows in. Eventually, the incoming flow rate equals the exit flow rate.

PROBLEM 4.1 (Contd.)

The following plot is obtained for the instantaneous rate of change of mass:

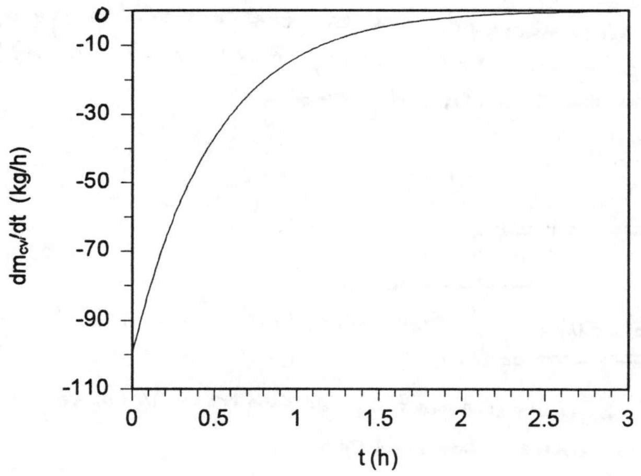

Initially, the amount of mass in the tank is decreasing at a rapid rate. Eventually the rate of change (decrease) of mass slows to zero.

The following plot is obtained for the amount of mass in the tank as a function of time:

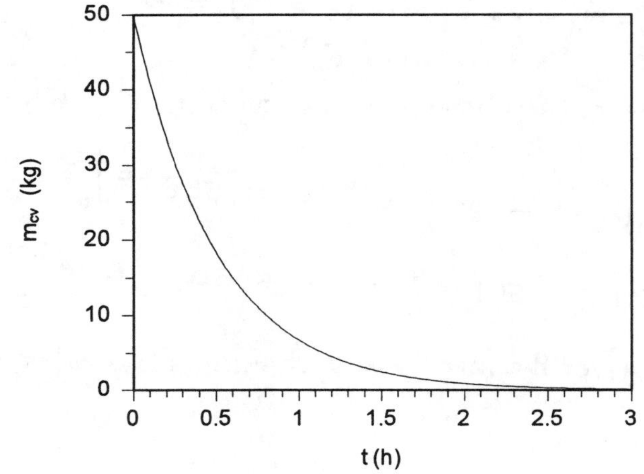

From this plot, we see that the tank initially contains 50 kg and empties.

(b) From the plot, we see that the tank is nearly empty at $t \approx 2.5$ h.

PROBLEM 4.2

<u>KNOWN</u>: Mass flows into and out of a control volume with known flow rates.

<u>FIND</u>: Plot the time rate of change of mass and the net change in the amount of mass versus time.

<u>SCHEMATIC AND GIVEN DATA</u>:

<u>ANALYSIS</u>: The rate of change of mass at time t is

$$\frac{dm_{cv}}{dt} = \dot{m}_i - \dot{m}_e \Rightarrow \frac{dm_{cv}}{dt} = 1.5 - 1.5(1 - e^{-.002t})\ \text{kg/s} \qquad (1)$$

$$= 1.5\, e^{-.002t}$$

Integrating the mass rate balance over time

$$\int dm_{cv} = \int_0^t (\dot{m}_i - \dot{m}_e)\,dt \qquad (2)$$

or

$$\Delta m_{cv}(t) = \int_0^t [1.5\, e^{-.002t'}]\,dt' = \left[\frac{1.5\, e^{-.002t'}}{(-0.002)}\right]_0^t$$

$$= 750(1 - e^{-.002t})$$

<u>*IT* program</u>

```
mdot_i = 1.5  // kg/s
mdot_e = 1.5 * (1 - exp(-.002*t))  // kg/s
dmcvdt = mdot_i - mdot_e
delta_m = Integral(dmcvdt,t)
```

Using the Explore button, sweep t from 0 to 3600 s in steps of 4.

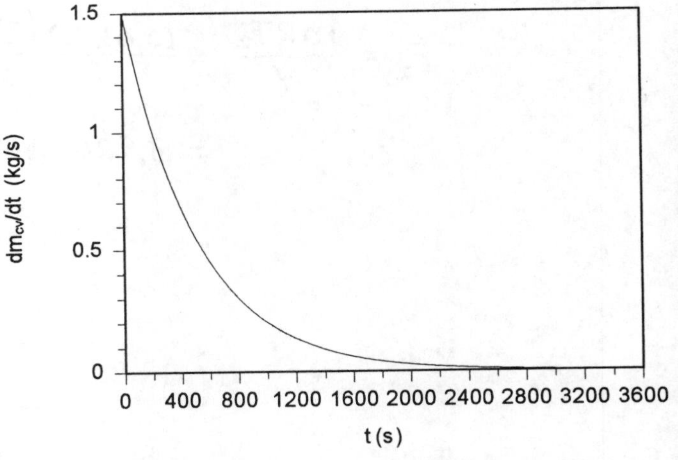

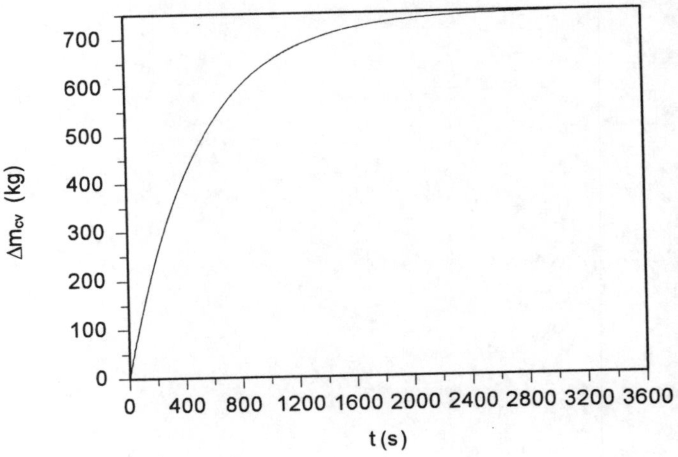

From the upper graph and Eq. (1) we see that initially the inlet flow rate is high and the exit flow rate is zero. Thus, dm_{cv}/dt starts high. As t increases, the exit flow rate increases, and dm_{cv}/dt approaches zero as $t \to 2400$ s.

The bottom graph shows that the net change in mass increases until it approaches 750 kg. At that point there is little change in mass (t > 2400 s).

PROBLEM 4.3

KNOWN: Air leaks into an initially evacuated tank at a constant mass flow rate.

FIND: Determine the pressure in the tank after 20 s.

SCHEMATIC & GIVEN DATA:

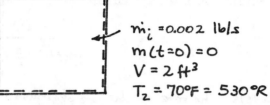

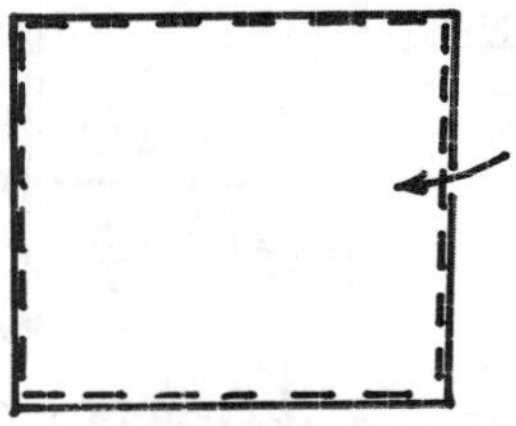

$\dot{m}_i = 0.002$ lb/s

$m(t=0) = 0$

$V = 2$ ft^3

$T_2 = 70°F = 530°R$

ASSUMPTIONS: (1) The control volume has one inlet and no exits. (2) The air behaves as an ideal gas.

ANALYSIS: The mass rate balance, Eq. 4.6, reduces to

$$\frac{dm_{cv}}{dt} = \dot{m}_i \Rightarrow dm_{cv} = \dot{m}_i \, dt$$

Integrating

$$\int_0^m dm_{cv} = \int_0^{20} \dot{m}_i \, dt = \int_0^{20} (0.002) \, dt$$

$$m(t=20) = (0.002)(20) = 0.04 \text{ lb}$$

Now, using the ideal gas model

$$P_2 = \frac{mRT_2}{V} = \frac{(0.04 \text{ lb})\left(\frac{1545}{28.97} \frac{\text{ft}\cdot\text{lbf}}{\text{lb}\cdot°\text{R}}\right)(530°\text{R})}{(2 \text{ ft}^3)} \left| \frac{1 \text{ ft}^2}{144 \text{ in}^2} \right|$$

$$= 3.926 \text{ lbf/in}^2 \leftarrow P_2$$

PROBLEM 4.4

KNOWN: A tank containing Refrigerant 134a develops a leak. Refrigerant escapes at a constant mass flow rate.

FIND: Determine the time at which half the mass has escaped and the pressure in the tank at that time.

SCHEMATIC & GIVEN DATA:

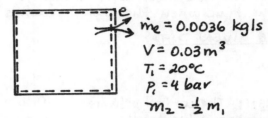

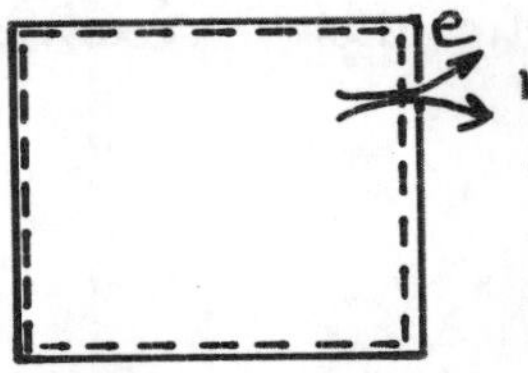

ASSUMPTIONS: (1) The control volume is shown on the accompanying diagram. (2) The temperature in the tank remains constant. (3) The volume is constant.

ANALYSIS: Applying the mass rate balance

$$\frac{dm_{cv}}{dt} = -\dot{m}_e = -0.0036 \text{ kg/s}$$

Integrating over time

$$\int_{m_{cv}(0)}^{m_{cv}(t)} dm_{cv} = -\int_0^t \dot{m}_e \, dt = -0.0036\, t$$

Solving for t

$$t = \frac{m_{cv}(t) - m_{cv}(0)}{(-0.0036)}$$

From Table A-12, $v_1 = 0.05397$ m³/kg. Thus, the initial mass is

$$m_{cv}(0) = \frac{V}{v_1} = 0.5559 \text{ kg}$$

and

$$m_{cv}(t) = \frac{1}{2} m_{cv}(0) = 0.27795 \text{ kg}$$

Thus

$$t = \frac{(0.27795 - 0.5559)\text{ kg}}{(-0.0036 \text{ kg/s})} = 77.2 \text{ s} \longleftarrow t$$

Now, to get p_2, determine v_2

$$v_2 = \frac{V}{m_{cv}(t)} = 0.10793 \text{ m}^3/\text{kg}$$

Interpolating in Table A-12 at $T_2 = 20°C$, $v_2 = 0.10793$ m³/kg

$$p_2 \approx 2.12 \text{ bar} \longleftarrow p_2$$

PROBLEM 4.5

KNOWN: Water enters at the top of a tank through a supply pipe at a constant mass flow rate and exits to a pump through a pipe in the bottom of the tank. The velocity of the water exiting to the pump varies with height of the water surface. An overflow pipe stands in the tank.

FIND: Plot (a) the height of the water surface, (b) the rate mass exits to the pump, (c) the rate mass exit through the overflow pipe, each versus time. Determine the time when the water reaches the top of the overflow pipe.

SCHEMATIC & GIVEN DATA:

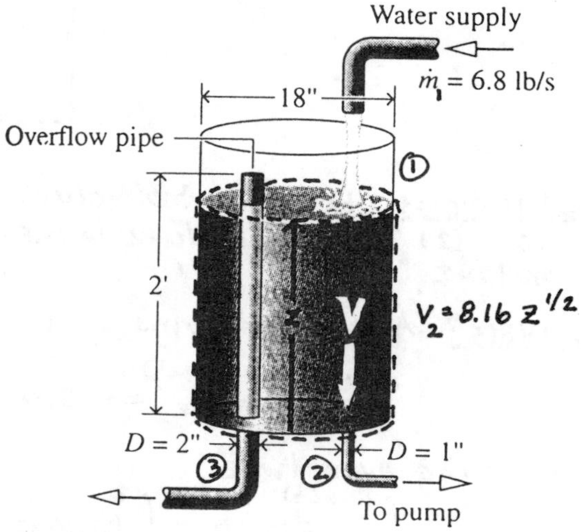

ASSUMPTIONS: (1) The control volume encloses the water in the tank. The control volume is at steady state when the water level reaches the top of the overflow pipe. (2) The water is incompressible, with $\rho = 1/v_{f\,@\,60°F} = 62.34$ lb/ft³. (3) Initially, the tank is empty.

ANALYSIS: (a) Before the water level reaches z = 2 ft, the mass rate balance is

$$\frac{dm_{cv}}{dt} = \dot{m}_1 - \dot{m}_2 \Rightarrow \rho A_{tank} \frac{dz}{dt} = \dot{m}_1 - \rho A_2 V_2$$

Evaluating the areas:

$$A_2 = \frac{\pi D_2^2}{4} = \frac{\pi (1/12 \text{ ft})^2}{4} = 5.454 \times 10^{-3} \text{ ft}^2$$

$$A_{tank} = \frac{\pi D_{tank}^2}{4} = \frac{\pi (18/12 \text{ ft})^2}{4} = 1.767 \text{ ft}^2$$

Thus

$$\left(62.34 \frac{\text{lb}}{\text{ft}^3}\right)(1.767 \text{ ft}^2)\frac{dz}{dt} = 6.8 \frac{\text{lb}}{\text{s}} - \left(62.34 \frac{\text{lb}}{\text{ft}^3}\right)(5.454\times10^{-3} \text{ ft}^2)\, 8.16\, z^{1/2}$$

$$\Rightarrow \frac{dz}{dt} = 0.06173 - 0.02519\, z^{1/2} \quad (z \text{ in ft}, t \text{ in s}) \qquad (*)$$

Equation (*) is a non-linear differential equation. By assumption (3), the initial condition is z = 0 at t = 0. The expression can be integrated using the integration feature of IT, as follows:

IT Program

```
f = 0.06173 - 0.02519 * z^0.5
der(z,t) = f
```

Using the Explore button, sweep t from 0 to 60 s in steps of 0.1.

The following graph can be obtained from the resulting data using IT:

PROBLEM 4.5 (Contd.)

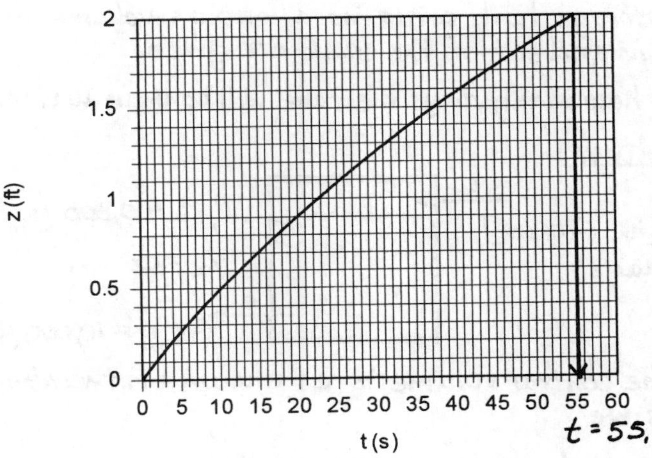

$t = 55.2$ s ← time to overflow

($z = 2$ ft)

(b) For $t < 55.2$ s, $\dot{m}_2$ varies with z as follows:

$$\dot{m}_2 = \rho A_2 V_2 = \rho A_2 (8.16 z^{1/2})$$

$$= (62.34)(5.454 \times 10^{-3})(8.16 z^{1/2}) = 2.7744 z^{1/2}$$

at $t = 55.2$ s; $z = 2$ ft and $\dot{m}_2 = 3.9236$ lb/s

For $t > 55.2$ s; $\dot{m}_2 = 3.9236$ lb/s (z remains constant at 2 ft) ← $\dot{m}_2$

(c) For $t < 55.2$ s; $\dot{m}_3 = 0$ (flow hasn't reached the overflow pipe inlet)

For $t > 55.2$ s; steady state. Thus

$$\frac{dm_{cv}}{dt}^{\,0} = \dot{m}_1 - \dot{m}_2 - \dot{m}_3 \Rightarrow \dot{m}_3 = \dot{m}_1 - \dot{m}_2$$

$$= 6.8 \text{ lb/s} - 3.9236 \text{ lb/s}$$

$$= 2.876 \text{ lb/s}$$ ← $\dot{m}_3$ ($t > 55.2$)

The following plot is constructed using IT:

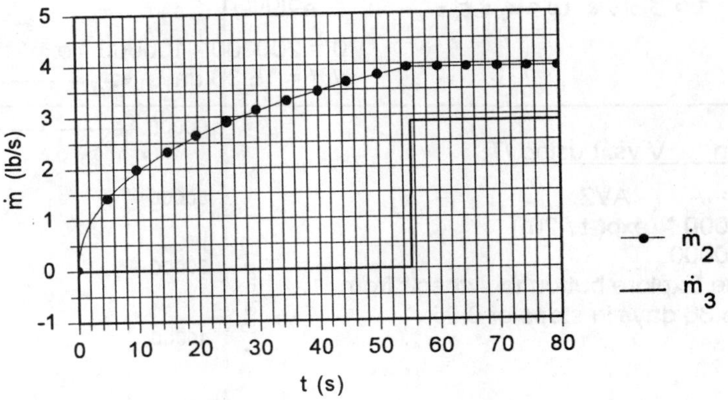

PROBLEM 4.6

KNOWN: A water storage tank contains a known volume of water. The volume flow rates in and out of the tank are given.

FIND: Determine how many days the tank will contain water.

SCHEMATIC & GIVEN DATA:

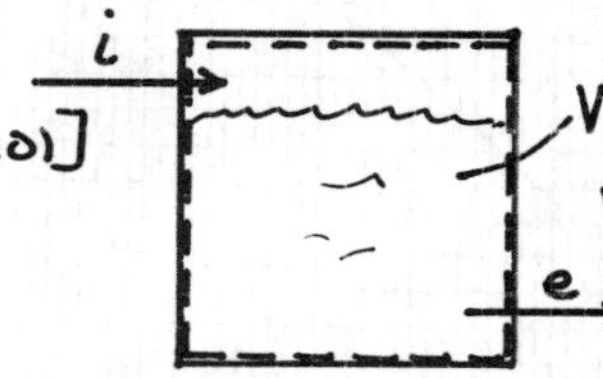

ASSUMPTIONS: (1) The control volume is as shown on the above sketch. (2) The water is incompressible.

ANALYSIS: The mass rate balance reduces to

$$\frac{dm_{cv}}{dt} = \dot{m}_i - \dot{m}_e$$

With $m_{cv} = \rho V$, and $\dot{m}_i = \rho(AV)_i$ and $\dot{m}_e = \rho(AV)_e$

$$\rho\frac{dV}{dt} = \rho(AV)_i - \rho(AV)_e$$

$$\Rightarrow \quad \frac{dV}{dt} = (AV)_i - (AV_e) = 5000[\exp(-t/20)] - 10{,}000$$

Integrating from $t=0$ to $t=t_f$

$$V_2 - V_1 = \int_0^{t_f} (5000[\exp(-t/20)] - 10{,}000)\,dt$$

$$= \left(\frac{5000}{(-1/20)}\exp(-t/20) - 10{,}000t\right)\Bigg|_0^{t_f}$$

① $$= [-100{,}000\exp(-t_f/20) - 1] - 10{,}000t_f \qquad (*)$$

With $V_1 = 100{,}000$ gal and for $V_2 = 0$

$$0 = 200{,}000 - 100{,}000\exp(-t_f/20) - 10{,}000t_f$$

Using IT to solve we get:

IT solution:

```
0 = 200000 - 100000 * exp(-tf/20) - 10000 * tf
// tf = 15.36 days  <-- tf
```

---

1. To obtain V vs. t using *IT*:

```
der(V,t) = AV1 - AV2
AV1 = 5000 * (exp(-t / 20))
AV2 = 10000
// Use the Explore button to sweep t from
// 0 to 15.36 days in steps of 0.05.
```

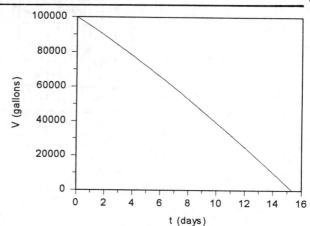

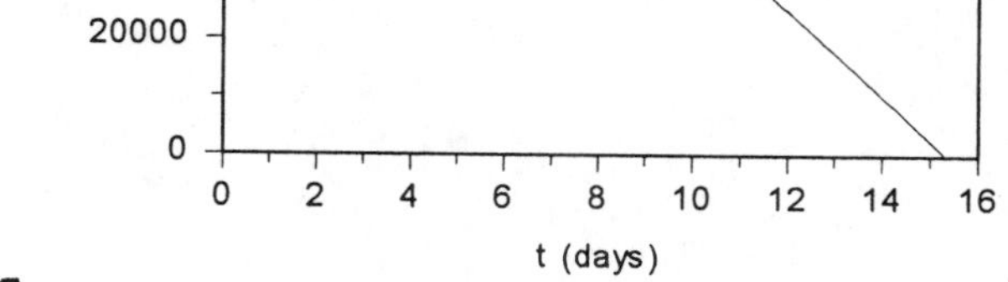

PROBLEM 4.7

KNOWN: A pipe carrying an incompressible liquid contains an expansion chamber.

FIND: (a) Develop an expression for the rate of change of liquid level in the chamber in terms of certain quantities. (b) Compare the relative magnitudes of the mass flow rates for specified values for dL/dt, the rate of change of liquid level.

SCHEMATIC & GIVEN DATA:

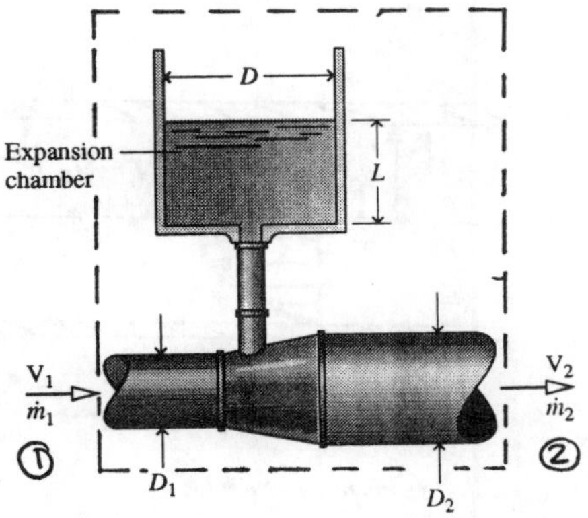

ASSUMPTIONS: 1. The control volume is shown on the accompanying schematic. 2. The liquid is modeled as incompressible. 3. Flow is one-dimensional at 1, 2.

ANALYSIS: (a) The mass rate balance for the control volume is

$$\frac{dm_{cv}}{dt} = \dot{m}_1 - \dot{m}_2$$

with

$$m_{cv} = \rho\left(\frac{\pi D^2}{4}\right)L$$

$$\dot{m}_1 = \rho\left(\frac{\pi D_1^2}{4}\right)V_1$$

$$\dot{m}_2 = \rho\left(\frac{\pi D_2^2}{4}\right)V_2$$

the mass rate balance becomes

$$\rho\left(\frac{\pi D^2}{4}\right)\frac{dL}{dt} = \rho\left(\frac{\pi D_1^2}{4}\right)V_1 - \rho\left(\frac{\pi D_2^2}{4}\right)V_2$$

or, solving for dL/dt and simpifying

$$\frac{dL}{dt} = \frac{D_1^2 V_1 - D_2^2 V_2}{D^2} \longleftarrow dL/dt$$

(b) The mass flow rate expressions indicate $\dot{m}_1 \sim D_1^2 V_1$, $\dot{m}_2 \sim D_2^2 V_2$.

Thus,

$$\frac{dL}{dt} > 0 \Rightarrow D_1^2 V_1 > D_2^2 V_2 \Rightarrow \dot{m}_1 > \dot{m}_2$$

$$\frac{dL}{dt} = 0 \Rightarrow D_1^2 V_1 = D_2^2 V_2 \Rightarrow \dot{m}_1 = \dot{m}_2$$

$$\frac{dL}{dt} < 0 \Rightarrow D_1^2 V_1 < D_2^2 V_2 \Rightarrow \dot{m}_1 < \dot{m}_2$$

PROBLEM 4.8

KNOWN: Velocity distributions are given for laminar and turbulent flow of an incompressible liquid in a circular pipe.

FIND: For each distribution, plot V/V_0 vs. r/R, derive expressions for the mass flow rate, average flow velocity, and specific kinetic energy. Determine the percent error if the specific kinetic energy is evaluated in terms of average velocity. Discuss.

SCHEMATIC & GIVEN DATA:

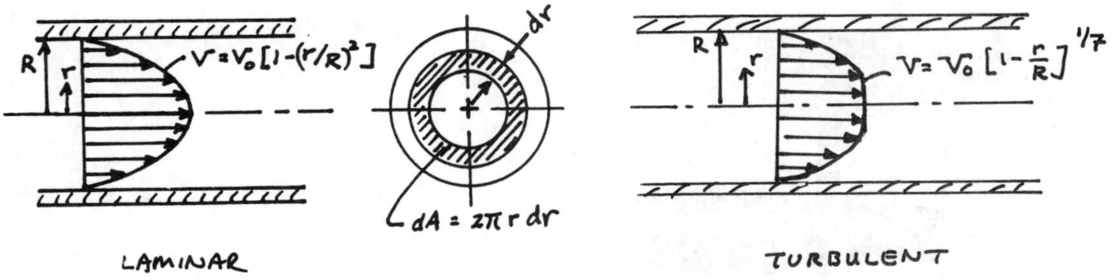

ASSUMPTIONS: The liquid is modeled as incompressible.

ANALYSIS: (a)

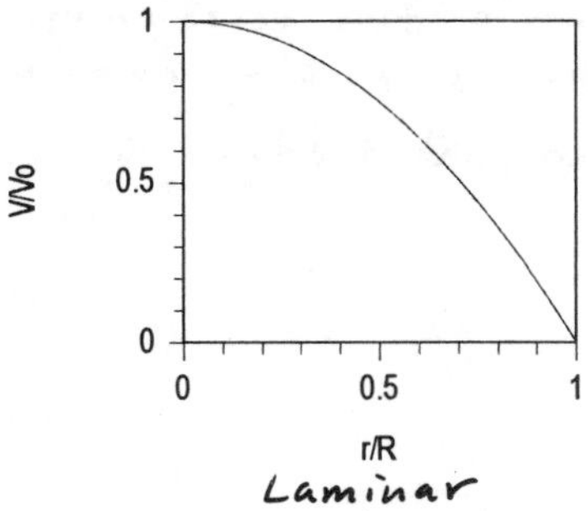

Laminar

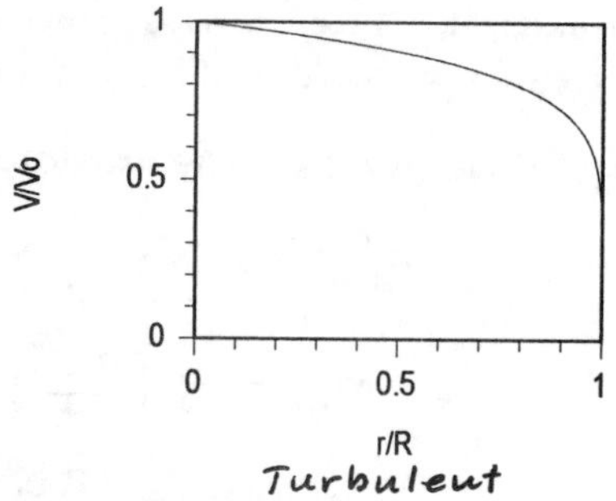

Turbulent

(b) Using Eq. 4.9, the mass flow rate for laminar flow is

$$\dot{m} = \int_A \rho V\,dA$$

$$= \int_0^R \rho V_0\left[1-\left(\frac{r}{R}\right)^2\right]2\pi r\,dr$$

$$= \rho V_0 2\pi \int_0^R \left[r - \frac{r^3}{R^2}\right]dr$$

$$= \rho V_0 2\pi \left[\frac{r^2}{2} - \frac{r^4}{4R^2}\right]\Bigg|_{r=0}^{r=R} = \rho V_0 2\pi\left[\frac{R^2}{2} - \frac{R^2}{4}\right]$$

$$= \rho V_0 \pi \frac{R^2}{2}$$

Using Eq. 4.11a, the average velocity is

$$V_{ave} = \frac{\dot{m}}{\rho A} = \frac{\rho V_0 \pi R^2/2}{\rho(\pi R^2)}$$

$$= \frac{V_0}{2}$$

PROBLEM 4.8 (Contd.)

Using Eq. 4.9, the mass flow rate for turbulent flow is

$$\dot{m} = \int_A \rho V dA$$

$$= \int_0^R \rho V_0 [1 - \frac{r}{R}]^{\frac{1}{7}} 2\pi r dr$$

$$= \rho V_0 2\pi \int_0^R r [1 - \frac{r}{R}]^{\frac{1}{7}} dr$$

To evaluate the integral, let $u = 1 - r/R$ and $dr = -R du$. Then

$$\dot{m} = \rho V_0 2\pi \int_1^0 R(1-u) u^{1/7} (-R du) = \rho V_0 2\pi [-R^2 \int_1^0 (u^{1/7} - u^{8/7}) du]$$

$$= \rho V_0 2\pi [-R^2 (\frac{7}{8} u^{8/7} - \frac{7}{15} u^{15/7})_{u=1}^{u=0}]$$

$$= \rho V_0 2\pi [R^2 \frac{49}{120}] = \rho V_0 \pi (\frac{49}{60}) R^2$$

Using Eq. 4.11a, the average velocity is

$$V_{ave} = \frac{\dot{m}}{\rho A} = \frac{\rho V_0 \pi (\frac{49}{60}) R^2}{\rho (\pi R^2)} = \frac{49}{60} V_0$$

(c) The specific kinetic energy (kinetic energy per unit of mass) carried through an area normal to the flow is

$$ke = \frac{\int_A \frac{V^2}{2} \rho V dA}{\dot{m}} = \frac{\rho \int_A \frac{V^3}{2} dA}{\rho \int_A V dA} = \frac{\int_A \frac{1}{2} V^3 dA}{\int_A V dA} = \frac{\int_A \frac{1}{2} V^3 dA}{V_{ave} A} \quad (*)$$

Forming the ratio of Eq. (*) to the specific kinetic energy calculated as $V_{ave}^2/2$, we have the kinetic energy coefficient (or kinetic energy correction factor):

$$\alpha = \frac{\int_A \frac{1}{2} V^3 dA}{\frac{1}{2} V_{ave}^3 A} = \frac{\int_A V^3 dA}{V_{ave}^3 A}$$

For Laminar Flow: $\alpha = 2$ → % ERROR = 50

(1) For Turbulent Flow: $\alpha = 1.058$ → % ERROR = 5.5

The flatter turbulent velocity profile adheres most closely to the idealization of one-dimensional flow.

---

1. For further discussion, see R. W. Fox and A. T. McDonald, Introduction to Fluid Mechanics, 5th ed., J. Wiley & Sons, New York, pp 354-356.

PROBLEM 4.9

KNOWN: Data are provided for a vegetable oil-filled spray can.

FIND: Determine the mass flow rate per spray, and the mass remaining in the can after a specified number of sprays.

SCHEMATIC & GIVEN DATA:

- 560 sprays, each lasting 0.25 s and having a mass of 0.25 g
- initial mass of vegetable oil in the can is 170 g

ASSUMPTION: The control volume is shown in the accompanying schematic.

ANALYSIS: (a) Since each spray has a duration of 0.25 s and consists of 0.25 g

$$\dot{m}_e = \frac{0.25 g}{0.25 s} = 1 \text{ g/s}$$

(b) The mass rate balance, Eq. 4.3, reduces to

$$m_{cv}(t+\Delta t) - m_{cv}(0) = \cancel{m_i}^{\,0} - m_e$$

where $m_{cv}(0)$ is the initial amount of mass within the can and m_e is the amount of mass that exits. Thus, with $m_{cv}(0) = 170$ g and

$$m_e = (560 \text{ sprays})\left(\frac{0.25 g}{\text{spray}}\right) = 140 g$$

The mass of vegetable oil remaining in the can after 560 sprays is

$$m_{cv} = m_{cv}(0) - m_e$$
$$= 170 g - 140 g = 30 g$$

PROBLEM 4.10

KNOWN: Air flows through a one-inlet, one-exit control volume. Data are known at the inlet and exit.

FIND: Determine (a) the mass flow rate, (b) the exit area.

SCHEMATIC & GIVEN DATA:

$p_1 = 10$ bar
$T_1 = 400$ K
$V_1 = 20$ m/s
$A_1 = 20\ cm^2 = 0.002\ m^2$

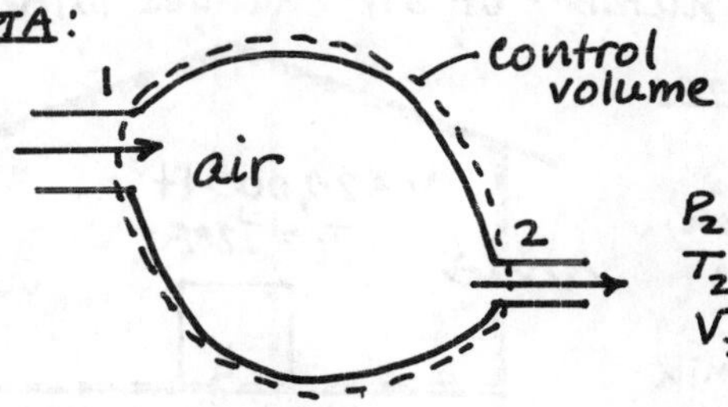

$p_2 = 6$ bar
$T_2 = 345.7$ K
$V_2 = 330.2$ m/s

ASSUMPTIONS: (1) The control volume is at steady state. (2) The air behaves as an ideal gas. (3) The flow is one-dimensional at the inlet and exit.

ANALYSIS: Beginning with the mass rate balance

$$\frac{dm_{cv}}{dt}^{\,0} = \dot{m}_1 - \dot{m}_2 \Rightarrow \dot{m}_1 = \dot{m}_2 \equiv \dot{m}$$

Using data at the inlet and the ideal gas equation of state

(a) $$\dot{m} = \rho_1 A_1 V_1 = \left(\frac{p_1}{RT_1}\right) A_1 V_1$$

$$= \frac{(10\ \text{bar})}{\left(\frac{8.314}{28.97}\frac{\text{kJ}}{\text{kg}\cdot\text{K}}\right)(400\text{K})} \left|\frac{10^5\ \text{N/m}^2}{1\ \text{bar}}\right| \left|\frac{1\ \text{kJ}}{10^3\ \text{N}\cdot\text{m}}\right| (0.002\ \text{m}^2)(20\text{m/s})$$

$$= 0.3484\ \text{kg/s} \longleftarrow \dot{m}$$

(b) From $\dot{m}_1 = \dot{m}_2$

$$\rho_1 A_1 V_1 = \rho_2 A_2 V_2$$

Thus

$$A_2 = \left(\frac{\rho_1}{\rho_2}\right)\left(\frac{V_1}{V_2}\right) A_1$$

with $\rho = p/RT$

$$A_2 = \left(\frac{p_1}{p_2}\right)\left(\frac{T_2}{T_1}\right)\left(\frac{V_1}{V_2}\right) A_1$$

$$= \left(\frac{10}{6}\right)\left(\frac{345.7}{400}\right)\left(\frac{20}{330.2}\right)(20\ \text{cm}^2)$$

$$= 1.745\ \text{cm}^2 \longleftarrow A_2$$

PROBLEM 4.11

KNOWN: Air enters a building through cracks around doors and windows and due to door openings. The internal volume of the building is known.

FIND: Estimate the number of air changes within the building per hour.

SCHEMATIC & GIVEN DATA:

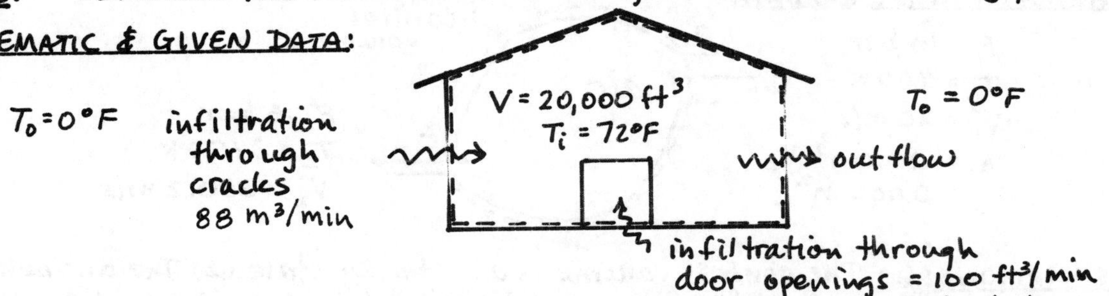

ASSUMPTIONS: (1) The control volume is at steady state. (2) The air behaves as an ideal gas. (3) The inside and outside air pressures are nearly equal.

ANALYSIS: At steady state, the mass balance reduces to

$$\dot{m}_{outflow} = \dot{m}_{cracks} + \dot{m}_{doors}$$

$$\rho_i (AV)_{outflow} = \rho_o [(AV)_{cracks} + (AV)_{doors}]$$

where ρ_i and ρ_o are the inside and outside densities, respectively. Assuming ideal gas behavior

$$\frac{\rho_o}{\rho_i} = \frac{\cancel{P_o}/\cancel{R}T_o}{\cancel{P_i}/\cancel{R}T_i} = \frac{T_o}{T_i}$$

Thus

$$(AV)_{outflow} = \left(\frac{T_o}{T_i}\right)[(AV)_{cracks} + (AV)_{doors}]$$

Inserting values

$$(AV)_{outflow} = \left(\frac{460°R}{532°R}\right)[88 \text{ ft}^3/\text{min} + 100 \text{ ft}^3/\text{min}] = 163 \text{ ft}^3/\text{min}$$

$$= 163 \text{ ft}^3/\text{min} = 9780 \text{ ft}^3/\text{h} \longleftarrow (AV)_{outflow}$$

$$\text{Air changes per hour} = \frac{9780 \text{ ft}^3/\text{h}}{20{,}000 \text{ ft}^3/\text{air change}}$$

$$\approx 0.5 \text{ air changes/h} \longleftarrow \text{air changes}$$

PROBLEM 4.12

KNOWN: Air enters a furnace operating at steady state and is to a duct system consisting of three ducts. Data are known at the inlet and each of the discharge ducts.

FIND: Determine (a) the mass flow rate entering the furnace, (2) the volumetric flow rate in each of the 6-in. exit ducts, (c) the velocity in the 12-in. exit duct.

SCHEMATIC & GIVEN DATA:

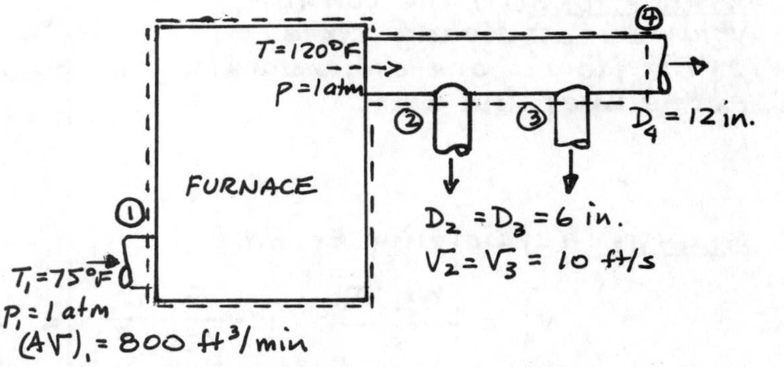

ASSUMPTIONS: (1) The control shown on the accompanying sketch is at steady state. (2) The air behaves as an ideal gas. (3) The temperature and pressure in each duct are the same as the temperature and pressure of the air delivered to the duct system.

ANALYSIS: (a) Using Eq. 4.11 b with the ideal gas equation

$$\dot{m}_1 = \frac{(AV)_1}{v_1} = \frac{p_1 (AV)_1}{R T_1}$$

$$= \frac{(1\,\text{atm})(800\,\text{ft}^3/\text{min})}{\left(\frac{1545}{28.97}\frac{\text{ft}\cdot\text{lbf}}{\text{lb}\cdot{}^\circ\text{R}}\right)(535\,{}^\circ\text{R})}\left|\frac{14.696\,\text{lbf/in.}^2}{1\,\text{atm}}\right|\left|\frac{144\,\text{in.}^2}{1\,\text{ft}^2}\right|\left|\frac{1\,\text{min}}{60\,\text{s}}\right|$$

$$= 0.989\,\text{lb/s} \leftarrow \dot{m}_1$$

(b) Since $D_2 = D_3$ and $V_2 = V_3$,

$$(AV)_2 = (AV)_3 = \left(\frac{\pi D^2}{4}\right) V = \left(\frac{\pi \left(\frac{6}{12}\right)^2 \text{ft}^2}{4}\right)\left(10\,\frac{\text{ft}}{\text{s}}\right)\left|\frac{60\,\text{s}}{1\,\text{min}}\right|$$

$$= 117.8\,\text{ft}^3/\text{min} \leftarrow (AV)_2, (AV)_3$$

(c) Applying the mass balance to get $\dot{m}_4$

$$\frac{dm_{cv}}{dt}\overset{0}{} = \dot{m}_1 - \dot{m}_2 - \dot{m}_3 - \dot{m}_4 \Rightarrow \dot{m}_4 = \dot{m}_1 - \dot{m}_2 - \dot{m}_3 \quad (*)$$

With Eq. 4.11 b

$$\dot{m}_2 = \dot{m}_3 = \frac{(AV)}{v} = \frac{p(AV)}{RT} = \frac{(14.696)(117.8)}{\left(\frac{1545}{28.97}\right)(580)}\left|\frac{144}{60}\right| = 0.1343\,\text{lb/s}$$

From (*)

$$\dot{m}_4 = 0.989 - 2(.1343) = 0.7204\,\text{lb/s}$$

Finally, from $\dot{m}_4 = \frac{A_4 V_4}{v_4}$

$$V_4 = \frac{v_4 \dot{m}_4}{A_4} = \frac{(RT_4)(\dot{m}_4)}{(p_4)\left(\frac{\pi D_4^2}{4}\right)} = \frac{\left(\frac{1545}{28.97}\right)(580)(0.7204)}{(14.696)\frac{\pi(12)^2}{4}} = 13.41\,\text{ft/s} \leftarrow V_4$$

PROBLEM 4.13

KNOWN: Refrigerant 22 flows through a refrigeration condenser. Data are known at the inlet and exit. The mass flow at the inlet is given.

FIND: Determine (a) the inlet velocity, (b) the diameter of the exit pipe.

SCHEMATIC & GIVEN DATA:

ASSUMPTIONS: (1) The control volume is at steady state. (2) The flow is one-dimensional at the inlet and exit.

$P_1 = 12$ bar
$T_1 = 50$ °C
$d_1 = 2.5$ cm
$\dot{m}_1 = 5$ kg/min

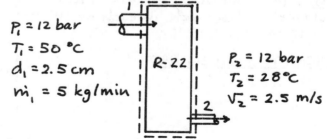

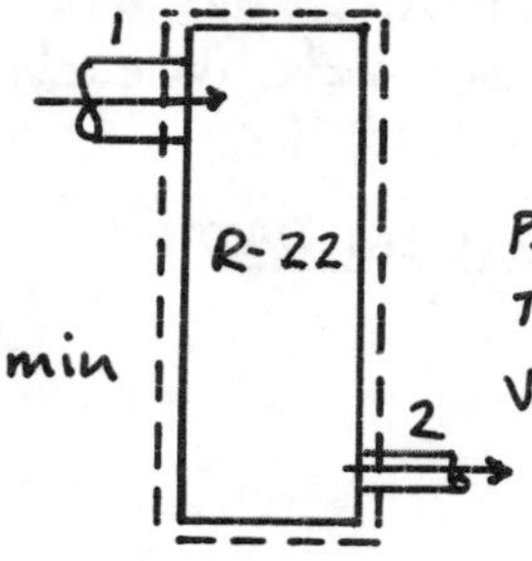

$P_2 = 12$ bar
$T_2 = 28$°C
$V_2 = 2.5$ m/s

ANALYSIS: (a) Solving Eq. 4.11b for V_1

$$V_1 = \frac{\dot{m}_1 v_1}{A_1} = \frac{\dot{m}_1 v_1}{\left(\frac{\pi d_1^2}{4}\right)} = \frac{4\dot{m}_1 v_1}{\pi d_1^2}$$

From Table A-9, $v_1 = 0.02204$ m³/kg. Thus

$$V_1 = \frac{(4)(5\text{ kg/min})(0.02204\text{ m}^3/\text{kg})}{\pi\,(2.5)^2\text{cm}^2}\left|\frac{1\text{ min}}{60\text{ s}}\right|\left|\frac{10^4\text{ cm}^2}{1\text{ m}^2}\right|$$

$= 3.74$ m/s ← V_1

(b) To find the diameter of the exit pipe, begin with mass rate balance

$$\cancelto{0}{\frac{dm_{cv}}{dt}} = \dot{m}_1 - \dot{m}_2 \Rightarrow \dot{m}_2 = \dot{m}_1$$

Thus, with $\dot{m}_2 = A_2 V_2 / v_2$

$$A_2 = \frac{\dot{m}_2 v_2}{V_2}$$

From Table A-8, $T_2 < T_{sat\,@\,12\,bar}$. Thus, the refrigerant is a sub-cooled liquid. From Table A-7

$$v_2 \approx v_{f\,@\,28°C} = 0.848\times10^{-3}\text{ m}^3/\text{kg}$$

Inserting values

$$A_2 = \frac{(5\text{ kg/min})(0.848\times10^{-3}\text{ m}^3/\text{kg})}{(2.5\text{ m/s})}\left|\frac{1\text{ min}}{60\text{ s}}\right| = 2.827\times10^{-5}\text{ m}^2$$

Finally, with $A = \pi d^2/4$

$$d_2 = \left(\frac{4A_2}{\pi}\right)^{1/2} = 0.006\text{ m} = 0.6\text{ cm}$$ ← d_2

PROBLEM 4.14

KNOWN: Data are given for steam flowing through a turbine with one inlet and two exits.

FIND: Determine the diameter of each exit duct.

SCHEMATIC & GIVEN DATA:

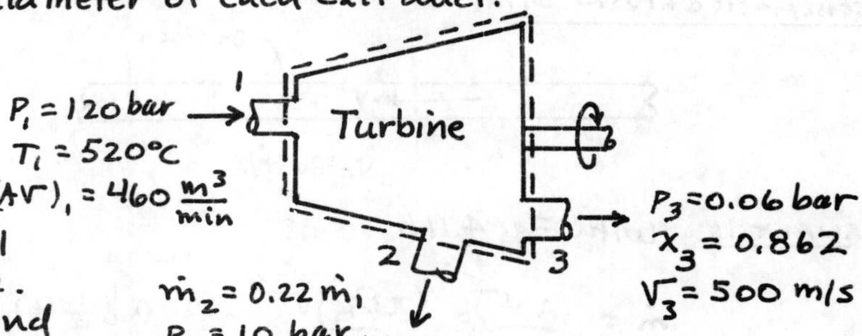

ASSUMPTIONS: (1) The control volume is at steady state. (2) The flow at the inlet and each exit is one-dimensional.

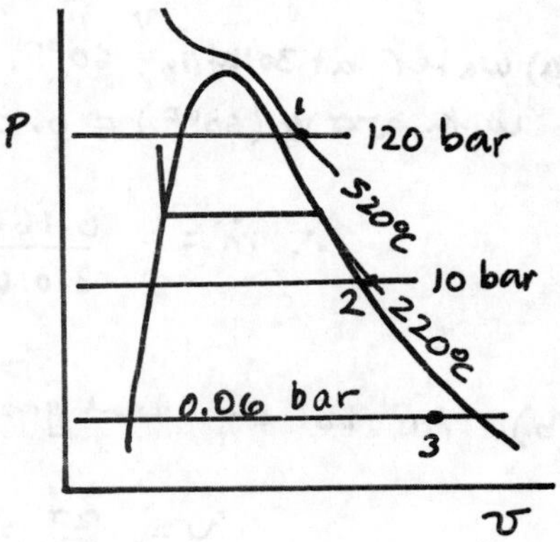

ANALYSIS: The mass flow rate at the inlet is

$$\dot{m}_1 = \frac{(AV)_1}{v_1}$$

From Table A-4, $v_1 = 0.02781\ m^3/kg$. Thus

$$\dot{m}_1 = \frac{(460\ m^3/min)}{(0.02781\ m^3/kg)}\left|\frac{1\ min}{60\ s}\right| = 275.7\ kg/s$$

and $\dot{m}_2 = 0.22\,\dot{m}_1 = 60.65$ kg/s

Applying the mass rate balance

$$\frac{dm_{cv}^{\,0}}{dt} = \dot{m}_1 - \dot{m}_2 - \dot{m}_3 \Rightarrow \dot{m}_3 = \dot{m}_1 - \dot{m}_2 = 215.1\ kg/s$$

Now, with $\dot{m} = AV/v$, and from Table A-4; $v_2 = 0.21675\ m^3/kg$

$$A_2 = \frac{\dot{m}_2 v_2}{V_2} = \frac{(60.65\ kg/s)(0.21675\ m^3/kg)}{(20\ m/s)} = 0.657\ m^2$$

Noting that $A = \pi d^2/4$

$$d_2 = \sqrt{\frac{4A_2}{\pi}} = 0.915\ m \longleftarrow \quad d_2$$

From Table A-3, at $P_3 = 0.06$ bar and $x_3 = 0.862$

$$v_3 = v_{f3} + x_3(v_{g3} - v_{f3})$$
$$= 1.0064\times10^{-3} + (.862)(23.739 - 1.0064\times10^{-3}) = 20.463\ m^3/kg$$

Thus

$$A_3 = \frac{\dot{m}_3 v_3}{V_3} = \frac{(215.1)(20.463)}{(500)} = 8.803\ m^2$$

and

$$d_3 = \sqrt{\frac{4A_3}{\pi}} = 3.35\ m \longleftarrow \quad d_3$$

PROBLEM 4.15

KNOWN: Data are provided for substances flowing through a pipe of known diameter.

FIND: Determine the mass flow rate for each of three specified substances.

SCHEMATIC & GIVEN DATA:

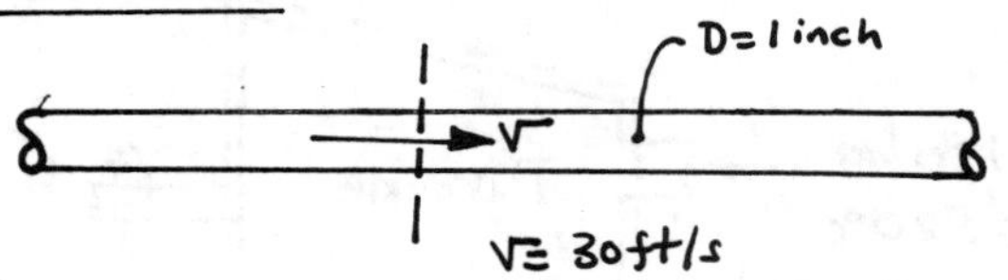

ANALYSIS: With Eq. 4.11b

$$\dot{m} = \frac{AV}{v} = \frac{(\pi D^2/4)V}{v} = \frac{\pi\left[\left(\frac{1}{12}\,ft\right)^2/4\right]\left[30\ ft/s\right]}{v} = \frac{0.1636\ ft^3/s}{v}$$

(a) water at 30 lbf/in², 60°F.

With $v \approx v_f(60°F) = 0.01604\ ft^3/lb$ (Table A-2E)

$$\therefore\ \dot{m} = \frac{0.1636\ ft^3/s}{0.01604\ ft^3/lb} = 10.2\ lb/s \qquad \text{(a)}$$

(b) Air as an ideal gas at 100 lbf/in², 100°F

$$v = \frac{RT}{p} = \frac{\left(\frac{1545}{28.97}\ \frac{ft\cdot lbf}{lb\cdot °R}\right)(560°R)}{(100\ lbf/in.^2)}\left|\frac{1\ ft^2}{144\ in^2}\right| = 2.074\ ft^3/lb$$

$$\therefore\ \dot{m} = \frac{0.1636}{2.074} = 0.079\ lb/s \qquad \text{(b)}$$

(c) R134a at 100 lbf/in², 100°F

$$v = 0.5086\ \frac{ft^3}{lb} \quad \text{from Table A-12E}$$

$$\therefore\ \dot{m} = \frac{0.1636}{0.5086} = 0.322\ lb/s \qquad \text{(c)}$$

PROBLEM 4.16

<u>KNOWN</u>: Data are known at the inlet and exit of an air compressor operating at steady state. Each unit of mass passing through undergoes a polytropic process.

<u>FIND</u>: Determine the velocity and temperature at the exit.

<u>SCHEMATIC & GIVEN DATA</u>:

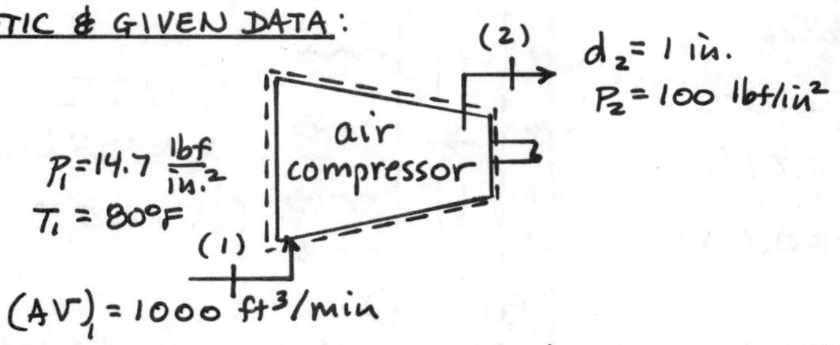

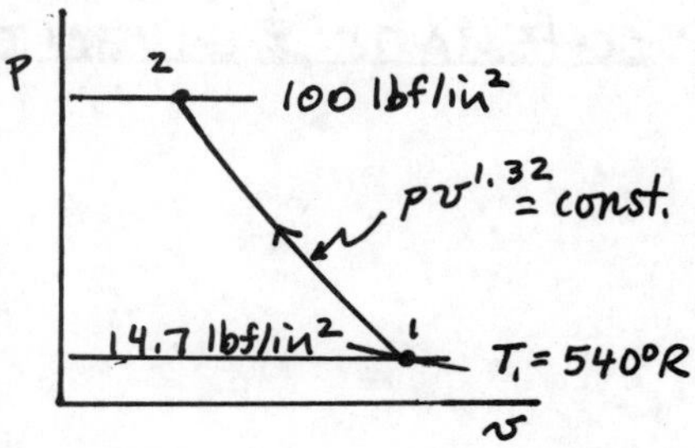

<u>ASSUMPTIONS</u>: (1) The control volume is at steady state. (2) Each unit of mass undergoes a polytropic process with $n = 1.32$. (3) The air behaves as an ideal gas, as can be verified by reference to the compressibility chart.

<u>ANALYSIS</u>: The exit velocity is determined by using the mass rate balance with Eq. 4.11b, as follows:

$$\cancelto{0}{\frac{dm_{cv}}{dt}} = \dot{m}_1 - \dot{m}_2 \Rightarrow \dot{m}_1 = \dot{m}_2 \Rightarrow \frac{(AV)_1}{v_1} = \frac{(AV)_2}{v_2} \qquad (*)$$

First, evaluate v_1 using the ideal gas equation

$$v_1 = \frac{RT_1}{P_1} = \frac{\left(\frac{1545}{28.97}\frac{\text{ft}\cdot\text{lbf}}{\text{lb}\cdot°\text{R}}\right)(540°\text{R})}{(14.7\ \text{lbf/in.}^2)}\left|\frac{1\ \text{ft}^2}{144\ \text{in}^2}\right| = 13.60\ \text{ft}^3/\text{lb}$$

Now, with the polytropic process expression

$$v_2 = \left(\frac{P_1}{P_2}\right)^{\frac{1}{1.32}} v_1 = \left(\frac{14.7}{100}\right)^{\frac{1}{1.32}}\left(13.60\ \frac{\text{ft}^3}{\text{lb}}\right) = 3.182\ \text{ft}^3/\text{lb}$$

The exit area is

$$A_2 = \frac{\pi d_2^2}{4} = \frac{\pi\left(\frac{1}{12}\right)^2 \text{ft}^2}{4} = 0.005454\ \text{ft}^2$$

Solving (*) for V_2 and inserting values

$$V_2 = \frac{(AV)_1 v_2}{A_2 v_1}$$

$$= \frac{(1000\ \text{ft}^3/\text{min})(3.182\ \text{ft}^3/\text{lb})}{(0.005454\ \text{ft}^2)(13.60\ \text{ft}^3/\text{lb})}\left|\frac{1\ \text{min}}{60\ \text{s}}\right| = 715\ \text{ft/s} \longleftarrow V_2$$

The exit temperature is determined using the ideal gas equation of state.

$$T_2 = \frac{P_2 v_2}{R} = \frac{(100\ \text{lbf/in}^2)(3.182\ \text{ft}^3/\text{lb})}{\left(\frac{1545}{28.97}\frac{\text{ft}\cdot\text{lbf}}{\text{lb}\cdot°\text{R}}\right)}\left|\frac{144\ \text{in}^2}{1\ \text{ft}^2}\right|$$

$$= 859°\text{R} = 399°\text{F} \longleftarrow T_2$$

PROBLEM 4.17

KNOWN: Data are given for air entering and exiting from a fan.

FIND: Determine at steady state (a) the mass flow rate, (b) the volumetric flow rate at the inlet, and (c) the inlet and exit velocities.

SCHEMATIC & GIVEN DATA:

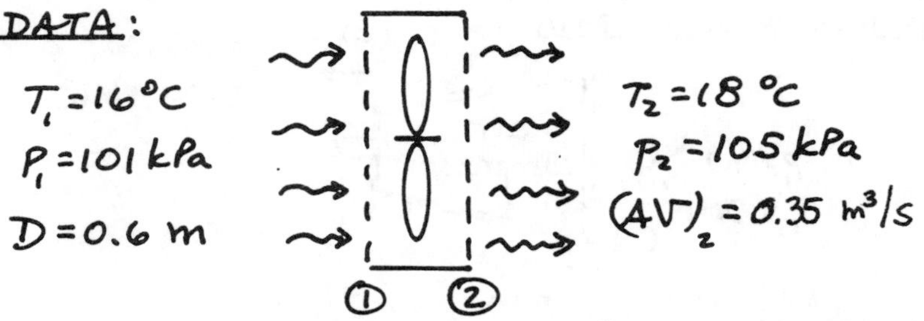

ASSUMPTIONS: (1) A control volume enclosing the fan is at steady state, (2) The air behaves as an ideal gas.

ANALYSIS:

(a) The mass balance reduces to: $\dot{m}_1 = \dot{m}_2$. Thus, using data at the exit of the fan

$$\dot{m} = \frac{(AV)_2}{v_2} = \frac{(AV)_2 P_2}{R T_2}$$

$$= \frac{(0.35\ \text{m}^3/\text{s})(105\ \text{kPa})}{\left(\frac{8.314}{28.97}\ \frac{\text{kJ}}{\text{kg}\cdot\text{K}}\right)(291\ \text{K})} \left|\frac{10^3\ \text{N/m}^2}{1\ \text{kPa}}\right| \left|\frac{1\ \text{kJ}}{10^3\ \text{N}\cdot\text{m}}\right|$$

$$= 0.44\ \text{kg/s} \longleftarrow$$

(b) At the inlet

$$(AV)_1 = \dot{m} v_1 = \dot{m}\left(\frac{RT_1}{P_1}\right) = 0.44 \left[\frac{\left(\frac{8.314}{28.97}\right)(289)}{(101)}\right]$$

$$= 0.361\ \text{m}^3/\text{s} \longleftarrow$$

(c) The cross-sectional area is

$$A = \frac{\pi D^2}{4} = \frac{\pi(.6\ \text{m})^2}{4} = 0.2827\ \text{m}^2$$

$$\Rightarrow \quad V_1 = \frac{(AV)_1}{A} = \frac{0.361\ \text{m}^3/\text{s}}{0.2827\ \text{m}^2} = 1.28\ \text{m/s} \longleftarrow$$

$$V_2 = \frac{(AV)_2}{A} = \frac{0.35}{0.2827} = 1.24\ \text{m/s} \longleftarrow$$

PROBLEM 4.18

<u>KNOWN</u>: Ammonia flows through a control volume at steady state. The control volume has one inlet and two exit, and data are known at each flow boundary.

<u>FIND</u>: Determine (a) the minimum inlet diameter so the ammonia velocity does not exceed 20 m/s. (b) the volumetric flow rate of the second exit stream.

<u>SCHEMATIC & GIVEN DATA</u>:

$P_1 = 14$ bar
$T_1 = 28$ °C
$\dot{m}_1 = 0.5$ kg/s

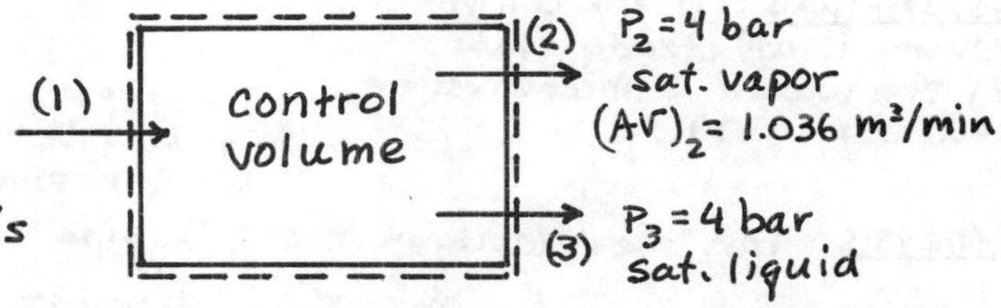

<u>ASSUMPTIONS</u>: (1) The control volume is at steady state. (2) The flow at the inlet is one-dimensional.

<u>ANALYSIS</u>: (a) To relate velocity and pipe diameter at the inlet, use Eq. 4.11b

$$V_1 = \frac{\dot{m}_1 v_1}{A_1} = \frac{\dot{m}_1 v_1}{\left(\frac{\pi d_1^2}{4}\right)}$$

Thus, velocity varies inversely with diameter. The minimum diameter corresponds to $V_1 = 20$ m/s.

To get v_1, note from Table A-14 that $T_1 = 28°C$ is less than T_{sat} at 14 bar. Hence, from Table A-13, $v_1 \approx v_{f@28°C} = 1.6714 \times 10^{-3}$ m³/kg, and

$$(d_1)_{min} = \sqrt{\frac{4 \dot{m}_1 v_1}{\pi V_1}} = \sqrt{\frac{(4)(0.5\ \text{kg/s})(1.6714 \times 10^{-3}\ \text{m}^3/\text{kg})}{\pi (20\ \text{m/s})}}$$

$$= 0.00729\ \text{m} = 0.729\ \text{cm} \longleftarrow (d_1)_{min}$$

(b) To find $(AV)_3$, begin with the mass rate balance

$$\frac{dm_{cv}}{dt}^{\nearrow 0} = \dot{m}_1 - \dot{m}_2 - \dot{m}_3 \Rightarrow \dot{m}_3 = \dot{m}_1 - \dot{m}_2$$

With $\dot{m} = (AV)/v$

$$(AV)_3 = v_3 \left[\dot{m}_1 - (AV)_2 / v_2 \right]$$

From Table A-14 at 4 bar ; $v_2 = 0.3094$ m³/kg and $v_3 = 1.5597 \times 10^{-3}$ m³/kg. Thus

$$(AV)_3 = \left(1.5597 \times 10^{-3} \frac{\text{m}^3}{\text{kg}}\right) \left[\left(0.5 \frac{\text{kg}}{\text{s}}\right)\left(\frac{60\ \text{s}}{1\ \text{min}}\right) - \frac{(1.036\ \text{m}^3/\text{min})}{(0.3094\ \text{m}^3/\text{kg})} \right]$$

$$= 0.0416\ \text{m}^3/\text{min} \longleftarrow (AV)_3$$

PROBLEM 4.19

KNOWN: A pump operating at steady state provides water through two exit pipes. Data are known at the inlet and each exit.

FIND: Determine the water velocity in each exit pipe.

SCHEMATIC & GIVEN DATA:

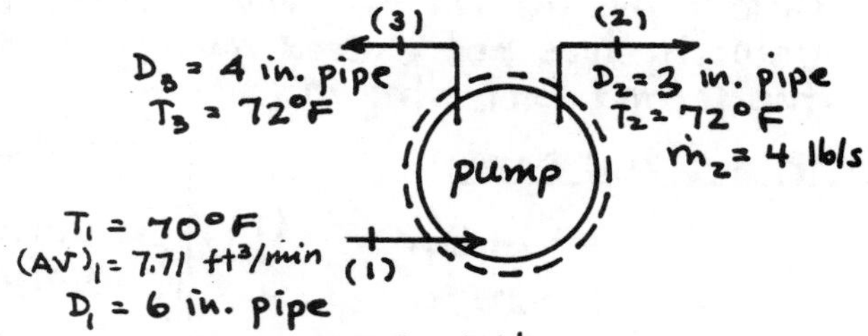

ASSUMPTIONS: (1) The control volume is at steady state. (2) The water is incompressible and $v \approx v_f(T)$.

ANALYSIS: For the velocity in the 3 in. pipe, use Eq. 4.11b

$$V_2 = \frac{\dot{m}_2 v_2}{A_2} = \frac{4\dot{m}_2 v_2}{\pi D_2^2}$$

From Table A-2E at 72°F; $v_2 \approx v_f$ = 0.01606 ft³/lb. Thus

$$V_2 = \frac{4\,(4\ \text{lb/s})(0.01606\ \text{ft}^3/\text{lb})}{\pi\,(3/12)^2\ \text{ft}^2} = 1.309\ \text{ft/s} \longleftarrow V_2$$

To find the velocity in the 4 in. pipe, use the mass rate balance to determine the mass flow rate. That is

$$\frac{dm_{cv}}{dt}^{\,0} = \dot{m}_1 - \dot{m}_2 - \dot{m}_3 \Rightarrow \dot{m}_3 = \dot{m}_1 - \dot{m}_2$$

At the inlet

$$\dot{m}_1 = \frac{A_1 V_1}{v_1} = \frac{(7.71\ \text{ft}^3/\text{min})}{(0.01605\ \text{ft}^3/\text{lb})}\left|\frac{1\ \text{min}}{60\ \text{s}}\right| = 8.01\ \text{lb/s}$$

where v_1 is obtained from Table A-2E. Finally,

$$\dot{m}_3 = \dot{m}_1 - \dot{m}_2 = 8.01 - 4 = 4.01\ \text{lb/s}$$

and

$$V_3 = \frac{4\dot{m}_3 v_3}{\pi D_3^2} = \frac{4\,(4.01\ \text{lb/s})(0.01606\ \text{ft}^3/\text{lb})}{\pi\,(4/12)^2\ \text{ft}^2}$$

$$= 0.738\ \text{ft/s} \longleftarrow V_3$$

PROBLEM 4.20

KNOWN: Separate streams of water and ethylene glycol (glycol) mix to form a single mixture that is half glycol by mass. Data are given for the inlet flows.

FIND: Determine (a) the molar and volumetric flow rates of the entering glycol. (b) the diameters of the supply pipe.

SCHEMATIC & GIVEN DATA:

Water:
$T_1 = 20°C$
$p_1 = 1$ bar
$\dot{n}_1 = 4.2 \frac{kmol}{min}$
$V_1 = V_2 = 2.5$ m/s

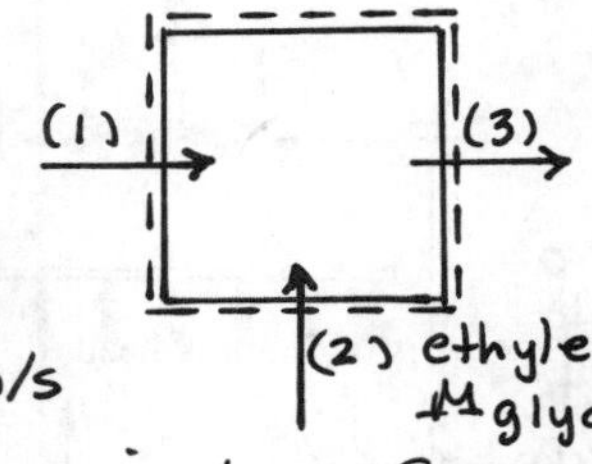

Mixture: 50% glycol by mass

(2) ethylene glycol
$M_{glycol} = 62.07$ kg/kmol
$\rho_{glycol} = 1.115\,\rho_{H_2O}$

ASSUMPTIONS: (1) The control volume is at steady state. (2) The water and glycol are each incompressible substances.

ANALYSIS: (a) To find the glycol flow rates, begin with a mass rate balance. At steady state: $\dot{m}_1 + \dot{m}_2 = \dot{m}_3$. For the mixture

$$\left.\begin{array}{l} 50\%\text{ glycol} \Rightarrow \dfrac{\dot{m}_2}{\dot{m}_3} = 0.5 \\ 50\%\text{ water} \Rightarrow \dfrac{\dot{m}_1}{\dot{m}_3} = 0.5 \end{array}\right\} \dot{m}_1 = \dot{m}_2$$

Now

$$\dot{m}_1 = \dot{n}_1 M_{H_2O} = \left(4.2\,\frac{kmol}{min}\right)\left(18.02\,\frac{kg}{kmol}\right) = 75.68\text{ kg/min}$$

$$\dot{n}_2 = \frac{\dot{m}_2}{M_{glycol}} = \frac{75.68\text{ kg/min}}{62.07\text{ kg/kmol}} = 1.219\text{ kmol/min} \longleftarrow \dot{n}_2$$

Also, with $\dot{m} = \rho(AV)$

$$(AV)_2 = \frac{\dot{m}_2}{\rho_{glycol}} = \frac{\dot{m}_2}{1.115\,\rho_{H_2O}}$$

From Table A-2, $v_{H_2O} \approx v_{f @ 20°C} = 1.0018 \times 10^{-3}$ m³/kg and $\rho_{H_2O} = 1/v_{H_2O} = 998.2$ kg/m³. Thus

$$(AV)_2 = \frac{75.68\text{ kg/min}}{1.115\,(998.2)\text{ kg/m}^3} = 0.068\text{ m}^3\text{/min} \longleftarrow (AV)_2$$

(b) The area of the water supply pipe is

$$A_1 = \frac{\dot{m}_1}{\rho_{H_2O} V_1} = \frac{(75.68\text{ kg/min})}{(998.2\text{ kg/m}^3)(2.5\text{ m/s})}\left|\frac{1\text{ min}}{60\text{ s}}\right|\left|\frac{10^4\text{ cm}^2}{1\text{ m}^2}\right| = 5.054\text{ cm}^2$$

With $A = \pi d^2/4$

$$d_1 = \sqrt{\frac{4A_1}{\pi}} = \sqrt{\frac{(4)(5.054\text{ cm}^2)}{\pi}} = 2.54\text{ cm} \longleftarrow d_1$$

Similarly for the glycol stream

$$A_2 = \frac{75.68}{(1.115)(998.2)(2.5)}\left|\frac{10^4}{60}\right| = 4.533\text{ cm}^2$$

$$d_2 = \sqrt{\frac{(4)(4.533)}{\pi}} = 2.4\text{ cm} \longleftarrow d_2$$

PROBLEM 4.21

KNOWN: Data are know for inlet and exit streams of an air conditioner cooling tower operating at steady state.

FIND: Determine the mass flow rate of the makeup water.

SCHEMATIC & GIVEN DATA:

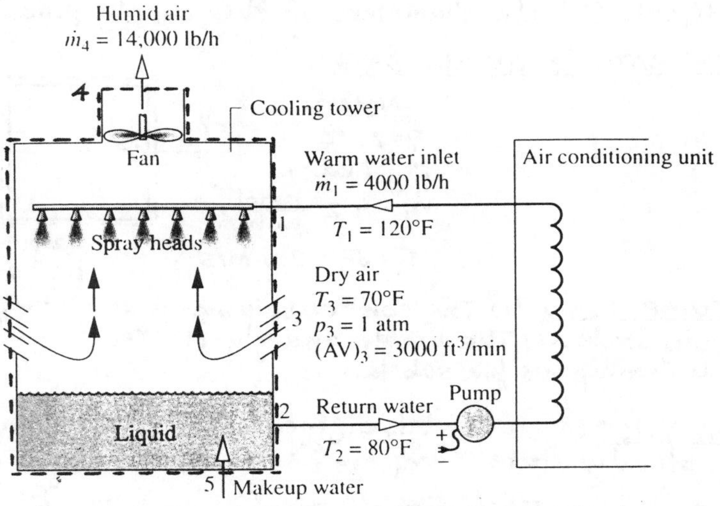

ASSUMPTIONS: (1) The control volume is at steady state. (2) The dry air stream at location 3 behaves as an ideal gas.

ANALYSIS: The mass rate balance for the control reduces as follows:

$$\frac{dm_{cv}}{dt}^{\,0} = \dot{m}_1 + \dot{m}_3 + \dot{m}_5 - \dot{m}_2 - \dot{m}_4$$

$$\therefore \quad \dot{m}_5 = \dot{m}_2 - \dot{m}_1 + \dot{m}_4 - \dot{m}_3$$

From the schematic, we see that circuit through the air conditioning unit is closed. Therefore, at steady state $\dot{m}_1 = \dot{m}_2$. Thus

$$\dot{m}_5 = \dot{m}_4 - \dot{m}_3 \qquad (*)$$

To get $\dot{m}_3$, use Eq. 4.11 b and the ideal gas equation

$$\dot{m}_3 = \frac{(AV)_3}{v_3} = \frac{p_3 (AV)_3}{R T_3}$$

$$= \frac{(1\text{ atm})(3000\text{ ft}^3/\text{min})}{\left(\frac{1545}{28.97}\frac{\text{ft}\cdot\text{lbf}}{\text{lb}\cdot{}^\circ\text{R}}\right)(530^\circ\text{R})} \left|\frac{14.696\text{ lbf/in}^2}{1\text{ atm}}\right|\left|\frac{144\text{ in}^2}{1\text{ ft}^2}\right|\left|\frac{60\text{ min}}{1\text{ h}}\right|$$

$$= 13480\text{ lb/h}$$

Finally, from (*)

$$\dot{m}_5 = 14{,}000 - 13{,}480 = 520\text{ lb/h} \longleftarrow \dot{m}_5$$

PROBLEM 4.22

KNOWN: Air passes through a control volume operating at steady state. Data are given at the inlet and exit, and the work input per kg of air flowing is known. The exit volumetric flow rate is known.

FIND: Determine the heat transfer rate.

SCHEMATIC & GIVEN DATA:

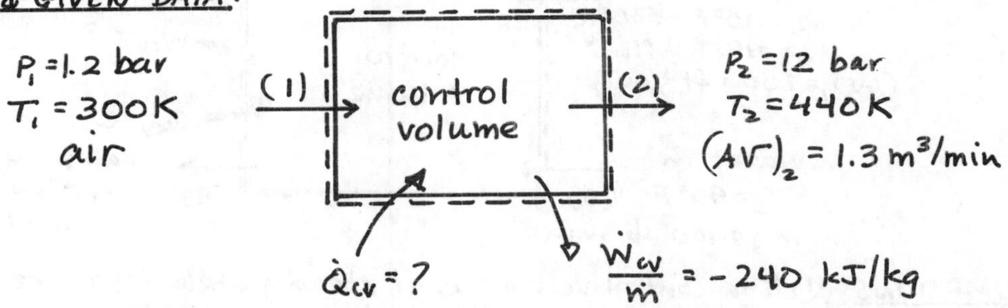

ASSUMPTIONS: (1) The control volume is at steady state. (2) The air behaves as an ideal gas. (3) Kinetic and potential energy effects can be neglected.

ANALYSIS: To determine the heat transfer rate, begin with the energy rate balance

$$\frac{dE_{cv}}{dt}^{\,0} = \dot{Q}_{cv} - \dot{W}_{cv} + \dot{m}_1\left(h_1 + \frac{V_1^2}{2} + gz_1\right) - \dot{m}_2\left(h_2 + \frac{V_2^2}{2} + gz_2\right)$$

With $\dot{m}_1 = \dot{m}_2 = \dot{m}$

$$\frac{\dot{Q}_{cv}}{\dot{m}} = \frac{\dot{W}_{cv}}{\dot{m}} + (h_2 - h_1) + \left(\frac{V_2^2 - V_1^2}{2}\right)^{0} + g(z_2 - z_1)^{0}$$

Using data from Table A-22

$$\frac{\dot{Q}_{cv}}{\dot{m}} = (-240 \text{ kJ/kg}) + (441.61 - 300.19) \text{ kJ/kg} = -98.58 \text{ kJ/kg}$$

The mass flow rate is

$$\dot{m} = \frac{(AV)_2}{v_2} = \frac{p_2 (AV)_2}{RT_2}$$

$$= \frac{(12 \text{ bar})(1.3 \text{ m}^3/\text{min})}{\left(\frac{8.314}{28.97}\frac{\text{kJ}}{\text{kg}\cdot\text{K}}\right)(440\text{K})}\left|\frac{1 \text{ min}}{60 \text{ s}}\right|\left|\frac{10^5 \text{ N/m}^2}{1 \text{ bar}}\right|\left|\frac{1 \text{ kJ}}{10^3 \text{ N}\cdot\text{m}}\right|$$

$$= 0.2059 \text{ kg/s}$$

Finally

$$\dot{Q}_{cv} = \left(\frac{\dot{Q}_{cv}}{\dot{m}}\right)\dot{m} = \left(-98.58 \frac{\text{kJ}}{\text{kg}}\right)\left(0.2059 \frac{\text{kg}}{\text{s}}\right)\left|\frac{1 \text{ kW}}{1 \text{ kJ/s}}\right|$$

① $= -20.3$ kW ← $\dot{Q}_{cv}$

1. The negative sign for the heat transfer denotes energy transfer from the control volume to the surroundings.

PROBLEM 4.23

KNOWN: Air and water pass through a control volume as separate streams. Data are known at the inlets and exits, and the control volume is at steady state.

FIND: Determine the power.

SCHEMATIC & GIVEN DATA:

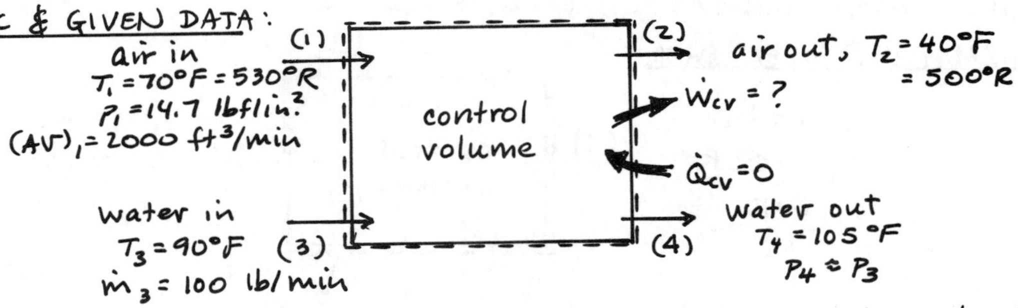

ASSUMPTIONS: (1) The control volume is at steady state. (2) There is no heat transfer: $\dot{Q}_{cv} = 0$. (3) Kinetic and potential energy effects can be neglected. (4) The air behaves as an ideal gas. (5) The water behaves as an incompressible substance, with negligible change in pressure.

ANALYSIS: To find the power, we apply mass and energy rate balances to the control volume. First, since the air and water streams are separate, by assumption 1

Air: $\frac{dm_{cv}}{dt}^{\,0} = \dot{m}_1 - \dot{m}_2 \Rightarrow \dot{m}_1 = \dot{m}_2 \equiv \dot{m}_a$

Water: $\frac{dm_{cv}}{dt}^{\,0} = \dot{m}_3 - \dot{m}_4 \Rightarrow \dot{m}_3 = \dot{m}_4 \equiv \dot{m}_w$

and

$$\frac{dE_{cv}}{dt}^{0} = \dot{Q}_{cv}^{\,0} - \dot{W}_{cv} + \dot{m}_1\left(h_1 + \frac{V_1^2}{2} + gz_1\right) - \dot{m}_2\left(h_2 + \frac{V_2^2}{2} + gz_2\right)$$
$$+ \dot{m}_3\left(h_3 + \frac{V_3^2}{2} + gz_3\right) - \dot{m}_4\left(h_4 + \frac{V_4^2}{2} g z_4\right)$$

Combining

$$0 = -\dot{W}_{cv} + \dot{m}_a\left[(h_1 - h_2) + \left(\frac{V_1^2 - V_2^2}{2}\right)^0 + g(z_1 - z_2)^0\right]$$
$$+ \dot{m}_w\left[(h_3 - h_4) + \left(\frac{V_3^2 - V_4^2}{2}\right)^0 + g(z_3 - z_4)^0\right]$$

or

$$\dot{W}_{cv} = \dot{m}_a (h_1 - h_2) + \dot{m}_w (h_3 - h_4)$$

The mass flow rate of air is found from $\dot{m}_a = (AV)_1/v_1$, or with $v_1 = RT_1/p_1$

$$\dot{m}_{air} = \frac{p_1 (AV)_1}{RT_1} = \frac{(14.7\ \text{lbf/in}^2)(2000\ \text{ft}^3/\text{min})}{\left(\frac{1545}{28.97}\frac{\text{ft}\cdot\text{lbf}}{\text{lb}\cdot°\text{R}}\right)(530°\text{R})}\left|\frac{144\ \text{in}^2}{1\ \text{ft}^2}\right| = 149.8\ \text{lb/min}$$

Now, from Table A-22E; $h_1 = 126.66$ Btu/lb and $h_2 = 119.48$ Btu/lb. Further, for the water; $h_3 \approx h_{f@70°F} = 58.07$ Btu/lb and $h_4 \approx h_{f@105°F} = 73.035$ Btu/lb. Thus

$$\dot{W}_{cv} = \left[\left(149.8\frac{\text{lb}}{\text{min}}\right)(126.66 - 119.48)\frac{\text{Btu}}{\text{lb}} + \left(100\frac{\text{lb}}{\text{min}}\right)(58.07 - 73.035)\frac{\text{Btu}}{\text{lb}}\right]\left|\frac{60\ \text{min}}{1\ \text{h}}\right|\left|\frac{1\ \text{hp}}{2545\ \text{Btu/h}}\right|$$

① $= -9.92$ hp ← $\dot{W}_{cv}$

1. The negative sign denotes energy transfer into the control volume by work.

PROBLEM 4.24

KNOWN: Steam flows through a nozzle with known conditions at the inlet and exit. The mass flow rate is given.

FIND: Determine (a) the exit velocity, (b) the inlet and exit flow areas.

SCHEMATIC & GIVEN DATA:

$P_1 = 30$ bar
$T_1 = 320°C$
$V_1 = 100$ m/s

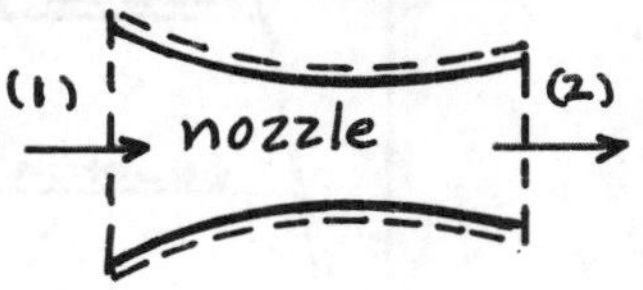

$P_2 = 10$ bar
$T_2 = 200°C$
$\dot{m} = 2$ kg/s

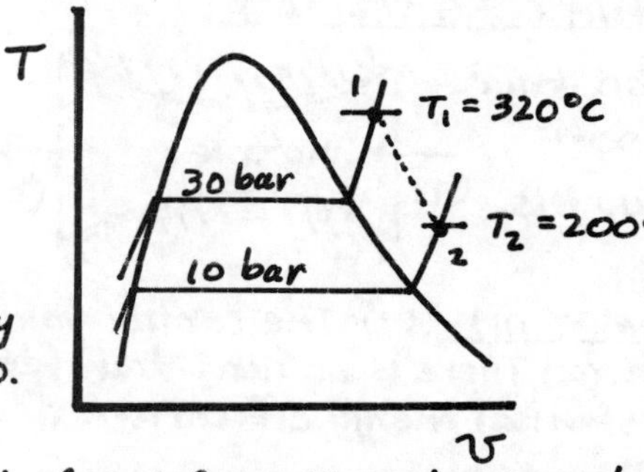

ASSUMPTIONS: (1) The control volume is at steady state. (2) Heat transfer is negligible and $\dot{W}_{cv} = 0$. (3) Potential energy effects are negligible.

ANALYSIS: (a) The velocity of steam at the exit is found from the steady-state energy balance

$$0 = \cancel{\dot{Q}_{cv}}^{0} - \cancel{\dot{W}_{cv}}^{0} + \dot{m}\left[(h_1 - h_2) + \left(\frac{V_1^2 - V_2^2}{2}\right) + g\cancel{(z_1 - z_2)}^{0}\right]$$

where $\dot{m}_1 = \dot{m}_2 \equiv \dot{m}$. Solving for V_2

$$V_2 = \sqrt{2(h_1 - h_2) + V_1^2}$$

From Table A-4, $h_1 = 3043.4$ kJ/kg and $h_2 = 2827.9$ kJ/kg. Thus

$$V_2 = \sqrt{2(3043.4 - 2827.9)\frac{\text{kJ}}{\text{kg}}\left|\frac{1\ \text{kg}\cdot\text{m/s}^2}{1\ \text{N}}\right|\left|\frac{10^3\ \text{N}\cdot\text{m}}{1\ \text{kJ}}\right| + (100^2)\,\text{m}^2/\text{s}^2}$$

$= 664.1$ m/s ← V_2

(b) To find the inlet and exit flow areas, use $\dot{m} = (AV)/v$. Solving

$$A_1 = \frac{\dot{m} v_1}{V_1} \quad \text{and} \quad A_2 = \frac{\dot{m} v_2}{V_2}$$

From Table A-4, $v_1 = 0.0850$ m³/kg and $v_2 = 0.2060$ m³/kg. Thus

$$A_1 = \frac{(2\ \text{kg/s})(0.0850\ \text{m}^3/\text{kg})}{(100\ \text{m/s})}\left|\frac{10^4\ \text{cm}^2}{1\ \text{m}^2}\right| = 17\ \text{cm}^2$$ ← A_1

and

$$A_2 = \frac{(2)(0.2060)}{(664.1)}\left|\frac{10^4}{1}\right| = 6.2\ \text{cm}^2$$ ← A_2

PROBLEM 4.25

KNOWN: Steam flows through a well-insulated nozzle with known conditions at the inlet and exit.

FIND: Determine the exit temperature.

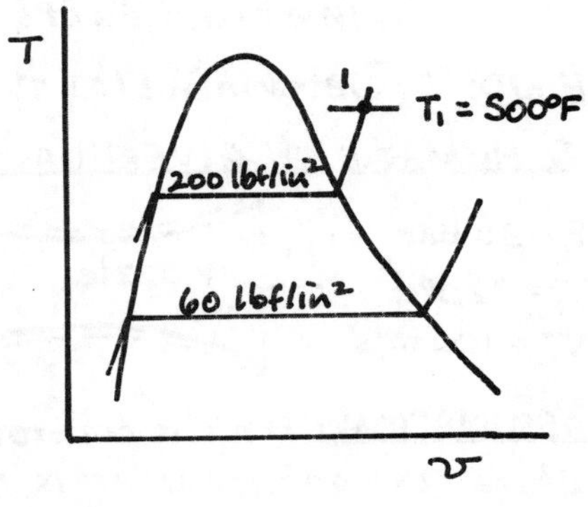

SCHEMATIC & GIVEN DATA:

$p_1 = 200$ lbf/in²
$T_1 = 500°F$
$V_1 = 200$ ft/s

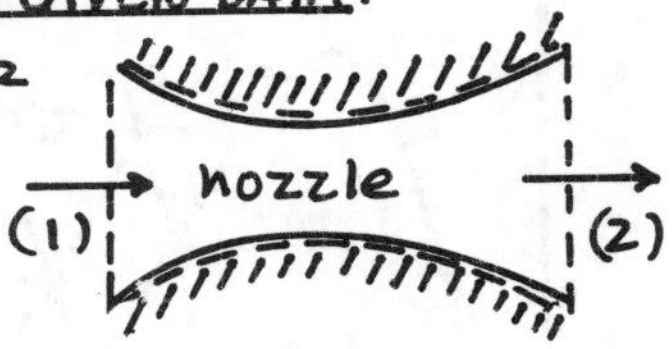

$p_2 = 60$ lbf/in²
$V_2 = 1700$ ft/s

ASSUMPTIONS: (1) The control volume is at steady state. (2) There is no heat transfer, and $\dot{W}_{cv} = 0$. (3) Potential energy effects are negligible.

ANALYSIS: The pressure is known at the exit. The state is fixed by determining h_2 using the steady-state energy balance

$$0 = \dot{Q}_{cv}^{\,0} - \dot{W}_{cv}^{\,0} + \dot{m}\left[(h_1 - h_2) + \left(\frac{V_1^2 - V_2^2}{2}\right) + g(z_1 - z_2)^{0}\right]$$

where $\dot{m}_1 = \dot{m}_2 \equiv \dot{m}$. Solving for h_2

$$h_2 = h_1 + \left(\frac{V_1^2 - V_2^2}{2}\right)$$

From Table A-4E, $h_1 = 1268.8$ Btu/lb. Thus

$$h_2 = (1268.8 \text{ Btu/lb}) + \left(\frac{200^2 - 1700^2}{2}\right)\frac{\text{ft}^2}{\text{s}^2}\left|\frac{1 \text{ lbf}}{32.2 \text{ lb}\cdot\text{ft/s}^2}\right|\left|\frac{1 \text{ Btu}}{778 \text{ ft}\cdot\text{lbf}}\right|$$

$$= 1211.92 \text{ Btu/lb}$$

Interpolating in Table A-4E at $p_2 = 60$ lbf/in², $h_2 = 1211.92$ Btu/lb gives

$$T_2 \approx 357.4\ °F \longleftarrow \underline{T_2}$$

PROBLEM 4.26

KNOWN: Methane gas flows through a nozzle with known inlet conditions and a specified range of exit velocities.

FIND: Plot exit temperature versus exit velocity.

SCHEMATIC & GIVEN DATA:

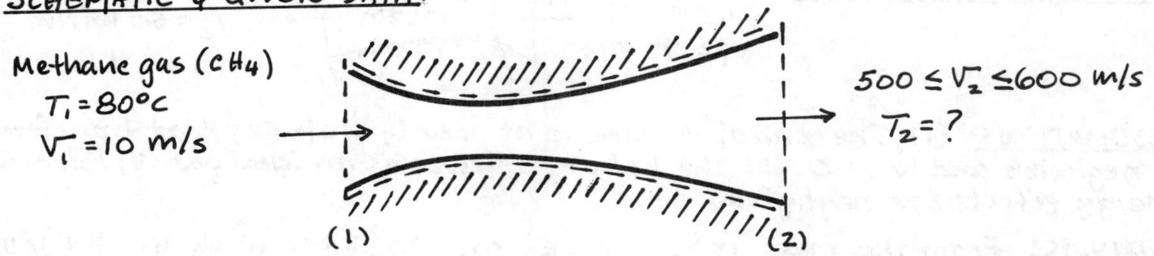

ASSUMPTIONS: (1) The control volume is at steady state. (2) For the control volume, $\dot{Q}_{cv} = \dot{W}_{cv} = 0$. (3) Potential energy effects can be ignored. (4) The methane behaves as an ideal gas.

ANALYSIS: Since $h = h(T)$ for the ideal gas, the exit temperature is determined by evaluating h_2 using steady state mass and energy balances:

$$0 = \dot{Q}_{cv}^{\,0} - \dot{W}_{cv}^{\,0} + \dot{m}\left[(h_1 - h_2) + \left(\frac{V_1^2 - V_2^2}{2}\right) + g(z_1 - z_2)^{0}\right]$$

where $\dot{m}_1 = \dot{m}_2 \equiv \dot{m}$. Thus, with $h = h(T)$

$$0 = h(T_1) - h(T_2) + \left(\frac{V_1^2 - V_2^2}{2}\right)\left|\frac{1\text{ N}}{1\text{ kg}\cdot\text{m/s}^2}\right|\left|\frac{1\text{ kJ}}{10^3\text{ N}\cdot\text{m}}\right| \qquad (*)$$

for h in kJ/kg and V in m/s.

Using IT, the functions $h(T)$ for methane (CH_4) as an ideal gas are accessed readily, and data for T_2 is obtained using the following code. The results are shown on the accompanying plot.

IT Code

```
T1 = 80  // °C
V1 = 10  // m/s
V2 = 550  // m/s

0 = (h1 - h2) + ((V1^2 - V2^2) / 2) * (1 / 10^3)  // kJ/kg
h1 = h_T("CH4", T1)
h2 = h_T("CH4", T2)

// Using the Explore button, sweep V2 from
// 500 to 600 in steps of 0.1.
```

(1) Result for V2 = 550: T2 = 13.63°C

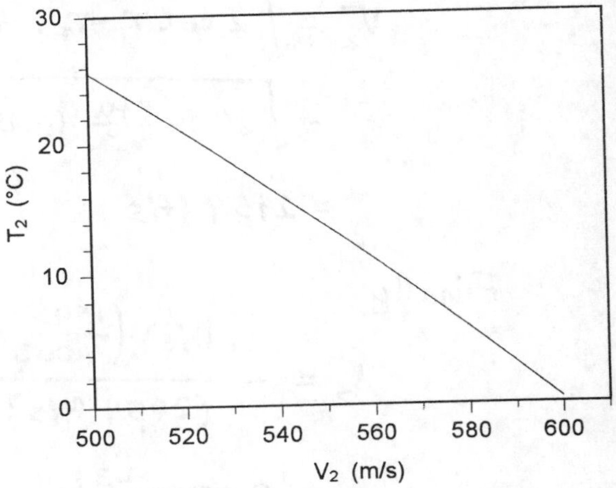

From (*), $h(T_2) = h(T_1) + \left(\frac{V_1^2 - V_2^2}{2}\right)$. Thus, as V_2 increases, $h(T_2)$ decreases. Therefore, T_2 decreases as expected.

1. A sample calculation using the c_p function from Table A-21 to evaluate the enthalpy change in (*) confirms this result. With IT, this integration is not necessary.

PROBLEM 4.27

KNOWN: Helium flows through a well-insulated nozzle with known conditions at the inlet and exit. The mass flow rate is given.

FIND: Determine the exit area.

SCHEMATIC & GIVEN DATA:

$T_1 = 600°R$
$V_1 = 175$ ft/s

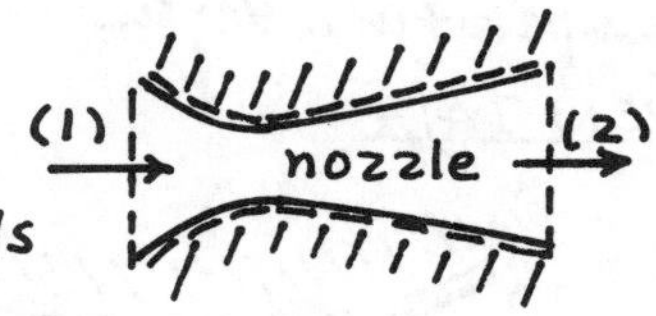

$T_2 = 460°R$
$p_2 = 50$ lbf/in²
$\dot{m} = 1$ lb/s

ASSUMPTIONS: (1) The control volume is at steady state. (2) Heat transfer is negligible and $\dot{W}_{cv} = 0$. (3) The helium behaves as an ideal gas. (4) Potential energy effects are negligible.

ANALYSIS: From the mass rate balance, $\dot{m}_1 = \dot{m}_2 \equiv \dot{m}$. With $\dot{m} = (AV)/v$

$$A_2 = \frac{\dot{m} v_2}{V_2}$$

Introducing the ideal gas equation of state

$$A_2 = \frac{\dot{m} R T_2}{V_2 \, p_2}$$

The exit velocity is found using the steady-state energy balance

$$0 = \dot{Q}_{cv}^{\,0} - \dot{W}_{cv}^{\,0} + \dot{m}\left[(h_1 - h_2) + \left(\frac{V_1^2 - V_2^2}{2}\right) + g(z_1 - z_2)^0\right]$$

From Table A-21, the specific heat c_p of helium is

$$c_p = (5/2)R = (5/2)\frac{(1.986 \text{ Btu/lbmol}\cdot°R)}{(4.003 \text{ lb/lbmol})} = 1.24 \text{ Btu/lb}\cdot°R$$

Since c_p is a constant $\Delta h = c_p \Delta T$, and

$$V_2 = \sqrt{2 c_p (T_1 - T_2) + V_1^2}$$

$$= \sqrt{2\left(1.24 \frac{\text{Btu}}{\text{lb}\cdot°R}\right)(600 - 460)°R \left|\frac{32.2 \text{ lb}\cdot\text{ft/s}^2}{1 \text{ lbf}}\right|\left|\frac{778 \text{ ft}\cdot\text{lbf}}{1 \text{ Btu}}\right| + (175^2)\frac{\text{ft}^2}{\text{s}^2}}$$

$= 2954$ ft/s

Finally

$$A_2 = \frac{(1 \text{ lb/s})\left(\frac{1545}{4.003}\frac{\text{ft}\cdot\text{lbf}}{\text{lb}\cdot°R}\right)(460°R)}{(2954 \text{ ft/s})(50 \text{ lbf/in}^2)}\left|\frac{1 \text{ ft}^2}{144 \text{ in}^2}\right|$$

$= 8.35 \times 10^{-3}$ ft² ← A_2

PROBLEM 4.28

KNOWN: Air flows through a nozzle with known conditions at the inlet and exit. Heat transfer occurs from the air to the surroundings.

FIND: Determine the exit velocity.

SCHEMATIC & GIVEN DATA:

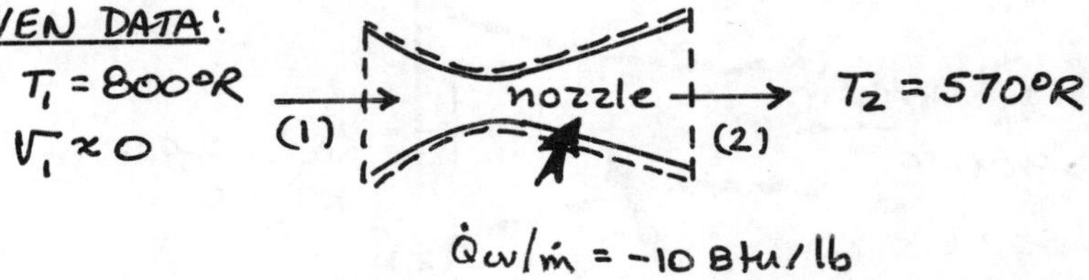

ASSUMPTIONS: (1) The control volume is at steady state. (2) For the control volume, $\dot{W}_{cv} = 0$. (3) The air behaves as an ideal gas. (4) The inlet kinetic energy and potential energy effects are negligible.

ANALYSIS: To determine the exit velocity, begin with the steady-state energy balance

$$0 = \dot{Q}_{cv} - \dot{W}_{cv}^{\,0} + \dot{m}\left[(h_1 - h_2) + \left(\frac{V_1^{2\,0} - V_2^2}{2}\right) + g(z_1 - z_2)^{0}\right]$$

Solving for V_2

$$V_2 = \sqrt{2\left[(h_1 - h_2) + \frac{\dot{Q}_{cv}}{\dot{m}}\right]}$$

From Table A-22E, $h_1 = 191.81$ Btu/lb and $h_2 = 136.26$ Btu/lb. Thus

$$V_2 = \sqrt{2\left[(191.81 - 136.26)\text{ Btu/lb} + (-10\text{ Btu/lb})\right]\left|\frac{778\text{ ft}\cdot\text{lbf}}{1\text{ Btu}}\right|\left|\frac{32.2\text{ ft}\cdot\text{lb/s}^2}{1\text{ lbf}}\right|}$$

$= 1511$ ft/s ← V_2

PROBLEM 4.29

KNOWN: Steam is decelerated in passing through a diffuser.

FIND: Determine the exit pressure.

SCHEMATIC & GIVEN DATA:

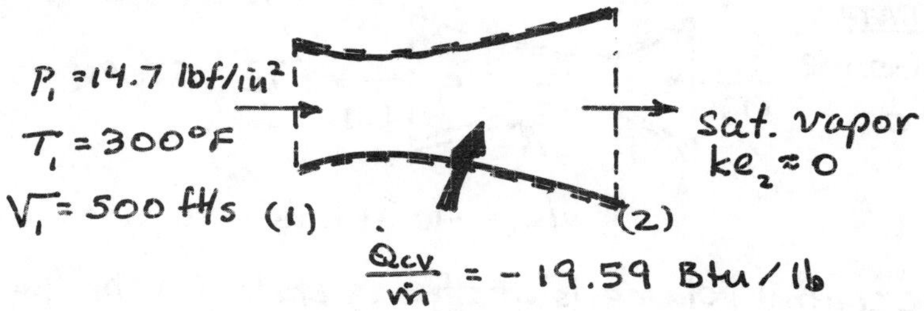

ASSUMPTIONS: (1) The control volume is at steady state. (2) For the control volume, $\dot{W}_{cv} = 0$. (3) The exit kinetic energy is negligible. (4) Potential energy effects are negligible.

ANALYSIS: Applying the energy balance at steady state to get h_2

$$0 = \dot{Q}_{cv} - \dot{W}_{cv}^{\nearrow 0} + \dot{m}\left[(h_1 - h_2) + \left(\frac{V_1^2 - V_2^2}{2}\right) + g(z_1 - z_2)\right]$$

where $\dot{m}_1 = \dot{m}_2 \equiv \dot{m}$. Thus

$$h_2 = \dot{Q}_{cv}/\dot{m} + h_1 + V_1^2/2$$

From Table A-4E at $p_1 = 14.7$ lbf/in², $T_1 = 300$°F ; $h_1 = 1192.6$ Btu/lb. Inserting values

$$h_2 = (-19.59 \text{ Btu/lb}) + (1192.6 \text{ Btu/lb}) + \left[\frac{(500^2)\text{ ft}^2/\text{s}^2}{2}\right]\left|\frac{1 \text{ lbf}}{32.2 \text{ ft}\cdot\text{lb/s}^2}\right|\left|\frac{1 \text{ Btu}}{778 \text{ ft}\cdot\text{lbf}}\right|$$

$= 1178$ Btu/lb

The pressure at the exit is the saturation pressure corresponding to $h_g = h_2 = 1178$ Btu/lb. From Table A-3E

$p_2 = 60$ lbf/in² ← p_2

PROBLEM 4.30

KNOWN: Data is provided for a diffuser at steady state, through which air is flowing.

FIND: Determine the ratio of the exit flow area to the inlet flow area, and the exit velocity.

SCHEMATIC & GIVEN DATA:

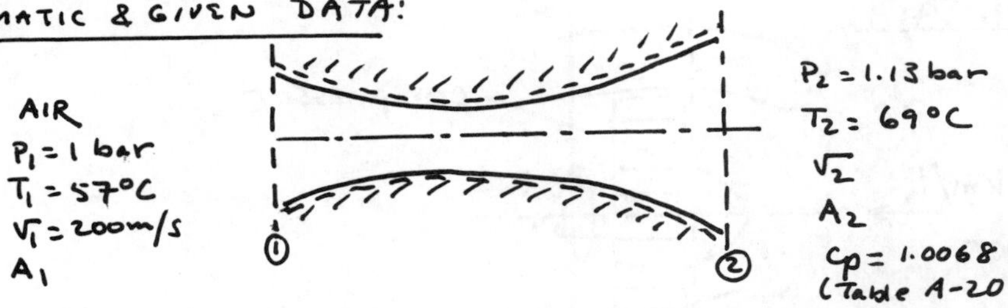

ASSUMPTIONS: 1. The control volume shown in the schematic is at steady state. 2. For the control volume, $\dot{Q}_{cv} = \dot{W}_{cv} = 0$, and potential energy effects can be ignored. 3. Air is modeled as an ideal gas with constant c_p.

ANALYSIS: The mass rate balance reads $\dot{m}_2 = \dot{m}_1$, or

$$\frac{A_2 V_2}{v_2} = \frac{A_1 V_1}{v_1} \Rightarrow \frac{A_2 V_2}{(RT_2/p_2)} = \frac{A_1 V_1}{(RT_1/p_1)} \Rightarrow \frac{A_2}{A_1} = \left(\frac{p_1}{p_2}\right)\left(\frac{T_2}{T_1}\right)\left(\frac{V_1}{V_2}\right) \quad (1)$$

The exit velocity, V_2, can be obtained from an energy rate balance:

$$0 = \cancel{\dot{Q}_{cv}} - \cancel{\dot{W}_{cv}} + \dot{m}\left[h_2 - h_1 + \frac{V_2^2 - V_1^2}{2} + g\cancel{(z_1 - z_2)}\right]$$

$$\Rightarrow \frac{V_2^2}{2} = \frac{V_1^2}{2} + h_1 - h_2 \Rightarrow V_2 = \sqrt{V_1^2 + 2(h_1 - h_2)}$$

Or, since c_p is constant,

$$V_2 = \sqrt{V_1^2 + 2c_p(T_1 - T_2)}$$

$$= \sqrt{\left(200\frac{m}{s}\right)^2 + 2\left(1.0068\frac{kJ}{kg \cdot K}\right)(-12\text{ K})\left|\frac{10^3 \text{ N}\cdot\text{m}}{\text{kJ}}\right|\left|\frac{1\text{ kg}\cdot\text{m/s}^2}{1\text{ N}}\right|}$$

$= 125.8$ m/s ← V_2

Returning to Eq. (1)

$$\frac{A_2}{A_1} = \left(\frac{1\text{ bar}}{1.13\text{ bar}}\right)\left(\frac{342\text{ K}}{330\text{ K}}\right)\left(\frac{200\text{ m/s}}{125.8\text{ m/s}}\right)$$

$= 1.458$ ← A_2/A_1

PROBLEM 4.31

KNOWN: Air enters the diffuser at the inlet of a jet engine and decelerates to zero velocity before entering the jet engine's compressor.

FIND: Determine the temperature of the air entering the compressor.

SCHEMATIC & GIVEN DATA:

$V_1 = 1000$ km/h
$P_1 = 0.6$ bar
$T_1 = 8$°C

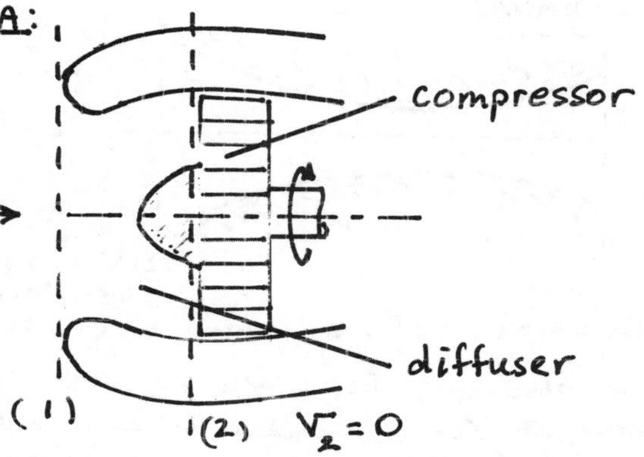

ASSUMPTIONS: (1) The control volume is at steady state. (2) Heat transfer is negligible and $\dot{W}_{cv} = 0$. (3) Potential energy effects are ① negligible. (4) The air behaves as an ideal gas.

ANALYSIS: Since $h = h(T)$ for an ideal gas, the exit temperature can be found by evaluating h_2. Beginning with the steady-state energy balance

$$0 = \dot{Q}_{cv}^{\nearrow 0} - \dot{W}_{cv}^{\nearrow 0} + \dot{m}\left[(h_1 - h_2) + \frac{(V_1^2 - V_2^{2\nearrow 0})}{2} + g(z_1 - z_3)^{\nearrow 0}\right]$$

where $\dot{m}_1 = \dot{m}_2 \equiv \dot{m}$, and using assumption (3)

$$0 = (h_1 - h_2) + \tfrac{1}{2}V_1^2$$

or

$$h_2 = h_1 + \tfrac{1}{2}V_1^2$$

From Table A-22; $h_1 = 281.3$ kJ/kg, and

$$h_2 = 281.3\,\frac{\text{kJ}}{\text{kg}} + \frac{1}{2}(1\times10^6)^2\,\frac{\text{m}^2}{\text{h}^2}\left|\frac{1\,\text{h}^2}{3600^2\,\text{s}^2}\right|\left|\frac{1\,\text{N}}{1\,\text{kg}\cdot\text{m/s}^2}\right|\left|\frac{1\,\text{kJ}}{10^3\,\text{N}\cdot\text{m}}\right|$$

$$= 319.9\ \text{kJ/kg}$$

Interpolating in Table A-22; $T_2 \approx 319.6$ K ← T_2

1. The applicability of the ideal gas model can be checked by reference to the compressibility chart.

PROBLEM 4.32

KNOWN: Ammonia passes through an insulated diffuser. Data are known at the inlet and exit.

FIND: Determine the exit temperature.

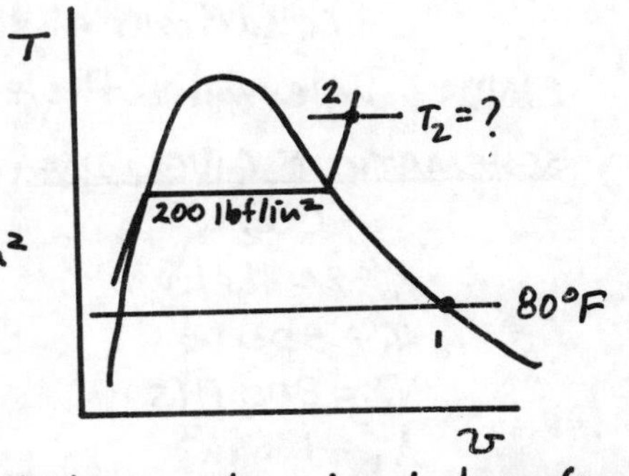

SCHEMATIC & GIVEN DATA:

Ammonia
$T_1 = 80°F$
sat. vapor
$V_1 = 1200$ ft/s

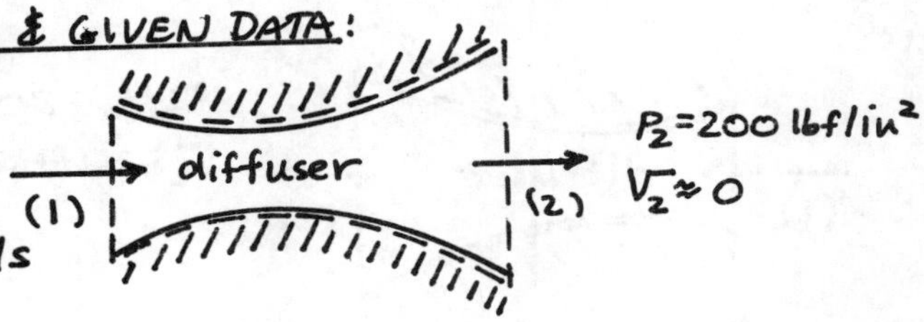

ASSUMPTIONS: (1) The control volume is at steady state. (2) The heat transfer is negligible and $\dot{W}_{cv} = 0$. (3) Potential energy effects and kinetic energy at exit can be neglected.

ANALYSIS: The exit pressure is given. The state is fixed by determining h_2 using a steady-state energy balance

$$0 = \dot{Q}_{cv}^{\nearrow 0} - \dot{W}_{cv}^{\nearrow 0} + \dot{m}\left[(h_1 - h_2) + \left(\frac{V_1^2 - V_2^2{}^{\nearrow 0}}{2}\right) + g(z_1 - z_2)^{\nearrow 0}\right]$$

where $\dot{m}_1 = \dot{m}_2 \equiv \dot{m}$. Solving for h_2

$$h_2 = h_1 + \frac{V_1^2}{2}$$

From Table A-13E, at $T_1 = 80°F$, $h_1 = h_g = 629.93$ Btu/lb. Thus

$$h_2 = 629.93 \text{ Btu/lb} + \frac{1}{2}(1200^2 \text{ ft}^2/\text{s}^2)\left|\frac{1 \text{ lbf}}{32.2 \text{ lb}\cdot\text{ft/s}^2}\right|\left|\frac{1 \text{ Btu}}{778 \text{ ft-lbf}}\right|$$

$$= 658.67 \text{ Btu/lb}$$

Interpolating in Table A-15E at $P_2 = 200$ lbf/in², $h_2 = 658.67$ Btu/lb gives

$$T_2 \approx 132.4°F \longleftarrow T_2$$

PROBLEM 4.33

KNOWN: Carbon dioxide flows through a well-insulated diffuser with known conditions at the inlet and exit.

FIND: Determine the exit temperature and pressure and the mass flow rate.

SCHEMATIC & GIVEN DATA:

CO_2

$p_1 = 20$ lbf/in²

$T_1 = 500°R$

$V_1 = 800$ ft/s

$A_1 = 1.4$ in²

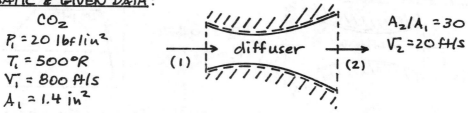

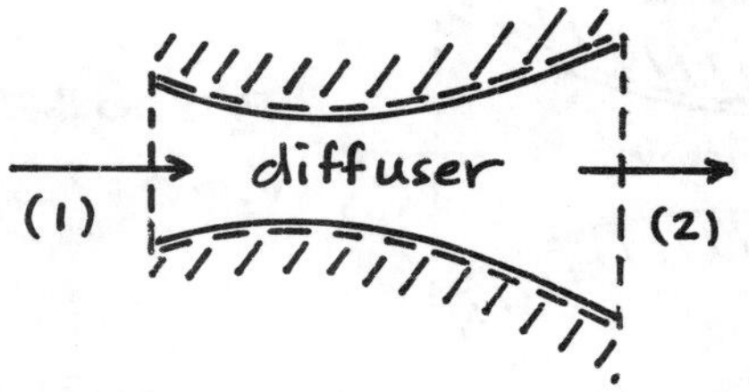

$A_2/A_1 = 30$

$V_2 = 20$ ft/s

ASSUMPTIONS: (1) The control volume is at steady state. (2) Heat transfer is negligible and $\dot{W}_{cv} = 0$. (3) Potential energy effects are negligible. (4) The carbon dioxide behaves as an ideal gas.

ANALYSIS: Since $h = h(T)$ for an ideal gas, the exit temperature can be found by evaluating h_2. Beginning with the steady-state energy balance

$$0 = \cancel{\dot{Q}_{cv}}^{0} - \cancel{\dot{W}_{cv}}^{0} + \dot{m}\left[(h_1 - h_2) + \left(\frac{V_1^2 - V_2^2}{2}\right) + g\cancel{(z_1 - z_2)}^{0}\right]$$

where $\dot{m}_1 = \dot{m}_2 \equiv \dot{m}$. Solving for h_2, and noting that $\bar{h} = h/M$

$$h_2 = h_1 + \left(\frac{V_1^2 - V_2^2}{2}\right) \Rightarrow \bar{h}_2 = \bar{h}_1 + \left(\frac{V_1^2 - V_2^2}{2}\right)M$$

where M is the molecular weight. From Table A-23E, $\bar{h}_1 = 3706.2$ Btu/lbmol and

$$\bar{h}_2 = 3706.2\,\frac{\text{Btu}}{\text{lbmol}} + \left(\frac{800^2 - 20^2}{2}\right)\frac{\text{ft}^2}{\text{s}^2}\left|\frac{1\ \text{lbf}}{32.2\ \text{lb}\cdot\text{ft/s}^2}\right|\left|\frac{44.01\ \text{lb}}{1\ \text{lbmol}}\right|\left|\frac{1\ \text{Btu}}{778\ \text{ft}\cdot\text{lbf}}\right|$$

$$= 4268\ \text{Btu/lbmol}$$

Interpolating in Table A-23E; $T_2 \approx 563.6°R$ ← T_2

To get p_2, begin with $\dot{m}_1 = \dot{m}_2$. Thus

$$\frac{A_1 V_1}{v_1} = \frac{A_2 V_2}{v_2}$$

with $pv = RT$

$$\frac{p_1}{RT_1}(A_1 V_1) = \frac{p_2}{RT_2}(A_2 V_2)$$

or

$$p_2 = p_1\left(\frac{T_2}{T_1}\right)\left(\frac{A_1}{A_2}\right)\left(\frac{V_1}{V_2}\right) = \left(20\,\frac{\text{lbf}}{\text{in}^2}\right)\left(\frac{563.6}{500}\right)\left(\frac{1}{30}\right)\left(\frac{800}{20}\right)$$

$$= 30.06\ \text{lbf/in}^2$$ ← p_2

Finally, the mass flow rate is

$$\dot{m} = \frac{(A_1 V_1)p_1}{RT_1} = \frac{(1.4\ \text{in}^2)(800\ \text{ft/s})(20\ \text{lbf/in}^2)}{\left(\frac{1545}{44.01}\,\frac{\text{ft}\cdot\text{lbf}}{\text{lb}\cdot°\text{R}}\right)(500°\text{R})}$$

$$= 1.276\ \text{lb/s}$$ ← $\dot{m}$

PROBLEM 4.34

KNOWN: Air expands through a turbine with known conditions at the inlet and exit. The power developed is known.

FIND: Determine the mass flow rate and the exit area.

SCHEMATIC & GIVEN DATA:

air

$p_1 = 10$ bar

$T_1 = 900$ K

$V_1 << V_2$

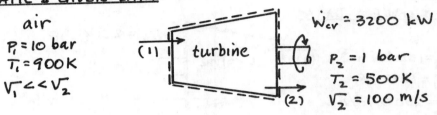

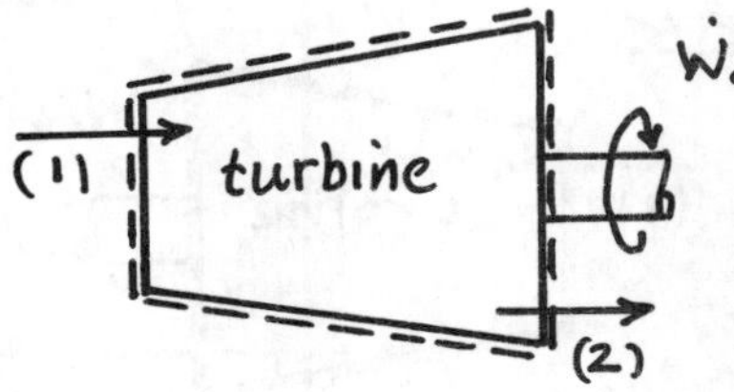

$\dot{W}_{cv} = 3200$ kW

$p_2 = 1$ bar

$T_2 = 500$ K

$V_2 = 100$ m/s

ASSUMPTIONS: (1) The control volume is at steady state. (2) Heat transfer is negligible. (3) Potential energy effects and kinetic energy at the inlet can ① be neglected. (4) The air behaves as an ideal gas.

ANALYSIS: Begin with a steady-state energy balance

$$0 = \dot{Q}_{cv}^{\nearrow 0} - \dot{W}_{cv} + \dot{m}\left[(h_1 - h_2) + \left(\frac{V_1^2 - V_2^2}{2}\right) + g(z_1 - z_2)^{\nearrow 0}\right]$$

where $\dot{m}_1 = \dot{m}_2 \equiv \dot{m}$. Solving for $\dot{m}$

$$\dot{m} = \frac{\dot{W}_{cv}}{(h_1 - h_2) - \frac{V_2^2}{2}}$$

From Table A-22; $h_1 = 932.93$ kJ/kg and $h_2 = 503.02$ kJ/kg. Thus

$$\dot{m} = \frac{(3200 \text{ kW}) \left|\frac{1 \text{ kJ/s}}{1 \text{ kW}}\right|}{(932.93 - 503.02) \text{ kJ/kg} - \left(\frac{100^2 \text{ m}^2/\text{s}^2}{2}\right)\left|\frac{1 \text{ N}}{\text{kg}\cdot\text{m/s}^2}\right|\left|\frac{1 \text{ kJ}}{10^3 \text{ N}\cdot\text{m}}\right|}$$

$= 7.53$ kg/s ← $\dot{m}$

The exit area is

$$A_2 = \frac{v_2 \dot{m}}{V_2} = \frac{R T_2 \dot{m}}{p_2 V_2}$$

$$= \frac{\left(\frac{8.314}{28.97} \frac{\text{kJ}}{\text{kg}\cdot\text{K}}\right)(500 \text{K})(7.53 \text{ kg/s})}{(1 \text{ bar})(100 \text{ m/s})}\left|\frac{1 \text{ bar}}{10^5 \text{ N/m}^2}\right|\left|\frac{10^3 \text{ N}\cdot\text{m}}{1 \text{ kJ}}\right|$$

$= 0.108 \text{ m}^2$ ← A_2

1. The applicability of the ideal gas model can be checked by reference to the compressibility chart.

PROBLEM 4.35

KNOWN: Air expands through a turbine with known conditions at the inlet and exit. The inlet volumetric flow rate and the power developed are given.

FIND: Determine the exit temperature.

SCHEMATIC & GIVEN DATA:

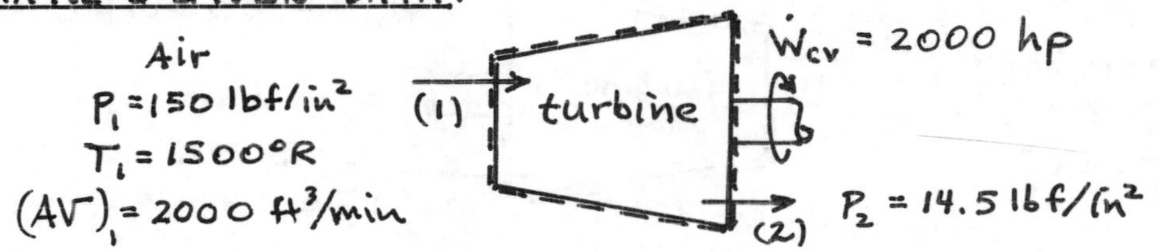

ASSUMPTIONS: (1) The control volume is at steady state. (2) Heat transfer is negligible. (3) Kinetic and potential energy effects are negligible. (4) The air behaves as an ideal gas.

ANALYSIS: Since $h = h(T)$ for an ideal gas, the exit temperature can be found by evaluating h_2. Beginning with the steady-state energy balance

$$0 = \dot{Q}_{cv}^{\,0} - \dot{W}_{cv} + \dot{m}\left[(h_1 - h_2) + \left(\frac{V_1^2 - V_2^2}{2}\right) + g(z_1 - z_2)\right]$$

where $\dot{m}_1 = \dot{m}_2 \equiv \dot{m}$, and with assumption (3)

$$0 = -\dot{W}_{cv} + \dot{m}(h_1 - h_2)$$

or

$$h_2 = -\dot{W}_{cv}/\dot{m} + h_1 \qquad (1)$$

The mass flow rate is evaluated next

$$\dot{m} = \frac{(AV)_1}{v_1} = \frac{(AV)_1 p_1}{RT_1}$$

$$= \frac{(2000 \text{ ft}^3/\text{min})(150 \text{ lbf/in}^2)}{\left(\frac{1545}{28.97}\frac{\text{ft}\cdot\text{lbf}}{\text{lb}\cdot°\text{R}}\right)(1500\ °\text{R})}\left|\frac{144 \text{ in}^2}{1 \text{ ft}^2}\right| = 540 \text{ lb/min}$$

Using data from Table A-22E for h_1 and inserting values into (1)

$$h_2 = -\frac{(2000 \text{ hp})}{(540 \text{ lb/min})}\left(\frac{1 \text{ h}}{60 \text{ min}}\right)\left(\frac{2545 \text{ Btu/h}}{1 \text{ hp}}\right) + 369.18 \text{ Btu/lb}$$

$$= 212.08 \text{ Btu/lb}$$

Interpolating in Table A-22E; $T_2 = 883.0$ °R ← T_2

1. The applicability of the ideal gas model can be checked by reference to the compressibility chart.

PROBLEM 4.36

KNOWN: A steam turbine operates at steady state with known inlet and exit conditions. The power developed and the heat transfer rate between the turbine and its surroundings are specified.

FIND: Determine the mass flow rate of steam.

SCHEMATIC & GIVEN DATA:

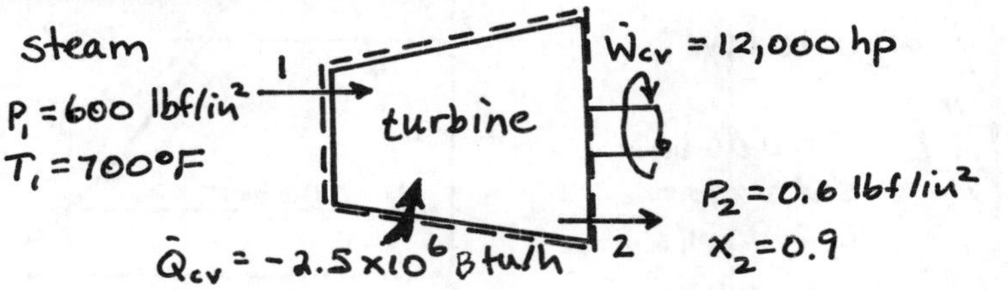

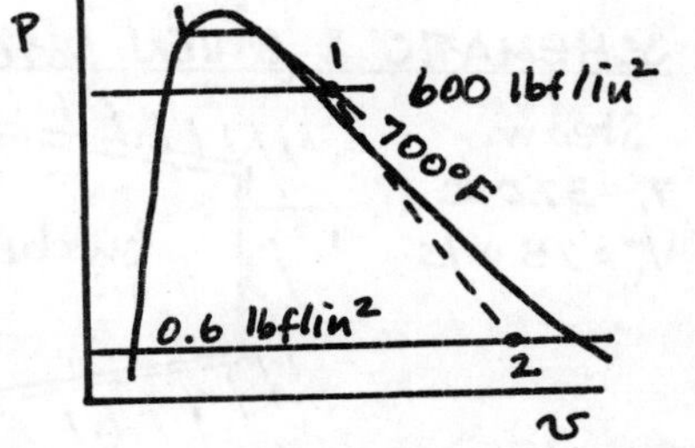

ASSUMPTIONS: (1) The control volume is at steady state. (2) Kinetic and potential energy changes from inlet to exit can be neglected.

ANALYSIS: To calculate the mass flow rate, begin with mass and energy rate balances for the one-inlet, one-exit control volume at steady state

$$0 = \dot{m}_1 - \dot{m}_2 \Rightarrow \dot{m}_1 = \dot{m}_2 \equiv \dot{m}$$

$$0 = \dot{Q}_{cv} - \dot{W}_{cv} + \dot{m}\left[(h_1 - h_2) + \left(\frac{V_1^2 - V_2^2}{2}\right)^0 + g(z_1 - z_2)^0\right]$$

Solving

$$\dot{m} = \frac{\dot{Q}_{cv} - \dot{W}_{cv}}{(h_2 - h_1)}$$

From Table A-4E, at $p_1 = 600$ lbf/in², $T_1 = 700°F$; $h_1 = 1350.6$ Btu/lb. From Table A-3E, at $p_2 = 0.6$ lbf/in²

$$h_2 = h_{f2} + x_2 h_{fg2}$$

$$= (53.27) + (.9)(1045.4) = 994.13 \text{ Btu/lb}$$

Thus

$$\dot{m} = \frac{(-2.5\times10^6 \text{ Btu/h}) - (12{,}000 \text{ hp})\left|\frac{2545 \text{ Btu/h}}{1 \text{ hp}}\right|}{(994.13 - 1350.6) \text{ Btu/lb}}$$

$$= 9.27\times10^4 \text{ lb/h} \leftarrow \dot{m}$$

PROBLEM 4.37

KNOWN: A well-insulated steam turbine operates at steady-state with known inlet and exit conditions. The power output and mass flow rate are specified.

FIND: Determine the inlet pressure.

SCHEMATIC & GIVEN DATA:

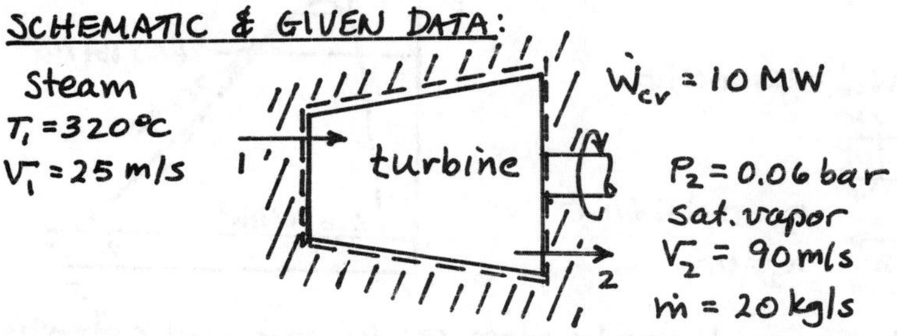

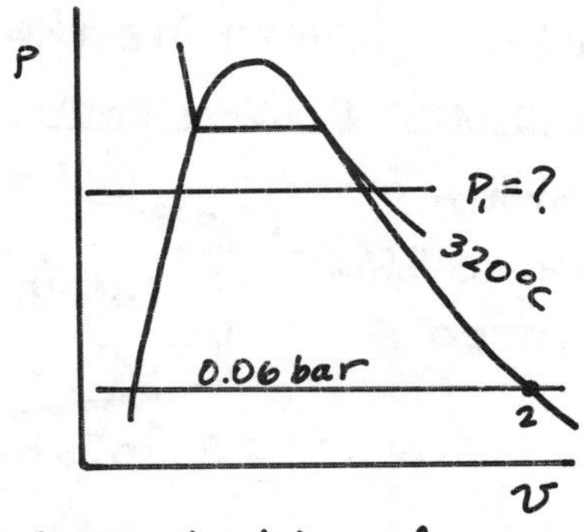

ASSUMPTIONS: (1) The control volume is at steady state. (2) Heat transfer can be neglected. (3) The potential energy change from inlet to exit can be neglected.

ANALYSIS: The temperature is known at the inlet. To fix the inlet state, h_1 is determined using mass and energy balances. For the control volume

$$\frac{dm_{cv}}{dt}^{\,0} = \dot{m}_1 - \dot{m}_2 \Rightarrow \dot{m}_1 = \dot{m}_2 \equiv \dot{m}$$

and

$$0 = \dot{Q}_{cv}^{\,0} - \dot{W}_{cv} + \dot{m}\left[(h_1 - h_2) + \left(\frac{V_1^2 - V_2^2}{2}\right) + g(z_1 - z_2)^{\,0}\right]$$

Solving for h_1

$$h_1 = \frac{\dot{W}_{cv}}{\dot{m}} + h_2 + \left(\frac{V_2^2 - V_1^2}{2}\right)$$

From Table A-3 for saturated vapor at $p_2 = 0.06$ bar; $h_2 = 2567.4$ kJ/kg. Thus

$$h_1 = \frac{(10\text{ MW})}{(20\text{ kg/s})}\left|\frac{10^3\text{ kW}}{1\text{ MW}}\right| + \left(2567.4\frac{\text{kJ}}{\text{kg}}\right) + \left(\frac{90^2 - 25^2}{2}\right)\frac{\text{m}^2}{\text{s}^2}\left|\frac{1\text{ N}}{1\text{ kg}\cdot\text{m/s}^2}\right|\left|\frac{1\text{ kJ}}{10^3\text{ N}\cdot\text{m}}\right|$$

$$= 3071.1\text{ kJ/kg}$$

Interpolating in Table A-4, with $T_1 = 320°C$ and $h_1 = 3071.1$ kJ/kg gives

$p_1 \approx 19.35$ bar ← p_1

PROBLEM 4.38

KNOWN: Data are provided for a turbine at steady state, through which N_2 is expanding.

FIND: Determine the power developed.

SCHEMATIC & GIVEN DATA:

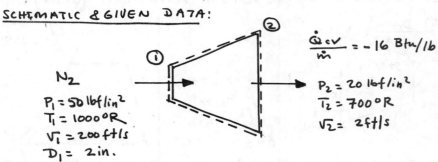

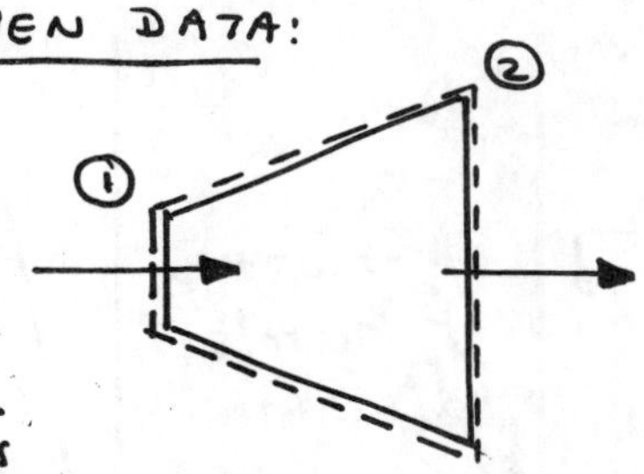

ASSUMPTIONS: 1. The control volume shown in the schematic is at steady state. 2. For the control volume, the change in potential energy from inlet to exit is negligible. 3. Nitrogen is modeled as an ideal gas.

ANALYSIS: Reducing the mass and energy rate balances,

$$0 = \dot{Q}_{cv} - \dot{W}_{cv} + \dot{m}\left[h_1 - h_2 + \frac{V_1^2 - V_2^2}{2} + \cancelto{0}{g(z_1 - z_2)}\right]$$

$$\Rightarrow \quad \dot{W}_{cv} = \dot{Q}_{cv} + \dot{m}\left[h_1 - h_2 + \frac{V_1^2 - V_2^2}{2}\right]$$

To find $\dot{m}$

$$\dot{m} = \frac{A_1 V_1}{v_1} = \frac{A_1 V_1}{(RT_1/P_1)} = \frac{(\pi D_1^2/4) V_1 P_1}{RT_1} = \frac{\left[\pi\left(\frac{2}{12}\text{ ft}\right)^2/4\right]\left[200\text{ ft/s}\right]\left[50\times144\text{ lbf/ft}^2\right]}{\left(\frac{1545}{28.01}\frac{\text{ft}\cdot\text{lbf}}{\text{lb}\cdot°\text{R}}\right)(1000°\text{R})}$$

$$= 0.57 \text{ lb/s}$$

Then, with enthalpy data from Table A-23E and $\dot{Q}_{cv} = (\dot{Q}_{cv}/\dot{m})(\dot{m})$

$$\dot{W}_{cv} = \left[-16\frac{\text{Btu}}{\text{lb}}\right]\left[0.57\frac{\text{lb}}{\text{s}}\right] + 0.57\frac{\text{lb}}{\text{s}}\left[\frac{(6977.9 - 4864.9)}{(28.01\text{ lb/lbmol})}\frac{\text{Btu}}{\text{lbmol}} + \right.$$

$$\left.\left[\frac{(200\text{ ft/s})^2 - (2\text{ ft/s})^2}{2}\right]\left|\frac{1\text{ lbf}}{32.2\text{ lbf-ft/s}^2}\right|\left|\frac{1\text{ Btu}}{778\text{ ft}\cdot\text{lbf}}\right|\right]$$

$$= 0.57\left[-16 + 75.44 + 0.8\right]\frac{\text{Btu}}{\text{s}}$$

$$= 34.34\frac{\text{Btu}}{\text{s}}\left|\frac{1\text{ hp}}{2545\text{ Btu/h}}\right|\left|\frac{3600\text{ s}}{1\text{ h}}\right| = 48.58\text{ hp} \quad \longleftarrow$$

PROBLEM 4.39

KNOWN: Steam expands through a well-insulated turbine. Inlet and exit conditions are known, and the exit diameter is given.

FIND: Plot the power developed by the turbine versus exit quality.

SCHEMATIC & GIVEN DATA:

$p_1 = 4$ MPa $= 40$ bar
$T_1 = 320°C$
$V_1 \approx 0$

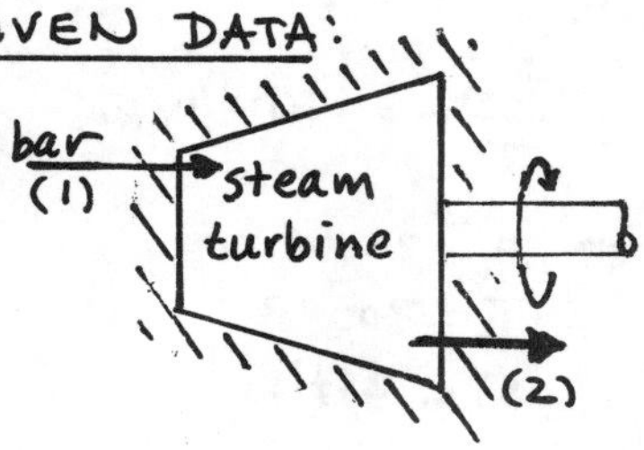

$p_2 = 0.07$ MPa
$V_2 = 90$ m/s
$d_2 = 0.6$ m
$0.9 \le x_2 \le 1.0$

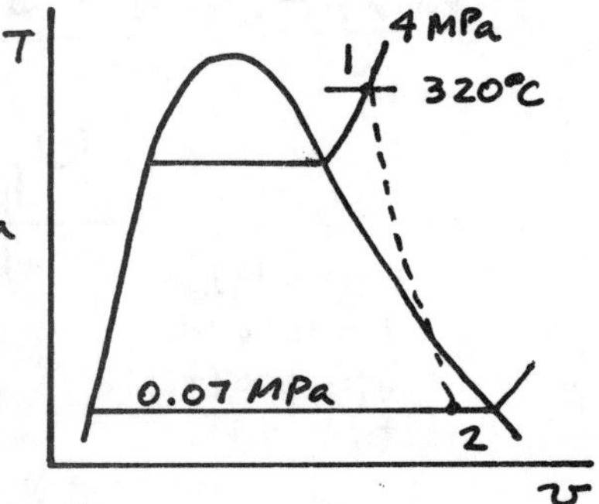

ASSUMPTIONS: (1) The control volume is at steady state. (2) Heat transfer is negligible. (3) Kinetic energy at the inlet and potential energy effects are negligible.

ANALYSIS: To determine the power, use the energy balance at steady state

$$0 = \dot{Q}_{cv}^{\,0} - \dot{W}_{cv} + \dot{m}\left[(h_1 - h_2) + \left(\frac{V_1^2 - V_2^2}{2}\right) + g(z_2 - z_1)\right]$$

where $\dot{m}_1 = \dot{m}_2 \equiv \dot{m}$. With assumption (3)

$$\dot{W}_{cv} = \dot{m}\left[(h_1 - h_2) - \frac{V_2^2}{2}\right] \qquad (*)$$

Evaluating the mass flow rate

$$\dot{m} = \frac{A_2 V_2}{v_2} = \frac{\pi d_2^2 V_2}{4 v_2} \qquad (**)$$

Sample Calculation: $\boxed{x_2 = 0.9}$ From Table A-4 at $p_1 = 4$ MPa $= 40$ bar, $T_1 = 320°C$; we get $h_1 = 3015.4$ kJ/kg. From Table A-3; $v_2 = 1.036 \times 10^{-3} + (.9)(2.365 - 1.036 \times 10^{-3})$ or $v_2 = 2.1286$ m³/kg. Similarly, $h_2 = 376.70 + (.9)(2283.3) = 2431.7$ kJ/kg. Now, from (**)

$$\dot{m} = \frac{\pi (.6)^2 \text{m}^2 (90 \text{ m/s})}{4(2.1286 \text{ m}^3/\text{kg})} = 11.95 \text{ kg/s}$$

Now, from (*)

$$\dot{W}_{cv} = (11.95 \tfrac{\text{kg}}{\text{s}})\left[(3015.4 - 2431.7)\frac{\text{kJ}}{\text{kg}} - \left(\frac{90^2}{2}\right)\frac{\text{m}^2}{\text{s}^2}\left|\frac{1 \text{ N}}{1 \text{ kg}\cdot\text{m/s}^2}\right|\left|\frac{1 \text{ kJ}}{10^3 \text{N}\cdot\text{m}}\right|\right]\left|\frac{1 \text{ kW}}{1 \text{kJ/s}}\right|$$

$$= 6927 \text{ kW}$$

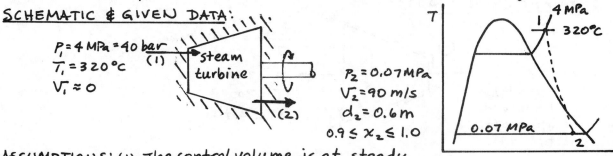

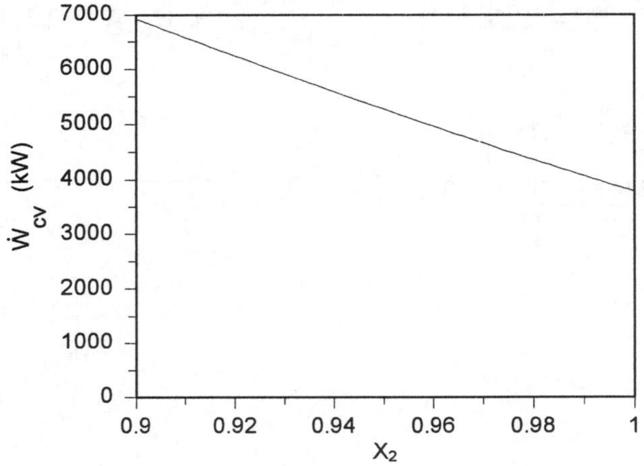

IT Code

```
p1 = 40  // bar
T1 = 320  // °C
p2 = 0.7  // bar
V2 = 90  // m/s
d2 = 0.6  // m
x2 = 0.9
h1 = h_PT("Water/Steam", p1, T1)
h2 = hsat_Px("Water/Steam", p2, x2)
v2 = vsat_Px("Water/Steam", p2, x2)
mdot = pi * d2^2 * V2 / (4 * v2)
0 = -Wdot + mdot * ((h1 - h2) - V2^2 / (2 * 1000))
// Using the Explore button, sweep x2 from
// 0.9 to 1.0 in steps of 0.001
```

PROBLEM 4.40

KNOWN: Water flows through a hydraulic turbine with known conditions at the inlet and exit. The power output is specified.

FIND: Determine the mass flow rate.

SCHEMATIC & GIVEN DATA:

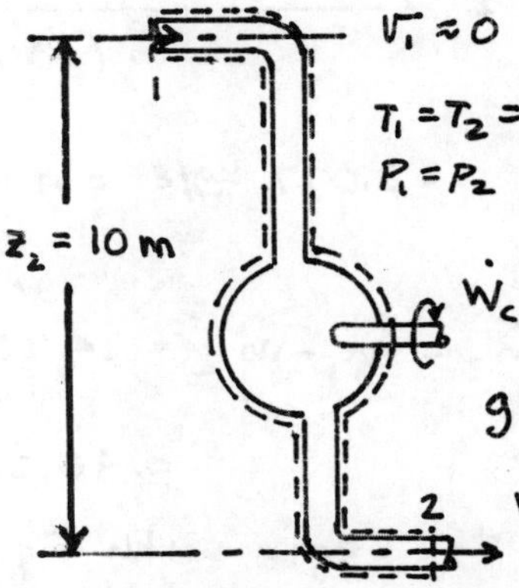

ASSUMPTIONS: (1) The control volume is at steady state. (2) Heat transfer is negligible. (3) Changes in temperature and pressure from inlet to exit are negligible. (4) Kinetic energy can be neglected at the inlet. (5) The acceleration of gravity is constant; $g = 9.81$ m/s².

ANALYSIS: To find the mass flow rate, begin with steady state mass and energy rate balances

$$0 = \dot{m}_1 - \dot{m}_2 \Rightarrow \dot{m}_1 = \dot{m}_2 \equiv \dot{m}$$

and

$$0 = \cancel{\dot{Q}_{cv}}^{0} - \dot{W}_{cv} + \dot{m}\left[\cancel{(h_1 - h_2)}^{0} + \left(\frac{\cancel{V_1^2}^{0} - V_2^2}{2}\right) + g(z_1 - z_2)\right]$$

where the enthalpy term is cancelled because of assumption (3). Solving

$$\dot{m} = \frac{\dot{W}_{cv}}{-\frac{V_2^2}{2} + g(z_1 - z_2)}$$

Inserting values

$$\dot{m} = \frac{(500 \text{ kW})\left|\frac{1 \text{ kJ/s}}{1 \text{ kW}}\right|}{\left[\left(-\frac{10^2}{2}\right)\frac{\text{m}^2}{\text{s}^2} + (9.81 \frac{\text{m}}{\text{s}^2})(10 \text{ m})\right]\left|\frac{1 \text{ N}}{1 \text{ kg} \cdot \text{m/s}^2}\right|\left|\frac{1 \text{ kJ}}{10^3 \text{ N} \cdot \text{m}}\right|}$$

$$= 10{,}400 \text{ kg/s} \leftarrow \dot{m}$$

PROBLEM 4.40 (Contd.)

Inserting values

$$\dot{m}_2 = \frac{-(11{,}400\ \text{kW})\left|\frac{1\ \text{kJ/s}}{1\ \text{kW}}\right| + (14.25\ \text{kg/s})(3230.9 - 2325.9)\frac{\text{kJ}}{\text{kg}}}{(2812.0 - 2325.8)\frac{\text{kJ}}{\text{kg}}}$$

$$= 3.077\ \text{kg/s} = 11{,}079\ \text{kg/h} \longleftarrow \dot{m}_2$$

From (*)

$$\dot{m}_3 = \dot{m}_1 - \dot{m}_2 = 14.25\ \frac{\text{kg}}{\text{s}}\ \frac{3600\ \text{s}}{1\ \text{h}} - 11{,}079\ \frac{\text{kg}}{\text{h}}$$

$$= 40{,}221\ \text{kg/h} \longleftarrow \dot{m}_3$$

(b) To find d_2, begin with Eq. 4.11 b

$$\dot{m}_2 = \frac{A_2 V_2}{v_2}$$

or

$$A_2 = \frac{\dot{m}_2 v_2}{V_2} = \frac{(3.077\ \text{kg/s})(0.4045\ \text{m}^3/\text{kg})}{(20\ \text{m/s})} = 0.06223\ \text{m}^2$$

where the value of v_2 is from Table A-4. Finally, with $A_2 = \pi d_2^2/4$

$$d_2 = \sqrt{\frac{4A_2}{\pi}} = \sqrt{\frac{(4)(0.6223\ \text{m}^2)}{\pi}} = 0.249\ \text{m} \longleftarrow d_2$$

PROBLEM 4.41

KNOWN: Steam passes through a well-insulated extraction turbine with known conditions at the inlet and exits. The power output is given.

FIND: Determine (a) the mass flow rate at each exit, and (b) the diameter of the duct where steam is extracted.

SCHEMATIC & GIVEN DATA:

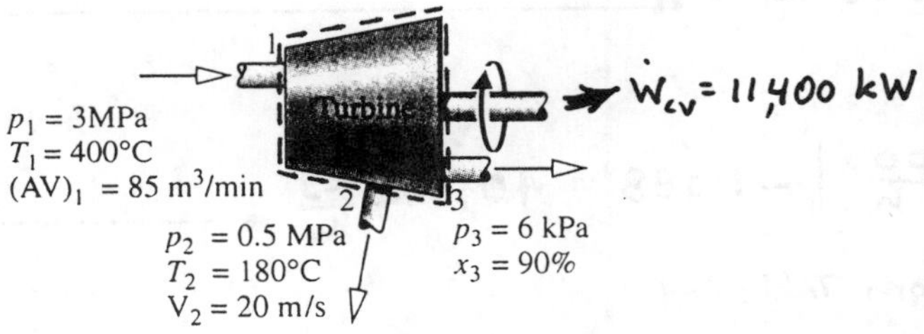

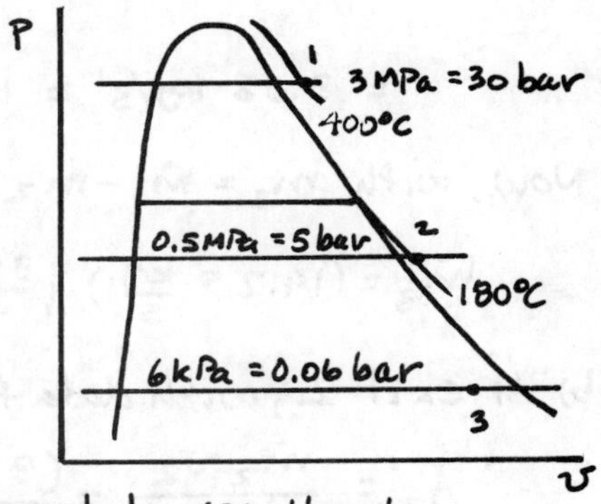

ASSUMPTIONS: (1) The control volume is at steady state. (2) Heat transfer is negligible. (3) Kinetic and potential energy effects can be neglected.

ANALYSIS: (a) To determine the mass flow rates $\dot{m}_2$ and $\dot{m}_3$, begin with a mass rate balance

$$\cancel{\frac{dm_{cv}}{dt}}^{0} = \dot{m}_1 - \dot{m}_2 - \dot{m}_3 \Rightarrow \dot{m}_3 = \dot{m}_1 - \dot{m}_2 \qquad (*)$$

Using Eq. 4.4b, with $v_1 = 0.0994$ m³/kg from Table A-4

$$\dot{m}_1 = \frac{(AV)_1}{v_1} = \frac{(85\ \text{m}^3/\text{min})}{(.0994\ \text{m}^3/\text{kg})}\left|\frac{1\ \text{min}}{60\ \text{s}}\right| = 14.25\ \text{kg/s}$$

With $\dot{m}_1$ known, (*) has two unknowns. Another relation is obtained using the energy rate balance.

$$\cancel{\frac{dE_{cv}}{dt}}^{0} = \cancel{\dot{Q}_{cv}}^{0} - \dot{W}_{cv} + \dot{m}_1\left(h_1 + \cancel{\frac{V_1^2}{2}} + \cancel{gz_1}\right) - \dot{m}_2\left(h_2 + \cancel{\frac{V_2^2}{2}} + \cancel{gz_2}\right) - \dot{m}_3\left(h_3 + \cancel{\frac{V_3^2}{2}} + \cancel{gz_3}\right)$$

where the indicated terms are deleted based on the assumptions. Thus

$$0 = -\dot{W}_{cv} + \dot{m}_1 h_1 - \dot{m}_2 h_2 - \dot{m}_3 h_3 \qquad (**)$$

With (*), we can reduce (**) as follows

$$0 = -\dot{W}_{cv} + \dot{m}_1 h_1 - \dot{m}_2 h_2 - (\dot{m}_1 - \dot{m}_2) h_3$$

$$= -\dot{W}_{cv} + \dot{m}_1 (h_1 - h_3) - \dot{m}_2 (h_2 - h_3)$$

Solving for $\dot{m}_2$

$$\dot{m}_2 = \frac{-\dot{W}_{cv} + \dot{m}_1 (h_1 - h_3)}{(h_2 - h_3)}$$

From Table A-4; $h_1 = 3230.9$ kJ/kg and $h_2 = 2812.0$ kJ/kg. The specific enthalpy at 3 is determined using data from Table A-3

$$h_3 = h_{f3} + x_3 h_{fg3} = 151.53 + (.9)2415.9 = 2325.8\ \text{kJ/kg}$$

PROBLEM 4.41 (Cont'd.)

Inserting values

$$\dot{m}_2 = \frac{-(11{,}400\ \text{kW})\left|\frac{1\ \text{kJ/s}}{1\ \text{kW}}\right| + (14.25\ \frac{\text{kg}}{\text{s}})(3230.9 - 2325.8)\frac{\text{kJ}}{\text{kg}}}{(2812.0 - 2325.8)\ \text{kJ/kg}}$$

$$= 3.08\ \text{kg/s} = 11088\ \text{kg/h} \longleftarrow \dot{m}_2$$

Now, with $\dot{m}_3 = \dot{m}_1 - \dot{m}_2$

$$\dot{m}_3 = (14.25\ \frac{\text{kg}}{\text{s}})\left|\frac{3600\ \text{s}}{1\ \text{h}}\right| - 11088 = 40{,}212\ \frac{\text{kg}}{\text{h}} \longleftarrow \dot{m}_3$$

(b) At exit 2, with data from Table A-4

$$A_2 = \frac{\dot{m}_2 v_2}{V_2} = \frac{(3.08\ \text{kg/s})(0.4045\ \text{m}^3/\text{kg})}{(20\ \text{m/s})} = 0.0623\ \text{m}^2$$

$$d_2 = \sqrt{\frac{4A_2}{\pi}} = 0.282\ \text{m} \longleftarrow d_2$$

PROBLEM 4.42

KNOWN: Steam passes through an extraction turbine operating at steady state with know inlet and exit conditions. The power output is specified.

FIND: Determine (a) the inlet mass flow rate, (b) the diameter of the extraction duct.

SCHEMATIC & GIVEN DATA:

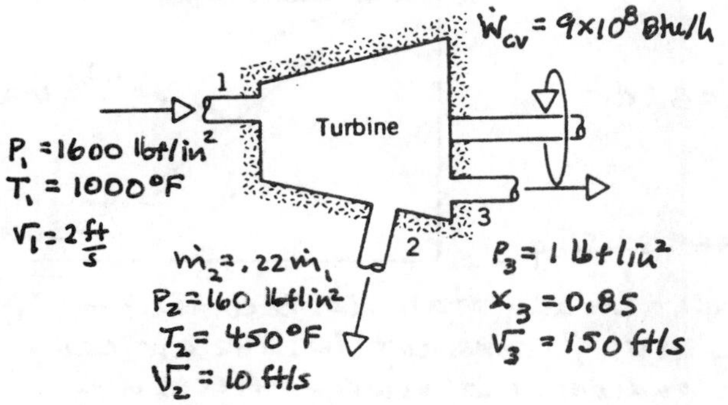

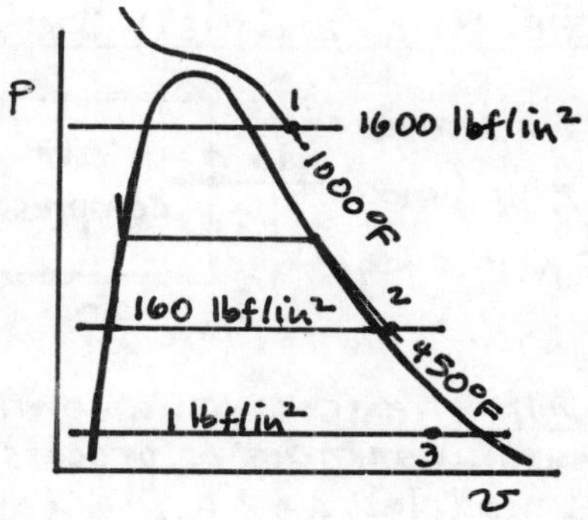

ASSUMPTIONS: 1. A control volume enclosing the turbine is at steady state. 2. For the control volume, heat transfer and potential energy effects are negligible.

ANALYSIS: (a) To find $\dot{m}_1$, apply mass rate balance : $\dot{m}_1 = \dot{m}_2 + \dot{m}_3$. Then, since $\dot{m}_2/\dot{m}_1 = 0.22$, we have $\dot{m}_3/\dot{m}_1 = 0.78$. Next, apply an energy rate balance:

$$0 = \dot{Q}_{cv}^{\,0} - \dot{W}_{cv} + \dot{m}_1\left[h_1 + \frac{V_1^2}{2}\right] - \dot{m}_2\left[h_2 + \frac{V_2^2}{2}\right] - \dot{m}_3\left[h_3 + \frac{V_3^2}{2}\right]$$

where the potential energy terms are omitted by assumption 2. Solving

$$\dot{m}_1 = \frac{\dot{W}_{cv}}{\left[h_1 + \frac{V_1^2}{2}\right] - \frac{\dot{m}_2}{\dot{m}_1}\left[h_2 + \frac{V_2^2}{2}\right] - \frac{\dot{m}_3}{\dot{m}_1}\left[h_3 + \frac{V_3^2}{2}\right]}$$

From Table A-4E, $h_1 = 1487$ Btu/lb, $h_2 = 1246.1$ Btu/lb. With Table A-3E data

$h_3 = h_{f3} + x_3(h_{g3} - h_{f3}) = 69.74 + 0.85\,(1036) = 950.3$ Btu/lb. Thus

$$\dot{m}_1 = \frac{(9. \times 10^8 \text{ Btu/h})}{\left[1487\frac{\text{Btu}}{\text{lb}} + \frac{(2\text{ ft/s})^2}{2}\left|\frac{1\text{ lbf}}{32.2\text{ lb}\cdot\text{ft/s}^2}\right|\left|\frac{1\text{ Btu}}{778\text{ ft}\cdot\text{lbf}}\right|\right] - 0.22\left[1246.1 + \frac{(10)^2}{(2)|32.2||778|}\right] - 0.78\left[950.3 + \frac{(150)^2}{(2)|32.2||778|}\right]}$$

$= 1.91 \times 10^6$ lb/h ← $\dot{m}_1$

(b) Using $v_2 = 3.228$ from Table A-4E

$$A_2 = \frac{\dot{m}_2 v_2}{V_2} = \frac{(.22)(1.91\times10^6 \text{ lb/h})(3.228 \text{ ft}^3/\text{lb})}{(10 \text{ ft/s})}\left|\frac{1\text{ h}}{3600\text{ s}}\right| = 37.68 \text{ ft}^2$$

Since $A_2 = \pi d_2^2/4$

$$d_2 = \sqrt{\frac{4A_2}{\pi}} = \sqrt{\frac{(4)(37.68\text{ ft}^2)}{\pi}} = 6.93\text{ ft}$$ ← d_2

PROBLEM 4.43

KNOWN: Air is compressed at steady state from a given initial state to a given final pressure. The mass flow is known, and each unit of mass undergoes a specified process in going from inlet to exit.

FIND: Determine the compressor power.

SCHEMATIC & GIVEN DATA:

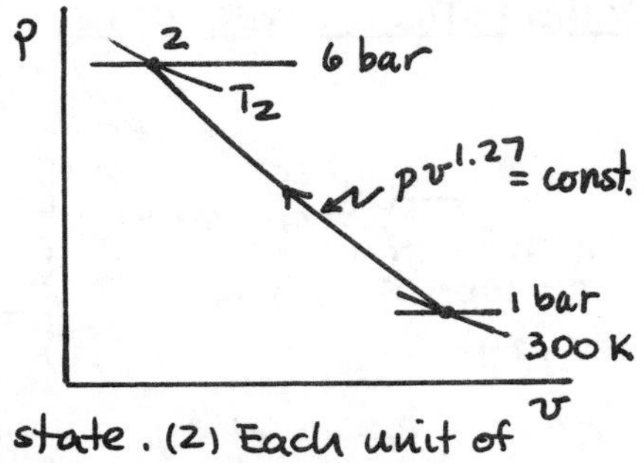

$T_1 = 300K$
$p_1 = 1$ bar
$\dot{m} = 4$ kg/s

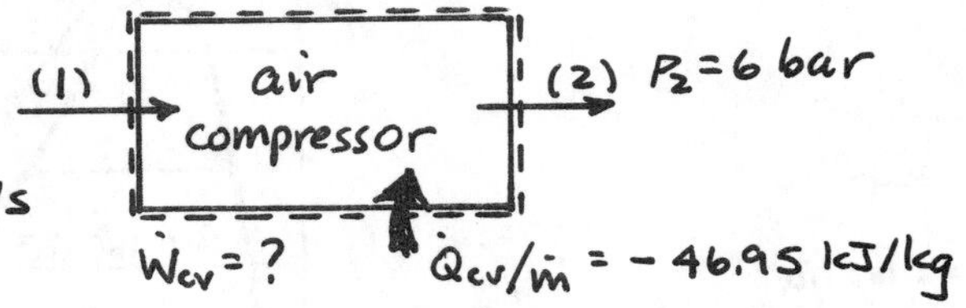

ASSUMPTIONS: (1) The control volume is at steady state. (2) Each unit of
① mass undergoes a process described by $pv^{1.27}$ = constant. (3) The air can be modeled as an ideal gas. (4) Kinetic and potential energy effects are negligible.

ANALYSIS: To determine the power, begin with mass and energy rate balances at steady state

$$0 = \dot{Q}_{cv} - \dot{W}_{cv} + \dot{m}\left[(h_1 - h_2) + \left(\frac{V_1^2 - V_2^2}{2}\right)^{0} + g(z_1 - z_2)^{0}\right]$$

where $\dot{m}_1 = \dot{m}_2 \equiv \dot{m}$ and the indicated terms are deleted by assumption 4. Rearranging and solving for $\dot{W}_{cv}$

$$\dot{W}_{cv} = \dot{m}\left[(\dot{Q}_{cv}/\dot{m}) + (h_1 - h_2)\right] \qquad (*)$$

Now, specific enthalpy h_1 is read from Table A-22 at 300K: $h_1 = 300.19 \frac{kJ}{kg}$.
To get T_2, we use Eq. 3.56 with $n = 1.27$

$$\frac{T_2}{T_1} = \left(\frac{p_2}{p_1}\right)^{\frac{n-1}{n}} \Rightarrow T_2 = \left(\frac{p_2}{p_1}\right)^{\frac{n-1}{n}} T_1 = \left(\frac{6}{1}\right)^{\frac{1.27-1}{1.27}} (300K) = 439.1\ K$$

Interpolating in Table A-22; $h_2 = 440.7$ kJ/kg. Inserting values in (*)

$$\dot{W}_{cv} = \left(4\frac{kg}{s}\right)\left[\left(-46.95\frac{kJ}{kg}\right) + (300.19 - 440.7)\frac{kJ}{kg}\right]\left|\frac{1\ kW}{1\ kJ/s}\right|$$

② $= -750$ kW ◀ $\dot{W}_{cv}$

---

1. The applicability of the ideal gas model can be checked by reference to the generalized compressibility chart.
2. The negative sign for power denotes energy transfer by work into the control volume, as expected.

PROBLEM 4.44

KNOWN: A well-insulated air compressor operates with known inlet and exit states. The inlet volumetric flow rate is also known.

FIND: Determine the compressor power and the exit volumetric flow rate.

SCHEMATIC & GIVEN DATA:

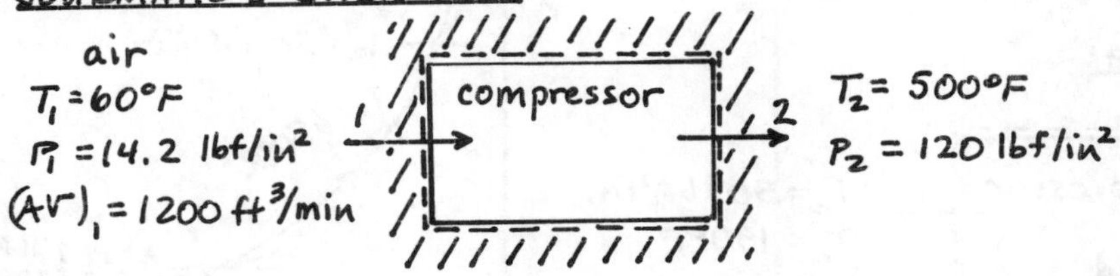

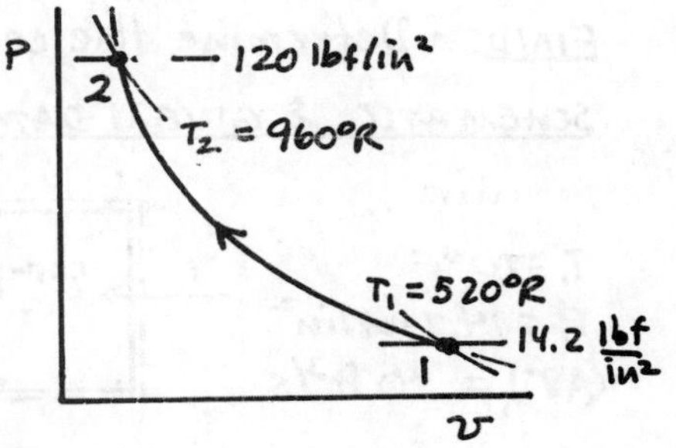

ASSUMPTIONS: (1) The control volume is at steady
① state. (2) Heat transfer is negligible. (3) The air behaves as an ideal gas. (4) Kinetic and potential energy changes from inlet to exit can be neglected.

ANALYSIS: To find the power, begin with steady state mass and energy balances

$$0 = \cancel{\dot{Q}_{cv}}^{0} - \dot{W}_{cv} + \dot{m}\left[(h_1 - h_2) + \cancel{\left(\frac{V_1^2 - V_2^2}{2}\right)}^{0} + g\cancel{(z_1 - z_2)}\right]$$

where $\dot{m}_1 = \dot{m}_2 \equiv \dot{m}$. Solving for $\dot{W}_{cv}$

$$\dot{W}_{cv} = \dot{m}\,[h_1 - h_2]$$

To evaluate $\dot{m}$, use Eq. 4.11b and the ideal gas equation of state

$$\dot{m} = \frac{(AV)_1}{v_1} = \frac{p_1 (AV)_1}{RT_1}$$

$$= \frac{(14.2\ \text{lbf/in}^2)(1200\ \text{ft}^3/\text{min})}{\left(\frac{1545}{28.97}\ \frac{\text{ft}\cdot\text{lbf}}{\text{lb}\cdot°\text{R}}\right)(520°\text{R})}\left|\frac{60\ \text{min}}{1\ \text{h}}\right|\left|\frac{144\ \text{in}^2}{1\ \text{ft}^2}\right| = 5309\ \text{lb/h}$$

Thus, with $h_1 = 124.27$ Btu/lb and $h_2 = 231.06$ Btu/lb from Table A-22E

$$\dot{W}_{cv} = \left(5309\ \frac{\text{lb}}{\text{h}}\right)(124.27 - 231.06)\frac{\text{Btu}}{\text{lb}}\left|\frac{1\ \text{hp}}{2545\ \text{Btu/h}}\right|$$

② $= -222.8$ hp ← $\dot{W}_{cv}$

The exit volumetric flow rate is

$$(AV)_2 = \dot{m}\, v_2 = \dot{m}\left(\frac{RT_2}{p_2}\right)$$

$$= (5309\ \text{lb/h})\left|\frac{1\ \text{h}}{60\ \text{min}}\right| \frac{\left(\frac{1545}{28.97}\ \frac{\text{ft}\cdot\text{lbf}}{\text{lb}\cdot°\text{R}}\right)(960°\text{R})}{(120\ \text{lbf/in}^2)}\left|\frac{1\ \text{ft}^2}{144\ \text{in}^2}\right|$$

$= 262$ ft³/min ← $(AV)_2$

1. The applicability of the ideal gas model can be checked by reference to the generalized compressibility chart.
2. The negative sign for power denotes energy transfer <u>into</u> the control volume, as expected.

PROBLEM 4.45

KNOWN: An air compressor operates with known inlet an exit conditions. The inlet volumetric flow rate and the heat transfer rate per unit mass of air flow are also specified.

FIND: Determine the compressor power.

SCHEMATIC & GIVEN DATA:

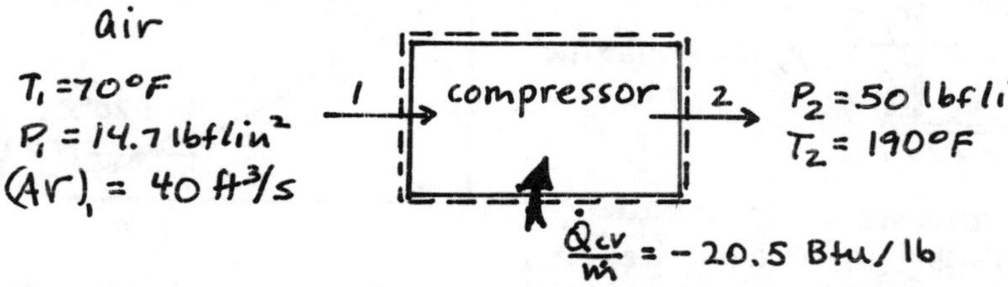

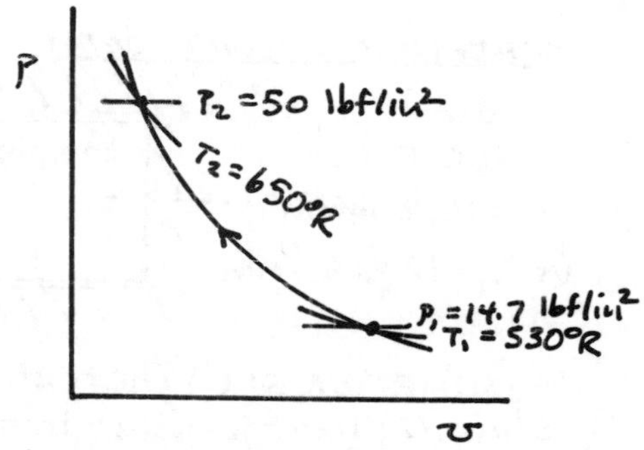

ASSUMPTIONS: (1) The control volume is at steady state. (2) Kinetic and ① potential energy changes from inlet to exit are negligible. (3) The air behaves as an ideal gas.

ANALYSIS: To find the compressor power, begin with steady state mass and energy rate balances

$$0 = \dot{Q}_{cv} - \dot{W}_{cv} + \dot{m}\left[(h_1 - h_2) + \left(\frac{V_1^2 - V_2^2}{2}\right)^{\nearrow 0} + g(z_1 - z_2)^{\nearrow 0}\right]$$

where $\dot{m}_1 = \dot{m}_2 \equiv \dot{m}$. Solving

$$\dot{W}_{cv} = \dot{m}\left[\left(\frac{\dot{Q}_{cv}}{\dot{m}}\right) + (h_1 - h_2)\right]$$

To evaluate $\dot{m}$, use Eq. 4.11b and the ideal gas equation of state

$$\dot{m} = \frac{(AV)_1}{v_1} = \frac{P_1 (AV)_1}{RT_1}$$

$$= \frac{(14.7\ \text{lbf/in}^2)(40\ \text{ft}^3/\text{s})}{\left(\frac{1545}{28.97}\frac{\text{ft}\cdot\text{lbf}}{\text{lb}\cdot°\text{R}}\right)(530°\text{R})}\left|\frac{144\ \text{in}^2}{1\ \text{ft}^2}\right| = 3.00\ \text{lb/s}$$

Interpolating in Table A-22E; $h_1 = 126.66$ Btu/lb and $h_2 = 155.5$ Btu/lb. Thus

$$\dot{W}_{cv} = \left(3.00\ \frac{\text{lb}}{\text{s}}\right)\left[(-20.5) + (126.66 - 155.5)\right]\frac{\text{Btu}}{\text{lb}}\left|\frac{3600\ \text{s}}{1\ \text{h}}\right|\left|\frac{1\ \text{hp}}{2545\ \text{Btu/h}}\right|$$

② $= -209.4$ hp ← $\dot{W}_{cv}$

---

1. The applicability of the ideal gas model can be checked by reference to the generalized compressibility chart.
2. The negative sign for power denotes energy transfer <u>into</u> the compressor, as expected.

PROBLEM 4.46

KNOWN: Data are provided for a R134a compressor operating at steady state.

FIND: Determine the power required by the compressor.

SCHEMATIC & GIVEN DATA:

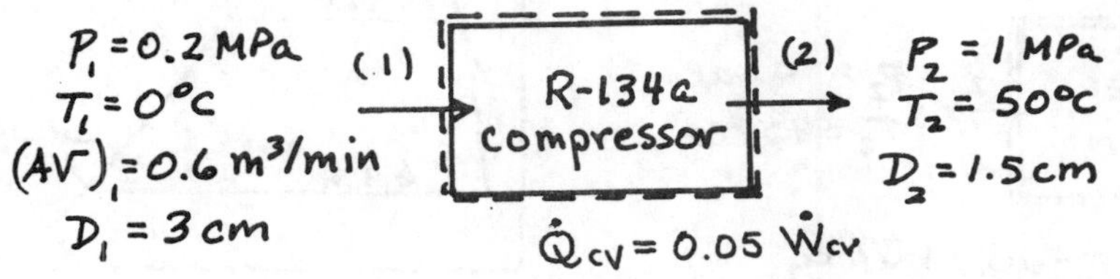

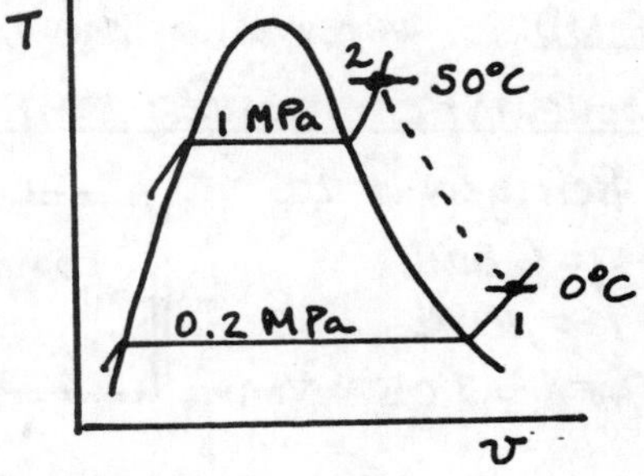

ASSUMPTIONS: (1) The control volume is at steady state. (2) Potential energy effects can be neglected.

ANALYSIS: Noting that $\dot{W}_{cv}$ is negative for a compressor, the rate of heat transfer from the compressor is given by $\dot{Q}_{cv} = 0.05\,\dot{W}_{cv}$. For steady-state operation, the mass and energy rate balances reduce to

$$0 = \dot{Q}_{cv} - \dot{W}_{cv} + \dot{m}\left[h_1 - h_2 + \frac{V_1^2 - V_2^2}{2} + g(z_1 - z_2)\right]$$

$$\Rightarrow \dot{W}_{cv} = (0.05\,\dot{W}_{cv}) + \dot{m}\left[h_1 - h_2 + \frac{V_1^2 - V_2^2}{2}\right] \Rightarrow \dot{W}_{cv} = \frac{\dot{m}\left[h_1 - h_2 + \frac{V_1^2 - V_2^2}{2}\right]}{0.95}$$

From Table A-12, $h_1 = 250.1$ kJ/kg, $h_2 = 280.19$ kJ/kg, $v_1 = 0.10438$ m³/kg, $v_2 = 0.02171$ m³/kg. The mass flow rate is

$$\dot{m} = \frac{(AV)_1}{v_1} = \frac{0.6\ \text{m}^3/\text{min}}{0.10438\ \text{m}^3/\text{kg}} = 5.748\ \text{kg/min} = 0.0958\ \text{kg/s}$$

Further

$$A_1 = \frac{\pi D_1^2}{4} = \frac{\pi (3\times 10^{-2}\ \text{m})^2}{4} = 7.0686\times 10^{-4}\ \text{m}^2$$

$$A_2 = 1.767\times 10^{-4}\ \text{m}^2$$

Thus, the velocities at the inlet and exit are, respectively

$$V_1 = \frac{(AV)_1}{A_1} = \frac{(0.6\ \text{m}^3/\text{min})}{(7.0686\times 10^{-4}\ \text{m}^2)}\left|\frac{1\ \text{min}}{60\ \text{s}}\right| = 14.15\ \text{m/s}$$

$$V_2 = \frac{\dot{m}\,v_2}{A_2} = \frac{(0.0958\ \text{kg/s})(0.02171\ \text{m}^3/\text{kg})}{(1.767\times 10^{-4}\ \text{m}^2)} = 11.77\ \text{m/s}$$

The required power is then

$$\dot{W}_{cv} = \frac{(0.0958\ \text{kg/s})\left[(250.1 - 280.19)\ \text{kJ/kg} + \left[\frac{(14.15\ \text{m/s})^2 - (11.77)^2}{2}\left|\frac{1\ \text{N}}{1\ \text{kg}\cdot\text{m/s}^2}\right|\left|\frac{1\ \text{kJ}}{10^3\ \text{N}\cdot\text{m}}\right|\right]\right]}{0.95}$$

$$= \frac{(0.0958)\left[-30.09 + (0.03)\right]\ \text{kJ/s}}{0.95} = -3.03\ \frac{\text{kJ}}{\text{s}}\left|\frac{1\ \text{kW}}{1\ \text{kJ/s}}\right|$$

$$= -3.03\ \text{kW}$$

PROBLEM 4.47

KNOWN: A Refrigerant 22 compressor has known conditions at the inlet and exit. The inlet volumetric flow rate and the compressor power per unit mass of refrigerant flowing are also specified.

FIND: Determine the heat transfer rate.

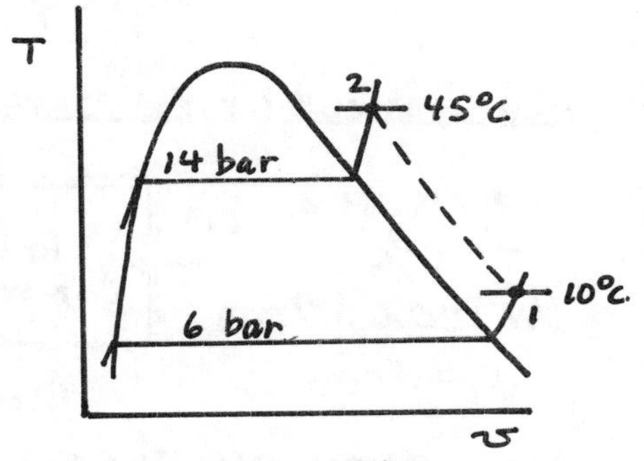

SCHEMATIC & GIVEN DATA:

Refrigerant 22
$p_1 = 6$ bar
$T_1 = 10°C$
$(AV)_1 = 2.05\ m^3/min$

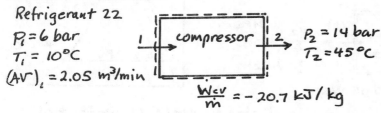

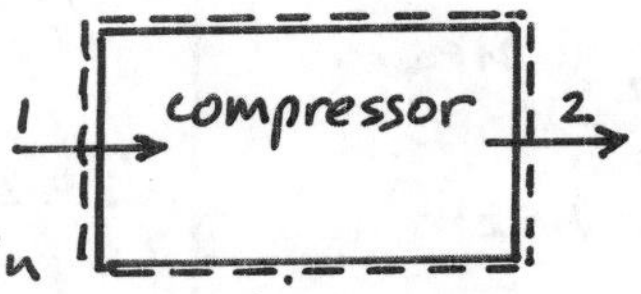

$p_2 = 14$ bar
$T_2 = 45°C$

$\frac{\dot{W}_{cv}}{\dot{m}} = -20.7\ kJ/kg$

ASSUMPTIONS: (1) The control volume is at steady state. (2) Kinetic and potential energy changes from inlet to exit can be neglected.

ANALYSIS: To find the heat transfer rate, begin with steady state energy and mass rate balances

$$0 = \dot{Q}_{cv} - \dot{W}_{cv} + \dot{m}\left[(h_1 - h_2) + \left(\frac{V_1^2 - V_2^2}{2}\right)^{0} + g(z_1 - z_2)^{0}\right]$$

where $\dot{m}_1 = \dot{m}_2 \equiv \dot{m}$. Solving

$$\dot{Q}_{cv} = \dot{m}\left[\left(\frac{\dot{W}_{cv}}{\dot{m}}\right) + (h_2 - h_1)\right]$$

To evaluate $\dot{m}$, use Eq. 4.11b and data for v_1 from Table A-9

$$\dot{m} = \frac{(AV)_1}{v_1} = \frac{2.05\ m^3/min}{0.04015\ m^3/kg} = 51.06\ kg/min$$

From Table A-9; $h_1 = 255.14$ kJ/kg, and interpolating; $h_2 \approx 268.3$ kJ/kg. Thus

$$\dot{Q}_{cv} = \left(51.06\ \frac{kg}{min}\right)\left[(-20.7) + (268.3 - 255.14)\right]\frac{kJ}{kg}\left(\frac{1\ min}{60\ s}\right)\left(\frac{1\ kW}{1\ kJ/s}\right)$$

$$= -6.42\ kW \longleftarrow \dot{Q}_{cv}$$

PROBLEM 4.48

KNOWN: Data are provided for a CO_2 compressor operating at steady state.

FIND: Determine the power required by the compressor.

SCHEMATIC & GIVEN DATA:

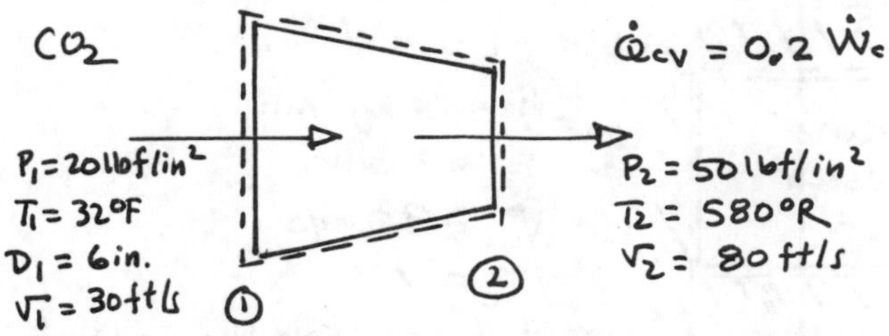

ASSUMPTIONS: 1. The control volume shown in the figure is at steady state. 2. Potential energy effects can be ignored. 3. CO_2 is modeled as an ideal gas.

ANALYSIS: Since the compressor requires a power input, the heat transfer from the compressor is given by $\dot{Q}_{cv} = 0.2\dot{W}_{cv}$. To evaluate $\dot{W}_{cv}$, reduce the mass and energy rate balances

$$0 = \dot{Q}_{cv} - \dot{W}_{cv} + \dot{m}\left[h_1 - h_2 + \frac{V_1^2 - V_2^2}{2} + g(\cancel{z_1 - z_2})^{0}\right]$$

With $\dot{Q}_{cv} = 0.2\dot{W}_{cv}$

$$\dot{W}_{cv} = \frac{\dot{m}\left[h_1 - h_2 + \frac{V_1^2 - V_2^2}{2}\right]}{0.8}$$

Using data at the inlet, the mass flow rate is

$$\dot{m} = \frac{A_1 V_1}{v_1} = \frac{A_1 V_1 p_1}{RT_1} = \frac{\left[\frac{\pi}{4}(0.5\ ft)^2\right]\left[30\ ft/s\right]\left[(20)(144)\ lbf/ft^2\right]}{\left(\frac{1545}{44.01}\ \frac{ft\cdot lbf}{lb\cdot °R}\right)(492°R)} = 0.982\ lb/s$$

From Table A-23E, $\bar{h}_2 = 4417.2$ kJ/kmol, $\bar{h}_1 = 3637.7$ kJ/kmol. Thus

$$\dot{W}_{cv} = \frac{(0.982\ lb/s)\left[\left(\frac{3637.7 - 4417.2}{44.01}\right)\frac{Btu}{lb} + \left[\frac{(30\ ft/s)^2 - (80)^2}{2}\right]\left|\frac{1\ lbf}{32.2\ lb\cdot ft/s^2}\right|\left|\frac{1\ Btu}{778\ ft\cdot lbf}\right|\right]}{0.8}$$

$$= \frac{(0.982)}{0.8}\left[-17.71 + (-0.11)\right]\frac{Btu}{s} = -21.87\ \frac{Btu}{s}$$

$$= -21.8\ \frac{Btu}{s}\left|\frac{3600\ s}{1\ h}\right|\left|\frac{1\ hp}{2545\ Btu/h}\right|$$

$$= -30.84\ hp \longleftarrow \dot{W}_{cv}$$

PROBLEM 4.49

KNOWN: Methane gas is compressed with negligible heat transfer. At the inlet, pressure, temperature, and velocity are given. Pressure and velocity are specified at the exit. The power input is known.

FIND: Determine the exit temperature.

SCHEMATIC & GIVEN DATA:

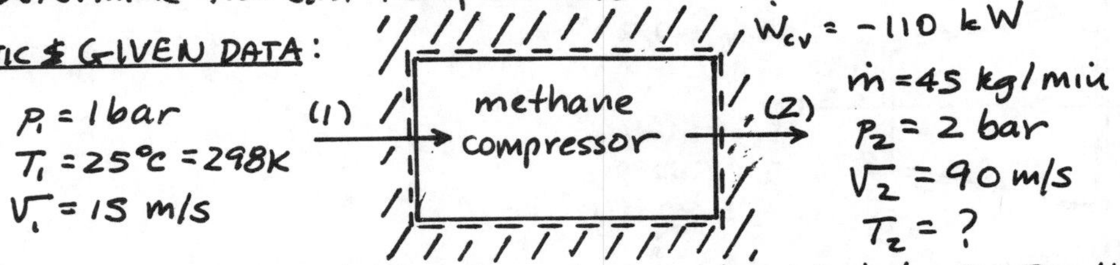

$\dot{W}_{cv} = -110$ kW
$p_1 = 1$ bar
$T_1 = 25°C = 298$K
$V_1 = 15$ m/s
(1)
methane
compressor
(2)
$\dot{m} = 45$ kg/min
$p_2 = 2$ bar
$V_2 = 90$ m/s
$T_2 = ?$

ASSUMPTIONS: (1) The control volume is at steady state. (2) For the control volume, $\dot{Q}_{cv} = 0$ and potential energy effects are negligible. (3) The methane
① gas can be modeled as an ideal gas.

ANALYSIS: By assumption (3), $h_2 = h(T_2)$. Thus, using energy and mass rate balances we can solve for h_2 and hence T_2. That is

$$0 = \overset{0}{\cancel{\dot{Q}_{cv}}} - \dot{W}_{cv} + \dot{m}\left[(h_1 - h_2) + \left(\frac{V_1^2 - V_2^2}{2}\right) + g(\overset{0}{\cancel{z_1 - z_2}})\right]$$

or

$$0 = -\dot{W}_{cv} + \dot{m}\left[(h_1 - h_2) + \left(\frac{V_1^2 - V_2^2}{2}\right)\right] \qquad (*)$$

The term $h_1 - h_2$ can be expressed in terms of $c_p(T)$, as follows

$$h_1 - h_2 = \int_{T_2}^{T_1} c_p(T)\, dT \qquad (**)$$

One alternative to evaluate the integral is use the function for c_p in Table A-21. The result would be a polynomial in which T_2 would be unknown. Using the result in conjunction with Eq. (*), an iterative procedure or an equation-solving computer program would be needed.

A simpler alternative is to use the $h(T)$ function for methane included in IT, as follows:

IT Code

```
    T1 = 25 + 273  // K
    V1 = 15  // m/s
    V2 = 90  // m/s
    mdot = 45 / 60  // kg/s
    Wdot = -110  // kW
②  0 = - Wdot + mdot * ((h1 - h2) + (V1^2 - V2^2) / (2 * 1000))
    h1 = h_T("CH4", T1)
③  h2 = h_T("CH4", T2)
```

IT Result

T2 = 359.9 K ← T_2

1. The applicability of the ideal gas model can be checked by reference to the compressibility chart.
2. Carefully note the unit conversions required in this expression.
3. This result compares very favorably with the result obtaine using the alternative method involving Table A-21 data.

PROBLEM 4.50

KNOWN: An ammonia compressor has known conditions at the inlet and exit. The compressor power and the heat transfer rate are also specified.

FIND: Determine the inlet volumetric flow rate using ammonia tables and ideal gas relationships. Discuss.

SCHEMATIC & GIVEN DATA:

Ammonia
$P_1 = 20$ lbf/in²
$T_1 = 0°F$

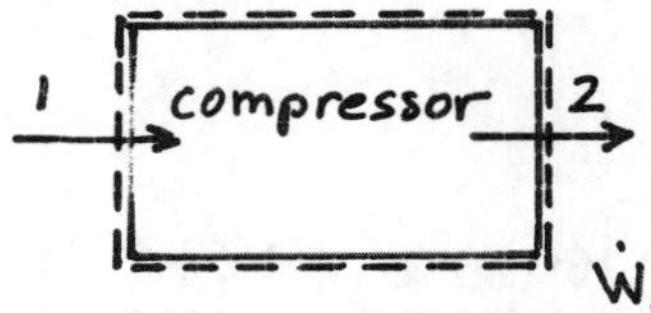

$P_2 = 250$ lbf/in²
$T_2 = 300°F$
$\dot{W}_{cv} = -10$ hp
$\dot{Q}_{cv} = -5000$ Btu/h

ASSUMPTIONS: (1) The control volume is at steady state. (2) Kinetic and potential energy changes from inlet to exit are negligible. (3) For the second part, the ammonia behaves as an ideal gas.

ANALYSIS: To begin, determine the mass flow rate by using the steady-state mass and energy balances

$$0 = \dot{Q}_{cv} - \dot{W}_{cv} + \dot{m}\left[(h_1 - h_2) + \left(\frac{V_1^2 - V_2^2}{2}\right)^{\nearrow 0} + g(z_1 - z_2)^{\nearrow 0}\right]$$

where $\dot{m}_1 = \dot{m}_2 \equiv \dot{m}$. Solving

$$\dot{m} = \frac{\dot{Q}_{cv} - \dot{W}_{cv}}{(h_2 - h_1)} \qquad (*)$$

From Table A-15E; $h_1 = 614.84$ Btu/lb and $h_2 = 760.39$ Btu/lb. Thus

$$\dot{m} = \frac{(-5000 \text{ Btu/h}) - (-10 \text{ hp})\left|\frac{2545 \text{ Btu/h}}{1 \text{ hp}}\right|}{(760.39 - 614.84)\text{ Btu/lb}}\left|\frac{1 \text{ h}}{60 \text{ min}}\right| = 2.342 \text{ lb/min}$$

Now, using $v_1 = 14.078$ ft³/lb from Table A-15E and Eq. 4-11b

$$(AV)_1 = \dot{m} v_1 = (2.342)(14.078) = 32.97 \text{ ft}^3/\text{min} \longleftarrow \underset{\text{(Ammonia table)}}{(AV)_1}$$

To evaluate $h_2 - h_1$ for an ideal gas, integrate the specific heat function $c_p(T)$ for ammonia from Table A-21E

$$h_2 - h_1 = \frac{\bar{R}}{M}\int_{T_1}^{T_2}(\alpha + \beta T + \gamma T^2 + \delta T^3 + \epsilon T^4)\,dT$$

$$= \frac{\bar{R}}{M}\left[\alpha(T_2 - T_1) + \frac{\beta}{2}(T_2^2 - T_1^2) + \frac{\gamma}{3}(T_2^3 - T_1^3) + \frac{\delta}{4}(T_2^4 - T_1^4) + \frac{\epsilon}{5}(T_2^5 - T_1^5)\right]$$

With $T_1 = 460°R$, $T_2 = 760°R$, and coefficient values from Table A-21E

$$h_2 - h_1 = \frac{(1.986 \text{ Btu/lbmol}\cdot°R)}{(17.04 \text{ lb/lbmol})}\left[1329 \text{ °R}\right]$$

$$= 154.9 \text{ Btu/lb}$$

Using this result with (*)

$$\dot{m} = \frac{(-5000) - (-10)(2545)}{(154.9)} \left| \frac{1}{60} \right| = 2.2 \text{ lb/min}$$

Incorporating the ideal gas equation of state into Eq. 4-11b

$$(AV)_1 = \dot{m} v_1 = \dot{m}\left(\frac{RT_1}{P_1}\right)$$

$$= \left(2.2 \frac{\text{lb}}{\text{min}}\right) \frac{\left(\frac{1545}{17.04} \cdot \frac{\text{ft}\cdot\text{lbf}}{\text{lb}\cdot{}^\circ\text{R}}\right)(460^\circ\text{R})}{(20 \text{ lbf/in}^2)} \left| \frac{1 \text{ ft}^2}{144 \text{ in}^2} \right|$$

$$= 31.86 \text{ ft}^3/\text{min} \longleftarrow (AV)_1 \text{ (ideal gas)}$$

Discussion: The % deviation in assuming ideal gass behavior is

$$\% \text{ deviation} = \left[\frac{(AV)_{tables} - (AV)_{ideal\ gas}}{(AV)_{tables}}\right] \times 100$$

$$= \left[\frac{32.97 - 31.86}{32.97}\right] \times 100 = 3.4\%$$

Thus, the ideal gas model is reasonably accurate. To explore this further, consider

$$\% \text{ deviation} = \left[\frac{\Delta h_{ideal\ gas} - \Delta h_{tables}}{\Delta h_{tables}}\right] \times 100$$

$$= \left[\frac{154.9 - 145.55}{145.55}\right] \times 100 = 6.4\%$$

The applicability of the ideal gas model can also be checked by determining the compressibility factor. For state 1

$$Z_1 = \frac{P_1 v_1}{RT_1} = \frac{(20)|144|(14.078)}{\left(\frac{1545}{17.04}\right)(460)} = 0.972$$

For state 2, $v_2 = 1.8191$ ft³/lb from Table A-15E. Thus

$$Z_2 = \frac{(250)|144|(1.8191)}{\left(\frac{1545}{17.04}\right)(760)} = 0.95$$

Both of these values are reasonably close to unity.

PROBLEM 4.51

KNOWN: Refrigerant 134a is compressed in a water-jacketed compressor between two known states. The inlet volumetric flow rate, input power, and cooling water temperature rise are given.

FIND: Determine the mass flow rate of the cooling water.

SCHEMATIC & GIVEN DATA:

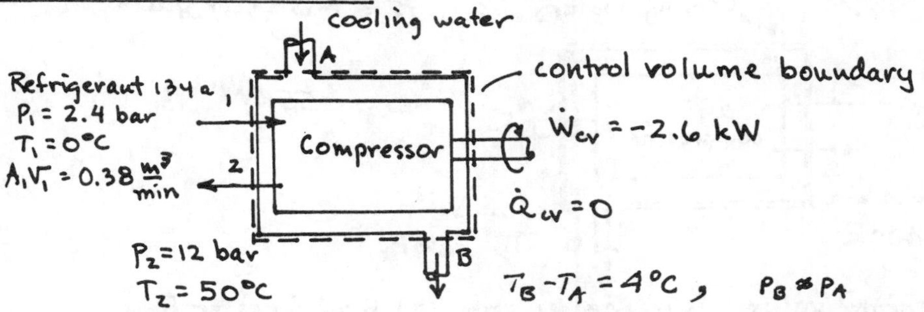

ASSUMPTIONS: (1) The control volume shown is at steady-state. (2) There is no heat transfer with the surroundings. (3) Kinetic and potential energy effects can be neglected. (4) The cooling water is incompressible with $c = 4.179$ kJ/kg·K from Table A-19.

ANALYSIS: The energy rate balance applied to the overall compressor and water jacket at steady state reduces to

$$0 = \dot{Q}_{cv}^{\nearrow 0} - \dot{W}_{cv} + \dot{m}_1\left(h_1 + \frac{V_1^2}{2} + gz_1\right) - \dot{m}_2\left(h_2 + \frac{V_2^2}{2} + gz_2\right) + \dot{m}_A\left(h_A + \frac{V_A^2}{2} + gz_A\right) - \dot{m}_B\left(h_B + \frac{V_B^2}{2} + gz_B\right)$$

where $\dot{Q}_{cv} = 0$ by assumption (2), and the underlined terms drop out by assumption (3). Since the water and refrigerant streams are separate

$$\dot{m}_1 = \dot{m}_2 \equiv \dot{m}_R$$
$$\dot{m}_A = \dot{m}_B \equiv \dot{m}_W$$

Thus

$$0 = -\dot{W}_{cv} + \dot{m}_R(h_1 - h_2) + \dot{m}_W(h_A - h_B)$$

Applying Eq. 3.20b for the water stream

$$h_A - h_B = c(T_A - T_B) + v(p_A - p_B)^{\nearrow 0}$$

Inserting this result and solving for $\dot{m}_W$

$$\dot{m}_W = \frac{-\dot{W}_{cv} + \dot{m}_R(h_1 - h_2)}{c(T_B - T_A)} \qquad (*)$$

From Table A-12; $h_1 = 248.89$ kJ/kg, $v_1 = 0.08574$ m³/kg and $h_2 = 275.52$ kJ/kg.

Evaluating $\dot{m}_R$

$$\dot{m}_R = \frac{(AV)_1}{v_1} = \frac{(0.38\ \text{m}^3/\text{min})}{(0.08574\ \text{m}^3/\text{kg})}\left|\frac{1\ \text{min}}{60\ \text{s}}\right| = 0.0739\ \text{kg/s}$$

Finally, inserting values in (*)

$$\dot{m}_W = \frac{-(-2.6\ \text{kW})\left|\frac{1\ \text{kJ/s}}{1\ \text{kW}}\right| + (0.0739\ \text{kg/s})(248.89 - 275.52)\frac{\text{kJ}}{\text{kg}}}{(4.179\ \text{kJ/kg·K})(4\ ^\circ\text{C})} = 0.0378\ \frac{\text{kg}}{\text{s}} \longleftarrow \dot{m}_W$$

PROBLEM 4.52

KNOWN: Data are provided for a water-jacketed air compressor operating at steady state.

FIND: Determine the cooling water temperature increase for a specified cooling water mass flow rate. Plot the cooling water temperature increase versus cooling water mass flow rate.

SCHEMATIC & GIVEN DATA:

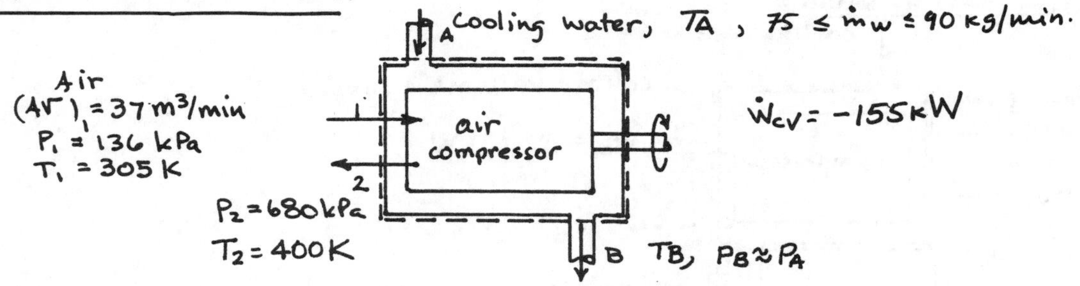

ASSUMPTIONS: (1) The control volume is at steady state. (2) Heat transfer from the outside of the cooling water jacket is negligible. (3) Kinetic and potential energy effects are negligible. (4) The air behaves as an ideal gas. (5) The cooling water is incompressible with $c_p = 4.179$ kJ/kg·K from Table A-19.

ANALYSIS: (a) The steady-state energy balance for the overall compressor is

$$0 = \dot{Q}_{cv}^{\,0} - \dot{W}_{cv} + \dot{m}_1\left(h_1 + \frac{V_1^2}{2} + gz_1\right) - \dot{m}_2\left(h_2 + \frac{V_2^2}{2} + gz_2\right) + \dot{m}_A\left(h_A + \frac{V_A^2}{2} + gz_A\right) - \dot{m}_B\left(h_B + \frac{V_B^2}{2} + gz_B\right)$$

where the indicated terms drop out by assumptions (2) and (3). Since the water and air streams are separate

$$\dot{m}_1 = \dot{m}_2 \equiv \dot{m}_{air}$$
$$\dot{m}_A = \dot{m}_B \equiv \dot{m}_w$$

Thus

$$0 = -\dot{W}_{cv} + \dot{m}_{air}(h_1 - h_2) + \dot{m}_w(h_A - h_B) \qquad (1)$$

With assumption 5, the enthalpy change of the cooling water can be found using Eq. 3.20b:

$$h_B - h_A = c(T_B - T_A) + v\cancel{(p_B - p_A)} = c(T_B - T_A)$$

Accordingly, Eq. (1) gives

$$T_B - T_A = \frac{\left[-\dot{W}_{cv} + \dot{m}_{air}(h_1 - h_2)\right]}{c\,\dot{m}_w}$$

Using the data at location 1 to evaluate $\dot{m}_{air}$

$$\dot{m}_{air} = \frac{(AV)_1}{v_1} = \frac{(AV)_1 p_1}{RT_1} = \frac{(37\ m^3/min)(136\ kPa)}{\left(\frac{8.314}{28.97}\frac{kJ}{kg\cdot K}\right)(305\ K)}\left|\frac{10^3\ N/m^2}{1\ kPa}\right|\left|\frac{1\ kJ}{10^3\ N\cdot m}\right|$$

$$= 57.49\ kg/min$$

Also, from Table A-22, $h_1 = 305.22$ kJ/kg, $h_2 = 400.48$ kJ/kg

Collecting results

$$T_B - T_A = \frac{\left[-(-155\ kJ/s)(60\ s/min) + (57.49\ kg/min)(305.22 - 400.48)\ kJ/kg\right]}{(4.179\ kJ/kg\cdot K)\ \dot{m}_w}$$

PROBLEM 4.52 (Contd.)

or

$$(T_B - T_A) = \frac{3823.5 \text{ kJ/min}}{(4.179 \text{ kJ/kg}\cdot\text{K})\, \dot{m}_w}$$

$$= \frac{914.93 \text{ kg}\cdot\text{K/min}}{\dot{m}_w} \qquad (*)$$

For the case of $\dot{m}_w = 82$ kg/min; $T_B - T_A = 11.2$ K ← ΔT_w (part a)

(b) Equation (*) can be plotted readily using a spreadsheet, plotting program, or IT. The IT plot follows:

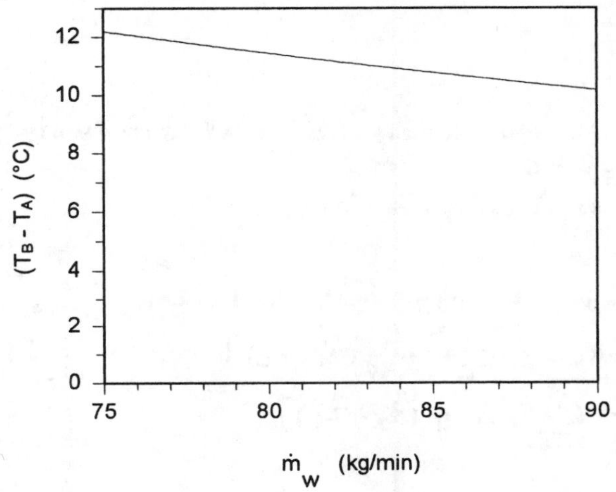

PROBLEM 4.53

KNOWN: Water is steadily pumped through a piping arrangement to a higher elevation where it is discharged with a known mass flow rate.

FIND: Determine the power required by the pump.

SCHEMATIC & GIVEN DATA:

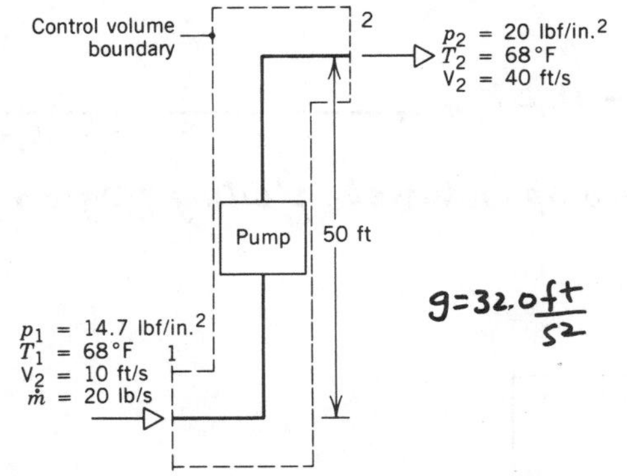

$g = 32.0 \frac{ft}{s^2}$

ASSUMPTIONS:
1. The control volume shown in the schematic is at steady state.
(1) 2. For the control volume, $\dot{Q}_{cv} \sim 0$.
(2) 3. The water is modeled as incompressible.
4. $g = 32.0\ ft/s^2$

ANALYSIS: Reducing mass and energy rate balances

$$0 = \dot{Q}_{cv}^{\,0} - \dot{W}_{cv} + \dot{m}\left[h_1 - h_2 + \frac{V_1^2 - V_2^2}{2} + g(z_1 - z_2)\right]$$

$$\Rightarrow \quad (-\dot{W}_{cv}) = \dot{m}\left[(h_2 - h_1) + \frac{V_2^2 - V_1^2}{2} + g(z_2 - z_1)\right]$$

With assumption 3 and Eq. 3.20b

$$(h_2 - h_1) = c(T_2 - T_1) + v(p_2 - p_1) = v(p_2 - p_1)$$

$$= \left(0.01605 \frac{ft^3}{lb}\right)(20 - 14.7)\frac{lbf}{in^2}\left|\frac{144\ in^2}{ft^2}\right|\left|\frac{1\ Btu}{778\ ft \cdot lbf}\right| = 0.0157 \frac{Btu}{lb}$$

where $v = v_f(68°F)$ from Table A-2E. The change in specific kinetic energy is

$$\frac{V_2^2 - V_1^2}{2} = \left[\frac{(40\ ft/s)^2 - (10\ ft/s)^2}{2}\right]\left|\frac{1\ lbf}{32.2\ lb \cdot ft/s^2}\right|\left|\frac{1\ Btu}{778\ ft \cdot lbf}\right| = 0.0299 \frac{Btu}{lb}$$

Finally

$$g(z_2 - z_1) = \left(32.0 \frac{ft}{s^2}\right)(50\ ft)\left|\frac{1\ lbf}{32.2\ lb \cdot ft/s^2}\right|\left|\frac{1\ Btu}{778\ ft \cdot lbf}\right| = 0.0639 \frac{Btu}{lb}$$

Collecting results

$$(-\dot{W}_{cv}) = \left(20 \frac{lb}{s}\right)\left[0.0157 + 0.0299 + 0.0639\right]\frac{Btu}{lb}$$

$$= 2.19\ Btu/s$$

$$= \left(2.19 \frac{Btu}{s}\right)\left|\frac{3600\ s}{1\ h}\right|\left|\frac{1\ hp}{2545\ Btu/h}\right| = 3.1\ hp \quad \leftarrow (-\dot{W}_{cv})$$

1. Since the water is at 68°F, only a small temperature difference with normal surroundings would be observed. Accordingly, heat transfer can be ignored.
2. Alternatively, Eq. 3.13 can be invoked to obtain the same result.

PROBLEM 4.54

KNOWN: Water flows through pumping system with known inlet and exit conditions. The power required by the pump is also specified.

FIND: Determine the mass flow rate.

SCHEMATIC & GIVEN DATA:

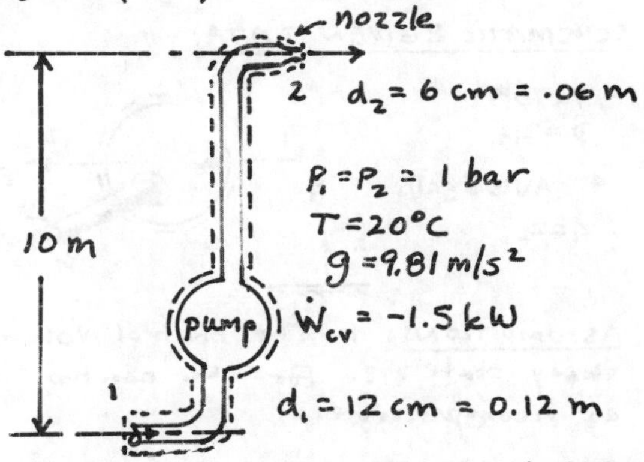

ASSUMPTIONS: (1) The control volume is at steady state. (2) Heat transfer is negligible. (3) The water behaves as an incompessible liquid. (4) The acceleration of gravity is constant at $g = 9.81$ m/s². (5) The temperature and pressure are nearly constant throughout.

ANALYSIS: To find $\dot{m}$, begin with steady-state mass and energy rate balances

$$0 = \dot{Q}_{cv}^{\,0} - \dot{W}_{cv} + \dot{m}\left[(h_1 - h_2)^{0} + \left(\frac{V_1^2 - V_2^2}{2}\right) + g(z_1 - z_2)\right]$$

$$= -\dot{W}_{cv} + \dot{m}\left[\left(\frac{V_1^2 - V_2^2}{2}\right) + g(z_1 - z_2)\right] \qquad (*)$$

where $\dot{m}_1 = \dot{m}_2 \equiv \dot{m}$, and the specific enthalpy term is eliminated based on assumption (3) and Eq. 3.20b.

From Eq. 4.11b, $V = \dot{m}v/A$, and (*) becomes

$$0 = -\dot{W}_{cv} + \dot{m}\left[\frac{(\dot{m}v/A_1)^2 - (\dot{m}v/A_2)^2}{2} + g(z_1 - z_2)\right]$$

$$= -\dot{W}_{cv} + \frac{\dot{m}^3 v^2}{2}\left(\frac{1}{A_1^2} - \frac{1}{A_2^2}\right) + \dot{m}g(z_1 - z_2) \qquad (**)$$

From $A = \pi d^2/4$; $A_1 = 0.01131$ m² and $A_2 = 0.002827$ m². Now, with $v \approx v_{f@20°C} = 1.0018 \times 10^{-3}$ m³/kg from Table A-2, we can insert values in (**)

$$0 = -(-1.5\text{ kW})\left|\frac{1\text{ kJ/s}}{1\text{ kW}}\right| + \dot{m}^3\frac{(1.0018\times10^{-3}\ \frac{\text{m}^3}{\text{kg}})^2}{2}\left(\frac{1}{0.01131^2} - \frac{1}{0.002827^2}\right)\frac{1}{\text{m}^4}\left|\frac{1\text{ N}}{1\text{ kg}\cdot\text{m/s}^2}\right|\left|\frac{1\text{ kJ}}{10^3\text{ N}\cdot\text{m}}\right| + \dot{m}(9.81\ \tfrac{\text{m}}{\text{s}^2})(-10\text{ m})\left|\frac{1\text{ N}}{1\text{ kg}\cdot\text{m/s}^2}\right|\left|\frac{1\text{ kJ}}{10^3\text{ N}\cdot\text{m}}\right|$$

or

$$0 = 1.5 - 5.8865\times10^{-5}\,\dot{m}^3 - 0.0981\,\dot{m} \quad (\text{where } \dot{m} \text{ is in kg/s})$$

This equation is cubic in $\dot{m}$. The solution is

$$\dot{m} = 13.74\text{ kg/s} \longleftarrow \dot{m}$$

PROBLEM 4.55

KNOWN: Data are provided for a water pump operating at steady state

FIND: Plot the pressure rise from inlet to exit versus volumetric flow.

SCHEMATIC & GIVEN DATA:

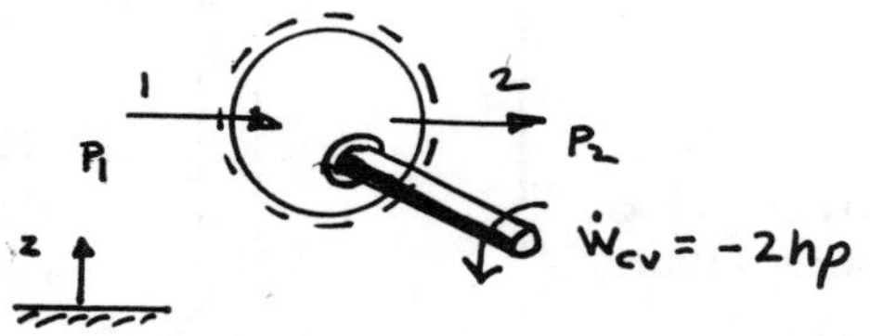

ASSUMPTIONS: 1. The control volume shown in the schematic is at
① steady state. 2. For the control volume, $\dot{Q}_{cv}=0$. 3. Water is modeled as incompressible.

ANALYSIS: The mass rate balance is $\dot{m}_2=\dot{m}_1$. Since the inlet and exit pipe diameters are the same and the water is incompressible, $V_1=V_2$; and so the kinetic term vanishes in the energy rate balance, which reads

$$0=\cancel{\dot{Q}_{cv}}-\dot{W}_{cv}+\dot{m}\left[h_1-h_2+\cancel{\frac{V_1^2-V_2^2}{2}}+\cancel{g(z_1-z_2)}\right]$$

② With Eq. 3.20b,

$$h_2-h_1=c\cancel{(T_2-T_1)}^{0}+v(p_2-p_1)$$

Thus

$$(-\dot{W}_{cv})=\dot{m}\left[v(p_2-p_1)\right] \Rightarrow p_2-p_1=\frac{(-\dot{W}_{cv})}{\dot{m}v}$$

with $\dot{m}=\frac{AV}{v} \Rightarrow AV=\dot{m}v$,

$$(p_2-p_1)=\frac{(-\dot{W}_{cv})}{(AV)}$$

$$=\frac{(2\,hp)\left|\frac{2545\,Btu/h}{1\,hp}\right|\left|\frac{1h}{3600s}\right|\left|\frac{778\,ft\cdot lbf}{1\,Btu}\right|\left|\frac{1\,ft^2}{144\,in^2}\right|}{(AV)(gal/s)\left|\frac{0.13368\,ft^3}{1\,gal}\right|}$$

$$=\frac{57.14}{(AV)}\ \frac{lbf}{in^2} \qquad (*)$$

Eq. (*) can be plotted readily using a spreadsheet, plotting program, or IT. The IT plot is shown at the right.

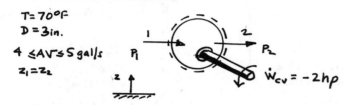

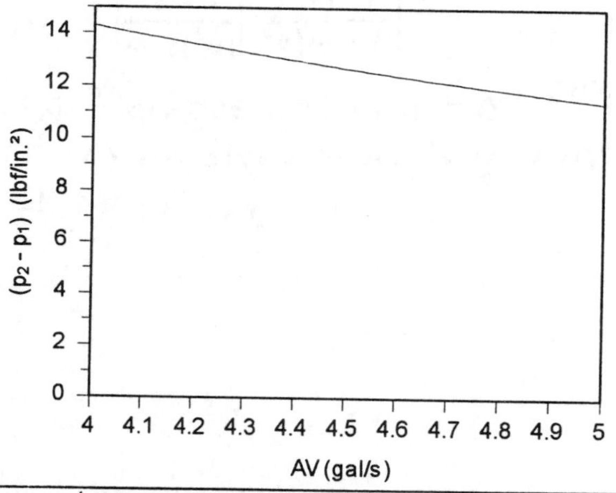

1. Since the water is at 70°F, only a small temperature difference with normal surroundings would be observed. So, heat transfer can be ignored.
2. Alternatively, Eq. 3.13 can be used to obtain the same result.

PROBLEM 4.56

KNOWN: An oil pump operating at steady state delivers oil with a known mass flow rate and pressure rise from inlet to exit.

FIND: If pumps are available in 1/4-horsepower increments, determine the horsepower rating of the pump needed for this application.

SCHEMATIC & GIVEN DATA:

$\rho_{oil} = 100 \frac{lb}{ft^3}$

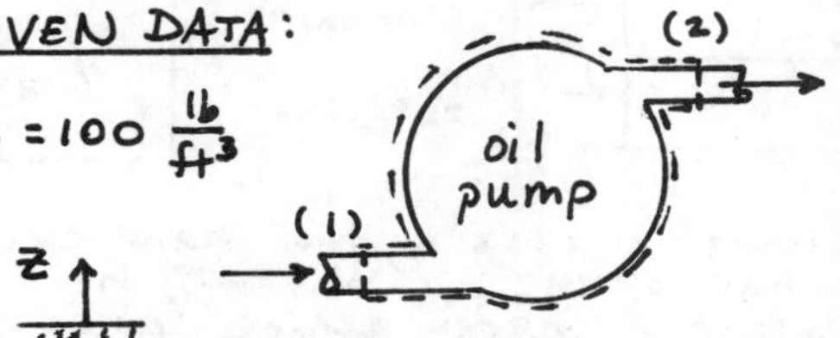

$\dot{m} = 12$ lb/s
$d_2 = 1$ in. $= (1/12)$ ft
$p_2 - p_1 = 40$ lbf/in.²
T is constant
$z_1 \approx z_2$

ASSUMPTIONS: (1) The control volume is at steady state. (2) For the control
① volume, $\dot{Q}_{cv} \approx 0$. (3) The oil is incompressible. (4) The potential energy change from inlet to exit and the inlet kinetic energy are negligible.

ANALYSIS: To determine the power required, begin with steady state forms of the mass and energy rate balances, as follows:

$$0 = \dot{Q}_{cv}^{\nearrow 0} - \dot{W}_{cv} + \dot{m}\left[(h_1 - h_2) + \left(\frac{V_1^2 - V_2^2}{2}\right) + g(z_1 - z_2)\right]$$

where $\dot{m}_1 = \dot{m}_2 = \dot{m}$ and assumptions (2) and (4) have been applied. Thus

$$\dot{W}_{cv} = \dot{m}\left[(h_1 - h_2) - \frac{V_2^2}{2}\right]$$

With Eq. 3.20b

② $$h_1 - h_2 = c(T_1 - T_2)^{\nearrow 0} + v(p_1 - p_2) = \frac{p_1 - p_2}{\rho}$$

Thus

$$\dot{W}_{cv} = \dot{m}\left[\frac{p_1 - p_2}{\rho} - \frac{V_2^2}{2}\right]$$

Using Eq. 4.11a with $A = \pi d^2/4$

$$V_2 = \frac{4\dot{m}}{\rho \pi d^2} = \frac{4\,(12\text{ lb/s})}{(100\text{ lb/ft}^3)\,\pi\,(1/12)^2\text{ ft}^2} = 22\text{ ft/s}$$

Inserting values and noting that $p_1 - p_2 = -(p_2 - p_1)$

$$\dot{W}_{cv} = \left(12\frac{lb}{s}\right)\left[\frac{-40\text{ lbf/in.}^2}{100\text{ lb/ft}^3}\left|\frac{144\text{ in.}^2}{1\text{ ft}^2}\right| - \frac{(22\text{ ft/s})^2}{2}\left|\frac{1\text{ lbf}}{32.2\text{ lb}\cdot\text{ft/s}^2}\right|\right]$$

$$= \left(-781.4\frac{\text{ft}\cdot\text{lbf}}{s}\right)\left|\frac{1\text{ hp}}{550\text{ ft}\cdot\text{lbf/s}}\right| = -1.42\text{ hp}$$

pump size = 1.5 hp ← hp rating

---

1. It is assumed that there is a small temperature difference between the pump and the surroundings, so heat transfer can be ignored.
2. Alternatively, Eq. 3.13 can be used to obtain the same result.

PROBLEM 4.57

KNOWN: Refrigerant 134a and air pass in separate streams through a heat exchanger at steady state, for which data are provided.

FIND: Determine the mass flow rate of the air.

SCHEMATIC & GIVEN DATA:

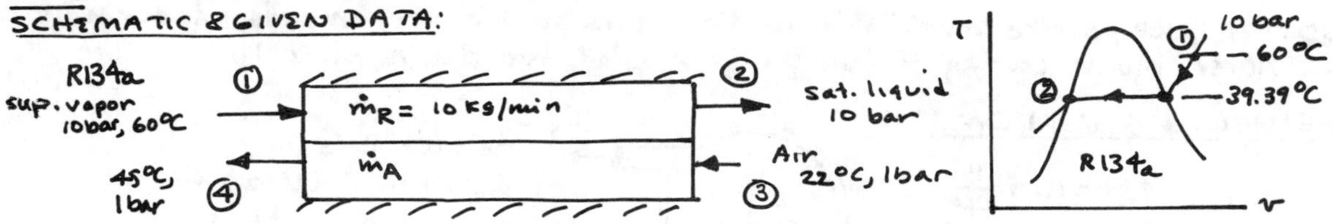

ASSUMPTIONS: 1. A control volume enclosing the heat exchanger is at steady state. 2. For the control volume, $\dot{W}_{cv} = 0$, heat transfer can be ignored, and kinetic/potential energy effects are negligible.
(1) 3. Air is modeled as an ideal gas.

ANALYSIS: Since the streams flow separately, the conservation of mass principle indicates at steady state: $\dot{m}_1 = \dot{m}_2 \equiv \dot{m}_R$ and $\dot{m}_3 = \dot{m}_4 \equiv \dot{m}_A$. An energy rate balance reads

$$0 = \dot{Q}_{cv} - \dot{W}_{cv} + \dot{m}_R\left[h_1 - h_2 + \frac{V_1^2 - V_2^2}{2} + g(z_1 - z_2)\right] - \dot{m}_A\left[h_3 - h_4 + \frac{V_3^2 - V_4^2}{2} + g(z_3 - z_4)\right]$$

$\Rightarrow$

$$\dot{m}_A = \dot{m}_R\left[\frac{h_1 - h_2}{h_4 - h_3}\right]$$

From Table A-12: $h_1 = 291.36$ kJ/kg. From Table A-11, $h_2 = 105.29$ kJ/kg. From Table A-22, $h_3 = 295.17$ kJ/kg, $h_4 = 318.28$ kJ/kg. Then

$$\dot{m}_A = 10\,\frac{\text{kg}}{\text{min}}\left[\frac{291.36 - 105.29}{318.28 - 295.17}\right] = 80.5\,\frac{\text{kg}}{\text{min}} \longleftarrow$$

---

1. The validity of the ideal gas model is readily checked using the generalized compressibility chart.

PROBLEM 4. 58

<u>KNOWN</u>: Ammonia and cooling water pass in separate streams through a condenser (heat exchanger). The volumetric flow rate of cooling water and other data are given at the inlets and exits.

<u>FIND</u>: Determine (a) the mass flow rate of ammonia, and (b) the rate of energy transfer from the condensing ammonia to the cooling water.

<u>SCHEMATIC & GIVEN DATA</u>:

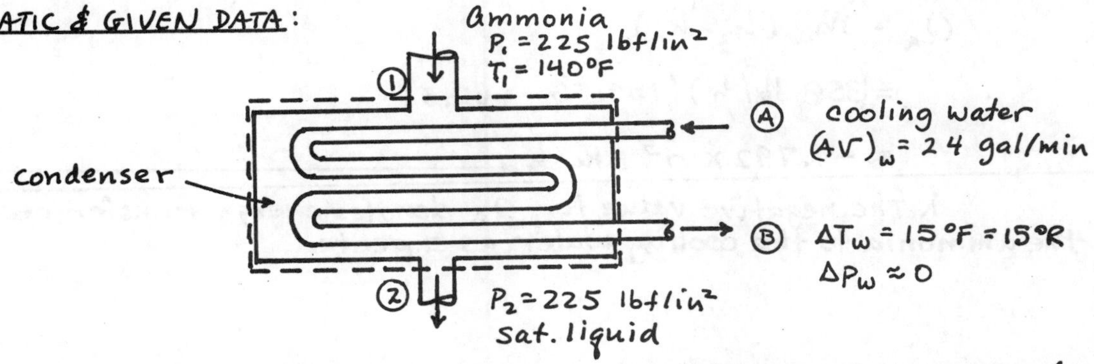

<u>ASSUMPTIONS</u>: (1) The control volume is at steady state. (2) Heat transfer from the outside of the condenser is negligible. (3) Kinetic and potential energy changes from inlet to exit are negligible. (4) The cooling water is modeled as an incompressible liquid with constant specific heat.

<u>ANALYSIS</u>: (a) Since the ammonia and cooling water are separate streams

<u>Ammonia</u>: $\dot{m}_1 = \dot{m}_2 \equiv \dot{m}_a$

<u>Water</u>: $\dot{m}_A = \dot{m}_B \equiv \dot{m}_w$

The mass flow rate of ammonia is found from the steady-state energy balance

$$0 = \cancel{\dot{Q}_{cv}}^{0} - \cancel{\dot{W}_{cv}}^{0} + \dot{m}_a\left[(h_1 - h_2) + \cancel{\left(\frac{V_1^2 - V_2^2}{2}\right)}^{0} + \cancel{g(z_1 - z_2)}^{0}\right] + \dot{m}_w\left[(h_A - h_B) + \cancel{\left(\frac{V_A^2 - V_B^2}{2}\right)}^{0} + \cancel{g(z_A - z_B)}^{0}\right]$$

Or

$$\dot{m}_a = \frac{\dot{m}_w (h_B - h_A)}{(h_1 - h_2)}$$

For the water, using Eq. 3.20b

$$h_B - h_A = c\,(T_B - T_A) + v\,\cancel{(p_B - p_A)}^{0}$$

and

$$\dot{m}_w = \frac{(AV)_w}{v_w}$$

From Table A-19E; $c \approx 1$ Btu/lb·°R, and $v = 1/\rho = 0.0161$ ft³/lb.
For the ammonia, $h_1 = 660.3$ Btu/lb from interpolating in Table A-15E and $h_2 = 159.58$ Btu/lb from Table A-14E. Inserting values

$$\dot{m}_a = \left(\frac{(24\ \text{gal/min})\left|\frac{0.13368\ \text{ft}^3}{1\ \text{gal}}\right|}{(0.0161\ \text{ft}^3/\text{lb})}\right)\frac{(1\ \text{Btu/lb}\cdot°\text{R})(15\ °\text{R})}{(660.3 - 159.58)\ \text{Btu/lb}}\left|\frac{60\ \text{min}}{1\ \text{h}}\right|$$

$$\doteq 358\ \text{lb/h} \longleftarrow \dot{m}_a$$

(b) For a control volume enclosing only the ammonia

$$0 = \dot{Q}_a - \dot{W}_a^{\,\nearrow 0} + \dot{m}_a \left[(h_1 - h_2) + \left(\frac{V_1^2 - V_2^2}{2} \right)^{\nearrow 0} + g(z_1 - z_2)^{\nearrow 0} \right]$$

where $\dot{Q}_a$ denotes the heat transfer rate for the ammonia only. Thus

$$\dot{Q}_a = \dot{m}_a (h_2 - h_1)$$

$$= (358 \text{ lb/h})(159.58 - 660.3)$$

(1) $$= -1.792 \times 10^5 \text{ Btu/h} \leftarrow \dot{Q}_a$$

1. The negative value for $\dot{Q}_a$ denotes energy transfer heat from the ammonia to the cooling water, as expected.

PROBLEM 4.59

KNOWN: A steam boiler is constructed by passing a direct current through the stainless steel pipe through which the steam is flowing.

FIND: Determine the required size of the power supply and the expected current draw.

SCHEMATIC & GIVEN DATA:

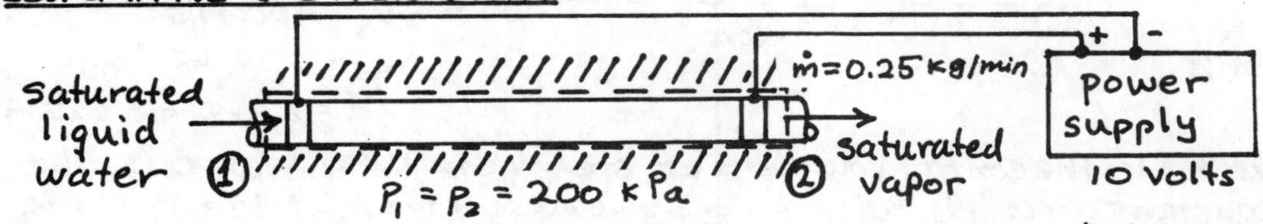

ASSUMPTIONS: (1) The control volume is at steady state. (2) Heat transfer can be neglected. (3) Kinetic and potential energy effects are negligible.

ANALYSIS: To find the power required start with the steady-state energy balance

$$0 = \dot{Q}_{cv}^{\,0} - \dot{W}_{cv} + \dot{m}\left[(h_1 - h_2) + \left(\frac{V_1^2 - V_2^2}{2}\right) + g(z_1 - z_2)\right]$$

where $\dot{m}_1 = \dot{m}_2 \equiv \dot{m}$. With assumption (3)

$$|\dot{W}_{cv}| = \dot{m}(h_2 - h_1)$$

From Table A-3 at 2bar, $h_2 - h_1 = h_{fg} = 2201.9$ kJ/kg. Thus

$$|\dot{W}_{cv}| = \left(0.25 \frac{kg}{min}\right)\left|\frac{1\ min}{60\ s}\right|\left(2201.9 \frac{kJ}{kg}\right)\left|\frac{1\ kW}{1\ kJ/s}\right|$$

$$= 9.175\ kW \longleftarrow$$

Then, with Eq. 2.21, the current is

$$i = \frac{|\dot{W}_{cv}|}{\mathcal{E}} = \left(\frac{9175\ W}{10\ volts}\right)\left(\frac{1\ Volt}{1\ W/amp}\right)$$

$$= 917.5\ amp \longleftarrow$$

PROBLEM 4.60

KNOWN: Carbon dioxide is heated as it flows through a constant area pipe. Inlet and exit conditions are known.

FIND: Determine the rate of heat transfer.

SCHEMATIC & GIVEN DATA:

carbon dioxide
$p_1 = 2$ bar
$T_1 = 300$ K
$V_1 = 100$ m/s

$p_2 = 0.9413$ bar
$V_2 = 400$ m/s
$c_p = 0.94$ kJ/kg·K
$d = 2.5$ cm $= 0.025$ m

ASSUMPTIONS: (1) The control volume is at steady state. (2) $\dot{W}_{cv} = 0$. (3) The duct has constant area. (4) Potential energy change from inlet to exit can be neglected. (5) The carbon dioxide can be modeled as an ideal gas with constant specific heats.

ANALYSIS: To fix the exit state, begin with the steady-state mass balance

$$\dot{m}_1 = \dot{m}_2 \equiv \dot{m}$$

$$\frac{A_1 V_1}{v_1} = \frac{A_2 V_2}{v_2} \Rightarrow v_2 = \left(\frac{V_2}{V_1}\right) v_1$$

with the ideal gas equation of state

$$\frac{RT_2}{p_2} = \left(\frac{V_2}{V_1}\right)\frac{RT_1}{p_1} \Rightarrow T_2 = \left(\frac{V_2}{V_1}\right)\left(\frac{p_2}{p_1}\right)T_1 = \left(\frac{400}{100}\right)\left(\frac{0.9413}{2}\right)(300\text{K})$$

$$= 564.8\text{K}$$

Evaluating the mass flow rate

$$\dot{m} = \frac{AV_1}{v_1} = \frac{\left(\frac{\pi d^2}{4}\right)V_1}{(RT_1/p_1)}$$

$$= \frac{\left(\frac{\pi(.025^2 \text{m}^2)}{4}\right)(100\text{ m/s})(2\text{ bar})}{\left(\frac{8.314}{44.01}\frac{\text{kJ}}{\text{kg}\cdot\text{K}}\right)(300\text{K})}\left|\frac{10^5\text{ N/m}^2}{1\text{ bar}}\right|\left|\frac{1\text{ kJ}}{10^3\text{ N}\cdot\text{m}}\right|$$

$$= 0.1732\text{ kg/s}$$

Using the steady-state energy balance to find the heat transfer rate

$$0 = \dot{Q}_{cv} - \dot{W}_{cv}^{\,0} + \dot{m}\left[(h_1 - h_2) + \left(\frac{V_1^2 - V_2^2}{2}\right) + g(z_1 - z_2)^{0}\right]$$

with $h_1 - h_2 = c_p(T_1 - T_2)$

$$\dot{Q}_{cv} = \dot{m}\left[c_p(T_2 - T_1) + \left(\frac{V_2^2 - V_1^2}{2}\right)\right]$$

$$= \left(0.1732\frac{\text{kg}}{\text{s}}\right)\left[\left(0.94\frac{\text{kJ}}{\text{kg}\cdot\text{K}}\right)(564.8-300)\text{K} + \left(\frac{400^2 - 100^2}{2}\right)\frac{\text{m}^2}{\text{s}^2}\left|\frac{1\text{ N}}{1\text{ kg}\cdot\text{m/s}^2}\right|\left|\frac{1\text{ kJ}}{10^3\text{ N}\cdot\text{m}}\right|\right]\left|\frac{1\text{ kW}}{1\text{ kJ/s}}\right|$$

$$= 56.1\text{ kW} \longleftarrow \dot{Q}_{cv}$$

PROBLEM 4.61

KNOWN: Data are provided for a feedwater heater at steady state.

FIND: Determine the ratio of the mass flow rates of the two incoming streams, $\dot{m}_2/\dot{m}_1$

SCHEMATIC & GIVEN DATA:

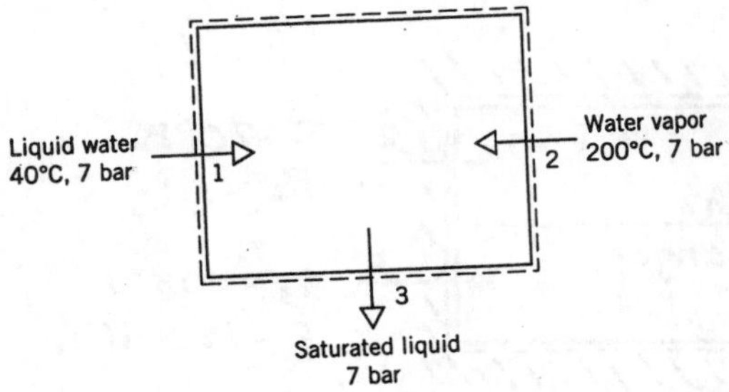

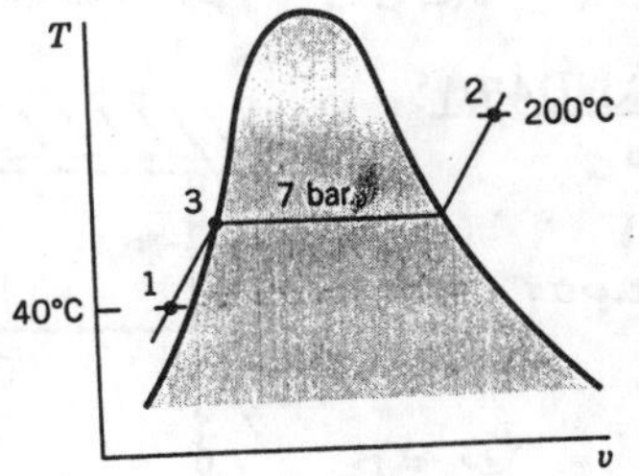

ASSUMPTIONS: 1. The control volume shown in the schematic is at steady state. 2. For the control volume, $\dot{W}_{cv} = 0$, and heat transfer with the surroundings can be ignored. Kinetic and potential energy effects also can be neglected.

ANALYSIS: At steady state the mass rate balance reads

$$0 = \dot{m}_1 + \dot{m}_2 - \dot{m}_3 \Rightarrow \dot{m}_3 = \dot{m}_1 + \dot{m}_2$$

The energy rate balance reduces to

$$0 = \cancel{\dot{Q}_{cv}} - \cancel{\dot{W}_{cv}} + \dot{m}_1\left(h_1 + \cancel{\frac{V_1^2}{2}} + \cancel{gz_1}\right) + \dot{m}_2\left(h_2 + \cancel{\frac{V_2^2}{2}} + \cancel{gz_2}\right) - \dot{m}_3\left(h_3 + \cancel{\frac{V_3^2}{2}} + \cancel{gz_3}\right)$$

$$\Rightarrow 0 = \dot{m}_1 h_1 + \dot{m}_2 h_2 - \dot{m}_3 h_3$$

Introducing $\dot{m}_3 = \dot{m}_1 + \dot{m}_2$

$$0 = \dot{m}_1 h_1 + \dot{m}_2 h_2 - (\dot{m}_1 + \dot{m}_2) h_3 \Rightarrow \frac{\dot{m}_1}{\dot{m}_2} = \frac{h_2 - h_3}{h_3 - h_1}$$

From Table A-4, $h_2 = 2844.8$ kJ/kg. From Table A-3, $h_3 = 697.22$ kJ/kg. With Eq. 3.14 and data from Table A-2

① $$h_1 \approx h_f(40°C) = 167.57 \text{ kJ/kg}$$

Thus

$$\frac{\dot{m}_1}{\dot{m}_2} = \frac{2844.8 - 697.22}{697.22 - 167.57} = 4.05$$

---

1. Using Eq. 3.13, $h_1 \approx 168.27$ kJ/kg and $\dot{m}_1/\dot{m}_2 = 4.06$

PROBLEM 4.62

KNOWN: Separate vapor and liquid streams of Refrigerant 134a pass in counter flow trough a well-insulated heat exchanger. Data are known at the inlets and exits.

FIND: Determine the exit temperature of the liquid stream.

SCHEMATIC & GIVEN DATA:

R-134a
$T_1 = 0°F$
sat. vapor

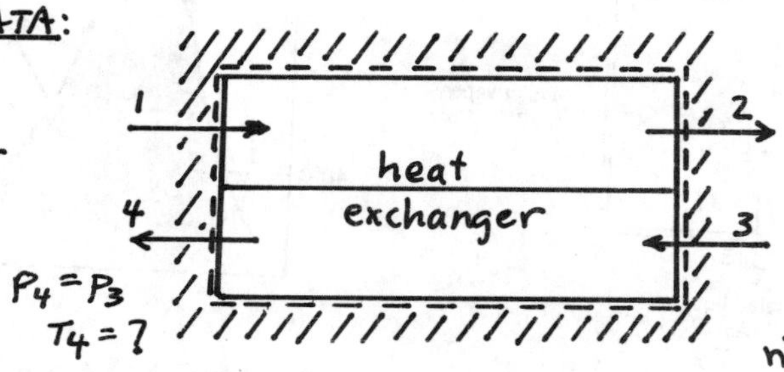

$T_2 = 20°F$
$P_2 = P_1$

$T_3 = 105°F$
$P_3 = 160$ lbf/in²
$\dot{m}_1 = \dot{m}_3$

ASSUMPTIONS: (1) The control volume is at steady state. (2) Heat transfer between the heat exchanger and the surroundings can be neglected and $\dot{W}_{cv} = 0$. (3) Kinetic and potential energy changes from inlet to exit are negligible.

ANALYSIS: To fix state 4, begin with steady state mass and energy balances to determine h_4

$$\dot{m}_1 = \dot{m}_2$$
$$\dot{m}_3 = \dot{m}_4$$

$$0 = \dot{Q}_{cv}^{\,0} - \dot{W}_{cv}^{\,0} + \dot{m}_1\left[(h_1 - h_2) + \left(\frac{V_1^2 - V_2^2}{2}\right)^0 + g(z_1 - z_2)^0\right] + \dot{m}_3\left[(h_3 - h_4) + \left(\frac{V_3^2 - V_4^2}{2}\right)^0 + g(z_3 - z_4)^0\right]$$

with $\dot{m}_1 = \dot{m}_3$

$$h_4 = (h_1 - h_2) + h_3$$

From Table A-10E; $h_1 = 101.75$ Btu/lb and $p_1 = 21.203$ lbf/in². Interpolating in Table A-12E; $h_2 = 105.76$ Btu/lb.

States 3 and 4 are both subcooled liquid states. The following approximations are reasonable

$$h_3 \approx h_{f@T_3}$$
$$h_4 \approx h_{f@T_4}$$

with $h_3 = 46.01$ Btu/lb from Table A-10E

$$h_4 = (101.75 - 105.76) + 46.01 = 42 \text{ Btu/lb}$$

Interpolating in Table A-10E

$$T_4 \approx 93.7\ °F \longleftarrow T_4$$

PROBLEM 4.63

KNOWN: Air and Refrigerant 22 pass in separate streams through a heat exchanger. Data are known at the inlet and exit of each stream.

FIND: Determine (a) the mass flow rate of refrigerant and (b) the rate of energy transfer from the air to the refrigerant.

SCHEMATIC & GIVEN DATA:

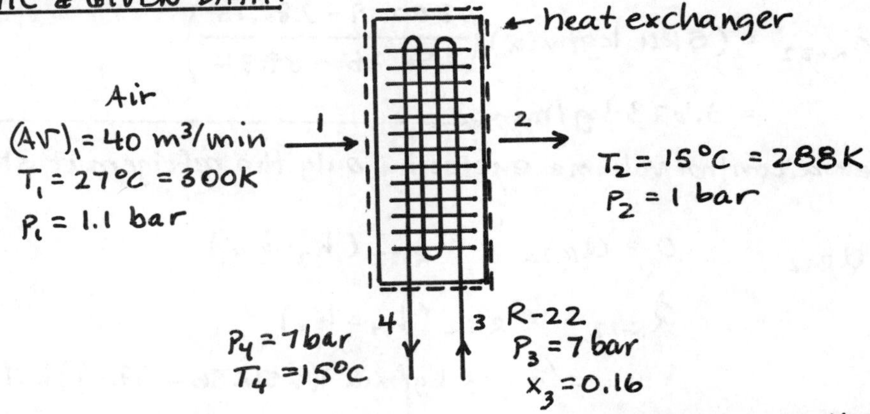

ASSUMPTIONS: (1) The control volume is at steady state. (2) Heat transfer from the outside of the heat exchanger is negligible, and $\dot{W}_{cv} = 0$. (3) Kinetic and potential effects can be neglected. (4) The air behaves as an ideal gas, as can be verified by reference to the compressibility chart.

ANALYSIS: (a) The mass flow rate of refrigerant is determined using steady-state mass and energy balances. First, since the air and refrigerant flow as separate streams

$$\dot{m}_1 = \dot{m}_2 \equiv \dot{m}_{air}$$
$$\dot{m}_3 = \dot{m}_4 \equiv \dot{m}_{R-22}$$

Thus, the energy rate balance reduces as follows

$$0 = \cancel{\dot{Q}_{cv}}^{0} - \cancel{\dot{W}_{cv}}^{0} + \dot{m}_{air}\left[(h_1 - h_2) + \cancel{\left(\frac{V_1^2 - V_2^2}{2}\right)}^{0} + \cancel{g(z_1 - z_2)}^{0}\right] + \dot{m}_{R-22}\left[(h_3 - h_4) + \cancel{\left(\frac{V_3^2 - V_4^2}{2}\right)}^{0} + \cancel{g(z_3 - z_4)}^{0}\right]$$

and

$$\dot{m}_{R-22} = \dot{m}_{air}\left(\frac{h_1 - h_2}{h_4 - h_3}\right)$$

The mass flow rate of air is found using data at the inlet and the ideal gas equation of state

$$\dot{m}_{air} = \frac{(AV)_1}{v_1} = \frac{p_1 (AV)_1}{RT_1}$$

$$= \frac{(1.1\ \text{bars})(40\ m^3/min)}{\left(\frac{8.314}{28.97}\frac{kJ}{kg\cdot K}\right)(300\ K)}\left|\frac{10^5\ N/m^2}{1\ bar}\right|\left|\frac{1\ kJ}{10^3\ N\cdot m}\right|$$

$$= 51.11\ kg/min$$

PROBLEM 4.63 (Cont'd)

From Table A-22; $h_1 = 300.19$ kJ/kg and $h_2 = 288.15$ kJ/kg. Further, using data from Table A-8

$$h_3 = h_{f3} + x_3 h_{fg3} = 58.04 + (.16)(195.60) = 89.34 \text{ kJ/kg}$$

And, from Table A-9; $h_4 = 256.86$ kJ/kg. Thus

$$\dot{m}_{R\text{-}22} = (51.11 \text{ kg/min})\left(\frac{300.19 - 288.15}{256.86 - 89.34}\right)$$

$$= 3.673 \text{ kg/min} \longleftarrow \dot{m}_{R\text{-}22}$$

(b) Consider a control volume enclosing only the refrigerant stream

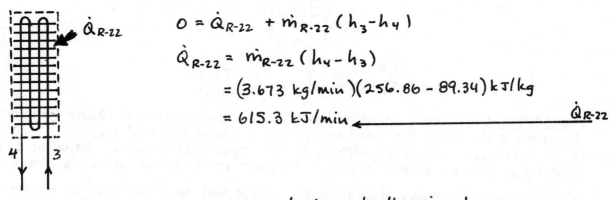

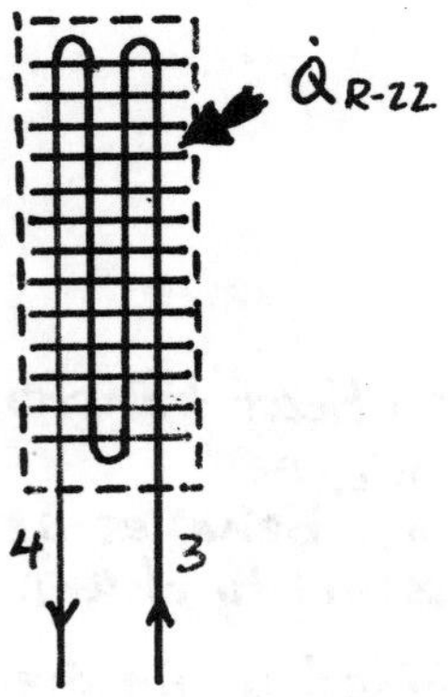

$$0 = \dot{Q}_{R\text{-}22} + \dot{m}_{R\text{-}22}(h_3 - h_4)$$

$$\dot{Q}_{R\text{-}22} = \dot{m}_{R\text{-}22}(h_4 - h_3)$$

$$= (3.673 \text{ kg/min})(256.86 - 89.34) \text{ kJ/kg}$$

$$= 615.3 \text{ kJ/min} \longleftarrow \dot{Q}_{R\text{-}22}$$

COMMENT: For a control volume enclosing only the air stream

$$\dot{Q}_{air} = \dot{m}_{air}(h_2 - h_1)$$

$$= (51.11 \text{ kg/min})(288.15 - 300.19) \text{ kJ/kg}$$

$$= -615.3 \text{ kJ/min}$$

Thus, $\dot{Q}_{R\text{-}22} = -\dot{Q}_{air}$, as expected.

PROBLEM 4.64

KNOWN: Refrigerant 134a flows through a horizontal pipe at steady state, for which operating data are provided.

FIND: Determine the exit temperature and velocity, and the inlet velocity.

SCHEMATIC & GIVEN DATA:

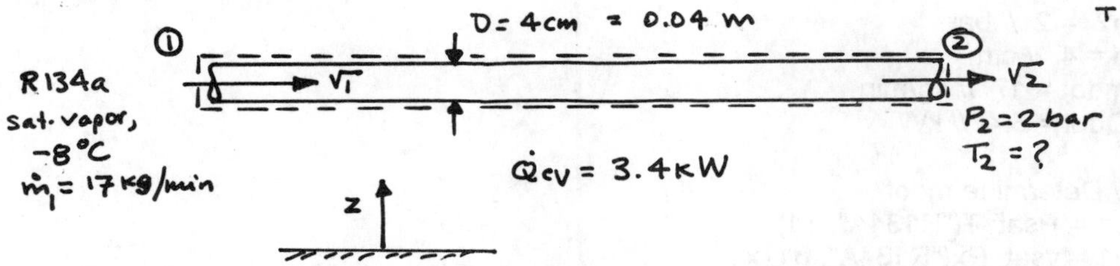

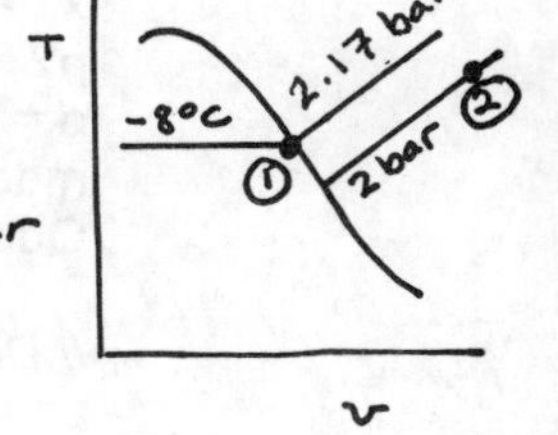

ASSUMPTIONS: 1. The control volume shown in the schematic is at steady state. 2. For the control volume, $\dot{W}_{cv} = 0$ and there is no change in potential energy from inlet to exit.

ANALYSIS: At steady state, the mass rate balance reduces to $\dot{m}_1 = \dot{m}_2 \equiv \dot{m}$. The inlet velocity is found as follows:

$$\dot{m} = \frac{A_1 V_1}{v_1} \Rightarrow V_1 = \frac{\dot{m} v_1}{A_1} = \frac{4 \dot{m} v_1}{\pi d^2}$$

with v_1 from Table A-10

$$V_1 = \frac{4(17\ \text{kg/min})\left|\frac{1\ \text{min}}{60\ \text{s}}\right|(0.0919\ \text{m}^3/\text{kg})}{\pi (0.04\ \text{m})^2} = 20.72\ \text{m/s} \leftarrow V_1$$

Similarly, V_2 is found from $\dot{m}_1 = \dot{m}_2$ as follows:

$$\dot{m}_1 = \dot{m}_2 \Rightarrow \frac{\cancel{A_1} V_1}{v_1} = \frac{A_2 V_2}{v_2} \Rightarrow V_2 = \frac{v_2}{v_1} V_1 \qquad (*)$$

In this expression, v_1 and V_1 are known. However, $v_2 = v(T_2, P_2)$ is unknown.

Another relation is obtained using the energy rate balance

$$0 = \dot{Q}_{cv} - \cancel{\dot{W}_{cv}}^{0} + \dot{m}\left[(h_1 - h_2) + \left(\frac{V_1^2 - V_2^2}{2}\right) + g\cancel{(z_1 - z_2)}^{0}\right]$$

$$= \dot{Q}_{cv} + \dot{m}\left[(h_1 - h_2) + \left(\frac{V_1^2 - V_2^2}{2}\right)\right]$$

Inserting known values, including h_1 from Table A-10

$$0 = (3.4\ \text{kW})\left|\frac{1\ \text{kJ/s}}{1\ \text{kW}}\right| + \left(17\ \frac{\text{kg}}{\text{min}}\right)\left|\frac{1\ \text{min}}{60\ \text{s}}\right|\left[(242.54 - h_2)\frac{\text{kJ}}{\text{kg}} + \left(\frac{(20.72\ \text{m/s})^2 - V_2^2}{2}\right)\left|\frac{1\ \text{N}}{1\ \text{kg}\cdot\text{m/s}^2}\right|\left|\frac{1\ \text{kJ}}{10^3\ \text{N}\cdot\text{m}}\right|\right] \qquad (**)$$

where $h_2 = h(T_2, P_2)$.

Equations (*) and (**) can be solved simultaneously by referring to Table A-12 for v_2 and h_2 as functions of T_2 and P_2. The process is iterative. To avoid iteration, IT can be used effectively, as follows:

IT Code

```
// Data
T1 = -8  // °C
x1 = 1
p2  = 2  // bar
d = 4  // cm
mdot = 17  // kg/min
Qdot = 3.4  // kW

// Determine mdot
p1 = Psat_T("R134A", T1)
v1 = vsat_Px("R134A", p1, x1)
h1 = hsat_Px("R134A", p1, x1)
mdot  = ((pi * (d / 100)^2  / 4) * V1 / v1) * 60

// Find exit state
0 = Qdot + (mdot / 60) * ((h1 - h2) + (V1^2 - V2^2) / (2 * 1000))
V2 / v2 = V1 / v1
h2 = h_PT("R134A", p2, T2)
v2 = v_PT("R134A", p2, T2)
```

IT Results

① $T_2 = 4.976°C \approx 5°C$ ← T_2
$V_2 = 24.08$ m/s ← V_2
$V_1 = 20.71$ m/s

---

1. These values can be verified using Eq.s (*) and (**) with data from Table A-12. The results compare very favorably.

PROBLEM 4.65

KNOWN: Water is heated as it passes through a solar collector.

FIND: Determine the mass flow rate of water and the number of gallons of heated water that 8 collectors can provide in 30 min.

SCHEMATIC & GIVEN DATA:

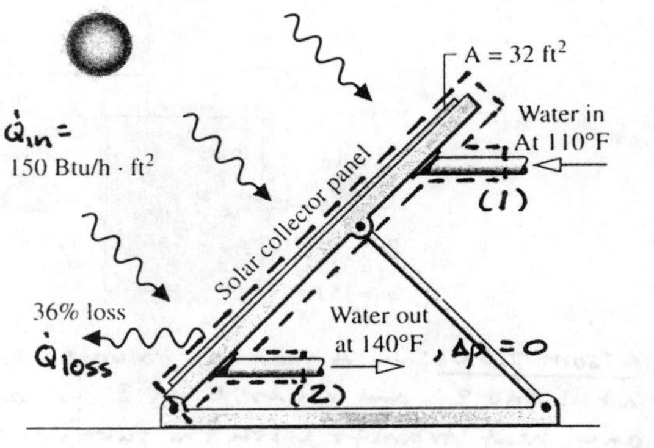

ASSUMPTIONS: (1) The control volume is at steady state. (2) For the control volume, $\dot{W}_{cv} = 0$ and kinetic and potential energy effects are negligible. (3) The water behaves as an incompressible substance with $c = 0.999$ Btu/lb·°R.

ANALYSIS: The mass flow rate is determined using the steady-state energy balance, as follows

$$0 = \dot{Q}_{cv} - \dot{W}_{cv}^{\,0} + \dot{m}\left[(h_1 - h_2) + \left(\frac{V_1^2 - V_2^2}{2}\right) + g(z_1 - z_2)\right]$$

where $\dot{m}_1 = \dot{m}_2 = \dot{m}$. With $\dot{Q}_{cv} = \dot{Q}_{in} - \dot{Q}_{loss}$ and assumption (3)

$$0 = \dot{Q}_{in} - \dot{Q}_{loss} + \dot{m}(h_1 - h_2)$$

① Using Eq. 3.20b, $h_1 - h_2 = c(T_1 - T_2) + v(p_1 - p_2)^{0}$. Thus

$$\dot{m} = \frac{\dot{Q}_{in} - \dot{Q}_{loss}}{c(T_2 - T_1)}$$

From the given data, $\dot{Q}_{in} = (150\ \text{Btu/h}\cdot\text{ft}^2)(32\ \text{ft}^2) = 4800$ Btu/h and $\dot{Q}_{loss} = (.36)\dot{Q}_{in} = 1728$ Btu/h. Inserting values

$$\dot{m} = \frac{(4800 - 1728)\ \text{Btu/h}}{(1\ \text{Btu/lb}\cdot°\text{R})(30°\text{R})}\left|\frac{1\ \text{h}}{60\ \text{min}}\right| = 1.71\ \text{lb/min} \longleftarrow \dot{m}$$

In 30 min., one collector can provide the following amount of 140°F water:

$$m = \int_{t_1}^{t_2} \dot{m}\,dt = \dot{m}\Delta t = \left(1.71\ \frac{\text{lb}}{\text{min}}\right)(30\ \text{min}) = 51.3\ \text{lb/collector}$$

Thus, the volume is

$$V = m v_2 = (51.3\ \text{lb/collector})\left(0.01629\ \frac{\text{ft}^3}{\text{lb}}\right)\left|\frac{1\ \text{gal}}{0.13368\ \text{ft}^3}\right| = 6.25\ \frac{\text{gal}}{\text{collector}}$$

where $v_2 \approx v_{f@140°F}$ from Table A-2E. Finally, for eight collectors

$$V_{tot} = (8\ \text{collectors})\left(6.25\ \frac{\text{gal}}{\text{collector}}\right) = 50\ \text{gal} \longleftarrow V_{tot}$$

---

1. Alternatively, Eq. 3.13 could be invoked to obtain the same result.

PROBLEM 4.66

KNOWN: Data are provided for a desuperheater operating at steady state.

FIND: (a) For a specified temperature for the entering liquid, determine the liquid mass flow rate. (b) Plot the liquid mass flow rate versus the liquid temperature.

SCHEMATIC & GIVEN DATA:

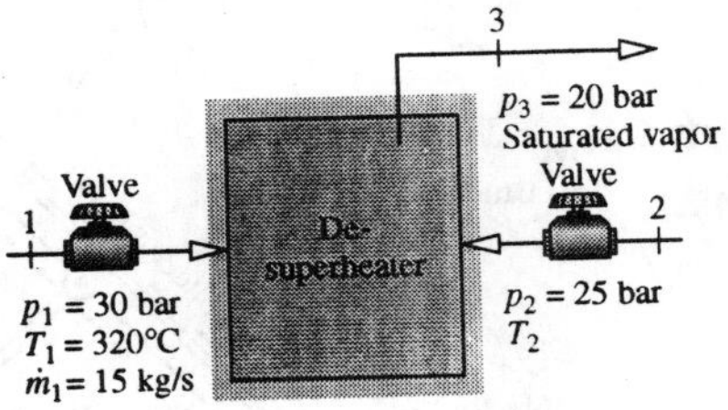

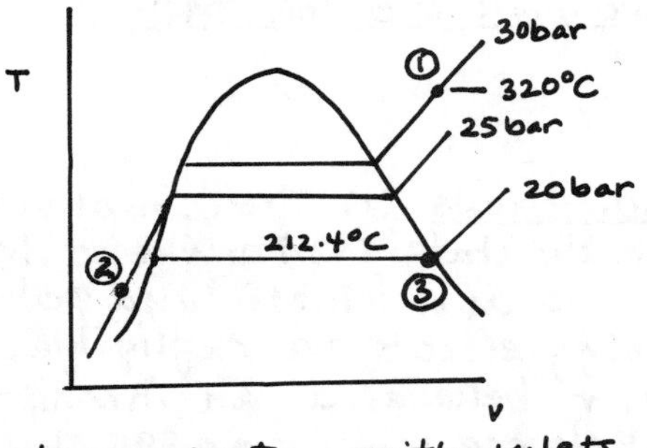

ASSUMPTIONS: 1. A control volume enclosing the desuperheater with inlets at 1 and 2 and an exit at 3 is at steady state. 2. For the control volume, $\dot{W}_{cv} = 0$ and heat transfer with the surroundings can be ignored. Kinetic and potential energy effects can be ignored.

ANALYSIS: The mass rate balance at steady state reads $\dot{m}_1 + \dot{m}_2 = \dot{m}_3$. The energy rate balance at steady state reduces as follows:

$$0 = \cancel{\dot{Q}_{cv}} - \cancel{\dot{W}_{cv}} + \dot{m}_1\left[h_1 + \cancel{\frac{V_1^2}{2}} + \cancel{gz_1}\right] + \dot{m}_2\left[h_2 + \cancel{\frac{V_2^2}{2}} + \cancel{gz_2}\right] - \dot{m}_3\left[h_3 + \cancel{\frac{V_3^2}{2}} + \cancel{gz_3}\right]$$

$$0 = \dot{m}_1 h_1 + \dot{m}_2 h_2 - (\dot{m}_1 + \dot{m}_2) h_3$$

Solving for $\dot{m}_2$

$$\dot{m}_2 = \dot{m}_1\left[\frac{h_3 - h_1}{h_2 - h_3}\right] \qquad (1)$$

(a) From Table A-4; h_1 = 3043.4 kJ/kg. From Table A-3; h_3 = 2799.5 kJ/kg. Further, at T_2 = 200°C, p_2 = 25 bar; Table A-5 gives h_2 = 852.8 kJ/kg. Thus

$$\dot{m}_2 = \left(15\ \frac{kg}{s}\right)\left[\frac{2799.5 - 3043.4}{852.8 - 2799.5}\right] = 1.88\ kg/s \longleftarrow \dot{m}_2 \text{ (part a)}$$

(b) The following IT code is used to develop data to construct a plot of $\dot{m}_2$ vs. T_2:

<u>IT Code</u>

```
p1 = 30  // bar
T1 = 320  // °C
mdot1 = 15  // kg/s
p2 = 25  // bar
T2 = 20  // °C
p3 = 20  // bar

h1 = h_PT("Water/Steam", p1, T1)
h2 = h_PT("Water/Steam", p2, T2)
h3 = hsat_Px("Water/Steam", p3, 1)

mdot2 = mdot1 * ((h3 - h1) / (h2 - h3))
```

①

IT Result for $T_2 = 200°C$

$h_1 = 3043$ kJ/kg
$h_2 = 852.4$ kJ/kg
$h_3 = 2799$ kJ/kg
mdot2 = 1.879 kg/s

This result compares very favorably with the result of part (a). Thus, with the computer solution validated, use the Explore button and sweep T_2 from 20 to 220°C in steps of 10. The resulting data are used to construct the following plot:

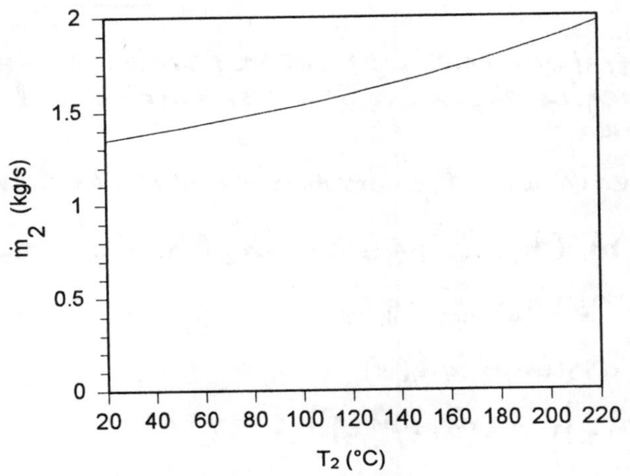

---

1. Note that IT uses the approximation of Eq. 3.14 for liquid enthalpies.

PROBLEM 4.67

KNOWN: An open feedwater heater operates with known inlet and exit conditions. The mass flow rate at one inlet is given.

FIND: Determine the mass flow rate at the second inlet.

SCHEMATIC & GIVEN DATA:

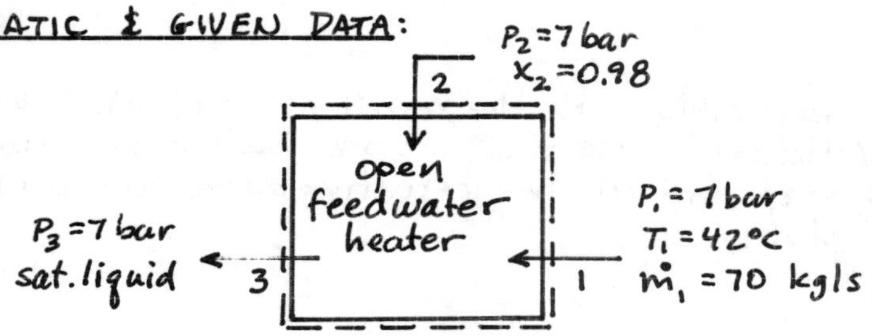

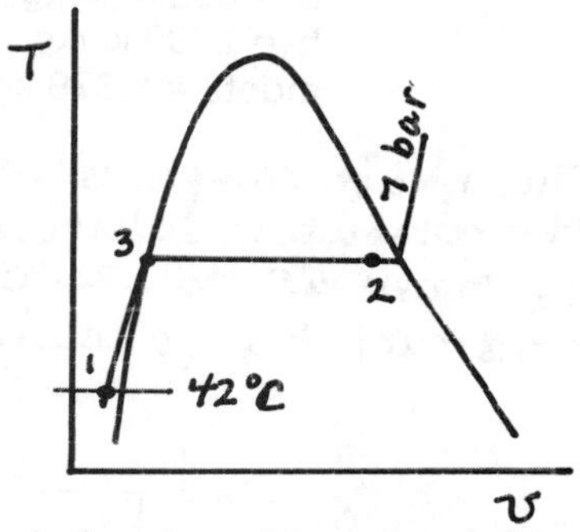

ASSUMPTIONS: (1) The control volume is at steady state. (2) Heat transfer with the surroundings is negligible, and $\dot{W}_{cv} = 0$. (3) Kinetic and potential energy effects are negligible.

ANALYSIS: To find $\dot{m}_2$, begin with the steady-state mass and energy balances

$$0 = \cancel{\dot{Q}_{cv}}^{0} - \cancel{\dot{W}_{cv}}^{0} + \dot{m}_1\left(h_1 + \frac{V_1^2}{2} + gz_1\right) + \dot{m}_2\left(h_2 + \frac{V_2^2}{2} + gz_2\right) - \dot{m}_3\left(h_3 + \frac{V_3^2}{2} + gz_3\right)$$

With $\dot{m}_1 + \dot{m}_2 = \dot{m}_3$ and assumption (3)

$$0 = \dot{m}_1 h_1 + \dot{m}_2 h_2 - (\dot{m}_1 + \dot{m}_2) h_3$$

or

$$\dot{m}_2 = \dot{m}_1 \left(\frac{h_1 - h_3}{h_3 - h_2}\right)$$

From Table A-3, $T_1 < T_{sat}$ at 7 bar. Hence, state 1 is compressed liquid. Using Eq. 3.14 and data from Table A-2 at 42°C

$$h_1 \approx h_f(T) + v_f(T)\left[p - p_{sat}(T)\right]$$
$$= 175.9 \text{ kJ/kg} + \left(1.0086 \times 10^{-3} \frac{\text{m}^3}{\text{kg}}\right)\left[7 - 0.08268\right]\text{bar} \left|\frac{10^5 \text{ N/m}^2}{1 \text{ bar}}\right|\left|\frac{1 \text{ kJ}}{10^3 \text{ N}\cdot\text{m}}\right|$$
$$= 176.6 \text{ kJ/kg}$$

Further, from Table A-3

$$h_2 = h_{f2} + x_2 h_{fg2} = 697.22 + (.98)(2066.3) = 2722.2 \text{ kJ/kg}$$
$$h_3 = 697.22 \text{ kJ/kg}$$

Finally

$$\dot{m}_2 = (70 \text{ kg/s})\left(\frac{176.6 - 697.22}{697.22 - 2722.2}\right)$$
$$= 18.0 \text{ kg/s} \leftarrow \dot{m}_2$$

PROBLEM 4.68

KNOWN: Two ducts carrying air in a ventilation system merge into one exit duct. Data are known at the inlets and exit.

FIND: Determine the exit temperature and the diameter of the exit duct.

SCHEMATIC & GIVEN DATA:

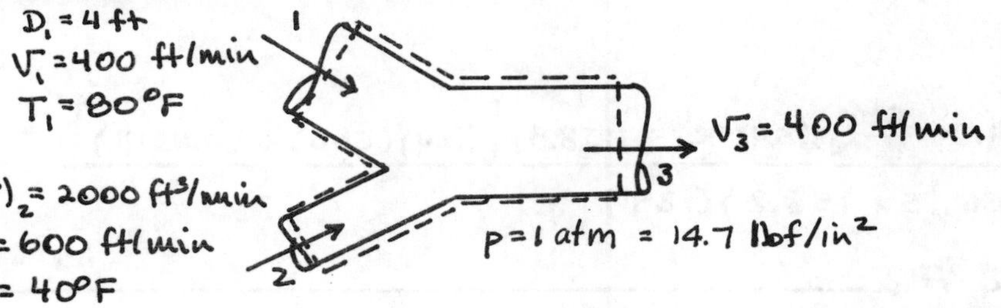

ASSUMPTIONS: (1) The control volume is at steady state. (2) Heat transfer with the surroundings is negligible, and $\dot{W}_{cv} = 0$. (3) Potential energy effects can be neglected. (4) The air behaves as an ideal gas with constant specific heats.

ANALYSIS: To find T_3, begin with steady-state mass and energy balances

$$0 = \cancel{\dot{Q}_{cv}}^{0} - \cancel{\dot{W}_{cv}}^{0} + \dot{m}_1\left(h_1 + \frac{V_1^2}{2} + gz_1\right) + \dot{m}_2\left(h_2 + \frac{V_2^2}{2} + gz_2\right) - \dot{m}_3\left(h_3 + \frac{V_3^2}{2} + gz_3\right)$$

and $\dot{m}_1 + \dot{m}_3 - \dot{m}_3 = 0 \Rightarrow \dot{m}_3 = \dot{m}_1 + \dot{m}_2$

Combining and incorporating assumption (3)

$$0 = \dot{m}_1\left[(h_1 - h_3) + \cancel{\left(\frac{V_1^2 - V_3^2}{2}\right)}^{0}\right] + \dot{m}_2\left[(h_2 - h_3) + \left(\frac{V_2^2 - V_3^2}{2}\right)\right]$$

Referring to Table A-20E, $c_p = 0.24$ Btu/lb·°R for the temperature range in this problem. Using $\Delta h = c_p \Delta T$

$$0 = \dot{m}_1\left[c_p(T_1 - T_3)\right] + \dot{m}_2\left[c_p(T_2 - T_3) + \left(\frac{V_2^2 - V_3^2}{2}\right)\right] \qquad (*)$$

The mass flow rates are evaluated using Eq. 4-11b and the ideal gas equation of state

$$\dot{m}_1 = \frac{A_1 V_1}{v_1} = \frac{\left(\frac{\pi D_1^2}{4}\right) V_1 p_1}{RT_1} = \frac{\left(\frac{\pi(4^2)\,\text{ft}^2}{4}\right)(400\ \text{ft/min})(14.7\ \text{lbf/in}^2)}{\left(\frac{1545}{28.97}\frac{\text{ft·lbf}}{\text{lb·°R}}\right)(540°\text{R})}\left|\frac{144\ \text{in}^2}{1\ \text{ft}^2}\right|$$

$$= 369.5\ \text{lb/min}$$

$$\dot{m}_2 = \frac{(AV)_2\, p_2}{RT_2} = \frac{(2000)(14.7)|144|}{\left(\frac{1545}{28.97}\right)(500)} = 158.8\ \text{lb/min}$$

Returning to (*)

$$T_3 = \frac{\dot{m}_1 c_p T_1 + \dot{m}_2\left[c_p T_2 + \left(\frac{V_2^2 - V_3^2}{2}\right)\right]}{(\dot{m}_1 + \dot{m}_2)c_p}$$

Evaluating the kinetic energy term

$$\frac{V_2^2 - V_3^2}{2} = \left(\frac{600^2 - 400^2}{2}\right)\frac{ft^2}{min^2}\left|\frac{1\ min^2}{3600\ s^2}\right|\left|\frac{1\ lbf}{32.2\ lb\cdot ft/s^2}\right|\left|\frac{1\ Btu}{778\ ft\cdot lbf}\right|$$

$$= 0.00111\ Btu/lb$$

Inserting values

① $$T_3 = \frac{(369.5\ lb/min)(.24\ \frac{Btu}{lb\cdot°R})(540°R) + (158.8)\left[(.24)(500) + (0.00111)\right]}{(369.5 + 158.8)(.24)}$$

$$= 528\ °R = 68\ °F \longleftarrow T_3$$

To get D_3, note that

$$\dot{m}_3 = \dot{m}_1 + \dot{m}_2 = 528.3\ lb/min$$

Thus

$$A_3 = \frac{v_3 \dot{m}_3}{V_3} = \frac{R T_3 \dot{m}_3}{p_3 V_3} = \frac{\left(\frac{1545}{28.97}\right)(528)(528.3)}{(14.7)|144|(400)} = 17.57\ ft^2$$

and

$$D_3 = \sqrt{\frac{4A_3}{\pi}} = 4.73\ ft \longleftarrow D_3$$

---

1. Note that if kinetic energy is neglected there is virtually no effect on the temperature at the exit.

PROBLEM 4.69

KNOWN: The electronic components of Example 4.8 are cooled by air flowing through the electronics enclosure.

FIND: Determine the largest average surface temperature of the components for which specified limits are met.

SCHEMATIC & GIVEN DATA:

Also See Fig. E4.8

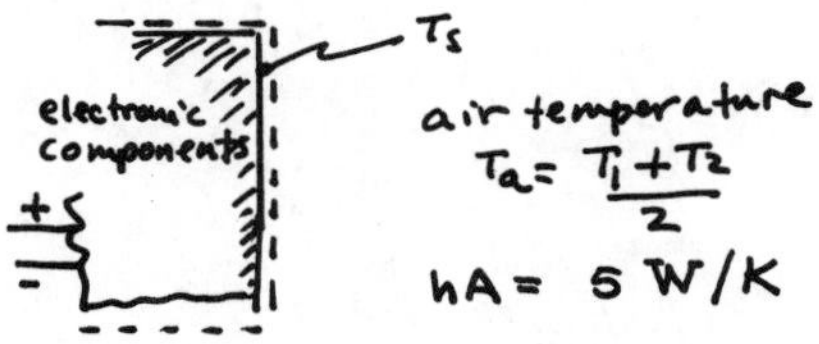

ASSUMPTIONS: 1. The electronic components form the system, which is at steady state.

ANALYSIS: The energy transfer from the electronic components to the air by convection is

$$\dot{Q}_c = hA[T_s - T_a]$$

(1) At steady state, the electric power provided to the electric components equals the energy removed by heat transfer: $\dot{Q}_c = 80\ W$. Also, $hA = 5\ W/K$. Solving for T_s, noting that $T_1 = 293K$ and $T_2 \leq 305K$ (22°C)

$$T_s = T_a + \frac{\dot{Q}_c}{hA}$$

$$T_s = \left(\frac{T_1 + T_2}{2}\right) + \frac{\dot{Q}_c}{hA}$$

(2) $$T_s \leq \left(\frac{293 + 305}{2}\right)K + \frac{80\ W}{5\ W/K} = 315\ K\ (42°C)$$

---

1. This can be obtained formally by applying an energy rate balance to a system consisting of the electronic components.

2. For reliability, excessive temperatures within the electronics are avoided by controlling the surface temperature T_s.

PROBLEM 4.70

KNOWN: Data are provided for an electronics enclosure cooled by an air flow induced by a fan.

FIND: Determine the volumetric flow rate of the air entering the fan.

SCHEMATIC & GIVEN DATA:

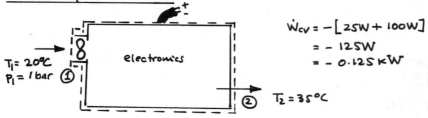

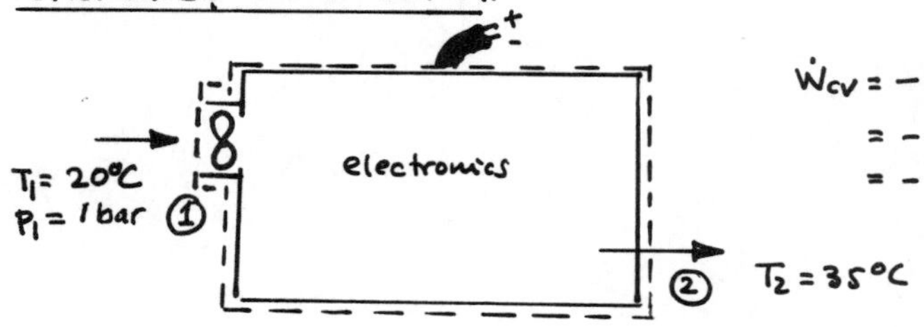

$\dot{W}_{cv} = -[25W + 100W]$
$= -125W$
$= -0.125 kW$

ASSUMPTIONS: 1. The control volume shown in the schematic is at steady state. 2. For the control volume, heat transfer with the surroundings and kinetic/potential energy effects can be ignored. 3. Air is modeled as an ideal gas.

ANALYSIS: At steady state $\dot{m}_1 = \dot{m}_2 \equiv \dot{m}$. The energy rate balance reads

$$0 = \cancel{\dot{Q}_{cv}} - \dot{W}_{cv} + \dot{m}\left[(h_1 - h_2) + \cancel{\frac{V_1^2 - V_2^2}{2}} + \cancel{g(z_1 - z_2)}\right]$$

or

$$\dot{m} = \frac{(-\dot{W}_{cv})}{h_2 - h_1}$$

With $\dot{m} = (AV)_1/v_1$ and using the ideal gas equation of state

$$\dot{m} = \frac{(AV)_1 p_1}{RT_1}$$

Collecting results and inserting data, including enthalpy values from Table A-22

$$(AV)_1 = \left(\frac{RT_1}{p_1}\right)\left[\frac{(-\dot{W}_{cv})}{h_2 - h_1}\right]$$

$$= \frac{\left(\frac{8314}{28.97}\frac{N \cdot m}{kg \cdot K}\right)(293 K)}{(10^5 N/m^2)}\left[\frac{0.125 kJ/s}{(308.2 - 293.2)\frac{kJ}{kg}}\right]$$

$$= 7 \times 10^{-3} \frac{m^3}{s}$$ ⟵

PROBLEM 4.71

KNOWN: Data are provided for a water-jacketed housing filled with electronic components, which is at steady state.

FIND: Determine the minimum cooling water mass flow rate to satisfy a limit on the temperature of the water exiting the enclosure.

SCHEMATIC & GIVEN DATA:

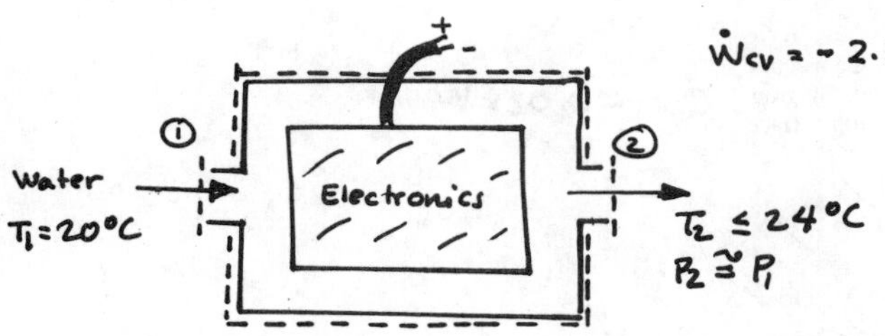

ASSUMPTIONS: 1. The control volume shown in the schematic is at steady state. 2. For the control volume, heat transfer with the surroundings can be ignored, as can kinetic/potential energy effects. 3. For the water
① entering and exiting the housing $h \approx h_f(T)$.

ANALYSIS: At steady state, $\dot{m}_1 = \dot{m}_2 \equiv \dot{m}$. An energy rate balance reads

$$0 = \cancel{\dot{Q}_{cv}} - \dot{W}_{cv} + \dot{m}\left[(h_1 - h_2) + \cancel{\left(\frac{V_1^2 - V_2^2}{2}\right)} + \cancel{g(z_1 - z_2)}\right]$$

giving

$$\dot{m} = \frac{(-\dot{W}_{cv})}{h_2 - h_1}$$

$$= \frac{(-\dot{W}_{cv})}{h_f(T_2) - h_f(T_1)}$$

Since $T_2 \leq 24°C$

$$\dot{m} \geq \frac{(-\dot{W}_{cv})}{h_f(24°C) - h_f(20°C)} = \frac{-(-2.5\ kJ/s)}{(100.70 - 83.96)\ kJ/kg}$$

$$\dot{m} \geq 0.149\ kg/s$$

where h_f values are from Table A-2.

---

1. Alternatively, Eq. 3.20b with c from Table A-19 can be used.

PROBLEM 4.72

KNOWN: Data are provided for electronic components mounted on a plate that are cooled by convection to the surroundings and water circulating through a tube bonded to the plate. Operation is at steady state.

FIND: Determine the tube diameter.

SCHEMATIC & GIVEN DATA:

$\dot{Q}_{cv} = -0.08\,kW$

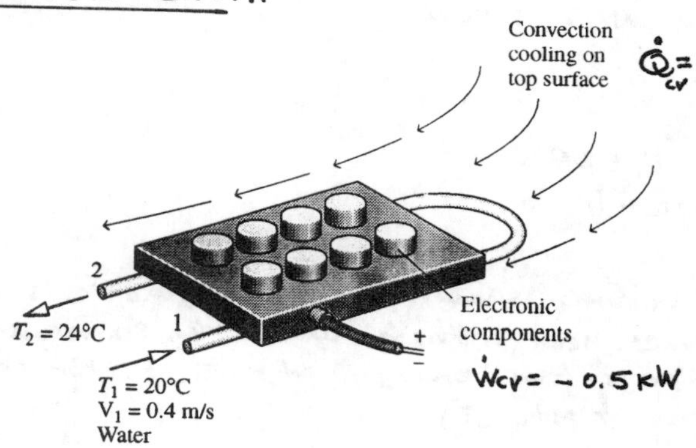

$\dot{W}_{cv} = -0.5\,kW$

ASSUMPTIONS: 1. A control volume encloses the plate-mounted electronic components with an inlet at 1 and an exit at 2. 2. The control volume ① is at steady state. 3. For the water entering and exiting, $h \approx h_f(T)$, $v \approx v_f(T)$. 4. Kinetic and potential energy effects can be ignored.

ANALYSIS: At steady state, $\dot{m}_1 = \dot{m}_2 \equiv \dot{m}$. Also,

$$\dot{m} = \frac{A_1 V_1}{v_1} = \frac{(\pi D^2/4) V_1}{v_f(T_1)}$$

An energy rate balance reads

$$0 = \dot{Q}_{cv} - \dot{W}_{cv} + \dot{m}\left[(h_1 - h_2) + \cancel{\left(\frac{V_1^2 - V_2^2}{2}\right)} + \cancel{g(z_1 - z_2)}\right]$$

or

$$\dot{m} = \frac{[\dot{Q}_{cv} - \dot{W}_{cv}]}{h_2 - h_1} = \frac{[\dot{Q}_{cv} - \dot{W}_{cv}]}{h_f(T_2) - h_f(T_1)}$$

Collecting results

$$D = \sqrt{\frac{4\, v_f(T_1)}{\pi V_1}\left[\frac{(\dot{Q}_{cv} - \dot{W}_{cv})}{h_f(T_2) - h_f(T_1)}\right]}$$

With data from Table A-2: $v_f(20°C) = (1.0018/10^3)\ m^3/kg$, $h_f(T_1) = 83.96$ kJ/kg, $h_f(T_2) = 100.7$ kJ/kg

$$D = \sqrt{\frac{(4)(1.0018/10^3)\ m^3/kg}{\pi\,(0.4\ m/s)}\left[\frac{(-0.08 - (-0.5))\ kJ/s}{(100.7 - 83.96)\ kJ/kg}\right]}$$

$$= 0.0089\ m \left|\frac{10^2\ cm}{m}\right|$$

$$= 0.89\ cm \quad \blacktriangleleft$$

1. Alternatively, the incompressible model can be used, with Eq. 3.20b and c from Table A-19.

PROBLEM 4.73

KNOWN: Data are provided for electronic components mounted on the inner surface of a horizontal duct. The components are cooled by air flowing through the cylinder and by convection from the outer surface.

FIND: Determine the minimum heat transfer from the outer surface for which a limit on the exiting air temperature is satisfied.

SCHEMATIC & GIVEN DATA:

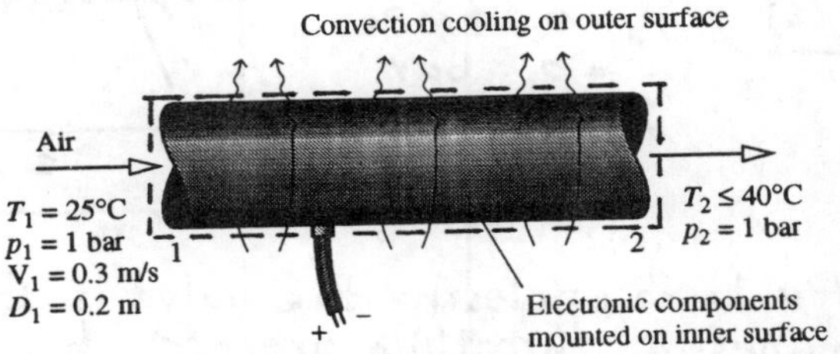

$\dot{W}_{cv} = -0.2 kW$

ASSUMPTIONS: 1. The control volume shown in the schematic is at steady state. 2. The air is modeled as an ideal gas. 3. Kinetic and potential energy effects can be ignored.

ANALYSIS: At steady state, $\dot{m}_1 = \dot{m}_2 = \dot{m}$, where

$$\dot{m}_1 = \frac{A_1 V_1}{v_1} = \frac{(\pi D_1^2/4)V_1}{RT_1/p_1}$$

$$= \frac{\pi(0.2m)^2(0.3m/s)(10^5 N/m^2)}{4\left(\frac{8314}{28.97}\frac{N\cdot m}{kg\cdot K}\right)(298K)} = 0.011\ \frac{kg}{s}$$

An energy rate balance is

$$0 = \dot{Q}_{cv} - \dot{W}_{cv} + \dot{m}\left[(h_1 - h_2) + \cancel{\frac{V_1^2 - V_2^2}{2}} + \cancel{g(z_1 - z_2)}\right]$$

or

$$\dot{Q}_{cv} = \dot{W}_{cv} + \dot{m}(h_2 - h_1)$$
$$= \dot{W}_{cv} + \dot{m}(h(T_2) - h(T_1))$$

Since $T_2 \leq 313 K (40°C)$

$$\dot{Q}_{cv} \leq (-0.2 kW) + \left(0.011 \frac{kg}{s}\right)\left[h(313K) - h(298K)\right]\left(\frac{kJ}{kg}\right)\left|\frac{1 kW}{1 kJ/s}\right|$$
$$\leq (-0.2kW) + \left(0.011 \frac{kg}{s}\right)\left[(313.3 - 298.2)\frac{kJ}{kg}\right]\left|\frac{1 kW}{1 kJ/s}\right|$$
$$\leq -0.034 kW \longleftarrow$$

The energy removed by convection must be at least 0.034 kW.

PROBLEM 4.74

KNOWN: Refrigerant 134a expands through a valve from a known pressure and temperature to a given final pressure.

FIND: Determine the exit quality.

SCHEMATIC & GIVEN DATA:

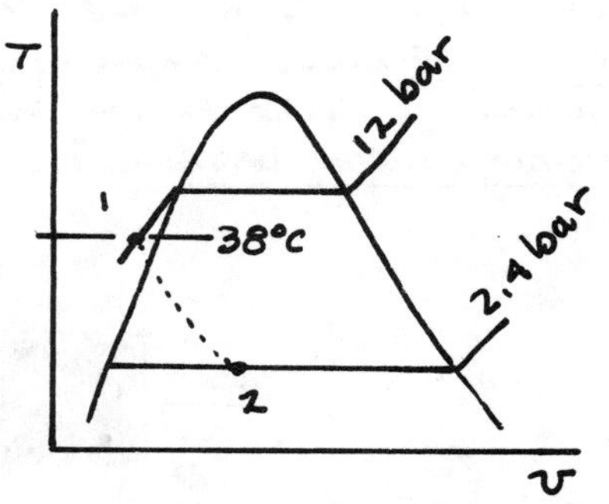

$p_1 = 1.2$ MPa $= 12$ bar
$T_1 = 38°C$

(1) → ⊗ → (2)

$p_2 = 0.24$ MPa $= 2.4$ bar
$x_2 = ?$

ASSUMPTIONS: (1) A control volume enclosing the valve is at steady state. (2) The refrigerant undergoes a throttling process; $h_1 = h_2$.

ANALYSIS: According to data from Table A-11, at $p_1 = 12$ bar; $T_{sat} = 46.32°C$. Since $T_1 < T_{sat}$, state 1 is in the compressed liquid region. For simplicity,
① we use Eq. 3.14 to evaluate h_1, as follows:

$$h_1 \approx h_f(T_1) = 103.21 \text{ kJ/kg}$$

where the value is obtained from Table A-10.

By assumption (2)

$$h_1 = h_2$$
$$= h_{f_2} + x_2 h_{fg2}$$

Solving and inserting data from Table A-11 at $p_2 = 2.4$ bar

$$x_2 = \frac{h_1 - h_{f2}}{h_{fg2}}$$

$$= \frac{103.21 \text{ kJ/kg} - 42.95 \text{ kJ/kg}}{201.14 \text{ kJ/kg}}$$

$$= 0.30 \ (30\%) \quad \leftarrow x_2$$

---

1. Here, we have ignored the effect of pressure on the specific enthalpy of liquid refrigerant. We could have used Eq. 3.13 to estimate the effect of pressure. In that case, h_1 would have been 104.0 kJ/kg, and the exit quality would have been $x_2 = 0.304$ (30.4%). Thus, using Eq. 3.14 is very accurate in this case, and the approximation of Eq. 3.14 is commonly used when refrigeration systems are analyzed.

PROBLEM 4.75

__KNOWN__: Ammonia expands through a valve from a known inlet state to a known exit pressure.

__FIND__: Determine the exit temperature.

__SCHEMATIC & GIVEN DATA__:

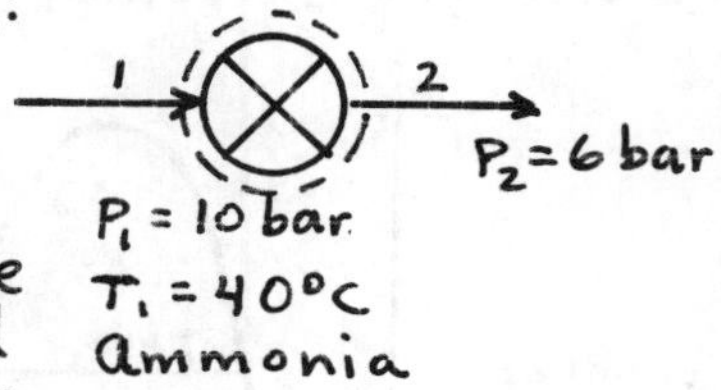

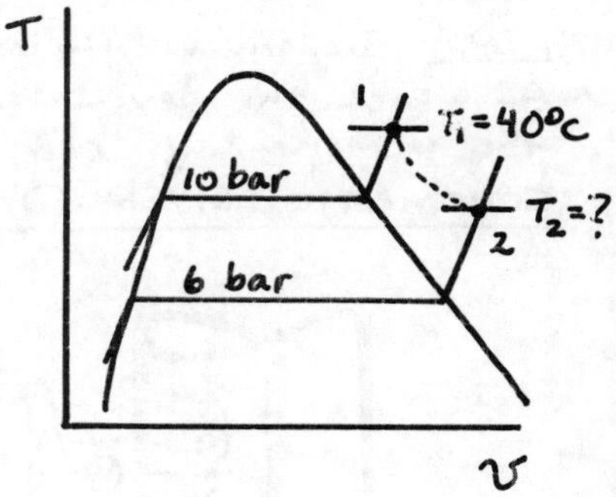

__ASSUMPTIONS__: (1) The control volume is at steady state. (2) Heat transfer is negligible and $\dot{W}_{cv} = 0$. (3) Kinetic and potential energy effects are negligible. (throttling process)

__ANALYSIS__: In accordance with the assumptions for a throttling process, $h_1 = h_2$. Using data from Table A-15

$$h_2 = h_1 = 1508.20 \text{ kJ/kg}$$

Interpolating in Table A-15 at $p_2 = 6$ bar, $h_2 = 1508.2$ kJ/kg

$$T_2 \approx 30.5°C \longleftarrow T_2$$

PROBLEM 4.76

KNOWN: Data are provided for a throttling calorimeter attached to a large pipe carrying a two-phase liquid-vapor mixture. Operation is at steady state.

FIND: Determine the range of throttling calorimeter exit temperatures for which the device can determine the quality of steam in the pipe, and the corresponding range of steam quality values.

SCHEMATIC & GIVEN DATA:

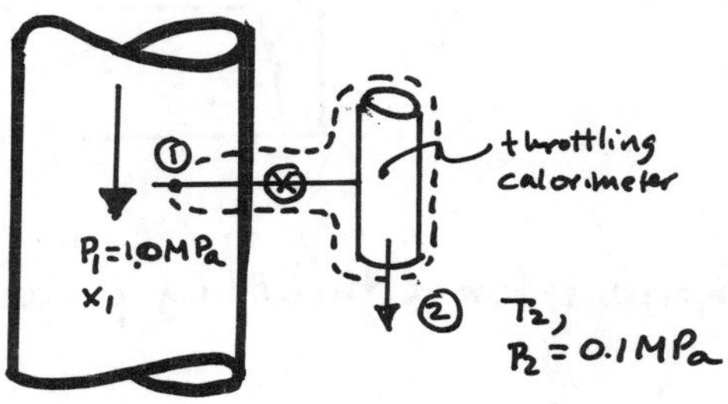

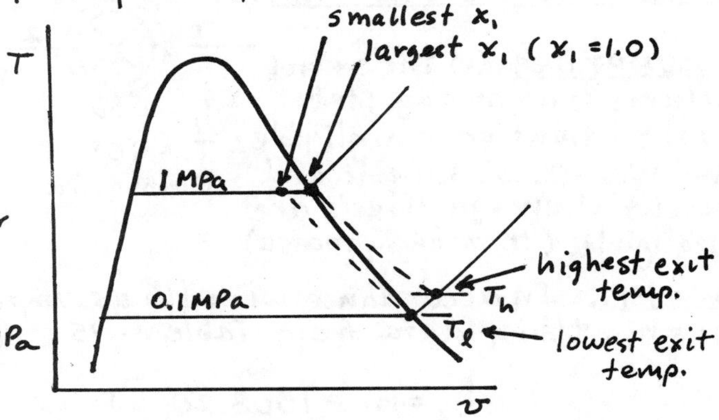

ASSUMPTIONS: 1. The control volume shown in the figure is at steady state. 2. The expansion through the calorimeter adheres to the throttling process model: $h_2 \approx h_1$, where $h_1 = h_{f1} + x_1(h_{g1} - h_{f1})$.

ANALYSIS: With assumption 2

$$h(T_2, P_2) = h_{f1} + x_1(h_{g1} - h_{f1}) \qquad (1)$$

where the state at the exit is fixed by T_2, P_2, and thus must be superheated vapor or, in the limit, saturated vapor. Also, the quality x_1 is required to be less than or, in the limit, equal to 1.0.

Referring to the T-v diagram, the highest exit temperature, T_h, corresponds to a steam quality of 1.0: That is, with data from Table A-3, Eq. (1) gives

$$h(T_h, P_2) = h_g(P_1) = 2778.1 \text{ kJ/kg}$$

Interpolating in Table A-4 at 0.1 MPa gives $T_h = 151°C$.

The lowest exit temperature, T_l, corresponds to saturated vapor exiting the calorimeter, for which $h_2 = 2675.5$ kJ/kg and $T_l = 99.6°C$. Eq. (1) gives

$$x_1 = \frac{h_2 - h_{f1}}{h_{g1} - h_{f1}} = \frac{2675.5 - 762.81}{2015.3} = 0.949$$

In summary

$$99.6 \le T_2 \le 151 \text{ °C}$$
$$0.949 \le x_1 \le 1.0$$

PROBLEM 4.77

<u>KNOWN</u>: Refrigerant 22 expands through a valve from a known inlet state to a known final pressure.

<u>FIND</u>: Determine the exit temperature and quality.

<u>SCHEMATIC & GIVEN DATA</u>:

<u>ASSUMPTIONS</u>: (1) The control volume is at steady state. (2) Heat transfer is negligible, and $\dot{W}_{cv} = 0$. (3) Kinetic and potential energy effects are negligible. (throttling process)

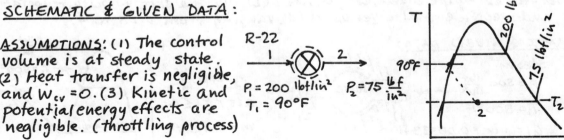

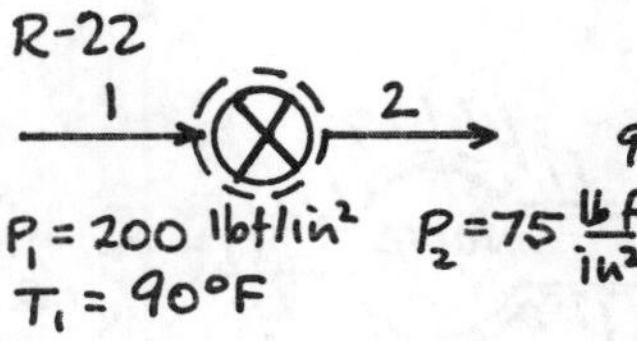

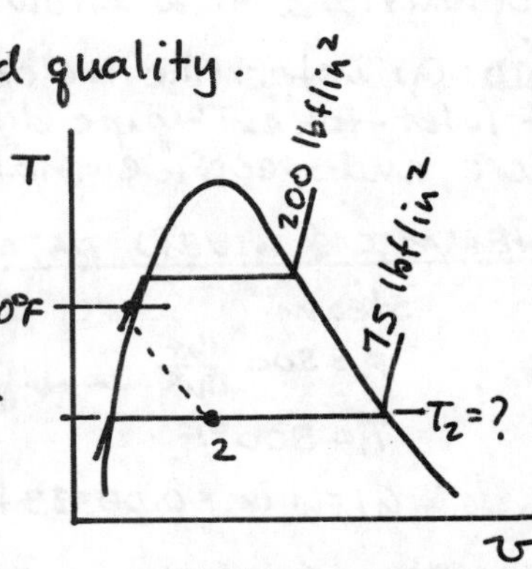

<u>ANALYSIS</u>: In accordance with the assumptions for a throttling process, $h_1 = h_2$. From Table A-8E; $T_1 < T_{sat}$ @ 200 lbf/in². Thus, the inlet state is in the compressed liquid region. Ignoring the effect of pressure on the enthalpy of the liquid, Table A-7E gives

$$h_1 \approx h_f(90°F) = 36.32 \text{ Btu/lb}$$

Since $h_2 = h_1 = h_{f2} + x_2 h_{fg2}$, data from Table A-8E at 75 lbf/in² give

$$x_2 = \frac{h_2 - h_{f2}}{h_{fg2}}$$

$$= \frac{36.32 - 19.99}{87.71}$$

$$= 0.186 \longleftarrow x_2$$

and

$$T_2 = 34.08°F \longleftarrow T_2$$

PROBLEM 4.78

KNOWN: Steam flows through a well-insulated valve from specified inlet conditions to a known exit pressure.

FIND: (a) Determine the exit velocity and exit temperature for a given ratio of inlet-to-exit pipe diameters, d_1/d_2. (b) Plot the exit velocity, temperature, and specific enthalpy versus d_1/d_2 ranging from 0.25 to 4.

SCHEMATIC & GIVEN DATA:

Steam
$p_1 = 500\ \frac{lbf}{in.^2}$
$T_1 = 500°F$
$d_1 = 1\ in. = 0.08333\ ft$

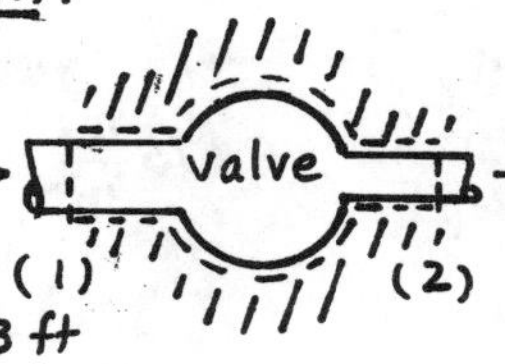

$p_2 = 200\ \frac{lbf}{in^2}$
$0.25 \le \frac{d_1}{d_2} \le 4$
$\dot{m} = 0.11\ lb/s$

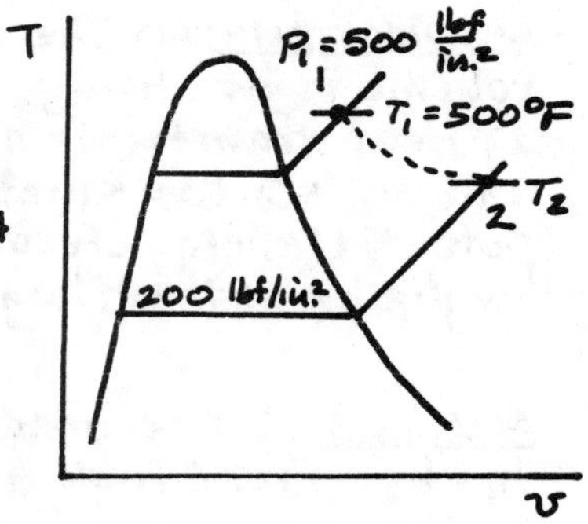

ASSUMPTIONS: (1) The control volume is at steady state.
(2) For the control volume, $\dot{Q}_{cv} = 0$ and $\dot{W}_{cv} = 0$.
(3) Potential energy effects are negligible.

ANALYSIS: (a) To determine the exit velocity, begin with the mass balance and Eq. 4.11b: $\dot{m}_1 = \dot{m}_2 \equiv \dot{m}$, and

$$\dot{m} = \frac{A_1 V_1}{v_1} = \frac{A_2 V_2}{v_2} \Rightarrow V_2 = \frac{A_1}{A_2} \cdot \frac{V_1}{v_1} \cdot v_2 = \left(\frac{d_1}{d_2}\right)^2 \cdot \frac{V_1}{v_1} \cdot v_2$$

Now, $V_1/v_1 = \dot{m}/A_1 = 4\dot{m}/\pi d_1^2$. Inserting values

$$\frac{V_1}{v_1} = \frac{(4)(.11\ lb/s)}{\pi(.08333\ ft)^2} = 20.17$$

Thus

$$V_2 = 20.17\ \left(\frac{d_1}{d_2}\right)^2 \cdot v(T_2, p_2) \qquad (1)$$

and, with $v_1 = 0.992\ ft^3/lb$ from Table A-4E

$$V_1 = (20.17)(0.992) = 20.01\ ft/s$$

From (1) we see that it is necessary to fix state 2 to evaluate V_2. Another relation is obtained from the energy balance at steady state.

$$0 = \cancel{\dot{Q}_{cv}}^{0} - \cancel{\dot{W}_{cv}}^{0} + \dot{m}\left[(h_1 - h_2) + \left(\frac{V_1^2 - V_2^2}{2}\right) + g(\cancel{z_1 - z_2}^{0})\right] \qquad (2)$$

or

$$V_2 = \sqrt{2(h_1 - h_2) + V_1^2}$$

with h_1 from Table A-4E

$$V_2 = \sqrt{2\left[1231.5 - h(T_2, p_2)\right]\frac{Btu}{lb}\left|\frac{1\ lbf}{32.2\ lb \cdot ft/s^2}\right|\left|\frac{1\ Btu}{778\ ft \cdot lbf}\right| + (20.01\ ft/s)^2} \qquad (3)$$

Equations (1) and (3) can be solved simultaneously using data from Table A-4E and an iterative process. The results are, for $d_1/d_2 = 0.25$

$$T_2 = 434.6\ °F$$

$$V_2 = 3.140\ ft/s$$

(b) The following IT code can be used to solve Eqs. (1) and (2) with the associated data for steam:

```
p1 = 500  // lbf/in.²
T1 = 500  // °F
mdot = 0.11  // lb/s
d1 = 1/12  // ft.
p2 = 200  // lbf/in.²
dratio = d1 / d2
dratio = .25

A1 = pi * d1^2 / 4
mdot = A1 * V1 / v1
V2 / V1 = (d1 / d2)^2 * (v2 / v1)

h1 = h_PT("Water/Steam", p1, T1)
v1 = v_PT("Water/Steam", p1, T1)
h2 = h_PT("Water/Steam", p2, T2)
v2 = v_PT("Water/Steam", p2, T2)

0 = (h1 - h2) + ((V1^2 - V2^2) / (2 * 32.2 * 778))
```

IT Results for the sample case of $d_1/d_2 = 0.25$

$T_2 = 434.2$ °F
$V_2 = 3.139$ ft/s
$h_2 = 1231$ Btu/lb
$V_1 = 20.01$ ft/s
$v_2 = 2.490$ ft³/lb

Using the Explore button, sweep dratio from 0.25 to 4 in steps of 0.25. Then, construct the following plots:

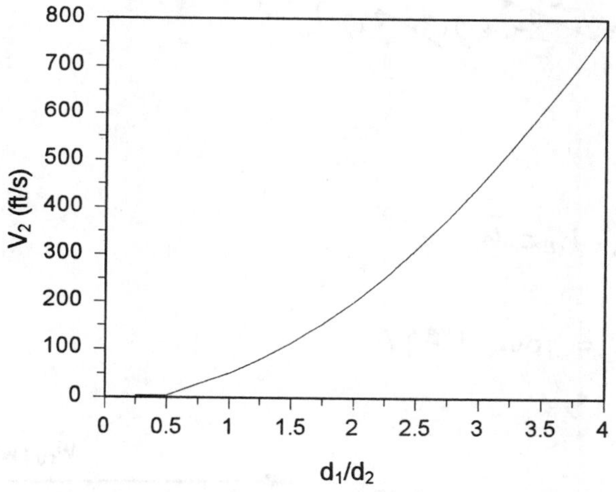

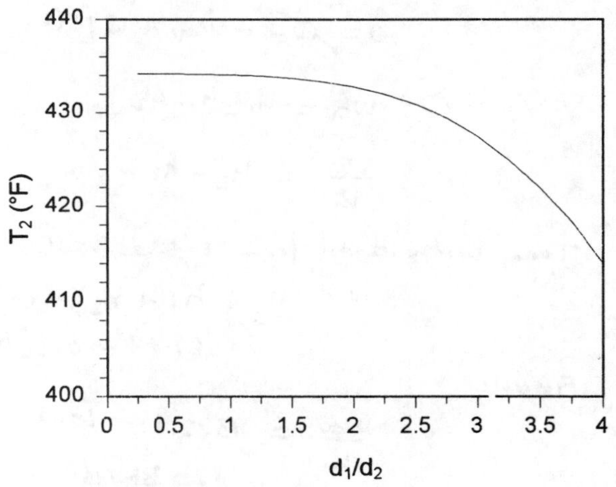

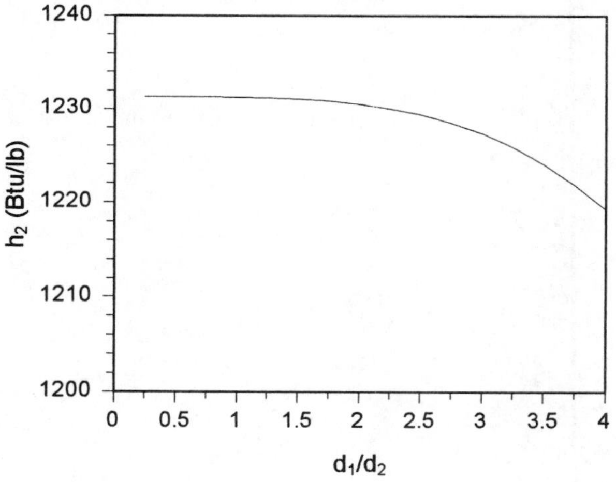

Note that as d_1/d_2 becomes large the throttling process assumptions (pp. 171-173) of negligible kinetic energy and $h_2 \approx h_1$ are not valid.

PROBLEM 4.79

KNOWN: Data are provided for a valve and turbine in series, each operating at steady state.

FIND: For the turbine, determine the temperature at the inlet and the power developed per unit mass of steam flowing.

SCHEMATIC & GIVEN DATA:

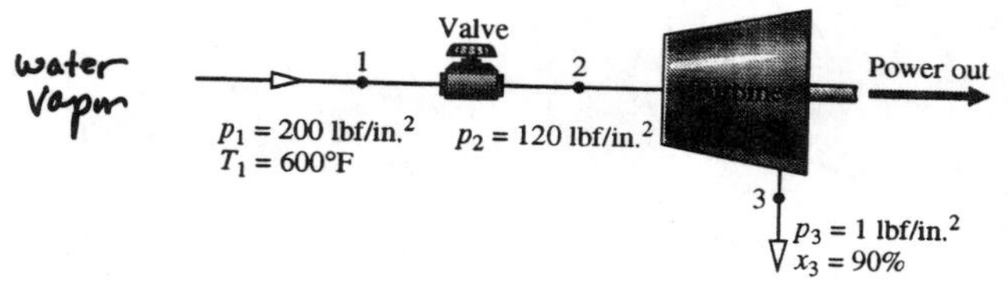

ASSUMPTIONS: 1. Control volumes at steady state enclose the valve and the turbine, respectively. 2. The expansion across the valve is a throttling process. 3. Heat transfer with the surroundings and kinetic/potential energy effects are negligible.

ANALYSIS: Considering the valve, since the expansion is a throttling process, we have $h_2 \approx h_1$. With data from Table A-4E, $h_1 = 1322.1$ Btu/lb. Then, interpolating at 120 lbf/in^2; $T_2 \approx 589°F$. ← T_2

Mass and energy rate balances applied to the turbine read

$\dot{m}_2 = \dot{m}_3 = \dot{m}$

$$0 = \dot{Q}_{cv} - \dot{W}_{cv} + \dot{m}\left[(h_1 - h_2) + \frac{V_1^2 - V_2^2}{2} + g(z_1 - z_2)\right]$$

$$\Rightarrow \dot{W}_{cv} = \dot{m}[h_1 - h_2]$$

$$\frac{\dot{W}_{cv}}{\dot{m}} = h_3 - h_2$$

Then, with data from Table A-3E and $h_2 = h_1$

$$h_3 = h_f + x_3(h_g - h_f)$$
$$= 69.74 + 0.9[1036] = 1002.1 \text{ Btu/lb}$$

Finally

$$\frac{\dot{W}_{cv}}{\dot{m}} = 1322.1 - 1002.1$$
$$= 320 \text{ Btu/lb}$$ ← $\dot{W}_{cv}/\dot{m}$

PROBLEM 4.80

KNOWN: Refrigerant 134a passes through a suction line heat exchanger and an evaporator. Data are given at various locations.

FIND: Determine the heat transfer rate for a control volume enclosing the evaporator.

SCHEMATIC & GIVEN DATA:

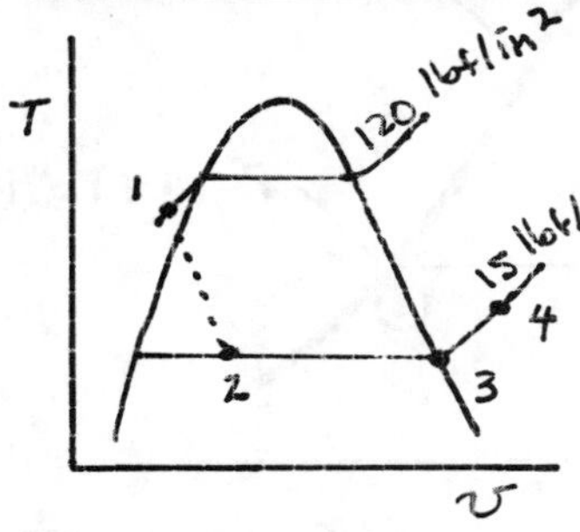

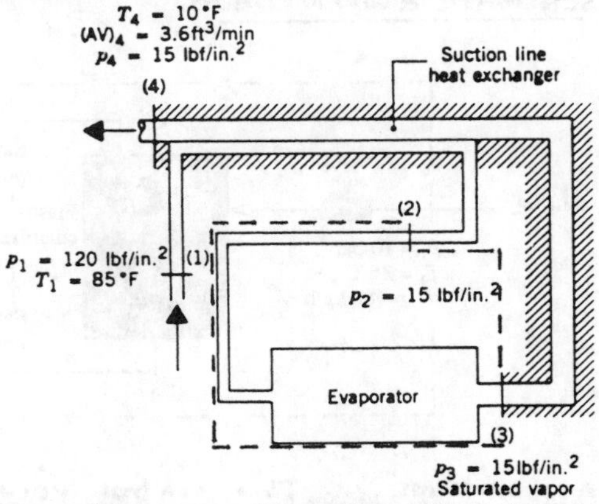

ASSUMPTIONS: (1) The control volume is at steady state. (2) Heat transfer is negligible between the suction line heat exchanger and its surroundings. (3) $\dot{W}_{cv} = 0$. (4) Kinetic and potential energy effects are negligible.

ANALYSIS: To find the evaporator heat transfer rate, begin with the steady-state energy and mass balances

$$0 = \dot{Q}_{cv} - \dot{W}_{cv}^{\,0} + \dot{m}\left[(h_2 - h_3) + \left(\frac{V_2^2 - V_3^2}{2}\right)^0 + g(z_2 - z_3)^0\right]$$

where $\dot{m}_1 = \dot{m}_2 = \dot{m}_3 = \dot{m}_4 \equiv \dot{m}$. Thus

$$\dot{Q}_{cv} = \dot{m}(h_3 - h_2)$$

From Table A-11E, $h_3 = h_{g@15\,lbf/in^2} = 99.66$ Btu/lb. To find h_2, write an energy balance for a control volume enclosing the heat exchanger

$$0 = \dot{m}(h_1 - h_2) + \dot{m}(h_3 - h_4)$$

or

$$h_2 = h_1 + h_3 - h_4$$

From Table A-12E, $h_4 = 104.38$ Btu/lb, and assuming $h_1 \approx h_f(T_1)$

$$h_2 = 38.99 + 99.66 - 104.38 = 34.27 \text{ Btu/lb}$$

Now, with $v_4 = 3.1680$ ft³/lb, the mass flow rate is

$$\dot{m} = \frac{(AV)_4}{v_4} = \frac{(3.6 \text{ ft}^3/\text{min})}{(3.1680 \text{ ft}^3/\text{lb})}\left|\frac{60 \text{ min}}{1 \text{ h}}\right| = 68.18 \text{ lb/h}$$

Finally, the evaporator heat transfer rate is

$$\dot{Q}_{cv} = \left(68.18 \frac{\text{lb}}{\text{h}}\right)(99.66 - 34.27) \text{ Btu/lb}$$

$$= 4458 \text{ Btu/h} \leftarrow \quad \dot{Q}_{cv}$$

PROBLEM 4.81

KNOWN: Data are provided for a flash chamber operating at steady state. Saturated vapor and saturated liquid streams exit at pressure p.

FIND: If p = 4 bar, determine the mass flow rates of the exiting streams. Plot the mass flow rates of the exiting streams versus pressure p, 1 ≤ p ≤ 9 bar.

SCHEMATIC & GIVEN DATA:

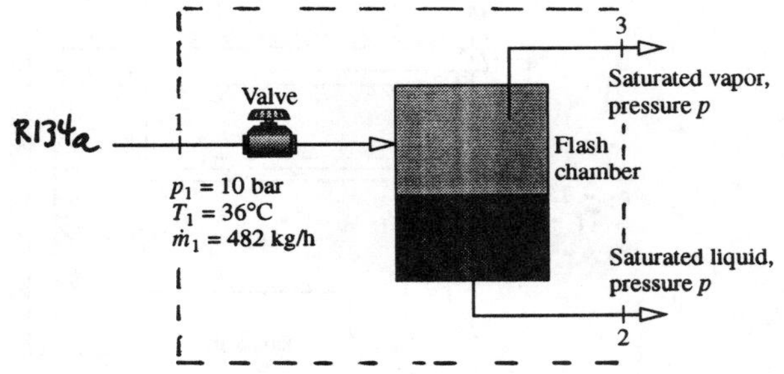

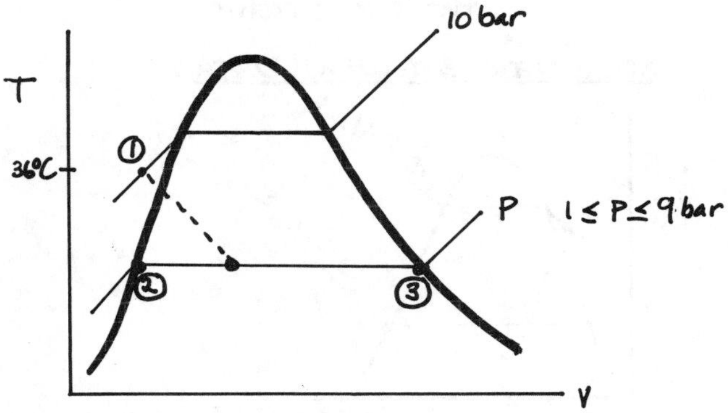

ASSUMPTIONS: 1. The control volume shown in the accompanying figure is at steady state. 2. For the control volume, $\dot{W}_{cv} = 0$, heat transfer with the surroundings is negligible, and kinetic/potential energy effects can be ignored. 3. For the liquid entering at 1, $h_1 \approx h_f(T_1)$.

ANALYSIS: For the control volume, the mass rate balance at steady state gives

$\dot{m}_1 = \dot{m}_2 + \dot{m}_3$, or $\dot{m}_3 = \dot{m}_1 - \dot{m}_2$.

An energy rate balance reduces to

$$0 = \dot{Q}_{cv} - \dot{W}_{cv} + \dot{m}_1\left[h_1 + \frac{V_1^2}{2} + gz_1\right] - \dot{m}_2\left[h_2 + \frac{V_2^2}{2} + gz_2\right] - \dot{m}_3\left[h_3 + \frac{V_3^2}{2} + gz_3\right]$$

$$\Rightarrow 0 = \dot{m}_1 h_1 - \dot{m}_2 h_2 - \dot{m}_3 h_3 \quad \text{or} \quad 0 = \dot{m}_1 h_1 - \dot{m}_2 h_2 - (\dot{m}_1 - \dot{m}_2) h_3$$

$$\Rightarrow \quad \dot{m}_2 = \dot{m}_1 \left[\frac{h_1 - h_3}{h_2 - h_3}\right]$$

(a) From Table A-10; $h_1 \approx h_f(36°C) = 100.25$ kJ/kg. Also, at p = 4 bar, Table A-11 gives; $h_2 = h_f(4\text{ bar}) = 62$ kJ/kg and $h_3 = h_g(4\text{ bar}) = 252.32$ kJ/kg. Thus

$$\dot{m}_2 = \left(482\ \frac{kg}{h}\right)\left[\frac{(100.25) - (252.32)}{(62) - (252.32)}\right] = 385.1\ \frac{kg}{h} \longleftarrow \dot{m}_2$$

From above

$$\dot{m}_3 = \dot{m}_1 - \dot{m}_2 = 482 - 385.1 = 96.9\ kg/h \longleftarrow \dot{m}_3$$

(b) The following IT code can be used to generate data for plots of $\dot{m}_2$ and $\dot{m}_3$ versus p:

IT Code

```
p1 = 10  // bar
T1 = 36  // °C
mdot1 = 482  // kg/h
p = 4  // bar
x2 = 0
x3 = 1
mdot1 = mdot2 + mdot3
0 = mdot1 * h1 - mdot2 * h2 - mdot3 * h3
h1 = h_PT("R134A", p1, T1)
h2 = hsat_Px("R134A", p, x2)
h3 = hsat_Px("R134A", p, x3)
```

PROBLEM 4.81 (Cont'd.)

Using $p = 4$ bars as a sample case

<u>*IT* Results (p = 4 bar)</u>
$\dot{m}_2 = 385.1$ kg/h
$\dot{m}_3 = 96.87$ kg/h
$h_1 = 100.3$ kJ/kg
$h_2 = 62$ kJ/kg
$h_3 = 252.3$ kJ/kg

These results compare very favorable with those of part(a). Now, using the Explore button, sweep p from 1 to 9 bar in steps of 0.1 bar. Then, the following plot can be constructed:

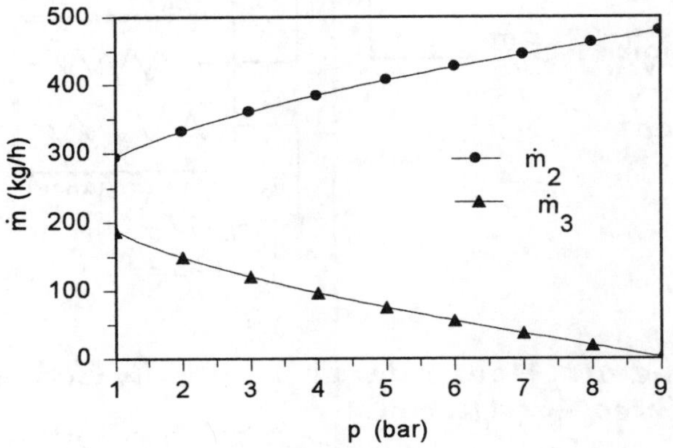

These plots imply that as the pressure in the flash chamber, p, is decreased a greater fraction of the incoming liquid flow "flashes" to saturated vapor.

PROBLEM 4.82

KNOWN: Air flows through two turbine stages and an inter-connecting heat exchanger. A separate hot air stream passes in counterflow through the heat exchanger. Data are known at various locations.

FIND: Determine the temperature of the main air stream exiting the heat exchanger and the power output of the second turbine.

SCHEMATIC & GIVEN DATA:

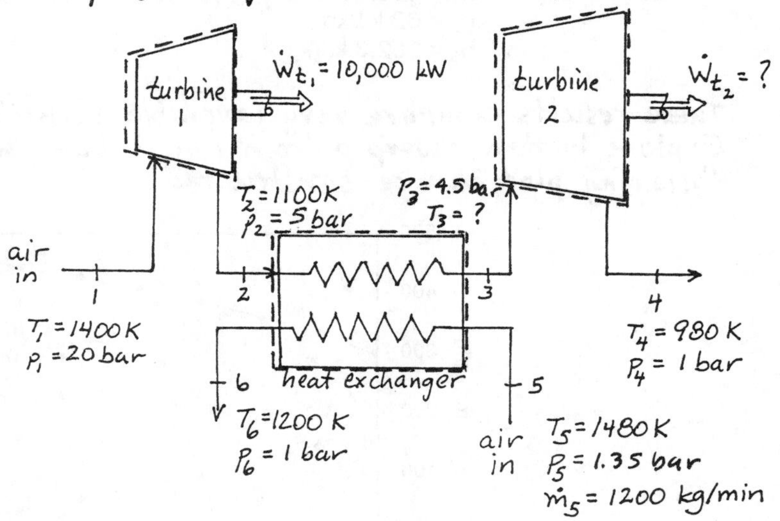

ASSUMPTIONS: (1) The control volumes are at steady state. (2) Heat transfer to the surroundings can be neglected. (3) Kinetic and potential energy effects are negligible. (4) The air behaves as an ideal gas. (5) For the heat exchanger, $W_{cv} = 0$.

ANALYSIS: First, find the air flow rate at 1. Begin with steady-state energy and mass balances for turbine 1

$$0 = \dot{Q}_{cv}^{\,\nearrow 0} - \dot{W}_{t_1} + \dot{m}_1\left[(h_1 - h_2) + \left(\frac{V_1^2 - V_2^2}{2}\right)^{\nearrow 0} + g(z_1 - z_2)^{\nearrow 0}\right]$$

and $$0 = \dot{m}_1 - \dot{m}_2 \Rightarrow \dot{m}_1 = \dot{m}_2$$

Solving for $\dot{m}_1$

$$\dot{m}_1 = \frac{\dot{W}_{t1}}{(h_1 - h_2)}$$

From Table A-22; $h_1 = 1515.42$ kJ/kg and $h_2 = 1161.07$ kJ/kg. Thus

$$\dot{m}_1 = \frac{(10{,}000 \text{ kW})}{(1515.42 - 1161.07)\text{ kJ/kg}}\left|\frac{1 \text{ kJ/s}}{1 \text{ kW}}\right| = 28.22 \text{ kg/s}$$

Turning next to the heat exchanger

$$\dot{m}_1 = \dot{m}_2 = \dot{m}_3 = \dot{m}_4$$
$$\dot{m}_5 = \dot{m}_6$$

$$0 = \dot{Q}_{cv}^{\,\nearrow 0} - \dot{W}_{cv}^{\,\nearrow 0} + \dot{m}_2\left[(h_2 - h_3) + \left(\frac{V_2^2 - V_3^2}{2}\right)^{\nearrow 0} + g(z_2 - z_3)^{\nearrow 0}\right] + \dot{m}_5\left[(h_5 - h_6) + \left(\frac{V_5^2 - V_6^2}{2}\right)^{\nearrow 0} + g(z_5 - z_6)^{\nearrow 0}\right]$$

$$0 = \dot{m}_2(h_2 - h_3) + \dot{m}_5(h_5 - h_6)$$

or $$h_3 = h_2 + \frac{\dot{m}_5}{\dot{m}_2}(h_5 - h_6)$$

Again, from Table A-22; $h_5 = 1611.79$ kJ/kg and $h_6 = 1277.79$ kJ/kg. Thus

PROBLEM 4.82 (Cont'd)

$$h_3 = 1161.07 + \left[\frac{(1200\ \text{kg/min})}{(28.22\ \text{kg/s})\left|60\ \text{s/1min}\right|}\right](1611.79 - 1277.79)$$

$$= 1397.8\ \text{kJ/kg}$$

Interpolating in Table A-22

$$T_3 = 1301.5\text{K} \longleftarrow T_3$$

Now, writing the steady-state energy balance for turbine 2

$$0 = \dot{Q}_{cv}^{\nearrow 0} - \dot{W}_{t2} + \dot{m}_3\left[(h_3 - h_4) + \left(\frac{V_3^2 - V_4^2}{2}\right)^{\nearrow 0} + g(z_3 - z_4)^{\nearrow 0}\right]$$

or

$$\dot{W}_{t2} = \dot{m}_3 (h_3 - h_4)$$

From Table A-22; $h_4 = 1023.25$ kJ/kg, and

$$\dot{W}_{t2} = (28.22\ \text{kg/s})(1397.8 - 1023.25)\ \text{kJ/kg}\left|\frac{1\ \text{kW}}{1\ \text{kJ/s}}\right|$$

$$= 10{,}570\ \text{kW} \longleftarrow \dot{W}_{t2}$$

PROBLEM 4.83

KNOWN: Steam is developed from a water stream by energy transfer from oven exhaust in a waste heat recovery steam generator. The steam then expands through a turbine. Data are known at various locations.

FIND: Determine the power developed by the turbine, evaluate it economically on an annual basis, and comment.

SCHEMATIC & GIVEN DATA:

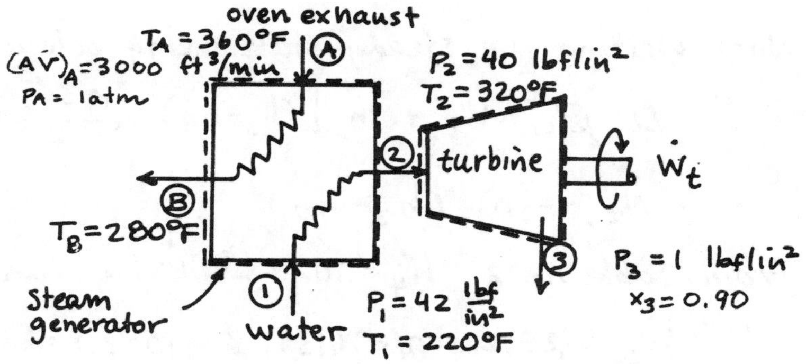

ASSUMPTIONS: (1) Both control volumes are at steady state. (2) Heat transfer is negligible between each control volume and its surroundings, and $\dot{W}_{cv} = 0$ for the heat exchanger. (3) Kinetic and potential energy effects are negligible. (4) The air behaves as an ideal gas, at a pressure of 14.7 lbf/in².

ANALYSIS: (a) The turbine power can be determined from an energy balance once the steam flow rate is determined. Begin with a steady-state energy balance on the steam generator

$$0 = \dot{m}_A - \dot{m}_B \Rightarrow \dot{m}_A = \dot{m}_B \equiv \dot{m}_{gas}$$

$$0 = \dot{m}_1 - \dot{m}_2 \Rightarrow \dot{m}_1 = \dot{m}_2 = \dot{m}_3 \equiv \dot{m}_{st}$$

$$0 = \dot{Q}_{cv}^{\,0} - \dot{W}_{cv}^{\,0} + \dot{m}_{gas}\left[(h_A - h_B) + \left(\frac{V_A^2 - V_B^2}{2}\right)^0 + g(z_A - z_B)^0\right] + \dot{m}_{st}\left[(h_1 - h_2) + \left(\frac{V_1^2 - V_2^2}{2}\right)^0 + g(z_1 - z_2)^0\right]$$

$$\dot{m}_{st} = \dot{m}_{gas}\frac{(h_A - h_B)}{(h_2 - h_1)}$$

The gas flow rate is

$$\dot{m}_{gas} = \frac{(AV)_A}{v_A} = \frac{P_A (AV)_A}{RT_A}$$

$$= \frac{(14.7\ \text{lbf/in}^2)(3000\ \text{ft}^3/\text{min})}{\left(\frac{1545}{28.97}\cdot\frac{\text{ft}\cdot\text{lbf}}{\text{lb}\cdot°\text{R}}\right)(820\ °\text{R})}\left|\frac{144\ \text{in}^2}{1\ \text{ft}^2}\right| = 145.2\ \text{lb/min}$$

From Table A-22E; $h_A = 196.69$ Btu/lb and $h_B = 177.23$ Btu/lb. For the water, from Table A-2E; $P_1 > P_{sat@220°F} \Rightarrow$ compressed liquid. Thus, $h_1 \approx h_{f@220°F} = 188.2$ Btu/lb. At 2, using Table A-4E; $h_2 = 1196.8$ Btu/lb. The flow rate is

$$\dot{m}_{st} = (145.2\ \text{lb/min})\frac{(196.69 - 177.23)}{(1196.8 - 188.2)} = 2.8\ \text{lb/min}$$

Turning now to the control volume enclosing the turbine

$$0 = \cancel{\dot{Q}_{cv}}^{0} - \dot{W}_t + \dot{m}_{st}\left[(h_2 - h_3) + \cancel{\left(\frac{V_2^2 - V_3^2}{2}\right)}^{0} + \cancel{g(z_2 - z_3)}^{0}\right]$$

$$\dot{W}_t = \dot{m}_{st}(h_2 - h_3)$$

Using data from Table A-3E

$$h_3 = h_f + x_3(h_{fg}) = 69.7 + 0.9(1036) = 1002.1 \text{ Btu/lb}$$

Thus

$$\dot{W}_t = \left(2.8 \frac{\text{lb}}{\text{min}}\right)(1196.8 - 1002.1)\frac{\text{Btu}}{\text{lb}}\left|\frac{60 \text{ min}}{1 \text{ h}}\right|\left|\frac{1 \text{ hp}}{2545 \text{ Btu/h}}\right|$$

$$= 12.85 \text{ hp}$$ ◄

(b) Evaluating power at 8 cents per kW·h, the value for 8000 h of operation annually is

$$\dot{\$} = (12.85 \text{ hp})\left|\frac{0.7457 \text{ kW}}{1 \text{ hp}}\right|\left(8000 \frac{\text{h}}{\text{year}}\right)\left(\frac{\$0.08}{\text{kW}\cdot\text{h}}\right)$$

$$= \$6133/\text{year}$$

It is unlikely that this value would cover the capital and operating costs associated with the overall system. Accordingly, even though the oven exhaust has a thermodynamic potential for use, the potential is not likely to be exploited by the proposed means owing to economic considerations.

PROBLEM 4.84

KNOWN: Operating data are provided for a residential heat pump at steady state.

FIND: Determine the rate of heat transfer between the compressor and the surrounding. Also determine the coefficient of performance.

SCHEMATIC & GIVEN DATA:

①

$\dot{m}_R = 4.6 \frac{kg}{min}$

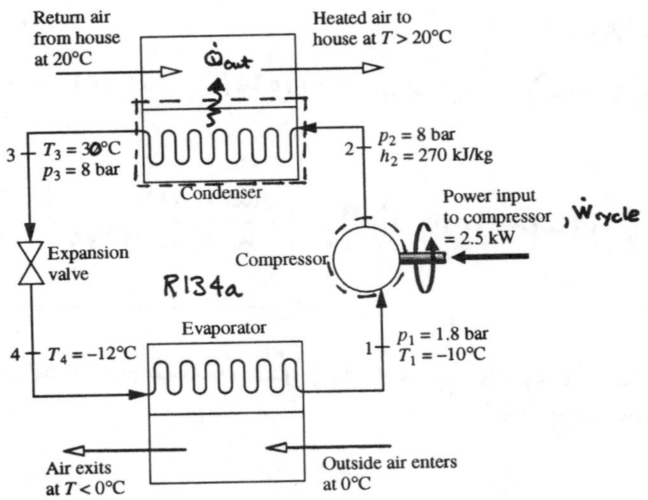

ASSUMPTIONS: 1. Control volumes enclosing the compressor and the refrigerant side of the condenser are at steady state. 2. Kinetic and potential energy effects can be ignored.

ANALYSIS: (a) For the compressor, the mass rate balance reads $\dot{m}_2 = \dot{m}_1 \equiv \dot{m}_R$. The energy rate balance is

$$0 = \dot{Q}_{cv} - \dot{W}_{cv} + \dot{m}_R\left[(h_1 - h_2) + \frac{V_1^2 - V_2^2}{2} + g(z_1 - z_2)\right]$$

$$\Rightarrow \quad \dot{Q}_{cv} = \dot{W}_{cv} + \dot{m}_R(h_2 - h_1)$$

From Table A-12, $h_1 = 242.06$ kJ/kg; $h_2 = 270$ kJ/kg (given). Thus

$$\dot{Q}_{cv} = \left(-2.5 \frac{kJ}{s}\right)\left|\frac{60\,s}{min}\right| + \left(4.6 \frac{kg}{min}\right)(270 - 242.06)\frac{kJ}{kg} = -21.48 \frac{kJ}{min} \longleftarrow$$

(b) To determine the coefficient of performance requires the energy supplied by the condensing refrigerant to the return air, $\dot{Q}_{out}$. Thus, an energy rate balance for the refrigerant side of the condenser is required:

$$0 = \dot{Q}_{cv} - \dot{W}_{cv} + \dot{m}_R\left[(h_2 - h_3) + \frac{V_2^2 - V_3^2}{2} + g(z_2 - z_3)\right] \Rightarrow \dot{Q}_{cv} = \dot{m}_R(h_3 - h_2).$$

Then, with $h_3 \cong h_f(T_3) = 91.49$ kJ/kg, $\dot{Q}_{cv} = (4.6)(91.49 - 270) = -821.15$ kJ/min.

Applying Eq. 2.47 with $\dot{W}_{cycle} = 2.5$ kW and $\dot{Q}_{out} = 821.15$ kJ/min

$$\gamma = \frac{821.15 \text{ kJ/min}}{\left(2.5 \frac{kJ}{s}\right)\left|\frac{60\,s}{min}\right|} = 5.47 \longleftarrow$$

---

1. In the first printing of the 4th edition, T_3 is given incorrectly as 32°C.

PROBLEM 4.85

KNOWN: Data are provided for a simple vapor power plant at steady state.

FIND: Determine the thermal efficiency and the mass flow rate of the cooling water through the condenser.

SCHEMATIC & GIVEN DATA:

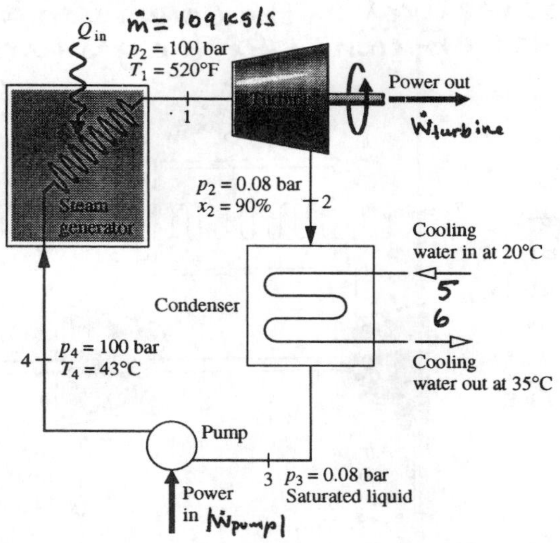

ASSUMPTIONS: 1. Control volumes enclosing each of the four components are at steady state. 2. For each control volume, kinetic/potential energy effects are negligible as are all stray heat transfers.

ANALYSIS: (a) For any power cycle, the thermal efficiency is the ratio of the net work developed to the heat added. In the present case

$$\eta = \frac{\dot{W}_{turbine} - |\dot{W}_{pump}|}{\dot{Q}_{in}}$$

Applying mass and energy rate balances to the turbine, pump, and steam generator, respectively, gives

$$\dot{W}_{turbine} = \dot{m}(h_1 - h_2).$$
$$|\dot{W}_{pump}| = \dot{m}(h_4 - h_3).$$
$$\dot{Q}_{in} = \dot{m}(h_1 - h_4).$$

Table A-4: $h_1 = 3425.1$ kJ/kg. Table A-3: $h_2 = h_f + x_2 h_{fg}$ or $h_2 = 173.88 + 0.9(2403.1) = 2336.7$ kJ/kg. Table A-3: $h_3 = 173.9$ kJ/kg. Table A-5: $h_4 = 188.9$ kJ/kg.

$$\Rightarrow \eta = \frac{\dot{m}(h_1 - h_2) - \dot{m}(h_4 - h_3)}{\dot{m}(h_1 - h_4)} = \frac{(h_1 - h_2) - (h_4 - h_3)}{h_1 - h_4} = \frac{(3425.1 - 2336.7) - (188.9 - 173.9)}{(3425.1 - 188.9)}$$

$$= 0.332 \ (33.2\%)$$

(b) Applying mass rate balances to the condenser: $\dot{m}_2 = \dot{m}_3 = 109$ kg/s, $\dot{m}_5 = \dot{m}_6 \equiv \dot{m}_w$. An energy rate balance for the condenser reduces with assumption 4 to read

$$0 = \dot{m}_2 [h_2 - h_3] + \dot{m}_{cw}[h_5 - h_6] \Rightarrow \dot{m}_{cw} = \dot{m}_2 \left[\frac{h_2 - h_3}{h_6 - h_5}\right]$$

With $h_5 \sim h_f(T_5)$, $h_6 \sim h_f(T_6)$ from Table A-2

$$\dot{m}_{cw} = 109 \frac{kg}{s} \left[\frac{2336.7 - 173.9}{146.68 - 83.96}\right] = 3759 \frac{kg}{s}$$

PROBLEM 4.86

KNOWN: A simple gas turbine power plant operates at steady state with a single stream of air as the working fluid. Data are known for the flow stream at various locations.

FIND: (a) Determine the power required by the compressor, and (b) define and evaluate a thermal efficiency for the gas turbine.

SCHEMATIC & GIVEN DATA:

ASSUMPTIONS: (1) Each control volume shown is at steady state. (2) Heat transfer is negligible for the control volumes enclosing the compressor and turbine. (3) For the heat exchanger, $\dot{W}_{cv} = 0$. (4) Kinetic and potential energy effects are negligible. (4) The air behaves as an ideal gas.

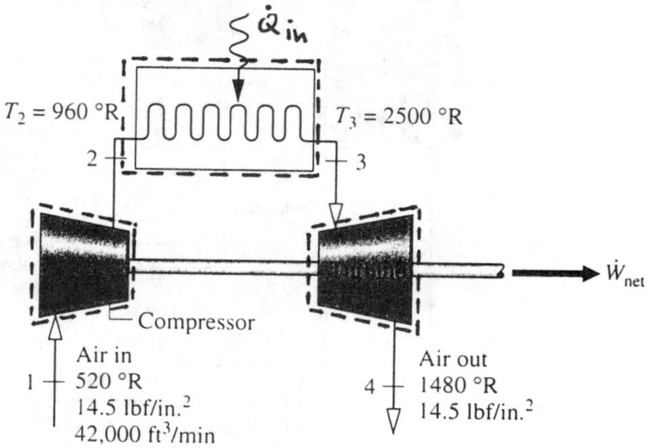

ANALYSIS: (a) The compressor power is determined by using energy and mass balances for the control volume enclosing the compressor. At steady state

$$0 = \cancel{\dot{Q}_{cv}} - \dot{W}_{comp} + \dot{m}\left[(h_1 - h_2) + \cancel{\left(\frac{V_1^2 - V_2^2}{2}\right)} + g\cancel{(z_1 - z_2)}\right]$$

or

$$\dot{W}_{comp} = \dot{m}(h_1 - h_2)$$

where $\dot{m}_1 = \dot{m}_2 \equiv \dot{m}$. To get $\dot{m}$

$$\dot{m} = \frac{(AV)_1}{v_1} = \frac{p_1 (AV)_1}{RT_1}$$

$$= \frac{(14.5\ \text{lbf/in.}^2)(42{,}000\ \text{ft}^3/\text{min})}{\left(\frac{1545}{28.97} \cdot \frac{\text{ft}\cdot\text{lbf}}{\text{lb}\cdot°\text{R}}\right)(520°\text{R})} \left|\frac{144\ \text{in.}^2}{1\ \text{ft}^2}\right| \left|\frac{60\ \text{min}}{1\ \text{h}}\right|$$

$$= 1.897 \times 10^5\ \text{lb/h}$$

Now, with specific enthalpies from Table A-22E; $h_1 = 124.27$ Btu/lb and $h_2 = 231.06$ Btu/lb. Thus

$$\dot{W}_{comp} = \left(1.897\times10^5 \frac{\text{lb}}{\text{h}}\right)(124.27 - 231.06)\frac{\text{Btu}}{\text{lb}} \left|\frac{1\ \text{hp}}{2545\ \text{Btu/h}}\right|$$

$$= -7960\ \text{hp} \longleftarrow \dot{W}_{comp}$$

(b) Comparing the operation of the gas turbine to Fig. 2.16, and Eq. 2.42, an appropriate thermal efficiency for the gas turbine is

$$\eta = \frac{\text{Net power developed}}{\text{Rate of heat transfer to the air passing through the heat exchanger}}$$

$$= \frac{\dot{W}_{net}}{\dot{Q}_{in}}$$

To determine $\dot{W}_{net}$, consider the control volume enclosing the turbine. With assumptions (1), (2), and (4)

$$0 = -(\dot{W}_{net} + |\dot{W}_{comp}|) + \dot{m}(h_3 - h_4)$$

or

$$\dot{W}_{net} = -|\dot{W}_{comp}| + \dot{m}(h_3 - h_4)$$

With data from Table A-22E; $h_3 = 645.78$ Btu/lb and $h_4 = 363.89$ Btu/lb. Thus

$$\dot{W}_{net} = -(7960\text{ hp}) + (1.897\times10^5 \tfrac{\text{lb}}{\text{h}})(645.78 - 363.89)\tfrac{\text{Btu}}{\text{lb}}\left|\frac{1\text{ hp}}{2545\text{ Btu/h}}\right|$$

$$= 13{,}052\text{ hp}$$

For $\dot{Q}_{in}$, consider the heat exchanger. With assumptions (1), (3), and (4)

$$0 = \dot{Q}_{in} - \cancel{\dot{W}_{cv}} + \dot{m}\left[(h_2 - h_3) + \cancel{\left(\frac{V_2^2 - V_3^2}{2}\right)} + \cancel{g(z_2 - z_3)}\right]$$

or

$$\dot{Q}_{in} = \dot{m}(h_3 - h_2)$$

Inserting data

$$\dot{Q}_{in} = (1.897\times10^5)(645.78 - 231.06)\left|\frac{1}{2545}\right|$$

$$= 30{,}912\text{ hp}$$

Finally

$$\eta = \frac{13{,}052}{30{,}912} = 0.422 \quad (42.2\%) \longleftarrow \eta$$

The following observations are made:

- In this case, 42.2% of the energy input by heat shows up as net power developed by the power plant. The second low of thermodynamics introduced in chapter 5 will allow us to compare this with ideal performance.
- Although the air doesn't undergo a cycle in passing through the gas turbine components, we could imagine that a cycle is completed in the surroundings as the air cools from 1480 °R back to 520°R.

PROBLEM 4.87

KNOWN: A pinhole leak develops in the wall of a rigid tank of known volume, and air from the surroundings enters at a known temperature and pressure. The temperature of the air within the tank remains constant at the temperature of the surroundings due to heat transfer from the tank contents.

FIND: Determine the heat transfer between the tank contents and surroundings if (a) the tank is initially evacuated, (b) the tank initially contains air at a specified condition.

SCHEMATIC & GIVEN DATA:

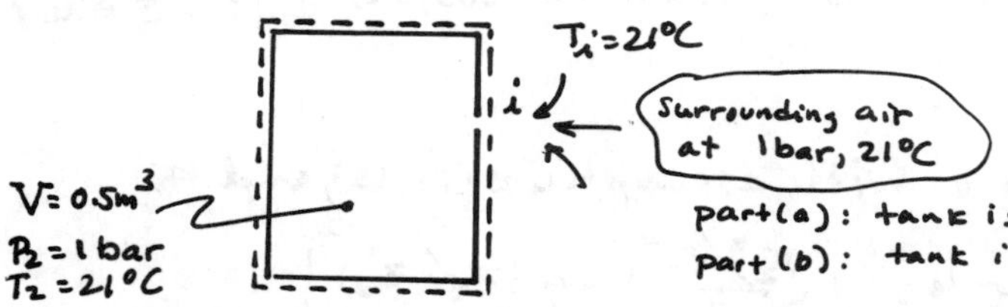

part (a): tank is initially evacuated
part (b): tank initially contains air at 21°C, 0.4 bar.

ASSUMPTIONS: 1. For the control volume shown in the accompanying figure, $\dot{W}_{cv} = 0$, and kinetic/potential energy effects are negligible. 2. The air is modeled as an ideal gas. 3. The condition of the air leaking into the tank at i remains constant at 1 bar, 21°C. 4. The temperature of the air within the tank is 21°C.

ANALYSIS: The mass rate balance for the control volume takes the form $dm_{cv}/dt = \dot{m}_i$. With indicated assumptions, the energy rate balance reduces to

$$\frac{dU_{cv}}{dt} = \dot{Q}_{cv} + \dot{m}_i h_i \Rightarrow \frac{dU_{cv}}{dt} = \dot{Q}_{cv} + h_i \frac{dm_{cv}}{dt}$$

With assumption 3, h_i remains constant. Thus, on integration

$$\Delta U_{cv} = Q_{cv} + \int_1^2 h_i\, dm_{cv} \Rightarrow m_2 u_2 - m_1 u_1 = Q_{cv} + h_i (m_2 - m_1)$$

Introducing $h_i = u_i + (pv)_i$, and collecting like terms

$$Q_{cv} = \underbrace{m_2(u_2 - u_i)}_{(1)} - \underbrace{m_1(u_1 - u_i)}_{(2)} - (m_2 - m_1)(pv)_i \qquad (1)$$

Since the specific internal energy of an ideal gas depends on temperature only, term (1) vanishes because $T_2 = T_i$. Term (2) vanishes in part (a) because $m_1 = 0$, and in part (b) because $T_1 = T_i$. Accordingly, Eq. (1) reduces in each part to the following working equation:

$$Q_{cv} = -(m_2 - m_1)(pv_i) \qquad (2a)$$

That is, the heat transfer from the tank removes the energy entering the tank at i by <u>flow work</u> (see Sec. 4.2.1). Since $pv_i = RT_i$, Eq. (2a) becomes alternatively

$$Q_{cv} = -(m_2 - m_1) RT_i \qquad (2b)$$

(a) <u>$m_1 = 0$</u>. Using the ideal gas equation of state, $m_2 = p_2 V/RT_2 = p_2 V/RT_i$. So

$$Q_{cv} = -p_2 V = -(1\,\text{bar}) \left| \frac{10^5\,N/m^2}{1\,\text{bar}} \right| (0.5\,m^3) \left| \frac{1\,kJ}{10^3\,N\cdot m} \right| = -50\,kJ \longleftarrow$$

(b) <u>$m_1 \neq 0$</u>. Since $m_1 = p_1 V/RT_1 = p_1 V/RT_i$, $m_2 = p_2 V/RT_i$,

$$Q_{cv} = -(p_2 - p_1)V = -(1\,\text{bar} - 0.4\,\text{bar}) \left| \frac{10^5\,N/m^2}{1\,\text{bar}} \right| (0.5\,m^3) \left| \frac{1\,kJ}{10^3\,N\cdot m} \right|$$

$$= -30\,kJ \longleftarrow$$

PROBLEM 4.88

KNOWN: A pinhole develops in the wall of an initially evacuated tank, and air enters from the surroundings until the pressure is 1 bar.

FIND: Determine the final temperature of the air in the tank in the absence of heat transfer between the tank contents and surroundings.

SCHEMATIC & GIVEN DATA:

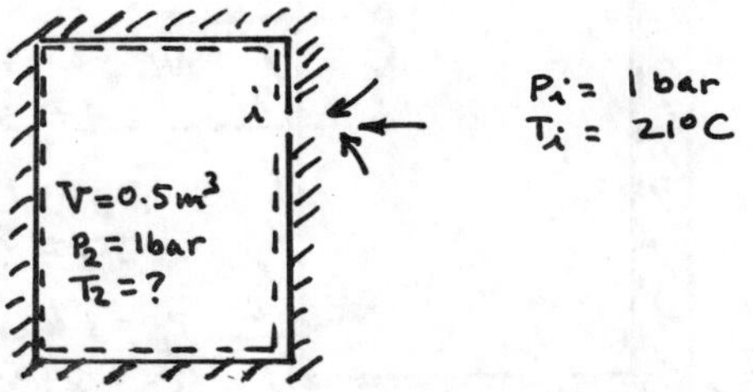

ASSUMPTIONS: 1. The control volume is shown in the accompanying figure. 2. For the control volume, $\dot{W}_{cv}=0$, $\dot{Q}_{cv}=0$, and kinetic/potential energy effects are negligible. 3. The air is modeled as an ideal gas. 4. The condition of the air entering the tank remain constant at 1 bar, 21°C.

ANALYSIS: The mass rate balance takes the form $dm_{cv}/dt = \dot{m}_i$. The energy rate balance reduces to

$$\frac{dU_{cv}}{dt} = \cancel{\dot{Q}_{cv}} - \cancel{\dot{W}_{cv}} + \dot{m}_i h_i$$

Combining the mass and energy rate balance, and integrating using assumption 4

$$\Delta U_{cv} = \int_1^2 \dot{m}_i h_i \, dt \Rightarrow \Delta U_{cv} = h_i \int_1^2 \dot{m}_i \, dt \Rightarrow \Delta U_{cv} = h_i(m_2 - m_1)$$

or since the tank is initially evacuated

$$m_2 u_2 - \cancel{m_1} u_1 = h_i(m_2 - \cancel{m_1}) \Rightarrow \quad u_2 = h_i \qquad (1)$$

From Table A-22 at $T_i = 294\,K\ (21°C)$, $h_i = 294.2\ kJ/kg$. Then,

① interpolating with $u_2 = 294.2\ kJ/kg$, $T_2 = 411\ K\ (138°C)$ ◄

---

1. To interpret the temperature increase of the air within the tank during the filling process, rewrite Eq. (1) with $h_i = u_i + (pv)_i$ to read

$$u_2 = u_i + \underline{(pv)_i}$$

where the underlined term is the flow work discussed in Sec. 4.2.1. Thus, the temperature of air in the tank increases during the process because work is done by the surroundings -- flow work -- on the matter entering the tank.

PROBLEM 4.89

KNOWN: An electric resistor transfers energy to air from the surroundings entering a rigid, well-insulated tank. The final pressure is known.

FIND: Determine the final temperature of air in the tank.

SCHEMATIC & GIVEN DATA:

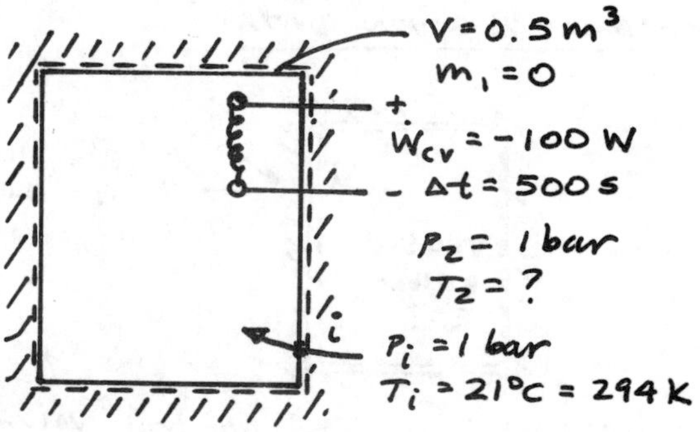

ASSUMPTIONS: (1) The control volume is shown in the accompanying figure. (2) For the control volume, $\dot{Q}_{cv} = 0$ and kinetic/potential energy effects are negligible. (3) The air is modeled as an ideal gas. (4) The condition of the air entering the tank remains constant. (5) For the resistor, $\Delta U_{resistor} = 0$.

ANALYSIS: The mass rate balance reduces to $dm_{cv}/dt = \dot{m}_i$, and the energy rate balance reduces to

$$\frac{dU_{cv}}{dt} = \cancel{\dot{Q}_{cv}} - \dot{W}_{cv} + \dot{m}_i h_i$$

Combining and integrating using assumption (4)

$$\Delta U_{cv} = -\int_{t_1}^{t_2} \dot{W}_{cv}\,dt + \int_{t_1}^{t_2} \dot{m}_i h_i\,dt$$

$$= -\dot{W}_{cv}\Delta t + h_i \int_{t_1}^{t_2} \dot{m}_i\,dt = -\dot{W}_{cv}\Delta t + h_i(m_2 - m_1)$$

With $\Delta U_{cv} = m_2u_2 - m_1u_1$

$$m_2u_2 - \cancel{m_1}u_1 = -\dot{W}_{cv}\Delta t + (m_2 - \cancel{m_1})h_i$$

or

$$m_2u_2 = -\dot{W}_{cv}\Delta t + m_2h_i \Rightarrow u_2 = \frac{-\dot{W}_{cv}\Delta t}{m_2} + h_i$$

By assumption (3), $m_2 = p_2V/RT_2$, and

$$u_2 = \left(\frac{RT_2}{p_2V}\right)(-\dot{W}_{cv}\Delta t) + h_i \qquad (1)$$

Inserting values

$$-\dot{W}_{cv}\Delta t = -(-100\ \text{W})(500\ \text{s})\left|\frac{1\ \text{kJ/s}}{10^3\ \text{W}}\right| = 50\ \text{kJ}$$

$$\frac{RT_2}{p_2V} = \frac{\left(\frac{8.314}{28.97}\frac{\text{kJ}}{\text{kg}\cdot\text{K}}\right)(T_2\ \text{in K})}{(1\ \text{bar})(0.5\ \text{m}^3)}\left|\frac{1\ \text{bar}}{10^5\ \text{N/m}^2}\right|\left|\frac{10^3\ \text{N}\cdot\text{m}}{1\ \text{kJ}}\right|$$

$$= (5.74\times10^{-3})\,T_2$$

From Table A-22; $h_i = 294.17$ kJ/kg. Thus

$$u_2 = \left(5.74\times10^{-3}\ \frac{1}{\text{kg}\cdot\text{K}}\right)(50\ \text{kJ})\,T_2 + 294.17\ \text{kJ/kg} \qquad (2)$$

Eq. (2) can be solved for T_2 using an iterative procedure with data for u_2 from Table A-22. The result is

$$T_2 = 665.3\ \text{K} = 392.3\ ^\circ\text{C}$$

PROBLEM 4.90

KNOWN: A small amount of air at a given state is injected into a container initially holding air at a different state.

FIND: Determine the final temperature and pressure of the air in the container.

SCHEMATIC & GIVEN DATA:

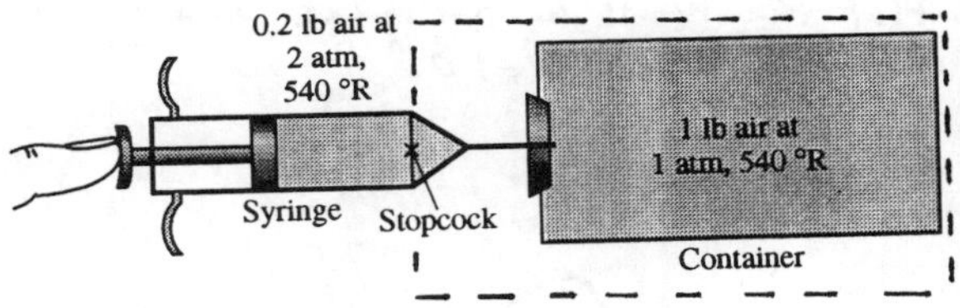

ASSUMPTIONS: 1. The control volume is shown in the accompanying schematic. 2. The state of the air remains constant until it passes the stopcock. 3. For the control volume, $\dot{W}_{cv} = 0$, heat transfer with the surroundings can be ignored, and kinetic/potential energy effects are negligible. 4. The ideal gas model applies for the air.

ANALYSIS: The mass rate balance reads $dm_{cv}/dt = \dot{m}_i$. The energy rate balance reduces with given assumptions to

① $$\frac{dU_{cv}}{dt} = \cancel{\dot{Q}_{cv}} - \cancel{\dot{W}_{cv}} + \dot{m}_i h_i \Rightarrow \frac{dU_{cv}}{dt} = h_i \frac{dm_{cv}}{dt}$$

With assumption 2, h_i remains constant. Thus, integration over time gives

$$\Delta U_{cv} = \int_1^2 h_i \, dm_{cv} \Rightarrow \Delta U_{cv} = h_i \Delta m_{cv} \Rightarrow m_2 u_2 - m_1 u_1 = h_i (m_2 - m_1)$$

where 1 and 2 denote the initial and final states within the container, respectively. That is, $m_1 = 1$ lb and $m_2 = 1\text{ lb} + 0.2\text{ lb} = 1.2$ lb. Solving for u_2, and inserting data from Table A-22E for u_1 and h_i

$$u_2 = \frac{m_1 u_1 + (m_2 - m_1) h_i}{m_2} = \frac{(1\text{ lb})[92.04\ \frac{\text{Btu}}{\text{lb}}] + (0.2\text{ lb})[129.06\ \frac{\text{Btu}}{\text{lb}}]}{1.2\text{ lb}} = 98.21\ \frac{\text{Btu}}{\text{lb}}$$

Interpolating in Table A-22E with u_2 gives $T_2 = 576°R$ (116°F).

Using the ideal gas equation of state, $P_2 = m_2 R T_2 / V$, where $V = m_1 R T_1 / P_1$, the final pressure is

$$P_2 = \left(\frac{m_2}{m_1}\right)\left(\frac{T_2}{T_1}\right) P_1 = \left(\frac{1.2\text{ lb}}{1\text{ lb}}\right)\left(\frac{576\text{ °R}}{540\text{ °R}}\right)\left(14.7\ \frac{\text{lbf}}{\text{in}^2}\right) = 18.8\text{ lbf/in}^2$$

---

1. The plunger does not directly enter the analysis -- it is located in the surroundings of the control volume. However, there is flow work where air enters the control volume. This is incorporated in the $(pv)_i$ term of the specific enthalpy at the inlet, h_i. As shown by the following analysis, the final temperature of the air in the container is higher than the initial temperature. The higher temperature is due to the flow work effect.

PROBLEM 4.91

KNOWN: A rigid tank of known volume initially containing a two-phase liquid-vapor mixture at a known state is heated, allowing saturated vapor to escape while maintaining constant pressure in the tank until a specified final quality is attained.

FIND: (a) If the final quality is 0.5, determine the mass in the tank and the amount of heat transfer. (b) Plot the mass in the tank and the heat transfer versus final quality ranging from 0.2 to 1.0.

SCHEMATIC & GIVEN DATA:

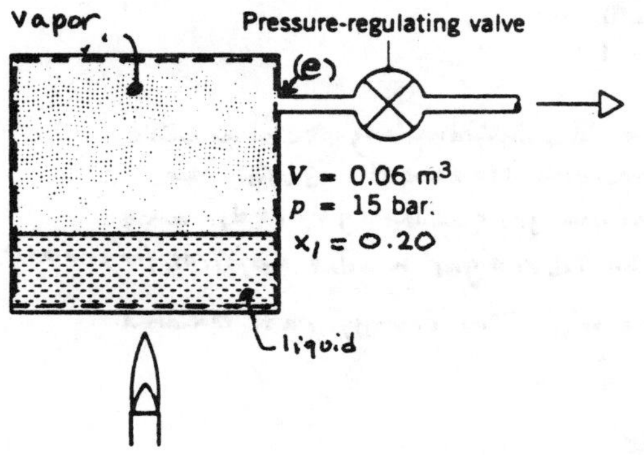

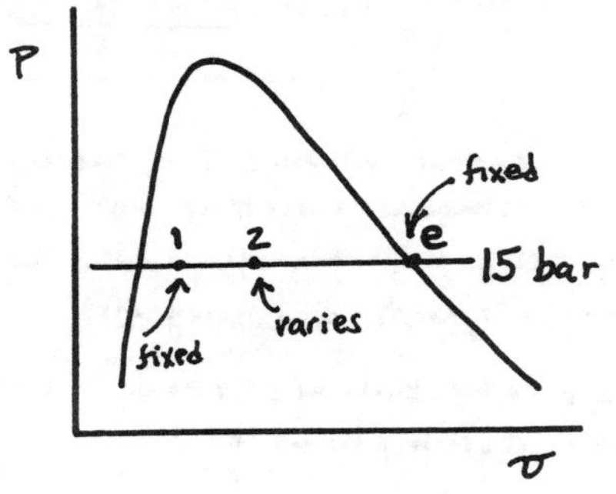

ASSUMPTIONS: 1. The control volume is shown in the schematic. 2. For the control volume, $\dot{W}_{cv} = 0$ and kinetic/potential energy effects are negligible. 3. At exit e the state remains constant, as shown in the p-v diagram.

ANALYSIS: The mass rate balance reads $dm_{cv}/dt = -\dot{m}_e$. With the indicated assumptions, the energy rate balance reduces to

$$\frac{dU_{cv}}{dt} = \dot{Q}_{cv} - \dot{m}_e h_e \Rightarrow \frac{dU_{cv}}{dt} = \dot{Q}_{cv} + h_e \frac{dm_{cv}}{dt}$$

Then, with assumption 3, integration over time results in

$$\Delta U_{cv} = Q_{cv} + h_e \int_1^2 dm_{cv} \Rightarrow m_2 u_2 - m_1 u_1 = Q_{cv} + h_e [m_2 - m_1] \text{ , or}$$

$$Q_{cv} = m_2 u_2 - m_1 u_1 - h_e [m_2 - m_1] \qquad (1)$$

where $h_e = 2792.2$ kJ/kg, and

$$v_1 = v_f + x_1(v_g - v_f)$$

$$= \left(\frac{1.1539}{10^3}\right) + 0.2\left[0.1318 - \frac{1.1539}{10^3}\right] = 0.02728 \frac{m^3}{kg} \Rightarrow m_1 = \frac{V}{v_1} = \frac{0.06\,m^3}{0.02728\,m^3/kg} = 2.199 \text{ kg}$$

$$u_1 = u_f + x_1(u_g - u_f) = 843.16 + 0.2[2594.5 - 843.16] = 1193.4 \text{ kJ/kg}$$

With the same values for v_f, v_g, u_f and u_g, $v_2 = v_f + x_2(v_g - v_f)$ and $u_2 = u_f + x_2(u_g - u_f)$. Also, $m_2 = V/v_2$.

(a) $x = 0.5$ Then,

$$v_2 = \left(\frac{1.1539}{10^3}\right) + 0.5\left[0.1318 - \frac{1.1539}{10^3}\right] = 0.06648 \frac{m^3}{kg} \Rightarrow m_2 = \frac{0.06\,m^3}{0.06648\,m^3/kg} = 0.903 \text{ kg.} \leftarrow m_2$$

$$u_2 = 843.16 - 0.5[2594.5 - 843.16] = 1718.8 \text{ kJ/kg}$$

$$\Rightarrow Q_{cv} = (0.903)(1718.8) - (2.199)(1193.4) - 2792.2(0.903 - 2.199)$$

$$= 2546.5 \text{ kJ} \qquad \leftarrow Q_{cv}$$

PROBLEM 4.91 (Cont'd.)

(b) Data are obtained to make the required plots using IT, as follows:

IT Code

```
V = 0.06  // m³
p = 15  // bar
x1 = 0.2
x2 = 0.5

m1 = V / v1
m2 = V / v2
Qcv = m2 * u2 - m1 * u1 - he * (m2 - m1)

he = hsat_Px("Water/Steam", p, 1)
u1 = usat_Px("Water/Steam", p, x1)
u2 = usat_Px("Water/Steam", p, x2)
v1 = vsat_Px("Water/Steam", p, x1)
v2 = vsat_Px("Water/Steam", p, x2)
```

IT Result (x_2 = 0.5)

Q_{cv} = 2547 kJ

The result for x_2=0.5 compares very favorably with the result of part (a). Now, using the Explore button, sweep x_2 from 0.2 to 1.0 in steps of 0.05. The following plots are constructed from the data:

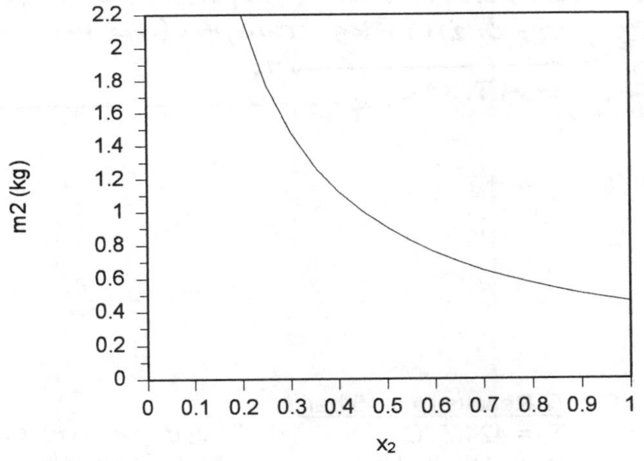

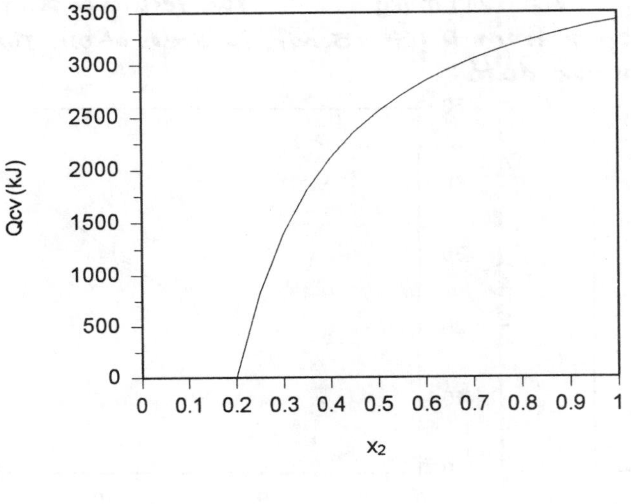

We see from the plots that the heat transfer increases rapidly with x_2 and that the mass in the tank drops rapidly as vapor escapes.

PROBLEM 4.92

KNOWN: An initially-evacuated, well-insulated tank is filled from a large steam line until the pressure in the tank attains a specified value, p.

FIND: (a) Determine the mass and temperature in the tank when p = 15 bar.
(b) Plot the mass and temperature in the tank versus p ranging from 0.1 to 15 bar.

SCHEMATIC & GIVEN DATA:

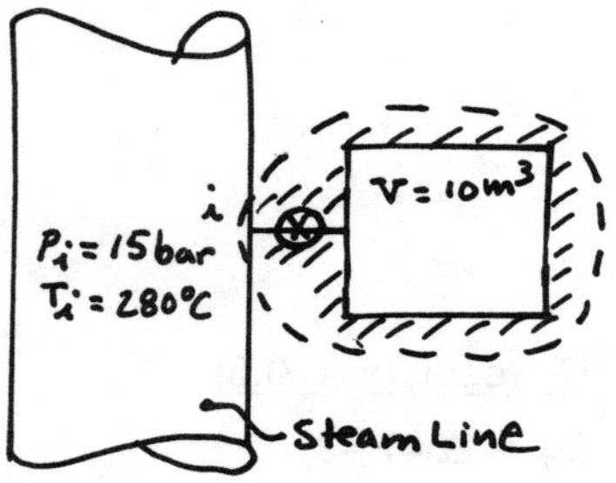

ASSUMPTIONS: 1. The control volume is shown in the schematic. 2. For the control volume, $\dot{W}_{cv} = 0$, $\dot{Q}_{cv} = 0$, and kinetic/potential energy effects are negligible. 3. At inlet i, the state remains constant.

ANALYSIS: The mass rate balance reads $dm_{cv}/dt = \dot{m}_i$. With the indicated assumptions, the energy rate balance reduces to

$$\frac{dU_{cv}}{dt} = \cancel{\dot{Q}_{cv}} - \cancel{\dot{W}_{cv}} + \dot{m}_i h_i \Rightarrow \frac{dU_{cv}}{dt} = h_i \frac{dm_{cv}}{dt}$$

Since h_i is constant (assumption 3), integration over time gives

$$\Delta U_{cv} = h_i \Delta m_{cv} \Rightarrow m_2 u_2 - \cancel{m_1 u_1}^{\,0} = h_i (m_2 - \cancel{m_1}^{\,0}) \Rightarrow \quad u_2 = h_i \qquad (1)$$

(a) p = 15 bar. From Table A-4, $h_i = 2992.7$ kJ/kg. Then, interpolation at 15 bar with $u_2 = 2992.7$ kJ/kg; $T_2 = 425°C$, $v_2 = 0.211$ m³/kg. Thus, the final mass is

$$m_2 = \frac{V}{v_2} = \frac{10\text{ m}^3}{0.211\text{ m}^3/\text{kg}} = 47.39\text{ kg}$$

← T_2 ← m_2

(b) *IT* is used as follows:

IT Code

```
V = 10  // m³
pline = 15  // bar
Tline = 280  // °C
p = 15  // bar
u2 = u_PT("Water/Steam", p, T2)
hi = h_PT("Water/Steam", pline, Tline)
u2 = hi
v2 = v_PT("Water/Steam", p, T2)
m2 = V / v2
```

IT Results (p = 15 bar)
T_2 = 424.7 °C
m_2 = 47.39 kg
v_2 = 0.211 m³/kg

These results compare very favorably with the results of part (a).

Now, using the Explore button, sweep p from 0.1 to 15 bar, in steps of 0.1. The following plots are constructed from the data:

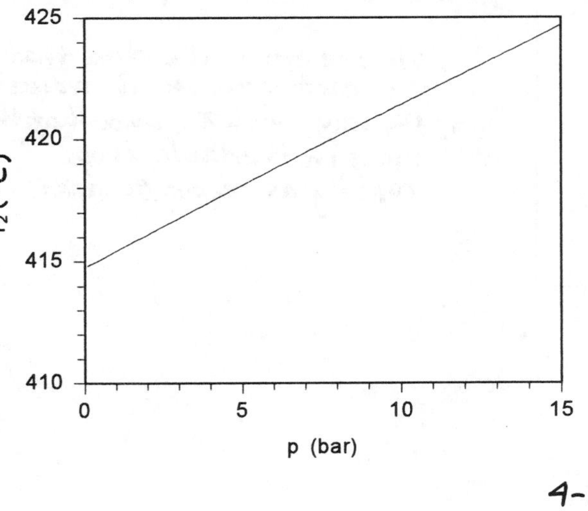

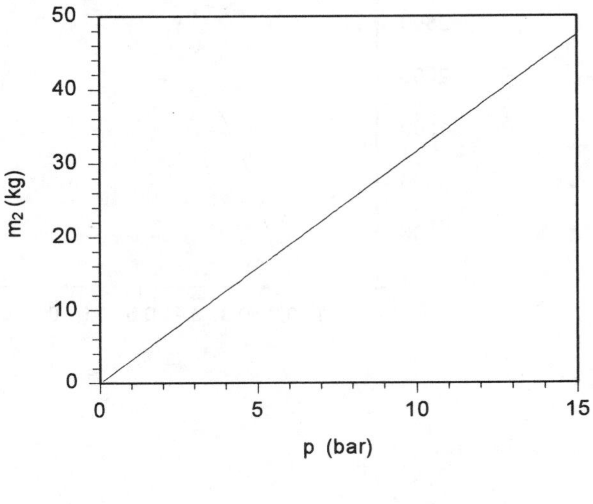

PROBLEM 4.93

KNOWN: A storage tank containing Refrigerant 22 develops a leak allowing saturated vapor to escape until the level of saturated liquid in the tank has dropped to one-half its original value.

FIND: Determine the mass of refrigerant that has escaped and the heat transfer.

SCHEMATIC & GIVEN DATA:

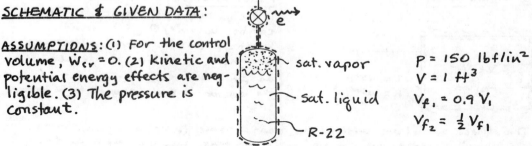

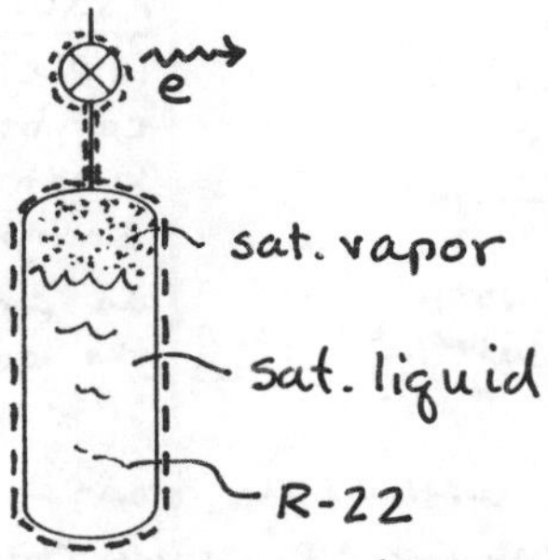

$p = 150$ lbf/in²
$V = 1$ ft³
$V_{f_1} = 0.9 V_1$
$V_{f_2} = \frac{1}{2} V_{f_1}$

ASSUMPTIONS: (1) For the control volume, $\dot{W}_{cv} = 0$. (2) Kinetic and potential energy effects are negligible. (3) The pressure is constant.

ANALYSIS: The mass rate balance takes the form $dm_{cv}/dt = -\dot{m}_e$. Integrating

$$\int_1^2 \dot{m}_e dt = m_1 - m_2 \quad \text{(mass escaped)}$$

To get the masses, use the given volumes and data from Table A-8E at a pressure of 150 lbf/in²

$$m_{f_1} = \frac{V_{f1}}{v_f} = \frac{0.9V}{v_f} = \frac{(0.9)(1 \text{ ft}^3)}{(0.01343 \text{ ft}^3/\text{lb})} = 67.01 \text{ lb}$$

$$m_{g_1} = \frac{V_{g1}}{v_g} = \frac{V - V_{f1}}{v_g} = \frac{1 - 0.9}{0.3649} = 0.274 \text{ lb}$$

$$m_1 = m_{f_1} + m_{g_1} = 67.28 \text{ lb}$$

Similarly for state 2

$$V_{f_2} = \tfrac{1}{2} V_{f_1} = 0.45 \text{ ft}^3, \quad V_{g_2} = 1 - 0.45 = 0.55 \text{ ft}^3$$

$$m_{f_2} = 33.51 \text{ lb}, \quad m_{g2} = 1.507 \text{ lb}; \quad m_2 = 35.017 \text{ lb}$$

and

$$\int_1^2 \dot{m}_e dt = 67.28 - 35.017 = 31.99 \text{ lb} \leftarrow \text{mass escaped}$$

With assumptions (1) and (2), the energy rate balance reduces to

$$\frac{dU_{cv}}{dt} = \dot{Q}_{cv} - \dot{m}_e h_e$$

Combining the mass and energy rate balances and integrating

$$Q_{cv} = m_2 u_2 - m_1 u_1 - h_e(m_2 - m_1)$$

From above, $x_1 = m_{g1}/m_1 = 0.00407$ and $x_2 = m_{g2}/m_2 = 0.04304$. Thus, with data from Table A-8E

$$u_1 = u_f + x_1(u_g - u_f) = 32.10 \text{ Btu/lb}; \quad u_2 = 34.78 \text{ Btu/lb}$$

and $h_e = h_g$ @ 150 lbf/in² $= 110.82$ Btu/lb

Thus

$$Q_{cv} = (35.017)(34.78) - (67.28)(32.10) - (110.82)(35.017 - 67.28)$$

$$= 2633.6 \text{ Btu}$$

PROBLEM 4.94

KNOWN: Helium is withdrawn slowly from a well-insulated tank of known volume until the pressure within the tank is p. The temperature within the tank is kept constant by an electrical resistor.

FIND: (a) When $p = 18\ lbf/in.^2$, determine the mass of helium withdrawn and the energy input to the resistor. (b) Plot the quantities of part (a) versus p ranging from 15 to 30 $lbf/in.^2$

SCHEMATIC & GIVEN DATA:

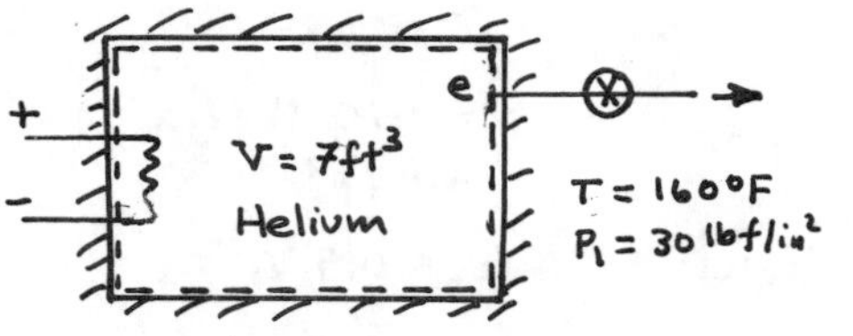

ASSUMPTIONS: 1. The control volume is shown in the schematic. 2. For the control volume, $\dot{Q}_{cv} = 0$, and kinetic/potential energy effects can be ignored. 3. Helium is modeled as an ideal gas. 4. The mass of the resistor is small enough to be ignored.

ANALYSIS: The mass of helium withdrawn over a time interval equals the difference between the initial amount of mass in the tank and the mass in the tank at the later time: (Amount withdrawn) = $m_1 - m_2$. Since $T_1 = T_2$, the ideal gas equation of state can be used to rewrite this as follows, where $P_2 = P$:

$$\begin{pmatrix}\text{Amount}\\ \text{withdrawn}\end{pmatrix} = \frac{P_1 V}{RT} - \frac{p V}{RT} = (P_1 - P)\frac{V}{RT} \qquad (1)$$

$$= (30-p)\left(\frac{lbf}{in.^2}\right)\left|\frac{144\ in.^2}{1\ ft^2}\right|\frac{(7\ ft^3)}{\left[\frac{1545}{4.003}\frac{ft\cdot lbf}{lb\cdot °R}\right](620°R)}$$

$$= (30-p)(4.21\times10^{-3})\ lb \qquad (2)$$

The mass rate balance is $dm_{cv}/dt = -\dot{m}_e$. The energy rate balance reduces to $dU_{cv}/dt = \dot{Q}_{cv}^{\,0} - \dot{W}_{cv} - \dot{m}_e h_e$. Introducing the mass rate balance, and noting that h_e remains constant because temperature is constant,

$$\frac{dU_{cv}}{dt} = -\dot{W}_{cv} + h_e\frac{dm_{cv}}{dt} \Rightarrow \Delta U_{cv} = -W_{cv} + h_e \Delta m_{cv} \Rightarrow (-W_{cv}) = m_2u_2 - m_1u_1 - (m_2 - m_1)h_e.$$

Since temperature is constant, $u_2 = u(T)$, $u_1 = u(T)$, $h_e = u(T) + (pv)_e$. Thus

① $(-W_{cv}) = (m_1 - m_2)(pv)_e \Rightarrow$ with the ideal gas equation of state, $(-W_{cv}) = (m_1 - m_2)RT$.

By Eq. (1) $(m_1 - m_2) = (P_1 - p)V/RT$, giving $(-W_{cv}) = (P_1 - p)V$. That is,

$$(-W_{cv}) = (30-p)\left(\frac{lbf}{in.^2}\right)\left|\frac{144\ in.^2}{1\ ft^2}\right|(7\ ft^3)\left|\frac{1\ Btu}{778\ ft\cdot lbf}\right| = 1.296(30-p)\ Btu \qquad (3)$$

(a) When $p = 18\ lbf/in.^2$, Eq. (2) gives (Amt. withdrawn) = 0.0505 kg, Eq. (3) gives $(-W_{cv}) = 15.55$ Btu.

(b) PLOTS

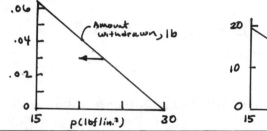

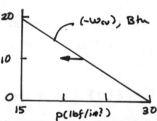

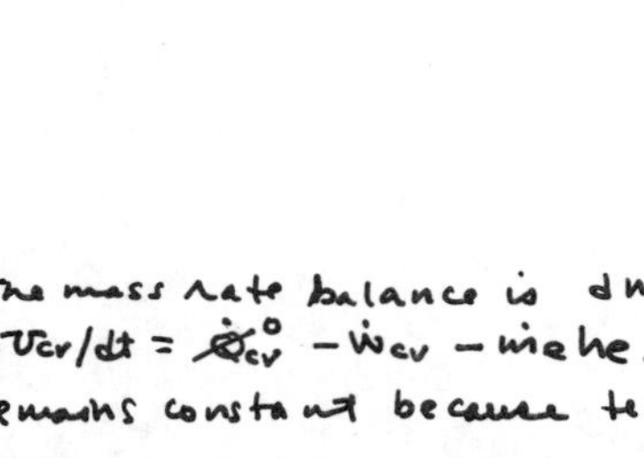

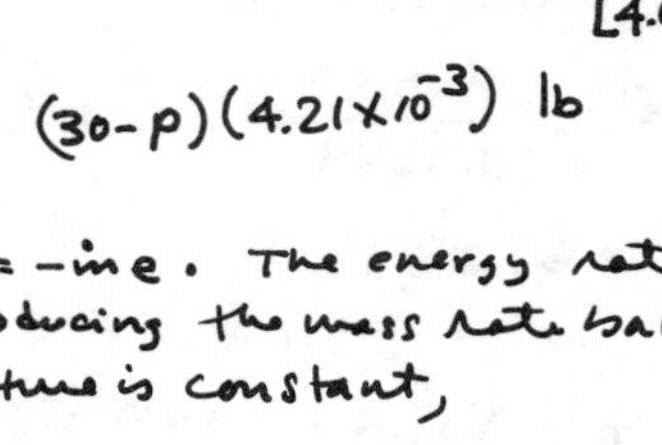

1. To maintain the temperature within the tank constant, the resistor must provide energy to the helium in the tank equal to the energy carried out by flow work at e.

PROBLEM 4.95

Good Problem. Not for a test.

KNOWN: An insulated tank containing carbon dioxide is connected to a supply line carrying carbon dioxide. A valve between the tank and the line is opened and gas flows into the tank until the pressure reaches that of the line. The valve is closed and eventually the tank contents cool back to their initial temperature.

FIND: Determine (a) the tank temperature when the valve closes and (b) the final pressure in the tank.

SCHEMATIC & GIVEN DATA:

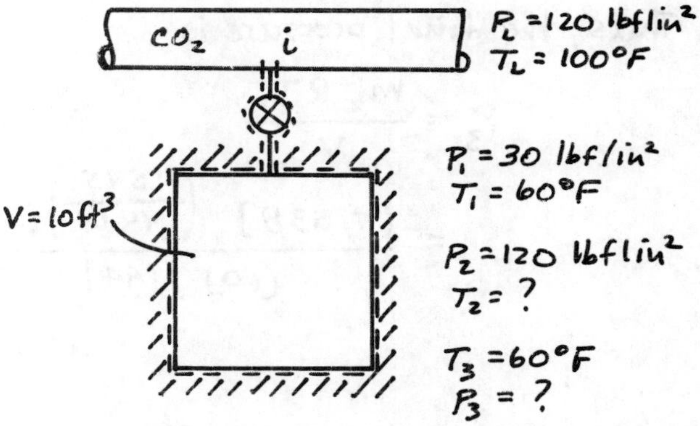

ASSUMPTIONS: (1) For the control volume shown, $\dot{W}_{cv} = 0$. (2) There is no heat transfer during the filling process. (3) Kinetic and potential energy effects are negligible. (4) The state of the CO_2 in the supply line remains constant. (5) The CO_2 behaves as an ideal gas.

On a Test... Give T_2 & ask for Q or make $m_1 = 0$ & ask for T_2. These do not req. iter. soln.

ANALYSIS: (a) Consider first the filling process. The mass rate balance takes the form $dm_{cv}/dt = \dot{m}_i$ and with assumptions (1), (2), and (3), the energy rate balance reduces to

$$\frac{dU_{cv}}{dt} = \cancel{\dot{Q}_{cv}}^{0} - \cancel{\dot{W}_{cv}}^{0} + \dot{m}_i h_i$$

The specific enthalpy in the supply line is constant. Combining the mass and energy rate balances, and integrating

$$m_2 u_2 - m_1 u_1 = h_i (m_2 - m_1)$$

Using the ideal gas equation of state

$$m_1 = \frac{P_1 V}{R T_1}, \quad m_2 = \frac{P_2 V}{R T_2}$$

Combining and rearranging gives

$$u_2 = h_i + \frac{T_2}{T_1} \cdot \frac{P_1}{P_2} (u_1 - h_i)$$

From Table A-23E

$$h_i = \left(\frac{4235.8 \text{ Btu/lbmol}}{44.01 \text{ lb/lbmol}}\right) = 96.25 \text{ Btu/lb}$$

$$u_1 = 2847.7 / 44.01 = 64.71 \text{ Btu/lb}$$

Inserting values

$$u_2 = 96.25 - (0.01516) T_2$$

Since u_2 depends on T_2, the final temperature can be found by iteration using data from Table A-23E and this expression. The result is

$$T_2 \approx 653°R = 193°F \longleftarrow T_2$$

Ref State: Gas P_1, T_1 : $\hat{u}_1 = 0$

$m_2 \hat{u}_2 = \hat{H}_{in} (m_2 - m_1)$

$\hat{u}_2 = \hat{H}_{in} (1 - m_1/m_2)$

$\hat{u}_2 = \hat{H}_{in} \left(1 - \frac{P_1/T_1}{P_2/T_2}\right)$

$C_v (T_2 - T_1) = \hat{H}_{in} \left(1 - \frac{P_1/T_1}{P_2/T_2}\right)$

$C_v T_2 - C_v T_1 = \hat{H}_{in} - \left(\frac{P_1 \hat{H}_{in}}{P_2 T_2}\right) \frac{1}{T_1}$

$a T_2^2 + b T + c = 0$

(b) To determine the pressure after the contents of the tank cool off, first find the mass in the tank after the valve closes

$$m_2 = \frac{P_2 V}{R T_2} = \frac{(120\ \text{lbf/in}^2)(10\ \text{ft}^3)}{\left(\frac{1545}{44.01}\ \frac{\text{ft}\cdot\text{lbf}}{\text{lb}\cdot°\text{R}}\right)(653\,°\text{R})}\left|\frac{144\ \text{in}^2}{1\ \text{ft}^2}\right|$$

$$= 7.538\ \text{lb}$$

Thus, the final pressure is

$$P_3 = \frac{m_2 R T_3}{V}$$

$$= \frac{(7.538)\left(\frac{1545}{44.01}\right)(520)}{(10)\,|144|} = 95.6\ \text{lbf/in}^2 \longleftarrow P_2$$

PROBLEM 4.96

KNOWN: A tank of known volume initially contains steam at a specified state. Steam is withdrawn slowly until the pressure drops to pressure p. The temperature is kept constant by heat transfer to the tank contents.

FIND: (a) Determine the heat tranfer for p = 1.5 MPa. (b) Plot the heat transfer versus p ranging from 0.5 to 6 MPa.

SCHEMATIC & GIVEN DATA:

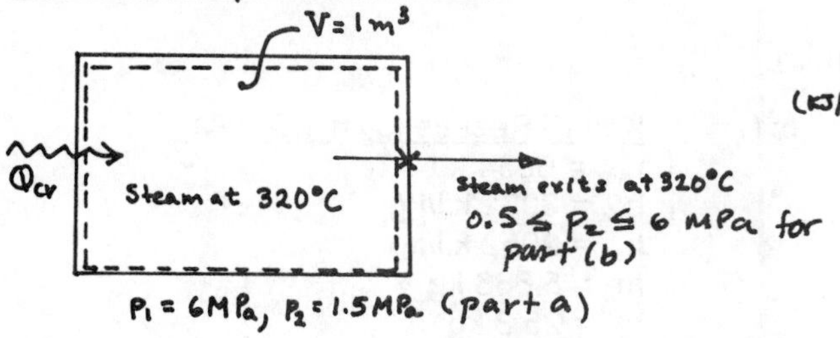

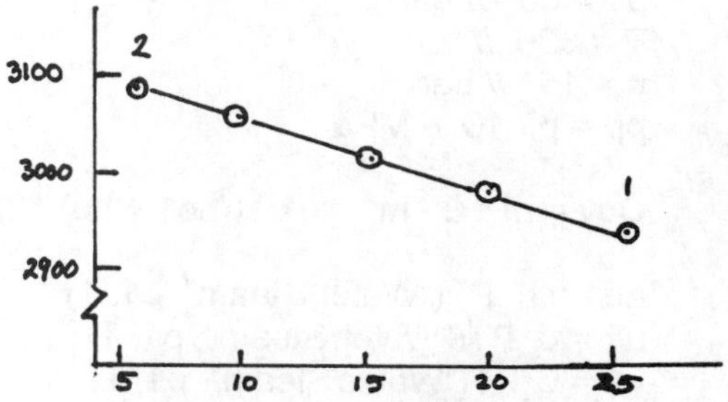

ASSUMPTIONS: (1) For the control volume shown in the figure, $\dot{W}_{cv} = 0$ and kinetic and potential energy effects can be ignored. (2) At each instant, pressure is uniform throughout the steam.

ANALYSIS: (a) The mass rate balance takes the form $dm_{cv}/dt = -\dot{m}_e$. Using the assumptions listed, the energy rate balance is

$$\frac{dU_{cv}}{dt} = \dot{Q}_{cv} - \dot{m}_e h_e$$

Introducing the mass rate balance and integrating

$$\Delta U_{cv} = Q_{cv} + \int_1^2 h_e dm_{cv} \Rightarrow Q_{cv} = m_2u_2 - m_1u_1 - \int_1^2 h_e dm \qquad (1)$$

where m denotes the mass contained within the tank. At any instant, $m = V/v$ where v is specific volume at that instant determined by 320°C and the tank pressure. Initially, $P_1 = 6$ MPa, so Table A-4 gives $v_1 = 0.03876\ m^3/kg$, $u_1 = 2720$ kJ/kg. Finally, $P_2 = 1.5$ MPa, so $v_2 = 0.1765\ m^3/kg$, $u_2 = 2817.1$ kJ/kg. Thus

$$m_1 = \frac{V}{v_1} = \frac{1\ m^3}{0.03876\ m^3/kg} = 25.8\ kg \quad , \quad m_2 = \frac{V}{v_2} = \frac{1\ m^3}{0.1765\ m^3/kg} = 5.67\ kg \qquad (2)$$

For each of several pressures in the interval 1.5 MPa < P < 6.0 MPa, the mass within the tank can be calculated in like manner and the specific enthalpy h_e determined from Table A-4 at 320°C and the specified pressure. These values allow the plot of h_e vs m given above to be constructed. The area under the line from 1 to 2 equals the term $-\int_1^2 h_e dm$ of Eq. (1). As this variation is very nearly linear, the average value of h_e can be used to evaluate this term:

$$-\int_1^2 h_e dm = \left(\frac{(h_e)_1 + (h_e)_2}{2}\right)(m_1 - m_2) = \left(\frac{3081.9 + 2952.6}{2}\right)(25.8 - 5.67)$$

$$= \left(3017.25\ \frac{kJ}{kg}\right)(20.13\ kg) = 60737\ kJ$$

Finally, inserting values into Eq. (1)

$$Q_{cv} = (5.67\ kg)\left(2817.1\ \frac{kJ}{kg}\right) - (25.8\ kg)\left(2720\ \frac{kJ}{kg}\right) + 60737\ kJ$$

$$= 6534\ kJ \longleftarrow Q_{cv}$$

PROBLEM 4.96 (cont'd.)

(b) Data are obtained to make the required plots using IT, as follows:

IT Code

```
V = 1 // m³
p1 = 60  // bar
T = 320  // °C
p = 15  // bar
pp = p / 10  // MPa

Qcv = m * u - m1 * u1 - ((he1 + he) / 2) * (m - m1)

he1 = h_PT("Water/Steam", p1, T)
u1 = u_PT("Water/Steam", p1, T)
v1 = v_PT("Water/Steam", p1, T)
m1 = V / v1
he = h_PT("Water/Steam", p,T)
u = u_PT("Water/Steam",p,T)
v = v_PT("Water/Steam", p, T)
m = V / v
```

Key *IT* Results (p_2 = 1.5 MPa)

Q_{cv} = 6534 kJ
h_{e2} = 3081 kJ/kg
h_{e1} = 2952 kJ/kg
m_2 = 5.666 kg
m_1 = 25.8 kg
u_2 = 2817 kJ/kg
u_1 = 2720 kJ/kg
v_2 = 0.1765 m³/kg
v1 = 0.03876 m³/kg

The results for p_2 = 1.5 MPa compare very favorably with the results of part (a). Now, using the Explore button, sweep p from 0.5 to 6 MPa in steps of 0.5. The following plot is constructed from the data:

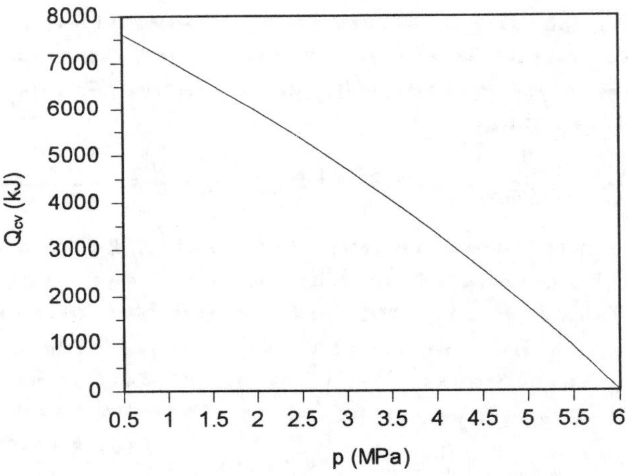

Note that the heat transfer increases as the pressure in the tank drops, as expected.

PROBLEM 4.97

<u>KNOWN:</u> Air escapes slowly from a tank. The initial state and final pressure are known. The that remains in the tank undergoes a polytropic process.

<u>FIND:</u> Determine the heat transfer for a control volume enclosing the tank.

<u>SCHEMATIC & GIVEN DATA:</u>

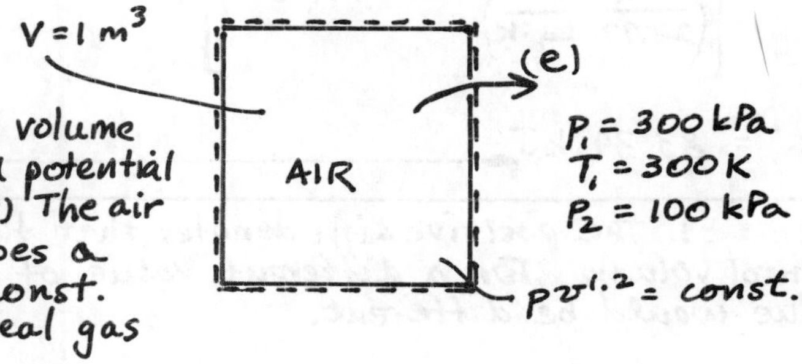

<u>ASSUMPTIONS:</u> (1) For the control volume shown, $\dot{W}_{cv} = 0$. (2) Kinetic and potential energy effects are negligible. (3) The air remaining in the tank undergoes a process described by $pv^{1.2} = \text{const.}$ (4) The air is modeled as an ideal gas with constant specific heats.

<u>ANALYSIS:</u> The mass rate balance takes the form $dm_{cv}/dt = -\dot{m}_e$. With the assumptions listed, the energy rate balance is

$$\frac{dU_{cv}}{dt} = \dot{Q}_{cv} - \dot{m}_e h_e \qquad (1)$$

Since the process occurs slowly, the state of the air in the tank is uniform at any time. Thus, $U_{cv} = mu$, and $h_e = u + RT$. Inserting these expressions into (1) along with the mass rate balance

$$m\frac{du}{dt} + \cancel{u\frac{dm}{dt}} = \dot{Q}_{cv} + (\cancel{u} + RT)\frac{dm}{dt}$$

Thus
$$\dot{Q}_{cv}dt = m\,du - RT\,dm = m\,c_v\,dT - RT\,dm \qquad (2)$$

Each term on the right side of (2) can be expressed in terms of pressure, as follows. First, from $pv^{1.2} = \text{constant}$

$$m = \frac{V}{v} = \left(\frac{p}{\text{const}}\right)^{\frac{1}{1.2}} V = c\,p^{\frac{1}{1.2}}$$

and
$$dm = c\left(\frac{1}{1.2}\right) p^{\left(\frac{1}{1.2}-1\right)} dp \qquad (3)$$

Further
$$RT = \frac{pV}{m} = \frac{pV}{c\,p^{1/1.2}} = \frac{V}{c}\,p^{\left(1-\frac{1}{1.2}\right)} \qquad (4)$$

and
$$dT = \frac{V}{Rc}\left(1 - \frac{1}{1.2}\right) p^{(-1/1.2)} dp \qquad (5)$$

Substituting (3), (4) and (5) into (2) gives

$$\dot{Q}_{cv}dt = \left[c\,p^{\frac{1}{1.2}}\right] c_v \left[\frac{V}{Rc}\left(1-\frac{1}{1.2}\right) p^{\left(-\frac{1}{1.2}\right)} dp\right] - \left[\frac{V}{c}\,p^{\left(1-\frac{1}{1.2}\right)}\right]\left[c\left(\frac{1}{1.2}\right) p^{\left(\frac{1}{1.2}-1\right)} dp\right]$$

$$= \left[\frac{c_v}{R}\left(1-\frac{1}{1.2}\right) - \left(\frac{1}{1.2}\right)\right] V\,dp$$

Integrating

$$Q_{cv} = \left[\frac{c_v}{R}\left(1-\frac{1}{1.2}\right) - \left(\frac{1}{1.2}\right)\right] V\,(p_2 - p_1)$$

PROBLEM 4.97 (Cont'd)

From Table A-20, $c_v \approx 0.717$ kJ/kg·K. Thus

$$Q_{cv} = \left[\frac{(0.717 \text{ kJ/kg·K})}{\left(\frac{8.314}{28.97}\frac{\text{kJ}}{\text{kg·K}}\right)}\left(1-\frac{1}{1.2}\right)-\left(\frac{1}{1.2}\right)\right](1\text{ m}^3)(100-300)\text{kPa}\left|\frac{10^3\text{ N/m}^2}{1\text{ kPa}}\right|\left|\frac{1\text{ kJ}}{10^3\text{ N·m}}\right|$$

① $= 83.39$ kJ ← Q_{cv}

1. The positive sign denotes that the energy transfer is into the control volume. For a different value of polytropic exponent, the value would be different.

PROBLEM 4.98

KNOWN: A well-insulated tank containing R-134a is connected to a supply line. As refrigerant is allowed to flow into the tank, a flexible bladder in the tank expands to maintain the refrigerant in the tank at constant pressure.

FIND: Determine the amount of mass admitted to the tank between the initial time and the instant when all the liquid in the tank is vaporized.

SCHEMATIC & GIVEN DATA:

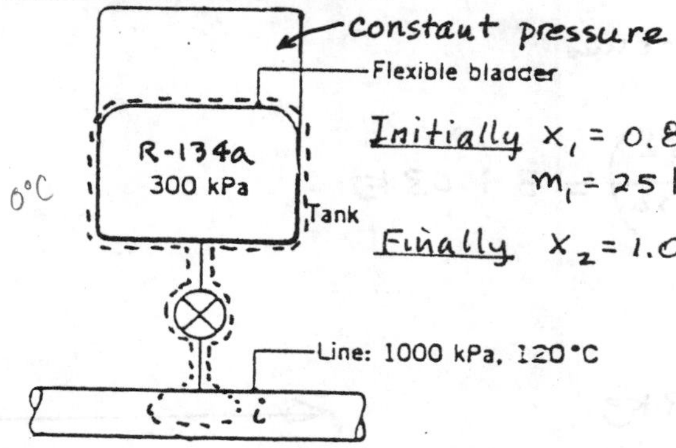

Initially $x_1 = 0.8$
$m_1 = 25$ kg
Finally $x_2 = 1.0$

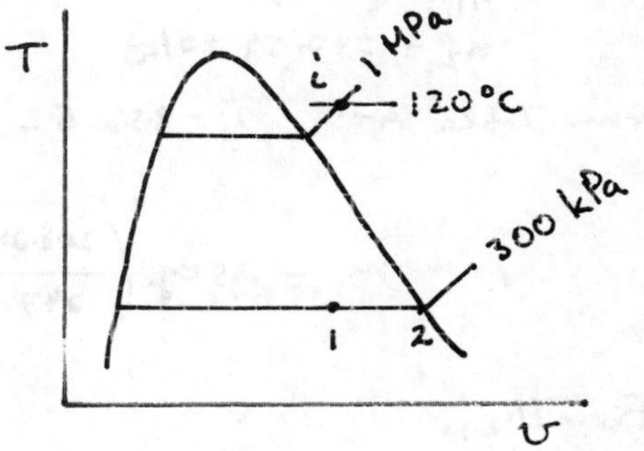

ASSUMPTIONS: (1) The control volume is shown with $\dot{Q}_{cv} = 0$. (2) Conditions in the supply line remain constant. (3) The pressure remains constant in the tank. (4) Kinetic and potential energy effects are negligible.

ANALYSIS: The mass rate balance takes the form; $dm_{cv}/dt = \dot{m}_i$. With the assumptions listed, the energy rate balance is

$$\frac{dU_{cv}}{dt} = \cancel{\dot{Q}_{cv}}^{\,0} - \dot{W}_{cv} + \dot{m}_i h_i$$

The specific enthalpy h_i is constant by assumption (2). Thus, combining the mass and energy rate balances and integrating

$$\Delta U_{cv} = -W_{cv} + \int_{t_1}^{t_2} \dot{m}_i h_i \, dt$$

Since h_i is constant

$$\Delta U_{cv} = -W_{cv} + h_i \int_{t_1}^{t_2} \dot{m}_i \, dt = -W_{cv} + h_i (m_2 - m_1) \qquad (1)$$

To evaluate the work, note that the pressure in the tank is constant. Thus

$$W_{cv} = \int p \, dV = p(V_2 - V_1) = p(m_2 v_2 - m_1 v_1) \qquad (2)$$

Combining (1) and (2), and noting that $\Delta U_{cv} = m_2 u_2 - m_1 u_1$

$$m_2 u_2 - m_1 u_1 = -p(m_2 v_2 - m_1 v_1) + h_i (m_2 - m_1)$$

or

$$m_2[(u_2 + p v_2) - h_i] = m_1[(u_1 + p v_1) - h_i]$$

$$m_2 [h_2 - h_i] = m_1 [h_1 - h_i]$$

PROBLEM 4.98 (Contd.)

Solving for m_2

$$m_2 = m_1\left(\frac{h_1 - h_i}{h_2 - h_i}\right)$$

Using data from Table A-11 at 3 bar: $h_f = 50.85$, $h_g = 247.59$ kJ/kg

$$h_1 = (50.85) + (0.81)[247.59 - 50.85] = 208.24 \text{ kJ/kg}$$
$$h_2 = 247.59 \text{ kJ/kg}$$

From Table A-12, $h_i = 356.52$ kJ/kg. Thus

$$m_2 = 25\text{ kg}\left(\frac{208.24 - 356.52}{247.59 - 356.52}\right) = 34.03 \text{ kg}$$

Finally,

$$\Delta m = m_2 - m_1$$
$$= 34.03 - 25 = 9.03 \text{ kg}$$ ←

PROBLEM 4.99

KNOWN: Air is admitted slowly into an insulated piston-cylinder assembly from a supply line until the volume inside the cylinder has doubled.

FIND: Plot the final temperature and mass inside the cylinder for a given range of supply temperatures.

SCHEMATIC & GIVEN DATA:

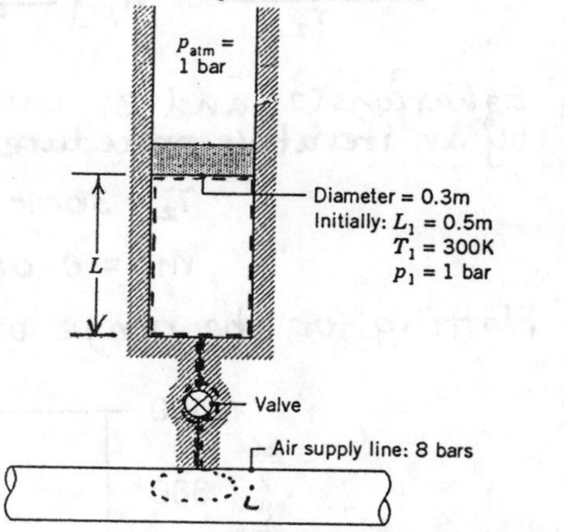

$300 \leq T_{supply} \leq 500K$

ASSUMPTIONS: (1) For the control volume shown, $\dot{Q}_{cv}=0$. (2) Kinetic and potential energy effects can be neglected. (3) The weight of the piston and friction between the piston and cylinder wall can be neglected. (4) The air behaves as an ideal gas.

ANALYSIS: The mass rate balance takes the form $dm_{cv}/dt = \dot{m}_i$. With the assumptions listed, the energy balance is

$$\frac{dU_{cv}}{dt} = \dot{Q}_{cv}^{\,\to 0} - \dot{W}_{cv} + \dot{m}_i h_i$$

The line conditions are constant, so h_i is constant. Combining the mass and energy rate balances and integrating

$$m_2u_2 - m_1u_1 = -W_{cv} + h_i(m_2 - m_1) \qquad (1)$$

To evaluate the work, note that the pressure in the cylinder is always atmospheric since the process is slow and since assumption (3) applies. Thus

$$W_{cv} = \int_{V_1}^{V_2} p\,dV = p(V_2 - V_1)$$

From the given data

$$V_1 = \left(\frac{\pi d_{pist}^2}{4}\right)L_1 = \left(\frac{\pi(.3^2)m^2}{4}\right)(0.5\,m) = 0.03534\,m^3$$

and $V_2 = 0.07068\ m^2$

$$W_{cv} = (1\,bar)(0.07068 - 0.03534)m^3 \left|\frac{10^5 N/m^2}{1\,bar}\right|\left|\frac{1\,kJ}{10^3\,N\cdot m}\right|$$

$$= 3.534\ kJ$$

From Table A-22, at $T_1 = 300K$; $u_1 = 214.07$ kJ/kg. Also, $h_i = h_i(T_{supply})$. For $T_{supply} = 300K$, $h_i = 300.19$ K.

Using the ideal gas model

$$m_1 = \frac{P_1V_1}{RT_1} = \frac{(1\,bar)(0.03534\,m^3)}{\left(\frac{8.314}{28.97}\frac{kJ}{kg\cdot K}\right)(300K)}\left|\frac{10^5 N/m^2}{1\,bar}\right|\left|\frac{1\,kJ}{10^3\,N\cdot m}\right| = 0.04105\ kg$$

and

$$m_2 = \frac{P_2V_2}{RT_2} = \frac{(1)(0.07068)|100|}{\left(\frac{8.314}{28.97}\right)(T_2)} = \frac{24.628}{T_2} \qquad (2)$$

Incorporating these results into (1) and rearranging

PROBLEM 4.99 (Cont'd)

$$m_2 u_2 - h_i (m_2 - m_1) = m_1 u_1 - W_{cv}$$

$$\frac{24.628\, u_2}{T_2} - h_i \left(\frac{24.628}{T_2} - 0.04105 \right) = 5.254 \qquad (3)$$

Equations (2) and (3) can be solved for given values of $h_i\,(T_{supply})$ by an iterative procedure and data from Table A-22. For $T_{supply} = 300\text{K}$,

$T_2 = 300\text{K}$

$m_2 = 0.0821$ kg

Plotting for the range of T_{supply} values

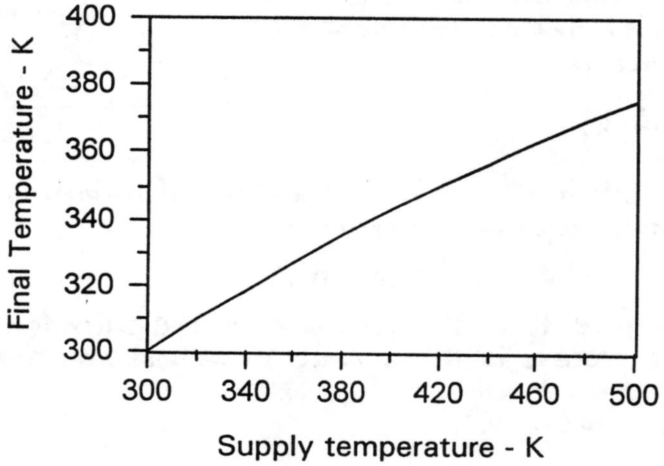

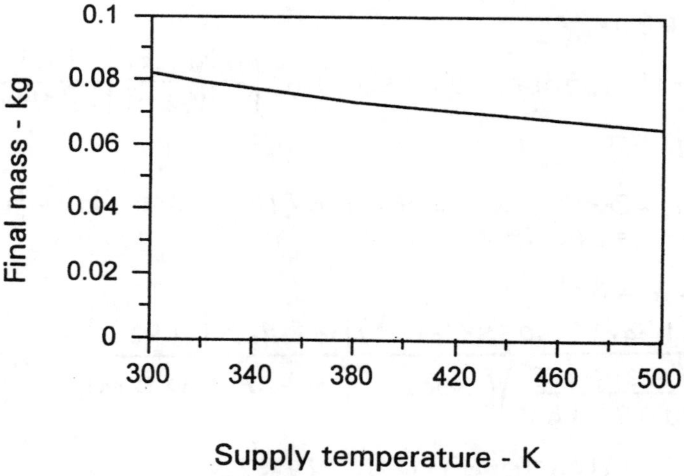

PROBLEM 4.100

KNOWN: A well-insulated piston-cylinder assembly is connected by a valve to an air supply. The supply conditions and the initial state of air inside the cylinder. Air is admitted slowly causing the piston to compress a spring. The initial and final volumes within the cylinder are known.

FIND: Plot the final pressure and final temperature within the cylinder versus spring constant k varying from 650 to 750 lbf/ft.

SCHEMATIC & GIVEN DATA:

Initially: $p_1 = 14.7\ lbf/in^2$
$T_1 = 80°F$
$V_1 = 0.1\ ft^3$

Finally: $V_2 = 0.4 ft^3$

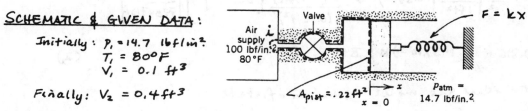

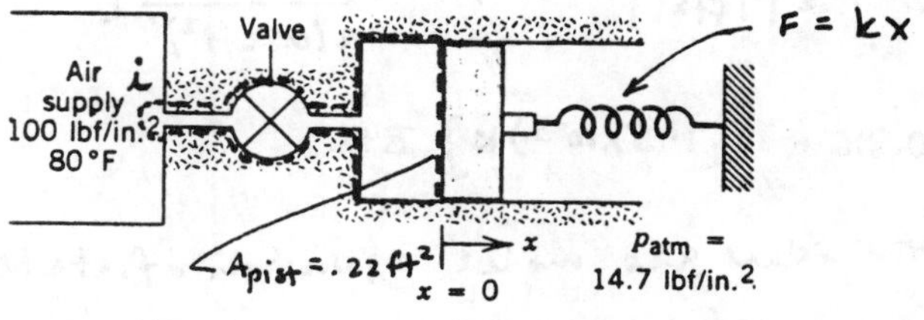

ASSUMPTIONS: (1) The control volume is shown on the accompanying diagram, with $\dot{Q}_{cv} = 0$. (2) Conditions in the air supply remain constant. (3) Kinetic and potential energy effects are negligible. (4) The air behaves as an ideal gas. (5) There is no friction between the piston and cylinder wall.

ANALYSIS: The final pressure is found by applying Newton's Law to the piston. Since air is admitted slowly; $\Sigma F_x = 0$. Thus

$$p A_{pist} = p_{atm} A_{pist} + kx$$

with $V - V_1 = x A_{pist}$, the pressure is

$$p = p_{atm} + \frac{k(V - V_1)}{A_{pist}^2} \quad (a)$$

At $V = V_2$

$$p_2 = 14.7 \frac{lbf}{in^2} + \frac{(k\ lbf/ft)(0.4 - 0.1) ft^3}{(.22^2)\ ft^4}\left(\frac{1\ ft^2}{144\ in^2}\right)$$

$$= \left[14.7 + 0.043k\right] lbf/in^2 \quad (1)$$

The mass rate balance takes the form; $dm_{cv}/dt = \dot{m}_i$. With assumptions listed, the energy rate balance is

$$\frac{dU_{cv}}{dt} = \cancel{\dot{Q}_{cv}}^{0} - \dot{W}_{cv} + \dot{m}_i h_i$$

The specific enthalpy h_i is constant by assumption (3). Thus, combining the mass and energy rate balances and integrating

$$\Delta U_{cv} = -W_{cv} - \int_1^2 h_i\, dm_{cv} \Rightarrow m_2 u_2 - m_1 u_1 = -W_{cv} + h_i(m_2 - m_1) \quad (2)$$

Since the process occurs slowly, $W_{cv} = \int p\,dV$. With Eq. (a) above

$$W_{cv} = \int_{V_1}^{V_2} \left[\left(p_{atm} - \frac{kV_1}{A_{pist}^2}\right) + \frac{kV}{A_{pist}^2}\right] dV$$

$$= \left(p_{atm} - \frac{kV_1}{A_{pist}^2}\right)(V_2 - V_1) + \frac{k(V_2^2 - V_1^2)}{2 A_{pist}^2}$$

PROBLEM 4.100 (Contd.)

$$\therefore W_{cv} = p_{atm}(V_2 - V_1) + \frac{k}{(A_{pist})^2}\left[\frac{V_2^2 - V_1^2}{2} - V_1(V_2 - V_1)\right]$$

$$= p_{atm}(V_2 - V_1) + \frac{k}{2(A_{pist})^2}\left[V_2 - V_1\right]^2$$

$$= \left[14.7\ \frac{lbf}{in^2}\left|\frac{144\ in.^2}{1\ ft^2}\right|(0.3\ ft^3) + \frac{k(lbf/ft)}{2(0.22\ ft^2)^2}\left[0.3\ ft^3\right]^2\right]\left|\frac{1\ Btu}{778\ ft\cdot lbf}\right|$$

$$= \left[0.816 + (1.195\times10^{-3})k\right]\ Btu \qquad (3)$$

Also, with the ideal gas model equation of state

$$m_1 = \frac{p_1 V_1}{RT_1} = \frac{(14.7\times144\ lbf/ft^2)(0.1\ ft^3)}{\left(\frac{1545}{28.97}\ \frac{ft\cdot lbf}{lb\cdot {}^\circ R}\right)(540^\circ R)} = 7.35\times10^{-3}\ lb$$

$$m_2 = \frac{p_2 V_2}{R T_2}, \text{ where } p_2 \text{ is given by Eq. (1)} \qquad (4)$$

Since $T_1 = T_i = 540^\circ R\ (80^\circ F)$, $u_1 = u(540^\circ R)$, $h_i = h(540^\circ R)$.
Accordingly, T_2 can be determined by solving Eq. (2), together with Eqs. (1), (3), (4) and known values of V_2, u_1, and h_i.
This scheme is implemented via the following IT code:

IT Code

```
p1 = 14.7  // lbf/in.²
T1 = 80  // °F
Ti = T1
V1 = 0.1  // ft²
V2 = 0.4  // ft²
K = 650 // lbf/ft

p2 = 14.7 + 0.043 * K
m2 * u2 - m1 * u1 = - Wcv + hi * (m2 - m1)
Wcv = 0.816 + 1.195E-3 * K
m1 = V1 / v1
m2 = V2 / v2

u1 = u_T("Air", T1)
v1 = v_TP("Air",T1,p1)
u2 = u_T("Air", T2)
v2 = v_TP("Air",T2,p2)
hi = h_T("Air", Ti)
```

Now, using the Explore button, sweep K from 650 to 750 lbf/ft in steps of 1, to get the following plots:

PROBLEM 4.100 (Contd.)

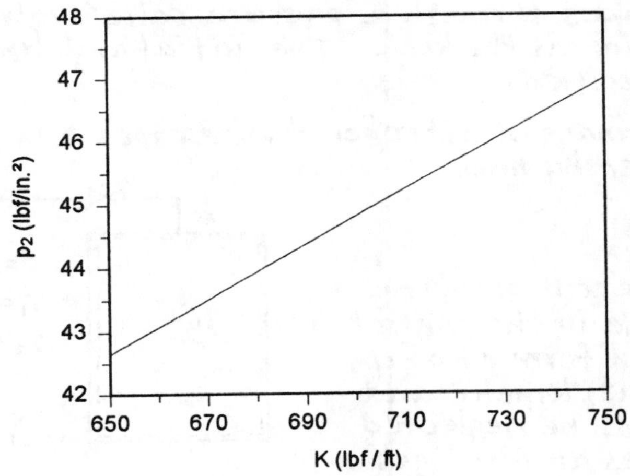

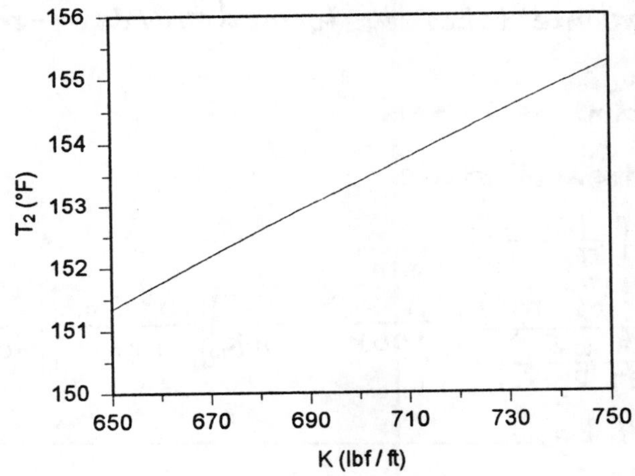

Note that the pressure plot is linear, as indicated by Eq.(1), and could have been readily plotted by hand. However, the temperature plot is non-linear and would be quite difficult to obtain without the use of IT or other software.

PROBLEM 4.101

KNOWN: Heat transfer occurs to nitrogen gas contained in a rigid tank. Gas escapes through a pressure relief valve, maintaining constant pressure in the tank. The initial and final temperatures are specified.

FIND: Determine the mass of nitrogen that escapes and the amount of energy transfer by heat.

SCHEMATIC & GIVEN DATA:

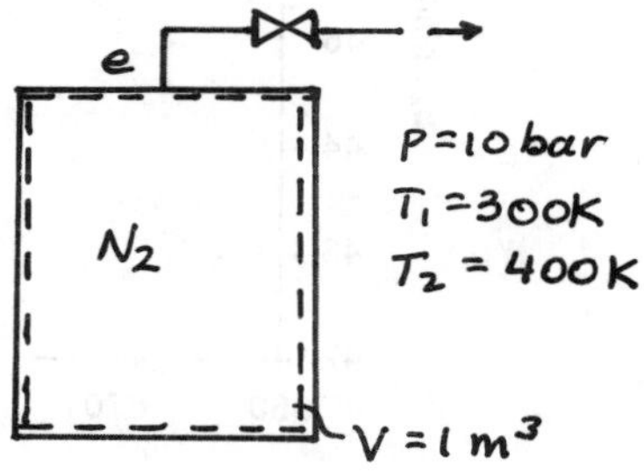

ASSUMPTIONS: (1) For the control volume shown, $\dot{W}_{cv}=0$. (2) The state in the control can be assumed to be uniform at any time during the process. (3) Kinetic and potential energy effects can be neglected. (4) The nitrogen behaves as an ideal gas with constant specific heats evaluated at 350K.

ANALYSIS: The mass rate balance takes the form $dm_{cv}/dt = -\dot{m}_e$. Thus the mass that escapes is

$$\int_1^2 \dot{m}_e dt = -\int_{m_1}^{m_2} dm_{cv} = m_1 - m_2$$

With the ideal gas equation of state

$$\int_1^2 \dot{m}_e dt = \frac{PV}{R}\left[\frac{1}{T_1} - \frac{1}{T_2}\right]$$

$$= \frac{(10\text{ bar})(1\text{ m}^3)}{\left(\frac{8.314}{28.97}\frac{\text{kJ}}{\text{kg}\cdot\text{K}}\right)}\left[\frac{1}{300\text{K}} - \frac{1}{400\text{K}}\right]\left|\frac{10^5\text{ N/m}^2}{1\text{ bar}}\right|\left|\frac{1\text{ kJ}}{10^3\text{ N}\cdot\text{m}}\right|$$

$$= 2.904\text{ kg} \longleftarrow \text{mass escaped}$$

With the assumptions listed, the energy rate balance is

$$\frac{dU_{cv}}{dt} = \dot{Q}_{cv} - \dot{m}_e h_e$$

Noting that $U_{cv} = mu$ and $h_e = u + RT$, and with the mass rate balance

$$m\frac{du}{dt} + u\frac{dm}{dt} = \dot{Q}_{cv} + (u+RT)\frac{dm}{dt}$$

$$\dot{Q}_{cv} = m\frac{du}{dt} - RT\frac{dm}{dt}$$

For an ideal gas, $du = c_v\, dT$. Also, with $m = PV/RT$

$$dm = \left(\frac{PV}{R}\right)\frac{dT}{(-T^2)}$$

Thus

$$\dot{Q}_{cv} = \left(\frac{PV}{RT}\right)c_v\frac{dT}{dt} + \left(\frac{PV}{RT}\right)R\frac{dT}{dt} = \frac{PV}{R}(c_v + R)\frac{1}{T}\frac{dT}{dt}$$

PROBLEM 4.101 (Cont'd.)

Noting that $c_v + R = c_p$

$$\dot{Q}_{cv}\,dt = \left(\frac{PVc_p}{R}\right)\frac{dT}{T}$$

From Table A-20; $c_p = 1.041$ kJ/kg·K. Integrating and inserting values

$$\int_1^2 \dot{Q}_{cv}\,dt = \left(\frac{PVc_p}{R}\right)\int_{T_1}^{T_2}\frac{dT}{T}$$

$$Q_{cv} = \left(\frac{PVc_p}{R}\right)\ln\frac{T_2}{T_1}$$

$$= \frac{(10\text{ bar})(1\text{ m}^3)(1.041\text{ kJ/kg·K})}{\left(\frac{8.314}{28.97}\frac{\text{kJ}}{\text{kg·K}}\right)}\ln\left(\frac{400}{300}\right)\left|\frac{10^5\text{ N/m}^2}{1\text{ bar}}\right|\left|\frac{1\text{ kJ}}{10^3\text{ N·m}}\right|$$

$$= 1043\text{ kJ} \longleftarrow Q_{cv}$$

PROBLEM 4.102

KNOWN: The air supply to an office is shut off overnight and the room temperature drops. In the morning, the thermostat is reset and a known volumetric flow rate of heated air is supplied.

FIND: Estimate the time it takes for the room to reach 70°F, and plot room temperature as a function of time.

SCHEMATIC & GIVEN DATA:

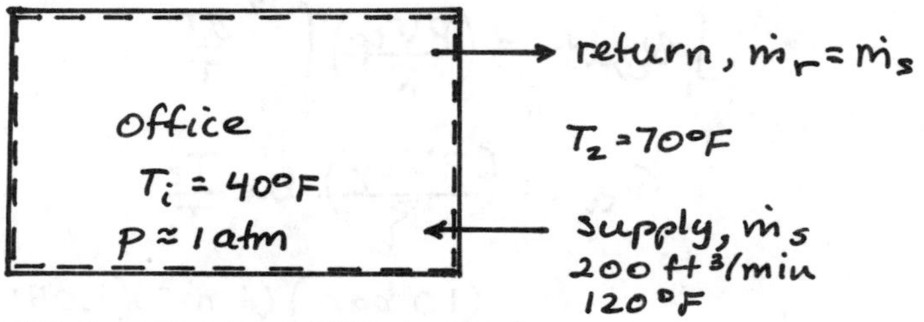

ASSUMPTIONS: (1) For the control volume shown, $\dot{Q}_{cv} = \dot{W}_{cv} = 0$. (2) The air behaves as an ideal gas with constant specific heats. (3) The pressure is taken as 1 atm everywhere. (4) Kinetic and potential energy effects can be neglected. (5) The room air is well-mixed.

ANALYSIS: The mass rate balance takes the form $dm/dt = \dot{m}_s - \dot{m}_r$. With $\dot{m}_s = \dot{m}_r$; $dm/dt = 0$. Thus, the mass in the room is constant with time. The energy rate balance reduces to

$$\frac{dU_{cv}}{dt} = \dot{Q}_{cv}^{\,0} - \dot{W}_{cv}^{\,0} + \dot{m}_s h_s - \dot{m}_r h_r$$

or, with $U_{cv} = mu$ and $h_r = h(T)$

$$m\frac{du}{dt} + u\frac{dm}{dt}^{0} = \dot{m}_s\left(h_s - h(T)\right)$$

Further, $du = c_v\,dT$ and $h_s - h(T) = c_p(T_s - T)$. Thus

$$m c_v \frac{dT}{dt} = \dot{m}_s c_p (T_s - T)$$

With $dT = -d(T_s - T)$ and $c_p/c_v = k$

$$-\frac{m\,d(T_s - T)}{k\,(T_s - T)} = \dot{m}_s\,dt \qquad (1)$$

Evaluating $\dot{m}_s$

$$\dot{m}_s = \frac{(AV)_s}{v_s} = \frac{p\,(AV)_s}{RT_s}$$

$$= \frac{(14.7\ lbf/in^2)(200\ ft^3/min)}{\left(\frac{1545\ ft\cdot lbf}{28.97\ lb\cdot °R}\right)(580°R)}\left|\frac{144\ in^2}{1\ ft^2}\right| = 13.69\ lb/min$$

and

$$m = \frac{pV}{RT} = \frac{(14.7)(20{,}000)\,|144|}{\left(\frac{1545}{28.97}\right)(515)} = 154\ lb$$

where the average of the initial and final temperatures is used to estimate the mass. Also, from A-20E; $k = 1.4$.

Integrating Eq. (1) from $t = 0$ ($T = T_i = 40°F$) to any time t

$$-\left(\frac{m}{k \cdot \dot{m}_s}\right) \ln\left(\frac{T_s - T}{T_s - T_i}\right) = t$$

or, solving for T

$$T = T_s - (T_s - T_i) \exp\left[\left(-\frac{\dot{m}_s k}{m}\right) t\right]$$

Inserting values

$$T = 120 - 80 \exp\left[(-0.1245) t\right] \quad (2)$$

Solving for t_2 when $T_2 = 70°F$ gives

$$t_2 = 3.79 \text{ min} \longleftarrow t_2$$

Eq.(2) can readily be plotted using software. The following plot was constructed using IT:

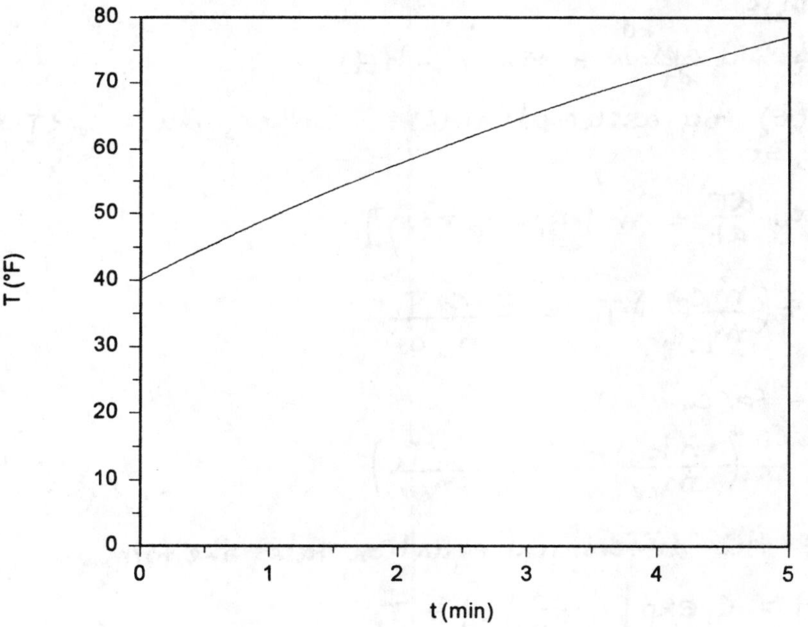

PROBLEM 4.103

KNOWN: Air flows through a well-insulated chamber. The initial conditions in the chamber, the supply conditions, and the inlet and exit mass flow rates are specified.

FIND: Determine the temperature and pressure of the air in the chamber as functions of time and plot for various mass flow rates.

SCHEMATIC & GIVEN DATA:

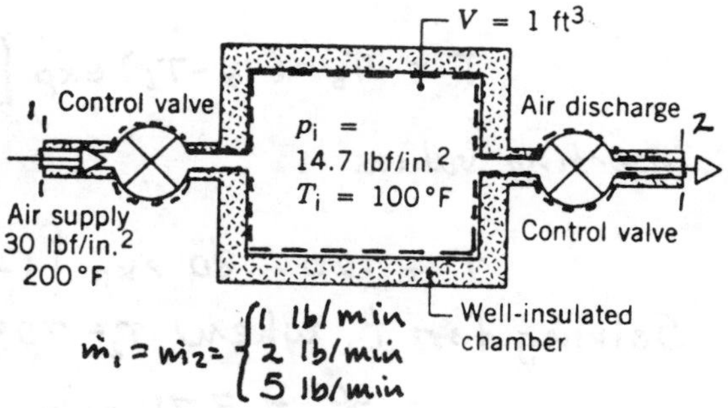

ASSUMPTIONS: (1) For the control volume shown, $\dot{Q}_{cv} = 0$ and $\dot{W}_{cv} = 0$. (2) The temperature and pressure in the chamber are uniform throughout at any time. (3) Kinetic and potential energy effects are negligible. (4) The air is modeled as an ideal gas with constant specific heats.

ANALYSIS: The mass rate balance takes the form $dm_{cv}/dt = \dot{m}_1 - \dot{m}_2$. With $\dot{m}_1 = \dot{m}_2 \equiv \dot{m}$, $dm_{cv}/dt = 0$. Thus, the amount of mass contained within the control volume is constant. Using assumptions listed, the energy balance is

$$\frac{dU_{cv}}{dt} = \cancel{\dot{Q}_{cv}}^{\,0} - \cancel{\dot{W}_{cv}}^{\,0} + \dot{m}(h_1 - h_2)$$

or, with $U_{cv} = m_{cv}u$

$$m_{cv}\frac{du}{dt} + u\cancel{\frac{dm_{cv}}{dt}}^{\,0} = \dot{m}[h_1 - h(t)]$$

where $h_2 = h(t)$ by assumption (2). Further, $du = c_v dT$ and $h(t) = c_p T(t)$, so

$$m_{cv} c_v \frac{dT}{dt} = \dot{m}[h_1 - c_p T(t)]$$

or

$$\frac{dT}{dt} + \left(\frac{\dot{m}c_p}{m_{cv}c_v}\right)T = \frac{\dot{m}c_p T_1}{m_{cv}c_v}$$

Introducing $k = c_p/c_v$

$$\frac{dT}{dt} + \left(\frac{\dot{m}k}{m_{cv}}\right)T = \left(\frac{\dot{m}kT_1}{m_{cv}}\right)$$

The solution of this differential equation takes the form

$$T(t) = c\exp\left[-\left(\frac{\dot{m}k}{m_{cv}}\right)t\right] + T_1$$

With $T(0) = T_i$; $c = T_i - T_1$, so

$$T(t) = (T_i - T_1)\exp\left[-\left(\frac{\dot{m}k}{m_{cv}}\right)t\right] + T_1 \qquad (1)$$

PROBLEM 4.103 (Cont'd.)

From the given data

$$m_{cv} = \frac{p_i V_i}{RT_i} = \frac{(14.7\ \text{lbf/in.}^2)(1\ \text{ft}^3)}{\left(\frac{1545}{28.97}\ \frac{\text{ft}\cdot\text{lbf}}{\text{lb}\cdot{}^\circ\text{R}}\right)(560^\circ\text{R})}\left|\frac{144\ \text{in.}^2}{1\ \text{ft}^2}\right| = 0.0709\ \text{lb}$$

From Table A-20E ; $k \approx 1.4$. Also, $T_i = 100°F$ and $T_1 = 200°F$. Thus, from (1)

$\underline{\dot{m} = 1\ \text{lb/s}}$ $\quad T(t) = -100\, e^{-0.3291\, t} + 200$

$\underline{\dot{m} = 2\ \text{lb/s}}$ $\quad T(t) = -100\, e^{-0.6582\, t} + 200$

$\underline{\dot{m} = 5\ \text{lb/s}}$ $\quad T(t) = -100\, e^{-1.6455\, t} + 200$

The pressure is $p = \frac{mR\,T(t)}{V}$. Thus

$$p = \left[\frac{(0.0709\ \text{lb})\ \frac{1545}{28.97}\ \frac{\text{ft}\cdot\text{lbf}}{\text{lb}\cdot{}^\circ\text{R}}}{(1\ \text{ft}^3)\left|\frac{144\ \text{in.}^2}{1\ \text{ft}^2}\right|}\right][T(t) + 460]$$

Finally

$\underline{\dot{m} = 1\ \text{lb/s}}$ $\quad p(t) = -2.6258\, e^{-0.3291\, t} + 17.3304$

$\underline{\dot{m} = 2\ \text{lb/s}}$ $\quad p(t) = -2.6258\, e^{-0.6582\, t} + 17.3304$

$\underline{\dot{m} = 5\ \text{lb/s}}$ $\quad p(t) = -2.6258\, e^{-1.6455\, t} + 17.3304$

Using IT to plot, we get

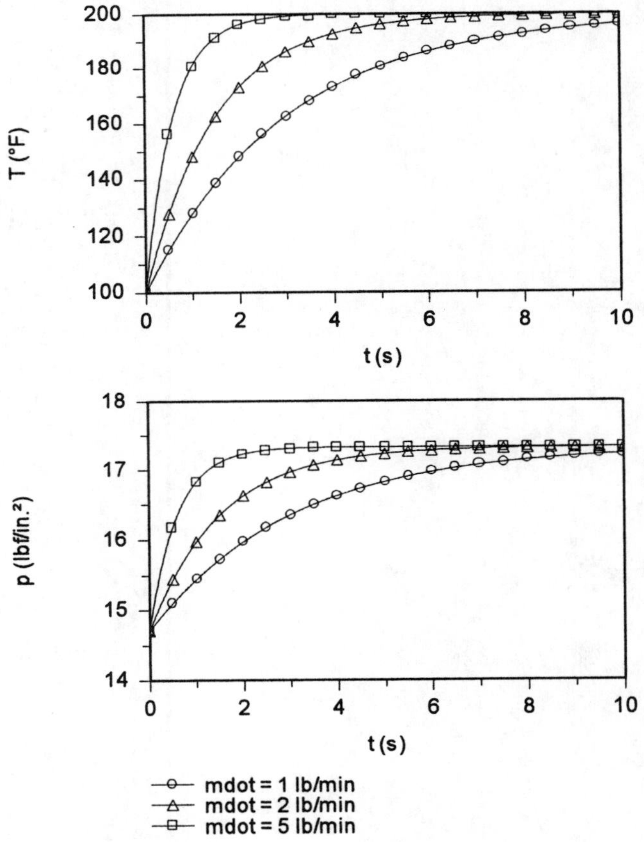

CHAPTER FIVE
THE SECOND LAW OF THERMODYNAMICS

Chapter 5 - Third and fourth edition problem correspondence

3rd	4th
5.1	---
5.2	---
---	5.1
5.3	5.2*
5.4	5.3
5.5	5.4*
5.6	---
5.7	---
5.8	---
5.9	5.5
5.10	5.6
5.11	---
5.12	5.7
---	5.8
5.13	5.9
5.14	5.10*
5.15	5.11
5.16	5.12
5.17	5.13a,b
5.18	5.13c,d
5.19	5.14
5.20	5.15
5.25	5.16
5.26	5.17
5.27	5.18
5.29	5.19*
5.28	5.20*
5.41	5.21*
5.50	5.22*
5.60	5.23*
---	5.24
5.21	5.25*
5.22	5.26
5.23	5.27*
5.24	5.28*
5.30	5.29*
5.31	5.30*
5.32	5.31*
5.33	5.32*
5.34	---
---	5.33
5.35	---
---	5.34
5.36	5.35*
5.37	5.36
5.38	5.37a
5.39	5.37b
---	5.38
---	5.39
5.40	5.40*
5.42	---
5.43	5.41
---	5.42
---	5.43
---	5.44
5.45	5.45*
5.47	5.46*
5.48	5.47
5.49	---
5.52	5.48*
5.53	---
5.56	5.49*
5.57	---
5.58	5.50*
5.59	5.51*
5.51	5.52a
5.61	5.52b
5.46	5.53*
5.54	5.54*
5.55	5.55*
5.44	5.56
---	5.57
---	5.58
---	5.59
---	5.60
5.62	5.61
5.63	5.62*
5.64	5.63*
5.65	---
5.66	5.64*
5.67	5.65*
---	5.66
5.68	5.67*
5.69	5.68*

PROBLEM 5.1

KNOWN: Operating data are provided for a heat pump.

FIND: Explain whether the heat pump is in violation of the Clausius statement.

SCHEMATIC & GIVEN DATA:

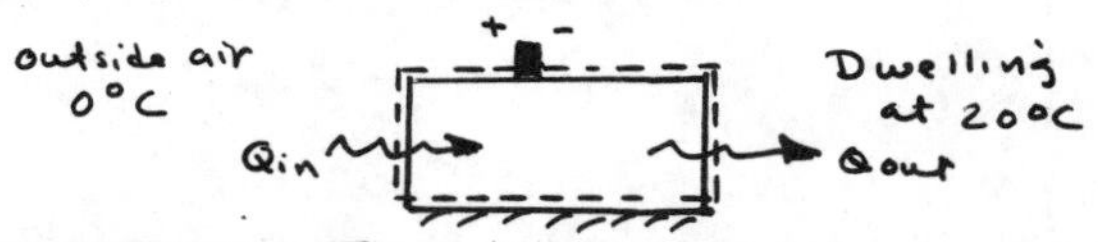

ASSUMPTION: The heat pump is a conventional one intended for dwelling heating from an electric power input.

ANALYSIS: Since the heat pump is driven by an electric power input, there is no violation of the Clausius statement, which refers only to cases where the sole effect would be a heat transfer from a colder to a hotter body.

PROBLEM 5.2

KNOWN: Air expands isothermally between specified states.

FIND: Evaluate the heat transfer and work for the process, and explain if a violation of the Kelvin-Planck statement occurs.

SCHEMATIC & GIVEN DATA:

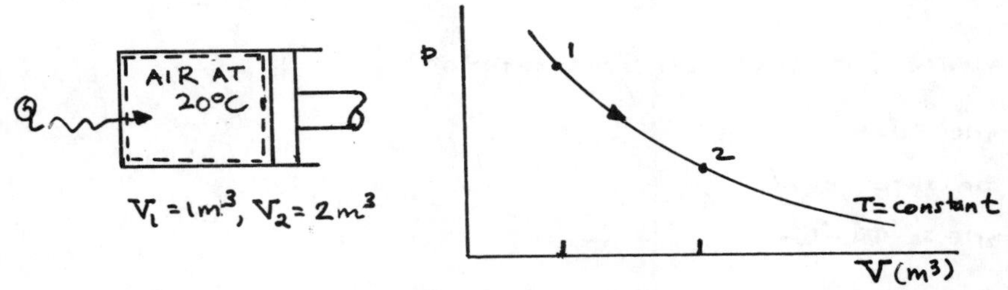

ASSUMPTIONS: (1) The system is 1 lb of air modeled as an ideal gas. (2) Temperature remains constant. (3) Kinetic and potential energy changes are absent. (4) The surroundings play the role of a thermal reservoir. (5) Volume change is the only work mode.

ANALYSIS: An energy balance for the system reads $U_2 - U_1 = Q - W$. However, as the internal energy of an ideal gas depends on temperature only, $U_2 = U_1$. Thus, the energy balance reduces to $Q = W$. To find W

$$W = \int_1^2 p\,dV = \int_1^2 \frac{mRT}{V} dV = mRT \ln \frac{V_2}{V_1}$$

$$\Rightarrow \quad \frac{W}{m} = \left(\frac{8.314}{28.97}\right)\left(\frac{kJ}{kg \cdot K}\right)(293\ K) \ln \frac{2\ m^3}{1\ m^3} = 58.3\ kJ/kg \quad \longleftarrow W/m, Q/m$$

The Kelvin-Planck statement of the second law refers to systems that undergo thermodynamic cycles while communicating thermally with a single reservoir. Here, the system undergoes a process—not a cycle. Accordingly, the Kelvin-Planck statement is not applicable, so no violation can be claimed. Similarly, the Clausius statement, which is equivalent to the Kelvin-Planck statement, is not applicable in this case. With the information provided, there is no apparent violation of the second law.

PROBLEM 5.3

KNOWN: A system undergoes a cycle in violation of the Kelvin-Planck statement of the second law.

FIND: Show that a violation of the Clausius statement of the second law is a consequence.

SCHEMATIC & GIVEN DATA:

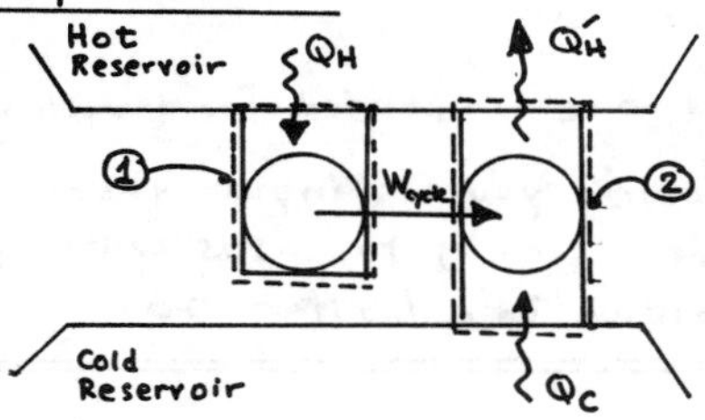

ASSUMPTIONS: (1) System 1 undergoes a cycle while receiving energy Q_H from the hot reservoir and developing work W_{cycle}. (This is in violation of the Kelvin-Planck statement of the second law.) (2) System 2 undergoes a cycle while removing energy Q_C from the cold reservoir and discharging energy Q_H' to the hot reservor. The work developed by System 1 is used to drive System 2.

ANALYSIS: An energy balance for system 1 reduces to

$$W_{cycle} = Q_H$$

For System 2 an energy balance reads

$$W_{cycle} = Q_H' - Q_C$$

Combining these expressions

$$Q_H' = Q_H + Q_C$$

Next, consider a combined system consisting of Systems 1 and 2:

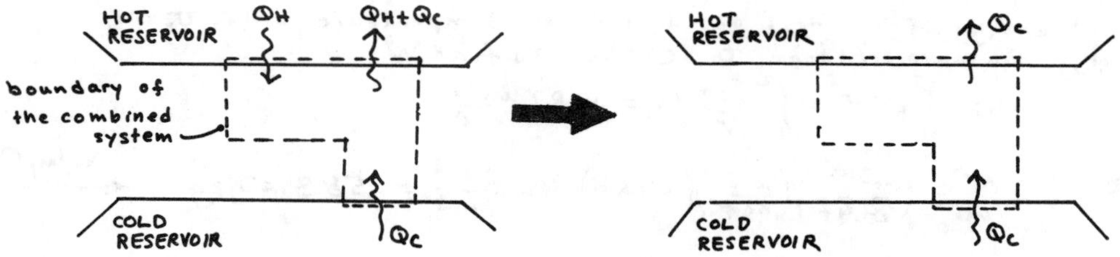

The sole result of the combined system is that a net amount of energy Q_C is transferred from the cold reservoir to the hot reservoir. This is a violation of the Clausius statement of the second law.

PROBLEM 5.4

KNOWN: A system undergoes a cycle while communicating with two reservoirs. The system develops work while receiving Q_C from the cold reservoir and discharging Q_H to the hot reservoir.

FIND: Evaluate the claimed performance using the second law.

SCHEMATIC & GIVEN DATA:

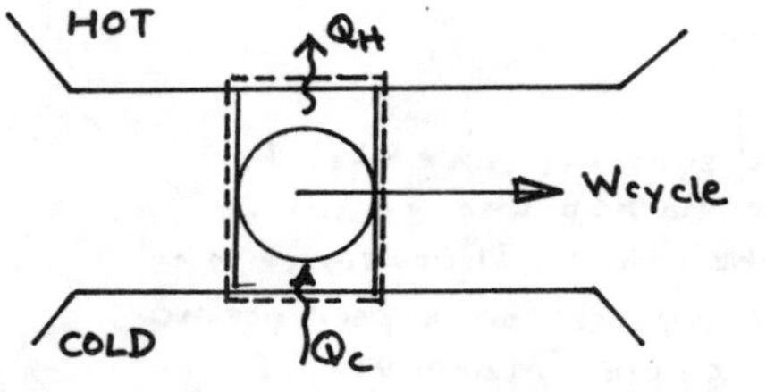

ASSUMPTION: The system shown in the accompanying figure undergoes a power cycle while communicating thermally with two reservoirs as shown.

ANALYSIS: An energy balance for the system reads

$$W_{cycle} = Q_C - Q_H$$

The claim can be evaluated using either the Clausius or the Kelvin-Plank statement:

(a) Clausius. As shown in the figure below, let the given cycle drive a heat pump cycle.

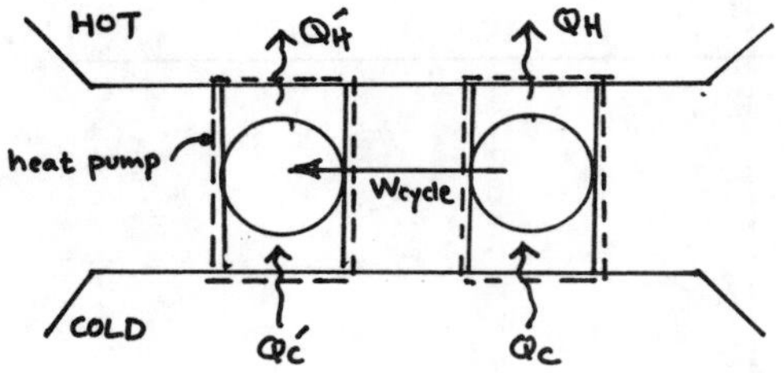

An energy balance for the heat pump reads

$$Q_H' = Q_C' + W_{cycle}$$

or, with $W_{cycle} = Q_C - Q_H$

$$Q_H' = Q_C' + Q_C - Q_H$$

Next, consider the combined system of the two systems together. The sole result of the combined system is that a net amount of energy $Q_C' + Q_C$ is transferred from the cold reservoir to the hot reservoir. This is a violation of the Clausius statement of the second law, so the original system cannot operate as assumed. The inventor's claim is invalid.

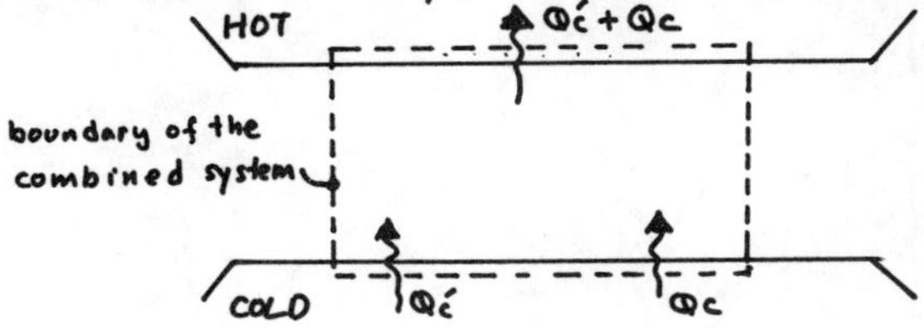

(b) Kelvin-Plank. As shown in the figure below, let Q_H be supplied to a system undergoing a power cycle.

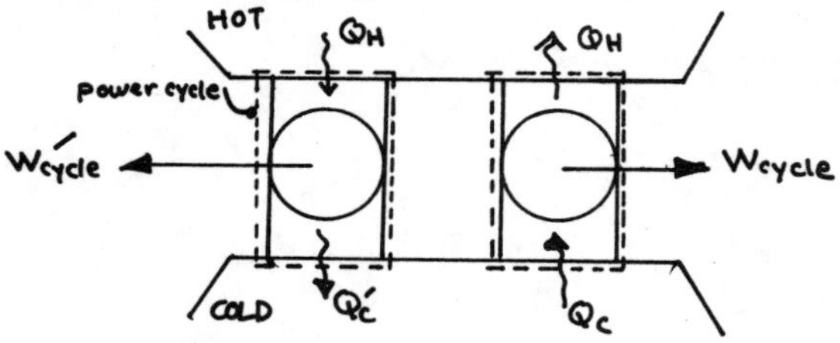

Next, consider the combined system formed by the two systems and the hot reservoir. Observe that the hot reservoir experiences no net change as it receives Q_H from one cycle and supplies Q_H to the other. Thus, since the combined system is made up of parts that undergo cycles or experience no net change, the combined system operates in a cycle. Moreover, it exchanges energy by heat transfer with a single reservoir: the cold reservoir, while producing a net amount of work. However, this is a violation of the Kelvin-Planck statement of the second law, so the original system cannot operate as assumed.

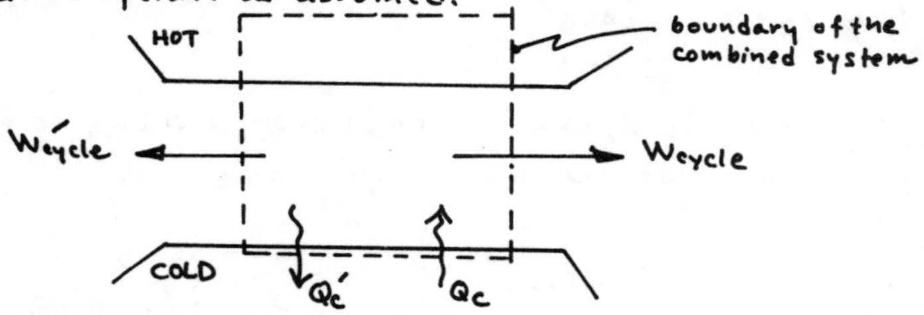

PROBLEM 5.5

KNOWN: Energy transfer by conduction from a hot reservoir to a cold reservoir takes place spontaneously.

FIND: Using the Kelvin-Planck statement of the second law, show that such a process is irreversible.

SCHEMATIC & GIVEN DATA:

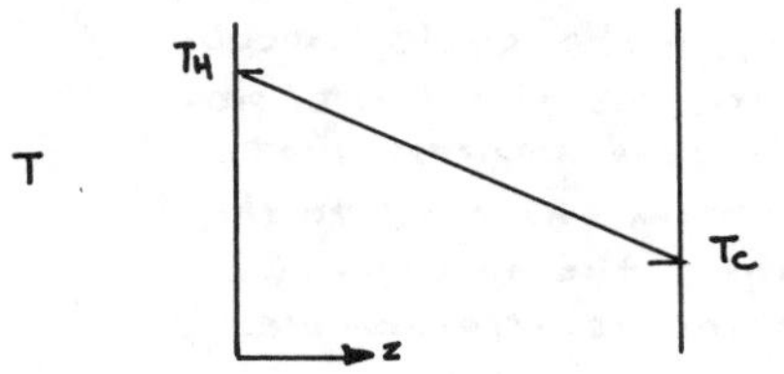

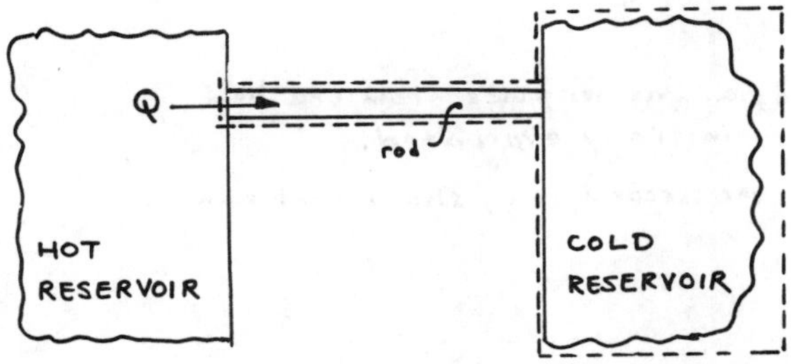

ASSUMPTIONS: (1) The system shown in the accompanying figure receives energy Q from the hot reservoir which passes by conduction through a rod at steady state to the cold reservoir. (2) A power cycle is available for use in demonstrating that the process is irreversible.

ANALYSIS: The objective in this proof by contradiction is to devise a system that undergoes a cycle which develops work while the system communicates thermally with a single reservoir, thereby violating the Kelvin-Planck statement.

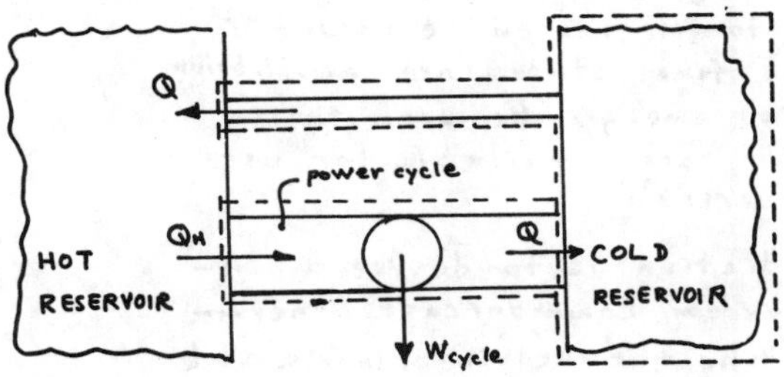

Note: Q represents the energy transferred from the cold reservoir to the hot reservoir without any other effect.

Note: Q represents the energy discharged to the cold reservoir from the power cycle.

The system shown above consists of the original system plus a system capable of undergoing a power cycle (assumption 2). This combined system undergoes a cycle as follows. (1) An amount of energy Q_H is transferred from the hot reservoir to the power cycle, producing work W_{cycle} and discharging energy Q to the cold reservoir. (2) An amount of energy Q is transferred from the cold reservoir to the hot reservoir without any other effects. (This would be possible only if the process described in assumption 1 were reversible.)

Since Q is added to the cold reservoir by the power cycle in the first

PROBLEM 5.5 (Contd.)

process of the cycle and the same amount of energy is removed from the cold reservoir in the second process of the cycle, this reservoir experiences no net change in its condition. The power cycle enclosed within the combined system also undergoes a cycle. Accordingly, the combined system undergoes a cycle in which work W_{cycle} is produced while exchanging energy by heat transfer with a single reservoir (the hot reservoir). Such a cycle violates the Kelvin-Planck statement of the second law, and thus is impossible. It follows that one, or both, of the processes making up the cycle executed by the combined system must be impossible. However, as the first process involving the power cycle can occur, it is the second process that must be impossible: energy Q cannot be transferred from the cold to the hot reservoir without other effects. By definition, then, the transfer of energy Q by conduction from the hot to the cold reservoir is irreversible.

PROBLEM 5.6

KNOWN: When an interconnecting valve is opened, a gas expands from one half of a tank to the other half which is initially evacuated.

FIND: Using the Kelvin-Planck statement of the second law, show that such a process is irreversible.

SCHEMATIC & GIVEN DATA:

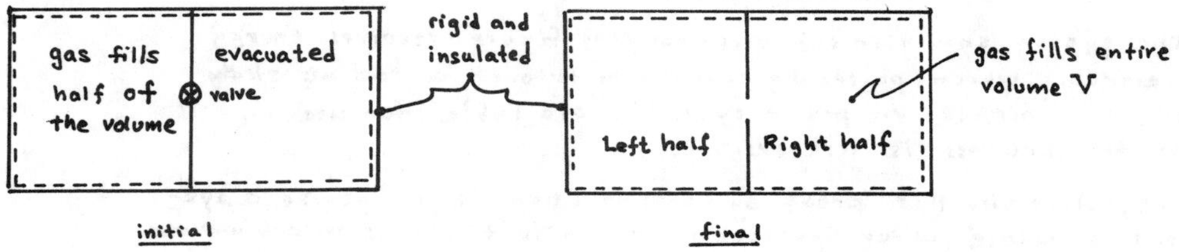

ASSUMPTIONS: (1) The system shown in the accompanying figure undergoes a process in which the gas expands spontaneously to fill the entire volume V. (2) During the process $Q = W = 0$. (3) The initial and final states are equilibrium states. There is no change in kinetic or potential energy between these states. (4) A turbine and a thermal reservoir are available for use in demonstrating that the process is irreversible.

ANALYSIS: The object in this proof by contradiction is to devise a system that undergoes a power cycle while the system communicates thermally with a single reservoir, thereby violating the Kelvin-Planck statement of the second law.

Before considering such a system, note that an energy balance for the spontaneous process is

$$U_{final} - U_{initial} = \cancel{Q}^{0} - \cancel{W}^{0}$$

where assumptions 2 and 3 have been used. Accordingly,

$$U_{final} = U_{initial}$$

That is, the internal energy of the system does not change in the free expansion.

PROBLEM 5.6 (Contd.)

As shown in the figure below, modify the system to include a turbine and introduce a thermal reservoir in the surroundings (assumption 4).

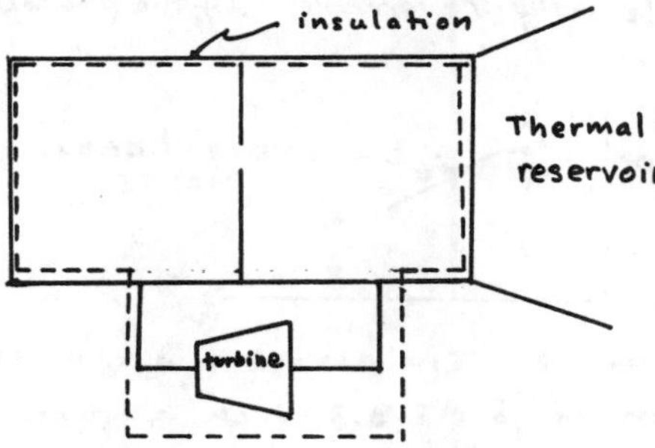

Starting with the gas filling the entire volume V, let the modified system undergo a cycle consisting of three processes.

Process 1. Let the reverse of the free expansion occur without any other effects. That is, the gas passes spontaneously from the right half of the tank until it fills only the left half. (This would be possible only if the process described in assumption 1 were reversible.)

Process 2. Let part of the gas expand through the turbine into the right half of the tank until the pressure in both halves is the same. In expanding through the turbine the gas does work so that its internal energy is decreased: $U < U_{initial}$.

Process 3. Remove part of the tank insulation and add energy by heat transfer from the thermal reservoir until the internal energy of the gas is restored to its initial value. Thus, a cycle is completed.

The net result of this cycle is to draw energy from a single reservoir by heat transfer and produce an equivalent amount of work. Such a cycle violates the Kelvin-Planck statement of the second law, and thus is impossible. Since both the heating of the gas by the reservoir (process 3) and the development of work as gas passes through the turbine (process 2) are possible, it can be concluded that it is process 1 that is impossible. Since process 1 is the reverse of the original free expansion, it follows that the original process is irreversible.

PROBLEM 5.7

KNOWN: A gas within a piston-cylinder assembly undergoes a quasiequilibrium process.

FIND: Determine if the process is internally reversible. If the process is reversible.

SCHEMATIC & GIVEN DATA:

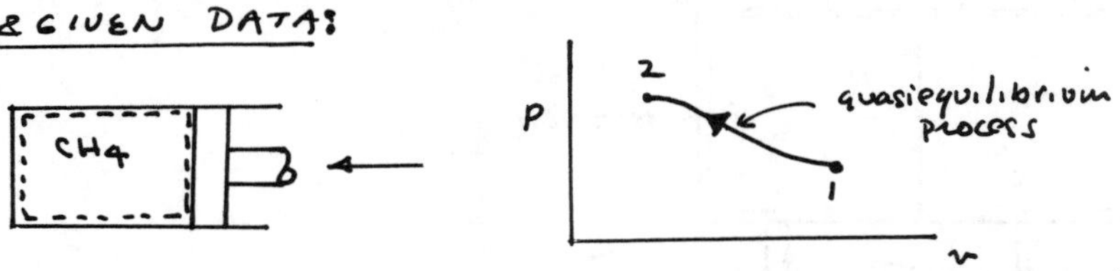

ASSUMPTIONS: 1. The gas is the system. 2. The process is a quasiequilibrium process.

ANALYSIS: Referring to the discussion of Sec. 5.3.3, such a process is internally reversible. The process would be reversible only if the surroundings were also free of irreversibilities as the gas undergoes the indicated process. A reversible process is one in which there are no internal or external irreversibilities. Thus, reversibility in this case can be decided only if further information about the surroundings is provided.

PROBLEM 5.8

KNOWN: Water within a piston-cylinder assembly cools isothermally at 100°C from saturated vapor to saturated liquid.

FIND: Determine if the process is internally reversible, is reversible.

SCHEMATIC & GIVEN DATA:

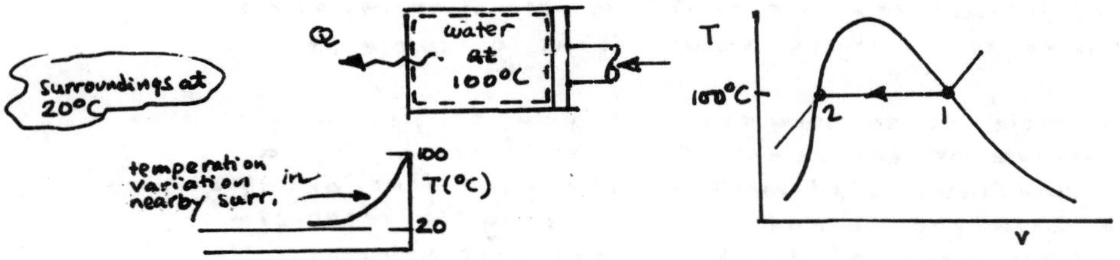

ASSUMPTIONS: 1. The system is the water in the piston-cylinder assembly. 2. The system undergoes a constant-temperature process from saturated vapor to saturated liquid.

ANALYSIS: Since temperature is constant during the process, the pressure also remains constant. As shown by the T-v diagram, the process is a sequence of equilibrium states, and thus is internally reversible.

The process is not reversible because there is a significant irreversibility in the surroundings — namely, the spontaneous heat transfer taking place between the water at 100°C and the surroundings at 20°C.

PROBLEM 5.9

KNOWN: A system undergoes a cycle reversibly while communicating thermally with a single reservoir.

FIND: Show that $W_{cycle} = 0$.

SCHEMATIC & GIVEN DATA:

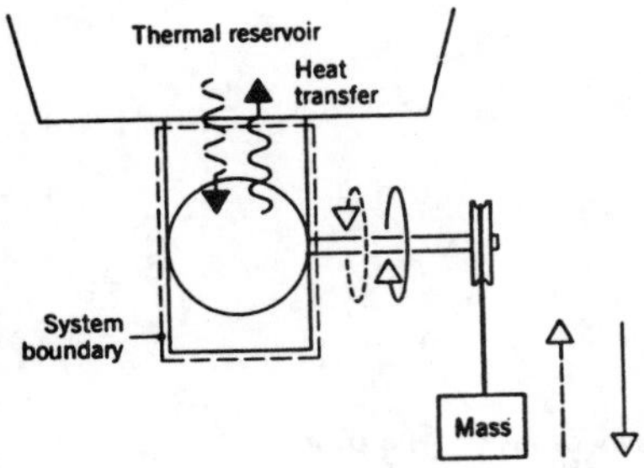

ASSUMPTIONS: (1) The system shown in the accompanying figure undergoes a reversible cycle. (2) During the cycle the system communicates thermally only with a single reservoir.

ANALYSIS: Let the system undergo one cycle. According to the Kelvin-Planck statement of the second law, the work for the cycle is zero or negative in value: $W_{cycle} \leq 0$.

As the cycle is reversible, it is possible to return both the system and its surroundings to their initial states. Accordingly, there would be no net change in the condition of the reservoir or the elevation of the mass used to store energy in the surroundings.

The above statements are consistent only if the sign of equal-
① ity is used: $W_{cycle} = 0$.

---

1. In Sec. 5.4.1 the converse is demonstrated: If the sign of equality applies: $W_{cycle} = 0$, then the cycle is reversible. Taken together, these two demonstrations establish the following proposition: If, and only if, the sign of equality applies in Eq. 5.1, the cycle is reversible.

PROBLEM 5.10

KNOWN: A power cycle I and a reversible power cycle R operate between the same two reservoirs. For these cycles, $\eta_I = \frac{2}{3}\eta_R$.

FIND: Show that cycle I must be irreversible.

SCHEMATIC & GIVEN DATA:

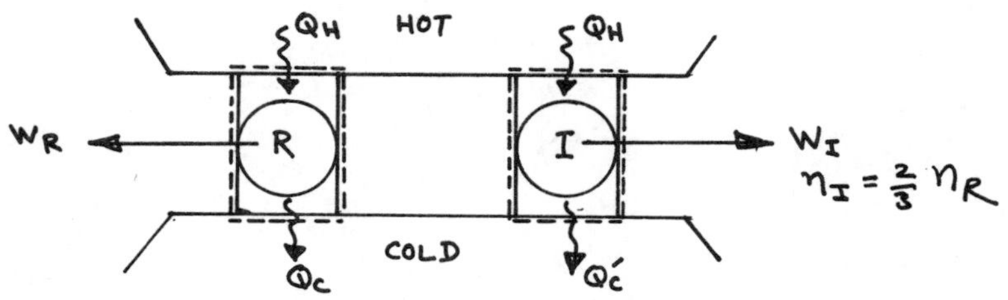

ASSUMPTIONS: (1) The system denoted by R in the accompanying figure undergoes a reversible power cycle while system I undergoes a power cycle such that $\eta_I = \frac{2}{3}\eta_R$. (2) Both cycles receive the same energy Q_H from the hot reservoir.

ANALYSIS: If $\eta_I = 2/3\,R$ and both cycles receive Q_H from the hot reservoir, $W_I = 2/3\,W_R < W_R$.

In this proof by contradiction, assume I is reversible. Then, if I operates in the opposite direction as a refrigeration (or heat pump) cycle, the magnitudes of the energy transfers Q_H, Q_C', and W_I would remain the same but would be oppositely directed as shown in the figure below. With I operating in the opposite direction, the hot reservoir would experience no net change since it would receive Q_H from I while passing Q_H to R.

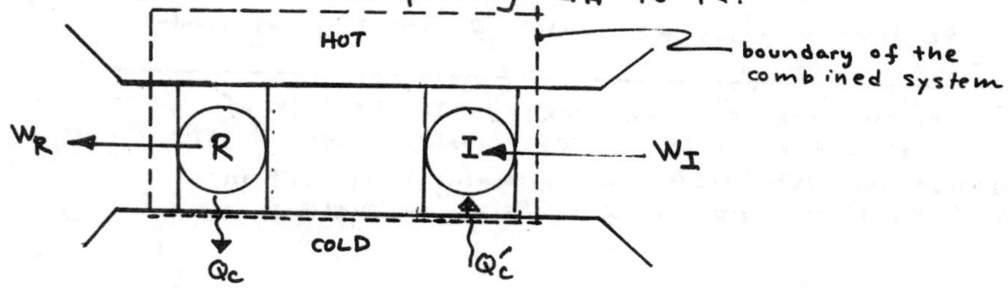

As the combined system shown in the figure above is made up of parts that execute cycles or experience no net change, the combined system operates in a cycle. Moreover, it exchanges energy by heat transfer with a single reservoir: the cold reservoir. Accordingly, the combined system must satisfy Eq. 5.1 expressed as $W_{cycle} = 0$, where the sign of equality is used because R is reversible and I has been assumed reversible. Evaluating W_{cycle} for the combined system in terms of the work amounts W_R and W_I, $W_{cycle} = W_R - W_I$, it follows that

$$W_I = W_R$$

As this conclusion is not in agreement with the requirement that $W_I < W_R$, it can be concluded that the hypothesis is false and I must be irreversible.

PROBLEM 5.11

KNOWN: A reversible power cycle R and an irreversible power cycle I operate between the same two reservoirs.

FIND: (a) If each cycle receives the same amount of energy from the hot reservoir, show that I discharges more energy to the cold reservoir.

(b) If each cycle develops the same net work, show that I receives more energy from the hot reservoir.

SCHEMATIC & GIVEN DATA:

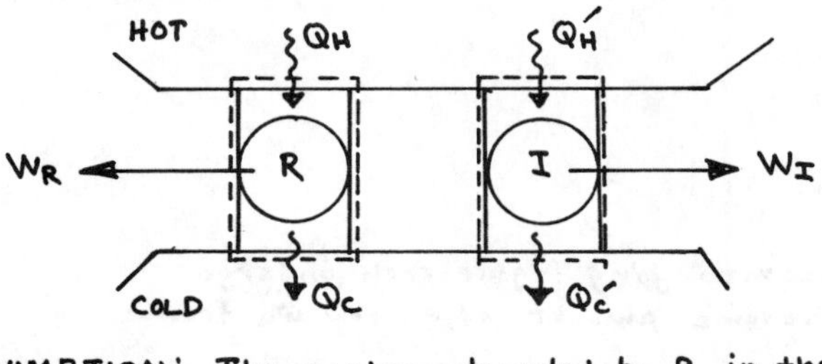

To show:

(a) $Q_H = Q_H' \Rightarrow Q_C' > Q_C$

(b) $W_R = W_I \Rightarrow Q_H' > Q_H$

ASSUMPTION: The system denoted by R in the accompanying figure undergoes a reversible power cycle while system I undergoes an irreversible power cycle.

ANALYSIS: (a) By the first Carnot Corollary, $\eta_R > \eta_I$. Since both cycles receive the same amount of energy Q_H, it follows that $W_R > W_I$.

An energy balance for R and I read, respectively

$$W_R = Q_H - Q_C$$

$$W_I = Q_H - Q_C'$$

Collecting results

$$Q_H - Q_C > Q_H - Q_C'$$

Accordingly

$$Q_C' > Q_C \longleftarrow$$

as was to be demonstrated.

Thus, not only do actual cycles develop less work they also discharge more energy by heat transfer to their surroundings, thereby increasing the effect of thermal pollution.

(b) Since $\eta_R > \eta_I$ and $W_R = W_I \equiv W$,

$$\frac{W}{Q_H} > \frac{W}{Q_H'} \Rightarrow Q_H' > Q_H$$

If the hot reservoir were maintained by, say, energy from the combustion of a fossil fuel, cycle I would have the greater fuel requirement. Also, note that cycle I also would have the greater energy discharge to the cold reservoir, so the comments of (a) would also be applicable.

PROBLEM 5.12

KNOWN: The Kelvin-Planck statement of the second law: $W_{cycle} \leq 0$ (single reservoir).

FIND: Show that all reversible power cycles operating between the same two reservoirs have the same thermal efficiency.

SCHEMATIC & GIVEN DATA:

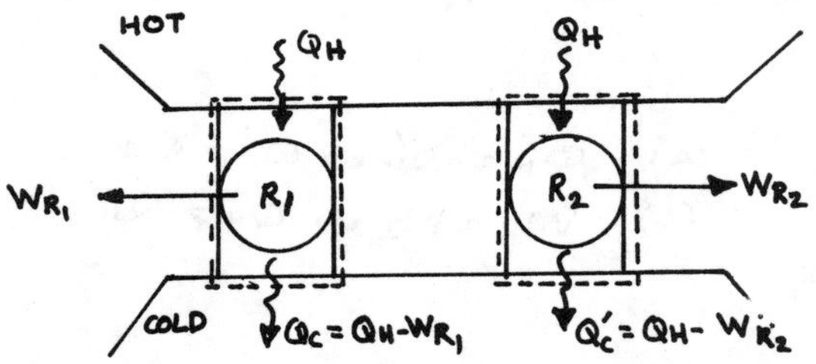

ASSUMPTION: The systems shown in the accompanying figure each undergo reversible power cycles while receiving the same amount of energy Q_H from the hot reservoir.

ANALYSIS: Let cycle R_1 operate as a reversible refrigeration (or heat pump cycle). The magnitudes of W_{R1}, Q_H, and Q_C remain the same but are now oppositely directed. Further, with R_1 working in the opposite direction, the hot reservoir would experience no net change in its condition since it would receive Q_H from R_1 while passing Q_H to R_2.

The demonstration is completed by considering the combined system consisting of the two cycles and the hot reservoir. Since its parts execute cycles or experience no net change, the combined system operates in a cycle. Moreover, it exchanges energy by heat transfer with a single reservoir: the cold reservoir. Accordingly, the combined system must satisfy the Kelvin-Planck statement expressed as

$$W_{cycle} = 0$$

where the equality is used because all parts of the combined system are free of irreversibilities. Evaluating W_{cycle} in terms of the work amounts W_{R_1} and W_{R_2}

$$W_{R_2} - W_{R_1} = 0$$

or

$$W_{R_2} = W_{R1}$$

Since each of the power cycles receives the same energy input Q_H, it follows that $\eta_{R_2} = \eta_{R_1}$ and this completes the demonstration.

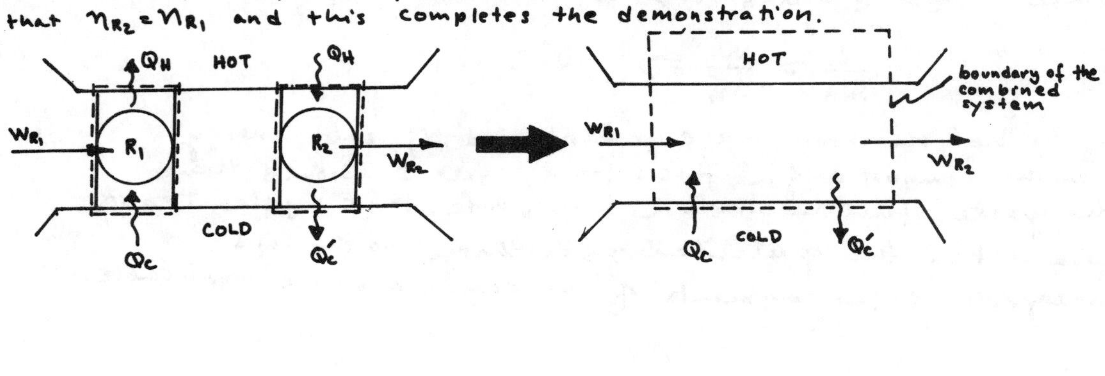

PROBLEM 5.13

KNOWN: The Kelvin-Planck statement of the second law: $W_{cycle} \le 0$ (single reservoir).

FIND: Show that (a) the coefficient of performance of an irreversible refrigeration cycle is less than the coefficient of performance of a reversible refrigeration cycle when both exchange energy by heat transfer with the same two reservoirs, (b) all reversible refrigeration cycles operating between the same two reservoirs have the same coefficient of performance. For parts (c), (d), see next page.

SCHEMATIC & GIVEN DATA:

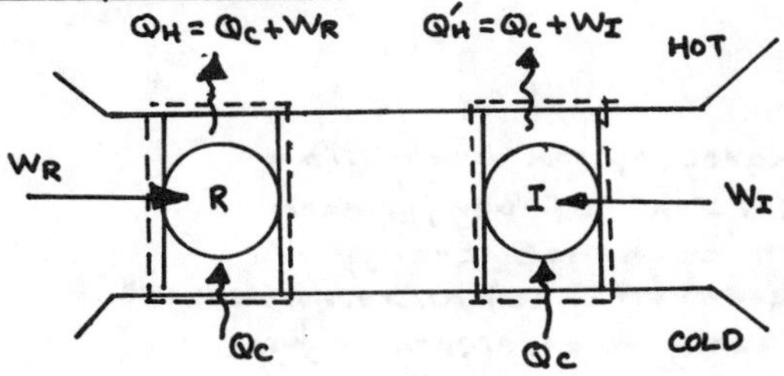

ASSUMPTIONS: The systems shown in the accompanying figure undergo refrigeration cycles. Each cycle removes energy Q_C from the cold reservoir. R accomplishes this reversibly while I is irreversible.

ANALYSIS: (a) Let R operate as a reversible power cycle. The magnitudes of W_R, Q_C and Q_H remain the same but are now oppositely directed. Further, with R working in the opposite direction, the cold reservoir would experience no net change in its condition since it would receive Q_C from R while passing Q_C to I.

The demonstration is completed by considering the combined system consisting of the two cycles and the cold reservoir. Since its parts execute cycles or experience no net change, the combined system operates in a cycle. Moreover, it exchanges energy by heat transfer with a single reservoir: the hot reservoir. Accordingly, the combined system must satisfy the Kelvin-Planck statement expressed as

$$W_{cycle} < 0$$

where the inequality is used because irreversible cycle I is included. Evaluating W_{cycle} in terms of the work amounts W_R and W_I

$$W_R - W_I < 0$$

or

$$W_R < W_I$$

Since each of the refrigeration cycles receives the same energy input Q_C, it follows that $\beta_R > \beta_I$ and this completes the demonstration.

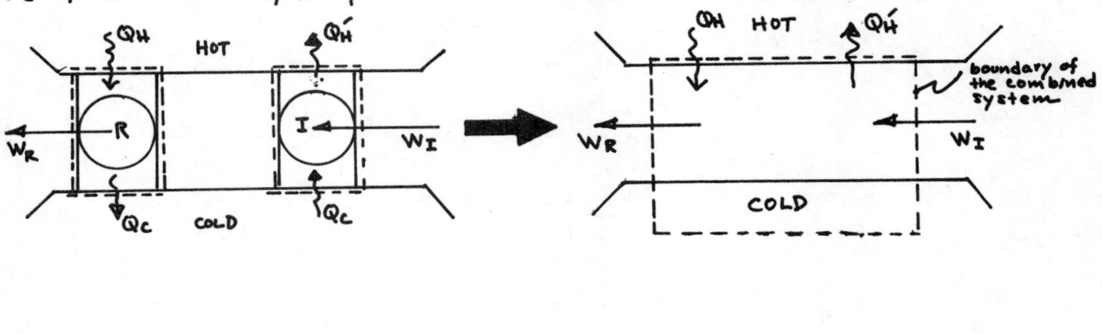

PROBLEM 5.13 (Contd.)

ANALYSIS (b) This proposition can be demonstrated in a parallel way by considering two reversible refrigeration cycles R_1 and R_2 operating between the same two reservoirs. Then, letting R_1 play the role of R and R_2 the role of I in the previous development, a combined system consisting of the two cycles and the hot reservoir may be formed which must satisfy $W_{cycle} = 0$, where the equality is used because all parts of the combined system are free of irreversibilities. Thus, it can be concluded that $W_{R_1} = W_{R_2}$, and therefore $\beta_{R_1} = \beta_{R_2}$.

Parts (c), (d):

FIND: Show that (c) the coefficient of performance of an irreversible heat pump cycle is less than the coefficient of performance a reversible heat pump cycle when both exchange energy by heat transfer with the same two reservoirs, (d) all reversible heat pump cycles operating between the same two reservoirs have the same coefficient of performance.

SCHEMATIC & GIVEN DATA:

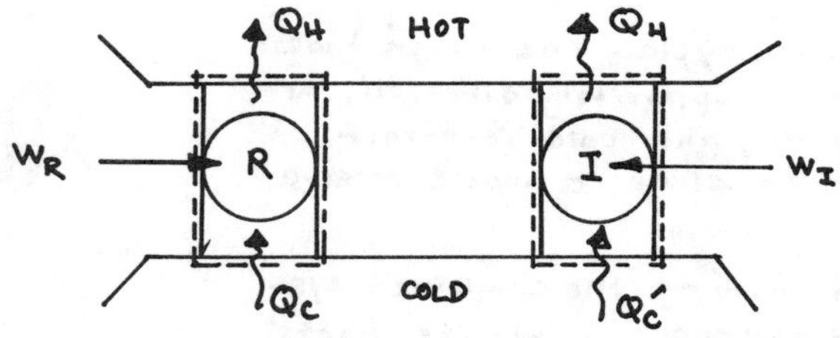

ASSUMPTIONS: (1) The systems shown in the accompanying figure undergo heat pump cycles. (2) Each cycle provides energy Q_H to the hot reservoir. R accomplishes this reversibly while I is irreversible. (3) Both cycles operate between the same two reservoirs.

ANALYSIS: (c) The demonstration parallels the approach used in part(a). However, in the present case the combined system would consist of the two cycles and the hot reservoir. (b). The demonstration parallels the approach used in part(b).

PROBLEM 5.14

KNOWN: The symbol Θ denotes temperature on Kelvin's logarithmic scale.

FIND: (a) Show that $\Theta = \ln T + C$, where T is temperature on the Kelvin scale. (b) Determine the range of temperature values on the logarithmic scale. (c) Obtain an expression for the thermal efficiency of a reversible power cycle operating between reservoirs at Θ_H and Θ_C on the logarithmic scale.

ANALYSIS: The two scales arise from different specifications of the function ψ in Eq. 5.5:

$$\left(\frac{Q_C}{Q_H}\right)_{\substack{rev\\cycle}} = \psi$$

That is

$$\left(\frac{Q_C}{Q_H}\right)_{\substack{rev\\cycle}} = \frac{T_C}{T_H} \qquad \text{Kelvin scale}$$

$$\left(\frac{Q_C}{Q_H}\right)_{\substack{rev\\cycle}} = \frac{\exp \Theta_C}{\exp \Theta_H} \qquad \text{Logarithmic scale}$$

(a) By comparison of the last two equations

$$\frac{T_C}{T_H} = \frac{\exp \Theta_C}{\exp \Theta_H} = \exp[\Theta_C - \Theta_H]$$

Thus

$$\ln T_C - \ln T_H = \Theta_C - \Theta_H$$

and therefore

$$\Theta = \ln T + C \quad \longleftarrow \text{(a)}$$

where C is a constant determining the level of temperature corresponding to zero on the logarithmic scale.

(b) Temperatures on the Kelvin scale vary from 0 to $+\infty$. With the relationship of part (a), temperatures on the logarithmic scale vary from $-\infty$ to $+\infty$. ← (b)

(c) Use of

$$\left(\frac{Q_C}{Q_H}\right)_{\substack{rev\\cycle}} = \frac{\exp \Theta_C}{\exp \Theta_H}$$

in Eq. 5.2 gives an expression for the thermal efficiency of a reversible power cycle while operating between reservoirs at temperatures Θ_H and Θ_C on the logarithmic scale

$$\eta = 1 - \frac{\exp \Theta_C}{\exp \Theta_H} \quad \longleftarrow \text{(c)}$$

PROBLEM 5.15

KNOWN: The gas temperature scale introduced in Sec. 1.6.3 and the Kelvin scale defined in Sec. 5.5.

FIND: Demonstrate that the two scales are identical.

SCHEMATIC & GIVEN DATA:

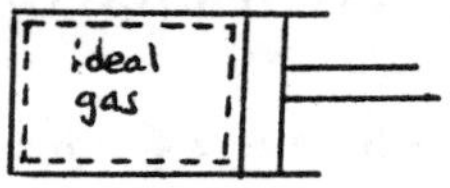

ASSUMPTIONS: (1) The system shown in the figure consists of a gas obeying $pv = RT'$, where T' denotes temperature on the ① gas scale. (2) The system undergoes a reversible cycle while communicating thermally with reservoirs at T_H' and T_C'. (3) The cycle consists of four processes in series: 1-2 isothermal at T_H', 2-3 adiabatic, 3-4 isothermal at T_C', 4-1 adiabatic. (4) Kinetic and potential energy effects are absent.

ANALYSIS: With assumption (4), an energy balance in differential form reads

$$du = \delta(Q/m) - \delta(W/m)$$

Then, since the processes are internally reversible, $\delta(W/m) = p\,dv$. Also, for the ideal gas, $du = c_v\,dT'$. Collecting results

$$c_v\,dT' = \delta(Q/m) - p\,dv \Rightarrow c_v\,dT' = \delta(Q/m) - \frac{RT'}{v}dv \Rightarrow \underline{c_v\,dT' = \delta(Q/m) - RT'\,d\ln v}$$

Process 1-2: The gas temperature is constant at T_H', so

$$\cancel{c_v\,dT'}^{\,0} = \delta(Q/m)_{1\text{-}2} - RT_H'\,d\ln v \Rightarrow (Q/m)_{1\text{-}2} = RT_H'\ln(v_2/v_1) \qquad (1)$$

Process 2-3: There is no heat transfer, and the gas temperature varies from T_H' to T_C', so

$$c_v\,dT' = \cancel{\delta(Q/m)}^{\,0} - RT'\,d\ln v \Rightarrow \frac{c_v\,dT'}{T'} = -R\,d\ln v \Rightarrow \int_{T_H'}^{T_C'} \frac{c_v\,dT'}{T'} = -R\ln\left(\frac{v_3}{v_2}\right) \qquad (2)$$

Process 3-4: The gas temperature is constant at T_C', and the heat rejected in the process is

$$(Q/m)_{3\text{-}4} = RT_C'\ln(v_4/v_3) \Rightarrow |(Q/m)_{3\text{-}4}| = RT_C'\ln\left(\frac{v_3}{v_4}\right) \qquad (3)$$

Process 4-1: There is no heat transfer, and the gas temperature varies from T_C' to T_H', so

$$\int_{T_C'}^{T_H'} \frac{c_v\,dT'}{T'} = -R\ln\left(\frac{v_1}{v_4}\right) \Rightarrow \int_{T_H'}^{T_C'} \frac{c_v\,dT'}{T'} = -R\ln\left(\frac{v_4}{v_1}\right) \qquad (4)$$

From Equations (2) and (4), $v_3/v_2 = v_4/v_1$ or $v_3/v_4 = v_2/v_1$. Using this result together with Equations (1) and (3) gives

$$\frac{|(Q/m)_{3\text{-}4}|}{(Q/m)_{1\text{-}2}} = \frac{T_C'}{T_H'}$$

which corresponds to Equation 5.6 underlying the Kelvin scale. This implies $T' \equiv T$, where T denotes temperature on the Kelvin scale.

---

1. The reversible cycle considered here corresponds to the Carnot cycle discussed in Sec. 5.7.

PROBLEM 5.16

KNOWN: A system undergoes a reversible power cycle while receiving and discharging energy by heat transfer with two thermal reservoirs.

FIND: To increase the thermal efficiency, determine whether it would be better to increase T_H or decrease T_C.

SCHEMATIC & GIVEN DATA:

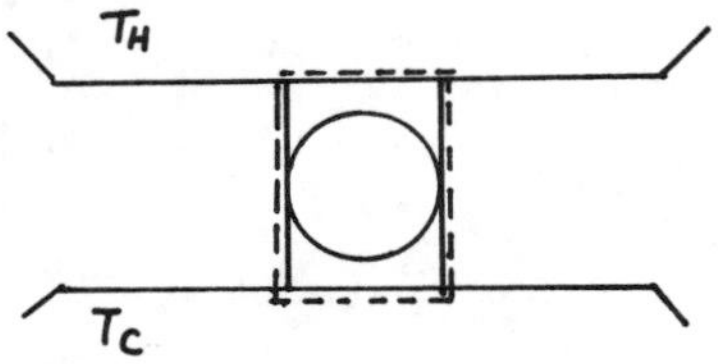

ASSUMPTION: The system shown in the accompanying figure undergoes a reversible power cycle while receiving and discharging energy by heat transfer with two thermal reservoirs.

ANALYSIS: Trends can be determined by differentiation of Eq. 5.8, which is applicable in this case:

$$\left(\frac{\partial \eta}{\partial T_H}\right)_{T_C} = \frac{T_C}{(T_H)^2} \Rightarrow \text{As } T_H \text{ increases, } \eta \text{ increases.}$$

$$\left(\frac{\partial \eta}{\partial T_C}\right)_{T_H} = -\frac{1}{T_H} \Rightarrow \text{As } T_C \text{ decreases, } \eta \text{ increases.}$$

Quantitative evaluations also can be had using Eq. 5.8,

$$\eta = 1 - \frac{T_C}{T_H}$$

If T_C is decreased by ϵ degrees, the thermal efficiency is increased:

$$\eta_{(-)} = 1 - \frac{(T_C - \epsilon)}{T_H} = \eta + \frac{\epsilon}{T_H} \qquad (1)$$

If T_H is increased by ϵ degrees, the thermal efficiency is increased:

$$\eta_{(+)} = 1 - \frac{T_C}{T_H + \epsilon}$$

$$= \frac{T_H - T_C + \epsilon}{T_H + \epsilon} = \frac{\eta T_H + \epsilon}{T_H + \epsilon} = \frac{\eta + \epsilon/T_H}{1 + \epsilon/T_H} \qquad (2)$$

Combining Eqs. (1), (2)

$$\eta_{(+)} = \frac{\eta_{(-)}}{1 + \epsilon/T_H} \Rightarrow \eta_{(+)} < \eta_{(-)}$$

Accordingly, it would be more beneficial to decrease T_C than increase T_H. The possibility of increasing thermal efficiency by reducing T_C below that of ① the environment is not practical, however, for maintaining T_C below the ambient temperature would require a refrigerator that would have to be supplied work to operate. An increase in thermal efficiency by increasing T_H is limited by the properties of the materials used to fabricate the system undergoing the cycle.

---

1. A discussion of the influence of the temperatures of heat addition and heat rejection on the thermal efficiency of practical power cycles is provided in Sec. 8.2.3.

PROBLEM 5.17

KNOWN: Two reversible power cycles are arranged in series. One cycle rejects energy by heat transfer to a reservoir at T while the other cycle receives energy by heat transfer from the reservoir at T.

FIND: Determine T when (a) the net work of the two cycles is equal, (b) the thermal efficiencies are equal.

SCHEMATIC & GIVEN DATA:

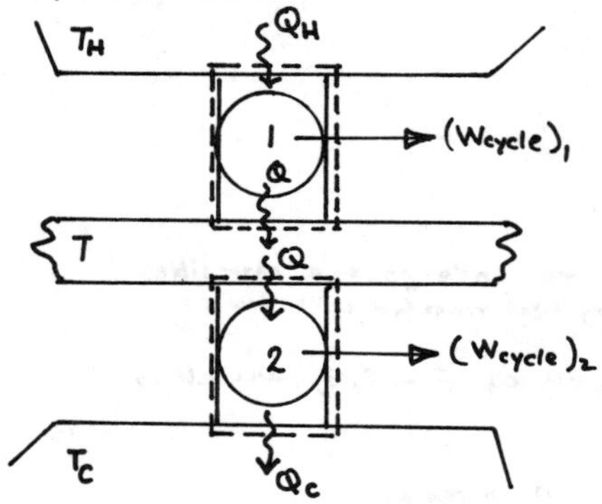

ASSUMPTION: The systems shown in the accompanying figure undergo reversible power cycles between thermal reservoirs.

ANALYSIS:

(a) Equal net work case. An energy balance for each system gives

$$(W_{cycle})_1 = Q_H - Q$$
$$(W_{cycle})_2 = Q - Q_C$$

In accord with the above assumption, Eq. 5.6 gives

$$\frac{Q}{Q_H} = \frac{T}{T_H} \quad , \quad \frac{Q_C}{Q} = \frac{T_C}{T}$$

Setting the work expressions equal and eliminating Q_H and Q_C with the above expressions

$$Q_H - Q = Q - Q_C$$
$$\frac{T_H}{T} Q - Q = Q - \frac{T_C}{T} Q$$
$$T = \frac{T_H + T_C}{2} \longleftarrow T$$

(b) Equal thermal efficiency case. In accord with the above assumption, Eq. 5.8 applies, giving

$$\eta_1 = 1 - \frac{T}{T_H}$$
$$\eta_2 = 1 - \frac{T_C}{T}$$

Setting the efficiencies equal and solving

$$T = \sqrt{T_C T_H} \longleftarrow T$$

PROBLEM 5.18

KNOWN: Systems undergo reversible power, refrigeration, and heat pump cycles while operating between the same two thermal reservoirs.

FIND: Obtain expressions for the coefficients of performance of the refrigeration and heat pump cycles in terms of η_{max}, the thermal efficiency of the power cycle.

SCHEMATIC & GIVEN DATA:

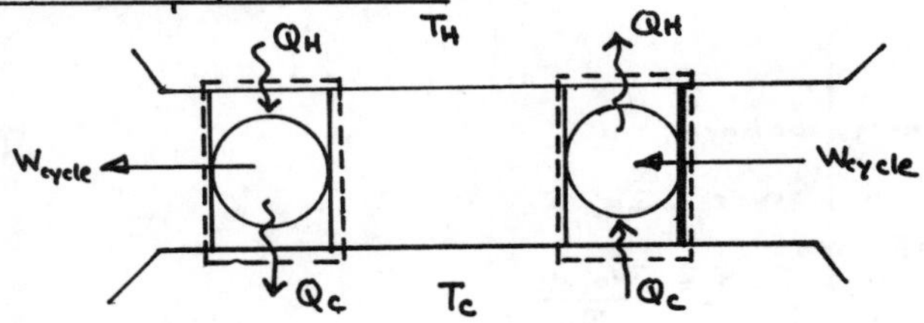

ASSUMPTION: The systems shown in the accompanying figure undergo reversible cycles.

ANALYSIS: In accord with the above assumption, Eqs. 5.8, 5.9, and 5.10 are applicable. For the power cycle

$$\eta_{max} = 1 - \frac{T_c}{T_H}$$

Thus

$$\frac{T_c}{T_H} = 1 - \eta_{max}$$

For the refrigeration cycle

$$\beta_{max} = \frac{T_c}{T_H - T_c} = \frac{T_c/T_H}{1 - T_c/T_H}$$

Accordingly

$$\beta_{max} = \frac{1 - \eta_{max}}{1 - (1 - \eta_{max})}$$

$$= \frac{1 - \eta_{max}}{\eta_{max}} \longleftarrow \beta_{max}$$

For the heat pump cycle

$$\gamma_{max} = \frac{T_H}{T_H - T_c} = \frac{1}{1 - T_c/T_H}$$

Accordingly

$$\gamma_{max} = \frac{1}{1 - (1 - \eta_{max})}$$

$$= \frac{1}{\eta_{max}} \longleftarrow \gamma_{max}$$

PROBLEM 5.19

KNOWN: Operating data are provided for a system undergoing a power cycle while receiving and discharging energy by heat transfer with two thermal reservoirs at specified temperatures.

FIND: For each of three sets of data, determine if any principles of thermodynamics are violated.

SCHEMATIC & GIVEN DATA:

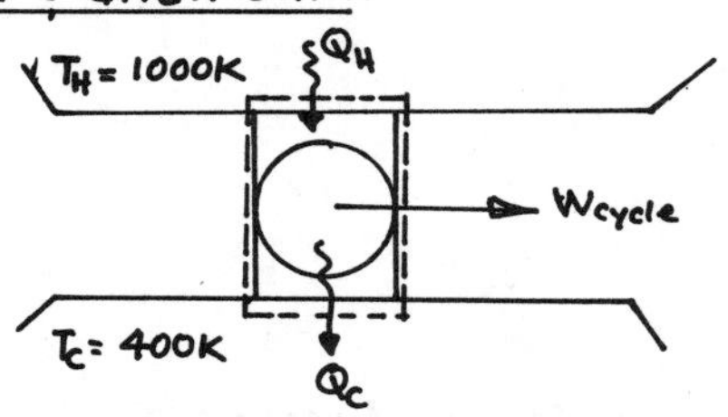

Energy balance:

$$W_{cycle} = Q_H - Q_C$$

Definition:

$$\eta = \frac{W_{cycle}}{Q_H}$$

ASSUMPTION: The system shown in the accompanying figure undergoes a power cycle.

ANALYSIS:

(a) $Q_H = 600\,kJ$, $W_{cycle} = 200\,kJ$, $Q_C = 400\,kJ$. These values satisfy the energy balance, and give

$$\eta = \frac{W_{cycle}}{Q_H} = \frac{200\,kJ}{600\,kJ} = 0.33\ (33\%).$$

The maximum thermal efficiency for any power cycle under the stated conditions is given by Eq. 5.8:

$$\eta_{MAX} = 1 - \frac{T_C}{T_H} = 1 - \frac{400K}{1000K} = 0.6\ (60\%)$$

Thus the second law is also satisfied. There is no apparent violation of the first and second laws.

(b) $Q_H = 400\,kJ$, $W_{cycle} = 240\,kJ$, $Q_C = 160\,kJ$. These values satisfy the energy balance, and give

$$\eta = \frac{240\,kJ}{400\,kJ} = 0.6\ (60\%)$$

As $\eta = \eta_{MAX}$, this power cycle must be reversible. There is no apparent violation of the first and second laws.

(c) $Q_H = 400\,kJ$, $W_{cycle} = 210\,kJ$, $Q_C = 180\,kJ$. With these values for Q_H and Q_C the energy balance gives

$$W_{cycle} = Q_H - Q_C = 400 - 180 = 220\,kJ$$

which does not agree with the given value. Accordingly, the conservation of energy principle is not satisfied. There is no need to consider the second law under such circumstances.

PROBLEM 5.20

KNOWN: Operating data are provided for a system undergoing a power cycle while receiving and discharging energy by heat tranfer with two thermal reservoirs at specified temperatures

FIND: For each of four sets of data, determine if the cycle is reversible, irreversible, or impossible.

SCHEMATIC & GIVEN DATA:

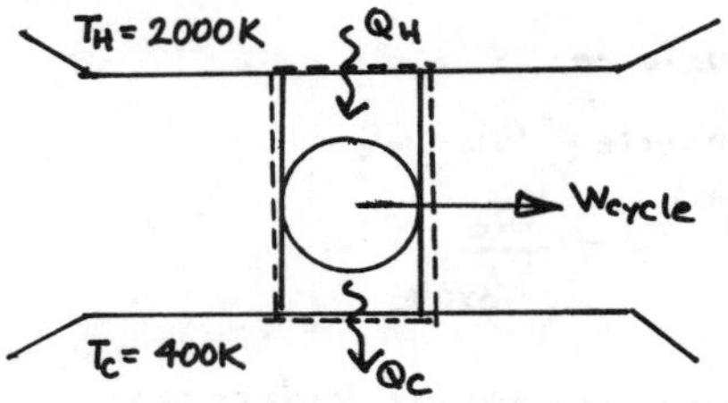

Energy balance:

$$W_{cycle} = Q_H - Q_C$$

Definition:

$$\eta = \frac{W_{cycle}}{Q_H}$$

ASSUMPTION: The system shown in the accompanying figure undergoes a power cycle.

ANALYSIS:

(a) $Q_H = 1200\text{ kJ}$, $W_{cycle} = 1020\text{ kJ}$. The maximum thermal efficiency for any power cycle under the stated conditions is given by Eq. 5.8:

$$\eta_{MAX} = 1 - \frac{T_C}{T_H} = 1 - \frac{400\text{K}}{2000\text{K}} = 0.8\ (80\%)$$

Using the given data

$$\eta = \frac{W_{cycle}}{Q_H} = \frac{1020\text{ kJ}}{1200\text{ kJ}} = 0.85\ (85\%)$$

Since $\eta > \eta_{MAX}$, this cycle is impossible. ◄ (a)

(b) $Q_H = 1200\text{ kJ}$, $Q_C = 240\text{ kJ}$. In this case, it is convenient to use the thermal efficiency in the following form:

$$\eta = 1 - \frac{Q_C}{Q_H} = 1 - \frac{240}{1200} = 0.8\ (80\%)$$

Since $\eta = \eta_{MAX}$, this cycle is reversible. ◄ (b)

(c) $W_{cycle} = 1400\text{ kJ}$, $Q_C = 600\text{ kJ}$. Expressing the thermal efficiency in terms of given quantities,

$$\eta = \frac{W_{cycle}}{W_{cycle} + Q_C} = \frac{1400}{1400 + 600} = 0.7\ (70\%)$$

Since $\eta < \eta_{MAX}$, this cycle is irreversible. ◄ (c)

(d) $\eta = 40\%$. Since $\eta < \eta_{MAX}$, this cycle is irreversible. ◄ (d)

PROBLEM 5.21

KNOWN: Operating data are provided for a system undergoing a refrigeration cycle while receiving and discharging energy by heat transfer with two thermal reservoirs at specified temperatures.

FIND: For each of four sets of data determine if the cycle is reversible, irreversible, or impossible.

SCHEMATIC & GIVEN DATA:

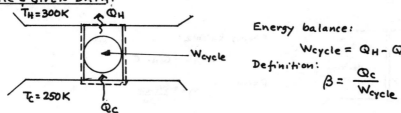

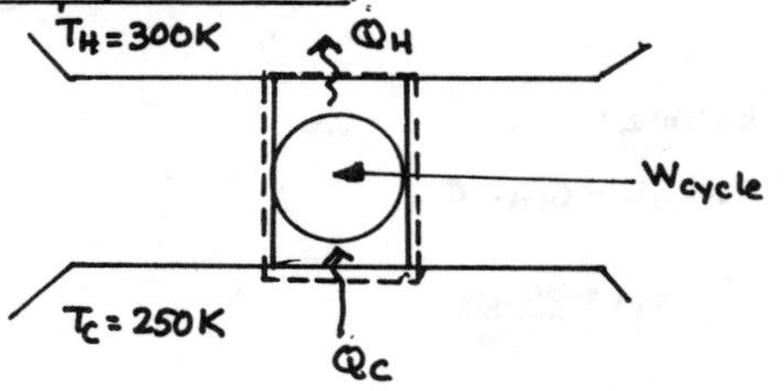

Energy balance:

$$W_{cycle} = Q_H - Q_C$$

Definition:

$$\beta = \frac{Q_C}{W_{cycle}}$$

ASSUMPTION: The system shown in the accompanying figure undergoes a refrigeration cycle.

ANALYSIS:

(a) $Q_C = 1000\ kJ,\ W_{cycle} = 400\ kJ$. The coefficient of performance is

$$\beta = \frac{Q_C}{W_{cycle}} = \frac{1000\ kJ}{400\ kJ} = 2.5$$

The maximum coefficient of performance under the stated conditions is given by Eq. 5.9:

$$\beta_{MAX} = \frac{T_C}{T_H - T_C} = \frac{250\ K}{(300-250)K} = 5.0$$

Accordingly, as $\beta < \beta_{max}$, the cycle is irreversible. ← (a)

(b) $Q_C = 1500\ kJ,\ Q_H = 1800\ kJ$. With the energy balance, the coefficient of performance can be written as

$$\beta = \frac{Q_C}{Q_H - Q_C} = \frac{1500\ kJ}{(1800-1500)\ kJ} = 5$$

As $\beta = \beta_{max}$, this cycle is reversible. ← (b)

(c) $Q_H = 1500\ kJ,\ W_{cycle} = 200\ kJ$. With the energy balance, the coefficient of performance can be written as

$$\beta = \frac{Q_H - W_{cycle}}{W_{cycle}} = \frac{1500-200}{200} = 6.5$$

As $\beta > \beta_{max}$, this cycle is impossible ← (c)

(d) $\beta = 4$. As $\beta < \beta_{max}$, this cycle is irreversible. ← (d)

PROBLEM 5.22

KNOWN: A reversible power cycle receiving Q_H from a hot reservoir drives a refrigeration cycle that removes Q_C from a cold reservoir.

FIND: Develop an expression for Q_C/Q_H in terms of T_H/T_0, T_C/T_0. Plot Q_C/Q_H versus T_H/T_0 and T_C/T_0, each for specified values of the other temperature ratio.

SCHEMATIC & GIVEN DATA:

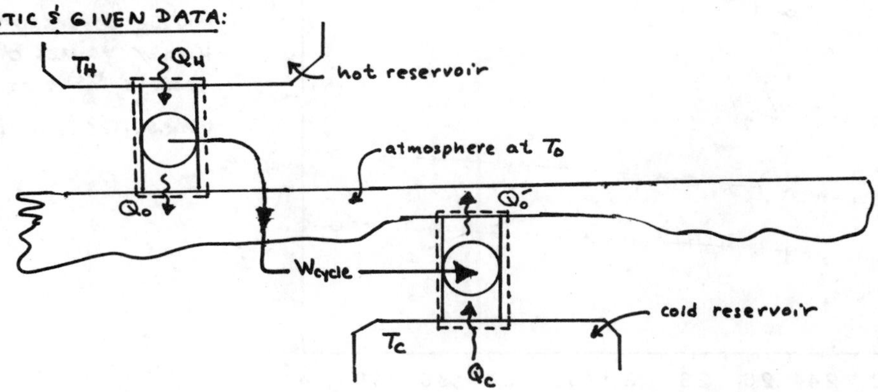

ASSUMPTIONS: (1) Each of the systems shown in the accompany figure undergo reversible cycles. (2) The atmosphere plays the role of a thermal reservoir.

ANALYSIS: (a) An energy balance for the power cycle gives

$$W_{cycle} = Q_H - Q_0$$

An energy balance for the refrigeration cycle gives

$$W_{cycle} = Q_0' - Q_C$$

Combining the last two expressions

$$Q_H - Q_0 = Q_0' - Q_C \qquad (1)$$

As each of the systems undergo reversible cycles between thermal reservoirs, Eq. 5.6 is applicable for each. That is

$$\frac{Q_0}{Q_H} = \frac{T_0}{T_H} \Rightarrow Q_0 = \frac{T_0}{T_H} Q_H \qquad (2)$$

$$\frac{Q_C}{Q_0'} = \frac{T_C}{T_0} \Rightarrow Q_0' = \frac{T_0}{T_C} Q_C \qquad (3)$$

Substituting Eqs. (2) and (3) into Eq. (1)

$$Q_H\left[1 - \frac{T_0}{T_H}\right] = Q_C\left[\frac{T_0}{T_C} - 1\right]$$

Solving for Q_C/Q_H

$$\frac{Q_C}{Q_H} = \frac{\left[1 - \frac{T_0}{T_H}\right]}{\left[\frac{T_0}{T_C} - 1\right]} = \frac{T_C}{T_H}\left[\frac{T_H - T_0}{T_0 - T_C}\right]$$

$$= \frac{(T_C/T_0)\left[\frac{T_H}{T_0} - 1\right]}{(T_H/T_0)\left[1 - \frac{T_C}{T_0}\right]} \longleftarrow$$

Sample calculations: when $T_H/T_0 = 2$, $T_C/T_0 = .9$, $Q_C/Q_H = 4.5$. When $T_H/T_0 = 4$, $T_C/T_0 = .9$, $Q_C/Q_H = 6.75$.

PROBLEM 5.22 (Contd.)

(b) Plots:

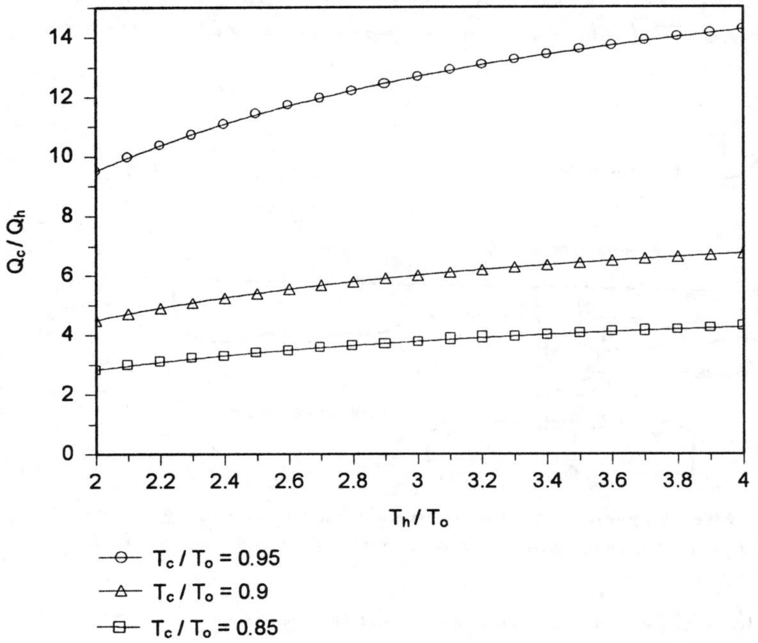

Discussion:
When T_c/T_o is specified, larger values of the ratio Q_c/Q_h correspond to larger values of T_H/T_o. That is, $\frac{Q_c}{Q_h} \uparrow$ as $\frac{T_H}{T_o} \uparrow$ when T_c/T_o is fixed.

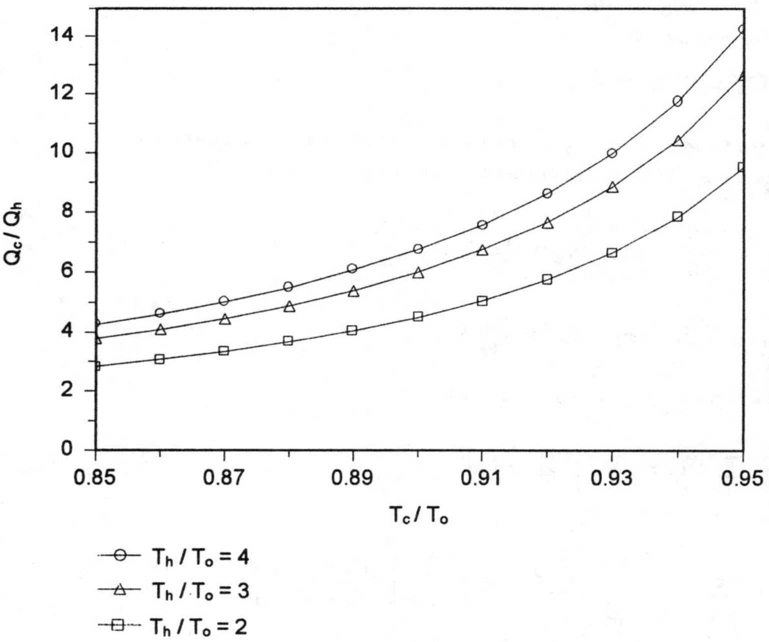

Discussion:
When T_H/T_o is specified, larger values of the ratio Q_c/Q_H correspond to larger values of T_c/T_o, which is limited by a value of unity. That is, $\frac{Q_c}{Q_H} \uparrow$ as $\frac{T_c}{T_o} \rightarrow 1.0$ when T_H/T_o is fixed.

Also note that for a fixed value of the ratio Q_c/Q_H – e.g. a value of 5 – the required value of T_H increases as T_c decreases.

PROBLEM 5.23

KNOWN: A reversible power cycle receiving Q_H from a hot reservoir at T_H drives a heat pump that discharges Q_H' to a reservoir at T_H'.

FIND: Develop an expression for Q_H'/Q_H in terms of T_H, T_C, T_H', T_C', and determine the relationship among the temperatures for the ratio to exceed unity. If $T_H' = T_C = T_0$, plot Q_H'/Q_H versus T_H/T_0 and T_C'/T_0, each for selected values of the other temperature ratio.

SCHEMATIC & GIVEN DATA:

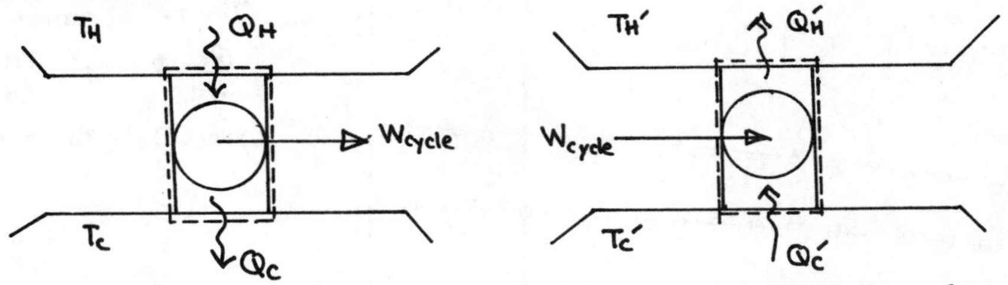

ASSUMPTION: Each of the systems shown in the accompanying figure undergo reversible cycles.

ANALYSIS: (a) An energy balance on the power cycle gives

$$W_{cycle} = Q_H - Q_C$$

An energy balance for the heat pump cycle gives

$$W_{cycle} = Q_H' - Q_C'$$

Combining

$$Q_H - Q_C = Q_H' - Q_C' \qquad (1)$$

As each of the systems undergo reversible cycles between thermal reservoirs, Eq. 5.6 is applicable for each. That is

$$\frac{Q_C}{Q_H} = \frac{T_C}{T_H} \Rightarrow Q_C = \frac{T_C}{T_H} Q_H \qquad (2)$$

$$\frac{Q_C'}{Q_H'} = \frac{T_C'}{T_H'} \Rightarrow Q_C' = \frac{T_C'}{T_H'} Q_H' \qquad (3)$$

Substituting Eqs. (2) and (3) into Eq. (1)

$$Q_H \left[1 - \frac{T_C}{T_H}\right] = Q_H' \left[1 - \frac{T_C'}{T_H'}\right]$$

Solving for Q_H'/Q_H

$$\frac{Q_H'}{Q_H} = \frac{\left[1 - \frac{T_C}{T_H}\right]}{\left[1 - \frac{T_C'}{T_H'}\right]} = \frac{T_H'[T_H - T_C]}{T_H[T_H' - T_C']} \longleftarrow \frac{Q_H'}{Q_H}$$

(b) $Q_H'/Q_H > 1$ when

$$\frac{T_C}{T_C'} < \frac{T_H}{T_H'} \longleftarrow$$

(c) If $T_H' = T_C = T_0$,

$$\frac{Q_H'}{Q_H} = \frac{T_0}{T_H}\left[\frac{T_H - T_0}{T_0 - T_C'}\right] \Rightarrow \frac{Q_H'}{Q_H} = \frac{\left[\frac{T_H}{T_0} - 1\right]}{\frac{T_H}{T_0}\left[1 - \frac{T_C'}{T_0}\right]}$$

PROBLEM 5.23 (Contd.)

Sample Calculations : $T_H/T_0 = 2$, $T_c'/T_0 = .9$; $Q_H'/Q_H = 5$. $T_H/T_0 = 4$, $T_c'/T_0 = .9$; $Q_H'/Q_H = 7.5$.

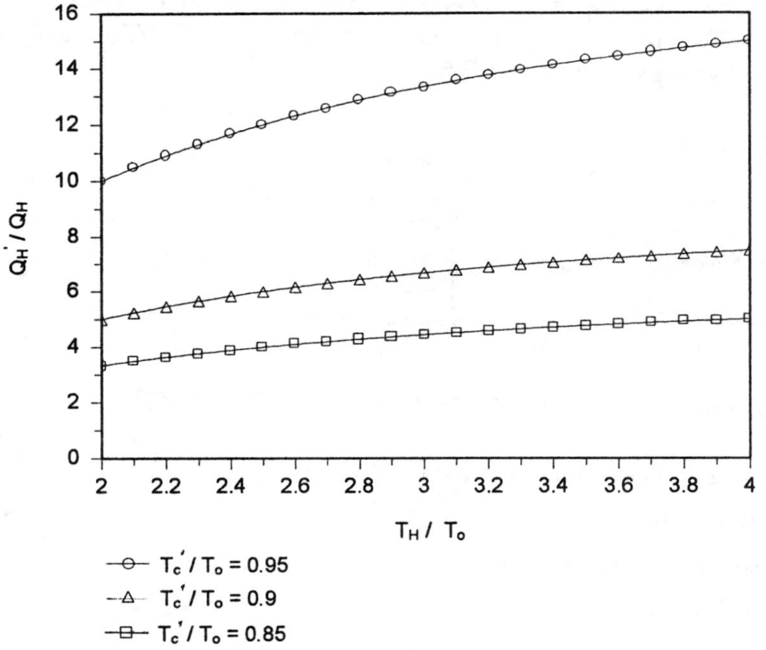

Discussion:

When T_c'/T_0 is fixed, the ratio Q_H'/Q_H increases as T_H/T_0 increases. That is,

$\frac{Q_H'}{Q_H} \uparrow$ as $\frac{T_H}{T_0} \uparrow$

when T_c'/T_0 is fixed.

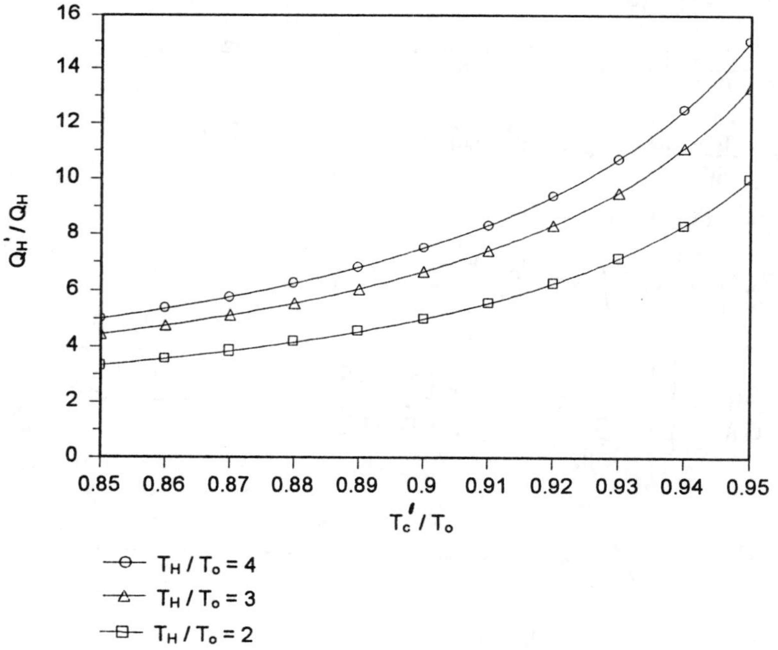

Discussion:

When T_H/T_0 is fixed, the ratio Q_H'/Q_H increases as T_c'/T_0 increases, where T_c'/T_0 is limited by a value of unity. That is,

$\frac{Q_H'}{Q_H} \uparrow$ as $\frac{T_c'}{T_0} \rightarrow 1.0$

When T_H/T_0 is fixed.

Also, note that for a fixed value of the ratio Q_H'/Q_H – e.g., a value of 5.0 – the required value of T_H increases as T_c decreases.

PROBLEM 5.24

KNOWN: Operating conditions specified for a system consisting of a power cycle driving a heat pump are shown on the schematic.

FIND: Obtain an expression for the maximum theoretical value of the ratio $(\dot{Q}_1+\dot{Q}_2)/\dot{Q}_s$ in terms of T_s/T_d and T_0/T_d. Plot the ratio $(\dot{Q}_1+\dot{Q}_2)/\dot{Q}_s$ versus (T_s/T_d) for specified values of T_0/T_d.

SCHEMATIC & GIVEN DATA:

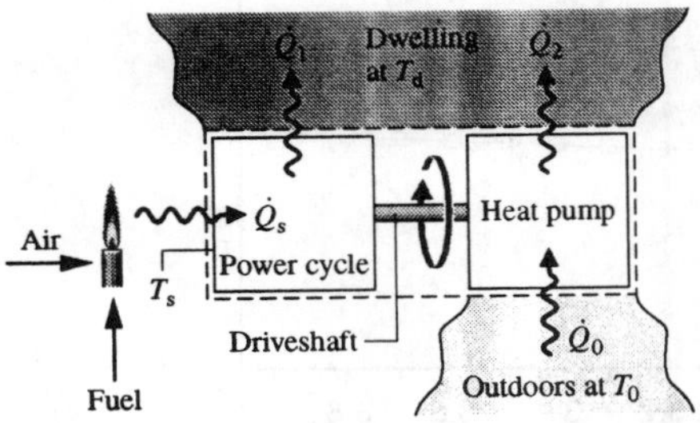

ASSUMPTIONS: 1. The system shown in the schematic is at steady state. 2. The maximum theoretical value for $(\dot{Q}_1+\dot{Q}_2)/\dot{Q}_s$ is achieved in the absence of internal irreversibilities. ①

ANALYSIS: (a) An energy rate balance for the system reduces to $\cancel{dE/dt}^{0} = \dot{Q} - \cancel{\dot{W}}^{0}$, or

$$0 = \dot{Q}_s + \dot{Q}_0 - \dot{Q}_1 - \dot{Q}_2 \Rightarrow \dot{Q}_0 = \dot{Q}_1 + \dot{Q}_2 - \dot{Q}_s \qquad (1)$$

With assumption 2, Eq. 5.6 can be invoked to write

$$\text{power cycle: } \frac{\dot{Q}_1}{T_d} = \frac{\dot{Q}_s}{T_s}, \quad \text{heat pump: } \frac{\dot{Q}_2}{T_d} = \frac{\dot{Q}_0}{T_0} \qquad (2)$$

The ratio $(\dot{Q}_1+\dot{Q}_2)/\dot{Q}_s$ can be expressed as follows, with Eqs. (1), (2):

From Eqs. (2), we have

$$\dot{Q}_1 = \frac{T_d}{T_s}\dot{Q}_s \quad \text{and} \quad \dot{Q}_0 = \frac{T_0}{T_d}\dot{Q}_2$$

Inserting these into Eq. (1) and solving for $\dot{Q}_2/\dot{Q}_s$

$$\frac{T_0}{T_d}\dot{Q}_2 = \frac{T_d}{T_s}\dot{Q}_s + \dot{Q}_2 - \dot{Q}_s \Rightarrow \left[\frac{T_0}{T_d} - 1\right]\dot{Q}_2 = \left[\frac{T_d}{T_s} - 1\right]\dot{Q}_s$$

$$\Rightarrow \dot{Q}_2 = \frac{T_d}{T_s}\left[\frac{T_d - T_s}{T_0 - T_d}\right]\dot{Q}_s$$

Then

$$\frac{\dot{Q}_1 + \dot{Q}_2}{\dot{Q}_s} = \frac{\frac{T_d}{T_s}\dot{Q}_s + \frac{T_d}{T_s}\left[\frac{T_d - T_s}{T_0 - T_d}\right]\dot{Q}_s}{\dot{Q}_s} = \frac{T_d}{T_s}\left[1 + \left[\frac{T_d - T_s}{T_0 - T_d}\right]\right]$$

$$= \frac{T_d}{T_s}\left[\frac{T_0 - T_s}{T_0 - T_d}\right]$$

$$= \frac{1}{(T_s/T_d)}\left[\frac{(T_s/T_d) - (T_0/T_d)}{1 - (T_0/T_d)}\right] \longleftarrow$$

Sample calculations:
$T_s/T_d = 2$, $T_0/T_d = 0.85$, $(\dot{Q}_1+\dot{Q}_2)/\dot{Q}_s = 3.83$
$T_s/T_d = 2$, $T_0/T_d = 0.9$, $(\dot{Q}_1+\dot{Q}_2)/\dot{Q}_s = 5.5$
$T_s/T_d = 2$, $T_0/T_d = 0.95$, $(\dot{Q}_1+\dot{Q}_2)/\dot{Q}_s = 10.5$

ROBLEM 5.24 (Contd.)

(b) PLOT

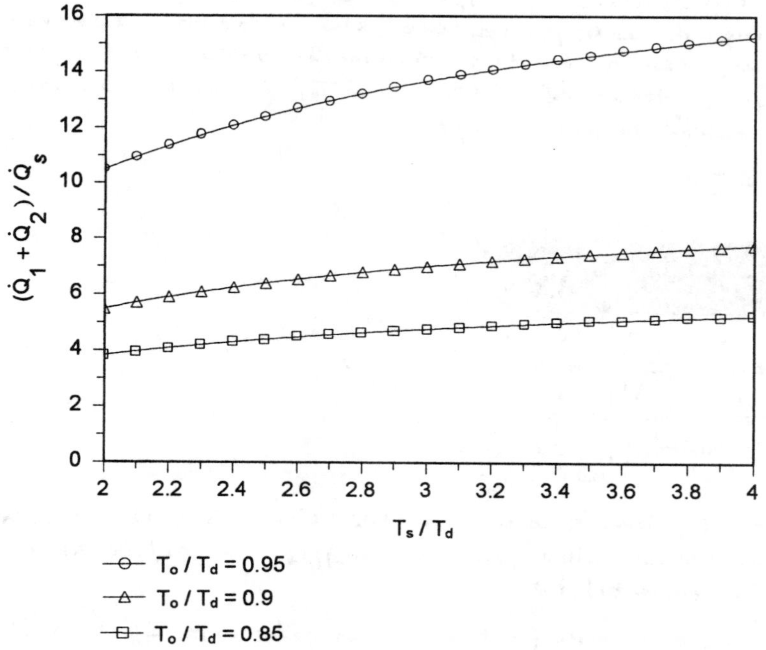

The plot shows that for fixed To/Td, the ratio $(\dot{Q}_1+\dot{Q}_2)/\dot{Q}_s$ increases as Ts/Td increases. Note that To/Td is limited by a value of unity: the dwelling temperature Td would never be less than the outdoors temperature. For example, when Td = 293 K (20°C), the three ratios shown in the plot correspond to outdoor temperatures To = 278 K (5°C), 264 K (−9°C), and 249 K (−24°C), respectively.

---

1. The presence of irreversibilities within the system is expected to exact a penalty. If two systems operating as shown in the schematic each receive $\dot{Q}_s$ from a fuel, and one system has internal irreversibilities and the other does not, it is in accord with intuition that the desirable outcome: $(\dot{Q}_1+\dot{Q}_2)$ would be greater in the ideal case than in the actual case. Using concepts from the present chapter, or from Chaps. 6 and 7 to follow, this conclusion can be demonstrated formally.

PROBLEM 5.25

KNOWN: A system undergoes a reversible cycle while receiving and discharging energy by heat transfer with hot and cold reservoirs.

FIND: Determine the thermal efficiency and the net work developed.

SCHEMATIC & GIVEN DATA:

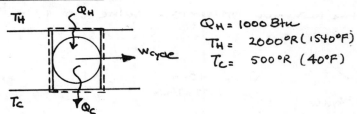

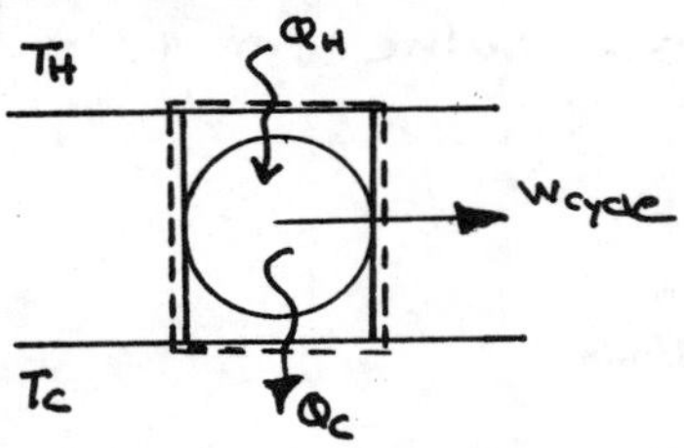

$Q_H = 1000$ Btu
$T_H = 2000°R\ (1540°F)$
$T_C = 500°R\ (40°F)$

ASSUMPTIONS: (1) The system is shown on the accompanying figure. (2) The system undergoes a reversible cycle while communicating thermally with the hot and cold reservoirs only.

ANALYSIS: With assumption 2, Eq. 5.6 is applicable, giving

$$\left(\frac{Q_C}{1000\ \text{Btu}}\right)_{\substack{rev\\cycle}} = \frac{500°R}{2000°R} \Rightarrow Q_C = 250\ \text{Btu}$$

For any thermodynamic cycle, the net work developed equals the net heat added:

$$W_{cycle} = Q_H - Q_C$$
$$= 1000 - 250 = 750\ \text{kJ} \longleftarrow W_{cycle}$$

For any power cycle the thermal efficiency is

① $$\eta = \frac{W_{cycle}}{Q_H} = \frac{750}{1000} = 0.75\ (75\%) \longleftarrow \eta$$

---

1. Alternatively, as the power cycle is reversible the thermal efficiency can be calculated using Eq. 5.8:

$$\eta = 1 - \frac{T_C}{T_H} = 1 - \frac{500}{2000} = 0.75\ (75\%)$$

Then,

$$W_{cycle} = \eta Q_H = (0.75)(1000\ \text{kJ}) = 750\ \text{kJ}$$

PROBLEM 5.26

KNOWN: Steady state operating data are provided for a system undergoing a power cycle while receiving and discharging energy by heat transfer with a hot and cold reservoir at temperatures T and 280 K, respectively

FIND: Determine the minimum theoretical value for T, in K.

SCHEMATIC & GIVEN DATA:

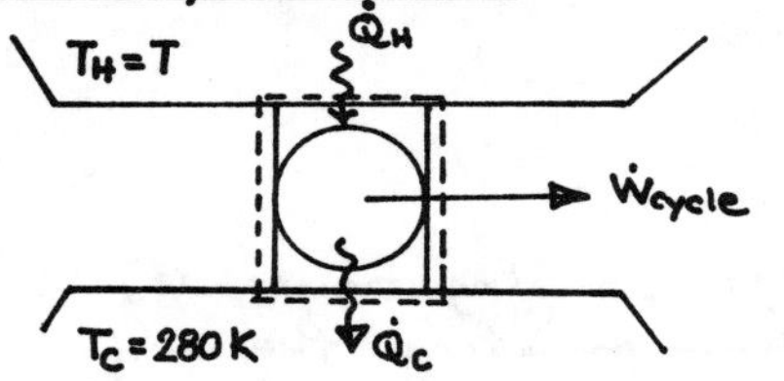

$\dot{W}_{cycle} = 40 kW$
$\dot{Q}_C = 1000 kJ/min$

ASSUMPTIONS: (1) The system is shown on the accompanying figure. (2) The system undergoes a power cycle. (3) The data provided is for operation at steady state.

ANALYSIS: An energy rate balance gives

① $$\dot{W}_{cycle} = \dot{Q}_H - \dot{Q}_C$$

or

$$\dot{Q}_H = \dot{W}_{cycle} + \dot{Q}_C$$
$$= 40 kW\left(\frac{1 kJ/s}{1 kW}\right) + 1000 \frac{kJ}{min}\left(\frac{min}{60 s}\right)$$
$$= 56.67 kJ/s$$

We know that $\eta \leq \eta_{max}$; that is

$$\eta = \frac{\dot{W}_{cycle}}{\dot{Q}_H} \leq 1 - \frac{T_C}{T}$$

Inserting values

$$\frac{40 kJ/s}{56.67 kJ/s} \leq 1 - \frac{280K}{T}$$

$$\Rightarrow \quad \frac{280}{T} \leq 1 - \frac{40}{56.67} = 0.294$$

$$\Rightarrow \quad \underbrace{952K}_{T_{min}} \leq T \quad \longleftarrow T_{min}$$

---

1. At steady state, the cycle energy balance and thermal efficiency can be expressed in terms of rates:

$$\dot{W}_{cycle} = \dot{Q}_H - \dot{Q}_C$$
$$\eta = \frac{\dot{W}_{cycle}}{\dot{Q}_H}$$

PROBLEM 5.27

KNOWN: Systems undergo reversible power cycles while receiving and discharging energy with thermal reservoirs. The thermal efficiency is the same for hot and cold reservoirs at 1000 and 500 K, respectively, as for hot and cold reservoirs at T and 1000 K.

FIND: Determine the temperature T.

SCHEMATIC & GIVEN DATA:

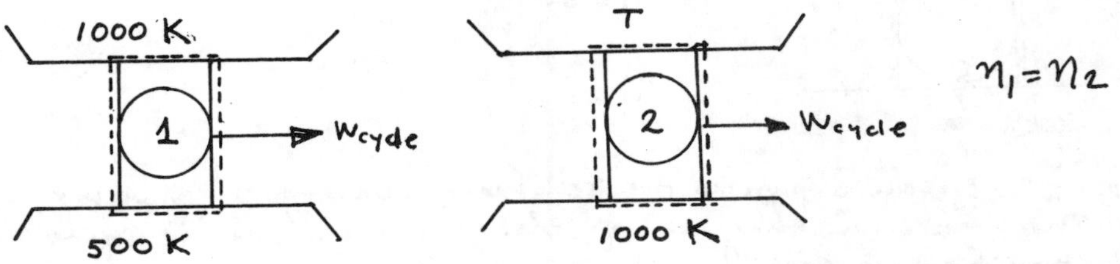

ASSUMPTION: The systems shown in the accompanying figure undergo reversible power cycles having the same thermal efficiency.

ANALYSIS: In accord with the above assumption, Eq. 5.8 is applicable, giving

$$\eta_1 = 1 - \frac{500}{1000}$$

$$\eta_2 = 1 - \frac{1000}{T}$$

Since $\eta_1 = \eta_2$

$$\frac{500}{1000} = \frac{1000}{T} \Rightarrow T = 2000\ K \leftarrow T$$

PROBLEM 5.28

KNOWN: A system undergoes a reversible power cycle whose thermal efficiency is 50% while receiving and discharging energy by heat transfer with reservoirs at temperatures 1800 K (1527°C) and T, respectively.

FIND: Determine T.

SCHEMATIC & GIVEN DATA:

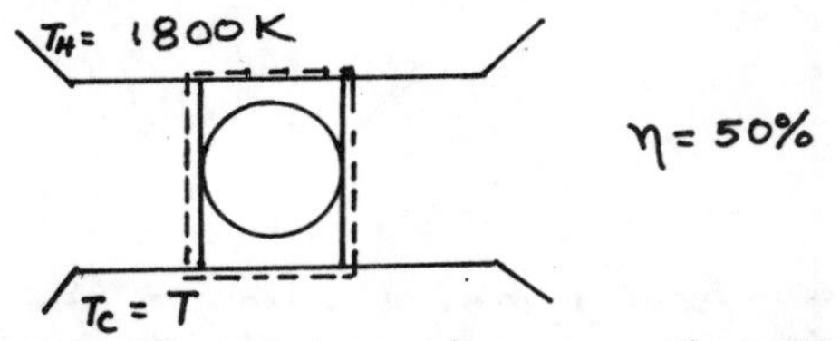

ASSUMPTION: The system shown on the accompanying figure undergoes a reversible power cycle.

ANALYSIS: In accord with the above assumption, Eq. 5.8 is applicable, giving

$$0.5 = 1 - \frac{T}{1800}$$

Solving for T

$$T = 900\ K \leftarrow T$$

PROBLEM 5.29

KNOWN: A system undergoes a cycle while receiving and discharging energy by heat transfer with two reservoirs of known temperatures.

FIND: For each of two specified values of thermal efficiency, determine if the claimed operation is feasible.

SCHEMATIC & GIVEN DATA:

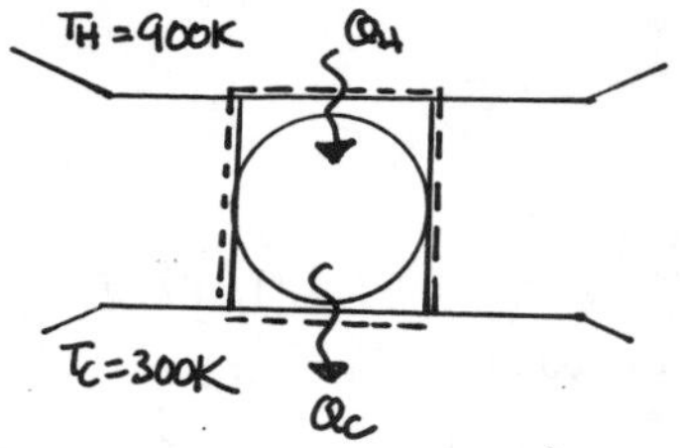

(a) $\eta = 66\%$
(b) $\eta = 50\%$

ASSUMPTION: The system shown in the schematic undergoes a power cycle.

ANALYSIS: The maximum thermal efficiency for any power cycle under the stated conditions is given by Eq. 5.8:

$$\eta_{MAX} = 1 - \frac{T_C}{T_H} = 1 - \frac{300}{900} = 0.667 \quad (66.7\%)$$

(a) $\eta = 66\%$. Since this claimed value is nearly the maximum theoretical value, the cycle must be nearly ideal in operation. Although not ruled out by the second law, it is unlikely for an actual power cycle to perform at such a level.

(b) $\eta = 50\%$. Since the claimed value is less than η_{max}, the cycle is feasible.

PROBLEM 5.30

KNOWN: A system undergoes a power cycle while receiving and discharging energy by heat transfer with two thermal reservoirs. Steady state operating data are provided.

FIND: Determine if the cycle can operate as claimed.

SCHEMATIC & GIVEN DATA:

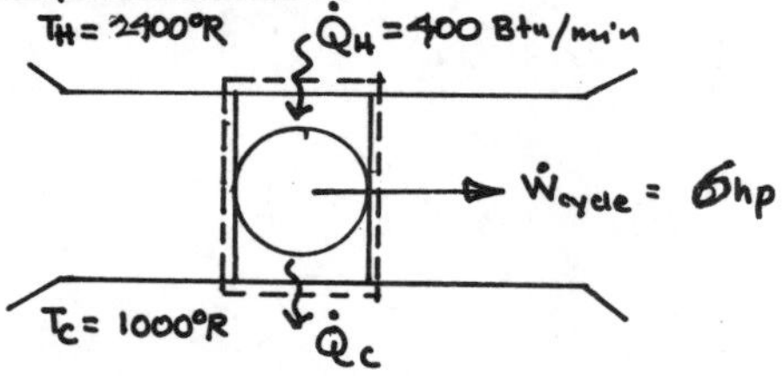

ASSUMPTIONS: (1) The system shown in the accompanying figure undergoes a power cycle. (2) The data provided is for operation at steady-state.

ANALYSIS: Using the given data, the thermal efficiency is

$$\eta = \frac{\dot{W}_{cycle}}{\dot{Q}_H} = \frac{(6\,hp)\left|\frac{2545\ Btu/h}{1\ hp}\right|\left|\frac{1\ h}{60\ min}\right|}{400\ \frac{Btu}{min}} = 0.636 \quad (63.6\%)$$

The maximum thermal efficiency for any power cycle under the stated conditions is given by Eq. 5.8:

$$\eta_{max} = 1 - \frac{1000}{2400} = 0.58 \quad (58\ \%)$$

Since the claimed value of thermal efficiency is greater than η_{MAX}, the claim cannot be valid.

PROBLEM 5.31

KNOWN: A system undergoes a power cycle while receiving and discharging energy by heat transfer from a source at 1500K and to cooling water at 300K, respectively. Steady state operating data are provided.

FIND: Determine the minimum number of cycles/min.

SCHEMATIC & GIVEN DATA:

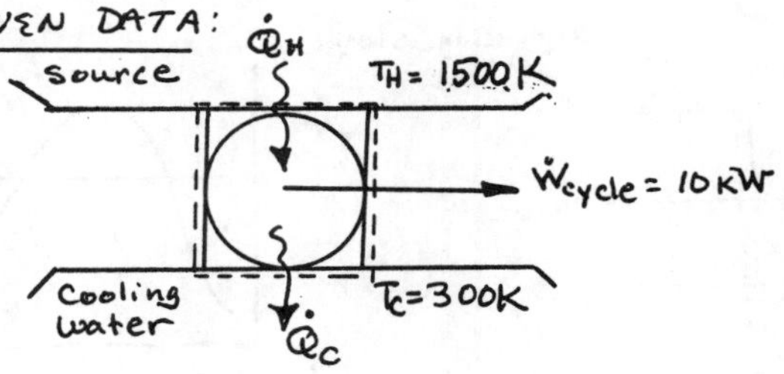

$$\frac{\dot{Q}_H}{n} = 10 \frac{kJ}{cycle}$$

where n = cycles/min

ASSUMPTIONS: (1) The system shown in the accompanying figure undergoes a power cycle. (2) The data provided is for operation at steady state. (3) The source and cooling water play the roles of hot and cold reservoirs, respectively.

ANALYSIS: The power developed must be less than, or equal to, the power that would be developed by a reversible power cycle operating between thermal reservoirs at the specified temperatures:

$$10\,kW \leq \left[1 - \frac{T_C}{T_H}\right]\dot{Q}_H$$

$$\leq \left[1 - \frac{300K}{1500K}\right]\left[10 \frac{kJ}{cycle}\right]\left[n \frac{cycles}{min}\right]\left[\frac{min}{60s}\right]\left[\frac{1\,kW}{1\,kJ/s}\right]$$

$$10\,kW \leq \left[\frac{(0.8)(10)\,n}{60}\right] kW$$

$$75 \leq n$$

Accordingly, the actual device would require 75, or more, cycles per minute.

PROBLEM 5.32

KNOWN: A system undergoes a power cycle while receiving energy by heat transfer from condensing steam and discharging energy by heat transfer to a lake at 70°F. For the cycle, $\eta = 40\%$.

FIND: Determine the lowest possible condensing steam temperature and the corresponding pressure.

SCHEMATIC & GIVEN DATA:

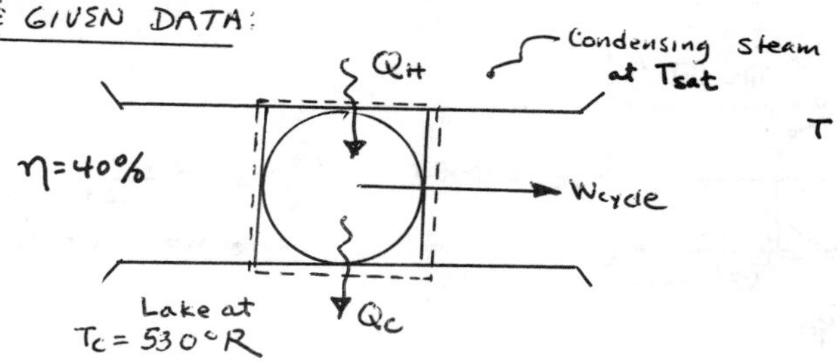

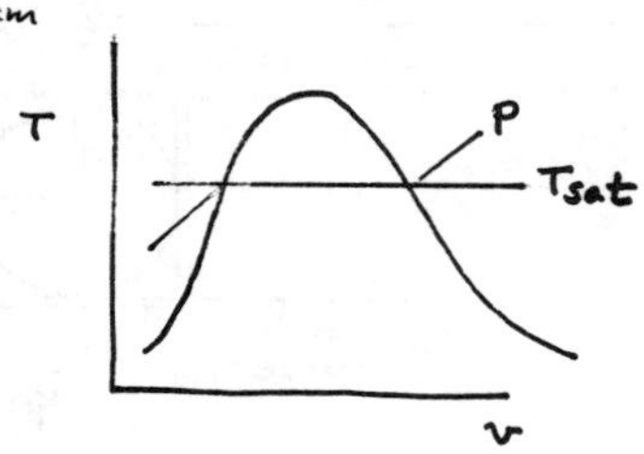

ASSUMPTIONS: (1) The system shown in the accompanying figure undergoes a power cycle. (2) The condensing steam and the lake play the roles of the hot and cold reservoirs, respectively.

ANALYSIS: Since P_{sat} decreases as T_{sat} decreases, the lowest possible condensing steam pressure corresponds to the lowest possible condensing steam temperature. Moreover, the thermal efficiency must be less than, or equal to, the thermal efficiency of a reversible power cycle operating between reservoirs at the specified temperatures:

$$0.40 \leq \left[1 - \frac{T_C}{T_H}\right] = 1 - \frac{530}{T_{sat}}$$

or

$$0.60 \geq \frac{530}{T_{sat}} \Rightarrow T_{sat} \geq 883°R \ (423°F).$$

Thus, the lowest possible saturation temperature is 423 °F. Interpolating in Table A-2E, the corresponding saturation pressure is 319 lbf/in.² ←

PROBLEM 5.33

KNOWN: Steady-state operating data are provided for a system undergoing a power cycle wh receiving and discharging energy by heat transfer at 900 and 70°F, respectively.

FIND: Determine the rate energy is discharged and compare with the minimum theoretical rate.

SCHEMATIC & GIVEN DATA:

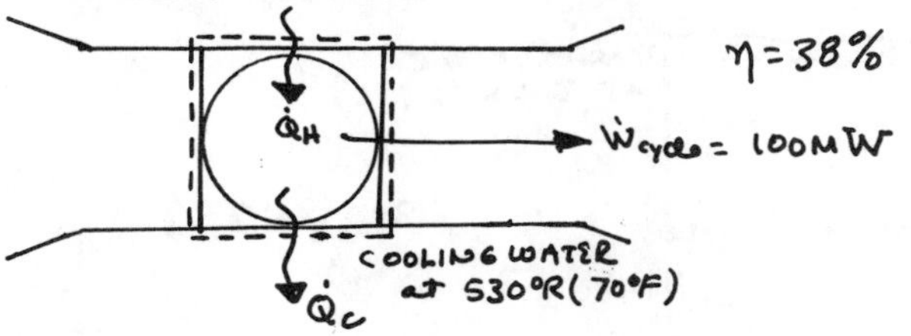

ASSUMPTIONS: 1. The system shown in the schematic undergoes a power cycle. 2. Operation is at steady state. 3. The steam and cooling water play the roles of hot and cold reservoirs, respectively.

ANALYSIS: (a) For any cycle at steady state, $\dot{W}_{cycle} = \dot{Q}_H - \dot{Q}_C$, or $\dot{Q}_c = \dot{Q}_H - \dot{W}_{cycle}$. Also, $\eta = \dot{W}_{cycle}/\dot{Q}_H \Rightarrow \dot{Q}_H = \dot{W}_{cycle}/\eta$. Collecting results

$$\dot{Q}_C = \dot{W}_{cycle}\left[\frac{1}{\eta} - 1\right] = 100 MW\left[\frac{1}{0.38} - 1\right] = 163.2 MW \left| \frac{3.413 \times 10^6 \ Btu/h}{1 MW} \right| = 5.57 \times 10^8 \frac{Btu}{h}$$

(b) The thermal efficiency must be less than, or equal to, the thermal efficiency of a reversible power cycle operating between reservoirs at $T_H = 1360°R$, $T_C = 530°R$:

$$\eta \leq \left[1 - \frac{T_C}{T_H}\right], \text{ where } \eta = \frac{\dot{W}_{cycle}}{\dot{W}_{cycle} + \dot{Q}_C}. \text{ That is}$$

① $$\left(\frac{100 MW}{100 MW + \dot{Q}_C}\right) \leq 1 - \frac{530}{1360} = 0.61 \Rightarrow \frac{100 MW}{0.61} \leq 100 MW + \dot{Q}_C \Rightarrow$$

$$\dot{Q}_C \geq 63.9 MW \left| \frac{3.413 \times 10^6 \ Btu/h}{1 MW} \right| = 2.18 \times 10^8 \frac{Btu}{h}$$

Thus, the minimum theoretical rate at which energy could be discharged to the cooling water is 2.18×10^8 Btu/h.

Comparing the values for $\dot{Q}_C$

$$\frac{\dot{Q}_C}{(\dot{Q}_C)_{MIN}} = \frac{5.57 \times 10^8 \ Btu/h}{2.18 \times 10^8 \ Btu/h} = 2.56$$

Thus, the actual heat discharge rate is over 2½ times the minimum theoretical rate. This illustrates that irreversibilities within the system during each cycle of operation contribute significantly to thermal pollution.

---

1. The actual thermal efficiency, 38%, is considerably less than the maximum value, 61%, owing to the effect of irreversibilities.

PROBLEM 5.34

KNOWN: A power cycle operates between surface ocean water at 27°C and water at a depth of 700 m where the temperature is 7°C.

FIND: Determine the maximum theoretical thermal efficiency for any such cycle and compare with the observed thermal efficiency of OTEC power plants

SCHEMATIC & GIVEN DATA:

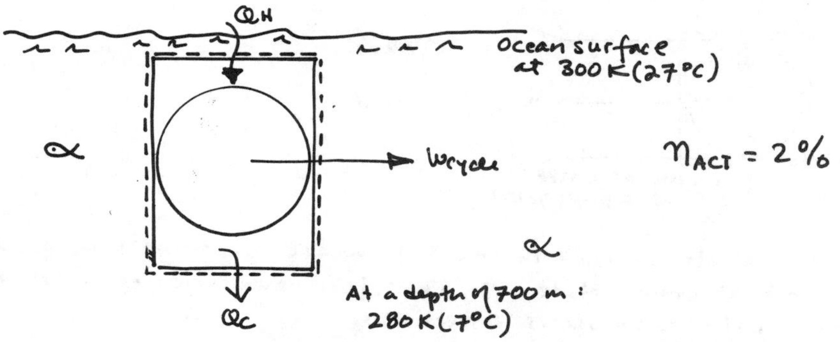

ASSUMPTIONS: 1. The system shown in the schematic undergoes a power cycle. 2. The ocean surface water and water at a depth of 700 m play the roles of hot and cold reservoirs, respectively.

ANALYSIS: (a) The thermal efficiency of any power cycle under these conditions must be less than, or equal to, the thermal efficiency of a reversible power cycle operating between reservoirs at $T_H = 300K$, $T_C = 280K$, respectively. That is

$$\eta \le \left[1 - \frac{T_C}{T_H}\right] = \left[1 - \frac{280}{300}\right] = 0.067 \quad (6.7\%) \Rightarrow \eta_{MAX} = 6.7\% \longleftarrow$$

(b) If $\eta_{act} = 2\%$, then

$$\frac{\eta_{act}}{\eta_{MAX}} = \frac{2\%}{6.7\%} = 0.3$$

That is, for the assumed conditions, actual OTEC cycles operate at 30% of the maximum theoretical value. This indicates that there may be some scope for improving actual performance, by a few percentage points, via conventional engineering measures. Still, the characteristically low value of η_{max} for OTEC plants suggests that power development by such means would not be competitive economically with other types of power plants in use today.

PROBLEM 5.35

KNOWN: Hot water at 167°C is available from an underground source. The local atmosphere is at 13°C.

FIND: Determine the maximum possible thermal efficiency for any power cycle operating between these temperatures.

SCHEMATIC & GIVEN DATA:

PROBLEM 5.35 (Contd.)

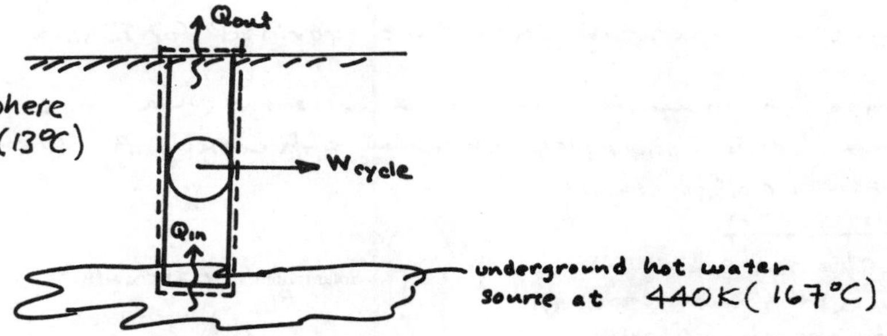

ASSUMPTIONS: (1) The system shown in the accompanying figure undergoes a power cycle. (2) The underground hot water source and the local atmosphere play the roles of the hot and cold reservoirs, respectively.

ANALYSIS: The maximum thermal efficiency corresponds to the case where $\dot{Q}_{in}$ is received from a thermal reservoir at 440K and $\dot{Q}_{out}$ is discharged to a thermal reservoir at 286 K, and the system operates without irreversibilities. That is, applying Eq. 5.8

$$\eta_{MAX} = 1 - \frac{286}{440} = 0.35\ (35\%) \longleftarrow$$

PROBLEM 5.36

KNOWN: A power cycle operates between the ground at a temperature of 55°F and the atmosphere at a temperature of −23°F. A thermal efficiency of 15% is claimed.

FIND: Evaluate this claim.

SCHEMATIC & GIVEN DATA:

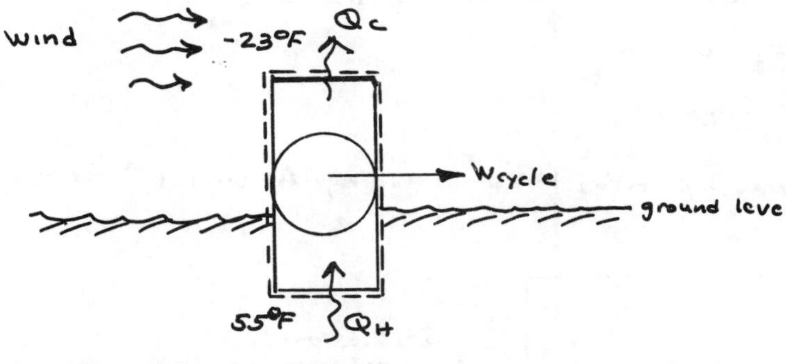

ASSUMPTIONS: (1) The system shown in the figure undergoes a power cycle. (2) The ground at 55°C and the atmosphere at −23°F play the roles of hot and cold reservoirs, respectively.

ANALYSIS: The maximum thermal efficiency any power cycle may have when operating between hot and cold reservoirs at T_H and T_C, respectively, is given by Eq. 5.8:

$$\eta_{MAX} = 1 - \frac{T_C}{T_H}$$

$$= 1 - \frac{(460-23)°R}{(460+55)°R} = 0.151 \quad (15.1\%)$$

The claimed operation is very close to ideal. Although not ruled out by the second law, an actual device is not likely to perform at such a level: Claim is doubtful.

PROBLEM 5.37

KNOWN: Steady-state operating data are provided for a solar-activated power plant.

FIND: Determine the minimum theoretical collector area required. Plot the collector area versus power plant thermal efficiency for specified values of the collector efficiency.

SCHEMATIC & GIVEN DATA:

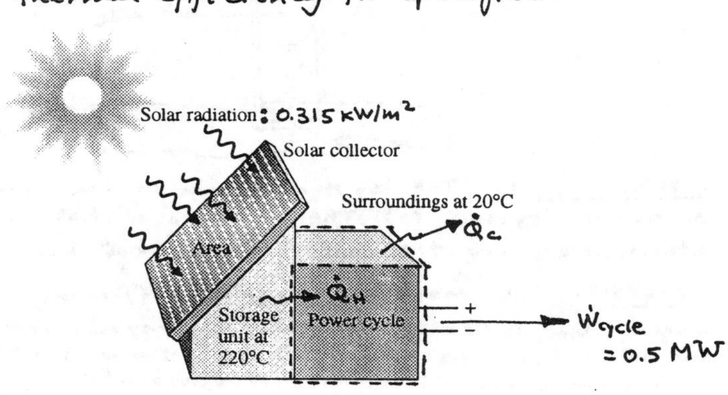

ASSUMPTIONS: 1. The system shown in the schematic is at steady state. All other components are also at steady state. 2. Some of the incident solar energy is lost through unavoidable heat transfer from the solar collector. All other components experience negligible stray heat transfer. 3. The storage unit and surroundings play roles of hot and cold reservoirs, respectively.

ANALYSIS: The rate energy is collected and supplied to the storage unit is evaluated as $(0.315\, A\, e)$ where A is the collector area and e is the collector efficiency. Further, at steady state the rate energy is supplied to the system undergoing the power cycle equals the rate energy enters the storage unit: $\dot{Q}_H = (0.315\, A\, e)$. The thermal efficiency of the power cycle is $\eta = \dot{W}_{cycle}/\dot{Q}_H$. Combining these expressions and solving for A

$$A = \frac{\dot{W}_{cycle}}{0.315\eta e} = \frac{(0.5\,\text{MW})\left|\frac{10^3\,\text{kW}}{1\,\text{MW}}\right|}{\left(0.315\frac{\text{kW}}{\text{m}^2}\right)\eta e} = \frac{1587\,\text{m}^2}{\eta e} \qquad (1)$$

(a) By inspection of Eq. (1), the collector area A is a minimum when η and e take on their respective maximum values: e = 1.0 and

$$\eta_{MAX} = 1 - \frac{T_C}{T_H} = 1 - \frac{293\,K}{493\,K} = 0.406.$$

Thus,

① $$A_{min} = \frac{1587\,m^2}{(0.406)(1)} = 3909\,m^2$$

(b) PLOT: For $0 < \eta \le \eta_{MAX} = 0.406$, the plot giving A versus η for e = 1.0, 0.75, and 0.5 is

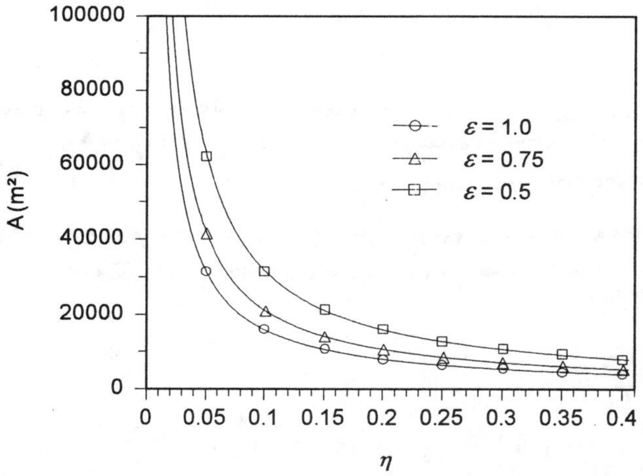

Discussion:
1. The collector area required decreases as η approaches η_{max}, as expected.
2. For fixed η, the collector area decreases as ϵ increases, also as expected.

---

1. For comparison, the area of a football field is about $4350 m^2$.

PROBLEM 5.38

KNOWN: Operating data are provided for a power plant that discharges energy by heat transfer to a river. For environmental reasons, the temperature of the river downstream of the power plant is restricted.

FIND: Estimate the maximum theoretical power that can be developed subject to the restriction on river temperature.

SCHEMATIC & GIVEN DATA

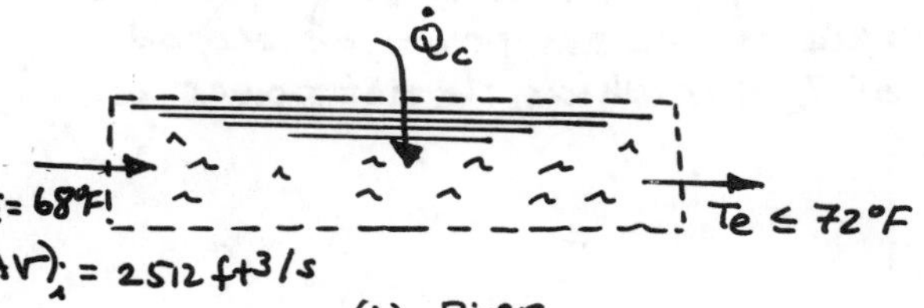

(A) River

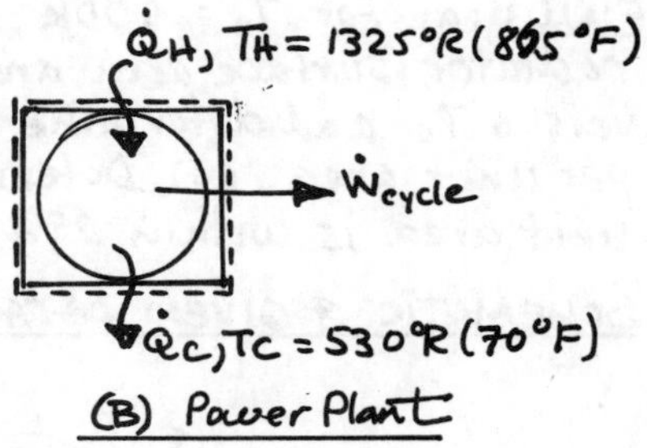

(B) Power Plant

ASSUMPTIONS: 1. The two systems shown in the schematic are at steady state. In each case, $\dot{Q}_C$ is in the direction of the arrow. 2. For system (A), effects of kinetic and potential energy can be ignored. 3. System (B) undergoes a power cycle that receives energy from a reservoir at T_H and discharges energy to a reservoir at T_C, which equals the average river temperature.

ANALYSIS: An energy rate balance for control volume (A) reduces to give

$$\dot{Q}_c = \dot{m}[h_e - h_i] = \frac{(AV)_i}{v_i}[h_e - h_i].$$

Then, with $v_i = v_f(T_i) = 0.01605\ ft^3/lb$, $h_i = h_f(T_i) = 36.09\ Btu/lb$, and $h_e \le h_f(72°F) = 40.09\ Btu/lb$ from Table A-2E

$$\dot{Q}_c \le \left(\frac{2512\ ft^3/s}{0.01605\ ft^3/lb}\right)(40.09 - 36.09)\left(\frac{Btu}{lb}\right) = 6.26 \times 10^5\ Btu/s$$

Thus, to meet the environmental constraint, the power cycle cannot discharge energy to the river at a rate greater than $(\dot{Q}_c)_{MAX} = 6.26 \times 10^5\ Btu/s$.

Turning to system (B), an energy balance reduces to $\dot{Q}_H = \dot{W}_{cycle} + \dot{Q}_c$. Moreover, the thermal efficiency must be less than, or equal to, the thermal efficiency of a reversible power cycle operating between reservoirs at T_H, T_C: $\eta \le [1 - \frac{T_C}{T_H}]$, where $\eta = \dot{W}_{cycle}/\dot{Q}_H$. Collecting results

$$\frac{\dot{W}_{cycle}}{\dot{W}_{cycle} + \dot{Q}_c} \le \left[1 - \frac{T_C}{T_H}\right] \Rightarrow \dot{W}_{cycle} \le \left[1 - \frac{T_C}{T_H}\right](\dot{W}_{cycle} + \dot{Q}_c)$$

$$\Rightarrow \quad \dot{W}_{cycle} \le \left[\frac{T_H}{T_C} - 1\right](\dot{Q}_c) \rightarrow \dot{W}_{cycle} \le \left[\frac{T_H}{T_C} - 1\right](\dot{Q}_c)_{MAX} \qquad (1)$$

Inserting values, Eq. (1) gives

$$\dot{W}_{cycle} \le \left[\frac{1325°R}{530°R} - 1\right]\left(6.26 \times 10^5\ \frac{Btu}{s}\right) = (1.5)\left(6.26 \times 10^5\ \frac{Btu}{s}\right) = 9.39 \times 10^5\ \frac{Btu}{s}$$

$$\le \left(9.39 \times 10^5\ \frac{Btu}{s}\right)\left|\frac{3600\ s}{1\ h}\right|\left|\frac{1\ MW}{3.413 \times 10^6\ Btu/h}\right| = 990.4\ MW$$

or

① $$(\dot{W}_{cycle})_{\substack{MAX \\ TH2O}} = 990.4\ MW$$

---

1. Owing to the effect of irreversibilities, an actual power plant would develop much less power than the maximum theoretical value determined here.

PROBLEM 5.39

KNOWN: A power cycle receives energy by heat transfer at a specified temperature and rejects energy by heat transfer at temperature T_c. The thermal efficiency is one-half that of a reversible cycle operating between reservoirs at the respective temperatures.

FIND: (a) For T_c = 400K, determine the net power developed per unit of radiator surface area and the thermal efficiency. (b) Plot these quantities versus T_c and determine the maximum value of the net power developed per unit area. (c) Determine the range of T_c for which the net power per unit area is within 2% of the maximum.

SCHEMATIC & GIVEN DATA:

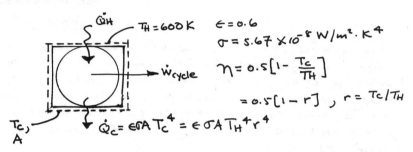

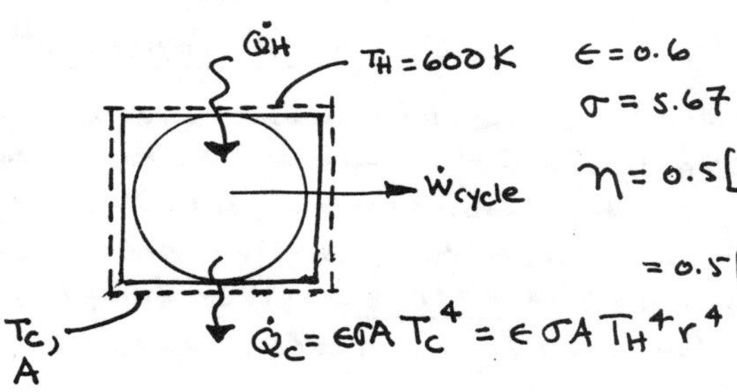

$\epsilon = 0.6$

$\sigma = 5.67 \times 10^{-8}\ W/m^2 \cdot K^4$

$\eta = 0.5\left[1 - \frac{T_c}{T_H}\right]$

$= 0.5[1 - r]$, $r = T_c/T_H$

ASSUMPTIONS: (1) The system shown in the schematic undergoes a power cycle and operates at steady state. (2) Heat transfer at T_c is only by thermal radiation, as indicated on the schematic. The surface receives no radiation. (3) For the power cycle, $\eta = 0.5\,\eta_{max}$.

ANALYSIS: The thermal efficiency of a power cycle operating at steady state is $\eta = \dot{W}_{cycle}/\dot{Q}_H$. Introducing the energy rate balance, $\dot{W}_{cycle} = \dot{Q}_H - \dot{Q}_c$

$$\eta = \frac{\dot{W}_{cycle}}{\dot{W}_{cycle} + \dot{Q}_c} \Rightarrow \dot{W}_{cycle} = \left(\frac{\eta}{1-\eta}\right)\dot{Q}_c$$

Now, from above $\eta = 0.5(1-r)$ and $\dot{Q}_c = \epsilon\sigma A T_H^4 r^4$. Thus

$$\dot{W}_{cycle} = \left(\frac{0.5(1-r)}{1 - 0.5(1-r)}\right)\epsilon\sigma A T_H^4 r^4$$

Simplifying and solving for $\dot{W}_{cycle}/A$

$$\dot{W}_{cycle}/A = (\epsilon\sigma T_H^4)\left[\frac{(1-r)}{(1+r)} r^4\right] \quad (1)$$

(a) T_c = 400K. In this case, r = 400/600 = 0.6667. Thus

$$\frac{\dot{W}_{cycle}}{A} = (0.6)\left(5.67\times10^{-8}\frac{W}{m^2\cdot K^4}\right)(600K)^4\left[\frac{(1-.6667)}{(1+.6667)}(.6667)^4\right]\left|\frac{1\,kW}{10^3\,W}\right|$$

$$= 0.174\ \frac{kW}{m^2} \longleftarrow \dot{W}_{cycle}/A$$

Also

$$\eta = 0.5(1-r) = 0.1667\ (16.67\%) \longleftarrow \eta$$

PROBLEM 5.39 (Cont'd)

(b) Plots The following plots are constructed using IT:

① 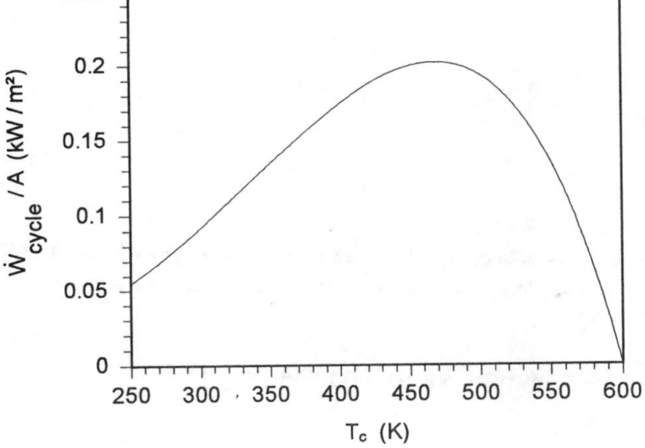

From the computer data, the maximum value for $\dot{W}_{cycle}/A$ is 0.2017 kW/m², and occurs at T_c = 468 K. This is confirmed by inspection of the graph.

② 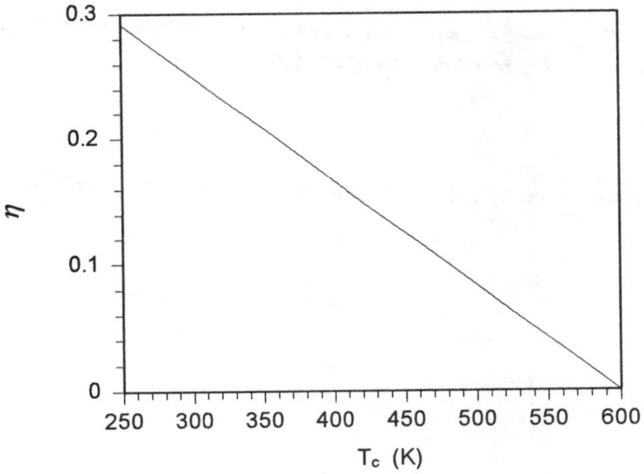

(c) For the net power per unit area within 2% of the maximum value,

$$0.98\,(\dot{W}_{cycle}/A)_{max} = (0.98)(0.2017) = 0.1977 \text{ kW/m}^2$$

By inspection of the computer data: $441 \text{ K} < T_c < 491 \text{ K}$ ← T_c range

Again, this is confirmed by inspection of the graph.

1. Alternatively, the value of r for which $\dot{W}_{cycle}/A$ is maximum could be obtained by differentiating (1) with respect to r, setting the resulting expression equal to zero, and solving for r. The result is $r = 0.781$, and $(\dot{W}_{cycle}/A)_{max} = 0.2017$. This confirms the computer solution.

2. Note that the thermal efficiency corresponding to $(\dot{W}_{cycle}/A)_{max}$ is 0.1095 (10.95%). In this application, maximizing η may not be the objective. The size of the radiator surface may be a more significant consideration.

PROBLEM 5.40

KNOWN: Operating data are claimed for a refrigeration cycle.

FIND: Evaluate the claim.

SCHEMATIC & GIVEN DATA:

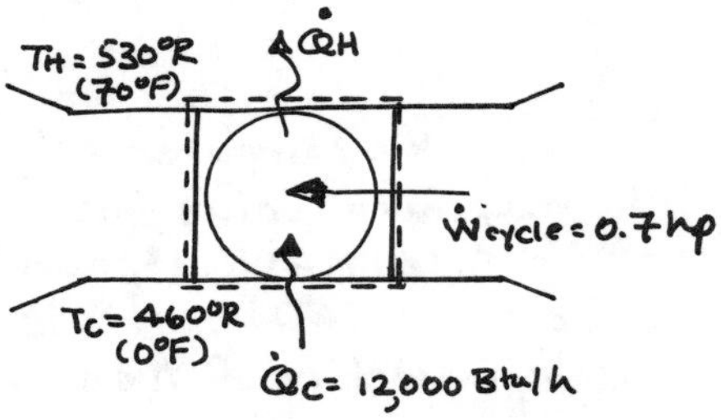

ASSUMPTION: The system shown in the schematic undergoes a refrigeration cycle while receiving energy by heat transfer from a reservoir at T_C and discharging energy by heat transfer to a reservoir at T_H.

ANALYSIS: Applying Eq. 2.45 together with claimed operating data, the coefficient of performance is

$$\beta = \frac{\dot{Q}_C}{\dot{W}_{cycle}} = \frac{12{,}000 \text{ Btu/h}}{0.7 \text{ hp} \left| \frac{2545 \text{ Btu/h}}{1 \text{ hp}} \right|} = 6.74$$

The maximum coefficient of performance any refrigeration cycle can have while operating between reservoirs at T_C, T_H is given by Eq. 5.9:

$$\beta_{max} = \frac{T_C}{T_H - T_C} = \frac{460°R}{(530-460)°R} = 6.57$$

Since the claimed value for β exceeds β_{max}, the inventor's claim is invalid.

PROBLEM 5.41

KNOWN: A tray of ice cubes is placed in a freezer having a coefficient of performance of 9.0 operating in a room at 32°C.

FIND: Determine if the cubes would remain frozen.

SCHEMATIC & GIVEN DATA:

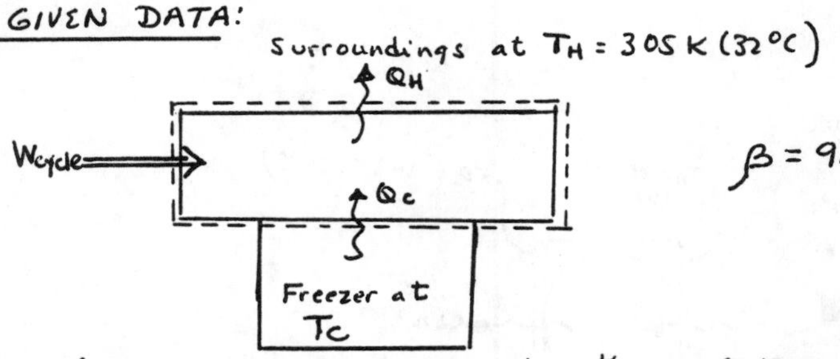

$\beta = 9.0$

ASSUMPTIONS: (1) The system shown in the accompanying figure undergoes a refrigeration cycle. (2) The freezer compartment and the surroundings play the roles of the cold and hot reservoirs, respectively.

ANALYSIS: As discussed in Sec. 5.4.3, the actual coefficient of performance must be less than, or equal to, the coefficient of performance of a reversible refrigeration cycle operating between reservoirs at the specified temperatures. Then, with Equation 5.9 and the given value of 9.0 for the coefficient of performance

$$9.0 \leq \frac{T_C}{T_H - T_C} = \frac{1}{(305/T_C) - 1}$$

or

$$\frac{305}{T_C} \leq 1 + \frac{1}{9}$$

giving

$$T_C \geqq 275 \text{ K}$$

However, to maintain the frozen cubes the freezer must be at 273 K, or less. Accordingly, the cubes would not remain frozen. ←

PROBLEM 5.42

KNOWN: Steady-state operating data are provided for a refrigerator.

FIND: Determine the maximum theoretical power that could be developed by a power cycle from the energy rejected from the refrigerator. Discuss.

SCHEMATIC & GIVEN DATA:

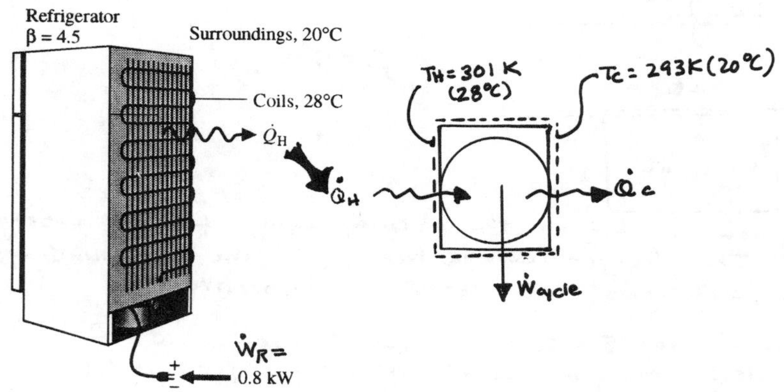

ASSUMPTION: 1. The system shown in the schematic undergoes a power cycle at steady state. 2. Data provided for the refrigerator also are for operation at steady state. 3. The coils and surroundings play the roles of hot and cold reservoirs, respectively.

ANALYSIS: For the power cycle, the thermal efficiency is $\eta = \dot{W}_{cycle}/\dot{Q}_H$, which must be less than, or equal to, the thermal efficiency of a reversible power cycle operating between reservoirs at T_H and T_C. Accordingly,

$$\dot{W}_{cycle} \leq \left[1 - \frac{T_C}{T_H}\right]\dot{Q}_H \qquad (1)$$

$\dot{Q}_H$ can be evaluated using data provided for the refrigerator: with Eq. 2.45,

$\beta = \frac{\dot{Q}_{in}}{\dot{W}_R}$. An energy rate balance for the refrigerator indicates that $\dot{Q}_{in} + \dot{W}_R = \dot{Q}_H$.

In these expressions, $\dot{Q}_{in}$ represents the energy provided to the circulating refrigerant from the food and other contents of the inside compartment of the refrigerator. $\dot{W}_R$ represents the power to the refrigerator (0.8 kW). Collecting results

$$\beta = \frac{\dot{Q}_H - \dot{W}_R}{\dot{W}_R} \Rightarrow \dot{Q}_H = \dot{W}_R(1+\beta) = 0.8\text{kW}(1+4.5) = 4.4\text{kW}$$

Returning to Eq. (1)

$$\dot{W}_{cycle} \leq \left[1 - \frac{293}{301}\right](4.4\text{kW}) = 0.12\text{kW}$$

This result shows that the refrigerator rejects energy at an appreciable rate, 4.4kW, but the thermodynamic value of this energy, as measured by the potential to develop power, is relatively small. Accordingly, there is little incentive for attempting to exploit this opportunity, and $\dot{Q}_H$ is simply discharged to the surroundings.

PROBLEM 5.43

KNOWN: A cryogenic sample must be kept at −195°F by a refrigeration system.

FIND: Determine the minimum theoretical power required by the refrigerator.

SCHEMATIC & GIVEN DATA:

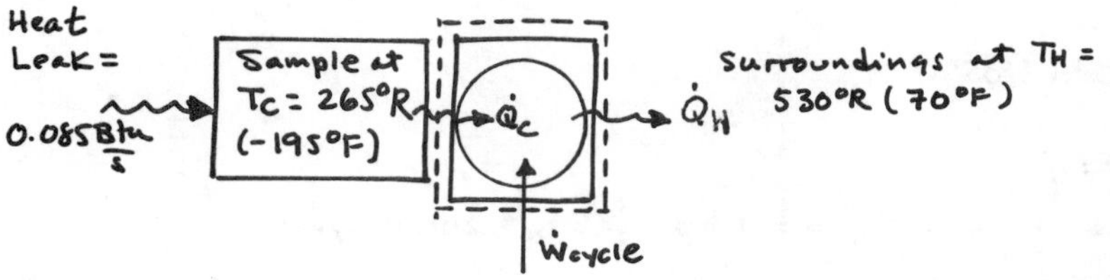

ASSUMPTIONS: 1. The system shown in the schematic undergoes a refrigeration cycle at steady state. 2. The sample is also at steady state. 3. The sample and surroundings play the role of cold and hot reservoirs, respectively.

ANALYSIS: The coefficient of performance of the refrigerator must be less than, or equal to, the coefficient of performance of a reversible refrigeration cycle operating between reservoirs at T_C, T_H. Thus, with Eq. 2.45 and Eq. 5.9,

$$\beta \le \beta_{MAX} \rightarrow \frac{\dot{Q}_C}{\dot{W}_{cycle}} \le \frac{T_C}{T_H - T_C} \rightarrow \dot{Q}_C\left[\frac{T_H - T_C}{T_C}\right] \le \dot{W}_{cycle}$$

At steady state, the rate energy must be removed from the sample equals the rate energy leaks into the sample. Thus, $\dot{Q}_C = 0.085$ Btu/s, and

$$0.085\,\frac{\text{Btu}}{\text{s}}\underbrace{\left[\frac{530-265}{265}\right]}_{=1} \le \dot{W}_{cycle} \quad \text{or} \quad \dot{W}_{cycle} \ge 0.085\,\frac{\text{Btu}}{\text{s}} \leftarrow \text{MIN. POWER}$$

PROBLEM 5.44

KNOWN: Steady-state operating data are provided for an ice maker.

FIND: Determine the maximum rate ice can be formed at 0°C, per kW of power input.

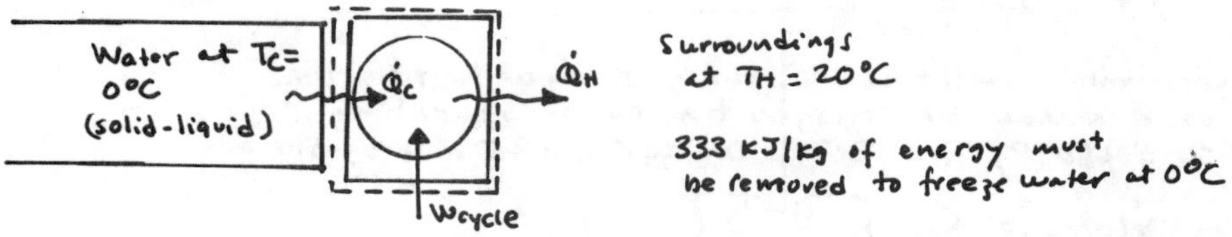

ASSUMPTIONS: 1. The system shown in the schematic undergoes a refrigeration cycle at steady state. 2. The water and the surroundings play the role of cold and hot reservoirs.

ANALYSIS: The coefficient of performance of the refrigeration cycle must be less than, or equal to, the coefficient of performance of a reversible refrigeration cycle operating between reservoirs at T_C, T_H. Thus, with Eq. 2.45 and Eq. 5.9,

$$\beta \le \beta_{MAX} \rightarrow \frac{\dot{Q}_C}{\dot{W}_{cycle}} \le \frac{T_C}{T_H - T_C}$$

At steady state, the energy removed while freezing water is $\dot{Q}_C = (333\text{ kJ/kg})\dot{M}$, where $\dot{M}$ is the rate water freezes, in kg/h. Collecting results

$$\frac{(333\text{ kJ/kg})\,\dot{M}(\text{kg/h})}{\dot{W}_{cycle}(\text{kJ/s})}\left|\frac{1\text{ h}}{3600\text{ s}}\right| \le \frac{273\text{ K}}{(293-273\text{ K})} \rightarrow \frac{\dot{M}}{\dot{W}_{cycle}} \le \left(\frac{273}{20}\right)\frac{|3600|}{333} = 147.6\,\frac{\text{kg/h}}{\text{kW}} \leftarrow \text{MAX. RATE.}$$

PROBLEM 5.45

KNOWN: A refrigerator maintains a refrigerated space at a specified temperature. Data for operation at steady state are provided.

FIND: Determine the coefficient of performance and the rate at which energy is discharged to the surroundings. Also, evaluate the minimum theoretical power input under the stated conditions.

SCHEMATIC & GIVEN DATA:

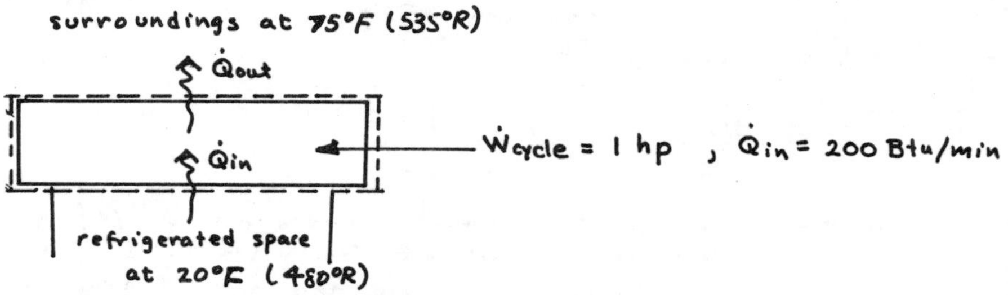

ASSUMPTIONS: (1) The system shown in the accompanying figure undergoes a refrigeration cycle. (2) The data are for operation at steady state. (3) The refrigerated space and the surroundings play the roles of cold and hot reservoirs, respectively.

ANALYSIS: (a) By definition of coefficient of performance

$$\beta = \frac{\dot{Q}_{in}}{\dot{W}_{cycle}} = \frac{(200\ \text{Btu/min})\left|60\ \text{min/h}\right|}{(1\ \text{hp})\left|\frac{2545\ \text{Btu/h}}{1\ \text{hp}}\right|} = 4.72 \longleftarrow \beta$$

From an energy balance

$$\dot{Q}_{out} = \dot{Q}_{in} + \dot{W}_{cycle}$$

$$= 200\ \frac{\text{Btu}}{\text{min}} + (1\ \text{hp})\left|\frac{2545\ \text{Btu/h}}{1\ \text{hp}}\right|\left|\frac{1\ \text{h}}{60\ \text{min}}\right|$$

$$= 242.4\ \frac{\text{Btu}}{\text{min}} \longleftarrow \dot{Q}_{out}$$

(b) The coefficient of performance must be less than, or equal to, the coefficient of performance of a reversible refrigeration cycle operating between reservoirs at $T_C = 480°R$, $T_H = 535°R$. With Eq. 2.45 and Eq. 5.9

$$\frac{\dot{Q}_{in}}{\dot{W}_{cycle}} \le \frac{T_C}{T_H - T_C} \Rightarrow \dot{W}_{cycle} \ge \dot{Q}_{in}\left[\frac{T_H - T_C}{T_C}\right]$$

$$\ge \left(200\ \frac{\text{Btu}}{\text{min}}\right)\left|\frac{60\ \text{min}}{1\ \text{h}}\right|\left|\frac{1\ \text{hp}}{2545\ \text{Btu/h}}\right|\left[\frac{535-480}{480}\right]$$

$$\ge 0.54\ \text{hp}$$

Thus, the minimum theoretical power required is 0.54 hp ← MIN. POWER

PROBLEM 5.46

KNOWN: A refrigerator maintains a refrigerated space at a specified temperature. Data for operation at steady state are provided.

FIND: Determine the power input required.

SCHEMATIC & GIVEN DATA:

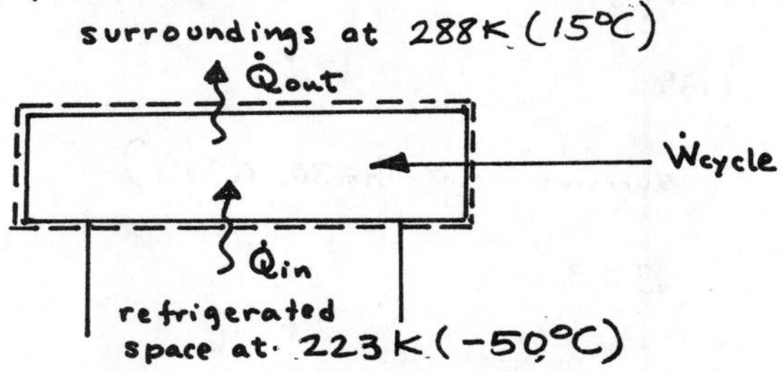

$\dot{Q}_{in}$ = 150 kJ/min

$\beta = 0.3\,\beta_{MAX}$

ASSUMPTIONS: (1) The system shown in the accompanying figure undergoes a refrigeration cycle. (2) The data are for operation at steady state. (3) The refrigerated space and the surroundings play the roles of the cold and hot reservoirs, respectively.

ANALYSIS: The coefficient of performance of a reversible refrigeration cycle operating between reservoirs at T_H = 288K and T_C = 223K is given by Eq. 5.9:

$$\beta_{MAX} = \frac{T_C}{T_H - T_C} = \frac{223\,K}{(288-223)\,K} = 3.43$$

Then

$$\beta = 0.3\,(3.43) = 1.03$$

Since $\beta = \dot{Q}_C/\dot{W}_{cycle}$,

$$\dot{W}_{cycle} = \frac{\dot{Q}_C}{\beta} = \frac{150\ kJ/min}{1.03}$$

$$= \left(145.6\ \frac{kJ}{min}\right)\left(\frac{1\ min}{60\ s}\right)\left(\frac{1\ kW}{1\ kJ/s}\right)$$

$$= 2.43\ kW$$ ◄

PROBLEM 5.47

KNOWN: A refrigeration cycle maintains a computer laboratory at 18°C on a day when the outside temperature is 30°C. Steady state operating data are provided.

FIND: Determine the power required, in kW, and compare with the minimum theoretical power required.

SCHEMATIC & GIVEN DATA

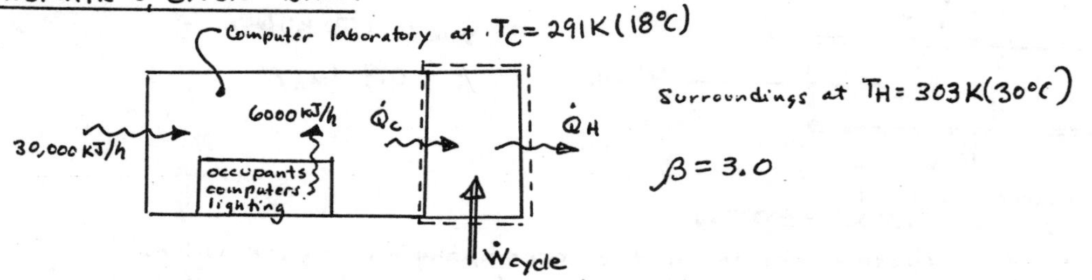

ASSUMPTIONS: (1) The system shown in the accompanying figure undergoes a refrigeration cycle. (2) All data provided are for operation at steady state. (3) The computer laboratory and surroundings play the roles of cold and hot reservoirs, respectively.

ANALYSIS: At steady state, the refrigeration cycle must remove energy from the computer laboratory at the same rate as energy enters from all sources: $\dot{Q}_c = (30,000 + 6000)\,kJ/h = 36,000\ kJ/h$. The coefficient of performance is given as 3.0. Thus

$$\beta = \frac{\dot{Q}_c}{\dot{W}_{cycle}} \Rightarrow 3.0 = \frac{36,000\,kJ/h}{\dot{W}_{cycle}} \Rightarrow \dot{W}_{cycle} = 12,000\ \frac{kJ}{h}$$

or

$$\dot{W}_{cycle} = \left(12000\ \frac{kJ}{h}\right)\left(\frac{h}{3600\,s}\right)\left(\frac{1\,kW}{1\,kJ/s}\right) = \mathbf{3.33\ kW} \longleftarrow \dot{W}_{cycle}$$

From Sec 5.4.3, we know that the actual coefficient of performance must be less than, or equal to, the coefficient of performance of a reversible refrigeration cycle operating between reservoirs at the specified temperatures. Then, with Equation 5.9

$$\frac{\dot{Q}_c}{\dot{W}_{cycle}} \le \frac{T_c}{T_H - T_c} = \frac{291}{303-291} = 24.25$$

Accordingly

$$\frac{36,000\ kJ/h}{24.25} \le \dot{W}_{cycle}$$

$$1484.5\ \frac{kJ}{h} \le \dot{W}_{cycle}$$

or

$$\dot{W}_{cycle} \ge \left(1484.5\ \frac{kJ}{h}\right)\left|\frac{1\,h}{3600\,s}\right|\left|\frac{1\,kW}{1\,kJ/s}\right| = 0.41\ kW \longleftarrow (\dot{W}_{cycle})_{min}$$

Forming the ratio of the actual power required to the minimum theoretical value

$$\frac{3.33\,kW}{0.41\,kW} = 8.12$$

it can be concluded that the actual requirement **is over 8 times greater** than the minimum theoretical requirement.

PROBLEM 5.48

KNOWN: Operating data at steady state are provided for a heat pump working between a dwelling at 70°F and (a) outdoor air at 32°F, (b) a pond at 40°F, (c) the ground at 55°F.

FIND: For each case determine the minimum theoretical power required.

SCHEMATIC & GIVEN DATA:

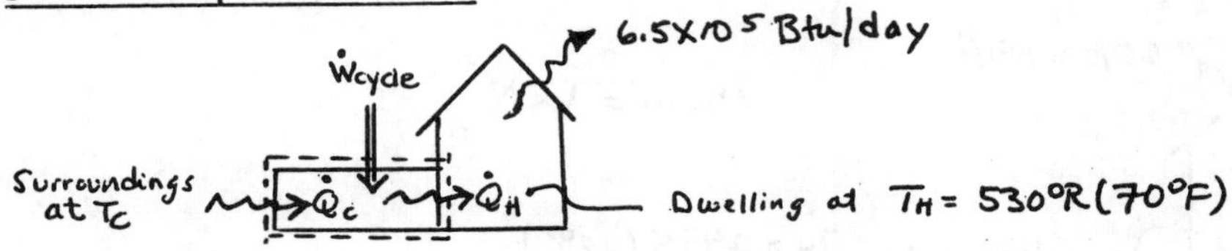

ASSUMPTIONS: (1) The system shown in the cycle undergoes a heat pump cycle. (2) All data are for operation at steady state. (3) The surroundings and dwelling play the roles of cold and hot reservoirs, respectively.

ANALYSIS: At steady state, the heat pump cycle must provide energy to the dwelling equal to the energy leaking through the walls and roof:

$$\dot{Q}_H = \left(6.5 \times 10^5 \frac{\text{Btu}}{\text{day}}\right)\left|\frac{1\ \text{day}}{24\ \text{h}}\right| = 2.71 \times 10^4 \frac{\text{Btu}}{\text{h}}$$

From Sec 5.4.3, we know that the coefficient of performance of the heat pump must be less than, or equal to, the coefficient of performance of a reversible heat pump operating between reservoirs at $T_H = 293\,K$ and T_C. Then, with Eq. 5.10

$$\frac{\dot{Q}_H}{\dot{W}_{cycle}} \le \frac{T_H}{T_H - T_C} \Rightarrow \dot{W}_{cycle} \ge \dot{Q}_H\left[\frac{T_H - T_C}{T_H}\right]$$

or

$$\dot{W}_{cycle} \ge \dot{Q}_H\left[1 - \frac{T_C}{T_H}\right]$$

Inserting values

$$\dot{W}_{cycle} \ge \left(2.71 \times 10^4 \frac{\text{Btu}}{\text{h}}\right)\left|\frac{1\ \text{hp}}{2545\ \text{Btu/h}}\right|\left[1 - \frac{T_C}{530}\right]$$

$$\ge (10.65\ \text{hp})\left[1 - \frac{T_C}{530}\right]$$

(a) $T_C = 492°R$

$$\dot{W}_{cycle} \ge (10.65\ \text{hp})\left[1 - \frac{492}{530}\right] = 0.76\ \text{hp}$$

(b) $T_C = 500°R$

$$\dot{W}_{cycle} \ge (10.65\ \text{hp})\left[1 - \frac{500}{530}\right] = 0.6\ \text{hp}$$

① (c) $T_C = 515°R$

$$\dot{W}_{cycle} \ge (10.65\ \text{hp})\left[1 - \frac{515}{530}\right] = 0.3\ \text{hp}$$

---

1. The required power decreases as the outside source temperature increases:

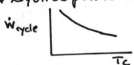

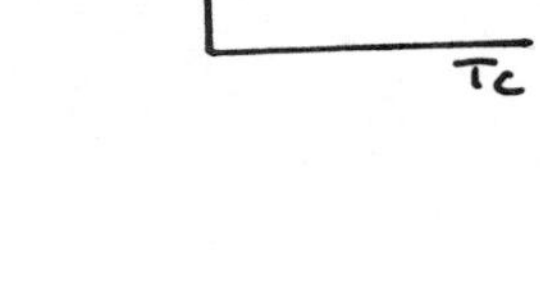

PROBLEM 5.49

KNOWN: Operating data are provided for a heat pump that provides heating for a building on a day when the outside temperature is 0°C and the building interior is to be kept at 20°C.

FIND: Determine if the heat pump would suffice.

SCHEMATIC & GIVEN DATA:

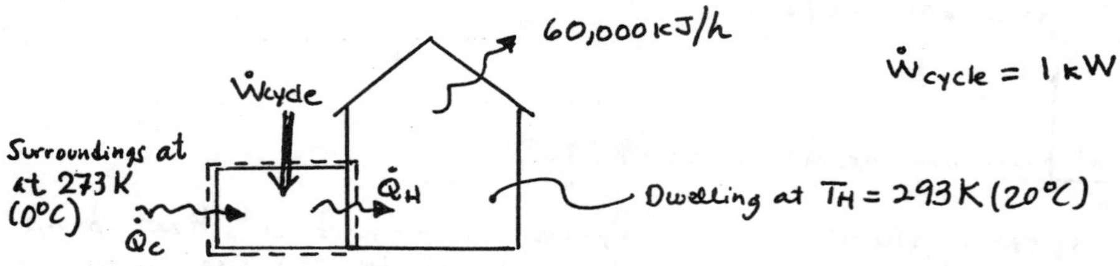

$\dot{W}_{cycle}$ = 1 kW

ASSUMPTIONS: (1) The system shown in the accompanying figure undergoes a heat pump cycle. (2) The data provided are for operation at steady state. (3) The surroundings and dwelling play the roles of cold and hot reservoirs, respectively.

ANALYSIS: At steady state, the heat pump cycle must provide energy to the dwelling equal to the energy leaking through the walls and roof: $\dot{Q}_H$ = 60,000 kJ/h.

From Sec. 5.4.3, we know that the coefficient of performance of the heat pump must be less than, or equal to, the coefficient of performance of a reversible heat pump operating between reservoirs at T_C = 273 K, T_H = 293 K. Then, with Eq. 5.10

$$\frac{\dot{Q}_H}{\dot{W}_{cycle}} \leq \frac{T_H}{T_H - T_C}$$

This equation can be used alternatively to determine if the heat pump would suffice: (a) Calculate $(\dot{Q}_H)_{MAX}$ from given values for $\dot{W}_{cycle}$, T_C, and T_H. (b) Calculate $(T_H)_{MAX}$ from known values of $\dot{Q}_H$, $\dot{W}_{cycle}$, T_C. (c) Calculate $(\dot{W}_{cycle})_{MIN}$ from known values of the remaining quantities. For the first alternative

$$\dot{Q}_H \leq \left(\frac{T_H}{T_H - T_C}\right)\dot{W}_{cycle} = \left(\frac{293}{20}\right) 1\,kW \left|\frac{1\,kJ/s}{1\,kW}\right| \left|\frac{3600\,s}{1\,h}\right| = 52{,}740\,\frac{kJ}{h}$$

$$\Rightarrow (\dot{Q}_H)_{MAX} = 52{,}740\,kJ/h$$

With alternative (b), $(T_H)_{MAX}$ = 290 K (17°C). With alternative (c), $(\dot{W}_{cycle})_{MIN}$ = 1.14 kW. With each approach, we conclude the heat pump would not suffice.

PROBLEM 5.50

KNOWN: Operating data are provided for a heat pump that maintains a dwelling at 70°F when the outside temperature is 40°F.

FIND: Determine the minimum theoretical power required, in horsepower.

SCHEMATIC & GIVEN DATA:

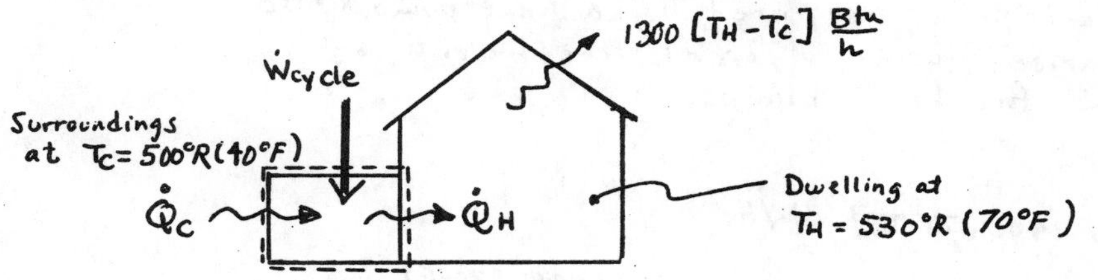

ASSUMPTIONS: (1) The system shown in the accompanying figure undergoes a heat pump cycle. (2) The data provided are for operation at steady state. (3) The dwelling and surroundings play the roles of hot and cold reservoirs, respectively.

ANALYSIS: At steady state the heat pump must provide energy to the dwelling equal to the energy leaking through the walls and roof:

$$\dot{Q}_H = 1300\,[T_H - T_C] = 1300\,[530 - 500] = 39{,}000\ \frac{\text{Btu}}{\text{h}}$$

From Sec. 5.4.3, we know that the coefficient of performance of the heat pump must be less than, or equal to, the coefficient of performance of a reversible heat pump operating between reservoirs at $T_H = 530°R$ and $T_C = 500°R$. Then, with Eq. 5.10

$$\frac{\dot{Q}_H}{\dot{W}_{cycle}} \leq \frac{T_H}{T_H - T_C}$$

Rearranging

$$\dot{Q}_H \left[\frac{T_H - T_C}{T_H}\right] \leq \dot{W}_{cycle}$$

Inserting known values

$$\left(39{,}000\ \frac{\text{Btu}}{\text{h}}\right)\left[\frac{530 - 500}{530}\right] \leq \dot{W}_{cycle}$$

$$2208\ \frac{\text{Btu}}{\text{h}} \leq \dot{W}_{cycle}$$

Or

$$\dot{W}_{cycle} \geq \left(2208\ \frac{\text{Btu}}{\text{h}}\right)\left(\frac{1\ \text{hp}}{2545\ \text{Btu/h}}\right) = 0.87\ \text{hp} \longleftarrow (\dot{W}_{cycle})_{min}$$

PROBLEM 5.51

KNOWN: Operating data are provided for a heat pump that maintains a dwelling at 68°F on a day when the outside temperature is 28°F.

FIND: Determine the power required, in kW, to provide the heating by electrical resistance elements and compare with the minimum theoretical power required by a heat pump. Also repeat the comparison using typical coefficient of performance data for heat pumps.

SCHEMATIC & GIVEN DATA:

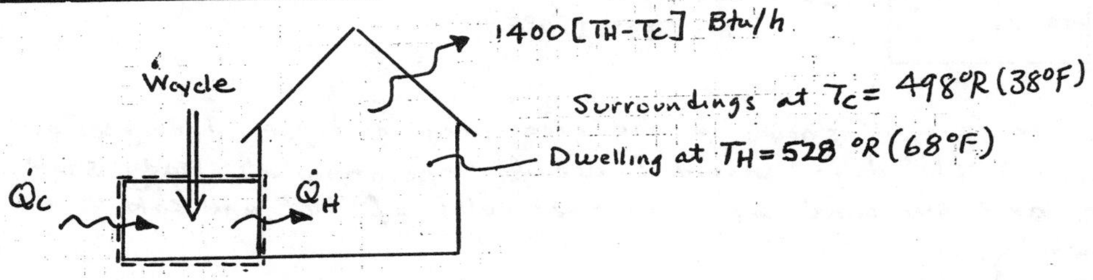

ASSUMPTIONS: (1) The system shown in the accompanying figure undergoes a heat pump cycle. (2) The data provided are for operation at steady state. (3) The dwelling and surroundings play the roles of hot and cold reservoirs, respectively.

ANALYSIS: At steady state, the electrical resistance elements, or the heat pump, must provide energy to the dwelling equal to the energy leaking through the walls and roof:

$$\dot{Q}_H = 1400[T_H - T_C] = 1400[528-498] = 42{,}000 \text{ Btu/h}$$

$$= (42{,}000 \text{ Btu/h})\left|\frac{1 \text{ hp}}{2545 \text{ Btu/h}}\right| = 16.5 \text{ hp}$$

Accordingly, the resistors would be provided with electrical power = 16.5 hp. ←

Turning next to the heat pump option, we know from Sec. 5.4.3 that the coefficient of performance of a heat pump must be less than, or equal to, the coefficient of performance of a reversible heat pump operating between reservoirs at T_C and T_H. Then, with Eq. 5.10

$$\frac{\dot{Q}_H}{\dot{W}_{cycle}} \leq \frac{T_H}{T_H - T_C} \Rightarrow \dot{W}_{cycle} \geq \dot{Q}_H\left[\frac{T_H - T_C}{T_H}\right]$$

Substituting known values

$$\dot{W}_{cycle} \geq (16.5 \text{ hp})\left[1 - \frac{498}{528}\right] = 0.94 \text{ hp}$$

Accordingly, the minimum theoretical power required = 0.94 hp ←

From manufacturer's data, an actual heat pump would have a coefficient of performance in the range 2.0–3.0. Using the higher value

$$\beta = \frac{\dot{Q}_H}{\dot{W}_{cycle}} \Rightarrow \dot{W}_{cycle} = \frac{\dot{Q}_H}{\beta} = \frac{16.5 \text{ hp}}{3.0} = 5.5 \text{ hp}$$ ←

←

Comparisons:

$$\frac{\text{All electric}}{\text{Ideal heat pump}} = \frac{16.5 \text{ hp}}{0.94 \text{ hp}} = 17.6, \quad \frac{\text{All electric}}{\text{Actual heat pump}} = \frac{16.5 \text{ hp}}{5.5 \text{ hp}} = 3$$

PROBLEM 5.52

KNOWN: Refrigeration and heat pump cycles operate reversibly between reservoirs at fixed T_H and a range of T_C values, respectively.

FIND: Plot (a) β_{max} given by Eq. 5.9 for the refrigeration cycle and (b) γ_{max} given by Eq. 5.10 for the heat pump cycle. Discuss practical implications in each case.

SCHEMATIC & GIVEN DATA:

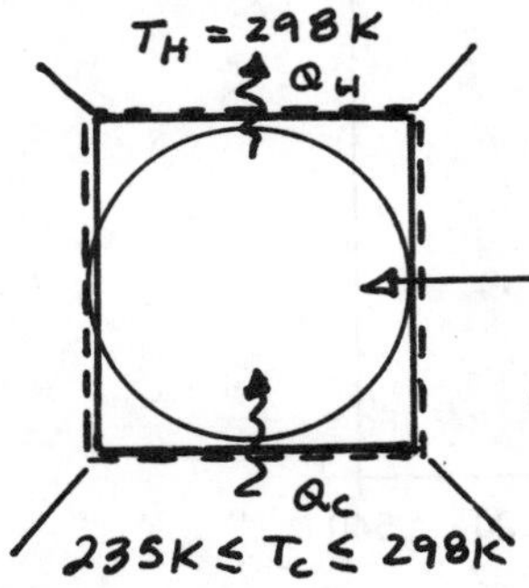

$$\beta_{max} = \frac{T_C}{T_H - T_C}$$

(a)

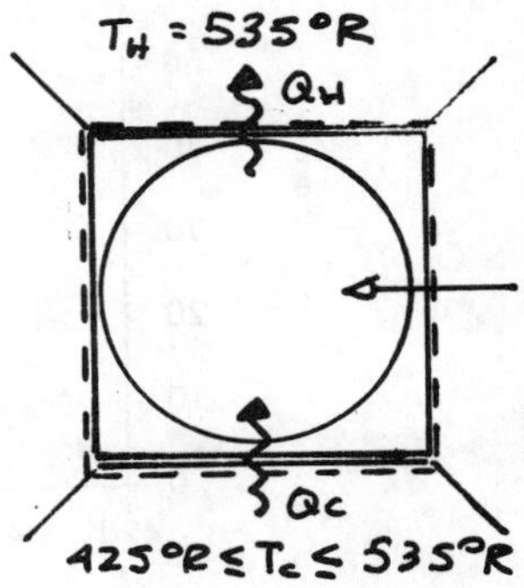

$$\gamma_{max} = \frac{T_H}{T_H - T_C}$$

(b)

ASSUMPTIONS: (1) Each system shown on the accompanying sketches undergoes a cycle. (2) The cycles are reversible.

ANALYSIS: (a) The coefficient of performance for the reversible refrigeration cycle operating between reservoirs at T_H and T_C is given by Eq. 5.9. With $T_H = 298K$, we get $\beta_{max} = T_C/(298 - T_C)$. The following plot is obtained using IT:

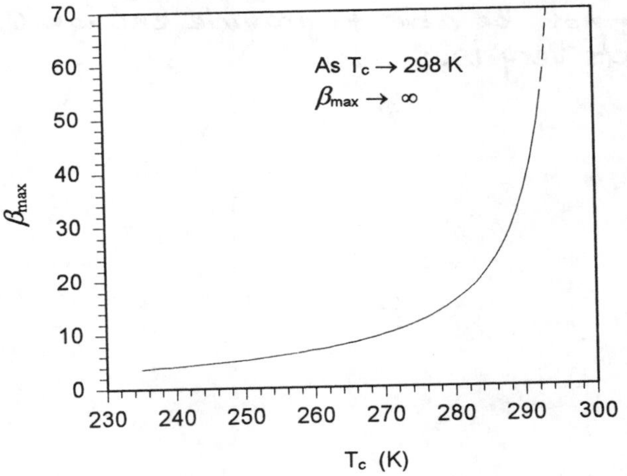

For an actual refrigeration system operating at given values of T_H and T_C, the coefficient of performance β would be less than β_{max}: $\beta < \beta_{max}$. Still, the trend shown in the plot is qualitatively correct: β_{max} decreases rapidly as T_C decreases for fixed T_H. Accordingly, as T_C decreases, more work input is needed per unit of energy Q_C removed from the region at T_C (cold reservoir). When electricity is used to provide the work input, there would normally a higher cost of operation associated with the lowering of T_C.

PROBLEM 5.52 (Contd.)

(b) The coefficient of performance of a reversible heat pump cycle operating between reservoirs at T_H and T_C is given by Eq. 5.10. With $T_H = 535$ °R, we get $\gamma_{max} = 535/(535 - T_C)$. The following plot is obtained using IT:

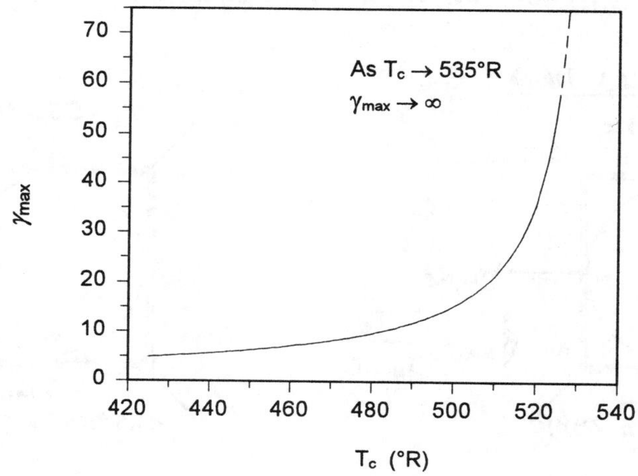

For an actual heat pump system operating given values of T_H and T_C, the coefficient of performance γ would be less than γ_{max}: $\gamma < \gamma_{max}$. Still, the trend shown on the plot is qualitatively correct: γ_{max} decreases rapidly as T_C decreases for fixed T_H. Accordingly, as T_C decreases, the energy delivered by heat transfer at T_H, Q_H, decreases for a given work input. Thus, when the heat pump is used to provide heat transfer to a dwelling, the system may not be able to provide enough Q_H when the outside temperture T_C gets very low.

PROBLEM 5.53

KNOWN: A refrigerator maintains a refrigerated space at a specified temperature. Data for operation at steady state are provided.

FIND: Determine the power input and compare with the power input for a reversible refrigeration cycle operating between reservoirs at the specified temperatures. Evaluate the daily actual and minimum theoretical operating costs.

SCHEMATIC & GIVEN DATA:

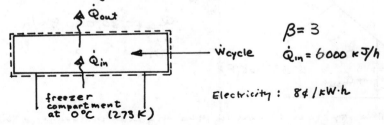

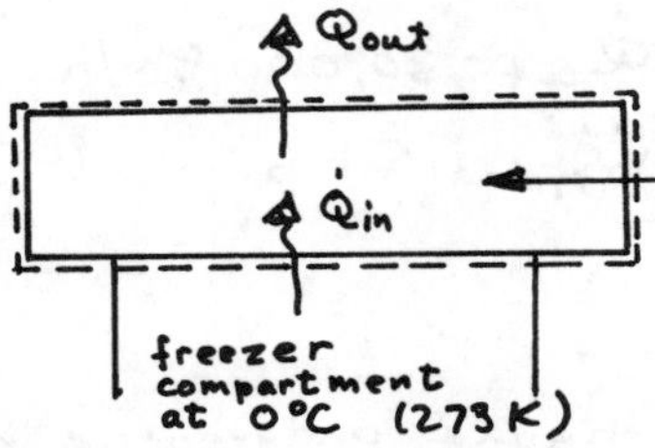

ASSUMPTIONS: (1) The system shown in the accompanying figure undergoes a refrigeration cycle. (2) The data are for operation at steady state. (3) The freezer compartment and the surroundings play the roles of cold and hot reservoirs, respectively.

ANALYSIS: (a) Upon rearrangement, the defining expression for the coefficient of performance gives

$$\dot{W}_{cycle} = \frac{\dot{Q}_{in}}{\beta} = \frac{6000 \text{ kJ/h}}{3} = 2000 \frac{\text{kJ}}{\text{h}} \longleftarrow \dot{W}_{cycle}$$

The coefficient of performance for a reversible refrigeration cycle operating between reservoirs at $T_H = 293$ K (20°C) and $T_C = 273$ K (0°C) is given by Eq. 5.9

$$\beta_{max} = \frac{273}{293-273} = 13.65$$

The power input for such a cycle is then

$$(\dot{W}_{cycle})_{min} = \frac{6000 \text{ kJ/h}}{13.65} = 440 \frac{\text{kJ}}{\text{h}} \longleftarrow (\dot{W}_{cycle})_{min}$$

The actual power input is over 4½ times the minimum theoretical power requirement.

(b)

Actual cost: $\dot{\$} = 2000 \frac{\text{kJ}}{\text{h}} \left|\frac{1 \text{ h}}{3600 \text{ s}}\right| \left|\frac{1 \text{ kW}}{1 \text{ kJ/s}}\right| \left|\frac{24 \text{ h}}{\text{day}}\right| \left(\frac{\$ 0.08}{\text{kW}\cdot\text{h}}\right) = \$1.07/\text{day}$

Min. Theo. Cost: $\dot{\$} = 440 \frac{\text{kJ}}{\text{h}} \left|\frac{1 \text{ h}}{3600 \text{ s}}\right| \left|\frac{1 \text{ kW}}{1 \text{ kJ/s}}\right| \left|\frac{24 \text{ h}}{\text{day}}\right| \left(\frac{\$ 0.08}{\text{kW}\cdot\text{h}}\right) = \$0.23/\text{day}$

PROBLEM 5.54

KNOWN: A heat pump provides heating to a dwelling. Operating data are provided.

FIND: Determine the actual operating cost and compare with the minimum theoretical operating cost for each day of operation.

SCHEMATIC & GIVEN DATA:

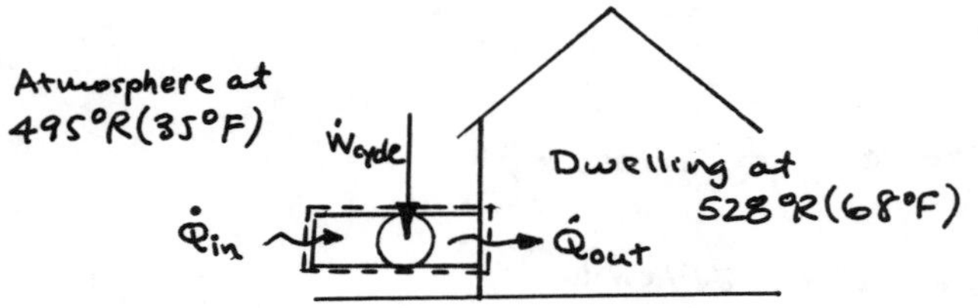

$\dot{Q}_{out} = 30{,}000$ Btu/h

$\dot{W}_{cycle} = 5$ hp

ASSUMPTIONS: (1) The system shown in the figure undergoes a heat pump cycle. (2) All data are for operation at steady state. (3) The atmosphere and the dwelling play the roles of cold and hot reservoirs, respectively. (4) The cost of electricity is 8 cents per kW·h.

ANALYSIS: Using given data, the actual operating cost is

$$\begin{pmatrix}\text{Cost per}\\ \text{day}\end{pmatrix} = (5\text{ hp})\left|\frac{1\text{ kW}}{1.341\text{ hp}}\right|\left|\frac{24\text{ h}}{1\text{ day}}\right|\left(\frac{0.08\ \$}{\text{kW}\cdot\text{h}}\right) = \$7.16/\text{day} \longleftarrow$$

The minimum theoretical operating cost corresponds to the minimum theoretical power requirement. Since $\beta < \beta_{MAX}$,

$$\Rightarrow \quad \frac{\dot{Q}_{out}}{\dot{W}_{cycle}} \leq \frac{T_H}{T_H - T_C}$$

or

$$\frac{30{,}000\text{ Btu/h}}{\dot{W}_{cycle}} \leq \frac{528\ °R}{(528 - 495)°R} = 16$$

$$\Rightarrow \quad \frac{30{,}000\text{ Btu/h}}{16} \leq \dot{W}_{cycle}$$

$$1875\ \frac{\text{Btu}}{\text{h}} \leq \dot{W}_{cycle}$$

The minimum cost is then

$$\begin{pmatrix}\text{minimum cost}\\ \text{per day}\end{pmatrix} = \left(1875\ \frac{\text{Btu}}{\text{h}}\right)\left(\frac{1\text{ kW}}{3413\text{ Btu/h}}\right)\left(\frac{24\text{ h}}{1\text{ day}}\right)\left(\frac{0.08\ \$}{\text{kW}\cdot\text{h}}\right)$$

$$= \$1.05/\text{day} \longleftarrow$$

The actual operating cost is about 7 times higher than the minimum theoretical cost.

PROBLEM 5.55

KNOWN: A heat pump maintains a dwelling at a specified temperature.

FIND: Determine the minimum theoretical cost.

SCHEMATIC & GIVEN DATA:

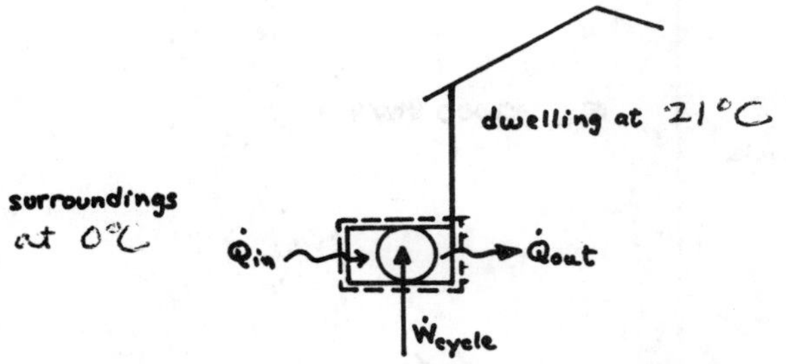

$\dot{Q}_{out} = 8\,kW$

cost of electricity = 0.08 $/kW·h

ASSUMPTIONS: (1) The system shown on the accompanying figure undergoes a heat pump cycle. (2) The data are for operation at steady state. (3) The dwelling and the surroundings play the roles of hot and cold reservoirs, respectively.

ANALYSIS: The minimum theoretical cost for any heat pump cycle under the stated conditions is the cost for a reversible cycle operating between reservoirs at $T_H = 294\,K\ (21°C)$ and $T_C = 273\,K\ (0°C)$. The power required by such a cycle can be obtained from

$$(\dot{W}_{cycle})_{min} = \frac{\dot{Q}_{out}}{\gamma_{max}}$$

where γ_{max} is

$$\gamma_{MAX} = \frac{294}{294-273} = 14$$

Accordingly,

$$(\dot{W}_{cycle})_{Min} = \frac{8\,KW}{14} = 0.57\,kW$$

Then, the minimum cost for one day of operation is

$$\text{Minimum theoretical cost: } \$ = (0.57\,kW)\left|\frac{24\,h}{day}\right|(\$\ 0.08/day)$$

$$= \$1.09/day \qquad \longleftarrow \text{MIN COST}$$

PROBLEM 5.56

KNOWN: A refrigerator maintains a freezer compartment at a specified temperature. Data for operation at steady state are provided.

FIND: Determine the minimum theoretical operating cost per day.

SCHEMATIC & GIVEN DATA:

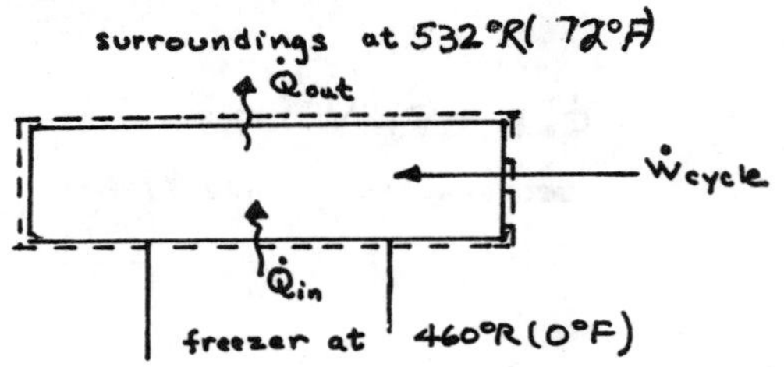

$\dot{Q}_{in} = 2000$ Btu/h

ASSUMPTIONS: (1) The system shown on the accompanying figure undergoes a refrigeration cycle. (2) The data provided are for operation at steady state. (3) The freezer compartment and the surroundings play the roles of cold and hot reservoirs, respectively. (4) The cost of electricity is 8 cents per kW·h.

ANALYSIS: The mimum cost of electricity corresponds to the minimum power input. Then, since $\beta \leq \beta_{max}$,

$$\frac{\dot{Q}_{in}}{\dot{W}_{cycle}} \leq \frac{T_C}{T_H - T_C}$$

or

$$\frac{2000 \text{ Btu/h}}{\dot{W}_{cycle}} \leq \frac{460°R}{(532-460)°R} = 6.39$$

$$\Rightarrow \quad \frac{2000 \text{ Btu/h}}{6.39} \leq \dot{W}_{cycle}$$

$$313 \frac{\text{Btu}}{\text{h}} \leq \dot{W}_{cycle}$$

Then

$$\left[\begin{matrix}\text{Minimum cost}\\ \text{per day}\end{matrix}\right] = \left[313 \frac{\text{Btu}}{\text{h}}\right]\left|\frac{24 \text{h}}{1 \text{ day}}\right|\left|\frac{1 \text{ kW}}{3413 \text{ Btu/h}}\right|\left[\frac{8 \text{ cents}}{\text{kW}\cdot\text{h}}\right]$$

$$= 17.6 \text{ cents/day} \quad \longleftarrow$$

PROBLEM 5.57

KNOWN: Operating data are provided for a heat pump receiving energy by heat transfer from the outdoor air. From well water.

FIND: For each case, determine the minimum theoretical daily cost to operate.

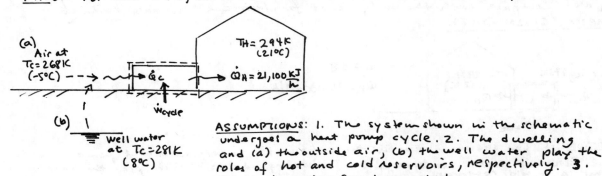

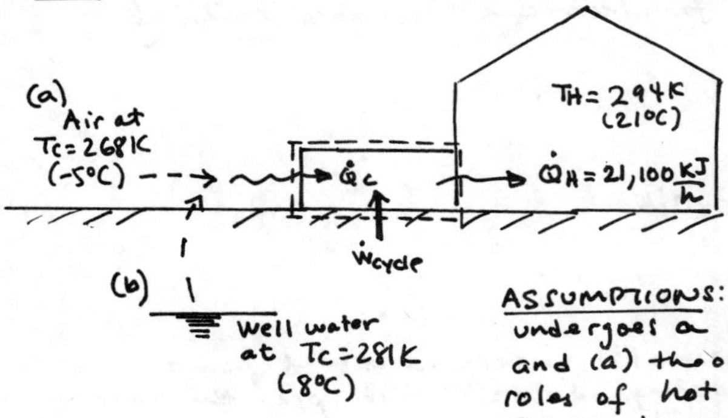

ASSUMPTIONS: 1. The system shown in the schematic undergoes a heat pump cycle. 2. The dwelling and (a) the outside air, (b) the well water play the roles of hot and cold reservoirs, respectively. 3. Electricity costs 8cents per kW·h.

ANALYSIS: The coefficient of performance of a heat pump is less than, or equal to, the coefficient of performance of a reversible heat pump operating between reservoirs at T_C, T_H. That is, $\gamma \le \gamma_{MAX}$. With Eqs. 2.47 and 5.10

$$\frac{\dot{Q}_H}{\dot{W}_{cycle}} \le \frac{T_H}{T_H - T_C} \Rightarrow \dot{Q}_H\left[1 - \frac{T_C}{T_H}\right] \le \dot{W}_{cycle}$$

Inserting values

$$\dot{W}_{cycle} \ge \left(21{,}100\ \frac{kJ}{h}\right)\left[1 - \frac{T_C}{294K}\right]\left|\frac{1h}{3600s}\right|\left|\frac{1kW}{1kJ/s}\right|$$

(a) Air at $T_C = 268K$

$$\dot{W}_{cycle} \ge (21{,}100)\left[1 - \frac{268}{294}\right]\left|\frac{1}{3600}\right| = 0.52kW$$

Costing:

$$\dot{\$} \ge (0.52kW)\left|\frac{24h}{1day}\right|\left(\frac{\$0.08}{kW\cdot h}\right) = \$1.00/day$$

(b) Well water at $T_C = 281K$

$$\dot{W}_{cycle} = (21{,}100)\left[1 - \frac{281}{294}\right]\left|\frac{1}{3600}\right| = 0.26kW$$

Costing:

$$\dot{\$} \ge (0.26kW)\left|\frac{24h}{day}\right|\left(\frac{\$0.08}{kW\cdot h}\right) = \$0.50/day$$

PROBLEM 5.58

KNOWN: Operating data are provided for a heat pump that maintains the temperature within a building at a specified value.

FIND: Determine the actual and minimum theoretical daily operating costs, and compare with the daily cost for electrical resistance heating.

SCHEMATIC & GIVEN DATA:

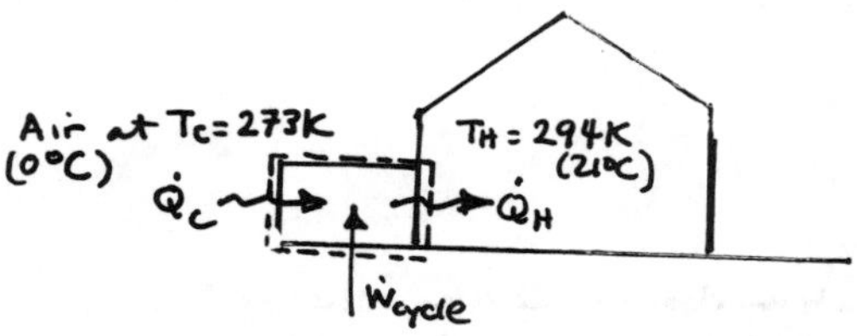

$\gamma = 3.8$

$\dot{Q}_H = 75{,}000$ kJ/h

ASSUMPTIONS: 1. The system shown in the schematic undergoes a heat pump cycle. 2. Electricity costs 8 cents/kW·h. 3. The dwelling and outside air play the roles of the hot and cold reservoirs.

ANALYSIS: (a) Using Eq. 2.47: $\gamma = \dot{Q}_H / \dot{W}_{cycle}$,

$$\dot{W}_{cycle} = \frac{\dot{Q}_H}{\gamma} = \frac{75{,}000 \text{ kJ/h}}{3.8} = 19{,}737 \frac{\text{kJ}}{\text{h}}$$

Costing:

$$\dot{\$} = \left(19{,}737 \frac{\text{kJ}}{\text{h}}\right)\left|\frac{1 \text{ h}}{3600 \text{ s}}\right|\left|\frac{1 \text{ kW}}{1 \text{ kJ/s}}\right|\left(\frac{24 \text{ h}}{\text{day}}\right)\left(\frac{\$0.08}{\text{kW}\cdot\text{h}}\right) = \$10.53/\text{day} \longleftarrow$$

For a reversible heat pump operating between reservoirs at T_C, T_H, Eq. 5.10 gives

$$\gamma_{MAX} = \frac{T_H}{T_H - T_C} = \frac{294}{294 - 273} = 14$$

Then

$$(\dot{W}_{cycle})_{MIN} = \frac{75000 \text{ kJ/h}}{14} = 5357 \frac{\text{kJ}}{\text{h}}$$

Costing:

$$(\dot{\$})_{MIN} = (5357)\left|\frac{1}{3600}\right|(24)(0.08) = \$2.86/\text{day} \longleftarrow$$

(b) For electrical-resistance heating, electricity would provide the full 75,000 kJ/h needed. Thus

Costing:

$$(\dot{\$})_{\substack{\text{electrical}\\\text{resistance}}} = (75{,}000)\left|\frac{1}{3600}\right|(24)(0.08) = \$40/\text{day} \longleftarrow$$

PROBLEM 5.59

KNOWN: Operating data are provided for a heat pump that maintains a dwelling at temperature T.

FIND: (a) When T = 20°C, determine the minimum theoretical daily operating cost. (b) Plot the minimum theoretical daily operating cost versus T.

SCHEMATIC & GIVEN DATA:

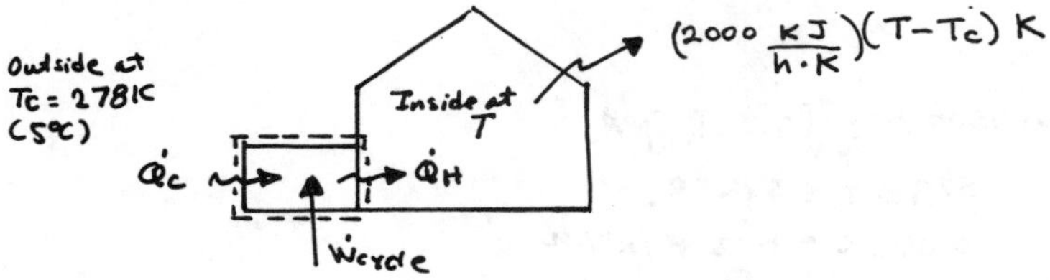

ASSUMPTIONS: 1. The system shown in the schematic undergoes a heat pump cycle. 2. The inside air and outside air play the roles of hot and cold reservoirs, respectively. 3. Electricity costs 8 cents per kW·h.

ANALYSIS: To maintain the dwelling at temperature T, the heat pump must provide energy to the dwelling at the rate

$$\dot{Q}_H = \left(2000 \frac{kJ}{h \cdot K}\right)(T - T_C) \quad , T_C = 278K \qquad (1)$$

The coefficient of performance must satisfy the following relationship, written using Eqs. 2.47 and 5.10

$$\frac{\dot{Q}_H}{\dot{W}_{cycle}} \leq \frac{T}{T - T_C} \longrightarrow \dot{W}_{cycle} \geq \left[\frac{T - T_C}{T}\right]\dot{Q}_H$$

Introducing Eq. (1)

$$\dot{W}_{cycle} \geq \left[\frac{T - T_C}{T}\right]\left[\left(2000 \frac{kJ}{h \cdot K}\right)(T - T_C)K\right] = \left(2000 \frac{kJ}{h \cdot K}\right)\frac{(T - T_C)^2}{T} \qquad (2)$$

where T, Tc are in K.

(a) When T = 293 K (20°C), Eq. (2) gives

$$\dot{W}_{cycle} \geq \left(2000 \frac{kJ}{h \cdot K}\right)\frac{(293 - 278)^2}{293} K = 1536 \frac{kJ}{h}$$

Thus, the minimum theoretical power required is 1536 kJ/h. The corresponding cost on a daily basis is

$$(\dot{\$})_{min} = \left(1536 \frac{kJ}{h}\right)\left|\frac{1h}{3600s}\right|\left|\frac{1kW}{1kJ/s}\right|\left|\frac{24h}{day}\right|\left(\frac{\$0.08}{kW \cdot h}\right) = \$0.82/day \longleftarrow$$

(b) For T ranging from 18 to 23°C, the cost on a daily basis is

$$(\dot{\$})_{min} = \left[2000 \frac{(T - T_C)^2}{T} \frac{kJ}{h}\right]\left|\frac{1h}{3600s}\right|\left|\frac{1kW}{1kJ/s}\right|\left|\frac{24h}{day}\right|\left(\frac{\$0.08}{kW \cdot h}\right) = 1.067\left[\frac{(T - T_C)^2}{T}\right](\$/day)$$

PLOT:

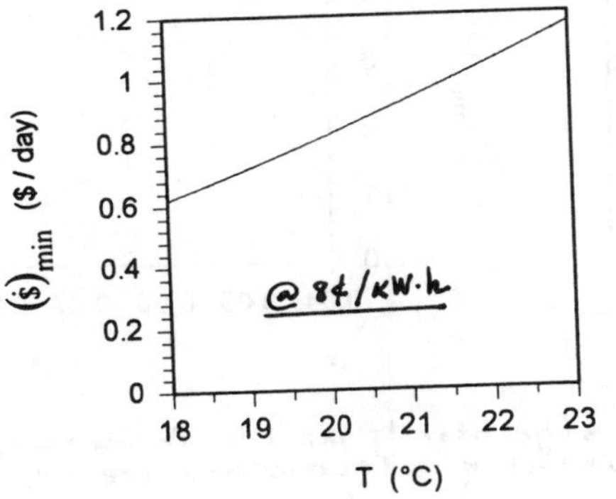

The minimum daily cost increases significantly as the inside temperature is required to be higher.

PROBLEM 5.60

KNOWN: Operating data are provided for a heat pump that maintains a dwelling at temperature T.

FIND: For a given electricity cost, plot the minimum theoretical operating cost versus T. For a given value of T, plot the minimum theoretical operating cost versus the cost of electricity in cents per kW·h.

SCHEMATIC & GIVEN DATA:

Outside at $T_C = 480°R$

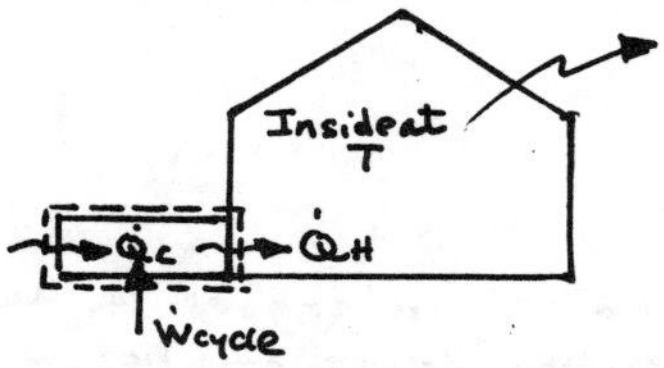

$1500 \frac{Btu}{h\cdot°R}[T - T_C]°R$

$528 \leq T \leq 532°R$

$0.04 \leq c \leq 0.12$ \$/kW·h

ASSUMPTIONS: 1. The system shown in the schematic undergoes a heat pump cycle. 2. The inside air and outside air play the roles of hot and cold reservoirs, respectively.

ANALYSIS: To maintain the dwelling at temperature T, the heat pump must provide energy to the dwelling at the rate

$$\dot{Q}_H = \left(1500 \frac{Btu}{h\cdot°R}\right)[T - T_C], \quad T_C = 480°R \tag{1}$$

The coefficient of performance must satisfy the following relationship, written using Eqs. 2.47 and 5.10

$$\frac{\dot{Q}_H}{\dot{W}_{cycle}} \leq \frac{T}{T - T_C} \Rightarrow \dot{W}_{cycle} \geq \left[\frac{T - T_C}{T}\right]\dot{Q}_H \Rightarrow (\dot{W}_{cycle})_{MIN} = \left[\frac{T - T_C}{T}\right]\dot{Q}_H$$

Introducing Eq. (1) and the unit cost of electricity, c, the minimum theoretical daily operating cost is

$$(\dot{\$})_{MIN} = \left[\frac{T - T_C}{T}\right]\left[\left(1500 \frac{Btu}{h\cdot°R}\right)[T - T_C]°R\right]\left(\frac{24\,h}{day}\right)\left|\frac{1\,kW\cdot h}{3413\,Btu}\right| c\left(\frac{\$}{kW\cdot h}\right)$$

$$= 10.55\left[\frac{(T - T_C)^2}{T}\right] c \left(\frac{\$}{day}\right) \tag{2}$$

<u>Sample calculation</u>: T = 530°R (70°F), c = \$0.08/kW·h. Then, $(\dot{\$})_{MIN} = 10.55\left[\frac{(530 - 480)^2}{530}\right](.08) =$ \$3.98/day.

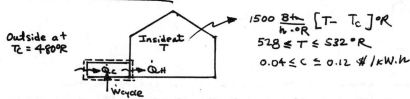

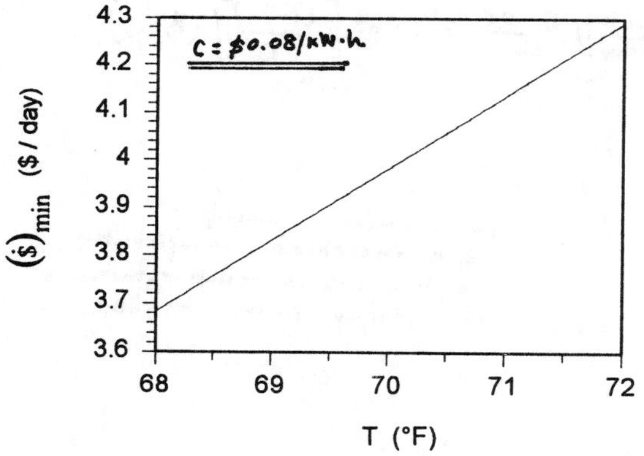

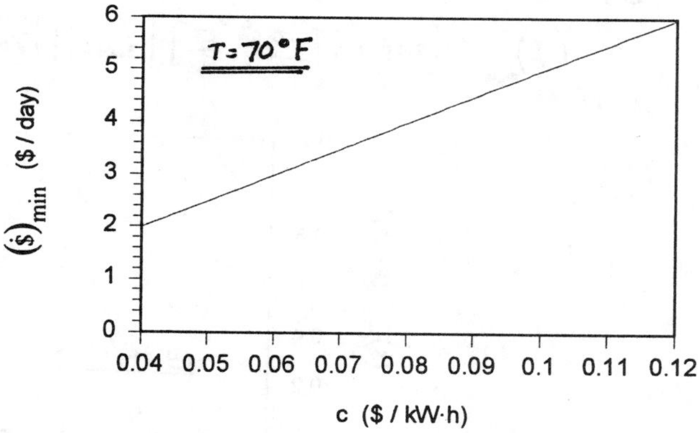

The minimum daily cost increases significantly as the inside temperature is required to be higher and/or as the unit cost of electricity increases.

PROBLEM 5.61

KNOWN: 0.5 kg of water executes a Carnot cycle for which property data are provided.

FIND: Sketch the cycle on p-v coordinates, evaluate the heat and work for each process, and evaluate the thermal efficiency.

SCHEMATIC AND GIVEN DATA:

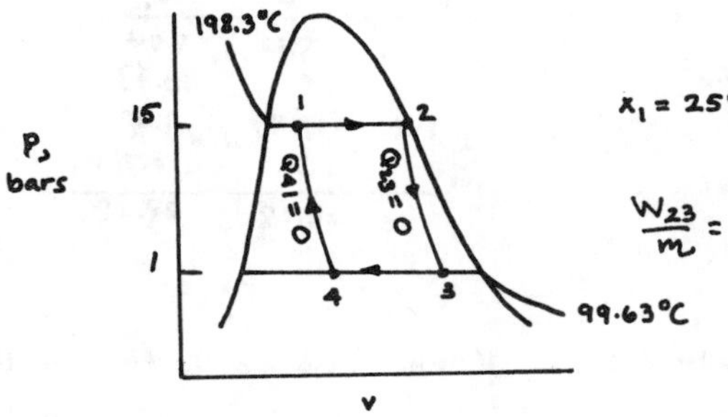

$x_1 = 25\%$

$\frac{W_{23}}{m} = 403.8 \frac{kJ}{kg}$

Summary:

Process	Q	W
1-2	730.3	73.5
2-3	0	201.9
3-4	−578.3	−43.3
4-1	0	−79.9

ASSUMPTION: The system shown in the figure undergoes a Carnot cycle.

Process 1-2: $\frac{W_{12}}{m} = \int_1^2 p\,dv = p(v_2 - v_1)$. An energy balance gives $\frac{Q_{12}}{m} = u_2 - u_1 + \frac{W_{12}}{m}$. Thus

$$\frac{Q_{12}}{m} = (u_2 - u_1) + p(v_2 - v_1)$$
$$= h_2 - h_1$$

with data from Table A-3. $v_2 = 0.1318\ m^3/kg$, $h_2 = 2792.2\ kJ/kg$, $u_2 = 2594.5\ kJ/kg$

$$v_1 = v_f + x(v_g - v_f) = 1.1539 \times 10^{-3} + 0.25(0.1318 - 1.1539 \times 10^{-3}) = 33.815 \times 10^{-3}\ m^3/kg$$

$$h_1 = h_f + x(h_g - h_f) = 844.84 + 0.25(1947.3) = 1331.67\ kJ/kg.$$

With these values, $Q_{12} = (0.5 kg)(2792.2 - 1331.67) kJ/kg = 730.3\ kJ$, and ← Q_{12}

$$W_{12} = (0.5 kg)(15 \text{ bars})\left(10^5 \frac{N/m^2}{bar}\right)\left[(131.8 - 33.815) \times 10^{-3} \frac{m^3}{kg}\right]\left(\frac{kJ}{10^3 N\cdot m}\right) = 73.5\ kJ$$ ← W_{12}

Process 2-3: $Q_{23} = 0$, $W_{23}/m = 403.8\ kJ/kg$. Thus $W_{23} = 201.9\ kJ$. ← Q_{23}, W_{23}
Using an energy balance and data from Table A-3, $u_3 = 2190.7\ kJ/kg$, giving $x_3 = 0.849$.

Process 3-4: As for process 1-2, $W_{34}/m = p(v_4 - v_3)$, $Q_{34}/m = (h_4 - h_3)$. Also, since the system undergoes a reversible cycle while communicating with reservoirs at $T_H = 471\ K$, $T_C = 373\ K$, Eq. 5.6 is applicable

$$\frac{|Q_{34}|}{Q_{12}} = \frac{373}{471} \Rightarrow h_3 - h_4 = \frac{373}{471}(1460.53) = 1156.6 \frac{kJ}{kg}$$

Thus, $h_4 = h_3 - 1156.6 = (417.46 + 0.849(2258)) - 1156.6 = 1177.9\ kJ/kg$. Using this value, $x_4 = (1177.9 - 417.46)/2258 = 0.337$. Then, $v_4 = 1.0432 \times 10^{-3} + 0.337[1.694 - 1.0432 \times 10^{-3}] = 0.5716\ m^3/kg$, $u_4 = 417.36 + 0.337(2506.1 - 417.36) = 1121.27\ kJ/kg$. Also, $v_3 = 1.438\ m^3/kg$.
Finally, $Q_{34} = -578.3\ kJ$, $W_{34} = -43.3\ kJ$. ← Q_{34}, W_{34}

Process 4-1: $Q_{41} = 0$. $W_{41}/m = u_4 - u_1 = 1121.27 - (843.16 + 0.25(1751.34)) = -159.73\ kJ/kg$.
So, $W_{41} = -79.9\ kJ$. ← Q_{41}, W_{41}

Thermal Efficiency.

Method #1: $$\eta = \frac{W_{cycle}}{Q_{in}} = \frac{73.5 + 201.9 - 43.3 - 79.9}{730.3} = 0.208\ (20.8\%)$$

Method #2: Equation 5.8
$$\eta_{MAX} = 1 - \frac{T_C}{T_H} = 1 - \frac{(99.63 + 273.15)}{(198.3 + 273.15)} = 1 - \frac{372.78}{471.45} = 0.208\ (20.8\%)$$

PROBLEM 5.62

KNOWN: 0.1 lb of water executes a Carnot cycle for which property data are provided.

FIND: Sketch the cycle on p-v coordinates, evaluate the heat and work for each process, and evaluate the thermal efficiency.

SCHEMATIC & GIVEN DATA:

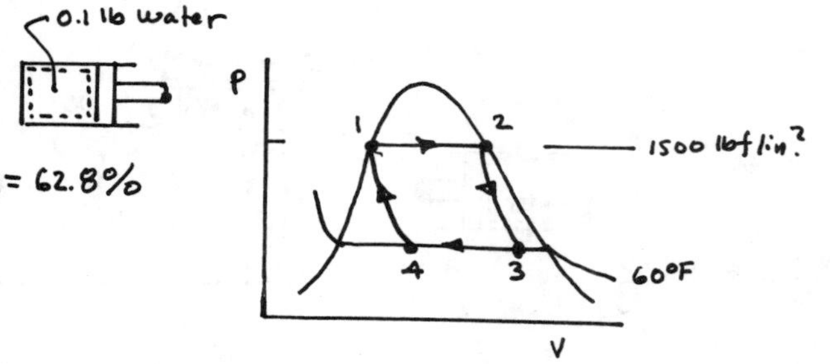

$x_3 = 62.8\%$

SUMMARY:

Process	Q	W
1-2	55.72	7.04
2-3	0	43.43
3-4	-27.44	-1.48
4-1	0	-20.71
Cycle	28.28	28.28

ASSUMPTION: 1. The system is shown in the schematic. 2. Volume change is the only work mode.

ANALYSIS: (b) Process 1-2: $W_{12}/m = \int_1^2 p\,dv = p(v_2 - v_1)$. Energy balance: $Q_{12}/m = (u_2 - u_1) + W_{12}/m$. Thus, $Q_{12}/m = (u_2 - u_1) + p(v_2 - v_1) = h_2 - h_1$. With data from Table A-3E, $Q_{12} = 0.1\text{lb}(557.2\ \frac{\text{Btu}}{\text{lb}}) = 55.72\ \text{Btu}$. $W_{12} = (0.1\text{lb})(1500 \times 144\ \text{lbf/ft}^2)(0.277 - 0.02346)(\text{ft}^3/\text{lb})\ \left|\frac{1\ \text{Btu}}{778\ \text{ft}\cdot\text{lbf}}\right| = 7.04\ \text{Btu}$.

Process 2-3: $Q_{23} = 0$. An energy balance reduces to $W_{23} = m(u_2 - u_3)$. From Table A-3, $u_2 = 1091.8$ Btu/lb. Table A-2: $u_3 = 28.08 + 0.628(1030.4 - 28.08) = 657.54$ Btu/lb. Then $W_{23} = 0.1(1091.8 - 657.54) = 43.43$ Btu.

Process 3-4: As for process 1-2: $W_{34} = mp(v_4 - v_3)$, $Q_{34} = m(h_4 - h_3)$. Since the system undergoes a reversible cycle while communicating with reservoirs at $T_H = 1056°R$ (sat. temp. for 1500 lbf/in²) and $T_C = 520°R$, Eq. 5.6 is applicable: $\frac{|Q_{34}|}{520} = \frac{Q_{12}}{1056} \Rightarrow Q_{34} = -27.44$ Btu.

Thus, $h_4 = \frac{Q_{34}}{m} + h_3 = -274.4 + (28.08 + 0.628(1059.6)) = 419.11$ Btu/lb.

Using this, the quality at state 4 can be determined: $x_4 = (h_4 - h_f)/h_{fg} = (419.11 - 28.08)/1059.6 = 0.369$.

Accordingly, $(v_4 - v_3) = (v_f + x_4 v_{fg}) - (v_f + x_3 v_{fg}) = v_{fg}(x_4 - x_3) = [1207 - 0.01604](0.369 - 0.628) = -312.61\ \text{ft}^3/\text{lb}$. Finally, $W_{34} = (0.1\ \text{lb})(0.2563 \times 144\ \text{lbf/ft}^2)(-312.61\ \text{ft}^3/\text{lb})\ |1\ \text{Btu}/778\ \text{ft}\cdot\text{lbf}| = -1.48$ Btu.

Process 4-1: $Q_{41} = 0$. Energy balance, $W_{41} = m(u_4 - u_1)$, $u_4 = 28.08 + 0.369(1030.4 - 28.08) = 397.94$ Btu/lb, $u_1 = 605$ Btu/lb. Then, $W_{41} = 0.1(397.94 - 605) = -20.71$ Btu

(c) The thermal efficiency is found using

$$\eta = 1 - \frac{T_C}{T_H} = 1 - \frac{520}{1056} = 0.508\ (50.8\%)$$

or

① $$\eta = \frac{W_{cycle}}{Q_{in}} = \frac{7.04 + 43.43 - 1.48 - 20.71}{55.72} = \frac{28.28}{55.72} = 0.508\ (50.8\%)$$

---

1. For any cycle, $W_{cycle} = Q_{cycle}$. This provides a check on calculations here:

$$28.28 = 55.72 + 0 - 27.44 + 0$$
$$= 28.28\ ✓$$

PROBLEM 5.63

KNOWN: One kg of air undergoes a Carnot cycle for which $\eta = 60\%$.

FIND: Determine the minimum and maximum temperatures, the pressure and volume at the beginning of the isothermal expansion, the work and heat transfer for each process, and sketch the cycle on p-v coordinates.

SCHEMATIC & GIVEN DATA:

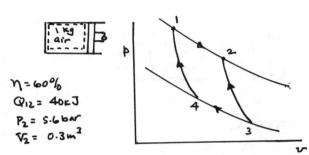

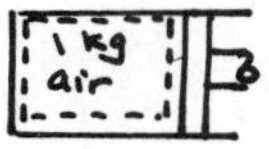

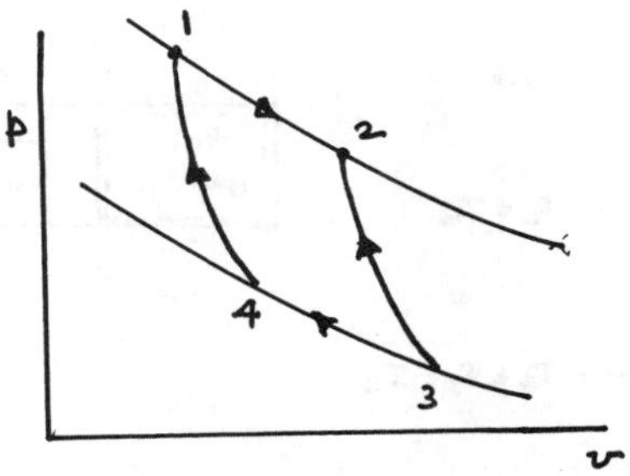

$\eta = 60\%$

$Q_{12} = 40\text{ kJ}$

$p_2 = 5.6\text{ bar}$

$V_2 = 0.3\text{ m}^3$

Summary:

Process	Q	W
1-2	40	40
2-3	0	256.7
3-4	−16	−16
4-1	0	−256.7
cycle	24	24

ASSUMPTIONS: 1. The system shown in the schematic consists of air modeled as an ideal gas. 2. Volume change is the only work mode.

ANALYSIS: (a) Using the ideal gas model equation of state, $T_2 = p_2 V_2 / mR$. Then

$$T_2 = \frac{(5.6\times10^5\,\text{N/m}^2)(0.3\,\text{m}^3)}{\left(\frac{8314}{28.97}\frac{\text{N}\cdot\text{m}}{\text{kg}\cdot\text{K}}\right)(1\,\text{kg})} = 585.4\,\text{K}.$$ Then, since $\eta = 1 - \frac{T_C}{T_H} \Rightarrow T_C = T_H(1-\eta)$. With $T_H = T_2$,

$T_C = 585.4\,\text{K}(1-0.6) = 234.2\,\text{K}$, where $T_C = T_3 = T_4$.

(b) For process 1-2, $Q_{12} = 40\text{ kJ}$ (given). An energy balance reads $m(u_2-u_1) = Q_{12} - W_{12}$, but since internal energy of an ideal gas depends on temperature and $T_1 = T_2$, $W_{12} = Q_{12}$. Further, $W_{12} = \int_1^2 p\,dV = \int_1^2 \frac{mRT_H}{V}dV = mRT_H \ln V_2/V_1$. Solving and inserting values

$$\ln\frac{V_2}{V_1} = \frac{W_{12}}{mRT_H} \Rightarrow \ln\frac{V_2}{V_1} = \frac{40\,\text{kJ}}{(1\,\text{kg})\left(\frac{8.314}{28.97}\frac{\text{kJ}}{\text{kg}\cdot\text{K}}\right)(585.4\,\text{K})} = 0.2381 \Rightarrow V_1 = 0.24\,\text{m}^3$$

Since $T_1 = T_2$, $p_1V_1 = mRT$, $p_2V_2 = mRT \Rightarrow p_2V_2 = p_1V_1$, $p_1 = p_2(V_2/V_1) = 5.6\,\text{bar}(0.3/0.24) = 7\,\text{bar}$.

(c) For process 2-3: $Q_{23} = 0$. An energy balance reduces to give $W_{23} = m(u_2 - u_3)$ with data from Table A-22, $W_{23} = (1\,\text{kg})(423.7 - 167.0) = 256.7\,\text{kJ}$. For process 3-4, $W_{34} = Q_{34}$ (as for process 1-2). Also, Eq. 5.6 is applicable:

$$\frac{|Q_{34}|}{T_C} = \frac{Q_{12}}{T_H} \Rightarrow |Q_{34}| = 0.4(40\,\text{kJ}) = 16\,\text{kJ} \Rightarrow Q_{34} = -16\,\text{kJ},\ W_{34} = -16\,\text{kJ}.$$

(1) Process 4-1, $Q_{41} = 0$. An energy balance reduces to give, $W_{41} = m(u_4 - u_1)$. Since $u_1 = u_2$, $u_4 = u_3$, $W_{41} = -256.7\,\text{kJ}$. The work and heat transfers are summarized in the table.

---

1. As checks on the calculations, note that (1) $W_{cycle} = Q_{cycle}$ (see summary above), and (2)

$$\eta = \frac{W_{cycle}}{Q_{in}} = \frac{24\,\text{kJ}}{40\,\text{kJ}} = 0.6$$

PROBLEM 5.64

KNOWN: A carnot cycle is executed by an ideal gas with constant specific heat ratio k.

FIND: Show that (a) $V_4 V_2 = V_1 V_3$, (b) $T_2/T_3 = (p_2/p_3)^{(k-1)/k}$, (c) $T_2/T_3 = (V_3/V_2)^{k-1}$

SCHEMATIC & GIVEN DATA.

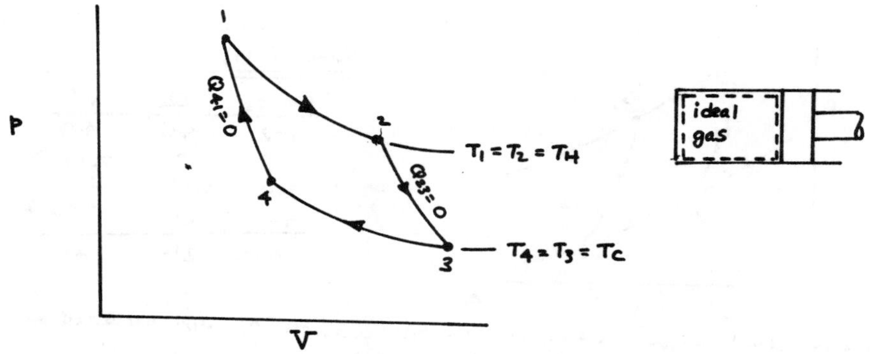

ASSUMPTIONS: (1) The system shown in the figure consists of an ideal gas. (2) The specific heat ratio k is constant (required in part (b) only). (3) The system undergoes a Carnot cycle.

ANALYSIS: (a) The thermal efficiency is

$$\eta = \frac{W_{cycle}}{Q_{in}} = \frac{W_{12} + W_{23}}{Q_{12}}$$

Since internal energy of an ideal gas depends on temperature only, an energy balance for process 1-2 reduces to $U_2 - U_1 = Q_{12} - W_{12}$, where $U_2 = U_1$. Thus, $Q_{12} = W_{12}$. Furthermore

$$W_{12} = \int_1^2 p\,dV = \int_1^2 \frac{mRT_H}{V}\,dV = mRT_H \ln\frac{V_2}{V_1}$$

Similarly, $W_{23} = mRT_C \ln V_4/V_3$. Collecting results

$$\eta = 1 - \frac{mRT_C \ln V_3/V_4}{mRT_H \ln V_2/V_1} = 1 - \left(\frac{\ln V_3/V_4}{\ln V_2/V_1}\right)\frac{T_C}{T_H}$$

However, for the Carnot cycle $\eta = 1 - T_C/T_H$, so it is necessary that

$$\frac{\ln(V_3/V_4)}{\ln(V_2/V_1)} = 1 \Rightarrow \ln\left(\frac{V_3}{V_4}\right) = \ln\left(\frac{V_2}{V_1}\right) \Rightarrow V_4 V_2 = V_3 V_1 \longleftarrow \quad (a)$$

(b) As process 2-3 is adiabatic, a energy balance in differential form reads

$$dU = \cancelto{0}{\delta Q} - \delta W$$

where $\delta W = p\,dV$ and with assumption 1 $dU = mc_v dT$. Collecting results and using $pV = mRT$ and $c_v = R/(k-1)$ (Eq. 3.47b)

$$\frac{1}{k-1} d\ln T = -d\ln V$$

Integration for constant k (assumption 2) gives

$$\ln\frac{T_3}{T_2} = -\ln\left(\frac{V_3}{V_2}\right)^{k-1} \Rightarrow \frac{T_3}{T_2} = \left(\frac{V_2}{V_3}\right)^{k-1} \longleftarrow \quad (c)$$

Finally, using $V = mRT/p$

$$\frac{T_3}{T_2} = \left[\frac{T_2}{T_3}\frac{p_3}{p_2}\right]^{k-1} \Rightarrow \frac{T_3}{T_2} = \left(\frac{p_3}{p_2}\right)^{\frac{k-1}{k}} \longleftarrow \quad (b)$$

PROBLEM 5.65

KNOWN: A quantity of N_2 undergoes a Carnot cycle.

FIND: Determine the work and heat transfer for each process, the thermal efficiency, and the pressures at the initial and final states of the isothermal compression.

SCHEMATIC & GIVEN DATA:

$k = 1.4$

$P_1 = 30$ bar

$P_2 = 15$ bar

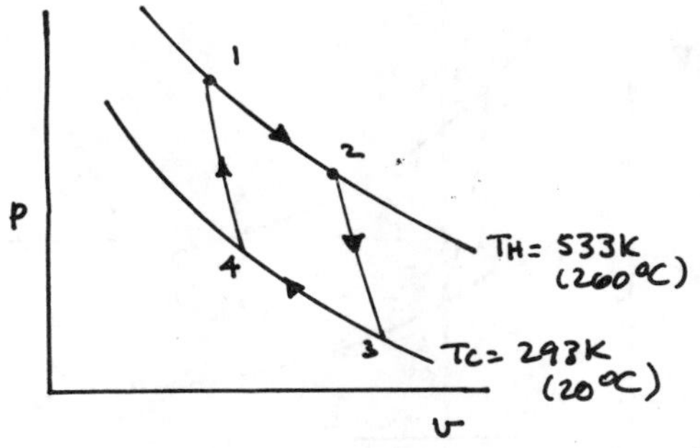

Summary:

Process	Q	W
1-2	109.7	109.7
2-3	0	178.1
3-4	-60.3	-60.3
4-1	0	-178.1
Cycle	49.4	49.4

ASSUMPTIONS: 1. The system consists of a quantity of N_2 modeled as an ideal gas with $k = 1.4$. 2. Volume change is the only work mode.

ANALYSIS: (a) Process 1-2: Since $u_2 = u_1$ for an ideal gas at fixed temperature, an energy balance reduces to give $Q_{12} = W_{12}$, where $W_{12} = \int_1^2 p\,dV = \int_1^2 \frac{mRT_H\,dV}{V}$, or $W_{12}/m = RT_H \ln(V_2/V_1)$. With $p_1V_1 = mRT_H$, $p_2V_2 = mRT_H$, we have

$$\frac{W_{12}}{m} = RT_H \ln\frac{p_1}{p_2} = \left(\frac{8.314}{28.01}\right)\left(\frac{kJ}{kg\cdot K}\right)(533K)\ln\left(\frac{30}{15}\right) = 109.7 \text{ kJ/kg}$$

Process 2-3: $Q_{23} = 0$. An energy balance gives $W_{23} = m(u_2 - u_3)$. For an ideal gas with constant k, Eq. 3.47b gives $c_v = \frac{R}{k-1}$. Then

$$\frac{W_{23}}{m} = c_v(T_2 - T_3) = \frac{R}{k-1}(T_H - T_C) = \left(\frac{8.314/28.01}{1.4-1}\right)(533 - 293) = 178.1 \text{ kJ/kg}$$

Process 3-4: Since $u_3 = u_4$, an energy balance gives $Q_{34} = W_{34}$, where

$$\frac{W_{34}}{m} = \int_3^4 \frac{RT_C}{V}dV = RT_C \ln\frac{V_4}{V_3} = RT_C \ln\frac{p_3}{p_4} = \left(\frac{8.314}{28.01}\right)(293)\ln\frac{1}{2} = -60.3 \text{ kJ/kg}$$

To find p_3, p_4, use part (b) of Problem 5.64:

$$\frac{p_3}{p_2} = \left(\frac{T_C}{T_H}\right)^{\frac{k}{k-1}} \Rightarrow p_3 = 15\text{bar}\left[\frac{293}{533}\right]^{\frac{1.4}{.4}} = 1.85 \text{bar}$$

① $$\frac{p_4}{p_1} = \left(\frac{T_C}{T_H}\right)^{\frac{k}{k-1}} \Rightarrow \frac{p_4}{p_1} = \frac{p_3}{p_2} \Rightarrow p_4 = \frac{p_1}{p_2}p_3 = 3.7 \text{bar}$$

← (c)

Process 4-1: $Q_{41} = 0$, and an energy balance gives $W_{41} = m(u_4 - u_1)$, or

$$\frac{W_{34}}{m} = c_v[T_4 - T_1] = \frac{R}{k-1}(T_4 - T_1) = \left(\frac{8.314/28.01}{1.4-1}\right)(293 - 533) = -178.1 \frac{kJ}{kg}$$

(b) The thermal efficiency is given in this case by

$$\eta = \frac{W_{cycle}}{Q_{in}} = \frac{109.7 + 178.1 - 60.3 - 178.1}{109.7} = \frac{49.4}{109.7} = .45 \ (45\%)$$

or using Eq. 5.8

$$\eta = 1 - \frac{T_C}{T_H} = 1 - \frac{293}{533} = 0.45 \ (45\%)$$

---

1. For any cycle, $Q_{cycle} = W_{cycle}$. See summary table above. This provides a check on the calculations.

PROBLEM 5.66

KNOWN: One-half pound of air executes a Carnot refrigeration cycle for which data are provided.

FIND: Determine (a) the pressure at each of the four principal states, (b) the work for each of the four processes, and (c) the coefficient of performance.

SCHEMATIC & GIVEN DATA:

$V_4 = 0.1\ m^3$
$Q_{12} = 7.5$ Btu
$k = 1.4$

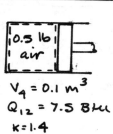

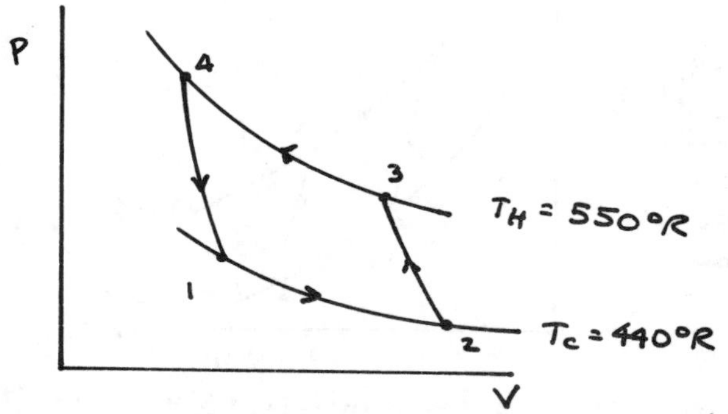

ASSUMPTIONS: (1) The system consists of air modeled as an ideal gas with $k = 1.4$. (2) Volume change is the only work mode.

ANALYSIS: From the solution to Problem 5.64; $T_4/T_1 = (P_4/P_1)^{k-1/k}$ and $V_4 V_2 = V_3 V_1$.

(a) Using the ideal gas equation

$$P_4 = \frac{mRT_4}{V_4} = \frac{(0.5\ \text{lb})\left(\frac{1545}{28.97}\ \frac{\text{ft}\cdot\text{lbf}}{\text{lb}\cdot°R}\right)(550°R)}{(0.1\ m^3)}\left|\frac{1\ \text{ft}^2}{144\ \text{in}^2}\right| = 1018\ \frac{\text{lbf}}{\text{in}^2} \longleftarrow P_4$$

Now, for process 4-1; $T_4/T_1 = (P_4/P_1)^{k-1/k}$, or

$$P_1 = \left(\frac{T_1}{T_4}\right)^{k/k-1} P_4 = \left(\frac{440}{550}\right)^{1.4/1.4-1}\left(1018\ \frac{\text{lbf}}{\text{in}^2}\right) = 466.2\ \frac{\text{lbf}}{\text{in}^2} \longleftarrow P_1$$

The volume at 1 is

$$V_1 = \frac{mRT_1}{P_1} = \frac{(0.5)\left(\frac{1545}{28.97}\right)(440)}{(466.2)}\left|\frac{1}{144}\right| = 0.1748\ \text{ft}^3$$

For process 1-2: $m(u_2 - u_1) = Q_{12} - W_{12}$. Since $T_1 = T_2 = T_c$, $u_2 = u_1$, and

$$Q_{12} = W_{12} = \int_1^2 p\,dV = \int_1^2 \frac{mRT}{V}dV = mRT_c \ln\frac{V_2}{V_1}$$

Solving for V_2/V_1 and inserting values

$$\frac{V_2}{V_1} = \exp\left[\frac{W_{12}}{mRT_c}\right] = \exp\left[\frac{(7.5\ \text{Btu})}{(0.5\ \text{lb})\left(\frac{1545}{28.97}\ \frac{\text{ft}\cdot\text{lbf}}{\text{lb}\cdot°R}\right)(440°R)}\left|\frac{778\ \text{ft}\cdot\text{lbf}}{1\ \text{Btu}}\right|\right]$$

$$= 1.6443$$

$$\therefore V_2 = (1.6443)(0.1748\ \text{ft}^3) = 0.2874\ \text{ft}^3$$

Thus

$$P_2 = \frac{(0.5)\left(\frac{1545}{28.97}\right)(440)}{(0.2874)}\left|\frac{1}{144}\right| = 283.5\ \frac{\text{lbf}}{\text{in}^2} \longleftarrow P_2$$

Next, from above

$$V_4V_2 = V_3V_1 \Rightarrow V_3 = \left(\frac{V_4}{V_1}\right)V_2 = \left(\frac{0.1}{0.1748}\right)(0.2874\ \text{ft}^3) = 0.1644\ \text{ft}^3$$

Finally

$$p_3 = \frac{(0.5)\left(\frac{1545}{28.97}\right)(550)}{(0.1644)}\left|\frac{1}{144}\right| = 619.5\ \frac{lbf}{in.^2} \longleftarrow p_3$$

(b) Process 1-2: $Q_{12} = W_{12} = 7.5$ Btu $\longleftarrow W_{12}$

process 2-3: $Q_{23} = 0 \Rightarrow W_{23} = -m(u_3 - u_2) = -m\,c_v(T_3 - T_2)$

With $c_v = R/(k-1)$, we get

$$W_{23} = -\frac{mR(T_3 - T_2)}{(k-1)} = -\frac{(0.5\ lb)\left(\frac{1.986}{28.97}\frac{Btu}{lb\cdot°R}\right)(550-440)°R}{(1.4-1)}$$

$= -9.426$ Btu $\longleftarrow W_{23}$

Process 3-4: As for process 1-2 in part (a)

$$W_{34} = \int_3^4 p\,dV = mRT_H \ln\frac{V_4}{V_3}$$

$$= (0.5\ lb)\left(\frac{1.986}{28.97}\frac{Btu}{lb\cdot°R}\right)(550°R)\ln\left(\frac{0.1}{0.1644}\right) = -9.372\ Btu \longleftarrow W_{34}$$

Process 4-1: $Q_{41} = 0 \Rightarrow W_{41} = -m(u_1 - u_4) = -m\,c_v(T_1 - T_4)$. Again

$$W_{41} = -\frac{mR(T_1 - T_4)}{(k-1)} = -\frac{(0.5)\left(\frac{1.986}{28.97}\right)(440-550)}{(1.4-1)} = 9.426\ Btu \longleftarrow W_{41}$$

(c) For the Carnot cycle

(1) $$\beta = \beta_{max} = \frac{T_C}{T_H - T_C} = \frac{440}{550-440} = 4 \longleftarrow \beta$$

---

1. Alternatively,

$$|W_{cycle}| = |W_{12} + W_{23} + W_{34} + W_{41}| = |-1.872\ Btu|$$

and, with $Q_{in} = Q_{12}$ for the cycle

$$\beta = \frac{Q_{in}}{|W_{cycle}|} = \frac{7.5}{1.872} = 4.006$$

The slight difference is due to round-off.

PROBLEM 5.67

KNOWN: Systems execute Carnot refrigeration cycles for which data are provided.

FIND: For each such cycle, evaluate the coefficient of performance.

SCHEMATIC & GIVEN DATA: See the solutions to the respective problems.

ASSUMPTIONS: See the solutions to the respective problems.

ANALYSIS: For each case, Eq. 5.9 is applicable. Accordingly, $\beta_{MAX} = T_C/(T_H - T_C)$. ①

(a) PROBLEM 5.61

$$\beta_{MAX} = \frac{373}{471-373} = 3.81$$

(b) PROBLEM 5.62

$$\beta_{MAX} = \frac{520}{1056-520} = 0.97$$

(c) PROBLEM 5.63

$$\beta_{MAX} = \frac{234.2}{585.4-234.2} = 0.67$$

(d) PROBLEM 5.65

$$\beta_{MAX} = \frac{293}{533-293} = 1.22$$

---

1. Using the result of Problem 5.18(a),

$$\beta_{MAX} = \frac{1-\eta_{MAX}}{\eta_{max}}$$

(a) $\beta_{MAX} = \frac{1-0.208}{0.208} = 3.81$

(b) $\beta_{MAX} = \frac{1-0.508}{0.508} = 0.97$

(c) $\beta_{MAX} = \frac{1-0.6}{0.6} = 0.67$

(d) $\beta_{MAX} = \frac{1-0.45}{0.45} = 1.22$

PROBLEM 5.68

KNOWN: Systems execute Carnot heat pump cycles for which data are provided.

FIND: For each cycle, evaluate the coefficient of performance.

SCHEMATIC & GIVEN DATA: See the solutions to the respective problems.

ASSUMPTIONS: See the solutions to the respective problems.

ANALYSIS: For each case, Eq. 5.10 is applicable. Accordingly, $\gamma_{MAX} = T_H/(T_H - T_C)$. ①

(a) PROBLEM 5.61

$$\gamma_{MAX} = \frac{471}{471-373} = 4.81$$

(b) PROBLEM 5.62

$$\gamma_{MAX} = \frac{1056}{1056-520} = 1.97$$

(c) PROBLEM 5.63

$$\gamma_{MAX} = \frac{585.4}{585.4-234.2} = 1.67$$

(d) PROBLEM 5.65

$$\gamma_{MAX} = \frac{533}{533-293} = 2.22$$

---

1. Using the result of Problem 5.18(b),

$$\gamma_{MAX} = \frac{1}{\eta_{MAX}}$$

(a) $\gamma_{MAX} = \frac{1}{0.208} = 4.81$

(b) $\gamma_{MAX} = \frac{1}{0.508} = 1.97$

(c) $\gamma_{MAX} = \frac{1}{0.6} = 1.67$

(d) $\gamma_{MAX} = \frac{1}{0.45} = 2.22$

CHAPTER SIX

USING ENTROPY

Chapter 6 - Third and fourth edition problem correspondence.

3rd	4th	3rd	4th	3rd	4th	3rd	4th
6.1	6.1*	6.37	6.46*	6.87	6.89*	6.122	6.132*
---	6.2	6.33	6.47	6.88	---	6.123	---
6.2	6.3*	6.39	6.48*	---	6.90	---	6.133
6.3	6.4*	6.40	6.49*	6.89	6.91	6.124	6.134
6.4	6.5	6.44	6.50*	6.90	---	6.125	6.135
6.5	6.6	6.45	6.51	---	6.92	6.127	6.136
6.6	6.7	6.46	6.52	6.91	6.93*	6.126	6.137*
6.18	6.8*	6.47	6.53*	6.92	6.94*	6.128	6.138
6.22	6.9*	6.48	6.54*	6.93	6.95*	6.129	6.139
6.38	6.10*	6.49	6.55*	6.94	6.96*	6.130	6.140*
6.41	6.11	6.50	6.56*	6.95	6.97*	6.131	6.141*
6.42	6.12*	6.51	6.57*	6.97	6.98*	6.132	6.142*
6.43	6.13	6.52	6.58*	6.98	6.99*	6.133	6.143*
6.66	6.14	6.53	6.59*	6.99	6.100*	6.134	6.144*
6.74	6.15	6.54	6.60*	6.100	6.101	6.135	6.145
6.75	6.16	6.55	6.61*	6.101	6.102*	6.136	6.146*
6.79	6.17*	6.56	6.62*	6.102	6.103*	6.137	6.147*
6.96	6.18	6.57	6.63*	6.103	6.104*	6.138	6.148*
6.7	6.19*	---	6.64	---	6.105	6.139	6.149
6.8	6.20*	---	6.65	---	6.106	6.140	6.150*
6.9	6.21*	6.58	6.66	---	6.107	6.141	6.151*
6.10	6.22*	6.59	6.67	---	6.108	6.142	6.152*
6.11	6.23*	6.60	6.68*	6.104	6.109*	6.143	6.153*
6.12	6.24*	6.70	6.69	6.105	6.110	6.144	6.154*
6.13	6.25*	6.62	6.70*	6.106	6.111	6.145	6.155*
6.14	6.26*	6.65	6.71	6.107	6.112*	6.146	6.156*
6.15	6.27*	6.61	6.72*	6.108	6.113	6.147	6.157*
6.16	6.28*	6.63	6.73*	6.109	6.114*	6.148	6.158*
6.17	6.29	6.64	6.74*	6.110	---	6.149	6.159*
6.23	6.30*	6.68	6.75	---	6.115	6.150	6.160*
6.19	6.31*	6.69	6.76*	6.111	6.116*	6.151	6.161*
6.20	---	6.71	6.77*	---	6.117	6.152	6.162*
---	6.32	6.72	6.78*	6.112	6.118*	6.153	6.163*
6.21	6.33*	6.67	6.79	6.113	6.119*	6.154	6.164*
6.24	6.34	6.73	6.80*	6.114	6.120*	6.156	6.165
6.25	6.35	6.76	---	6.115	6.121*	6.157	6.166
6.26	6.36*	6.77	---	6.116	6.122*	6.158	6.167
6.27	6.37*	6.78	6.81	---	6.123	6.155	6.168*
6.28	6.38*	6.80	---	---	6.124	6.159	6.169*
6.29	6.39*	---	6.82	---	6.125	6.160	6.170*
6.30	6.40*	6.81	6.83	---	6.126	6.161	6.171*
6.31	6.41*	6.82	6.84	6.117	6.127	6.162	6.172*
6.34	6.42*	6.83	6.85	6.118	6.128*	6.163	6.173*
6.35	6.43	6.84	6.86*	6.119	6.129	6.164	6.174*
6.32	6.44*	6.86	6.87*	6.120	6.130	6.165	6.175
6.36	6.45*	6.85	6.88*	6.121	6.131*		

* Revised

PROBLEM 6.1

KNOWN: A system undergoes a power cycle for which data are provided.

FIND: Using Eq. 6.2, evaluate σ_{cycle} if η is 75, 50, and 25%. Discuss.

SCHEMATIC & GIVEN DATA:

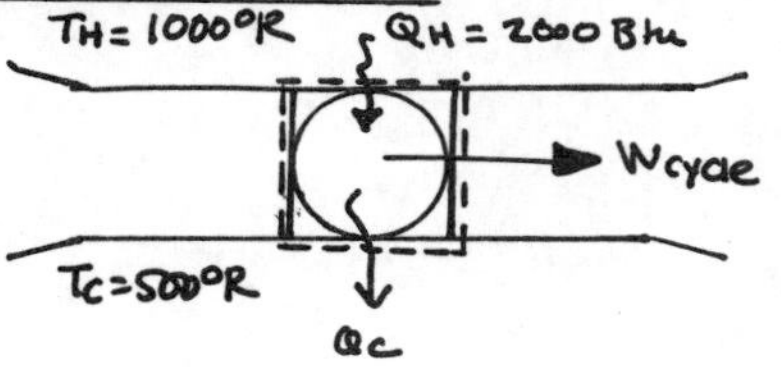

ASSUMPTION: The system shown in the accompanying figure undergoes a power cycle while receiving Q_H at T_H and discharging Q_C at T_C.

ANALYSIS: For any cycle, $\eta = 1 - Q_C/Q_H \Rightarrow Q_C = (1-\eta)Q_H$. Then, Eq. 6.2 gives

$$\sigma_{cycle} = -\oint\left(\frac{\delta Q}{T}\right)_b = -\left[\frac{Q_H}{T_H} - \frac{Q_C}{T_C}\right] = -Q_H\left[\frac{1}{T_H} - \frac{(1-\eta)}{T_C}\right]$$

(a) $\eta = 75\%$

$$\sigma_{cycle} = -(2000\ \text{Btu})\left[\frac{1}{1000°R} - \frac{(1-0.75)}{500°R}\right] = -1\ \frac{\text{Btu}}{°R}$$

① Since σ_{cycle} must be positive or zero in value, this case is impossible.

(b) $\eta = 50\%$

$$\sigma_{cycle} = -(2000\ \text{Btu})\left[\frac{1}{1000°R} - \frac{(1-.50)}{500°R}\right] = 0$$

Since $\sigma_{cycle} = 0$, this case corresponds to internally reversible operation.

(c) $\eta = 25\%$

$$\sigma_{cycle} = -(2000\ \text{Btu})\left[\frac{1}{1000°R} - \frac{(1-0.25)}{500°R}\right] = +1\ \frac{\text{Btu}}{°R}$$

In this case, irreversibilities are present within the system.

---

1. Using Eq. 5.8, the maximum thermal efficiency any cycle can have while operating between reservoirs at T_H and T_C is

$$\eta_{MAX} = 1 - \frac{T_C}{T_H} = 1 - \frac{500}{1000} = 0.5\ (50\%)$$

PROBLEM 6.2

KNOWN: A system undergoes a reversible power cycle while communicating thermally with three reservoirs at specified temperatures. The values for the energy transfer by heat between two of the reservoirs and the system are also specified.

FIND: Determine the thermal efficiency.

SCHEMATIC & GIVEN DATA:

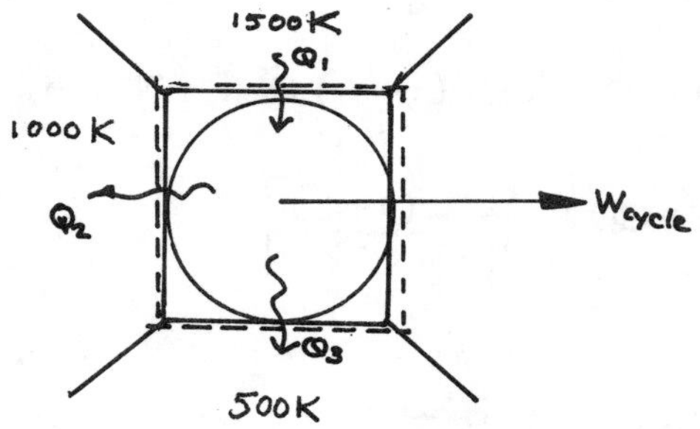

$Q_1 = 750\text{ kJ}$

$Q_3 = 100\text{ kJ}$

ASSUMPTION: 1. The system shown in the accompanying figure undergoes a power cycle with no internal irreversibilities. 2. All heat transfers take place at the indicated temperatures.

ANALYSIS: The thermal efficiency is

$$\eta = \frac{W_{cycle}}{Q_{in}} = \frac{W_{cycle}}{Q_1}$$

An energy balance gives

$$W_{cycle} = Q_1 - Q_2 - Q_3 = 750 - Q_2 - 100$$
$$= 650 - Q_2$$

Since the cycle involves no irreversibilities, and heat transfers are at T_1, T_2, T_3, Eq. 6.2 reads

$$\frac{Q_1}{T_1} - \frac{Q_2}{T_2} - \frac{Q_3}{T_3} = -\sigma_{cycle}^{\,0}$$

$$\Rightarrow \quad 0 = \frac{750}{1500} - \frac{Q_2}{1000} - \frac{100}{500} \Rightarrow Q_2 = 300\text{ kJ}$$

Thus

$$W_{cycle} = 650 - 300 = 350\text{ kJ}$$

and

$$\eta = \frac{350}{750} = 0.467 \quad (46.7\%)$$

PROBLEM 6.3.

KNOWN: A reversible power cycle R and an irreversible cycle I operate between the same two reservoirs.

FIND: (a) Evaluate σ_{cycle} for I in terms of W_R, W_I, and the temperature T_C.
(b) Demonstrate that $W_I < W_R$, $Q_C' > Q_C$.

SCHEMATIC & GIVEN DATA:

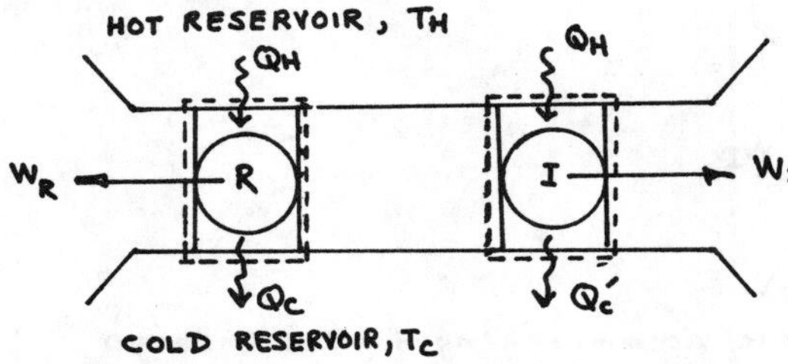

ASSUMPTIONS: (1) The systems shown in the accompanying figure undergo power cycles. R is reversible and I is irreversible. (2) Each system receives Q_H at T_H from the hot reservoir and discharges energy at T_C to the cold reservoir.

ANALYSIS: (a) For cycle I, Equation 6.2 takes the form

$$\sigma_{cycle} = -\oint\left(\frac{\delta Q}{T}\right)_b = -\left[\frac{Q_H}{T_H} - \frac{Q_C'}{T_C}\right]$$

Energy balances for cycles R and I give, respectively

$$\begin{aligned} Q_H &= W_R + Q_C \\ Q_H &= W_I + Q_C' \end{aligned} \quad \Longrightarrow \quad Q_C' = Q_C + W_R - W_I$$

Inserting this result into the expression for σ_{cycle}

$$\sigma_{cycle} = -\left[\left(\frac{Q_H}{T_H} - \frac{Q_C}{T_C}\right) - \left(\frac{W_R - W_I}{T_C}\right)\right]$$

Since R is reversible, Eq. 5.6 applies and the first term on the right vanishes, leaving

$$\sigma_{cycle} = \frac{W_R - W_I}{T_C} \quad \longleftarrow \quad \text{(a)}$$

(b) When irreversibilities are present, $\sigma_{cycle} > 0$. Thus, Eq. (a) reads

$$\frac{W_R - W_I}{T_C} > 0 \Rightarrow W_R > W_I$$

The energy balance result: $Q_C' = Q_C + W_R - W_I$ then indicates that $Q_C' > Q_C$. Accordingly, the effect of irreversibilities is to reduce the desired outcome: W_{cycle} and increase the heat rejected, which for actual power cycles increases the thermal pollution effect.

PROBLEM 6.4

KNOWN: A reversible refrigeration cycle R and an irreversible refrigeration cycle I operate between the same two reservoirs.

FIND: Show that $W_I > W_R$ and $Q_H' > Q_H$

SCHEMATIC & GIVEN DATA:

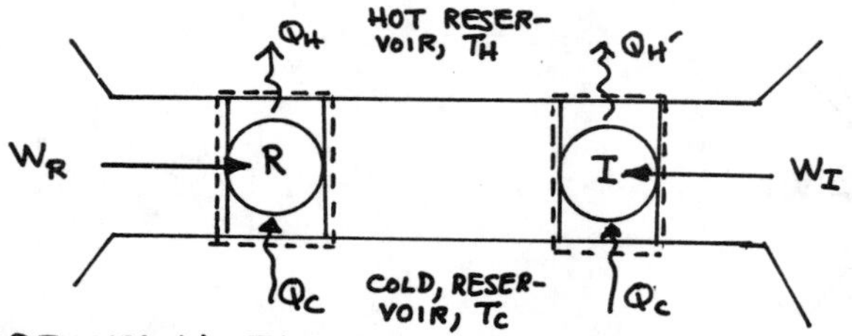

ASSUMPTIONS: (1) The systems shown in the accompanying figure undergo refrigeration cycles. R is reversible and I is irreversible. (2) Each system receives Q_C at T_C from the cold reservoir and discharges energy at T_H to the hot reservoir.

ANALYSIS: For cycle I, Eq. 6.2 takes the form

$$\sigma_{cycle} = -\oint\left(\frac{\delta Q}{T}\right)_b = -\left[\frac{Q_C}{T_C} - \frac{Q_H'}{T_H}\right]$$

Energy balances for cycles R and I give, respectively

$$Q_H = W_R + Q_C$$
$$Q_H' = W_I + Q_C \quad \Longrightarrow \quad Q_H' = Q_H + W_I - W_R$$

Inserting this result into the expression for σ_{cycle}

$$\sigma_{cycle} = -\left[\left(\frac{Q_C}{T_C} - \frac{Q_H}{T_H}\right) - \left(\frac{W_I - W_R}{T_H}\right)\right]$$

Since R is reversible, Eq. 5.6 applies and the first term on the right vanishes leaving

$$\sigma_{cycle} = \frac{W_I - W_R}{T_H}$$

Since cycle I irreversible, $\sigma_{cycle} > 0$ (Sec. 6.1). Thus, $W_I - W_R > 0$, or $W_I > W_R$. The energy balance result then gives $Q_H' > Q_H$. ◄

COMMENT: The current demonstration uses the Clausius inequality developed in Sec. 6.1. The conclusion that $W_I > W_R$ also can be demonstrated using the Kelvin-Planck statement of the second law, as illustrated in the solution to Problem 5.13a.

PROBLEM 6.5

KNOWN: A system undergoes a reversible power cycle while receiving Q_1, Q_2 from hot reservoirs at T_1, T_2 and discharging Q_3 to a cold reservoir at T_3.

FIND: Obtain $\eta = f(T_1/T_3, T_2/T_3, q)$, where $q = Q_2/Q_1$, and discuss the limiting behavior as $q \to 0$, $q \to \infty$, $T_1 \to \infty$.

SCHEMATIC & GIVEN DATA:

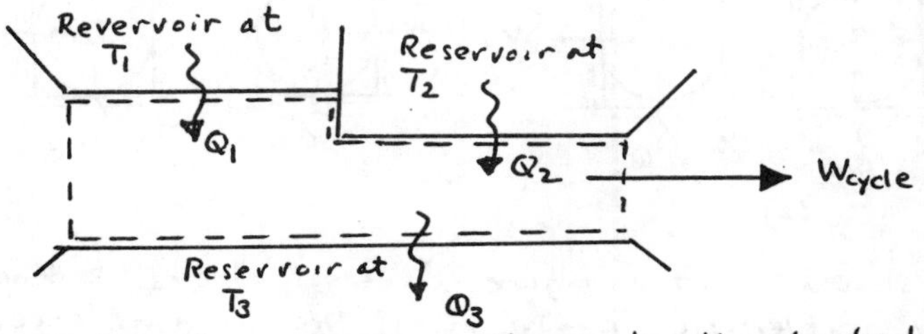

ASSUMPTIONS: (1) The system shown by the dashed line undergoes a power cycle. (2) There are no internal irreversibilities or irreversibilities assocated with heat transfer between the system and reservoirs. (3) The only energy transfers between system and surroundings are indicated on the sketch.

ANALYSIS: (a) An energy balance for the cycle reads $W_{cycle} = Q_1 + Q_2 - Q_3$. Equation 6.2 takes the form

$$\frac{Q_1}{T_1} + \frac{Q_2}{T_2} - \frac{Q_3}{T_3} = -\sigma_{cycle}^{\,0} \Rightarrow Q_3 = \frac{T_3}{T_1} Q_1 + \frac{T_3}{T_2} Q_2$$

The thermal efficiency can be expressed, then, as

$$\eta = \frac{W_{cycle}}{Q_1 + Q_2} = \frac{Q_1 + Q_2 - Q_3}{Q_1 + Q_2} = 1 - \frac{Q_3}{(Q_1 + Q_2)} = 1 - \frac{\left[\frac{T_3}{T_1} Q_1 + \frac{T_3}{T_2} Q_2\right]}{(Q_1 + Q_2)}$$

or

$$\eta = 1 - \frac{1}{(T_1/T_3)}\left[\frac{1}{1+q}\right] - \frac{1}{(T_2/T_3)}\left[\frac{q}{1+q}\right], \quad q = Q_2/Q_1 \longleftarrow \eta$$

(b)

$$\lim_{q \to 0} \eta = 1 - \frac{1}{(T_1/T_3)} = 1 - \frac{T_3}{T_1}$$ (Corresponds to the case of Eq. 5.8)

$$\lim_{q \to \infty} \eta = 1 - \frac{1}{T_2/T_3} = 1 - \frac{T_3}{T_2}$$ (Corresponds to the case of Eq. 5.8)

$$\lim_{T_1 \to \infty} \eta = 1 - \frac{1}{T_2/T_3}\left(\frac{q}{1+q}\right) = 1 - \frac{T_3}{T_2}\left[\frac{q}{1+q}\right]$$ (Although one reservoir might be at an extremely high temperature, the value of η would be strictly less than unity.)

PROBLEM 6.6

KNOWN: Operating details are provided for reversible and irreversible cycles.

FIND: (a) For power cycles, show $T_H' > T_H$, (b) For refrigeration cycles, show $T_C' > T_C$, (c) For heat pump cycles, show $T_H' < T_H$.

SCHEMATIC & GIVEN DATA:

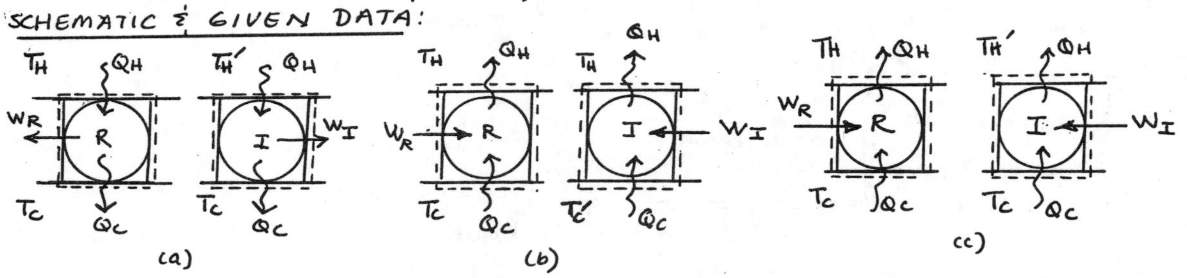

ASSUMPTIONS: (1) The systems shown in the figure undergo cycles. R denotes a reversible cycle and I denotes an irreversible cycle, (2) The systems receive and discharge energy by heat transfer at the indicated temperatures. (3) The only energy transfers taking place are shown on the figures.

ANALYSIS: (a) Applying Eq. 6.2 to R and I, respectively

R: $\frac{Q_H}{T_H} - \frac{Q_C}{T_C} = -\sigma_{cycle}^{\,0} \Rightarrow Q_C = \frac{T_C}{T_H} Q_H$

I: $\frac{Q_H}{T_H'} - \frac{Q_C}{T_C} = -\sigma_{cycle}$

$$\frac{Q_H}{T_H'} - \left(\frac{T_C}{T_H} Q_H\right)\frac{1}{T_C} = -\sigma_{cycle}$$

$$Q_H\left[\frac{1}{T_H'} - \frac{1}{T_H}\right] = -\sigma_{cycle} \qquad \longleftarrow \text{(a)}$$

$$\Rightarrow \quad T_H' > T_H$$

(b) Similarly

R: $\frac{Q_C}{T_C} - \frac{Q_H}{T_H} = 0 \Rightarrow Q_H = \frac{T_H}{T_C} Q_C$

I: $\frac{Q_C}{T_C'} - \frac{Q_H}{T_H} = -\sigma_{cycle}$

$$\frac{Q_C}{T_C'} - \left(\frac{T_H}{T_C} Q_C\right)\frac{1}{T_H} = -\sigma_{cycle}$$

$$Q_C\left[\frac{1}{T_C'} - \frac{1}{T_C}\right] = -\sigma_{cycle} \qquad \longleftarrow \text{(b)}$$

$$\Rightarrow \quad T_C' > T_C$$

(c) Similarly

R: $\frac{Q_C}{T_C} - \frac{Q_H}{T_H} = 0 \Rightarrow Q_C = \frac{T_C}{T_H} Q_H$

I: $\frac{Q_C}{T_C} - \frac{Q_H}{T_H'} = -\sigma_{cycle}$

$$\left(\frac{T_C}{T_H} Q_H\right)\frac{1}{T_C} - \frac{Q_H}{T_H'} = -\sigma_{cycle} \qquad \longleftarrow \text{(c)}$$

$$Q_H\left[\frac{1}{T_H} - \frac{1}{T_H'}\right] = -\sigma_{cycle}$$

$$\Rightarrow \quad T_H' < T_H$$

PROBLEM 6.7

KNOWN: A system undergoes a cycle while receiving $\dot{Q}_s$, $\dot{Q}_o$ at T_s, T_o, respectively, and delivering $\dot{Q}_u$ at T_u.

FIND: Obtain an expression for the maximum theoretical value of $\dot{Q}_u$ in terms of $\dot{Q}_s$, T_s, T_u, T_o.

SCHEMATIC & GIVEN DATA:

① 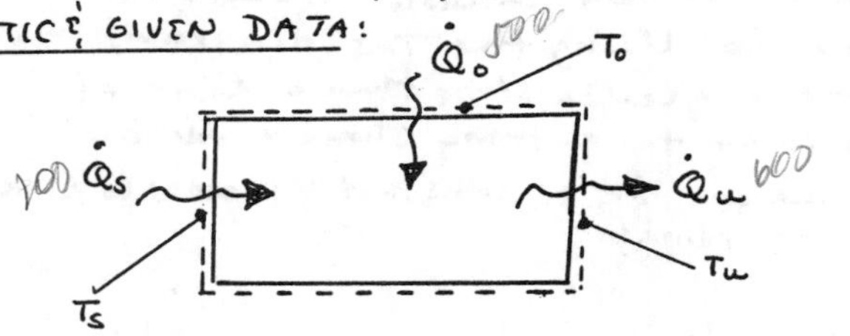

$T_s > T_u > T_o$

ASSUMPTIONS: (1) The system shown in the figure undergoes a cycle. (2) The only energy transfers experienced are shown on the figure. (3) $T_o < T_u < T_s$.

② ANALYSIS: An energy balance reduces to $\dot{Q}_u = \dot{Q}_s + \dot{Q}_o$. Equation 6.2 takes the form

$$\frac{\dot{Q}_s}{T_s} + \frac{\dot{Q}_o}{T_o} - \frac{\dot{Q}_u}{T_u} = -\dot{\sigma}_{cycle}$$

$$\Delta S = 0 = \int \frac{\delta Q}{T} + S_{gen}$$

Eliminating $\dot{Q}_o$ between these expressions

$$\frac{\dot{Q}_s}{T_s} + \frac{\dot{Q}_u - \dot{Q}_s}{T_o} - \frac{\dot{Q}_u}{T_u} = -\dot{\sigma}_{cycle}$$

$$\Rightarrow \quad \dot{Q}_s\left[\frac{1}{T_s} - \frac{1}{T_o}\right] + \dot{Q}_u\left[\frac{1}{T_o} - \frac{1}{T_u}\right] = -\dot{\sigma}_{cycle}$$

$$\Rightarrow \quad \dot{Q}_u = \frac{\dot{Q}_s\left[1 - \frac{T_o}{T_s}\right] - T_o\dot{\sigma}_{cycle}}{\left[1 - \frac{T_o}{T_u}\right]}$$

$$\frac{Q_H}{T_s} + \frac{Q_{surr}}{T_o} - \frac{Q_c}{T_h} = -S_{gen}$$

$$\frac{Q_H}{T_s} + \frac{Q_c - Q_H}{T_o} - \frac{Q_c}{T_h} = -S_{gen}$$

$$Q_c\left(\frac{1}{T_o} - \frac{1}{T_h}\right) + Q_H\left(\frac{1}{T_s} - \frac{1}{T_o}\right) = (-S_{gen})$$

As $T_s > T_o$ and $T_u > 0$, the terms in brackets are positive. The maximum theoretical value of $\dot{Q}_u$ would be realized as $\dot{\sigma}_{cycle} \to 0$:

$$Q_c = \frac{Q_H(1 - T_o/T_s) - T_o S_{gen}}{1 - T_o/T_u}$$

$$(\dot{Q}_u)_{MAX} = \dot{Q}_s\left[\frac{1 - \frac{T_o}{T_s}}{1 - \frac{T_o}{T_u}}\right]$$

$$Q_c\left(\frac{1}{T_o} - \frac{1}{T_h}\right) = -S_{gen} + Q\left(\frac{1}{T_o} - 1/T_s\right)$$

1. This device corresponds to an absorption heat pump -- See Secs. 10.5, 6.
2. The energy balance shows that $\dot{Q}_u > \dot{Q}_s$

PROBLEM 6.8 No format – True/false

(a) TRUE. Entropy is a property. Thus, its change in value between two states is independent of the process -- see Sec. 1.3 for discussion.

(b) FALSE. When T is constant, Eq. 6.18 reduces to

$$s_2 - s_1 = \overbrace{\int_{T_1}^{T_2} \frac{c_v(T)}{T} dT}^{0} + R \ln \frac{v_2}{v_1}$$

. In a compression, $v_2 < v_1$; so, $s_2 - s_1 < 0$.

(c) TRUE. As shown by Eqs. 6.18, 6.19, the specific entropy of an ideal gas depends on T, v or T, p respectively. Also, s can be expressed in terms of p, v.

(d) FALSE. See Eq. 6.12a.

(e) TRUE. See Eq. 6.24

PROBLEM 6.9

KNOWN: A system consisting of an ideal gas with constant specific heat ratio k undergoes constant volume, constant pressure, and constant temperature processes from the same initial state.

FIND: (a) Show that the entropy change is greater for the constant pressure process than for the constant volume process. , Sketch the processes on p-v, T-s coordinates. (b) Using the T-s diagram, show that a line of constant v has a greater slope than a line of constant p. (c) Show that the ratio of the entropy change for a constant T process to the entropy change for a constant v process is (1-k).
Sketch the processes on p-v and T-s coordinates.

SCHEMATIC & GIVEN DATA:

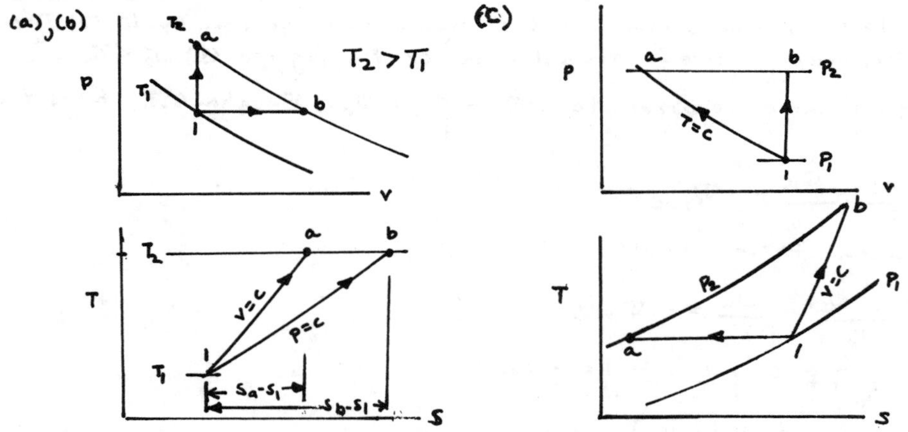

ASSUMPTION: The system consists of an ideal gas with constant k.

ANALYSIS: (a) For process 1-a, specific volume is constant. For process 1-b, pressure is constant. Thus, Eqs. 6.22 and 6.23 reduce, respectively, to give

$$s_a - s_1 = c_v \ln \frac{T_2}{T_1} \qquad\qquad \frac{s_b - s_1}{s_a - s_1} = \frac{c_p \ln T_2/T_1}{c_v \ln T_2/T_1}$$

$$s_b - s_1 = c_p \ln \frac{T_2}{T_1} \qquad\qquad = \frac{c_p}{c_v} = k \longleftarrow$$

As $k > 1$, $(s_b - s_1) > (s_a - s_1)$.

(b) Here, we compare, at state 1 $(\partial T/\partial s)_v$ to $(\partial T/\partial s)_p$. Since $(\partial T/\partial s)$ at fixed v (or fixed p) $= \lim_{\Delta s \to 0} \frac{\Delta T}{\Delta s}$, it is evident from the T-s diagram that a constant specific volume line passing through state 1 has a greater slope than a constant pressure line. ←

PROBLEM 6.9 (Contd.)

For process 1-a, temperature is constant. For process 1-b, volume is constant. Thus, Eqs. 6.19 and 6.18 reduce, respectively, to give

$$s_a - s_1 = -R \ln \frac{p_2}{p_1} \quad \text{(entropy decreases)}$$

$$s_b - s_1 = c_v \ln \frac{T_b}{T_1} \quad \text{(entropy increases)}$$

Using Eq. 3.44, the first of these can be written as

$$s_a - s_1 = -(c_p - c_v) \ln \frac{p_2}{p_1}$$

Using the ideal gas equation of state with $v_b = v_1$, the second can be written as

$$s_b - s_1 = c_v \ln \frac{p_2}{p_1}$$

Forming the ratio

$$\frac{s_a - s_1}{s_b - s_1} = \frac{-(c_p - c_v) \ln p_2/p_1}{c_v \ln p_2/p_1} = 1 - k \longleftarrow$$

PROBLEM 6.10 No Format. True/false

(a) FALSE. See discussion of Sec. 5.1.

(b) FALSE. For any process, internally reversible or otherwise, $\Delta E = Q - W$. Accordingly, the magnitude and direction of Q is not determined by W alone -- unless $\Delta E = 0$.

(c) FALSE. See Eq. 6.29 and the accompanying discussion.

(d) FALSE. See Eq. 6.27 and the accompanying discussion.

(e) FALSE. See Eq. 6.28 and the accompanying discussion.

(f) TRUE. See Sec. 6.4, p. 250 for discussion.

(g) FALSE. See Sec. 6.5.5

PROBLEM 6.11

KNOWN: A fixed mass of water, initially a saturated liquid, undergoes a constant temperature, constant pressure process to a saturated vapor condition.

FIND: (a) Derive expressions for W, Q in terms of m and steam table properties. (b) Show that the process is internally reversible.

SCHEMATIC & GIVEN DATA:

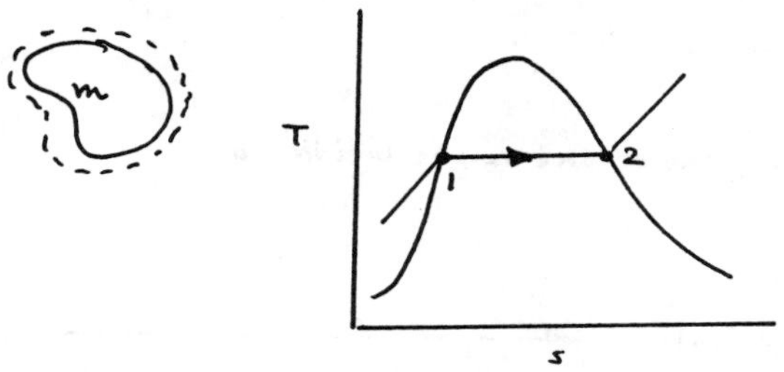

ASSUMPTIONS: (1) The system is a fixed mass of water. (2) Kinetic and potential energy effects are absent. (3) In the process, p = constant, T = constant. (4) Volume change is the only work mode.

ANALYSIS: (a) With assumption (4) and since the pressure remains constant,

$$W = m\int_1^2 p\,dv = mp(v_2 - v_1) = mp[v_g - v_f] \longleftarrow W$$

With assumption (2) and the above expression for W, an energy balance gives

$$m[u_2 - u_1] = Q - W \Rightarrow Q = m[u_g - u_f] + mp(v_g - v_f)$$

$$= m[(u_g + pv_g) - (u_f + pv_f)]$$

$$= m[h_g - h_f] \longleftarrow Q$$

(b) An entropy balance reduces to

$$m(s_2 - s_1) = \frac{Q}{T} + \sigma$$

Introducing the expression for Q

$$m(s_g - s_f) = \frac{m(h_g - h_f)}{T} + \sigma$$

From the discussion of Sec. 6.3.1, Eq. 6.14, $s_g - s_f = (h_g - h_f)/T$, and so

$\sigma = 0 \Rightarrow$ the process is internally reversible.

PROBLEM 6.12

KNOWN: A quantity of air undergoes a process in which the air is both stirred and heated from a reservoir.

FIND: Explain how the process can be conducted with (a) minimum entropy production, (b) maximum entropy production.

SCHEMATIC & GIVEN DATA:

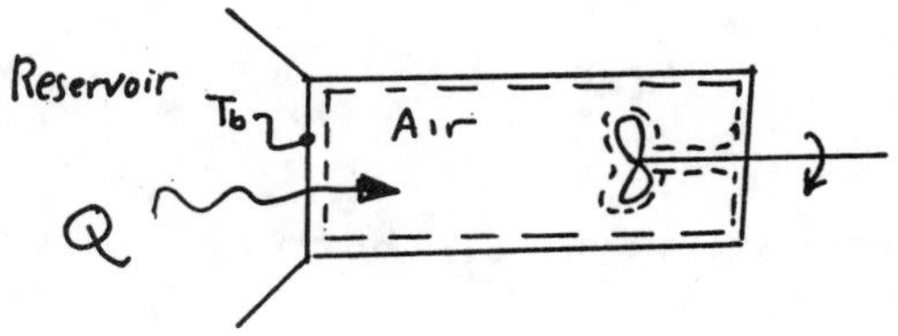

ASSUMPTIONS (1) The system consists of the air only. (2) The temperature of the air at the place where energy is received from the reservoir by heat transfer is T_b. (3) The air is modeled as an ideal gas. (4) The volume of the air is constant during the process and there is no significant change in kinetic or potential energy.

ANALYSIS: The change of state can be brought about by a combination of heat and work. Q and W are related by an energy balance:

$$\Delta U = Q - W \Rightarrow m\int_{T_1}^{T_2} c_v(T)\,dT = Q - W$$

where assumptions 3 and 4 have been used.

Since Q is received at T_b, the closed system entropy balance reads

$$\Delta S = \frac{Q}{T_b} + \sigma \Rightarrow \sigma = \Delta S - \frac{Q}{T_b}$$

Then, with Eq 6.18 and assumption 4

$$\Delta S = m\left[\int_{T_1}^{T_2} \frac{c_v(T)}{T}\,dT + \cancel{R\ln\frac{V_2}{V_1}}^{\,0}\right]$$

Collecting results

$$\sigma = \underbrace{m\int_{T_1}^{T_2} \frac{c_v(T)}{T}\,dT}_{\text{(strictly positive)}} - \underbrace{\frac{Q}{T_b}}_{\text{(positive or zero)}}$$

Accordingly, σ increases as Q decreases, and conversely. This suggests

- Maximum σ corresponds to $Q = 0$: the process takes place by stirring only.
- Minimum σ corresponds to heating to the fullest extent, followed by stirring, as required. Observe that some stirring would be required if T_2 were greater than the reservoir temperature.

PROBLEM 6.13

KNOWN: A closed system and its surroundings make up an isolated system.

FIND: Four each of four cases answer true or false, and explain.

ANALYSIS: From the discussion of Sec. 6.5.5

(energy) $\quad \Delta E]_{sys} + \Delta E]_{surr} = 0 \quad$ (1)

(entropy) $\quad \Delta S]_{sys} + \Delta S]_{surr} = \sigma_{isol} \; (\geq 0) \quad$ (2)

(a) False — Equation (2) allows, but does not require, both entropy changes to be positive.

(b) True — Equation (2) allows $\Delta S]_{system}$ to decrease while $\Delta S]_{surr}$ decreases, and conversely — provided the sum is nonnegative: provided $\sigma_{isol} \geq 0$.

(c) False — If $\Delta S)_{sys} = \Delta S)_{surr} = 0$, it follows that σ_{isol} must be zero, and $\sigma_{isol} = 0$ is an allowed outcome.

(d) False — If both $\Delta S]_{sys}$ and $\Delta S]_{surr}$ decrease, it follows that $\sigma_{isol} < 0$, which is not an allowed outcome.

PROBLEM 6.14

KNOWN: An isolated system of total mass m is formed by mixing equal masses of the same liquid initially at temperatures T_1 and T_2.

FIND: (a) Show that the amount of entropy produced is $\sigma = mc \ln[(T_1+T_2)/2(T_1 T_2)^{1/2}]$ and (b) that σ must be positive.

SCHEMATIC & GIVEN DATA:

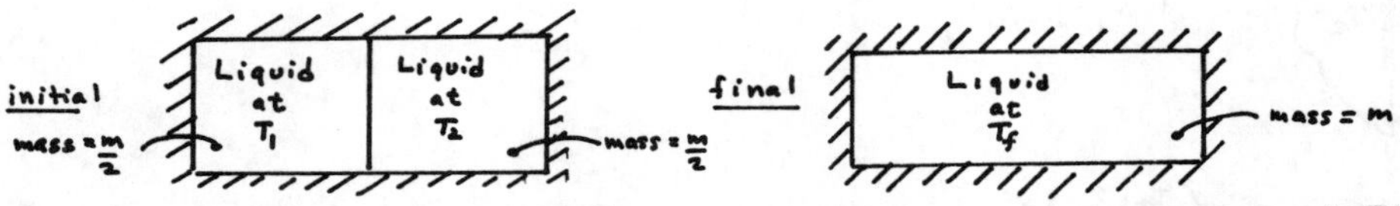

ASSUMPTIONS: (1) The system consists of the total mass of liquid. (2) The system is isolated. (3) The liquid is incompressible with constant specific heat c

ANALYSIS: (a) The final temperature T_f can be evaluated from an energy balance: $\Delta U = \cancel{Q}^0 - \cancel{W}^0$. Thus $\Delta U = 0$, or

$$m\,u(T_f) - \left[\frac{m}{2}u(T_1) + \frac{m}{2}u(T_2)\right] = 0$$

$$\Rightarrow \frac{m}{2}[u(T_f) - u(T_1)] + \frac{m}{2}[u(T_f) - u(T_2)] = 0$$

Since the liquid is incompressible with constant specific heat c

$$\frac{m}{2}c[T_f - T_1] + \frac{m}{2}c[T_f - T_2] = 0$$

$$\Rightarrow T_f = \frac{T_1 + T_2}{2}$$

An entropy balance gives

$$\Delta S = \cancel{\int_1^2 \left(\frac{\delta Q}{T}\right)_b}^{\,0} + \sigma$$

or

$$\sigma = m s_f - \left[\frac{m}{2}s_1 + \frac{m}{2}s_2\right]$$

$$= \frac{m}{2}\left[(s_f - s_1) + (s_f - s_2)\right]$$

Using Eq. 6.28

$$\sigma = \frac{m}{2}c\left[\ln\frac{T_f}{T_1} + \ln\frac{T_f}{T_2}\right] = \frac{m}{2}c\ln\left[\frac{T_f^2}{T_1 T_2}\right]$$

$$= mc\ln\left[\frac{T_f}{(T_1 T_2)^{1/2}}\right]$$

$$= mc\ln\left[\frac{T_1 + T_2}{2(T_1 T_2)^{1/2}}\right] \longleftarrow \text{(a)}$$

(b) $\sigma \geq 0$ when $\ln\left[\frac{T_1+T_2}{2(T_1T_2)^{1/2}}\right] \geq 0$

$$\Rightarrow \frac{T_1+T_2}{2(T_1T_2)^{1/2}} \geq 1 \quad \text{or} \quad T_1 + T_2 \geq 2(T_1T_2)^{1/2}$$

Squaring both sides

$$(T_1+T_2)^2 \geq 4(T_1T_2) \quad \text{or} \quad T_1^2 + 2T_1T_2 + T_2^2 \geq 4T_1T_2$$

$$\Rightarrow T_1^2 - 2T_1T_1 + T_2^2 \geq 0$$

$$\Rightarrow (T_1 - T_2)^2 \geq 0$$

The inequality is satisfied for either $T_1 > T_2$ or $T_2 > T_1$. The equality applies only when $T_1 = T_2$ ⟵ (b)

PROBLEM 6.15

KNOWN: The temperature within a rod initially in contact with hot and cold walls at its ends is linear with position. The rod is insulated overall and eventually comes to a final equilibrium state where the temperature is T_f.

FIND: Evaluate the final temperature and the amount of entropy produced.

SCHEMATIC & GIVEN DATA:

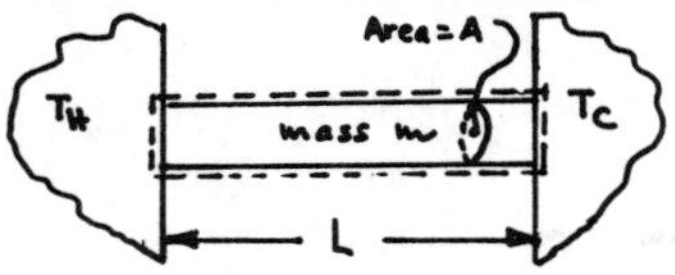

$T_C < T_H$

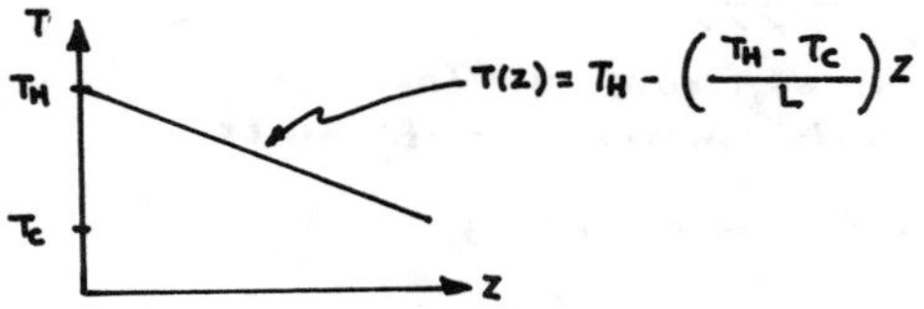

ASSUMPTIONS: (1) The system is the rod which is insulated on its lateral surface. (2) The rod is modeled as incompressible with constant specific heat c. (3) Initially, the temperature within the rod varies linearly from T_H to T_C.

ANALYSIS: The final temperature can be determined using an energy balance which reduces to give $\Delta U = \cancel{Q}^0 - \cancel{W}^0$ or $\Delta U = 0$. Each element of rod dz changes temperature from T(z) to the final temperature T_f, and thus contributes to the change in internal energy

$$dU = dm\, c\,(T_f - T(z))$$
$$= (\rho A\, dz)\, c\,(T_f - T(z))$$

Accordingly

$$\Delta U = \int_0^L (\rho A dz) c\,(T_f - T(z))$$
$$= \rho A c \int_0^L \left[T_f - T_H + \left(\frac{T_H - T_C}{L}\right) z\right] dz$$
$$= \rho A c \left[(T_f - T_H) z + \left(\frac{T_H - T_C}{L}\right) \frac{z^2}{2}\right]_0^L$$
$$= \rho A c L \left[(T_f - T_H) + \frac{(T_H - T_C)}{2}\right]$$

Since $\Delta U = 0$, $T_f = (T_H + T_C)/2$. ← T_f

To find the entropy production, an entropy balance reduces to give $\Delta S = \cancel{\int_1^2 \frac{\delta Q}{T}}^0 + \sigma$ or $\sigma = \Delta S$. With Eq. 6.24, the entropy change of an element of rod dz is

$$dS = dm\, c \ln \frac{T_f}{T(z)}$$
$$= (\rho A dz)\, c \ln \frac{T_f}{T(z)}$$

Accordingly

$$\sigma = \rho A c \int_0^L (\ln T_f - \ln T(z))\, dz = \rho A c \left[(\ln T_f) L - \int_0^L (\ln T(z))\, dz\right]$$

Using the given temperature distribution, the variable of integration can be changed from z to T:

$$dT = -\left(\frac{T_H - T_C}{L}\right) dz \quad \Rightarrow \quad dz = -\left(\frac{L}{T_H - T_C}\right) dT$$

With this, the integral can be expressed as

PROBLEM 6.15 (Contd.)

$$\int_0^L (\ln T(z))\,dz = \int_{T_H}^{T_C} (\ln T)\left(\frac{-L}{T_H - T_C}\right) dT$$

$$= \frac{L}{(T_H - T_C)} \int_{T_C}^{T_H} \ln T \, dT$$

$$= \frac{L}{(T_H - T_C)} \Big[T \ln T - T \Big]_{T_C}^{T_H}$$

$$= \frac{L}{T_H - T_C} \Big[(T_H \ln T_H - T_H) - (T_C \ln T_C - T_C) \Big]$$

$$= L \left[\frac{T_H \ln T_H}{T_H - T_C} - \frac{T_C \ln T_C}{T_H - T_C} - 1 \right]$$

Collecting results

$$\sigma = \rho A c \left[(\ln T_f) L - L \left[\frac{T_H \ln T_H}{T_H - T_C} - \frac{T_C \ln T_C}{T_H - T_C} - 1 \right] \right]$$

$$= mc \left[1 + \ln T_f + \frac{T_C \ln T_C}{T_H - T_C} - \frac{T_H \ln T_H}{T_H - T_C} \right] \longleftarrow \sigma$$

PROBLEM 6.16

KNOWN: A system undergoes a cycle while receiving Q_H at T_H' and discharging Q_C at T_C'. Q_H and Q_C are with hot and cold reservoirs at T_H and T_C, respectively.

FIND: (a) Determine an expression for W_{cycle} in terms of Q_H, T_C', T_H', and σ. (b) State the relationship of T_H' to T_H and T_C' to T_C. (c) Obtain an expression for W_{cycle} when there are (i) no internal irreversibilities, (ii) no irreversibilities.

SCHEMATIC & GIVEN DATA:

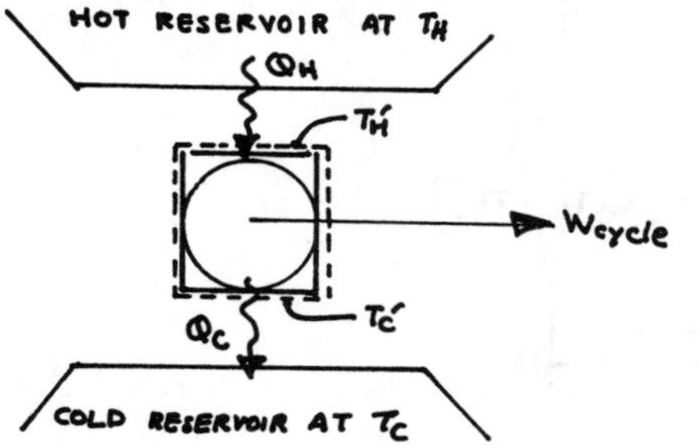

ASSUMPTIONS: The system shown in the accompanying figure undergoes a power cycle while receiving Q_H at T_H' and discharging Q_C at T_C'. There are no other heat transfers.

ANALYSIS: (a) An energy balance gives

$$W_{cycle} = Q_H - Q_C \qquad (1)$$

An entropy balance gives

$$\Delta S_{cycle}^{\,0} = \frac{Q_H}{T_H'} - \frac{Q_C}{T_C'} + \sigma_{cycle} \qquad (2)$$

where σ is the amount of entropy produced within the system per cycle and $\Delta S = 0$ becouse the system undergoes a cycle. Solving Eq. (2) for Q_C and substituting the result into Eq. (1)

$$W_{cycle} = Q_H\left[1 - \frac{T_C'}{T_H'}\right] - T_C'\sigma_{cycle} \qquad (3)$$ ← (a)

(b) For heat transfer to occur from the hot reservoir to the system $T_H \geq T_H'$. For heat transfer to occur from the system to the cold reservoir $T_C' \geq T_C$. ← (b)

(c) If there were no irreversibilities within the system during the cycle, the term σ_{cycle} in Eq. (3) would vanish leaving

$$W_{cycle} = Q_H\left[1 - \frac{T_C'}{T_H'}\right] \qquad (4)$$ ← (c-i)

External irreversibilities are associated with heat transfer between the reservoirs and the system. If these are also absent, $T_H' = T_H$ and $T_C' = T_C$, and Eq. (4) becomes

$$W_{cycle} = Q_H\left[1 - \frac{T_C}{T_H}\right] \qquad (5)$$ ← (c-ii)

which is the maximum theoretical work that can be obtained (Sec. 5.6).

PROBLEM 6.17

KNOWN: A system undergoes a power cycle while receiving energy by heat transfer from a body initially at T_H and discharging energy by heat transfer to a similar body initially at T_C. Work is developed until the temperature of each body is T'.

FIND: Obtain expressions for the minimum theoretical final temperature T' and for the maximum theoretical work that can be developed. Also obtain an expression for the minimum theoretical work input that would be required by a refrigeration cycle to restore the two bodies to their respective initial temperatures.

SCHEMATIC & GIVEN DATA:

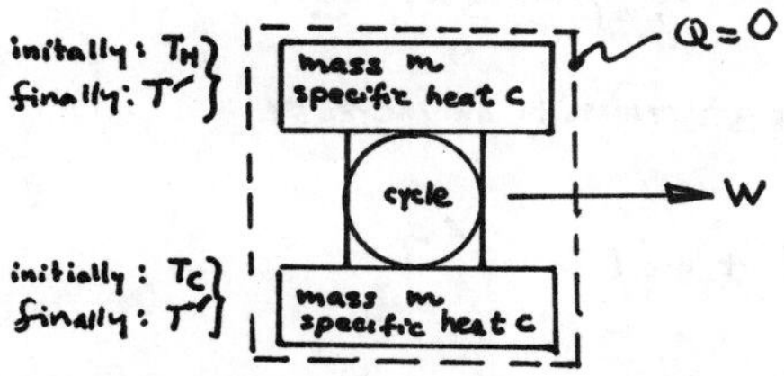

ASSUMPTIONS: (1) The system shown in the accompany figure is composed of three subsystems: An incompressible body of mass m and specific heat c initially at T_H. An incompressible body of mass m and specific heat c initially at T_C. A system that undergoes a power cycle. (2) For the overall system, $Q = 0$.

ANALYSIS: (a) An energy balance gives $\Delta U = \cancel{Q}^{0} - W \Rightarrow$

$$W = -[\Delta U]_{HOT} + \cancel{\Delta U]_{cycle}}^{0} + \Delta U]_{COLD}] = -[mc(T' - T_H) + mc(T' - T_C)]$$

or

$$W = mc[T_H + T_C - 2T'] \qquad (1)$$

An entropy balance gives $\Delta S = \cancel{\int_1^2 \left(\frac{\delta Q}{T}\right)_b}^{0} + \sigma$. Then, with Eq. 6.24

$$\sigma = [\Delta S]_{HOT} + \cancel{\Delta S]_{cycle}}^{0} + \Delta S]_{COLD}] = \left[mc \ln \frac{T'}{T_H} + mc \ln \frac{T'}{T_C}\right]$$

or

$$\sigma = mc \ln\left[\frac{(T')^2}{T_H T_C}\right]$$

Solving for T'

$$T' = [T_H T_C \exp(\sigma/mc)]^{1/2} \qquad (2)$$

From this, it can be concluded that the minimum value of T' corresponds to the case of no internal irreversibilities: $\sigma = 0$. Then

$$(T')_{min} = [T_H T_C]^{1/2} \qquad (3)$$

PROBLEM 6.17 (Continued)

(b) Substituting Eq. (2) into Eq. (1)

$$W = mc\left[T_H + T_C - 2\left[T_H T_C \exp(\sigma/mc)\right]^{1/2}\right]$$

Since $\sigma \geq 0$, it follows that W is maximum when $\sigma = 0$, giving

$$W_{MAX} = mc\left[T_H + T_C - 2(T_H T_C)^{1/2}\right] \longleftarrow$$

(c) In this part, the system shown in the schematic undergoes a refrigeration cycle that must be driven by a <u>work input</u>. The analysis very closely parallels that given above in parts (a), (b):

As $Q = 0$, the energy added by the <u>work input</u> results in an increase in the system internal energy: $-W = \Delta U$, or

$$-W = \Delta U]_{HOT} + \Delta U]_{cycle}^{0} + \Delta U]_{COLD} = \left[mc(T_H - T') + mc(T_C - T')\right]$$

$$\Rightarrow \quad -W = mc\left[T_H + T_C - 2T'\right] \qquad (4)$$

An entropy balance gives $\Delta S = \int_1^2 \left(\frac{\delta Q}{T}\right)_b^{0} + \sigma$. With Eq. 6.24

$$\sigma = \Delta S]_{HOT} + \Delta S]_{cycle}^{0} + \Delta S]_{COLD} = mc \ln \frac{T_H}{T'} + mc \ln \frac{T_C}{T'}$$

$$= mc \ln\left[\frac{T_H T_C}{(T')^2}\right]$$

Solving for T'

$$T' = \left[\frac{T_H T_C}{\exp(\sigma/mc)}\right]^{1/2} \qquad (5)$$

Substituting Eq. (5) into Eq. (4)

$$-W = mc\left[T_H + T_C - 2\left[\frac{T_H T_C}{\exp(\sigma/m)}\right]^{1/2}\right]$$

As $\sigma \geq 0$, it follows that $(-W)$ is a minimum when $\sigma = 0$, giving

$$-W_{min} = mc\left(T_H + T_C - 2(T_H T_C)^{\frac{1}{2}}\right)$$

or

$$W_{min} = mc\left[2(T_H T_C)^{1/2} - T_H - T_C\right] \longleftarrow$$

which corresponds to the result of part (b), except here a <u>work input</u> is evaluated.

PROBLEM 6.18

KNOWN: An insulated mixing chamber at steady state receives liquid streams of the same substance at $\dot{m}_1, T_1$ and $\dot{m}_2, T_2$. A single stream exits at $\dot{m}_3, T_3$.

FIND: (a) Evaluate T_3 in terms of $T_1, T_2, \dot{m}_1/\dot{m}_3$. (b) Evaluate $\dot{\sigma}/\dot{m}_3$ in terms of the specific heat c, T_1/T_2, $\dot{m}_1/\dot{m}_3$. (c) For fixed c and T_1/T_2, determine the value of $\dot{m}_1/\dot{m}_3$ for which $\dot{\sigma}/\dot{m}_3$ is a maximum.

SCHEMATIC & GIVEN DATA:

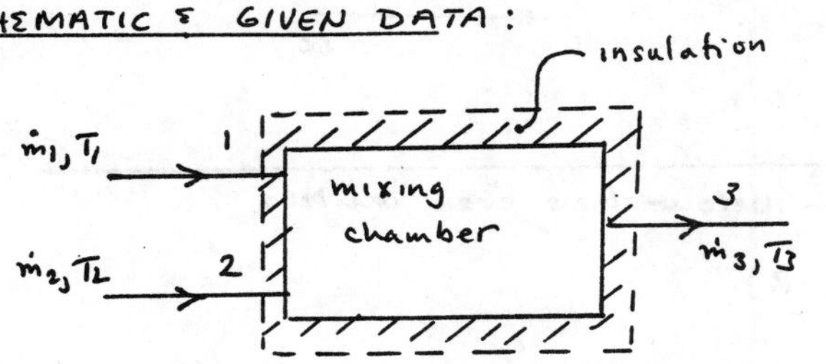

ASSUMPTIONS: (1) The control volume shown in the figure is at steady state. (2) For the control volume, $\dot{Q}_{cv} = \dot{W}_{cv} = 0$, and all kinetic and potential energy effects are negligible. (3) The liquid streams can be modeled as incompressible with constant specific heat c and negligible effects of pressure.

ANALYSIS: (a) At steady state, a mass balance reads $\dot{m}_3 = \dot{m}_1 + \dot{m}_2$. An energy balance becomes, with assumptions(2), $0 = \dot{m}_1 h_1 + \dot{m}_2 h_2 - \dot{m}_3 h_3$. Combining these equations gives

$$0 = \dot{m}_1 h_1 + (\dot{m}_3 - \dot{m}_1)h_2 - \dot{m}_3 h_3$$

$$0 = \dot{m}_1 (h_1 - h_2) + \dot{m}_3 (h_2 - h_3)$$

① Or with assumption (3) and Eq. 3.20b

$$0 = \dot{m}_1 c[T_1 - T_2] + \dot{m}_3 c[T_2 - T_3] \Rightarrow T_3 = T_2 + \left[\frac{\dot{m}_1}{\dot{m}_3}\right](T_1 - T_2) \quad \longleftarrow \text{(a)}$$

(b) An entropy balance reduces at steady state to give

$$0 = \sum \frac{\dot{Q}_j}{T_j}^{\,0} + \dot{m}_1 s_1 + \dot{m}_2 s_2 - \dot{m}_3 s_3 + \dot{\sigma} \quad \text{or} \quad 0 = \dot{m}_1 s_1 + (\dot{m}_3 - \dot{m}_1)s_2 - \dot{m}_3 s_3 + \dot{\sigma}$$

Rearranging, and using Eq. 6.2

$$0 = \dot{m}_1(s_1 - s_2) + \dot{m}_3(s_2 - s_3) + \dot{\sigma} \quad \text{or} \quad 0 = \dot{m}_1 c \ln\frac{T_1}{T_2} + \dot{m}_3 c \ln\frac{T_2}{T_3} + \dot{\sigma}$$

Solving

② $$\frac{\dot{\sigma}}{\dot{m}_3} = c\left[\frac{\dot{m}_1}{\dot{m}_3}\ln\frac{T_2}{T_1} + \ln\frac{T_3}{T_2}\right]$$

Using the result of part(a) this becomes

$$\frac{\dot{\sigma}}{\dot{m}_3} = c\left[\frac{\dot{m}_1}{\dot{m}_3}\ln\frac{T_2}{T_1} + \ln\left[\frac{T_2 + \left(\frac{\dot{m}_1}{\dot{m}_3}\right)(T_1 - T_2)}{T_2}\right]\right] = c\left[\frac{\dot{m}_1}{\dot{m}_3}\ln\frac{T_2}{T_1} + \ln\left[1 + \left(\frac{\dot{m}_1}{\dot{m}_3}\right)\left(\frac{T_1}{T_2}\right) - \frac{\dot{m}_1}{\dot{m}_3}\right]\right]$$

Letting $x = \dot{m}_1/\dot{m}_3$, this can be expressed as

$$\frac{\dot{\sigma}}{\dot{m}_3} = c\ln\left[\left(\frac{T_2}{T_1}\right)^x\left[1 + x\left(\frac{T_1}{T_2}\right) - x\right]\right] = c\ln\left[\left(\frac{T_1}{T_2}\right)^{-x}\left[1 + x\left[\left(\frac{T_1}{T_2}\right) - 1\right]\right]\right] \quad \longleftarrow \text{(b)}$$

(c) Taking the derivative of $\dot{\sigma}/\dot{m}_3$ with respect to x while holding c and T_2/T_1 constant,

$$\left.\frac{\partial(\dot{\sigma}/\dot{m}_3)}{\partial x}\right]_{c, T_1/T_2} = \frac{-\left(\frac{T_1}{T_2}\right)^{-x}\ln\left(\frac{T_1}{T_2}\right)\left[1 + x\left[\left(\frac{T_1}{T_2}\right) - 1\right]\right] + \left(\frac{T_1}{T_2}\right)^{-x}\left[\left(\frac{T_1}{T_2}\right) - 1\right]}{\left(\frac{T_1}{T_2}\right)^{-x}\left[1 + x\left[\left(\frac{T_1}{T_2}\right) - 1\right]\right]}$$

PROBLEM 6.18 (Contd.)

Setting this to zero

$$-\ln\left(\frac{T_1}{T_2}\right)\left[1 + x\left[\left(\frac{T_1}{T_2}\right)-1\right]\right] + \left[\left(\frac{T_1}{T_2}\right) - 1\right] = 0$$

Solving

③ $$x = \frac{\ln\left(\frac{T_1}{T_2}\right) + \left[1 - \left(\frac{T_1}{T_2}\right)\right]}{\ln\left(\frac{T_1}{T_2}\right)\left[1 - \left(\frac{T_1}{T_2}\right)\right]} \longleftarrow \quad (c)$$

---

1. In applying Eq. 3.20b the underlined term below has been omitted

$$h_2 - h_1 = c(T_2 - T_1) + \underline{v(p_2 - p_1)}$$

Since the specific volume of liquids are small, a substantial pressure difference $(p_2 - p_1)$ would be required before the underlined term becomes significant. Effects of pressure are neglected in assumption 3.

2 when $\dot{m}_1/\dot{m}_3 = 0$, Eq (a) gives $T_3 = T_2$, then

$$\frac{\dot{\sigma}}{\dot{m}_3} = c\left[\left(\frac{\dot{m}_1}{\dot{m}_3}\right)\ln\frac{T_2}{T_1} + \ln\frac{T_3}{T_2}\right] = 0$$

And when $\dot{m}_1/\dot{m}_3 = 1$, Eq. (a) gives $T_3 = T_1$, then

$$\frac{\dot{\sigma}}{\dot{m}_3} = c\left[\left(\frac{\dot{m}_1}{\dot{m}_3}\right)\ln\frac{T_2}{T_1} + \ln\frac{T_3}{T_2}\right] = c\left[\ln\frac{T_2}{T_1} + \ln\frac{T_1}{T_2}\right] = 0$$

In these cases there is no mixing, and thus no entropy production.

3. To verify that this locates a maximum, form the derivative $\partial^2(\dot{\sigma}/\dot{m}_3)/\partial x^2$ and evaluate the resulting expression with the expression for x here to show the second derivative is negative.

PROBLEM 6.19

FIND: Determine the specific entropy, in kJ/kg·K, of water at the following states:

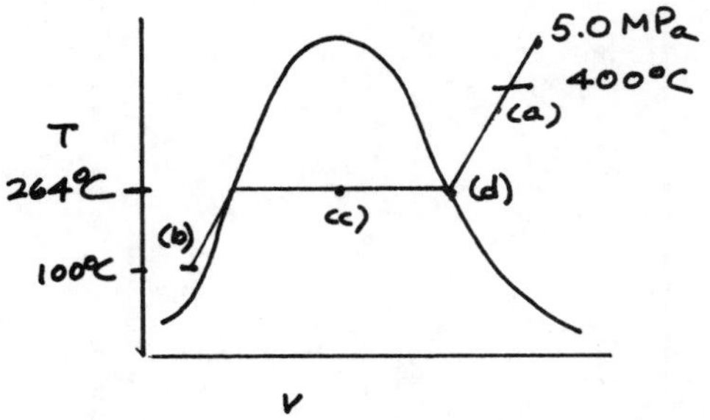

(a) 5.0 MPa, 400°C. Interpolation in Table A-4: $s = 6.6549$ kJ/kg·K. ← (a)

(b) 5.0 MPa, 100°C. Table A-5: $s = 1.303$ kJ/kg·K. ← (b)

(c) 5.0 MPa, $u = 1872.5$ kJ/kg. Table A-3 gives $u_f = 1147.8$ kJ/kg, $u_g = 2597.1$ kJ/kg.
Thus

$$x = \frac{1872.5 - 1147.8}{2597.1 - 1147.8} = 0.5$$

Then

$$s = s_f + x(s_g - s_f) = 2.9202 + 0.5[5.9734 - 2.9202] = 4.4468 \text{ kJ/kg·K} \leftarrow (c)$$

(d) 5.0 MPa, Saturated vapor. Table A-3: $s_g = 5.9734$ kJ/kg·K ← (d)

IT results. Note at a liquid state, such as (b), IT returns entropy using Eq. 6.7. See Methodology Update — p. 244.

IT Code

```
p = 50  // bar

Ta = 400  // °C
sa = s_PT("Water/Steam", p, Ta)

Tb = 100
sb = s_PT("Water/Steam", p, Tb)

uc = 1872.5  // kJ/kg
uc = usat_Px("Water/Steam", p, xc)
sc = ssat_Px("Water/Steam", p, xc)

xd = 1
sd = ssat_Px("Water/Steam", p, xd)
```

IT Results

(a) sa = 6.645 kJ/kg·K
(b) sb = 1.308 kJ/kg·K
(c) xc = 0.5001
    sc = 4.447 kJ/kg·K
(d) sd = 5.973 kJ/kg·K

PROBLEM 6.20

FIND: Determine the specific entropy, in Btu/lb·°R, of water at the following states:

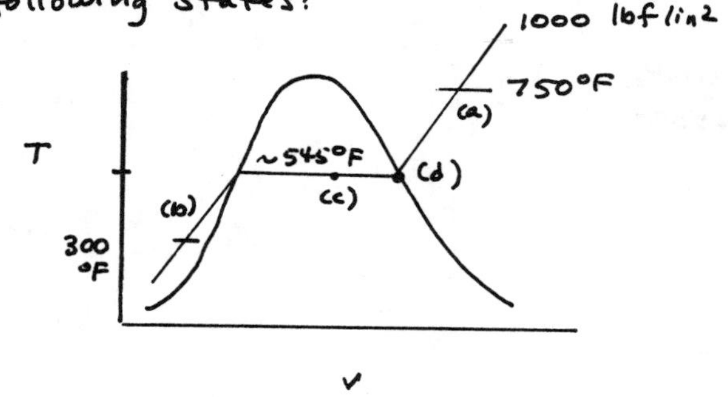

(a) 1000 lbf/in², 750°F. Interpolation in Table A-4E: $s = 1.54$ Btu/lb·°R. ← (a)

(b) 1000 lbf/in², 300°F. Table A-5E: $s = 0.43552$ Btu/lb·°R. ← (b)

(c) 1000 lbf/in², $h = 932.4$ Btu/lb. Table A-3E gives $h_f = 542.4$ Btu/lb, $h_{fg} = 650.0$, $h_g = 1192.4$ Btu/lb. Thus

$$x = \frac{932.4 - 542.4}{650} = 0.6$$

Then,

$$s = s_f + x(s_g - s_f)$$

$$= 0.7432 + 0.6(0.6471) = 1.1315 \text{ Btu/lb·°R}.$$ ← (c)

(d) 1000 lbf/in², saturated vapor. Table A-3E. $s_g = 1.3903$ Btu/lb·°R ← (d)

IT Results. Note that at liquid states, such as (b), IT returns entropy using Eq. 6.7. See Methodology Update - p. 244.

IT Code

```
p = 1000  // lbf/in.²

Ta = 750  // °F
sa = s_PT("Water/Steam", p, Ta)

Tb = 300
sb = s_PT("Water/Steam", p, Tb)

hc = 932.4  // Btu/lb
hc = hsat_Px("Water/Steam", p, xc)
sc = ssat_Px("Water/Steam", p, xc)

xd = 1
sd = ssat_Px("Water/Steam", p, xd)
```

IT Results

(a) sa = 1.541 Btu/lb·°R
(b) sb = 0.4369 Btu/lb·°R
(c) xc = 0.6001
sc = 1.131 Btu/lb·°R
(d) sd = 1.39 Btu/lb·°R

PROBLEM 6.21

FIND: Determine the change in specific entropy, in kJ/kg·K, between the specified states.

(a) Water. $P_1 = 10$ MPa, $T_1 = 400°C$, $P_2 = 10$ MPa, $T_2 = 100°C$.

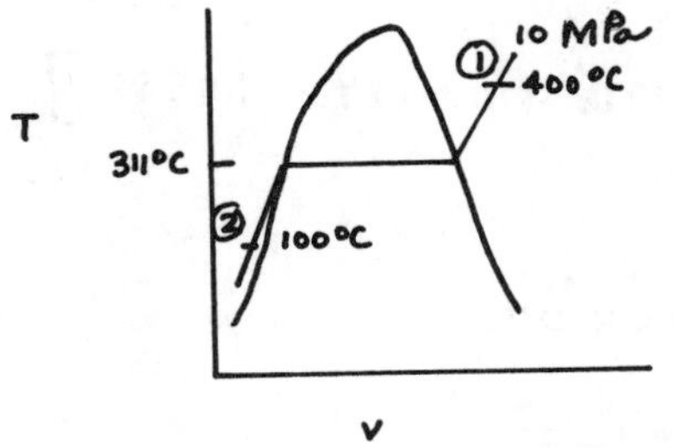

Table A-4, $s_1 = 6.2120$ kJ/kg·K

Table A-5, $s_2 = 1.2992$ kJ/kg·K

$s_2 - s_1 = 1.2992 - 6.2120$

$= -4.9128$ kJ/kg·K ← (a)

(b) R134a. $h_1 = 111.44$ kJ/kg, $T_1 = -40°C$, saturated vapor at $P_2 = 5$ bar.

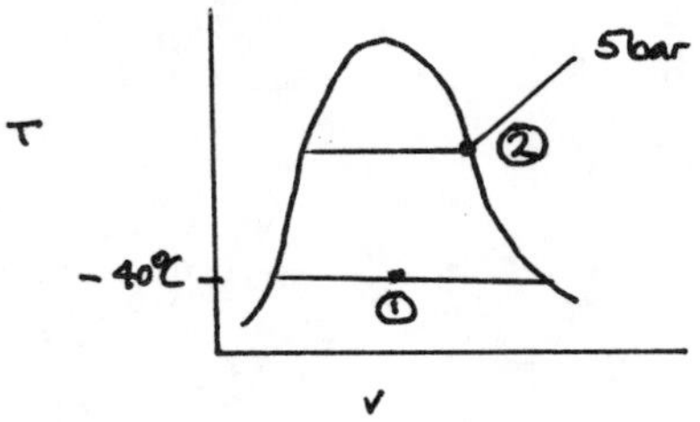

Table A-10 at −40°C

$h_f = 0$, $h_{fg} = 222.88$, $h_g = 222.88$ kJ/kg

Thus,

$$x_1 = \frac{111.44 - 0}{222.88} = 0.5$$

Then

$s_1 = s_f + x_1(s_g - s_f) = 0 + 0.5[0.9560 - 0]$

$= 0.478$ kJ/kg·K

Then, with s_2 from Table A-11

$s_2 - s_1 = 0.9117 - 0.478 = 0.4337$ kJ/kg·K ← (b)

(c) Air as an ideal gas. $T_1 = 280$ K (7°C), $P_1 = 2$ bar, $T_2 = 600$ K (327°C), $P_2 = 1$ bar.

With Eq. 6.21a and $s°$ data from Table A-22

$$s_2 - s_1 = s°(T_2) - s°(T_1) - \frac{\bar{R}}{M} \ln \frac{P_2}{P_1}$$

$$= (2.40902 - 1.63279)\frac{kJ}{kg·K} - \left(\frac{8.314}{28.97}\frac{kJ}{kg·K}\right)\ln\frac{1}{2}$$

$$= 0.97515 \frac{kJ}{kg·K}$$ ← (c)

(d) Hydrogen (H_2) as an ideal gas. $T_1 = 1000$ K (727°C), $P_1 = 1$ bar, $T_2 = 298$ K (25°C), $P_2 = 3$ bar

Table A-21 provides the following expression for $\bar{c}_p(T)$

$$\frac{\bar{c}_p}{\bar{R}} = \alpha + \beta T + \gamma T^2 + \delta T^3 + \varepsilon T^4$$

Inserting this into Eq. 6.19, written on a molar basis, and integrating

$$\frac{\bar{s}_2 - \bar{s}_1}{\bar{R}} = \int_{T_1}^{T_2}\left(\frac{\alpha + \beta T + \gamma T^2 + \delta T^3 + \varepsilon T^4}{T}\right) dT - \ln\frac{p_2}{p_1}$$

$$= \alpha \ln\frac{T_2}{T_1} + \beta(T_2 - T_1) + \frac{\gamma}{2}(T_2^2 - T_1^2) + \frac{\delta}{3}(T_2^3 - T_1^3) + \frac{\varepsilon}{4}(T_2^4 - T_1^4) - \ln\frac{p_2}{p_1}$$

The values for the constants α, β, γ, δ, and ε for H_2 are obtained from Table A-21. Hence

$$\frac{\bar{s}_2-\bar{s}_1}{\bar{R}} = 3.057 \ln\frac{298}{1000} + 2.677\frac{(298-1000)}{10^3} - \frac{5.810}{2}\left[\frac{298^2-1000^2}{10^6}\right] + \frac{5.521}{3}\left[\frac{298^3-1000^3}{10^9}\right] - \frac{1.812}{4}\left[\frac{298^4-1000^4}{10^{12}}\right] - \ln\frac{3}{1}$$

$$\bar{s}_2-\bar{s}_1 = 8.314\,\frac{kJ}{kmol\cdot K}\left[-3.701 - 1.8793 + 2.647 - 1.7916 + 0.4494 - 1.0986\right]$$

$$= -44.68\,\frac{kJ}{kmol\cdot K}$$

With the molecular weight for H_2 from Table A-1, $M = 2.016$

$$s_2 - s_1 = \frac{\bar{s}_2-\bar{s}_1}{M} = \frac{-44.68\ kJ/kmol\cdot K}{2.016\ kg/kmol} = -22.163\,\frac{kJ}{kg\cdot K}$$

IT SOLUTION:

IT Code

```
Ta1 = 400 // °C
pa1 = 100  // bar
Ta2 = 100
pa2 = 100
sa1 = s_PT("Water/Steam", pa1, Ta1)
sa2 = s_PT("Water/Steam", pa2, Ta2)
dels_a = sa2 - sa1

hb1 = 111.44  // kJ/kg
Tb1 = -40
xb2 = 1
pb2 = 5
pb1 = Psat_T("R134A", Tb1)
hb1 = hsat_Px("R134A", pb1, xb1)
sb1 = ssat_Px("R134A", pb1, xb1)
sb2 = ssat_Px("R134A", pb2, xb2)
dels_b = sb2 - sb1

Tc1 = 7
pc1 = 2
Tc2 = 327
pc2 = 1
dels_c = s_TP("Air", Tc2, pc2) -  s_TP("Air", Tc1, pc1)

Td1 = 727
pd1 = 1
Td2 = 25
pd2 = 3
dels_d = s_TP("H2", Td2, pd2) -  s_TP("H2", Td1, pd1)
```

IT Results

(a) Δs = -4.903 kJ/kg·K
(b) Δs = 0.4338 kJ/kg·K
(c) Δs = 0.9749 kJ/kg·K
(d) Δs = -22.16 kJ/kg·K

PROBLEM 6.22

FIND: Determine the change in specific entropy, in Btu/lb·°R.

(a) Water. $p_1 = 1000$ lbf/in², $T_1 = 800°F$, $p_2 = 1000$ lbf/in², $T_2 = 100°F$.

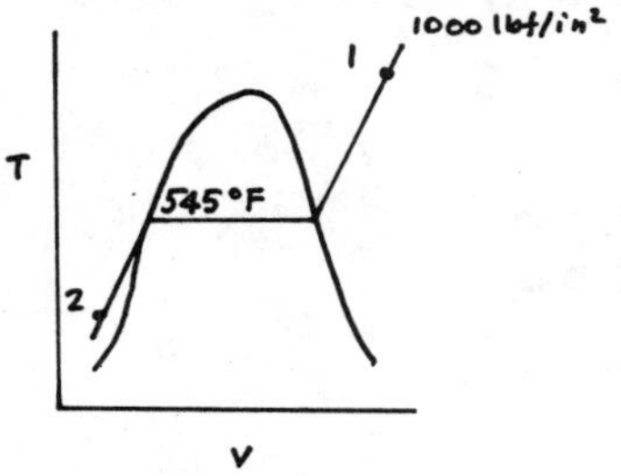

Table A-4E. $s_1 = 1.5665$ Btu/lb·°R

Table A-5E. $s_2 = 0.12901$ Btu/lb·°R

$s_2 - s_1 = -1.43749$ Btu/lb·°R ← (a)

(b) R-134a. $h_1 = 47.91$ Btu/lb, $T_1 = -40°F$, saturated vapor at $p_2 = 40$ lbf/in².

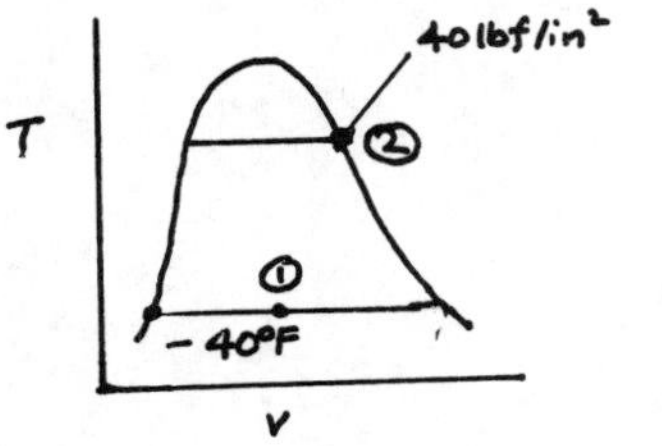

Table A-10E at −40°F: $h_f = 0$, $h_{fg} = 95.82$ Btu/lb

$$\therefore \; x_1 = \frac{47.91 - 0}{95.82} = 0.5$$

And

$$s_1 = s_f + x_1(s_g - s_f) = 0 + 0.5(0.2283 - 0) = 0.11415 \text{ Btu/lb·°R.}$$

Table A-11E, $s_2 = 0.2197$ Btu/lb·°R. Then

$s_2 - s_1 = 0.2197 - 0.11415 = 0.1056$ Btu/lb·°R ← (b)

(c) Air as an ideal gas. $T_1 = 40°F$, $p_1 = 2$ atm, $T_2 = 420°F$, $p_2 = 1$ atm.
With Eq. 6.25a and $s°$ data from Table A-22E at 500°R and 880°R

$$s_2 - s_1 = s°(T_2) - s°(T_1) - \frac{\bar{R}}{M} \ln \frac{p_2}{p_1}$$

$$= (0.71886 - 0.58233)\left(\frac{\text{Btu}}{\text{lb·°R}}\right) - \left(\frac{1.986}{28.97}\frac{\text{Btu}}{\text{lb·°R}}\right) \ln \frac{1}{2}$$

$= 0.18405$ Btu/lb·°R ← (c)

(d) Carbon dioxide as an ideal gas. $T_1 = 820°F$, $p_1 = 1$ atm, $T_2 = 77°F$, $p_2 = 3$ atm.
With Eq. 6.25b and $\bar{s}°$ data from Table A-23E at 1280°R and 537°R

$$\bar{s}_2 - \bar{s}_1 = \bar{s}°(T_2) - \bar{s}°(T_1) - \bar{R} \ln \frac{p_2}{p_1}$$

$$= (51.032 - 60.044)\left(\frac{\text{Btu}}{\text{lbmol·°R}}\right) - \left(1.986 \frac{\text{Btu}}{\text{lbmol·°R}}\right) \ln \frac{3}{1}$$

$$= -11.1938 \frac{\text{Btu}}{\text{lbmol·°R}}$$

Then, with the molecular weight of CO_2 from Table A-1E

$$s_2 - s_1 = \frac{\bar{s}_2 - \bar{s}_1}{M}$$

$$= \frac{-11.1938 \frac{\text{Btu}}{\text{lbmol·°R}}}{44.01 \text{ lb/lbmol}}$$

$= -0.25435$ Btu/lb·°R ← (d)

PROBLEM 6.22 (Contd). IT Solution ...

IT Code

```
p = 1000  // lbf/in.²
Ta1 = 800  // °F
Ta2 = 100
sa1 = s_PT("Water/Steam", p, Ta1)
sa2 = s_PT("Water/Steam", p, Ta2)
dels_a = sa2 - sa1

hb1 = 47.91  // Btu/lb
Tb1 = -40
xb1 = 1
pb2 = 40
pb1 = Psat_T("R134A", Tb1)
hb1 = hsat_Px("R134A", pb1, xb1)
sb1 = ssat_Px("R134A", pb1, xb1)
sb2 = ssa_Px("R134A", pb2, xb2)
dels_b = sb2 - sb1

Tc1 = 40
pc1 = 2 * 14.696 // lbf/in.²
Tc2 = 420
pc2 = 1 * 14.696
sc1 = s_TP("Air", Tc1, pc1)
sc2 = s_TP("Air", Tc2, pc2)
dels_c = sc2 - sc1

Td1 = 820
pd1 = 1 * 14.696
Td2 = 77
pd2 = 3 * 14.696
sd1 = s_TP("CO2", Td1, pd1)
sd2 = s_TP("CO2", Td2, pd2)
dels_d = sd2 - sd1
```

IT Results

(a) $\Delta s = -1.437$ Btu/lb·°R
(b) $\Delta s = 0.1056$ Btu/lb·°R
(c) $\Delta s = 0.184$ Btu/lb·°R
(d) $\Delta s = -0.2545$ Btu/lb·°R

PROBLEM 6.23

KNOWN: Ammonia undergoes a process between two specified states.

FIND: Determine the temperature at the final state, in °C, and the final specific enthalpy, in kJ/kg.

SCHEMATIC & GIVEN DATA:

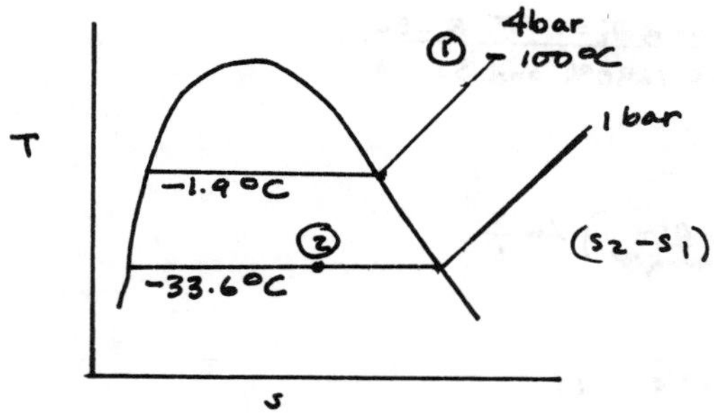

$(s_2 - s_1) = -3.1378$ kJ/kg·K

ASSUMPTION: The system is 1 kg of NH_3.

ANALYSIS: From Table A-15, $s_1 = 6.1169$ kJ/kg·K. Then

$s_2 = s_1 + (s_2 - s_1) = 6.1169 - 3.1378 = 2.9791$ kJ/kg·K.

Since $s_f < s_2 < s_g$ at 1 bar, Table A-14, state 2 is in the two-phase liquid-vapor regime. Then, $T = -33.6°C$ and

$$x_2 = \frac{2.9791 - 0.1191}{5.839 - 0.1191} = 0.5$$

$\Rightarrow \quad h_2 = h_f + x_2(h_{fg}) = 28.18 + 0.5(1370.23) = 713.3$ kJ/kg

PROBLEM 6.24

KNOWN: One lb of water undergoes a process with no change in specific entropy between two specified states.

FIND: Determine x or T at the final state, as appropriate, using alternative methods.

SCHEMATIC & GIVEN DATA:

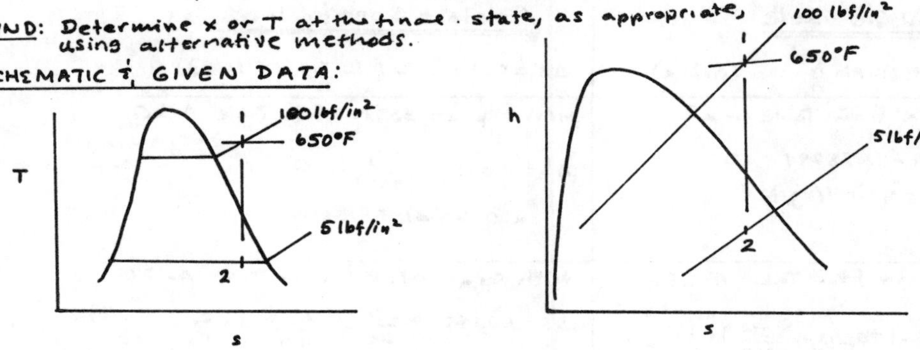

ASSUMPTION: The one lb of water is the system.

ANALYSIS: (a) From Table A-4E, $s_1 = 1.78075$ Btu/lb·°R. Then, with s_f and s_g at 5 lbf/in² from Table A-3E

$$x_2 = \frac{s_2 - s_f}{s_g - s_f} = \frac{1.78075 - 0.2349}{1.8441 - 0.2349} = 0.961$$

(b) By inspection of Figure A-8E at $p_1 = 100$ lbf/in², $T_1 = 650$°F, $s_1 = 1.781$ Btu/lb·°R. Then, with $s_2 = s_1$ and $p_2 = 5$ lbf/in², $x_2 = 0.96$.

(c) *IT* Code

```
T1 = 650  // °F
p1 = 100  // lbf/in.²
p2 = 5
s1 = s_PT("Water/Steam", p1, T1)
s2 = ssat_Px("Water/Steam", p2, x2)
s2 = s1
```

IT Results

$s_1 = s_2 = 1.781$ Btu/lb·°R
$x_2 = 0.9609$

PROBLEM 6.25

FIND: Determine the change in specific entropy, in kJ/kg·K, using the ideal gas model and three alternative approaches.

(1) case	Ideal Gas Table $\Delta s = s^\circ(T_2) - s^\circ(T_1) - R \ln \frac{P_2}{P_1}$ (6.21 a)	Constant Specific Heat $\Delta s = c_p \ln \frac{T_2}{T_1} - R \ln \frac{P_2}{P_1}$ (6.23)
(a) Air $P_1 = 100$ kPa $P_2 = 100$ kPa $T_1 = 293$ K $T_2 = 373$ K	With s° data from Table A-22 $\Delta s = 1.92119 - 1.678298 - 0$ $= 0.24289$ kJ/kg·K	With c_p at 333 K from Table A-20 $\Delta s = 1.007 \ln \frac{373}{293} - 0$ $= 0.2431$ kJ/kg·K ← (a)
(b) Air $P_1 = 1$ bar $P_2 = 3$ bar $T_1 = 300$ K $T_2 = 650$ K	With s° data from Table A-22 $\Delta s = 2.49364 - 1.70203 - \frac{8.314}{28.97} \ln \frac{3}{1}$ $= 0.47632$ kJ/kg·K	With c_p at 475 K from Table A-20 $\Delta s = 1.0245 \ln \frac{650}{300} - \frac{8.314}{28.97} \ln \frac{3}{1}$ $= 0.47684$ kJ/kg·K ← (b)
(c) CO_2 $P_1 = 150$ kPa $P_2 = 300$ kPa $T_1 = 303$ K $T_2 = 573$ K	With $\bar{s}^\circ$ data from Table A-23 and M from Table A-1 $\Delta s = \frac{241.033 - 214.284 - 8.314 \ln \frac{300}{150}}{44.01}$ $= 0.4769$ kJ/kg·K	With c_p at 438 K from Table A-20 $\Delta s = .9686 \ln \frac{573}{303} - \frac{8.314}{44.01} \ln \frac{300}{150}$ $= 0.4862$ kJ/kg·K ← (c)
(d) CO $T_1 = 300$ K $T_2 = 500$ K $v_1 = 1.1$ m³/kg $v_2 = 0.75$ m³/kg	With $pv = RT$ $\frac{P_2}{P_1} = \left(\frac{T_2}{T_1}\right)\left(\frac{v_1}{v_2}\right)$ $= \left(\frac{500}{300}\right)\left(\frac{1.10}{0.75}\right) = 2.444$ With $\bar{s}^\circ$ data from Table A-23 and M from Table A-1 $\Delta s = \frac{212.719 - 197.723 - 8.314 \ln 2.444}{28.01}$ $= 0.2701$ kJ/kg·K	With c_p at 400 K from Table A-20 $\Delta s = 1.047 \ln \frac{500}{300} - \frac{8.314}{28.01} \ln 2.444$ $= 0.2696$ kJ/kg·K ← (d)
(e) N_2 $P_1 = 2$ MPa $P_2 = 1$ MPa $T_1 = 800$ K $T_2 = 300$ K	With $\bar{s}^\circ$ data from Table A-23 and M from Table A-1 $\Delta s = \frac{191.682 - 220.907 - 8.314 \ln \frac{1}{2}}{28.02}$ $= -0.8373$ kJ/kg·K	With c_p at 550 K from Table A-20 $\Delta s = 1.065 \ln \frac{300}{800} - \frac{8.314}{28.02} \ln \frac{1}{2}$ $= -0.8389$ kJ/kg·K ← (e)

IT Results

(a) $\Delta s = 0.2429$ kJ/kg·K
(b) $\Delta s = 0.4762$ kJ/kg·K
(c) $\Delta s = 0.4764$ kJ/kg·K
(d) $\Delta s = 0.2702$ kJ/kg·K
(e) $\Delta s = -0.838$ kJ/kg·K

1. The constant c_p value is evaluated from Table A-20 at the average temperature. Alternatively, the arithmetic average of the c_p values at the two end temperatures can be used.

PROBLEM 6.25 (Contd.)

IT Code

```
pa1 = 1  // bar
pa2 = 1
Ta1 = 293 // K
Ta2 = 373
sa1 = s_TP("Air", Ta1, pa1)
sa2 = s_TP("Air", Ta2, pa2)
dels_a = sa2 - sa1

pb1 = 1
pb2 = 3
Tb1 = 300
Tb2 = 650
sb1 = s_TP("Air", Tb1, pb1)
sb2 = s_TP("Air", Tb2, pb2)
dels_b = sb2 - sb1

pc1 = 1.5
pc2 = 3
Tc1 = 303
Tc2 = 573
sc1 = s_TP("CO2", Tc1, pc1)
sc2 = s_TP("CO2", Tc2, pc2)
dels_c = sc2 - sc1

Td1 = 300
Td2 = 500
vd1 = 1.1 // m³/kg
vd2 = 0.75
vd1 = v_TP("CO",Td1,pd1)
vd2 = v_TP("CO",Td2,pd2)
sd1 = s_TP("CO", Td1, pd1)
sd2 = s_TP("CO", Td2, pd2)
dels_d = sd2 - sd1

pe1 = 20
pe2 = 10
Te1 = 800
Te2 = 300
se1 = s_TP("N2", Te1, pe1)
se2 = s_TP("N2", Te2, pe2)
dels_e = se2 - se1
```

PROBLEM 6.26

FIND: Determine the change in specific entropy, in Btu/lbmol·°R, using the ideal gas model and three alternative approaches.

①

case	Ideal Gas Table	Constant Specific Heat
	$\Delta\bar{s} = \bar{s}^\circ(T_2) - \bar{s}^\circ(T_1) - \bar{R}\ln\frac{p_2}{p_1}$ (6.21b)	$\Delta s = \bar{c}_p \ln\frac{T_2}{T_1} - \bar{R}\ln\frac{p_2}{p_1}$ (6.23)
(a) Air p_1 = 1 atm p_2 = 1 atm T_1 = 500°R T_2 = 860°R	With s° data from Table A-22E and M = 28.97 lb/lbmol $\Delta\bar{s} = 28.97(0.71323 - 0.58233) - 1.986 \ln 1$ (→ 0) = 3.7922 Btu/lbmol·°R	With c_p at 220°F from Table A-20E and M = 28.97 lb/lbmol $\Delta\bar{s} = (28.97)(0.241)\ln\frac{860}{500} - 1.986\ln 1$ (→ 0) = 3.7864 Btu/lbmol·°R ← (a)
(b) Air p_1 = 20 lbf/in² p_2 = 60 lbf/in² T_1 = 560°R T_2 = 760°R	With s° data from Table A-22E and M = 28.97 lb/lbmol $\Delta\bar{s} = 28.97(.68312 - .6095) - 1.986 \ln\frac{60}{20}$ = -0.0491 Btu/lbmol·°R	With c_p at 200°F from Table A-20E and M = 28.97 lb/lbmol $\Delta\bar{s} = (28.97)(0.241)\ln\frac{760}{560} - 1.986\ln\frac{60}{20}$ ← (b) = -0.0498 Btu/lbmol·°R
(c) CO_2 p_1 = 1atm p_2 = 3atm T_1 = 500°R T_2 = 960°R	With $\bar{s}^\circ$ data from Table A-23E $\Delta\bar{s} = 56.765 - 50.408 - 1.986\ln\frac{3}{1}$ = 4.1752 Btu/lbmol·°R	With c_p at 270°F from Table A-20E and M = 44.01 lb/lbmol from Table A-1E $\Delta\bar{s} = (44.01)(0.225)\ln\frac{960}{500} - 1.986\ln\frac{3}{1}$ ← (c) = 4.2776 Btu/lbmol·°R
(d) CO_2 T_1 = 660°R T_2 = 860°R v_1 = 20 ft³/lb v_2 = 15 ft³/lb	With pv = RT $\frac{p_2}{p_1} = \left(\frac{T_2}{T_1}\right)\left(\frac{v_1}{v_2}\right) = \left(\frac{860}{660}\right)\left(\frac{20}{15}\right) = 1.7374$ With $\bar{s}^\circ$ data from Table A-23E $\Delta\bar{s} = 55.389 - 52.934 - 1.986\ln 1.7374$ = 1.558 Btu/lbmol·°R	With c_p at 300°F from Table A-20E and M = 44.01 lb/lbmol from Table A-1E $\Delta\bar{s} = (44.01)(0.229)\ln\frac{860}{660} - 1.986\ln 1.7374$ = 1.571 Btu/lbmol·°R ← (d)
(e) N_2 p_1 = 2atm p_2 = 1atm T_1 = 1260°R T_2 = 660°R	With $\bar{s}^\circ$ data from Table A-23E $\Delta\bar{s} = 47.178 - 51.771 - 1.986\ln\frac{1}{2}$ = -3.2164 Btu/lbmol·°R	With c_p at 500°F from Table A-20E and M = 28.02 lb/lbmol from Table A-1E $\Delta\bar{s} = (28.02)(0.254)\ln\frac{660}{1260} - 1.986\ln\frac{1}{2}$ ← (e) = -3.2255 Btu/lbmol·°R

IT Results

(a) Δs = 3.79 Btu/lbmol·°R
(b) Δs = -0.04884 Btu/lbmol·°R
(c) Δs = 4.176 Btu/lbmol·°R
(d) Δs = 1.557 Btu/lbmol·°R
(e) Δs = -3.217 Btu/lbmol·°R

1. The constant c_p value is evaluated from Table A-20 at the average temperature. Alternatively, the arithmetic average of the c_p values at the two end states can be used.

PROBLEM 6.26 (Contd.)

<u>*IT* Code</u>

```
pa1 = 1  // atm
pa2 = 1
Ta1 = 500  // °R
Ta2 = 860
sa1 = s_TP("Air", Ta1, pa1)
sa2 = s_TP("Air", Ta2, pa2)
dels_a = sa2 - sa1

pb1 = 20 / 14.696  // atm
pb2 = 60 / 14.696
Tb1 = 560
Tb2 = 760
sb1 = s_TP("Air", Tb1, pb1)
sb2 = s_TP("Air", Tb2, pb2)
dels_b = sb2 - sb1

pc1 = 1
pc2 = 3
Tc1 = 500
Tc2 = 960
sc1 = s_TP("CO2", Tc1, pc1)
sc2 = s_TP("CO2", Tc2, pc2)
dels_c = sc2 - sc1

Td1 = 660
Td2 = 860
vd1 = 20 * 44.01  // ft³/lbmol
vd2 = 15 * 44.01
vd1 = v_TP("CO2",Td1,pd1)
vd2 = v_TP("CO2",Td2,pd2)
sd1 = s_TP("CO2", Td1, pd1)
sd2 = s_TP("CO2", Td2, pd2)
dels_d = sd2 - sd1

pe1 = 2
pe2 = 1
Te1 = 1260
Te2 = 660
se1 = s_TP("N2", Te1, pe1)
se2 = s_TP("N2", Te2, pe2)
dels_e = se2 - se1
```

PROBLEM 6.27

FIND: Determine the indicated property for a process where $s_2 = s_1$.

(a) Water. $P_1 = 14.7\ lbf/in^2$, $T_1 = 500°F$, $P_2 = 100\ lbf/in^2$. Find T_2.

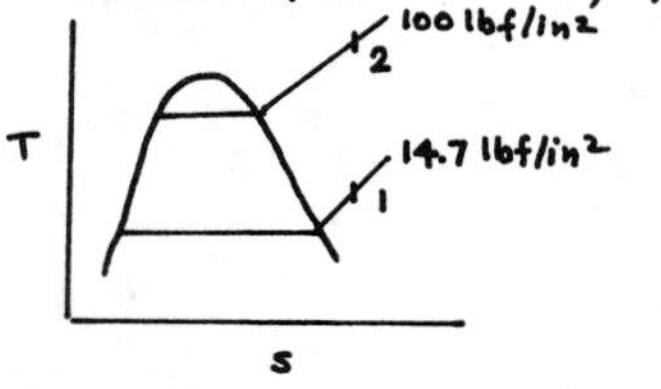

From Table A-4E, $s_1 = 1.9263$ Btu/lb·°R.
Interpolating with $s_2 = s_1$ in Table A-4E at 100 lbf/in²,
$T_2 = 1017\ °F$. ← (a)

(b) Water. $T_1 = 10°C$, $x_1 = 0.75$. Saturated vapor at state 2. Find P_2.

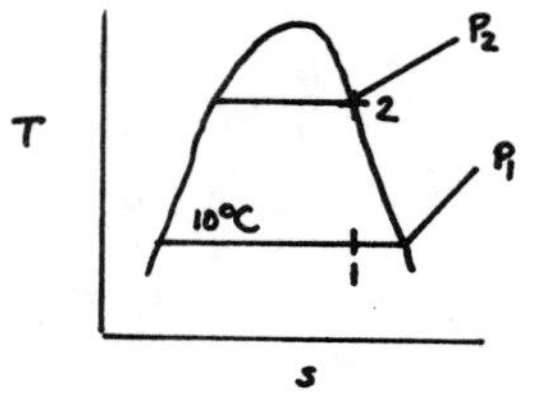

Using data from Table A-2

$$s_1 = s_f + x_1(s_g - s_f) = 0.1510 + 0.75(8.9008 - 0.1510)$$
$$= 6.71335\ kJ/kg\cdot K$$

Then, interpolating with $s_2 = s_1$ in Table A-3, $P_2 = 6.897$ bar. ← (b)

(c) Air as an ideal gas. $T_1 = 300K\ (27°C)$, $P_1 = 1.5$ bar, $T_2 = 400K\ (127°C)$. Find P_2.

Since $s_2 = s_1$, Eq. 6.25a gives

$$\ln \frac{P_2}{P_1} = \frac{s°(T_2) - s°(T_1)}{R}$$

With s° data from Table A-22

$$\ln \frac{P_2}{P_1} = \frac{1.99194 - 1.70203}{(8.314/28.97)}$$

$$\Rightarrow \quad \frac{P_2}{P_1} = 2.746 \quad \Rightarrow \quad P_2 = 4.119\ \text{bar}$$ ← (c)

(d) Air as an ideal gas. $T_1 = 560°R\ (100°F)$, $P_1 = 3$ atm, $P_2 = 2$ atm. Find T_2.

Since $s_2 = s_1$, Eq. 6.21a gives

$$s°(T_2) = s°(T_1) + R \ln \frac{P_2}{P_1}$$

With s° at T_1 from Table A-22E

$$s°(T_2) = 0.60950 + \frac{1.986}{28.97} \ln \frac{2}{3} = 0.5817\ \text{Btu/lb}\cdot°R.$$

Interpolating with $s°(T_2)$ in Table A-22E, $T_2 = 498.7°R\ (39°F)$. ← (d)

(e) R134a. $T_1 = 20°C$, $P_1 = 5$ bar, $P_2 = 1$ bar. Find v_2

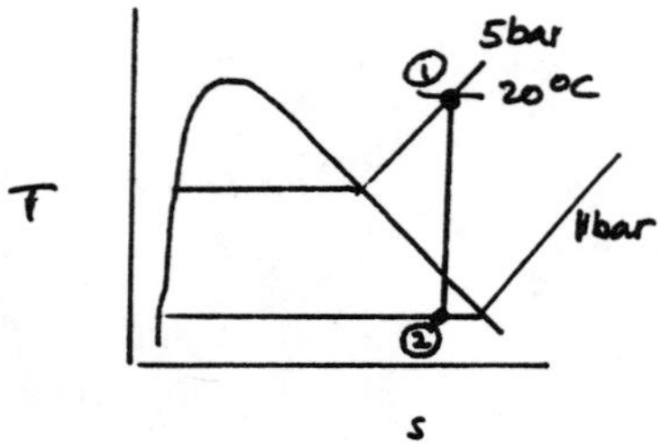

From Table A-12, $s_1 = 0.9264$ kJ/kg·K.
Then, with data from Table A-11 and $s_2 = s_1$

$$x_2 = \frac{0.9264 - 0.0678}{0.9395 - 0.0678} = 0.985$$

And

$$v_2 = v_f + x_2(v_g - v_f)$$
$$= \frac{0.7258}{10^3} + 0.985\left[0.1917 - \frac{0.7258}{10^3}\right]$$
$$= 0.188\ m^3/kg$$ ← (e)

PROBLEM 6.27 (Contd.)

IT Code

```
/* (a)  Water
p1 = 14.7  // lbf/in.²
T1 = 500  // °F
p2 = 100
s1 = s_PT("Water/Steam", p1, T1)
s2 = s_PT("Water/Steam", p2, T2)
s2 = s1

(b)  Water
T1 = 10  // °C
x1 = 0.75
x2 = 1
p1 = Psat_T("Water/Steam", T1)
s1 = ssat_Px("Water/Steam", p1, x1)
s2 = ssat_Px("Water/Steam", p2, x2)
s2 = s1

(c)  Air
T1 = 27  // °C
p1 = 1.5  // bar
T2 = 127
s1 = s_TP("Air", T1, p1)
s2 = s_TP("Air", T2, p2)
s2 = s1

(d)  Air
T1 = 100  // °F
p1 = 3  // atm
p2 = 2
s1 = s_TP("Air", T1, p1)
s2 = s_TP("Air", T2, p2)
s2 = s1

(e)  R-134a
*/
T1 = 20  // °C
p1 = 5  // bar
p2 = 1
s1 = s_PT("R134A", p1, T1)
x2 = x_sP("R134A", s2, p2)
v2 = vsat_Px("R134A", p2, x2)
s2 = s1
```

IT Results

(a) s = 1.926Btu/lb·°R; T_2 = 1016°F
(b) s = 6.712 kJ/kg·K; p_2 = 6.903 bar
(c) p_2 = 4.117 bar
(d) T_2 = 38.75°F
(e) s = 0.9264 kJ/kg·K; x_2 = 0.985, v_2 = 0.1888 m³/kg

PROBLEM 6.28

KNOWN: One kg of O_2 undergoes a process between specified states.

FIND: Using three methods, determine the change in specific entropy: (a) Eq. 6.19 with $\bar{c}_p$ from Table A-21, (b) Eq. 6.21b with $\bar{s}^\circ$ from Table A-23 (c) Eq. 6.23 with c_p at 900 K from Table A-20, (d) using IT.

SCHEMATIC & GIVEN DATA:

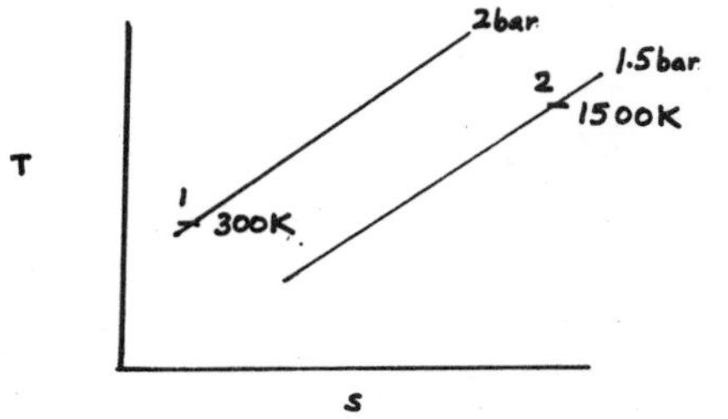

ASSUMPTION: The system consists of one kg of O_2, which behaves as an ideal gas.

(a) Using Eq. 6.19

$$s_2 - s_1 = \int_{T_1}^{T_2} \frac{c_p(T)}{T} dT - R \ln \frac{p_2}{p_1}$$

Table A-21 gives

$$\frac{\bar{c}_p}{\bar{R}} = \alpha + \beta T + \gamma T^2 + \delta T^3 + \epsilon T^4$$

where for O_2

$$\alpha = 3.626,\ \beta = -1.878 \times 10^{-3},\ \gamma = 7.056 \times 10^{-6},\ \delta = -6.764 \times 10^{-9},\ \epsilon = 2.156 \times 10^{-12}$$

Thus

$$\int_{T_1}^{T_2} \frac{c_p(T)}{T} dT = \frac{\bar{R}}{M} \int_{T_1}^{T_2} \left(\frac{\alpha + \beta T + \gamma T^2 + \delta T^3 + \epsilon T^4}{T} \right) dT = \frac{\bar{R}}{M} \left(\alpha \ln T + \beta T + \frac{\gamma T^2}{2} + \frac{\delta T^3}{3} + \frac{\epsilon T^4}{4} \right) \Bigg|_{T_1}^{T_2}$$

$$= \frac{8.314}{32} \left[3.626 \ln \frac{1500}{300} - \frac{1.878}{10^3}(1500-300) + \frac{7.056}{2(10^6)}(1500^2 - 300^2) - \frac{6.764}{3(10^9)}(1500^3 - 300^3) + \frac{2.156}{4(10^{12})}(1500^4 - 300^4) \right]$$

$$= 1.657 \text{ kJ/kg·K}$$

Finally

$$s_2 - s_1 = \int_{T_1}^{T_2} \frac{c_p}{T} dT - R \ln \frac{p_2}{p_1} = 1.657 - \frac{8.314}{32} \ln \frac{1.5}{2} = 1.7317 \text{ kJ/kg·K} \longleftarrow \text{(a)}$$

(b) Using Eq. 6.21b with $\bar{s}^\circ$ from Table A-23

$$s_2 - s_1 = \frac{\bar{s}_2 - \bar{s}_1}{M} = \frac{\bar{s}^\circ(T_2) - \bar{s}^\circ(T_1) - \bar{R} \ln p_2/p_1}{M}$$

$$= \frac{257.965 - 205.213 - 8.314 \ln 1.5/2}{32} = 1.7235 \text{ kJ/kg·K} \longleftarrow \text{(b)}$$

(c) Using Eq. 6.23 with c_p at 900 K from Table A-20

$$s_2 - s_1 = c_p \ln \frac{T_2}{T_1} - R \ln \frac{p_2}{p_1}$$

$$= 1.074 \ln \frac{1500}{300} - \frac{8.314}{32} \ln \frac{1.5}{2} = 1.8033 \text{ kJ/kg·K} \longleftarrow \text{(c)}$$

(d)

IT Code

```
T1 = 300 // K
p1 = 2  // bar
T2 = 1500  // K
p2 = 1.5  // bar

s1 = s_TP("O2", T1, p1)
s2 = s_TP("O2", T2, p2)
dels = s2 - s1
```

IT Result

Δs = 1.724 kJ/kg·K

PROBLEM 6.29

KNOWN: Two kilograms of water undergo a specified process.

FIND: Determine the entropy change if the process is (a) irreversible, (b) internally reversible.

SCHEMATIC & GIVEN DATA:

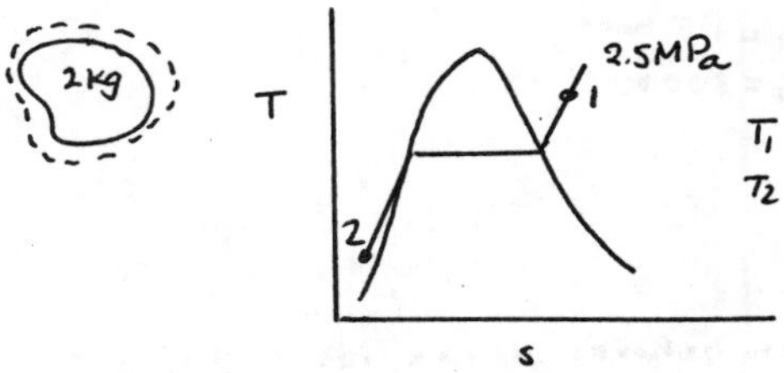

$T_1 = 400°C$
$T_2 = 100°C$

ASSUMPTION: The system consists of 2kg of water.

ANALYSIS: Since entropy is a property, the change in entropy between the two states is the same regardless of the nature of the process.

s_1, Table A-4: $7.02415 \frac{kJ}{kg \cdot K}$ (interpolation)

s_2, Table A-5: $1.3050 \frac{kJ}{kg \cdot K}$

Thus

$$\Delta s = s_2 - s_1 = (1.3050 - 7.02415)\frac{kJ}{kg \cdot K} = -5.7192 \frac{kJ}{kg \cdot K}$$

and for two kg

$$\Delta S = (2kg)\left(-5.7192 \frac{kJ}{kg \cdot K}\right) = -11.4384 \frac{kJ}{K}$$ ◄

PROBLEM 6.30

KNOWN: Water undergoes a process from a liquid state at 80°C, 5MPa to sat. liq. at 40°C

FIND: Determine the change in specific entropy using four alternative methods.

ANALYSIS: (a) Tables A-2 and A-5. $s_1 = 1.0720$ kJ/kg·K, $s_2 = 0.5725$, $\Delta s = -0.4995$ kJ/kg·K.

(b) Sat. Liq. data from Table A-2. $s_1 \approx s_f(T_1) = 1.0753$ kJ/kg·K, s_2 (as above),
$\Delta s = -0.5028$ kJ/kg·K.

(c) Incompressible liquid model with $c = 4.18 \frac{kJ}{kg \cdot K}$ from Table A-19 (at Tave): with Eq. 6.24

$$\Delta s = c \ln \frac{T_2}{T_1} = 4.18 \frac{kJ}{kg \cdot K} \ln \frac{313}{353} = -0.5027 \text{ kJ/kg} \cdot K$$

① (d) *IT* Code

```
T1 = 80 // °C
p1 = 50 // bar = 5 MPa
T2 = 40 // °C
x2 = 0

s1 = s_PT("Water/Steam", p1, T1)
p2 = Psat_T("Water/Steam", T2)
s2 = ssat_Px("Water/Steam", p2, x2)
dels = s2 - s1
```

IT Results

$s_1 = 1.077$
$s_2 = 0.5711$
$\Delta s = -0.506$

---

1. As noted in the Methodology Update on p. 244, IT returns liquid entropy data using Eq. 6.7.

PROBLEM 6.31

<u>KNOWN</u>: 0.1 kmol of CO undergoes a process between known states.

<u>FIND</u>: Determine (a) the heat transfer and (b) the change in entropy.

<u>SCHEMATIC & GIVEN DATA</u>:

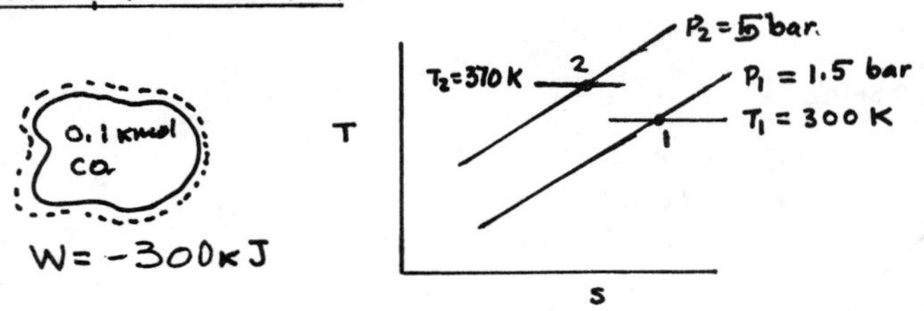

<u>ASSUMPTIONS</u>: (1) The system consists of the CO which behaves as an ideal gas. (2) There are no changes in kinetic and potential energy between the two end states.

<u>ANALYSIS</u>: (a) With assumption 2, an energy balance gives

$$Q = W + \Delta U$$
$$= W + n[\bar{u}_2 - \bar{u}_1]$$

With $\bar{u}$ data from Table A-23

$$Q = -300\ \text{kJ} + (0.1\ \text{kmol})[7689 - 6229]\ \text{kJ/kmol} = -154\ \text{kJ} \longleftarrow Q$$

(b) With Eq. 6.21b and $\bar{s}^o$ data from Table A-23

$$\Delta S = n[\bar{s}^o(T_2) - \bar{s}^o(T_1) - \bar{R}\ln p_2/p_1] = 0.1[203.842 - 197.723 - 8.314 \ln 5/1.5]$$
$$= -0.3891\ \text{kJ/K} \longleftarrow \Delta S$$

PROBLEM 6.32

<u>KNOWN</u>: CH_4 is compressed from 298K, 1 bar to temperature T, 2 bar in a process for which $s_2 = s_1$

<u>FIND</u>: Determine T.

<u>SCHEMATIC & GIVEN DATA</u>:

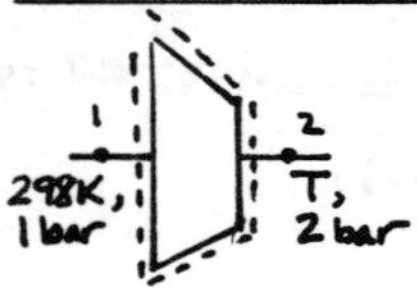

<u>ASSUMPTION</u>: CH_4 is modeled as an ideal gas.

① <u>ANALYSIS</u> Using IT

IT code

```
T1 = 298  // K
p1 = 1  // bar
p2 = 2  // bar

s1 = s_TP("CH4", T1, p1)
s2 = s_TP("CH4", T2, p2)
s2 = s1
```

IT Result

T_2 = 348.7 K ◁———

---

1. Alternatively, using Eq. 6.19 with $\bar{c}_p(T)$ from Table A-21

$$\bar{s}_2 - \bar{s}_1 = \int_{T_1}^{T_2} \left(\frac{\bar{c}_p}{T}\right) dT - \bar{R}\ln\left(\frac{p_2}{p_1}\right)$$

$$0 = \bar{R}\left[\alpha \ln T + \beta T + \frac{\gamma}{2}T^2 + \frac{\delta}{3}T^3 + \frac{\epsilon}{4}T^4\right]_{298}^{T} - \bar{R}\ln\left(\frac{p_2}{p_1}\right)$$

Using an equation solver, such as IT, this equation can be solved for T.

PROBLEM 6.33

KNOWN: A quantity of air undergoes a thermodynamic cycle consisting of three processes.

FIND: Evaluate the change in entropy for each process and sketch the cycle on p-v coordinates.

SCHEMATIC & GIVEN DATA:

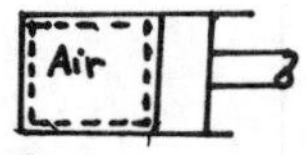

$p_1 = 0.1\,MPa$

$V_1 = 0.02\,m^3$

$p_2 = 0.42\,MPa$

$m = 2.42 \times 10^{-2}\,kg$

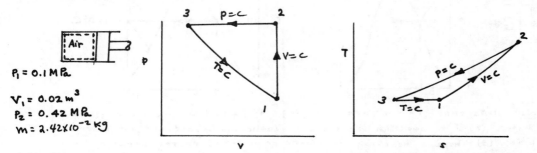

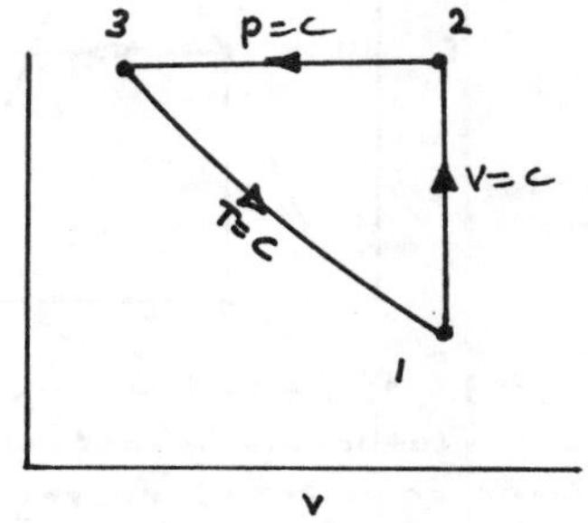

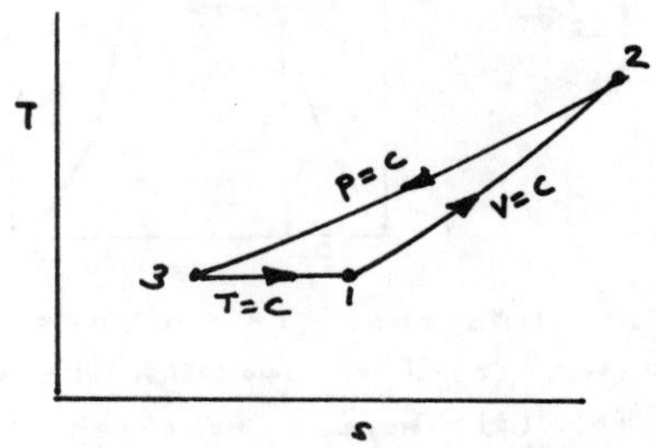

ASSUMPTIONS: (1) As shown in the accompanying figure, the system consists of the quantity of air. (2) The air behaves as an ideal gas with $c_p = 1\,kJ/kg \cdot K$.

ANALYSIS: The entropy changes can be evaluated using Eqs. 6.22 and 6.23. First, some preliminary results are calculated. Using the ideal gas equation of state

$$T_1 = \frac{p_1 V_1}{mR} = \frac{(0.1 \times 10^6\,N/m^2)(0.02\,m^3)}{(2.42 \times 10^{-2}\,kg)\left(\frac{8314}{28.97}\,\frac{N \cdot m}{kg\,K}\right)} = 288\,K$$

Also, since $V_1 = V_2$

$$\left.\begin{matrix} p_1 V = mRT_1 \\ p_2 V = mRT_2 \end{matrix}\right\} \Rightarrow T_2 = \frac{p_2}{p_1} T_1 = \left(\frac{0.42}{0.10}\right)(288) = 1210\,K$$

With Eq. 3.44

$$c_v = c_p - R = 1\,\frac{kJ}{kg \cdot K} - \frac{8.314}{28.97}\,\frac{kJ}{kg \cdot K} = 0.713\,kJ/kg \cdot K$$

Process 1-2: With $v_2 = v_1$, Eq. 6.22 reduces so

$$S_2 - S_1 = m c_v \ln \frac{T_2}{T_1} = (0.024\,kg)\left(0.713\,\frac{kJ}{kg \cdot K}\right) \ln \frac{1210}{288}$$

$$= 0.0246\,kJ/K \longleftarrow S_2 - S_1$$

Process 2-3: With $p_1 = p_2$, Eq. 6.23 reduces so

$$S_3 - S_2 = m c_p \ln \frac{T_3}{T_2} = m c_p \ln \frac{T_1}{T_2}$$

$$= (0.024\,kg)\left(1.0\,\frac{kJ}{kg \cdot K}\right) \ln \frac{288}{1210}$$

$$= -0.0344\,kJ/K \longleftarrow S_3 - S_2$$

Process 3-1: Since $T_3 = T_1$, Eq. 6.23 reduces so

$$S_1 - S_3 = -mR \ln \frac{p_1}{p_3} = -mR \ln \frac{p_1}{p_2}$$

$$= -(0.024)\left(\frac{8.314}{28.97}\right) \ln \frac{0.1}{0.42}$$

$$= 0.0099\,kJ/K \longleftarrow S_1 - S_3$$

PROBLEM 6.34

KNOWN: One kg of water undergoes an isothermal process between two specified states.

FIND: Determine the heat transfer and the work. Sketch the process on p-v and T-s coordinates

SCHEMATIC & GIVEN DATA:

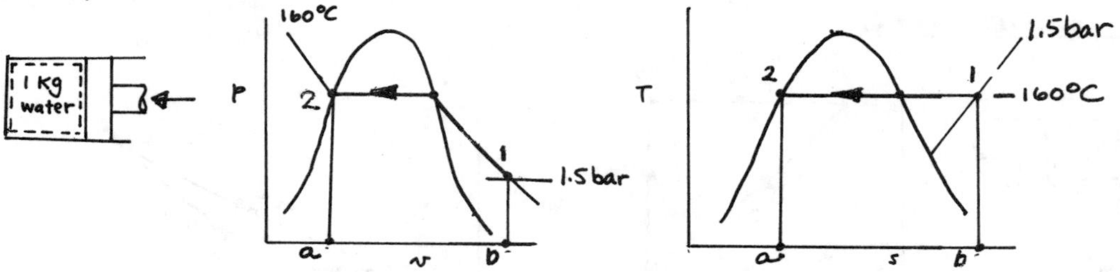

ASSUMPTIONS: (1) As shown in the accompanying figure, the system consists of one kg of water. (2) The compression takes place isothermally and without internal irreversibilities. (3) There is no change in kinetic or potential energy between the end states.

ANALYSIS: Using assumption 2, Eq. 6.25 becomes

$$Q = \int_1^2 T dS = m T [s_2 - s_1]$$

Then, with data from Tables A-2 and A-4

$$Q = (1\,kg)(433\,K)(1.9427 - 7.4665)\frac{kJ}{kg \cdot K} = -2391.8\,kJ \longleftarrow Q$$

The magnitude of the heat transfer is represented by area 1-2-a-b-1 on the T-s diagram.

The energy balance reduces to give $W = Q - m(u_2 - u_1)$. With data from Tables A-2 and A-4

$$W = -2391.8 - (1\,kg)[674.86 - 2595.2]\frac{kJ}{kg} = -471.5\,kJ \longleftarrow$$

Alternatively, $W = \int_1^2 p dV$. The magnitude of the work is represented by area 1-2-a-b-1 on the p-v diagram.

PROBLEM 6.35

KNOWN: Two kg of water undergo an isothermal process.

FIND: Determine the heat transfer and the work, each in kJ.

SCHEMATIC & GIVEN DATA:

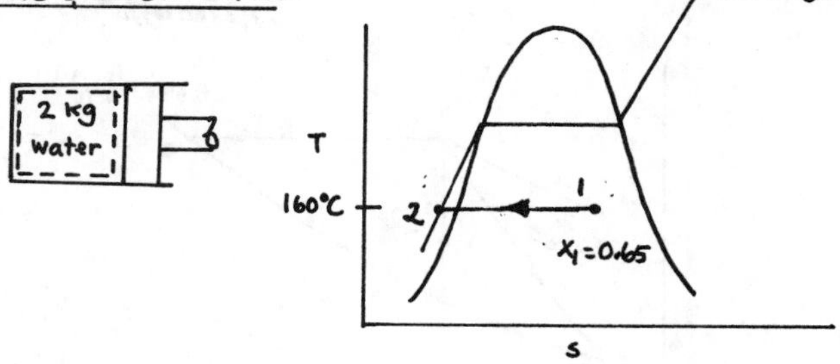

ASSUMPTIONS: (1) As shown in the accompanying figure, the system consists of 2 kg of water. (2) The process takes place isothermally and without internal irreversibilities. (3) There is no change in kinetic or potential energy between the end states.

ANALYSIS: Using assumption 2, Eq. 6.25 becomes

$$Q = \int_1^2 T\,dS = mT(s_2 - s_1)$$

With data from Table A-2 at 160°C, $s_2 \approx s_f(160°C) = 1.9427$ kJ/kg·K, and

$$s_2 = s_f + x_2(s_g - s_f) = 1.9427 + 0.65(6.7502 - 1.9427)$$
$$= 5.068 \text{ kJ/kg·K}$$

so

$$Q = (2\text{kg})(433\text{K})(1.9427 - 5.068) = -2706.5 \text{ kJ} \longleftarrow Q$$

An energy balance reduces to give

$$W = Q - m(u_2 - u_1)$$

With data from Table A-2 at 160°C, $u_2 \approx u_f(160°C) = 674.86$ kJ/kg, and

$$u_1 = u_f + x_1(u_g - u_f) = 674.86 + 0.65(2568.4 - 674.86) = 1905.7 \text{ kJ/kg}$$

so

$$W = -2706.5 + 2(1905.7 - 674.86)$$
$$= -244.8 \text{ kJ} \longleftarrow W$$

PROBLEM 6.36

KNOWN: Air undergoes an internally reversible process between two specified states.

FIND: For each of three cases, determine the heat transfer and the work, per unit of mass, and show the process on p-v and T-s coordinates.

SCHEMATIC & GIVEN DATA:

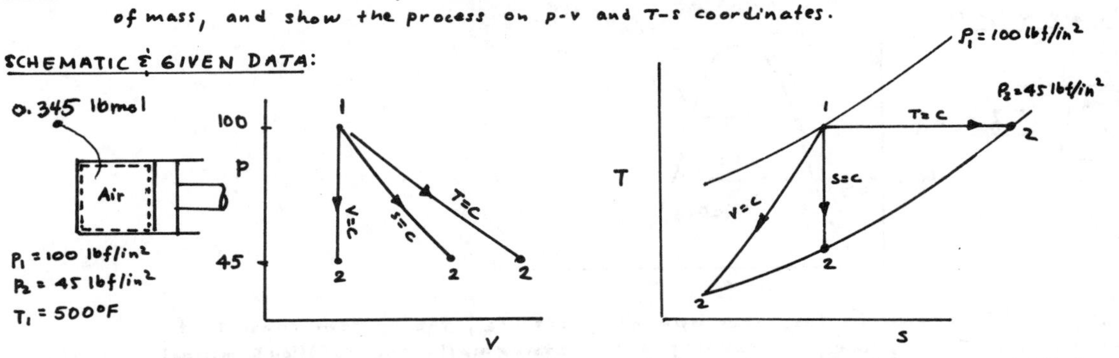

ASSUMPTIONS: (1) The system consists of air which behaves as an ideal gas. (2) Each process is internally reversible. (3) There is no change in kinetic or potential energy between the end states.

ANALYSIS: (a) Isothermal process. With assumption (2), Eq. 6.25 is applicable. Then, as temperature is constant

$$Q = \int_1^2 T\,dS = m\,T(s_2 - s_1)$$

With Eq. 6.21a and $m = nM$

$$Q = m\,T(s_2 - s_1) = mT\left[-R \ln \frac{p_2}{p_1}\right]$$

$$= \left((0.345)(28.97)\ \text{lb}\right)(960°R)\left(-\frac{1.986}{28.97}\frac{\text{Btu}}{\text{lb}\cdot°R}\right) \ln \frac{45}{100} = +525.2\ \text{Btu}$$ ← Q

Since internal energy depends on temperature alone for an ideal gas, $\Delta U = 0$ because $T_1 = T_2$. Thus, an energy balance reduces with assumption 3 to give $W = Q = +525.2$ Btu. ← W

(b) Adiabatic Process. Since adiabatic, $Q = 0$. Then, an energy balance gives ← Q

$$W = -m(u_2 - u_1)$$

This requires that state 2 be fixed. However, as the process is adiabatic and internally reversible, it is an isentropic process, so $s_2 = s_1$. Eq. 6.21a gives

$$s°(T_2) = s°(T_1) + R \ln \frac{p_2}{p_1}$$

With data from Table A-22E

$$s°(T_2) = 0.74030 + \frac{1.986}{28.97} \ln \frac{45}{100} = 0.68556 \Rightarrow u_2 = 131.36\ \text{Btu/lb}$$

At $T_1 = 960°R$, $u_1 = 165.26$ Btu/lb. Finally, with $m = nM = 9.99$ kg

$$W = -(9.99)(131.36 - 165.26) = +338.7\ \text{Btu}$$ ← W

(c) Constant volume process. Since volume is constant, $W = 0$. Then, an energy balance gives $Q = m(u_2 - u_1)$. To find u_2, first determine T_2 using the ideal gas equation of state ← W

$$\left.\begin{array}{l} p_1 v = RT_1 \\ p_2 v = RT_2 \end{array}\right\} \Rightarrow T_2 = T_1\left(\frac{p_2}{p_1}\right) = 960\left(\frac{45}{100}\right) = 432°R$$

Then, from Table A-22E $u_2 = 73.57$ Btu/lb, and

$$Q = (9.99)(73.57 - 165.26) = -916\ \text{Btu}$$ ← Q

PROBLEM 6.37

KNOWN: A gas expands from a specified initial state to a specified final pressure isothermally and without internal irreversibilities.

FIND: Determine the heat transfer and work, per unit of mass, for (a) R134a (b) Air as an ideal gas.

SCHEMATIC & GIVEN DATA:

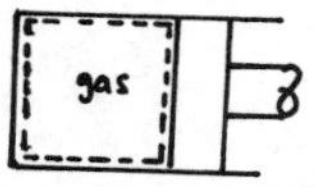

$T = 333\,K\ (60°C)$

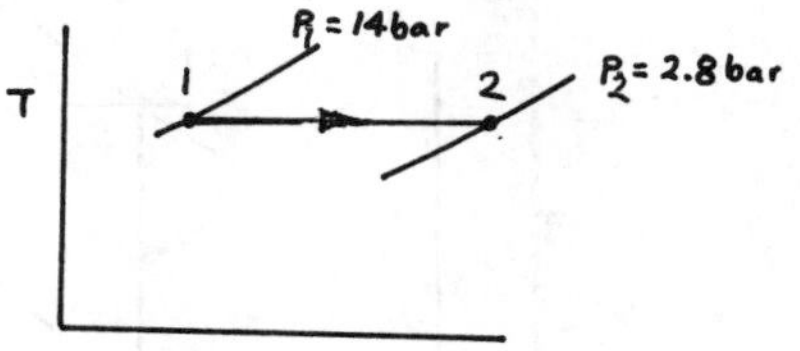

ASSUMPTIONS: (1) As shown in the accompanying figure, the system is the gas. (2) The expansion takes place isothermally and without internal irreversibilities. (3) There is no change in kinetic or potential energy between the end states. (4) Air is modeled as an ideal gas.

ANALYSIS: Using assumption 2, Eq. 6.25 becomes

$$Q = \int_1^2 T\,dS = mT(s_2 - s_1) \implies Q/m = T(s_2 - s_1)$$

With assumption 2, an energy balance gives

$$W = Q - \Delta U \implies W/m = Q/m - (u_2 - u_1)$$

(a) R134a. Table A-12: $u_1 = 262.17$ kJ/kg, $s_1 = 0.9297$ kJ/kg·K, $u_2 = 277.23$, $s_2 = 1.1079$ kJ/kg·K.

$$\Rightarrow \quad Q/m = 333\,K\left(1.1079 - 0.9297\right)\frac{kJ}{kg\cdot K} = 59.34\ kJ/kg \quad \longleftarrow \quad Q/m$$

$$W/m = 59.34 - (277.23 - 262.17) = 44.28\ kJ/kg \quad \longleftarrow \quad W/m$$

(b) AIR. Since $T_1 = T_2$, Eq. 6.21a reduces to read

$$s_2 - s_1 = -R \ln p_2/p_1 = -\left(\frac{8.314}{28.97}\frac{kJ}{kg\cdot K}\right)\ln\frac{2.8}{14} = +0.46188\ \frac{kJ}{kg\cdot K}$$

and

$$Q/m = (333)(0.46188) = 153.81\ \frac{kJ}{kg} \quad \longleftarrow \quad Q/m$$

Since internal energy depends on temperature alone for an ideal gas, $\Delta U = 0$ because $T_1 = T_2$. The energy balance then gives

$$W/m = Q/m - (\cancel{u_2 - u_1})^{\,0}$$

$$= Q/m = 153.81\ kJ/kg \quad \longleftarrow \quad W/m$$

PROBLEM 6.38

KNOWN: A gas expands from a specified initial state to a specified final pressure adiabatically and without internal irreversibilities.

FIND: Determine the work, per unit of mass, for (a) R134a, (b) air as an ideal gas.

SCHEMATIC & GIVEN DATA:

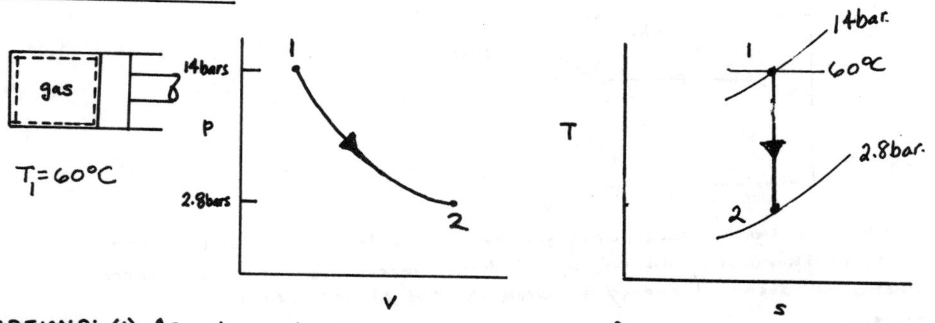

ASSUMPTIONS: (1) As shown in the accompanying figure, the system is the gas. (2) The expansion takes place adiabatically and without internal irreversibilities. (3) There is no change in kinetic or potential energy between the end states. (4) Air is modeled as an ideal gas.

ANALYSIS: Since the process is both adiabatic and internally reversible, it is isentropic (Sec. 6.4). This together with the specified final pressure fixes state 2. With assumption 3, an energy balance for the adiabatic process reduces to give $W = -m(u_2 - u_1)$, or $W/m = -(u_2 - u_1)$.

(a) R134a. Table A-12: $u_1 = 262.17$ kJ/kg, $s_1 = 0.9297$ kJ/kg·K. Then, with $s_2 = s_1$ and data from Table A-12 at 2.8 bar, interpolation gives $u_2 = 228.82$ kJ/kg. Finally,

$$\frac{W}{m} = -(228.82 - 262.17)\frac{kJ}{kg} = 33.25 \frac{kJ}{kg} \longleftarrow \text{(a)}$$

(b) AIR. With $s_2 = s_1$, Eq. 6.21a reduces to

$$s°(T_2) = s°(T_1) + R \ln P_2/P_1$$

With $s°(T_1)$ from Table A-22

$$s°(T_2) = 1.80685 + \frac{8.314}{28.97} \ln \frac{2.8}{14} = 1.345$$

Using this value, interpolation in Table A-22 gives $u_2 = 149.8$ kJ/kg. Also, $u_1 = 237.8$ kJ/kg. Finally

$$\frac{W}{m} = -(149.8 - 237.8) = 88 \text{ kJ/kg} \longleftarrow \text{(b)}$$

PROBLEM 6.39

KNOWN: 0.1 lb of helium is compressed from a specified initial state to a specified final pressure without irreversibilities.

FIND: Determine the work and the change in entropy if the process is (a) isothermal, (b) polytropic with $n = 1.3$, (c) adiabatic.

SCHEMATIC & GIVEN DATA:

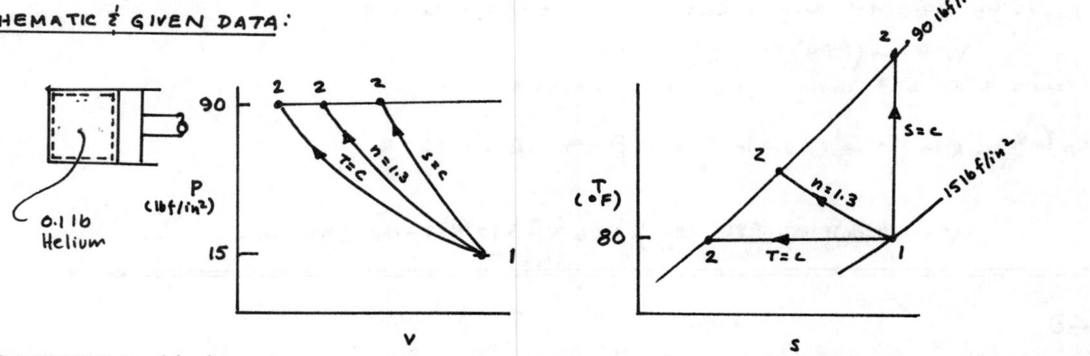

ASSUMPTIONS: (1) As shown in the accompanying figure, the system is the gas. (2) The gas is modeled as an ideal gas. (3) The compression is internally reversible.

ANALYSIS: (a) Isothermal. Using the ideal gas equation of state

$$W = \int_1^2 p\,dV = \int_1^2 \frac{mRT\,dV}{V} = mRT \ln \frac{V_2}{V_1}$$

Since $T_1 = T_2$, $V_2/V_1 = p_1/p_2$, so with the molecular weight from Table A-1E

$$W = mRT \ln \frac{p_1}{p_2} = (0.1\,\text{lb})\left(\frac{1.986}{4.003}\ \frac{\text{Btu}}{\text{lb}\cdot{}^\circ\text{R}}\right)(540^\circ\text{R}) \ln \frac{15}{90} = -48.0\ \text{Btu} \longleftarrow W$$

Since $T_1 = T_2$, Eq. 6.21a reduces to give

$$\Delta S = -mR \ln \frac{p_2}{p_1} = -(0.1\,\text{lb})\left(\frac{1.986}{4.003}\ \frac{\text{Btu}}{\text{lb}\cdot{}^\circ\text{R}}\right)\ln \frac{90}{15} = -0.089\ \frac{\text{Btu}}{{}^\circ\text{R}} \longleftarrow \Delta S$$

(b) Polytropic, $n = 1.3$. The process is described by $p = C/V^n$ where C is a constant (Sec. 2.2.3). Thus,

$$W = \int_1^2 p\,dV = \int_1^2 \frac{C}{V^n}\,dV = \frac{p_2V_2 - p_1V_1}{1-n} = \frac{mR(T_2 - T_1)}{1-n}$$

where in the last step the ideal gas equation of state has been used. Introducing Eq. 3.56

$$W = \frac{mRT_1}{1-n}\left[\left(\frac{p_2}{p_1}\right)^{(n-1)/n} - 1\right] = \frac{(0.1\,\text{lb})\left(\frac{1.986}{4.003}\ \frac{\text{Btu}}{\text{lb}\cdot{}^\circ\text{R}}\right)(540^\circ\text{R})}{1-1.3}\left[\left(\frac{90}{15}\right)^{0.3/1.3} - 1\right]$$

$$= -45.73\ \text{Btu} \longleftarrow W$$

For a monatomic gas such as helium $c_p/R = 2.5$ (Table A-21), so Eq. 6.23 is appropriate for evaluating Δs. The temperature at state 2 is obtained using Eq. 3.56 as 817°R. Thus

$$\Delta S = m\left[c_p \ln \frac{T_2}{T_1} - R \ln \frac{p_2}{p_1}\right] = (0.1\,\text{lb})\left(\frac{1.986}{4.003}\ \frac{\text{Btu}}{\text{lb}\cdot{}^\circ\text{R}}\right)\left[2.5 \ln \frac{817}{540} - \ln \frac{90}{15}\right]$$

$$= -0.0375\ \frac{\text{Btu}}{{}^\circ\text{R}} \longleftarrow \Delta S$$

PROBLEM 6.39 (Contd.)

(c) Adiabatic. As the process is both adiabatic and internally reversible, it is isentropic (Sec. 6.4). Thus $\Delta S = 0$. ← ΔS

Since $Q = 0$, an energy balance reduces with assumption 4 to give $W = -m(u_2 - u_1)$. Using Eq. 3.44 with $c_p = 2.5R$, $c_v = 1.5R$. Thus

$$W = -m(1.5R)(T_2 - T_1)$$

To find T_2, rewrite Eq. 6.23 using $s_2 = s_1$ and $c_p = 2.5R$

$$0 = c_p \ln\frac{T_2}{T_1} - R\ln\frac{p_2}{p_1} \Rightarrow 2.5\ln\frac{T_2}{T_1} = \ln\frac{p_2}{p_1} \Rightarrow T_2 = 1106\,°R$$

Finally

$$W = -(0.1 lb)(1.5)\left(\frac{1.986}{4.003}\frac{Btu}{lb\cdot°R}\right)(1106\,°R - 540\,°R) = -42.12\ Btu \quad \leftarrow W$$

PROBLEM 6.40

KNOWN: Air undergoes an internally reversible compression between specified states during which pV^n = constant, where $n = 1.27$.

FIND: Determine (a) p_2, (b) the work and heat transfer, (c) the entropy change.

SCHEMATIC & GIVEN DATA:

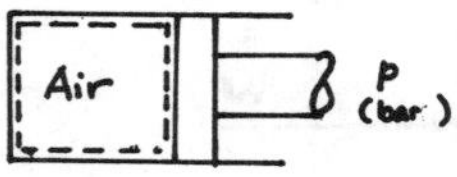

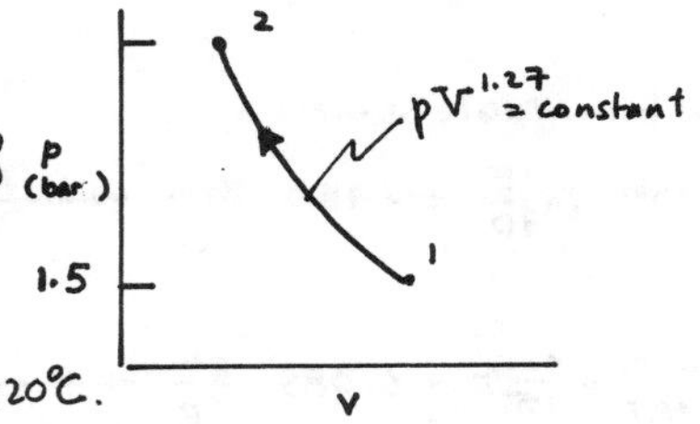

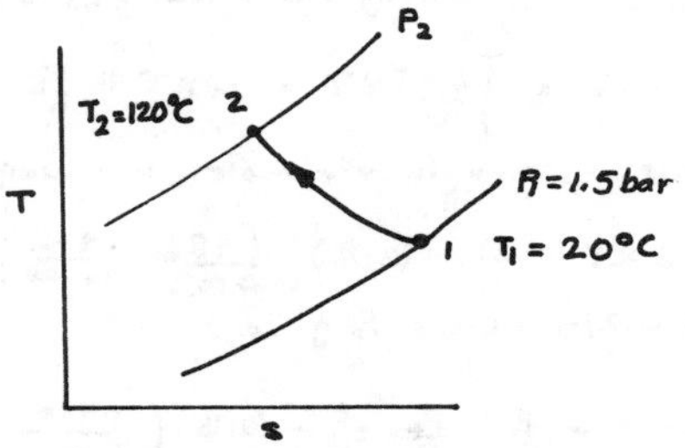

$V_1 = 1\ m^3$
$p_1 = 1.5$ bar
$T_1 = 20°C$, $T_2 = 120°C$.

ASSUMPTIONS: (1) As shown in the accompanying figure, the system consists of the air which is modeled as an ideal gas. (2) The compression is internally reversible and described by pV^n = constant. (3) There is no change in kinetic or potential energy between the end states.

ANALYSIS: (a) With Eq 3.56,

$$p_2 = p_1\left(\frac{T_2}{T_1}\right)^{n/(n-1)} = (1.5\,bar)\left(\frac{393}{293}\right)^{\frac{1.27}{.27}} = 5.97\ bar \quad \leftarrow$$

(b) The mass of air present can be found using the ideal gas equation of state

$$m = \frac{p_1V_1}{RT_1} = \frac{(1.5\times10^5\,N/m^2)(1\,m^3)}{\left(\frac{8314}{28.97}\frac{N\cdot m}{kg\cdot K}\right)(293\,K)} = 1.784\ kg$$

The work is

$$W = \int_1^2 p\,dV = \int_1^2 \frac{c}{V^n}dV = \frac{p_2V_2 - p_1V_1}{1-n} = \frac{mR(T_2 - T_1)}{1-n} = \frac{(1.784\,kg)\left(\frac{8.314}{28.97}\frac{kJ}{kg}\right)(100\,K)}{1 - 1.27}$$

$$= -189.6\ kJ \quad \leftarrow W$$

With assumption 3, an energy balance reduces to give $Q = m(u_2 - u_1) + W$. Evaluating u_1 and u_2 from Table A-22

$$Q = (1.784\,kg)(281.1 - 209.06)\,kJ/kg + (-189.6\,kJ) = -61.1\ kJ \quad \leftarrow Q$$

(c) The entropy change can be found with $s°$ data from Table A-22 and Eq. 6.21a

$$\Delta S = m\left[s°(T_2) - s°(T_1) - R\ln\frac{p_2}{p_1}\right] = 1.784\left[1.9740 - 1.6783 - \frac{8.314}{28.97}\ln\frac{5.97}{1.5}\right] = -0.1797\ \frac{kJ}{K} \quad \leftarrow \Delta S$$

PROBLEM 6.41

KNOWN: Air undergoes two internally reversible processes in series.

FIND: Sketch the processes on p-v and T-s diagrams. Determine the temperatures at states 2 and 3, and the net work.

SCHEMATIC & GIVEN DATA:

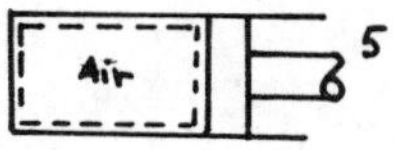

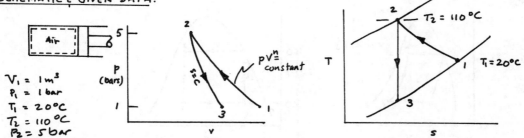

$V_1 = 1\ m^3$
$p_1 = 1$ bar
$T_1 = 20°C$
$T_2 = 110°C$
$p_2 = 5$ bar

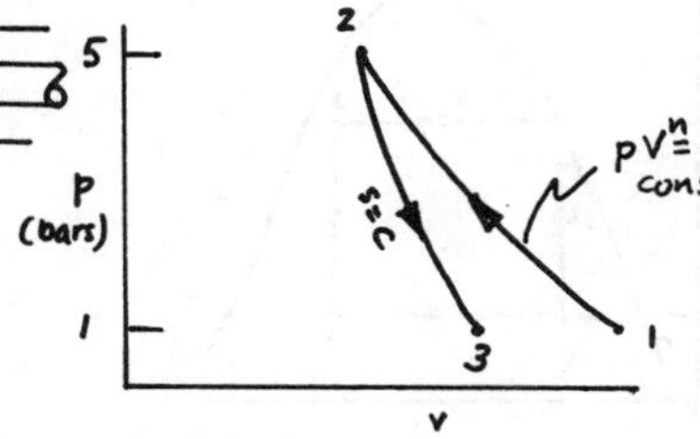

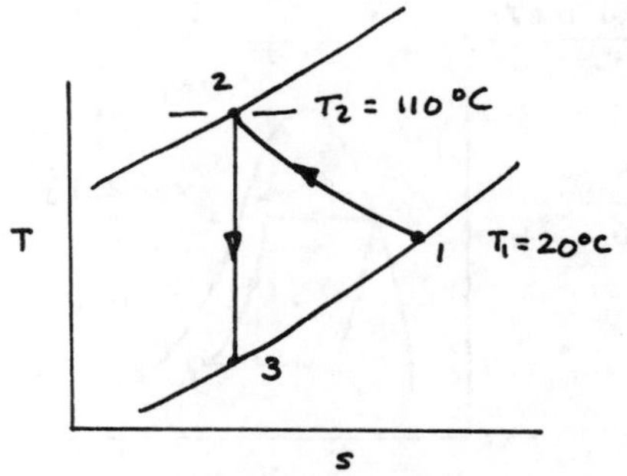

ASSUMPTIONS: (1) As shown in the accompanying figure, the system is the air. (2) The air is modeled as an ideal gas. (3) Each process is internally reversible. Process 1-2 is described by pV^n = constant and process 2-3 is adiabatic. (4) There is no change in kinetic or potential energy between the end states of either process.

ANALYSIS: (b) Using Eq. 3.56.

$$\frac{T_2}{T_1} = \left(\frac{P_2}{P_1}\right)^{\frac{(n-1)}{n}} \Rightarrow \frac{n-1}{n} = \frac{\ln(T_2/T_1)}{\ln(P_2/P_1)} = \frac{\ln(383/293)}{\ln(5/1)} \Rightarrow n = 1.2 \longleftarrow n$$

(c) Since process 2-3 is adiabatic and internally reversible, it is isentropic (Sec. 6.4). Thus, with $s_3 = s_2$, Eq. 6.21a gives

$$s°(T_3) = s°(T_2) + R \ln \frac{P_3}{P_2} = 1.94791 + \frac{8.314}{28.97} \ln \frac{1}{5} = 1.48602$$

where $s°(T_2)$ is from Table A-22. Interpolating with $s°(T_3)$ in Table A-22, $T_3 = 242$ K. $\longleftarrow T_3$

(d) The mass can be determined using the ideal gas equation of state

$$m = \frac{P_1 V_1}{R T_1} = \frac{(10^5\ N/m^2)(1\ m^3)}{\left(\frac{8314}{28.97}\frac{N\cdot m}{kg\cdot K}\right)(293\ K)} = 1.189\ kg$$

The work for process 1-2 is

$$W_{12} = \int_1^2 p\,dV = \int_1^2 \frac{C}{V^n}\,dV = \frac{P_2V_2 - P_1V_1}{1-n} = \frac{mR(T_2 - T_1)}{1-n} = \frac{(1.189\ kg)\left(\frac{8.314}{28.97}\frac{kJ}{kg\cdot K}\right)(383-293)\ K}{1-1.2}$$

$$= -153.55\ kJ$$

For process 2-3 an energy balance reduces with assumption 4 to give, $W_{23} = -m(u_3 - u_2)$. With data from Table A-22

$$W_{23} = -\ 1.189\ kg\,(172.56 - 273.86)\ \frac{kJ}{kg} = +\ 120.45\ kJ$$

The net work is

$$W = W_{12} + W_{23} = (-153.55) + (120.45) = -33.1\ kJ \longleftarrow W$$

PROBLEM 6.42

KNOWN: 0.1 kg of water executes a Carnot power cycle.

FIND: Sketch the cycle on p-v and T-s coordinates. Determine the heat added, net work, and thermal efficiency.

SCHEMATIC & GIVEN DATA:

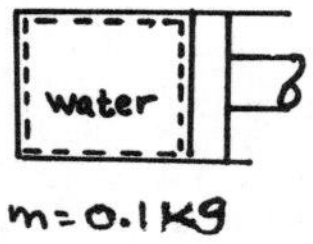

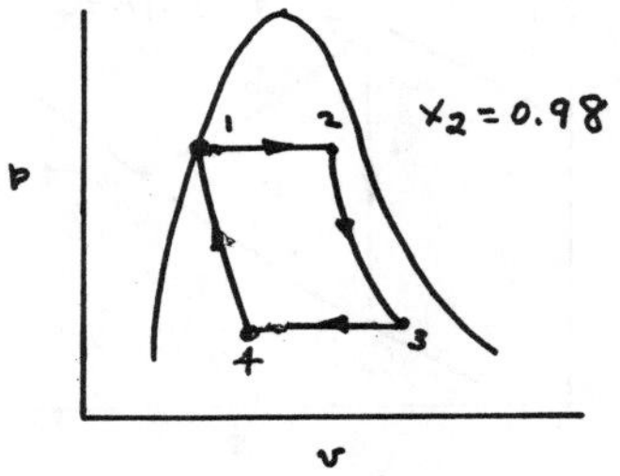

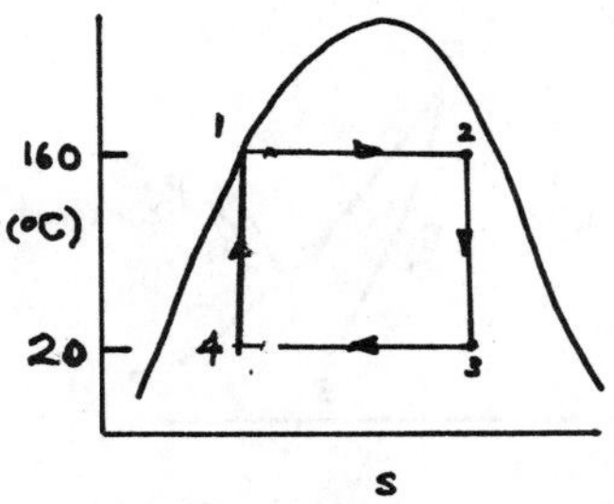

ASSUMPTION: The system shown in the accompanying figure executes a Carnot cycle.

ANALYSIS: Since each process is internally reversible, Eq. 6.29 is applicable.

$$Q_{12} = \int_1^2 T\,dS = m\,T_1(s_2 - s_1) \qquad \text{(heat added)}$$

$$Q_{34} = \int_3^4 T\,dS = m\,T_3(s_4 - s_3) = -m\,T_3(s_2 - s_1)$$

$$Q_{23} = Q_{41} = 0$$

For any cycle $W_{cycle} = Q_{cycle}$. Thus $W_{cycle} = m[T_1(s_2 - s_1) - T_3(s_2 - s_1)]$, or

$$W_{cycle} = m(T_1 - T_3)(s_2 - s_1) \qquad \text{(net work)}$$

From Table A-2, $s_1 = 1.9427$ kJ/kg·K. To find s_2, use the known value for x_2 with data from Table A-2

$$s_2 = s_f + x_2(s_g - s_f) = 1.9427 + 0.98(6.7502 - 1.9427) = 6.654 \text{ kJ/kg·K}$$

Then

$$Q_{12} = (0.1\text{ kg})(433\text{ K})(6.654 - 1.9427)\frac{\text{kJ}}{\text{kg·K}} = 203.99 \text{ kJ}$$

$$W_{cycle} = (0.1)(140)(6.654 - 1.9427) = 65.96 \text{ kJ}$$

① $$\eta = \frac{W_{cycle}}{Q_{12}} = \frac{65.96}{203.99} = 0.323 \quad (32.3\%)$$

---

1. Alternatively, using Eq. 5.8

$$\eta_{MAX} = 1 - \frac{T_C}{T_H} = 1 - \frac{293}{433} = 0.323$$

PROBLEM 6.43

KNOWN: Data is provided for 0.1 Kg of air undergoing a Carnot power cycle.

FIND: Determine (a) the pressures at the end of the isothermal expansion, adiabatic expansion, and isothermal compression. the net work, and (c) the thermal efficiency.

SCHEMATIC & GIVEN DATA:

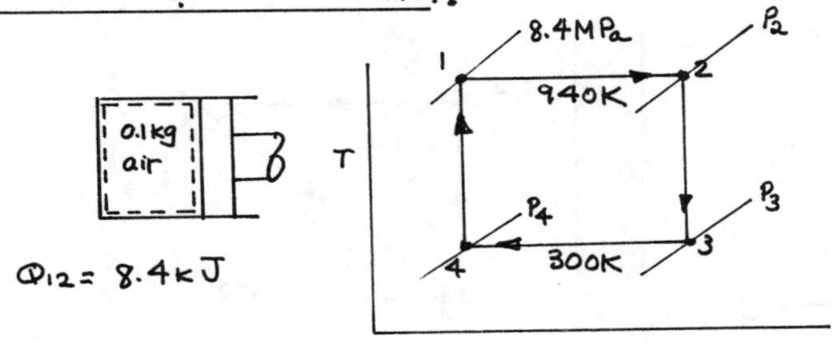

ASSUMPTIONS: (1) The system is the 0.1 kg of air. (2) The air is modeled as an ideal gas. (3) The system undergoes a Carnot power cycle.

ANALYSIS: (a) As discussed in Sec. 6.4, for process 1-2 $Q_{12} = m\,T_H\,(s_2 - s_1)$. And for an ideal gas at fixed temperature, Eq. 6.25 reduces to read $s_2 - s_1 = -R \ln P_2/P_1$. Combining these $Q_{12} = -mRT_H \ln P_2/P_1$. Solving

$$\ln \frac{P_2}{P_1} = -\frac{Q_{12}}{mRT_H} = -\frac{8.4\text{ kJ}}{(0.1\text{ kg})\left(\frac{8.314}{28.97}\frac{\text{kJ}}{\text{kmol}\cdot\text{K}}\right)(940\text{K})} = -0.3114$$

giving $P_2/P_1 = 0.7324$, $P_2 = 6.15$ MPa. ← P_2

For process 2-3 and data from Table A-22

$$0 = s^\circ(T_3) - s^\circ(T_2) - R \ln \frac{P_3}{P_2} \Rightarrow \ln \frac{P_3}{P_2} = \frac{s^\circ(T_3) - s^\circ(T_2)}{R} = \frac{1.70203 - 2.89748}{(8.314/28.97)}$$

giving $\ln P_3/P_2 = -4.1655$, $P_3/P_2 = 0.0155$, $P_3 = 0.095$ MPa. ← P_3

Similarly, for process 4-1

$$\ln \frac{P_1}{P_4} = \frac{s^\circ(T_1) - s^\circ(T_4)}{R} = +4.1655,\quad \frac{P_1}{P_4} = 64.425,\quad P_4 = 0.13\text{ MPa}$$ ← P_4

(b) For any cycle, $W_{net} = \eta\, Q_{in}$. Here, η corresponds to η_{MAX} and $Q_{in} = Q_{12}$.

$$\therefore W_{net} = \underbrace{\left[1 - \frac{300}{940}\right]}_{0.681\ (68.1\%)} 8.4\text{ kJ} = 5.7\text{ kJ}$$ ← W_{net}, ← η

Alternatively $W_{net} = Q_{net} = Q_{12} + Q_{34}$. From an energy balance, since T is constant, $Q_{34} = W_{34}$, where

$$W_{34} = m\int_3^4 p\,dv = m\int_3^4 \frac{RT}{v}\,dv = mRT \ln \frac{v_4}{v_3} = mRT \ln \frac{P_3}{P_4}$$

(where $v_4/v_3 = P_3/P_4$)

$$= (0.1)\left(\frac{8.314}{28.97}\right)(300) \ln \frac{0.095}{0.13} = -2.7\text{ kJ}$$

So $Q_{net} = 8.4 - 2.7 = 5.7$ kJ, and $W_{net} = 5.7$ kJ

PROBLEM 6.44

KNOWN: Air undergoes three internally reversible processes in series between specified states.

FIND: Sketch the processes on p-v and T-s coordinates. Determine the temperature at state 3. Also find the net work per unit of mass and the thermal efficiency.

SCHEMATIC & GIVEN DATA:

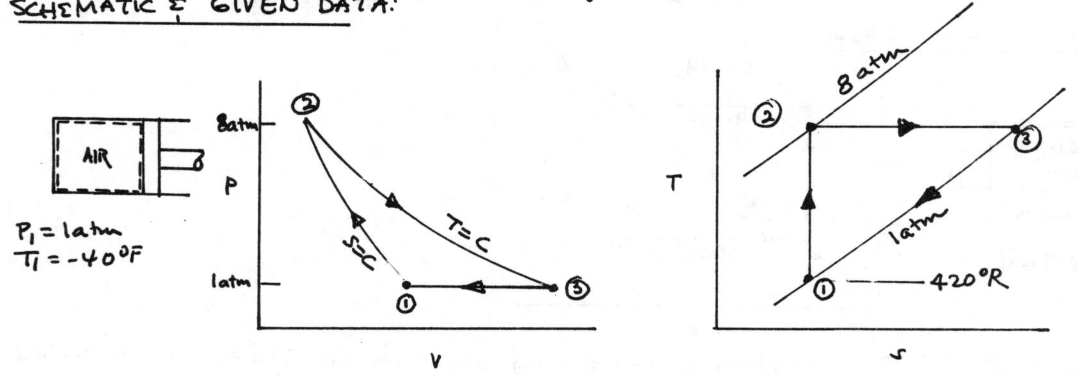

ASSUMPTIONS: (1) As shown in the accompanying figure, the system is the air. (2) The air is modeled as an ideal gas. (3) Each process is internally reversible. (4) There is no change in kinetic or potential energy for either of the two processes.

ANALYSIS: As process 1-2 is adiabatic and internally reversible, it is an isentropic process (Sec. 6.4): $s_2 = s_1$. Thus, Eq. 6.21b reduces to give, with data from Table A-22E,

$$s°(T_2) = s°(T_1) + R \ln \frac{P_2}{P_1} = 0.54058 + \frac{1.986}{28.97} \ln 8 = 0.6831 \text{ Btu/lb·°R}$$

Thus, Table A-22E gives $T_2 = 760°R$, $u_2 = 129.99$ Btu/lb. Also, $u_1 = 71.52$ Btu/lb. Therefore, W_{12} can be evaluated using an energy balance:

$$\frac{W_{12}}{m} = \cancel{\frac{Q_{12}}{m}}^{0} - (u_2 - u_1) = -(129.99 - 71.52) \text{ Btu/lb} = -58.5 \text{ Btu/lb}$$

Since process 2-3 is isothermal, $T_3 = T_2 = 760°R$. ◄ T_3

Also,

$$\frac{W_{23}}{m} = \int_2^3 p\,dv = \int_2^3 \frac{RT}{v} dv = RT \ln \frac{v_3}{v_2} = RT \ln \frac{P_2}{P_3}$$

$$\left[\begin{matrix} P_2 v_2 = RT_2 \\ P_3 v_3 = RT_3 \end{matrix} \Rightarrow \frac{v_3}{v_2} = \frac{P_2}{P_3}\right]$$

$$= \left(\frac{1.986}{28.97}\right)(760) \ln (8)$$

$$= +108.3 \text{ Btu/lb}$$

PROBLEM 6.44 (Contd.)

Since process 3-1 is at constant pressure

$$\frac{W_{31}}{m} = \int_3^1 p\,dv = p(v_1 - v_3) = R(T_1 - T_3)$$

$$= \left(\frac{1.986}{28.97}\ \frac{\text{Btu}}{\text{lb}\cdot°\text{R}}\right)(420 - 760)°\text{R} = -23.3\ \text{Btu/lb}$$

The net work is then,

① $$\frac{W_{cycle}}{m} = \frac{W_{12}}{m} + \frac{W_{23}}{m} + \frac{W_{31}}{m} = -58.5 + 108.3 - 23.3$$
$$= 26.5\ \text{Btu/lb}$$ ← W_{net}/m

By inspection of the T-s diagram, using Eq. 6-25, it can be concluded that $Q_{23} > 0$ (entropy increases) and $Q_{31} < 0$ (entropy decreases). Thus, the heat addition for the cycle occurs during process 2-3:

$$\frac{Q_{23}}{m} = \int_2^3 T\,ds = T(s_3 - s_2)$$
$$= T\Big(\underbrace{s^\circ(T_3) - s^\circ(T_2)}_{=0\ (T_2 = T_3)} - R\ln p_3/p_2\Big)$$
$$= -760°\text{R}\left(\frac{1.986}{28.97}\ \frac{\text{Btu}}{\text{lb}\cdot°\text{R}}\right)\ln\frac{1}{8}$$
$$= 108.3\ \text{Btu/lb}$$

Finally, the thermal efficiency is

$$\eta = \frac{W_{cycle}/m}{Q_{23}/m} = \frac{26.5}{108.3} = 0.245\ (24.5\%)$$ ← η

---

1. Alternatively, $Q_{cycle} = W_{cycle}$, where

$$Q_{cycle} = \overset{0}{\cancel{Q_{12}}} + Q_{23} + Q_{31}$$

An energy balance reduces to give $Q_{31} = u_1 - u_3 + p(v_1 - v_3)$, or

$$\frac{Q_{31}}{m} = h_1 - h_3 = 100.32 - 182.08 = -81.76\ \text{Btu/lb}.$$

Then, with Q_{23}/m determined below we get

$$\frac{Q_{cycle}}{m} = 108.3 - 81.76 = 26.5\ \text{Btu/lb}.$$

which checks the value given here for W_{cycle}/m.

PROBLEM 6.45

KNOWN: Air as an ideal gas undergoes a cycle consisting of three internally reversible processes.

FIND: Sketch the cycle on p-v and T-s coordinates. Determine temperature T_3 and the thermal efficiency or coefficient of performance, as appropriate.

SCHEMATIC & GIVEN DATA:

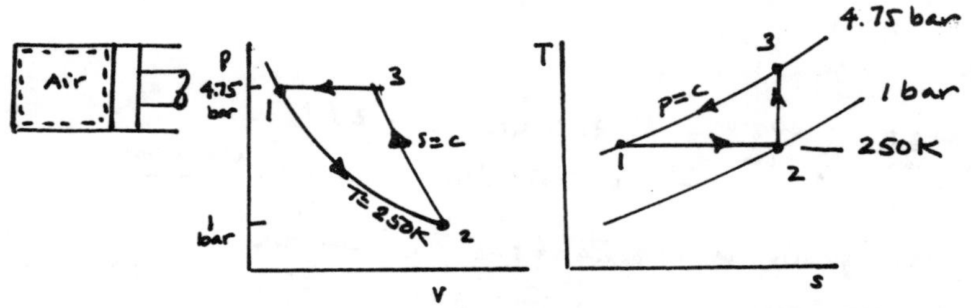

ASSUMPTIONS: 1. As shown in the schematic, the system is the air. 2. Air is modeled as an ideal gas. 3. All processes are internally reversible. 4. Kinetic & potential energy changes can be ignored.

ANALYSIS: (a) To find T_3, consider process 2-3, which has $s_2 = s_2$. Then, with Eq. 6.21a and data from Table A-22

$$s^\circ(T_3) = s^\circ(T_2) + R \ln \frac{P_3}{P_2} = 1.51917 + \frac{8.314}{28.97} \ln \frac{4.75}{1} = 1.9663 \frac{kJ}{kg \cdot K} \rightarrow T_3 = 390\,K \longleftarrow$$

(b) Study of the areas under the processes on the p-v diagram indicates that there is a net work input requirement. Accordingly, the cycle can be a refrigeration cycle but not a power cycle. Study of the areas under the processes of the T-s diagram suggest that process 1-2 involves a heat addition (entropy increases) and process 3-1 involves a heat rejection (entropy decreases). Thus,

① $\beta = \dfrac{Q_{12}}{|W_{cycle}|}$, where $W_{cycle} = W_{12} + W_{23} + W_{31}$. Alternatively,

$W_{cycle} = Q_{cycle} = Q_{12} + \cancel{Q_{23}} + Q_{31}$. Then, $\beta = \dfrac{Q_{12}}{|Q_{12} + Q_{31}|}$.

Using Eq. 6.25 with Eq. 6.21a

$$\frac{Q_{12}}{m} = \int_1^2 T\,ds = T(s_2 - s_1) = T[\underbrace{s^\circ(T_2) - s^\circ(T_1)}_{=0 \text{ since } T \text{ is const.}} - R \ln \frac{P_2}{P_1}] = +RT \ln \frac{P_1}{P_2}$$

$$= \left(\frac{8.314}{28.97}\right)(250) \ln \frac{4.75}{1.00}$$

$$= 111.79\ kJ/kg$$

An energy balance for process 3-1 reads: $(u_1 - u_3) = \frac{Q_{31}}{m} - \frac{W_{31}}{m}$, where

$\frac{W_{31}}{m} = \int_3^1 p\,dv = p(v_1 - v_3)$. Thus, $\frac{Q_{31}}{m} = (u_1 - u_3) + p(v_1 - v_3)$, which can be rewritten concisely as $\frac{Q_{31}}{m} = h_1 - h_3$. With data from Table A-22

$\frac{Q_{31}}{m} = 250.05 - 390.88 = -140.83 \frac{kJ}{kg}$. Finally, $Q_{cycle} = (111.79) + (-140.83) = -29.04 \frac{kJ}{kg}$

$$\Rightarrow \beta = \frac{111.79}{29.04} = 3.85$$

---

1. From the discussion of Sec. 2.6.3 recall that β requires the net work into the system to achieve the refrigeration effect. Thus, $|W_{cycle}|$ is used here.

PROBLEM 6.46

KNOWN: Air as an ideal gas undergoes a cycle consisting of three internally reversible processes.

FIND: Sketch the cycle on p-v, T-s coordinates, determine T_2, and evaluate the thermal efficiency or coefficient of performance, as appropriate.

SCHEMATIC & GIVEN DATA:

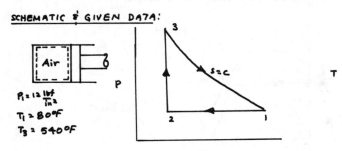

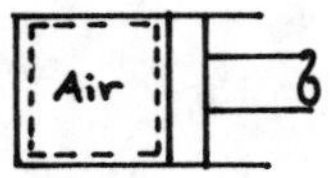

$p_1 = 12\ \frac{lbf}{in^2}$

$T_1 = 80°F$

$T_3 = 540°F$

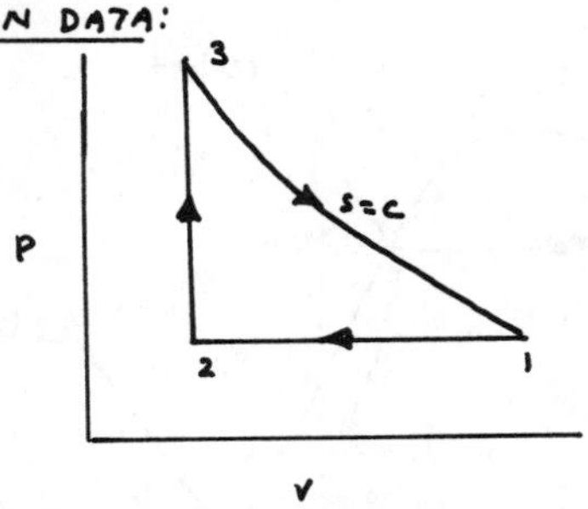

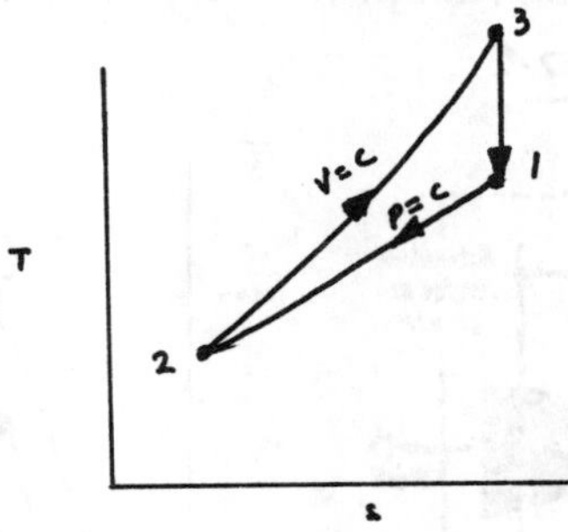

ASSUMPTIONS: (1) As shown in the accompanying figure, the system is the air. (2) The air is modeled as an ideal gas. (3) All processes are internally reversible. (4) There is no change in kinetic or potential energy between states 1, 2, 3.

ANALYSIS: As the processes occur in a clockwise sense, the cycle would be a power cycle. The heat added can be evaluated using an energy balance with assumption 4

$$\frac{Q_{23}}{m} = u_3 - u_2 + \cancelto{0}{\frac{W_{23}}{m}} = u_3 - u_2 \qquad (1)$$

Similarly the heat rejected is

$$\frac{Q_{12}}{m} = u_2 - u_1 + \frac{W_{12}}{m} = u_2 - u_1 + \int_1^2 p\,dv = u_2 - u_1 + p_1(v_2 - v_1) = h_2 - h_1 \qquad (2)$$

To evaluate these requires T_2. Since process 2-3 is at constant volume, the ideal gas equation of state gives $p_3/p_2 = T_3/T_2$. Also, with $s_3 = s_1$, Eq. 6.2a gives with $p_2 = p_1$

$$0 = s°(T_3) - s°(T_1) - R \ln p_3/p_1 \Rightarrow 0 = s°(T_3) - s°(T_1) - R \ln p_3/p_2$$

Combining these results, and using data from Table A-22E

$$\ln \frac{T_3}{T_2} = \frac{s°(T_3) - s°(T_1)}{R} = \frac{0.63395 - 0.60078}{(1.986/28.97)} \Rightarrow T_2 = 382.2\ °R \longleftarrow$$

Using Eqs. (1), (2) with data from Table A-22E

$$\frac{Q_{23}}{m} = u_3 - u_2 = 105.78 - 65.08 = 40.7\ Btu/lb\ , \quad \frac{Q_{12}}{m} = 91.28 - 129.06 = -37.78\ Btu/lb$$

The thermal efficiency is

$$① \quad \eta = \frac{W_{cycle}}{Q_{23}} = \frac{Q_{12} + Q_{23} + \cancelto{0}{Q_{31}}}{Q_{23}} = \frac{\overbrace{(-37.78) + (40.7)}^{2.92}}{40.7} = 0.072 \quad (7.2\%)$$

---

1. For the cycle, $W_{cycle} = Q_{cycle}$, where $W_{cycle} = W_{12} + \cancelto{0}{W_{23}} + W_{31}$. An energy balance gives $W_{31}/m = u_3 - u_1$. Also, $\frac{W_{12}}{m} = \int_1^2 p\,dv = p(v_2 - v_1)$, which can be expressed with the ideal gas equation of state as $W_{12}/m = R(T_2 - T_1)$. Then,

$$\frac{W_{31}}{m} = 105.78 - 92.04 = 13.74, \quad \frac{W_{12}}{m} = \frac{1.986}{28.97}(382.2 - 540) = -10.82\ Btu/lb$$

So, $(W_{cycle}/m) = 13.74 - 10.82 = 2.92\ Btu/lb$, which agrees with the value of Q_{cycle}/m, as expected.

PROBLEM 6.47

KNOWN: The schematic and steady state operating data are provided for a vapor power plant.

FIND: (a) sketch the cycle on T-s coordinates, (b) determine the thermal efficiency and compare with the thermal efficiency of a Carnot cycle operating between the same maximum and minimum temperatures.

SCHEMATIC & GIVEN DATA:

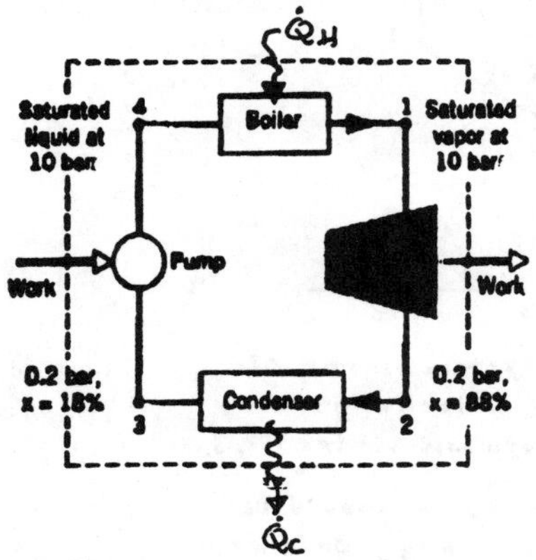

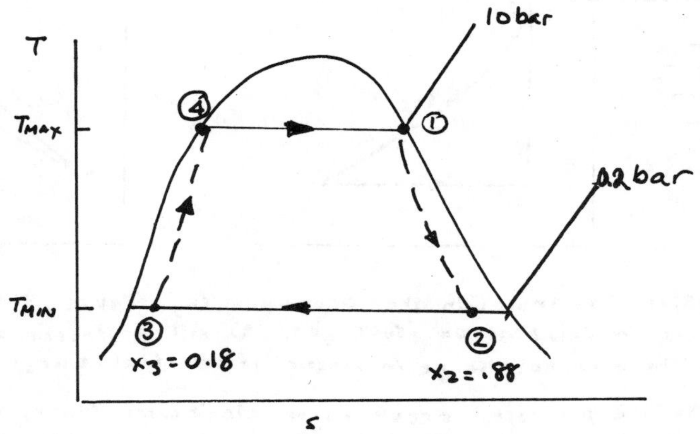

ASSUMPTIONS: (1) The overall system shown in the figure and thus each component is at steady state. (2) The only significant heat transfers occur with the two reservoirs. (3) Kinetic and potential energy effects can be ignored. (3) The maximum and minimum temperatures correspond, respectively, to the saturation temperatures at 10 bars and 1bar.

ANALYSIS: The thermal efficiency is

$$\eta = \frac{\dot{W}_{net}}{\dot{Q}_H} = 1 - \frac{\dot{Q}_c}{\dot{Q}_H}$$

The heat transfer rates are found by reducing mass and energy balances for the boiler and condenser: $\dot{Q}_H = \dot{m}(h_1 - h_4)$, $\dot{Q}_c = \dot{m}(h_2 - h_3)$. Thus

$$\eta = 1 - \left[\frac{h_2 - h_3}{h_1 - h_4}\right].$$

With data from Table A-3, $h_1 - h_4 = 2015.3$ kJ/kg. Also, $h_2 = h_f + x_2(h_g - h_f)$ and $h_3 = h_f + x_3(h_g - h_f)$. So, $h_2 - h_3 = (x_2 - x_3)(h_g - h_f)$. That is,

$$h_2 - h_3 = (0.88 - 0.18)(2358.3) = 1650.8 \text{ kJ/kg}$$

Accordingly,

$$\eta = 1 - \left[\frac{1650.8}{2015.3}\right] = 0.181 \quad (18.1\%)$$ ← η

For a Carnot cycle

$$\eta_{MAX} = 1 - \frac{T_{MIN}}{T_{MAX}} = 1 - \frac{(60.06 + 273.15)}{(179.9 + 273.15)} = 0.265 \quad (26.5\%)$$ ← η_{MAX}

PROBLEM 6.48

KNOWN: A closed system undergoes a process in which Q occurs at T_b.

FIND: For each of several cases, determine whether the entropy change is positive, negative, zero, or indeterminate.

SCHEMATIC & GIVEN DATA:

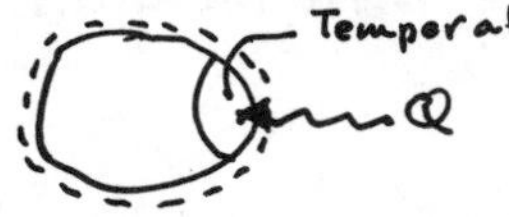

ASSUMPTION: The system shown in the figure interacts thermally with its surroundings only at a place on its boundary where temperature is T_b.

ANALYSIS: In this case, Eq. 6.27 takes the form $\Delta S = \frac{Q}{T_b} + \sigma$.

(a) $\sigma = 0$, $Q > 0$ $\Rightarrow$ $\Delta S = \frac{Q}{T_b} + \cancel{\sigma}^{0} > 0$

(b) $\sigma = 0$, $Q = 0$ $\Rightarrow$ $\Delta S = \cancel{\frac{Q}{T_b}}^{0} + \cancel{\sigma}^{0} = 0$

(c) $\sigma = 0$, $Q < 0$ $\Rightarrow$ $\Delta S = \frac{Q}{T_b} + \cancel{\sigma}^{0} < 0$

(d) $\sigma > 0$, $Q > 0$ $\Rightarrow$ $\Delta S = \frac{Q}{T_b} + \sigma > 0$

(e) $\sigma > 0$, $Q = 0$ $\Rightarrow$ $\Delta S = \cancel{\frac{Q}{T_b}}^{0} + \sigma > 0$

(f) $\sigma > 0$, $Q < 0$ $\Rightarrow$ $\Delta S = \underbrace{Q/T_b}_{\ominus} + \underbrace{\sigma}_{\oplus}$ Indeterminate: ΔS may be positive or negative depending on the relative magnitudes of the two terms.

PROBLEM 6.49

KNOWN: Closed systems undergo specified processes.

FIND: For each case, indicate whether the entropy change is +, −, 0, or indeterminate.

ANALYSIS: (a) Water undergoing an adiabatic process.

$$\Delta S = \cancel{\int \left(\frac{\delta Q}{T}\right)_b}^{0} + \sigma \Rightarrow \Delta S = \sigma.$$ Since $\sigma \geq 0$, ΔS can be + or 0 depending on the nature of the process. Indeterminate.

(b) Nitrogen heated internally reversibly.

$$\Delta S = \int \left(\frac{\delta Q}{T}\right)_b + \cancel{\sigma}^{0}.$$ As entropy is transferred in the direction of heat transfer, the entropy would increase.

(c) R134a stirred adiabatically.

$$\Delta S = \cancel{\int \left(\frac{\delta Q}{T}\right)_b}^{0} + \sigma.$$ Since stirring takes place, there would be entropy production. Thus, entropy would increase.

(d) CO_2 cooled isothermally.

$$\Delta S = \int \left(\frac{\delta Q}{T}\right)_b + \sigma \Rightarrow \Delta S = \underbrace{\frac{Q}{T_b}}_{\ominus} + \underbrace{\sigma}_{+\text{ or }0}.$$ Indeterminate.

(e) Ideal gas undergoing a constant-pressure process, $T_2 > T_1$.
Using Eq. 6.21(a),

$$\Delta S = s^{\circ}(T_2) - s^{\circ}(T_1) - R\cancel{\ln p_2/p_1}^{0}.$$ If $T_2 > T_1$, $s^{\circ}(T_2) > s^{\circ}(T_1)$. So, $\Delta S > 0$. Increase.

(f) Ideal gas undergoing a constant-temperature process, $p_2 < p_1$.
Using Eq. 6.21(a),

$$\Delta S = \underbrace{s^{\circ}(T_2) - s^{\circ}(T_1)}_{=0} - R\ln p_2/p_1 = -R\ln p_2/p_1 > 0$$ when $p_2 < p_1$. Increase

PROBLEM 6.50

KNOWN: One lb of O_2 expands isothermally between specified states while receiving energy through a thin intervening wall from a reservoir.

FIND: (a) For the O_2 as the system, evaluate W, Q, σ. (b) For the O_2 plus the thin wall as the system, evaluate σ and compare with the result of part (a).

SCHEMATIC & GIVEN DATA:

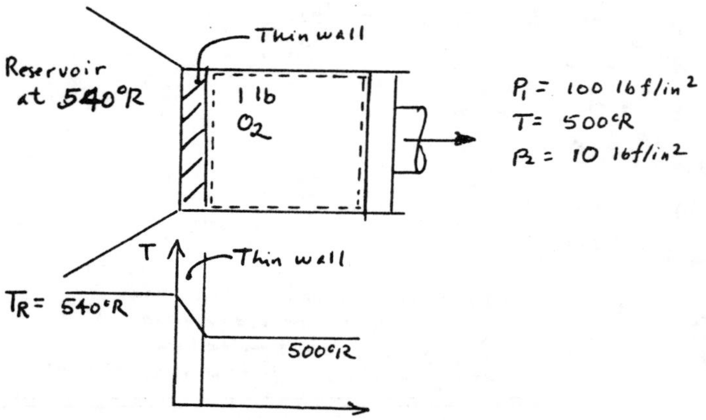

ASSUMPTIONS: (1) In part (a) the system is the O_2 only. In part (b) the system is the O_2 plus the thin wall. (2) O_2 is modeled as an ideal gas. (3) There is no change in kinetic or potential energy. (4) The state of the thin wall does not change.

ANALYSIS: With assumptions (2), (3) an energy balance reduces to $\Delta U \overset{0}{=} Q - W$, where $\Delta U = 0$ because $u(T)$ for an ideal gas. Thus $Q = W$. To find W, use the ideal gas equation of state to write

$$W = m\int_1^2 p\,dv = m\int_1^2 \frac{RT}{v}\,dv = mRT\ln\frac{V_2}{V_1} = mRT\ln\frac{P_1}{P_2}$$

$$= -(1\,lb)\left(\frac{1.986}{32}\frac{Btu}{lb\cdot°R}\right)(500°R)\ln\left(\frac{100}{10}\right) = 71.45\ Btu$$ ← W, Q

An entropy balance reads

$$\Delta S = \frac{Q}{T} + \sigma \Rightarrow \sigma = \Delta S - \frac{Q}{T}$$

From Eq. 6.21b, $\Delta S = -mR\ln P_2/P_1$. Also, from above $Q = mRT\ln P_1/P_2$. Thus

$$\sigma = -mR\ln\frac{P_2}{P_1} - \frac{mRT\ln\frac{P_1}{P_2}}{T} \equiv 0$$ (the process is internally reversible) ← σ

(b) For the enlarged system of O_2 plus thin wall an entropy balance reads

$$\Delta S = \frac{Q}{T_R} + \sigma \Rightarrow \sigma = \Delta S - \frac{Q}{T_R}$$

where by assumption (4) ΔS is the same as in part (a): $-mR\ln\frac{P_2}{P_1}$. Introducing $Q = mRT\ln\frac{P_1}{P_2}$ and simplifying

$$\sigma = -mR\ln\frac{P_2}{P_1} - \frac{mRT\ln\frac{P_1}{P_2}}{T_R} = mR\ln\frac{P_1}{P_2}\left[1 - \frac{T}{T_R}\right]$$

$$= (1\,lb)\left(\frac{1.986}{32}\frac{Btu}{lb\cdot°R}\right)\ln\frac{100}{10}\left[1 - \frac{500}{540}\right] = 0.011\ Btu/°R$$ ← σ

Discussion: The enlarged system includes an irreversibility not present in the O_2 alone: heat transfer through a finite temperature difference, and so a nonzero value of σ is determined for the enlarged system.

PROBLEM 6.51

KNOWN: A quantity of water expands adiabatically from a specified state to a specified pressure.

FIND: Plot the work done versus the amount of entropy produced.

SCHEMATIC & GIVEN DATA:

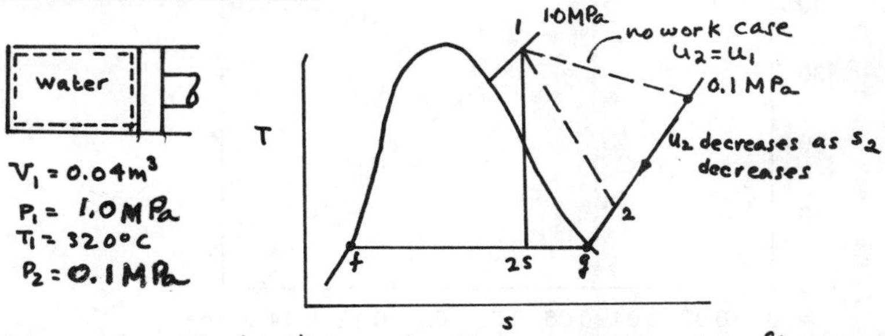

ASSUMPTIONS: (1) As shown in the accompanying figure, the system is the water. (2) The expansion occurs adiabatically. (3) There is no change in kinetic or potential energy between the end states.

ANALYSIS: With assumptions 2 and 3 an energy balance reduces to

$$W = \cancel{Q}^{0} - \Delta U = m(u_1 - u_2)$$

With data from Table A-4, $u_1 = 2826.1$ kJ/kg, $v_1 = 0.2678\ m^3/kg$; $s_1 = 7.1962$ kJ/kg·K. So,

$$W = \left(\frac{0.04\ m^3}{0.2678\ m^3/kg}\right)(2826.1 - u_2)\frac{kJ}{kg} = 0.1494(2826.1 - u_2)\ kJ \qquad (1)$$

An entropy balance reduces to read, $m(s_2 - s_1) = \sigma$. Thus

$$\sigma = (0.1494)(s_2 - 7.1962)\ \frac{kJ}{K} \qquad (2)$$

The T-s diagram shows that u_2 decreases as s_2 decreases. Thus W is a maximum and $\sigma = 0$, when $s_2 = s_1$.

Case $s_2 = s_1$: The specific internal energy at this state is the smallest allowed value: u_{2s}. To determine u_{2s}, first evaluate the quality with data from Tables A-3 and A-4

$$x_{2s} = \frac{s_{2s} - s_f}{s_g - s_f} = \frac{s_1 - s_f}{s_g - s_f} = \frac{7.1962 - 1.3026}{7.3594 - 1.3026} = 0.9731$$

$$\Rightarrow \quad u_{2s} = u_f + x_{2s}(u_g - u_f) = 417.36 + 0.9731(2506.1 - 417.36) = 2449.9\ kJ/kg$$

So

$$W_{MAX} = 0.1494(2826.1 - 2449.9) = 56.20\ kJ \qquad \longleftarrow W_{MAX}$$

Case $u_2 = u_1$: In the limit as $u_2 \to u_1$, $W \to 0$, and the entropy production approaches a maximum. Interpolating in Table A-4 at 0.1 MPa with $u_2 = u_1$ gives $s_2 = 8.25$ kJ/kg·K. Then

$$\sigma_{MAX} = (0.1494)(8.25 - 7.1962) = 0.1574\ \frac{kJ}{K} \qquad \longleftarrow \sigma_{max}$$

PROBLEM 6.51 (Contd.)

PLOT: W vs. σ

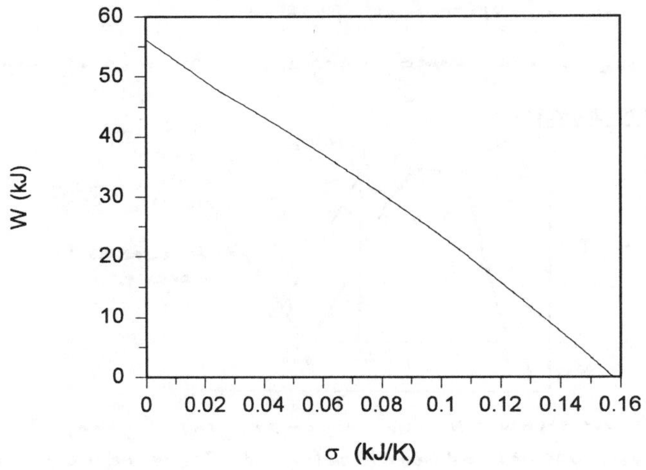

PROBLEM 6.52

KNOWN: Two lb of steam expand adiabatically from a specified state to a specified pressure.

FIND: Plot the work done versus the amount of entropy produced.

SCHEMATIC & GIVEN DATA:

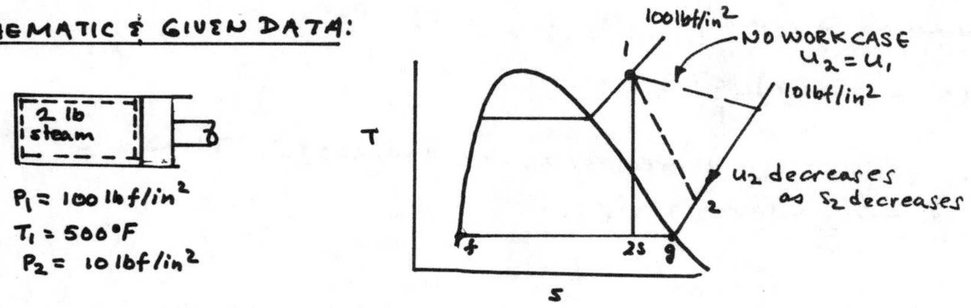

ASSUMPTIONS: (1) As shown in the accompanying figure, the system is the water. (2) The expansion occurs adiabatically. (3) There is no change in kinetic or potential energy between the end states.

ANALYSIS: (a) The work can be found from an energy balance using assumptions 2 and 3

$$W = \cancel{Q}^{0} - \Delta U = m(u_1 - u_2)$$

From Table A-4E at 100 lbf/in², 500°F, $u_1 = 1175.7$ Btu/lb, $s_1 = 1.7085$ Btu/lb·°R. Thus

$$W = (2\,\text{lb})(1175.7 - u_2)\ \frac{\text{Btu}}{\text{lb}} \qquad (1)$$

An entropy balance reduces to read $m(s_2 - s_1) = \sigma$. Thus

$$\sigma = (2\,\text{lb})(s_2 - 1.7085)\ \frac{\text{Btu}}{\text{lb}\cdot°\text{R}} \qquad (2)$$

The T-s diagram shows that u_2 decreases as s_2 decreases. ~~Thus~~ W is a maximum and $\sigma = 0$ when $s_2 = s_1$.

PROBLEM 6.52 (Contd.)

Case $s_2 = s_1$. To determine u_{2s}, first evaluate the quality x_{2s} with data from Tables A-3E and A-4E

$$x_{2s} = \frac{s_{2s} - s_f}{s_g - s_f} = \frac{s_1 - s_f}{s_g - s_f} = \frac{1.7085 - 0.2836}{1.5041} = 0.9473$$

Then

$$u_{2s} = u_f + x_{2s}(u_g - u_f) = 161.2 + 0.9473(1072.2 - 161.2) = 1024.19 \text{ Btu/lb}$$

Finally

$$W_{max} = 2(1175.7 - 1024.19) = 303.02 \text{ Btu} \longleftarrow W_{max}$$

Case $u_2 = u_1$. In the limit as $u_2 \to u_1$, $W \to 0$, and the entropy production approaches a maximum.

Interpolation in Table A-4E at 10 lbf/in² with $u_2 = u_1$, gives $s_2 = 1.9597$ Btu/lb·°R. Then

$$\sigma_{MAX} = 2(1.9597 - 1.7085) = 0.5024 \text{ Btu/°R} \longleftarrow \sigma_{MAX}$$

PLOT: W vs σ:

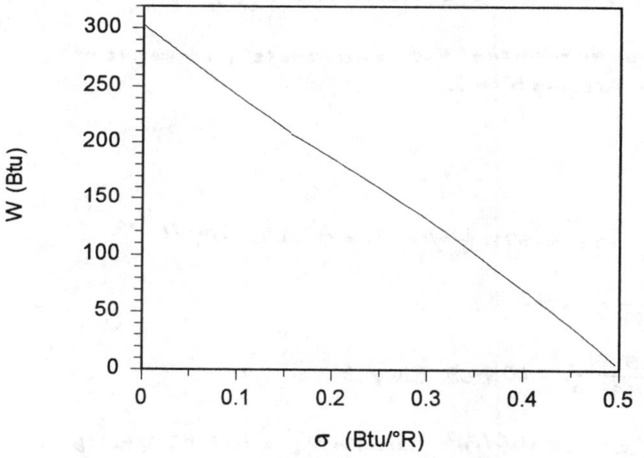

PROBLEM 6.53

KNOWN: R134a expands adiabatically from a specified initial state to a specified pressure. The work developed is measured.

FIND: Determine if the value for work can be correct.

SCHEMATIC & GIVEN DATA:

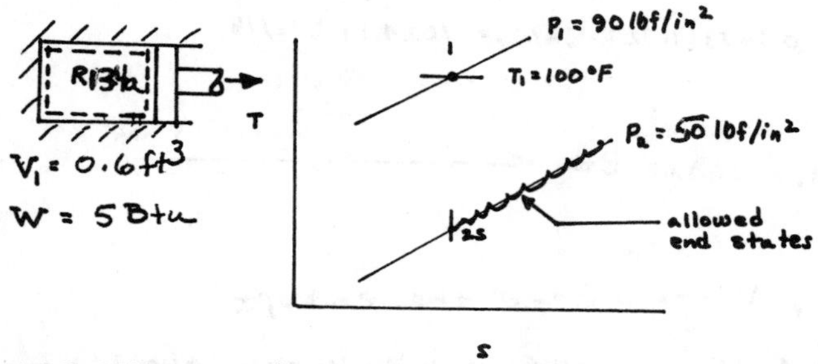

ASSUMPTIONS: (1) As shown in the figure the R134a is the system. (2) The expansion occurs adiabatically and with no change in kinetic or potential energy between the end states.

ANALYSIS: The allowed end states can be determined using an entropy balance which reduces to give

$$\Delta S = \int_1^2 \left(\frac{\delta Q}{T}\right)_b^{\,0} + \sigma$$

or

$$s_2 - s_1 = \sigma/m \geq 0 \Rightarrow \text{only allowed end states have } s_2 \geq s_1.$$

Next, use an energy balance to determine the end state that corresponds to the given value for W. With assumption 2

$$\Delta U = \cancel{Q}^{\,0} - W$$

or

$$u_2 - u_1 = -W/m$$

From Table A-12E, $u_1 = 108.82$ Btu/lb, $v_1 = 0.5751$ ft³/lb, $s_1 = 0.2295$ Btu/lb·°R. Thus

$$m = \frac{V_1}{v_1} = \frac{0.6 \text{ ft}^3}{0.5751 \text{ ft}^3/\text{lb}} = 1.043 \text{ lb}$$

and

$$u_2 = 108.82 - \frac{5 \text{ Btu}}{1.043 \text{ lb}} = 104.03 \text{ Btu/lb}$$

Turning to Table A-12E, interpolation at 50 lbf/in² with $u_2 = 104.03$ Btu/lb gives $s_2 = 0.2306$ Btu/lb·°R. Since $s_2 > s_1$, the measured value for W can be correct.

PROBLEM 6.54

KNOWN: One lb of air undergoes a process between two specified states.
FIND: Determine if the process can occur adiabatically.

SCHEMATIC & GIVEN DATA:

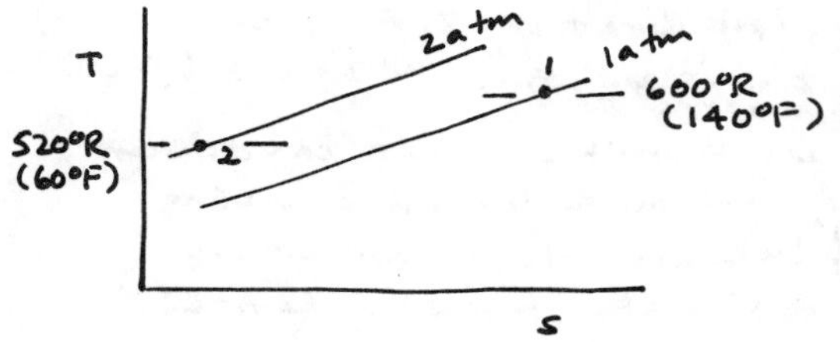

ASSUMPTIONS: 1. The system is the one lb of air. 2. Air is modeled as an ideal gas.

ANALYSIS: Whether the process can occur adiabatically can be determined using an entropy balance together with s^o data from Table A-22E and Eq. 6.21a. The entropy balance reads

$$\Delta S = \underline{\int_1^2 \left(\frac{\delta Q}{T}\right)_b} + \sigma$$

If the process takes place adiabatically, the underlined term vanishes, giving $\Delta S = \sigma$. Since σ cannot be negative, ΔS cannot be negative. Using Eq. 6.21a and s^o from Table A-22E

$$\Delta S = m\left[s^o(T_2) - s^o(T_1) - R \ln \frac{p_2}{p_1}\right] = (1\text{ lb})\left[0.59172 - 0.62607 - \frac{1.986}{28.97}\ln\frac{2}{1}\right]\frac{\text{Btu}}{\text{lb}\cdot°\text{R}}$$

$$= -0.0819 \text{ Btu/°R}$$

Since ΔS is negative, an adiabatic process between these states cannot occur.

PROBLEM 6.55

KNOWN: One kg of R134a undergoes a process between specified states.
FIND: Determine the entropy change for the process. Can the process occur with $Q = 0$?

SCHEMATIC & GIVEN DATA:

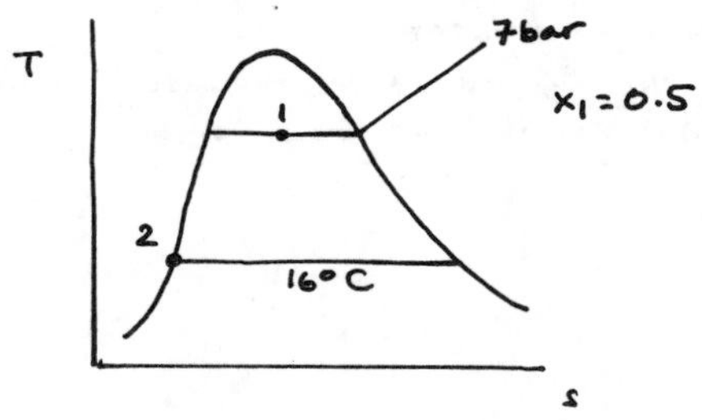

ASSUMPTION: The system is the one kg of R134a.

ANALYSIS: The entropy change can be determined using data from Tables A-10, 11.

$$s_1 = s_f + x_1(s_g - s_f) = 0.3242 + 0.5(0.9080 - 0.3242)$$
$$= 0.6161 \text{ kJ/kg}\cdot\text{K}$$
$$s_2 = 0.2735 \text{ kJ/kg}\cdot\text{K}$$

Thus, $\Delta S = (1\text{ kg})[0.2735 - 0.6161]\text{ kJ/kg}\cdot\text{K} = -0.3426 \text{ kJ/K}$

To consider the possibility of an adiabatic process, begin with an entropy balance:

$$\Delta S = \underline{\int_1^2 \left(\frac{\delta Q}{T}\right)_b} + \sigma$$

For an adiabatic process, the underlined term vanishes, giving $\Delta S = \sigma$. Since σ cannot be negative, ΔS also cannot be negative. Accordingly, the indicated process cannot occur adiabatically.

PROBLEM 6.56

KNOWN: Air is compressed between specified states.

FIND: Determine if the process can occur adiabatically. If so, evaluate the work. If not, determine the direction of the heat transfer.

SCHEMATIC & GIVEN DATA:

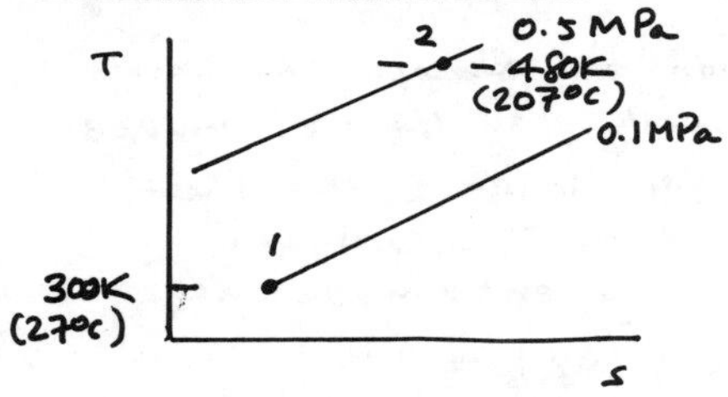

ASSUMPTIONS: 1. The system is the quantity of air under consideration. 2. Air is modeled as an ideal gas. 3. Ignore kinetic and potential energy.

ANALYSIS: Whether the process can occur adiabatically can be determined using an entropy balance together with Eq. 6.21a and s^o data from Table A-22:

The entropy balance reads, $\Delta S = \int \left(\frac{\delta Q}{T}\right)_b + \sigma$. In an adiabatic process, the underlined term vanishes, giving $\Delta S = \sigma$. Since σ cannot be negative, ΔS cannot be negative. Then, with Eq. 6.21a and s^o data,

$$\Delta s = s^o(T_2) - s^o(T_1) - R \ln \frac{p_2}{p_1} = 2.1776 - 1.70203 - \frac{8.314}{28.97} \ln 5 = +0.0137 \text{ kJ/kg·K}$$

Accordingly, an adiabatic process between these states is allowed. For any such process, the energy balance reads, $\Delta U = \cancel{Q} - W$, or

$W = -m(u_2 - u_1)$, $W/m = u_1 - u_2 = 214.07 - 344.7 = -130.6$ kJ/kg. ⟵

PROBLEM 6.57

KNOWN: Air is compressed in a process between two specified states.

FIND: Determine if the process can occur adiabatically. If adiabatic, evaluate the work. If not adiabatic, determine the direction of heat transfer.

SCHEMATIC & GIVEN DATA:

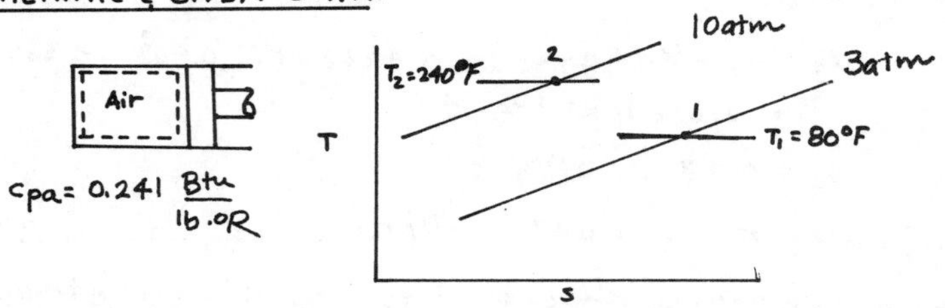

$c_{pa} = 0.241 \frac{\text{Btu}}{\text{lb·°R}}$

ASSUMPTIONS: (1) As shown in the accompanying figure, the system is the air. (2) Air is modeled as an ideal gas with $c_p = 0.241$ Btu/lb·°R.

ANALYSIS: Whether the process can occur adiabatically can be determined using an entropy balance together with Eq. 6.23 which gives

$$\Delta s = c_p \ln \frac{T_2}{T_1} - R \ln \frac{p_2}{p_1} = 0.241 \ln \frac{700}{540} - \frac{1.986}{28.97} \ln \frac{10}{3} = -0.02 \text{ Btu/lb·°R}$$

An entropy balance then reads

$$m[-0.02] = \int_1^2 \left(\frac{\delta Q}{T}\right)_b + \sigma$$

Since $\sigma \geq 0$ for all processes, the entropy transfer term must be negative which implies that energy is <u>removed</u> by heat transfer during the process. ⟵

PROBLEM 6.58

KNOWN: R134a is compressed adiabatically from a specified initial state to a state where the volume is specified.

FIND: Determine if the pressure at the final state can be (a) 200 lbf/in², (b) 300 lbf/in²

SCHEMATIC & GIVEN DATA:

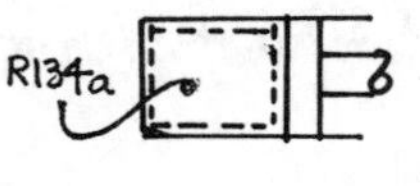

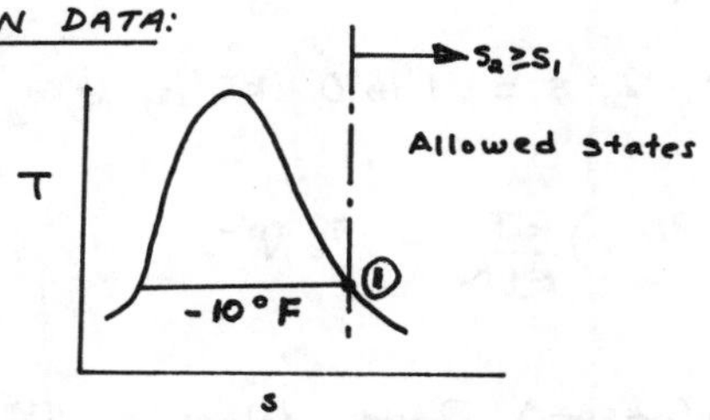

$m = 1.0$ lb

$V_2 = 0.8$ ft³

ASSUMPTIONS: (1) As shown in the accompanying figure, the system is the R134a (2) The compression occurs adiabatically.

ANALYSIS: The allowed end states can be determined using an entropy balance which reduces to give

$$\Delta S = \int_1^2 \left(\frac{\delta Q}{T}\right)_b^{\,0} + \sigma$$

or

$$s_2 - s_1 = \sigma/m \geq 0 \Rightarrow \text{ only allowed states have } s_2 \geq s_1.$$

From Table A-10E at -10°F, $s_1 = 0.2236$ Btu/lb·°R. At state 2,

$$v_2 = \frac{V_2}{m} = \frac{0.8 \text{ ft}^3}{1 \text{ lb}} = 0.8 \text{ ft}^3/\text{lb}.$$

(a) $p_2 = 60$ lbf/in²:

From Table A-12E at $p_2 = 60$ lbf/in², $v_2 = 0.8$ ft³/lb, we see that $s_2 < s_1$, and so this state is not allowed. No. ◄

(b) $p_2 = 70$ lbf/in²:

From Table A-12E at $p_2 = 70$ lbf/in², $v_2 = 0.8$ ft³/lb, we see that $s_2 > s_1$, and so this state is allowed. Yes. ◄

PROBLEM 6.59

KNOWN: Two kg of R12 is compressed between two specified states, during which the temperature of the R134a is 60°C ± 0.1°C.

FIND: Determine the minimum theoretical heat transfer from the refrigerant.

SCHEMATIC & GIVEN DATA:

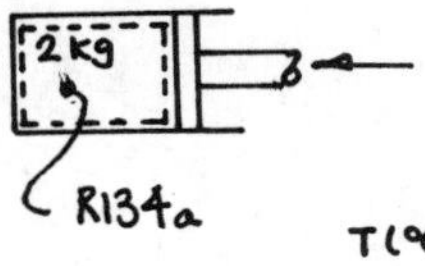

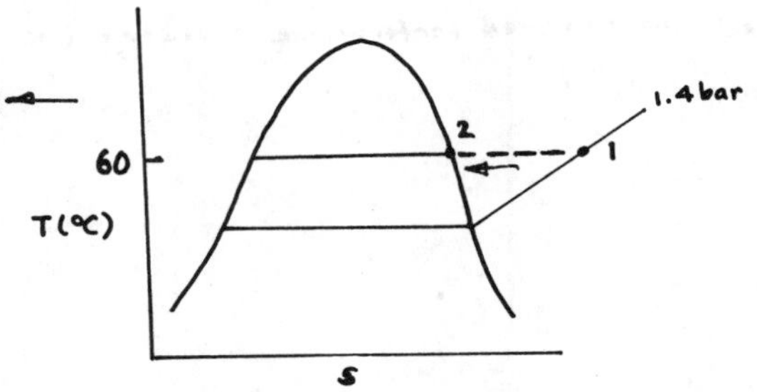

ASSUMPTIONS: (1) The R134a is the closed system. (2) The temperature at which heat transfer occurs is $T_b = 60$°C.

ANALYSIS: An entropy balance for the process reads

$$\Delta S = \int_1^2 \left(\frac{\delta Q}{T}\right)_b + \sigma = \frac{Q}{T_b} + \sigma$$

PROBLEM 6.59 (Continued)

Thus

$$Q = T_b \Delta S - T_b \sigma$$

With data from Tables A-10 and A-12, $s_1 = 1.1690$ kJ/kg·K, $s_2 = 0.8973$ kJ/kg·K.

Then

$$Q = (333\,K)(2\,kg)(0.8973 - 1.1690)\frac{kJ}{kg\,K} - T_b \sigma$$

$$= -180.95\,kJ - \underline{T_b \sigma}$$

As $\sigma \geq 0$, the underlined term is positive (or zero). Thus, $Q_{min} = -180.95$ kJ ←

PROBLEM 6.60

KNOWN: A device operating at steady state receives 1 kW by heat transfer at 167°C and generates electricity. There are no other energy transfers.

FIND: Determine if the claimed performance of the device violates any principles of thermodynamics.

SCHEMATIC & GIVEN DATA:

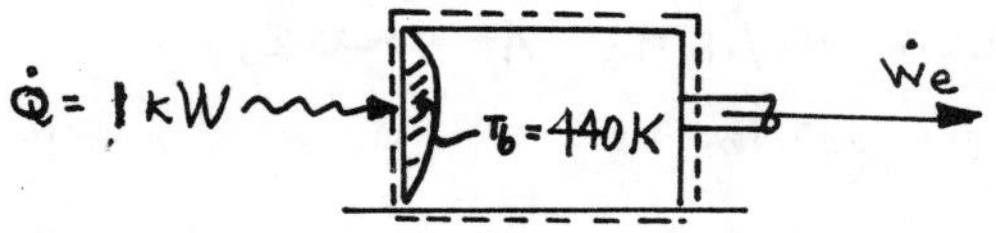

ASSUMPTIONS: (1) The system shown in the accompanying figure operates at steady state. (2) The system receives energy by heat transfer at $T_b = 440K$ (167°C).

ANALYSIS: An energy rate balance for the device reduces with assumption 1 to give

$$\cancel{\frac{dE}{dt}}^{0} = \dot{Q} - \dot{W}_e \Rightarrow \dot{W}_e = \dot{Q} = 1\,kW$$

This result is in accord with the conservation of energy principle. (If $\dot{W}_e$ had any other value, there would be a violation of this principle.)

With assumptions 1 and 2, entropy rate balance reduces to give

$$\cancel{\frac{dS}{dt}}^{0} = \frac{\dot{Q}}{T_b} + \dot{\sigma} \Rightarrow \dot{\sigma} = -\frac{\dot{Q}}{T_b}$$

Thus

$$\dot{\sigma} = -\frac{(1\,kW)}{440\,K} = 2.27 \times 10^{-3}\,\frac{kW}{K}$$

However, as entropy must be produced the claimed performance violates the second law of thermodynamics.

PROBLEM 6.61

KNOWN: Data is provided for a quantity of water rapidly compressed.

FIND: Determine whether the reported value for the work is feasible.

SCHEMATIC & GIVEN DATA:

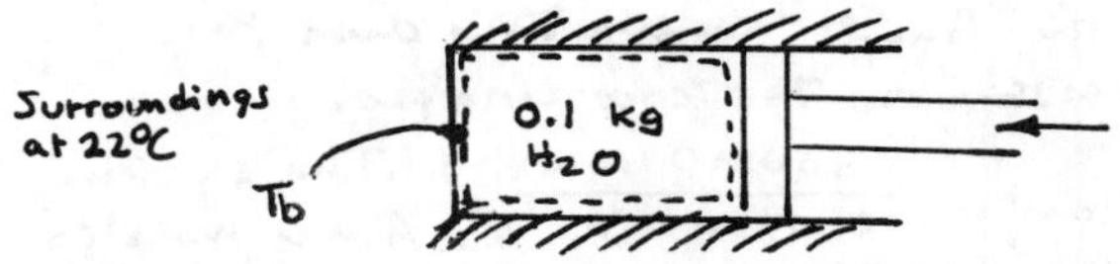

$W = -17.5\,kJ$

$P_1 = 3\,bar$, $T_1 = 200°C$

$P_2 = 15\,bar$, $T_2 = 210°C$

ASSUMPTIONS: 1. The water is the closed system. 2. Heat transfer with the surroundings occurs at T_b, which cannot be less than 295 K (22°C) and is most likely in the interval from 473 K (200°C) to 483 K (210°C). Kinetic and potential energy effects can be ignored.

ANALYSIS: An energy balance reduces to $\Delta U = Q - W$, or with data from Table A-4

$$Q = m(u_2 - u_1) + W = (0.1\,kg)[2617.8 - 2650.7]\frac{kJ}{kg} + (-17.5\,kJ) = -20.79\,kJ$$

An entropy balance takes the form,

$$\Delta S = \frac{Q}{T_b} + \sigma \Rightarrow \sigma = m[s_2 - s_1] - \frac{Q}{T_b}$$

With Eq. 6.21a and data from Table A-4

$$\sigma = (0.1\,kg)\left[(6.5067 - 7.3115)\frac{kJ}{kg\cdot K}\right] - \frac{(-20.79)}{T_b}$$

$$\sigma = -0.0805\frac{kJ}{kg} + \frac{20.79\,kJ}{T_b}$$

For $T_b \geq 295\,K$, $\sigma < 0$. Thus, the value for W cannot be correct.

PROBLEM 6.62

KNOWN: Operating data are provided for a gearbox at steady state.

FIND: Determine the rate of heat transfer and the rate of entropy production. Can the heat transfer be achieved by free convection?

SCHEMATIC & GIVEN DATA:

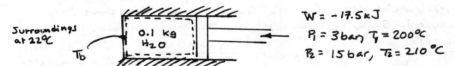

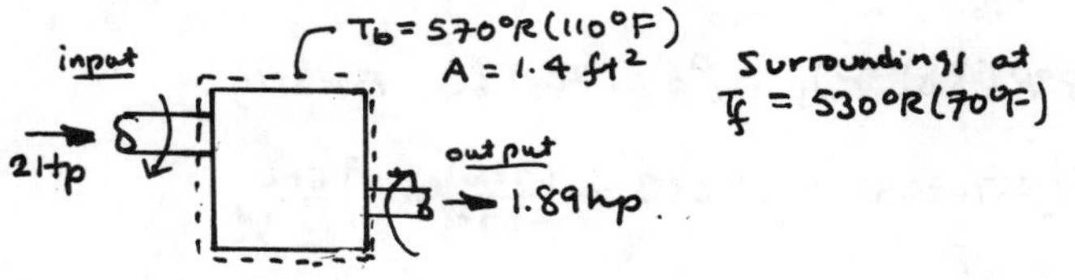

ASSUMPTIONS: 1. The system shown in the schematic is at steady state. 2. Heat transfer occurs at temperature T_b and is by convection.

ANALYSIS: At steady state an energy rate balance reduces to $\frac{dE}{dt}\overset{0}{} = \dot{Q} - \dot{W}$, or

$$\dot{Q} = \dot{W} \Rightarrow \dot{Q} = [1.89\,hp - 2\,hp]\left|\frac{2545\,Btu/h}{1\,hp}\right|\left|\frac{1\,h}{3600\,s}\right| = -7.78\times10^{-2}\,\frac{Btu}{s}$$

At steady state an entropy rate balance reduces to $\frac{dS}{dt}\overset{0}{} = \frac{\dot{Q}}{T_b} + \dot{\sigma}$, or

$$\dot{\sigma} = -\frac{\dot{Q}}{T_b} = -\frac{(-7.78\times10^{-2}\,Btu/s)}{570°R} = 1.36\times10^{-4}\,\frac{Btu}{s\cdot°R}$$

Using Eq. 2.34 written as $\dot{Q} = -hA[T_b - T_f]$, where the minus sign is needed for consistency with the sign for $\dot{Q}$ in the energy rate balance. Solving for the heat transfer coefficient,

$$h = \frac{-\dot{Q}}{A[T_b - T_f]} = -\frac{(-7.78\times10^{-2}\,Btu/s)\left|\frac{3600\,s}{h}\right|}{(1.4\,ft^2)(570 - 530)°R} = 5\,\frac{Btu}{h\cdot ft^2\cdot°R}$$

. Checking Table 2.1 with this value for h suggests that forced convection would be required.

PROBLEM 6.63

KNOWN: One lb of air is compressed adiabatically from 40°F, 1atm to 5 atm.

FIND: If the air is compressed without internal irreversibilities, find the final temperature and the work. If the work is 20% greater than the value for no internal irreversibilities, find the final temperature and the amount of entropy produced. Show both processes on T-s coordinates.

SCHEMATIC & GIVEN DATA:

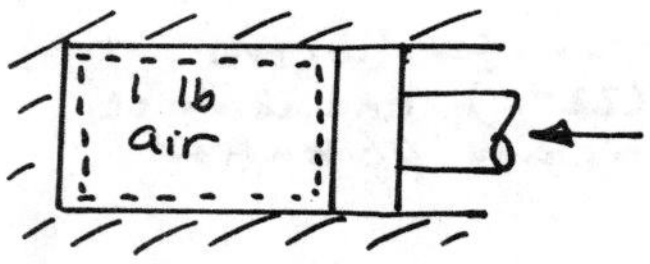

Q = 0

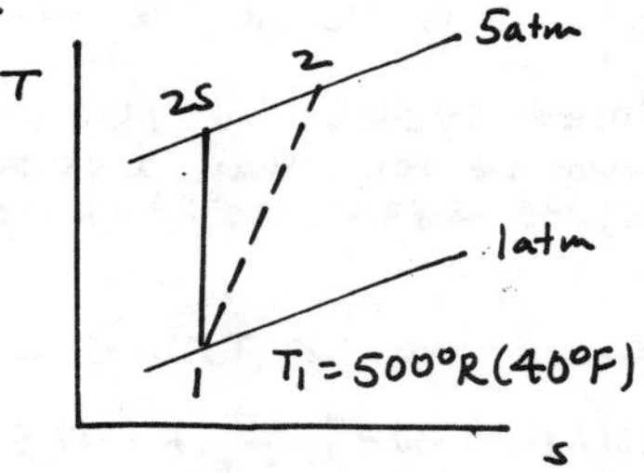

ASSUMPTIONS: 1. The system is the air. 2. Air is modeled as an ideal gas. 3. There is no heat transfer nor significant kinetic/potential energy effects. 4. For part (a), $\sigma = 0$.

ANALYSIS: With assumption 3, the energy balance reduces to $\Delta U = \cancel{Q}^{\,0} - W$, which gives the following expression for the work input:

$(-W) = m(u_2 - u_1)$. The entropy balance reduces to $\Delta S = \cancel{\int\left(\frac{\delta Q}{T}\right)_b}^{\,0} + \sigma$, which gives $\sigma = m(s_2 - s_1)$.

(a) $\sigma = 0$. Reducing the above entropy balance gives $s_2 = s_1$. Thus, the process is 1-2s on the above T-s diagram. The corresponding expression for work is $(-W)_s = m(u_{2s} - u_1)$. From Table A-22E, $u_1 = 85.2$ Btu/lb. To find u_{2s}, first apply Eq. 6.21a with $s_2 = s_1$, obtaining

$$s^o_{2s} = s^o_1 + R \ln \frac{P_2}{P_1} \Rightarrow s^o_{2s} = 0.58233 + \frac{1.986}{28.97} \ln 5 = 0.69266$$

Then, interpolation in Table A-22E with s^o_{2s} gives $u_{2s} = 135.3$ Btu/lb, $T_{2s} = 791°R\ (331°F)$.

Finally,

$$(-W)_s = (1\text{ lb})(135.3 - 85.2)\frac{\text{Btu}}{\text{lb}} = 50.1 \text{ Btu}$$

(b) The actual work input is 20% greater than in part (a): $(-W) = 1.2(50.1) = 60.1$ Btu.

Thus, the expression for $(-W)$ can be rearranged to give

$$u_2 = \frac{(-W)}{m} + u_1 = \left(\frac{60.1 \text{ Btu}}{1 \text{ lb}}\right) + 85.2\frac{\text{Btu}}{\text{lb}} = 145.3\frac{\text{Btu}}{\text{lb}}$$

Interpolating in Table A-22E, $T_2 = 848°R\ (388°F)$, $s^o_2 = 0.70963$. Finally,

$$\sigma = m\left[s^o_2 - s^o_1 - R\ln\frac{P_2}{P_1}\right] = (1\text{ lb})\left[0.70963 - 0.58233 - \frac{1.986}{28.97}\ln 5\right]\frac{\text{Btu}}{\text{lb}\cdot°R}$$

$$= 0.017 \frac{\text{Btu}}{°R}.$$

PROBLEM 6.64

KNOWN: Data are provided for a silicon chip at steady state.

FIND: Determine the rate of entropy production and identify the principal source of irreversibility.

SCHEMATIC & GIVEN DATA: See Fig. E 2.5.

ASSUMPTIONS: 1. The chip is a closed system at steady state. 2. There is no heat transfer between the chip and the substrate. All heat transfer is by convection from the top surface at T_b.

ANALYSIS: An entropy rate balance at steady state reads

$$\cancel{\frac{dS}{dt}}^{0} = \frac{\dot{Q}}{T_b} + \dot{\sigma} \Rightarrow \dot{\sigma} = -\frac{\dot{Q}}{T_b} \quad \text{where } \dot{Q} = -0.225\text{ W (from Example 2.5).}$$

Thus

$$\dot{\sigma} = \frac{-(-0.225)}{353\text{K}} = 6.37\times10^{-4}\ \frac{\text{watt}}{\text{K}}\left|\frac{1\text{ kW}}{10^3\text{ W}}\right| = 6.37\times10^{-7}\ \frac{\text{kW}}{\text{K}} \longleftarrow$$

The principal source of internal irreversibility is electric current flow through a resistance.

PROBLEM 6.65

KNOWN: Data are provided for an electric water heater.

FIND: For each of two systems determine the amount of entropy produced. Compare.

SCHEMATIC & GIVEN DATA:

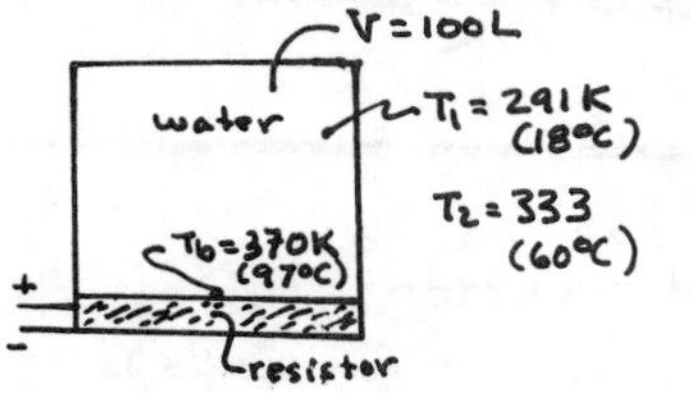

ASSUMPTIONS: 1. In part (a) the system is the water alone. In part (b) the system is the water plus the resistor. 2. The water is modeled as incompressible. 3. Heat transfer from the outer surface of the water heater can be ignored. 4. The states of the tank wall and the resistor are constant.

(a) For the water as the system, the entropy balance reduces to $\Delta S = \frac{Q}{T_b} + \sigma$. The heat transfer to the water from the resistor is found from an energy balance on the water: $\Delta U = Q - \cancel{W}$, which with Eq. 3.20a takes the form

$Q = \Delta U \Rightarrow Q = m[c\,\Delta T]$. With Eq. 6.24, $\Delta S = mc\ln T_2/T_1$. Collecting results

$$\sigma = \Delta S - \frac{Q}{T_b} \Rightarrow \sigma = mc\left[\ln\frac{T_2}{T_1} - \frac{(T_2-T_1)}{T_b}\right]$$

Using the known volume and $v = v_f(T_{ave}) = 1.008\times10^{-3}\ \text{m}^3/\text{kg}$ from Table A-2 and $c = 4.18$ kJ/kg·K from Table A-19

$$m = \frac{V}{v} = \frac{(100\text{L})\,|10^{-3}\text{m}^3/\text{L}|}{1.008\times10^{-3}\ \text{m}^3/\text{kg}} = 99.2\text{ kg},\quad \sigma = (99.2\text{kg})\left(4.18\frac{\text{kJ}}{\text{kg}\cdot\text{K}}\right)\left[\ln\frac{333}{291} - \frac{(42)}{370}\right] = 8.83\ \frac{\text{kJ}}{\text{K}} \quad \text{(a)}$$

(b) For the water plus the resistor as the system, the entropy balance reduces to

$\Delta S = \cancel{\int\left(\frac{\delta Q}{T}\right)_b}^{0} + \sigma \Rightarrow \sigma = \Delta S$. With assumption 4, ΔS can be identified with the water only: $\Delta S = mc\ln\frac{T_2}{T_1}$. Thus

$$\sigma = mc\ln\frac{T_2}{T_1} = (99.2\text{kg})\left(4.18\frac{\text{kJ}}{\text{K}\cdot\text{K}}\right)\ln\frac{333}{291} = 55.9\ \frac{\text{kJ}}{\text{K}} \longleftarrow \text{(b)}$$

The entropy production in the enlarged system is greater because there is an additional source of irreversibility — namely, electric current flow through the resistor. For the water alone, the irreversibility is associated with internal heat transfer from the resistor through the water as it is heated.

PROBLEM 6.66

KNOWN: At steady state the outer surface of a 15-W curling iron is at 90°C.

FIND: Determine the rate of heat transfer and the rate of entropy production.

SCHEMATIC & GIVEN DATA:

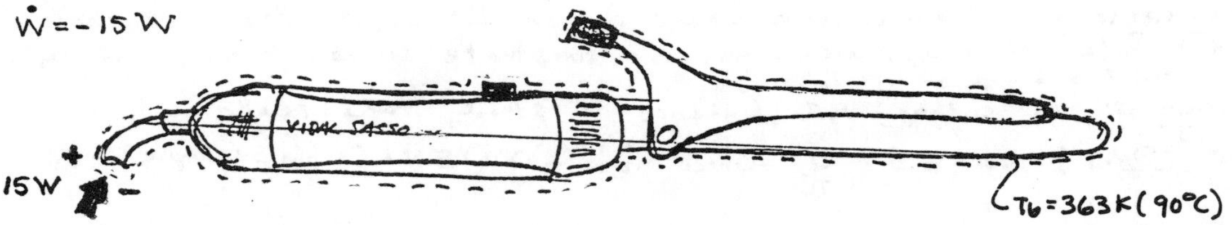

ASSUMPTIONS: (1) The system shown in the figure is at steady state. (2) Heat transfer occurs at $T_b = 363$ K (90°C) only.

ANALYSIS: An energy rate balance reduces to read

$$\frac{dE}{dt}^{0} = \dot{Q} - \dot{W} \Rightarrow \dot{Q} = \dot{W} = -15 \text{ W} \longleftarrow \dot{Q}$$

An entropy rate balance reduces to read

$$\frac{dS}{dt}^{0} = \frac{\dot{Q}}{T_b} + \dot{\sigma} \Rightarrow \dot{\sigma} = -\frac{\dot{Q}}{T_b} = \frac{15 \text{ W}}{363 \text{K}} = 0.04 \frac{\text{W}}{\text{K}} \longleftarrow \dot{\sigma}$$

PROBLEM 6.67

KNOWN: Data are provided for a thermally insulated resistor.

FIND: Plot the temperature and amount of entropy produced versus t for $0 \le t \le 3$s.

SCHEMATIC & GIVEN DATA:

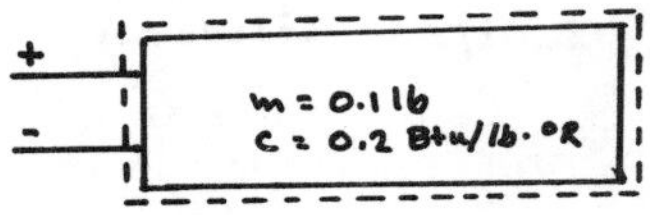

R = 30 ohm
I = 6 amps
$T_1 = 70$°F

ASSUMPTIONS: (1) The system shown in the accompanying figure is modeled as an incompressible substance. (2) Q = 0.

ANALYSIS An energy balance reduces to give

$$\Delta U = \cancel{Q}^{0} - W \Rightarrow m(u_2 - u_1) = -W$$

With assumption 1

$$mc(T_2 - T_1) = -W \Rightarrow T_2 = T_1 - W/mc$$

Electrical work is done on the system. Using given data

$$W = -I^2 R \Delta t = -(6 \text{ amps})^2 (30 \text{ ohm})(\Delta t)\left(\frac{\text{W}}{(\text{amps})^2 \cdot \text{ohm}}\right)$$

$$= -1080 \, \Delta t \text{ (W·s)}$$

Accordingly

$$T_2 = 530°\text{R} - \frac{(-1080 \, \Delta t \text{ (W·s)})}{(0.1 \text{ lb})(0.2 \text{ Btu/lb·°R})}\left(\frac{3.413 \text{ Btu/h}}{1 \text{ W}}\right)\left(\frac{1 \text{ h}}{3600 \text{ s}}\right)$$

$$= (530 + 51.2 \, \Delta t) \text{ °R} \qquad (1)$$

An entropy balance reduces to give

$$\Delta S = \cancel{\int_1^2 \left(\frac{\delta Q}{T}\right)_b} + \sigma \Rightarrow \sigma = \Delta S = mc \ln \frac{T_2}{T_1} \quad \text{(Eq. 6.24)}$$

PROBLEM 6.67 (Contd.)

Introducing Eq.(1) the expression for σ becomes

$$\sigma = (0.1\ lb)\left(0.2\ \frac{Btu}{lb \cdot °R}\right) \ln\left(1 + \frac{51.2\Delta t}{530}\right) = 0.02\ \frac{Btu}{°R} \ln\left(1 + \frac{51.2\Delta t}{530}\right) \quad (2)$$

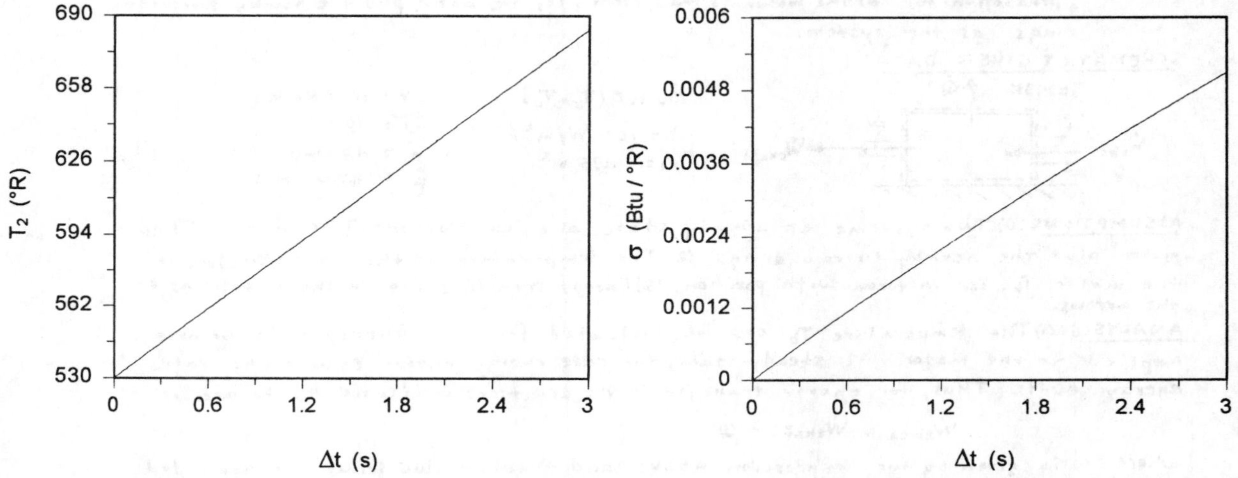

PROBLEM 6.68

KNOWN: Data are provided for an electric motor at steady state.

FIND: For the motor, determine the rates of heat transfer and entropy production.

SCHEMATIC & GIVEN DATA:

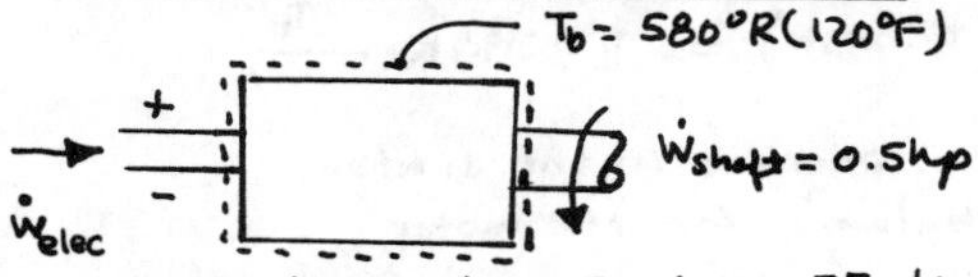

Motor draws 4 amps at 120 Volts

ASSUMPTIONS: 1. The system shown in the accompanying figure operates at steady state.
2. The power quantities indicated are in the directions of the respective arrows.

ANALYSIS: At steady state, the energy rate balance reduces to give

$$\frac{dE}{dt}^{0} = \dot{Q} - \dot{W} \Rightarrow \dot{Q} = \dot{W} \quad \text{where} \quad \dot{W} = \dot{W}_{shaft} - \dot{W}_{elec}. \text{ That is}$$

$$\dot{W} = 0.5hp \left|\frac{2545\ Btu/h}{1hp}\right| - \left[(4amps)(120volts)\left|\frac{1W}{amp\cdot volt}\right|\left|\frac{3.413\ Btu/h}{1W}\right|\right] = -365.7\ \frac{Btu}{h}$$

$$\Rightarrow \dot{Q} = -365.7\ Btu/h.$$

At steady state, an entropy rate balance reduces to give

$$\frac{dS}{dt}^{0} = \frac{\dot{Q}}{T_b} + \dot{\sigma} \Rightarrow \dot{\sigma} = -\frac{\dot{Q}}{T_b} = -\frac{(-365.7\ Btu/h)}{580\ °R} = 0.631\ \frac{Btu}{h\cdot °R}$$

PROBLEM 6.69

KNOWN: Operating data are provided for an electric motor at steady state.

FIND: (a) Determine the motor outer surface temperature. Evaluate the rate of entropy production for (b) the motor as the system, (c) the motor plus the nearby surroundings as the system.

SCHEMATIC & GIVEN DATA:

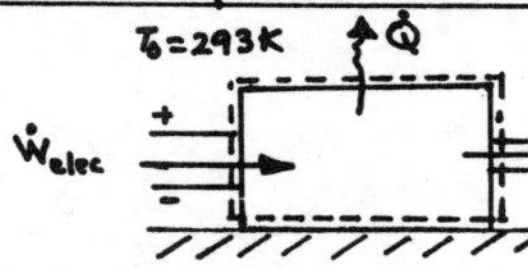

$\dot{Q} = hA(T_b - T_0)$

$h = 100\ W/m^2 \cdot K$

$A = 0.195\ m^2$

$\omega = 1000\ RPM$

$\mathcal{T} = 16\ N \cdot m$

$i = 10\ amp$

$\mathcal{E} = 220\ Volts$

ASSUMPTIONS: (1) Two systems at steady state are considered: The motor. The motor plus the nearby surroundings. (2) The temperature at the outer surface of the motor, T_b, is uniform with position. (3) Energy transfers are in the directions of the arrows.

ANALYSIS: (a) The temperature T_b can be evaluated from an energy rate balance applied to the motor. At steady-state, the rate energy enters equals the rate energy exits. Thus, for energy transfer in the directions indicated by the arrows

$$\dot{W}_{elect} = \dot{W}_{shaft} + \dot{Q}$$

where $\dot{Q}$ is given by the expression above and the other two terms are evaluated using Eq. 2.20 and 2.21:

$$\dot{W}_{shaft} = \mathcal{T}\omega = (16\ N \cdot m)\left(1000\ \frac{rev}{min}\right)\left(\frac{2\pi\ rad}{rev}\right)\left(\frac{min}{60\ s}\right)\left(\frac{1\ kW}{10^3\ N \cdot m}\right) = 1.676\ kW$$

$$\dot{W}_{elec} = \mathcal{E}i = (220\ volts)(10\ amp)\left[\frac{1\ watt/amp}{1\ volt}\right]\left(\frac{kW}{10^3\ W}\right) = 2.2\ kW$$

Thus

$$2.2\ kW = 1.676\ kW + \left(100\ \frac{W}{m^2 K}\right)(0.195\ m^2)\left(\frac{kW}{10^3}\right)(T_b - 293K) \Rightarrow T_b = 319.9K \leftarrow T_b$$

(b) With assumption 2, and taking in account that $\dot{Q}$ is positive in the direction of the arrow on the sketch above, an entropy rate balance for the motor as the system reduces at steady state to give

$$\cancelto{0}{\frac{dS}{dt}} = \frac{(-\dot{Q})}{T_b} + \dot{\sigma} \Rightarrow \dot{\sigma} = \frac{\dot{Q}}{T_b}$$

From part (a), $\dot{Q} = \dot{W}_{elect} - \dot{W}_{shaft} = 0.524\ kW$, so

$$\dot{\sigma} = \frac{0.524\ kW}{319.9\ K} = 1.64 \times 10^{-3}\ \frac{kW}{K} \leftarrow \dot{\sigma}$$

(c) For a system consisting of the motor plus the nearby surroundings, the heat transfer takes place at $T_0 = 293K$. At steady state an entropy rate balance reduces to give for this enlarged system

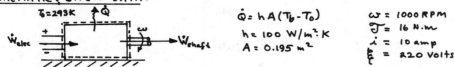

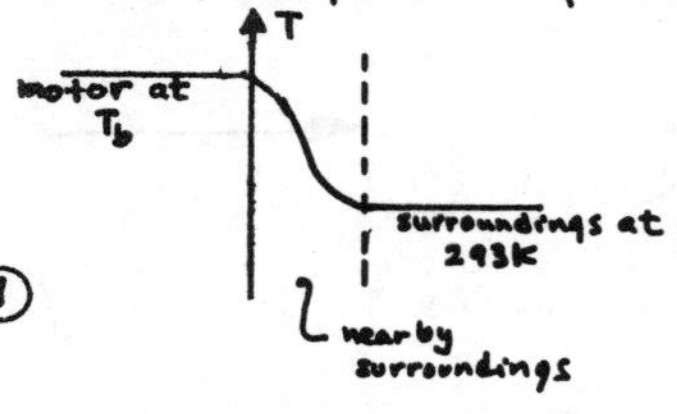

①

$$\dot{\sigma} = \frac{\dot{Q}}{T_0}$$

$$= \frac{0.524\ kW}{293\ K} = 1.79 \times 10^{-3}\ \frac{kW}{K} \leftarrow \dot{\sigma}$$

---

1. The rate of entropy production is greater for the enlarged system because it includes an additional source of irreversibility associated with heat transfer from the motor at T_b to the surroundings at T_0.

PROBLEM 6.70

KNOWN: Operating data are provided for a closed tank fitted with a paddle wheel and containing a slurry.

FIND: Determine the rate of entropy production for systems consisting of (a) the tank and its contents, (b) the tank and its nearby surroundings.

SCHEMATIC & GIVEN DATA:

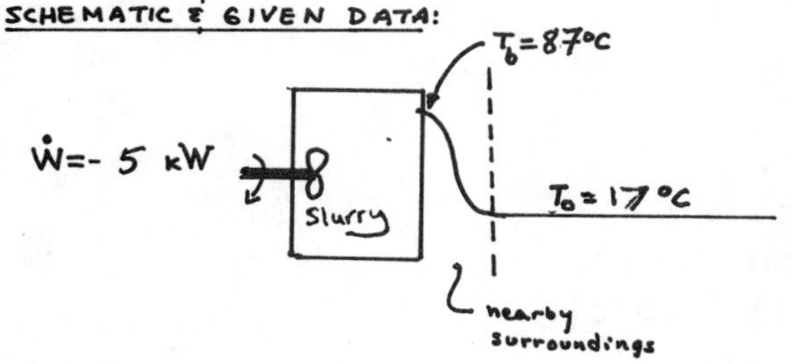

ASSUMPTIONS: (1) Two systems are under consideration. One comprises the tank and the slurry within the tank. The other includes the tank, the slurry, and as shown in the sketch the nearby surroundings. (2) Each system operates at steady state, while receiving a power input from a paddle wheel.

ANALYSIS: An energy rate balance reduces at steady state to give

$$\cancel{\frac{dE}{dt}}^{0} = \dot{Q} - \dot{W} \Rightarrow \dot{Q} = \dot{W} = -5 \text{ kW}$$

(a) Tank and slurry as the system. An entropy rate balance reduces at steady state to give

$$\cancel{\frac{dS}{dt}}^{0} = \frac{\dot{Q}}{T_b} + \dot{\sigma}$$

$$\Rightarrow \quad \dot{\sigma} = -\frac{\dot{Q}}{T_b} = \frac{-(-5 \text{ kW})}{360 \text{ K}} = 0.014 \text{ kW/K}$$

(b) Tank, slurry, and nearby surroundings as the system. An entropy rate balance reduces at steady state to give

① $$\cancel{\frac{dS}{dt}}^{0} = \frac{\dot{Q}}{T_0} + \dot{\sigma} \Rightarrow \dot{\sigma} = \frac{-(-5 \text{ kW})}{290 \text{ K}} = 0.017 \frac{\text{kW}}{\text{K}}$$

---

1. The entropy production rate is greater in part (b) than in part (a) because the enlarged system of part (b) has an additional source of irreversibility: the irreversibility associated with heat transfer from the tank to the surroundings.

PROBLEM 6.71

KNOWN: A system consisting of a fixed volume of hydrogen gas is brought from a specified initial state to a specified final temperature by heat transfer from a reservoir.

FIND: For the hydrogen, find Q, ΔS, and σ. For the reservoir, find ΔS. Compare ΔS values and discuss.

SCHEMATIC & GIVEN DATA:

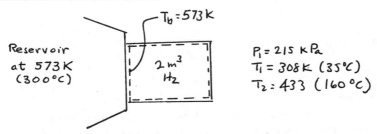

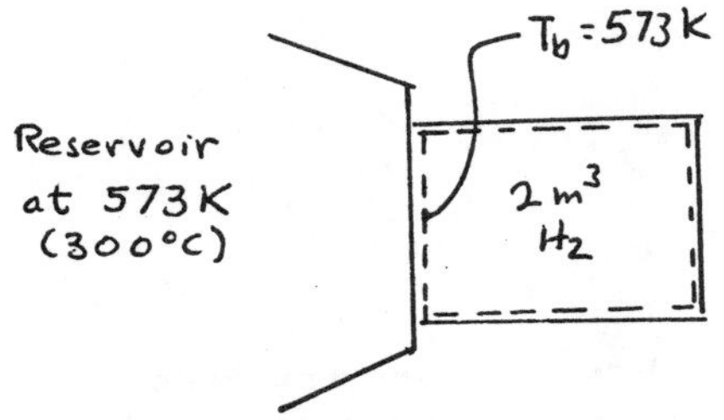

$P_1 = 215$ kPa
$T_1 = 308$K (35°C)
$T_2 = 433$ (160°C)

ASSUMPTIONS: (1) The hydrogen gas is the system. (2) The hydrogen gas is modeled as an ideal gas. (3) Heat transfer takes place at $T_b = 573K$. (4) Kinetic and potential energy effects are absent.

ANALYSIS: An energy balance reduces to give $\Delta U = Q - W^{\,0}$, or

$$Q = m(u_2 - u_1)$$

Inspection of Table A-20 indicates that c_v varies only slightly for H_2 over the range 308 to 433 K. Taking the value at $T_{ave} = 371$ K and evaluating m using the ideal gas equation of state

$$Q = \left(\frac{P_1 V}{R T_1}\right) c_v [T_2 - T_1] = \left[\frac{(215 \times 10^3\ N/m^2)(2m^3)}{\left(\frac{8314}{2.016}\frac{N \cdot m}{kg \cdot K}\right)(308K)}\right]\left(10.32\ \frac{kJ}{kg \cdot K}\right)[433 - 308]K = 436.7\ kJ \longleftarrow$$

To find ΔS, use Eq. 6.22

$$\Delta S = m\left[c_v \ln\frac{T_2}{T_1} + R \ln\frac{v_2}{v_1}^{\,0}\right] = \left[\frac{(215 \times 10^3)(2)}{\left(\frac{8314}{2.016}\right)(308)}\right](10.32)\ln\left(\frac{433}{308}\right) = 1.19\ \frac{kJ}{K} \longleftarrow$$

With assumption 3, an entropy balance reduces to

$$\Delta S = \frac{Q}{T_b} + \sigma \Rightarrow \sigma = \Delta S - \frac{Q}{T_b} = \left(1.19 - \frac{436.7}{573}\right) = 0.428\ \frac{kJ}{K} \longleftarrow$$

Taking the reservoir as the system, the entropy balance reads

$$\Delta S)_{res} = \frac{Q)_{res}}{T_b} + \sigma_{res}^{\,0} = \frac{(-436.7)}{573} = -0.762\ \frac{kJ}{K}$$

Entropy is transferred from the reservoir accompanying heat transfer, and so the entropy of the reservoir decreases by the amount carried out. For the hydrogen, entropy is carried in accompanying heat transfer. In addition, entropy is produced owing to irreversibilities. Accordingly, the entropy change for the hydrogen exceeds the amount carried in by the amount produced.

PROBLEM 6.72

KNOWN: Data are provided on an isolated system consisting of an aluminum vessel and two quantities of liquid water.

FIND: Determine (a) the final temperature when the system has come to equilibrium, (b) the entropy change for the aluminum vessel and each of the liquid masses, (c) the amount of entropy produced.

SCHEMATIC & GIVEN DATA:

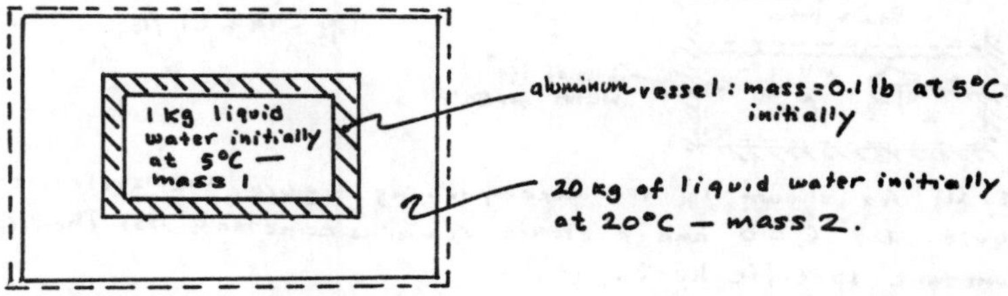

ASSUMPTIONS: (1) The system shown in the accompanying figure is isolated. (2) The aluminum vessel and each of the liquid masses can be modeled as incompressible with constant specific heat.

ANALYSIS: From Table A-19, the specific heat of aluminum is $c = 0.9$ kJ/kg·K and of liquid water $c_w = 4.18$ kJ/kg·K.

(a) With assumption 1, $Q = W = 0$ and an energy balance reduces to

$$\Delta U = \cancel{Q}^{\,0} - \cancel{W}^{\,0}$$

$$(\Delta U)_1 + (\Delta U)_2 + (\Delta U)_{copper} = 0$$

With assumption 2

$$m_1 c_w [T_f - T_1] + m_2 c_w [T_f - T_2] + mc[T_f - T_1] = 0$$

where m_1 and T_1 are the mass and initial temperature of liquid mass 1, m_2 and T_2 are the mass and initial temperature of liquid mass 2, and m is the mass of the vessel.

Solving for the final temperature, T_f

$$T_f = \frac{m_1 c_w T_1 + m_2 c_w T_2 + mc\,T_1}{m_1 c_w + m_2 c_w + mc}$$

$$= \frac{(1)(4.18)(278) + (20)(4.18)(293) + (0.1)(0.9)(278)}{(1)(4.18) + (20)(4.18) + (0.1)(0.9)} = 292.3 \text{ K} \quad (19°\text{C}) \longleftarrow T_f$$

(b) Entropy changes. Using Eq. 6.24 (assumption 2)

$$(\Delta S)_1 = m_1 c_w \ln\frac{T_f}{T_1} = (1)(4.18)\ln\frac{292.3}{278} = 0.20967 \frac{\text{kJ}}{\text{K}} \longleftarrow \Delta S$$

$$(\Delta S)_2 = m_2 c_w \ln\frac{T_f}{T_2} = (20)(4.18)\ln\frac{292.3}{293} = -0.19996 \frac{\text{kJ}}{\text{K}}$$

$$(\Delta S)_{copper} = mc\ln\frac{T_f}{T_1} = (0.1)(0.9)\ln\frac{292.3}{278} = 0.00451 \frac{\text{kJ}}{\text{K}}$$

(c) Since there is no heat transfer to or from the system an entropy balance gives

$$\Delta S = \cancel{\int_1^2 \left(\frac{\delta Q}{T}\right)_b}^{\,0} + \sigma \Rightarrow \sigma = (\Delta S)_1 + (\Delta S)_2 + (\Delta S)_{Aluminum}$$

$$= 0.0142 \frac{\text{kJ}}{\text{K}} \longleftarrow \sigma$$

PROBLEM 6.73

KNOWN: A system initially consisting of 0.4 lb of ice at 32°F and 2 lb of liquid water at 80°F attains an equilibrium state adiabatically at a constant pressure of 1 atm.

FIND: Determine the final temperature and the amount of entropy produced.

SCHEMATIC & GIVEN DATA:

initially:

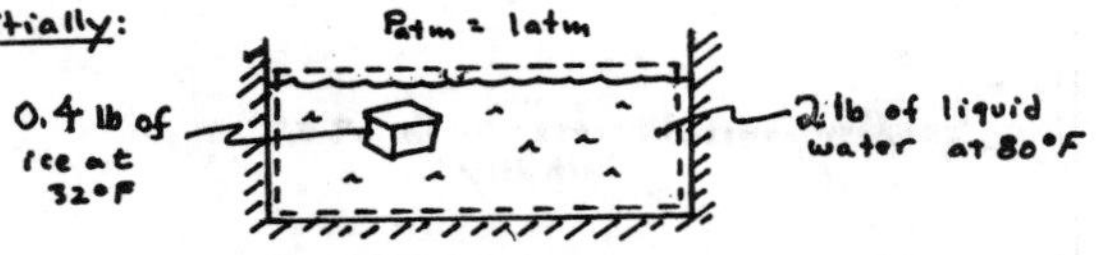

$h_{if} = 144$ Btu/lb

ASSUMPTIONS: (1) As shown in the accompanying figure, the system consists of the ice and liquid. (2) Q = 0 and pressure remains constant. (3) The liquid is incompressible with constant specific heat.

ANALYSIS: (a) The final temperature can be determined using an energy balance. Thus, $\Delta U = \cancel{Q}^0 - W$, where $W = \int p\,dV = p\Delta V$ since pressure is constant. So

$$\Delta U = -p\Delta V \Rightarrow \Delta H = 0 \text{ for the constant pressure process}$$

If all the ice does not melt, the final temperature would be 32°F. To investigate this, let T_f be the final temperature when all of the ice is regarded to have melted. Then with c = 1.0 Btu/lb·°R for liquid water from Table A-19E

$$\left\{\Delta H]_{ice} + \Delta H]_{\substack{melt \\ 32°F \to T_f}}\right\} + \Delta H]_{\substack{liquid \\ 80°F \to T_f}} = 0$$

$$\left\{(0.4\text{lb})\left(144\,\frac{\text{Btu}}{\text{lb}}\right) + (0.4\text{lb})\left(1\,\frac{\text{Btu}}{\text{lb}\cdot°\text{R}}\right)(T_f - 492)°\text{R}\right\} + (2\text{lb})\left(1\,\frac{\text{Btu}}{\text{lb}\cdot°\text{R}}\right)(T_f - 540)°\text{R} = 0$$

$$\Rightarrow T_f = 508°\text{R} \quad (48°\text{F}) \longleftarrow T_f$$

As $T_f > 32°F$, it can be concluded that all the ice melts.

(b) An entropy balance reduces to give

$$\Delta S = \cancel{\int_1^2 \left(\frac{\delta Q}{T}\right)_b}^0 + \sigma \Rightarrow \sigma = \Delta S$$

Thus

$$\sigma = \left\{\Delta S]_{\substack{ice \\ at\,32°F}} + \Delta S]_{\substack{melt \\ 32°F \to T_f}}\right\} + \Delta S]_{\substack{Liquid \\ 80°F \to T_f}}$$

The two liquid terms can be evaluated using Eq. 6.24. The entropy change for the melting ice can be found using the TdS equation, Eq 6.12b, which at constant pressure and temperature reduces to give

$$ds = \frac{dh}{T} \Rightarrow s_f - s_i = \frac{h_f - h_i}{T} = \frac{144\text{ Btu/lb}}{492°\text{R}} = 0.2927\text{ Btu/lb}\cdot°\text{R}$$

Thus

$$\sigma = (0.4\text{lb})\left(0.2927\,\frac{\text{Btu}}{\text{lb}\cdot°\text{R}}\right) + (0.4\text{lb})\left(1\,\frac{\text{Btu}}{\text{lb}\cdot°\text{R}}\right)\ln\frac{508}{492} + (2\text{lb})\left(1\,\frac{\text{Btu}}{\text{lb}}\right)\ln\left(\frac{508}{540}\right) \longleftarrow \sigma$$

$$= (0.1171) + (0.0128) + (-0.1222) = 0.0077\text{ Btu/°R}$$

PROBLEM 6.74

KNOWN: A system initially consisting of a rivet and a two-phase solid-liquid mixture of water attains an equilibrium state adiabatically at constant pressure.

FIND: Determine the final temperature and the amount of entropy produced.

SCHEMATIC & GIVEN DATA:

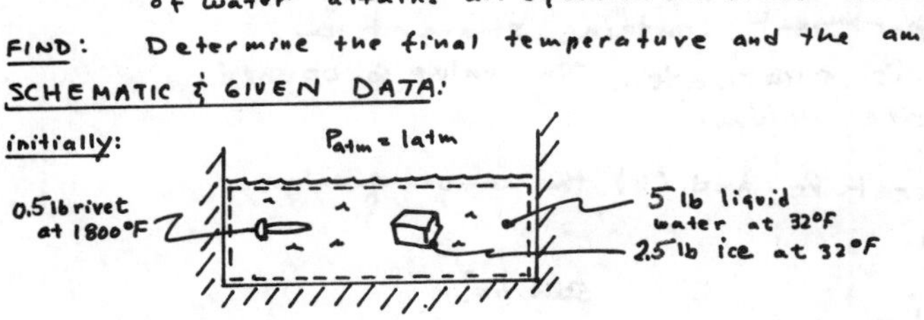

$h_{if} = 144$ Btu/lb

$c_r = 0.12$ Btu/lb°R

ASSUMPTIONS: (1) As shown in the accompanying figure, the system consists of the ice, liquid, and rivet. (2) $Q = 0$ and pressure remains constant. (3) The rivet and liquid can be regarded as incompressible with constant specific heats.

ANALYSIS: (a) The final temperature can be determined using an energy balance. Thus, $\Delta U = \cancel{Q}^{0} - W$, where $W = \int_1^2 p\,dV = p\Delta V$ since pressure is constant. So

$$\Delta U = -p\Delta V \Rightarrow \Delta H = 0 \text{ for the constant pressure process.}$$

If all the ice does not melt, the final temperature would be 32°F. To investigate this, let T_f be the final temperature when all of the ice is regarded to have melted. Then, with $c = 1.0$ Btu/lb·°R for liquid water from Table A-19E

$$\left\{ \Delta H\big]_{\substack{ice \\ at\ 32°F}} + \Delta H\big]_{\substack{melt \\ 32°F \to T_f}} \right\} + \Delta H\big]_{\substack{liquid \\ 32°F \to T_f}} + \Delta H\big]_{\substack{rivet \\ 1800°F \to T_f}} = 0$$

$$0 = \left\{(2.5\,lb)\left(144\,\frac{Btu}{lb}\right) + (2.5\,lb)\left(1\,\frac{Btu}{lb\cdot°R}\right)(T_f - 492)\right\} + (5\,lb)\left(1\,\frac{Btu}{lb\cdot°R}\right)(T_f - 492) + (0.5\,lb)\left(0.12\,\frac{Btu}{lb\cdot°R}\right)(T_f - 2260)$$

$$\Rightarrow T_f = 458.4\,°R$$

Since $T_f < 32°F$, it can be concluded that not all of the ice melts and the final temperature is 32°F. ◄——— T_f

(b) An entropy balance reduces to give $\Delta S = \int_1^2 \left(\frac{\delta Q}{T}\right)_b^{\,0} + \sigma$ or $\sigma = \Delta S$. Since not all of the ice melts it is necessary to determine the amount that does melt. Let x denote the amount of ice that melts. Then since the liquid remains at 32°F, the energy balance gives

$$\Delta H\big]_{\substack{ice \\ at\ 32°F}} + \Delta H\big]_{\substack{rivet \\ 1800°F \to 32°F}} = 0$$

or

$$(x\,lb)\left(144\,\frac{Btu}{lb}\right) + (0.5\,lb)\left(0.12\,\frac{Btu}{lb\cdot°R}\right)(492 - 2260)\,°R = 0 \Rightarrow x = 0.737\,lb$$

Returning to the calculation of the entropy production,

$$\sigma = \Delta S\big]_{ice} + \Delta S\big]_{rivet} + \cancel{\Delta S\big]_{liq}}^{\,0}$$

The entropy change of the rivet can be found using Eq. 6.24. Similarly, $\Delta S = 0$ for the liquid by Eq. 6.24 since the temperature remains constant. The entropy change for the melting ice can be found using the TdS equation, Eq. 6.12b, which at constant pressure and temperature reduces to give

$$ds = \frac{dh}{T} \Rightarrow s_f - s_i = \frac{h_f - h_i}{T} = \frac{144\,Btu/lb}{492\,°R} = 0.2927\,Btu/lb\cdot°R$$

Thus

$$\sigma = (0.737\,lb)\left(0.2927\,\frac{Btu}{lb\cdot°R}\right) + (0.5\,lb)\left(0.12\,\frac{Btu}{lb\cdot°R}\right)\ln\left(\frac{492}{2260}\right)$$

$$= (0.2157) + (-0.0915) = 0.1242\,Btu/°R \quad ◄——— \sigma$$

PROBLEM 6.75

KNOWN: An insulated vessel is divided into equal-sized compartments connected by a valve. Initially one compartment contains steam at a known state and the other is evacuated. The valve is opened and the steam fills the entire volume.

FIND: Determine (a) the final temperature and (b) the amount of entropy produced per unit mass of steam.

SCHEMATIC & GIVEN DATA:

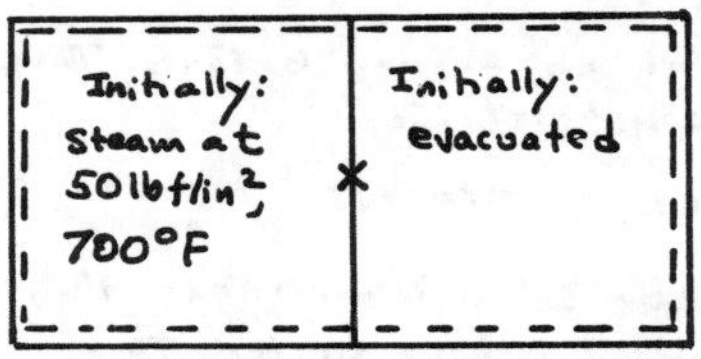

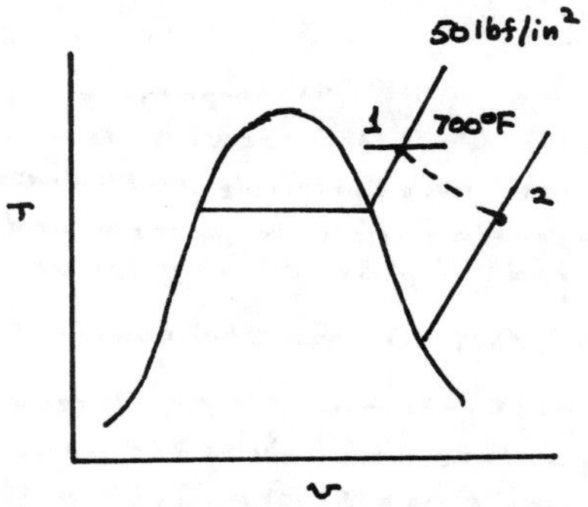

ASSUMPTIONS: (1) The system is shown by the dashed line in the figure above. (2) For the system, Q = 0, W = 0, and kinetic/potential energy effects can be ignored.

ANALYSIS: To fix the final state requires the values of two independent intensive properties: u_2 and v_2. Thus, since the steam expands to fill twice the initial volume occupied by the steam, $V_2 = 2V_1$. And from an energy balance

$$\Delta U = \cancel{Q}^{0} - \cancel{W}^{0} \Rightarrow u_2 = u_1$$

An entropy balance reduces to

$$\Delta S = \cancel{\int_1^2 \left(\frac{\delta Q}{T}\right)_b}^{0} + \sigma \Rightarrow \frac{\sigma}{m} = (s_2 - s_1) \qquad (1)$$

As the use of the pair of independent properties u_2, v_2 to fix the state with tabular steam table data is cumbersome, *IT* is employed to evaluate σ/m using Eq. (1):

IT Code

```
p1 = 50  // lbf/in.²
T1 = 700  // °F

u2 = u1
v2 = 2 * v1
u1 = u_PT("Water/Steam", p1, T1)
v1 = v_PT("Water/Steam", p1, T1)
u2 = u_PT("Water/Steam", p2, T2)
v2 = v_PT("Water/Steam", p2, T2)

sigma / m = s2 - s1
m = 1  // lb
s1 = s_PT("Water/Steam", p1, T1)
s2 = s_PT("Water/Steam", p2, T2)
```

IT Results

T_2 = 697.4 °F ← T_2
p_2 = 25.01 lbf/in.²
s_1 = 1.881 Btu/lb·°R
s_2 = 1.957 Btu/lb·°R
u_2 = 1255 Btu/lb
v_2 = 27.48 ft³/lb
σ/m = 0.07611 Btu/lb·°R ← σ/m

PROBLEM 6.76

KNOWN: Two insulated tanks containing air at known states are connected by a valve. The valve is opened and the two quantities of air mix.

FIND: Determine (a) the final temperature, (b) the final pressure, and (c) the amount of entropy produced.

SCHEMATIC & GIVEN DATA:

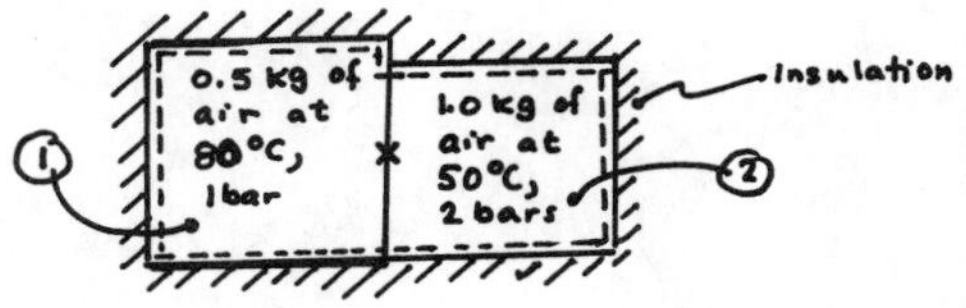

$c_v = 0.72 \frac{kJ}{kg \cdot K}$

ASSUMPTIONS: (1) As shown in the accompany figure, the system is the total quantity of air. (2) Air is modeled as an ideal gas. (3) $Q = W = 0$ and there is no change in kinetic or potential energy between the initial and final states.

ANALYSIS: (a) An energy balance reduces with assumption 3 to give $\Delta U = \cancel{Q}^0 - \cancel{W}^0$ or $\Delta U = 0$. That is

$$\Delta U = (m_1+m_2)u(T_f) - [m_1 u(T_1) + m_2 u(T_2)] = 0$$

or

$$0 = m_1[u(T_f) - u(T_1)] + m_2[u(T_f) - u(T_2)] \Rightarrow 0 = m_1 \cancel{c_v}[T_f - T_1] + m_2 \cancel{c_v}[T_f - T_2]$$

$$\Rightarrow \quad T_f = \frac{m_1 T_1 + m_2 T_2}{(m_1+m_2)} = \frac{(0.5\,kg)(353K) + (1.0\,kg)(323K)}{(1.5\,kg)} = 333\,K\ (60°C) \quad \longleftarrow (a)$$

(b) Using the ideal gas equation of state

$$P_f = \frac{(m_1+m_2)\,R\,T_f}{(V_1+V_2)}$$

where

$$V_1 = \frac{m_1 R T_1}{P_1} = \frac{(0.5)\left(\frac{8.314}{28.97}\right)(353)}{(10^5)} = 0.507\,m^3$$

$$V_2 = \frac{m_2 R T_2}{P_2} = \frac{(1)(8314/28.97)(323)}{2(10^5)} = 0.463\,m^3$$

Thus

$$P_f = \frac{(1.5)\left(\frac{8314}{28.97}\right)(333)}{(0.507 + 0.463)(10^5)} = 1.478\,bars \quad \longleftarrow (b)$$

(c) An entropy balance reduces to give $\Delta S = \cancel{\int_1^2 \left(\frac{\delta Q}{T}\right)_b}^0 + \sigma$, or

$$\sigma = \Delta S = \{(m_1+m_2)s_f - (m_1 s_1 + m_2 s_2)\} = m_1(s_f - s_1) + m_2(s_f - s_2)$$

$$= m_1\left[c_p \ln\frac{T_f}{T_1} - R\ln\frac{P_f}{P_1}\right] + m_2\left[c_p \ln\frac{T_f}{T_2} - R\ln\frac{P_f}{P_2}\right]$$

Using Eq. 3.44, $c_p = c_v + R = 0.72 + (8.314/28.97) = 1.007\,kJ/kg\cdot K$. Then

$$\sigma = (0.5\,kg)\left[\left(1.007\frac{kJ}{kg\cdot K}\right)\ln\frac{333}{353} - \frac{8.314}{28.97}\ln\frac{1.478}{1}\right] + (1)\left[(1.007)\ln\frac{333}{323} - \frac{8.314}{28.97}\ln\frac{1.478}{2}\right]$$

$$= [(-0.0854) + (0.1175)]\frac{kJ}{K}$$

$$= 0.0321\,kJ/K \quad \longleftarrow (c)$$

PROBLEM 6.77

KNOWN: An insulated cylinder is initially divided into halves by a piston. On either side of the piston is a gas at a known state. The piston is released and equilibrium is attained

FIND: Determine the final pressure, final temperature, and the amount of entropy produced.

SCHEMATIC & GIVEN DATA:

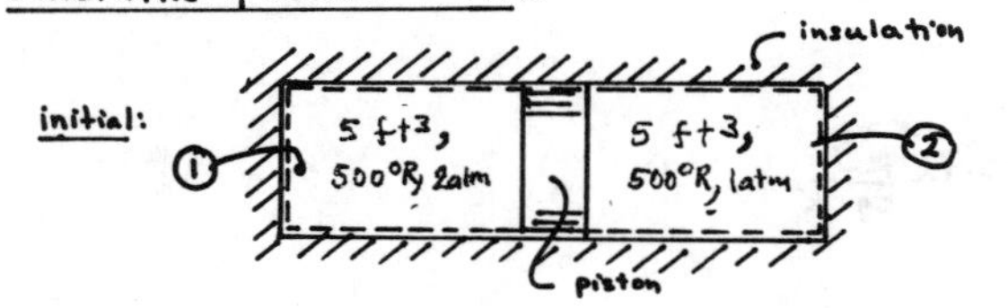

ASSUMPTIONS: (1) The system consists of both quantities of gas and the piston. But the piston experiences no net change in state. (2) The piston moves freely in the cylinder and is thermally conducting. (3) The gas is modeled as an ideal gas. (4) $Q = 0$.

ANALYSIS: (a) An energy balance reduces to give $\Delta U = \cancel{Q}^0 - \cancel{W}^0 \Rightarrow \Delta U = 0 \Rightarrow$

$$\Delta U]_1 + \Delta U]_2 + \cancel{\Delta U]_{piston}}^0 = 0 \Rightarrow m_1[u(T_f) - u(500°R)] + m_2[u(T_f) - u(500°R)] = 0$$

$$\Rightarrow u(T_f) = u(500°R) \Rightarrow T_f = 500°R \longleftarrow T_f$$

(b) Using the ideal gas equation of state, and noting that the total mass of gas occupies the same total volume at the final state as initially: $V = 10\ ft^3$,

$$P_f = \frac{(m_1 + m_2)RT_f}{V}$$

where

$$m_1 = \frac{P_1(V/2)}{RT_1}, \quad m_2 = \frac{P_2(V/2)}{RT_2}$$

Thus, since $T_1 = T_2 = T_f$

$$P_f = \frac{\left(\frac{P_1(V/2)}{RT} + \frac{P_2(V/2)}{RT}\right)RT}{V} = \frac{P_1 + P_2}{2} = 1.5\ atm \longleftarrow P_f$$

(c) An entropy balance reduces to give $\Delta S = \cancel{\int_1^2 \left(\frac{\delta Q}{T}\right)_b}^0 + \sigma$ or

$$\sigma = \Delta S \Rightarrow \sigma = \Delta S]_1 + \Delta S]_2 + \cancel{\Delta S]_{piston}}^0$$

where

$$\Delta S]_1 = m_1[\underbrace{s°(T_f) - s°(T_1)}_{=0\ (T_f = T_1)} - R\ln\frac{P_f}{P_1}] = -Rm_1\ln\frac{P_f}{P_1}$$

$$\Delta S]_2 = m_2[\underbrace{s°(T_f) - s°(T_2)}_{=0\ (T_f = T_2)} - R\ln\frac{P_f}{P_2}] = -Rm_2\ln\frac{P_f}{P_2}$$

Thus

$$\sigma = -R\left[m_1\ln\frac{P_f}{P_1} + m_2\ln\frac{P_f}{P_2}\right] = -R\left[\frac{P_1(V/2)}{RT}\ln\frac{P_f}{P_1} + \frac{P_2(V/2)}{RT}\ln\frac{P_f}{P_2}\right]$$

$$= -\frac{V/2}{T}\left[P_1\ln\frac{P_f}{P_1} + P_2\ln\frac{P_f}{P_2}\right]$$

$$= -\frac{5\ ft^3}{500°R}\left[(2\ atm)\ln\frac{1.5}{2} + (1\ atm)\ln\frac{1.5}{1}\right]\left|\frac{14.7\ lbf/in^2}{1\ atm} \times \frac{144\ in^2}{1\ ft^2}\right|$$

$$= \left(3.596\ \frac{ft\cdot lbf}{°R}\right)\left|\frac{1\ Btu}{778\ ft\cdot lbf}\right|$$

① $$= 4.6 \times 10^{-3}\ Btu/°R \longleftarrow \sigma$$

---

1. Provided the ideal gas model is assumed, the same numerical results are obtained for any gas.

PROBLEM 6.78

KNOWN: An insulated tank is divided into two compartments by a piston. On either side of the piston is water vapor at known states. The piston is released and equilibrium is attained.

FIND: Determine (a) the final pressure, (b) the final temperature, and (c) the amount of entropy produced.

SCHEMATIC & GIVEN DATA:

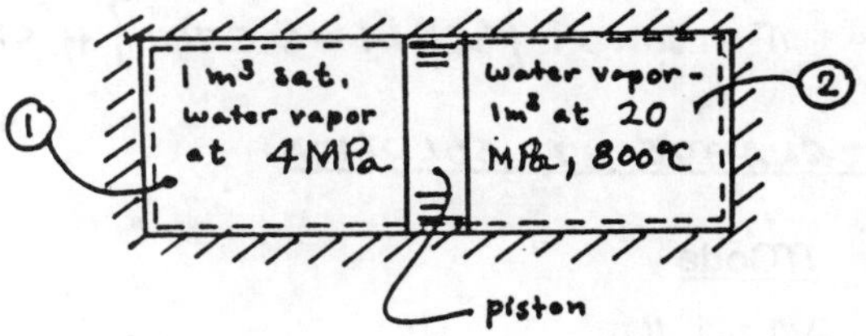

ASSUMPTIONS: (1) The closed system consists of both quantities of water vapor and the piston, but the piston experiences no change of state. (2) The piston moves freely and is thermally conducting. (3) For the system, $Q=0$, $W=0$, $\Delta KE = \Delta PE = 0$.

ANALYSIS: (a),(b) To fix the final state of the water vapor, we begin with an energy balance. That is, $\Delta U + \cancel{\Delta KE} + \cancel{\Delta PE} = \cancel{Q} - \cancel{W} \Rightarrow \Delta U = 0$. Thus

$$\Delta U]_1 + \Delta U]_2 + \cancel{\Delta U]_{piston}}^{=0} = 0 \Rightarrow m_1(\Delta u)_1 + m_2(\Delta u)_2 = 0 \qquad (1)$$

For each side, $m = V/v$. With data from Table A-3 at the initial state

$$m_1 = \frac{V_1}{v_1} = \frac{1\ m^3}{0.04978\ m^3/kg} = 20.09\ kg$$

Similarly, for side 2 initially Table A-4 gives $v_2 = 0.02385\ m^3/kg$. Thus

$$m_2 = \frac{V_2}{v_2} = \frac{1\ m^3}{0.02385\ m^3/kg} = 41.93\ kg$$

It follows from assumption 2 that at equilibrium, the states of water vapor on either side of the piston are the same. Thus, from (1)

$$m_1(u - u_1) + m_2(u - u_2) = 0$$

From Table A-3, $u_1 = 2602.3$ kJ/kg and from Table A-4, $u_2 = 3592.7$ kJ/kg. Inserting values and solving for u we get

$$20.09(u - 2602.3) + 41.93(u - 3592.7) = 0 \Rightarrow u = 3271.9\ kJ/kg$$

The overall system volume is constant: 2 m³, and the total mass of water vapor is fixed. Thus, at equilibrium $v = (2\ m^3)/(20.09 + 41.93)\ kg = 0.03225\ m^3/kg$. The equilibrium state on both sides of the piston or fixed by $u = 3271.9$ kJ/kg and $v = 0.03225\ m^3/kg$. Interpolation in Table A-4 with u and v is very inconvenient. Instead, we use IT to get

$$p = 121.5\ bar \qquad \text{(a) (b)}$$
$$T = 622.6\ °C$$
$$s = 6.861\ kJ/kg\cdot K$$

PROBLEM 6.78 (Contd.)

(c) The entropy produced can be determined from an entropy balance which reduces with assumptions (1) and (3) to give

$$\sigma = \Delta S \Rightarrow \sigma = \Delta S]_1 + \Delta S]_2 + \Delta S]_{piston}^{\nearrow 0} \Rightarrow \sigma = m_1(\Delta s)_1 + m_2(\Delta s)_2$$

From Table A-3, s_1 = 6.0701 kJ/kg·K, and from Table A-4, s_2 = 7.0544 kJ/kg·K. Thus

$$\sigma = 20.09\,[6.861 - 6.0701] + 41.93\,[6.861 - 7.0544] = 7.78 \text{ kJ/kg·K} \leftarrow \sigma$$

ALTERNATIVE IT SOLUTION

IT Code

```
V1 = 1  // m³
x1 = 1
p1 = 40  // bar
V2 = 1  // m³
p2 = 200  // bar
T2 = 800  // °C

m1 = V1 / v1
v1 = vsat_Px("Water/Steam", p1, x1)
m2 = V2 / v2
v2 = v_PT("Water/Steam", p2, T2)


m1 * (u - u1) + m2 * (u - u2) = 0
u1 = usat_Px("Water/Steam", p1, x1)
u2 = u_PT("Water/Steam", p2, T2)
v = (V1 + V2) / (m1 + m2)

u = u_PT("Water/Steam", p, T)
v = v_PT("Water/Steam", p, T)
```

IT Results - Alternative Solution

```
sigma = m1 * (s - s1) + m2 * (s - s2)
s1 = ssat_Px("Water/Steam",00 p1, x1)
s2 = s_PT("Water/Steam", p2, T2)
s = s_PT("Water/Steam",p, T)
```

PROBLEM 6.79

KNOWN: A system consisting of air at a specified state undergoes constant volume processes in which the temperature increases in each of two ways.

FIND: (a) When the air is stirred adiabatically, determine the entropy produced. (b) When the air is heated by a reservoir at temperature T, plot the entropy produced versus T. Compare and discuss.

SCHEMATIC & GIVEN DATA:

case a:

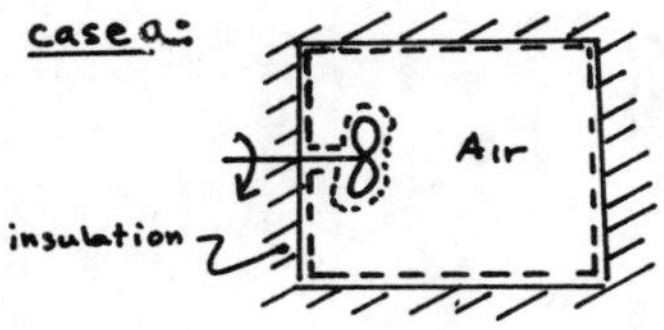

$T_1 = 300\,K$
$p_1 = 1\,bar$
$T_2 = 500\,K$

Case b:

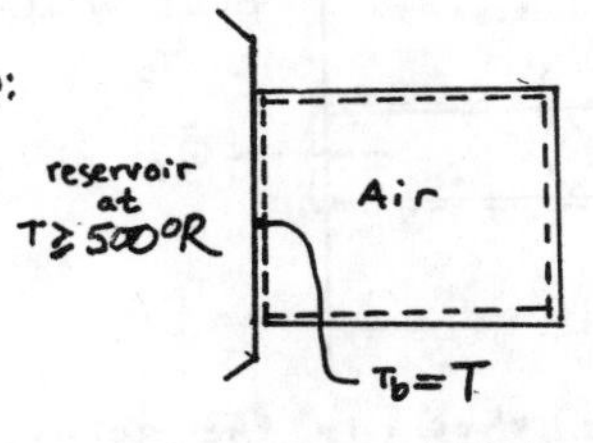

ASSUMPTIONS: (1) The system consists of the air. (2) Air is modeled as an ideal gas. (3) In case (a), $Q = 0$, $W \neq 0$. In case (b), $Q \neq 0$, $W = 0$.

ANALYSIS: The change in entropy of the air is required in the evaluation of σ in each case. Thus, with data from Table A-22 and the ideal gas equation which gives $p_2/p_1 = T_2/T_1$

$$s_2 - s_1 = s^\circ(T_2) - s^\circ(T_1) - R \ln\frac{p_2}{p_1} = 2.21952 - 1.70203 - \frac{8.314}{28.97}\ln\frac{500}{300} = 0.3709\ \text{kJ/kg}\cdot\text{K}$$

CASE a: An entropy balance reduces to give

$$\Delta S = \int_1^2 \left(\frac{\delta Q}{T}\right)_b^{\,0} + \sigma_a \Rightarrow \frac{\sigma_a}{m} = s_2 - s_1 = 0.3709\ \text{kJ/kg}\cdot\text{K} \quad \text{(a)}$$

CASE b: An entropy balance reduces to give

$$\Delta S = \frac{Q}{T_b} + \sigma_b \Rightarrow \frac{\sigma_b}{m} = (s_2 - s_1) - \frac{Q/m}{T_b}$$

To find Q, write an energy balance: $\Delta U = Q - W^{\,0}$. Thus, with data from Table A-22

$$\frac{Q}{m} = u(T_2) - u(T_1) = 359.49 - 214.07 = 145.42\ \text{kJ/kg}$$

Thus

$$\frac{\sigma_b}{m} = 0.3709 - \frac{145.42}{T} \quad \text{(b)}$$

Sample calculation: when $T = 500\,K$, $\sigma_b/m = 0.08\ \text{kJ/kg}\cdot\text{K}$.

Equation (b) is plotted below. Comparing the two cases, we get

$$(\sigma_b/m) = (\sigma_a/m) - (145.42/T).$$

Thus, the entropy produced by heating is always less than the entropy produced by stirring. The entropy produced by heating approaches the entropy produced stirring at higher reservoir temperatures (mathematically, as $T \to \infty$). See the accompanying plot.

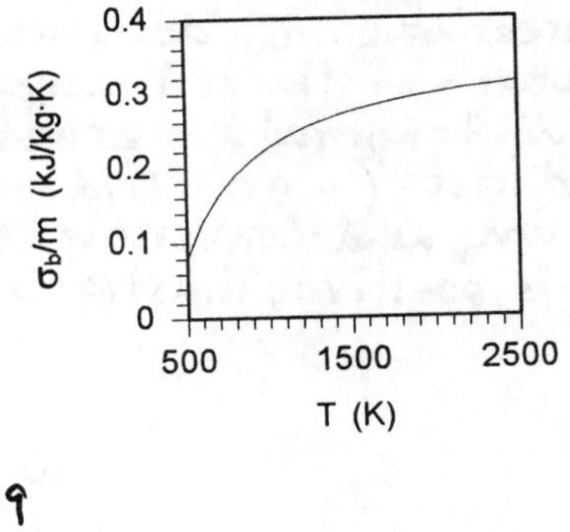

PROBLEM 6.80

KNOWN: Energy is conducted steadily through a copper rod from a hot wall to a cold wall. The rate of heat transfer, temperatures, and geometrical parameters are specified.

FIND: (a) Determine an expression for the rate of entropy production within the rod in terms of specified quantities, and (b) plot $\dot{Q}_H$ and $\dot{\sigma}$ versus L for a given set of numerical values for these quantities.

SCHEMATIC & GIVEN DATA:

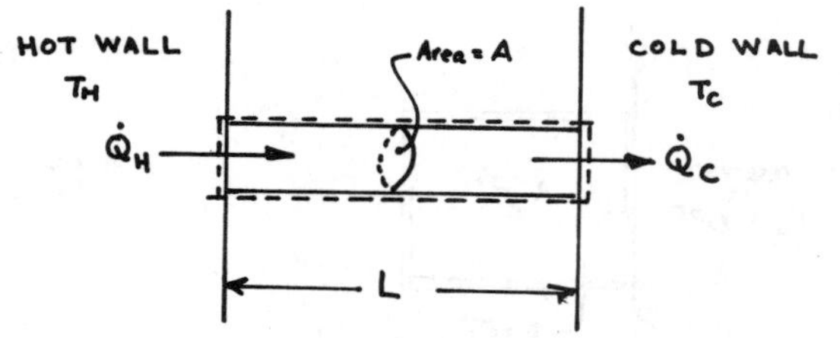

$$\dot{Q}_H = \kappa A (T_H - T_C)/L \qquad (1)$$

$T_H = 600\text{ K}$ (327°C)
$T_C = 350\text{ K}$ (77°C)
$\kappa = 0.4\text{ kW/m}\cdot\text{K}$
$A = 0.1\text{ m}^2$

ASSUMPTIONS: (1) As shown in the accompanying figure, the system is the copper rod. (2) The system is at steady state. (3) The rod is insulated on its lateral surface. (4) An expression for evaluating $\dot{Q}_H$ is provided.

ANALYSIS: (a) At steady state an entropy rate balance reduces to give

$$\cancelto{0}{\frac{dS}{dt}} = \frac{\dot{Q}_H}{T_H} - \frac{\dot{Q}_C}{T_C} + \dot{\sigma} \Rightarrow \dot{\sigma} = \frac{\dot{Q}_C}{T_C} - \frac{\dot{Q}_H}{T_H}$$

Noting that the energy transfers are positive in the directions of the arrows, an energy rate balance gives $\dot{Q}_C = \dot{Q}_H$. Collecting results

$$\dot{\sigma} = \dot{Q}_H \left[\frac{1}{T_C} - \frac{1}{T_H}\right] = \frac{\kappa A (T_H - T_C)^2}{L\, T_H T_C} \qquad (2)$$

(b) If $T_H = 600\text{K}$ (327°C), $T_C = 350\text{K}$ (77°C), $\kappa = 0.4\text{ kW/m}\cdot\text{K}$, $A = 0.1\text{ m}^2$, the variations of $\dot{Q}$ and $\dot{\sigma}$ with L are

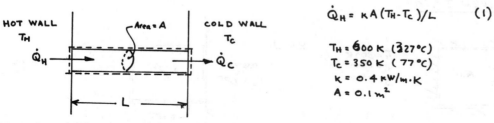

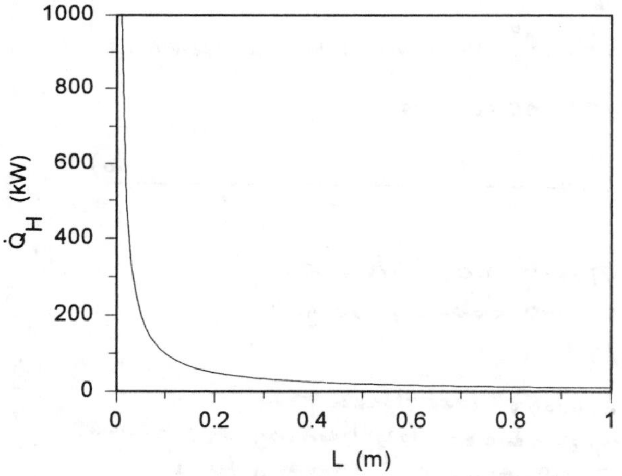

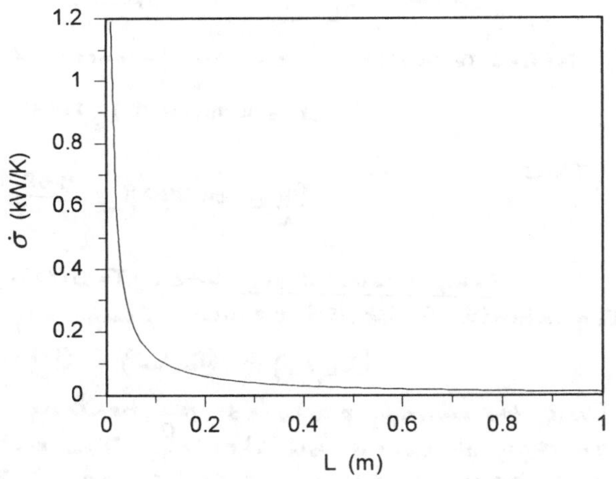

The plots show that both heat transfer and entropy production rate decrease rapidly with increasing L. $\dot{Q}_H$ decreases due to greater resistance to heat transfer as the length of the rod increases. The variation of $\dot{\sigma}$ is associated with the changing temperature gradient, dT/dx, in the rod. Higher temperature gradient ($L \to 0$; $dT/dx \to \infty$) corresponds to high rates of entropy production, and conversely. As a final point, Eq. (2) above indicates that $\dot{\sigma}$ is positive, illustrating the irreversible nature of heat tranfer.

PROBLEM 6.81

KNOWN: A system undergoes a cycle while receiving energy by heat transfer from a tank of water.

FIND: Determine the minimum theoretical volume of the water for the cycle to produce net work of 1.5×10^5 Btu.

SCHEMATIC & GIVEN DATA:

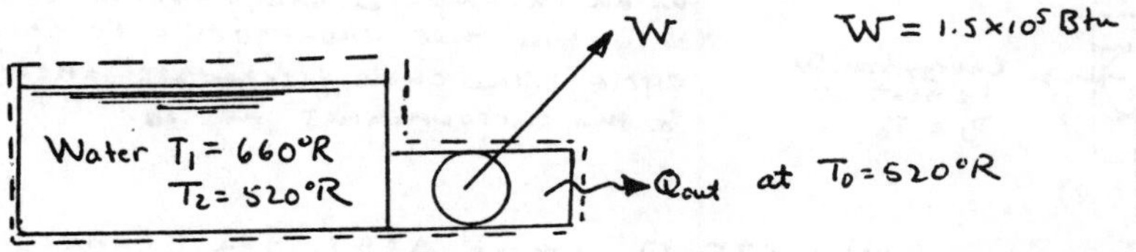

ASSUMPTIONS: (1) The system is composed of the tank of water plus a subsystem that undergoes a thermodynamic cycle. (2) The only energy transfers between system and cycle are shown on the figure. (3) Kinetic and potential energy effects are absent. (4) The liquid water is modeled as incompressible with constant specific heat c.

ANALYSIS: As the water temperature is reduced from T_1 to T_2 the system undergoes a multiplicity of thermodynamic cycles. The energy balance reads

$$\Delta U = -Q_{out} - W$$

where ΔU is the internal energy change for the water only because the subsystem undergoes cycles. Then with assumption (4)

$$mc[T_2 - T_1] = -Q_{out} - W \Rightarrow W = mc[T_1 - T_2] - Q_{out} \qquad (1)$$

With Eq. 6.24, an entropy balance yields

$$\Delta S = \frac{-Q_{out}}{T_0} + \sigma \Rightarrow mc\ln\frac{T_2}{T_1} = -\frac{Q_{out}}{T_0} + \sigma$$

$$\Rightarrow -Q_{out} = T_0 mc\ln\frac{T_2}{T_1} - T_0\sigma \qquad (2)$$

Eliminating $(-Q_{out})$ between Eqs. (1), (2) gives

$$W = mc[T_1 - T_2] + T_0 mc\ln\frac{T_2}{T_1} - T_0\sigma \qquad (3)$$

Solving Eq. (3) for m, and noting that $T_2 = T_0$, gives

$$m = \frac{W + T_0\sigma}{c\left[T_1 - T_0 + T_0\ln\frac{T_0}{T_1}\right]}$$

Since $\sigma \geq 0$, the minimum theoretical mass of water is

$$m_{MIN} = \frac{W}{c\left[T_1 - T_0 + T_0\ln\frac{T_0}{T_1}\right]} = \frac{(1.5\times10^5\ \text{Btu})}{\left(1\ \frac{\text{Btu}}{16.012}\right)\left(660 - 520 + 520\ln\frac{520}{660}\right)°\text{R}}$$

$$= 9359\ \text{lb}$$

where c is from Table A-19E. Then, with $v = v_f$ at 130°F from Table A-2E, the minimum volume of water is

$$V_{MIN} = m_{min}v = (9359\ \text{lb})\left(0.0163\ \frac{\text{ft}^3}{\text{lb}}\right)\left(\frac{1\ \text{gal}}{0.13368\ \text{ft}^3}\right)$$

$$= 1141.2\ \text{gal}$$

PROBLEM 6.82

KNOWN: The temperature of an incompressible substance of mass m and specific heat c is reduced from T_0 to T ($< T_0$) by a refrigeration cycle.

FIND: Plot (W_{min}/mcT_0) versus T/T_0, where W_{min} is the minimum theoretical work input required.

SCHEMATIC & GIVEN DATA:

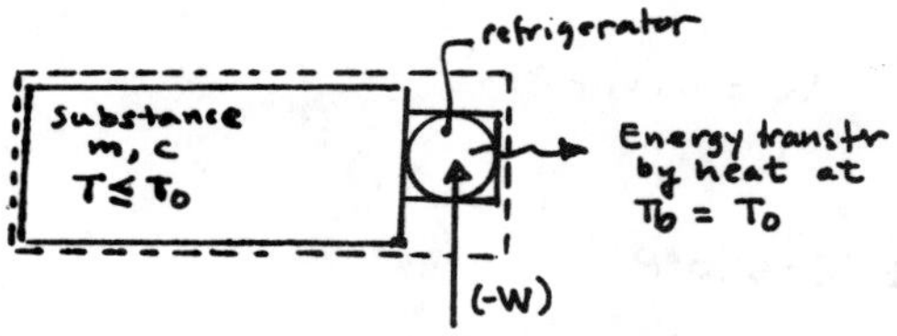

ASSUMPTIONS: 1. The system shown in the accompanying figure is composed of an incompressible substance and a system that undergoes a refrigeration cycle. 2. The cycle discharges energy to the surroundings at T_0.

ANALYSIS: An energy balance gives $(-W) = \Delta U - Q$, where ΔU is the change in internal energy of the substance because $(\Delta U)_{cycle} = 0$. With Eq. 3.20a

$$(-W) = mc[T - T_0] - Q \quad (1)$$

An entropy balance reads $\Delta S = (Q/T_0) + \sigma$, where ΔS is the change in entropy of the substance because $(\Delta S)_{cycle} = 0$. With Eq. 6.24

$$mc \ln\left(\frac{T}{T_0}\right) = \frac{Q}{T_0} + \sigma \Rightarrow Q = T_0 mc \ln\left(\frac{T}{T_0}\right) - T_0 \sigma \quad (2)$$

Collecting Eqs. (1), (2)

$$(-W) = mc[T - T_0] - T_0 mc \ln\left(\frac{T}{T_0}\right) + T_0 \sigma$$

Since $\sigma \geq 0$, the minimum work input corresponds to $\sigma = 0$, giving

$$W_{min} = mc\left[T - T_0 - T_0 \ln\frac{T}{T_0}\right] \Rightarrow \left(\frac{W_{min}}{mcT_0}\right) = \frac{T}{T_0} - 1 - \ln\frac{T}{T_0} \quad (3)$$

PLOT:

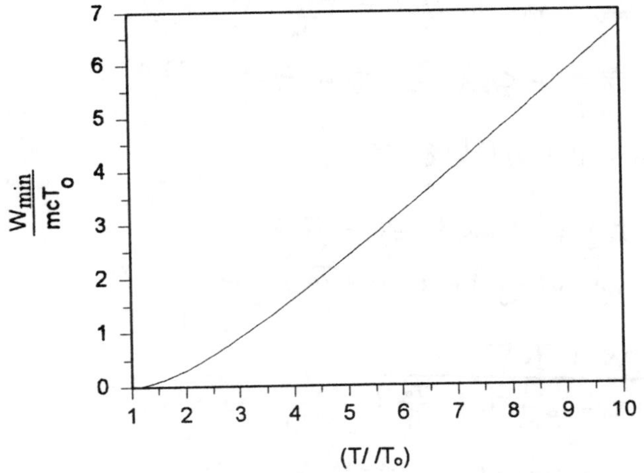

PROBLEM 6.83

KNOWN: A can of soft drink is cooled by a system undergoing a refrigeration cycle that receives energy by heat transfer from the soft drink and discharges energy by heat transfer to the surroundings.

FIND: Determine the minimum theoretical work required.

SCHEMATIC & GIVEN DATA:

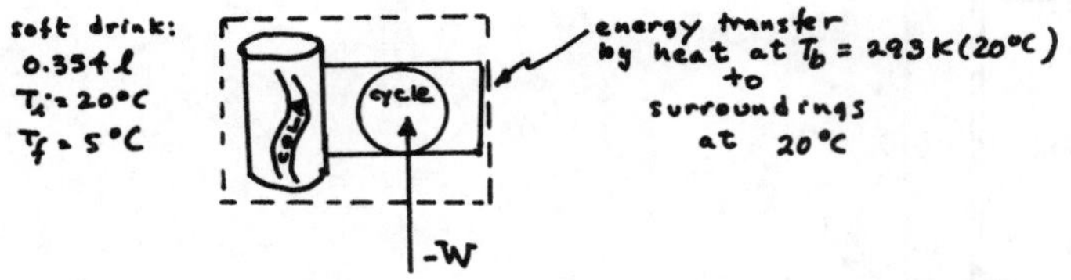

ASSUMPTIONS: (1) The system shown in the accompanying figure is composed of two subsystems: The can of soft drink and a system that undergoes a refrigeration cycle. (2) The cycle discharges energy by heat transfer at T_b = 293 K. (3) The soft drink can be modeled as an incompressible liquid with the properties of water. The aluminum can itself can be ignored.

ANALYSIS: An energy balance gives $(-W) = -Q + \Delta U$ where ΔU is the change in internal energy of the liquid because $\Delta U]_{cycle} = 0$ and the can is ignored. Thus, with Eq. 3.20a the work input is

$$(-W) = -Q + mc(T_f - T_i) \tag{1}$$

An entropy balance takes the form

$$\Delta S = \frac{Q}{T_b} + \sigma$$

where ΔS is the entropy change of the liquid because $\Delta S]_{cycle} = 0$ and the can is ignored. Thus, on solving for Q and using Eq. 6.24

$$Q = T_b \Delta S - T_b \sigma = T_b\left(mc \ln \frac{T_f}{T_i}\right) - T_b \sigma \tag{2}$$

Substituting Eq. (2) into Eq. (1)

$$(-W) = mc\left[T_b \ln \frac{T_i}{T_f} + T_f - T_i\right] + T_b \sigma$$

Since $\sigma \geq 0$, the work input $(-W)$ is a minimum when $\sigma = 0$, giving

$$(-W)_{min} = mc\left[T_b \ln \frac{T_i}{T_f} + T_f - T_i\right]$$

With $c \approx 4.2$ kJ/kg·K from Table A-19 and $v \approx 1$ cm³/g from Table A-2

$$(-W)_{min} = \left[(0.354\,l)\left(\frac{10^{-3}\,m^3}{l}\right)\left(1\frac{g}{cm^3}\right)\left(\frac{10^6\,cm^3}{m^3}\right)\left(\frac{kg}{10^3\,g}\right)\right]\left(4.2\frac{kJ}{kg\cdot K}\right)\left[293 \ln\frac{293}{278} + 278 - 293\right]K$$

$$= 0.591 \text{ kJ}$$

PROBLEM 6.84

KNOWN: A turbine is located between two tanks, initially one tank is filled with steam and the other is evacuated. Steam is allowed to flow through the turbine until equilibrium is established.

FIND: Determine the maximum theoretical work that can be developed.

SCHEMATIC & GIVEN DATA:

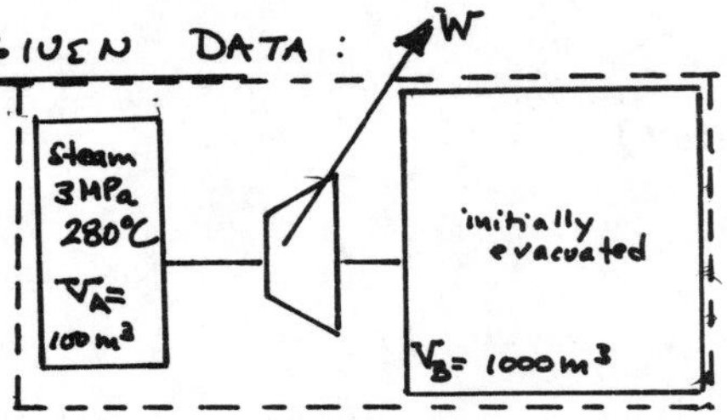

ASSUMPTIONS: (1) The system is shown by the dashed line above. (2) Heat transfer can be ignored. (3) The initial and final states are equilibrium states. At the final state the mass of steam contained within the turbine and interconnecting piping can be ignored.

ANALYSIS: An energy balance reduces to $\Delta U = \cancel{Q} - W$, or $W = m(u_1 - u_2)$, where $m = \frac{V_A}{v_1}$. Thus, with steam table data, $u_1 = 2709.9$ kJ/kg, $v_1 = 0.0771$ m³/kg, $s_1 = 6.4462$ kJ/kg·K,

$$W = \left(\frac{100\ m^3}{0.0771\ m^3/kg}\right)(2709.9 - u_2)\frac{kJ}{kg} = 1297(2709.9 - u_2)\ kJ \qquad (1)$$

The final specific volume is

$$v_2 = \frac{V_A + V_B}{m} = v_1\left[\frac{V_A + V_B}{V_A}\right] = (0.0771)\left(\frac{1100}{100}\right) = 0.8481\ \frac{m^3}{kg} \qquad (2)$$

An entropy balance reduces to $\Delta S = \int_1^2 \left(\frac{\delta Q}{T}\right)_b^{\,0} + \sigma \rightarrow m(s_2 - s_1) = \sigma$. (3)

By inspection of Eq.(1), the maximum value of W is attained when u_2 assumes the smallest allowed value at the value of v_2 given by Eq.(2). Since
① u and s change in the same direction at fixed v, the smallest allowed value for u_2 corresponds to the smallest allowed value for s_2. Since $\sigma \geq 0$, Eq.(3) indicates that the smallest value of s_2 corresponds to $\sigma = 0$: $s_2 = s_1$. The final state of the steam is then fixed by: $v_2 = .8481$ m³/kg, $s_2 = 6.4462$ kJ/kg·K.

As this pair of independent properties is cumbersome to use with tabular steam table data, IT is employed to obtain $u_2 = 2270$ kJ/kg. Returning to Eq.(1)

$$W_{MAX} = 1297(2709.9 - 2270) = 5.71 \times 10^5\ kJ$$

The final state is a two-phase liquid-vapor mixture with $T_2 = 117°C$, $P_2 = 1.815$ bar, $x_2 = 0.8741$.

When the entire problem is solved using IT, we get

IT Solution:

```
p1 = 30 bar
T1 = 280 °C
VA = 100 m³
VB = 1000 m³

v1 = v_PT("Water/Steam", p1, T1)
u1 = u_PT("Water/Steam", p1, T1)
s1 = s_PT("Water/Steam", p1, T1)
m = VA / v1

W = m * (u1 - u2)
v2 = (VA + VB) / m
s2 = s1

v2 = vsat_Px("Water/Steam", p2, x2)
s2 = ssat_Px("Water/Steam", p2, x2)
u2 = usat_Px("Water/Steam", p2, x2)
T2 = Tsat_P("Water/Steam", p2)
```

Note: These are the IT results when the entire problem is solved using IT

T2	117.2
W	5.707E5
s2	6.445
u2	2269
v2	0.8483

---

1. When v is fixed, the first Tds equation — Eq. 6.12a — reads $Tds = du$. Thus, changes in u and s occur in the same direction.

PROBLEM 6.85

KNOWN: A gas flows through a one inlet, one exit control volume operating at steady state. Heat transfer $\dot{Q}_{cv}$ takes place only at temperature T_b.

FIND: For each of several cases, determine if the change in specific entropy from inlet to exit is positive, negative, or indeterminate

SCHEMATIC & GIVEN DATA:

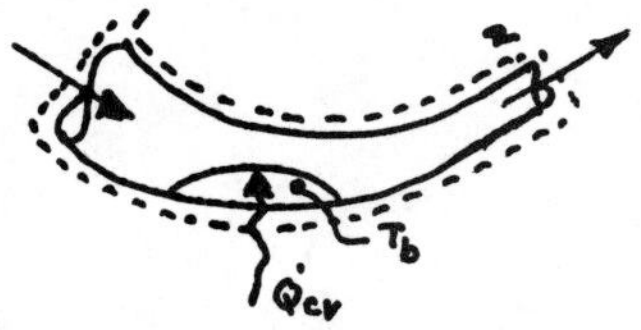

ASSUMPTIONS: (1) The control volume shown in the accompanying figure operates at steady state. (2) Heat transfer occurs only where the temperature is T_b.

ANALYSIS: At steady-state Eq 6.42 reduces to Eq. 6.44. Then, since heat transfer takes place only at T_b and there is a single inlet and single exit, the entropy rate balance becomes

$$0 = \frac{\dot{Q}_{cv}}{T_b} + \dot{m}(s_1 - s_2) + \dot{\sigma}_{cv}$$

Solving for the change in specific entropy from inlet to exit

$$s_2 - s_1 = \frac{\dot{Q}_{cv}}{\dot{m}T_b} + \frac{\dot{\sigma}_{cv}}{\dot{m}}$$

(a) $\dot{\sigma}_{cv} = 0$, $\dot{Q}_{cv} = 0$

$$s_2 - s_1 = \cancel{\frac{\dot{Q}_{cv}}{\dot{m}T_b}}^{\,0} + \cancel{\frac{\dot{\sigma}_{cv}}{\dot{m}}}^{\,0} = 0$$

<u>$s_2 = s_1$</u>

(b) $\dot{\sigma}_{cv} = 0$, $\dot{Q}_{cv} < 0$:

$$s_2 - s_1 = \underbrace{\frac{\dot{Q}_{cv}}{\dot{m}T_b}}_{\text{negative}} + \cancel{\frac{\dot{\sigma}_{cv}}{\dot{m}}}^{\,0} < 0$$

<u>s decreases</u>

(c) $\dot{\sigma}_{cv} = 0$, $\dot{Q}_{cv} > 0$:

$$s_2 - s_1 = \underbrace{\frac{\dot{Q}_{cv}}{\dot{m}T_b}}_{\text{positive}} + \cancel{\frac{\dot{\sigma}_{cv}}{\dot{m}}}^{\,0} > 0$$

<u>s increases</u>

(d) $\dot{\sigma}_{cv} > 0$, $\dot{Q}_{cv} < 0$

$$s_2 - s_1 = \underbrace{\frac{\dot{Q}_{cv}}{\dot{m}T_b}}_{\text{negative}} + \underbrace{\frac{\dot{\sigma}_{cv}}{\dot{m}}}_{\text{positive}}$$

<u>indeterminate without additional information about $\dot{Q}_{cv}/\dot{m}$, T_b, $\dot{\sigma}_{cv}/\dot{m}$</u>

(e) $\dot{\sigma}_{cv} > 0$, $\dot{Q}_{cv} \geq 0$

$$s_2 - s_1 = \underbrace{\frac{\dot{Q}_{cv}}{\dot{m}T_b}}_{\text{positive or zero}} + \underbrace{\frac{\dot{\sigma}_{cv}}{\dot{m}}}_{\text{positive}} > 0$$

<u>s increases</u>

PROBLEM 6.86

KNOWN: An insulated turbine at steady state has steam entering at 3 MPa, 500°C, 70 m/s and exiting at 0.3 MPa, 140 m/s. The work developed is claimed to be (a) 667 kJ/kg, (b) 619 kJ/kg.

FIND: Determine if the claim can be correct.

SCHEMATIC & GIVEN DATA:

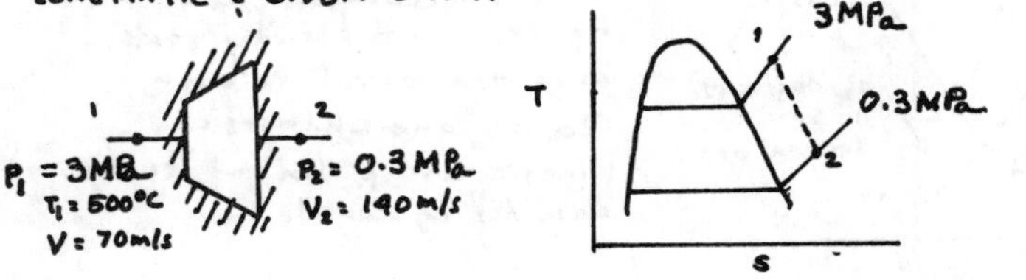

ASSUMPTIONS: (1) The turbine operates at steady-state and is well-insulated. (2) Potential energy effects can be ignored.

ANALYSIS: At steady-state the entropy rate balance reduces with the mass rate balance: $\dot{m}_1 = \dot{m}_2 = \dot{m}$ to give

$$0 = \cancel{\sum \frac{\dot{Q}_j}{T_j}}^{\,0} + \dot{m}(s_1 - s_2) + \dot{\sigma}_{cv}$$

Thus,

$$s_2 - s_1 = \frac{\dot{\sigma}_{cv}}{\dot{m}} \geq 0$$

Accordingly, the exiting specific entropy must be greater than, or equal to, the entering specific entropy.

From Table A-4 at 3MPa, 500°C, $s_1 = 7.2338$ kJ/kg·K. The exit specific entropy that corresponds to the claimed work value can be determined using an energy rate balance to fix state 2. Thus, at steady-state

$$0 = \cancel{\dot{Q}_{cv}}^{\,0} - \dot{W}_{cv} + \dot{m}\left(h_1 - h_2 + \frac{V_1^2 - V_2^2}{2} + \cancel{g(z_1 - z_2)}^{\,0}\right)$$

Solving for h_2, and introducing h_1 from Table A-4: $h_1 = 3456.5$ kJ/kg

$$h_2 = -\frac{\dot{W}_{cv}}{\dot{m}} + h_1 + \frac{V_1^2 - V_2^2}{2} = -\frac{\dot{W}_{cv}}{\dot{m}} + 3456.5 + \left[\frac{(70)^2 - (140)^2}{2}\left(\frac{m}{s}\right)^2\right]\left|\frac{1\ N}{1\ kg\cdot m/s^2}\right|\left|\frac{1\ kJ}{10^3\ N\cdot m}\right|$$

$$= -\frac{\dot{W}_{cv}}{\dot{m}} + 3449.2\ \frac{kJ}{kg} \qquad (1)$$

(a) $\dot{W}_{cv}/\dot{m} = 667$ kJ/kg. Eq. (1) gives $h_2 = 2782.2$ kJ/kg. Interpolating in Table A-4 at 0.3 MPa with h_2 gives $s_2 = 7.1276$ kJ/kg·K. Since this value is less than s_1, the claimed value cannot be correct.

(b) $\dot{W}_{cv}/\dot{m} = 619$ kJ/kg. Eq. (1) gives $h_2 = 2830.2$ kJ/kg. Interpolating in Table A-4 at 0.3 MPa with h_2 gives $s_2 = 7.2335$ kJ/kg·K. Since this value gives $(\dot{\sigma}_{cv}/\dot{m}) \approx 0$, the claimed value for the work per unit mass of steam flowing is unlikely in an actual steam turbine.

PROBLEM 6.87

KNOWN: Steady state operating data are provided for a steam turbine.

FIND: Determine if either or both of two alternative values for the power developed by the turbine can be correct.

SCHEMATIC & GIVEN DATA:

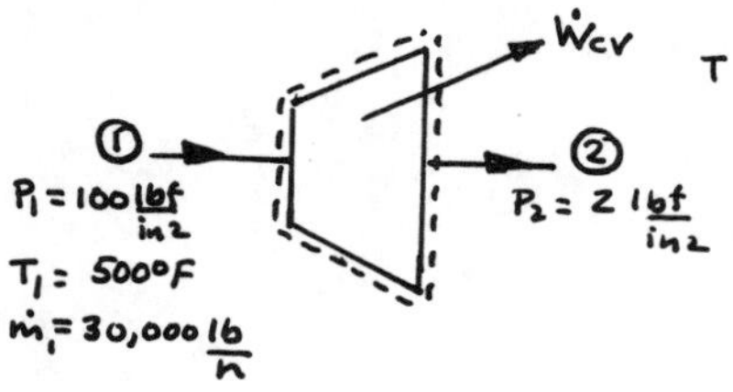

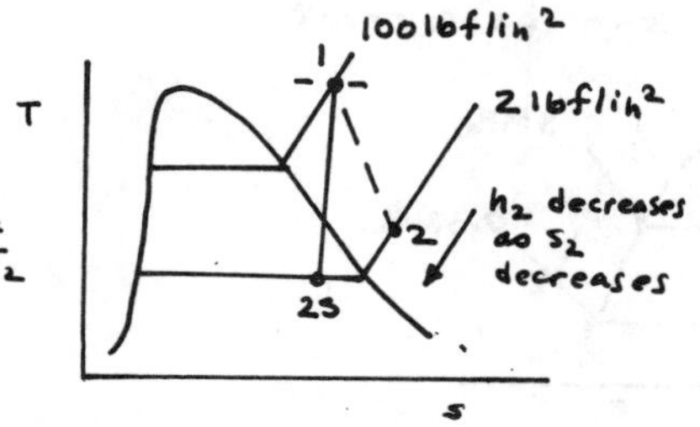

ASSUMPTIONS: 1. The control volume shown in the accompanying figure is at steady state. 2. For the control volume, $\dot{Q}_{cv} = 0$ and changes in kinetic and potential energy can be ignored.

ANALYSIS: The mass rate balance reduces to $\dot{m}_1 = \dot{m}_2 \equiv \dot{m}$. Then, with assumption 2 the energy rate balance reduces to give

$$\dot{W}_{cv} = \dot{m}(h_1 - h_2) \qquad (1)$$

Since $\dot{Q}_{cv} = 0$, the entropy rate balance reduces at steady state to give

$$0 = \sum \frac{\dot{Q}_j}{T_j}^{\,0} + \dot{m}(s_1 - s_2) + \dot{\sigma}_{cv} \Rightarrow s_2 - s_1 = \dot{\sigma}_{cv}/\dot{m} \qquad (2)$$

Method 1. Eq. (1) indicates that $\dot{W}_{cv}$ increases as h_2 decreases. Inspection of Table A-4E shows that h_2 decreases as s_2 decreases. Thus, the maximum theoretical value of $\dot{W}_{cv}$ corresponds to the minimum theoretical value for h_2, which corresponds to the minimum allowed value for s_2. Since $\dot{\sigma}_{cv} \geq 0$, Eq. (2) shows that the minimum value for s_2 corresponds to $\dot{\sigma}_{cv} = 0$ — that is, $s_2 = s_1$.

From Table A-4E, $h_1 = 1279.1$ Btu/lb, $s_1 = 1.7085$ Btu/lb·°R. The state s2 falls in the two-phase region. With values from Table A-3E

$$x_{2s} = \frac{s_{2s} - s_f}{s_g - s_f} = \frac{1.7085 - 0.175}{1.7448} = 0.879 \Rightarrow h_{2s} = 94.02 + (0.879)(1022.1) = 992.4 \text{ Btu/lb}$$

Then, Eq. (1) gives

$$(\dot{W}_{cv})_{MAX} = \dot{m}(h_1 - h_{2s}) = \left(30{,}000 \frac{lb}{h}\right)\left[(1279.1 - 992.4)\frac{Btu}{lb}\right]\left|\frac{1 \text{ hp}}{2545 \text{ Btu/h}}\right| = 3380 \text{ hp}$$

Of the two alternatives, only 3080 hp can be correct.

Method 2. The given power values can be used to determine $\dot{\sigma}_{cv}$. Thus, if $\dot{W}_{cv} = 3800$ hp, Eq. (1) gives on rearrangement

$$h_2 = h_1 - \frac{\dot{W}_{cv}}{\dot{m}} = 1279.1 - \left(\frac{3800 \text{ hp}}{30{,}000 \text{ lb/h}}\right)\left|\frac{2545 \text{ Btu/h}}{1 \text{ hp}}\right| = 956.7 \text{ Btu/lb}$$

Then, with data from Table A-3E

$$x_2 = \frac{h_2 - h_f}{h_{fg}} = \frac{956.7 - 94.02}{1022.1} = 0.844 \Rightarrow s_2 = 0.175 + 0.844(1.7448) = 1.6476 \frac{Btu}{lb \cdot °R}$$

With Eq. (2),

$$\dot{\sigma}_{cv}/\dot{m} = s_2 - s_1 = (1.6476 - 1.7085) = -0.061 \text{ Btu/lb·°R}.$$

But since $\dot{\sigma}_{cv}/\dot{m}$ cannot be negative, the power value 3800 hp cannot be correct.

Using the value 3080 hp, the same approach gives $\dot{\sigma}_{cv}/\dot{m} > 0$, and thus this value is allowed.

PROBLEM 6.88

KNOWN: Air enters and exits an insulated turbine at specified pressures and temperatures.

FIND: Determine the work developed per kg of air flowing, and whether the expansion is internally reversible, irreversible, or impossible.

SCHEMATIC & GIVEN DATA:

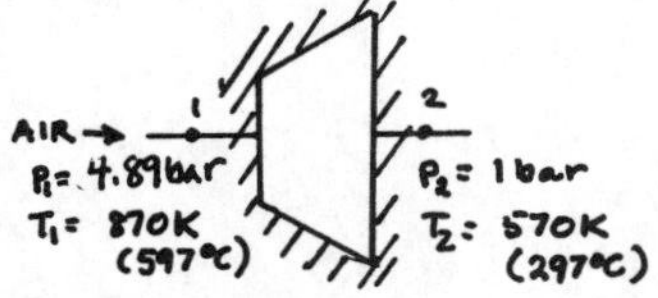

ASSUMPTIONS: (1) The turbine operates at steady state and is well-insulated. (2) Changes in kinetic and potential energy from inlet to exit can be neglected. (3) Air is modeled

① as an ideal gas.

ANALYSIS: An energy rate balance at steady state for a control volume enclosing the turbine reduces with a mass rate balance: $\dot{m}_1 = \dot{m}_2 = \dot{m}$ to give

$$0 = \dot{Q}_{cv}^{\,0} - \dot{W}_{cv} + \dot{m}\left(h_1 - h_2 + \frac{V_1^2 - V_2^2}{2}^{\,0} + g(z_1 - z_2)^{\,0}\right)$$

or

$$\frac{\dot{W}_{cv}}{\dot{m}} = h_1 - h_2$$

Then, with data from Table A-22

$$\frac{\dot{W}_{cv}}{\dot{m}} = 899.4 - 575.59 = 323.8 \text{ kJ/kg}$$ ◄

An entropy rate balance at steady state gives

$$0 = \sum \frac{\dot{Q}_j}{T_j}^{\,0} + \dot{m}(s_1 - s_2) + \dot{\sigma}_{cv}$$

or

$$\frac{\dot{\sigma}_{cv}}{\dot{m}} = s_2 - s_1$$

With Eq 6.25a and data from Table A-22

$$\frac{\dot{\sigma}_{cv}}{\dot{m}} = s^\circ(T_2) - s^\circ(T_1) - R \ln P_2/P_1$$

$$= 2.35531 - 2.81064 - \frac{8.314}{28.97} \ln \frac{1}{4.89} = 1.73 \times 10^{-4} \text{ kJ/kg}\cdot\text{K}$$

Since $\dot{\sigma}_{cv}/\dot{m}$ is positive, the expansion is irreversible. However, the small value suggests that the expansion approaches ideality.

---

1. From Table A-1, the critical pressure of air is $p_c = 37.7$ bar and the critical temperature is $T_c = 133$ K. Thus, at the inlet and exit, respectively

$$p_{R1} = \frac{4.89}{37.7} = 0.13 \qquad p_{R2} = \frac{1}{37.7} = 0.027$$

$$T_{R1} = \frac{870}{133} = 6.54 \qquad T_{R2} = \frac{570}{133} = 4.29$$

Referring to Fig A-1, these states fall into the region where the ideal gas model is appropriate.

PROBLEM 6.89

KNOWN: Steady-state operating data are provided for an R134a compressor.

FIND What can be said about the direction of heat transfer between the compressor and its surroundings?

SCHEMATIC & GIVEN DATA:

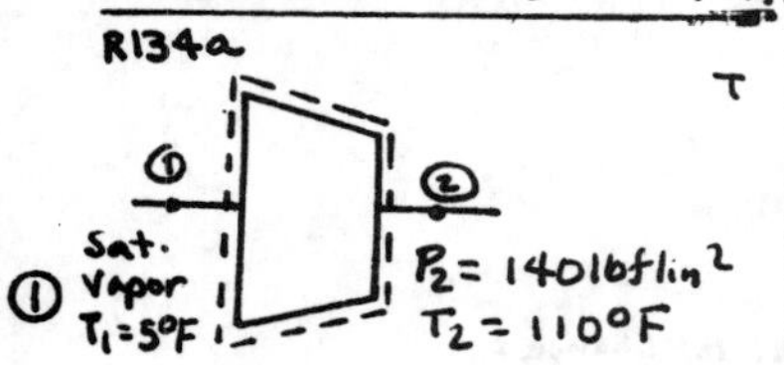

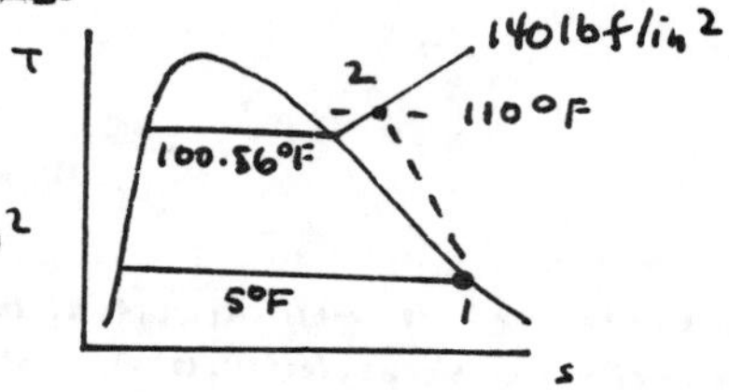

ASSUMPTIONS: 1. The control volume shown in the figure operates at steady state. 2. Kinetic and potential energy effects can be ignored.

ANALYSIS: Reducing mass and energy balances

$$\frac{\dot{Q}}{\dot{m}} = \frac{\dot{W}}{\dot{m}} + (h_2 - h_1)$$

With data from Tables A-10E, 12E

$$\frac{\dot{Q}}{\dot{m}} = \frac{\dot{W}}{\dot{m}} + (117.53 - 102.47) = \underbrace{\frac{\dot{W}}{\dot{m}}}_{\ominus} + 15.06 \ \frac{\text{Btu}}{\text{lb}}$$

Since $\frac{\dot{W}}{\dot{m}}$ must be minus, the direction of $\dot{Q}/\dot{m}$ is determined by the magnitude of $\dot{W}/\dot{m}$, which is unknown. Turning next to an entropy balance

$$0 = \sum \frac{\dot{Q}_j}{T_j} + \dot{m}(s_1 - s_2) + \dot{\sigma}_{cv} \Rightarrow \sum \frac{\dot{Q}_j}{T_j} = \dot{m}(s_2 - s_1) - \dot{\sigma}_{cv}$$

With table data

$$\frac{1}{\dot{m}} \sum \frac{\dot{Q}_j}{T_j} = s_2 - s_1 - (\dot{\sigma}_{cv}/\dot{m})$$

$$= (0.2206 - 0.2219) - (\dot{\sigma}_{cv}/\dot{m})$$

$$= (-0.0013 - \dot{\sigma}_{cv}/\dot{m}) \ \text{Btu/lb·°R} \quad < 0$$

Thus, entropy is carried out of the control volume by heat transfer. The direction of heat transfer is from the control volume to the surroundings.

---

1. In the first printing of the 4th edition of the book, T_2 was given incorrectly as 120°F.

PROBLEM 6.90

KNOWN: Methane gas enters and exits a compressor operating at steady state at specified temperatures and pressures.

FIND: Determine the rate of entropy production.

SCHEMATIC & GIVEN DATA:

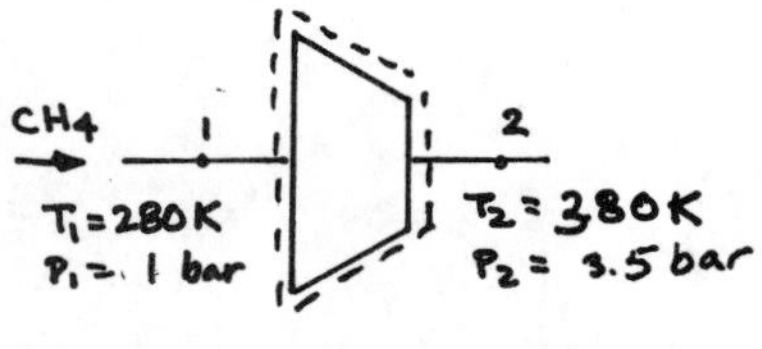

ASSUMPTIONS: (1) The control volume operates at steady state. (2) The methane is modeled as an ideal gas. (3) Heat transfer with the surroundings is negligible.

ANALYSIS: With assumption 3, the entropy rate balance reduces at steady state to

$$0 = \sum \frac{\dot{Q}_j}{T_j}^{0} + \dot{m}(s_1 - s_2) + \dot{\sigma}_{cv} \Rightarrow \dot{\sigma}_{cv} = \dot{m}(s_2 - s_1) \qquad (1)$$

① The change in specific entropy is evaluated using Eq. 6.19 and $\bar{c}_p(T)$ from Table A-21

$$s_2 - s_1 = \int_{T_1}^{T_2} \frac{c_p(T)}{T} dT - R \ln \frac{p_2}{p_1} = R \int_{T_1}^{T_2} \frac{(\alpha + \beta T + \gamma T^2 + \delta T^3 + \epsilon T^4)}{T} dT - R \ln \frac{p_2}{p_1}$$

$$= R \int_{T_1}^{T_2} \left(\frac{\alpha}{T} + \beta + \gamma T + \delta T^2 + \epsilon T^3\right) dT - R \ln \frac{p_2}{p_1}$$

$$= R \left[\alpha \ln T + \beta T + \frac{\gamma T^2}{2} + \frac{\delta T^3}{3} + \frac{\epsilon T^4}{4}\right]_{T_1}^{T_2} - R \ln \frac{p_2}{p_1}$$

$$= \left(\frac{8.314}{16.04} \frac{kJ}{kg \cdot K}\right) \left\{\left[3.826 \ln \frac{380}{280} - \frac{3.979}{10^3}(380 - 280) + \frac{24.558}{2(10^6)}(\overline{380}^2 - \overline{280}^2) - \frac{22.733}{3(10^9)}(\overline{380}^3 - \overline{280}^3) + \frac{6.963}{4(10^{12})}(\overline{380}^4 - \overline{280}^4)\right] - \ln \frac{3.5}{1}\right\}$$

$$= \left(\frac{8.314}{16.04} \frac{kJ}{kg \cdot K}\right) [1.357 - 1.253] = 0.0539 \frac{kJ}{kg \cdot K}$$

With Eq. (1)

$$\dot{\sigma}_{cv}/\dot{m} = 0.0539 \frac{kJ}{kg \cdot K} \longleftarrow$$

---

1. Alternatively, using IT

```
T1 = 280  // K
p1 = 1  // bar
T2 = 380
p2 = 3.5

dels = s2 - s1
s1 = s_TP("CH4", T1, p1)
s2 = s_TP("CH4", T2, p2)

/* Result
dels = 0.05407 kJ/kg·K

*/
```

PROBLEM 6.91

KNOWN: Steady-state operating data are provided for a well-insulated device carrying air at a known mass flow rate.

FIND: Determine the direction of flow and the power input or output.

SCHEMATIC & GIVEN DATA:

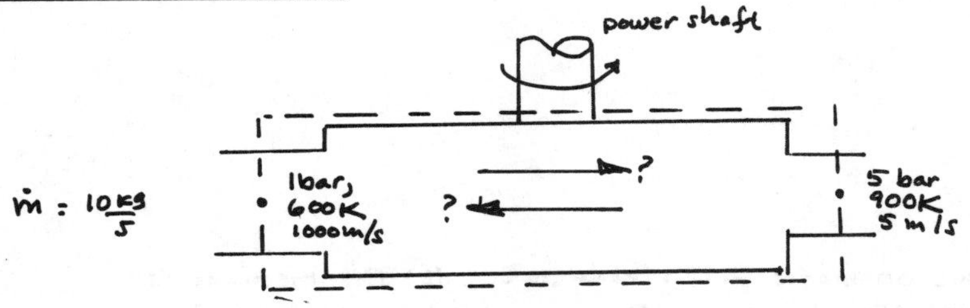

ASSUMPTIONS: (1) The control volume shown in the figure is at steady state. (2) For the control volume, $\dot{Q}_{cv} = 0$ and potential energy effects are negligible. (3) Air is modeled as an ideal gas.

ANALYSIS: Directionality is a second law aspect (Sec. 5.1). Thus, assume a flow direction and apply an entropy balance. Regarding the flow to be from left to right

$$0 = \cancel{\sum \frac{\dot{Q}_j}{T_j}}^{0} + \dot{m}(s_i - s_e) + \dot{\sigma}_{cv}$$

With ideal gas relations and data from Table A-22

$$\frac{\dot{\sigma}_{cv}}{\dot{m}} = s_e - s_i = s^\circ(T_e) - s^\circ(T_i) - \frac{\bar{R}}{M} \ln \frac{p_e}{p_i}$$

(T_e = 900K, T_i = 600K, p_e = 5 bar, p_i = 1 bar)

$$= 2.84856 - 2.40902 - \frac{8.314}{28.97} \ln 5 = -0.0223 \frac{kJ}{kg \cdot K}$$

As $\dot{\sigma}_{cv}$ cannot be negative, the direction of flow must be from right to left. ←

Applying mass and energy rate balances with data from Table A-22

$$\dot{W}_{cv} = \dot{m}\left[h_i - h_e + \frac{V_i^2 - V_e^2}{2}\right]$$

(h_i: 900K, h_e: 600K, V_i: 5 m/s, V_e: 1000 m/s)

$$= \left(10\frac{kg}{s}\right)\left\{(932.93 - 607.02)\frac{kJ}{kg} + \frac{[(5)^2 - (1000)^2]}{2}\left(\frac{m^2}{s^2}\right)\left(\frac{1 N}{1 kg \cdot m/s^2}\right)\left[\frac{1 kJ}{10^3 N \cdot m}\right]\right\}$$

$$= \left(10\frac{kg}{s}\right)\left\{325.91\frac{kJ}{kg} - 499.99\frac{kJ}{kg}\right\}$$

$$= -1740.8\left(\frac{kJ}{s}\right)\left(\frac{1 kW}{1 kJ/s}\right)$$

$$= -1740.8 \text{ kW} \longleftarrow$$

PROBLEM 6.92

KNOWN: Nitrogen (N_2) gas enters an insulated control volume operating at steady state at 21°C and pressure p. Half exits at 82°C, 1 bar and half exits at −40°C, 1 bar. For the control volume, $\dot{W}_{cv} = 0$.

FIND: Determine the minimum allowed value for p.

SCHEMATIC & GIVEN DATA:

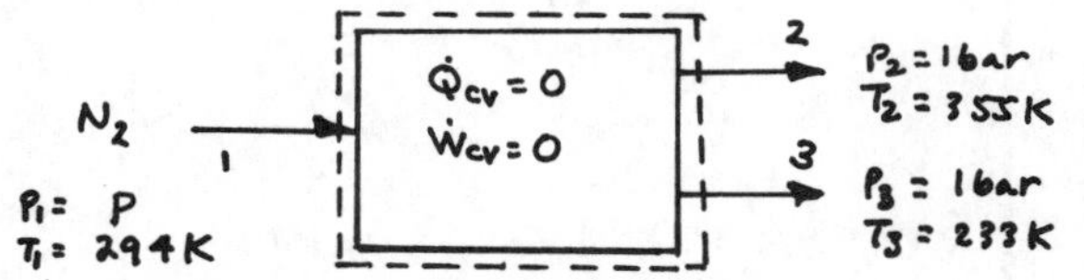

$\dot{m}_2 = \dot{m}_3 = \frac{1}{2}\dot{m}_1$

$c_p = 1.04$ kJ/kg·K

ASSUMPTIONS: (1) The control volume shown in the accompanying figure operates at steady state. (2) The effects of kinetic and potential energy are negligible. (3) Nitrogen is modeled as an ideal gas with constant c_p.

ANALYSIS: The given information must satisfy the conservation of energy principle. Thus, with $\dot{Q}_{cv} = \dot{W}_{cv} = 0$ and assumptions 1 and 2 an energy rate balance reduces to give

$$0 = \cancel{\dot{Q}_{cv}}^{0} - \cancel{\dot{W}_{cv}}^{0} + \dot{m}_1 h_1 - \dot{m}_2 h_2 - \dot{m}_3 h_3$$

or with $\dot{m}_2 = \dot{m}_3 = \dot{m}_1/2$

$$0 = h_1 - \frac{1}{2}h_2 - \frac{1}{2}h_3$$

As c_p is constant, $h = c_p(T - T_{ref})$, where T_{ref} is an arbitrary reference temperature. With this for h_1, h_2, and h_3 the energy rate balance gives

$$0 = T_1 - \frac{1}{2}T_2 - \frac{1}{2}T_3$$

Introducing the given temperatures

$$0 = 294 - \frac{355}{2} - \frac{233}{2} \Rightarrow 0 = 294 - 177.5 - 116.5$$

Thus, the conservation of energy principle is satisfied.

Turning next to the second law, an entropy rate balance reduces to give

$$0 = \cancel{\sum \frac{\dot{Q}_j}{T_j}}^{0} + \dot{m}_1 s_1 - \dot{m}_2 s_2 - \dot{m}_3 s_3 + \dot{\sigma}_{cv}$$

With $\dot{m}_2 = \dot{m}_3 = \dot{m}_1/2$

$$0 = \frac{1}{2}(s_1 - s_2) + \frac{1}{2}(s_1 - s_3) + \frac{\dot{\sigma}_{cv}}{\dot{m}}$$

Introducing Eq 6.23 with $p_2 = p_3$ and $p_1 = p$

$$0 = \frac{1}{2}\left[c_p \ln\frac{T_1}{T_2} - R\ln\frac{p}{p_2}\right] + \frac{1}{2}\left[c_p \ln\frac{T_1}{T_3} - R\ln\frac{p}{p_2}\right] + \frac{\dot{\sigma}_{cv}}{\dot{m}}$$

$$\Rightarrow \ln\frac{p}{p_2} = \frac{c_p}{2R}\ln\left(\frac{T_1^2}{T_2 T_3}\right) + \frac{1}{R}\left(\frac{\dot{\sigma}_{cv}}{\dot{m}}\right) = \left[\frac{1.04 \text{ kJ/kg·K}}{2(8.314/28.01)\text{ kJ/kg·K}}\right]\ln\left(\frac{(294)^2}{355 \times 233}\right) + \frac{1}{R}\left(\frac{\dot{\sigma}_{cv}}{\dot{m}}\right)$$

$$= 0.0771 + \frac{1}{R}\left(\frac{\dot{\sigma}_{cv}}{\dot{m}}\right)$$

Since $p_2 = 1$ bar

$$\Rightarrow \quad p = (1.08 \text{ bar})\exp\left(\frac{\dot{\sigma}_{cv}/\dot{m}}{R}\right)$$

where $\dot{\sigma}_{cv}/\dot{m} \geq 0$. Thus, in the limit as $\dot{\sigma}_{cv}/\dot{m} \to 0$, $\to$ 1.08 bar ← min. pressure

PROBLEM 6.93

KNOWN: Air enters an insulated control volume operating at steady state at 20°C, 3 bar. A stream exits at 60°C, 2.7 bar and another stream exits at 0°C, 2.7 bar. For the control volume $\dot{W}_{cv} = 0$.

FIND: Decide whether the device can operate as described.

SCHEMATIC & GIVEN DATA:

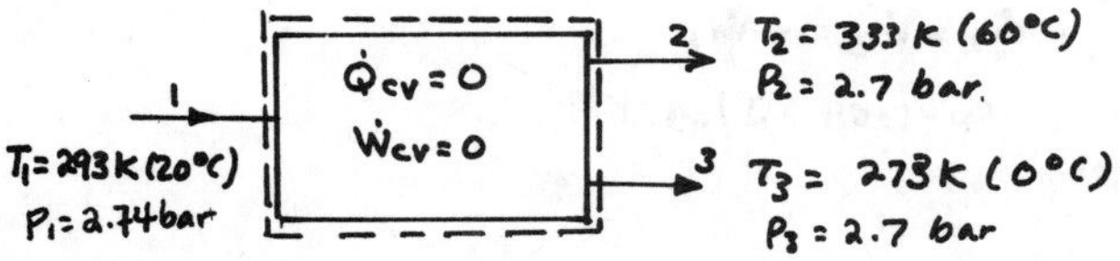

ASSUMPTIONS: (1) The control volume shown in the accompanying figure operates at steady state. (2) The effects of kinetic and potential energy are negligible. (3) Air is modeled as an ideal gas.

ANALYSIS: The conservation of mass and energy principles and the second law must be satisfied for the device to operate as described.

At steady state, the mass rate balance reduces to give

$$\dot{m}_1 = \dot{m}_2 + \dot{m}_3 \quad (1)$$

At steady state, the energy rate balance reduces with $\dot{Q}_{cv} = \dot{W}_{cv} = 0$ and assumption 2 to give

$$0 = \cancelto{0}{\dot{Q}_{cv}} - \cancelto{0}{\dot{W}_{cv}} + \dot{m}_1 h_1 - \dot{m}_2 h_2 - \dot{m}_3 h_3$$

or

$$0 = \dot{m}_1 h_1 - \dot{m}_2 h_2 - \dot{m}_3 h_3 \quad (2)$$

Letting $y = \dot{m}_2/\dot{m}_1$, Eqs. (1) and (2) combine to yield

$$y = \frac{h_1 - h_3}{h_2 - h_3}$$

with data from Table A-22

$$y = \frac{293.2 - 273.1}{333.3 - 273.1} = 0.334$$

Accordingly, the conservation of mass and energy principles require that the mass flow rates be in definite proportions: $\dot{m}_2/\dot{m}_1 = 0.334$, $\dot{m}_3/\dot{m}_1 = 0.666$.

An entropy rate balance at steady state reduces to give

$$0 = \cancelto{0}{\sum_j \frac{\dot{Q}_j}{T_j}} + \dot{m}_1 s_1 - \dot{m}_2 s_2 - \dot{m}_3 s_3 + \dot{\sigma}_{cv}$$

Thus

$$\frac{\dot{\sigma}_{cv}}{\dot{m}_1} = \frac{\dot{m}_3}{\dot{m}_1} s_3 + \frac{\dot{m}_2}{\dot{m}_1} s_2 - s_1$$

$$= (1-y) s_3 + y\, s_2 - s_1$$

$$= (s_3 - s_1) + y(s_2 - s_3)$$

Then, with data from Table A-22, and using Eq. 6.21a

$$\frac{\dot{\sigma}_{cv}}{\dot{m}} = \left(1.6073 - 1.6783 - \frac{8.314}{28.97} \ln \frac{2.7}{2.74}\right) + 0.334\left(1.8069 - 1.6073 - \cancelto{0}{\frac{8.314}{28.97} \ln \frac{2.7}{2.7}}\right)$$

$$= (-0.0668 + 0.0667) < 0$$

The device cannot operate as described.

PROBLEM 6.94

KNOWN: Steam enters and exits a compressor operating at steady state at specified temperatures and pressures. The ambient temperature is also specified. The required power input is claimed to be 4hp.

FIND: Determine whether this claim can be correct.

SCHEMATIC & GIVEN DATA:

$T_0 = 530°R\ (70°F)$

$\dot{W}_{cv} = -4\text{hp}$ (claim)

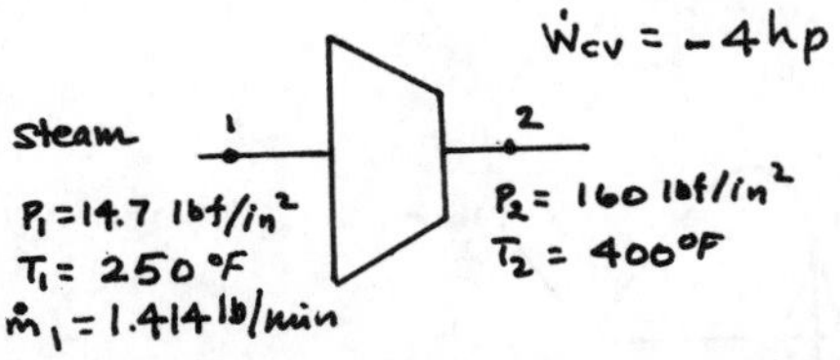

ASSUMPTIONS: (1) Two control volumes at steady state are considered. One encloses the compressor. The other encloses the compressor plus a portion of the nearby surroundings so heat transfer occurs at the ambient temperature T_0. (2) Changes in kinetic and potential energy from inlet to exit can be ignored.

ANALYSIS: An entropy rate balance is used to determine if the claimed value for the power input can be correct. But first consider a control volume enclosing the compressor, and apply an energy rate balance to obtain the heat transfer. Thus with $\dot{m}_1 = \dot{m}_2 = \dot{m}$

$$0 = \dot{Q}_{cv} - \dot{W}_{cv} + \dot{m}\left(h_1 - h_2 + \frac{V_1^2 - V_2^2}{2} + g(z_1 - z_2)\right)$$

with assumption 2 and data from Table A-4E

$$\frac{\dot{Q}_{cv}}{\dot{m}} = \frac{\dot{W}_{cv}}{\dot{m}} + h_2 - h_1 \Rightarrow \frac{\dot{Q}_{cv}}{\dot{m}} = \frac{(-4\text{hp})\left|\frac{2545\ Btu/h}{1\ hp}\right|\left|\frac{1h}{60min}\right|}{(1.414\ lb/min)} + 1217.8 - 1168.8 = -71\ \frac{Btu}{lb}$$

As the variation of temperature over the surface of this control volume is not specified, the entropy transfer accompanying heat transfer cannot be evaluated without further information. Accordingly, consider an entropy rate balance for an enlarged control volume enclosing the compressor plus a portion of the nearby surroundings so heat transfer occurs at the ambient temperature T_0. That is

$$0 = \frac{\dot{Q}_{cv}}{T_0} + \dot{m}(s_1 - s_2) + \dot{\sigma}_{cv}$$

where $\dot{\sigma}_{cv}$ represents the rate of entropy production within the enlarged control volume. With data from Table A-4E

$$\frac{\dot{\sigma}_{cv}}{\dot{m}} = -\frac{\dot{Q}_{cv}/\dot{m}}{T_0} + s_2 - s_1 = -\frac{(-71\ Btu/lb)}{530°R} + 1.5911\ \frac{Btu}{lb \cdot °R} - 1.7832\ \frac{Btu}{lb \cdot °R}$$

① $$= -0.0581\ Btu/lb \cdot °R$$

As $\dot{\sigma}_{cv} \geq 0$, the claimed value for the work input cannot be correct. ←

---

1. Note that the calculated value for $\dot{\sigma}_{cv}/\dot{m}$ is not for the compressor alone, but for an enlarged control volume of compressor plus nearby surroundings. For further discussion, see Sec. 6.5.3.

PROBLEM 6.95

KNOWN: Operating data are provided for ammonia throttled across a valve.

FIND: Determine the rate of entropy production. If the valve were relaced by a powery-recovery turbine, determine the maximum theoretical power that could be developed. Would you recommend the use of such a turbine?

SCHEMATIC & GIVEN DATA:

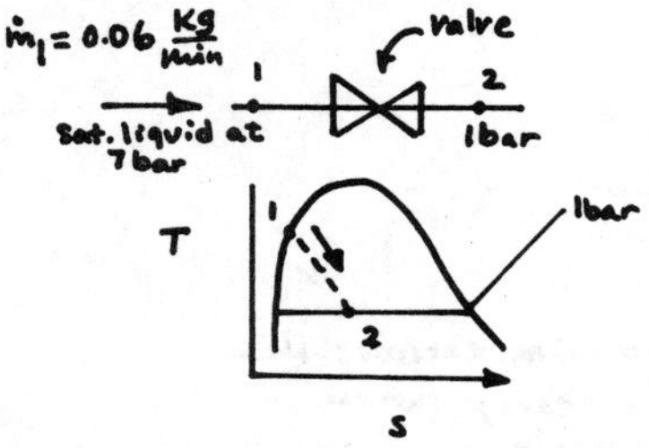

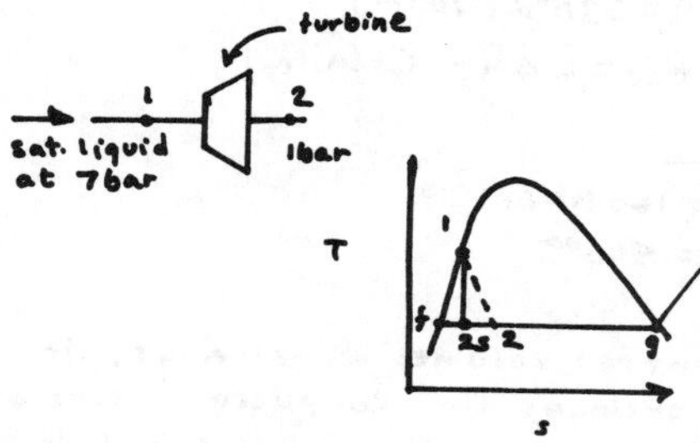

ASSUMPTIONS: (1) Control volumes enclosing the valve and enclosing the turbine operate at steady state. (2) $\dot{Q}_{cv} = 0$, and kinetic and potential energy changes from inlet to exit can be neglected.

ANALYSIS: Considering the valve first, the expansion is a throttling process for which $h_2 = h_1$ (Sec. 4.3). Thus, state 2 is fixed by $p_2 = 1$ bar and $h_2 = h_1$. The rate of entropy production can be evaluated from an entropy rate balance which a steady state reduces with $\dot{m}_1 = \dot{m}_2 = \dot{m}$ to

$$0 = \sum_j \frac{\dot{Q}_j}{T_j}^{\,0} + \dot{m}(s_1 - s_2) + \dot{\sigma}_{cv} \Rightarrow \dot{\sigma}_{cv} = \dot{m}(s_2 - s_1) \qquad (1)$$

From Table A-14 at 7bar, $s_1 = 0.9394$ kJ/kg·K. At state 2, $h_2 = h_1 = 244.69$ kJ/kg. Then

$$x_2 = \frac{h_2 - h_f}{h_g - h_f} = \frac{244.69 - 28.18}{1370.23} = 0.158$$

Then

$$s_2 = s_f + x_2(s_g - s_f) = 0.1191 + 0.158(5.8391 - .1191) = 1.0229 \text{ kJ/kg·K}.$$

Thus, with Eq. (1)

$$\dot{\sigma}_{cv} = \left(0.06 \frac{kg}{min}\right)\left|\frac{1\,min}{60\,s}\right|(1.0229 - 0.9394)\frac{kJ}{kg\cdot K}\left|\frac{1\,kW}{1\,kJ/s}\right| = 8.35\times10^{-5}\ \frac{kW}{K} \longleftarrow$$

If the valve is replaced by a turbine operating at steady state, an energy rate balance reduces with assumption 2 to give

$$\dot{W}_{cv} = \dot{m}(h_1 - h_2)$$

An entropy rate balance gives Eq. (1) above. As $h_1 = 244.69$ kJ/kg, the work developed is

$$\dot{W}_{cv} = \left(\frac{0.06}{60}\frac{kg}{s}\right)(244.69 - h_2)\frac{kJ}{kg} \qquad (2)$$

Accordingly, the power developed <u>increases</u> as h_2 <u>decreases</u>. Since $s_2 \geq s_1$, the smallest allowed value is h_{2s}, corresponding to the state where $s_{2s} = s_1$, as shown on the sketch above. The quality at state 2s is

$$x_{2s} = \frac{s_{2s} - s_f}{s_g - s_f} = \frac{0.9394 - 0.1191}{5.8391 - 0.1191} = 0.1434 \Rightarrow h_{2s} = h_f + x_{2s}(h_{fg}) = 28.18 + (0.1434)(1370.23) = 224.67\ \frac{kJ}{kg}$$

Then Eq. (2) gives

$$(\dot{W}_{cv})_{max} = \left(\frac{0.06}{60}\frac{kg}{s}\right)(244.69 - 224.67)\frac{kJ}{kg}\left|\frac{1\,kW}{1\,kJ/s}\right| = 0.02\,kW \longleftarrow$$

This is a small potential to develop power. Moreover, the affect of irreversibilities in an actual power-recovery turbine would be to reduce the actual power developed below the calculated value. Generally, for this type of application a power-recovery turbine would not be recommended.

PROBLEM 6.96

KNOWN: Steady-state operating data are provided for an air turbine.

FIND: Plot the work developed and entropy production, each per unit mass of air flowing, versus the turbine exit temperature.

SCHEMATIC & GIVEN DATA:

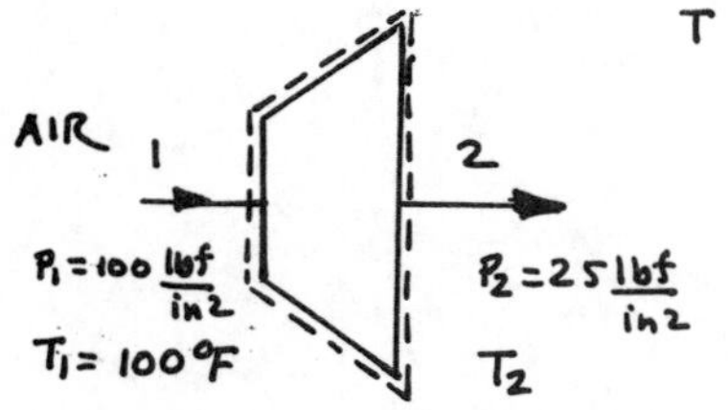

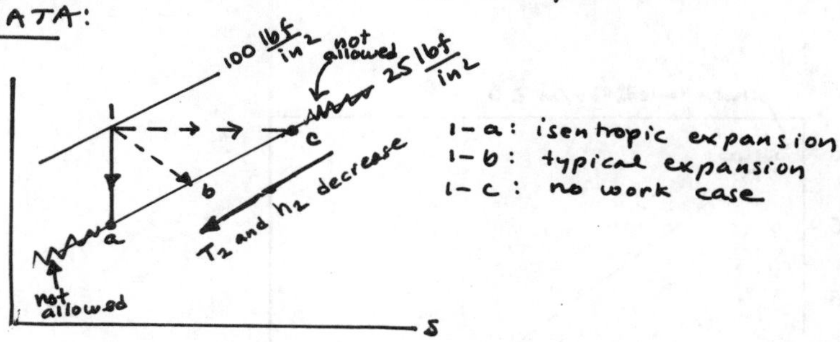

ASSUMPTIONS: 1. The control volume shown in the accompanying figure is at steady state. 2. For the control volume, $\dot{Q}_{cv} = 0$ and kinetic/potential energy effects are negligible. 3. Air is modeled as an ideal gas.

ANALYSIS: At steady state, mass, energy and entropy rate balances reduce with stated assumptions to give

$$\frac{\dot{W}_{cv}}{\dot{m}} = h_1 - h_2 \quad , \quad \frac{\dot{\sigma}_{cv}}{\dot{m}} = s_2 - s_1$$

Since $\dot{\sigma}_{cv}/\dot{m} \geq 0$, we have $s_2 \geq s_1$. Inspection of the T-s diagram shows that the lowest allowed exit temperature corresponds to $s_2 = s_1$, for which $\dot{\sigma}_{cv} = 0$. Since specific enthalpy at the exit depends only on the exit temperature (ideal gas model), process 1-a also corresponds to the maximum theoretical value for $\dot{W}_{cv}/\dot{m}$.

The maximum exit temperature corresponds to the case of no work: $h_2 = h_1$ (a throttling process). Since $h(T)$ for an ideal gas, this case -- process 1-c -- has $T_2 = T_1$.

SAMPLE CALCULATIONS: Process 1-a: Eq. 6.21a reduces to give with Table A-22E data,

$$s°(T_a) = s°(T_1) + R \ln p_2/p_1 = 0.6095 + (1.986/28.97)\ln(25/100) = 0.51446\ \frac{Btu}{lb \cdot °R} \Rightarrow T_a = 377°R\ (-83°F),$$

$h_a = 89.95$ Btu/lb. Then $(\dot{W}_{cv}/\dot{m})_a = 133.86 - 89.95 = 43.91$ Btu/lb. Process 1-c: $T_2 = T_1$. Thus, with Eq. 6.21a,

$$(\dot{\sigma}_{cv}/\dot{m})_c = \underbrace{s°(T_2) - s°(T_1)}_{=0} - R \ln p_2/p_1 = -\left(\frac{1.986}{28.97}\right)\ln\left(\frac{25}{100}\right) = 0.095\ \frac{Btu}{lb \cdot °R}.$$

IT Code

```
T1 = 100 // °F
p1 = 100 // lbf/in.²
p2 = 25
T2 = 100

mdot = 1
Wdot / mdot = h1 - h2

h1 = h_T("Air", T1)
h2 = h_T("Air", T2)
sigmadot / mdot = s2 - s1
s1 = s_TP("Air", T1, p1)
s2 = s_TP("Air", T2, p2)
```

PROBLEM 6.96 (Contd.)

Data for the following plots are generated by sweeping T_2 from -100 to 120 in steps of 1 °F:

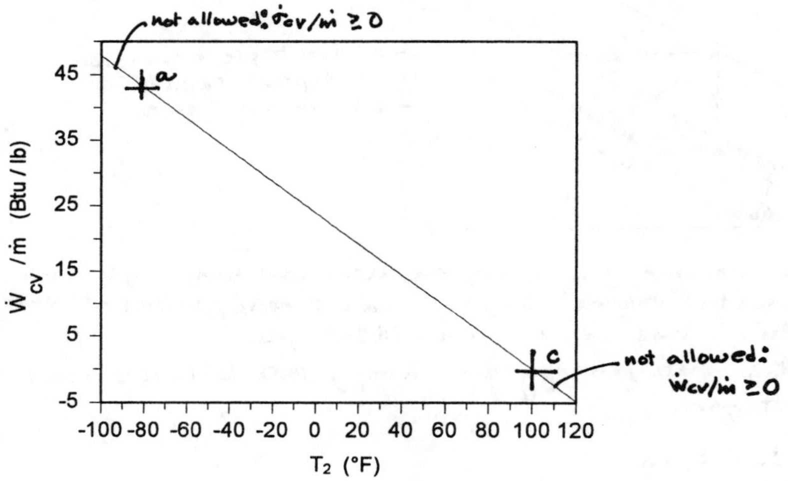

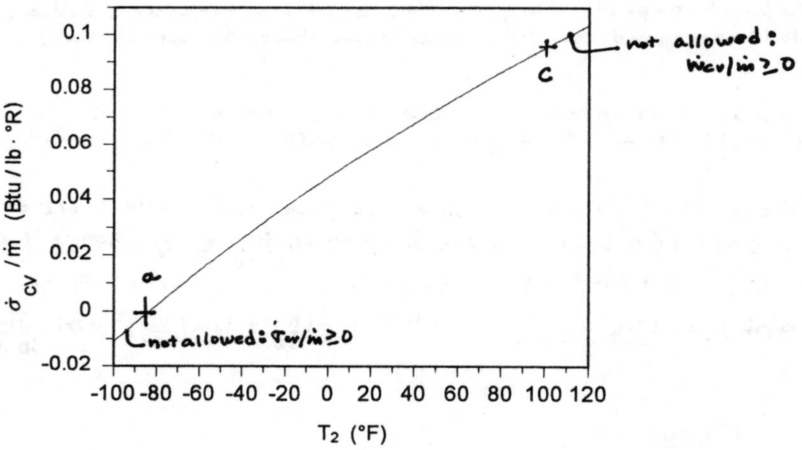

PROBLEM 6.97

KNOWN: Air enters an insulated diffuser operating at steady state at 1 bar, −3°C, 260 m/s and exits at 130 m/s.

FIND: Determine (a) the temperature of the air at the exit, (b) the maximum attainable exit pressure.

SCHEMATIC & GIVEN DATA:

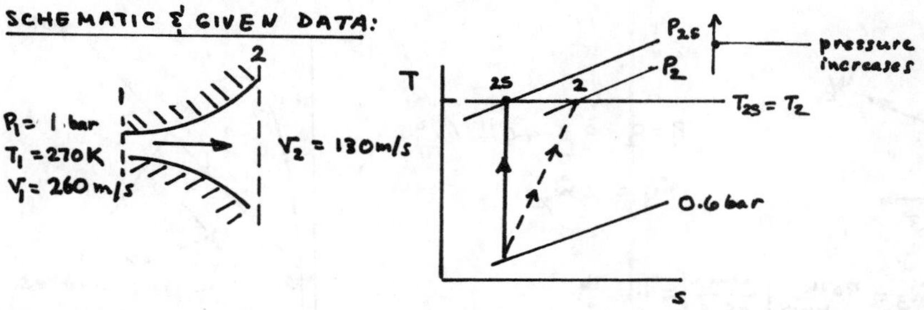

ASSUMPTIONS: (1) The diffuser operates at steady state and is insulated. (2) The change in potential energy from inlet to exit can be neglected. (3) Air is modeled as an ideal gas.

ANALYSIS: At steady state $\dot{m}_1 = \dot{m}_2 = \dot{m}$ and with stated assumptions an energy rate balance reduces to

$$0 = \cancel{\dot{Q}_{cv}}^{0} - \cancel{\dot{W}_{cv}}^{0} + \dot{m}\left(h_1 - h_2 + \frac{V_1^2 - V_2^2}{2} + \cancel{g(z_1 - z_2)}^{0}\right)$$

or

$$h_2 = h_1 + \frac{V_1^2 - V_2^2}{2}$$

with h_1 from Table A-22 at 270 K

$$h_2 = 270.11 + \left(\frac{260^2 - 130^2}{2}\right)\left(\frac{m^2}{s^2}\right)\left(\frac{1\,N}{1\,kg\cdot m/s^2}\right)\left(\frac{kJ}{10^3\,N\cdot m}\right)$$
$$= 295.46 \text{ kJ/kg}$$

Table A-22 then gives T_2 = 295 K (22°C). ← T_2

An entropy rate balance reduces to read

$$\frac{\dot{\sigma}_{cv}}{\dot{m}} = s_2 - s_1$$

Since $\dot{\sigma}_{cv} \geq 0$, $s_2 \geq s_1$. For fixed exit temperature T_2 (and thus fixed exit velocity V_2), the accompanying T-s diagram shows that the maximum allowed exit pressure corresponds to the case $s_2 = s_1$. To determine this pressure, P_{2s}, use Eq. 6.21a to write

$$0 = s^\circ(T_{2s}) - s^\circ(T_1) - R\ln\frac{P_{2s}}{P_1} \Rightarrow \text{with } T_{2s} = T_2 \quad \ln\frac{P_{2s}}{P_1} = \frac{s^\circ(T_2) - s^\circ(T_1)}{R}$$

with data from Table A-22 at T_1 = 270 K, T_2 = 295 K

$$\ln\frac{P_{2s}}{P_1} = \frac{1.68515 - 1.59634}{(8.314/28.97)} \Rightarrow P_{2s} = 1.363\,P_1 = 1.363 \text{ bar}$$ ← $(P_2)_{MAX}$

PROBLEM 6.98

KNOWN: Test data are provided for a new type of engine operating at steady state.

FIND: Determine the maximum theoretical rate power can be developed.

SCHEMATIC & GIVEN DATA:

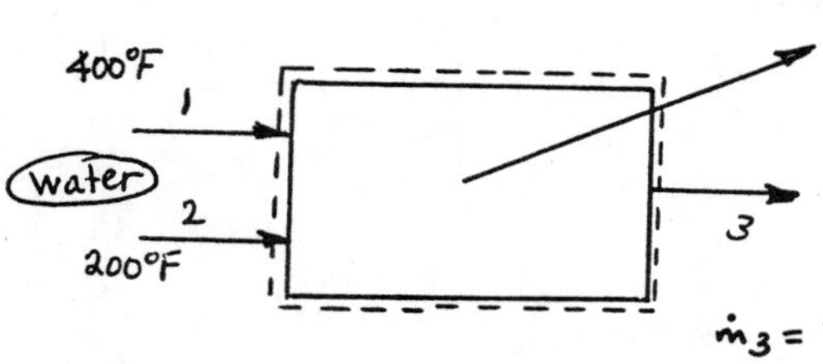

$P_1 = P_2 = P_3 = 40\,lbf/in^2$

$\dot{m}_1 = 2\dot{m}_2$

$\dot{m}_3 = 90 \frac{lb}{min}\left|\frac{1\,min}{60\,s}\right| = 1.5\,\frac{lb}{s}$

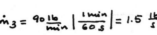

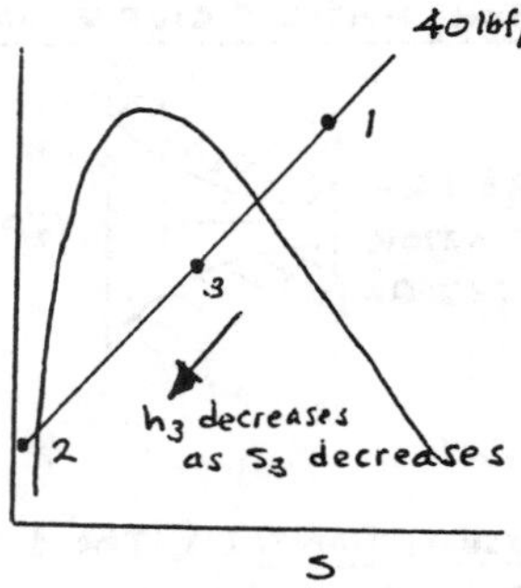

ASSUMPTIONS: (1) The control volume shown above is at steady state. (2) For the control volume, $\dot{Q}_{cv} = 0$, and kinetic and potential energy effects are negligible.

ANALYSIS: At steady state, a mass rate balance gives $\dot{m}_3 = \dot{m}_1 + \dot{m}_2$. Then, since $\dot{m}_3 = 1.5\,lb/s$ and $\dot{m}_1 = 2\dot{m}_2$, $\dot{m}_1 = 1\,lb/s$ and $\dot{m}_2 = 0.5\,lb/s$.

Using given assumptions, an energy rate balance reduces at steady state to give

$$\dot{W}_{cv} = \dot{m}_1 h_1 + \dot{m}_2 h_2 - \dot{m}_3 h_3$$

With data from Tables A-2E, 4E, using $h_2 \approx h_f(T_2)$

$$\dot{W}_{cv} = \left(1\,\frac{lb}{s}\right)\left(1236.4\,\frac{Btu}{lb}\right) + \left(0.5\,\frac{lb}{s}\right)\left(168.1\,\frac{Btu}{lb}\right) - \left(1.5\,\frac{lb}{s}\right)(h_3)$$

$$= (1320.5 - 1.5\,h_3)\,\frac{Btu}{s} \qquad (1)$$

From Eq.(1) it can be concluded that $\dot{W}_{cv}$ increases as h_3 decreases. The above Mollier diagram shows that the smallest allowed value for h_3 corresponds to the smallest allowed value for s_3.

To determine the smallest allowed value for s_3, apply an entropy rate balance at steady state: $0 = \dot{m}_1 s_1 + \dot{m}_2 s_2 - \dot{m}_3 s_3 + \dot{\sigma}_{cv}$

$$\Rightarrow s_3 = \frac{\dot{m}_1}{\dot{m}_3} s_1 + \frac{\dot{m}_2}{\dot{m}_3} s_2 + \frac{\dot{\sigma}_{cv}}{\dot{m}_3} \Rightarrow s_3 = \left(\frac{2}{3}\right)s_1 + \left(\frac{1}{3}\right)s_2 + \frac{\dot{\sigma}_{cv}}{\dot{m}_3}$$

Since $\dot{\sigma}_{cv} \geq 0$, the smallest allowed value for s_3, and thus h_3, corresponds to $\dot{\sigma}_{cv} = 0$. With data from Tables A-2E and 4E, and using $s_2 = s_f(T_2)$

$$(s_3)_{min} = \left(\frac{2}{3}\right)s_1 + \left(\frac{1}{3}\right)s_2 = \left(\frac{2}{3}\right)(1.7606) + \left(\frac{1}{3}\right)(0.2940) = 1.2717\,Btu/lb\cdot°R$$

Using this value, state 3 is located in the two-phase region with quality

$$(x_3)_{min} = \frac{1.2717 - 0.3921}{1.2845} = 0.685 \Rightarrow (h_3)_{min} = 236.16 + 0.685(933.8) = 875.8\,Btu/lb$$

① Finally, with Eq.(1), $(\dot{W}_{cv})_{MAX} = [1320.5 - 1.5(875.8)]\left(\frac{Btu}{s}\right)\left|\frac{3600\,s}{1\,h}\right|\left|\frac{1\,hp}{2545\,Btu/h}\right| = 9.62\,hp$ ←

1. From Eq.(1), the case $\dot{W}_{cv} = 0$ corresponds to $h_3 = 880.3\,Btu/lb$, or $x_3 = 0.69$, which differs only slightly from $(x_3)_{min}$. Accordingly, with such a small interval between maximum power and no power output, the effect of irreversibilities within the engine is expected to erode significantly the power that would be obtained.

PROBLEM 6.99

KNOWN: Steady state operating data are provided for a device.

FIND: Determine the maximum theoretical pressure achievable at the exit and discuss.

SCHEMATIC & GIVEN DATA:

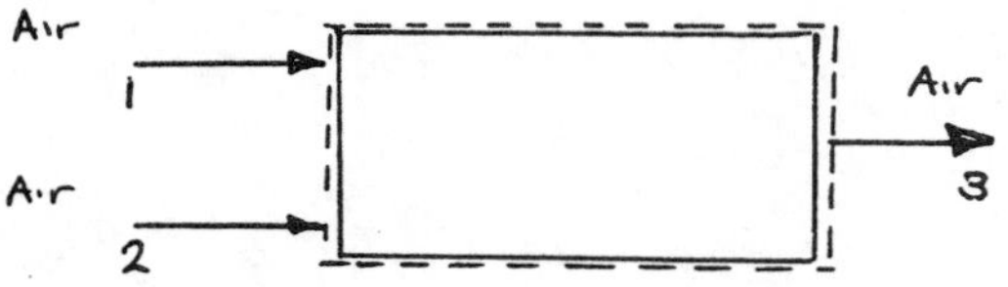

$p_1 = p_2 = 1$ atm
$T_1 = 600°R$, $\dot{m}_1 = 100$ lb/h
$T_2 = 2000°R$, $\dot{m}_2 = 120$ lb/h

ASSUMPTIONS: (1) The control volume shown above is at steady state. (2) For the control volume, $\dot{Q}_{cv} = \dot{W}_{cv} = 0$, and kinetic and potential energy effects are negligible. (3) The air is modeled as an ideal gas.

ANALYSIS: From a mass rate balance at steady state, $\dot{m}_3 = 220$ lb/h. An energy rate balance reduces with assumptions (1) and (2) to give $0 = \dot{m}_1 h_1 + \dot{m}_2 h_2 - \dot{m}_3 h_3$. Solving for h_3, and inserting data from Table A-22E

$$h_3 = \frac{\dot{m}_1 h_1 + \dot{m}_2 h_2}{\dot{m}_3} = \frac{(100)(143.47) + (120)(504.71)}{220} = 340.51 \ \frac{\text{Btu}}{\text{lb}}$$

Using h_3, Table A-22E gives $s_3^\circ = 0.8343$ Btu/lb·°R. From the same table $s_1^\circ = 0.62607$, $s_2^\circ = 0.93205$.

Next, applying an entropy rate balance, $0 = \dot{m}_1 s_1 + \dot{m}_2 s_2 - \dot{m}_3 s_3 + \dot{\sigma}$. Or, since $\dot{m}_3 = \dot{m}_1 + \dot{m}_2$, this gives

$$0 = \dot{m}_1 s_1 + \dot{m}_2 s_2 - (\dot{m}_1 + \dot{m}_2) s_3 + \dot{\sigma} \Rightarrow 0 = \dot{m}_1 (s_1 - s_3) + \dot{m}_2 (s_2 - s_3) + \dot{\sigma}$$

Then, using Eq 6.21a

$$0 = \dot{m}_1 \left[s_1^\circ - s_3^\circ - \frac{\bar{R}}{M} \ln \frac{p_1}{p_3} \right] + \dot{m}_2 \left[s_2^\circ - s_3^\circ - \frac{\bar{R}}{M} \ln \frac{p_2}{p_3} \right] + \dot{\sigma}$$

Solving

$$\dot{m}_1 \ln \frac{p_3}{p_1} + \dot{m}_2 \ln \frac{p_3}{p_2} = \frac{\dot{m}_1 [s_3^\circ - s_1^\circ] + \dot{m}_2 [s_3^\circ - s_2^\circ] - \dot{\sigma}}{\bar{R}/M}$$

Since $p_2 = p_1$, this becomes

$$\ln \frac{p_3}{p_1} = \frac{\dot{m}_1 (s_3^\circ - s_1^\circ) + \dot{m}_2 (s_3^\circ - s_2^\circ) - \dot{\sigma}}{(\dot{m}_1 + \dot{m}_2)\, \bar{R}/M}$$

As $\dot{\sigma} \geq 0$, the maximum theoretical value for the ratio p_3/p_1 corresponds to $\dot{\sigma} = 0$. Thus, with known values

$$\ln \left(\frac{p_3}{p_1} \right)_{MAX} = \frac{(100)(0.8343 - 0.62607) + (120)(.8343 - 0.93205)}{(220)(1.986/28.97)} = 0.6029$$

or

$$\left(\frac{p_3}{p_1} \right)_{MAX} = 1.827 \Rightarrow (p_3)_{MAX} = 1.827 \text{ atm} \quad \longleftarrow \text{MAX}$$

A pressure rise might be brought about by including a diffuser section as a component:

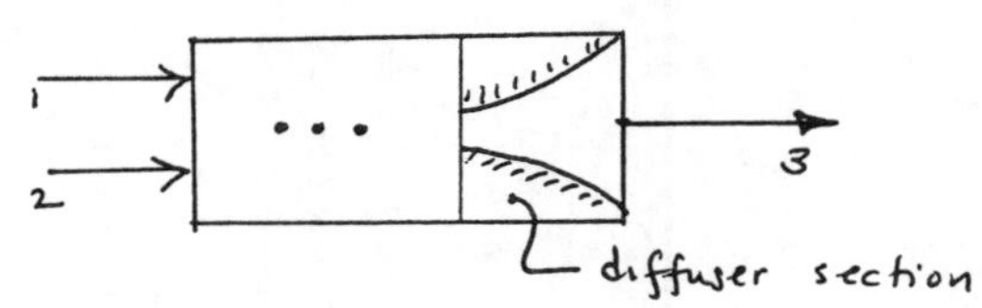

PROBLEM 6.100

KNOWN: Steady-state operating data are provided for a device to deliver energy by heat transfer at a specified temperature.

FIND: Determine the maximum theoretical amount of energy that could be delivered per unit mass of steam entering the device.

SCHEMATIC & GIVEN DATA:

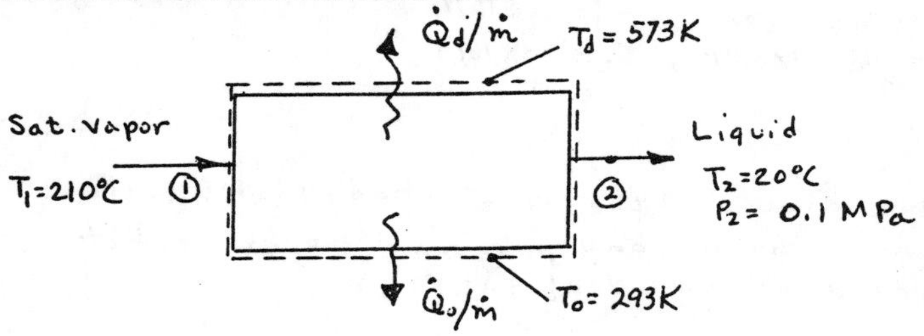

ASSUMPTIONS: (1) The control volume shown is at steady state. (2) For the control volume $\dot{W}_{cv} = 0$, all heat transfers are in the directions indicated, and kinetic and potential energies are negligible.

ANALYSIS: From Table A-2, $h_1 = 2798.5$ kJ/kg, $s_1 = 6.3585$ kJ/kg·K, $h_2 \approx h_f(T_2) = 83.96$ kJ/kg, $s_2 \approx s_f(T_2) = 0.2966$ kJ/kg·K.

By combining mass and energy balances at steady state, using indicated assumptions, $0 = \dot{Q}_{cv} + \dot{m}(h_1 - h_2)$, where $\dot{Q}_{cv} = -[\dot{Q}_d + \dot{Q}_o]$. Thus

$$\frac{\dot{Q}_o}{\dot{m}} = (h_1 - h_2) - \frac{\dot{Q}_d}{\dot{m}} \qquad (1)$$

At steady state, an entropy rate balance reduces to

$$0 = -\frac{\dot{Q}_o}{T_o} - \frac{\dot{Q}_d}{T_d} + \dot{m}(s_1 - s_2) + \dot{\sigma} \qquad (2)$$

Next, eliminate $\dot{Q}_o$ between Eqs. (1), (2) to obtain

$$0 = -\left[\frac{(h_1-h_2) - \dot{Q}_d/\dot{m}}{T_o}\right] - \frac{\dot{Q}_d/\dot{m}}{T_d} + (s_1 - s_2) + \frac{\dot{\sigma}}{\dot{m}} \Rightarrow \frac{\dot{Q}_d}{\dot{m}}\left[\frac{1}{T_o} - \frac{1}{T_d}\right] - \frac{(h_1-h_2)}{T_o} + (s_1 - s_2) + \frac{\dot{\sigma}}{\dot{m}} = 0$$

Solving

$$\frac{\dot{Q}_d}{\dot{m}} = \frac{\left(\frac{h_1 - h_2}{T_o}\right) + (s_2 - s_1) - \dot{\sigma}/\dot{m}}{\left[\frac{1}{T_o} - \frac{1}{T_d}\right]}$$

As $\dot{\sigma} \geq 0$, the maximum value corresponds to $\dot{\sigma} = 0$. That is

$$\left(\frac{\dot{Q}_d}{\dot{m}}\right)_{MAX} = \frac{\left(\frac{2798.5 - 83.96}{293}\right) + (0.2966 - 6.3585)}{\left[\frac{1}{293} - \frac{1}{573}\right]} = 1921 \text{ kJ/kg} \quad \longleftarrow \text{MAX}$$

PROBLEM 6.101

KNOWN: A thermally activated device for chilling water requires no power input to operate. The chilled water exits at temperature T.

FIND: Plot the minimum theoretical heat addition required per lb of chilled water versus T.

SCHEMATIC & GIVEN DATA:

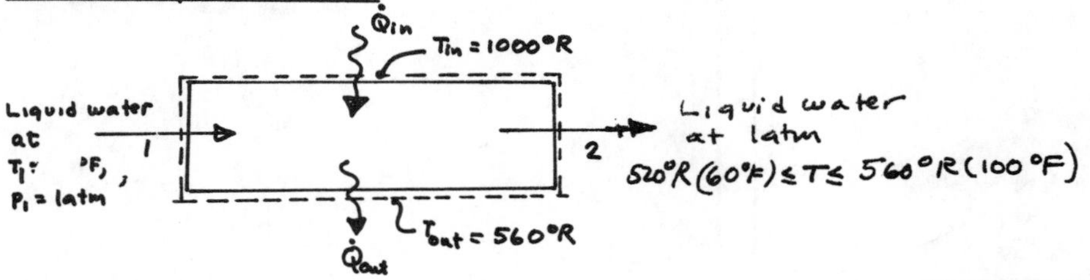

ASSUMPTIONS: (1) The control volume shown in the accompanying figure operates at steady state with $\dot{W}_{cv} = 0$. (2) Changes in kinetic and potential energy from inlet to exit can be neglected. (3) The water can be modeled as incompressible with constant specific heat c.

ANALYSIS: At steady state $\dot{m}_1 = \dot{m}_2 = \dot{m}$, and an energy rate balance reduces with listed assumptions to give

$$0 = \dot{Q}_{in} - \dot{Q}_{out} - \dot{W}_{cv}^{\,0} + \dot{m}(h_1 - h_2) \Rightarrow \frac{\dot{Q}_{in}}{\dot{m}} = \frac{\dot{Q}_{out}}{\dot{m}} + h_2 - h_1 \qquad (1)$$

At steady state an entropy rate balance reduces to give

$$0 = \frac{\dot{Q}_{in}}{T_{in}} - \frac{\dot{Q}_{out}}{T_{out}} + \dot{m}(s_1 - s_2) + \dot{\sigma}_{cv}$$

Solving for $\dot{Q}_{out}$

$$\frac{\dot{Q}_{out}}{\dot{m}} = \left(\frac{T_{out}}{T_{in}}\right)\left(\frac{\dot{Q}_{in}}{\dot{m}}\right) + T_{out}(s_1 - s_2) + T_{out}\frac{\dot{\sigma}_{cv}}{\dot{m}}$$

and substituting into Eq. (1)

$$\left[1 - \frac{T_{out}}{T_{in}}\right]\frac{\dot{Q}_{in}}{\dot{m}} = T_{out}(s_1 - s_2) + (h_2 - h_1) + T_{out}\dot{\sigma}_{cv}/\dot{m}$$

or, upon rearrangement

$$\frac{\dot{Q}_{in}}{\dot{m}} = \frac{T_{out}(s_1 - s_2) + (h_2 - h_1)}{\left[1 - \frac{T_{out}}{T_{in}}\right]} + \frac{T_{out}\dot{\sigma}_{cv}/\dot{m}}{\left[1 - \frac{T_{out}}{T_{in}}\right]} \qquad (1)$$

Evaluating $(h_2 - h_1)$ using Eq. 3.20b and $(s_2 - s_1)$ using Eq. 6.24, the minimum heat addition corresponds to $\dot{\sigma}_{cv} = 0$ in Eq. (1), giving

$$\left(\frac{\dot{Q}_{in}}{\dot{m}}\right)_{min} = \frac{c\left[T_{out}\ln\left(\frac{T_1}{T}\right) + (T - T_1)\right]}{\left[1 - \frac{T_{out}}{T_{in}}\right]}$$

, where c = 1 Btu/lb·°R (Table A-19E), $T_{in} = 1000°R$, $T_{out} = 560°R$, $T_1 = 560°R$, and $520 \leq T \leq 560°R$.

Sample calculation: T = 520 °R, $(\dot{Q}_{in}/\dot{m})_{min}$ = 3.41 Btu/lb.

Plot on next page

PROBLEM 6.101 (Contd.)

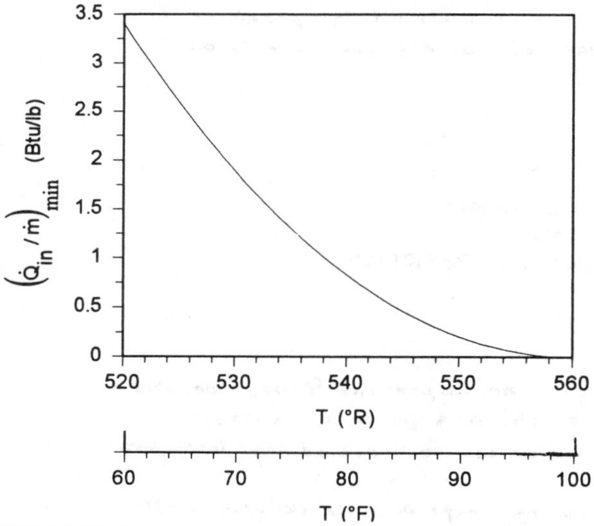

PROBLEM 6.102

KNOWN: A device is proposed for developing power using "waste heat".

FIND: Evaluate the maximum theoretical power that can be developed.

SCHEMATIC & GIVEN DATA:

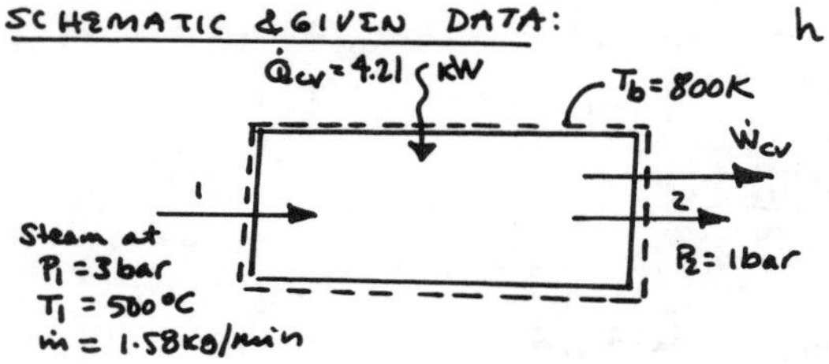

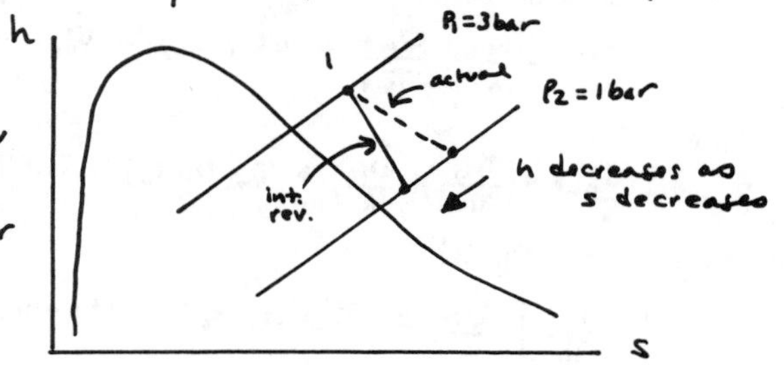

ASSUMPTIONS: 1. The control volume shown in the accompanying schematic operates at steady state. 2. Kinetic/potential energy effects can be ignored. 3. Heat transfer occurs only at T_b.

ANALYSIS: At steady state, the mass and energy rate balances reduce to give

① $$\dot{W}_{cv} = \dot{Q}_{cv} + \dot{m}(h_1 - h_2) = (4.21\,\text{kW}) + \left(\frac{1.58}{60}\frac{\text{kg}}{\text{s}}\right)\left(3486\frac{\text{kJ}}{\text{kg}} - h_2\right)\left|\frac{1\,\text{kW}}{1\,\text{kJ/s}}\right| \quad (1)$$

where h_1 is from Table A-4. Eq. (1) shows that $\dot{W}_{cv}$ increases as h_2 decreases. To determine the smallest allowed value for h_2, consider an entropy rate balance, which reduces at steady state to

$$0 = \frac{\dot{Q}_{cv}}{T_b} + \dot{m}(s_1 - s_2) + \dot{\sigma}_{cv} \Rightarrow s_2 = s_1 + \frac{\dot{Q}_{cv}}{\dot{m}T_b} + \frac{\dot{\sigma}_{cv}}{\dot{m}} \quad (2)$$

The h-s diagram shows that h_2 decreases as s_2 decreases. The smallest s_2 value allowed corresponds to $(\dot{\sigma}_{cv}/\dot{m}) = 0$ in Eq. (2):

$$(s_2)_{min} = s_1 + \frac{\dot{Q}_{cv}}{\dot{m}T_b} = 8.3251 + \frac{4.21\,\text{kJ/s}}{\left(\frac{1.58}{60}\frac{\text{kg}}{\text{s}}\right)(800\,\text{K})} = 8.5249\,\frac{\text{kJ}}{\text{kg}\cdot\text{K}}$$

where s_1 is from Table A-4. The corresponding value for specific enthalpy, $(h_2)_{min}$, is obtained by interpolation in Table A-4: $(h_2)_{min} = 3266$ kJ/kg. Then, Eq. (1) gives

$(\dot{W}_{cv})_{MAX} = (4.21) + (1.58/60)(3486 - 3266) = 10\,\text{kW}$ ◀

---

1. In principle, power is developed from two inputs: "waste" heat and steam. Note that $\dot{W}_{cv} = 0$ corresponds to $h_2 = 3645.9$ kJ/kg.

PROBLEM 6.103

KNOWN: Steady-state operating data are provided for a gas turbine power plant.

FIND: For the given conditions, determine the maximum theoretical value for the net power that can be developed.

SCHEMATIC & GIVEN DATA:

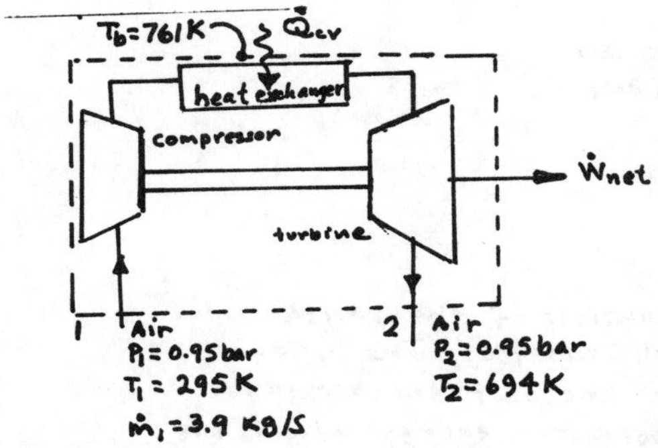

ASSUMPTIONS: 1. The control volume shown in the accompanying schematic operates at steady state. 2. Heat transfer to the control volume occurs only at T_b. 3. Kinetic/potential energy effects can be ignored. 4. Air is modeled as an ideal gas.

ANALYSIS: Mass and energy rate balances reduce at steady state to give

$$0 = \dot{Q}_{cv} - \dot{W}_{cv} + \dot{m}\left[(h_1 - h_2) + \cancel{\frac{V_1^2 - V_2^2}{2}} + \cancel{g(z_1 - z_2)}\right]$$

$$\Rightarrow \quad \dot{W}_{cv} = \dot{Q}_{cv} + \dot{m}\left[h_1 - h_2\right] \qquad (1)$$

An entropy rate balance at steady state gives

$$0 = \frac{\dot{Q}_{cv}}{T_b} + \dot{m}(s_1 - s_2) + \dot{\sigma}_{cv}$$

Solving for $\dot{Q}_{cv}$

$$\dot{Q}_{cv} = \dot{m}\,T_b(s_2 - s_1) - T_b\dot{\sigma}_{cv}$$

Inserting this into Eq. (1)

$$\dot{W}_{cv} = \underline{\dot{m}\left[(h_1 - h_2) + T_b(s_2 - s_1)\right]} - T_b\dot{\sigma}_{cv} \qquad (2)$$

The underlined term of Eq. (2) is fixed by conditions on the boundary of the control volume. The last term involving entropy production is greater than, or equal to, zero, depending on irreversibilities within the control volume. The maximum value of $\dot{W}_{cv}$ corresponds to $\dot{\sigma}_{cv} = 0$:

$$(\dot{W}_{cv})_{MAX} = \dot{m}\left[(h_1 - h_2) + T_b(s_2 - s_1)\right]$$

With Eq. 6.21a and data from Table A-22

$$(\dot{W}_{cv})_{MAX} = \dot{m}\left[h(T_1) - h(T_2) + T_b\left[s^o(T_2) - s^o(T_1) - \bar{R}\cancel{\ln\frac{P_2}{P_1}}^{\,0}\right]\right]$$

$$= \left(3.9\,\frac{kg}{s}\right)\left[(295.17 - 706.8)\frac{kJ}{kg} + 761K\left(2.5635 - 1.68515\right)\frac{kJ}{kg\cdot K}\right]$$

$$= 1001.5\,\frac{kJ}{s}\left|\frac{1\,MW}{10^3\,kJ/s}\right|$$

$$= 1\,MW \longleftarrow$$

PROBLEM 6.104

KNOWN: Operating data are provided for an electrical resistor located in an insulated duct carrying air.

FIND: (a) For the resistor as the system, determine the rate of entropy production.
(b) For a control volume enclosing the air in the duct and the resistor, determine the volumetric flow rate and the rate of entropy production.

SCHEMATIC & GIVEN DATA:

resistor: 30 ohm
current: 15 amp

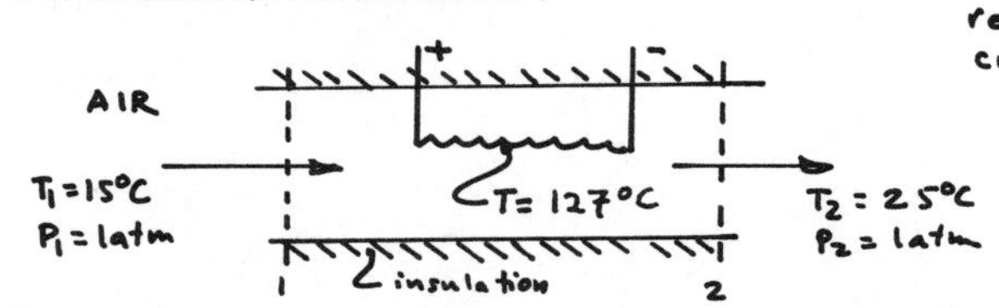

ASSUMPTIONS: (1) For part (a), the closed system consists of the resistor only. Heat transfer from this system takes place at $T_b = 400\,K\,(127°C)$. The system is at steady state. (2) For part (b), the control volume encloses the air in the duct and the resistor. $\dot{Q}_{cv} = 0$ and kinetic and potential energy effects are negligible. The control volume is at steady state. Air is modeled as an ideal gas.

ANALYSIS: (a) An energy rate balance for a closed system at steady state reduces to give $\dot{Q}_{cv} = \dot{W}_{cv}$. In this case, electrical work is done on the system. With given data

$$\dot{W}_{cv} = -(i)^2 R = -(15\,amp)^2 (30\,ohm)\left[\frac{kW}{10^3 (amp)^2 (ohm)}\right] = -6.75\,kW$$

At steady state an entropy rate balance becomes for heat transfer at T_b only

$$\cancel{\frac{dS}{dt}}^{0} = \frac{\dot{Q}_{cv}}{T_b} + \dot{\sigma} \Rightarrow \dot{\sigma} = -\frac{\dot{Q}_{cv}}{T_b}$$

Finally

$$\dot{\sigma} = -\frac{(-6.75\,kW)}{400\,K} = 0.0169\,\frac{kW}{K} \longleftarrow \dot{\sigma}$$

(b) For the control volume at steady state the energy rate balance reduces with listed assumptions to

$$0 = \cancel{\dot{Q}_{cv}}^{0} - \dot{W}_{cv} + \dot{m}\left[h_1 - h_2 + \cancel{\left(\frac{V_1^2 - V_2^2}{2}\right)}^{0} + \cancel{g(z_1 - z_2)}^{0}\right]$$

Solving for $\dot{m}$ and using enthalpies from Table A-22

$$\dot{m} = \frac{(-\dot{W}_{cv})}{h_2 - h_1} = \frac{-(-6.75\,kW)(1\,kJ/s / kW)}{[298.2 - 288.2]\,kJ/kg} = 0.675\,\frac{kg}{s} \longleftarrow \dot{m}$$

The volumetric flow rate entering the duct is then

$$(AV)_1 = \dot{m} v_1 \Rightarrow (AV)_1 = \dot{m}\left(\frac{RT_1}{P_1}\right) = \left(0.675\,\frac{kg}{s}\right)\left(\frac{8314}{28.97}\,\frac{N \cdot m}{kg \cdot K}\right)(288\,K)\left(\frac{1}{1.01325 \times 10^5\,N/m^2}\right)$$

$$= 0.55\,m^3/s \longleftarrow (AV)_1$$

An entropy rate balance reduces at steady state to

$$0 = \cancel{\sum \frac{\dot{Q}_j}{T_j}} + \dot{m}(s_1 - s_2) + \dot{\sigma}_{cv} \Rightarrow \dot{\sigma}_{cv} = \dot{m}(s_2 - s_1)$$

With Eq. 6.21a

$$s_2 - s_1 = s°(T_2) - s°(T_1) - R\,\cancel{\ln P_2/P_1}^{0}$$

Thus with data from Table A-22

$$\dot{\sigma}_{cv} = \dot{m}(s°(T_2) - s°(T_1)) = \left(0.695\,\frac{kg}{s}\right)(1.69528 - 1.66103)\,\frac{kJ}{kg \cdot K}$$

$$= 0.0238\,kW/K \longleftarrow \dot{\sigma}_{cv}$$

COMMENT: In part (a), $\dot{\sigma}$ accounts for the irreversibility of electrical current flow through a resistance. In part (b), $\dot{\sigma}_{cv}$ includes both the electrical irreversibility and the irreversibility of heat transfer from the resistor to the air.

PROBLEM 6.105

KNOWN: Operating data are provided for an electronics enclosure at steady state.

FIND: For the control volume of Example 4.8 determine the rate of entropy production when air exits at 32°C.

SCHEMATIC & GIVEN DATA: See Fig. E4.8

ASSUMPTIONS: 1. See Example 4.8. 2. Ignore the change in pressure from inlet to exit.

ANALYSIS: At steady state, mass and entropy rate balances reduce to give

$$0 = \sum_j \frac{\dot{Q}_j}{T_j}^{\,0} + \dot{m}(s_1 - s_2) + \dot{\sigma}_{cv}$$

$$\Rightarrow \quad \dot{\sigma}_{cv} = \dot{m}(s_2 - s_1)$$

Introducing $\dot{m}$ from Example 4.8 and evaluating $(s_2 - s_1)$ with Eq. 6.23

$$\dot{\sigma}_{cv} = \left[\frac{(-\dot{W}_{cv})}{c_p(T_2 - T_1)}\right]\left(c_p \ln\frac{T_2}{T_1} - R \ln\frac{p_2}{p_1}^{\,0}\right) \Rightarrow \dot{\sigma}_{cv} = \frac{(-\dot{W}_{cv})}{(T_2 - T_1)} \ln\frac{T_2}{T_1}$$

$$\Rightarrow \quad \dot{\sigma}_{cv} = \frac{(98\,W)}{(305-293)\,K} \ln\frac{305}{293} = 0.328\,\frac{W}{K}$$ ←

PROBLEM 6.106

KNOWN: Operating data are provided for an electronics enclosure at steady state.

FIND: For the control volume shown in the schematic determine the rate of entropy production.

SCHEMATIC & GIVEN DATA:

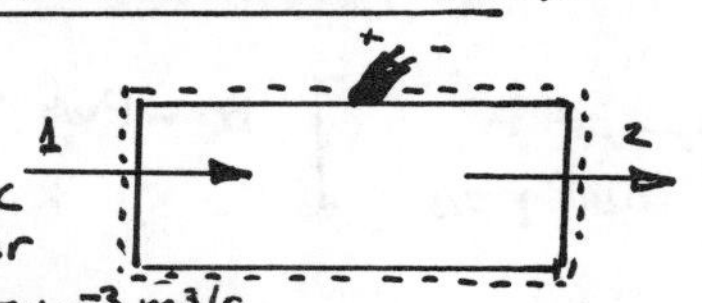

$T_1 = 20°C$, $p_1 = 1\,bar$ → $T_2 = 35°C$, $p_2 \sim p_1$

$(AV)_1 = 7 \times 10^{-3}\,m^3/s$ (from Solution to Prob. 4.70)

ASSUMPTIONS: 1. The control volume shown in the schematic is at steady state. 2. For the control volume, $\dot{Q}_{cv} = 0$ and $p_2 \sim p_1$. 3. Air is modeled as an ideal gas.

ANALYSIS: Mass and entropy rate balances reduce at steady state to give

$$0 = \sum_j \frac{\dot{Q}_j}{T_j} + \dot{m}(s_1 - s_2) + \dot{\sigma}_{cv} \Rightarrow \quad \dot{\sigma}_{cv} = \dot{m}(s_2 - s_1)$$

or with Eq. 6.21a and data from Table A-22

$$\dot{\sigma}_{cv} = \frac{(AV)_1}{v_1}\left[s^\circ(T_2) - s^\circ(T_1) - R \ln\frac{p_2}{p_1}^{\,0}\right]$$

$$= \frac{(AV)_1}{(RT_1/p_1)}\left[s^\circ(T_2) - s^\circ(T_1)\right] = \left[\frac{(7\times10^{-3}\,m^3/s)(10^5\,N/m^2)}{\left(\frac{8314}{28.97}\,\frac{N\cdot m}{kg\cdot K}\right)(293\,K)}\right]\left[1.7284 - 1.6783\right]\frac{kJ}{kg\cdot K}\left|\frac{1\,kW}{1\,kJ/s}\right|$$

$$= 4.15 \times 10^{-4}\,\frac{kW}{K}$$ ←

PROBLEM 6.107

KNOWN: Steady-state operating data are provided for a water-jacketed electronics enclosure.

FIND: Determine the rate of entropy production when water exits at 24°C.

SCHEMATIC & GIVEN DATA:

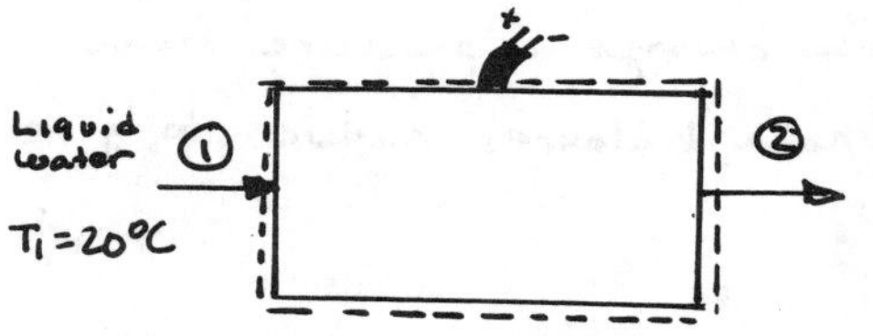

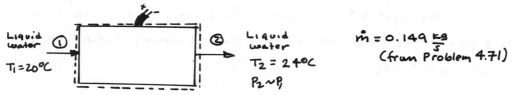

$\dot{m} = 0.149 \frac{kg}{s}$ (from Problem 4.71)

ASSUMPTIONS: 1. The control volume shown in the schematic is at steady state. 2. For the control volume, $\dot{Q}_{cv} = 0$ 3. For the liquid water, $s \approx s_f(T)$.

ANALYSIS: Mass and entropy rate balances reduce at steady state to give

$$0 = \sum_j \frac{\dot{Q}_j}{T_j}^{\,0} + \dot{m}(s_1 - s_2) + \dot{\sigma}_{cv} \Rightarrow \dot{\sigma}_{cv} = \dot{m}\left[s_f(T_2) - s_f(T_1)\right]$$

With data from Table A-2

$$\dot{\sigma}_{cv} = \left(0.149 \frac{kg}{s}\right)\left[0.3534 - 0.2966\right]\frac{kJ}{kg \cdot K}\left|\frac{1\ kW}{1\ kJ/s}\right| = 8.5 \times 10^{-3} \frac{kW}{K} \longleftarrow$$

PROBLEM 6.108

KNOWN: Steady-state operating data are provided for the electronics-laden cylinder of Problem 4.73.

FIND: Determine the rate of entropy production when air exits at 40°C.

SCHEMATIC & GIVEN DATA:

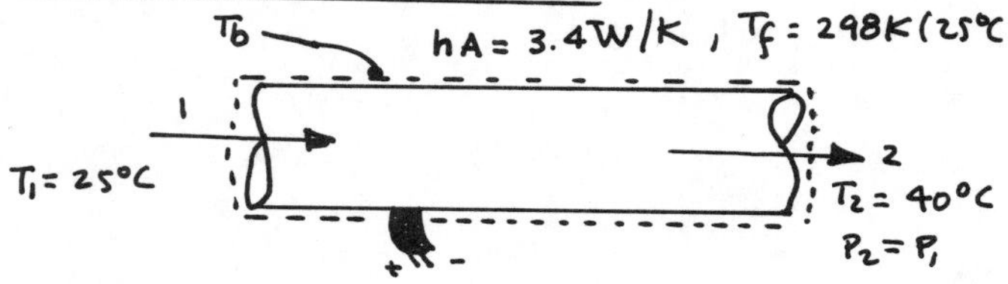

$\dot{Q}_{cv} = -34 W$, $\dot{m} = 0.011\ kg/s$ — from Prob. 4.73

ASSUMPTIONS: 1. The control volume shown in the figure is at steady state. 2. Heat transfer by convection from the cylinder to the surroundings occurs at temperature T_b. 3. Air is modeled as an ideal gas.

ANALYSIS: Mass and energy rate balances reduce at steady state to give

$$0 = \frac{\dot{Q}_{cv}}{T_b} + \dot{m}[s_1 - s_2] + \dot{\sigma}_{cv} \Rightarrow \dot{\sigma}_{cv} = -\frac{\dot{Q}_{cv}}{T_b} + \dot{m}(s_2 - s_1)$$

Using Eq. 6.21a

$$\dot{\sigma}_{cv} = -\frac{\dot{Q}_{cv}}{T_b} + \dot{m}\left[s^\circ(T_2) - s^\circ(T_1) - R \ln P_2/P_1^{\,0}\right]$$

T_b can be evaluated using Eq. 2-34, written as $\dot{Q}_{cv} = -hA[T_b - T_f]$, where the minus sign is used to conform with the sign convention for heat transfer in the energy and entropy balances. Thus

$$T_b = \left(\frac{-\dot{Q}_{cv}}{hA}\right) + T_f = \left(\frac{34\ W}{3.4\ W/K}\right) + 298K = 308K$$

Then, with data from Table A-22

$$\dot{\sigma}_{cv} = -\frac{(-34W)}{308K} + \left(0.011 \frac{kg}{s}\right)\left[1.7446 - 1.6953\right]\frac{kJ}{kg \cdot K}\left|\frac{1\ W}{1\ J/s}\right|\left|\frac{10^3 J}{1\ kJ}\right| = 0.65 \frac{W}{K} \longleftarrow$$

PROBLEM 6.109

KNOWN: Operating data are provided for a turbine operating at steady state from which steam is extracted at an intermediate pressure.

FIND: (a) For x=99% at the turbine exit, determine the power developed and the rate of entropy production. (b) Plot the quantities of part (a) versus x_3 ranging from 90 to 100%.

SCHEMATIC & GIVEN DATA:

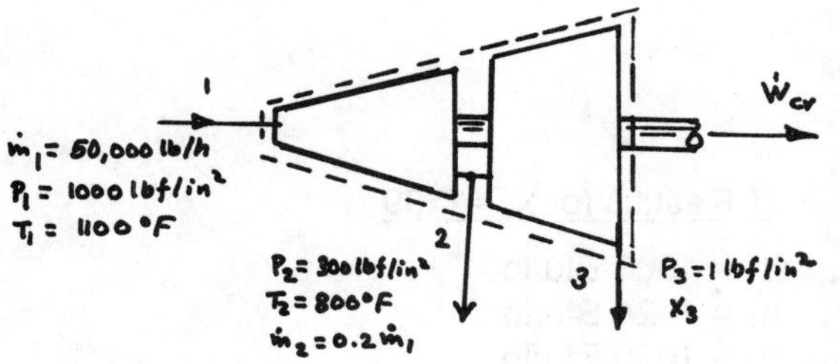

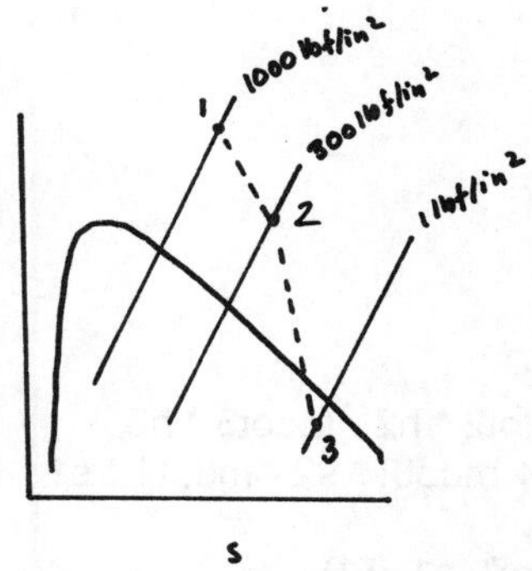

ASSUMPTIONS: (1) The control volume shown on the accompanying figure operates at steady state. (2) $\dot{Q}_{cv} = 0$ and kinetic and potential energy effects are negligible.

ANALYSIS: At steady state a mass rate balance reduces to $\dot{m}_1 = \dot{m}_2 + \dot{m}_3$. Then, with $\dot{m}_2 = 0.1\,\dot{m}_1$, it follows that $\dot{m}_3 = 0.8\,\dot{m}_1$.

With assumption 2 an energy rate balance at steady state reduces to give

$$\dot{W}_{cv} = \dot{m}_1 h_1 - \dot{m}_2 h_2 - \dot{m}_3 h_3$$

$$= \dot{m}_1 [h_1 - 0.2 h_2 - 0.8 h_3] \qquad (1)$$

An entropy rate balance at steady state takes the form

$$0 = \sum \frac{\dot{Q}_j}{T_j}^{\,0} + \dot{m}_1 s_1 - \dot{m}_2 s_2 - \dot{m}_3 s_3 + \dot{\sigma}_{cv}$$

or

$$\dot{\sigma}_{cv} = \dot{m}_3 s_3 + \dot{m}_2 s_2 - \dot{m}_1 s_1$$

$$= \dot{m}_1 [0.8 s_3 + 0.2 s_2 - s_1] \qquad (2)$$

(a) From Table A-4E, $h_1 = 1562.9$ Btu/lb, $h_2 = 1421$ Btu/lb, $s_1 = 1.6908$ Btu/lb·°R, $s_2 = 1.7187$ Btu/lb·°R. For $x_3 = 0.99$, using data from Table A-3E at 1 lbf/in².

$$h_3 = h_f + x_3 (h_g - h_f) = 69.74 + 0.99(1036) = 1095.38 \text{ Btu/lb}$$

$$s_3 = s_f + x_3 (s_g - s_f) = 0.1327 + 0.99(1.8453) = 1.9595 \text{ Btu/lb·°R}$$

Substituting values in Eq. (1)

$$\dot{W}_{cv} = 50{,}000 \frac{\text{lb}}{\text{h}} \left[1562.9 - 0.2(1421) - 0.8(1095.38) \right] \frac{\text{Btu}}{\text{lb}}$$

$$= 2.012 \times 10^7 \text{ Btu/h} \quad \longleftarrow$$

Eq. (2) gives

$$\dot{\sigma}_{cv} = 50{,}000 \frac{\text{lb}}{\text{h}} \left[0.8(1.9595) + 0.2(1.7187) - 1.6908 \right] \frac{\text{Btu}}{\text{lb·°R}} = 11{,}027 \frac{\text{Btu}}{\text{h·°R}} \quad \longleftarrow$$

PROBLEM 6.109 (Cont'd.)

(b) Data for the required plots are generated using *IT*, as follows:

IT Code

```
mdot1 = 50000  // lb/h
p1 = 1000  // lbf/in.²
T1 = 1100  // °F
p2 = 300
T2 = 800
mdot2 = 0.2 * mdot1
p3 = 1
x3 = 0.99

mdot1 = mdot2 + mdot3
Wdot = mdot1 * h1 - mdot2 * h2 - mdot3 * h3
sigmadot = mdot3 * s3 + mdot2 * s2 - mdot1 * s1

h1 = h_PT("Water/Steam", p1, T1)
h2 = h_PT("Water/Steam", p2, T2)
h3 = hsat_Px("Water/Steam", p3, x3)
s1 = s_PT("Water/Steam", p1, T1)
s2 = s_PT("Water/Steam", p2, T2)
s3 = ssat_Px("Water/Steam", p3, x3)
```

IT Results for $x_3 = 0.99$

$h_1 = 1563$ Btu/lb
$h_2 = 1421$ Btu/lb
$h_3 = 1095$ Btu/lb
$s_1 = 1.691$ Btu/lb·°R
$s_2 = 1.719$ Btu/lb·°R
$s_3 = 1.959$ Btu/lb·°R
$\dot{W}_{cv} = 2.012 \times 10^7$ Btu/h
$\dot{\sigma}_{cv} = 1.103 \times 10^4$ Btu/lb·°R

PLOTS:

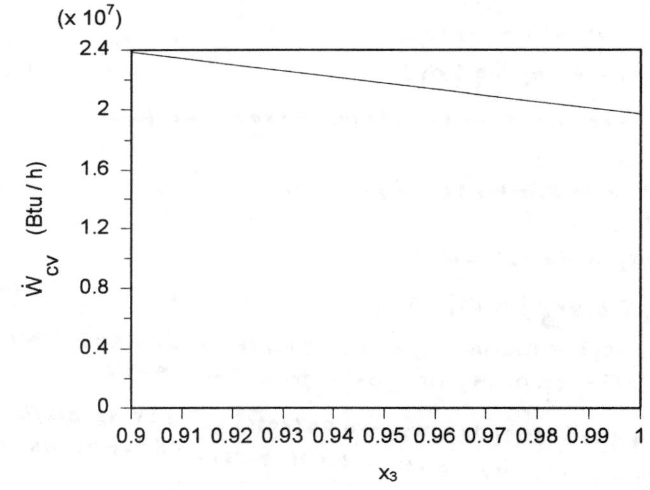

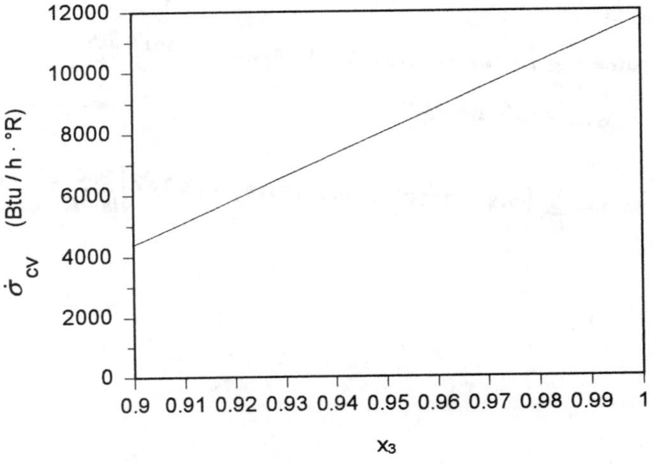

PROBLEM 6.110

KNOWN: Steady state operating data are provided for an insulated steam turbine. A two-phase liquid-vapor mixture with quality x exits the turbine.

FIND: Plot $\dot{W}_{cv}$ and $\dot{\sigma}_{cv}$ versus x

SCHEMATIC & GIVEN DATA:

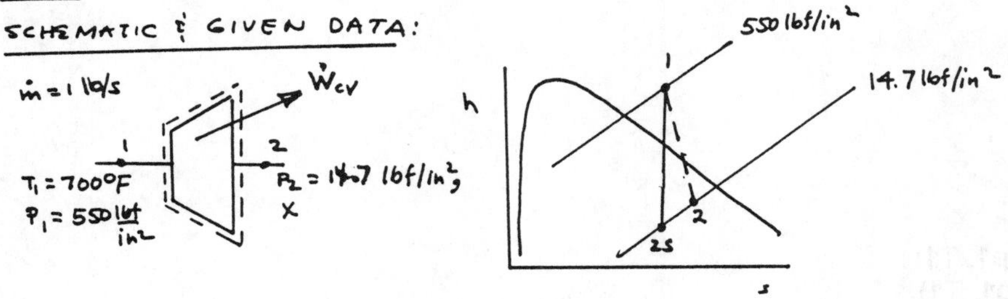

ASSUMPTIONS: (1) The control volume shown in the figure is at steady state. (2) At the exit the steam is a two-phase liquid-vapor mixture with quality x. (3) For the control volume $\dot{Q}_{cv} = 0$ and kinetic/potential energy changes can be ignored.

ANALYSIS: Reducing mass, energy, and entropy rate balances

$$\dot{W}_{cv} = \dot{m}(h_1 - h_2), \quad \dot{\sigma}_{cv} = \dot{m}(s_2 - s_1)$$

From steam table data, $h_1 = 1353.7$ Btu/lb, $s_1 = 1.5987$ Btu/lb·°R, $h_2 = h_f + x\,h_{fg} = 180.2 + x(970.4)$, $s_2 = s_f + x\,s_{fg} = 0.3121 + x(1.4446)$. Then

$$\dot{W}_{cv} = \left(1\ \frac{lb}{s}\right)(1353.7 - 180.2 - 970.4\,x)\left(\frac{Btu}{lb}\right) = [1173.5 - 970.4\,x]\ \frac{Btu}{s} \quad (3)$$

$$\dot{\sigma}_{cv} = \left(1\ \frac{lb}{s}\right)(0.3121 + 1.4446\,x - 1.5987)\left(\frac{Btu}{lb\cdot°R}\right) = [1.4446\,x - 1.2866]\ \frac{Btu}{s\cdot°R} \quad (4)$$

Sample calculation: Since $\dot{\sigma}_{cv} \geq 0$, the smallest allowed value for x is x_{2s} corresponding to $\dot{\sigma}_{cv} = 0$. Setting Eq. (4) to zero and solving; $x_{2s} = 0.8906$. The corresponding value of $\dot{W}_{cv}$ from Eq. (3) is 309.3 Btu/s.

The following plots of Eq. (3) and (4) are made using IT, although other plotting software such as a spreadsheet could be used.

Plots:

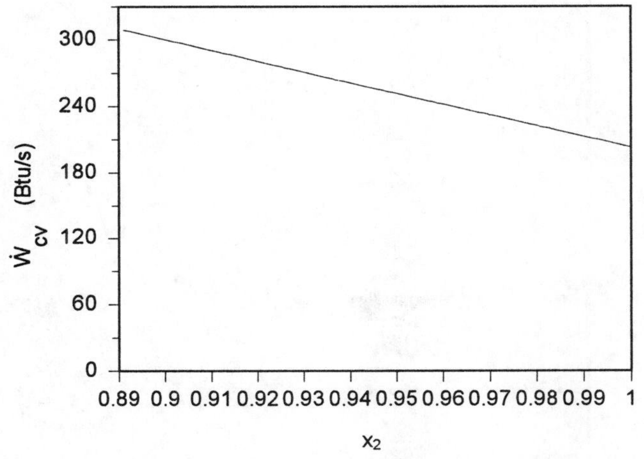

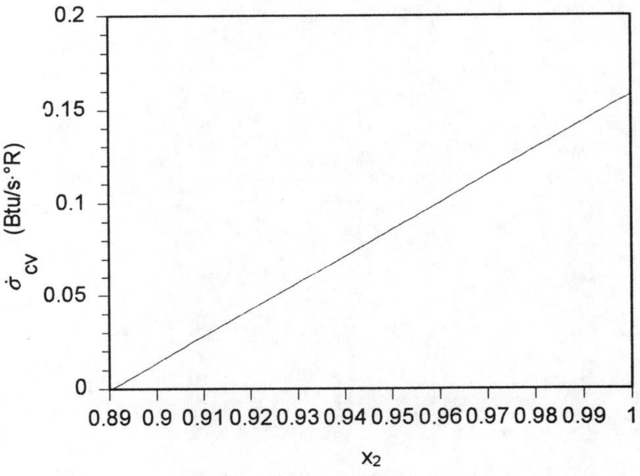

The plots show that greater entropy production (irreversibility) corresponds to less power output for the given operating conditions, as expected.

PROBLEM 6.110 (Contd.)

```
mdot = 1  // lb/s
T1 = 700  // °F
p1 = 550  // lbf/in.²
p2 = 14.7
x2 = 0.8906

Wdot = mdot * (h1 - h2)
sigmadot = mdot * (s2 - s1)

h1 = h_PT("Water/Steam", p1, T1)
s1 = s_PT("Water/Steam", p1, T1)
h2 = hsat_Px("Water/Steam", p2, x2)
s2 = ssat_Px("Water/Steam", p2, x2)
```

IT Results for $x_2 = 1$

$h_1 = 1353$ Btu/lb
$h_2 = 1150$ Btu/lb
$s_1 = 1.599$ Btu/lb·°R
$s_2 = 1.756$ Btu/lb·°R
$\dot{W}_{cv} = 203.2$ Btu/s
$\dot{\sigma}_{cv} = 0.1579$ Btu/s·°R

IT Results for $x_2 = 0.8906$

$h_1 = 1353$ Btu/lb
$h_2 = 1044$ Btu/lb
$s_1 = 1.599$ Btu/lb·°R
$s_2 = 1.598$ Btu/lb·°R
$\dot{W}_{cv} = 309.3$ Btu/s
$\dot{\sigma}_{cv} = 0$

PROBLEM 6.111

KNOWN: Steady-state operating data are provided for steam flowing through a pipe.

FIND: Determine (a) the velocity of the steam at the pipe exit, (b) the rate of heat transfer from the pipe, (c) the rate of entropy production for each of two specifications of the control volume.

SCHEMATIC & GIVEN DATA:

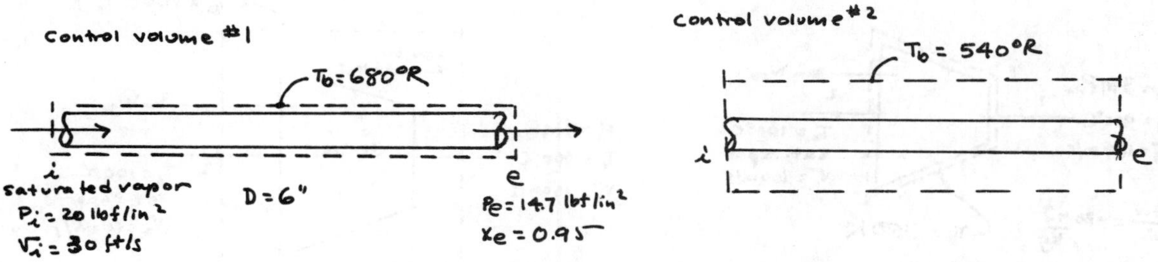

ASSUMPTIONS: (1) The two control volumes shown above are at steady state. (2) For the control volumes, heat transfer with the surroundings occurs at temperature T_b only; $\dot{W}_{cv} = 0$; and potential energy effects are absent.

ANALYSIS: Consider control volume #1. At steady state a mass rate balance reads $\dot{m}_i = \dot{m}_e$. Thus

$$\frac{A_i V_i}{v_i} = \frac{A_e V_e}{v_e} \Rightarrow V_e = \frac{v_e}{v_i} V_i = \left(\frac{25.46\ \text{ft}^3/\text{lb}}{20.09\ \text{ft}^3/\text{lb}}\right)\left(30\frac{\text{ft}}{\text{s}}\right) = 38.02\ \text{ft/s} \longleftarrow V_e$$

Steam table data: $v_i = 20.09$ ft³/lb, $v_e = v_f + x_e v_{fg} = 0.01672 + 0.95(26.8 - .01672) = 25.46$ ft³/lb

with these data the mass flow rate is

$$\dot{m} = \frac{A_i V_i}{v_i} = \left(\frac{\pi D^2}{4}\right)\left(\frac{V_i}{v_i}\right) = \frac{\pi}{4}(0.5\ \text{ft})^2\left(\frac{30\ \text{ft/s}}{20.09\ \text{ft}^3/\text{lb}}\right) = 0.293\ \frac{\text{lb}}{\text{s}}$$

Mass and energy balances reduce to give

$$\dot{Q}_{cv} = \dot{m}\left(h_e - h_i + \frac{V_e^2 - V_i^2}{2}\right) = \left(0.293\frac{\text{lb}}{\text{s}}\right)\left[(1102 - 1156.4)\frac{\text{Btu}}{\text{lb}} + \frac{(38.02\ \text{ft/s})^2 - (30\ \text{ft/s})^2}{2\left(32.2\ \frac{\text{lb}\cdot\text{ft}}{\text{lbf}\cdot\text{s}^2}\right)}\left|\frac{\text{Btu}}{778\ \text{ft}\cdot\text{lbf}}\right|\right] = -15.94\ \frac{\text{Btu}}{\text{s}} \longleftarrow \dot{Q}_{cv}$$

Steam table data: $h_i = 1156.4$ Btu/lb, $h_e = h_f + x_e h_{fg} = 180.15 + 0.95(970.4) = 1102$ Btu/lb.

Mass and entropy balances reduce to give

$$\dot{\sigma}_{cv} = -\frac{\dot{Q}_{cv}}{T_b} + \dot{m}(s_e - s_i)$$

Steam table data: $s_i = 1.7320$ Btu/lb·°R, $s_e = s_f + x_e s_{fg} = 0.3121 + 0.95(1.4446) = 1.6845$ Btu/lb·°R.

Control volume #1, $T_b = 680°R$:

$$\dot{\sigma}_{cv} = -\frac{(-15.94\ \text{Btu/s})}{680°\text{R}} + \left(0.293\frac{\text{lb}}{\text{s}}\right)(1.6845 - 1.7320)\frac{\text{Btu}}{\text{lb}\cdot°\text{R}} = 0.0095\ \frac{\text{Btu}}{\text{s}\cdot°\text{R}} \longleftarrow (\dot{\sigma}_{cv})_1$$

Control volume #2, $T_b = 540°R$:

$$\dot{\sigma}_{cv} = -\frac{(-15.94\ \text{Btu/s})}{540°\text{R}} + (0.293)(1.6845 - 1.7320) = 0.0156\ \frac{\text{Btu}}{\text{s}\cdot°\text{R}} \longleftarrow (\dot{\sigma}_{cv})_2$$

COMMENT: For control volume #1 there is a pressure decrease as the steam flows through the pipe. For this control volume the pressure drop is the most significant source of irreversibility. For control volume #2, in addition to the pressure drop irreversibility, there is the heat transfer from the outer surface of the pipe to the surroundings. Accordingly, the entropy production is greater for C.V. #2 than for C.V. #1. For control volume #1, the heat transfer effect is an *external* irreversibility.

PROBLEM 6.112

KNOWN: Steady state operating data are provided for a steam turbine.

FIND: Determine the work developed, and for each of two choices of the control volume the entropy produced.

SCHEMATIC & GIVEN DATA:

CONTROL VOLUME #1:

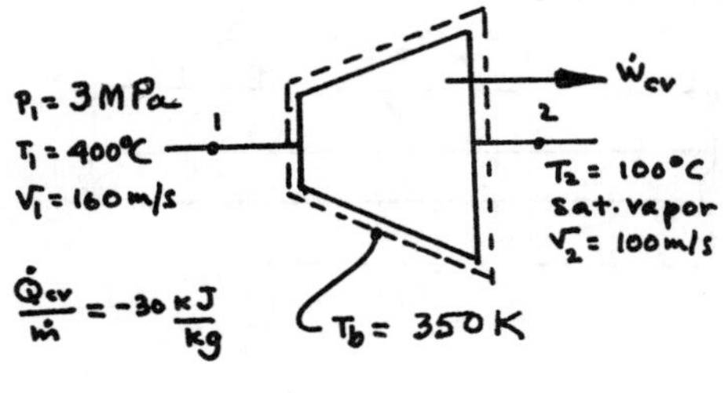

CONTROL VOLUME #2:

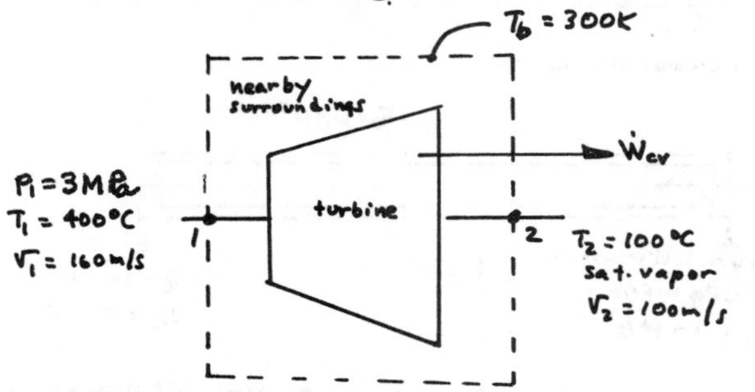

ASSUMPTIONS: (1) The control volumes shown above are at steady state. (2) For the control volumes heat transfer takes place only at temperature T_b. (3) Potential energy changes can be ignored.

ANALYSIS: For control volume #1, mass and energy rate balances reduce to read

$$\frac{\dot{W}_{cv}}{\dot{m}} = \frac{\dot{Q}_{cv}}{\dot{m}} + (h_1 - h_2) + \frac{V_1^2 - V_2^2}{2} \qquad (1)$$

With data from Table A-4, $h_1 = 3230.9$ kJ/kg, $s_1 = 6.9212$ kJ/kg·K, and from Table A-2 $h_2 = 2676.1$ kJ/kg, $s_2 = 7.3549$ kJ/kg·K. Thus

$$\frac{\dot{W}_{cv}}{\dot{m}} = -30\frac{kJ}{kg} + (3230.9 - 2676.1)\frac{kJ}{kg} + \frac{(160)^2 - (100)^2 (m^2/s^2)}{2}\left|\frac{1N}{1 kg\cdot m/s^2}\right|\left|\frac{1 kJ}{10^3 N\cdot m}\right|$$

$$= (-30 + 554.8 + 7.8)\frac{kJ}{kg} = +532.6\frac{kJ}{kg} \longleftarrow (\dot{W}_{cv}/\dot{m})$$

An entropy rate balance at steady state reduces with assumption 2

$$0 = \frac{\dot{Q}_{cv}}{T_b} + \dot{m}(s_1 - s_2) + \dot{\sigma}_{cv}$$

$$\frac{\dot{\sigma}_{cv}}{\dot{m}} = \frac{(-\dot{Q}_{cv}/\dot{m})}{T_b} + (s_2 - s_1) \qquad (2)$$

$$= \frac{-(-30 \text{ kJ/kg})}{350K} + (7.3549 - 6.9212)\frac{kJ}{kg\cdot K} = 0.5194 \frac{kJ}{kg\cdot K} \longleftarrow (\dot{\sigma}_{cv}/\dot{m})_1$$

Equation (2) is applicable to control volume #2, except now $T_b = 300K$. Thus

$$\frac{\dot{\sigma}_{cv}}{\dot{m}} = \frac{-(-30 \text{ kJ/kg})}{300K} + (7.3549 - 6.9212)\frac{kJ}{kg\cdot K} = 0.5337\frac{kJ}{kg\cdot K} \longleftarrow (\dot{\sigma}_{cv}/\dot{m})_2$$

COMMENT: The entropy production in control volume #2 is greater because this enlarged control volume includes an additional source of irreversibility: heat transfer from the outer surface of the turbine to the surroundings. For control volume #1 the heat transfer effect is an <u>external</u> irreversibility.

PROBLEM 6.113

KNOWN: Steady state operating data are provided for an air turbine.

FIND: Determine the work developed, and for each of two choices of the control volume the entropy produced.

SCHEMATIC & GIVEN DATA:

Control Volume #1

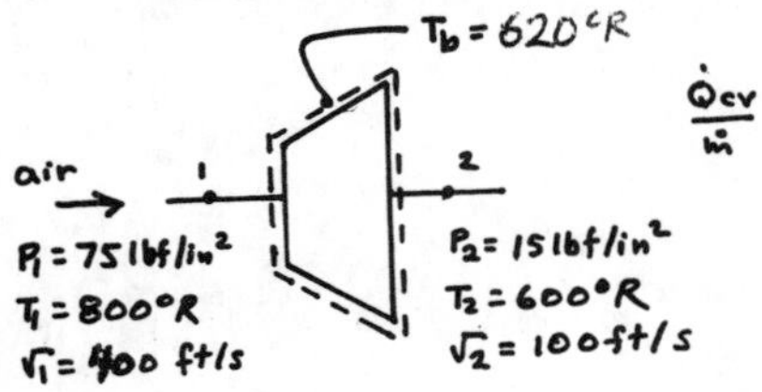

$\frac{\dot{Q}_{cv}}{\dot{m}} = -10\ \frac{Btu}{lb}$

Control Volume #2

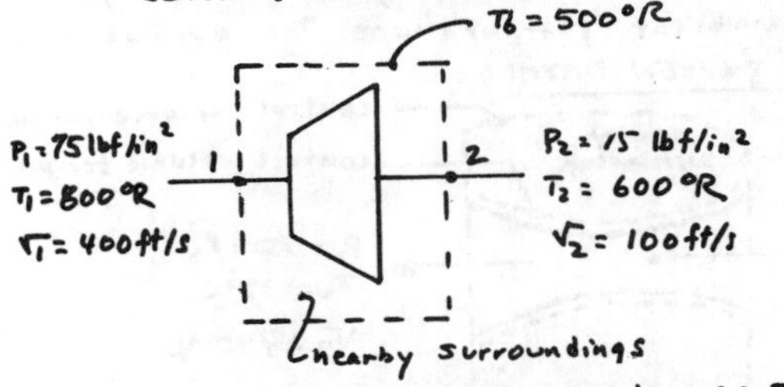

ASSUMPTIONS: (1) The control volumes shown above are at steady state. (2) For the control volumes heat transfer takes place only at temperature T_b. (3) Potential ① energy changes can be ignored. (4) Air is modeled as an ideal gas.

ANALYSIS: For control volume #1 mass and energy balances reduce to read

$$\frac{\dot{W}_{cv}}{\dot{m}} = \frac{\dot{Q}_{cv}}{\dot{m}} + (h_1 - h_2) + \left(\frac{V_1^2 - V_2^2}{2}\right) \qquad (1)$$

With data from Table A-22E $h_1 = 191.81$ Btu/lb, $s_1^\circ = 0.69558$ Btu/lb·°R, $h_2 = 143.47$ Btu/lb, $s_2^\circ = 0.62607$ Btu/lb·°R. Thus

$$\frac{\dot{W}_{cv}}{\dot{m}} = -10\ \frac{Btu}{lb} + (191.81 - 143.47)\frac{Btu}{lb} + \left(\frac{(400)^2 - (100)^2}{2}\right)\left(\frac{ft^2}{s^2}\right)\left|\frac{lbf}{32.2\ lb\cdot ft/s^2}\right|\left|\frac{Btu}{778\ ft\cdot lbf}\right|$$

$$= -10 + 48.34 + 2.99 = +41.33\ Btu/lb \longleftarrow (\dot{W}_{cv}/\dot{m})$$

An entropy balance at steady state reduces with assumption 2

$$0 = \frac{\dot{Q}_{cv}}{T_b} + \dot{m}(s_1 - s_2) + \dot{\sigma}_{cv}$$

or

$$\frac{\dot{\sigma}_{cv}}{\dot{m}} = \frac{(-\dot{Q}_{cv}/\dot{m})}{T_b} + (s_2 - s_1) \qquad (2)$$

$$= \frac{(-\dot{Q}_{cv}/\dot{m})}{T_b} + (s_2^\circ - s_1^\circ - R \ln p_2/p_1)$$

$$= -\frac{(-10\ Btu/lb)}{620°R} + \left(0.62607 - 0.69558 - \frac{1.986}{28.97}\ln\frac{15}{75}\right)\frac{Btu}{lb\cdot°R} = 0.057\ \frac{Btu}{lb\cdot°R} \longleftarrow (\dot{\sigma}_{cv}/\dot{m})_1$$

Equation (2) is applicable to control volume #2, except now $T_b = 500°R$. Thus

$$\frac{\dot{\sigma}_{cv}}{\dot{m}} = -\frac{(-10)}{500} + \left(0.62607 - 0.69558 - \frac{1.986}{28.97}\ln\frac{15}{75}\right) = 0.061\ \frac{Btu}{lb\cdot°R} \longleftarrow (\dot{\sigma}_{cv}/\dot{m})_2$$

COMMENT:

The entropy production in control volume #2 is greater because this enlarged control volume includes an additional source of irreversibility: heat transfer from the outer surface of the turbine to the surroundings. For control volume #1 the heat transfer effect is an external irreversibility.

---

1. From Table A-1E for air $p_c = 37.2$ atm, $T_c = 239°R$. Thus

$$p_{R1} = \frac{75/14.7}{37.2} = 0.137, \quad T_{R1} = \frac{800}{239} = 3.35$$

$$p_{R2} = \frac{15/14.7}{37.2} = 0.027, \quad T_{R2} = \frac{600}{239} = 2.51$$

Inspection of Fig. A-1 indicates that the ideal gas model is appropriate at these states.

PROBLEM 6.114

KNOWN: Operating data are provided for a nozzle at steady state through which O_2 is passing.

FIND: (a) For a control volume enclosing the nozzle only, evaluate $\dot{Q}_{cv}/\dot{m}$, and the change in specific entropy from inlet to exit. Indicate the additional data required to evaluate $\dot{\sigma}_{cv}/\dot{m}$. (b) For an enlarged control volume including enough of the nearby surroundings that heat transfer occurs at the ambient temperature T_0, evaluate $\dot{\sigma}_{cv}/\dot{m}$.

SCHEMATIC & GIVEN DATA:

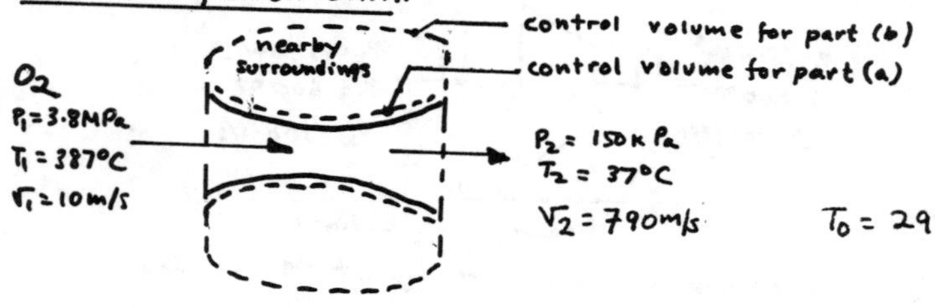

$T_0 = 293 K\ (20°C)$

ASSUMPTIONS: (1) Each of the control volumes shown in the accompanying figure operates at steady state. (2) The change in potential energy from inlet to exit ① is negligible. (3) O_2 can be modeled as an ideal gas.

ANALYSIS: (a) At steady state $\dot{m}_1 = \dot{m}_2 = \dot{m}$, and an energy rate balance reduces to

$$0 = \dot{Q}_{cv} - \cancel{\dot{W}_{cv}}^{0} + \dot{m}\left(h_1 - h_2 + \frac{V_1^2 - V_2^2}{2} + \cancel{g(z_1 - z_2)}^{0}\right)$$

or

$$\frac{\dot{Q}_{cv}}{\dot{m}} = h_2 - h_1 + \frac{V_2^2 - V_1^2}{2}$$

With M = 32 from Table A-1 and data from Table A-23, $\bar{h}_1 = 19{,}870$ kJ/kmol, $\bar{s}_1^\circ = 229.430$ kJ/kmol·K, $\bar{h}_2 = 9{,}030$ kJ/kmol, $\bar{s}_2^\circ = 206.177$ kJ/kmol·K. Thus

$$\frac{\dot{Q}_{cv}}{\dot{m}} = \frac{(9030 - 19870)\text{ kJ/kmol}}{32\text{ kg/kmol}} + \left(\frac{(790)^2 - (10)^2}{2}\right)\left(\frac{m^2}{s^2}\right)\left|\frac{1\,N}{1\,kg\cdot m/s^2}\right|\left|\frac{1\,kJ}{10^3\,N\cdot m}\right|$$

$$= -338.75\text{ kJ/kg} + 312\text{ kJ/kg} = -26.75\text{ kJ/kg} \longleftarrow (\dot{Q}_{cv}/\dot{m})$$

With Eq. 6.2 b

$$s_2 - s_1 = \frac{1}{M}\left[\bar{s}_2^\circ - \bar{s}_1^\circ - \bar{R}\ln\frac{P_2}{P_1}\right] = \frac{\left[206.177 - 229.430 - 8.314\ln\frac{0.15}{3.8}\right]}{32}$$

$$= +0.1131\text{ kJ/kg·K} \longleftarrow (s_2 - s_1)$$

To evaluate $\dot{\sigma}_{cv}/\dot{m}$ would require information about the temperature on the boundary of the control volume and the rate of heat transfer at each temperature.

(b) For the enlarged control volume the heat transfer takes place only at $T_0 = 293$ K, and the entropy rate balance reduces to

$$0 = \frac{\dot{Q}_{cv}}{T_0} + \dot{m}(s_1 - s_2) + \dot{\sigma}_{cv} \Rightarrow \dot{\sigma}_{cv}/\dot{m} = -\frac{\dot{Q}_{cv}/\dot{m}}{T_0} + s_2 - s_1$$

Using previously calculated values

$$\dot{\sigma}_{cv}/\dot{m} = -\frac{(-26.75\text{ kJ/kg})}{293\text{ K}} + 0.1131\frac{kJ}{kg\cdot K} = 0.2044\frac{kJ}{kg\cdot K} \longleftarrow \dot{\sigma}_{cv}/\dot{m}$$

---

1. From Table A-1, for O_2 $p_c = 50.5$ bar, $T_c = 154$ K. Then,

$$p_{R1} = \frac{3.8}{5.05} = 0.752 \qquad T_{R1} = \frac{660}{154} = 4.29$$

Inspection of Fig A-1 indicates that the ideal gas model is appropriate at this state. Similarly, the ideal gas model is appropriate at State 2, as can be verified.

PROBLEM 6.115

KNOWN: Steady-state operating data are provided for an aircompressor.

FIND: (a) Determine the heat transfer rate and change in entropy from inlet to exit for a control volume enclosing the compressor only. Discuss. (b) For an enlarged control volume, determine the rate of entropy production.

SCHEMATIC & GIVEN DATA:

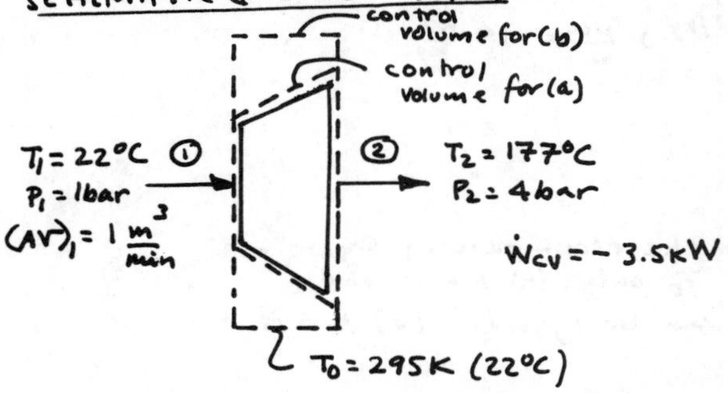

ASSUMPTIONS:

1. The control volumes shown in the accompanying figure are at steady state.
2. Kinetic and potential energy effects can be ignored.
3. Air is modeled as an ideal gas.

ANALYSIS: (a) At steady state, the mass rate balance reads, $\dot{m}_2 = \dot{m}_1 \equiv \dot{m}$, where

$$\dot{m} = \frac{(AV)_1}{v_1} = \frac{(AV)_1}{(RT_1/P_1)} = \frac{(1\, m^3/min)(10^5 N/m^2)}{\left(\frac{8314}{28.97}\frac{N\cdot m}{kg\cdot K}\right)(295K)} = 1.18\, kg/min$$

The energy rate balance at steady state reduces to give on rearrangement

$$\dot{Q}_{cv} = \dot{W}_{cv} + \dot{m}(h_2 - h_1) = -3.5kW + \left(1.18\frac{kg}{min}\right)\left|\frac{1 min}{60 s}\right|\left[451.8 - 295.17\right]\frac{kJ}{kg}\left|\frac{1 kW}{1 kJ/s}\right|$$
$$= -0.42 kW$$

where enthalpy data is from Table A-22. The entropy change is obtained with Eq. 6.21a and $s°$ data from Table A-22:

$$s_2 - s_1 = s°(T_2) - s°(T_1) - R\ln\frac{P_2}{P_1} = \left[2.11161 - 1.68515 - \frac{8.314}{28.97}\ln 4\right] = 0.0286\frac{kJ}{kg\cdot K}$$

To evaluate the entropy production for the control volume of part (a) would require information about the temperature at which heat transfer occurs on the boundary of the control volume.

(b) For the enlarged control volume of part (b), heat transfer takes place at $T_0 = 295K$. An entropy rate balance at steady state reads

$$0 = \frac{\dot{Q}_{cv}}{T_0} + \dot{m}(s_1 - s_2) + \dot{\sigma}_{cv} \Rightarrow \dot{\sigma}_{cv} = -\frac{\dot{Q}_{cv}}{T_0} + \dot{m}(s_2 - s_1)$$

with results from part (a)

$$\dot{\sigma}_{cv} = \frac{-(-0.42 kW)}{295K} + \left(\frac{1.18}{60}\frac{kg}{s}\right)\left[0.0286\frac{kJ}{kg\cdot K}\right]\left|\frac{1 kW}{1 kJ/s}\right|$$
$$= 1.98\times10^{-3}\frac{kW}{K}$$

PROBLEM 6.116

KNOWN: Operating data are provided for an air compressor at steady state.

FIND: Determine the temperature of the air exiting the compressor and the entropy production rate per kg of air flowing.

SCHEMATIC & GIVEN DATA:

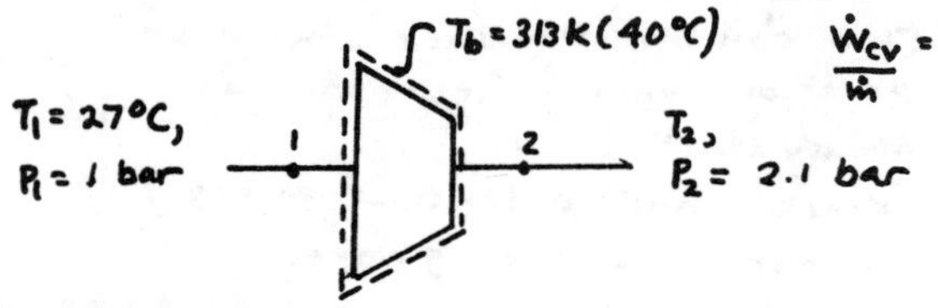

$\dot{W}_{cv}/\dot{m} = -94.6$ kJ/kg , $\dot{Q}_{cv}/\dot{m} = -14$ kJ/kg

ASSUMPTIONS: (1) The control volume shown in the accompanying figure is at steady state. (2) Heat transfer takes place at T_b only. (3) Kinetic and potential energy changes from inlet to exit can be ignored. (4) Air is modeled as an ideal gas.

ANALYSIS: At steady state $\dot{m}_1 = \dot{m}_2 = \dot{m}$, and an energy rate balance reduces with assumption 3

$$0 = \dot{Q}_{cv} - \dot{W}_{cv} + \dot{m}\left(h_1 - h_2 + \cancel{\frac{V_1^2 - V_2^2}{2}}^{\,0} + \cancel{g(z_1 - z_2)}^{\,0}\right)$$

or

$$h_2 = h_1 + \frac{\dot{Q}_{cv}}{\dot{m}} - \frac{\dot{W}_{cv}}{\dot{m}} = (300.19 - 14 + 94.6)\frac{kJ}{kg} = 380.79 \frac{kJ}{kg}$$

where h_1 is from Table A-22. From the same table with $h_2 = 380.79 \frac{kJ}{kg}$, $T_2 = 380K$. ← T_2

With assumption 2, an entropy rate balance at steady state takes the form

$$0 = \frac{\dot{Q}_{cv}}{T_b} + \dot{m}(s_1 - s_2) + \dot{\sigma}_{cv}$$

or

$$\frac{\dot{\sigma}_{cv}}{\dot{m}} = \frac{(-\dot{Q}_{cv}/\dot{m})}{T_b} + s_2 - s_1$$

$$= \left(\frac{-\dot{Q}_{cv}/\dot{m}}{T_b}\right) + \left(s_2^\circ - s_1^\circ - R \ln \frac{P_2}{P_1}\right)$$

$$= \left(\frac{14 \text{ kJ/kg}}{313 \text{ K}}\right) + \left(1.94001 - 1.70203 - \frac{8.314}{28.97} \ln \frac{2.1}{1}\right) \frac{kJ}{kg \cdot K}$$

$$= (0.0447 + 0.0251) \frac{kJ}{kg \cdot K}$$

$$= 0.0698 \frac{kJ}{kg \cdot K}$$ ← $\dot{\sigma}_{cv}/\dot{m}$

PROBLEM 6.117

KNOWN: Operating data are provided for a duct system at steady state.

FIND: Determine the rate of entropy production.

SCHEMATIC & GIVEN DATA:

Data from the solution to Problem 4.68: $\dot{m}_1 = 369.5$ lb/min, $\dot{m}_2 = 158.8$ lb/min, $T_3 = 528°R$.

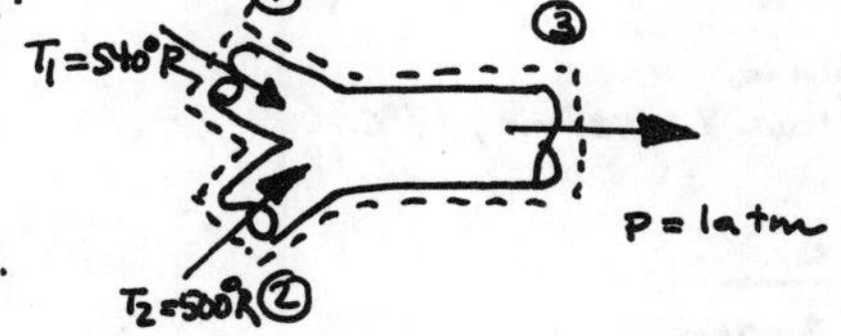

ASSUMPTIONS: 1. The control volume shown in the schematic is at steady state. 2. For the control volume, $\dot{Q}_{cv} = \dot{W}_{cv} = 0$. 3. Air is modeled as an ideal gas with constant $c_p = 0.24$ Btu/lb·°R.

ANALYSIS: An entropy rate balance at steady state reduces to read

$$0 = \sum_j \frac{\dot{Q}_j}{T_j}^{\,0} + \dot{m}_1 s_1 + \dot{m}_2 s_2 - \dot{m}_3 s_3 + \dot{\sigma}_{cv}$$

Since $\dot{m}_3 = \dot{m}_1 + \dot{m}_2$ at steady state

(1) $$\dot{\sigma}_{cv} = (\dot{m}_1 + \dot{m}_2) s_3 - \dot{m}_1 s_1 - \dot{m}_2 s_2$$

$$= \dot{m}_1 (s_3 - s_1) + \dot{m}_2 (s_3 - s_2)$$

Introducing Eq. 6.23

$$\dot{\sigma}_{cv} = \dot{m}_1 \left[c_p \ln \frac{T_3}{T_1} - R \ln \frac{p_3}{p_1}^{\,0} \right] + \dot{m}_2 \left[c_p \ln \frac{T_3}{T_2} - R \ln \frac{p_3}{p_2}^{\,0} \right]$$

$$= c_p \left[\dot{m}_1 \ln \frac{T_3}{T_1} + \dot{m}_2 \ln \frac{T_3}{T_2} \right]$$

$$= 0.24 \frac{\text{Btu}}{\text{lb}\cdot°\text{R}} \left[\left(369.5 \frac{\text{lb}}{\text{min}}\right) \ln \frac{528}{540} + \left(158.8 \frac{\text{lb}}{\text{min}}\right) \ln \frac{528}{500} \right]$$

$$= 0.084 \frac{\text{Btu}}{\text{min}\,°\text{R}}$$ ←

---

1. When using IT to perform this calculation, note that IT would return s_1, s_2 and s_3 directly for air as an ideal gas.

PROBLEM 6.118

KNOWN: Operating data are provided for a water-jacketed air compressor.

FIND: Determine the power required and the rate of entropy production.

SCHEMATIC & GIVEN DATA:

cooling water in, $\dot{m}_{cw} = 8.6$ kg/min, $17°C$

AIR
$P_1 = 1$ bar
$T_1 = 20°C$
$(AV)_1 = 9$ m³/min

$P_2 = 5$ bar
$T_2 = 160°C$

cooling water out, $25°C$

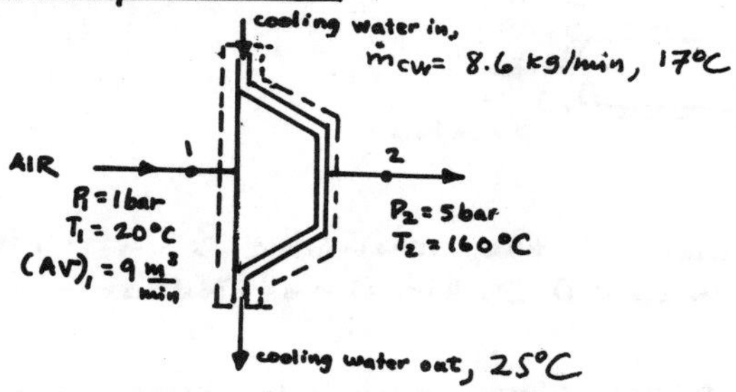

ASSUMPTIONS: (1) The control volume shown in the figure operates at steady state with negligible kinetic and potential energy effects and $\dot{Q}_{cv} = 0$. (2) Air is modeled as an ideal gas. (3) The cooling water is modeled as incompressible with constant specific heat c and a negligible change in pressure.

① **ANALYSIS:** At steady state $\dot{m}_1 = \dot{m}_2 = \dot{m}$ and the rates of cooling water entering and exiting are equal. An energy rate balance reduces with listed assumptions to give

$$\dot{W}_{cv} = \dot{m}[h_1 - h_2] + \dot{m}_{cw}[h_{cw,in} - h_{cw,out}]$$

With the ideal gas equation of state

$$\dot{m} = \frac{(AV)_1}{v_1} = \frac{P_1(AV)_1}{RT_1} = \frac{(10^5 \text{ N/m}^2)(9 \text{ m}^3/\text{min})}{\left(\frac{8314}{28.97} \frac{\text{N·m}}{\text{kg·K}}\right)(293 \text{ K})} = 10.703 \frac{\text{kg}}{\text{min}}$$

Then, with specific enthalpies h_1 and h_2 from Table A-22, and a specific heat value for liquid water from Table A-19, $c = 4.19$ kJ/kg·K, together with Eq. 3.20b: $\Delta h = c\Delta T + v\Delta p^{\to 0}$, where Δp is dropped by assumption 3, the power is

$$\dot{W}_{cv} = (10.703 \tfrac{\text{kg}}{\text{min}})(293.2 - 434.5)\tfrac{\text{kJ}}{\text{kg}} + (8.6 \tfrac{\text{kg}}{\text{min}})(4.19 \tfrac{\text{kJ}}{\text{kg·K}})(290 - 298)\text{ K}$$

$$= (-1512.3 - 288.3)\left(\tfrac{\text{kJ}}{\text{min}}\right)\left|\tfrac{\text{min}}{60\text{s}}\right|\left|\tfrac{\text{kW}}{\text{kJ/s}}\right| = -30.01 \text{ kW} \quad \longleftarrow \dot{W}_{cv}$$

At steady state an entropy rate balance reduces to give

$$0 = \sum \frac{\dot{Q}_j}{T_j}^0 + \dot{m}(s_1 - s_2) + \dot{m}_{cw}(s_{cw,in} - s_{cw,out}) + \dot{\sigma}_{cv}$$

Or, with Eq. 6.21a and Eq. 6.24

$$\dot{\sigma}_{cv} = \dot{m}\left(s_2° - s_1° - R\ln\frac{P_2}{P_1}\right) + \dot{m}_{cw}\, c\ln\frac{T_{cw,out}}{T_{cw,in}}$$

$$= (10.703 \tfrac{\text{kg}}{\text{min}})\left(2.07234 - 1.6783 - \tfrac{8.314}{28.97}\ln 5\right)\tfrac{\text{kJ}}{\text{kg·K}} +$$

$$(8.6 \tfrac{\text{kg}}{\text{min}})(4.19 \tfrac{\text{kJ}}{\text{kg·K}})\ln\tfrac{298}{290}$$

$$= (-0.72617 + 0.9801)\left(\tfrac{\text{kJ/min}}{\text{K}}\right)\left|\tfrac{\text{min}}{60\text{s}}\right|\left|\tfrac{\text{kW}}{\text{kJ/s}}\right|$$

$$= 4.23 \times 10^{-3} \text{ kW/K} \quad \longleftarrow$$

---

1. Alternatively, for the cooling water saturated liquid data could be used: $h \approx h_f(T)$, $s \approx s_f(T)$.

PROBLEM 6.119

KNOWN: Operating data are provided for a counterflow heat exchanger at steady state with NH_3 flowing on one side and air flowing on the other side.

FIND: Determine the mass flow rate of the ammonia and the rate of entropy production within the heat exchanger.

SCHEMATIC & GIVEN DATA:

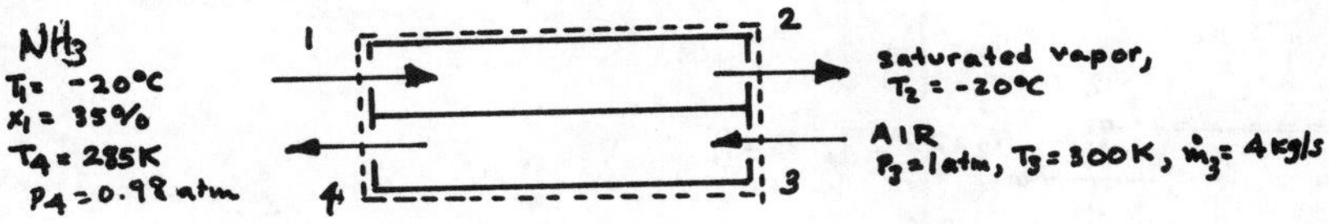

ASSUMPTIONS: (1) The control volume shown in the accompanying figure is at steady state. (2) $\dot{Q}_{cv} = 0$ and kinetic and potential energy effects are negligible. (3) Air is modeled as an ideal gas.

ANALYSIS: At steady state the inlet and exit mass flow rates of each side of the heat exchanger are equal: $\dot{m}_1 = \dot{m}_2 = \dot{m}_R$, $\dot{m}_3 = \dot{m}_4 = \dot{m}$, and the entropy rate balance reduces to give

$$0 = \sum \frac{\dot{Q}_j}{T_j}^{\,0} + \dot{m}_R(s_1 - s_2) + \dot{m}(s_3 - s_4) + \dot{\sigma}_{cv} \qquad (1)$$

This requires the mass flow rate $\dot{m}_R$ which can be evaluated from an energy rate balance at steady state with assumption 2:

$$0 = \dot{Q}_{cv}^{\,0} - \dot{W}_{cv}^{\,0} + \dot{m}_R(h_1 - h_2) + \dot{m}(h_3 - h_4)$$

or upon rearrangement

$$\dot{m}_R = \frac{\dot{m}(h_3 - h_4)}{(h_2 - h_1)}$$

The enthalpies h_3 and h_4 are obtained from Table A-22: $h_3 = 300.19$ kJ/kg, $h_4 = 285.14$ kJ/kg. The enthalpies h_1 and h_2 are obtained from Table A-13 at -20°C:

$$h_2 - h_1 = h_g - (h_f + x_1(h_g - h_f)) = (1 - x_1)(h_g - h_f) = 0.65(1329.1 \text{ kJ/kg}) = 863.9 \text{ kJ/kg}$$

Thus

$$\dot{m}_R = (4 \text{ kg/s})\frac{(300.19 - 285.14) \text{ kJ/kg}}{863.9 \text{ kJ/kg}} = 0.07 \text{ kg/s} \longleftarrow$$

Upon rearrangement Eq. (1) gives

$$\dot{\sigma}_{cv} = \dot{m}_R(s_2 - s_1) + \dot{m}(s_4 - s_3)$$

With Eq. 6.21a and data from Table A-22

$$s_4 - s_3 = s°(T_4) - s°(T_3) - R \ln P_4/P_3 = (1.65055 - 1.70203) - \frac{8.314}{28.97} \ln \frac{0.98}{1}$$
$$= -0.0457 \text{ kJ/kg·K}$$

With data from Table A-13 at -20°C

$$s_2 - s_1 = s_g - (s_f + x_1(s_g - s_f)) = (1 - x_1)(s_g - s_f) = 0.65(5.6144 - 0.3642) = 3.4126 \frac{\text{kJ}}{\text{kg·K}}$$

Finally,

$$\dot{\sigma}_{cv} = \left(0.07 \frac{\text{kg}}{\text{s}}\right)\left(3.4126 \frac{\text{kJ}}{\text{kg·K}}\right) + \left(4 \frac{\text{kg}}{\text{s}}\right)\left(-0.0457 \frac{\text{kJ}}{\text{kg·K}}\right) \longleftarrow$$
$$= 0.056 \text{ kW/K}$$

PROBLEM 6.120

KNOWN: Operating data are provided for a counterflow heat exchanger at steady state with liquid water flowing on one side and R134a flowing on the other side.

FIND: Determine the mass flow rate of the water stream and the rate of entropy production.

SCHEMATIC & GIVEN DATA:

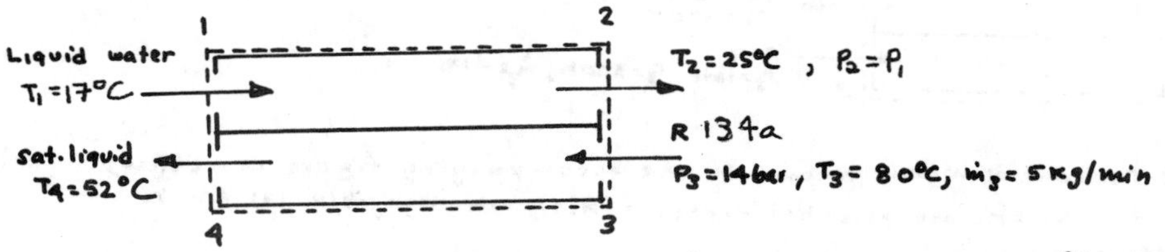

ASSUMPTIONS: (1) The control volume shown in the accompanying figure is at steady state. (2) $\dot{Q}_{cv} = 0$ and kinetic and potential energy effects are negligible. (3) Liquid water is modeled as incompressible with constant specific
① heat c and negligible pressure drop.

ANALYSIS: At steady state the inlet and exit mass flow rates of each side of the heat exchanger are equal: $\dot{m}_1 = \dot{m}_2 = \dot{m}$, $\dot{m}_3 = \dot{m}_4 = \dot{m}_R$, and the energy rate balance reduces with assumption 2 to

$$0 = \cancelto{0}{\dot{Q}_{cv}} - \cancelto{0}{\dot{W}_{cv}} + \dot{m}(h_1 - h_2) + \dot{m}_R(h_3 - h_4)$$

Thus

$$\dot{m} = \frac{\dot{m}_R(h_3 - h_4)}{(h_2 - h_1)}$$

With assumption 3, Eq. 3.20 gives $(h_2 - h_1) = c(T_2 - T_1) + \cancelto{0}{v(P_2 - P_1)}$. With $c = 4.2$ kJ/kg·K from Table A-19, $h_2 - h_1 = (4.2 \text{ kJ/kg·K})(8 \text{ K}) = 33.6$ kJ/kg. Then, with data from Tables A-10, 12

$$\dot{m} = \frac{(5 \text{ kg/min})(307.1 - 124.6)\text{ kJ/kg}}{33.6 \text{ kJ/kg}} = 27.2 \text{ kg/min} \quad \longleftarrow \dot{m}$$

An entropy rate balance at steady state reduces to give

$$0 = \cancelto{0}{\sum \frac{\dot{Q}_j}{T_j}} + \dot{m}(s_1 - s_2) + \dot{m}_R(s_3 - s_4) + \dot{\sigma}_{cv}$$

or

$$\dot{\sigma}_{cv} = \dot{m}(s_2 - s_1) + \dot{m}_R(s_4 - s_3)$$

Introducing Eq. 6.24 and data from Tables A-10 and A-12

$$\dot{\sigma}_{cv} = \dot{m}\, c \ln\frac{T_2}{T_1} + \dot{m}_R(s_4 - s_3)$$

$$= \left(27.2 \frac{\text{kg}}{\text{min}}\right)\left(4.2 \frac{\text{kJ}}{\text{kg·K}}\right) \ln\frac{298}{290} + \left(5 \frac{\text{kg}}{\text{min}}\right)(0.4432 - 0.9997)\frac{\text{kJ}}{\text{kg·K}}$$

$$= (3.1073 - 2.7825)\ \frac{\text{kJ/min}}{\text{K}}$$

$$= \left(0.3248\ \frac{\text{kJ/min}}{\text{K}}\right)\left|\frac{1 \text{ min}}{60 \text{ s}}\right|\left|\frac{1 \text{ kW}}{1 \text{ kJ/s}}\right|$$

$$= 5.4 \times 10^{-3}\ \frac{\text{kW}}{\text{K}} \quad \longleftarrow \dot{\sigma}_{cv}$$

---

1. Alternatively, for the liquid water stream saturated liquid data can be used: $h \approx h_f(T)$, $s \approx s_f(T)$.

PROBLEM 6.121

KNOWN: Operating data are provided for an open feedwater heater at steady state. The exiting stream is at pressure p.

FIND: (a) If $p = 0.7$ MPa, determine the ratio of the mass flow rates of the two incoming streams and the rate of entropy production within the heater.
(b) Plot the quantities of part (a) versus p.

SCHEMATIC & GIVEN DATA:

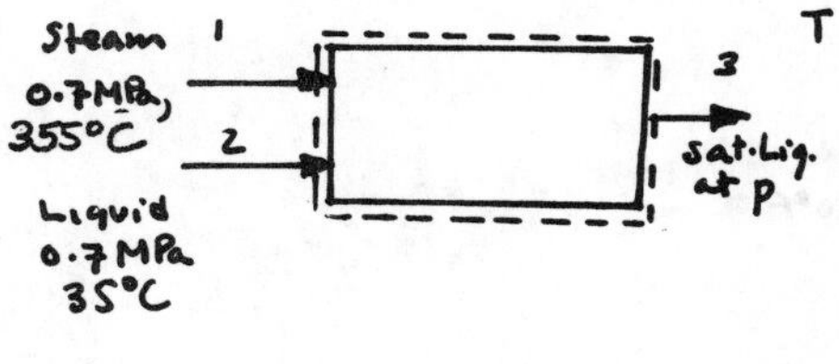

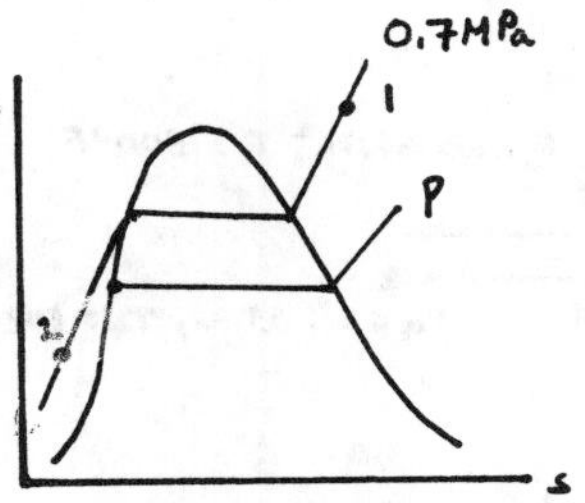

ASSUMPTIONS:
1. The control volume shown in the schematic is at steady state.
2. For the control volume, $\dot{Q}_{cv} = \dot{W}_{cv} = 0$. Kinetic and potential energy effects can be ignored.

ANALYSIS: At steady state, the mass rate balance gives $\dot{m}_3 = \dot{m}_1 + \dot{m}_2$, and the energy rate balance reduces as follows

$$0 = \dot{Q}_{cv}^{\,0} - \dot{W}_{cv}^{\,0} + \dot{m}_1 h_1 + \dot{m}_2 h_2 - (\dot{m}_1 + \dot{m}_2) h_3 \Rightarrow \frac{\dot{m}_1}{\dot{m}_2} = \frac{h_3 - h_2}{h_1 - h_3} \qquad (1)$$

An entropy rate balance reduces to

$$0 = \sum_j \frac{\dot{Q}_j}{T_j}^{\,0} + \dot{m}_1 s_1 + \dot{m}_2 - (\dot{m}_1 + \dot{m}_2) s_3 + \dot{\sigma}_{cv}$$

or

$$\frac{\dot{\sigma}_{cv}}{\dot{m}_3} = s_3 - \left[\frac{\dot{m}_1}{\dot{m}_1 + \dot{m}_2}\right] s_1 - \left[\frac{\dot{m}_2}{\dot{m}_1 + \dot{m}_2}\right] s_2 \Rightarrow \frac{\dot{\sigma}_{cv}}{\dot{m}_3} = s_3 - \left(\frac{r}{r+1}\right) s_1 - \left(\frac{1}{r+1}\right) s_2 \qquad (2)$$

where $r = \dot{m}_1/\dot{m}_2$.

(a) $p = 0.7$ MPa. With $h_2 = h_f(T_2)$ and $s_2 = s_f(T_2)$, together with data from Tables A-2, 3, 4 Eq. (1) and Eq. (2) give, respectively

① $$r = \frac{(697.22 - 146.68)\ \text{kJ/kg}}{(3174.2 - 697.22)\ \text{kJ/kg}} = 0.2223 \longleftarrow$$

$$\frac{\dot{\sigma}_{cv}}{\dot{m}_3} = 1.9922 - \left(\frac{0.2223}{1.2223}\right)(7.4896) - \left(\frac{1}{1.2223}\right)(0.5053) = 0.217\ \frac{\text{kJ}}{\text{kg}\cdot\text{K}} \longleftarrow$$

(b)

② 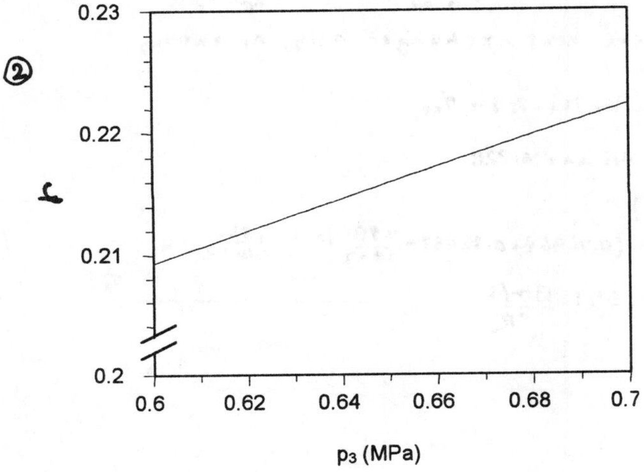

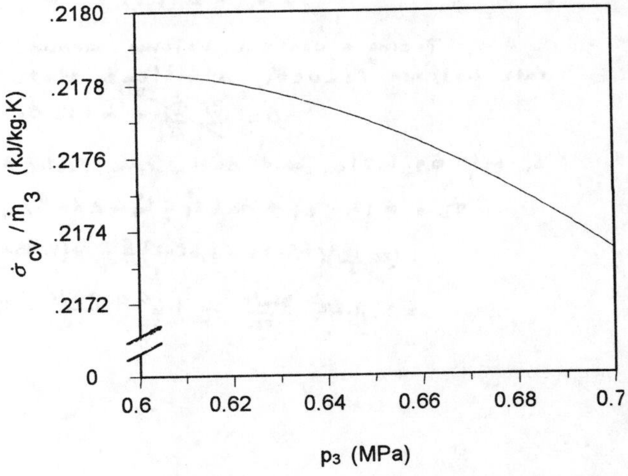

1. Values determined using Eqs. (1), (2) are sensitive to roundoff.
2. The entropy production varies only slightly over this pressure range.

PROBLEM 6.122

KNOWN: Steam expands through a turbine and then passes through a counterflow heat exchanger in which the other stream is air. Operating data at steady state are provided.

FIND: Determine the mass flow rate of the air and the rates of entropy production in the turbine and heat exchanger, respectively.

SCHEMATIC & GIVEN DATA:

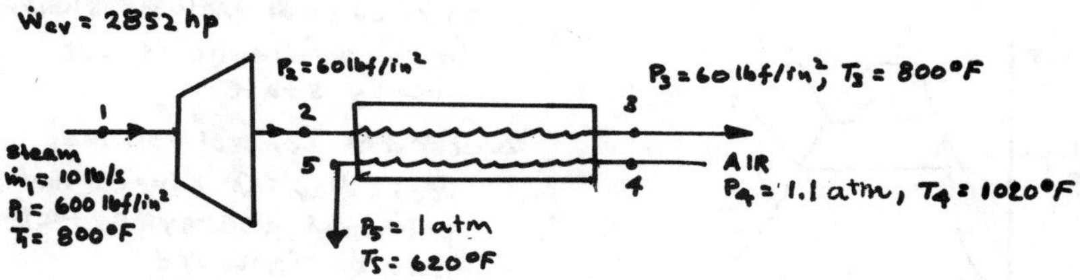

ASSUMPTIONS: (1) The turbine and heat exchanger operate at steady state. (2) There is no significant heat transfer between either component and its surroundings. (3) Kinetic and potential energy effects can be ignored. (4) Air is modeled as an ideal gas.

ANALYSIS: At steady state $\dot{m}_1 = \dot{m}_2 = \dot{m}_3 = \dot{m}$ and $\dot{m}_4 = \dot{m}_5 = \dot{m}_a$. Taking a control volume about the turbine and heat exchanger, an energy rate balance at steady state reduces to give with assumption 3

$$0 = \cancelto{0}{\dot{Q}_{cv}} - \dot{W}_{cv} + \dot{m}(h_1 - h_3) + \dot{m}_a(h_4 - h_5)$$

Thus

$$\dot{m}_a = \frac{\dot{W}_{cv} + \dot{m}(h_3 - h_1)}{h_4 - h_5} = \frac{2852\,\text{hp}\left|\frac{2545\,\text{Btu/h}}{\text{hp}}\right|\left|\frac{\text{h}}{3600\,\text{s}}\right| + (10\,\tfrac{\text{lb}}{\text{s}})(1431.2 - 1407.6)}{363.89 - 260.97}$$

$$= 21.88\ \text{lb/s} \longleftarrow \dot{m}_a$$

where enthalpies are from Tables A-4E and A-22E.

Taking a control volume about the turbine only, an entropy rate balance at steady state reduces to $\dot{\sigma}_{cv} = \dot{m}(s_2 - s_1)$. To determine $\dot{\sigma}_{cv}$ requires state 2 to be fixed. This can be accomplished using an energy rate balance for the same control volume:

$$0 = \cancel{\dot{Q}_{cv}} - \dot{W}_{cv} + \dot{m}(h_1 - h_2) \Rightarrow h_2 = h_1 - \dot{W}_{cv}/\dot{m}$$

Thus

$$h_2 = 1407.6\ \text{Btu/lb} - \frac{(2852)\,|2545/3600|\ \text{Btu/s}}{10\ \text{lb/s}} = 1205.98\ \text{Btu/lb}$$

State 2 is fixed by $p_2 = 60\ \text{lbf/in}^2$, $h_2 = 1205.98$ Btu/lb. Interpolating in Table A-4E, $s_2 = 1.6802$ Btu/lb·°R. Thus, for the turbine

$$\dot{\sigma}_{cv} = \dot{m}(s_2 - s_1) = 10\,\tfrac{\text{lb}}{\text{s}}\,(1.6802 - 1.6348)\,\frac{\text{Btu}}{\text{lb}\cdot{}^\circ\text{R}} = 0.459\ \frac{\text{Btu/s}}{{}^\circ\text{R}} \longleftarrow \dot{\sigma}_{cv}$$

Taking a control volume around the heat exchanger only, an entropy rate balance reduces at steady state to

$$0 = \cancelto{0}{\sum \frac{\dot{Q}_j}{T_j}} + \dot{m}(s_2 - s_3) + \dot{m}_a(s_4 - s_5) + \dot{\sigma}_{cv}$$

or with Eq. 6.21a and data from Tables A-4E and A-22E

$$\dot{\sigma}_{cv} = \dot{m}(s_3 - s_2) + \dot{m}_a(s_5^\circ - s_4^\circ - R\ln p_5/p_4)$$

$$= (10\,\tfrac{\text{lb}}{\text{s}})(1.9022 - 1.6802)\frac{\text{Btu}}{\text{lb}\cdot{}^\circ\text{R}} + (21.88\,\tfrac{\text{lb}}{\text{s}})\left(0.76964 - 0.85062 - \frac{1.986}{28.97}\ln\frac{1}{1.1}\right)\frac{\text{Btu}}{\text{lb}\cdot{}^\circ\text{R}}$$

$$= 2.22\ \frac{\text{Btu/s}}{{}^\circ\text{R}} - 1.629\ \frac{\text{Btu/s}}{{}^\circ\text{R}} = 0.591\ \frac{\text{Btu/s}}{{}^\circ\text{R}} \longleftarrow \dot{\sigma}_{cv}$$

PROBLEM 6.123

KNOWN: Steady state operating data are provided for two waste-heat recovery systems: (a) Example 4.10, (b) Problem 4.83.

FIND: For each system, determine the rates of entropy production for the steam generator and the turbine. Discuss.

SCHEMATIC & GIVEN DATA:

(a) See Fig. E 4.10

(b) See Fig. P 4.83. From the solution, $\dot{m}_A = 145.2$ lb/min, $\dot{m}_1 = 2.8$ lb/min.

ASSUMPTIONS: 1. Control volumes enclosing the steam generator and the turbine are at steady state. 2. For each control volume, $\dot{Q}_{cv} = 0$. 3. The combustion products of part (a) can be modeled as air as an ideal gas. The ideal gas model also applies to the air of part (b).

ANALYSIS: At steady state the entropy rate balance for the control volume enclosing the turbine reduces to give

(a) $\dot{\sigma}_t = \dot{m}(s_5 - s_4)$ (b) $\dot{\sigma}_t = \dot{m}(s_3 - s_2)$

An entropy rate balance for the steam generator gives

(a) $\dot{\sigma}_{HX} = \dot{m}_1(s_2 - s_1) + \dot{m}_3(s_4 - s_3)$ (b) $\dot{\sigma}_{HX} = \dot{m}_A(s_B - s_A) + \dot{m}_1(s_2 - s_1)$

(a) From the solution to Example 4.10, $h_4 = 1213.6$ Btu/lb, $p_4 = 40$ lbf/in^2; thus Table A-4E gives $s_4 = 1.7334$ Btu/lb·°R. With data from Table A-3E

$s_5 = s_f + x_5(s_g - s_f) = 0.1327 + 0.93(1.8453) = 1.8488$ Btu/lb·°R. Thus,

$$\dot{\sigma}_t = \left(275 \frac{\text{lb}}{\text{min}}\right)(1.8488 - 1.7334)\frac{\text{Btu}}{\text{lb}\cdot°\text{R}}\left|\frac{1\text{ min}}{60\text{ s}}\right| = 0.529 \frac{\text{Btu}}{\text{s}\cdot°\text{R}} \longleftarrow$$

With $s°$ data from Table A-22E and $s_3 = s_f(T_3)$ from Table A-2E

$$\dot{\sigma}_{HX} = \left[\left(9230.6 \frac{\text{lb}}{\text{min}}\right)\left[0.67002 - 0.71323 - R\ln \frac{p_2}{p_1}\right]\frac{\text{Btu}}{\text{lb}\cdot°\text{R}} + 275(1.7334 - 0.1331)\right]\left|\frac{1\text{ min}}{60\text{ s}}\right|$$

(with the $R \ln p_2/p_1$ term struck out as 0)

$$= 0.687 \frac{\text{Btu}}{\text{s}\cdot°\text{R}} \longleftarrow$$

The steam generator contributes most to inefficient operation.

(b) With data from Table A-4E and A-3E, $s_2 = 1.7121$ Btu/lb·°R, $s_3 = s_f + x_3 s_{fg}$, $s_3 = 0.1327 + 0.9(1.8453) = 1.7935$ Btu/lb·°R. Thus,

$$\dot{\sigma}_t = \left(2.8 \frac{\text{lb}}{\text{min}}\right)(1.7935 - 1.7121)\frac{\text{Btu}}{\text{lb}\cdot°\text{R}}\left|\frac{1\text{ min}}{60\text{ s}}\right| = 3.8 \times 10^{-3} \frac{\text{Btu}}{\text{s}\cdot°\text{R}} \longleftarrow$$

With $s°$ data from Table A-22E, noting that $p_B = p_A$, and $s_1 = s_f(T_1)$ from Table A-2E,

$$\dot{\sigma}_{HX} = \left[\left(145.2 \frac{\text{Btu}}{\text{min}}\right)\left[0.67665 - 0.70160\right]\frac{\text{Btu}}{\text{lb}\cdot°\text{R}} + 2.8\left[1.7121 - 0.3241\right]\right]\left|\frac{1\text{ min}}{60\text{ s}}\right|$$

$$= 4.4 \times 10^{-3} \frac{\text{Btu}}{\text{s}\cdot°\text{R}} \longleftarrow$$

The steam generator contributes most to inefficient operation.

PROBLEM 6.124

KNOWN: Air passes through a compressor and heat exchanger. Data for the various flow streams are known.

FIND: Determine the compressor power and the mass flow rate of the cooling water. Also, evaluate the rates of entropy production in the compressor and heat exchanger.

SCHEMATIC & GIVEN DATA:

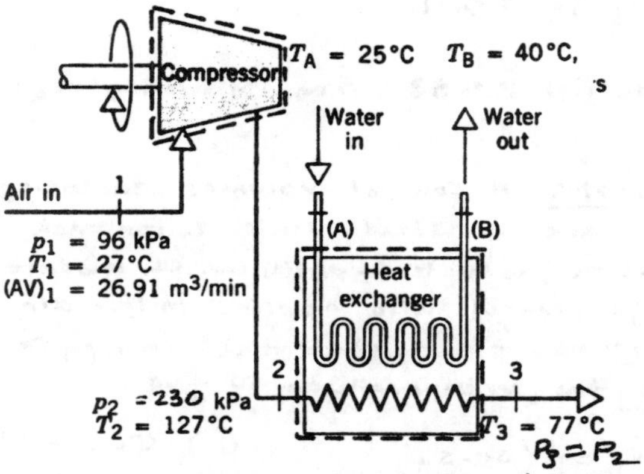

ASSUMPTIONS: (1) Both control volumes shown are at steady state. (2) Heat transfer is negligible, and $\dot{W}_{cv} = 0$ for the heat exchanger. (3) Kinetic and potential energy effects are negligible. (4) The air behaves as an ideal gas.

ANALYSIS: (a) The compressor power is found from the steady-state energy balance for the compressor control volume

$$0 = \cancelto{0}{\dot{Q}_{cv}} - \dot{W}_{cv} + \dot{m}_{air}\left[(h_1 - h_2) + \left(\frac{V_1^2 - V_2^2}{2}\right) + g(z_1 - z_2)\right]$$

where $\dot{m}_1 = \dot{m}_2 \equiv \dot{m}_{air}$. With assumption (3)

$$\dot{W}_{cv} = \dot{m}_{air}(h_1 - h_2)$$

The mass flow rate of air is

$$\dot{m}_{air} = \frac{(AV)_1}{v_1} = \frac{P_1 (AV)_1}{RT_1} = \frac{(96\text{ kPa})(26.91\text{ m}^3/\text{min})}{\left(\frac{8.314}{28.97}\frac{\text{kJ}}{\text{kg}\cdot\text{K}}\right)(300\text{ K})}\left|\frac{10^3\text{ N/m}^2}{1\text{ kPa}}\right|\left|\frac{1\text{ kJ}}{10^3\text{ N}\cdot\text{m}}\right| = 30\frac{\text{kg}}{\text{min}}$$

Thus

$$\dot{W}_{cv} = \left(30\frac{\text{kg}}{\text{min}}\right)\left(\frac{1\text{ min}}{60\text{ s}}\right)(300.19 - 400.98)\left|\frac{1\text{ kW}}{1\text{ kJ/s}}\right| = -50.4\text{ kW} \longleftarrow \dot{W}_{cv}$$

where the specific enthalpies are from Table A-22.

To find the cooling water flow rate, note that the air and water pass through the heat exchanger as separate streams. Thus

$$\dot{m}_2 = \dot{m}_3 \equiv \dot{m}_{air} \quad \text{and} \quad \dot{m}_A = \dot{m}_B \equiv \dot{m}_{cw}$$

With assumptions (2) and (3), the steady-state energy balance for the heat exchanger control volume reduces to

$$0 = \cancelto{0}{\dot{Q}_{cv}} - \cancelto{0}{\dot{W}_{cv}} + \dot{m}_{air}(h_2 - h_3) + \dot{m}_{cw}(h_A - h_B)$$

or

$$\dot{m}_{cw} = \left(\frac{h_2 - h_3}{h_B - h_A}\right)\dot{m}_{air}$$

From Table A-22, $h_3 = 350.49$ kJ/kg. For the cooling water, $h = h_f(T)$:

$h_A = 104.89$ kJ/kg, $h_B = 167.57$ kJ/kg. Thus, the cooling water mass flow rate is

$$\dot{m}_{cw} = \left(\frac{400.98 - 350.49}{167.57 - 104.89}\right)\left(30\frac{\text{kg}}{\text{min}}\right)\left|\frac{1\text{ min}}{60\text{ s}}\right| = 0.403\text{ kg/s} \longleftarrow$$

PROBLEM 6.1.24 (Contd.)

(b) Reducing mass and entropy rate balances for the compressor, and using Eq. 6.21a

$$\dot{\sigma}_C = \dot{m}_{Air}(s_2 - s_1)$$

$$= \dot{m}_{Air}\left(s^\circ(T_2) - s^\circ(T_1) - R \ln P_2/P_1 \right)$$

$$= \left(30 \frac{kg}{min}\right)\left|\frac{1 min}{60 s}\right|\left(1.99194 - 1.70203 - \frac{8.314}{28.97} \ln\left(\frac{230}{96}\right)\right] \frac{kJ}{kg \cdot K}\left|\frac{1 kW}{1 kJ/s}\right|$$

$$= 0.0196 \frac{kW}{K} \longleftarrow$$

where s° values are from Table A-22.

Reducing mass and entropy rate balances for the heat exchanger

$$\dot{\sigma}_{HX} = \dot{m}_{Air}[s_3 - s_2] + \dot{m}_{cw}[s_B - s_A]$$

$$= \dot{m}_{Air}\left[s^\circ(T_3) - s^\circ(T_2) - R \cancelto{0}{\ln \frac{P_3}{P_2}} \right] + \dot{m}_{cw}\left[s_f(T_B) - s_f(T_A) \right]$$

With data from Tables A-2, 22

$$\dot{\sigma}_{HX} = \left[\left(0.5 \frac{kg}{s}\right)\left[1.85708 - 1.99194\right] \frac{kJ}{kg \cdot K} + 0.403\,(0.5725 - 0.3674)\right]\left|\frac{1 kW}{1 kJ/s}\right|$$

$$= 0.0152 \frac{kW}{K} \longleftarrow$$

PROBLEM 6.125

KNOWN: Operating data are provided for two turbine stages and an interconnecting heat exchanger.

FIND: Determine the rates of entropy production for the two turbines and the heat exchanger. Place in rank order.

SCHEMATIC & GIVEN DATA:

See Fig. P 4.82

From the solution to Problem 4.82, $T_3 = 1301.5\,K$, $\dot{m}_1 = 28.22\,kg/s$.

ASSUMPTIONS: 1. Control volumes enclosing each of the turbines and the heat exchanger are at steady state. 2. For each control volume, $\dot{Q}_{cv}=0$. 3. Air is modeled as an ideal gas.

ANALYSIS: Mass and entropy rate balances reduce to give for the first turbine

$$\dot{\sigma}_{t1} = \dot{m}_1(s_2 - s_1) = \dot{m}_1\left(s^\circ(T_2) - s^\circ(T_1) - R\ln p_2/p_1\right)$$

$$= \left(28.22\,\frac{kg}{s}\right)\left[3.07732 - 3.3620 - \frac{8.314}{28.97}\ln\frac{5}{20}\right]\frac{kJ}{kg\cdot K}\left|\frac{1\,kW}{1\,kJ/s}\right|$$

$$= 3.1936\ kW/K$$

where Eq. 6.21a and data from Table A-22 have been used. Likewise for turbine 2

$$\dot{\sigma}_{t2} = \dot{m}_1(s_4 - s_3) = \dot{m}_1\left[s^\circ(T_4) - s^\circ(T_3) - R\ln p_4/p_3\right]$$

$$= (28.22)\left[2.94468 - 3.27481 - \frac{8.314}{28.97}\ln\frac{1}{4.5}\right] = 2.8649\ kW/K$$

Mass and entropy rate balances for the interconnecting heat exchanger give

$$\dot{\sigma}_{HX} = \dot{m}_1[s_3 - s_2] + \dot{m}_5[s_6 - s_5]$$

$$= \dot{m}_1\left[s^\circ(T_3) - s^\circ(T_2) - R\ln\frac{p_3}{p_2}\right] + \dot{m}_5\left[s^\circ(T_6) - s^\circ(T_5) - R\ln\frac{p_6}{p_5}\right]$$

$$= 28.22\left[3.27481 - 3.07732 - \frac{8.314}{28.97}\ln\frac{4.5}{5}\right] +$$

$$\left(\frac{1200}{60}\right)\left[3.17868 - 3.42892 - \frac{8.314}{28.97}\ln\frac{1}{1.35}\right]$$

$$= 6.4265 - 3.2783 = 3.1482\ kW/K$$

In rank order: Turbine 1, heat exchanger, turbine 2.

PROBLEM 6.126

KNOWN: Steady-state operating data are provided for a simple vapor power plant.

FIND: Determine the rates of entropy production for the turbine, condenser, and pump. Place in rank order.

SCHEMATIC & GIVEN DATA:

- See Figure P 4.85
- From the solution to Problem 4.85, $\dot{m}_{cw} = 3759$ kg/s

ASSUMPTIONS: 1. Control volumes enclosing the turbine, condenser, and pump are at steady state. 2. For each control volume, $\dot{Q}_{cv} = 0$.

ANALYSIS: Mass and entropy rate balances for the turbine give with data from Tables A-3 and A-4

$$\dot{\sigma}_t = \dot{m}(s_2 - s_1) = \left(109\,\frac{kg}{s}\right)[7.4651 - 6.6622]\,\frac{kJ}{kg\cdot K}\left|\frac{1\,kW}{1\,kJ/s}\right| = 87.51\,\frac{kW}{K} \leftarrow$$

$$s_2 = s_f + x_2(s_g - s_f) = 0.5926 + 0.9(8.2287 - 0.5926) = 7.4651\ kJ/kg\cdot K.$$

For the pump, and data from Tables A-3,5

$$\dot{\sigma}_p = \dot{m}[s_4 - s_3] = (109)[0.6061 - 0.5926] = 1.47\,\frac{kW}{K} \leftarrow$$

Mass and entropy rate balances for the condenser give

$$\dot{\sigma}_{cond} = \dot{m}[s_3 - s_2] + \dot{m}_{cw}[s_6 - s_5]$$

Using $s \approx s_f(T)$ for the cooling water and data from Table A-2

$$\dot{\sigma}_{cond} = (109)[0.5926 - 7.4651] + (3759)[0.5053 - 0.2966]$$

$$= -749.1 + 784.5 = 35.4\ kW/K \leftarrow$$

In rank order, turbine, condenser, pump.

PROBLEM 6.127

KNOWN: Steam contained in a large tank at a known state passes from the tank through a turbine into a small vessel until a specified final condition is attained in the vessel.

FIND: Determine the amount of entropy produced. Repeat for the case where no work is developed by the turbine.

SCHEMATIC & GIVEN DATA: See Figure E4.12.

ASSUMPTIONS: See assumptions listed in Example 4.12.

ANALYSIS: The mass rate balance reduces to

$$\frac{dm_{cv}}{dt} = \dot{m}_i$$

An entropy rate balance takes the form

$$\frac{dS_{cv}}{dt} = \sum \frac{\dot{Q}_j}{T_j}^{\,0} + \dot{m}_i s_i + \dot{\sigma}_{cv}$$

Combining the mass and entropy rate balances

$$\frac{dS_{cv}}{dt} = s_i \frac{dm_{cv}}{dt} + \dot{\sigma}_{cv}$$

Integrating

$$\Delta S_{cv} = \int_1^2 s_i\, dm_{cv} + \sigma_{cv} \Rightarrow \Delta S_{cv} = s_i \,\Delta m_{cv} + \sigma_{cv}$$

In accordance with assumption 3, the specific entropy of the steam entering the control volume is constant at the value corresponding to the state in the large tank.

Since the small vessel is initially evacuated, the terms ΔS_{cv} and Δm_{cv} reduce to the entropy and mass within the vessel at the end of the process. That is

$$\Delta S_{cv} = m_2 s_2 - m_1 s_1^{\,0} \quad , \quad \Delta m_{cv} = m_2 - m_1^{\,0}$$

Collecting results and solving for the amount of entropy produced

$$\sigma_{cv} = m_2 (s_2 - s_i) \tag{1}$$

At 15 bar, 320°C, Table A-4 gives $s_i = 6.9938$ kJ/kg·K. At 15 bar, 400°C, $s_2 =$ 7.2690 kJ/kg·K. The solution to Example 4.12 provides m_2. Thus, Eq.(1) gives

$$\sigma_{cv} = (2.96\ \text{kg})(7.2690 - 6.9938)\frac{\text{kJ}}{\text{kg}\cdot\text{K}} = 0.8146\ \frac{\text{kJ}}{\text{K}} \longleftarrow \sigma_{cv}$$

In the case where no work is developed, the solution to Example 4.12 gives the final temperature of the steam in the vessel as 477°C. Thus, interpolation in Table A-4 at 15 bar gives $s_2 = 7.5024$ kJ/kg·K, $v_2 = 0.22784$ m³/kg. Finally, substituting into Eq.(1)

$$\sigma_{cv} = \left(\frac{0.6\ \text{m}^3}{0.22784\ \text{m}^3/\text{kg}}\right)(7.5024 - 6.9938)\frac{\text{kJ}}{\text{kg}\cdot\text{K}}$$

$$= 1.3394\ \text{kJ/K} \longleftarrow \sigma_{cv}$$

As there is less control of the entering stream in this case, the amount of entropy produced is greater.

PROBLEM 6.128

KNOWN: Steam contained in a large vessel at a specified state is allowed to flow into an initially evacuated tank until a specified pressure p is attained in the tank.

FIND: (a) For p = 100 lbf/in², determine the final temperature of the steam in the tank and the amount of entropy produced. (b) Plot the quantities of part (a) versus p ranging from 10 to 100 lbf/in².

SCHEMATIC & GIVEN DATA:

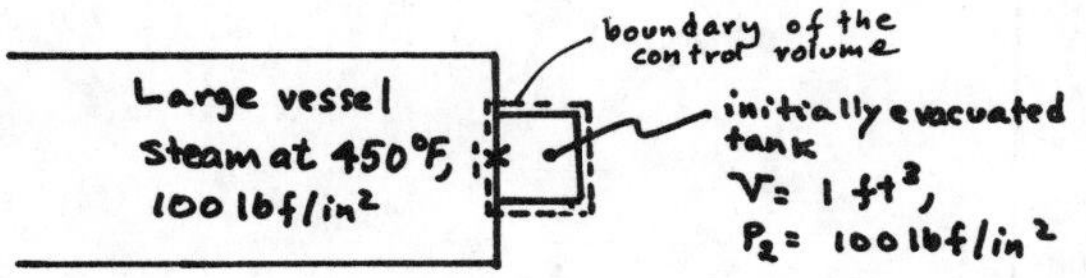

ASSUMPTIONS: (1) For the control volume shown in the accompanying figure, $\dot{Q}_{cv} = 0$ and kinetic and potential energy effects are negligible. (2) The state of the steam in the large vessel remains constant.

ANALYSIS: The mass rate balance and the energy rate balance reduce, respectively, to

$$\frac{dm_{cv}}{dt} = \dot{m}_i$$

$$\frac{dU_{cv}}{dt} = \dot{m}_i h_i$$

Combining these expressions and integrating

$$\frac{dU_{cv}}{dt} = h_i \frac{dm_{cv}}{dt} \Rightarrow \Delta U_{cv} = \int_1^2 h_i\, dm_{cv} \Rightarrow \Delta U_{cv} = h_i \Delta m_{cv}$$

In accordance with assumption 2, the specific enthalpy of the steam entering the control volume is constant at the value corresponding to the state in the large vessel. Since the tank is initially evacuated, the terms ΔU_{cv} and Δm_{cv} reduce to the internal energy and mass within the tank at the end of the process: $\Delta U_{cv} = m_2 u_2$, $\Delta m_{cv} = m_2$. Accordingly, the energy balance reduces simply to

$$m_2 u_2 = m_2 h_i \Rightarrow u_2 = h_i \qquad (1)$$

Thus, the final state of the steam in the tank is fixed by pressure p and $u_2 (= h_i)$.

An entropy rate balance for the control volume reduces to

$$\frac{dS_{cv}}{dt} = \sum \frac{\dot{Q}_j}{T_j}^{\,0} + \dot{m}_i s_i + \dot{\sigma}_{cv}$$

Introducing the above mass rate balance and integrating

$$\frac{dS_{cv}}{dt} = s_i \frac{dm_{cv}}{dt} + \dot{\sigma}_{cv} \Rightarrow \Delta S_{cv} = \int_1^2 s_i\, dm_{cv} + \sigma_{cv} \Rightarrow \Delta S_{cv} = s_i \Delta m_{cv} + \sigma_{cv}$$

As for h_i, s_i remains constant during the process. Then, since the tank is initially evacuated, $\Delta S_{cv} = m_2 s_2$, $\Delta m_{cv} = m_2$. Inserting these expressions and solving for the amount of entropy produced is

$$\sigma_{cv} = m_2 (s_2 - s_i) = \left(\frac{V}{v_2}\right)(s_2 - s_i) \qquad (2)$$

(a) <u>p = 100 lbf/in²</u> From Table A-4E at 100 lbf/in², 450°F, $h_i = 1253.6$ Btu/lb. Then, with $u_2 = 1253.6$ Btu/lb, interpolation in Table A-4E at 100 lbf/in² gives $T_2 = 702°F$, $s_2 = 1.8042$ Btu/lb·°R, $v_2 = 6.847$ ft³/lb ←

PROBLEM 6.128 (Contd.)

With Eq. (2) and s_i from Table A-4E at 100 lbf/in.², 450°F

$$\sigma_{cv} = \frac{V}{v_2}(s_2 - s_i) = \left(\frac{1\ ft^3}{6.847\ ft^3/lb}\right)(1.8042 - 1.6812)\frac{Btu}{lb \cdot °R} = 0.018\ \frac{Btu}{°R} \leftarrow$$

(b) PLOTS

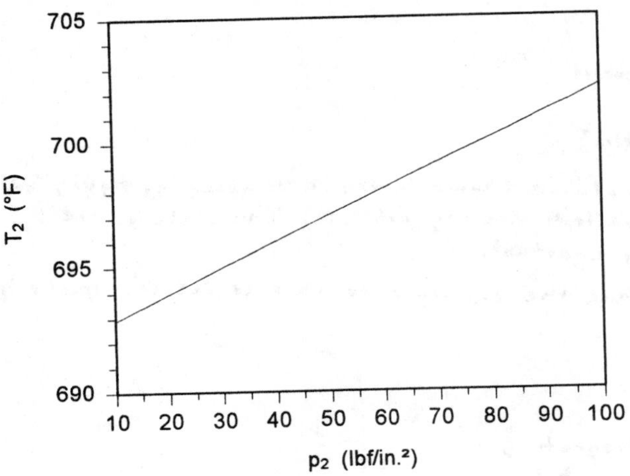

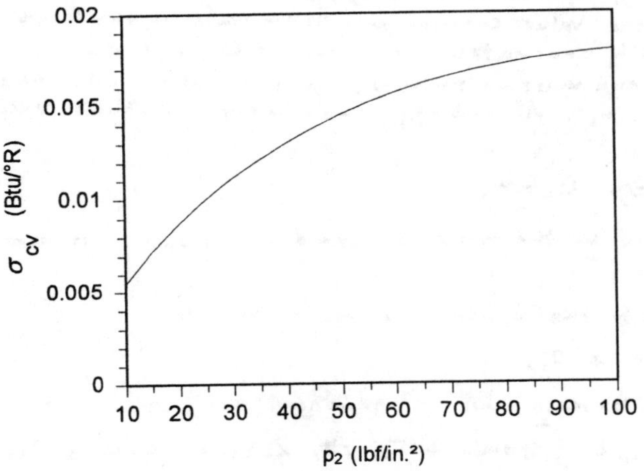

PROBLEM 6.129

KNOWN: Air flows from a large supply line into an initially evacuated, insulated tank until the tank pressure is p.

FIND: Plot the tank temperature, mass of air within the tank, and the amount of entropy produced versus p for p<10 bar.

SCHEMATIC & GIVEN DATA:

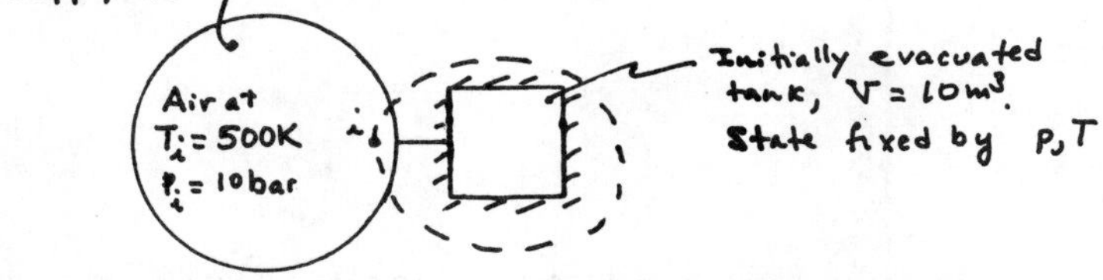

ASSUMPTIONS: (1) The control volume is shown in the above figure. (2) For the control volume, $\dot{Q}_{cv} = 0$, and kinetic/potential energy effects are negligible. (3) The state of the air within the supply line remains constant. (4) The mass of air within the piping connecting the supply line and tank can be ignored. (5) Air is modeled as an ideal gas with constant specific heat ratio k.

ANALYSIS: Paralleling the application of mass and energy balances in Example 4.12 but with $W_{cv} = 0$, the result is $u(T) = h(T_i)$, where T_i is the temperature of the air in the supply line and T is the temperature within the tank. For an ideal gas $h(T_i) = u(T_i) + RT_i$, so $u(T) = u(T_i) + RT_i$. Using ideal gas relations (assumption 5): $u(T) - u(T_i) = c_v[T - T_i]$, $R = c_p - c_v$, this gives

$$T = kT_i \tag{1}$$

That is, the temperature of the air within the tank is a constant independent of the pressure p.

Using the ideal gas equation of state, the mass of the air within the tank is

$$m = \frac{pV}{RT} = \frac{p}{p_i}\left[\frac{p_i V}{R(kT_i)}\right] = r\left[\frac{(10\times10^5\,\text{N/m}^2)(10\,\text{m}^3)}{\left(\frac{8314}{28.97}\frac{\text{N}\cdot\text{m}}{\text{kg}\cdot\text{K}}\right)(1.39)(500\,\text{K})}\right] = (50.14\,r)\,\text{kg} \tag{2}$$

where $r = p/p_i$ and $k = 1.39$ from Table A-20.

An entropy balance reads

$$\frac{dS_{cv}}{dt} = \dot{m}_i s_i + \dot{\sigma}_{cv} \xrightarrow{\text{(integration)}} [m\,s(T,p) - 0] = \int \dot{m}_i s_i + \sigma_{cv}$$

$$= m\,s(T_i, p_i) + \sigma_{cv}$$

That is, with ideal gas relations

$$\sigma_{cv} = m[s(T,p) - s(T_i,p_i)] = m\left[c_p \ln\frac{T}{T_i} - R\ln\frac{p}{p_i}\right]$$

($c_p = kR/(k-1)$ (Eq. 3.47a); m: Eq. (2); T/T_i: Eq. (1); p/p_i: r)

$$\Rightarrow \sigma_{cv} = (50.14\,r)R\left[\frac{k}{k-1}\ln k - \ln r\right] = (50.14\,r)\left(\frac{8.314}{28.97}\right)\left[\frac{1.39}{0.39}\ln 1.39 - \ln r\right]\frac{\text{kJ}}{\text{K}}$$

$$\sigma_{cv} = 14.39\,r\,[1.174 - \ln r]\,\frac{\text{kJ}}{\text{K}} \tag{3}$$

SAMPLE CALCULATION: If p = 10 bar, r = 1. Eq. (1), T = 1.39 (500 K) = 695 K. Eq (2) gives, m = 50.14 kg. Eq (3) gives $\sigma_{cv} = 16.89\ \frac{\text{kJ}}{\text{K}}$.

PROBLEM 6.129 (Cont'd.)

To generate data for the required plots, we use *IT*, as follows:

IT Code

```
p_i = 10  // bar
Ti = 500  // K
V = 10  // m³

cp = cp_T("Air", Ti)
cv = cv_T("Air", Ti)
k = cp / cv

T = k * Ti

r = p / p_i
r = 1
v = v_TP("Air",T,p)
m = V / v

sigma_cv = m * (cp * ln(T / Ti) - R * ln(p / p_i))
R = 8.314 / 28.97
```

IT Results for p = 10 bar (r = 1)

c_p = 1.029 kJ/kg·K
k = 1.387
v = 0.199 m³/kg
m = 50.26 kg
T = 693.3 K
σ_{cv} = 16.91 kJ/kg·K

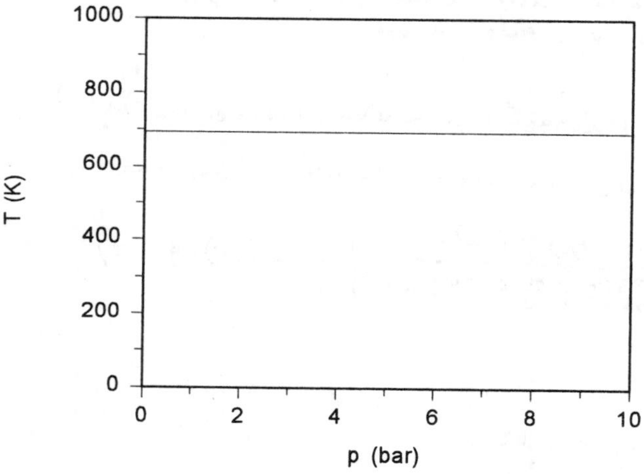

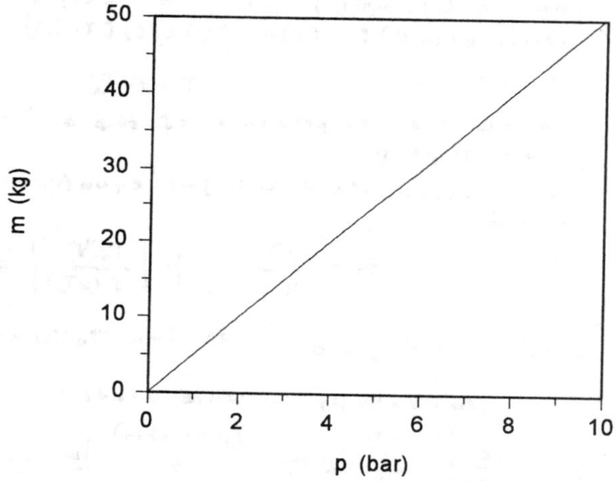

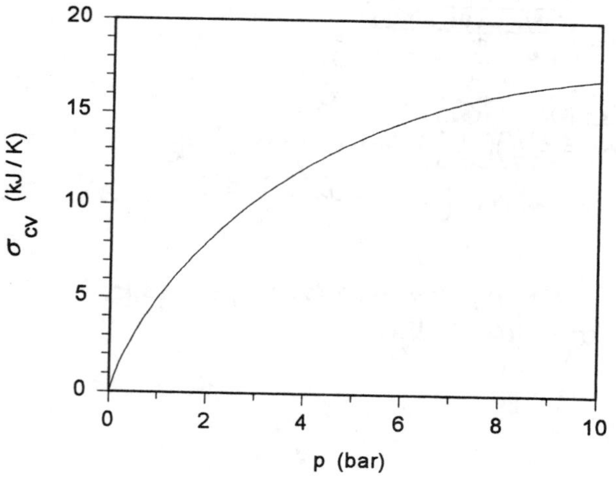

Observe:

- The temperature in the tank is constant.
- The mass increases directly with tank pressure.
- There is more entropy produced as the pressure increases, as expected.

PROBLEM 6.130

KNOWN: A tank initially filled with air is evacuated by pumping out the air.

FIND: Determine the minimum theoretical work required.

SCHEMATIC & GIVEN DATA:

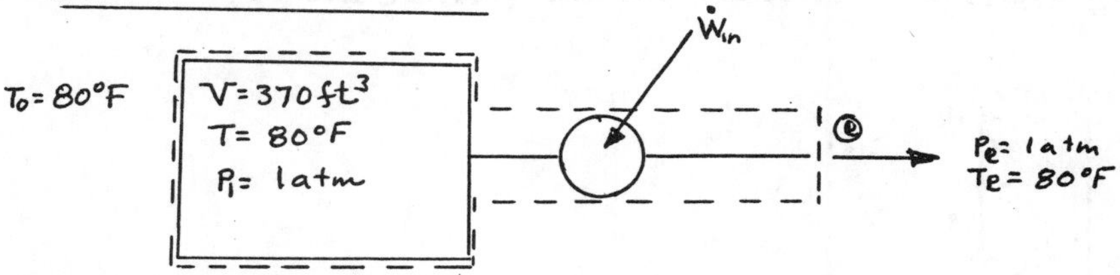

ASSUMPTIONS: (1) The control volume is shown in the figure. (2) Heat transfer from the surroundings maintains the temperature of the air in the tank at 80°F. (2) Kinetic and potential energy effects can be ignored. (3) The air is modeled as an ideal gas.

ANALYSIS: A mass balance reduces to $dm/dt = -\dot{m}_e$. An energy balace reads $dU/dt = \dot{Q} + \dot{W}_{in} - \dot{m}_e h_e$. Combinining these expressions

$$\frac{dU}{dt} = \dot{Q} + \dot{W}_{in} + h_e \frac{dm}{dt} \Rightarrow \Delta U = Q + W_{in} + h_e \Delta m \quad (1)$$

where h_e is determined by T_e and thus is constant.

An entropy balance reads $dS/dt = \dot{Q}/T_0 - \dot{m}_e s_e + \dot{\sigma}$. With $-\dot{m}_e = dm/dt$

$$\frac{dS}{dt} = \frac{\dot{Q}}{T_0} + s_e \frac{dm}{dt} + \dot{\sigma} \Rightarrow \Delta S = \frac{Q}{T_0} + s_e \Delta m + \sigma \quad (2)$$

where s_e is determined by T_e, P_e and thus is constant.

Eliminating Q between Eqs. (1), (2)

$$W_{in} = \Delta U - T_0 \Delta S + T_0 s_e \Delta m - h_e \Delta m + T_0 \sigma$$

$$= [\cancel{m_2}^{0} u_2 - m_1 u_1] - T_0(\cancel{m_2}^{0} s_2 - m_1 s_1) + T_0 s_e[\cancel{m_2}^{0} - m_1] - h_e(\cancel{m_2}^{0} - m_1) + T_0 \sigma$$

$$= m_1 T_0 (s_1 - s_e) + m_1 (h_e - u_1) + T_0 \sigma \quad (3)$$

Since the states are the same initially and at the exit, $s_1 = s_e$. Also, with $h_e = u_e + P_e v_e$ $(h_e - u_1) = (u_e - u_1) + P_e v_e = P_e v_e$, since $T_1 = T_e$. With these, Eq. (3) reduces to

$$W_{in} = m_1 (P_e v_e) + T_0 \sigma = P_e V + T_0 \sigma \quad (4)$$

The specific volume at the exit is the same as the initial specific volume in the tank, and so $V = m_1 v_e$.

Finally, since $\sigma \geq 0$, the minimum theoretical value corresponds to $\sigma = 0$:

$$(W_{in})_{MIN} = P_e V$$

$$= \left(14.7 \times 144 \frac{lbf}{ft^2}\right)(370 ft^3)\left(\frac{1 Btu}{778 ft \cdot lbf}\right)$$

$$= 1006.7 \text{ Btu}$$

← MIN

PROBLEM 6.131

KNOWN: CO_2 contained in a piston cylinder is compressed isentropically from a specified initial state to a specified final pressure.

FIND: Using four alternative approaches, evaluate the final pressure and the work.

SCHEMATIC & GIVEN DATA:

$V_1 = 0.10\ m^3$
$p_1 = 0.3\ bar$
$T_1 = 400K$
$T_2 = 560K$

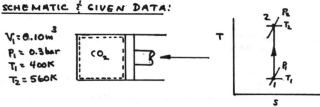

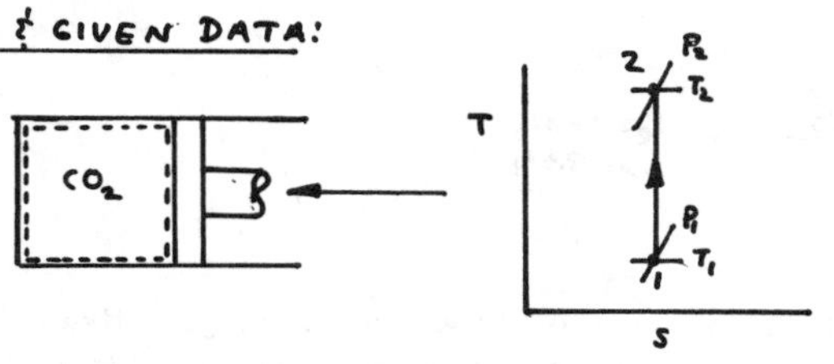

ASSUMPTIONS: (1) As shown in the figure, the system is the CO_2 gas. (2) CO_2 is modeled as an ideal gas. (3) There are no significant kinetic or potential energy effects. 4 The process is isentropic, and thus Q = 0.

ANALYSIS: An energy balance reduces to $\Delta U = \cancel{Q}^{0} - W$, or $W = -\Delta U$.

$$W = -n(\bar{u}_2 - \bar{u}_1) \qquad (1)$$

With the ideal gas equation of state

$$n = \frac{p_1 V_1}{\bar{R} T_1} = \frac{(0.3\times10^5\ N/m^2)(0.1\ m^3)}{(8314\ \frac{N\cdot m}{kmol\cdot K})(400K)} = 9.02\times10^{-4}\ kmol$$

(a) Table A-27. As $\Delta s = 0$, Eq. 6.21b gives

$$0 = \bar{s}^\circ(T_2) - \bar{s}^\circ(T_1) - \bar{R}\ln\frac{p_2}{p_1} \Rightarrow \ln\frac{p_2}{p_1} = \frac{\bar{s}^\circ(T_2) - \bar{s}^\circ(T_1)}{\bar{R}} = \frac{239.962 - 225.225}{8.314} \Rightarrow p_2 = 1.766\ bar \longleftarrow p_2$$

where $\bar{s}^\circ$ values are from Table A-23, with $\bar{u}$ values from the same soure, Eq. (1) gives

$$W = -(9.02\times10^{-4}\ kmol)(15{,}751 - 10{,}046)\frac{kJ}{kmol} = -5.146\ kJ \longleftarrow W$$

(c) k from Table A-20 at 480 K: k = 1.233. The pressures and temperatures are related by Eq. 6.45. Thus

$$\frac{p_2}{p_1} = \left(\frac{T_2}{T_1}\right)^{k/k-1} \Rightarrow \frac{p_2}{p_1} = \left(\frac{560}{400}\right)^{1.233/.233} \Rightarrow p_2 = 1.78\ bar \longleftarrow p_2$$

Using Eq. 3.47b on a molar basis, Eq. (1) becomes

$$W = -\frac{n\bar{R}}{k-1}(T_2 - T_1) = -\frac{(9.02\times10^{-4}\ kmol)(8.314\ kJ/kmol\cdot K)(560-400)K}{0.233} = -5.15\ kJ \longleftarrow W$$

(d) k from Table A-20 at 300 K: k = 1.288. Then, as in part (b)

① $$\frac{p_2}{p_1} = \left(\frac{T_2}{T_1}\right)^{k/k-1} \Rightarrow \frac{p_2}{p_1} = \left(\frac{560}{400}\right)^{1.288/.288} \Rightarrow p_2 = 1.351\ bar \longleftarrow p_2$$

$$W = -\frac{n\bar{R}}{k-1}(T_2 - T_1) = -\frac{(9.02\times10^{-4}\ kmol)(8.314\ kJ/kmol\cdot K)(560-400)K}{0.288} = -4.166\ kJ \longleftarrow W$$

(b) *IT* Results

n = 0.0009021 kmol
v_1 = 110.9 m³/kmol
p_2 = 1.766 bar
W = -5.144 kJ

IT Code

```
V1 = 0.1  // m³
p1 = 0.3  // bar
T1 = 400  // K
T2 = 560  // K

W = - n * (u2 - u1)
u1 = u_T("CO2", T1)   // kJ/kmol
u2= u_T("CO2", T2)
v1 = v_TP("CO2",T1,p1)   // m³/kmol
n = V1 / v1

s2 = s1
s1 = s_TP("CO2", T1, p1)
s2 = s_TP("CO2", T2, p2)
```

1. In the present case, the approach of part (d) gives results departing significantly from those obtained with the other approaches

PROBLEM 6.132

KNOWN: Air expands isentropically through a turbine at steady state from a specified inlet state to a specified pressure.

FIND: Using three alternative approaches determine the exit pressure and the work developed per unit of air flowing.

SCHEMATIC & GIVEN DATA:

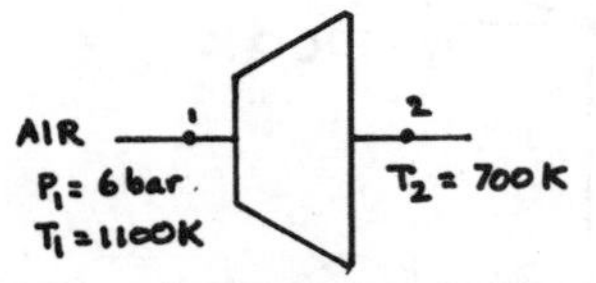

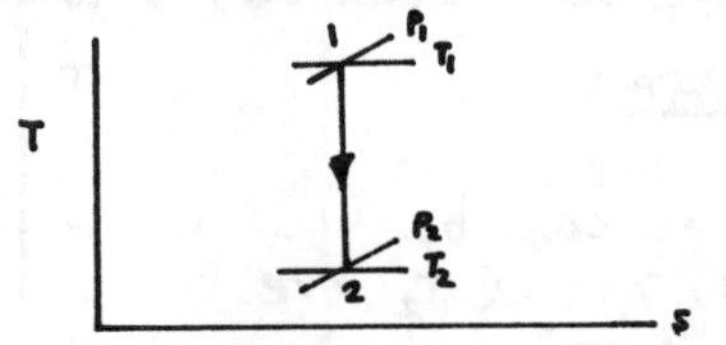

ASSUMPTIONS: (1) The turbine is at steady state and the expansion is isentropic. (2) Kinetic and potential energy changes from inlet to exit can be neglected. (3) Air is modeled as an ideal gas.

ANALYSIS: At steady state $\dot{m}_1 = \dot{m}_2 = \dot{m}$ and an energy rate balance reduces to give

$$\frac{\dot{W}_{cv}}{\dot{m}} = h_1 - h_2 \qquad (1)$$

(a) Table A-22. With h_1 and h_2 from the table, Eq. (1) gives

$$\frac{\dot{W}_{cv}}{\dot{m}} = (1161.07 - 713.27)\frac{kJ}{kg} = 447.8 \text{ kJ/kg} \longleftarrow (\dot{W}_{cv}/\dot{m})$$

The pressures and temperatures are related by Eq. 6.43, which with data from the table gives

$$p_2 = p_1\left(\frac{p_{r2}}{p_{r1}}\right) = 6 \text{ bar}\left(\frac{28.8}{167.1}\right) = 1.034 \text{ bar} \longleftarrow p_2$$

(c) k from Table A-20 at 900K: k = 1.344. With Eq. 3.47a, Eq. (1) becomes

$$\frac{\dot{W}_{cv}}{\dot{m}} = \frac{kR}{k-1}(T_1 - T_2)$$

$$= \frac{(1.344)(8.314/28.97)(kJ/kg\cdot K)(1100-700)K}{0.344} = 448.5 \text{ kJ/kg} \longleftarrow (\dot{W}_{cv}/\dot{m})$$

The pressures and temperatures are related by Eq. 6.45. Thus

$$\frac{p_2}{p_1} = \left(\frac{T_2}{T_1}\right)^{k/k-1} = \left(\frac{700}{1100}\right)^{1.344/.344} \Rightarrow p_2 = 1.026 \text{ bar} \longleftarrow p_2$$

(d) k from Table A-20 at 300 K: k = 1.4. Then, as in part (b)

① $$\frac{\dot{W}_{cv}}{\dot{m}} = \frac{kR}{k-1}(T_1 - T_2) = \frac{(1.4)(8.314/28.97)(1100-700)}{0.4} = 401.8 \text{ kJ/kg} \longleftarrow (\dot{W}_{cv}/\dot{m})$$

$$\frac{p_2}{p_1} = \left(\frac{T_2}{T_1}\right)^{k/k-1} = \left(\frac{700}{1100}\right)^{1.4/.4} \Rightarrow p_2 = 1.233 \text{ bar} \longleftarrow p_2$$

(b) *IT* Results

h1 = 1161 kJ/kg
h2 = 713 kJ/kg
$\dot{W}_{cv}$ = 447.6 kJ/kg
p2 = 1.034 bar

IT Code

```
p1 = 6 // bar
T1 = 1100 // K
T2 = 700 // K

Wdot_cv = mdot * (h1 - h2)
mdot = 1
h1 = h_T("Air", T1)
h2 = h_T("Air", T2)

s2 = s1
s1 = s_TP("Air", T1, p1)
s2 = s_TP("Air", T2, p2)
```

1. In the present case, the approach of part (d) gives results departing significantly from those obtained with the other approaches.

PROBLEM 6.133

KNOWN: Methane gas undergoes an isentropic expansion from a given initial state to a final state at T, p.

FIND: Using the ideal gas model, and $c_p(T)$ data from Table A-21, determine (a) p when T = 500K and (b) T when p = 1 bar. Check using IT.

SCHEMATIC AND GIVEN DATA:

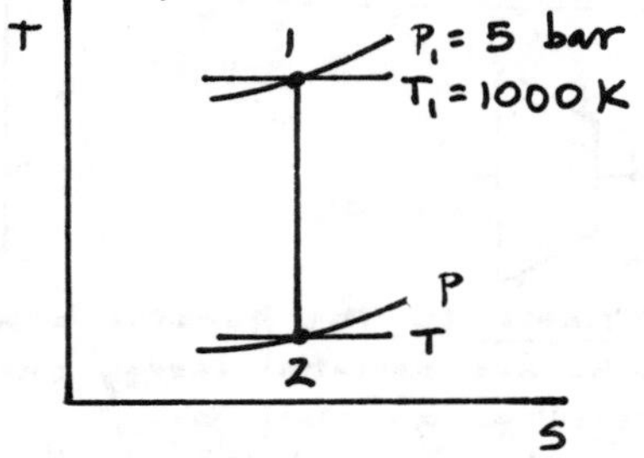

ASSUMPTIONS: (1) The methane can be modeled as an ideal gas. (2) T_1 and T_2 are in the range for which the $c_p(T)$ data in Table A-22 are applicable.

ANALYSIS: When expressed on a molar basis, Eq. 6.19 becomes for $\Delta s = 0$

$$0 = \int_{T_1}^{T} \frac{\bar{c}_p(T)}{T} dT - \bar{R} \ln \frac{p}{T_1}$$

where the integration is between state 1 (T_1, p_1) and a second state expressed a (T, p). With $\bar{c}_p$ from Table A-21: $\bar{c}_p = \bar{R}[\alpha + \beta T + \gamma T^2 + \delta T^3 + \epsilon T^4]$ this expression becomes

$$\ln \frac{p}{p_1} = \int_{T_1}^{T} [\frac{\alpha}{T} + \beta + \gamma T + \delta T^2 + \epsilon T^3] dT$$

or

$$\ln \frac{p}{p_1} = \alpha \ln\left(\frac{T}{T_1}\right) + \beta(T - T_1) + \frac{\gamma}{2}(T^2 - T_1^2) + \frac{\delta}{3}(T^3 - T_1^3) + \frac{\epsilon}{4}(T^4 - T_1^4) \quad (1)$$

(a) and (b): Eq. (1) can be solved for p when T is given or for T when p is given. The following IT code illustrates this, using data for the coefficients $\alpha, \beta, \ldots$ from Table A-21:

```
p1 = 5  // bar
T1 = 1000  // K

ln (p / p1) = alpha*ln(T/T1) + beta*(T-T1) + (gamma/2)*(T^2-T1^2) +
(delta/3)*(T^3-T1^3) + (epsilon/4)*(T^4-T1^4)
alpha = 3.826
beta = -3.979E-03
gamma = 24.558E-06
delta = -22.733E-09
epsilon = 6.962E-12
```

Results: (a) T = 500 K
// p = 0.03825 bar ← (a) p

(b) p = 1 bar
// T = 821.1 K ← (b) T

(c) To check these results, we use the s = s_TP("CH4", T, P) functions of it, as follows:

```
s = s1
s1 = s_TP("CH4", T1, p1)
s = s_TP("CH4", T, p)
```

① Results: (c) T = 500 K; p = 0.03818 bar ← (c)
p = 1 bar; T = 821.2 K

---

1. Note that the results using the internal functions of IT and using data from Table A-21 agree quite well in this case.

PROBLEM 6.134

KNOWN: Air as an ideal gas with constant specific heat ratio k expands isentropically from p_1, T_1, V_1 to a pressure p_2.

FIND: (a) Develop an expression for the exit velocity V_2 in terms of k, R, V_1, T_1, p_1, p_2.
(b) Plot V_2 versus p_2/p_1 for selected values of k, V_1, and T_1

SCHEMATIC & GIVEN DATA:

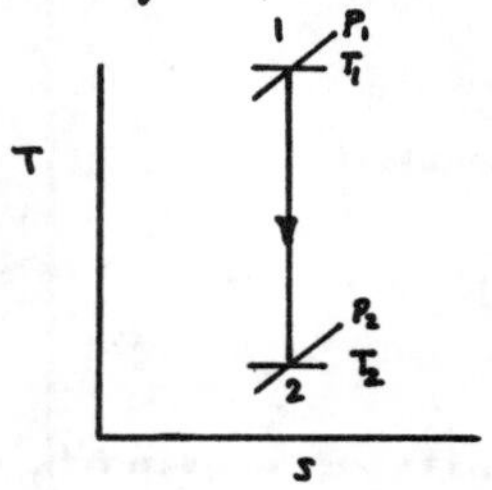

ASSUMPTIONS: (1) The expansion through the nozzle is isentropic. (2) The nozzle operates at steady state. (3) The gas is modeled as an ideal gas with constant specific heat ratio k. (4) Potential energy effects are negligible.

ANALYSIS: (a) At steady state $\dot{m}_1 = \dot{m}_2 = \dot{m}$, and an energy rate balance reduces to

$$0 = \dot{Q}_{cv}^{\,0} - \dot{W}_{cv}^{\,0} + \dot{m}\left(h_1 - h_2 + \frac{V_1^2 - V_2^2}{2} + g(z_1 - z_2)^{0}\right)$$

or

$$0 = h_1 - h_2 + \frac{V_1^2 - V_2^2}{2}$$

With Eq. 3.47a

$$\frac{V_2^2}{2} = \frac{V_1^2}{2} + \frac{kR}{k-1}(T_1 - T_2)$$

$$= \frac{V_1^2}{2} + \frac{kRT_1}{k-1}\left(1 - \frac{T_2}{T_1}\right)$$

For constant k and no change in specific entropy, Eq. 6.45 relates the temperatures and pressures, giving

$$V_2 = \sqrt{V_1^2 + 2\frac{kRT_1}{k-1}\left[1 - \left(\frac{p_2}{p_1}\right)^{(k-1)/k}\right]} \longleftarrow V_2$$

Sample calculation $V_1 = 0$, $T_1 = 1000\,K$, $k = 1.4$, $(p_2/p_1) = 0.1$

$$V_2 = \sqrt{\frac{(2)(1.4)}{1.4-1}\,\frac{8314}{28.97}\left(\frac{N\cdot m}{kg\cdot K}\right)(1000\,K)\left[\frac{1\,kg\cdot m/s^2}{1\,N}\right]\left[1 - (0.1)^{0.4/1.4}\right]} = 984\,m/s$$

(b) PLOT:

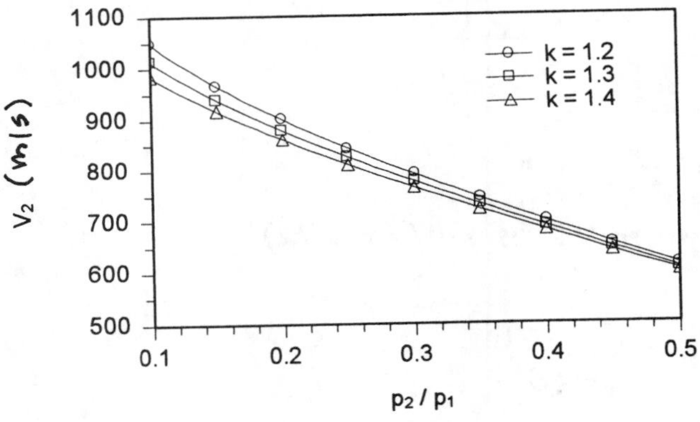

The plots show that the exit velocity increases as p_2/p_1 decreases. For a specified value of p_2/p_1, V_2 increases as k decreases.

PROBLEM 6.135

KNOWN: An ideal gas undergoes a polytropic process from T_1, p_1 to T_2.

FIND: (a) Derive an expression for $s_2 - s_1$ in terms of n, R, T_1, T_2 and s°. (b) Determine n when $\Delta s = 0$.

SCHEMATIC & GIVEN DATA:

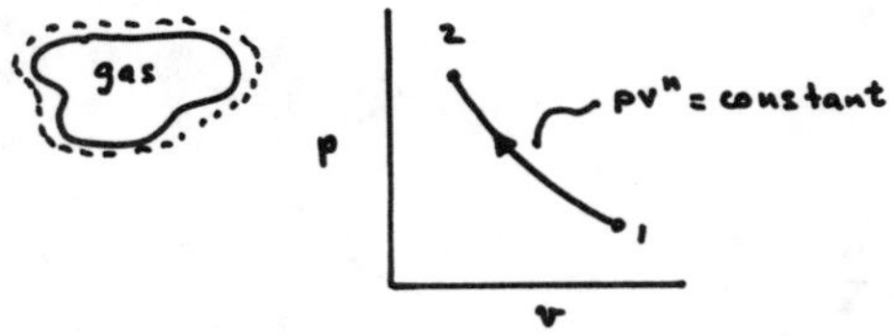

ASSUMPTIONS: (1) The system consists of a quantity of a gas that is modeled as an ideal gas. (2) The process is a polytropic process.

ANALYSIS: (a) With Eq. 6.21a

$$s_2 - s_1 = s^\circ(T_2) - s^\circ(T_1) - R \ln \frac{p_2}{p_1}$$

For a polytropic process of an ideal gas (Eq. 3.56)

$$\frac{T_2}{T_1} = \left(\frac{p_2}{p_1}\right)^{(n-1)/n}$$

Thus

$$s_2 - s_1 = s^\circ(T_2) - s^\circ(T_1) - R \ln \left(\frac{T_2}{T_1}\right)^{\frac{n}{n-1}} \longleftarrow \text{(a)}$$

(b) If $\Delta s = 0$, the expression of part (a) gives

$$\ln \left(\frac{T_2}{T_1}\right)^{\frac{n}{n-1}} = \frac{s^\circ(T_2) - s^\circ(T_1)}{R}$$

$$\frac{n}{n-1} \ln \left(\frac{T_2}{T_1}\right) = \frac{s^\circ(T_2) - s^\circ(T_1)}{R}$$

or

$$\frac{n}{n-1} = \alpha \quad \text{where} \quad \alpha = \frac{s^\circ(T_2) - s^\circ(T_1)}{R \ln T_2/T_1}$$

① $$\Rightarrow \quad n = \frac{\alpha}{\alpha - 1} = \frac{s^\circ(T_2) - s^\circ(T_1)}{s^\circ(T_2) - s^\circ(T_1) - R \ln T_2/T_1} \longleftarrow \text{(b)}$$

---

1. For the special case c_p = constant,

$$s^\circ(T_2) - s^\circ(T_1) = \int_{T_1}^{T_2} \frac{c_p}{T} dT = c_p \ln \frac{T_2}{T_1}$$

Thus

$$\alpha = \frac{c_p \ln T_2/T_1}{R \ln T_2/T_1} = \frac{c_p}{R}$$

With Eq. 3.47a, $c_p = kR/k-1$, so

$$\alpha = \frac{k}{k-1}$$

and

$$n = \frac{k/(k-1)}{\left(\frac{k}{k-1}\right) - 1} = \frac{k}{k-(k-1)} = k$$

which is in accord with the discussion of Fig. 6.11 (Sec. 6.7.2).

PROBLEM 6.136

KNOWN: Data are provided for a rigid, insulated tank initially filled with water vapor. A leak develops and steam slowly excapes until the pressure becomes 0.15 MPa

FIND: Determine (a) the final temperature of the water within the tank and (b) the amount of mass that exits.

SCHEMATIC & GIVEN DATA:

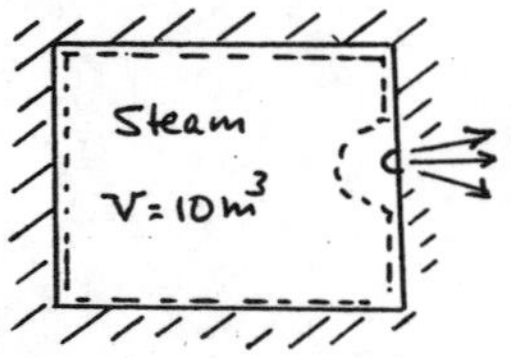

Initially $P_1 = 0.7$ MPa, $T_1 = 240°C$
$P_2 = 0.15$ MPa

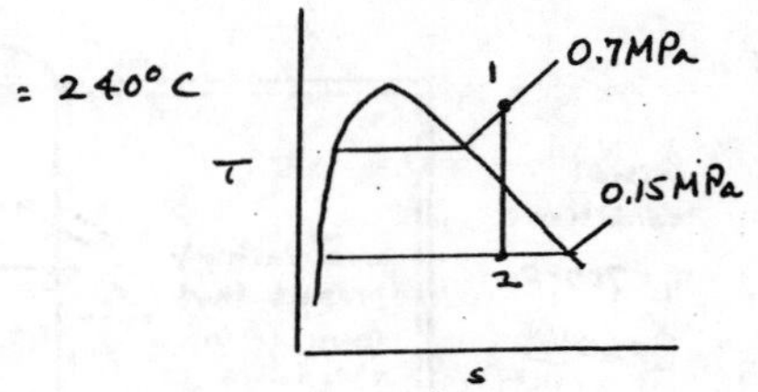

ASSUMPTIONS: (1) The system consists of the mass initially in the tank that remains in the tank. (2) For the system, $\dot{Q} = 0$ and the effects of kinetic and potential energy can be ignored. (3) Irreversibilities within the tank can be ignored as the air slowly escapes.

ANALYSIS: The solution closely follows that of Example 6.10, leading to the result that the mass within the system undergoes an isentropic expansion: $s_2 = s_1$.

With values from Table A-4, $s_1 = 7.0641$ kJ/kg·K, $v_1 = 0.3292$ m³/kg. Since $s_2 = s_1$, state 2 is located in the two phase liquid-vapor region with quality

$$x_2 = \frac{s_2 - s_f}{s_g - s_f} = \frac{7.0641 - 1.4336}{7.2233 - 1.4336} = 0.9725$$

(a) The final temperature is the saturation temperature corresponding to $P_2 = 0.15$ MPa: $T_2 = 111.4°C$. ← (a)

(b) The mass that exits $= m_1 - m_2$, or

$$\Delta m = m_1 - m_2 = V\left[\frac{1}{v_1} - \frac{1}{v_2}\right]$$

where

$$v_2 = \left(\frac{1.0528}{10^3}\right) + 0.9725\left[1.159 - \frac{1.0528}{10^3}\right] = 1.127 \frac{m^3}{kg}$$

So

$$\Delta m = 10 m^3 \left[\frac{1}{0.3292} - \frac{1}{1.127}\right]\frac{kg}{m^3} = 21.51 \text{ kg}$$ ← (b)

PROBLEM 6.137

KNOWN: A rigid, insulated tank is initially filled with water vapor at a known state. Water vapor slowly escapes until vapor at pressure p remains.

FIND: (a) For p = 14.7 lbf/in.², determine the final temperature and mass.
(b) Plot final temperature and mass versus p ranging from 14.7 to 60 lbf/in.²

SCHEMATIC & GIVEN DATA:

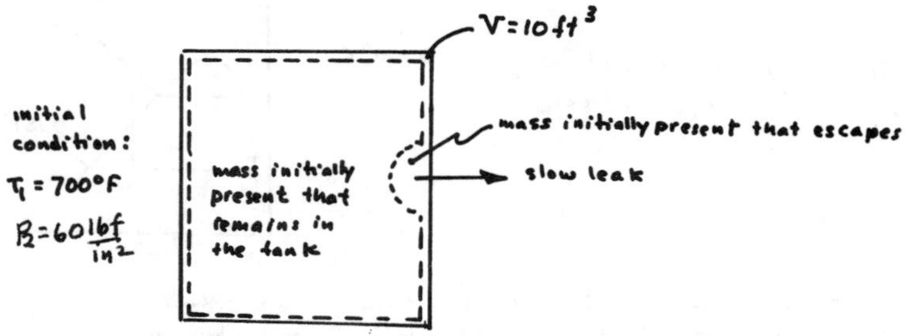

ASSUMPTIONS: (1) As shown in the figure, the closed system is the mass initially in the tank that remains in the tank. (2) No significant heat transfer occurs between system and surroundings. (3) Irreversibilities within the tank can be ignored as the water vapor slowly escapes.

Analysis: For the closed system under consideration, there are no significant irreversibilities (assumption 3), and no heat transfer occurs (assumption 2). For this system an entropy balance reduces to

$$m\Delta s = \int_1^2 \left(\frac{\delta Q}{T}\right)_b^{\,0} + \sigma^{\,0} \Rightarrow s_2 = s_1$$

Thus, the final state within the tank is fixed by $p_2 = p$, $s_2 = s_1$. The final mass in the tank is $m_2 = V/v_2$.

(a) $p = 14.7\ \text{lbf/in.}^2$ From Table A-4E, $s_1 = 1.8609\ \text{Btu/lb}\cdot{}^\circ\text{R}$. Then, with $s_2 = s_1$ at 14.7 lbf/in.², interpolation in Table A-4E gives $T_2 = 377.4^\circ\text{F}$ ←
and $v_2 = 33.73\ \text{ft}^3/\text{lb}$, so $m_2 = 10\ \text{ft}^3/(33.73\ \text{ft}^3/\text{lb}) = 0.296\ \text{lb}$. ←
The initial mass is $m_1 = V/v_1 = (10/11.44) = 0.874\ \text{lb}$.

(b) PLOTS

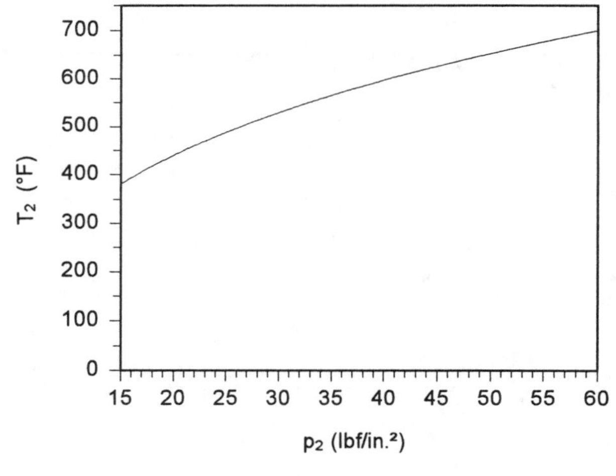

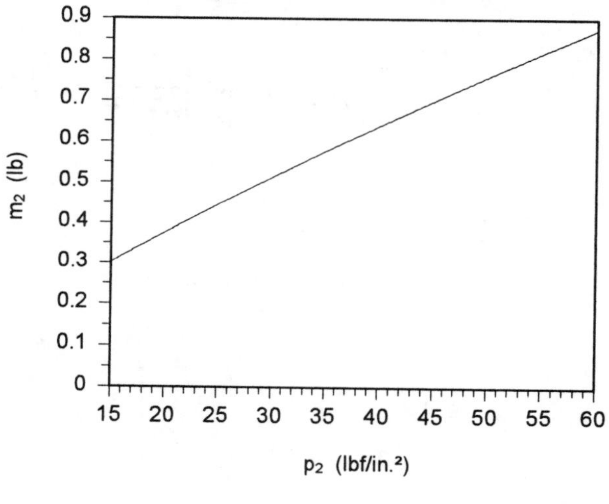

PROBLEM 6.138

KNOWN: A tank initially filled with air is allowed to discharge through a turbine until the pressure in the tank becomes atmospheric

FIND: Determine the maximum theoretical work that could be developed.

SCHEMATIC & GIVEN DATA:

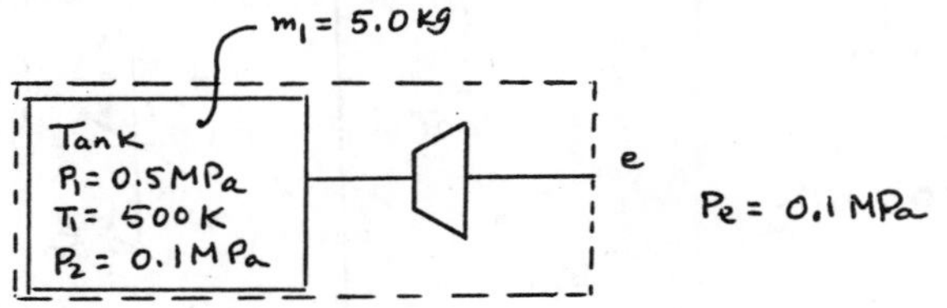

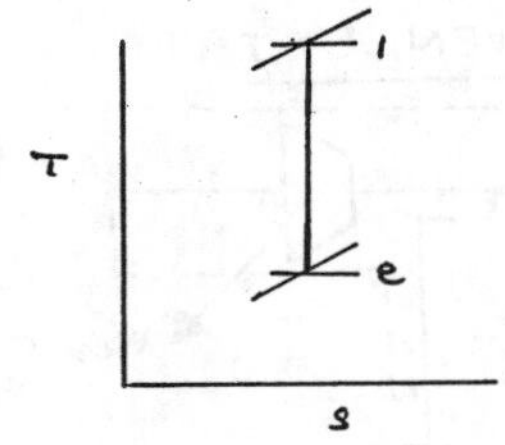

ASSUMPTIONS: (1) The control volume is shown above. (2) For the control volume, $\dot{Q}_{cv} = 0$ and kinetic/potential energy effects are negligible. (3) The air is modeled as an ideal gas. (4) The control volume is free of irreversibilities.

ANALYSIS: A mass rate balance reduces to $dm/dt = -\dot{m}_e$. Using this, an energy rate balance reads

$$\frac{dU}{dt} = \cancel{\dot{Q}_{cv}}^{0} - \dot{W}_{cv} - \dot{m}_e h_e \quad , \quad \frac{dU}{dt} = -\dot{W}_{cv} + h_e \frac{dm}{dt}$$

or

$$\dot{W}_{cv} = -\frac{dU}{dt} + h_e \frac{dm}{dt} \Rightarrow W_{cv} = -\Delta U + \int h_e \, dm \qquad (1)$$

We expect that the maximum theoretical work would be developed in the absence of irreversibilities within the tank and the turbine (assumption (4)). The data given are recognized as corresponding to those for Example 6.10. The solution to Example 6.10 shows that a typical unit of mass remaining in the tank would undergo an isentropic expansion from T_1, P_1 to T_2, P_2. Moreover, each unit of mass passing through the turbine expands isentropically. Accordingly, if P, T denote the pressure and temperature within the tank at a particular instant

$$\frac{P_r(T)}{P_r(T_1)} = \frac{P}{P_1} \qquad (2)$$

But P, T would also correspond to the condition of the mass entering the turbine. So

$$\frac{P_r(T_e)}{P_r(T)} = \frac{P_e}{P} \qquad (3)$$

Combining Eqs. (2), (3)

$$\frac{P_r(T_e)}{P_r(T_1)} = \frac{P_e}{P_1} \rightarrow \frac{P_r(T_e)}{P_r(T_1)} = \frac{P_2}{P_1} \quad \text{since } P_e = P_2$$

Then, with the result of Example 6.10, $T_e = T_2 = 317$ K.

Returning to Eq. (1), h_e is fixed by T_e, and thus a constant:

$$(W_{cv})_{MAX} = -\Delta U + h_e \Delta m$$

$$= (m_1 u_1 - m_2 u_2) + h_e (m_2 - m_1)$$

From Example 6.10, $m_2 = 1.58$ kg. With data from Table A-22

$$(W_{cv})_{MAX} = (5\,\text{kg})(359.49\,\tfrac{\text{kJ}}{\text{kg}}) - (1.58\,\text{kg})(226.28\,\tfrac{\text{kJ}}{\text{kg}}) + (317.28\,\tfrac{\text{kJ}}{\text{kg}})(1.58-5)\,\text{kg}$$

$$= 1797.45\,\text{kJ} - 357.52\,\text{kJ} - 1085.1\,\text{kJ}$$

$$= 354.83\,\text{kJ} \quad \longleftarrow \text{MAX}$$

PROBLEM 6.139

KNOWN: A bottle initially filled with air is allowed to discharge through a turbine, developing work, until the pressure in the bottle becomes 3 atm.

FIND: Determine the volume of the bottle when irreversibilities are absent.

SCHEMATIC & GIVEN DATA:

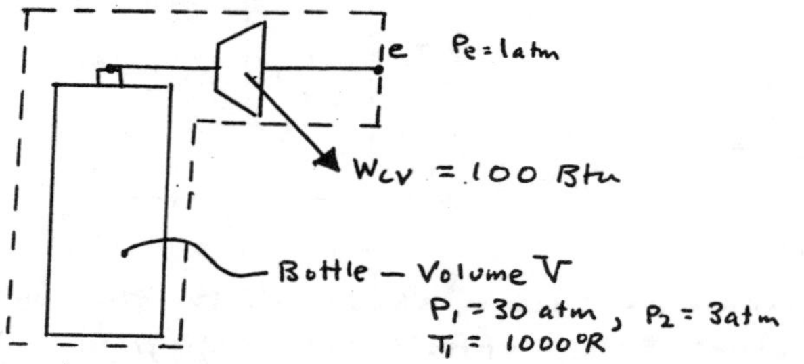

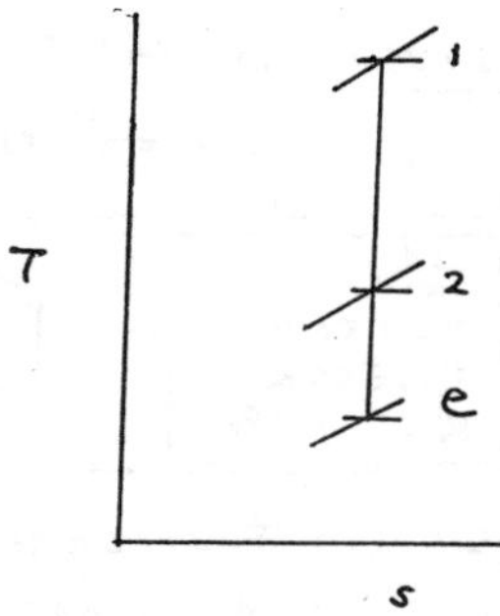

ASSUMPTIONS: (1) The control volume is shown in the figure. (2) For the control volume $\dot{Q}_{cv} = 0$ and kinetic/potential energy effects are negligible. (3) The air is modeled as an ideal gas. (4) Irreversibilities are absent within the control volume.

① (in margin beside assumption 3)

ANALYSIS: A mass rate balance reduces to $dm/dt = -\dot{m}_e$. Using this an energy rate balance reads

$$\frac{dU}{dt} = \dot{Q}_{cv}^{\,0} - \dot{W}_{cv} - \dot{m}_e h_e, \quad \frac{dU}{dt} = -\dot{W}_{cv} + h_e \frac{dm}{dt} \Rightarrow W_{cv} = -\Delta U + \int h_e\, dm \qquad (1)$$

The plan is to evaluate the bottle volume in the absence of irreversibilities (assumption (4)). Following the reasoning of Example 6.10, a typical unit of mass remaining in the bottle would undergo an isentropic expansion from P_1, T_1 to P_2, T_2. Moreover, each unit of mass passing through the turbine expands isentropically. Accordingly, if P, T denote the pressure and temperature within the tank at a particular instant,

$$\frac{P_r(T)}{P_r(T_1)} = \frac{P}{P_1} \qquad (2)$$

But P, T would also correspond to the condition of the mass entering the turbine, so

$$\frac{P_r(T_e)}{P_r(T)} = \frac{P_e}{P} \qquad (3)$$

Combining Eqs. (2), (3), and using data from Table A-22E

$$\frac{P_r(T_e)}{P_r(T_1)} = \frac{P_e}{P_1} \Rightarrow P_r(T_e) = \left(\frac{1\,\text{atm}}{30\,\text{atm}}\right)(12.3) = 0.41 \Rightarrow h_e = 90.98\ \text{Btu/lb}.$$

Since h_e is fixed, Eq. (1) becomes, with the ideal gas equation of state ($m = pV/RT$),

$$W_{cv} = -\Delta U + h_e(\Delta m) \Rightarrow W_{cv} = (m_1 u_1 - m_2 u_2) + h_e(m_2 - m_1)$$

$$= \left(\frac{P_1 V}{RT_1}\right)u_1 - \left(\frac{P_2 V}{RT_2}\right)u_2 + h_e\left[\frac{P_2 V}{RT_2} - \frac{P_1 V}{RT_1}\right]$$

$$= \frac{P_1 V}{R}\left[\frac{u(T_1)}{T_1} - \left(\frac{P_2}{P_1}\right)\frac{u(T_2)}{T_2} + h_e\left[\left(\frac{P_2}{P_1}\right)\frac{1}{T_2} - \frac{1}{T_1}\right]\right] \qquad (4)$$

With data from Table A-22E, $u(T_1) = 172.43$ Btu/lb, $P_r(T_2) = P_2/P_1\, P_r(T_1) = (3/30)(12.3) = 1.23$, giving $T_2 = 522°R$, $u(T_2) = 88.93$ Btu/lb. Evaluating Eq. (4)

$$100\ \text{Btu} = \frac{V(30 \times 14.7 \times 144\ \text{lbf/ft}^2)}{(1545/28.97)(\text{ft}\cdot\text{lbf/lb}\cdot°R)}\left[\frac{172.43}{1000} - \left(\frac{3}{30}\right)\frac{88.93}{522} + 90.98\left[\left(\frac{3}{30}\right)\frac{1}{522} - \frac{1}{1000}\right]\right]\frac{\text{Btu}}{\text{lb}\cdot°R} \qquad (5)$$

Solving gives $V = 1.03\ \text{ft}^3$. ←

---

1. Checking the compressibility chart with $T_{R1} = 1000/239 = 4.18$, $P_{R2} = 3/37.2 = 0.081$, it can be concluded that the ideal gas model is appropriate at these states.

PROBLEM 6.140

KNOWN: Steady state operating data are provided for an air turbin fitted with a diffuser at its exit.

FIND: Determine the pressure and temperature at the turbine exit and the rate of entropy production in the turbine. Show the processes on a T-s diagram.

SCHEMATIC & GIVEN DATA:

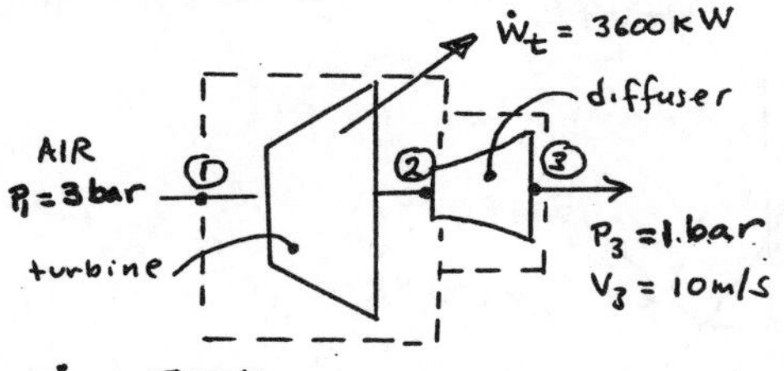

$\dot{m}_1 = 18$ kg/s
$T_1 = 800°C$
$V_1 = 100$ m/s
$V_2 = 150$ m/s

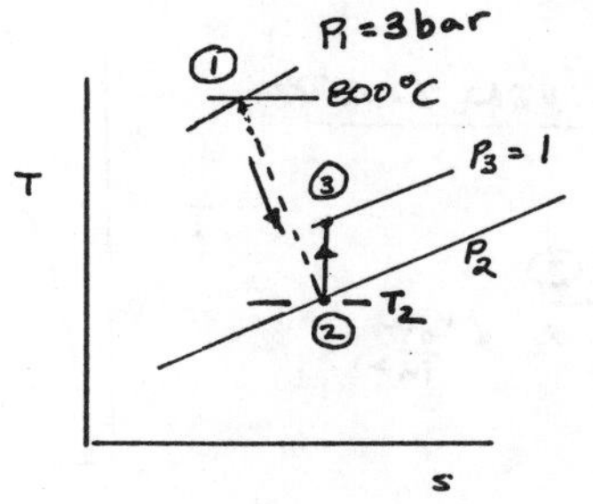

ASSUMPTIONS: (1) The overall control volume consists of two sub control volumes, each at steady state. (2) The turbine operates adiabatically. (3) Flow through the diffuser is isentropic. (4) Potential energy effects can be ignored. (4) Air is modeled as an ideal gas.

ANALYSIS: (a) For the isentropic process of air from 2 to 3, $P_2/P_3 = P_r(T_2)/P_r(T_3)$. Accordingly, to find P_2 requires $P_r(T_2)$ and $P_r(T_3)$.

Mass and energy rate balances for the turbine reduce to give

$$0 = \dot{Q}_{cv}^{\,0} - \dot{W}_t + \dot{m}\left(h_1 - h_2 + \frac{V_1^2 - V_2^2}{2}\right) \Rightarrow h_2 = -\frac{\dot{W}_t}{\dot{m}} + h_1 + \frac{(V_1^2 - V_2^2)}{2}$$

with h_1 from Table A-22

$$h_2 = -\frac{3600\ \text{kJ/s}}{18\ \text{kg/s}} + 1129.83\frac{\text{kJ}}{\text{kg}} + \frac{[(100)^2 - (150)^2](\text{m}^2/\text{s}^2)}{2}\left|\frac{1\ \text{N}}{1\ \text{kg}\cdot\text{m/s}^2}\right|\left|\frac{1\ \text{kJ}}{10^3\ \text{N}\cdot\text{m}}\right| = 923.6\ \frac{\text{kJ}}{\text{kg}}$$

Then, interpolation in Table A-22 gives $P_{r2} = 72.66$, $T_2 = 892\text{K}\ (619°C)$. ← T_2

Mass and energy rate balances for the diffuser reduce to give

$$0 = \dot{Q}_{cv} - \dot{W}_{cv} + \dot{m}\left(h_2 - h_3 + \frac{V_2^2 - V_3^2}{2}\right) \Rightarrow h_3 = h_2 + \frac{V_2^2 - V_3^2}{2}$$

or $$h_3 = 923.6 + \frac{(150)^2 - (10)^2}{2000} = 934.8\ \text{kJ/kg}.$$

Interpolating in Table A-22, $P_{r3} = 75.85$. Then

$$P_2 = P_3\frac{P_{r2}}{P_{r3}} = (1\ \text{bar})\left(\frac{72.66}{75.85}\right) = 0.958\ \text{bar}$$ ← P_2

(b) Mass and entropy rate balances reduce to give for the turbine

$$\dot{\sigma}_t = \dot{m}(s_2 - s_1)$$

With Eq. 6.21a this becomes

$$\dot{\sigma}_t = \dot{m}\left[s^\circ(T_2) - s^\circ(T_1) - R\ln P_2/P_1\right]$$

$$= (18\ \text{kg/s})\left[2.8381 - 3.0485 - \frac{8.314}{28.97}\ln\left(\frac{0.958}{3}\right)\right]\frac{\text{kJ}}{\text{kg}\cdot\text{K}}\left|\frac{1\ \text{kW}}{1\ \text{kJ/s}}\right|$$

$$= 2.11\ \frac{\text{kW}}{\text{K}}$$ ← $\dot{\sigma}_t$

~~PR~~OBLEM 6.141

KNOWN: Steam at a specified state enters a turbine and expands adiabatically with a specified mass flow rate to a given exit pressure.

FIND: (a) Determine the maximum theoretical power that can be developed and the corresponding exit temperature.

(b) For a given exit temperature, determine the isentropic turbine efficiency.

SCHEMATIC & GIVEN DATA:

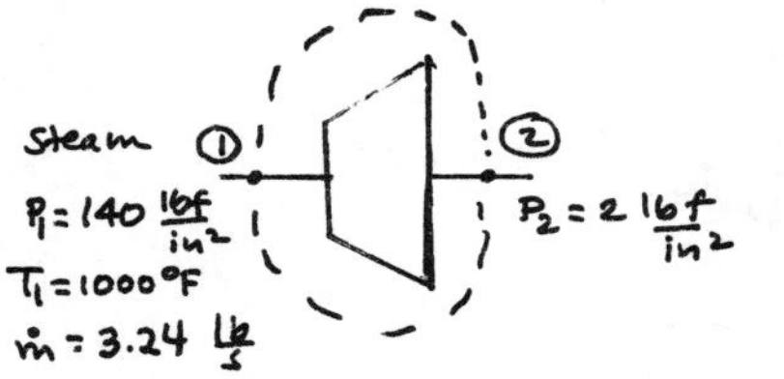

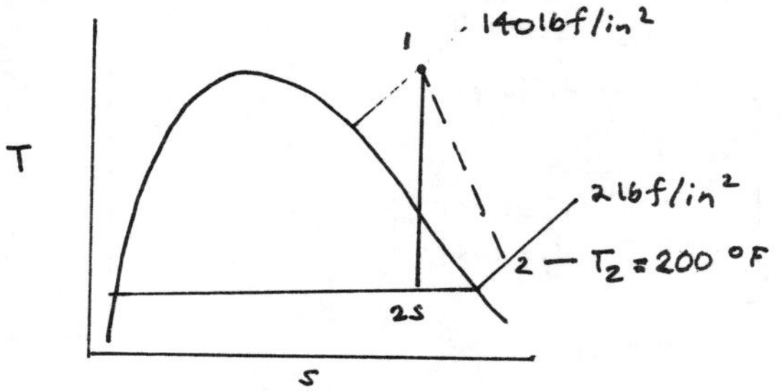

ASSUMPTIONS: (1) The control volume shown in the figure is at steady state. (2) For the control volume, $\dot{Q}_{cv} = 0$, and kinetic/potential energy effects are negligible.

ANALYSIS: Mass and energy rate balances reduce to give $\dot{W}_{cv} = \dot{m}(h_1 - h_2)$.

(a) As discussed in Sec. 6.8, the maximum theoretical power is obtained for an isentropic expansion: $s_{2s} = s_1$. With data from Table A-3E, 4E, $h_1 = 1531.0$ Btu/lb, $s_1 = 1.8827$ Btu/lb·°R,

$$x_{2s} = \frac{1.8827 - 0.1750}{1.7448} = 0.9787$$

$$h_{2s} = h_f + x_{2s} h_{fg} = 94.02 + 0.9787(1022.1) = 1094.3\ \text{Btu/lb}, \quad T_{2s} = 126°F$$

Then

$$(\dot{W}_{cv})_{max} = \left(3.24\frac{lb}{s}\right)(1531 - 1094.3)\left(\frac{Btu}{lb}\right)\left|\frac{3600\,s}{h}\right|\left|\frac{1\ hp}{2545\ Btu/h}\right| = 2000\ hp \longleftarrow (\dot{W}_{cv})_{max}$$

(b) The isentropic turbine efficiency, given by Eq. 6.48, takes the form

$$\eta_t = \frac{h_1 - h_2}{h_1 - h_{2s}} = \frac{(1531 - 1149.7)\ Btu/lb}{(1531 - 1094.3)\ Btu/lb} = 0.873\ (87.3\%) \longleftarrow \eta_t$$

where h_2 is obtained from the steam tables at 2 lbf/in², 200°F.

PROBLEM 6.142

KNOWN: Steam at a specified state enters a turbine and expands adiabatically to a specified pressure. The work developed per unit mass of steam is measured.

FIND: Determine the work developed by the turbine per unit mass of steam flowing and the isentropic turbine efficiency.

SCHEMATIC & GIVEN DATA:

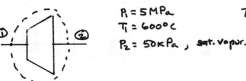

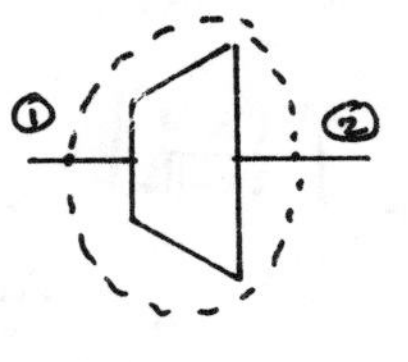

$P_1 = 5\ MPa$
$T_1 = 600°C$
$P_2 = 50\ kPa$, sat. vapor.

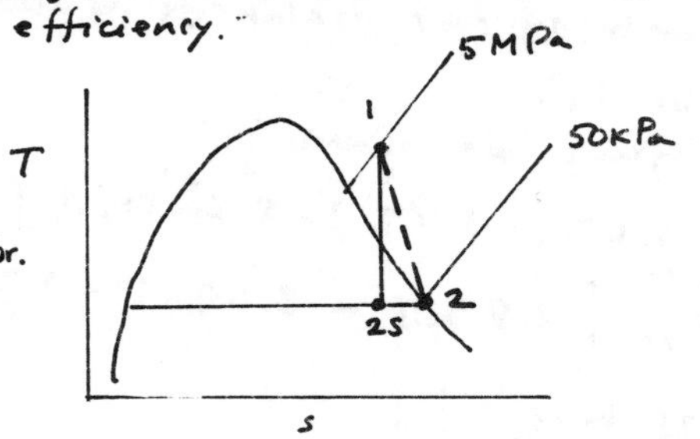

ASSUMPTIONS: (1) The control volume shown in the figure is at steady state. (2) For the control volume, $\dot{Q}_{cv} = 0$, and kinetic/potential energy effects are negligible.

ANALYSIS: With the given assumptions, the isentropic turbine efficiency, Eq. 6.48, takes the form

PROBLEM 6.142 (Continued)

$$\eta_t = \frac{\dot{W}_t/\dot{m}}{h_1 - h_{2s}}, \text{ where } \frac{\dot{W}_t}{\dot{m}} = h_1 - h_2$$

To find h_{2s}, Table A-4 gives $s_1 = 7.2683$ kJ/kg·K. Then with data from Table A-3 at 50 kPa

$$x_{2s} = \frac{s_{2s} - s_f}{s_g - s_f} = \frac{7.2683 - 1.0910}{7.5939 - 1.0910} = 0.95, \quad h_{2s} = h_f + x_{2s} h_{fg} = 340.49 + 0.95(2305.4) = 2530.6 \frac{kJ}{kg}$$

Then, with $h_1 = 3666.4$ kJ/kg, $h_2 = 2645.9$ kJ/kg, $\dot{W}_t/\dot{m} = 1020.5$ kJ/kg ← $\dot{W}_t/\dot{m}$

$$\eta_t = \frac{1020.5 \text{ kJ/kg}}{(3666.4 - 2530.6) \text{ kJ/kg}} = 0.898 \ (89.8\%)$$ ← η_t

---

PROBLEM 6.143

KNOWN: Hydrogen (H_2) at a specified state expands through an insulated turbine at steady state to a specified pressure. The isentropic turbine efficiency is given.

FIND: Determine the temperature at the turbine exit.

SCHEMATIC & GIVEN DATA:

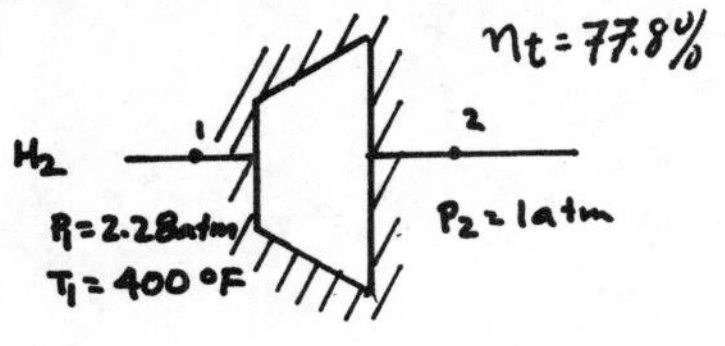

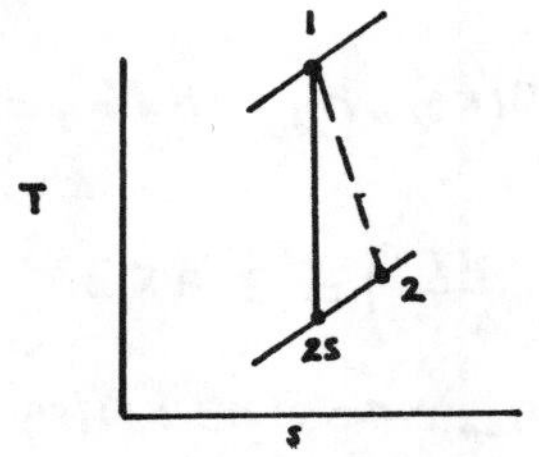

ASSUMPTIONS: (1) The turbine is well insulated and at steady state. (2) Kinetic and potential energy effects are negligible. (3) H_2 is modeled as an ideal gas with constant c_p.

ANALYSIS: The isentropic turbine efficiency is

$$\eta_t = \frac{\dot{W}_{cv}/\dot{m}}{(\dot{W}_{cv}/\dot{m})_s}$$

With the specified assumptions, mass and energy rate balances allow this to be expressed as

$$\eta_t = \frac{h_1 - h_2}{h_1 - h_{2s}} = \frac{c_p[T_1 - T_2]}{c_p[T_1 - T_{2s}]} \quad (1)$$

where T_{2s} can be determined using Eq. 6.45:

$$T_{2s} = T_1 \left[\frac{P_2}{P_1}\right]^{(k-1)/k} \quad (2)$$

As the H_2 expands through the turbine its temperature decreases -- see T-s diagram. Table A-20E shows that for $T \leq 400$°F, the value of k is closely 1.4. Thus, Eq. (2) gives

$$T_{2s} = 860°R \left[\frac{1}{2.28}\right]^{0.4/1.4} = 679.6 °R$$

Then, Eq. (1) gives on rearrangement

$$T_2 = T_1 - \eta_t (T_1 - T_{2s}) = 860°R - 0.778(860 - 679.6)°R = 719.6°R$$

$\Rightarrow T_2 = 260$°F ←

PROBLEM 6.144

KNOWN: Air at a specified temperature and pressure enters an insulated turbine operating at steady state and expands to a specified temperature and pressure.

FIND: Determine the work developed per unit mass flowing and the isentropic turbine efficiency.

SCHEMATIC & GIVEN DATA:

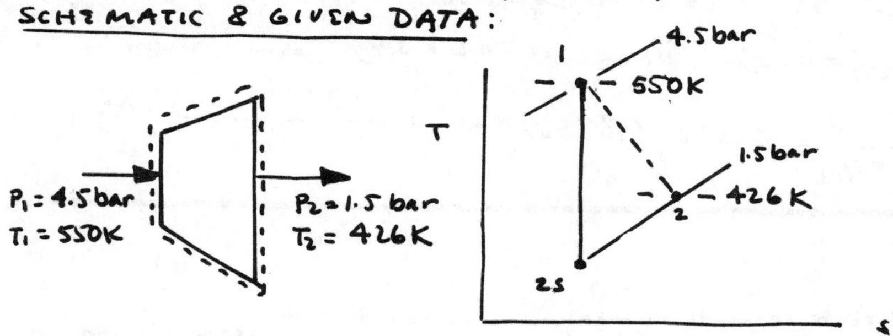

ASSUMPTIONS: 1. The turbine is insulated and operates at steady state. 2. Kinetic and potential energy effects are negligible. 3. Air is modeled as an ideal gas.

ANALYSIS: Table A-22 gives $h_1 = 554.74$ kJ/kg, $p_{r1} = 11.86$, and $h_2 = 427.36$ kJ/kg. To find h_{2s} use

$$\frac{p_{r2}}{p_{r1}} = \frac{p_2}{p_1} \Rightarrow p_{r2} = p_{r1}\left(\frac{p_2}{p_1}\right) = 11.86\left(\frac{1.5}{4.5}\right) = 3.953$$

Then, interpolating in Table A-22 with p_{r2}, $h_{2s} = 405.28$ kJ/kg.

The work developed per unit of mass flowing is

$$\frac{\dot{W}_t}{\dot{m}} = h_1 - h_2 = 554.74 - 427.36 = 127.38 \text{ kJ/kg} \quad \leftarrow$$

The isentropic turbine efficiency is

$$\eta_t = \frac{\dot{W}_t/\dot{m}}{h_1 - h_{2s}} = \frac{127.38}{554.74 - 405.28} = 0.852\ (85.2\%) \quad \leftarrow$$

PROBLEM 6.145

KNOWN: Steady state operating data are provided for a well-insulated steam turbine having two stages in series.

FIND: Show the principal states on a T-s diagram. Determine the state at the exit of the second stage. Determine the work developed by each stage, per unit of mass of steam flowing.

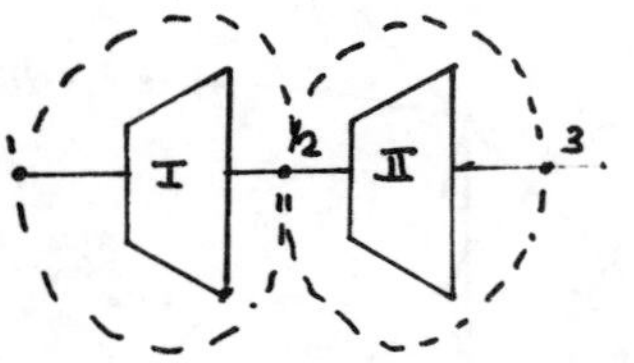

$T_1 = 700°F$
$P_1 = 550\ lbf/in^2$
$P_2 = 200\ lbf/in^2$
$P_3 = 14.7\ lbf/in^2$
$\eta_I = 88\%$
$\eta_{II} = 92\%$

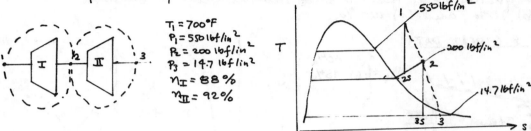

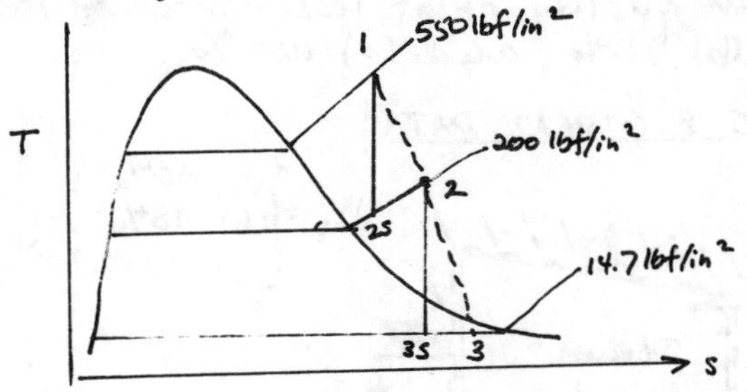

ASSUMPTIONS: (1) Each control volume shown in the figure is at steady state. (2) For each control volume, $\dot{Q}_{cv} = 0$, and kinetic/potential energy effects can be ignored.

ANALYSIS: Steam table data give $h_1 = 1353.7$ Btu/lb, $s_1 = 1.5987$ Btu/lb·°R. Interpolating at 200 lbf/in² with $s_{2s} = s_1$, $h_{2s} = 1245.2$ Btu/lb. Then, using the isentropic turbine efficiency of the first stage

$$\eta_I = 0.88 = \frac{h_1 - h_2}{h_1 - h_{2s}} \Rightarrow h_2 = h_1 - 0.88(h_1 - h_{2s}) = 1353.7 - 0.88(1353.7 - 1245.2) = 1258.2\ \frac{Btu}{lb}$$

Interpolating at 200 lbf/in² with h_2 gives $s_2 = 1.6127$ Btu/lb·°R. Using $s_{3s} = s_2$,

$$x_{3s} = \frac{s_{3s} - s_f}{s_g - s_f} = \frac{1.6127 - 0.31212}{1.4446} = 0.9, \quad h_{3s} = h_f + x_{3s} h_{fg} = 180.15 + 0.9(970.4) = 1053.5\ Btu/lb.$$

Then, using the isentropic turbine efficiency of the second stage

$$\eta_{II} = 0.92 = \frac{h_2 - h_3}{h_2 - h_{3s}} \Rightarrow h_3 = h_2 - 0.92(h_2 - h_{3s}) = 1258.2 - 0.92(1258.2 - 1053.5) = 1069.9\ \frac{Btu}{lb}$$

Using h_3

$$x_3 = \frac{h_3 - h_f}{h_{fg}} = \frac{1069.9 - 180.15}{970.4} = 0.917\ (91.7\%)$$

The state at the exit of Stage I is fixed by two of P_2, h_2, s_2. The state at the exit of Stage II is fixed by two of P_3, h_3, x_3. ← States 2, 3

The work developed for each of the stages is obtained by reducing mass and energy rate balances to give

$$(\dot{W}_{cv}/\dot{m})_I = h_1 - h_2 = 1353.7 - 1258.2 = 95.5\ Btu/lb$$

$$(\dot{W}_{cv}/\dot{m})_{II} = h_2 - h_3 = 1258.2 - 1069.9 = 188.3\ Btu/lb.$$

← ($\dot{W}_{cv}/\dot{m}$)

PROBLEM 6.146

KNOWN: Steam at a known temperature enters an insulated turbine and expands to a known pressure. The isentropic efficiency is specified.

FIND: Determine the range of turbine inlet pressures that insures a turbine exit steam quality of at least 90% for isentropic turbine efficiencies of (a) 80%, (b) 90%, and (c) 100%.

SCHEMATIC & GIVEN DATA:

$\eta_t = $ (a) 80%, (b) 90%, (c) 100%

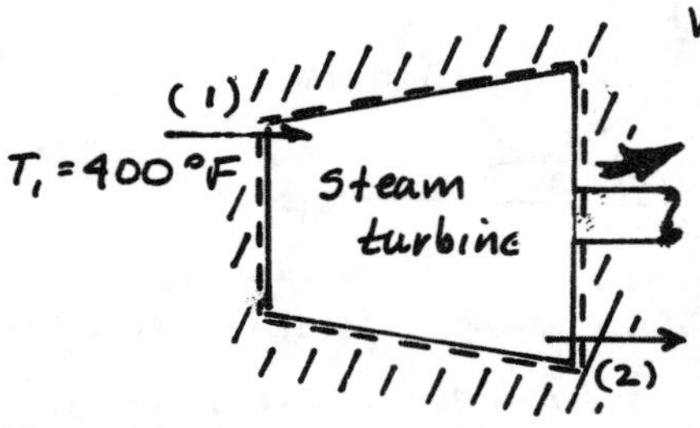

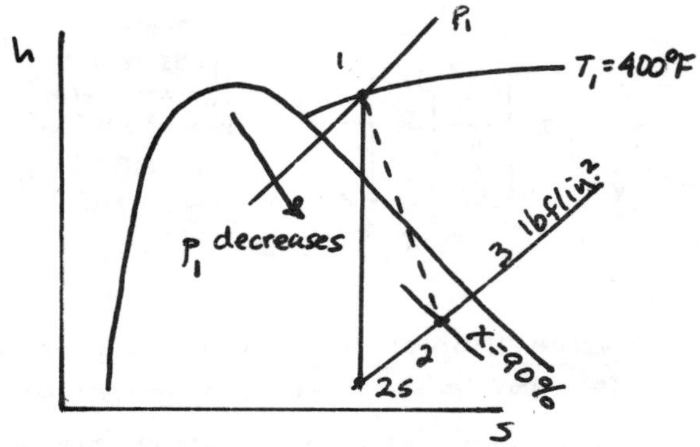

ASSUMPTIONS: (1) The turbine is at steady state, and $\dot{Q}_{cv} = 0$. (2) Kinetic and potential energy effects are negligible.

ANALYSIS: As shown on the h-s diagram, p_1 decreases as s_2 increases. Thus, the maximum allowed turbine inlet pressure corresponds to the case where $x_2 = 0.9$. For any pressure less than this, $x_2 > 0.9$.

Thus, with $x_2 = 0.9$ and data for h_f and h_{fg} at 3 lbf/in.², h_2 is found from

$$h_2 = h_f + 0.9\, h_{fg} \quad (1)$$

and

$$h_{2s} = h_f + x_{2s}\, h_{fg} \quad (2)$$

thus

$$\eta_t = \frac{h_1 - h_2}{h_1 - h_{2s}} = \frac{h_1 - h_f - 0.9\, h_{fg}}{h_1 - h_f - x_{2s}\, h_{fg}} \quad (3)$$

For a given value of p_1, there is a corresponding value of h_1 and x_{2s}. Thus, for that p_1 there is a value of η_t that can be calculated from (3). Using *IT*, an iterative numerical process is used to determine the unique combination of p_1 and x_{2s} and h_1 values to give a desired value of η_t:

IT Code

```
T1 = 400  // °F
p2 = 3  // lbf/in.²
x2 = 0.9

h2 = hsat_Px("Water/Steam", p2, x2)
h2s = hsat_Px("Water/Steam", p2, x2s)

eta = 80
h1 - h2 = (eta/100) * (h1 - h2s)

h1 = h_PT("Water/Steam", p1, T1)
s1 = s_PT("Water/Steam", p1, T1)
s2s = ssat_Px("Water/Steam", p2, x2s)
s1 = s2
```

IT Results

(a) η = 80%; $(p_1)_{max}$ = 116.1 lbf/in.² ← (a)

(b) η = 90%; $(p_1)_{max}$ = 80.04 lbf/in.² ← (b)

(c) η = 100%; $(p_1)_{max}$ = 57.86 lbf/in.² ← (c)

PROBLEM 6.147

KNOWN: Steady-state operating data are provided for a throttling valve in parallel with a steam turbine.

FIND: Determine the power developed by the turbine and the mass flow rate through the valve. Determine the rates of entropy production for the turbine, valve, and mixing chamber. Locate key states on an h-s diagram.

SCHEMATIC & GIVEN DATA:

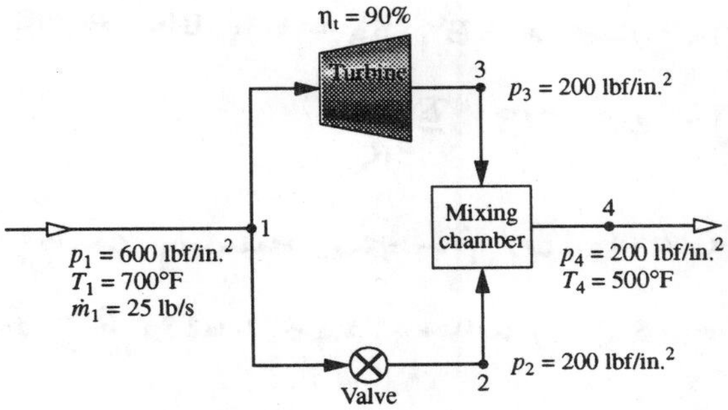

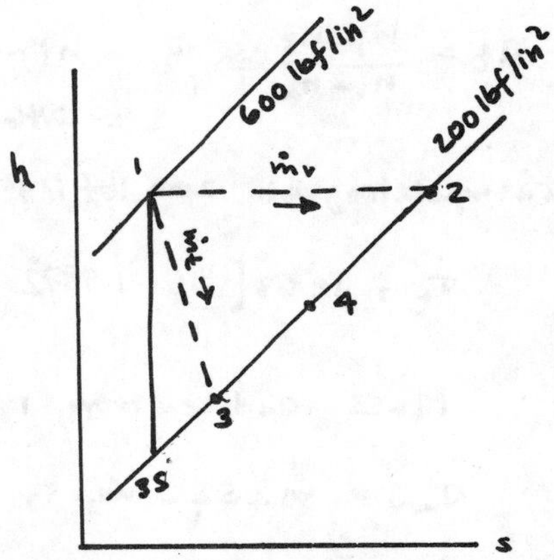

ASSUMPTIONS: 1. Control volumes enclosing the turbine, valve, and mixing chamber are at steady state. For each, $\dot{Q}_{cv} = 0$. 2. The expansion across the valve is a throttling process. 3. Kinetic and potential energy effects can be ignored.

ANALYSIS: (a) Considering an overall control volume at steady state including the turbine, valve, and mixing chamber, mass and energy rate balances reduce to give $\dot{W}_t = \dot{m}[h_1 - h_4]$. With data from Table A-4E

$$\dot{W}_t = \left(25\ \frac{lb}{s}\right)[1350.6 - 1268.8]\ \frac{Btu}{lb}\left|\frac{3600\ s}{1\ h}\right|\left|\frac{1\ hp}{2545\ Btu/h}\right| = 2893\ hp \quad \longleftarrow (a)$$

(b) Then mass and energy rate balances for a control volume enclosing only the turbine give $\dot{W}_t = \dot{m}_t[h_1 - h_3]$. Using the isentropic turbine efficiency $\eta_t = (h_1 - h_3)/(h_1 - h_{3s}) \Rightarrow (h_1 - h_3) = \eta_t(h_1 - h_{3s})$. So

$$\dot{W}_t = \dot{m}_t \eta_t (h_1 - h_{3s}) \Rightarrow \dot{m}_t = \frac{\dot{W}_t}{\eta_t(h_1 - h_{3s})}$$

. With $s_1 = s_{3s} = 1.5872$ Btu/lb·°R,

Table A-4E gives $h_{3s} = 1234.86$ Btu/lb. Then

$$\dot{m}_t = \frac{(25)[1350.6 - 1268.8]}{(0.9)[1350.6 - 1234.86]} = 19.63\ lb/s$$

The mass flow rate through the valve is then $\dot{m} = \dot{m}_t + \dot{m}_v$ or $\longleftarrow$ (b)

$\dot{m}_v = \dot{m} - \dot{m}_t = 25 - 19.63 = 5.37$ lb/s

(c) Mass and entropy rate balances for the valve give

$$\dot{\sigma}_v = \dot{m}_v[s_2 - s_1]$$

Since $h_2 \sim h_1$, state 2 is fixed by $h_2 = 1350.6$ Btu/lb, $P_2 = 200$ lbf/in², giving $s_2 = 1.7024$ Btu/lb·°R. Also, $s_1 = 1.5872$ Btu/lb·°R. So

$$\dot{\sigma}_v = \left(5.37\ \frac{lb}{s}\right)[1.7024 - 1.5872]\ \frac{Btu}{lb\cdot°R} = 0.6186\ \frac{Btu}{s\cdot°R}$$

PROBLEM 6.147 (Contd.)

Mass and entropy rate balances for the turbine give

$\dot{\sigma}_t = \dot{m}_t [s_3 - s_1]$

To find s_3, use the isentropic turbine efficiency to get

$$\eta_t = \frac{h_1 - h_3}{h_1 - h_{3s}} \Rightarrow h_3 = h_1 - \eta_t(h_1 - h_{3s}) = 1350.6 - .9(1350.6 - 1234.86)$$
$$= 1246.4 \text{ Btu/lb}$$

Interpolating at 200 lbf/in² in Table A-4E, $s_3 = 1.6$ Btu/lb·°R. Thus

$$\dot{\sigma}_t = 19.63[1.6 - 1.5872] = 0.2513 \frac{\text{Btu}}{\text{s}\cdot°\text{R}}$$

Mass and entropy rate balances for the mixing chamber give

$\sigma_{mix} = \dot{m}_4 s_4 - \dot{m}_t s_3 - \dot{m}_V s_2$, where $s_4 = 1.6239 \frac{\text{Btu}}{\text{lb}\cdot°\text{R}}$ from Table A-4E.

Thus

$$\sigma_{mix} = (25)(1.6239) - (19.63)(1.6) - (5.37)(1.7024)$$
① $$= 0.0476 \frac{\text{Btu}}{\text{s}\cdot°\text{R}}$$

---

1. The total rate of entropy production: $\dot{\sigma}_V + \dot{\sigma}_V + \dot{\sigma}_{mix} = 0.9175$ Btu/s·°R. Alternatively, mass and entropy rate balances for an overall control volume reduce to
$$\sigma_{overall} = \dot{m}[s_4 - s_1] = 25[1.6239 - 1.5872] = 0.9175 \text{ Btu/s}\cdot°\text{R}$$
which is in agreement with the above-calculated sum, as expected.

Also, note that an energy balance on the mixing chamber reduces to $\dot{m}_4 h_4 = \dot{m}_3 h_3 + \dot{m}_2 h_2$. Inserting previously determined values, the equation is satisfied, providing a check on the data.

PROBLEM 6.148

KNOWN: Steam at 60 lbf/in², 350°F, 10 ft/s enters an insulated nozzle at steady state and exits at 35 lbf/in². The isentropic nozzle efficiency is 94%.

FIND: Determine the velocity at the nozzle exit.

SCHEMATIC & GIVEN DATA:

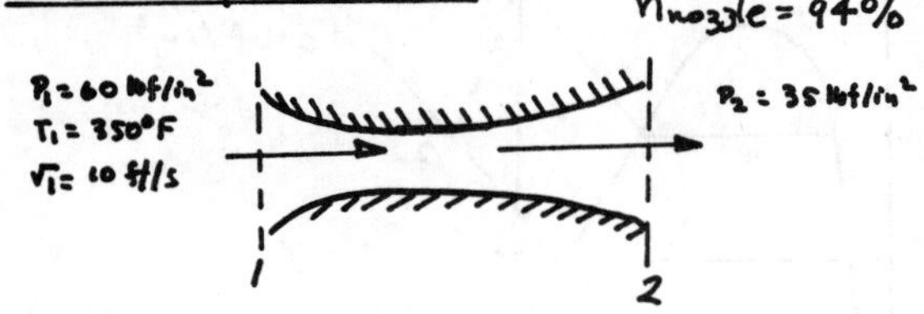

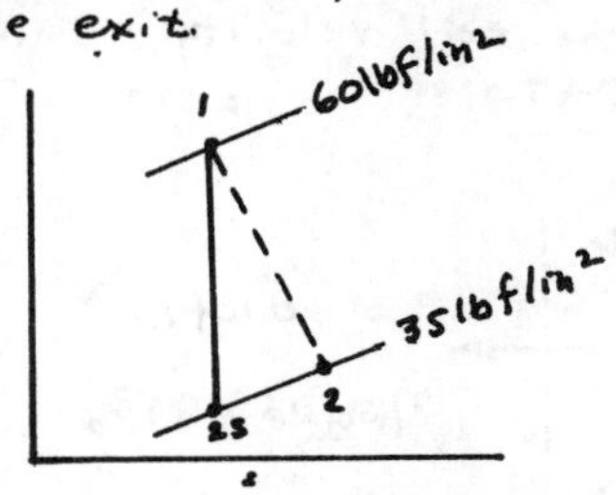

ASSUMPTIONS: (1) The nozzle is insulated and at steady state. (2) Potential energy effects are negligible.

ANALYSIS: By Eq. 6.49, the isentropic nozzle efficiency is

$$\eta_{nozzle} = \frac{V_2^2/2}{V_{2s}^2/2} \qquad (1)$$

where $V_{2s}^2/2$ is the exit kinetic energy per unit of mass for an isentropic expansion from the given inlet state to the specified exit pressure.

An energy rate balance for the nozzle in which the expansion is isentropic reduces at steady state to give

$$0 = \cancelto{0}{\dot{Q}_{cv}} - \cancelto{0}{\dot{W}_{cv}} + \dot{m}\left(h_1 - h_{2s} + \frac{V_1^2 - V_{2s}^2}{2} + \cancelto{0}{g(z_1 - z_2)}\right)$$

Thus

$$\frac{V_{2s}^2}{2} = \frac{V_1^2}{2} + h_1 - h_{2s} \qquad (2)$$

From Table A-4E $h_1 = 1208.2$ Btu/lb, $s_1 = 1.6830$ Btu/lb·°R. And with $s_{2s} = s_1$ Table A-3E gives

$$x_{2s} = \frac{s_{2s} - s_f}{s_g - s_f} = \frac{1.6830 - 0.3809}{1.6873 - 0.3809} = 0.9967$$

Accordingly

$$h_{2s} = h_f + x_{2s} h_{fg} = 228.04 + 0.9967(939.3) = 1164.24 \text{ Btu/lb}$$

Substituting values into Eq. (2)

$$\frac{V_{2s}^2}{2} = \frac{(10 \text{ ft/s})^2}{2} + (1208.2 - 1164.24)\left(\frac{\text{Btu}}{\text{lb}}\right)\frac{778 \text{ ft·lbf}}{\text{Btu}}\;\frac{32.2 \text{ lb·ft/s}^2}{\text{lbf}}$$

$$= 1101318 \, (\text{ft/s})^2 \Rightarrow V_{2s} = 1484 \text{ ft/s}$$

Returning to Eq. (1)

$$V_2 = \sqrt{\eta_{nozzle}}\, V_{2s} = \sqrt{0.94}\,(1484 \text{ ft/s}) = 1439 \text{ ft/s} \longleftarrow$$

PROBLEM 6.149

KNOWN: Steady state operating data are provided for an insulated nozzle.

FIND: Determine the exit velocity.

SCHEMATIC & GIVEN DATA:

Steam

$p_1 = 100 \text{ lbf/in}^2$

$T_1 = 500°F$

$V_1 = 100 \text{ ft/s}$

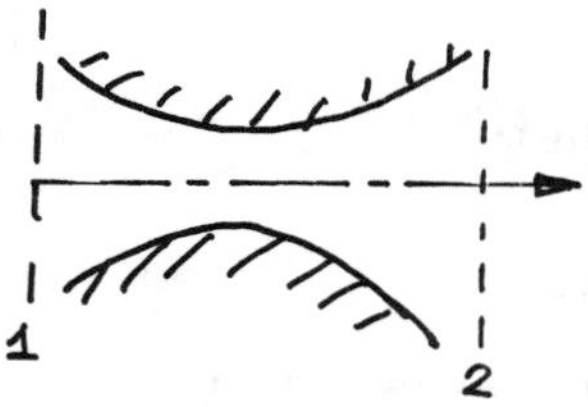

$p_2 = 40 \text{ lbf/in}^2$

$\eta_{nozzle} = 95\%$

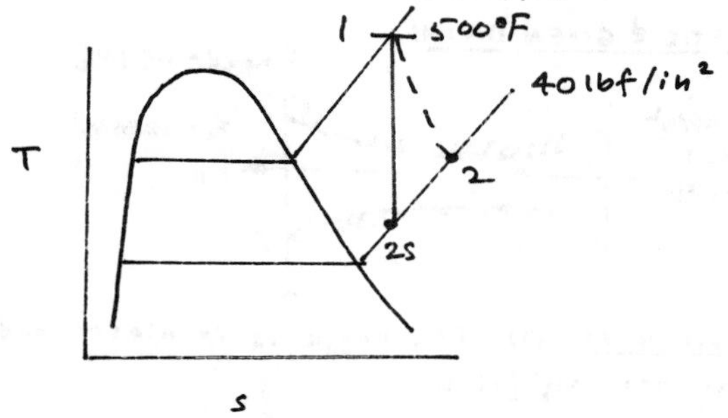

ASSUMPTIONS: (1) The control volume shown in the figure is at steady state. (2) For the control volume, $\dot{Q}_{cv} = 0$, and potential energy effects are negligible.

ANALYSIS: The isentropic nozzle efficiency, Eq. 6.49, yields

$$V_2 = \sqrt{\eta_{Nozzle}}\; V_{2s} \tag{1}$$

where V_{2s} is the exit velocity for an isentropic expansion. Considering the isentropic expansion, reduction of mass and energy balances gives

$$\frac{V_{2s}^2}{2} = \frac{V_1^2}{2} + h_1 - h_{2s} \tag{2}$$

With data from Table A-4E, $h_1 = 1279.1$ Btu/lb, $s_1 = 1.7085$ Btu/lb·°R. Then, interpolating at 40 lbf/in² with $s_{2s} = s_1$, $h_{2s} = 1194$ Btu/lb.

Substituting values in Eq. (2)

$$\frac{V_{2s}^2}{2} = \frac{\left(100\frac{ft}{s}\right)^2}{2} + (1279.1 - 1194)\left(\frac{Btu}{lb}\right)\left|\frac{778 \text{ ft·lbf}}{1 \text{ Btu}}\right|\left|\frac{32.2 \text{ lb·ft/s}^2}{1 \text{ lbf}}\right|$$

giving

$$V_{2s} = 2067.3 \frac{ft}{s} \Rightarrow V_2 = \sqrt{0.95}\,(2067.3) = 2015 \text{ ft/s} \longleftarrow V_2$$

PROBLEM 6.150

KNOWN: Air enters an insulated nozzle at steady state at a specified state and mass flow rate and exits at a specified pressure with a known velocity.

FIND: Determine the exit area and the isentropic nozzle efficiency.

SCHEMATIC & GIVEN DATA:

AIR

$p_1 = 80 \text{ lbf/in}^2$

$T_1 = 120°F$

$V_1 = 10 \text{ ft/s}$

$\dot{m}_1 = 0.4 \text{ lb/s}$

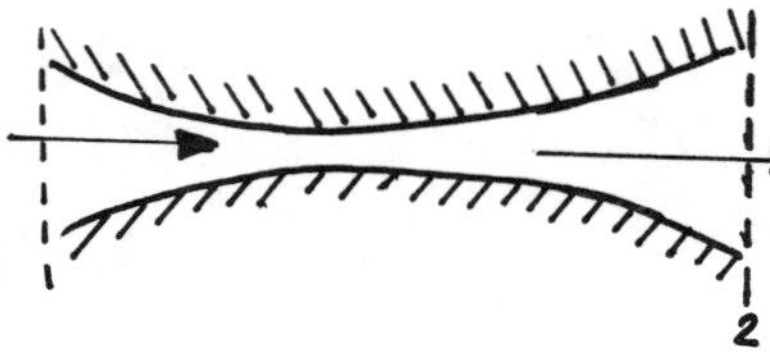

$p_2 = 50 \text{ lbf/in}^2$

$V_2 = 914 \text{ ft/s}$

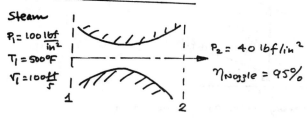

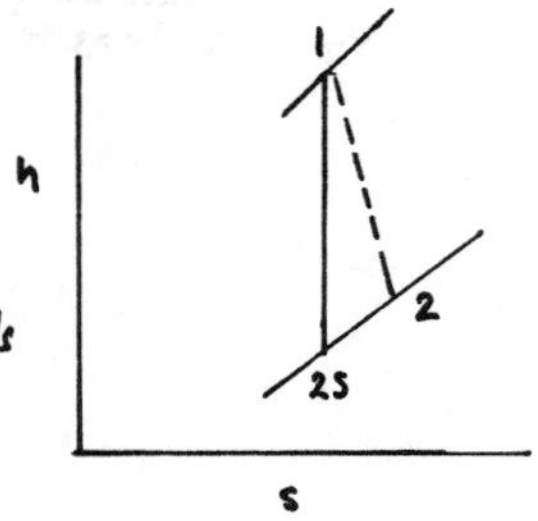

ASSUMPTIONS: (1) The nozzle is insulated and at steady state. (2) Potential energy effects are negligible. (3) Air is modeled as an ideal gas.

ANALYSIS: The isentropic nozzle efficiency is given by

$$\eta_{nozzle} = \frac{V_2^2}{V_{2s}^2} \tag{1}$$

PROBLEM 6.150 (Continued)

where V_{2s} is the exit velocity for an isentropic expansion from the specified inlet state to the specified exit pressure. The velocity V_{2s} can be found from an energy rate balance a steady state

$$0 = \cancel{\dot{Q}_{cv}}^{0} - \cancel{\dot{W}_{cv}}^{0} + \dot{m}\left(h_1 - h_{2s} + \frac{V_1^2 - V_{2s}^2}{2} + \cancel{g(z_1 - z_2)}^{0}\right) \qquad (2)$$

Thus

$$V_{2s} = \sqrt{V_1^2 + 2(h_1 - h_{2s})}$$

From Table A-22E, $h_1 = 138.66$ Btu/lb, $p_{r1} = 1.78$. Using Eq. 6.43

$$\frac{p_{2s}}{p_1} = \frac{p_r(T_{2s})}{p_r(T_1)} \Rightarrow p_r(T_{2s}) = \left(\frac{50}{80}\right)(1.78) = 1.1125$$

Returning to Table A-13E, the corresponding value for h_{2s} is $h_{2s} = 121.13$ Btu/lb. Thus

$$V_{2s} = \sqrt{(10\,ft/s)^2 + 2(138.66 - 121.13)(Btu/lb)\left(778\,\frac{ft\cdot lbf}{Btu}\right)\left(\frac{32.2\,lb\cdot ft/s^2}{lbf}\right)}$$

$$= 937.2\ ft/s$$

Finally, Eq. (1) gives $\eta_{nozzle} = \dfrac{(914)^2/2}{(937.2)^2/2} = 0.951\ (95.1\%)$ ← η_{nozzle}

The exit area can be evaluated from the mass flow rate using the ideal gas equation of state: $\dot{m} = A_2 V_2 / v_2$ where $v_2 = RT_2/p_2$. The exit temperature, T_2, can be evaluated from an energy rate balance:

$$0 = \cancel{\dot{Q}_{cv}}^{0} - \cancel{\dot{W}_{cv}}^{0} + \dot{m}\left(h_1 - h_2 + \frac{V_1^2 - V_2^2}{2} + \cancel{g(z_1 - z_2)}^{0}\right) \qquad (3)$$

Thus

① $$h_2 = h_1 + \frac{V_1^2 - V_2^2}{2} = 138.66 + \frac{(10)^2 - (914)^2}{2(32.2)(778)} = 121.99\ Btu/lb$$

Interpolating in Table A-22E, $T_2 = 510.5°R$. Accordingly

$$A_2 = \frac{\dot{m} v_2}{V_2} = \frac{\dot{m} R T_2}{p_2 V_2} = \frac{(0.4\,lb/s)\left(\frac{1545}{28.97}\,\frac{ft\cdot lbf}{lb\cdot °R}\right)(510.5°R)}{((50)(144)\,lbf/ft^2)(913.5\,ft/s)} = 1.66 \times 10^{-3}\ ft^2 \quad \leftarrow A_2$$

1. The energy rate balances Eqs. (2) and (3) are for the isentropic and actual expansions, respectively.

PROBLEM 6.151

KNOWN: Argon enters an insulated nozzle operating at steady state at a specified pressure, temperature, and velocity and expands to a known pressure and velocity.

FIND: Determine the exit temperature, the isentropic nozzle efficiency, and the rate of entropy production per unit mass flowing.

SCHEMATIC & GIVEN DATA:

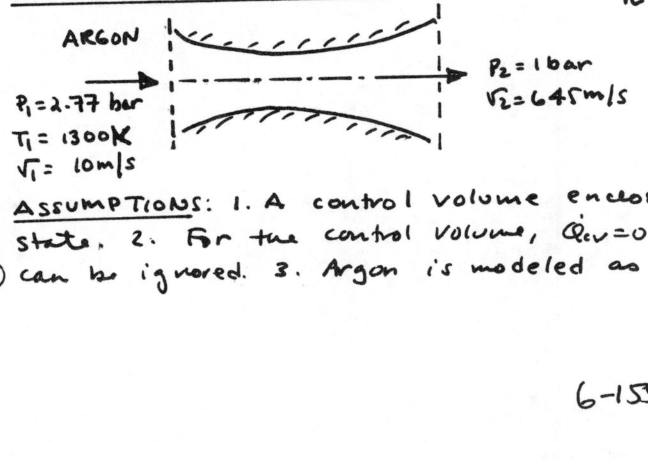

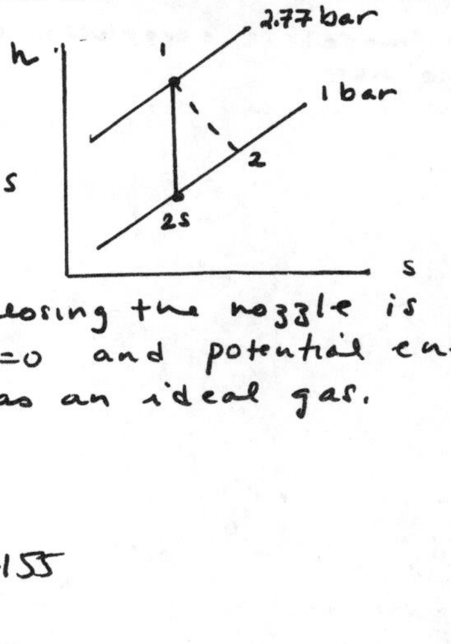

ASSUMPTIONS: 1. A control volume enclosing the nozzle is at steady state. 2. For the control volume, $\dot{Q}_{cv} = 0$ and potential energy effects ① can be ignored. 3. Argon is modeled as an ideal gas.

PROBLEM 6.151 (Contd.)

ANALYSIS: (a) Mass and energy rate balances reduce to give

$$0 = \cancel{\dot{Q}_{cv}} - \cancel{\dot{W}_{cv}} + \dot{m}\left[h_1 - h_2 + \frac{V_1^2 - V_2^2}{2} + \cancel{g(z_1 - z_2)}\right] \Rightarrow 0 = h_1 - h_2 + \frac{V_1^2 - V_2^2}{2}$$

Or since c_p is constant for Argon as an ideal gas: $c_p = 5/2\, R$ (Table A-21)

$$0 = c_p[T_1 - T_2] + \frac{V_1^2 - V_2^2}{2} \Rightarrow T_2 = T_1 + \frac{V_1^2 - V_2^2}{2c_p} = (1300\,K) + \frac{[(10)^2 - (645)^2]\,m^2/s^2}{2\left(\frac{5}{2}\right)\left(\frac{8314}{39.94}\frac{N\cdot m}{kg\cdot K}\right)}\left|\frac{1\,N}{1\,kg\cdot m/s^2}\right|$$

➧ $T_2 = 900\,K$.

(b) The isentropic nozzle efficiency is given by Eq. 6.49. To find V_{2s} requires the temperature at state 2s. Since $c_p = 5/2\, R$, $k = 1.666$ (Eq. 3.47a). Equation 6.45 gives

$$T_{2s} = T_1\left[\frac{p_2}{p_1}\right]^{(k-1)/k} = 1300\left[\frac{1}{2.77}\right]^{\frac{0.666}{1.666}} \Rightarrow T_{2s} = 865\,K.$$

Then, applying mass and energy rate balances to the isentropic expansion, we get

$$0 = c_p[T_1 - T_{2s}] + \frac{V_1^2 - V_{2s}^2}{2} \Rightarrow \frac{V_{2s}^2}{2} = \frac{V_1^2}{2} + c_p[T_1 - T_{2s}]$$

or

$$\frac{V_{2s}^2}{2} = \frac{(10\,m/s)^2}{2} + \frac{5}{2}\left[\frac{8314}{39.94}\frac{N\cdot m}{kg\cdot K}\right][1300 - 865]K\left|\frac{1\,kg\cdot m/s^2}{1\,N}\right|$$

$$= 226{,}426 (m/s)^2 \quad (V_{2s} = 672.9\,m/s)$$

Then

$$\eta_{nozzle} = \frac{(645)^2/2}{226{,}426} = 0.919 \quad (91.9\%)$$ ⟵

(c) Reducing mass and entropy rate balances gives with Eq. 6.23

$$\frac{\dot{\sigma}_{cv}}{\dot{m}} = s_2 - s_1 \Rightarrow \frac{\dot{\sigma}_{cv}}{\dot{m}} = c_p \ln\frac{T_2}{T_1} - R\ln\frac{p_2}{p_1}$$

with $c_p = 5/2\, R$

$$\frac{\dot{\sigma}_{cv}}{\dot{m}} = R\left[\frac{5}{2}\ln\frac{T_2}{T_1} - \ln\frac{p_2}{p_1}\right] = \left(\frac{8.314}{39.94}\frac{kJ}{kg\cdot K}\right)\left[\frac{5}{2}\ln\frac{900}{1300} - \ln\frac{1}{2.77}\right]$$

$$= 0.0207\,\frac{kJ}{kg\cdot K}$$ ⟵

---

1. Checking the generalized compressibility chart with

$$p_{R1} = \frac{p_1}{p_c} = \frac{2.77\,bar}{48.6\,bar} = 0.057 \qquad p_{R2} = \frac{p_2}{p_c} = \frac{1}{48.6} = 0.021$$

the ideal gas model is acceptable over a wide range of temperatures, including those here.

PROBLEM 6.152

KNOWN: R134a enters a compressor operating adiabatically at steady state as a saturated vapor and exits at a known pressure.

FIND: (a) Determine the minimum theoretical work input per unit mass flowing and the corresponding exit temperature. (b) For a specified exit temperature, determine the isentropic compressor efficiency.

SCHEMATIC & GIVEN DATA:

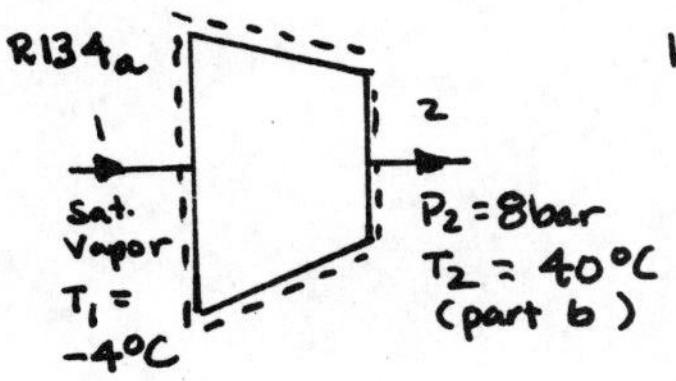

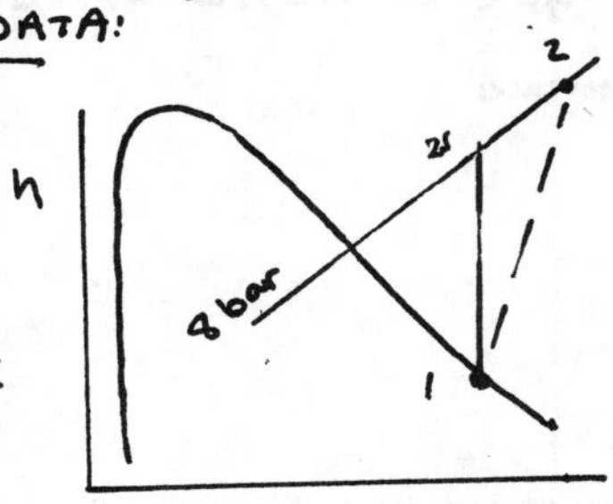

ASSUMPTIONS: 1. The control volume shown in the schematic operates at steady state with no appreciable heat transfer with the surroundings.
2. Kinetic and potential energy effects can be ignored.

ANALYSIS: (a) The mass, energy, and entropy rate balances reduce to give

$$\left(\frac{-\dot{W}_{cv}}{\dot{m}}\right) = h_2 - h_1, \quad \frac{\dot{\sigma}_{cv}}{\dot{m}} = s_2 - s_1$$

The work input decreases as h_2 decreases. The entropy balance and h-s diagram show that the smallest allowed value for h_2 corresponds to state 2s, where $s_{2s} = s_1$. From Table A-10, at −4°C, $h_1 = 244.9$ kJ/kg, $s_1 = 0.9213$ kJ/kg·K. Then, interpolation in Table A-12 at 8 bar with $s_{2s} = s_1$ gives, $T_{2s} = 35.5°C$ ←
and $h_{2s} = 268.7$ kJ/kg. Accordingly

$$\left(\frac{-\dot{W}_{cv}}{\dot{m}}\right)_{MIN} = h_{2s} - h_1 = (268.7 - 244.9)\frac{kJ}{kg} = 23.8\frac{kJ}{kg}$$ ←

(b) The isentropic compressor efficiency is given by Eq. 6.50, which reduces to

$$\eta_c = \frac{h_{2s} - h_1}{h_2 - h_1} = \frac{23.8}{273.66 - 244.9} = \frac{23.8}{28.76} = 0.828 \quad (82.8\%)$$ ←

where h_2 is from Table A-12 at 8 bar, 40°C.

PROBLEM 6.153

KNOWN: Air enters an insulated compressor operating at steady state at a specified state and mass flow rate and exits at a known pressure.

FIND: (a) Determine the minimum theoretical power input and the corresponding exit temperature. (b) For a specified exit temperature, determine the power input and the isentropic compressor efficiency.

SCHEMATIC & GIVEN DATA:

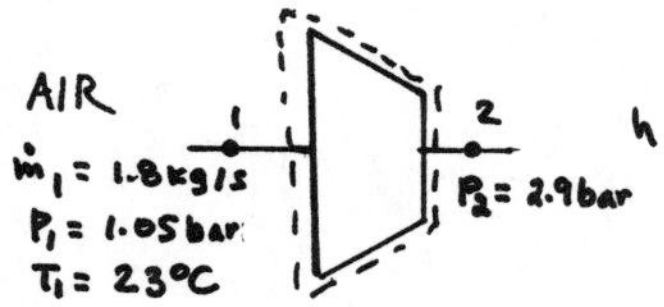

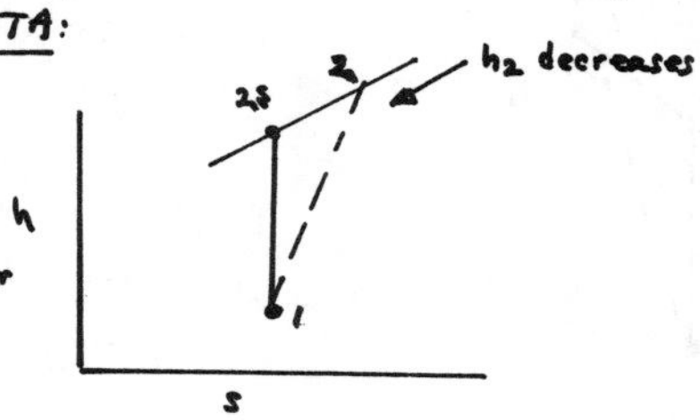

ASSUMPTIONS: (1) The compressor operates at steady state and $\dot{Q}_{cv} = 0$. (2) Kinetic and potential energy effects are negligible. (3) Air is modeled as an ideal gas. ①

ANALYSIS: (a) At steady state $\dot{m}_1 = \dot{m}_2 = \dot{m}$, and an energy rate balance reduces to

$$\left(-\frac{\dot{W}_{cv}}{\dot{m}}\right) = h_2 - h_1 \qquad (1)$$

An entropy rate balance at steady state gives, $s_2 - s_1 = \dot{\sigma}_{cv}/\dot{m} \geq 0$. The work input decreases as h_2 decreases. The h-s diagram and the entropy balance shows that the smallest allowed value for h_2 corresponds to state 2s, where $s_1 = s_{2s}$. That is

$$\left(-\frac{\dot{W}_{cv}}{\dot{m}}\right)_{min} = h_{2s} - h_1$$

To find h_{2s} use p_r data from Table A-22 and

$$\frac{p_r(T_{2s})}{p_r(T_1)} = \frac{p_{2s}}{p_1} \Rightarrow p_r(T_{2s}) = p_r(T_1)\left(\frac{p_{2s}}{p_1}\right) = 1.3226\left(\frac{2.9}{1.05}\right) = 3.6529$$

Interpolating in Table A-22, $T_{2s} = 395.3\,K$ (122°C), $h_{2s} = 396.22$ kJ/kg. Thus, with h_1 from Table A-22 ← T_{2s}

$$\left(-\frac{\dot{W}_{cv}}{\dot{m}}\right)_{min} = 396.22 - 296.17 = 100.05\ \frac{kJ}{kg}$$

$$\Rightarrow \quad (-\dot{W}_{cv})_{min} = 180.09\ kW$$ ← $(-\dot{W}_{cv})_{min}$

(b) $T_2 = 420\,K$ (147°C). With $h_2 = 421.26$ kJ/kg from Table A-22, Eq. (1) gives

$$\left(-\frac{\dot{W}_{cv}}{\dot{m}}\right) = h_2 - h_1 = 421.26 - 296.17 = 125.09\ \frac{kJ}{kg}$$

$$\Rightarrow \quad (-\dot{W}_{cv}) = 225.16\ kW$$ ← $(-\dot{W}_{cv})$

With Eq. 6.50

$$\eta_c = \frac{180.09}{225.16} = 0.8 \quad (80\%)$$ ← η_c

---

1. Checking the compressibility chart with

$$p_{R1} = \frac{1.05\ bar}{37.7\ bar} = 0.028, \quad p_{R2} = \frac{2.9}{37.7} = 0.077$$

indicates that the ideal gas model is valid over a wide range of temperature including those here.

PROBLEM 6.154

KNOWN: R134a enters a compressor at steady state at a specified state and exits at a specified pressure. The isentropic compressor efficiency is known.

FIND: Determine the exit temperature and the work input per unit mass of R134a flowing.

SCHEMATIC & GIVEN DATA:

$\eta_c = 0.75$

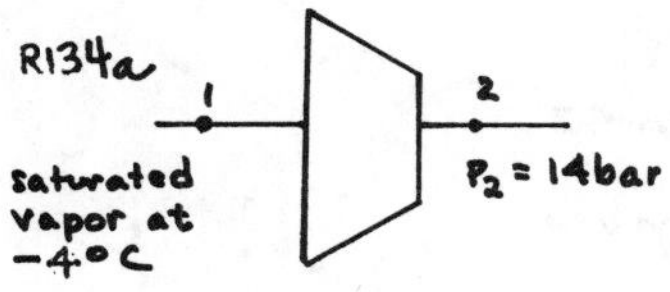

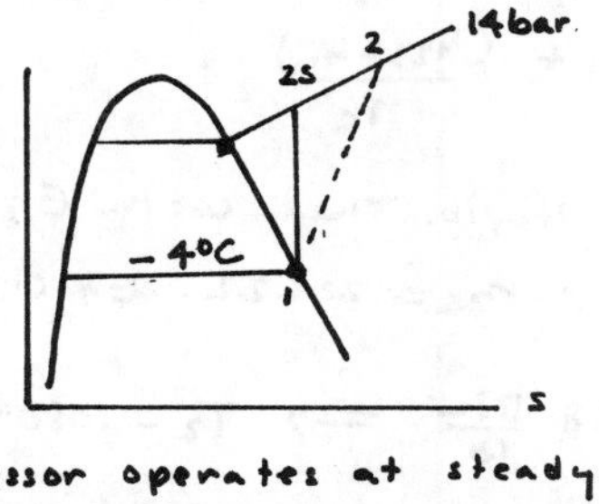

ASSUMPTIONS: (1) The compressor operates at steady state. (2) Heat transfer between the compressor and the surroundings can be ignored as can changes in kinetic and potential energy from inlet to exit.

ANALYSIS: At steady state $\dot{m}_1 = \dot{m}_2 = \dot{m}$ and an energy rate balance reduces to

$$\frac{-\dot{W}_{cv}}{\dot{m}} = h_2 - h_1 \qquad (1)$$

Similarly, for an isentropic compression

$$\left(\frac{-\dot{W}_{cv}}{\dot{m}}\right)_s = h_{2s} - h_1 \qquad (2)$$

State 2s is fixed by $P_2 = 14$ bar and $S_{2s} = S_1$. From Table A-10, at −4°C, $s_1 = 0.9213$ kJ/kg·K, $h_1 = 244.9$ kJ/kg. Interpolation in Table A-12 at 14 bar, $h_{2s} = 280.33$ kJ/kg. Thus

$$\left(-\frac{\dot{W}_{cv}}{\dot{m}}\right)_s = 280.33 - 244.9 = 35.43 \ \frac{kJ}{kg}$$

The isentropic compressor efficiency is given by Eq. 6.50. Thus

$$\left(-\frac{\dot{W}_{cv}}{\dot{m}}\right) = \frac{(-\dot{W}_{cv}/\dot{m})_s}{\eta_c} = \left(\frac{35.43\,kJ/kg}{0.75}\right) = 47.24\ \frac{kJ}{kg} \longleftarrow (-\dot{W}_{cv}/\dot{m})$$

Equation (1) allows h_2 to be evaluated:

$$h_2 = h_1 + (-\dot{W}_{cv}/\dot{m}) = 244.9 + 47.24 = 292.1\ kJ/kg$$

Interpolating in Table A-12 at 14 bar with h_2, $T_2 = 67.4$ °C $\longleftarrow T_2$

PROBLEM 6.155

KNOWN: Steady-state operating data are provided for a compressor.

FIND: Determine the temperature of the air at the compressor exit and the rate of entropy production per unit of mass flowing.

SCHEMATIC & GIVEN DATA:

$\eta_c = 71.9\%$

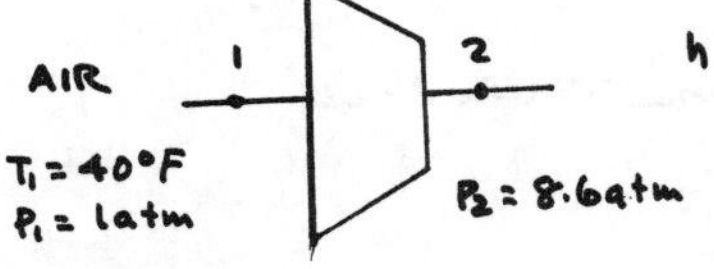

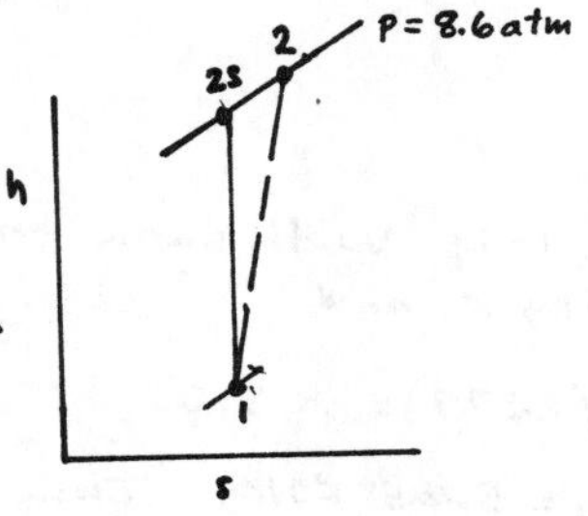

ASSUMPTIONS: (1) The compressor operates at steady state. (2) Heat transfer between the compressor and the surroundings can be ignored as can changes in kinetic and potential energy from inlet to exit. (3) Air can be modeled as an

① ideal gas.

PROBLEM 6.155 (Contd.)

ANALYSIS: At steady state mass and energy rate balances give for the actual and isentropic compressions, respectively

$$\left(-\frac{\dot{W}_{cv}}{\dot{m}}\right) = h_2 - h_1 \quad , \quad \left(-\frac{\dot{W}_{cv}}{\dot{m}}\right)_s = h_{2s} - h_1$$

The isentropic compressor efficiency is given by Eq. 6.50:

$$\eta_c = \frac{(-\dot{W}_{cv}/\dot{m})_s}{(-\dot{W}_{cv}/\dot{m})} = \frac{h_{2s}-h_1}{h_2-h_1} \Rightarrow h_2 = h_1 + \frac{(h_{2s}-h_1)}{\eta_c} \qquad (1)$$

From Table A-22E at 500°R, $h_1 = 119.48$ Btu/lb. Then, with Eq. 6.43

$(p_r)_{2s} = p_{r1}\left[\frac{p_2}{p_1}\right] = 1.0590\left[\frac{8.6}{1}\right] = 9.1074 \Rightarrow h_{2s} = 221.22$. Eq. (1) gives

$$h_2 = 119.48 + \frac{(221.22 - 119.48)}{0.719} = 260.98 \ \frac{\text{Btu}}{\text{lb}} \Rightarrow T_2 = 1080°R\ (620°F) \leftarrow$$

Mass and entropy rate balances reduce to give with Eq. 6.21a

$$\frac{\dot{\sigma}_{cv}}{\dot{m}} = s°(T_2) - s°(T_1) - R \ln p_2/p_1 = 0.76964 - 0.58233 - \frac{1.986}{28.97}\ln\frac{8.6}{1}$$

$$\Rightarrow \dot{\sigma}_{cv}/\dot{m} = 0.0398 \text{ Btu/lb·°R} \leftarrow$$

1. The validity of the ideal gas model can be checked using generalized compressibility data.

PROBLEM 6.156

KNOWN: Steady state operating data are provided for an air compressor.

FIND: Determine the power input and rate of entropy production using (a) Table A-22, (b) IT, (c) constant $k = 1.39$.

SCHEMATIC & GIVEN DATA:

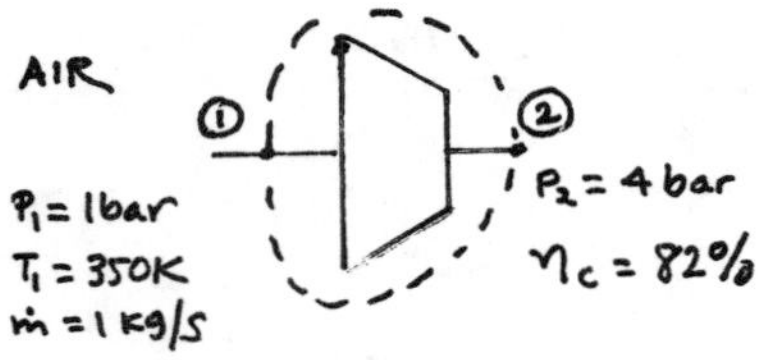

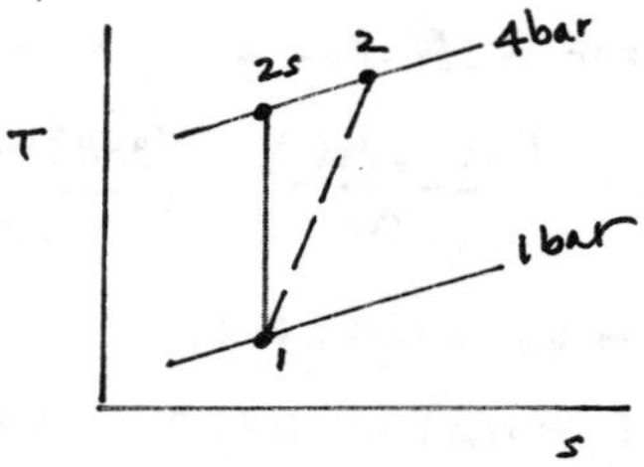

ASSUMPTIONS: (1) The control volume shown in the figure is at steady state. (2) For the control volume, $\dot{Q}_{cv} = 0$ and kinetic/potential energy effects can be ignored. (4) Air is modeled as an ideal gas. For solution method (c) the specific heat ratio is also constant, $k = 1.39$.

ANALYSIS: Reduction of mass, energy, and entropy balances with the given assumptions yields

$$\dot{W}_{cv} = \dot{m}(h_1 - h_2) \qquad (1)$$

$$\dot{\sigma}_{cv} = \dot{m}(s_2 - s_1) \qquad (2)$$

(a) Air Table Analysis. Consider process 1-2s. With data from Table A-22
$h_1 = 350.49$ kJ/kg, $s°(T_1) = 1.85708$ kJ/kg·K, and

$$p_{r,2s} = \frac{p_2}{p_1} p_{r,1} = \left(\frac{4 \text{ bars}}{1 \text{ bar}}\right)(2.379) = 9.516$$

Interpolation in Table A-22 gives $h_{2s} = 520.98$ kJ/kg. Then, using the isentropic compressor efficiency

$$\eta_c = \frac{h_{2s}-h_1}{h_2-h_1} \Rightarrow h_2 = h_1 + \frac{h_{2s}-h_1}{\eta_c} = 350.49 + \frac{(520.98-350.49)}{0.82} = 558.4 \text{ kJ/kg}$$

PROBLEM 6.156 (Continued)

Then, interpolating in Table A-22 with h_2 gives $s^\circ(T_2) = 2.3247$ kJ/kg·K. Substituting values in Eqs.(1), (2)

$$\dot{W}_{cv} = \dot{m}(h_1 - h_2) = \left(1\frac{kg}{s}\right)(350.49 - 558.4)\left(\frac{kJ}{kg}\right)\left|\frac{1\,kW}{1\,kJ/s}\right| = -207.9\,kW$$

$$\dot{\sigma}_{cv} = \dot{m}(s_2 - s_1) = \dot{m}(s^\circ(T_2) - s^\circ(T_1) - R\ln p_2/p_1) = \left(1\frac{kg}{s}\right)\left(2.3247 - 1.8571 - \frac{8.314}{28.97}\ln 4\right)\left(\frac{kJ}{kg\cdot K}\right)$$

$$= 0.0698\frac{kJ/s}{K} = 0.0698\frac{kW}{K}$$

(b) *IT Code*

```
p1 = 1 // bar
T1 = 350 // K
mdot = 1 // kg/s
p2 = 4 // bar
eta = 0.82

Wdotcv = mdot * (h1 - h2)
sigmadot = mdot * (s2 - s1)

h1 = h_T("Air", T1)
s1 = s_TP("Air", T1, p1)

h2 = h1 + (h2s - h1) / eta
h2s = h_T("Air", T2s)
s2s = s_TP("Air", T2s, p2)
s2s = s1

h2 = h_T("Air", T2)
s2 = s_TP("Air", T2, p2)
```

IT Results

$\eta = 0.82$
$T_2 = 553.5$ K
$T_{2s} = 517.4$ K
$h_1 = 350.3$ kJ/kg
$h_2 = 558.3$ kJ/kg
$h_{2s} = 520.9$ kJ/kg

$\dot{W}_{cv} = -208$ kW

$\dot{\sigma}_{cv} = 0.06993$ kW/K

① (c) Ideal gas, k = 1.39 Analysis. Applying Eq. 6.45 to process 1-2s

$$T_{2s} = T_1\left[\frac{p_2}{p_1}\right]^{\frac{k-1}{k}} = 350\,K\left[\frac{4\,bars}{1\,bar}\right]^{\frac{1.39-1}{1.39}} = 516.4\,K$$

When k is constant, c_p is constant, and so the isentropic compressor efficiency gives

$$\eta_c = \frac{h_{2s} - h_1}{h_2 - h_1} = \frac{c_p[T_{2s} - T_1]}{c_p[T_2 - T_1]} \Rightarrow T_2 = T_1 + \frac{(T_{2s} - T_1)}{\eta_c} \Rightarrow T_2 = 350 + \frac{(516.4 - 350)}{.82}$$

$$T_2 = 552.9\,K$$

With Eq. 3.47a

$$c_p = \frac{kR}{k-1} = \frac{1.39(8.314/28.97)\,kJ/kg\cdot K}{1.39 - 1} = 1.02\,kJ/kg\cdot K$$

Accordingly, Eqs. (1), (2) give

$$\dot{W}_{cv} = \dot{m}(h_1 - h_2) = \dot{m}c_p(T_1 - T_2) = \left(1\frac{kg}{s}\right)\left(1.02\frac{kJ}{kg\cdot K}\right)[350 - 552.9]\,K$$

$$= -207\frac{kJ}{s} = -207\,kW$$

$$\dot{\sigma}_{cv} = \dot{m}(s_2 - s_1) = \dot{m}\left[c_p\ln\frac{T_2}{T_1} - R\ln\frac{p_2}{p_1}\right] = \left(1\frac{kg}{s}\right)\left[\left(1.02\frac{kJ}{kg\cdot K}\right)\ln\frac{552.9}{350} - \left(\frac{8.314}{28.97}\frac{kJ}{kg\cdot K}\right)\ln 4\right]$$

$$= 0.0685\frac{kJ/s}{K} = 0.0685\frac{kW}{K}$$

---

1. Inspection of Table A-20 indicates that k = 1.39 is a suitable constant value for the interval from T_1 to T_2. Thus, good agreement is realized between the results of this part and those of parts a) and b).

PROBLEM 6.157

KNOWN: A compressor operating at steady state takes in atmospheric air at a known mass flow rate and discharges air at a given pressure.

FIND: Plot the power required and the compressor exit temperature, each versus compressor isentropic efficiency ranging from 70% to 100%.

SCHEMATIC & GIVEN DATA:

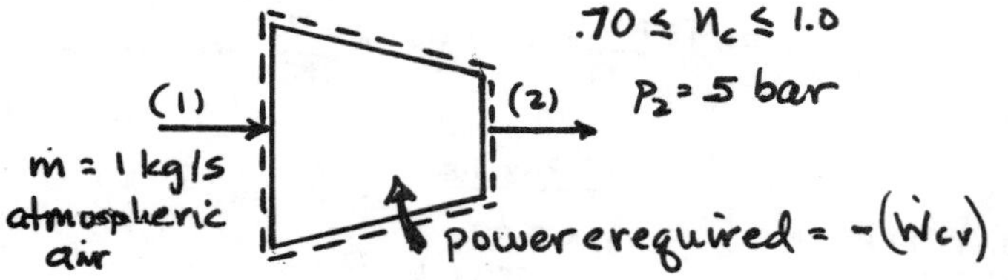

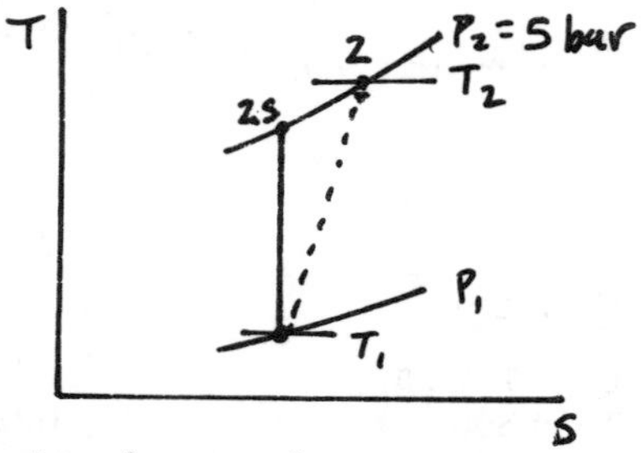

ASSUMPTIONS: (1) The control volume shown is at steady state. (2) For the control volume, $\dot{Q}_{cv} = 0$. (3) Kinetic and potential energy changes from inlet to exit can be ignored. (4) Since the incoming air is atmospheric, we assume $p_1 = 1$ bar and $T_1 = 20°C$.

ANALYSIS: The mass and energy rate balances reduce to give

$$\text{power required} = -(\dot{W}_{cv}) = \dot{m}(h_2 - h_1) \qquad (1)$$

the enthalpy h_1 can be determined since p_1 and T_1 are known from assumption 4. The enthapy h_2 is found using compressor efficiency η_c, as follows:

$$h_2 = h_1 + \frac{(h_{2s} - h_1)}{\eta_c} \qquad (2)$$

where h_{2s} corresponds to an isentropic compression from state 1 to p_2.

Sample Calculation

A sample calculation will be performed using data from Table A-22. From the table; $h_1 = 293.2$ kJ/kg and $p_{r1} = 1.27652$. Then, from Eq. 6.43

$$p_r(T_{2s}) = 1.27652\left(\frac{5}{1}\right) = 6.3826$$

Interpolating in Table A-22; $h_{2s} = 464.8$ kJ/kg. Accordingly, for $\eta = 0.7$

$$h_2 = 293.2 + \frac{(464.8 - 293.2)}{0.7} = 538.3 \text{ kJ/kg}$$

and

$$\begin{pmatrix}\text{power}\\ \text{required}\end{pmatrix} = \left(1\ \frac{kg}{s}\right)(538.3 - 293.2)\ \frac{kJ}{kg}\left|\frac{1\ kW}{1\ kJ/s}\right| = 245.1\ kW$$

The software IT is used to generate data for the required plots, as follows:

PROBLEM 6.157 (cont'd.)

IT Code

```
T1 = 20  // °C
p1 = 1  // bar
mdot = 1  // kg/s
p2 = 5  // bar
eta = 70  // %

Wdotcv = mdot * (h1 - h2)
power = - Wdotcv

h2 = h1 + (h2s - h1)*100 / eta
h1 = h_T("Air", T1)
s1 = s_TP("Air", T1, p1)
h2s = h_T("Air", T2s)
s2s = s_TP("Air", T2s, p2)
s2s = s1
h2 = h_T("Air", T2)
```

IT Results for η = 70%

h_1 = 293.2 kJ/kg
h_{2s} = 464.9 kJ/kg
h_2 = 538.6 kJ/kg
T_2 = 261.4 °C
T_{2s} = 189.6 °C
$-\left(\dot{W}_{cv}\right)$ = 245.4 kW

These results compare very favorably with those obtained in the sample calculation from table data.

PLOTS:

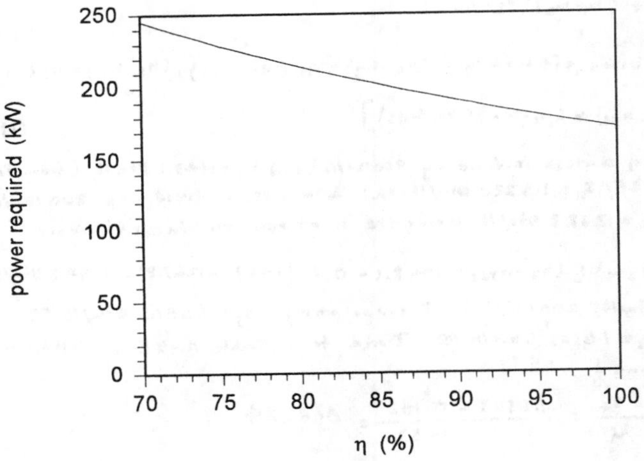

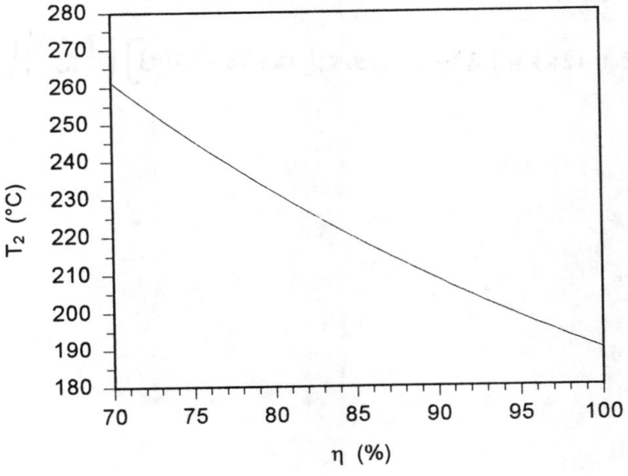

Note: As η_c decreases, the required power increases as does the exit temperature. This is expected.

PROBLEM 6.158

KNOWN: Operating data are provided for a two stage turbine at steady state from which steam is extracted between the stages.

FIND: Determine the power output.

SCHEMATIC & GIVEN DATA:

$\eta_t = 85\%$ $\eta_t = 85\%$

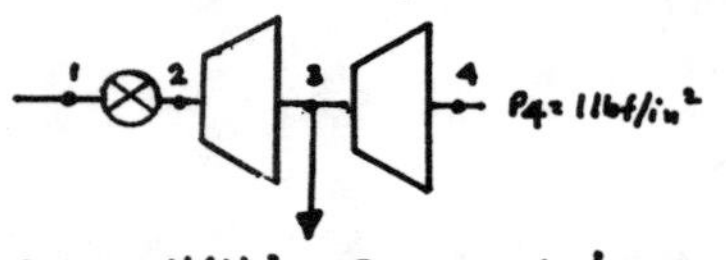

$P_1 = 1000\,lbf/in^2$ $P_2 = 900\,lbf/in^2$

$T_1 = 900\,°F$ $P_3 = 200\,lbf/in^2$

$\dot{m}_1 = 30\,lb/s$ $\dot{m}_3 = 4.5\,lb/s$

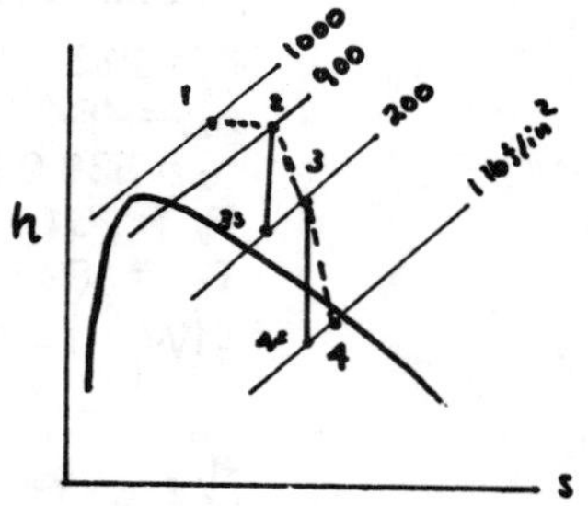

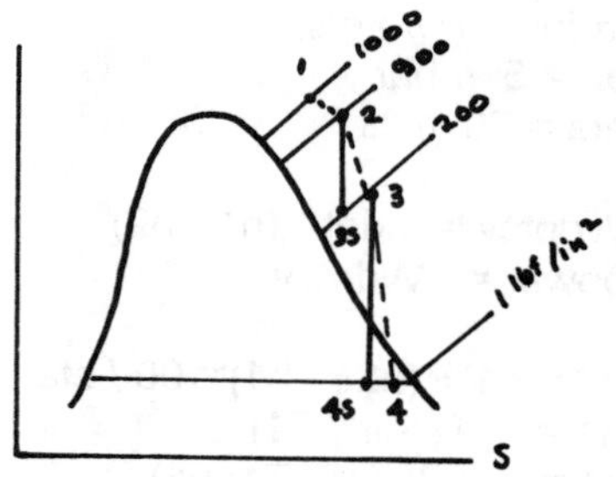

ASSUMPTIONS: (1) The turbine operates at steady state. (2) The expansion through each stage occurs adiabatically. (3) The expansion through the valve is a throttling process. (4) Kinetic and potential energy effects are negligible.

ANALYSIS: The total power developed is the sum of the power developed by the two stages. With the listed assumptions, mass and energy rate balances reduce to give

$$\dot{W}_t = \dot{m}_1 (h_2 - h_3) + (\dot{m}_1 - \dot{m}_3)(h_3 - h_4)$$

Introducing the isentropic turbine efficiency, $(h_2 - h_3) = \eta_t (h_2 - h_{3s})$, $(h_3 - h_4) = \eta_t (h_3 - h_{4s})$. Thus

$$\dot{W}_t = \eta_t \left[\dot{m}_1 (h_2 - h_{3s}) + (\dot{m}_1 - \dot{m}_3)(h_3 - h_{4s}) \right] \qquad (1)$$

With assumption 3, $h_2 = h_1$. So, state 2 is fixed by $900\,lbf/in^2$ and $h_2 = 1448.1\,Btu/lb$ (from Table A-4E). Interpolating in Table A-4E, $s_2 = 1.6228\,Btu/lb \cdot °R$. State 3s is fixed by $200\,lbf/in^2$ and $s_{3s} = s_2$. Interpolating, $h_{3s} = 1267.8\,Btu/lb$. Using the isentropic turbine efficiency

$$h_3 = h_2 - \eta_t (h_2 - h_{3s}) = 1448.1 - 0.85(1448.1 - 1267.8) = 1294.8\,Btu/lb$$

State 3 is fixed by $h_3 = 1294.8\,Btu/lb$, $200\,lbf/in^2$. Interpolating, $s_3 = 1.6503\,Btu/lb \cdot °R$. State 4s is fixed by $1\,lbf/in^2$, $s_3 = 1.6503\,Btu/lb \cdot °R$. Data from Table A-3E at $1\,lbf/in^2$ allows the quality to be determined

$$x_{4s} = \frac{s_{4s} - s_f}{s_g - s_f} = \frac{1.6503 - 0.1327}{1.8453} = 0.8224$$

Thus

$$h_{4s} = h_f + x_{4s} h_{fg} = 69.74 + 0.8224(1036) = 921.7\,Btu/lb$$

Inserting values into Eq. (1)

$$\dot{W}_t = 0.85 \left[(30\,lb/s)[1448.1 - 1267.8] \frac{Btu}{lb} + (25.5)[1294.8 - 921.7] \right] \left| \frac{3600\,s}{h} \right| \left| \frac{1\,hp}{2545\,Btu/h} \right|$$

$$= 17{,}943\,hp$$

PROBLEM 6.159

KNOWN: Operating data are provided for a gas turbine power plant operating at steady state.

FIND: Determine the net power developed for each of two cases: (a) the turbine and compressor are without internal irreversibilities, (b) the compressor and turbine efficiencies are 82 and 85%, respectively.

SCHEMATIC & GIVEN DATA:

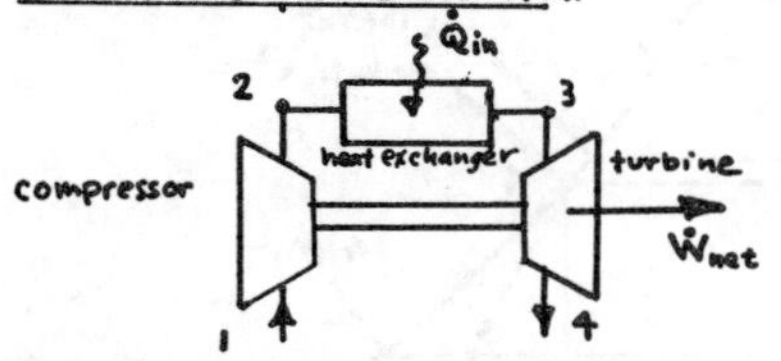

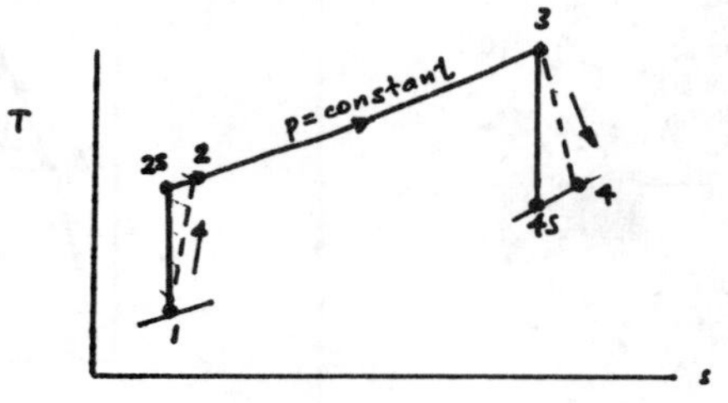

$p_1 = p_4 = 0.95$ bar, $p_2 = p_3 = 5.7$ bars
$T_1 = 22°C$, $T_3 = 1100$ K
$\dot{m}_1 = 5$ kg/s

ASSUMPTIONS: (1) The gas turbine power plant operates at steady state. (2) The compressor and turbine operate adiabatically. (3) Kinetic and potential energy effects are negligible. (4) Air is modeled as an ideal gas.

ANALYSIS: (a) $\eta_t = \eta_c = 100\%$. In this case, mass and energy rate balances at steady state give the following expression for the net work developed per unit mass flowing

$$\frac{\dot{W}_{net}}{\dot{m}} = \underbrace{(h_3 - h_{4s})}_{\text{turbine output}} - \underbrace{(h_{2s} - h_1)}_{\text{compressor input}}$$

From Table A-22, $h_1 = 295.17$ kJ/kg, $p_{r1} = 1.3068$, $h_3 = 1161.07$ kJ/kg, $p_{r3} = 167.1$. Then

$$p_{r2} = p_{r1}\left[\frac{p_2}{p_1}\right] = 1.3068\left[\frac{5.7}{0.95}\right] = 7.8408 \Rightarrow h_{2s} = 493.03 \text{ kJ/kg}$$

$$p_{r4} = p_{r3}\left[\frac{p_4}{p_3}\right] = 167.1\left[\frac{0.95}{5.7}\right] = 27.85 \Rightarrow h_{4s} = 706.51 \text{ kJ/kg}$$

Thus

$$\frac{\dot{W}_{net}}{\dot{m}} = (1161.07 - 706.51) - (493.03 - 295.17) = 454.56 - 197.86 = 256.7 \frac{\text{kJ}}{\text{kg}}$$

$$\Rightarrow \dot{W}_{net} = \left(5\frac{\text{kg}}{\text{s}}\right)\left[256.7\frac{\text{kJ}}{\text{kg}}\right]\left|\frac{1\text{ kW}}{1\text{ kJ/s}}\right| = 1284 \text{ kW} \quad \longleftarrow$$

(b) $\eta_c = 0.82$, $\eta_t = 0.85$. In this case the net work developed per unit mass flowing is

$$\frac{\dot{W}_{net}}{\dot{m}} = \underbrace{(h_3 - h_4)}_{\text{turbine output}} - \underbrace{(h_2 - h_1)}_{\text{compressor input}}$$

Using the compressor and turbine efficiencies

$$(h_3 - h_4) = \eta_t(h_3 - h_{4s}), \quad h_2 - h_1 = \frac{(h_{2s} - h_1)}{\eta_c}$$

Thus

① $$\frac{\dot{W}_{net}}{\dot{m}} = \eta_t(h_3 - h_{4s}) - \frac{(h_{2s} - h_1)}{\eta_c}$$

$$= 0.85(454.56) - \frac{197.86}{.82} = 386.38 - 241.29 = 145.09 \frac{\text{kJ}}{\text{kg}}$$

$$\Rightarrow \dot{W}_{net} = (5\text{ kg/s})\left[145.09\frac{\text{kJ}}{\text{kg}}\right]\left|\frac{1\text{ kW}}{1\text{ kJ/s}}\right| = 725 \text{ kW} \quad \longleftarrow$$

---

1. The effect of irreversibilities is to decrease the work developed by the turbine while increasing the work required by the compressor. Overall, the net work developed is significantly decreased.

PROBLEM 6.160

KNOWN: Steady state operating data are provided for a valve, flash chamber, and turbine in series.

FIND: Determine the power developed by the turbine. Also, determine the rates of entropy production for each of the three components, and compare.

SCHEMATIC & GIVEN DATA:

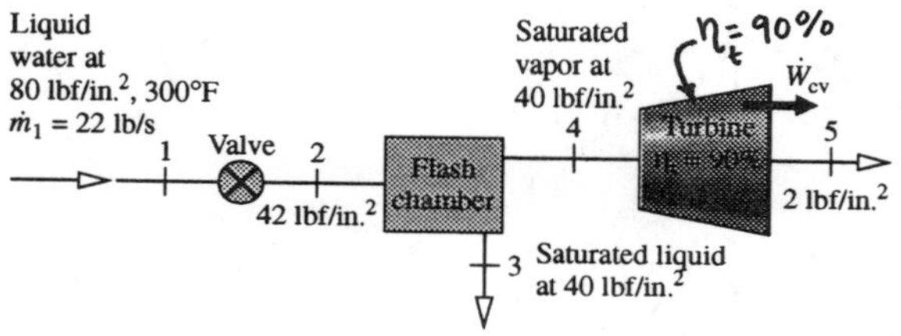

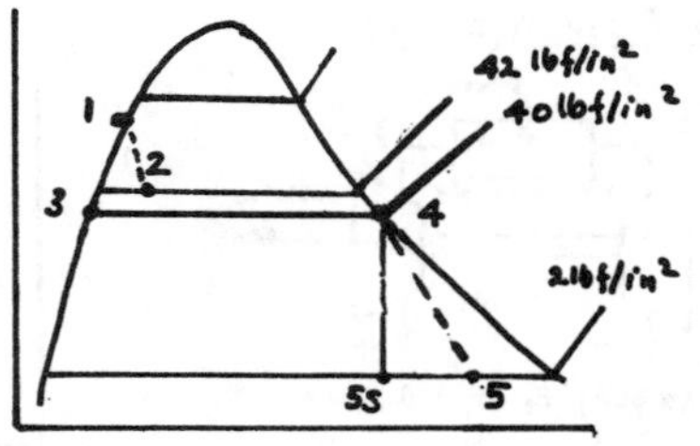

ASSUMPTIONS: (1) Each component operates at steady state with negligible heat transfer between each component and its surroundings. (2) Kinetic and potential energy effects can be ignored. (3) The expansion across the valve is a throttling process.

ANALYSIS: At steady state, mass rate balances reduce to give $\dot{m}_2 = \dot{m}_1$, $\dot{m}_3 + \dot{m}_4 = \dot{m}_2$, $\dot{m}_4 = \dot{m}_5$. An entropy rate balance for a control volume enclosing the flash chamber takes the form

$$0 = \sum \frac{\dot{Q}_j}{T_j}^{\,0} + \dot{m}_2 s_2 - \dot{m}_3 s_3 - \dot{m}_4 s_4 + \dot{\sigma}_{cv}$$

or

$$\dot{\sigma}_{cv} = \dot{m}_4 s_4 + \dot{m}_3 s_3 - \dot{m}_2 s_2 \qquad \text{[flash chamber]} \qquad (1)$$

Similarly, for the valve and turbine, the rates of entropy production are

$$\dot{\sigma}_{cv} = \dot{m}_1 (s_2 - s_1) \qquad \text{[valve]} \qquad (2)$$

$$\dot{\sigma}_{cv} = \dot{m}_4 (s_5 - s_4) \qquad \text{[turbine]} \qquad (3)$$

To evaluate Eqs. (1)–(3) requires $\dot{m}_3$, $\dot{m}_4$, s_1, s_2, s_3, s_4, and s_5. These will now be evaluated in turn.

Using the mass rate balance together with listed assumptions, an energy rate balance for a control volume enclosing the flash chamber reduces to

$$0 = \dot{m}_2 h_2 - \dot{m}_3 h_3 - \dot{m}_4 h_4$$

$$0 = \dot{m}_1 h_2 - \dot{m}_3 h_3 - (\dot{m}_1 - \dot{m}_3) h_4$$

Across the valve $h_2 = h_1$, so on solving the above equation

$$\dot{m}_3 = \dot{m}_1 \frac{[h_1 - h_4]}{[h_3 - h_4]} = \left(22 \frac{\text{lb}}{\text{s}}\right) \left[\frac{269.7 - 1170}{236.16 - 1170}\right] = 21.21 \text{ lb/s}$$

where $h_1 \approx h_f(T_1)$ from Table A-2E and h_3 and h_4 are from Table A-3E. Then, $\dot{m}_4 = \dot{m}_2 - \dot{m}_3 = 22 - 21.21 = 0.79$ lb/s.

From Table A-2E, $s_1 \approx s_f(T_1) = 0.4372$ Btu/lb·°R. State 2 is fixed by 42 lbf/in² and $h_2 = h_1 = 269.7$ Btu/lb. Thus, with data from Table A-3E

$$x_2 = \frac{h_2 - h_f}{h_{fg}} = \frac{269.7 - 239.1}{931.8} = 0.0328$$

Thus

$$s_2 = s_f + x_2 s_{fg} = 0.3961 + 0.0328(1.2767) = 0.438 \text{ Btu/lb·°R}$$

From Table A-3E at 40 lbf/in², $s_3 = 0.3921$ Btu/lb·°R, $s_4 = 1.6767$ Btu/lb·°R. State 5 is fixed by 2 lbf/in² and h_5. Using the turbine isentropic efficiency,

$$\eta_t = \frac{h_4 - h_5}{h_4 - h_{5s}} \qquad (4)$$

where h_{5s} is the specific entropy at the turbine exit for an isentropic expansion from state 4 to 2 lbf/in². The value of h_{5s} is determined by 2 lbf/in² and $s_{5s} = s_4$.

Thus, with data from Table A-3E at 2 lbf/in²

$$x_{5s} = \frac{s_{5s} - s_f}{s_{fg}} = \frac{1.6767 - 0.175}{1.7448} = 0.861 \Rightarrow h_{5s} = h_f + x_{5s} h_{fg}$$
$$= 94.02 + (0.861)(1022.1) = 974.05 \text{ Btu/lb}$$

Then, with data from Table A-3E at 2 lbf/in²

$$x_{5s} = \frac{s_{5s} - s_f}{s_{fg}} = \frac{1.6767 - 0.175}{1.7448} = 0.861, \quad h_{5s} = h_f + x_{5s} h_{fg} = 94.02 + 0.861(1022.1) = 974.05 \text{ Btu/lb}$$

Solving Eq. (4) for h_5 and inserting values

$$h_5 = h_4 - \eta_t (h_4 - h_{5s}) = 1170 - 0.9(1170 - 974.05) = 993.65 \text{ Btu/lb}$$

Using this together with data from Table A-3E at 2 lbf/in²

$$x_5 = \frac{h_5 - h_f}{h_{fg}} = \frac{993.65 - 94.02}{1022.1} = 0.88$$

$$s_5 = s_f + x_5 s_{fg} = 0.175 + 0.88(1.7448) = 1.7104 \text{ Btu/lb·°R}.$$

Finally, substituting values into Eqs. (1)-(3)

- flash chamber: $\dot{\sigma}_{cv} = (0.79)[1.6767] + (21.21)[0.3921] - (22)[0.438] = 0.0054 \dfrac{\text{Btu}}{\text{s·°R}}$ ←
- valve: $\dot{\sigma}_{cv} = (22)[0.438 - 0.4372] = 0.0176 \dfrac{\text{Btu}}{\text{s·°R}}$ ←
- turbine: $\dot{\sigma}_{cv} = (0.79)[1.7104 - 1.6767] = 0.0266 \dfrac{\text{Btu}}{\text{s·°R}}$ ←

Mass and energy rate balances reduce for the turbine to give

$$\dot{W}_t = \dot{m}_4[h_4 - h_5] = \left(0.79 \frac{\text{lb}}{\text{s}}\right)[1170 - 993.65] = 139.3 \frac{\text{Btu}}{\text{s}} \leftarrow$$

The overall rate of entropy production is obtained by summing the individual contributions:

$$\dot{\sigma}_{overall} = (0.0054 + 0.0176 + 0.0266) = 0.0496 \frac{\text{Btu}}{\text{s·°R}}$$

When expressed as percentages of the overall entropy production rate, the individual contributions are

①

Component	Contribution to overall, %
flash chamber	10.9
valve	35.5
turbine	53.6

---

1. Although the turbine contributes most to the overall entropy production a significant reduction in the turbine's contribution might not be feasible because the turbine is operating near its practical limit: $\eta_t = 90\%$. Attention would then turn to the next most important contributor — the valve. For example, the valve irreversibility could be reduced by replacing the valve with a power-recovery turbine. However, such a change would be governed by economic considerations.

PROBLEM 6.161

KNOWN: Data are provided for an air compressor operating at steady state.

FIND: For each of three cases, evaluate the work and heat transfer per unit mass of air flowing. Discuss.

SCHEMATIC & GIVEN DATA:

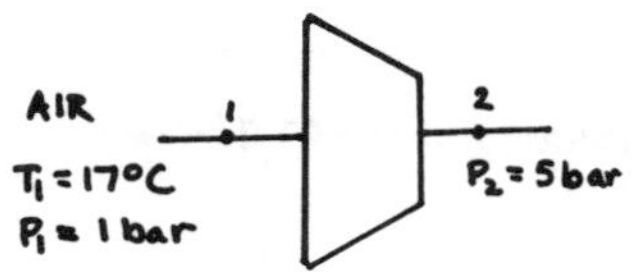

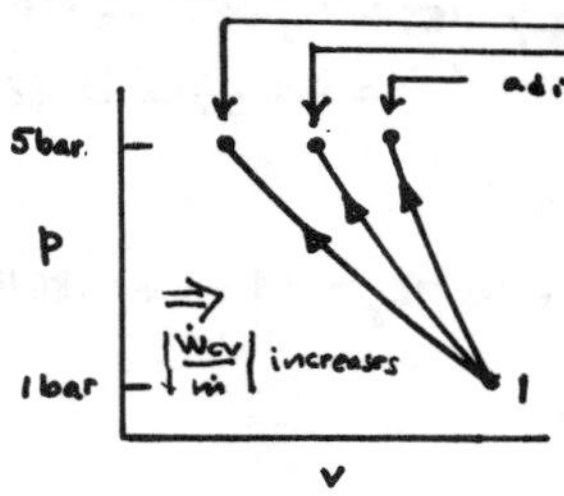

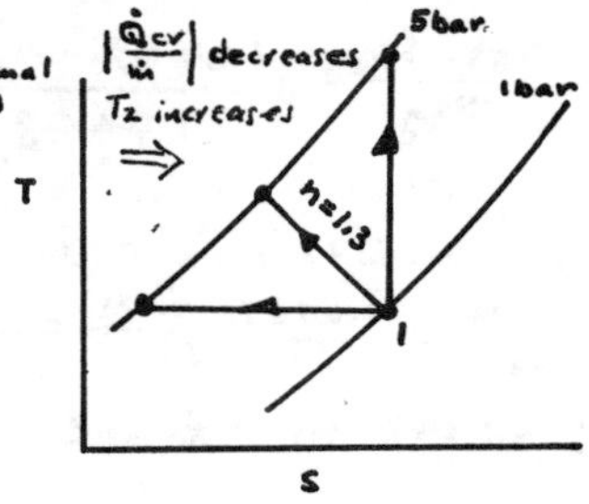

ASSUMPTIONS: (1) The air compressor operates at steady state without internal irreversibilities. (2) Kinetic and potential energy effects are negligible. (3) Air is modeled as an ideal gas.

ANALYSIS: Following the discussions of Eqs. 6.51 and 6.53 of Sec. 6.9, the areas below the process curves of the T-s diagram represent the magnitudes of the heat transfer per unit of mass flowing. The areas behind the process curves of the p-v diagram represent the magnitudes of the work per unit of mass flowing.

(a) Isothermal. An energy rate balance reduces at steady state to give

$$0 = \frac{\dot{Q}_{cv}}{\dot{m}} - \frac{\dot{W}_{cv}}{\dot{m}} + (h_1 - h_2) + \cancel{\frac{V_1^2 - V_2^2}{2}} + \cancel{g(z_1 - z_2)}$$

As the specific enthalpy of an ideal gas depends on temperature alone, and temperature is constant, $h_2 = h_1$. The energy rate balance then becomes simply $\dot{Q}_{cv}/\dot{m} = \dot{W}_{cv}/\dot{m}$.

Since irreversibilities are absent, and $\Delta ke = \Delta pe \approx 0$, the work per unit of mass flowing is given by Eq. 6.53b. Introducing $v = RT/p$ and integrating

$$\frac{\dot{W}_{cv}}{\dot{m}} = -\int_1^2 v\,dp = -\int_1^2 \frac{RT}{p}\,dp = -RT \ln P_2/P_1 = -\left(\frac{8.314}{28.97}\frac{kJ}{kg\cdot K}\right)(290K)\ln\frac{5}{1} = -133.95\ \frac{kJ}{kg} \quad \longleftarrow (a)$$

(b) Polytropic with n=1.3. As in part (a), the work per unit of mass flowing is given by Eq. 6.53b. Introducing pv^n = constant, and integrating, Eq. 6.55 results. Then, with the ideal gas equation of state, Eq. 6.57a is obtained:

$$\frac{\dot{W}_{cv}}{\dot{m}} = -\frac{nR}{(n-1)}(T_2 - T_1)$$

For a polytropic process of an ideal gas (Eq. 3.56)

$$T_2 = T_1\left(\frac{P_2}{P_1}\right)^{(n-1)/n} = (290K)\left(\frac{5}{1}\right)^{0.3/1.3} = 420.4\ K$$

Thus

$$\frac{\dot{W}_{cv}}{\dot{m}} = -\frac{(1.3)}{(0.3)}\left(\frac{8.314}{28.97}\right)(420.4 - 290) = -162.17\ \frac{kJ}{kg}$$

An energy rate balance at steady state reduces to give

$$\frac{\dot{Q}_{cv}}{\dot{m}} = \frac{\dot{W}_{cv}}{\dot{m}} + h_2 - h_1 = -162.17 + 421.67 - 290.16 = -30.66\ \frac{kJ}{kg} \quad \longleftarrow (b)$$

where enthalpy data is from Table A-22.

(c) Adiabatic. If the process is adiabatic and internally reversible, it is isentropic. An energy rate balance reduces with $\dot{Q}_{cv} = 0$ and assumption 2 to give

$$\frac{\dot{W}_{cv}}{\dot{m}} = h_1 - h_2$$

To find h_2, use P_r data from Table A-22

$$P_{r2} = P_{r1}(P_2/P_1) = 1.2311\,(5/1) = 6.1555 \Rightarrow h_2 = 460.1\ kJ/kg$$

Finally

$$\frac{\dot{W}_{cv}}{\dot{m}} = 290.16 - 460.1 = -169.94\ kJ/kg \quad \longleftarrow (c)$$

Only work and heat magnitudes are associated with areas on the p-v and T-s diagrams, respectively. Thus, the areas behind the curves on the p-v diagram increase from case (a) to case (b) to case (c) in accordance with the magnitudes calculated for $\dot{W}_{cv}/\dot{m}$. The areas below the curves on the T-s diagram decrease from case (a) to case (b) to case (c) in accordance with the magnitudes for $\dot{Q}_{cv}/\dot{m}$.

PROBLEM 6.162

KNOWN: Data are provided for an air compressor operating at steady state.

FIND: For each of three cases, evaluate the work and heat transfer per unit mass of air flowing.

SCHEMATIC & GIVEN DATA:

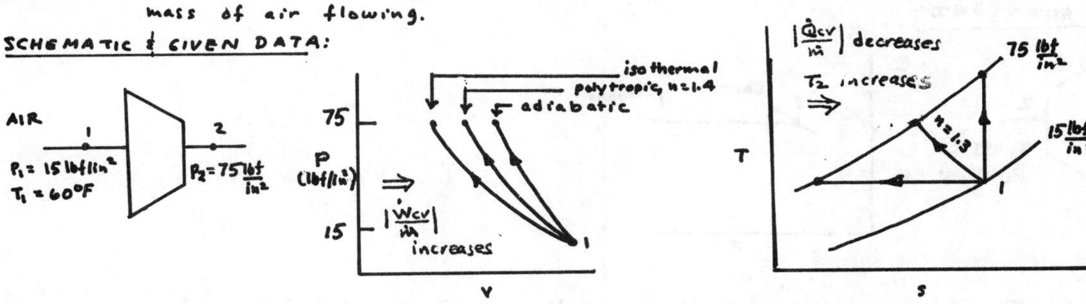

ASSUMPTIONS: (1) The air compressor operates at steady state without internal irreversibilities. (2) Kinetic and potential energy effects are negligible. (3) Air is modeled as an ideal gas.

ANALYSIS: Following the discussions of Eqs. 6.51 and 6.53 of Sec. 6.9, the areas below the process curves of the T-s diagram represent the magnitudes of the heat transfer per unit of mass flowing. The areas behind the process curves of the p-v diagram represent the magnitudes of the work per unit of mass flowing.

(a) Isothermal. An energy rate balance reduces at steady state to give

$$0 = \frac{\dot{Q}_{cv}}{\dot{m}} - \frac{\dot{W}_{cv}}{\dot{m}} + (h_1 - h_2) + \cancelto{0}{\frac{V_1^2 - V_2^2}{2}} + \cancelto{0}{g(z_1 - z_2)}$$

As the specific enthalpy of an ideal gas depends on temperature alone, and temperature is constant, $h_2 = h_1$. The energy rate balance then becomes simply $\dot{Q}_{cv}/\dot{m} = \dot{W}_{cv}/\dot{m}$.

Since irreversibilities are absent, and $\Delta ke = \Delta pe$, the work per unit of mass flowing is given by Eq. 6.53b. Introducing $v = RT/p$ and integrating

$$\frac{\dot{W}_{cv}}{\dot{m}} = -\int_1^2 v\,dp = -\int_1^2 \frac{RT}{p}\,dp = -RT \ln\frac{P_2}{P_1} = -\left(\frac{1.986}{28.97}\ \frac{Btu}{lb \cdot °R}\right)(520°R) \ln\frac{75}{15} = -57.37\ \frac{Btu}{lb} \quad \longleftarrow (a)$$

(b) Polytropic with n=1.3. As in part (a), the work per unit of mass flowing is given by Eq. 6.53b. Introducing pv^n = constant, and integrating, Eq. 6.55 results. Then, with the ideal gas equation of state, Eq. 6.57a. is obtained:

$$\frac{\dot{W}_{cv}}{\dot{m}} = -\frac{nR}{(n-1)}(T_2 - T_1)$$

For a polytropic process of an ideal gas (Eq. 3.56)

$$T_2 = T_1\left(\frac{P_2}{P_1}\right)^{(n-1)/n} = 520\left(\frac{75}{15}\right)^{0.3/1.3} = 753.9\ °R$$

Thus

$$\frac{\dot{W}_{cv}}{\dot{m}} = -\frac{(1.3)}{(0.3)}\left(\frac{1.986}{28.97}\right)(753.9 - 520) = -69.48\ \frac{Btu}{lb}$$

An energy rate balance at steady state reduces to give

$$\frac{\dot{Q}_{cv}}{\dot{m}} = \frac{\dot{W}_{cv}}{\dot{m}} + h_2 - h_1 = -69.48 + 180.61 - 124.27 = -13.14\ Btu/lb. \quad \Big> (b)$$

where enthalpy data is from Table A-22E.

(c) Adiabatic. If the process is adiabatic and internally reversible, it is isentropic. An energy balance reduces with $\dot{Q}_{cv}$ and assumption 2 to give

$$\frac{\dot{W}_{cv}}{\dot{m}} = h_1 - h_2$$

To find h_2, use p_r data from Table A-22E

$$p_{r2} = p_{r1}(P_2/P_1) = 1.2147(75/15) = 6.0735 \Rightarrow h_2 = 197.06\ Btu/lb$$

Finally

$$\frac{\dot{W}_{cv}}{\dot{m}} = 124.27 - 197.06 = -72.79\ Btu/lb \quad \longleftarrow (c)$$

Only work and heat magnitudes are associated with areas on the p-v and T-s diagrams, respectively. Thus, the areas behind the curves on the p-v diagram increase from case (a) to case (b) to case (c) in accordance with the magnitudes calculated for $\dot{W}_{cv}/\dot{m}$. The areas below the curves on the T-s diagram decrease from case (a) to case (b) to case (c) in accordance with the magnitudes for $\dot{Q}_{cv}/\dot{m}$.

PROBLEM 6.153

KNOWN: Operating data are provided for an air compressor at steady state.

FIND: Determine the power required and the rate of heat transfer.

SCHEMATIC & GIVEN DATA:

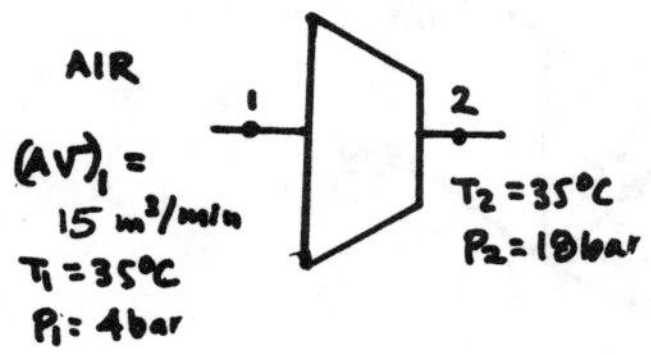

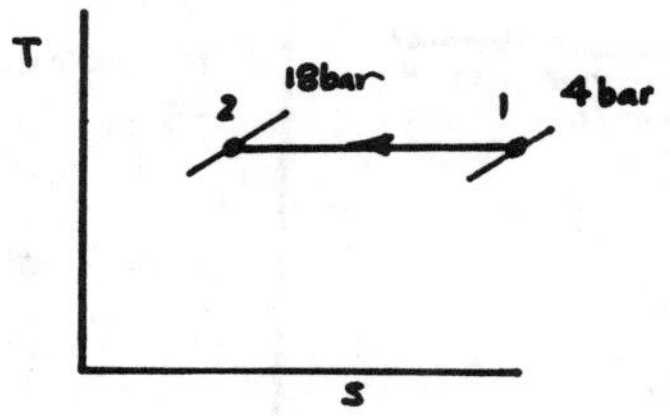

ASSUMPTIONS: (1) The compressor is at steady state. (2) The process is isothermal and internally reversible. (3) Kinetic and potential energy effects are negligible. (4) Air is modeled as an ideal gas.

ANALYSIS: At steady state $\dot{m}_1 = \dot{m}_2 = \dot{m}$, and an energy rate balance reduces to

$$0 = \frac{\dot{Q}_{cv}}{\dot{m}} - \frac{\dot{W}_{cv}}{\dot{m}} + (h_1 - h_2) + \cancel{\frac{V_1^2 - V_2^2}{2}} + \cancel{g(z_1 - z_2)}$$

As the enthalpy of an ideal gas depends on temperature only, and temperature remains constant, $h_2 = h_1$. Thus

$$\frac{\dot{Q}_{cv}}{\dot{m}} = \frac{\dot{W}_{cv}}{\dot{m}}$$

As the process is internally reversible, there are two solution approaches that can be taken: (1) Equation 6.51 can be integrated giving

$$\frac{\dot{Q}_{cv}}{\dot{m}} = T(s_2 - s_1)$$

$$= T\big(\underbrace{s^\circ(T_2) - s^\circ(T_1)}_{=0 \text{ since } T_1 = T_2} - R \ln P_2/P_1\big) = -RT \ln P_2/P_1$$

(2) Equation 6.53b can be integrated, giving

$$\frac{\dot{W}_{cv}}{\dot{m}} = -\int_1^2 v\,dp = -\int_1^2 \frac{RT}{p}\,dp = -RT \ln \frac{P_2}{P_1}$$

The mass flow rate is

$$\dot{m} = \frac{(AV)_1}{v_1} = \frac{P_1 (AV)_1}{RT_1} = \frac{(4\times10^5 \frac{N}{m^2})(15 \frac{m^3}{min})\left|\frac{min}{60 s}\right|}{\left(\frac{8314}{28.97}\frac{N\cdot m}{kg\cdot K}\right)(308 K)} = 1.131 \frac{kg}{s}$$

Thus

$$\dot{Q}_{cv} = \dot{W}_{cv} = -\dot{m} RT \ln \frac{P_2}{P_1} = -\left(1.131 \frac{kg}{s}\right)\left(\frac{8.314}{28.97}\frac{kJ}{kg\cdot K}\right)(308 K)\left(\ln \frac{18}{4}\right)\left|\frac{kW}{kJ/s}\right|$$

$$= -150.4 \text{ kW} \longleftarrow \dot{Q}_{cv}, \dot{W}_{cv}$$

PROBLEM 6.164

KNOWN: Steady state operating data are provided for a R134a compressor.

FIND: Determine the power required and rate of heat transfer.

SCHEMATIC & GIVEN DATA:

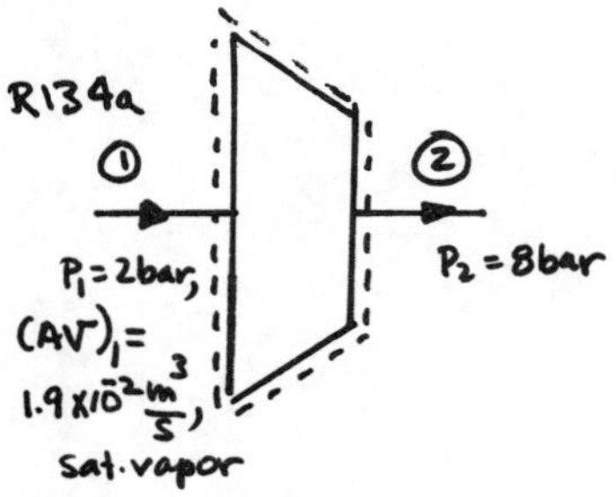

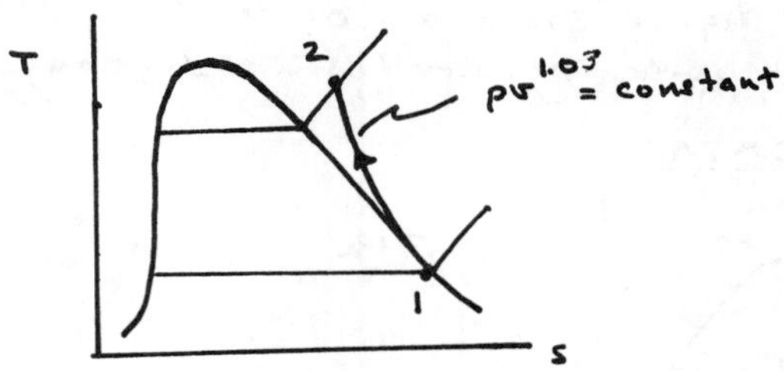

ASSUMPTIONS: 1. The control volume shown in the accompanying sketch is at steady state. 2. For the control volume, kinetic and potential energy effects can be ignored. 3. The compression is described by $pv^{1.03}$ = constant. The process is internally reversible.

ANALYSIS: (a) Using Eq. 6.55, the power is

$$(\dot{W}_{cv})_{\substack{int\\rev}} = -\dot{m}\left(\frac{n}{n-1}\right)\left[p_2v_2 - p_1v_1\right]$$

To find v_2, write $p_2v_2^{1.03} = p_1v_1^{1.03} \Rightarrow v_2 = \left(\frac{p_1}{p_2}\right)^{\frac{1}{1.03}} v_1$, where $v_1 = 0.0993$ m³/kg from Table A-11. Thus

$$v_2 = \left(\frac{2}{8}\right)^{\frac{1}{1.03}}\left(0.0993\ \frac{m^3}{kg}\right) = 0.02585\ \frac{m^3}{kg}.$$

Also,

$$\dot{m} = \frac{(AV)_1}{v_1} = \frac{1.9\times10^{-2}\ m^3/s}{0.0993\ m^3/kg} = 0.191\ kg/s.$$

Finally

$$(\dot{W}_{cv})_{\substack{int\\rev}} = -\left(0.191\ \frac{kg}{s}\right)\left[\frac{1.03}{1.03-1}\right]\left[\left(8\times10^5\frac{N}{m^2}\right)\left(0.02585\frac{m^3}{kg}\right) - (2\times10^5)(0.0993)\right]\left|\frac{1\ kJ}{10^3\ N\cdot m}\right|\left|\frac{1\ kW}{1\ kJ/s}\right|$$

$$= -5.38\ kW \quad \longleftarrow$$

(b) Mass and energy rate balances reduce to give

$$0 = \dot{Q}_{cv} - \dot{W}_{cv} + \dot{m}\left[(h_1-h_2) + \frac{V_1^2-V_2^2}{2} + g(z_1-z_2)\right]$$

$$\Rightarrow \quad \dot{Q}_{cv} = \dot{W}_{cv} + \dot{m}\left[h_2 - h_1\right]$$

State 2 is fixed by 8 bar, v_2 = 0.02585 m³/kg. Thus, from Table A-12, h_2 = 266.66 kJ/kg. At state 1, h_1 = 241.3 kJ/kg from Table A-11.

$$\dot{Q}_{cv} = -5.38\ kW + \left(0.191\ \frac{kg}{s}\right)\left[266.66 - 241.3\right]\frac{kJ}{s}\left|\frac{1\ kW}{1\ kJ/s}\right| \quad \longleftarrow$$

$$= 0.54\ kW$$

PROBLEM 6.165

KNOWN: At steady state, water vapor is compressed isentropically to 3 MPa from the saturated vapor state at 0.1 MPa and liquid water is pumped isentropically to 3 MPa from the saturated liquid state at 0.1 MPa.

① FIND: Compare the work required in each case.

SCHEMATIC & GIVEN DATA:

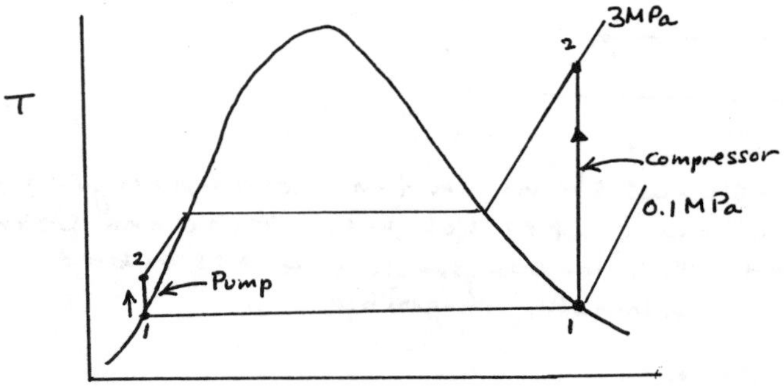

ASSUMPTIONS: (1) In each case, a control volume enclosing the device is under consideration. (2) For each control volume, $\dot{Q}_{cv} = 0$, kinetic/potential energy effects are negligible, and operation is at steady state.

ANALYSIS: Reducing mass and energy rate balances

$$-\frac{\dot{W}_{cv}}{\dot{m}} = h_2 - h_1$$

Compressor: From Table A-4, $h_1 = 2675.5$ kJ/kg, $s_1 = 7.3594$ kJ/kg·K, $h_2 = 3556.71$ kJ/kg. Thus

$$-\frac{\dot{W}_{cv}}{\dot{m}} = 3556.71 - 2675.5 = 881.21 \ \frac{kJ}{kg} \quad \longleftarrow$$

PUMP: From Table A-3, $h_1 = 417.46$ Btu/lb, $s_1 = 1.3026$ kJ/kg·K. Then, interpolating in Table A-5 (double interpolation), $h_2 = 420.49$ kJ/kg. Thus

$$-\frac{\dot{W}_{cv}}{\dot{m}} = 420.49 - 417.46 = 3.03 \ \frac{kJ}{kg} \quad \longleftarrow$$

Comparison:

$$\% = \left(\frac{\text{Pump Work}}{\text{Compressor Work}}\right)100 = \frac{(3.03)(100)}{881.21} = 0.34\% \quad \longleftarrow$$

---

1. In chapters 8 and 9 the significance of pump work and compressor work and their relation to each other will be apparent in the study of Rankine and Brayton power cycles. Also, see the discussion of Eq. 6.53b in Sec. 6.9.

PROBLEM 6.166

KNOWN: Steady state operating data are provided for an electrically-driven pump that draws water from a pond.

FIND: Estimate the hourly cost of operating the pump.

SCHEMATIC & GIVEN DATA:

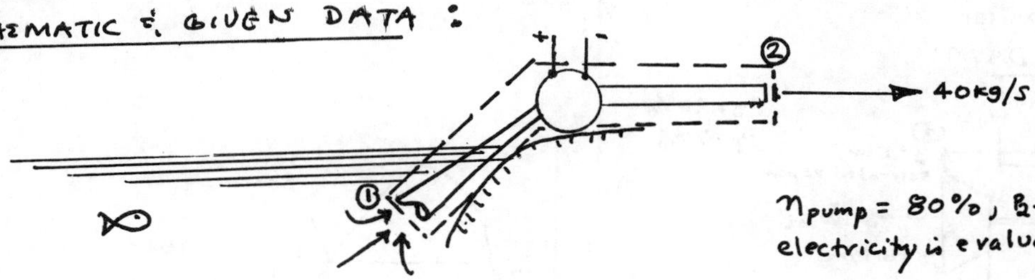

$\eta_{pump} = 80\%$, $P_2 - P_1 = 3$ bar
electricity is evaluated at 8¢/kW·h

ASSUMPTIONS: (1) The control volume shown in the figure is at steady state. (2) For the control volume, the changes in kinetic energy and potential energy from the inlet to the exit are ignored. (3) Liquid water is modeled as incompressible.

ANALYSIS: Using the pump efficiency, the power required to operate the pump is

$$\dot{W}_{pump} = \frac{(\dot{W}_{int\,rev})}{\eta_p}$$

where the numerator is provided by Eq. 6.53a or, with the assumptions listed, by Eq. 6.53c. Thus

$$\dot{W}_{pump} = -\frac{\dot{m}v(P_2 - P_1)}{\eta_p}$$

With v from Table A-2, $v \approx 10^{-3}$ m³/kg,

$$\dot{W}_{pump} = -\frac{\left(40\frac{kg}{s}\right)\left(10^{-3}\frac{m^3}{kg}\right)\left(3\times10^5\frac{N}{m^2}\right)}{0.80}\left|\frac{1\,kJ}{10^3\,N\cdot m}\right|\left|\frac{1\,kW}{1\,kJ/s}\right|$$

$$= -15\,kW \quad \text{(minus denotes input)}$$

Then, the hourly cost of operating the pump is

$$\begin{pmatrix}\text{hourly}\\ \text{cost}\end{pmatrix} = (15\,kW)(1\cdot h)\left(0.08\frac{\$}{kW\cdot h}\right)$$

$$= \$1.20/h$$

PROBLEM 6.167

KNOWN: Steady state operating data are provided for a pump, a boiler, and a turbine in series.

FIND: Determine in kJ per kg of steam flowing (a) the pump work, (b) the net work developed by the turbine, and (c) the heat transfer to the boiler.

SCHEMATIC & GIVEN DATA:

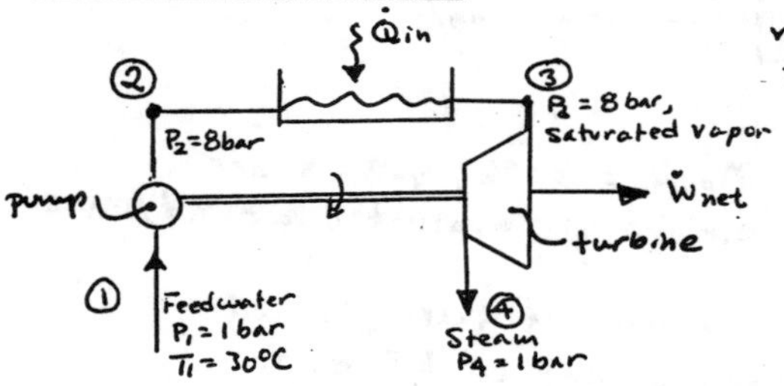

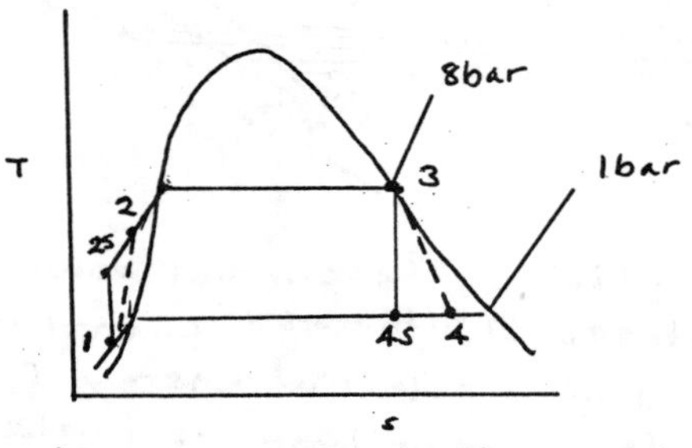

ASSUMPTIONS: (1) Control volumes enclosing the pump, the boiler, and the turbine are at steady state. For the pump and turbine, $\dot{Q}_{cv}=0$. (2) All kinetic and potential energy effects are negligible.

ANALYSIS: For the pump, **Eq. 6.53b** can be invoked to evaluate the work in the internally reversible process. With data from Table A-2, $h_1 \approx h_f(T_1) = 125.79$ kJ/kg, $v_1 \approx v_f(T_1) = 1.0043 \times 10^{-3}$ m³/kg. Then

$$\left(\frac{\dot{W}_P}{\dot{m}}\right)_{\substack{int\\rev}} \approx -v_1 \Delta p = \left(\frac{1.0043}{10^3}\frac{m^3}{kg}\right)(8-1)\times 10^5 \frac{N}{m^2}\left|\frac{1\,kJ}{10^3\,N\cdot m}\right| = -0.7\frac{kJ}{kg}$$

Then using the given isentropic pump efficiency

$$\eta_p = \frac{(-\dot{W}_P/\dot{m})_{s=c}}{(-\dot{W}_p/\dot{m})} \Rightarrow \frac{\dot{W}_P}{\dot{m}} = \frac{-0.7\,kJ/kg}{0.7} = -1\frac{kJ}{kg}$$ ◄ PUMP

Mass and energy rate balances reduce to give, $0 = \dot{Q}_{cv}^{\,0} - \dot{W}_P + \dot{m}(h_1 - h_2)$. Thus $h_2 = (-\dot{W}_P/\dot{m}) + h_1 = 126.79$ kJ/kg.

For the boiler, mass and energy rate balances reduce to give

$$\frac{\dot{Q}_{in}}{\dot{m}} = h_3 - h_2 = 2769.1 - 126.79 = 2642.3\ kJ/kg$$ ◄ BOILER

where h_3 is from Table A-3.

For the turbine, mass and energy rate balances reduce to give

$$\frac{\dot{W}_t}{\dot{m}} = h_3 - h_4$$

Or on introducing the isentropic turbine efficiency

$$\frac{\dot{W}_t}{\dot{m}} = \eta_t(h_3 - h_{4s})$$

From Table A-3, $s_3 = s_{4s} = 6.6628$ kJ/kg·K. The quality at 4s is

$$x_{4s} = \frac{6.6628 - 1.3026}{7.3594 - 1.3026} = 0.885$$

Then, $h_{4s} = 417.46 + 0.885(2258) = 2415.8$ kJ/kg. Accordingly

$$\frac{\dot{W}_t}{\dot{m}} = 0.9\,(2769.1 - 2415.8) = 318\ kJ/kg$$

Of this amount, 1 kJ/kg is required by the pump, leaving $\frac{\dot{W}_{net}}{\dot{m}} = 317\frac{kJ}{kg}$ ◄ Net

PROBLEM 6.168

KNOWN: Steady-state operating data are provided for a hydraulic turbine-generator.

FIND: In the absence of internal irreversibilities, determine the value of the power produced on a daily basis.

SCHEMATIC & GIVEN DATA:

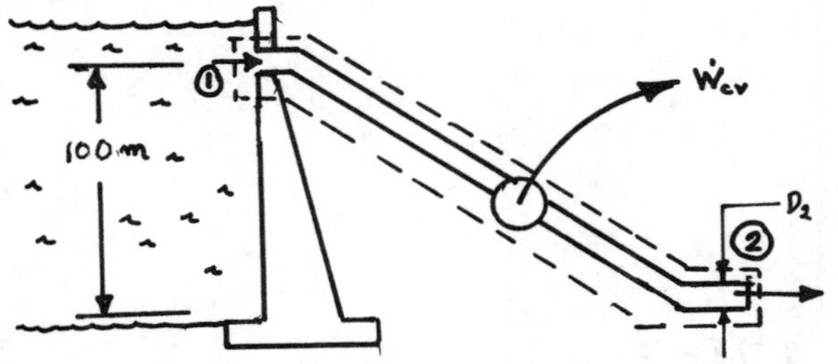

$p_1 = 1.05$ bar, $V_1 = 1$ m/s
$p_2 = 1$ bar, $V_2 = 10$ m/s, $T_2 = 15$°C, $D_2 = 1.2$ m
$g = 9.8$ m/s²
electricity is evaluated at 8¢/kW·h

ASSUMPTIONS: 1. The control volume shown in the figure is at steady state. 2. Internal irreversibilities are ignored. 3. Liquid water is modeled as incompressible.

ANALYSIS: Using assumption 2 indicating that internal irreversibilities are ignored, Eq. 6.53a can be used to evaluate the power developed. Since water is modeled as incompressible, we get

$$\left(\frac{\dot{W}_{cv}}{\dot{m}}\right)_{\substack{int\\rev}} = -v(p_2 - p_1) + \frac{V_1^2 - V_2^2}{2} + g(z_1 - z_2)$$

With $v = v_f(15°C) = 1.0009 \times 10^{-3}$ m³/kg and given data

$$\left(\frac{\dot{W}_{cv}}{\dot{m}}\right)_{\substack{int\\rev}} = -\left(\frac{1.0009}{10^3}\right)\left(\frac{m^3}{kg}\right)\left[(1-1.05)\times 10^5 \frac{N}{m^2}\right] + \left[\frac{(1\,m/s)^2 - (10\,m/s)^2}{2}\right]\left|\frac{1\,N}{1\,kg\cdot m/s^2}\right| + \left(9.8\frac{m}{s^2}\right)(100\,m)\left|\frac{1\,N}{kg\cdot m/s^2}\right|$$

$$= \left(+5\frac{N\cdot m}{kg}\right) + \left(-49.5\frac{N\cdot m}{kg}\right) + \left(980\frac{N\cdot m}{kg}\right) = 935.5\frac{N\cdot m}{kg}\left|\frac{1\,kJ}{10^3\,N\cdot m}\right| = 0.936\frac{kJ}{kg}$$

The mass flow rate is

$$\dot{m} = \frac{A_2 V_2}{v_2} = \frac{\pi D_2^2 V_2}{4 v_2} = \frac{\pi(1.2\,m)^2(10\,m/s)}{4(1.0009\times 10^{-3})\,m^3/kg} = 11{,}300\,\frac{kg}{s}$$

so

$$(\dot{W}_{cv})_{\substack{int\\rev}} = \left(11{,}300\frac{kg}{s}\right)\left(0.936\frac{kJ}{kg}\right)\left|\frac{1\,kW}{1\,kJ/s}\right| = 10{,}577\,kW$$

① Daily Value = $(10{,}577\,kW)\left(\frac{24\,h}{day}\right)\left(\frac{\$0.08}{kW\cdot h}\right) = \$20{,}308/day$

---

1. Owing to the effect of irreversibilities, the actual daily revenue would be <u>less</u> than the calculated value. Also, the daily value of the power developed is <u>not</u> the profit. The plant owners would have a considerable capital investment to recoup, and a variety of operating and maintenance costs to bear.

PROBLEM 6.169

KNOWN: Steady-state operating data are provided for a hydraulic turbine.

FIND: For a turbine power output of 1 MW, estimate the minimum required mass flow rate.

SCHEMATIC & GIVEN DATA:

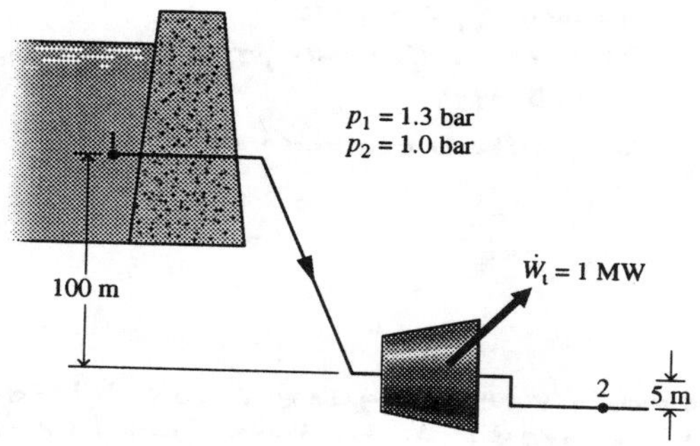

$g = 9.8\ m/s^2$

$D_2 = D_1$

ASSUMPTIONS: 1. A one-inlet, one-exit control volume with inlet at 1 and exit at 2, and enclosing the turbine, is at steady state.
2. Liquid water is modeled as incompressible, with $v = 10^{-3}\ m^3/kg$ (Table A-19).

ANALYSIS: We expect that the effect of internal irreversibilities would be a lower power output in the actual case than in the ideal case — that is, we expect

$$(\dot{W}_t)_{act} \le (\dot{W})_{\substack{int\\rev}}$$

Introducing Eq. 6.53a with assumption 2

$$(\dot{W}_t)_{act} \le \dot{m}\left[-v(p_2-p_1) + g(z_1-z_2) + \cancel{\frac{V_1^2-V_2^2}{2}}\right] \qquad (1)$$

where the kinetic energy term drops out because $D_1 = D_2$, v is constant, and $\dot{m}_1 = \dot{m}_2$. From (1), we have

$$\dot{m} \ge \left[\frac{(\dot{W}_t)_{act}}{-v(p_2-p_1)+g(z_1-z_2)}\right] = \frac{(1\ MW)\left|\frac{10^6\ N\cdot m/s}{1\ MW}\right|}{\left[-(10^{-3}\frac{m^3}{kg})(-0.3\times10^5\frac{N}{m^2})\right] + \left[(9.8\frac{m}{s^2})(105\ m)\right]\left|\frac{1\ N}{1\ kg\cdot m/s^2}\right|}$$

$$\dot{m} \ge 944.3\ kg/s$$

The minimum required mass flow rate is 944.3 kg/s. ←

PROBLEM 6.170

KNOWN: Operating data are provided for liquid water flowing through a pipe.

FIND: In the absence of internal irreversibilities, determine the pressure required at the pipe inlet. Discuss.

SCHEMATIC & GIVEN DATA:

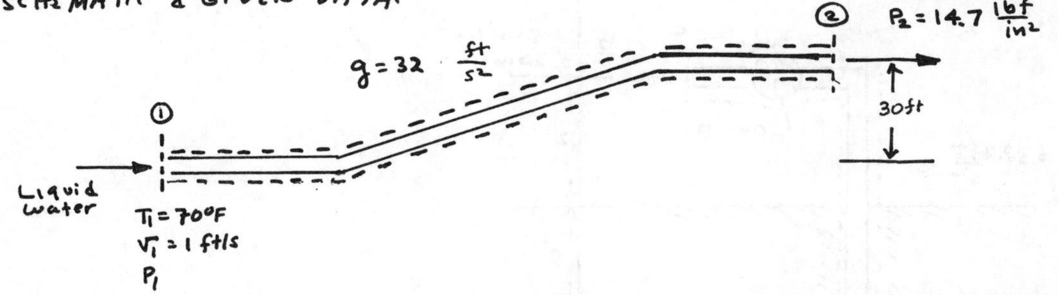

ASSUMPTIONS: 1. A one-inlet, one-exit control volume encloses the pipe. 2. The control volume is at steady state. 3. Liquid water is modeled as incompressible with $v = v_f(T_1) = 0.01605\ ft^3/lb$ (Table A-2E). 4. Internal irreversibilities are ignored.

ANALYSIS: Since internal irreversibilities are ignored and there is no power input or output, Eq. 6.54 applies

① $$\int_1^2 v\,dp + \frac{V_2^2 - V_1^2}{2} + g(z_2 - z_1) = 0$$

$$\Rightarrow \quad v(P_2 - P_1) + \frac{V_2^2 - V_1^2}{2} + g(z_2 - z_1) = 0$$

Solving for P_1

$$P_1 = P_2 + \frac{\left(\frac{V_2^2 - V_1^2}{2}\right) + g(z_2 - z_1)}{v}$$

$$= 14.7\ \frac{lbf}{in^2} + \frac{\frac{1}{2}\left[(20)^2 - (1)^2\right]\left(\frac{ft}{s}\right)^2 + \left(32\ \frac{ft}{s^2}\right)(30\ ft)}{\left(0.01605\ \frac{ft^3}{lb}\right)} \left|\frac{1\ lbf}{32.2\ \frac{lb \cdot ft}{s^2}}\right| \left|\frac{1\ ft^2}{144\ in^2}\right|$$

$$= 14.7\ \frac{lbf}{in^2} + 15.58\ \frac{lbf}{in^2} = 30.28\ \frac{lbf}{in^2}$$

To overcome the effects of internal irreversibilities, such as friction between the flowing water and pipe wall, we expect the pressure at the inlet to be greater than calculated.

---

1. This is the Bernoulli equation.

PROBLEM 6.171

KNOWN: Steady-state operating data are provided for a pump drawing water from underground and delivering it above ground.

FIND: In the absence of internal irreversibilities, determine the power required by the pump. Discuss.

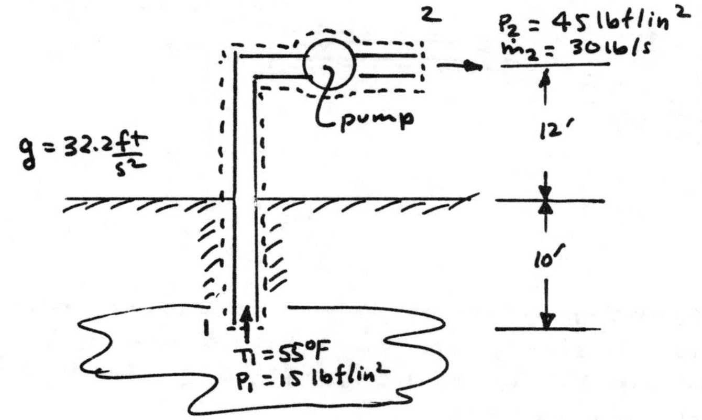

ASSUMPTIONS: 1. The control volume shown in the schematic is at steady state. 2. Internal irreversibilities are absent. 3. Kinetic energy effects can be ignored. 4. Liquid water is modeled as incompressible with $v = v_f(T_1) = 0.01603$ ft³/lb (Table A-2E).

ANALYSIS: In the absence of internal irreversibilities, Eq. 6.53a applies. Then, with assumptions 3, 4, we get

$$\left(\frac{\dot{W}_{cv}}{\dot{m}}\right)_{\substack{int\\rev}} = -v(P_2 - P_1) + g(z_1 - z_2) + \cancelto{0}{\frac{(V_1^2 - V_2^2)}{2}}$$

$$= -\left(0.01603 \frac{ft^3}{lb}\right)(45 - 15)\frac{lbf}{in^2}\left|\frac{144\,in^2}{ft^2}\right| + \left(32.2\frac{ft}{s^2}\right)(-22\,ft)\left|\frac{1\,lbf}{32.2\,lb\cdot ft/s^2}\right|$$

$$= -69.25\,\frac{ft\cdot lbf}{lb} - 22\,\frac{ft\cdot lbf}{lb} = -91.25\,\frac{ft\cdot lbf}{lb}$$

$$\Rightarrow \quad (\dot{W}_{cv})_{\substack{int\\rev}} = \left(30\frac{lb}{s}\right)\left(-91.25\frac{ft\cdot lbf}{lb}\right)\left|\frac{1\,hp}{550\,ft\cdot lbf/s}\right|$$

$$= -4.98\,hp \quad \rightarrow \quad \text{Power input} = 4.98\,hp \quad \leftarrow$$

To overcome the effects of internal irreversibilities, such as friction between the flowing liquid and pipe wall, we expect the power input to the pump to be greater than determined here.

PROBLEM 6.172

KNOWN: Steady state operating data are provided for a 3-hp pump.

FIND: Determine if it would be possible to pump 1000 gal in 10 min or less.

SCHEMATIC & GIVEN DATA:

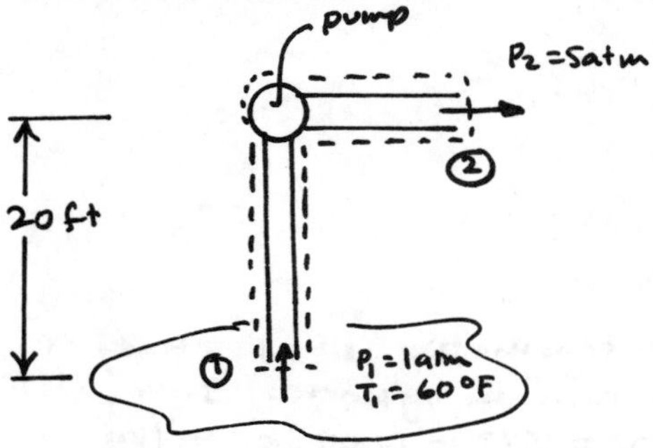

$\dot{W}_{cv} = -3\text{hp}$

$g = 32.2\,\text{ft/s}^2$

ASSUMPTIONS: 1. The control volume shown in the schematic is at steady state. 2. For the control volume, kinetic energy effects can be ignored. 3. Liquid water is modeled as incompressible with $v = v_f(T_1) = 0.01604\,\text{ft}^3/\text{lb}$ (Table A-2E).

ANALYSIS: To pump 1000 gal in 10 minutes, or less, would require a volumetric flow rate (AV)

$$(AV) \geq \left(\frac{1000\text{ gal}}{10\text{ min}}\right)\left|\frac{0.13368\text{ ft}^3}{\text{gal}}\right|\left|\frac{1\text{ min}}{60\text{ s}}\right| = 0.2228\,\frac{\text{ft}^3}{\text{s}}$$

which corresponds to a mass flow rate

$$\dot{m} \geq \frac{0.2228\,\text{ft}^3/\text{s}}{0.01604\,\text{ft}^3/\text{lb}} = 13.89\,\frac{\text{lb}}{\text{s}} \qquad (1)$$

In the absence of internal irreversibilities, the power input required would be given by the following expression obtained from Eq. 6.53a using assumptions 2, 3

$$(-\dot{W}_{cv})_{\substack{int\\rev}} = \dot{m}\left[v(p_2 - p_1) + g(z_2 - z_1) + \cancelto{0}{\frac{(V_2^2 - V_1^2)}{2}}\right]$$

Then, with (1)

$$(-\dot{W}_{cv})_{\substack{int\\rev}} \geq \left(13.89\,\frac{\text{lb}}{\text{s}}\right)\left[\left(0.01604\,\frac{\text{ft}^3}{\text{lb}}\right)\left(4 \times 14.7 \times 144\,\frac{\text{lbf}}{\text{ft}^2}\right) + \left(32.2\,\frac{\text{ft}}{\text{s}^2}\right)[20\text{ ft}]\left|\frac{1\text{ lbf}}{32.2\text{ lb}\cdot\text{ft/s}^2}\right|\right]$$

$$\geq \left(13.89\,\frac{\text{lb}}{\text{s}}\right)\left[135.81\,\frac{\text{ft}\cdot\text{lbf}}{\text{lb}} + 20\,\frac{\text{ft}\cdot\text{lbf}}{\text{lb}}\right]\left|\frac{1\text{ hp}}{550\text{ ft}\cdot\text{lbf/s}}\right|$$

$$\geq 3.9\text{ hp}$$

To overcome the effect of internal irreversibilities, such as friction between the flowing liquid and pipe wall, we expect the <u>actual</u> power input to the pump to be greater than calculated here. Since, only a 3-hp pump is available, the desired flow rate cannot be accomodated.

PROBLEM 6.173

KNOWN: Steady state operating data are provided for a 4-kW pump.

FIND: Determine if it would be possible for the pump to deliver water at a pressure of 10 bar.

SCHEMATIC & GIVEN DATA:

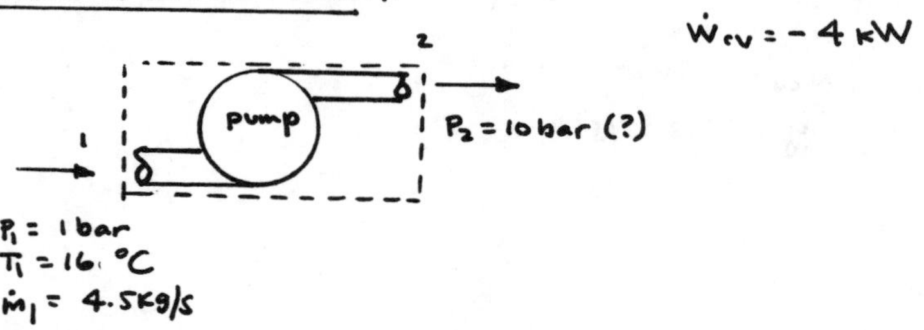

ASSUMPTIONS: 1. The control volume shown in the schematic is at steady state. 2. Kinetic and potential energy changes can be ignored. 3. Liquid water is modeled as incompressible with $v = v_f(T_1) = 1.0011 \times 10^{-3}$ m³/kg.

ANALYSIS: In the absence of internal irreversibilities, the power input required to deliver water at $P_2 = 10$ bar would be given by the following expression obtained from Eq. 6.53a using assumptions 3, 4

$$(-\dot{W}_{cv})_{\substack{int \\ rev}} = \dot{m}\, v (P_2 - P_1)$$

$$= \left(4.5 \frac{kg}{s}\right)\left(\frac{1.0011}{10^3}\right)\frac{m^3}{kg}\left(9 \times 10^5 \frac{N}{m^2}\right)\left|\frac{1\ kW}{10^3\ N \cdot m/s}\right|$$

$$= 4.05\ kW$$

To overcome the effect of internal irreversibilities, such as friction between the flowing water and the pump walls, we expect the actual power required by the pump to be greater than calculated here. Since only a 4-kW pump is available, the desired exit pressure cannot be accomodated.

PROBLEM 6.174

KNOWN: Steady-state operating data are provided for a pump that delivers water at an elevation above the pump inlet.

FIND: Determine the maximum theoretical elevation at which the water can be delivered.

SCHEMATIC & GIVEN DATA:

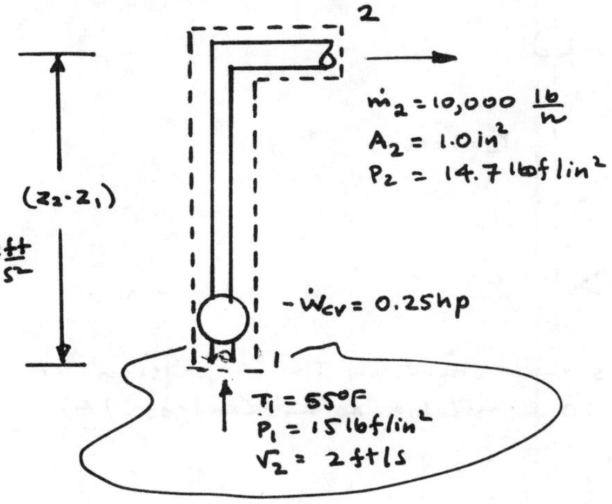

ASSUMPTIONS: 1. The control volume shown in the schematic is at steady state. 2. Liquid water is modeled as incompressible with $v = v_f(T_1) = 0.01733\ \frac{ft^3}{lb}$ (Table A-2E).

ANALYSIS: Using data at 2, the velocity V_2 can be found:

$$\dot{m}_2 = \frac{A_2 V_2}{v}$$

$$\Rightarrow V_2 = \frac{\dot{m}_2 v}{A} = \frac{\left(10{,}000\frac{lb}{h}\right)\left|\frac{1h}{3600s}\right|\left(0.01733\frac{ft^3}{lb}\right)}{(1.0\ in^2)\left|\frac{1\ ft^2}{144\ in^2}\right|}$$

$$= 6.9\ ft/s$$

Owing to the effect of internal irreversibilities, such as friction between the flowing water and pipe wall, the actual power input would be greater than in the ideal case — that is, we expect

$$(-\dot{W}_{cv})_{act} \geq (-\dot{W}_{cv})_{\substack{int\\rev}}$$

Introducing Eq. 6.53a with assumption 2

$$(-\dot{W}_{cv})_{act} \geq \dot{m}\left[v(p_2 - p_1) + g(z_2 - z_1) + \frac{(V_2^2 - V_1^2)}{2}\right]$$

$$\Rightarrow (z_2 - z_1) \leq \frac{1}{g}\left[\frac{(-\dot{W}_{cv})_{act}}{\dot{m}} + v(p_1 - p_2) - \left(\frac{V_2^2 - V_1^2}{2}\right)\right]$$

$$\leq \left(\frac{1}{32.2\ ft/s^2}\right)\left[\frac{(0.25hp)\left|\frac{550\ ft\cdot lbf/s}{1hp}\right|}{(10^4\ lb/h)\left|\frac{1h}{3600s}\right|} + \left(0.01733\frac{ft^3}{lb}\right)\left(0.3 \times 144\ \frac{lbf}{ft^2}\right) - \frac{\left[(6.9)^2 - (2)^2\right]\frac{ft^2}{s^2}}{2}\left|\frac{1\ lbf}{32.2\ lb\cdot\frac{ft}{s^2}}\right|\right]$$

$$\leq \left(\frac{1}{32.2\ ft/s^2}\right)\left[(49.5 + 0.75 - 0.67)\ \frac{ft\cdot lbf}{lb}\right]\left|\frac{32.2\ lb\cdot ft/s^2}{1\ lbf}\right|$$

$$\leq 49.58\ ft$$

The maximum theoretical elevation at the exit is 49.58 ft.

PROBLEM 6.175

KNOWN: Data are provided for CO expanding through a nozzle operating at steady state.

FIND: Determine the velocity at the nozzle exit and the rate of heat transfer per kg of CO flowing.

SCHEMATIC & GIVEN DATA:

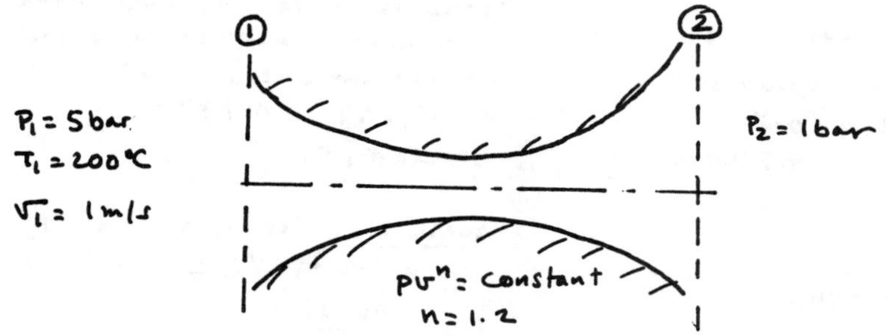

ASSUMPTIONS: (1) The nozzle is at steady state. (2) The expansion is a polytropic process with n=1.2. (3) CO is modeled as an ideal gas. (4) Potential energy effects are ignored.

ANALYSIS: (a) As a polytropic process is internally reversible, Eq. 6.53a is applicable. Then, with $\dot{W}_{cv} = 0$, the Bernoulli equation, Eq. 6.54 results:

$$\int_1^2 v\,dp + \frac{V_2^2 - V_1^2}{2} + \cancel{g(z_2 - z_1)}^{\,0} = 0 \tag{1}$$

Using the relationship pv^n = constant, the integral can be performed (Sec. 6.9), giving

$$\int_1^2 v\,dp = \frac{n}{n-1}(p_2 v_2 - p_1 v_1)$$

$$= \frac{nR}{n-1}(T_2 - T_1) \tag{2}$$

where the ideal gas equation of state has been used to obtain the last expression. For a polytropic process of an ideal gas (Eq. 3.56)

$$\frac{T_2}{T_1} = \left(\frac{p_2}{p_1}\right)^{(n-1)/n} \tag{3}$$

Collecting Eqs. (1)-(3)

$$V_2 = \sqrt{V_1^2 + 2\frac{nRT_1}{n-1}\left[1 - (p_2/p_1)^{(n-1)/n}\right]} \tag{4}$$

$$= \sqrt{\left(1\frac{m}{s}\right)^2 + 2\left(\frac{1.2}{1.2-1}\right)\left(\frac{8314}{28}\frac{N\cdot m}{kg\cdot K}\right)(473K)\left|\frac{1\ kg\cdot m/s^2}{1N}\right|\left[1 - \left(\frac{1}{5}\right)^{0.2/1.2}\right]} = 629.7\ \frac{m}{s} \longleftarrow V_2$$

(b) An energy balance reduces to give

$$\frac{\dot{Q}_{cv}}{\dot{m}} = h_2 - h_1 + \frac{V_2^2 - V_1^2}{2}$$

With Eq. (3), $T_2 = T_1\left(p_2/p_1\right)^{\frac{n-1}{n}} = (473)\left(\frac{1}{5}\right)^{\frac{0.2}{1.2}} = 362K$

Then with data from Table A-23

$$\frac{\dot{Q}_{cv}}{\dot{m}} = \frac{(10531 - 13797)}{28}\frac{kJ}{kg} + \left[\frac{(629.7)^2 - (1)^2}{2}\left(\frac{m^2}{s^2}\right)\right]\left|\frac{1N}{1kg\cdot m/s^2}\right|\left|\frac{1\ kJ}{10^3 N\cdot m}\right| = 81.7\ \frac{kJ}{kg} \longleftarrow \dot{Q}_{cv}/\dot{m}$$

CHAPTER SEVEN

EXERGY (AVAILABILITY) ANALYSIS

Chapter 7 - Third and fourth edition problem correspondence

3rd	4th	3rd	4th	3rd	4th	3rd	4th
7.1	---	---	7.30	---	7.56	---	7.92
7.2	7.1*	---	7.31	7.57	7.57*	7.82	7.93*
7.3	---	7.30	7.32	7.58	7.58*	7.83	---
---	7.2	---	7.33	7.59	7.59*	7.84	7.94*
---	7.3	---	7.34	7.60	7.60*	7.85	7.95*
7.4	7.4*	7.32	---	7.61	7.61*	7.86	7.96
7.5	---	7.31	7.35*	7.62	7.62*	7.87	7.97
7.7	7.5	---	7.36	7.63	7.63*	7.88	7.98
7.6	---	7.33	7.37	7.64	7.64*	7.89	---
---	7.6	7.34	---	7.65	7.65*	7.92	7.99
7.9	7.7*	---	7.38	7.66	7.66	7.90	7.100*
7.8	7.8*	---	7.39	7.67	7.67*	7.91	---
---	7.9	7.36	7.40	---	7.68	---	7.101
---	7.10	7.35	7.41*	---	7.69	7.93	7.102
7.10	---	7.37	---	---	7.70	7.94	7.103*
7.11	---	---	7.42	7.68	7.71*	7.95	7.104*
7.12	7.11*	7.38	---	7.69	7.72	7.96	---
7.13	7.12*	---	7.43	7.70	7.73*	---	7.105
7.14	7.13	7.39	---	---	7.74	7.97	7.106*
7.16	7.14*	7.40	---	---	7.75	7.98	7.107*
7.15	7.15	7.41	---	7.71	7.76*	7.99	7.108*
7.17	7.16	7.42	---	7.72	---	7.100	7.109*
---	7.17	7.44	---	---	7.77	7.101	7.110
---	7.18	---	7.44	7.73	---	---	7.111
7.18	---	7.43	7.45*	---	7.78	---	7.112
---	7.19	7.46	7.46	7.74	7.79	---	7.113
7.19	---	7.45	7.47	7.75	7.80*	---	7.114
7.20	7.20	7.47	---	7.76	7.81	---	7.115
7.21	7.21	7.48	7.48*	7.77	---	---	7.116
---	7.22	7.49	7.49*	7.78	7.82	7.102	7.117
7.22	---	7.50	7.50	7.79	7.83	---	7.118
7.23	7.23*	7.51	7.51*	---	7.84	7.103	7.119
7.24	7.24	---	7.52	---	7.85	7.104	7.120
7.25	7.25	7.52	---	---	7.86	7.105	7.121
---	7.26	---	7.53	---	7.87	---	7.122
7.26	---	7.53	---	---	7.88	---	7.123
7.27	7.27*	7.54	7.54*	---	7.89	---	7.124
7.29	7.28	7.55	7.55*	7.80	7.90		
7.28	7.29*	7.56	---	7.81	7.91*		

*Revised

PROBLEM 7.1

KNOWN: A system consists of 5 kg of water at 10°C, 1 bar. The system is at rest and zero elevation relative to the exergy reference environment.

FIND: Determine the exergy.

SCHEMATIC & GIVEN DATA:

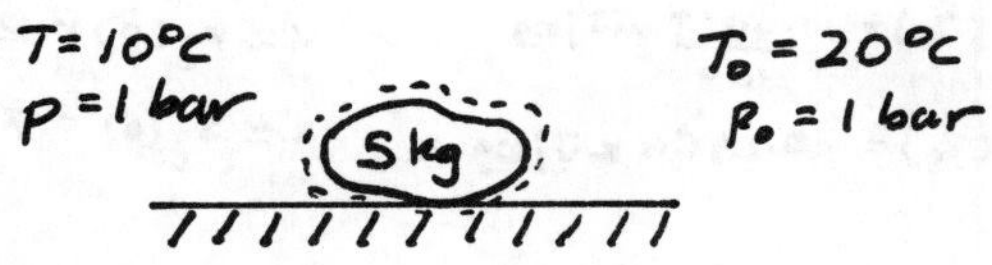

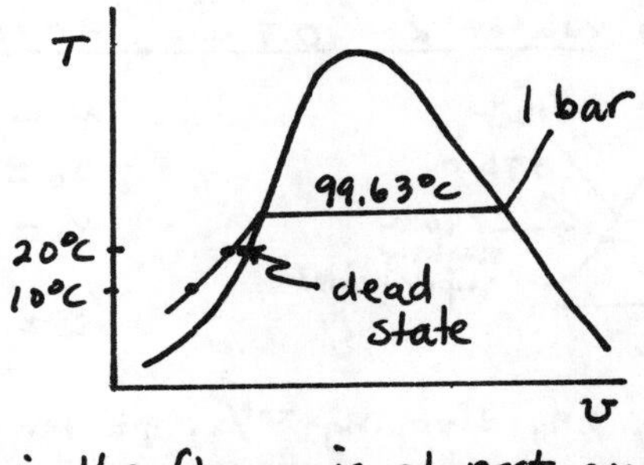

ASSUMPTION: The closed system shown in the figure is at rest and zero elevation relative to the environment for which $T_0 = 20°C$, $p_0 = 1 \text{ bar}$.

ANALYSIS: Equation 7.2 gives

$$E = m\left[(u-u_0) + p_0(v-v_0) - T_0(s-s_0) + \frac{V^2}{2}^{\,0} + gz^{\,0}\right]$$

From Table A-2, using the approximations of sec. 3.3.6: $v \approx v_f(T)$ and $u \approx u_f(T)$ with the appoximation of Eq. 6.7: $s \approx s_f(T)$

$v_0 = 1.0018 \times 10^{-3}\ \text{m}^3/\text{kg}$ $\qquad v = 1.0004 \times 10^{-3}\ \text{m}^3/\text{kg}$

$u_0 = 83.95\ \text{kJ/kg}$ $\qquad u = 42.00\ \text{kJ/kg}$

$s_0 = 0.2966\ \text{kJ/kg·K}$ $\qquad s = 0.1510\ \text{kJ/kg}$

Thus

$$E = (5\text{ kg})\left[(42.00-83.95)\frac{\text{kJ}}{\text{kg}} + (1\text{ bar})\left(\frac{1.0004-1.0018}{10^3}\right)\frac{\text{m}^3}{\text{kg}}\left|\frac{10^5\text{ N/m}^2}{1\text{ bar}}\right|\left|\frac{1\text{ kJ}}{10^3\text{ N·m}}\right| - (293\text{ K})(0.1510-0.2966)\frac{\text{kJ}}{\text{kg·K}}\right]$$

$$= (5\text{ kg})\left[-41.95 - 1.4\times10^{-4} + 42.66\right]\frac{\text{kJ}}{\text{kg}}$$

$$= 3.55\text{ kJ} \quad \longleftarrow \quad E$$

PROBLEM 7.2

KNOWN: A state is specified for each of three substances.

FIND: Determine the exergy.

ASSUMPTIONS: 1. Ignore the effects of motion and gravity. 2. $T_0 = 20°C$, $p_0 = 1$ bar.

ANALYSIS: (a) Water at 0.7 bar, 90°C.

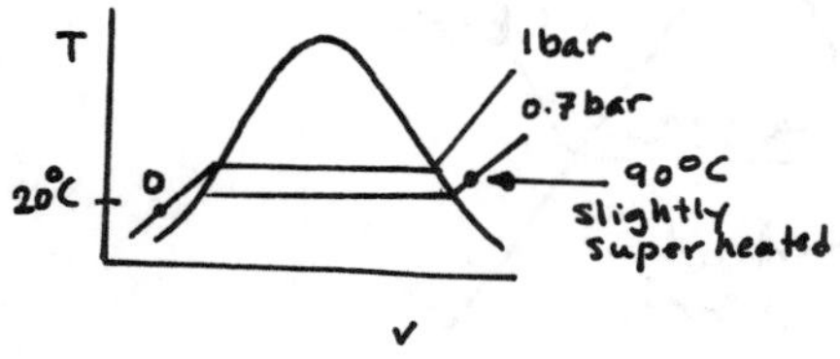

Tables A-2,3: $T_0 = 20°C$, $p_0 = 1$ bar

$u_0 \approx u_f(T_0) = 83.95$ kJ/kg

$v_0 \approx v_f(T_0) = \frac{1.0018}{10^3}$ m³/kg

$s_0 \approx s_f(T_0) = 0.2966$ kJ/kg·K

T 90°C, p = 0.7bar

$u \approx u_g(p) = 2494.5$

$v \approx v_g(p) = 2.365$

$s \approx s_g(p) = 7.4797$

Then, with Eq. 7.9, dropping $V^2/2$, gz terms

$$e = (u - u_0) + p_0(v - v_0) - T_0(s - s_0)$$

$$= (2494.5 - 83.95)\frac{kJ}{kg} + \left(10^5 \frac{N}{m^2}\right)\left(2.365 - \frac{1.0018}{10^3}\right)\frac{m^3}{kg}\left|\frac{1\,kJ}{10^3\,N\cdot m}\right| - (293K)(7.4797 - 0.2966)\frac{kJ}{kg\cdot K}$$

$$= [(2410.55) + (236.4) - (2104.65)]\ kJ/kg$$

$$= 542.27\ \frac{kJ}{kg} \Rightarrow E = (1kg)\left(542.27\frac{kJ}{kg}\right) = 542.27\ kJ \longleftarrow$$

(b) R134a at 0.7bar, 90°C

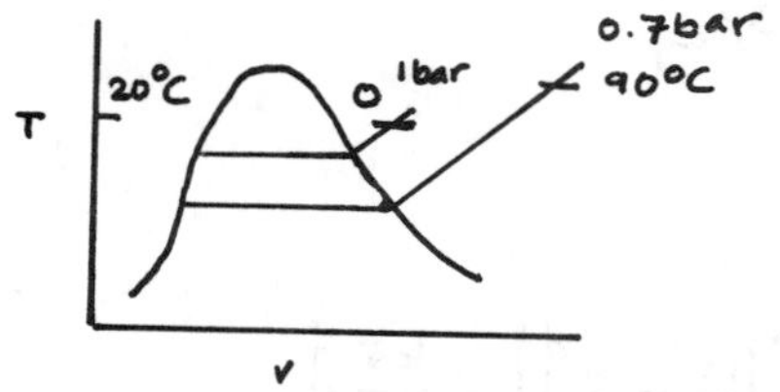

Table A-12:

$u_0 = 246.67$ kJ/kg

$v_0 = 0.23349$ m³/kg

$s_0 = 1.0829$ kJ/kg·K

$u = 305.5$

$v = 0.441$

$s = 1.3118$

$$e = (u - u_0) + p_0(v - v_0) - T_0(s - s_0)$$

$$= (305.5 - 246.67) + 10^5(0.441 - 0.23349)\left|\frac{1}{10^3}\right| - 293(1.3118 - 1.0829)$$

$$= (58.83) + (20.75) - (67.07) = 12.51\ kJ/kg \Rightarrow E = 12.51\ kJ \longleftarrow$$

(c) Air as an ideal gas with c_p constant.

Constant $c_p \Rightarrow$ constant c_v. Evaluating c_v at T_{ave} from Table A-20: $c_v = 0.72 \frac{kJ}{kg\cdot K}$

$c_p = 1.007 \frac{kJ}{kg\cdot K}$

$$e = (u - u_0) + p_0(v - v_0) - T_0(s - s_0)$$

$$= c_v[T - T_0] + p_0\left[\frac{RT}{p} - \frac{RT_0}{p_0}\right] - T_0\left[c_p \ln\frac{T}{T_0} - R\ln\frac{p}{p_0}\right]$$

$$= c_v[T - T_0] + R\left[T\left(\frac{p_0}{p}\right) - T_0\right] - T_0\left[c_p \ln\frac{T}{T_0} - R\ln\frac{p}{p_0}\right]$$

$$= \left(0.72\frac{kJ}{kg\cdot K}\right)[70K] + \frac{8.314}{28.97}\frac{kJ}{kg\cdot K}\left[363K\left(\frac{1}{0.7}\right) - 293K\right]$$

$$- 293K\left[1.007\frac{kJ}{kg\cdot K}\ln\frac{363}{293} - \frac{8.314}{28.97}\frac{kJ}{kg\cdot K}\ln\left(\frac{0.7}{1}\right)\right]$$

$$= [(50.4) + (64.74) - (93.2)]\frac{kJ}{kg} = 21.94\frac{kJ}{kg} \Rightarrow E = 21.94\ kJ \longleftarrow$$

PROBLEM 7.3

KNOWN: Water at each of three alternative states.

FIND: Determine the specific exergy.

SCHEMATIC & GIVEN DATA:

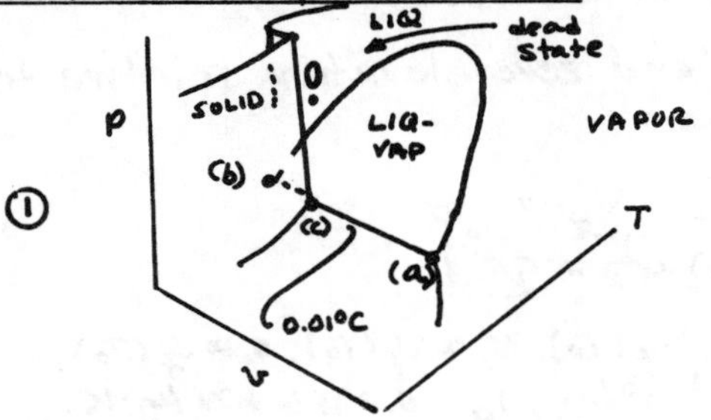

dead state: $T_0 = 20°C$, $p_0 = 1$ bar

$u_0 \sim u_f(20°C) = 83.95$ kJ/kg

$v_0 \sim v_f(20°C) = (1.0018/10^3)$ m³/kg

$s_0 \sim s_f(20°C) = 0.2966$ kJ/kg·K

ASSUMPTION: 1. At the dead state $v \sim v_f(T_0)$, $u \sim u_f(T_0)$, $s \sim s_f(T_0)$. $T_0 = 20°C$, $p_0 = 1$ bar.
2. Ignore the effects of motion and gravity.

(a) Saturated vapor at 0.01 °C. With Eq. 7.9, dropping $V^2/2$, gz terms

$$e = (u - u_0) + p_0(v - v_0) - T_0(s - s_0)$$, with data from Table A-2:

$$= (2375.3 - 83.95)\frac{kJ}{kg} + \left(10^5 \frac{N}{m^2}\right)\left[206.136 - \frac{1.0018}{10^3}\right]\frac{m^3}{kg}\left|\frac{1 kJ}{10^3 N\cdot m}\right| - 293[9.1562 - 0.2966]\frac{kJ}{kg\cdot K}$$

$$= [(2291.35) + 20,613.5 - (2595.86)]\frac{kJ}{kg} = 20,309 \text{ kJ/kg} \longleftarrow$$

(b) Saturated liquid at 0.01 °C. With data from Table A-2:

$$e = (0 - 83.95) + (10^5)\left[\frac{1.002}{10^3} - \frac{1.0018}{10^3}\right]\left|\frac{1}{10^3}\right| - 293(0 - 0.2966)$$

$$= [(-83.95) + (\sim 0) + (86.90)]\frac{kJ}{kg} = 2.95 \text{ kJ/kg} \longleftarrow$$

(c) Saturated solid at 0.01 °C. With data from Table A-5:

$$e = ((-333.4) - 83.95) + (10^5)\left[\frac{1.0908}{10^3} - \frac{1.0018}{10^3}\right]\left|\frac{1}{10^3}\right| - 293((-1.221) - 0.2966)$$

$$= [(-417.35) + (0.01) + (444.66)]\frac{kJ}{kg} = 27.32 \text{ kJ/kg} \longleftarrow$$

1. The "further" a state is from the dead state, the greater the value for exergy. Thus, state (a) is most distant, state (c) is closer, and state (b) is closest. The calculated exergy values confirm this rule.

PROBLEM 7.4

KNOWN: Systems at specified states are defined.

FIND: Determine the specific exergy for 1 kg of each substance.

ASSUMPTIONS: The closed systems are at rest and zero elevation relative to the environment for which $T_0 = 20°C$, $p_0 = 1 bar$.

ANALYSIS: Equation 7.2 gives

$$E = m[(u-u_0)+p_0(v-v_0)-T_0(s-s_0)+\cancel{\frac{V^2}{2}}^0+\cancel{gz}^0]$$

(a) Saturated water vapor at 100°C. Using $u_0 \approx u_f(T_0)$, $v_0 \approx v_f(T_0)$, $s_0 \approx s_f(T_0)$, Table A-2 gives $u_0 = 83.95$ kJ/kg, $v_0 = 1.0018\times10^{-3}$ m³/kg, $s_0 = 0.2966$ kJ/kg·K. Table A-2 also gives $u = 2506.5$ kJ/kg, $v = 1.673$ m³/kg, $s = 7.3549$ kJ/kg·K. Thus

$$E = (1\ kg)\left[(2506.5-83.95)\frac{kJ}{kg}+(1\ bar)\left(\frac{1.673-1.0018}{10^3}\right)\frac{m^3}{kg}\left|\frac{10^5\ N/m^2}{1\ bar}\right|\left|\frac{1\ kJ}{10^3\ N\cdot m}\right|-(293K)(7.3549-0.2966)\frac{kJ}{kg\cdot K}\right]$$

$$= (1\ kg)(2422.55+167.2-2068.08)\frac{kJ}{kg} = 521.7\ kJ \quad \text{(a)}$$

(b) Saturated liquid water at 5°C. The values of u_0, v_0, s_0 are the same as in part(a). From Table A-2, $u = 20.97$ kJ/kg, $v = 1.0001\times10^{-3}$ m³/kg, $s = 0.0761$ kJ/kg·K. Thus

$$E = (1)\left[(20.97-83.95)+(1)\left(\frac{1.0018-1.0001}{10^3}\right)\left|\frac{10^5}{1}\right|\left|\frac{1}{10^3}\right|-(293)(0.0761-0.2966)\right]$$

1 $$= (1\ kg)(-62.98+1.7\times10^{-4}+64.61)\frac{kJ}{kg} = 1.63\ kJ \quad \text{(b)}$$

(c) Ammonia at −10°C, 1 bar. From Table A-15

at 20°C, 1 bar	at −10°C, 1 bar
$v_0 = 1.4153$ m³/kg	$v = 1.2621$ m³/kg
$u_0 = 1374.27$ kJ/kg	$u = 1324.33$ kJ/kg
$s_0 = 6.2816$ kJ/kg·K	$s = 6.0467$ kJ/kg·K

$$E = (1)\left[(1324.33-1374.27)+(1)(1.2621-1.4153)\left|\frac{10^5}{1}\right|\left|\frac{1}{10^3}\right|-(293)(6.0467-6.2816)\right]$$

$$= (1\ kg)(-49.94-15.32+68.83)\frac{kJ}{kg} = 3.57\ kJ \quad \text{(c)}$$

---

1. Even though $T < T_0$, the value of exergy is positive, as expected

PROBLEM 7.5

KNOWN: A balloon filled with helium is at a specified state.

FIND: Determine the specific exergy.

SCHEMATIC & GIVEN DATA:

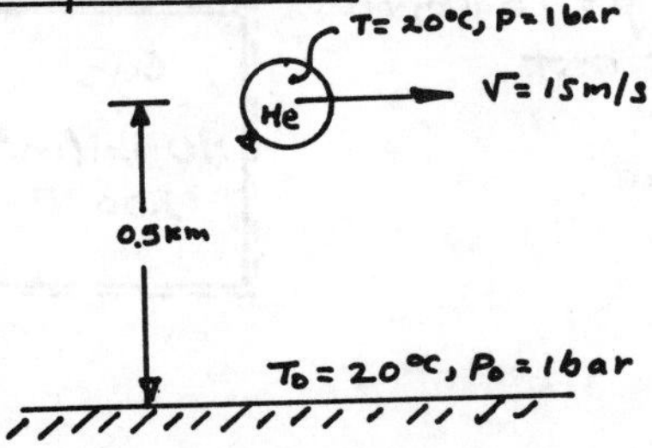

ASSUMPTIONS: (1) The closed system is the helium within the baloon. (2) Helium is modeled as an ideal gas. (3) $g = 9.807\ m/s^2$.

ANALYSIS: The specific exergy is given by Eq. 7.9:

$$e = u - u_0 + p_0(v - v_0) - T_0(s - s_0) + \frac{V^2}{2} + gz$$

For a monatomic gas such as Helium the specific heats c_p and c_v are constant*, so with ideal gas relations

$$e = \underline{c_v[T - T_0] + p_0\left[\frac{RT}{p} - \frac{RT_0}{p_0}\right] - T_0\left[c_p \ln\frac{T}{T_0} - R\ln\frac{p}{p_0}\right]} + \frac{V^2}{2} + gz$$

Since $T = T_0$ and $p = p_0$, the underlined term vanishes, leaving

$$e = \frac{V^2}{2} + gz$$

$$= \left[\frac{(15\ m/s)^2}{2} + (9.807\ m/s^2)(500\ m)\right]\left|\frac{1\ N}{1\ kg \cdot \frac{m}{s^2}}\right|\left|\frac{kJ}{10^3\ N \cdot m}\right|$$

$$= 5.016\ \frac{kJ}{kg} \longleftarrow a$$

---

* See Tables A-21

PROBLEM 7.6

KNOWN: A vessel contains carbon dioxide (CO_2).

FIND: (a) Determine specific exergy e at $p = 90$ lbf/in.², $T = 200°F$. (b) Plot e versus pressure for $T = 80°F$. (c) Plot e versus T for $p = 15$ lbf/in.²

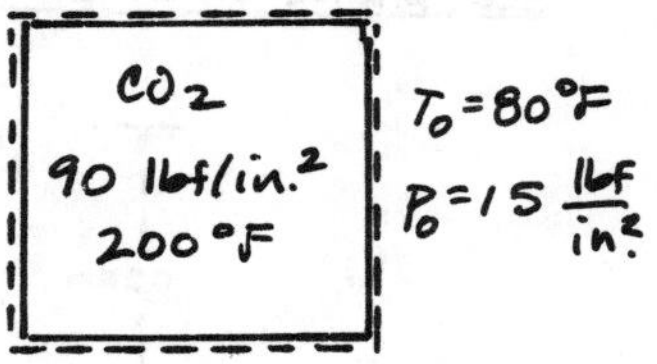

ASSUMPTIONS: (1) The CO_2 is a closed system at rest and zero elevation relative to the reference environment at $T_0 = 80°F$, $p_0 = 15$ lbf/in.² (2) The CO_2 is modeled as an ideal gas.

ANALYSIS: Equation 7.9 reduces to give

$$e = (u - u_0) + p_0(v - v_0) - T_0(s - s_0) \qquad (1)$$

(a) When using gas table data, $u - u_0 = \dfrac{\bar{u}(T) - \bar{u}(T_0)}{M}$, $s - s_0 = \dfrac{\bar{s}°(T) - \bar{s}°(T_0) - \bar{R} \ln p/p_0}{M}$

Thus, with ideal gas relations, (1) becomes

$$e = \frac{1}{M}\left\{[\bar{u}(T) - \bar{u}(T_0)] + p_0\left[\frac{\bar{R}T}{p} - \frac{\bar{R}T_0}{p_0}\right] - T_0\left[\bar{s}°(T) - \bar{s}°(T_0) - \bar{R}\ln\frac{p}{p_0}\right]\right\}$$

$$= \frac{1}{M}\left\{[\bar{u}(T) - \bar{u}(T_0)] + \bar{R}T\left[\frac{p_0}{p} - \frac{T_0}{T}\right] - T_0\left[\bar{s}°(T) - \bar{s}°(T_0) - \bar{R}\ln\frac{p}{p_0}\right]\right\}$$

Using data from Table A-23E

$$\bar{u}(T) - \bar{u}(T_0) = 3854.6 - 2984.4 = 870.2 \text{ Btu/lbmol}$$

$$\bar{R}T\left[\frac{p_0}{p} - \frac{T_0}{T}\right] = \left(1.986 \frac{\text{Btu}}{\text{lbmol}\cdot°\text{R}}\right)(660°\text{R})\left[\frac{15}{90} - \frac{540}{660}\right] = -853.98 \text{ Btu/lbmol}$$

$$T_0\left[\bar{s}°(T) - \bar{s}°(T_0) - \bar{R}\ln p/p_0\right] = (540°\text{R})\left[52.934 - 51.082 - (1.986)\ln\frac{90}{15}\right]\frac{\text{Btu}}{\text{lbmol}\cdot°\text{R}}$$

$$= -921.47 \text{ Btu/lbmol}$$

Finally

$$e = \left(\frac{1}{44.01 \frac{\text{lb}}{\text{lbmol}}}\right)\left\{(870.2 - 853.98 + 921.47)\right\}\frac{\text{Btu}}{\text{lbmol}} = 21.31 \frac{\text{Btu}}{\text{lb}} \longleftarrow e$$

(b)(c) The following IT code is used to generate data for the required plots. The evaluations are based on Eq.(1) above and internal functions in IT for u, v, and s of CO_2 as an ideal gas.

IT Code

```
T = 200  // °F
p = 90  // lbf/in.²
To = 80  // °F
po = 15  // lbf/in.²

e = (u - uo) + (po * (144 / 778)) * (v - vo) - (To + 460) * (s - so)
u = u_T("CO2", T)
uo = u_T("CO2", To)
v = v_TP("CO2",T,p)
vo = v_TP("CO2",To,po)
s = s_TP("CO2", T, p)
so = s_TP("CO2", To, po)
```

IT Result for p = 90 lbf/in.², T = 200 °F

① e = 21.32 Btu/lb

PROBLEM 7.6 (Cont'd.)

PLOTS:

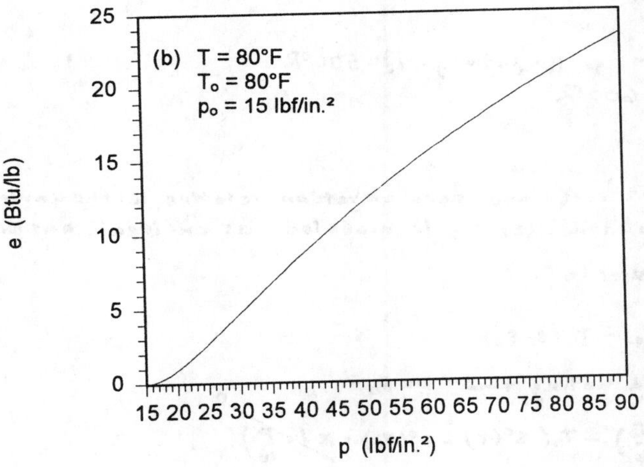

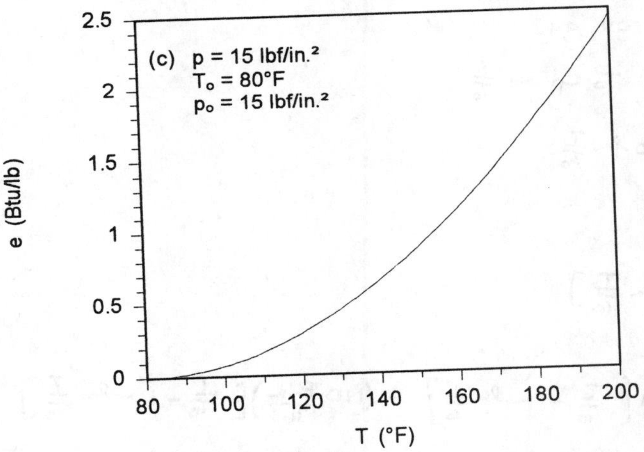

---

1. The IT result compares very favorably with the ideal gas table result in part (a).

PROBLEM 7.7

KNOWN: Oxygen (O_2) at T, 1atm fills a balloon at rest on the surface of the earth where the ambient temperature and pressure are known.

FIND: Plot specific exergy versus T.

SCHEMATIC & GIVEN DATA:

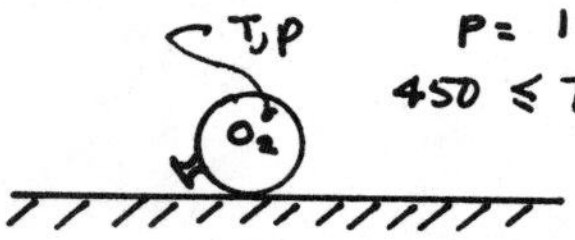

$p = 1\text{atm}$, $p_0 = 1\text{atm}$, $T_0 = 500°R$

$450 \le T \le 600°R$

ASSUMPTIONS: (1) The O_2 is at rest and zero elevation relative to the environ-
① ment for which $T_0 = 500°R$, $p_0 = 1\text{atm}$. (2) O_2 is modeled as an ideal gas with constant c_p.

ANALYSIS: Equation 7.9 reduces to

$$e = u - u_0 + p_0(v - v_0) - T_0(s - s_0)$$

With ideal gas relations, and noting that $p = p_0$

$$a = u(T) - u(T_0) + p_0\left(\frac{RT}{p} - \frac{RT_0}{p_0}\right) - T_0\left(s°(T) - s°(T_0) - R\ln\frac{p}{p_0}\right)$$

(the term $R\ln\frac{p}{p_0}$ is struck through and marked 0)

$$= u(T) - u(T_0) + R(T - T_0) - T_0\left(s°(T) - s°(T_0)\right) \qquad (1)$$

By inspection of Table A-20E, over the range $460 < T < 600K$, $c_p \cong 0.22\,\text{Btu/lb}\cdot°R$. Thus, Eq.(1) can be expressed as

$$e = c_v[T - T_0] + R[T - T_0] - T_0\,c_p\ln\frac{T}{T_0}$$

$$= \underbrace{(c_v + R)}_{c_p\ (\text{Eq. } 3.44)}[T - T_0] - T_0\,c_p\ln\frac{T}{T_0}$$

or

$$e = T_0\,c_p\left[\frac{T}{T_0} - 1 - \ln\frac{T}{T_0}\right]$$

$$= \underbrace{\left(0.22\,\frac{\text{Btu}}{\text{lb}\cdot°R}\right)(500°R)}_{110\,\text{Btu/lb}}\left[\frac{T}{T_0} - 1 - \ln\frac{T}{T_0}\right] = \left(110\,\frac{\text{Btu}}{\text{lb}}\right)\left[\frac{T}{T_0} - 1 - \ln\frac{T}{T_0}\right] \qquad (2)$$

SAMPLE CALCULATION: $T = 600°R$, $e = (110\,\text{Btu/lb})\left[\frac{600}{500} - 1 - \ln\frac{600}{500}\right] = 1.945\,\text{Btu/lb}$

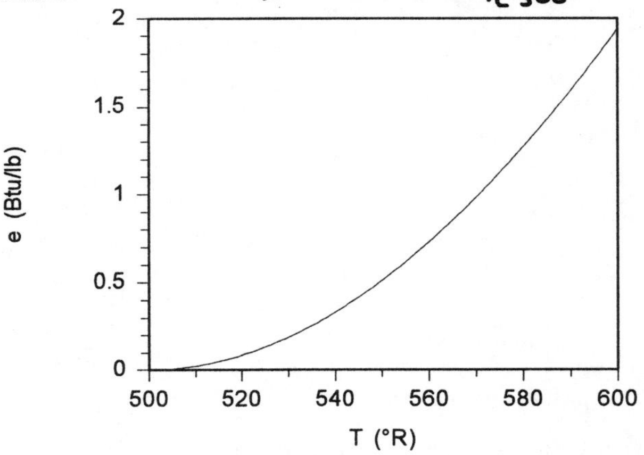

1. Alternatively, $(u - u_0)$ and $(s - s_0)$ can be evaluated using data from Table A-23E.

PROBLEM 7.8

KNOWN: A vessel contains a known amount of air at pressure p and 200°F.

FIND: Plot the specific exergy versus p ranging from 0.5 to 2 atm.

SCHEMATIC & GIVEN DATA:

$T = 200°F$

$0.5 \le p \le 2$ atm

$T_0 = 60°F$

$p_0 = 1$ atm

ASSUMPTIONS: (1) The air is a closed system, at rest and zero elevation relative to the reference environment which is at $T_0 = 60°F$, $p_0 = 1$ atm. (2) The air is modeled as an ideal gas.

ANALYSIS: Equation 7.9 reduces to $e = (u - u_0) + p_0(v - v_0) - T_0(s - s_0)$ (1)

SAMPLE CALCULATION: This calculation will be done using gas table data. That is, $u - u_0 = u(T) - u(T_0)$, $s - s_0 = s°(T) - s°(T_0) - R \ln p/p_0$. With the ideal gas equation of state

$$e = u(T) - u(T_0) + p_0\left(\frac{RT}{p} - \frac{RT_0}{p_0}\right) - T_0\left[s°(T) - s°(T_0) - R \ln p/p_0\right]$$

Using data from Table A-22E, for the case $T = 200°F$, $p = 0.5$ atm

$$u(T) - u(T_0) = 112.67 - 88.62 = 24.05 \text{ Btu/lb}$$

$$p_0 R\left(\frac{T}{p} - \frac{T_0}{p_0}\right) = (1 \text{ atm})\left(\frac{1.986}{28.97} \frac{\text{Btu}}{\text{lb}\cdot°R}\right)\left(\frac{660°R}{0.5 \text{ atm}} - \frac{520°R}{1 \text{ atm}}\right) = 54.843 \text{ Btu/lb}$$

$$T_0\left[s°(T) - s°(T_0) - R \ln p/p_0\right] = (520°R)\left[(0.64902 - 0.59172) - \left(\frac{1.986}{28.97}\right) \ln\left(\frac{0.5}{1}\right)\right] \frac{\text{Btu}}{\text{lb}\cdot°R}$$

$$= 54.505 \text{ Btu/lb}$$

Finally

$$e = 24.05 + 54.843 - 54.505 = 24.388 \text{ Btu/lb}$$

Using *IT* to generate the data required, the evaluations are based on Eq. (1) above and internal *IT* functions for u, v, and s, as follows:

IT Code

```
T = 100  // °F
p = 0.5  // atm
To = 60 // °F
po = 1  // atm

e = (u - uo) + po * (v - vo) * (14.696*144/778) - (To + 460) * (s - so)
u = u_T("Air", T)
uo = u_T("Air", To)
v = v_TP("Air",T,p)
vo = v_TP("Air",To,po)
s = s_TP("Air", T, p)
so = s_TP("Air", To, po)
```

IT Result for p = 0.5 atm, T = 200°F

(1) e = 24.37 Btu/lb

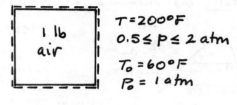

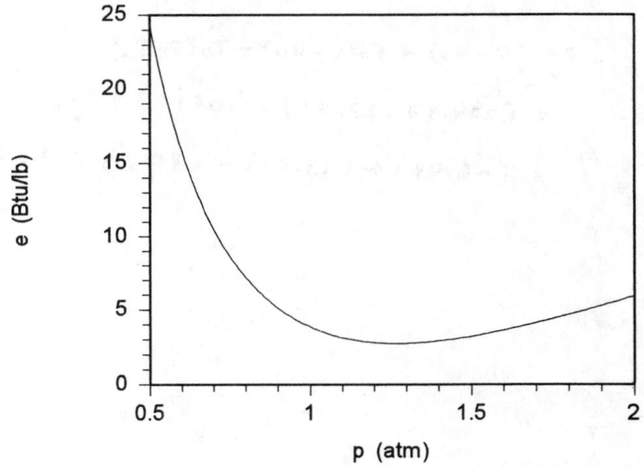

1. The *IT* result is in close agreement with the sample calculation based on Table A-22E, as expected.

PROBLEM 7.9

KNOWN: A state is specified for each of three substances.

FIND: Determine the specific exergy

ASSUMPTIONS: 1. Ignore the effects of motion and gravity. 2. $T_0 = 0°C$, $P_0 = 1$ bar.

ANALYSIS: (a) Ammonia at 0.6 bar, −10°C.

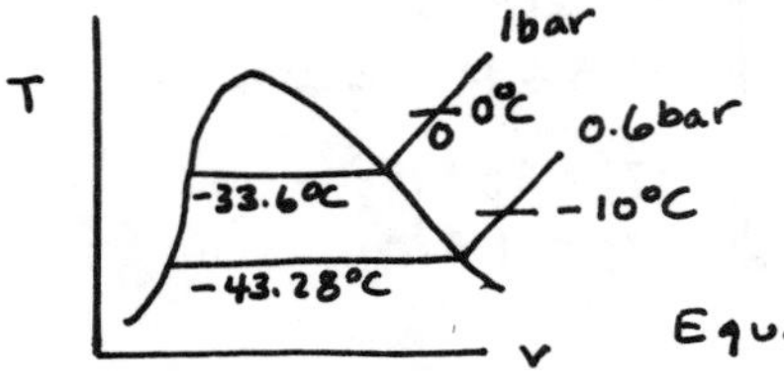

Table A-15

$u_0 = 1341$ kJ/kg $u = 1327.37$

$v_0 = 1.3136$ m³/kg $v = 2.1188$

$s_0 = 6.1281$ kJ/kg·K $s = 6.3077$

Equation 7.9 reduces with assumption 1 to give

$$e = (u - u_0) + P_0(v - v_0) - T_0(s - s_0)$$

$$= (1327.37 - 1341)\frac{kJ}{kg} + \left(10^5 \frac{N}{m^2}\right)\left[2.1188 - 1.3136\right]\frac{m^3}{kg}\left|\frac{1\ kJ}{10^3\ N\cdot m}\right| - 273K\left[6.3077 - 6.1281\right]\frac{kJ}{kg\cdot K}$$

$$= \left[(-13.63) + (80.52) - (49.03)\right]\frac{kJ}{kg} = 17.86\ kJ/kg$$ ←

(b) R22 at 0.6 bar, −10°C

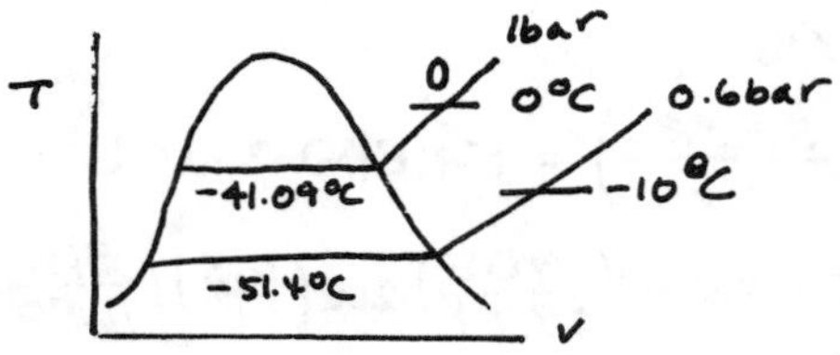

Table A-9

$u_0 = 232.36$ kJ/kg $u = 227.65$

$v_0 = 0.25747$ m³/kg $v = 0.41608$

$s_0 = 1.1035$ kJ/kg·K $s = 1.1314$

$$e = (u - u_0) + P_0(v - v_0) - T_0(s - s_0)$$

$$= (227.65 - 232.36) + (10^5)\left[0.41608 - 0.25747\right]\left|\frac{1}{10^3}\right| - 273(1.1314 - 1.1035)$$

$$= (-4.71) + (15.86) - (7.62) = 3.53\ kJ/kg$$ ←

(c) R134a at 0.6 bar, −10°C

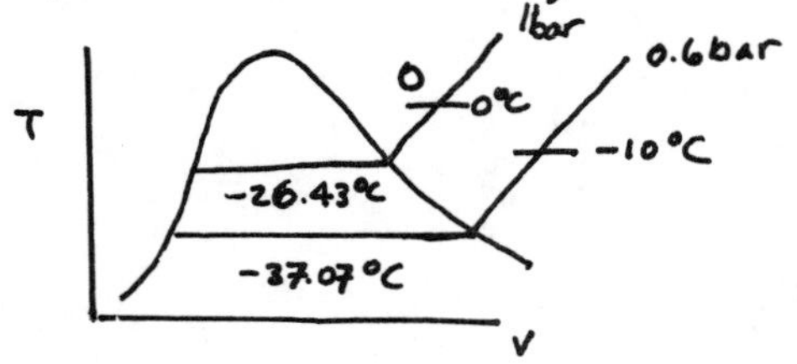

Table A-12

$u_0 = 231.41$ kJ/kg $u = 224.97$

$v_0 = 0.21587$ m³/kg $v = 0.34992$

$s_0 = 1.0227$ m³/kg $s = 1.0371$

$$e = (u - u_0) + P_0(v - v_0) - T_0(s - s_0)$$

$$= \left[224.97 - 231.41\right] + 10^5\left[0.34992 - 0.21587\right]\left|\frac{1}{10^3}\right| - 273(1.0371 - 1.0227)$$

$$= (-6.44) + (13.41) - (3.93) = 3.04\ kJ/kg$$ ←

PROBLEM 7.10

KNOWN: A two-phase solid-vapor mixture of H_2O is at a known temperature. The mass of each phase is the same.

FIND: Determine the specific exergy.

SCHEMATIC & GIVEN DATA:

$$x = \frac{m_g}{m_g + m_i} = 0.5$$

$T = -10°C$

$T_0 = 20°C$, $p_0 = 1\ atm$

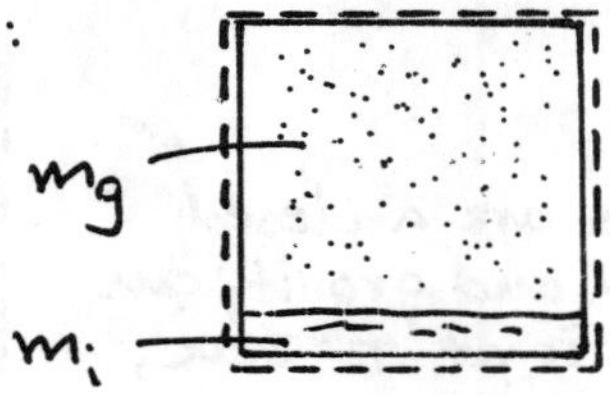

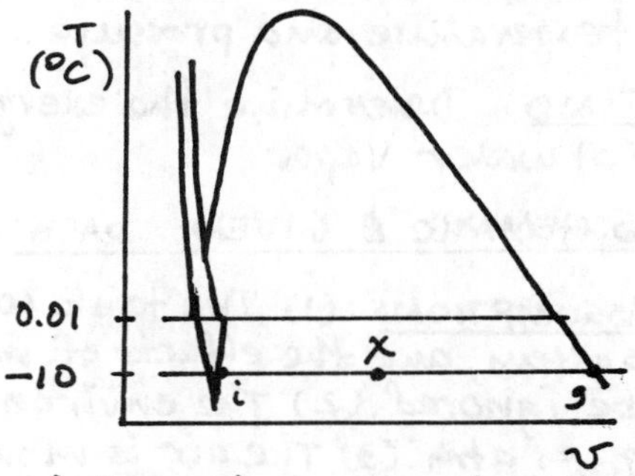

ASSUMPTIONS: (1) The two-phase solid-vapor mixture is a closed system at rest and zero elevation relative to the reference environment at $T_0 = 20°C$, $p_0 = 1\ atm$.

ANALYSIS: Equation 7.9 reduces to

$$e = (u - u_0) + p_0(v - v_0) - T_0(s - s_0) \qquad (1)$$

Using $u_0 \approx u_f(T_0)$, $v_0 \approx v_f(T_0)$, $s_0 \approx s_f(T_0)$ and data from Table A-2

$$u_0 = 83.95\ kJ/kg,\quad v_0 = 1.0018 \times 10^{-3}\ m^3/kg,\quad s_0 = 0.2966\ kJ/kg \cdot K$$

Defining "quality" x as above, and using data from Table A-6

$$u = u_i + x(u_g - u_i) = -354.09 + (.5)(2715.5) = 1003.66\ kJ/kg$$

$$v = v_i + x(v_g - v_i) = 1.0891 \times 10^{-3}\ \frac{m^3}{kg} + (0.5)(466.7 - 1.0891 \times 10^{-3})\ \frac{m^3}{kg}$$

$$= 233.35\ m^3/kg$$

$$s = s_i + x(s_g - s_i) = -1.299 + (0.5)(10.781) = 4.0915\ kJ/kg \cdot K$$

Inserting values in (1)

$$e = (1003.66 - 83.95)\frac{kJ}{kg} + \left(101.325\ \frac{kN}{m^2}\right)(233.35 - 1.0018 \times 10^{-3})\frac{m^3}{kg}\left|\frac{1\ kJ}{1\ kN \cdot m}\right| - (273 - 10)\,K\,(4.0915 - 0.2966)\frac{kJ}{kg \cdot K}$$

① $$= 919.71 + 23644 - 998.1 = 23{,}566\ kJ/kg \qquad e$$

---

1. The large value indicates that a two-phase solid-vapor mixture of H_2O has a great deal of exergy relative to the dead state of 20°C, 1 atm.

PROBLEM 7.11

KNOWN: The contents of a storage tank of known volume are at a given temperature and pressure.

FIND: Determine the exergy if the contents are (a) air as an ideal gas, (b) water vapor.

SCHEMATIC & GIVEN DATA:

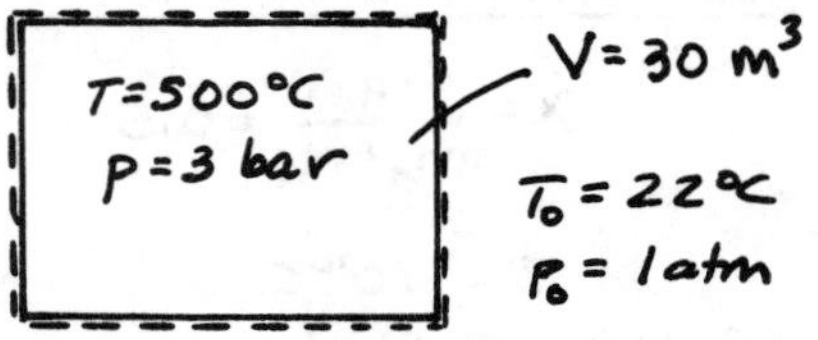

ASSUMPTIONS: (1) The tank contents are a closed system, and the effects of motion and gravity can be ignored. (2) The environment is at $T_0 = 22°C$, $p_0 = 1$ atm. (3) The air is modeled as an ideal gas.

ANALYSIS: With Eq. 7.9 and assumption (1)

$$E = m\left[(u-u_0) + p_0(v - v_0) - T_0(s-s_0)\right] \quad (1)$$

(a) Introducing ideal gas relations, Eq. (1) becomes

$$E = m\left[u(T) - u(T_0) + RT\left(\frac{p_0}{p} - \frac{T_0}{T}\right) - T_0\left(s^\circ(T) - s^\circ(T_0) - R\ln p/p_0\right)\right]$$

Using the ideal gas equation of state

$$m = \frac{pV}{RT} = \frac{(3\text{ bar})(30\text{ m}^3)}{\left(\frac{8.314}{28.97}\frac{\text{kJ}}{\text{kg}\cdot\text{K}}\right)(773\text{K})}\left|\frac{10^5\text{ N/m}^2}{1\text{ bar}}\right|\left|\frac{1\text{ kJ}}{10^3\text{ N}\cdot\text{m}}\right| = 40.57\text{ kg}$$

Now, with data from Table A-22

$$E = (40.57\text{kg})\left[(570.49-210.49)\frac{\text{kJ}}{\text{kg}} + \left(\frac{8.314}{28.97}\frac{\text{kJ}}{\text{kg}\cdot\text{K}}\right)(773\text{K})\left(\frac{1.01325}{3} - \frac{295}{773}\right) - (295\text{K})\left\{(2.6802-1.68515) - \frac{8.314}{28.97}\ln\left(\frac{3}{1.01325}\right)\right\}\right]$$

$= 6030$ kJ ← (a)

(b) For water, $u_0 \approx u_f(T_0)$, $v_0 \approx v_f(T_0)$, $s_0 \approx s_f(T_0)$. Thus, From Table A-2

$u_0 = 92.32$ kJ/kg, $v_0 = 1.0022\times10^{-3}$ m³/kg, $s_0 = 0.3251$ kJ/kg·K

Table A-4 gives

$u = 3130.0$ kJ/kg, $v = 1.187$ m³/kg, $s = 8.3251$ kJ/kg·K

The mass is: $m = V/v = (30\text{ m}^3)/(1.187\text{ m}^3/\text{kg}) = 25.27$ kg. Inserting values in Eq. (1)

$$E = (25.27\text{ kg})\left[(3130.0-92.32)\frac{\text{kJ}}{\text{kg}} + (1.01325\times10^5\frac{\text{N}}{\text{m}^2})(1.187-1.0022\times10^{-3})\frac{\text{m}^3}{\text{kg}}\left|\frac{1\text{ kg}\cdot\text{m/s}^2}{1\text{N}}\right|\left|\frac{1\text{ kJ}}{10^3\text{ N}\cdot\text{m}}\right| - (295\text{ K})(8.3251-0.3251)\frac{\text{kJ}}{\text{kg}\cdot\text{K}}\right]$$

$= 20{,}162$ kJ ← (b)

PROBLEM 7.12

KNOWN: Air as an ideal gas is stored in a closed vessel at pressure p, Volume V, and temperature T_0.

FIND: (a) Obtain an expression for the exergy of the air in a specified form. (b) Use the result of (a) to plot V versus p/p_0.

SCHEMATIC & GIVEN DATA:

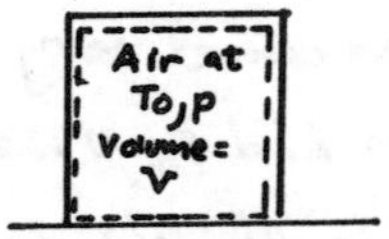

ASSUMPTIONS: (1) The effects of motion and gravity can be ignored. (2) Air is modeled as an ideal gas.

ANALYSIS: (a) Ignoring motion and gravity, Eq. 7.2 takes the form

$$E = \cancel{(U - U_0)}^{0} + p_0(V - V_0) - T_0(S - S_0)$$

The first term drops out because the internal of an ideal gas depends on temperature only and $T = T_0$. Also,

$$S - S_0 = m\left[\cancel{\int_{T_0}^{T} \frac{c_p}{T} dT}^{0} - R \ln \frac{p}{p_0}\right] = -mR \ln \frac{p}{p_0}$$

Accordingly

$$E = p_0 V \left[1 - \frac{V_0}{V} + \frac{mRT_0}{p_0 V} \ln \frac{p}{p_0}\right]$$

Since $T = T_0$, the ideal gas equation of state: $p = mRT_0/V$ gives $V_0/V = p/p_0$, so

$$E = p_0 V \left[1 - \frac{p}{p_0} + \frac{p}{p_0} \ln \frac{p}{p_0}\right] \longleftarrow$$

(b) using the above expression

$$V = \frac{E}{p_0\left[1 - \frac{p}{p_0} + \frac{p}{p_0} \ln \frac{p}{p_0}\right]} = \frac{(1\ \text{kW}\cdot\text{h})\left|\frac{3600\ \text{kJ}}{1\ \text{kW}\cdot\text{h}}\right|\left|\frac{10^3\ \text{N}\cdot\text{m}}{\text{kJ}}\right|}{10^5 \frac{\text{N}}{\text{m}^2}\left[1 - \frac{p}{p_0} + \frac{p}{p_0} \ln \frac{p}{p_0}\right]}$$

$$= \frac{36\ \text{m}^3}{\left[1 - \frac{p}{p_0} + \frac{p}{p_0} \ln \frac{p}{p_0}\right]}$$

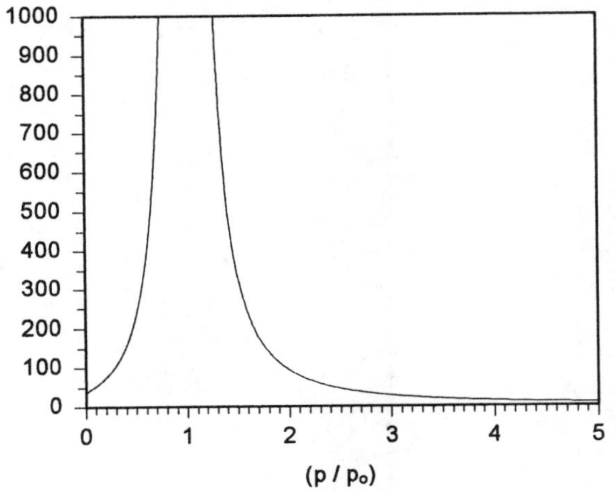

$\lim_{p/p_0 \to 0} V \to 36\ ft^3$

$\lim_{p/p_0 \to 1} V \to \infty$

$\lim_{p/p_0 \to \infty} V \to 0$

PROBLEM 7.12 (Contd.)

For a fixed exergy value, the tank size depends on the pressure:

- As $p/p_0 \to \infty$, the required tank volume would become very small owing to the large difference between p and p_0. That is, with a large pressure difference not much air would be needed for a specified exergy value.

- As $p/p_0 \to 1$, the required tank volume would become very large owing to the small difference between p and p_0. That is, with a small pressure difference a large quantity of air would be needed for a specified exergy value.

- As $p/p_0 \to 0$, the tank volume approaches a constant as the tank holds less and less air. In the limit the tank would be evacuated, illustrating that an evacuated space has a nonzero value for exergy. We can think of this as the minimum theoretical work that would be required to create the evacuated space. See the 4th bulleted item in the discussion of Sec. 7.2.4.

PROBLEM 7.13

KNOWN: An ideal gas is stored in a closed vessel at p, T.

FIND: (a) If $T=T_0$, derive $e = f(p, p_0, T_0, R)$

(b) If $p=p_0$, derive $e = f(T, T_0, c_p)$

ASSUMPTIONS: (1) The gas obeys the ideal gas model. In part (b), c_p is constant as well. (2) The effects of motion and gravity can be ignored.

ANALYSIS:

(a) With assumptions (1), (2), Eq. 7.9 becomes

$$e = \cancelto{0}{(u-u_0)} + p_0(v-v_0) - T_0(s-s_0)$$

The first term vanishes because u of an ideal gas depends only on temperature, and $T = T_0$. Also,

$$s - s_0 = \cancelto{0}{(s^\circ(T) - s^\circ(T_0))} - R \ln \frac{p}{p_0}$$

Collecting results

$$e = p_0(v-v_0) + RT_0 \ln \frac{p}{p_0}$$

With $v = RT_0/p$ and $v_0 = RT_0/p_0$

$$e = p_0\left[\frac{RT_0}{p} - \frac{RT_0}{p_0}\right] + RT_0 \ln \frac{p}{p_0}$$

$$= RT_0\left[\frac{p_0}{p} - 1 + \ln \frac{p}{p_0}\right] \longleftarrow$$

(b) With assumption (2), Eq. 7.9 becomes

$$e = u - u_0 + p_0(v-v_0) - T_0(s-s_0)$$

Introducing ideal gas relations, with $p = p_0$

$$e = u(T) - u(T_0) + p_0\left[\frac{RT}{p_0} - \frac{RT_0}{p_0}\right] - T_0\left[\int_{T_0}^{T} \frac{c_p}{T}dT - R \cancelto{0}{\ln \frac{p_0}{p_0}}\right]$$

If c_p is constant, so is c_v and the two differ by the gas constant R. Thus

$$e = c_v[T - T_0] + R[T - T_0] - T_0 c_p \ln \frac{T}{T_0}$$

$$= \underbrace{(c_v + R)}_{c_p}[T - T_0] - T_0 c_p \ln \frac{T}{T_0}$$

$$= c_p T_0\left[\frac{T}{T_0} - 1 - \ln \frac{T}{T_0}\right] \longleftarrow$$

PROBLEM 7.14

KNOWN: An evacuated container of known volume is under consideration.

FIND: For the space inside the tank, determine the exergy.

SCHEMATIC & GIVEN DATA:

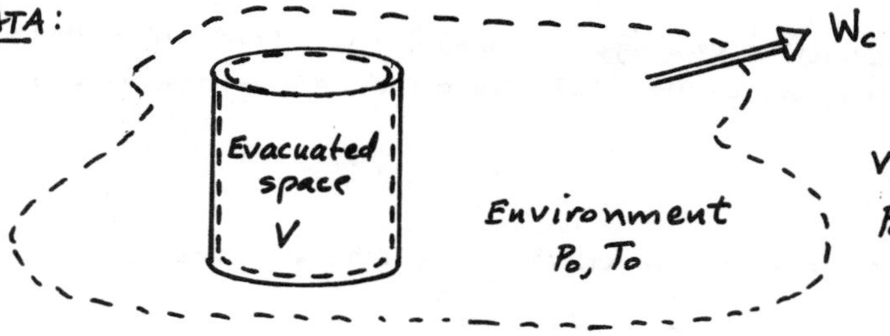

$V = 10\ ft^3$

$p_0 = 14.7\ lbf/in.^2$

ASSUMPTIONS: (1) A combined system is considered consisting of the evacuated space (a closed system) and the environment. (2) The volume of the combined system is constant. (3) For the combined system, $Q = 0$. (4) For the environment, p_0 and T_0 are constant. (5) The effects of gravity and motion are ignored.

ANALYSIS: Consider a process where the evacuated space collapses and work is developed by the combined system. An energy balance for the combined system reduces with assumption 3 to give

$$W_c = -\Delta E_c$$

Since $\Delta U = 0$ for the evacuated space, $\Delta E_c = \Delta U_e$. With Eq. 7.1

$$W_c = -[T_0 \Delta S_e - p_0 \Delta V_e]$$

By assumption 2, $\Delta V_e = -\Delta V = -(\cancel{V_0}^{\,0} - V) = V$. Further, an entropy balance for the combined system gives

$$\Delta S_c = \sigma_c \Rightarrow \cancel{\Delta S}^{\,0} + \Delta S_e = \sigma_c$$

Collecting results

$$W_c = p_0 V - T_0 \sigma_c$$

Since $\sigma_c \geq 0$, the maximum theoretical work, or exergy, is obtained when $\sigma_c = 0$. Thus

$$E = p_0 V \qquad (1)$$

Inserting values

$$E = \left(14.7 \frac{lbf}{in.^2}\right)(10\ ft^3) \left|\frac{144\ in.^2}{1\ ft^2}\right| \left|\frac{1\ Btu}{778\ ft \cdot lbf}\right| = 27.21\ Btu \longleftarrow E$$

PROBLEM 7.15

KNOWN: Equal molar amounts of CO_2 and He are at the same T, P.

FIND: Determine which gas has the greater exergy value, $\bar{e}$.

ASSUMPTIONS: (1) Each gas obeys the ideal gas model with constant $\bar{c}_v$. (2) There are no significant effects of motion and gravity.

ANALYSIS: With assumption (2), Eq. 7.9 reduces to give on a molar basis

① $$\bar{e} = \bar{u} - \bar{u}_0 + p_0(\bar{v} - \bar{v}_0) - T_0(\bar{s} - \bar{s}_0)$$

Then, with assumption (1)

$$\bar{e} = \bar{c}_v(T - T_0) + \bar{R}T_0\left[\frac{T}{T_0}\cdot\frac{p_0}{p} - 1\right] - T_0\left[\bar{c}_p \ln\frac{T}{T_0} - \bar{R}\ln\frac{p}{p_0}\right]$$

$$= \bar{c}_v(T - T_0) - T_0\bar{c}_p \ln\frac{T}{T_0} + \bar{R}T_0\left\{\left[\frac{T}{T_0}\cdot\frac{p_0}{p} - 1\right] + \ln\frac{p}{p_0}\right\}$$

Applying this to each of CO_2 and He, and subtracting the resulting equations gives

$$\bar{e}_{CO_2} - \bar{e}_{He} = \underbrace{\left[\bar{c}_{v,CO_2} - \bar{c}_{v,He}\right]}_{\equiv(\bar{c}_{p,CO_2} - \bar{c}_{p,He})}(T - T_0) - \left[\bar{c}_{p,CO_2} - \bar{c}_{p,He}\right]T_0 \ln\frac{T}{T_0}$$

$$= \left[\bar{c}_{p,CO_2} - \bar{c}_{p,He}\right]\left[T - T_0 - T_0\ln\frac{T}{T_0}\right]$$

By inspection of Figure 3.13, $\bar{c}_{p,CO_2} > \bar{c}_{p,He}$, giving

② $$\bar{e}_{CO_2} > \bar{e}_{He}$$ ◄

---

1. Here only the thermomechanical component is considered. The chemical contribution of Section 13.6 is not included in the present discussion.
2. But note that the molecular weights are much different: $M_{CO_2} = 44.01$, $M_{He} = 4.003$.

PROBLEM 7.16

KNOWN: NH_3 vapor, initially at T_0, p_0, is cooled at fixed volume to a specified temperature.

FIND: For the ammonia, evaluate Q/m and Δe.

SCHEMATIC & GIVEN DATA:

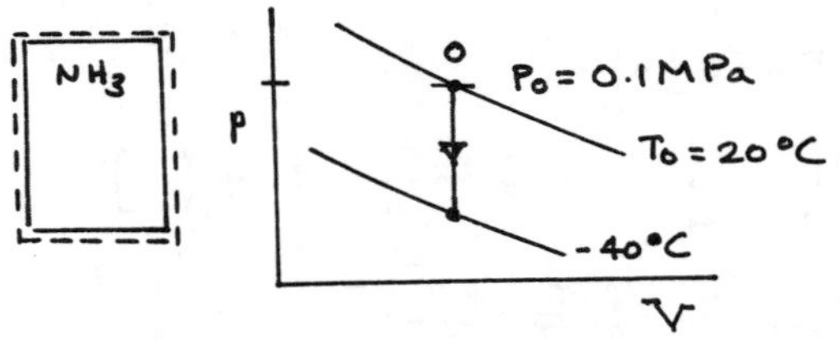

ASSUMPTIONS: (1) The system is the NH_3 vapor, as shown in the figure. (2) For the system, $W=0$ and there are no significant kinetic/potential energy effects. (3) For the exergy reference environment, $p_0 = 0.1\,MPa$, $T_0 = 20°C$

ANALYSIS: An energy balance reduces to read $Q/m = u_2 - u_1$. With data from Table A-15 $v_1 = 1.4153\ m^3/kg$, $u_1 = 1374.27\ kJ/kg$, $s_1 = 6.2816\ kJ/kg\cdot K$. Then with $v_2 = v_1$ and data from Table A-13

$$x_2 = \frac{v_2 - v_f}{v_g - v_f} = \frac{1.4153 - (1.4493/10^3)}{1.5524 - (1.4493/10^3)} = 0.912$$

and

$$u_2 = u_f + x_2 u_{fg} = -0.10 + 0.912(1277.3) = 1164.8\ kJ/kg$$

$$s_2 = s_f + x_2 s_{fg} = 0 + 0.912(5.9557) = 5.4316\ kJ/kg\cdot K$$

Then

$$\frac{Q}{m} = 1164.8 - 1374.27 = -209.5\ kJ/kg \longleftarrow Q/m$$

To find Δe, use Eq. 7.9 to obtain

$$\Delta e = \Delta u + p_0 \cancel{\Delta v}^{\,0} - T_0 \Delta s$$

$$= -209.5\ \frac{kJ}{kg} - (293\,K)(5.4316 - 6.2816)\frac{kJ}{kg\cdot K}$$

$$= +39.6\ kJ/kg \longleftarrow \Delta a$$

COMMENT: Here, despite the removal of energy by heat transfer, the exergy increases. The exergy increases because the system is brought <u>from</u> the dead state <u>to</u> a state where T, p differ from T_0, p_0. The exergy at <u>any</u> state other than the dead state is strictly positive. For further comment, see the discussion of Figure 7.4.

PROBLEM 7.17

KNOWN: Two kg of water undergo a process between specified states.

FIND: Determine the exergy at the initial and final states and the change in exergy.

SCHEMATIC & GIVEN DATA:

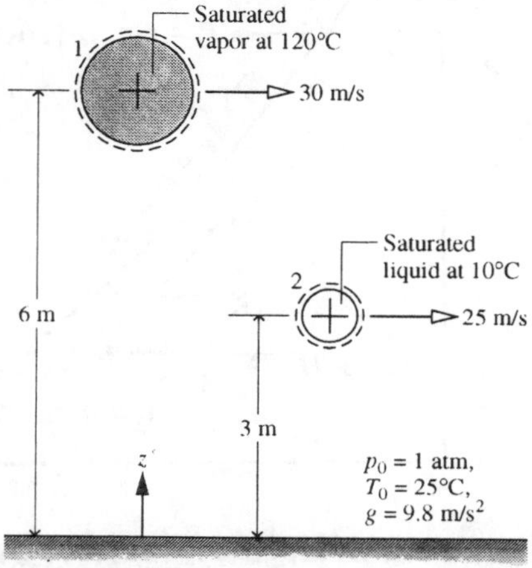

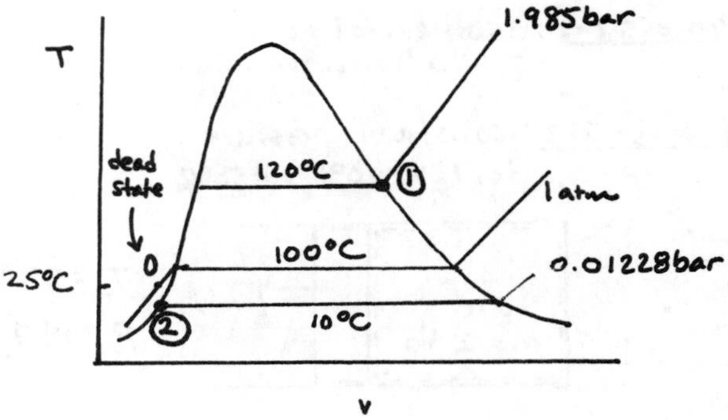

ASSUMPTIONS: 1. The water is a closed system at equilibrium states initially and finally. 2. The velocities and elevations are measured relative to the environment. 3. $T_0 = 25°C$, $p_0 = 1 atm$, $v_0 \sim v_f(T_0)$, $u_0 \sim u_f(T_0)$, $s_0 \sim s_f(T_0)$.

ANALYSIS: The exergy at the initial and final states can be calculated using Eq. 7.9:

① $$E = m\left[(u-u_0) + p_0(v-v_0) - T_0(s-s_0) + \frac{V^2}{2} + gz\right]$$

(a) For saturated vapor at 120°C, Table A-2 gives $v = 0.8919\ m^3/kg$, $u = 2529.3\ kJ/kg$, $s = 7.1296\ kJ/kg\cdot K$. Table A-2 gives $v_0 = 1.0029\times10^{-3}\ m^3/kg$, $u_0 = 104.88\ kJ/kg$, $s_0 = 0.3674\ kJ/kg\cdot K$. Thus,

$$E_1 = (2\ kg)\left\{(2529.3 - 104.88)\frac{kJ}{kg} + \left(1.013\times10^5\frac{N}{m^2}\right)\left[0.8919 - \frac{1.0029}{10^3}\right]\frac{m^3}{kg}\left|\frac{1\ kJ}{10^3\ N\cdot m}\right| - (298\ K)[7.1296 - 0.3674]\frac{kJ}{kg\cdot K} + \left[\frac{(30\ m/s)^2}{2} + (9.8\ m/s^2)(6\ m)\right]\left|\frac{1\ N}{1\ kg\cdot m/s^2}\right|\left|\frac{1\ kJ}{10^3\ N\cdot m}\right|\right\}$$

$$= (2)[2424.42 + 90.25 - 2015.14 + 0.45 + 0.06]\ kJ = 1000.1\ kJ \longleftarrow$$

② (b) For saturated liquid at 10°C, $v = 1.0004\times10^{-3}\ m^3/kg$, $u = 42\ kJ/kg$, $s = 0.151\ kJ/kg\cdot K$. We get $E_2 = 3.9\ kJ$. ⟵

③ (c) The change in exergy is $\Delta E = E_2 - E_1 = 3.9 - 1000.1 = -996.2\ kJ$ ⟵

---

1. The kinetic and potential energies measured relative to the environment contribute their full magnitudes to the value of exergy, for in principle each could be completely converted to work were the system brought to rest at zero elevation relative to the environment.

2. Exergy is a measure of the departure of the state of the system from that of the environment. At all states, $E \geq 0$. This applies when $T > T_0$, $p > p_0$, as in part (a), and when $T < T_0$, $p < p_0$, as in part (b).

3. Alternatively, Eq. 7.10 can be used. This requires dead state property values only for T_0, p_0. In parts (a), (b) u_0, v_0 and s_0 are also required; so more computation is needed with that approach.

PROBLEM 7.18

KNOWN: Air undergoes two specified processes in series.

FIND: (a) Represent each process on a p-v diagram and indicate the dead state. (b) Determine the change in exergy for each process.

SCHEMATIC & GIVEN DATA:

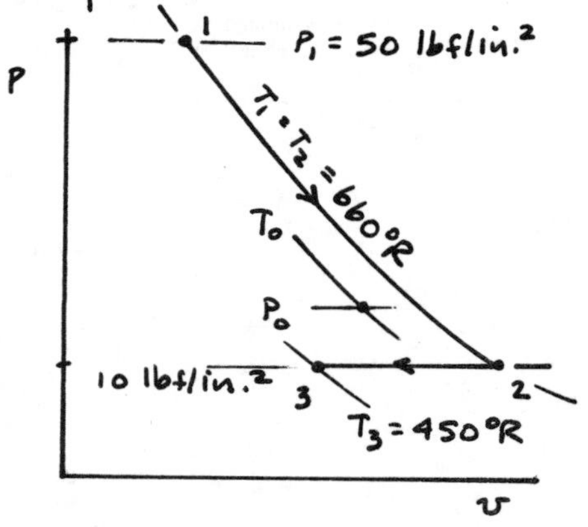

Process 1-2: isothermal to $p_2 = 10$ lbf/in.²

Process 2-3: constant pressure to $T_3 = -10°F = 450°R$

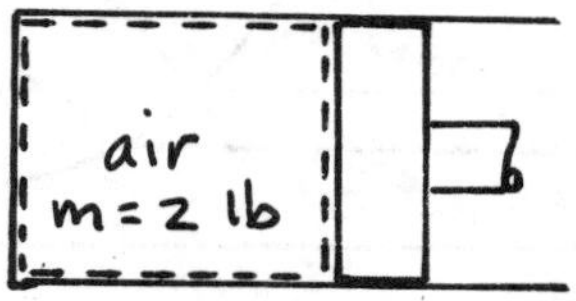

$T_0 = 77°F = 537°R$
$p_0 = 14.7$ lbf/in.²

ASSUMPTIONS: (1) The air is the closed system. (2) The effects of motion and gravity can be ignored. (3) The air is modeled as an ideal gas. (4) For the environment, $T_0 = 77°F$, $p_0 = 14.7$ lbf/in.²

ANALYSIS: (a) From the given information, $p_1 = 50$ lbf/in.², $T_1 = 660°R$. Thus

$$v_1 = \frac{RT_1}{p_1} = \frac{\left(\frac{1545}{28.97}\frac{\text{ft}\cdot\text{lbf}}{\text{lb}\cdot°R}\right)(660°R)}{(50\ \text{lbf/in.}^2)}\left|\frac{1\ \text{ft}^2}{144\ \text{in.}^2}\right| = 4.889\ \text{ft}^3/\text{lb}$$

Similarly, $p_2 = 10$ lbf/in.², $T_2 = T_1 = 660°R \Rightarrow v_2 = 24.44$ ft³/lb and $p_3 = p_2 = 10$ lbf/in.² and $T_3 = 450°R \Rightarrow v_3 = 16.67$ ft³/lb.

(b) Using Eq. 7.10 and ideal gas relations

$$E_2 - E_1 = m\left[\cancel{u(T_2) - u(T_1)}^{\,0} + p_0(v_2 - v_1) - T_0\left(\cancel{s°(T_2) - s°(T_1)}^{\,0} - R\ln p_2/p_1\right)\right]$$

$$E_3 - E_2 = m\left[u(T_3) - u(T_2) + p_0(v_3 - v_2) - T_0\left(s°(T_3) - s°(T_2) - \cancel{R\ln p_3/p_2}^{\,0}\right)\right]$$

with data from Table A-22E as needed

$$E_2 - E_1 = (2\ \text{lb})\left[\left(14.7\frac{\text{lbf}}{\text{in}^2}\right)(24.44 - 4.889)\frac{\text{ft}^3}{\text{lb}}\left|\frac{144\ \text{in}^2}{1\ \text{ft}^2}\right|\left|\frac{1\ \text{Btu}}{778\ \text{ft}\cdot\text{lbf}}\right| - (537°R)\left(-\frac{1.986}{28.97}\frac{\text{Btu}}{\text{lb}\cdot°R}\ln\frac{10}{50}\right)\right]$$

$= -12.11$ Btu ← $E_2 - E_1$

$$E_3 - E_2 = (2)\left[(76.645 - 112.67) + (14.7)(16.67 - 24.44)\left|\frac{144}{778}\right| - (537)(.55704 - .64902)\right]$$

$= -15.55$ Btu ← $E_3 - E_2$

PROBLEM 7.19

KNOWN: 5 kg of air fill a rigid tank. The air is cooled from a known initial state to a specified final state.

FIND: Locate the initial, final, and dead states on a T-v diagram. Evaluate the heat transfer and the exergy change, and interpret the sign of the exergy change using the p-v diagram.

SCHEMATIC & GIVEN DATA:

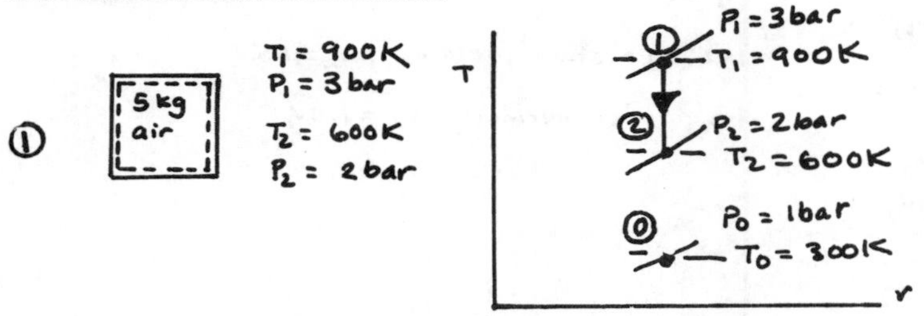

ASSUMPTIONS: 1. The air is the closed system. 2. For the system, W = 0 and the effects of motion and gravity can be ignored. 3. Air is modeled as an ideal gas. 4. $T_0 = 300K$, $P_0 = 1$ bar.

ANALYSIS: (a) Since volume and mass are fixed, the air undergoes a constant specific volume cooling process, as shown in the T-v diagram.

(b) Reducing an energy balance and using data from Table A-22

$$m(u_2-u_1) = Q - \cancel{W}^0 \Rightarrow Q = m(u_2-u_1) = (5\text{kg})[434.78 - 674.58]\frac{kJ}{kg} = -1199\text{ kJ} \leftarrow$$

(c) Using Eq. 7.10, the change in exergy is

$$E_2 - E_1 = m[(u_2-u_1) + \underbrace{p_0(v_2-v_1)}_{=0 \text{ since } v_2 = v_1} - T_0(s_2-s_1)] = m[(u_2-u_1) - T_0(s_2-s_1)]$$

where the kinetic/potential energy terms have been dropped by assumption 2. With Eq. 6.21a, this becomes

$$E_2 - E_1 = m[(u_2-u_1) - T_0(s^\circ(T_2) - s^\circ(T_1) - R\ln\frac{P_2}{P_1})]$$

$$= (5\text{kg})[(434.78 - 674.58)\frac{kJ}{kg} - 300K(2.40902 - 2.84856 - \frac{8.314}{28.97}\ln\frac{2}{3})\frac{kJ}{kg\cdot K}]$$

$$= -1683.8\text{ kJ}$$

The T-v diagram shows that the state of the system is brought "closer" to the dead state as the air is cooled. Since the value of exergy is a measure of the departure of the state of the system from the state of the environment, exergy decreases in the process from 1 to 2.

---

1. Since $pv = RT$, $\frac{v}{R} = \frac{T}{p}$. Then, using given data we get

$$\frac{v_1}{R} = \frac{T_1}{P_1} = \frac{900K}{3\text{bar}} = 3, \quad \frac{v_2}{R} = \frac{T_2}{P_2} = \frac{600K}{2\text{ bar}} = 3, \quad \frac{v_0}{R} = \frac{T_0}{P_0} = \frac{300K}{1\text{ bar}} = 3$$

Thus, states 1, 2, and 0 fall as shown on the T-v diagram.

PROBLEM 7.20

KNOWN: One lb of steam initially at 200 lbf/in² and 500°F undergoes two different processes.

FIND: For each process determine the change in exergy.

SCHEMATIC & GIVEN DATA:

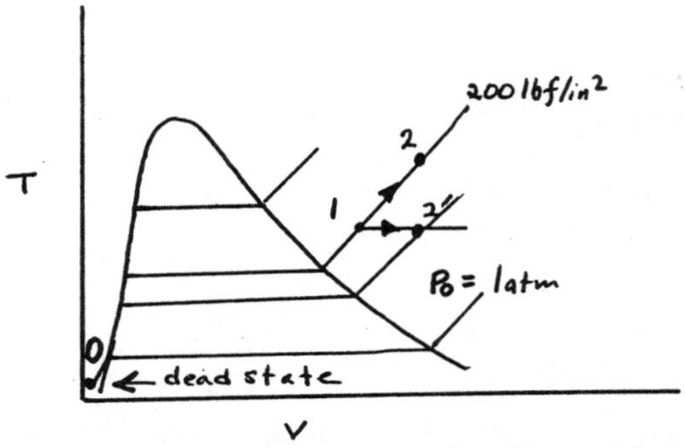

$T_1 = 500°F$

1-2: constant pressure, $V_2 = 1.5V_1$

1-2': isothermal, $V_2 = 1.5V_1$

ASSUMPTIONS: (1) The system is the 1 lb of steam. (2) For the environment, $T_0 = 60°F$, $p_0 = 1atm$. (3) Ignore the effects of motion and gravity.

ANALYSIS: Equation 7.10 reduces to give $E_2 - E_1 = m[u_2 - u_1 + p_0(V_2 - V_1) - T_0(s_2 - s_1)]$. From Table A-4E $u_1 = 1168$ Btu/lb, $v_1 = 2.724$ ft³/lb, $s_1 = 1.6239$ Btu/lb·°R. For each process $v_2 = 1.5v_1 = 4.086$ ft³/lb.

(a) Constant pressure process: Interpolating in Table A-4E at 200 lbf/in² with $v_2 = 4.086$ ft³/lb, $u_2 = 1339.96$ Btu/lb, $s_2 = 1.8155$ Btu/lb·°R.

$$E_2 - E_1 = (1\text{ lb})\left[(1339.96 - 1168)\frac{\text{Btu}}{\text{lb}} + \left(14.7 \times 144 \frac{\text{lbf}}{\text{ft}^2}\right)(4.086 - 2.724)\frac{\text{ft}^3}{\text{lb}}\left(\frac{\text{Btu}}{778\text{ ft}\cdot\text{lbf}}\right) - 520°R(1.8155 - 1.6239)\frac{\text{Btu}}{\text{lb}\cdot°R}\right]$$

① $$= (1\text{ lb})[171.96 + 3.71 - 99.63]\frac{\text{Btu}}{\text{lb}} = 76.04\text{ Btu} \quad \text{(a)}$$

(b) Isothermal process: Interpolating in Table A-4E at 500°F with $v_2 = 4.086$ ft³/lb, $u_2 = 1173$ Btu/lb, $s_2 = 1.6719$ Btu/lb·°R

$$E_2 - E_1 = (1)\left[(1173 - 1168) + (14.7)(144)[1.362]/778 - 520(1.6719 - 1.6239)\right]$$

② $$= (1)[5 + 3.71 - 24.96] = -16.25\text{ Btu} \quad \text{(b)}$$

---

1. Here, pressure is kept constant and temperature increases relative to T_0. Exergy increases in the process.

2. Here, temperature is kept constant and pressure decreases relative to P_0. Exergy decreases in the process.

PROBLEM 7.21

KNOWN: Data is provided for a flywheel braked to rest.

FIND: Determine the final temperature of the brake lining after braking to rest and the maximum theoretical rotational speed that could be attained by the flywheel using the energy stored in the flywheel.

SCHEMATIC & GIVEN DATA:

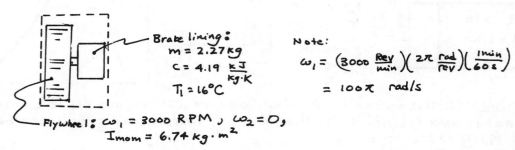

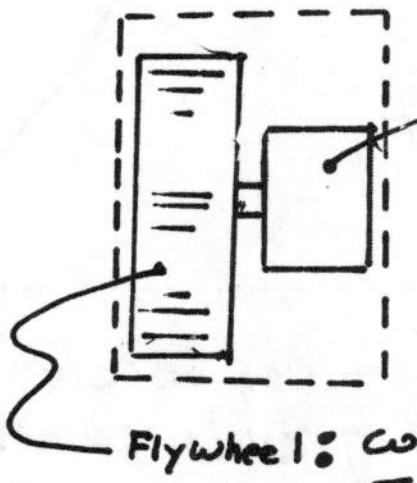

Note:

$$\omega_1 = \left(3000\ \frac{\text{Rev}}{\text{min}}\right)\left(2\pi\ \frac{\text{rad}}{\text{rev}}\right)\left(\frac{1\ \text{min}}{60\ \text{s}}\right) = 100\pi\ \text{rad/s}$$

ASSUMPTIONS: (1) For the system shown in the figure $W = Q = 0$ and $\Delta PE = 0$. (2) The brake lining is modeled as an incompressible substance with constant specific heat c. (3) For the environment, $T_0 = 16°C$.

ANALYSIS: (a) With $W = Q = \Delta PE = 0$, an energy balance reduces to read

$$\underset{\text{(brake lining)}}{\Delta U} + \underset{\text{(flywheel)}}{\Delta KE} = 0$$

or

$$mc[T_2 - T_1] + \left[0 - \tfrac{1}{2}\omega_1^2 I_{mom}\right] = 0$$

Thus

$$T_2 - T_1 = \frac{\frac{1}{2}\omega_1^2 I_{mom}}{mc} = \frac{\frac{1}{2}\left(100\pi\ \frac{\text{rad}}{\text{s}}\right)^2 (6.74\ \text{kg}\cdot\text{m}^2)\left(\frac{1\ \text{N}}{1\ \text{kg}\cdot\text{m/s}^2}\right)}{(2.27\ \text{kg})\left(4.19\ \frac{\text{kJ}}{\text{kg}\cdot\text{K}}\right)\left(\frac{10^3\ \text{N}\cdot\text{m}}{1\ \text{kJ}}\right)}$$

$$= 35\ \text{K}$$

The final temperature of the brake lininning is then 324 K (51°C). ←

(b) As witnessed by the energy balance, energy is stored in the brake lining. ① However, in determining the maximum possible rotational speed achievable, it is the exergy stored, and not the energy stored, that is significant.

At the final state, the flywheel makes no contribution to the system exergy because the flywheel is at rest. The brake lining does contribute exergy because now $T_2 \neq T_0$. The final exergy is then found using incompressible substance relations as

$$E_2 = m[(u_2 - u_0) + \cancelto{0}{p_0(v - v_0)} - T_0(s_2 - s_0)] = m[c(T_2 - T_0) - T_0 c \ln(T_2/T_0)]$$

$$= (2.27\ \text{kg})\left[4.19\ \frac{\text{kJ}}{\text{kg}\cdot\text{K}}\right]\left[35\ \text{K} - 289 \cdot \ln\left(\frac{324}{289}\right)\right] = 18.67\ \text{kJ}$$

In principle, the exergy at the final state can be used to bring the flywheel into motion as the brake lining is restored to its initial temperature. Accordingly, the maximum theoretical rotational speed, ω_{max}, must satisfy $E_2 = \frac{1}{2}\omega_{MAX}^2 I_{mom}$, which is the maximum kinetic energy achievable. Then,

$$\omega_{MAX} = \sqrt{\frac{2E_2}{I_{mom}}} = \sqrt{\frac{(2)(18.67\ \text{kJ})}{6.74\ \text{kg}\cdot\text{m}^2}\left|\frac{10^3\ \text{N}\cdot\text{m}}{\text{kJ}}\right|\left|\frac{1\ \text{kg}\cdot\text{m/s}^2}{1\ \text{N}}\right|\left|\left(\frac{60\ \text{s}}{1\ \text{min}}\right)^2\right|}\ \left|\frac{1\ \text{rev}}{2\pi\ \text{rad}}\right| = 711\ \text{RPM} \leftarrow$$

1. Initially, the brake lining contributes nothing to the system exergy because the lining is at temperature $T_1 = T_0$. The initial system exergy equals the kinetic energy of the flywheel:

$$E_1 = \tfrac{1}{2}\omega_1^2 I_{mom} = \tfrac{1}{2}\left(100\pi\ \frac{\text{rad}}{\text{s}}\right)^2 (6.74\ \text{kg}\cdot\text{m}^2)\left|\frac{1\ \text{kJ}}{10^3\ \text{N}\cdot\text{m}}\right|\left|\frac{1\ \text{N}}{1\ \text{kg}\cdot\text{m/s}^2}\right| = 332.61\ \text{kJ}.$$

PROBLEM 7.22

KNOWN: One kilogram of R-134a cools at constant pressure from an initial state to saturated liquid.

FIND: Determine the work, heat transfer, and amounts of exergy transfer accompanying heat and work.

SCHEMATIC & GIVEN DATA:

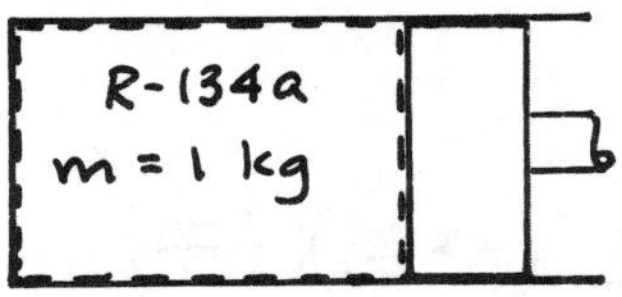

$p_1 = 6$ bar
$T_1 = 80°C$
p = const.
$x_2 = 0$

$T_0 = 20°C$
$p_0 = 1$ bar

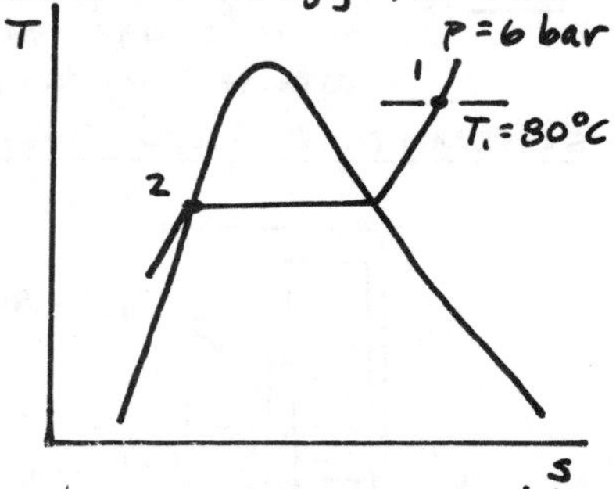

ASSUMPTIONS: (1) The R-134a is the closed system. (2) The process occurs with no internal irreversibilities. (3) The effects of motion and gravity can be ignored. (4) $T_0 = 20°C$, $p_0 = 1$ bar.

ANALYSIS: To determine the work, we use Eq. 2.17

$$W = \int_1^2 p\,dV = mp(v_2 - v_1)$$

With data from Tables A-12 and A-11, respectively, $v_1 = 0.04469\ m^3/kg$ and $v_2 = 0.8196 \times 10^{-3}\ m^3/kg$

$$W = (1\ kg)(6\ bar)(0.8196 \times 10^{-3} - 0.04469)\frac{m^3}{kg}\left|\frac{10^5\ N/m^2}{1\ bar}\right|\left|\frac{1\ kJ}{10^3\ N\cdot m}\right| = -26.32\ kJ \leftarrow W$$

The heat transfer is determined by using the energy balance, which reduces to

$$Q = m(u_2 - u_1) + W$$

From Table A-12, $u_1 = 291.86$ kJ/kg, and From Table A-11, $u_2 = 78.99$ kJ/kg. Thus

$$Q = (1\ kg)(78.99 - 291.86)\ kJ/kg + (-26.32\ kJ) = -239.2\ kJ \leftarrow Q$$

Now, the exergy transfer accompanying work, E_w, is

$$E_W = W - p_0 \Delta V = W - mp_0(v_2 - v_1)$$

$$= (-26.32\ kJ) - (1)(1)(0.8196 \times 10^{-3} - 0.04469)\left|\frac{10^5}{10^3}\right| = -21.93\ kJ \leftarrow E_W$$

The exergy transfer accompanying heat transfer, is , is

$$E_Q = \int_1^2 \left(1 - \frac{T_0}{T_b}\right)\delta Q = Q - T_0 \int_1^2 \frac{\delta Q}{T_b}$$

From an entropy balance: $m(s_2 - s_1) = \int_1^2 \frac{\delta Q}{T_b} + \sigma^{\nearrow 0}$. Thus

$$E_Q = Q - mT_0(s_2 - s_1)$$

With data from the appropriate tables

$$E_Q = (-239.2\ kJ) - (1\ kg)(293\ K)(0.2999 - 1.0938)\frac{kJ}{kg\cdot K}$$

$$= -6.587\ kJ \leftarrow E_Q$$

PROBLEM 7.23

KNOWN: 1 kg of air is heated at constant pressure with no internal irreversibilities.

FIND: Determine the work, heat transfer, and the amounts of exergy transfer accompanying work and heat transfer.

SCHEMATIC & GIVEN DATA:

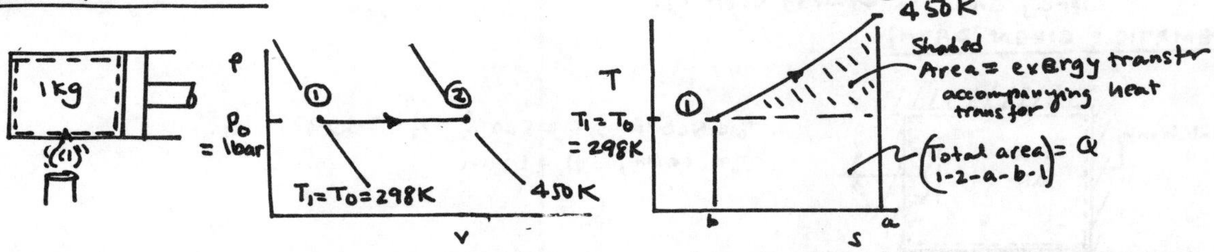

ASSUMPTIONS: 1. The closed system is 1 kg of air, as shown in the schematic. 2. The process occurs at constant pressure without internal irreversibilities. 3. Air is modeled as an ideal gas. 4. $T_0 = 298K (25°C)$, $p_0 = 1$ bar 4. The effects of motion and gravity can be ignored.

ANALYSIS: Using the ideal gas equation of state, the work is

$$W = \int_1^2 p\,dV = mp(v_2 - v_1) = mR(T_2 - T_1) = (1\,kg)\left(\frac{8.314}{28.97}\frac{kJ}{kg \cdot K}\right)(450 - 298)\,K = 43.62\,kJ \leftarrow$$

Using Eq. 7.13, the accompanying exergy transfer is

$$\begin{bmatrix}\text{Exergy transfer} \\ \text{accompanying work}\end{bmatrix} = W - p_0 \Delta V = mp(v_2 - v_1) - mp_0(v_2 - v_1)$$

$$= m(p - p_0)(v_2 - v_1) = 0, \text{ since } p = p_0. \leftarrow$$

Reducing an energy balance for the process: $\Delta U = Q - W \Rightarrow Q = \Delta U + W$.
Or, with data from Table A-22

① $$Q = m(u_2 - u_1) + W = (1\,kg)[322.62 - 212.64]\frac{kJ}{kg} + 43.62$$

$$= 153.6\,kJ \leftarrow$$

Using Eq. 7.12, the accompanying exergy transfer is

$$\begin{bmatrix}\text{Exergy transfer} \\ \text{accompanying heat}\end{bmatrix} = \int_1^2 \left[1 - \frac{T_0}{T_b}\right] \delta Q = Q - T_0 \underbrace{\int_1^2 \left(\frac{\delta Q}{T}\right)}_{= S_2 - S_1 \text{, since the process is without internal irrevs.}}$$

$$= Q - T_0[S_2 - S_1]$$

$$= Q - mT_0[s°(T_2) - s°(T_1) - R \ln p_2/p_1]$$ (with $\ln p_2/p_1 \to 0$)

$$= 153.6\,kJ - (1\,kg)(298\,K)[2.11161 - 1.69528]\frac{kJ}{kg \cdot K}$$

$$= 29.53\,kJ \leftarrow$$

---

1. Alternatively, for the constant pressure process

$$Q = m[u_2 - u_1] + mp(v_2 - v_1)$$

$$= m[h_2 - h_1] = 1\,kg(451.8 - 298.2)\frac{kJ}{kg} = 153.6\,kJ$$

PROBLEM 7.24

KNOWN: Air contained in a closed, rigid, insulated tank is stirred by a paddle wheel.

FIND: Determine the change in exergy, the transfer of exergy accompanying work, and the exergy destruction.

SCHEMATIC & GIVEN DATA:

$T_0 = 500°R$, $T_1 = 500°R$, $T_2 = 700°R$
$p_0 = 1 atm$, $p_1 = 1 atm$

ASSUMPTIONS: (1) For the system shown in the accompanying figure, $Q = 0$, and kinetic and potential energy effects can be ignored. (2) Air is modeled as an ideal gas. (3) For the environment $T_0 = 500°R$, $p_0 = 1 atm$.

ANALYSIS: With Eq. 7.10 the change in exergy is

$$E_2 - E_1 = m\left[(u_2 - u_1) + p_0\cancelto{0}{(V_2 - V_1)} - T_0(s_2 - s_1)\right]$$

$$= m\left[u(T_2) - u(T_1) - T_0\left(s^\circ(T_2) - s^\circ(T_1) - R \ln p_2/p_1\right)\right] \qquad (1)$$

This requires p_2. Using the ideal gas equation of state: $p_2/p_1 = T_2/T_1$. Then, with data from Table A-22E

$$E_2 - E_1 = (1\ lb)\left[119.58 - 85.2 - 500\left(0.66321 - 0.58233 - \frac{1.986}{28.97}\ln\frac{700}{500}\right)\right] = 5.47\ Btu \longleftarrow$$

To find W, reduce the energy balance to read $W = m(u_1 - u_2)$, or

$$W = (1\ lb)\left[85.2 - 119.58\right] = -34.38\ Btu$$

Then, since $\Delta V = 0$ in the process, Eq. 7.13 reads

$$\begin{bmatrix}\text{exergy transfer}\\\text{accompanying work}\end{bmatrix} = W - \cancelto{0}{p_0 \Delta V} = -34.38\ Btu \longleftarrow$$

① Using the exergy balance, Eq. 7.11, with Eq. 7.14: $E_d = T_0\sigma$, we get

$$E_d = \cancelto{0}{\int_1^2\left(1 - \frac{T_0}{T_b}\right)\delta Q} - \left[W - \cancelto{0}{p_0\Delta V}\right] - \left[E_2 - E_1\right]$$

$$= -W - \left[E_2 - E_1\right]$$

② $$= -(-34.38\ Btu) - \left[5.47\ Btu\right] = 28.91\ Btu \longleftarrow$$

---

1. Alternatively, $E_d = T_0\sigma$, where σ is the amount of entropy produced, which is obtained from an entropy balance. Since the tank involves no heat transfer, the entropy balance reduces to give $\sigma = S_2 - S_1$. Thus, in this case, $E_d = T_0(S_2 - S_1)$.

2. In this case, the destruction of exergy can be traced to the shearing action of the paddle wheel on the air in the tank.

PROBLEM 7.25

KNOWN: Argon contained in a closed, rigid, insulated tank is stirred by a paddle wheel.

FIND: Determine the work and the exergy destruction.

SCHEMATIC & GIVEN DATA:

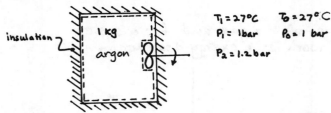

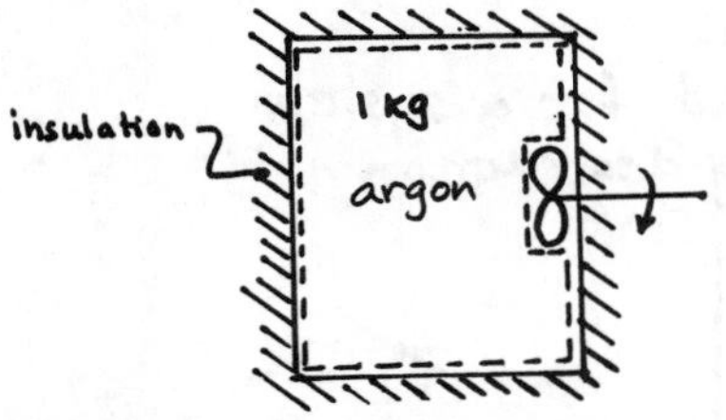

$T_1 = 27°C$ $T_0 = 27°C$
$P_1 = 1\,bar$ $P_0 = 1\,bar$
$P_2 = 1.2\,bar$

ASSUMPTIONS: (1) For the system shown in the accompanying figure, Q=0, and kinetic and potential energy effects can be ignored. (2) Argon is modeled as an ideal gas. (3) For the environment, $T_0 = 300K$, $P_0 = 1\,bar$.

ANALYSIS: With assumption 1 an energy balance reduces to

$$\Delta U = \cancel{Q}^{0} - W \Rightarrow W = -m\left(u(T_2) - u(T_1)\right) \quad (1)$$

To find T_2, use the ideal gas equation of state to write

$$\left.\begin{array}{l} P_1 V = mRT_1 \\ P_2 V = mRT_2 \end{array}\right\} : \quad T_2 = T_1\left(\frac{P_2}{P_1}\right) = 300K\left(\frac{1.2}{1}\right) = 360K$$

The value of $\bar{c}_p$ is constant for monatomic gases (see Table A-21): $\bar{c}_p = 5/2\bar{R}$.
Then, with $\bar{c}_p - \bar{c}_v = R$, $\bar{c}_v = 3/2\bar{R}$. Equation (1) gives

$$W = -m\, c_v (T_2 - T_1)$$

$$= -(1\,kg)\left((1.5)\left(\frac{8.314}{39.94}\frac{kJ}{kg\cdot K}\right)\right)(360-300)K = -18.73\,kJ \longleftarrow$$

The exergy destruction can be found from an exergy balance or using $E_d = T_0\sigma$, where σ is the amount of entropy produced, obtained from an entropy balance: Using the second of these approaches

$$\Delta S = \cancel{\int_1^2 \left(\frac{\delta Q}{T}\right)_b}^{0} + \sigma \Rightarrow \sigma = m(s_2 - s_1)$$

or

$$\sigma = m\left[\underbrace{c_p}_{\frac{5}{2}R} \ln \underbrace{\frac{T_2}{T_1}}_{=P_2/P_1} - R \ln \frac{P_2}{P_1}\right]$$

$$\Rightarrow \sigma = m\left(1.5\frac{\bar{R}}{M}\right)\ln\frac{P_2}{P_1}$$

$$= (1\,kg)(1.5)\left(\frac{8.314}{39.94}\frac{kJ}{kg\cdot K}\right)\ln 1.2 = 0.057\,\frac{kJ}{K}$$

So,

$$E_d = T_0\sigma = (300\,K)\left(0.057\,\frac{kJ}{K}\right) = 17.1\,kJ \longleftarrow$$

PROBLEM 7.26

KNOWN: A known amount of carbon dioxide gas is contained in a rigid, insulated vessel of known volume and at a specified initial pressure. An electric resistor in the vessel transfers energy to the gas at a known rate for a specified period of time.

FIND: Determine the exergy change of the gas and, for a system consisting of the resistor and gas, the work and exergy destruction.

SCHEMATIC & GIVEN DATA:

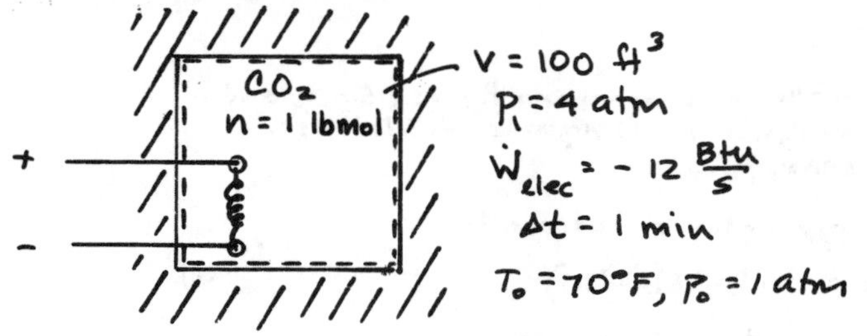

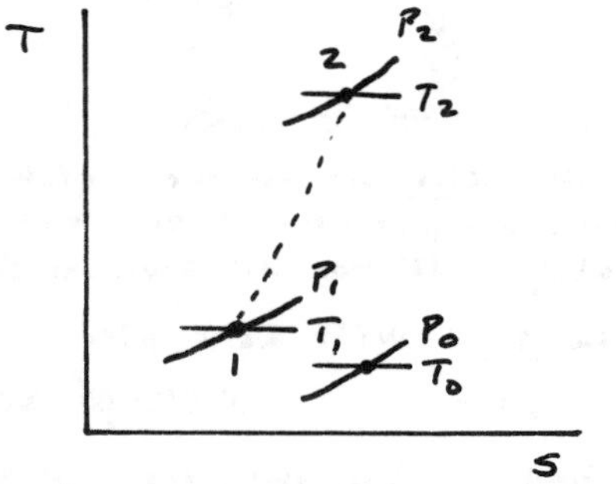

ASSUMPTIONS: (1) The system consists of the gas and the resistor. (2) The volume is constant and kinetic and potential energy effects are neglected. (3) The resistor is of negligible mass and thus undergoes no change of state. (4) The carbon dioxide is modeled as an ideal gas. (5) For the system, Q=0.

ANALYSIS: First, fix both states, as follows:

$$T_1 = \frac{p_1 V}{n\bar{R}} = \frac{(4\text{ atm})(100\text{ ft}^3)}{(1\text{ lbmol})(1545\,\frac{\text{ft}\cdot\text{lbf}}{\text{lbmol}\,°\text{R}})}\left|\frac{14.696\text{ lbf/in}^2}{1\text{ atm}}\right|\left|\frac{144\text{ in}^2}{1\text{ ft}^2}\right| = 548\,°\text{R}$$

To determine T_2, begin with an energy balance: $\Delta U = Q - W$. With $\Delta U_{res} = 0$,

$$n(\bar{u}_2 - \bar{u}_1) = \cancelto{0}{Q} - W$$

where $\bar{u}_2$ and $\bar{u}_1$ are obtained from Table A-23E. To get W, use Eq. 2.14

$$W = \int_1^2 \dot{W}dt = \dot{W}_{elec}\,\Delta t = (-12\,\tfrac{\text{Btu}}{\text{s}})(1\text{ min})\left|\frac{60\text{ s}}{1\text{ min}}\right| = -720\text{ Btu}$$

Thus, solving for $\bar{u}_2$ and inserting data

$$\bar{u}_2 = -\frac{W}{n} + u_1 = -\frac{(-720\text{ Btu})}{(1\text{ lbmol})} + 3040.1\text{ Btu/lbmol} = 3760.1\text{ Btu/lbmol}$$

Interpolating in Table A-23E, $T_2 = 647.5\,°\text{R}$. Thus

$$p_2 = \frac{n\bar{R}T_2}{V} = \frac{(1)(1545)(647.5)}{(100)}\left|\frac{1}{144}\right|\left|\frac{1}{14.696}\right| = 4.727\text{ atm}$$

Now, the change in exergy of the gas is obtained using Eq. 7.10 and ideal gas relations

$$E_2 - E_1 = n\left[\bar{u}(T_2) - \bar{u}(T_1) + p_0(\cancelto{0}{V_2 - V_1}) - T_0\left(\bar{s}°(T_2) - \bar{s}°(T_1) - \bar{R}\ln p_2/p_1\right)\right]$$

Inserting data interpolated from Table A-23E

$$E_2 - E_1 = (1\text{ lbmol})\left[(3760.1 - 3040.1)\frac{\text{Btu}}{\text{lbmol}} - (530°\text{R})\left(52.750 - 51.2124 - 1.986\ln\frac{4.727}{4}\right)\frac{\text{Btu}}{\text{lbmol}\,°\text{R}}\right]$$

$$= 80.85\text{ Btu} \longleftarrow E_2 - E_1$$

The exergy transfer accompanying work, E_W, is

$$E_W = W - p_0 \Delta V^{\nearrow 0} = -720 \text{ Btu} \longleftarrow E_W$$

In this case, all of the energy transfer by work is "available" in the sense that the exergy transfer accompanying work <u>equals</u> the energy transfer.

The exergy destruction, E_d, is determined by solving the exergy balance, Eq. 7.11, for E_d

$$E_2 - E_1 = \int_1^2 \left(1 - \frac{T_0}{T_b}\right)\delta Q^{\nearrow 0} - (W - p_0 \Delta V^{\nearrow 0}) - E_d$$

or

$$E_d = -(E_2 - E_1) - (W) = -(80.85 \text{ Btu}) - (-720 \text{ Btu})$$

(1) $$= 639.15 \text{ Btu} \longleftarrow E_d$$

---

1. Note that the exergy input by work (720 Btu) results in only 80.85 Btu being stored as increased exergy of the gas. The rest, 639.15 Btu, is destroyed by irreversibilities.

PROBLEM 7.27

KNOWN: A rigid insulated tank has two equal-volume compartments separated by a valve. Initially, one is evacuated and the other holds CH_4 gas at a known condition. The valve is opened and the gas fills the entire volume.

FIND: Determine the final temperature and pressure, and evaluate the exergy destruction. Discuss.

SCHEMATIC & GIVEN DATA:

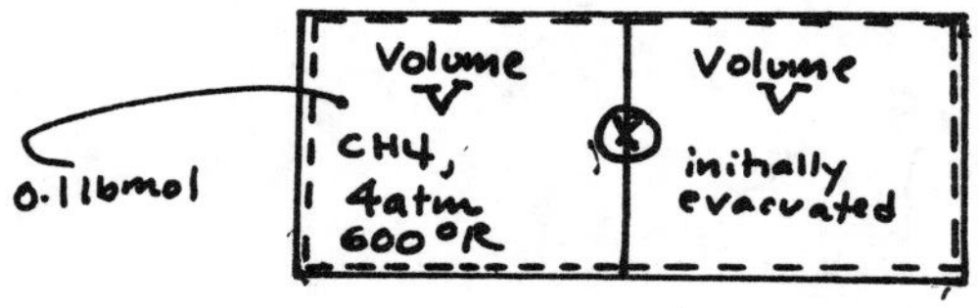

ASSUMPTIONS: 1. The closed system consists of both volumes, as shown in the schematic. 2. For the system, $Q = W = 0$ and effects of motion and gravity can be ignored. 3. CH_4 is modeled as an ideal gas. 4. For the environment, $T_0 = 530°R$, $p_0 = 1\,atm$.

ANALYSIS: With assumptions 1, 2 the energy balance reduces to $\Delta U = \cancel{Q}^{0} - \cancel{W}^{0} \Rightarrow \Delta U$. Since the specific internal energy of an ideal gas depends on temperature alone: $T_2 = T_1 = 600°R$ ←

Using the ideal gas equation of state

initially: $p_1 V = m R T_1$

finally: $p_2 (2V)\ m R T_2$

but $T_2 = T_1$, giving

$$p_2(2V) = p_1 V \Rightarrow p_2 = \tfrac{1}{2} p_1 = 2\,atm$$ ←

The exergy destruction can be found from an exergy balance or using $E_d = T_0 \sigma$, where σ is obtained from an entropy balance. That is,

$$\Delta S = \cancel{\int_1^2 \left(\frac{\delta Q}{T}\right)_b}^{\,0} + \sigma$$

$$\Rightarrow \quad \sigma = \Delta S = n\left[\underbrace{\bar{s}^\circ(T_2) - \bar{s}^\circ(T_1)}_{=0 \text{ since } T_2 = T_1} - \bar{R} \ln p_2/p_1\right]$$

$$\sigma = -n\bar{R}\ln\frac{p_2}{p_1} = -(0.1\,lbmol)\left(1.986\,\frac{Btu}{lbmol \cdot °R}\right)\ln\left(\tfrac{1}{2}\right)$$

$$= 0.1377\ Btu/°R$$

Finally

$$E_d = T_0 \sigma$$

$$= (530°R)\left(0.1377\,\frac{Btu}{°R}\right)$$

$$= 72.98\ Btu$$ ←

Exergy is destroyed in this case because the CH_4 undergoes an unrestrained, and thus irreversible, expansion to a lower pressure.

PROBLEM 7.28

KNOWN: One kg of R134a is compressed adiabatically from a specified initial state to a specified final pressure and temperature.

FIND: Determine the work and the exergy destruction.

SCHEMATIC & GIVEN DATA:

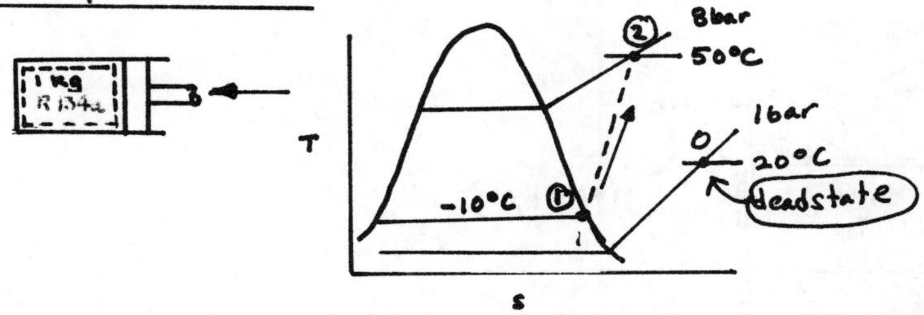

ASSUMPTIONS: (1) The system shown in the accompanying figure undergoes an adiabatic process between the indicated states. Changes in kinetic and potential energy can be ignored. (2) For the environment, $T_0 = 20°C$, $p_0 = 1\,bar$.

ANALYSIS: An energy balance reduces with assumption 1 to

$$\Delta U = \cancel{Q}^{0} - W \Rightarrow W = -m(u_2 - u_1)$$

With data from Tables A-10 and A-12

$$W = -(1\,kg)(261.62 - 221.48)\,\frac{kJ}{kg}$$
$$= -40.14\,kJ \longleftarrow W$$

To find the exergy destruction, use $E_d = T_0\sigma$, where σ is the amount of entropy produced, obtained from an entropy balance:

$$\Delta S = \cancel{\int_1^2 \left(\frac{\delta Q}{T}\right)_b}^{0} + \sigma$$

With data from Tables A-10 and 12

$$\sigma = m(s_2 - s_1) = (1\,kg)[0.9711 - 0.9253]\,\frac{kJ}{kg \cdot K}$$
$$= 0.0458\,\frac{kJ}{K}$$

Finally

$$E_d = (293\,K)(0.0458) = 13.42\,kJ \longleftarrow I$$

PROBLEM 7.29

KNOWN: Two solid blocks of mass m and specific heat c are brought into contact and attain thermal equilibrium.

FIND: Derive an expression for the exergy destruction; demonstrate that the exergy destruction cannot be negative, and discuss.

SCHEMATIC & GIVEN DATA:

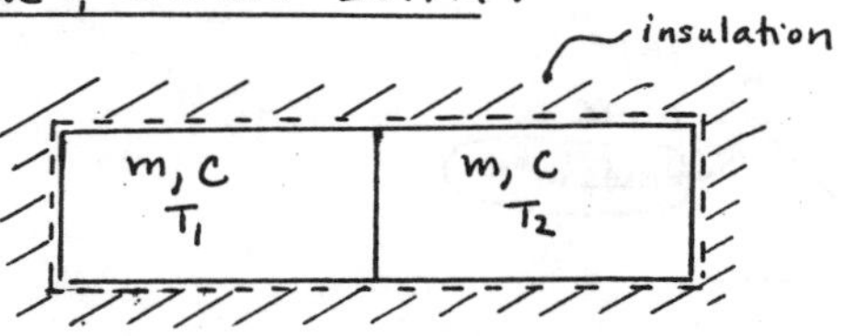

ASSUMPTIONS: (1) As shown in the figure, the two solid blocks form a closed system. (2) The blocks are modeled as incompressible with constant specific heat c. (3) For the system, Q = W = 0 and there are no kinetic and potential energy effects.

ANALYSIS: (a) Using assumptions (3), the energy balance reduces to

$$\Delta U = \cancel{Q}^{0} - \cancel{W}^{0} \Rightarrow \Delta U = 0 \Rightarrow [(2m)u(T_f) - [mu(T_1) + mu(T_2)]] = 0$$

where T_f is the final temperature at equilibrium. Accordingly

$$[2u(T_f) - u(T_1) - u(T_2)] = 0 \quad \Rightarrow \quad [u(T_f) - u(T_1)] + [u(T_f) - u(T_2)] = 0$$

$$\Rightarrow c[T_f - T_1] + c[T_f - T_2] = 0 \Rightarrow T_f = \frac{T_1 + T_2}{2}.$$

The amount of exergy destroyed is $E_d = T_0\sigma$, where σ is the amount of entropy produced, obtained from an entropy balance:

$$\Delta S = \cancel{\int\left(\frac{\delta Q}{T}\right)_b}^{0} + \sigma \Rightarrow [(2m)s(T_f) - [ms(T_1) + ms(T_2)]] = \sigma$$

or

$$\sigma = m[[s(T_f) - s(T_1)] + [s(T_f) - s(T_2)]]$$

$$= mc\ln\frac{T_f}{T_1} + mc\ln\frac{T_f}{T_2}$$

$$= mc\ln\left[\frac{T_f^2}{T_1T_2}\right] = mc\ln\left[\frac{(T_1+T_2)^2}{4T_1T_2}\right]$$

Finally

$$E_d = mcT_0\ln\left[\frac{(T_1+T_2)^2}{4T_1T_2}\right] \longleftarrow$$

(b) The value of E_d would be negative only if $\left[\frac{(T_1+T_2)^2}{4T_1T_2}\right]$ were less than unity. Considering this possibility...

$$\frac{(T_1+T_2)^2}{4T_1T_2} < 1 \Rightarrow (T_1+T_2)^2 < 4T_1T_2 \Rightarrow T_1^2 + 2T_1T_2 + T_2^2 < 4T_1T_2$$

or

$$T_1^2 - 2T_1T_2 + T_2^2 < 0 \Rightarrow (T_1 - T_2)^2 < 0$$

Since this inequality cannot be satisfied, we can conclude that $E_d \geq 0$.

(c) The exergy destruction in this case can be traced to the spontaneous heat transfer that takes place <u>within</u> the two blocks as they come to thermal equilibrium.

PROBLEM 7.30

KNOWN: A hot metal bar is quenched by immersing it in a tank of water.

FIND: Determine the exergy destruction.

SCHEMATIC & GIVEN DATA:

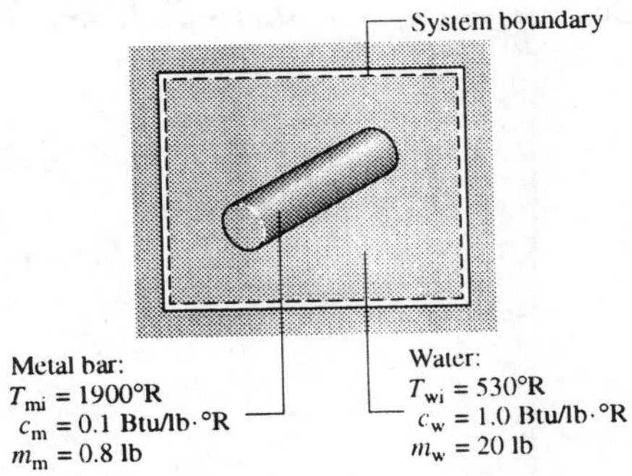

ASSUMPTIONS: 1. As shown in the schematic, the metal bar and water form a ① closed system. 2. For the system, $Q = W = 0$ and there are no effect of motion or gravity. 3. The metal bar and water are each modeled as incompressible. 4. $T_0 = 537°R\,(77°F)$.

ANALYSIS: An energy balance for the system reduces to give

$$\Delta U)_{metal} + \Delta U)_{water} = 0 \qquad (1)$$

An exergy balance reduces to give

$$\Delta E = \int_1^2 \left[1 - \frac{T_0}{T_b}\right]\delta Q^{\,0} - [W - p_0 \Delta V]^{\,0} - E_d \Rightarrow E_d = -\Delta E$$

Since exergy is an extensive property, $\Delta E = \Delta E)_{metal} + \Delta E)_{water}$. Then, with Eq. 7.10

$$E_d = -\left[[\Delta U + p_0 \Delta V^{\,0} - T_0 \Delta S]_{metal} + [\Delta U + p_0 \Delta V^{\,0} - T_0\, \Delta S]_{water}\right]$$

Using Eq. (1) this becomes

② $$E_d = T_0 [\Delta S)_{metal} + \Delta S)_{water}] \qquad (2)$$

The term in square brackets is the amount of entropy produced, which is evaluated in Example 6.5 as 0.0864 Btu/°R. Thus

$$E_d = 537°R\,(0.0864\ Btu/°R) = 46.4\ Btu \quad \longleftarrow$$

---

1. With the indicated assumptions, the system is <u>isolated</u>. Eq. (1) indicates that the total energy of the system remains constant. Eq. (2) indicates that the exergy of the system does not remain constant because exergy is destroyed.

2. Alternatively, Eq. (2) can be expressed as $E_d = T_0 \sigma$, where σ is the amount of entropy produced within the system.

PROBLEM 7.31

KNOWN: There is heat transfer through a wall for which the temperatures at the inner and outer surfaces are known.

FIND: Determine the rates of exergy transfer accompanying heat transfer into and out of the wall and the rate of exergy destruction. Discuss.

SCHEMATIC & GIVEN DATA:

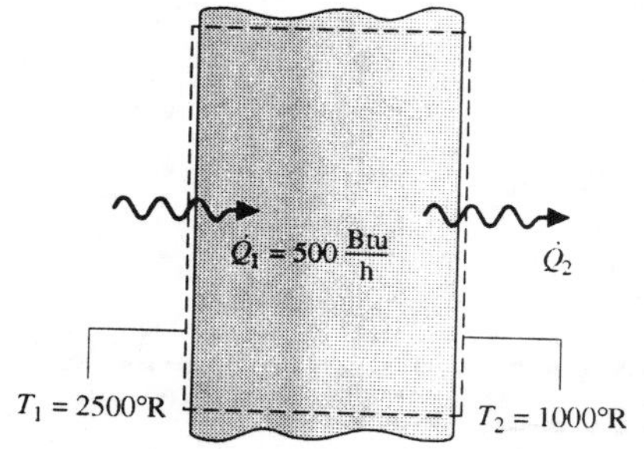

ASSUMPTIONS: 1. The system shown in the schematic is at steady state with $\dot{W} = 0$ 2. $T_0 = 500°R$

ANALYSIS: At steady state an energy rate balance reduces to

$$\frac{dE}{dt}^{0} = \dot{Q} - \dot{W}^{0} \Rightarrow \dot{Q} = 0 \Rightarrow \dot{Q}_2 = \dot{Q}_1 = 500\ \text{Btu/h}.$$ The directions of $\dot{Q}_1, \dot{Q}_2$ are indicated by the arrows on the schematic.

Since T_1 and T_2 are are uniform over the inner and outer surfaces, the rates of exergy transfer accompanying heat transfer are, respectively

$$\begin{bmatrix}\text{rate of exergy transfer} \\ \text{in accompanying} \\ \text{heat transfer}\end{bmatrix} = \left[1 - \frac{T_0}{T_1}\right]\dot{Q}_1$$

$$= \left[1 - \frac{500}{2500}\right]\left(500 \frac{\text{Btu}}{\text{h}}\right) = 400 \frac{\text{Btu}}{\text{h}} \quad \leftarrow$$

$$\begin{bmatrix}\text{rate of exergy transfer} \\ \text{out accompanying} \\ \text{heat transfer}\end{bmatrix} = \left[1 - \frac{T_0}{T_2}\right]\dot{Q}_2$$

$$= \left[1 - \frac{500}{1000}\right]\left(500 \frac{\text{Btu}}{\text{h}}\right) = 250 \frac{\text{Btu}}{\text{h}} \quad \leftarrow$$

Reducing the exergy rate balance, Eq. 7.17, at steady state

$$\frac{dE}{dt}^{0} = \left[1 - \frac{T_0}{T_1}\right]\dot{Q}_1 - \left[1 - \frac{T_0}{T_1}\right]\dot{Q}_2 - \left[\dot{W}^{0} - p_0 \frac{dV}{dt}^{0}\right] - \dot{E}_d$$

$$\Rightarrow \quad \dot{E}_d = \left[1 - \frac{T_0}{T_1}\right]\dot{Q}_1 - \left[1 - \frac{T_0}{T_1}\right]\dot{Q}_2 = 400 \frac{\text{Btu}}{\text{h}} - 250 \frac{\text{Btu}}{\text{h}} = 150 \frac{\text{Btu}}{\text{h}} \quad \leftarrow$$

① Exergy is destroyed because heat transfer occurs spontaneously through a finite temperature difference.

---

1. The term $\left[1 - \frac{T_0}{T_1}\right]\dot{Q}_1$ is the power that could be obtained in principle by supplying $\dot{Q}_1$ to a reversible power cycle operating between T_1 and T_0. Likewise, $\left[1 - \frac{T_0}{T_2}\right]\dot{Q}_2$ is the power that could be obtained in principle by supplying $\dot{Q}_2$ to a reversible power cycle operating between T_2 and T_0. The difference between these — $\dot{E}_d$ — represents the opportunity to develop power that is irrevocably lost in the spontaneous conduction process.

PROBLEM 7.32

KNOWN: A gear box operates at steady state. Performance data are provided.

FIND: Determine the power delivered along the output shaft, the rate of exergy transfer accompanying heat, and the rate of exergy destruction, each in Btu/h. Also express each as a percentage of the input power.

SCHEMATIC & GIVEN DATA:

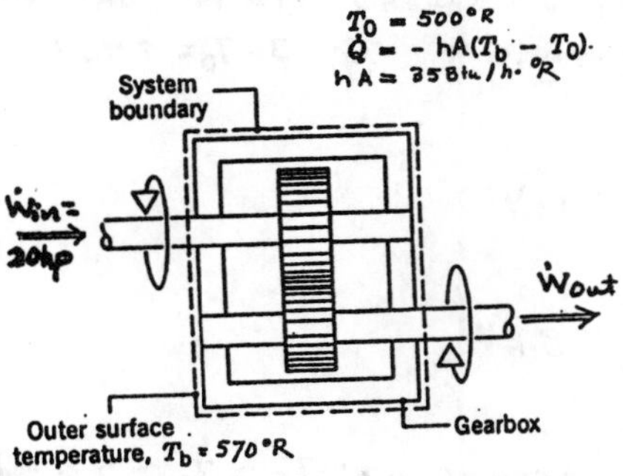

ASSUMPTIONS: (1) The system shown in the figure is at steady state. (2) The temperature of the outer surface of the gearbox and the temperature of the surroundings are each uniform. (3) For the environment $T_0 = 500°R$.

ANALYSIS: An energy rate balance reduces at steady state to give: $\dot{W} = \dot{Q}$. Then, inserting given information

$$\dot{W}_{out} = \dot{Q} + \dot{W}_{in} = -hA[T_b - T_0] + \dot{W}_{in}$$

$$= -\left(35 \frac{Btu}{h\,°R}\right)(570-500)°R + (20 hp)\left(\frac{2545 Btu/h}{1 hp}\right)$$

$$= -2450 \frac{Btu}{h} + 50{,}900 \frac{Btu}{h} = 48{,}450 \frac{Btu}{h} \longleftarrow$$

The rate of exergy transfer accompanying heat is

$$\left[1 - \frac{T_0}{T_b}\right]\dot{Q} = \left[1 - \frac{500}{570}\right]\left[-2450 \frac{Btu}{h}\right] = -300.9 \frac{Btu}{h} \longleftarrow$$

The rate of exergy destruction can be found by reducing the exergy rate balance at steady state to obtain

$$\dot{E}_d = \left[1 - \frac{T_0}{T_b}\right]\dot{Q} - \dot{W}$$

$$= -300.9 \frac{Btu}{h} - \left[48{,}450 - 50{,}900\right] \frac{Btu}{h} = 2149.1 \frac{Btu}{h} \longleftarrow$$

Expressing values as percentages of the input power

$$\frac{\dot{W}_{out}}{\dot{W}_{in}} = \left(\frac{48{,}450}{50{,}900}\right)(100) = 95.19\%$$

① $$\frac{\left|\left[1 - \frac{T_0}{T_b}\right]\dot{Q}\right|}{\dot{W}_{in}} = \left(\frac{300.9}{50{,}900}\right)(100) = 0.59\%$$

$$\frac{\dot{E}_d}{\dot{W}_{in}} = \left(\frac{2149.1}{50{,}900}\right)(100) = 4.22\%$$

$$= 100\%$$

1. This represents the true thermodynamic value of the heat loss. An energy analysis overstates its significance: $\frac{|\dot{Q}|}{\dot{W}_{in}} = \left(\frac{2450}{50{,}900}\right)(100) = 5\%$

PROBLEM 7.33

KNOWN: Steady-state data are provided for a gearbox.

FIND: Develop a full exergy accounting of the power input and compare with the results of Example 7.4.

SCHEMATIC & GIVEN DATA: See Fig. E 7.4

ASSUMPTIONS: 1. The gearbox and a portion of the surroundings are taken as the closed system. 2. The system is at steady state and heat transfer occurs on the boundary at temperature T_f. 3. $T_0 = 293$ K.

ANALYSIS: As in Example 7.4

$$\begin{bmatrix}\text{Rate exergy is transferred} \\ \text{in via the high-speed} \\ \text{shaft}\end{bmatrix} = 60 \text{ kW}$$

$$\begin{bmatrix}\text{Rate exergy is transferred} \\ \text{out via the low-speed} \\ \text{shaft}\end{bmatrix} = 58.8 \text{ kW}$$

Since heat transfer from the system occurs at $T_f = T_0 = 293$ K

$$\begin{bmatrix}\text{Rate exergy is transferred} \\ \text{out accompanying heat} \\ \text{transfer}\end{bmatrix} = \left[1 - \frac{T_0}{T_f}\right]\dot{Q} = 0$$

The rate of exergy destruction is then obtained from an exergy rate balance as

$$\dot{E}_d = (60 - 58.8) \text{ kW} = 1.2 \text{ kW}$$

or, alternatively, from $\dot{E}_d = T_0 \dot{\sigma}$, where $\dot{\sigma}$ is obtained from the solution to Example 6.4b:

$$\dot{E}_d = (293 \text{ K})\left(4.1 \times 10^{-3} \frac{\text{kW}}{\text{K}}\right) = 1.2 \text{ kW}$$

Exergy balance sheet in terms of exergy magnitudes on a rate basis:

- Rate of exergy in:
  - high-speed shaft — 60 kW (100%)
- Disposition of the exergy:
  - Rate of exergy out
    - low-speed shaft — 58.8 kW (98%)
    - heat transfer — 0 (0%)
  - Rate of exergy destruction — 1.2 (2%)

Total: 60.0 kW (100%)

In the present case the rate of exergy destruction is greater than in Example 7.4 because an additional source of irreversibility is included in the system: the irreversibility associated with internal heat transfer from the gearbox to the surroundings.

The analysis of Example 7.4 provides a sharper picture of gearbox performance because it distinguishes between the effects of exergy destruction within the gearbox and the exergy transferred from the gearbox associated with heat transfer. The present analysis lumps those effects.

PROBLEM 7.34

KNOWN: Air at a known initial state is contained in a closed, rigid tank. The air receives a specified energy transfer by heat through a wall separating the gas from a thermal reservoir at 900°R.

FIND: (a) For the air as the system, determine the exergy change, the exergy transfer accompanying heat transfer, and the exergy destruction. (b) Evaluate the exergy destruction for an enlarged system that includes the wall and discuss.

SCHEMATIC & GIVEN DATA:

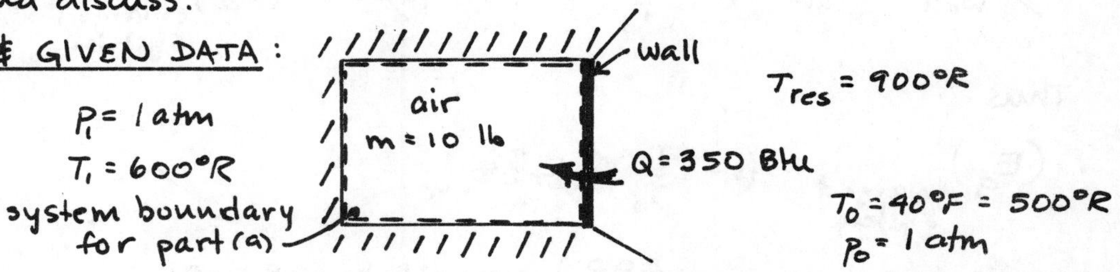

ASSUMPTIONS: (1) Closed systems are under consideration for which the effects of motion and gravity can be ignored. (2) The volume is constant and there is no work. (3) The air behaves as an ideal gas. (4) The state of the wall remains unchanged. (5) The air undergoes an internally reversible process.

ANALYSIS: (a) First, we apply the energy balance in order fix state 2. That is

$$\Delta KE^{\,0} + \Delta PE^{\,0} + \Delta U = Q - W^{\,0} \Rightarrow m(u_2 - u_1) = Q$$

Solving for u_2 and inserting data from Table A-22E

$$u_2 = \frac{Q}{m} + u_1 = \frac{350 \text{ Btu}}{10 \text{ lb}} + 102.34 \text{ Btu/lb} = 137.34 \text{ Btu/lb}$$

Interpolating in Table A-22E; $T_2 = 802.1°R$, $s°(T_2) = 0.69622$ Btu/lb·°R.

The exergy change of the air is

$$\Delta E_{air} = m\left[u(T_2) - u(T_1) + p_0(V_2 - V_1)^{0} - T_0\left(s°(T_2) - s°(T_1) - R \ln p_2/p_1\right)\right]$$

Noting that $p_2/p_1 = T_2/T_1$ and with $s°(T_1)$ from Table A-22E

$$\Delta E_{air} = (10 \text{ lb})\left[(137.34 - 102.34)\frac{\text{Btu}}{\text{lb}} - (500°R)\left(0.69622 - 0.62607 - \left(\frac{1.986}{28.97}\right)\ln\frac{802.1}{600}\right)\frac{\text{Btu}}{\text{lb}\cdot°R}\right]$$

$= 98.76$ Btu ← ΔE_{air}

Denoting the exergy transfer accompanying heat transfer as $E_{Q,air}$, the exergy balance gives

$$\Delta E_{air} = E_{Q,air} - (W^{0} - p_0 \Delta V^{0}) - E_{d,air}^{\,0}$$

Thus

$E_{d,air} = 0$ ←

$E_{Q,air} = 98.76$ ←

PROBLEM 7.34 (Cont'd.)

(b) Applying the exergy balance to the enlarged system that includes the air and the wall

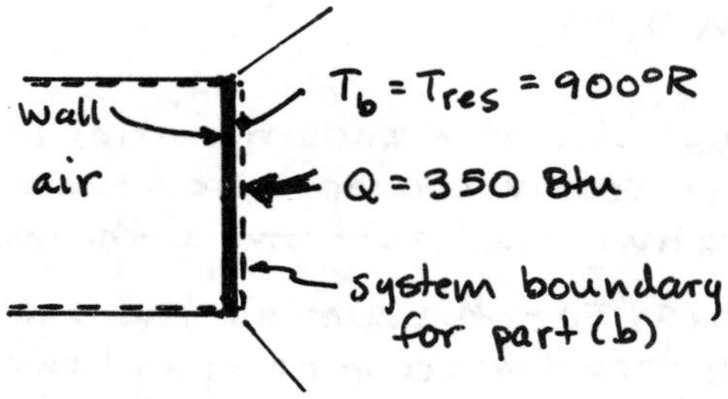

$$\Delta E_{wall}^{\,0} + \Delta E_{air} = \left(1 - \frac{T_0}{T_b}\right)Q - (W^{0} - p_0 \Delta V^{0}) - (E_d)_{\substack{enlarged\\ system}}$$

Thus

$$(E_d)_{\substack{enlarged\\ system}} = \left(1 - \frac{T_0}{T_b}\right)Q - \Delta E_{air}$$

$$= \left(1 - \frac{500}{900}\right)(350\ \text{Btu}) - (98.76\ \text{Btu})$$

$$= 155.56 - 98.76 = 56.80\ \text{Btu}$$ ← (b)

Discussion

From the analysis of the enlarged system, the exergy transfer accompanying heat transfer to the wall is 155.56 Btu. This occurs at the reservoir temperature of 900°R. The exergy transfer from the wall to the air, $E_{Q,air}$, occurs at a much lower temperature, and amounts to only 98.76 Btu. The difference is accounted for by the exergy destruction associated with the temperature gradient across the wall.

PROBLEM 7.35

KNOWN: Steady-state operating data are provided for each of two systems.

FIND: For each system, evaluate the rates of exergy transfer accompanying heat and work, and the rate of exergy destruction. Discuss.

SCHEMATIC & GIVEN DATA:

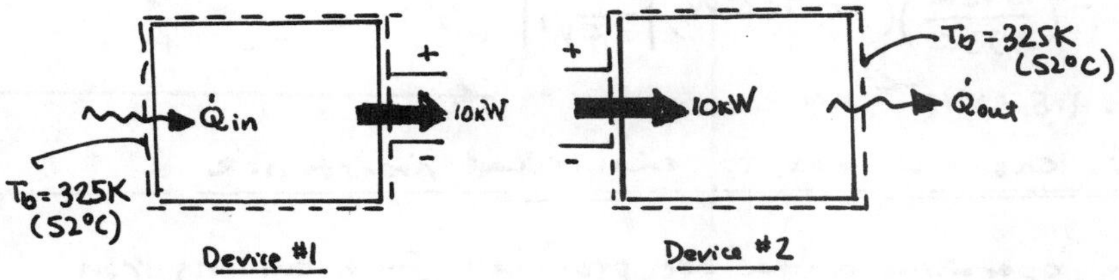

ASSUMPTIONS: 1. Each of the systems shown in the schematic is at steady state. 2. The only energy transfers are those occuring in the directions of the arrows. 3. $T_0 = 300K$.

ANALYSIS: Energy rate balances give

Device 1: $\frac{dE}{dt}^{0} = \dot{Q}_{in} - \dot{W}_{out} \Rightarrow \dot{Q}_{in} = \dot{W}_{out} = 10kW$

Device 2: $\frac{dE}{dt}^{0} = \dot{W}_{in} - \dot{Q}_{out} \Rightarrow \dot{Q}_{out} = \dot{W}_{in} = 10kW$

Exergy Analysis:

Device 1: $\begin{bmatrix} \text{rate of exergy transfer} \\ \text{in accompanying heat} \\ \text{transfer} \end{bmatrix} = \left[1 - \frac{T_0}{T_b}\right]\dot{Q}_{in} = \left[1 - \frac{300}{325}\right](10kW) = 0.77kW$ ←

$\begin{bmatrix} \text{rate of exergy transfer} \\ \text{out accompanying power} \end{bmatrix} = 10kW$ ←

$\dot{E}_d = 0.77kW - 10kW = -9.23\ kW$ ←

Device 1 cannot operate as indicated because $\dot{E}_d$ cannot be negative. ←

Device 2: $\begin{bmatrix} \text{Rate of exergy transfer} \\ \text{in accompanying power} \end{bmatrix} = 10kW$ ←

$\begin{bmatrix} \text{Rate of exergy transfer} \\ \text{out accompanying heat transfer} \end{bmatrix} = \left[1 - \frac{T_0}{T_b}\right]\dot{Q}_{out} = \left[1 - \frac{300}{325}\right](10kW) = 0.77kW$ ←

$\dot{E}_d = 10kW - 0.77kW = +9.23\ kW$ ←

Device 2 can operate as indicated. However, the very high exergy destruction rate, which is 92% of the exergy input, suggests that there is considerable scope for improving performance. Other uses for the exergy input also might be considered. ←

PROBLEM 7.36

KNOWN: Steady-state operating data are provided for a silicon chip.

FIND: Determine the rate of exergy destruction. Discuss.

SCHEMATIC & GIVEN DATA: See Fig. E2.5

ASSUMPTIONS: 1. The chip is a closed system at steady state. 2. No heat transfer occurs between the chip and substrate. 3. $T_0 = 293K$.

ANALYSIS: In this case it is convenient to evaluate the rate of exergy destruction using $\dot{E}_d = T_0\dot{\sigma}$, where $\dot{\sigma}$ is the rate of entropy production.

PROBLEM 7.36 (Contd.)

An entropy rate balance for the chip reduces to give

$$\cancel{\frac{dS}{dt}}^0 = \frac{\dot{Q}}{T_b} + \dot{\sigma} \Rightarrow \dot{\sigma} = -\frac{\dot{Q}}{T_b}$$

Then, with data from Example 2.5

$$\dot{E}_d = -\frac{T_0}{T_b}\dot{Q} = -\left(\frac{293K}{353K}\right)(-0.225W)\left|\frac{1kW}{10^3W}\right|$$

$$= 1.87 \times 10^{-4} \text{ kW}$$

Exergy destruction in this case is due to electrical resistance.

PROBLEM 7.37.

KNOWN: Steady-state operating data are provided for a curling iron.

FIND: Evaluate the rates of exergy destruction and exergy transfer accompanying heat transfer. Express each as a percentage of the electrical power supplied.

SCHEMATIC & GIVEN DATA: See solution to Problem 6.66.

ASSUMPTIONS: 1. The system shown in the figure is at steady state. 2. Heat transfer occurs at T_b (= 363 K (90°C)) only. 3. T_0 = 293 K.

ANALYSIS: At steady state, an energy rate balance reduces to give

$$\cancel{\frac{dE}{dt}}^0 = \dot{Q} - \dot{W} \Rightarrow \dot{Q} = \dot{W} = -15 \text{ W}$$

Then

$$\begin{bmatrix}\text{rate of exergy transfer}\\ \text{out accompanying heat}\end{bmatrix} = \left[1 - \frac{T_0}{T_b}\right]\dot{Q} = \left[1 - \frac{293}{363}\right](-15W) = -2.89 \text{ W}$$
$$= -2.89 \times 10^{-3} \text{ kW}$$

The rate of exergy destruction can be found using $\dot{E}_d = T_0\dot{\sigma}$, where $\dot{\sigma}$ is the rate of entropy production, or by reducing an exergy rate balance:

$$\cancel{\frac{dE}{dt}}^0 = \left[1 - \frac{T_0}{T_b}\right]\dot{Q} - \left[\dot{W} - \cancel{p_0\frac{dV}{dt}}^0\right] - \dot{E}_d$$

$$\Rightarrow \quad \dot{E}_d = \left[1 - \frac{T_0}{T_b}\right]\dot{Q} - \dot{W}$$

$$= (-2.89 \text{ W}) - (-15 \text{ W}) = 12.11 \text{ W} \quad \Rightarrow \quad \dot{E}_d = 1.211 \times 10^{-2} \text{ kW}$$

When expressed as percentages of the electrical power supplied

$$\left|\frac{\left[1 - \frac{T_0}{T_b}\right]\dot{Q}}{\dot{W}}\right|(100) = \left(\frac{2.89}{15}\right)(100) = 19.3\%$$

① $$\frac{\dot{E}_d}{|\dot{W}|}(100) = \left(\frac{12.11}{15}\right)(100) = 80.7\%$$

1. The significant exergy destruction in this case indicates that good use is not being made of the valuable electrical power input.

PROBLEM 7.38

KNOWN: Two kilograms of a two-phase liquid-vapor mixture of water undergo two different processes at constant volume between the same end states: (a) adiabatic process, stirring with a paddle wheel, (b) heat transfer from a thermal reservoir at 900K.

FIND: For each case, determine the change in exergy, the net amounts of exergy transfer by work and heat, and the amount of exergy destruction.

SCHEMATIC & GIVEN DATA:

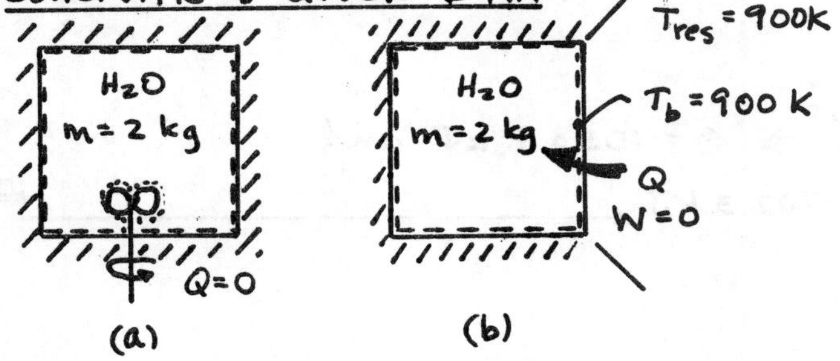

$T_1 = 300°C$
$x_1 = 0.5$
$x_2 = 1.0$

$T_0 = 300K$
$p_0 = 1$ bar

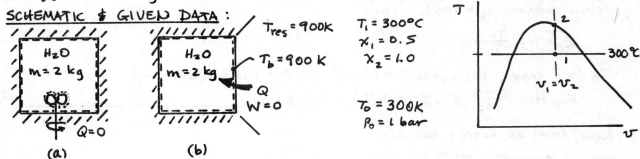

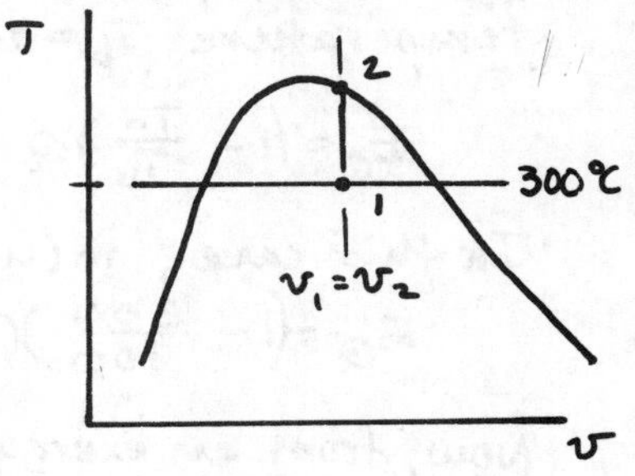

ASSUMPTIONS: (1) The water is a closed system. (2) The volume is constant. (3) In case (a); Q = 0, in case (b); W = 0. (4) Kinetic and potential energy effects are negligible. (5) For the environment, $T_0 = 300K$, $p_0 = 1 bar$.

ANALYSIS: For both cases, the initial and final states are the same. Let us begin by obtaining data at each state. From Table A-2 at 300°C

$$v_1 = v_{f1} + x_1(v_{g1} - v_{f1}) = 1.4036 \times 10^{-3} + (.5)(0.02167 - 1.4036 \times 10^{-3}) = 0.01154 \text{ m}^3/\text{kg}$$

$$u_1 = u_{f1} + x_1(u_{g1} - u_{f1}) = 1332.0 + (.5)(2563.0 - 1332.0) = 1947.5 \text{ kJ/kg}$$

$$s_1 = s_{f1} + x_1(s_{g1} - s_{f1}) = 3.2534 + (.5)(5.7045 - 3.2534) = 4.479 \text{ kJ/kg·K}$$

By assumptions (1) and (2), $v_2 = v_1 = v_g(T_2)$. From Table A-2, $T_2 = 336.8$ °C and

$$u_2 = 2474.2 \text{ kJ/kg}, \quad s_2 = 5.3673 \text{ kJ/kg·K}$$

(a) In this case, the volume is constant. Thus

$$\Delta E = m[(u_2 - u_1) + p_0(\cancel{v_2 - v_1})^{0} - T_0(s_2 - s_1)]$$

$$= (2 \text{ kg})[(2474.2 - 1947.5) - (300)(5.3673 - 4.479)] \frac{\text{kJ}}{\text{kg}} = 520.4 \text{ kJ} \longleftarrow \Delta E$$

$E_W = (W - p_0 \cancel{\Delta V}^{0}) = W$. From an energy balance, $m(u_2 - u_1) = \cancel{Q}^{0} - W$. Thus

$$W = -m(u_2 - u_1) = -(2 \text{ kg})(2474.2 - 1947.5) \frac{\text{kJ}}{\text{kg}} = -1053.4 \text{ kJ} \longleftarrow E_W$$

The exergy transfer accompanying heat transfer, denoted E_Q, is

$$E_Q = \int_1^2 \left(1 - \frac{T_0}{T_b}\right) \cancel{\delta Q}^{0} = 0 \longleftarrow E_Q$$

Now, from an exergy balance

$$\Delta E = \cancel{E_Q}^{0} - E_W - E_d$$

$$E_d = -E_W - \Delta E = -(-1053.4 \text{ kJ}) - (520.4 \text{ kJ})$$

$$= 533 \text{ kJ} \longleftarrow E_d$$

(b) Since the change of state is the same, the change in exergy is the same as in part (a). That is

$$\Delta E = 520.4 \text{ kJ} \longleftarrow \Delta E$$

Since the work is zero, $E_w = 0 \longleftarrow E_w$

The exergy transfer accompanying heat transfer is evaluated at the boundary temperature, $T_b = 900\text{K}$. Thus

$$E_Q = \left(1 - \frac{T_0}{T_b}\right)Q$$

In this case, $m(u_2 - u_1) = Q - \cancel{W}^{0} \Rightarrow Q = 1053.4 \text{ kJ}$ and

$$E_Q = \left(1 - \frac{300}{900}\right)(1053.4 \text{ kJ}) = 702.3 \text{ kJ} \longleftarrow E_Q$$

Now, from an exergy balance

$$\Delta E = E_Q - \cancel{E_W}^{0} - E_d$$

$$E_d = E_Q - \Delta E = 702.3 - 520.4 = 181.9 \text{ kJ} \longleftarrow E_d$$

<u>Discussion</u>

The exergy destruction for case (b) is significantly less than for case (a). Since the exergy change is the same in both cases, case (b) would allow the exergy change to be brought about by less exergy transfer from the surroundings.

PROBLEM 7.39

KNOWN: Operating data are provided for a water heater.

FIND: Determine the exergy destruction and exergy transfer for (a) the water alone as the system (b) the water and resistor as the system. Compare and discuss.

SCHEMATIC & GIVEN DATA:

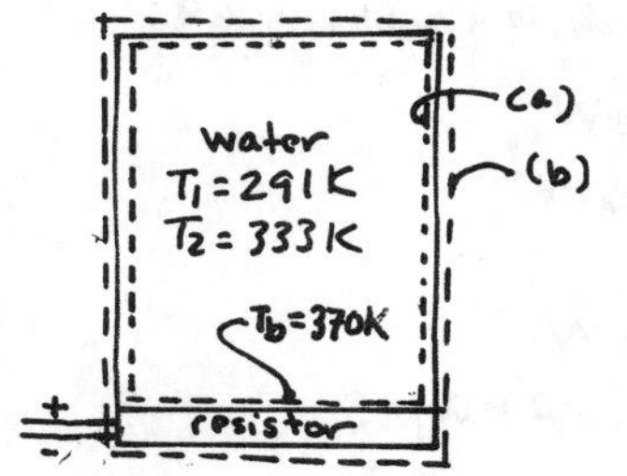

From solution to Problem 6.65...

(a) $\sigma = 8.83 \frac{kJ}{K}$

(b) $\sigma = 55.9 \frac{kJ}{kg}$

For water, $m = 99.2$ kg, $c = 4.18$ kJ/kg·K

ASSUMPTIONS: 1. In part (a) the system is the water only. In part (b) the system is the water plus the resistor. 2. The water is modeled as incompressible. 3. Heat transfer from the outer surface of the water heater can be ignored. 4. The states of the tank wall and the resistor are constant. 5. $T_0 = 291 K$.

ANALYSIS: (a) For system (a), the exergy is transfer is by heat transfer from the resistor to the water:

$$\begin{bmatrix}\text{exergy transfer}\\ \text{accompanying}\\ \text{heat transfer}\end{bmatrix} = \left[1 - \frac{T_0}{T_b}\right]Q = \left[1 - \frac{291}{370}\right](17{,}416 kJ) = 3719 kJ \quad \leftarrow$$

where $Q = \Delta U$ from an energy balance, or $Q = mc(T_2 - T_1) = (99.2 kg)(4.18 \frac{kJ}{kg \cdot K})(42 K) = 17{,}416 kJ$

The exergy destruction can be obtained conveniently here as $T_0\sigma$:

$$E_d = T_0\sigma = (291 K)(8.83 \frac{kJ}{K}) = 2570 kJ \quad \leftarrow$$

(b) For system (b), the exergy transfer is by electrical work. An overall energy balance reduces with assumptions 3, 4 to give $(-W) = \Delta U$, or $(-W) = 17{,}416$ kJ. As before the exergy destruction can be evaluated as

$$E_d = T_0\sigma = (291 K)(55.9 \frac{kJ}{kg}) = 16{,}267 kJ$$

DISCUSSION:

The exergy destruction in the enlarged system (b) is greater because there is an additional source of irreversibility — namely, electric current flow through the resistor. For the water alone, system (a), the irreversibility is associated with internal heat transfer from the resistor through the water as it is heated.

Observe that the increase in the exergy stored in the water equals, in each case, the difference between the exergy input and the exergy destroyed:

① System (a) $\Delta E = 3719 - 2570 = 1149$ kJ. System (b) $\Delta E = 17{,}416 - 16{,}267 = 1149$ kJ

Alternatively, ΔE for the water can be obtained using Eq. 7.10 with assumption 2:

$$\Delta E = \Delta U + p_0 \Delta V^{\,0} - T_0 \Delta S = mc[T_2 - T_1] - T_0 mc \ln\frac{T_2}{T_1} = mc\left[(T_2 - T_1) - T_0 \ln\frac{T_2}{T_1}\right]$$

$$\Rightarrow \Delta E = (99.2 kg)(4.18 \frac{kJ}{kg \cdot K})\left[42 K - 291 K \ln\frac{333}{291}\right] = 1148 K,$$ which agrees to within roundoff.

1. The percentage of the electrical input that is stored in the water: $\left(\frac{1149}{17{,}416}\right)(100) = 6.6\%$ suggests that such water heaters do not make effective use of the electrical power supplied to them.

PROBLEM 7.40

KNOWN: Steady-state operating data are provided for an electric motor.

FIND: Evaluate the rate of exergy destruction and the rate of exergy transfer accompanying heat. Express each as a percentage of the electrical power supplied.

SCHEMATIC & GIVEN DATA:

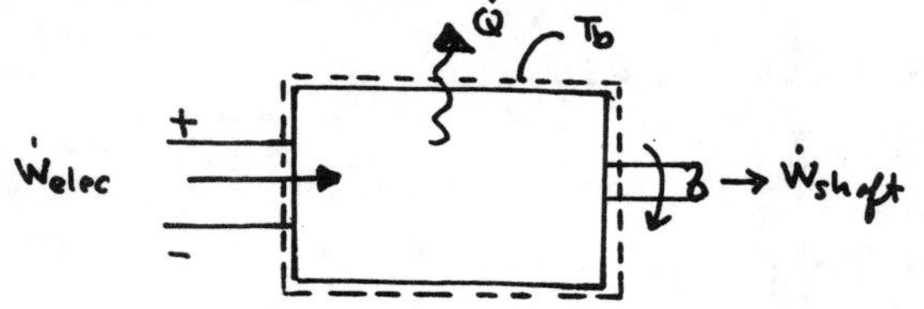

From the solution to Problem 6.69:

$\dot{W}_{elect} = 2.2\text{ kW}$

$\dot{W}_{shaft} = 1.676\text{ kW}$

$T_b = 319.9\text{ K}$

$\dot{Q} = 0.524\text{ kW}$

$\dot{\sigma} = 1.64 \times 10^{-3}\text{ kW/K}$

ASSUMPTIONS: 1. The system shown in the schematic is at steady state. 2. Energy transfers are in the directions of the arrows. 3. Heat transfer occurs at temperature T_b. 4. $T_0 = 293\text{ K}$.

ANALYSIS: With assumption 3

$$\begin{bmatrix}\text{rate of exergy transfer}\\ \text{accompanying heat transfer}\end{bmatrix} = \left[1 - \frac{T_0}{T_b}\right]\dot{Q}$$

$$= \left[1 - \frac{293}{319.9}\right](0.524\text{ kW}) = 0.044\text{ kW} \longleftarrow$$

The rate of exergy destruction can be found by reducing an exergy rate balance or using $\dot{E}_d = T_0\dot{\sigma}$. Thus

$$\dot{E}_d = T_0\dot{\sigma} = (293\text{ K})\left(1.64\times10^{-3}\ \frac{\text{kW}}{\text{K}}\right) = 0.481\text{ kW} \longleftarrow$$

Exergy accounting:

- rate exergy enters electrically : 2.2 kW (100%)
- disposition of the exergy :
  - exergy out via shaft 1.676 kW (76.18%)
  - exergy out via heat 0.044 kW (2.0%)
  - exergy destroyed 0.481 kW (21.86%) } ←

PROBLEM 7.41

KNOWN: Energy is conducted from a thermal reservior through a cylindrical rod at steady state to another thermal reservoir at a lower temperature.

FIND: Plot the rate of conduction through the rod, the rates of exergy transfer accompanying heat transfer into and out of the rod, and the rate of exergy destruction, each versus the rod length, L.

SCHEMATIC & GIVEN DATA:

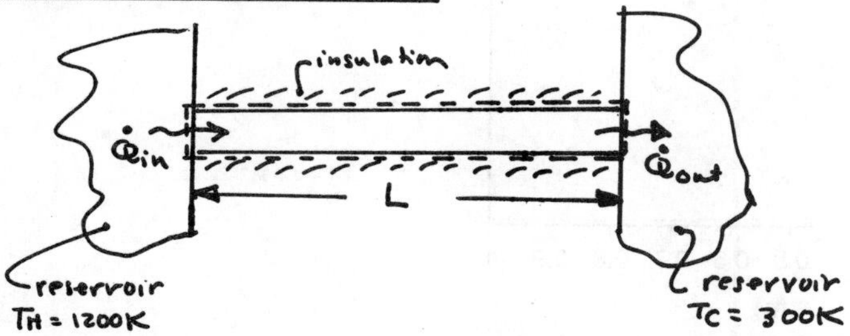

D = 2cm
k = 0.4 kW/m·K
0.01 ≤ L ≤ 1.0 m

ASSUMPTIONS: 1. The system shown in the schematic is at steady state. 2. Energy transfer is in the directions of the arrows only. 3. $T_0 = 300K$.

ANALYSIS: An energy rate balance reduces to give $\frac{dE}{dt}^{0} = \dot{Q} - \dot{W}^{0} \Rightarrow \dot{Q} = 0$, or $\dot{Q}_{in} = \dot{Q}_{out}$. Using Eq. 2.32, the common heat transfer rate by conduction is

$$\dot{Q}_{in} = \dot{Q}_{out} = kA\left[\frac{T_H - T_C}{L}\right] = \left(0.4 \frac{kW}{m \cdot K}\right)\left(\frac{\pi}{4}(.02m)^2\right)\left(\frac{1200-300}{L}\right)K$$

$$= \left(\frac{0.1131}{L}\right) kW \qquad (1)$$

The rates of exergy transfer accompanying heat transfer are, respectively

$$\begin{bmatrix}\text{rate of exergy}\\ \text{transfer in}\end{bmatrix} = \left[1 - \frac{T_0}{T_H}\right]\dot{Q}_{in} = \left[1 - \frac{300}{1200}\right]\dot{Q}_{in} = 0.75\,\dot{Q}_{in} \qquad (2)$$

$$\begin{bmatrix}\text{rate of exergy}\\ \text{transfer out}\end{bmatrix} = \left[1 - \frac{T_0}{T_C}\right]\dot{Q}_{out} = \left[1 - \frac{300}{300}\right]\dot{Q}_{out} = 0 \qquad (3)$$

The rate of exergy destruction can be obtained by reducing an exergy rate balance, which in this case corresponds to finding the difference between the exergy entering and exiting the rod. Thus,

$$\dot{E}_d = \begin{bmatrix}\text{rate of exergy}\\ \text{transfer in}\end{bmatrix} - \begin{bmatrix}\text{rate of exergy}\\ \text{transfer out}\end{bmatrix} = 0.75\,\dot{Q}_{in} \qquad (4)$$

Sample Calculation: L = 1m, $\dot{Q}_{in}$ = 0.1131 kW, [Rate of exergy transfer in] = $\dot{E}_d$ = 0.0848 kW.

PLOTS:

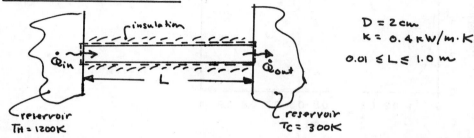

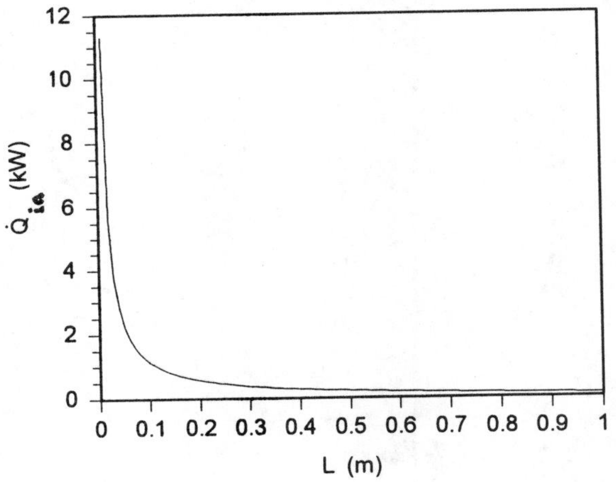

PROBLEM 7.41 (Cont'd.)

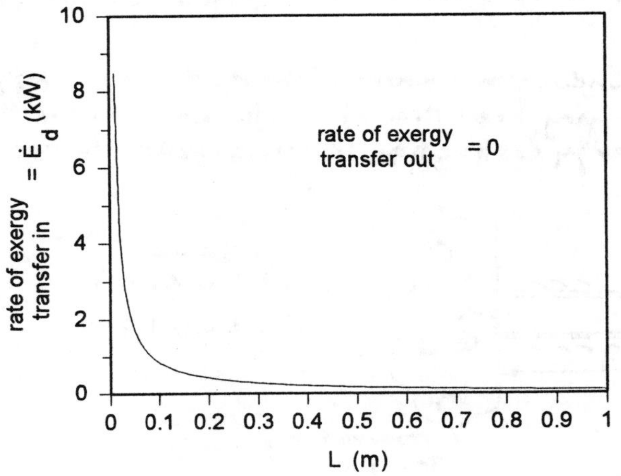

<u>Discussion</u>: As the length of the rod increases, the heat transfer rate and the exergy destruction rate decrease rapidly. Since exergy can be viewed as having economic value (secs. 7.6, 7.7), a reduction in exergy destruction can be viewed as a cost saving. However, the cost of the rod increases with length, so an economic tradeoff is inherent in selecting a value for L.

PROBLEM 7.42

KNOWN: One lb of O_2 expands isothermally while receiving energy by heat transfer through a wall separating the O_2 from a thermal reservoir.

FIND: (a) For the O_2 as the system, evaluate W, Q, the exergy transfers accompanying work and heat transfer, and the exergy destruction. (b) Evaluate the exergy destruction for a system consisting of the wall and O_2. Discuss.

SCHEMATIC & GIVEN DATA:

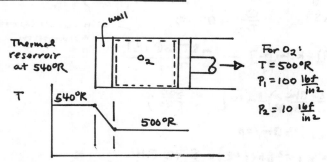

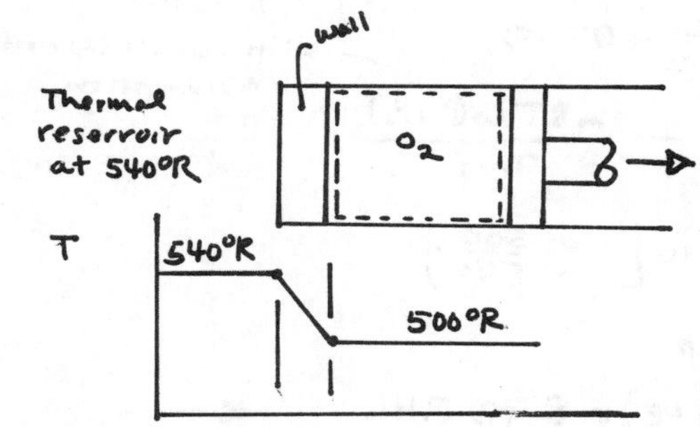

For O_2: $T = 500°R$, $P_1 = 100\ \frac{lbf}{in^2}$, $P_2 = 10\ \frac{lbf}{in^2}$

ASSUMPTIONS: 1. In part (a), the system is the O_2 only. In part (b), the system is the O_2 plus the wall. 2. O_2 is modeled as an ideal gas. 3. The state of the wall does not change. 4. Effects of motion and gravity are ignored. 5. $T_0 = 540°R$, $p_0 = 15\ \frac{lbf}{in^2}$.

ANALYSIS: (a) An energy balance reduces to $\Delta U = Q - W$. But $\Delta U = 0$ for an ideal gas at fixed temperature, giving $Q = W$. (1)

To find W, use the ideal gas model with

$$W = \int_1^2 p\,dV = \int_1^2 \frac{mRT}{V} dV = mRT \ln \frac{V_2}{V_1}.$$ Since $P_1V_1 = mRT$, $P_2V_2 = mRT$,

$$V_2/V_1 = \frac{P_1}{P_2} \Rightarrow \quad W = mRT \ln \frac{P_1}{P_2}. \quad (2)$$

An exergy balance takes the form

$$\Delta E = \underbrace{\left[1 - \frac{T_0}{T}\right] Q}_{\text{Exergy transfer with heat}} - \underbrace{[W - P_0 \Delta V]}_{\text{Exergy transfer with work}} - \underbrace{E_d}_{\text{Exergy destruction}}$$

Alternatively

$$(\Delta U + P_0 \Delta V - T_0 \Delta S) = \left[1 - \frac{T_0}{T}\right] Q - [W - P_0 \Delta V] - E_d$$

With $\Delta U = 0$, $Q = W$, and cancellation of $P_0\Delta V$ on both sides, we get

① $$E_d = T_0\left[\Delta S - \frac{Q}{T}\right]$$

$$= T_0\left[m[\underbrace{s°(T_2) - s°(T_1)}_{=0 \text{ since } T = \text{constant}} - R \ln \frac{P_2}{P_1}] - \frac{mRT \ln P_1/P_2}{T}\right] \equiv 0 \quad \leftarrow$$ (Q from Eqs. (1), (2))

Accordingly, the O_2 undergoes an internal reversible process.

With Eqs. (1), (2)

$$Q = W = (1\,lb)\left(\frac{1.986}{32} \frac{Btu}{lb \cdot °R}\right)(500°R) \ln\left(\frac{100}{10}\right) = 71.452\ Btu \quad \leftarrow$$

Using the ideal gas equation of state and introducing Eq. 2

$$\begin{bmatrix}\text{Exergy transfer} \\ \text{associated with work}\end{bmatrix} = W - P_0 \Delta V$$

$$= mRT \ln \frac{P_1}{P_2} - P_0\left[\frac{mRT}{P_2} - \frac{mRT}{P_1}\right] = mRT\left[\ln \frac{P_1}{P_2} - \frac{P_0}{P_2} + \frac{P_0}{P_1}\right]$$

$$= (1\,lb)\left(\frac{1.986}{32} \frac{Btu}{lb \cdot °R}\right)(500°R)\left[\ln 10 - \frac{15}{10} + \frac{15}{100}\right] = 29.56\ Btu \quad \leftarrow$$

PROBLEM 7.42 (Contd.)

$$\begin{bmatrix}\text{Exergy transfer} \\ \text{associated with heat}\end{bmatrix} = \left[1 - \frac{T_0}{T}\right]Q$$

② ③ $$= \left[1 - \frac{540}{500}\right](71.452 \text{ Btu}) = -5.716 \text{ Btu} \quad \longleftarrow$$

(b) For an enlarged system of O_2 plus wall, $E_d = T_0\sigma$, where σ is the amount of entropy produced within the enlarged system. Using assumption 3,

$$\Delta S = \frac{Q}{T_{res}} + \sigma \quad \text{or} \quad \cancelto{0}{(\Delta S)_{wall}} + (\Delta S)_{O_2} = \frac{Q}{T_{res}} + \sigma \Rightarrow$$

with Eq. (1), (2) and assumption 3

$$\sigma = (\Delta S)_{O_2} - \frac{Q}{T_{res}} = m\left[\underbrace{s^\circ(T_2) - s^\circ(T_1)}_{=0} - R \ln\frac{P_2}{P_1}\right] - \frac{mRT\ln(P_1/P_2)}{T_{res}}$$

or

④ ⑤ $$\sigma = mR\ln(P_1/P_2)\left[1 - \frac{T}{T_{res}}\right] = (1 \text{ lb})\left(\frac{1.986}{32}\frac{\text{Btu}}{\text{lb}\cdot{}^\circ\text{R}}\right)\ln 10\left[1 - \frac{500}{540}\right]$$

$$= 1.059 \times 10^{-2} \text{ Btu}/{}^\circ\text{R}$$

Finally, $E_d = T_0\sigma = 540^\circ\text{R}(1.059 \times 10^{-2} \text{ Btu}/{}^\circ\text{R}) = 5.72 \text{ Btu} \quad \longleftarrow$

Exergy is destroyed within the enlarged system owing to spontaneous heat transfer through a finite temperature difference through the wall.

---

1. The term in brackets is the entropy production within the O_2, as can be seen by applying the entropy balance to the system of part (a). For this system, we get $\sigma = 0$.

2. Since heat transfer to the system occurs at $T < T_0$, the accompanying exergy transfer is oppositely directed. See the discussion of Sec. 7.3.2.

3. Inserting calculated values into exergy balance

$$\Delta E = (-5.716) - (29.56) - (0) = -35.28 \text{ Btu}$$

This also can be determined using $\Delta E = \Delta U + p_0\Delta V - T_0\Delta S$, as can be readily checked.

4. $\sigma \to 0$ as $T_{res} \to T$.

5. Since the O_2 undergoes an internally reversible process, the entropy production (and exergy destruction) can be evaluated by study of the wall. That is, an entropy balance for the wall reduces with assumption 3 to give

$$\cancelto{0}{(\Delta S)_{wall}} = \frac{Q}{T_{res}} - \frac{Q}{T} + \sigma$$

$$\Rightarrow \quad \sigma = Q\left[\frac{1}{T} - \frac{1}{T_{res}}\right] = 71.452 \text{ Btu}\left[\frac{1}{500} - \frac{1}{540}\right]\left(\frac{1}{{}^\circ\text{R}}\right)$$

$$= 1.059 \times 10^{-2} \text{ Btu}/{}^\circ\text{R}$$

which agrees with the result reported here.

PROBLEM 7.43

KNOWN: A system undergoes a refrigeration cycle while receiving Q_c at temperature T_c and discharging Q_H at T_H, where $T_H > T_c$. Q_c and Q_H are the only heat transfers.

FIND: (a) Show that W_{cycle} cannot be zero. (b) Obtain a specified expression for coefficient of performance β. (c) Determine β_{max}.

SCHEMATIC & GIVEN DATA:

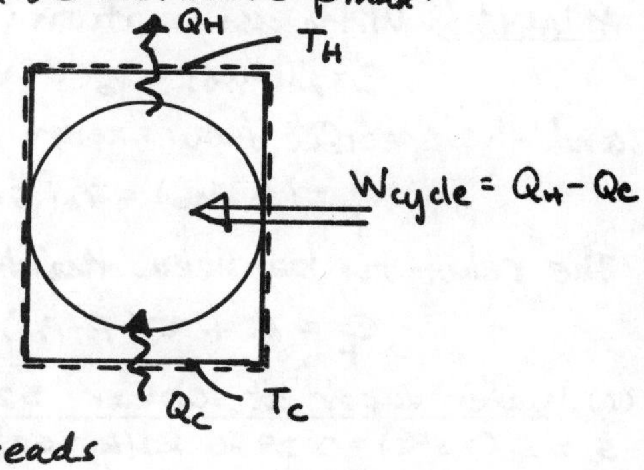

ASSUMPTIONS: (1) The system shown undergoes a refrigeration cycle. (2) Q_c and Q_H are the only heat transfers and are in the directions of the arrows. (3) T_c and T_H are constant and $T_H > T_c$. (4) The environment temperature is T_0.

ANALYSIS: (a) An exergy balance for the cycle reads

$$\cancel{\Delta E}\Big]_{cycle}^{0} = \left[1 - \frac{T_0}{T_c}\right]Q_c - \left[1 - \frac{T_0}{T_H}\right]Q_H - \left[W_{cycle} - p_0\cancel{\Delta V}^{0}\right]_{cycle} - E_d$$

where $\Delta E = \Delta V = 0$ for a cycle. Introducing the energy balance, $Q_H = W_{cycle} + Q_c$ we get

$$0 = \left[1 - \frac{T_0}{T_c}\right]Q_c - \left[1 - \frac{T_0}{T_H}\right](W_{cycle} + Q_c) - W_{cycle} - E_d$$

$$= \left[(\cancel{1} - \frac{T_0}{T_c}) - (\cancel{1} - \frac{T_0}{T_H})\right]Q_c + \left[\cancel{1} - (\cancel{1} - \frac{T_0}{T_H})\right]W_{cycle} - E_d$$

$$= T_0\left[\frac{1}{T_H} - \frac{1}{T_c}\right]Q_c + \frac{T_0}{T_H}W_{cycle} - E_d \qquad (1)$$

Solving for E_d, and setting W_{cycle} to zero

$$E_d = \underbrace{T_0}_{pos.}\underbrace{\left[\frac{1}{T_H} - \frac{1}{T_c}\right]}_{neg.}\underbrace{Q_c}_{pos} + \frac{T_0}{T_H}\cancel{W_{cycle}}^{0} \Rightarrow E_d < 0 \quad \text{impossible!}$$

Thus W_{cycle} cannot be zero. ← (a)

(b) Solving (1) for $\beta = Q_c/W_{cycle}$

$$\left[\frac{T_H - T_c}{T_H T_c}\right]Q_c = \frac{W_{cycle}}{T_H} - E_d/T_0$$

$$\beta = \frac{Q_c}{W_{cycle}} = \left[\frac{T_c}{T_H - T_c}\right]\left[1 - \frac{T_H E_d}{T_0 W_{cycle}}\right] = \left[\frac{T_c}{T_H - T_c}\right]\left[1 - \frac{T_H E_d}{T_0(Q_H - Q_c)}\right] \qquad \text{(b)}$$

(c) From the result of part (b), β increases as $E_d \to 0$. Thus, when $E_d = 0$

$$\beta_{max} = \frac{T_c}{T_H - T_c}$$

as expected.

PROBLEM 7.44

KNOWN: Various substances are at specified states at the inlet to a control volume.

FIND: In each case, determine the specific exergy and specific flow exergy.

ASSUMPTIONS: (1) The velocity is relative to an exergy reference environment for which $T_0 = 20°C$, $p_0 = 1$ bar. (2) The effects of gravity are negligible.

ANALYSIS: With assumption (2), the specific exergy is

$$e = (u - u_0) + p_0(v - v_0) - T_0(s - s_0) + \frac{V^2}{2} \quad (1)$$

and the specific flow exergy is

$$e_f = (h - h_0) - T_0(s - s_0) + \frac{V^2}{2} \quad (2)$$

The relation between the two is

$$e_f = e + v(p - p_0) \quad (3)$$

(a) Water vapor at 100 bar, 520°C, 100 m/s. With $h_0 \approx h_f(20°C) = 83.96$ kJ/kg and $s_0 \approx s_f(25°C) = 0.2966$ kJ/kg·K from Table A-2, and $h = 3425.1$ kJ/kg, $s = 6.6622$ kJ/kg·K, and $v = 0.03394$ m³/kg from Table A-4, we get

$$e_f = (3425.1 - 83.96)\frac{kJ}{kg} - (293\,K)(6.6622 - 0.2966)\frac{kJ}{kg \cdot K} + \left(\frac{100^2}{2}\right)\frac{m^2}{s^2}\left|\frac{1\,N}{1\,kg \cdot m/s^2}\right|\left|\frac{1\,kJ}{10^3\,N \cdot m}\right|$$

$$= 1481\ kJ/kg \longleftarrow e_f$$

From (3) above

$$e = e_f - v(p - p_0) = 1481\frac{kJ}{kg} - \left(0.03394\frac{m^3}{kg}\right)(100 - 1)\,bar\left|\frac{10^5\,N/m^2}{1\,bar}\right|\left|\frac{1\,kJ}{10^3\,N \cdot m}\right|$$

$$= 1145\ kJ/kg \longleftarrow e$$

(b) Ammonia at 3 bar, 0°C, 5 m/s. With data from Table A-15, $h_0 = 1515.8$ kJ/kg, $s_0 = 6.2816$ kJ/kg·K, $h = 1454.43$ kJ/kg, $s = 5.5409$ kJ/kg·K, $v = 0.42382$ m³/kg. Thus

$$e_f = (1454.43 - 1515.8) - (293)(5.5409 - 6.2816) + \left(\frac{5^2}{2}\right)\left|\frac{1}{10^3}\right| = 155.67\ kJ/kg \longleftarrow e_f$$

and

$$e = e_f - v(p - p_0) = 155.67 - (0.42382)(3 - 1)\left|\frac{10^5}{10^3}\right| = 70.91\ kJ/kg \longleftarrow e$$

(c) Nitrogen as an ideal gas at 527°C (800 K), 50 bar, 200 m/s. Using ideal gas relations, Eq. (2) becomes

$$e_f = \left(\frac{1}{M}\right)\left[\bar{h}(T) - \bar{h}(T_0) - T_0\left\{\bar{s}^\circ(T) - \bar{s}^\circ(T_0) - \bar{R}\ln\frac{p}{p_0}\right\}\right] + \frac{V^2}{2}$$

With data from Table A-23

$$e_f = \frac{1}{28.01}\left[(23{,}714 - 8521) - (293)\left\{(220.907 - 190.805) - (8.314)\ln\left(\frac{50}{1}\right)\right\}\right] + \left(\frac{200^2}{2}\right)\left|\frac{1}{10^3}\right|$$

$$= 587.8\ kJ/kg \longleftarrow e_f$$

Now, from (3) above

$$e = e_f - v(p - p_0) = e_f - RT\left[1 - \frac{p_0}{p}\right]$$

$$= 587.8\frac{kJ}{kg} - \left(\frac{8.314}{28.01}\frac{kJ}{kg \cdot K}\right)(800\,K)\left[1 - \frac{1}{50}\right] = 355.1\ kJ/kg \longleftarrow e$$

PROBLEM 7.45

KNOWN: Steam at a specified state is under consideration.

FIND: Determine the specific exergy and the specific flow exergy.

SCHEMATIC & GIVEN DATA:

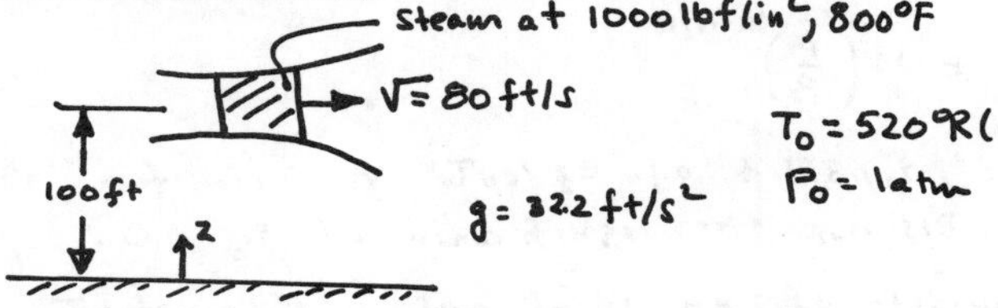

ASSUMPTIONS: The velocity and elevation are each relative to the exergy reference environment for which $T_0 = 520°R\ (60°F)$, $p_0 = 1\,\text{atm}$, $v_0 \approx v_f(T_0)$, $u_0 \approx u_f(T_0)$, $h_0 \approx h_f(T_0)$, $s_0 \approx s_f(T_0)$.

ANALYSIS: The specific exergy is found from Eq. 7.9:

$$e = (u - u_0) + p_0(v - v_0) - T_0(s - s_0) + \frac{V^2}{2} + gz$$

With data from Tables A-2E, 4E

$$e = (1261.2 - 28.08)\frac{\text{Btu}}{\text{lb}} + \left(14.7 \times 144 \frac{\text{lbf}}{\text{ft}^2}\right)(0.688 - 0.016)\frac{\text{ft}^3}{\text{lb}}\left|\frac{1\ \text{Btu}}{778\ \text{ft}\cdot\text{lbf}}\right|$$

$$- 520°R\,(1.5665 - 0.05555)\frac{\text{Btu}}{\text{lb}\cdot°R} + \left[\frac{(80\ \text{ft/s})^2}{2} + (32.2\ \text{ft/s}^2)(100\ \text{ft})\right]\left|\frac{1\ \text{lbf}}{32.2\ \text{lb}\cdot\text{ft/s}^2}\right|\left|\frac{1\ \text{Btu}}{778\ \text{ft}\cdot\text{lbf}}\right|$$

$$= 1233.12 + 1.83 - 785.69 + 0.26 = 449.52\ \text{Btu/lb} \longleftarrow$$

① The specific flow exergy is found from Eq. 7.20 and data from Tables A-2E, 4E:

$$e_f = (h - h_0) - T_0(s - s_0) + \frac{V^2}{2} + gz$$

$$= (1388.5 - 28.08) - 785.69 + 0.26 = 574.99\ \text{Btu/lb} \longleftarrow$$

---

1. Alternatively, Eq. 7.30 can be used to write

$$e_f = e + v(p - p_0) = 449.52\frac{\text{Btu}}{\text{lb}} + \left(0.688\frac{\text{ft}^3}{\text{lb}}\right)(1000 - 14.7)\frac{\text{lbf}}{\text{in}^2}\left|\frac{144\ \text{in}^2}{\text{ft}^2}\right|\left|\frac{\text{Btu}}{778\ \text{ft}\cdot\text{lbf}}\right|$$

$$= (449.52 + 125.47)\frac{\text{Btu}}{\text{lb}} = 574.99\frac{\text{Btu}}{\text{lb}}$$

PROBLEM 7.46

KNOWN: An ideal gas with constant specific heat ratio k.

FIND: Show that in the absence of motion and gravity effects

$$\frac{e_f}{c_p T_0} = \frac{T}{T_0} - 1 - \ln\frac{T}{T_0} + \ln\left(\frac{p}{p_0}\right)^{(k-1)/k}$$

and develop plots for $k = 1.2, 1.3, 1.4$ of e_f/c_pT_0 vs. T/T_0 for $p/p_0 = 0.25, 0.5, 1, 2, 4$. Discuss the significance of $e_f < 0$.

ASSUMPTIONS: (1) The system consists of an ideal gas with constant specific heat ratio. (2) The effects of motion and gravity are not significant.

ANALYSIS: With assumption (2) and ideal gas relationships, Eq. 7.20 gives

$$e_f = h - h_0 - T_0(s - s_0)$$

$$= c_p[T - T_0] - T_0\left[c_p \ln\frac{T}{T_0} - R\ln\frac{p}{p_0}\right]$$

where $R = c_p\left[\frac{k-1}{k}\right]$ (Eq. 3.47a)

Thus

$$\frac{e_f}{c_p T_0} = \frac{T}{T_0} - 1 - \ln\frac{T}{T_0} + \ln\left[\frac{p}{p_0}\right]^{(k-1)/k}$$

(a) See plots on the next page.

(b) Discussion.

The specific exergy is

$$e = (u - u_0) + p_0(v - v_0) - T_0(s - s_0)$$

and the specific flow exergy is

$$e_f = h - h_0 - T_0(s - s_0)$$

$$= (u + pv) - (u_0 + p_0 v_0) - T_0(s - s_0) = (u - u_0) + (pv - p_0 v_0) - T_0(s - s_0)$$

Subtracting e from e_f gives

$$e_f = e + v(p - p_0)$$

Thus, although the specific exergy e cannot be negative, e_f can take on negative values when $p < p_0$, as shown in the plot below. Physically, from the development of Sec. 7.4, the flow exergy is the sum of the exergy transfers accompanying mass and accompanying flow work. The value of e_f can be negative, then, when these contributions have opposite signs: the flow work contribution is opposite to the direction of flow.

PROBLEM 7.46 (Cont'd.)

PLOTS:

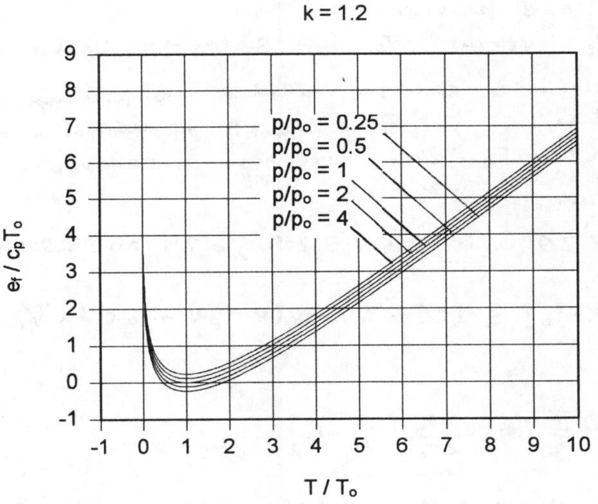

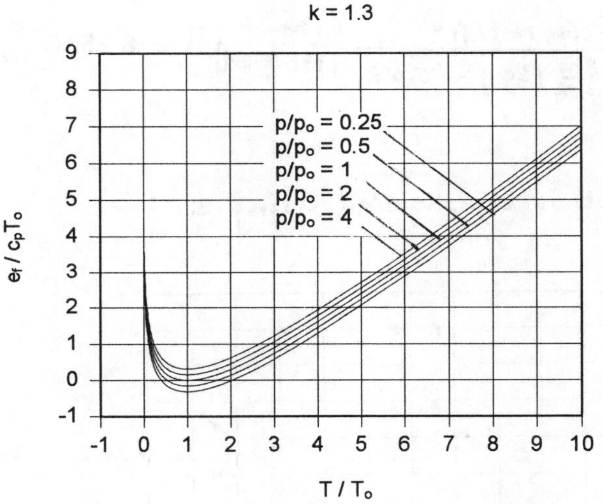

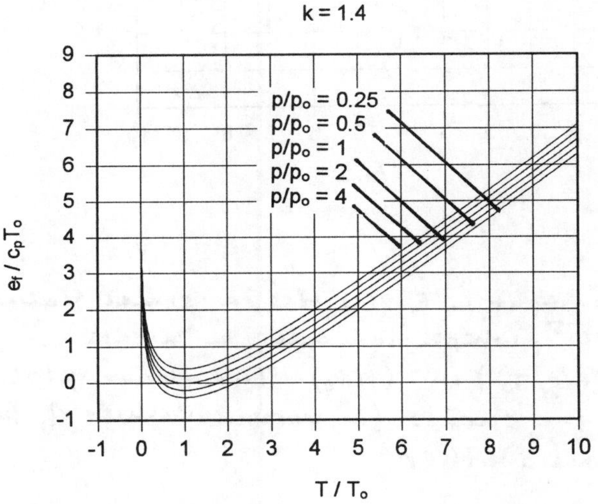

PROBLEM 7.47

KNOWN: A geothermal source provides a stream of liquid water at $T(\geq T_0)$ and pressure p.

FIND: Plot e_f/cT_0 versus T/T_0 for selected values of p/p_0.

ASSUMPTIONS: (1) The water can be modeled as incompressible with constant specific heat c. (2) The effects of motion and gravity can be neglected. (3) For the exergy reference environment, $T_0 = 60°F$, $p_0 = 1\,atm$.

ANALYSIS: With Eq. 7.20, using Eqs. 3.20b, 6.24 and assumption 2

$$e_f = (h-h_0) - T_0(s-s_0) = (c(T-T_0) + v(p-p_0)) - T_0 c \ln T/T_0$$

or

$$\frac{e_f}{cT_0} = \frac{T}{T_0} - 1 - \ln\frac{T}{T_0} + \underline{\frac{vp_0}{cT_0}}\left[\frac{p}{p_0} - 1\right] \qquad (1)$$

Selecting c and ρ from Table A-19E typical for geothermal applications, the underlined term of Eq. (1) is

$$\left[\frac{vp_0}{cT_0}\right] = \left[\frac{(14.7 \times 144)\,lbf/ft^2}{(1.01\,\frac{Btu}{lb \cdot °R})(60\,\frac{lb}{ft^3})(520°R)}\left|\frac{1\,Btu}{778\,ft \cdot lbf}\right|\right] = 8.63 \times 10^{-5}$$

The following plot is constructed using IT:

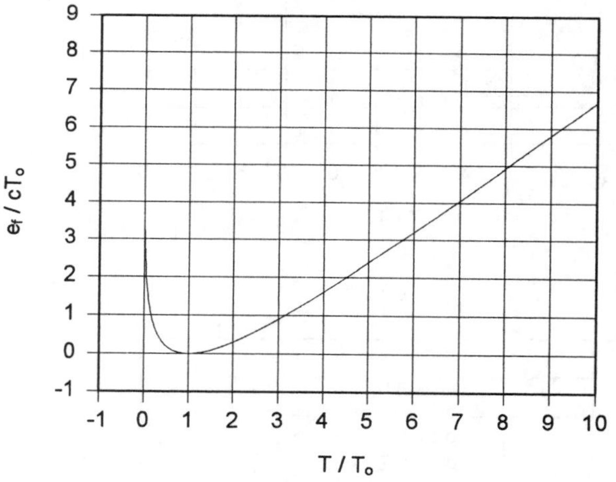

Note that the constant in Eq. (1) has a small value: $[vp_0/cT_0] = 8.63 \times 10^{-5}$, which indicates a weak dependence of (e_f/c_pT_0) on (p/p_0). The above plot reflects this, as the curves for various values of (p/p_0) fall on top of one another.

PROBLEM 7.48

KNOWN: The state of a flowing gas is defined by h, s, V, z.

FIND: Determine the maximum theoretical work, per unit of mass flowing, that could be developed as the stream is reduced to the dead state.

SCHEMATIC & GIVEN DATA:

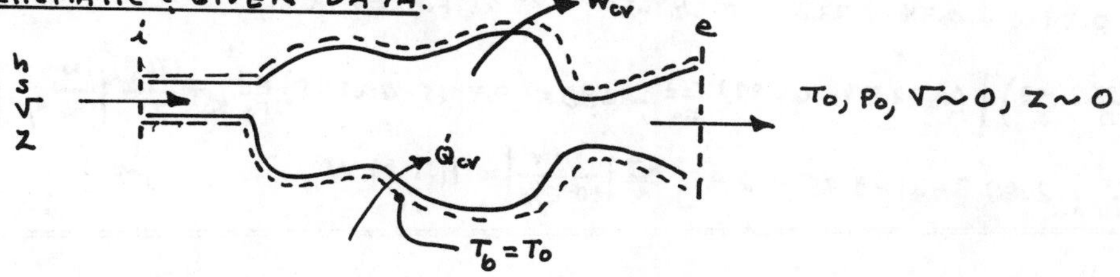

ASSUMPTIONS: (1) The control volume shown in the accompanying figure is at steady state. (2) Heat transfer occurs only at temperature $T_b = T_0$. (3) The velocity and elevation are relative to the environment, which is at T_0, p_0.

ANALYSIS: At steady state the exergy rate balance, Eq. 7.31, reduces to

$$\cancel{\frac{dE_{cv}}{dt}}^{\,0} = \underbrace{\left[1 - \frac{T_0}{T_b}\right]\dot{Q}_{cv}}_{=0 \text{ since } T_b = T_0} - \left[\dot{W}_{cv} - p_0 \cancel{\frac{dV}{dt}}^{\,0}\right] + \dot{m}(e_{fi} - e_{fe}) - \dot{E}_d$$

$e_{fe} = 0$ since the stream is reduced to the dead state at the exit

Thus

$$\frac{\dot{W}_{cv}}{\dot{m}} = e_{fi} - \frac{\dot{E}_d}{\dot{m}}$$

The maximum theoretical value is obtained when $\dot{E}_d = 0$: when there are no internal irreversibilities. Thus

$$\left(\frac{\dot{W}_{cv}}{\dot{m}}\right)_{max} = (h - h_0) - T_0(s - s_0) + \frac{V^2}{2} + gz \qquad \longleftarrow \left(\dot{W}_{cv}/\dot{m}\right)_{max}$$

Comparing this expression to Eq. 7.20, it can be concluded that the specific flow exergy can be interpreted as the maximum theoretical work, per unit of mass flowing, that could be developed as the stream is brought to the dead state while allowing heat transfer only at T_0. This interpretation complements those given in Sec. 7.3.2.

PROBLEM 7.49

KNOWN: Steam exits a turbine at a specified state and mass flow rate.

FIND: Determine the maximum theoretical power that could be developed by reducing the exiting stream to the dead state while allowing heat transfer only at temperature T_0.

SCHEMATIC & GIVEN DATA:

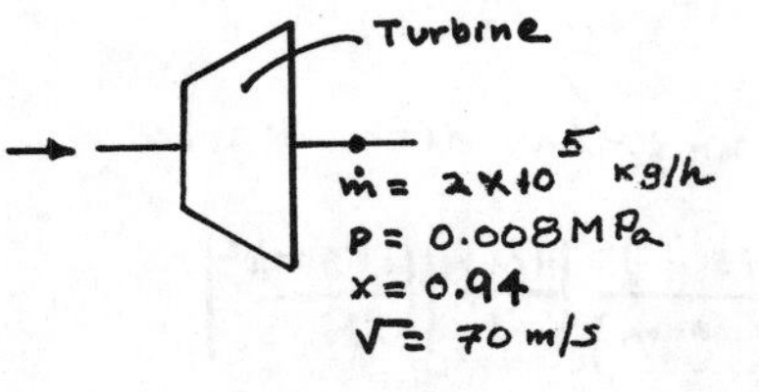

ACTUAL

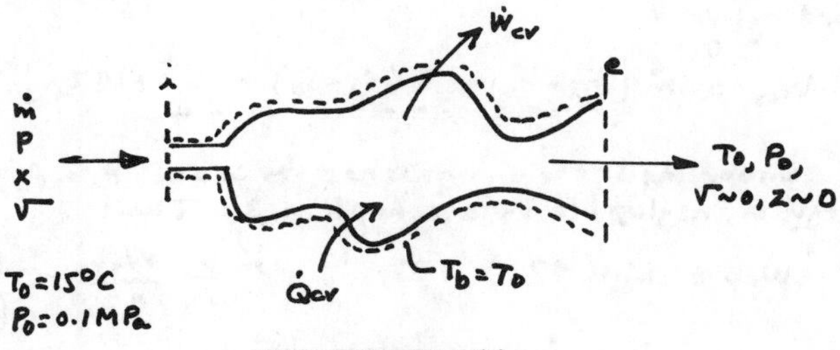

HYPOTHETICAL

ASSUMPTIONS: (1) The hypothetical control volume is at steady state and heat transfer occurs only at temperature $T_b = T_0$. (2) The velocity is relative to the environment for which temperature is T_0 and pressure is p_0. (3) The effect of gravity can be neglected.

ANALYSIS: The solution to Problem 7.48 shows that the maximum theoretical power that could be developed by any one-inlet, one-exit control volume at steady state that would reduce the stream to the dead state at the exit while allowing heat transfer only at T_0 is

PROBLEM 7.49 (Contd)

$$(\dot{W}_{cv})_{max} = \dot{m}\, e_f = \dot{m}\left[(h-h_0) - T_0(s-s_0) + \frac{V^2}{2}\right]$$

Then, with $h_0 \sim h_f(T_0) = 62.99$ kJ/kg, $s_0 \sim s_f(T_0) = 0.2245$ kJ/kg·K, and

$$h = h_f + x h_{fg} = 173.88 + 0.94(2403.1) = 2432.79 \text{ kJ/kg}$$

$$s = s_f + x s_{fg} = 0.5926 + 0.94(8.2287 - 0.5926) = 7.7705 \text{ kJ/kg·K}$$

$$(\dot{W}_{cv})_{max} = \left(\frac{2\times10^5}{3600}\frac{kg}{s}\right)\left[(2432.79 - 62.99)\frac{kJ}{kg} - 288K(7.7705 - 0.2245)\frac{kJ}{kg\cdot K} + \frac{(70\frac{m}{s})^2}{2}\left|\frac{1N}{1kg\cdot m/s^2}\right|\left|\frac{1kJ}{10^3 N\cdot m}\right|\right]$$

$$= \left(\frac{2\times10^5}{3600}\right)\left[2369.8 - 2173.25 + 2.45\right]\frac{kJ}{s}\left|\frac{1MW}{10^3 kJ/s}\right| = 11.1\ MW \quad \longleftarrow$$

PROBLEM 7.50

KNOWN: Water drawn from a lake flows through a hydraulic turbine-generator to a pond located 1 km below the lake.

FIND: Determine the minimum mass flow rate required to generate electricity at a rate of 1 MW.

SCHEMATIC & GIVEN DATA:

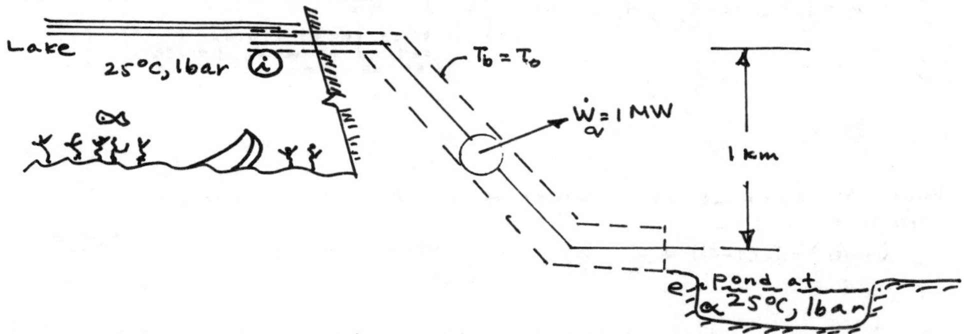

ASSUMPTIONS: 1. The control volume shown in the schematic is at steady state, and heat transfer occurs only at $T_b = T_0$. 2. At the exit e, the state corresponds to the dead state: T_0, p_0, $V_0 \sim 0$, $z_0 \sim$, $h_e = h_0$, $s_e = s_0$. 3. At the inlet i, kinetic energy can be ignored. 4. For the environment, $T_0 = 25°C$, $p_0 = 1$ bar, $g = 9.81$ m/s².

ANALYSIS: The actual power developed must satisfy, $\dot{W}_{cv} \leq (\dot{W}_{cv})_{max}$. Using the result of Problem 7.48, the maximum power that could be developed by any control volume of the type under consideration is

① $$(\dot{W}_{cv})_{max} = \dot{m}\, e_{fi} = \dot{m}\left[(h_i - h_0) - T_0(s_i - s_0) + \frac{V_i^2}{2} + g z_i\right]$$

Accordingly

$$\dot{W}_{cv} \leq \dot{m}\left[\underline{(h_i - h_0) - T_0(s_i - s_0) + \frac{V_i^2}{2}} + g z_i\right]$$

The underlined term vanishes because $p_i = p_0$, $T_i = T_0$ and the inlet kinetic energy is negligible by assumption 3. Thus,

$$\dot{W}_{cv} \leq \dot{m}(g z_i) \quad \text{or} \quad \dot{m} \geq \frac{\dot{W}_{cv}}{g z_i} = \frac{(10^6\ J/s)}{(9.81\frac{m}{s^2})(1000m)}\left|\frac{1N\cdot m}{J}\right|\left|\frac{1kg\cdot m/s^2}{1N}\right|$$

$\Rightarrow \dot{m} \geq 101.9$ kg/s $\qquad \dot{m}_{min} = 101.9$ kg/s $\longleftarrow$

---

1. Alternatively, the analysis could invoke Eq. 6.53a.

PROBLEM 7.51

KNOWN: Data are provided for the throttling of steam across a valve.

FIND: (a) Determine the flow exergy rates at the valve inlet and exit and the rate of exergy destruction. (b) Evaluate the annual cost associated with exergy destruction.

SCHEMATIC & GIVEN DATA:

$\dot{m} = 2\ \frac{lb}{s}$
$T_1 = 500°F$
$P_1 = 500\ lbf/in^2$
$P_2 = 400\ \frac{lbf}{in^2}$

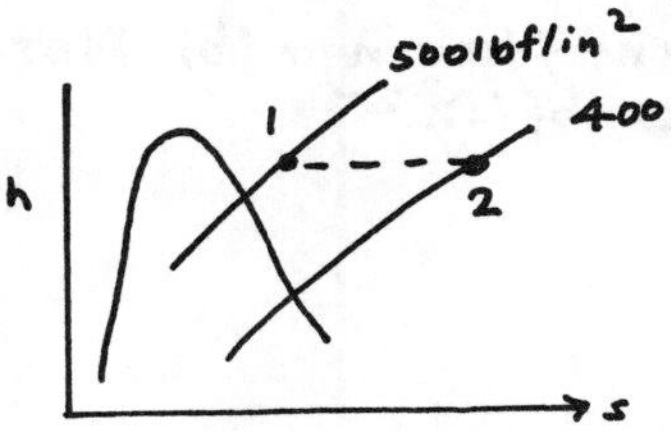

ASSUMPTIONS: 1. The expansion across the valve is a throttling process: $h_2 \approx h_1$. 2. For the environment, $T_0 = 537°R\ (77°F)$, $P_0 = 1\ atm$. 3. Exergy is valued at \$0.08/kW·h. 4. 8000 operating hours annually. 5. Ignore the effects of motion and gravity.

ANALYSIS: (a) With assumption 1, $h_2 \approx h_1$. Thus, state 2 is fixed by P_2, h_2. From Table A-4E, $h_1 = 1231.5$ Btu/lb, $s_1 = 1.4923$ Btu/lb·°R, $s_2 = 1.5135$ Btu/lb·°R. Table A-2E gives $h_0 \sim h_f(T_0) = 45.09$ Btu/lb, $s_0 \sim s_f(T_0) = 0.08775$ Btu/lb·°R. Then, with assumption 5, Eq. 7.20 gives

$$e_{f1} = h_1 - h_0 - T_0(s_1 - s_0) = (1231.5 - 45.09) - 537(1.4923 - 0.08775) = 432.2\ Btu/lb$$

$$e_{f2} = h_2 - h_0 - T_0(s_2 - s_0) = (1231.5 - 45.09) - 537(1.5135 - 0.08775) = 420.8\ Btu/lb$$

The associated rates are

$$\dot{E}_{f1} = \dot{m}\, e_{f1} = \left(2\ \frac{lb}{s}\right)\left(432.2\ \frac{Btu}{lb}\right) = 864.4\ \frac{Btu}{s} \quad \leftarrow$$

$$\dot{E}_{f2} = \dot{m}\, e_{f2} = (2)(420.8) = 841.6\ \frac{Btu}{s}$$

An exergy rate balance reduces to give

$$0 = \sum_j \left(1 - \frac{T_0}{T_j}\right)\dot{Q}_j^{\,0} - \dot{W}^{0} + (\dot{E}_{f1} - \dot{E}_{f2}) - \dot{E}_d$$

$$\Rightarrow \quad \dot{E}_d = \dot{E}_{f1} - \dot{E}_{f2} = (864.4 - 841.6) = 22.8\ \frac{Btu}{s} \quad \leftarrow$$

(b) Evaluating exergy at \$0.08/kW·h, the cost associated with the exergy destroyed for 8000 hours of operation annually is

$$\dot{\$} = \left(22.8\ \frac{Btu}{s}\right)\left|\frac{3600\,s}{h}\right|\left|\frac{1\,kW}{3413\,Btu/h}\right|\left(\frac{8000\,h}{year}\right)\left(\frac{\$0.08}{kW\cdot h}\right)$$

① $= \$15{,}392/year \quad \leftarrow$

---

1. The irreversibility associated with the expansion across the valve can be reduced by replacing the valve with a power-recovery turbine. However, the costs of the turbine and its operation would have to considered carefully before such a replacement would be approved.

PROBLEM 7.52

KNOWN: Steam at a known state enters a valve operating at steady state and undergoes a throttling process to pressure p_2.

FIND: (a) For $p = 500$ lbf/in.^{2}, determine the exit temperature and the exergy destruction per unit of steam flowing. (b) Plot these quantities versus p ranging from 500 to 1000 lbf/in.^{2}

SCHEMATIC & GIVEN DATA:

$p_1 = 1000$ lbf/in.^{2} $\quad 500 \le p_2 \le 1000$ lbf/in.^{2}

$T_1 = 600°F$

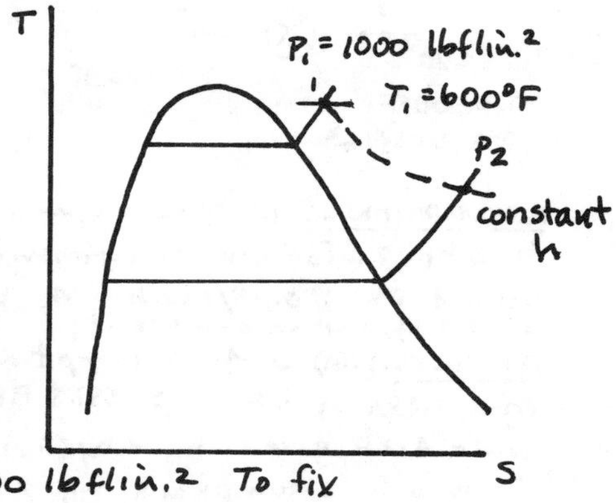

ASSUMPTIONS: (1) The control volume shown is at steady state. (2) The steam undergoes a throttling process in passing through the valve. (3) For the reference environment, $T_0 = 70°F$, $p_0 = 14.7$ lbf/in.^{2}

ANALYSIS: State 1 is fixed by $T_1 = 600°F$, $p_1 = 1000$ lbf/in.^{2} To fix state 2, we use assumption (2) to get

$$h_1 = h_2 \qquad (1)$$

Thus, with p_2 and h_2, state 2 is fixed as well and T_2 can be determined.

To get $\dot{E}_d/\dot{m}$, we begin with an exergy balance at steady state

$$0 = \sum_j \left(1 - \frac{T_0}{T_j}\right)\dot{Q}_j^{\,0} - \dot{W}^{\,0} + \dot{m}\left[e_{f1} - e_{f2}\right] - \dot{E}_d$$

or

$$\dot{E}_d/\dot{m} = e_{f1} - e_{f2} = (h_1 - h_2)^{0} - T_0(s_1 - s_2) \qquad (2)$$

① (a) From Table A-4E; $h_1 = 1248.8$ Btu/lb and $s_1 = 1.4450$ Btu/lb·°R. Interpolating in Table A-4E at $p_2 = 500$ lbf/in.^{2}, $h_2 = h_1 = 1248.8$ Btu/lb we get

$$T_2 = 524.6°F \qquad \text{(a) } T_2$$

$$s_2 = 1.5098 \text{ Btu/lb°R}$$

Now, from Eq. (2) above

$$\dot{E}_d/\dot{m} = -T_0(s_1 - s_2)$$

$$= -(530°R)(1.4450 - 1.5098)\ \frac{\text{Btu}}{\text{lb·°R}}$$

$$= 34.34 \text{ Btu/lb} \qquad \text{(a) } \dot{E}_d/\dot{m}$$

(b) The data to construct the required plots are obtained using IT, as follows:

PROBLEM 7.52 (Cont'd.)

IT Code

```
p1 = 1000 // lbf/in.²
T1 = 600 // °F
p2 = 500 // lbf/in.²
To = 530 // °R

h1 = h_PT("Water/Steam", p1, T1)
h2 = h_PT("Water/Steam", p2, T2)
h1 = h2
s1 = s_PT("Water/Steam", p1, T1)
s2 = s_PT("Water/Steam", p2, T2)

Ed = - To * (s1 - s2)
```

IT Results for p_2 = 500 lbf/in.²

h_1 = 1249 Btu/lb
s_1 = 1.445 Btu/lb·°R
s_2 = 1.51 Btu/lb·°R
$\dot{E}_d/\dot{m}$ = 34.43 Btu/lb
T_2 = 523.8 °F

PLOTS:

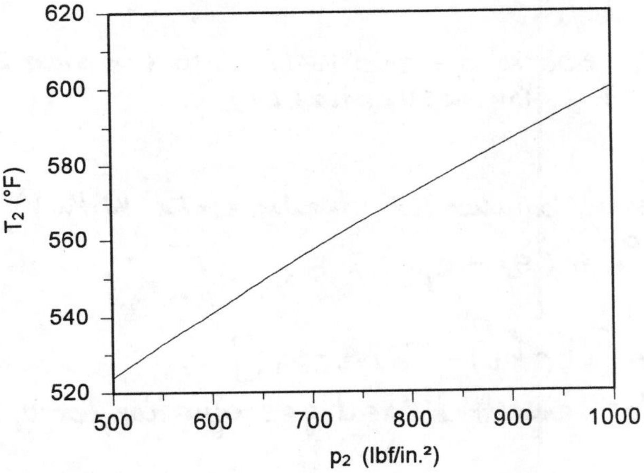

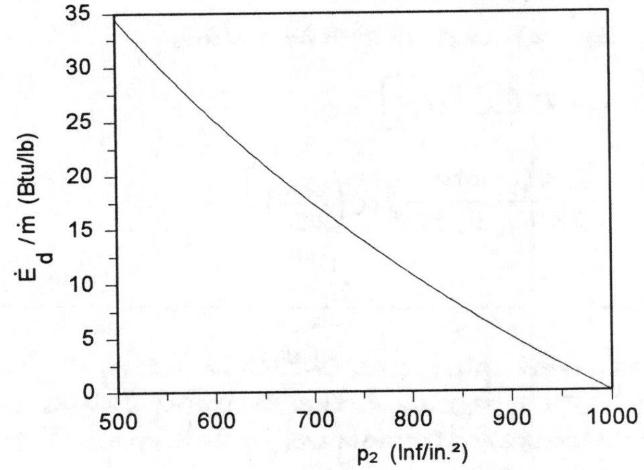

Discussion

- As seen on the accompanying T-s diagram, the exit temperature decreases as p_2 decreases for fixed h. Thus, the plot of exit temperature exhibits expected behavior.
- The rate of exergy destruction per unit mass of steam flowing increases with decreasing p_2. This is also expected based on greater entropy production and hence increasing s_2. (See the T-s diagram.)

1. In early printing of the 4th edition, h_1 is incorrectly given in Table A-4E as 1238.8 Btu/lb. The value used here (in the solution) is the correct value.

PROBLEM 7.53

KNOWN: Air at a given state enters a valve operating at steady state with a known volumetric flow rate. The air undergoes a throttling process to exit pressure p_2.

FIND: (a) Determine the rate of exergy destruction for $p_2 = 15$ lbf/in.² (b) Plot the exergy destruction rate versus p_2 ranging from 15 to 200 lbf/in.²

SCHEMATIC & GIVEN DATA:

$p_1 = 200$ lbf/in.²

$T_1 = 800°R$

$(AV)_1 = 100$ ft³/min

$15 \le p_2 \le 200$ lbf/in.²

$p_0 = 15$ lbf/in.²

$T_0 = 530°R$

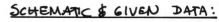

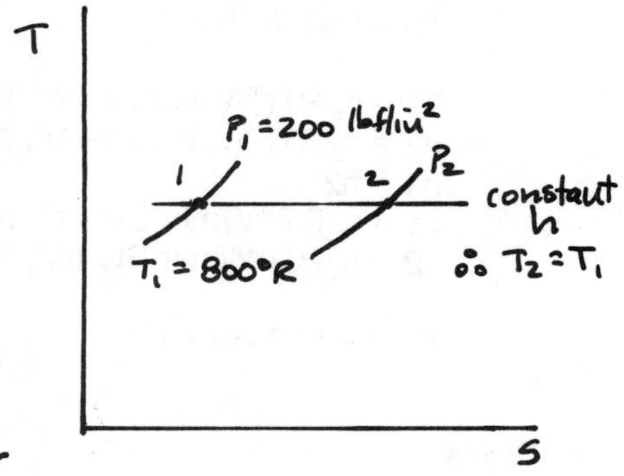

ASSUMPTIONS: (1) The control volume shown is at steady state. (2) The air undergoes a throttling process in passing through the valve. (3) The air is modeled as an ideal gas. (4) For the reference environment, $T_0 = 530°R$, $p_0 = 15$ lbf/in.²

ANALYSIS: State 1 is fixed by $T_1 = 800°R$, $p_1 = 200$ lbf/in.² To fix state 2, we use assumption (2) to get $h_1 = h_2$. By assumption (3)

$$T_2 = T_1 \quad (1)$$

To get $\dot{E}_d$, begin with an exergy balance at steady state. With $\dot{m}_1 = \dot{m}_2 \equiv \dot{m}$

$$0 = \sum_j \left[1 - \frac{T_0}{T_j}\right]\dot{Q}_j^{\,0} - \dot{W}_j^{\,0} + \dot{m}(e_{f1} - e_{f2}) - \dot{E}_d$$

or

$$\dot{E}_d = \dot{m}(e_{f1} - e_{f2}) = \dot{m}\left[(h_1 - h_2)^0 - T_0(s_1 - s_2)\right] \quad (2)$$

To get $\dot{m}$, we use $\dot{m} = (AV)_1/v_1$ and the ideal gas equation for v_1.

(a) Evaluating $\dot{m}$

$$\dot{m} = \frac{p_1 (AV)_1}{RT_1} = \frac{(200\ \text{lbf/in.}^2)(100\ \text{ft}^3/\text{min})}{\left(\frac{1545}{28.97}\ \frac{\text{ft}\cdot\text{lbf}}{\text{lb}\cdot°R}\right)(800°R)}\left|\frac{144\ \text{in.}^2}{1\ \text{ft}^2}\right| = 67.5\ \text{lb/min}$$

Using ideal gas relations with Eq. (2) and inserting values

$$\dot{E}_d = -\dot{m}T_0\left[(s^\circ(T_1) - s^\circ(T_2))^0 - R\ln p_1/p_2\right]$$

$$= -\left(67.5\ \frac{\text{lb}}{\text{min}}\right)(530°R)\left[-\frac{1.986}{28.97}\ \frac{\text{Btu}}{\text{lb}\cdot°R}\ln\left(\frac{200}{15}\right)\right]$$

$$= 6353\ \text{Btu/min} \qquad \text{(a) } \dot{E}_d$$

(b) The data to construct the required plot are obtained using IT, as follows. For the IT solution, $v_1(T_1, p_1)$ and the entropy values $s_1(T_1, p_1)$ and $s_2(T_2, p_2)$ are evaluated directly using internal IT property functions.

PROBLEM 7.53 (Cont'd.)

IT Code

```
p1 = 200  // lbf/in.²
T1 = 800  // °R
p2 = 15  // lbf/in.²
AV1 = 100  // ft³/min
To = 530  // °R

T2 = T1
v1 = v_TP("Air",T1,p1)
s1 = s_TP("Air", T1, p1)
s2 = s_TP("Air", T2, p2)

mdot = AV1 / v1
Edot = - mdot * To * (s1 - s2)
```

IT Results for p_2 = 15 lbf/in.²

v_1 = 1.481 ft³/lb
s_1 = 0.5167 Btu/lb·°R
s_2 = 0.6941 Btu/lb·°R
$\dot{m}$ = 67.5 lb/min
$\dot{E}_d$ = 6347 Btu/min

PLOT:

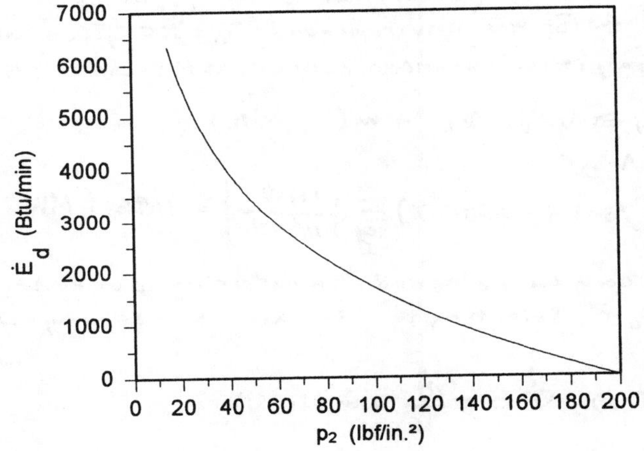

From the plot, we see that the rate of exergy destruction increases with decreasing p_2. This is expected based on greater entropy production and hence increasing s_2. (See the T-s diagram.)

PROBLEM 7.54

KNOWN: Steady-state operating data are provided for a steam turbine.

FIND: Determine the power developed. For each of two choices for the control volume, evaluate the rate of exergy destruction. Discuss.

SCHEMATIC & GIVEN DATA:

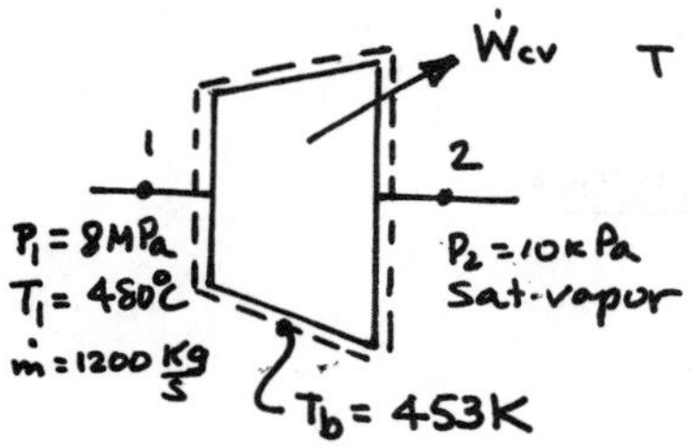

Control Volume (a)

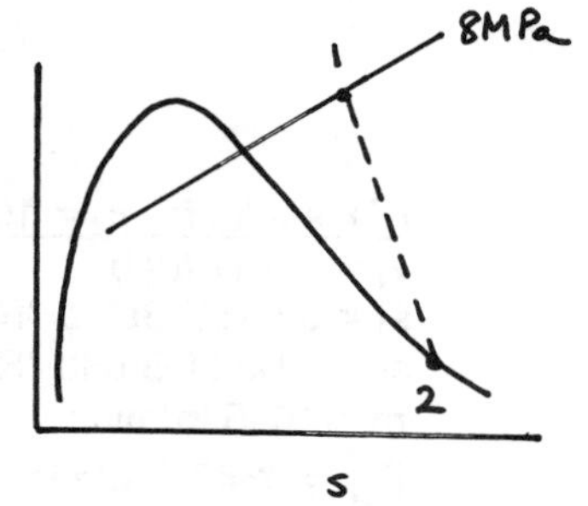

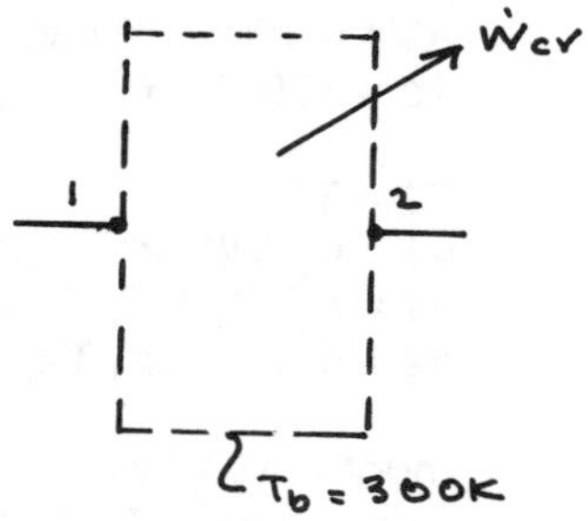

Control Volume (b)

ASSUMPTIONS: 1. The control volumes shown in the schematic are at steady state. 2. In each case, heat transfer occurs at T_b. 3. Effects of motion and gravity can be ignored. 4. For the environment, $T_0 = 300$ K, $p_0 = 100$ kPa.

ANALYSIS: (a) Mass and energy rate balances reduce at steady state to give

$$0 = \dot{Q}_{cv} - \dot{W}_{cv} + \dot{m}(h_1 - h_2) \Rightarrow \dot{W}_{cv} = \dot{Q}_{cv} + \dot{m}(h_1 - h_2).$$

Then, with data from Tables A-3, 4

$$\dot{W}_{cv} = (-30\,\text{MW}) + \left(1200\,\frac{\text{kg}}{\text{s}}\right)[3348.4 - 2584.7]\,\frac{\text{kJ}}{\text{kg}}\left|\frac{1\,\text{MW}}{10^3\,\text{kJ/s}}\right| = 886.4\,\text{MW} \leftarrow$$

The rate of exergy destruction can be obtained by reducing an exergy rate balance or using $\dot{E}_d = T_0\dot{\sigma}$. Selecting the second, an entropy rate balance reads

$$0 = \frac{\dot{Q}}{T_b} + \dot{m}(s_1 - s_2) + \dot{\sigma} \Rightarrow \dot{\sigma} = -\frac{\dot{Q}}{T_b} + \dot{m}(s_2 - s_1)$$

Then,

$$\dot{\sigma} = -\left(\frac{-30\,\text{MW}}{453\,\text{K}}\right) + \left(1200\,\frac{\text{kg}}{\text{s}}\right)[8.1502 - 6.6586]\,\frac{\text{kJ}}{\text{kg}\cdot\text{K}}\left|\frac{1\,\text{MW}}{10^3\,\text{kJ/s}}\right|$$

$$= 1.8561\,\frac{\text{MW}}{\text{K}} \Rightarrow \dot{E}_d = (300\,\text{K})\left(1.8561\,\frac{\text{MW}}{\text{K}}\right) = 556.8\,\text{MW}$$

(b) For the enlarged control volume, the entropy rate balance reads

$$\dot{\sigma} = -\frac{\dot{Q}}{T_b} + \dot{m}(s_2 - s_1)$$

$$= -\frac{(-30\,\text{MW})}{300\,\text{K}} + (1200)[8.1502 - 6.6586]\frac{1}{|10^3|} = 1.8899\,\frac{\text{MW}}{\text{K}}$$

① $$\Rightarrow \dot{E}_d = (300\,\text{K})\left(1.8899\,\frac{\text{MW}}{\text{K}}\right) = 567\,\text{MW}$$

The exergy destruction in control volume (b) is greater because there is an additional source of irreversibility — namely, the irreversibility related to heat transfer from the outer surface of the turbine to the ambient. For control volume (a), the heat transfer effect is an *external* irreversibility.

---

1. Using values obtained in the analysis, the net rate exergy is supplied to the turbine by the steam, $(\dot{E}_{f1} - \dot{E}_{f2})$, is 1453.4 MW. Since $[\dot{E}_d / (\dot{E}_{f1} - \dot{E}_{f2})] = [567/1453.4] = 0.39$ (39%), there is considerable scope in this case for improving thermodynamic performance.

PROBLEM 7.55

KNOWN: Steady state operating data are provided for an air turbine.

FIND: DETERMINE, per unit of mass flowing, the work developed and, for each of two choices of control volume, the exergy destruction. Discuss.

SCHEMATIC & GIVEN DATA:

(a)

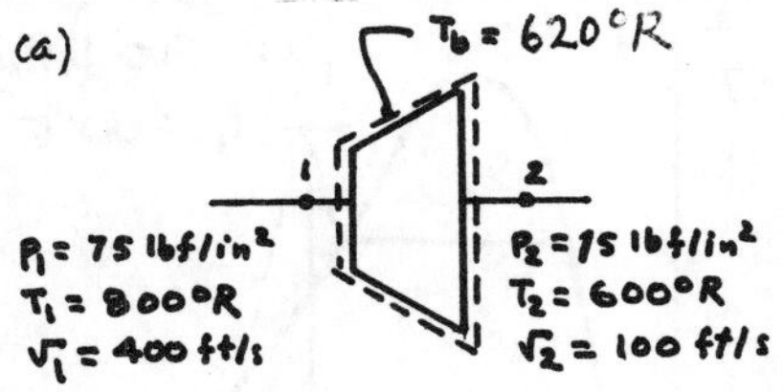

$\frac{\dot{Q}_{cv}}{\dot{m}} = -2 \frac{Btu}{lb}$

$T_0 = 40°F,\ p_0 = 15$ lbf/in²

(b)

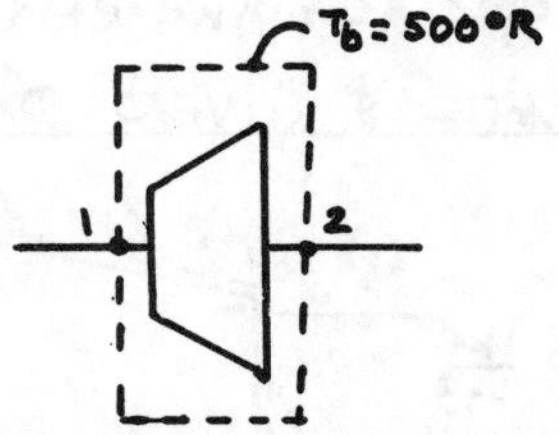

ASSUMPTIONS: (1) The control volumes in the accompanying figure are at steady state. (2) Heat transfer takes place at temperature T_b. (3) There is no effect of gravity. (4) For the environment, $T_0 = 500°R$, $p_0 = 1$ atm. (5) Air is modeled as an ideal gas.

ANALYSIS: At steady state an energy rate balance reduces to give

$$0 = \dot{Q}_{cv} - \dot{W}_{cv} + \dot{m}\left[(h_1 - h_2) + \left(\frac{V_1^2 - V_2^2}{2}\right) + g(z_1 - z_2)^{0}\right]$$

or

$$\frac{\dot{W}_{cv}}{\dot{m}} = \frac{\dot{Q}_{cv}}{\dot{m}} + (h_1 - h_2) + \left(\frac{V_1^2 - V_2^2}{2}\right) \qquad (1)$$

With data from Table A-22E

$$\frac{\dot{W}_{cv}}{\dot{m}} = -2\frac{Btu}{lb} + (191.81 - 143.47)\frac{Btu}{lb} + \frac{[(400)^2 - (100)^2](ft^2/s^2)}{(2)\ 32.2 \frac{lb \cdot ft}{lbf \cdot s^2}\ 778 \frac{ft \cdot lbf}{Btu}}$$

$$= -2\frac{Btu}{lb} + 48.34\frac{Btu}{lb} + 2.99\frac{Btu}{lb} = 49.33\frac{Btu}{lb} \longleftarrow \dot{W}_{cv}/\dot{m}$$

Control Volume (a)

The exergy destruction can be evaluated by reducing an exergy rate balance. Thus

$$\frac{\dot{E}_d}{\dot{m}} = \left[1 - \frac{T_0}{T_b}\right]\left(\frac{\dot{Q}_{cv}}{\dot{m}}\right) - \frac{\dot{W}_{cv}}{\dot{m}} + \left[(h_1 - h_2) - T_0(s_1 - s_2) + \frac{V_1^2 - V_2^2}{2}\right]$$

Introducing Eq. (1) and simplifying

$$\frac{\dot{E}_d}{\dot{m}} = T_0\left[(s_2 - s_1) - \frac{\dot{Q}_{cv}/\dot{m}}{T_b}\right] \qquad (2)$$

Or, for an ideal gas (assumption 5)

$$\frac{\dot{E}_d}{\dot{m}} = T_0\left[\left(s°(T_2) - s°(T_1) - \frac{\bar{R}}{M}\ln\frac{p_2}{p_1}\right) - \frac{\dot{Q}_{cv}/\dot{m}}{T_b}\right] \qquad (3)$$

With data from Table A-22E

$$\frac{\dot{E}_d}{\dot{m}} = 500°R\left[\left(0.62607 - 0.69558 - \frac{1.986}{28.97}\ln\frac{15}{75}\right)\frac{Btu}{lb \cdot °R} - \frac{(-2\ Btu/lb)}{620}\right]$$

$$= 500°R\left[0.04082 + 0.0032\right]\frac{Btu}{lb \cdot °R} = 22.01\ Btu/lb \longleftarrow \dot{I}_{cv}/\dot{m}$$

Control Volume (b). Eqs. (2)-(3) remain applicable. Thus

$$\frac{\dot{E}_d}{\dot{m}} = 500°R\left[0.04082 - \frac{(-2\ Btu/lb)}{500°R}\right]$$

$$= 22.41\frac{Btu}{lb} \longleftarrow (\dot{I}_{cv}/\dot{m})$$

Discussion: The irreversibility is greater for control volume (b) because the enlarged control volume has an additional source of irreversibility: the irreversibility associated with heat transfer from the turbine's surface to the surroundings. For control volume (a), the heat transfer effect is an external irreversibility.

PROBLEM 7.56

KNOWN: An insulated steam turbine operating at steady state receives steam at a known state and exhausts at a given pressure.

FIND: Plot the exergy destruction rate per unit of steam flowing versus isentropic turbine efficiency ranging from 70 to 100%.

SCHEMATIC & GIVEN DATA:

$p_1 = 400\ \frac{lbf}{in^2}$

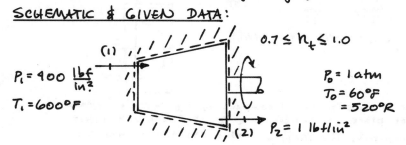

$T_1 = 600°F$

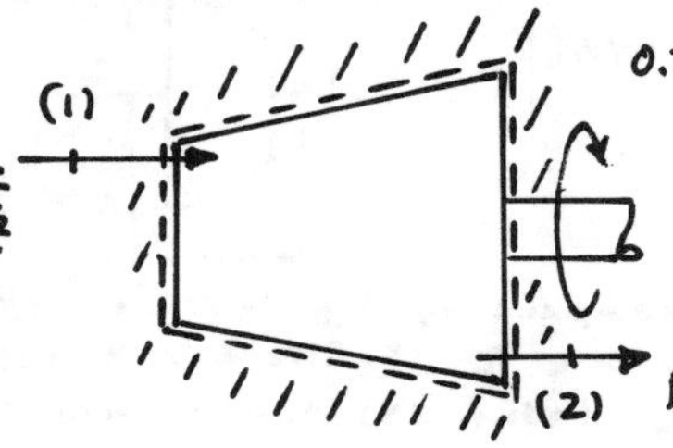

$0.7 \le \eta_t \le 1.0$

$p_0 = 1$ atm

$T_0 = 60°F = 520°R$

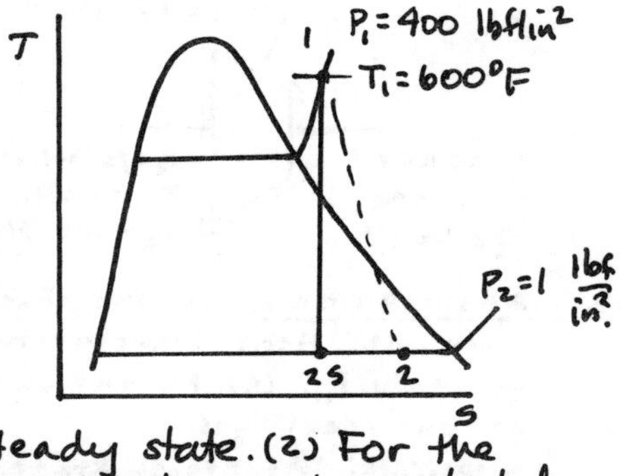

ASSUMPTIONS: (1) The control volume shown is at steady state. (2) For the turbine, $\dot{Q}_{cv} = 0$ and kinetic and potential energy effects can be neglected. (3) For the exergy reference environment, $T_0 = 520°R$, $p_0 = 1$ atm.

ANALYSIS: To determine the exergy destruction rate, we begin with mass and entropy rate balances which reduce to give

$$0 = \sum_j \left(\frac{\dot{Q}_j}{T_j}\right)^{0} + \dot{m}(s_1 - s_2) + \dot{\sigma}_{cv} \Rightarrow \dot{\sigma}_{cv}/\dot{m} = (s_2 - s_1)$$

With $\dot{E}_d = T_0 \dot{\sigma}_{cv}$

$$\dot{E}_d/\dot{m} = T_0(s_2 - s_1) \qquad (1)$$

Since $p_1 = 400$ lbf/in.², $T_1 = 600°F$, state 1 is in the superheated vapor region and the state is fixed. To fix state 2, we use the isentropic turbine efficiency

$$\eta_t = \frac{h_1 - h_2}{h_1 - h_{2s}} \Rightarrow h_2 = h_1 - \eta_t(h_1 - h_{2s}) \qquad (2)$$

where h_{2s} is determined using $p_2 = 1$ lbf/in.² and $s_{2s} = s_1$. Then, with h_2 known from Eq.(2) and p_2, s_2 can be determined.

Sample calculation: CASE: $\eta_t = 0.7$ From Table A-4E, $h_1 = 1306.6$ Btu/lb and $s_1 = 1.5892$ Btu/lb·°R. With $s_{2s} = s_1 = 1.5892$

$$x_{2s} = \frac{s_{2s} - s_{f2}}{s_{fg2}} = \frac{1.5892 - 0.1327}{1.8453} = 0.7893$$

and $h_{2s} = h_{f2} + x_{2s} h_{fg2} = 69.74 + (.7893)(1036.0) = 887.45$ Btu/lb

From (2)

$$h_2 = h_1 - \eta_t(h_1 - h_{2s}) = 1306.6 - (.7)(1306.6 - 887.45) = 1013.2 \text{ Btu/lb}$$

Finally, with $p_2 = 1$ lbf/in.²; $x_2 = \frac{h_2 - h_{f2}}{h_{fg2}} = 0.9107$ and $s_2 = s_{f2} + x_2 s_{fg2} = 1.8132\ \frac{Btu}{lb·°R}$

Thus

$$\dot{E}_d/\dot{m} = T_0(s_2 - s_1) = (520°R)(1.8132 - 1.5892)\frac{Btu}{lb·°R}$$

$= 116.5$ Btu/lb ← $\dot{E}_d/\dot{m}$ ($\eta_t = 70\%$)

PROBLEM 7.56 (Cont'd.)

The data to construct the required plot are obtained using IT, as follows:

IT Code

```
p1 = 400  // lbf/in.²
T1 = 600  // °F
p2 = 1  // lbf/in.²
eff = 0.7
To = 60 + 460  // °R

h1 = h_PT("Water/Steam", p1, T1)
s1 = s_PT("Water/Steam", p1, T1)
s2s = s1
h2s = h_Ps("Water/Steam", p2, s2s)

h2 = h1 - eff * (h1 - h2s)
Ed = To * (s2 - s1)
h2 = h_Ps("Water/Steam", p2, s2)
```

IT Results for η_t = 70%

h_1 = 1306 Btu/lb
h_2 = 1013 Btu/lb
h_{2s} = 887.3 Btu/lb
s_1 = 1.589 Btu/lb·°R
s_2 = 1.813 Btu/lb·°R
s_{2s} = 1.589 Btu/lb·°R
$\dot{E}_d / \dot{m}$ = 116.5 Btu/lb

PLOT:

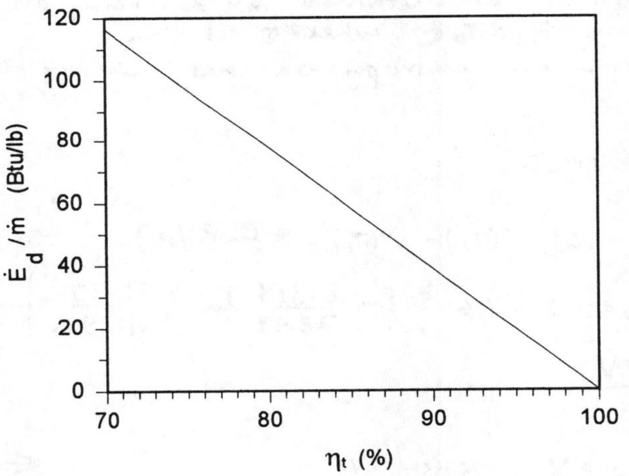

From the plot we see that lower values of isentropic turbine efficiency correspond to increased exergy destruction, as expected.

PROBLEM 7.57

KNOWN: Steady-state operating data are provided for an air compressor.

FIND: Determine the power required by the compressor and the rate of exergy destruction. Express the exergy destruction as a percentage of the power input.

SCHEMATIC & GIVEN DATA:

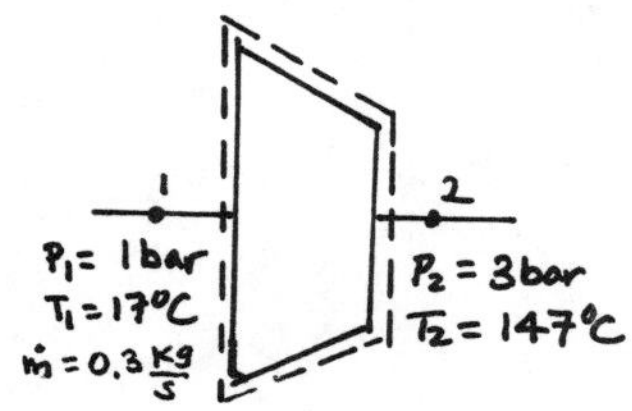

ASSUMPTIONS:

1. The control volume shown in the schematic is at steady state, $\dot{Q}_{cv} = 0$.
2. Effects of motion and gravity can be ignored.
3. Air is modeled as an ideal gas. (1)
4. $T_0 = 290$ K, $p_0 = 1$ bar

ANALYSIS: Mass and energy rate balances reduce to give

$$0 = \dot{Q}_{cv}^{\,0} - \dot{W}_{cv} + \dot{m}\left[(h_1 - h_2) + \frac{V_1^2 - V_2^2}{2}^{\,0} + g(z_1 - z_2)^{\,0}\right]$$

with data from Table A-22

$$\dot{W}_{cv} = \dot{m}(h_1 - h_2) = \left(0.3\,\frac{kg}{s}\right)\left[290.16 - 421.26\right]\frac{kJ}{kg}\left|\frac{1\,kW}{1\,kJ/s}\right| = -39.33\ kW \quad \leftarrow$$

The rate of exergy destruction can be obtained by reducing an exergy rate balance or using $\dot{E}_d = T_0\dot{\sigma}_{cv}$, where $\dot{\sigma}$ is the rate of entropy production from an entropy balance. Using the entropy approach,

$$0 = \sum_j \frac{\dot{Q}_j}{T_j}^{\,0} + \dot{m}(s_1 - s_2) + \dot{\sigma}_{cv}$$

$$\Rightarrow \quad \dot{\sigma}_{cv} = \dot{m}(s_2 - s_1) = \dot{m}\left[s^\circ(T_2) - s^\circ(T_1) - R\ln p_2/p_1\right]$$

$$= \left(0.3\,\frac{kg}{s}\right)\left[2.04142 - 1.66802 - \frac{8.314}{28.97}\ln\frac{3}{1}\right]\frac{kJ}{kg\cdot K}\left|\frac{1\,kW}{1\,kJ/s}\right|$$

$$= 0.0174\ \frac{kW}{K}$$

Then,

$$\dot{E}_d = (290\,K)\left(0.0174\,\frac{kW}{K}\right) = 5.05\ kW \quad \leftarrow$$

Forming a ratio

$$\left(\frac{\dot{E}_d}{-\dot{W}_{cv}}\right) = \frac{5.05\ kW}{39.33\ kW} = 0.128\ (12.8\%) \quad \leftarrow$$

---

1. Can be checked using the generalized compressibility chart.

PROBLEM 7.58

KNOWN: Operating data are provided for an insulated compressor at steady state.

FIND: Determine (a) the exit temperature, (b) the power input, and (c) the exergy destruction rate expressed as a percentage of the power input.

SCHEMATIC & GIVEN DATA:

R-22

$T_1 = -15°C$
$P_1 = 200$ kPa $= 2$ bar
$\dot{m} = 90$ kg/h

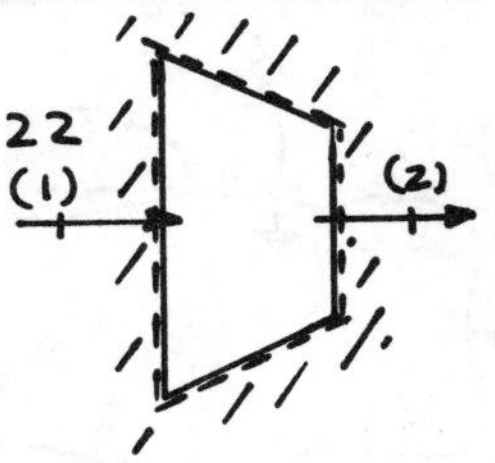

$P_2 = 1$ MPa $= 10$ bar
$\eta_c = 0.85$

$T_0 = 20°C = 293$ K
$P_0 = 1$ bar

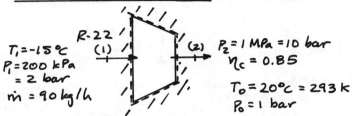

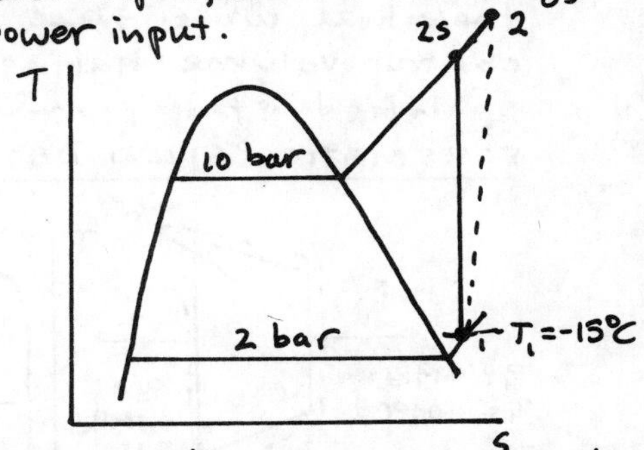

ASSUMPTIONS: (1) The control volume shown is at steady state. (2) For the control volume, $\dot{Q}_{cv} = 0$ and kinetic and potential energy effects can be neglected. (3) For the exergy reference environment, $T_0 = 293$K, $p_0 = 1$ bar.

ANALYSIS: (a) To find the exit temperature, we begin with the isentropic compressor efficiency

$$\eta_c = \frac{h_{2s} - h_1}{h_2 - h_1} \Rightarrow h_2 = h_1 + \frac{h_{2s} - h_1}{\eta_c}$$

From Table A-9, $h_1 = 246.47$ kJ/kg, $s_1 = 0.9952$ kJ/kg·K. Interpolating in Table A-9 with $p_2 = 10$ bar and $s_{2s} = s_1 = 0.9952$; $h_{2s} = 288.84$ kJ/kg. Thus

$$h_2 = 246.47 + \frac{(288.84 - 246.47)}{0.85} = 296.32 \text{ kJ/kg}$$

Interpolating again in Table A-9 with $p_2 = 10$ bar, $h_2 = 296.32$ kJ/kg

$$T_2 = 71.6\ °C \longleftarrow T_2$$

$$s_2 = 1.0172 \text{ kJ/kg·K}$$

(b) Applying energy and mass rate balances with assumptions (1) and (2)

$$\dot{m}_1 = \dot{m}_2 \equiv \dot{m}$$

$$0 = \cancelto{0}{\dot{Q}_{cv}} - \dot{W}_{cv} + \dot{m}\left[(h_1 - h_2) + \cancelto{0}{\left(\frac{V_1^2 - V_2^2}{2}\right)} + g\cancelto{0}{(z_1 - z_2)}\right]$$

Solving for the power input, $(-\dot{W}_{cv})$, and inserting values

$$-(\dot{W}_{cv}) = \dot{m}(h_2 - h_1)$$

$$= \left(90 \frac{kg}{h}\right)\left|\frac{1 h}{3600 s}\right|(296.32 - 246.47)\frac{kJ}{kg}\left|\frac{1 kW}{1 kJ/s}\right| = 1.246 \text{ kW} \longleftarrow (-\dot{W}_{cv})$$

(c) The exergy destruction rate is conveniently evaluated using $\dot{E}_d = T_0 \dot{\sigma}_{cv}$, where $\dot{\sigma}_{cv}$ is obtained from and entropy balance

$$0 = \cancelto{0}{\sum_j \left(\frac{\dot{Q}}{T}\right)_j} + \dot{m}(s_1 - s_2) + \dot{\sigma}_{cv} \Rightarrow \dot{E}_d = T_0 \dot{m}(s_2 - s_1)$$

Thus

$$\dot{E}_d = (293 K)\left(90 \frac{kg}{h}\right)(1.0172 - 0.9952)\frac{kJ}{kg·K}\left|\frac{1 h}{3600 s}\right|\left|\frac{1 kW}{1 kJ/s}\right| = 0.1612 \text{ kW}$$

Finally

$$\% = \left(\frac{0.1612}{1.246}\right)(100) = 12.4\% \longleftarrow \%$$

PROBLEM 7.59

KNOWN: Steady-state operating data are provided for a steam turbine.

FIND: (a) Determine the work developed and exergy destruction, each per unit mass of steam flowing. (b) Determine the maximum theoretical work that could be developed by any one-inlet, one-exit control volume having steam entering and exiting at the specified states, while allowing heat transfer only at T_0. Discuss.

SCHEMATIC & GIVEN DATA:

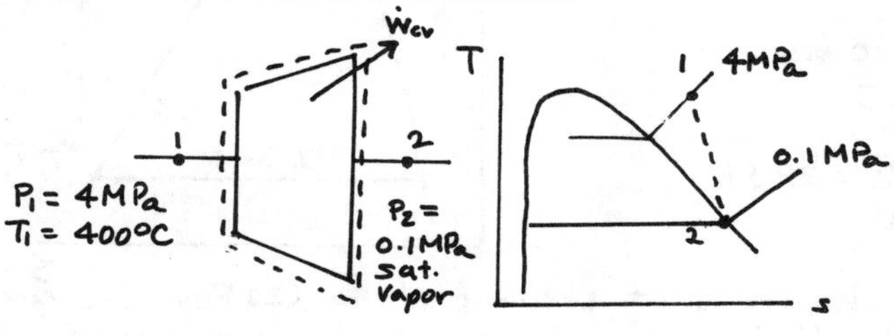

ACTUAL

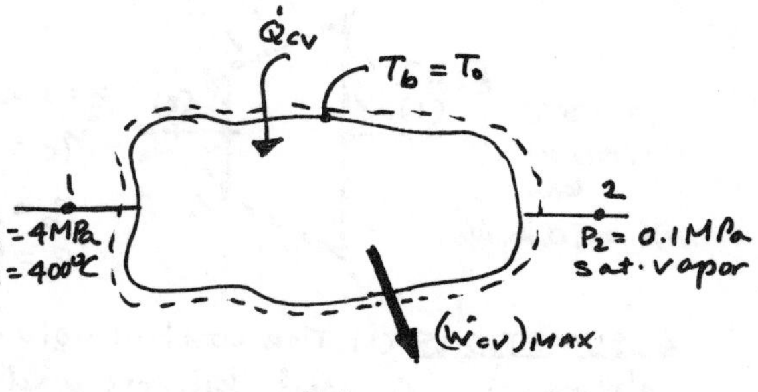

HYPOTHETICAL

ASSUMPTIONS: 1. The control volumes shown in the schematic are at steady state. 2. In the actual case, $\dot{Q}_{cv} = 0$. In the hypothetical case, heat transfer occurs only at $T_b = T_0$. 3. The effects of motion and gravity can be ignored. 4. $T_0 = 300K (27°C)$, $P_0 = 0.1 MPa$.

ANALYSIS: (a) Reducing mass and energy rate balances, we get

$$\frac{\dot{W}_{cv}}{\dot{m}} = h_1 - h_2 = (3213.6 - 2675.5)\frac{kJ}{kg} = 538.1 \text{ kJ/kg} \quad \leftarrow$$

Reducing mass and entropy rate balances

$$\frac{\dot{\sigma}}{\dot{m}} = s_2 - s_1 = (7.3594 - 6.7690)\frac{kJ}{kg \cdot K} = 0.5904 \frac{kJ}{kg \cdot K}$$

Then

$$\frac{\dot{E}_d}{\dot{m}} = T_0\left(\frac{\dot{\sigma}}{\dot{m}}\right) = 300K(.5904)\frac{kJ}{kg \cdot K} = 177.12 \frac{kJ}{kg} \quad \leftarrow$$

(b) Mass and exergy rate balances reduce for the hypothetical case to give

$$0 = \underbrace{\left[1 - \frac{T_0}{T_b}\right]}_{=0 \text{ since } T_b = T_0}\dot{Q}_{cv} - \dot{W}_{cv} + \dot{m}(e_{f1} - e_{f2}) - \dot{E}_d$$

$$\Rightarrow \quad \frac{\dot{W}_{cv}}{\dot{m}} = (e_{f1} - e_{f2}) - \frac{\dot{E}_d}{\dot{m}} \implies \left(\frac{\dot{W}_{cv}}{\dot{m}}\right)_{MAX} = e_{f1} - e_{f2}$$

That is, the maximum power is obtained when there are no internal irreversibilities.

Then

$$\left(\frac{\dot{W}_{cv}}{\dot{m}}\right)_{MAX} = (h_1 - h_2) - T_0(s_1 - s_2) = (3213.6 - 2675.5) - 300(6.7690 - 7.3594)$$

$$= 538.1 + 177.12 = 715.22 \text{ kJ/kg} \quad \leftarrow$$

Combining the principal results for (a), (b)

$$\left(\frac{\dot{W}_{cv}}{\dot{m}}\right)_{MAX} = \left(\frac{\dot{W}_{cv}}{\dot{m}}\right) + \left(\frac{\dot{E}_d}{\dot{m}}\right)$$

That is, the difference between the two work amounts is the exergy destruction for the actual turbine expansion.

PROBLEM 7.60

KNOWN: Operating data are provided for an insulated turbine at steady state.

FIND: Determine (a) the work developed and the exergy destruction, each per unit mass of air flowing, and (b) the maximum work that could be developed by any one-inlet, one-exit control volume at steady state with air entering and exiting at the specified states while allowing heat transfer only at temperature T_0.

SCHEMATIC & GIVEN DATA:

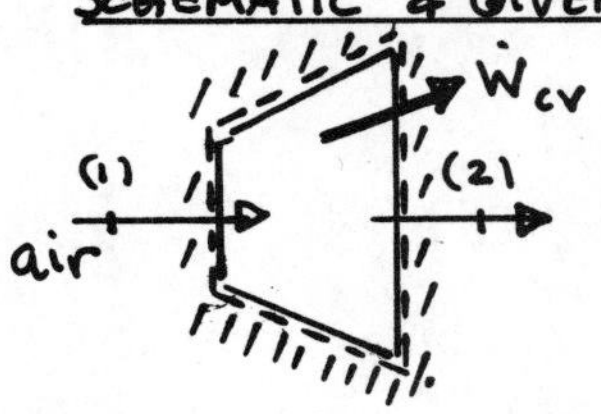

$p_1 = 8\,bar$, $T_1 = 500K$, $V_1 = 150\,m/s$

$p_2 = 1\,bar$, $T_2 = 320K$, $V_2 = 10\,m/s$

$T_0 = 300K$, $p_0 = 1\,bar$

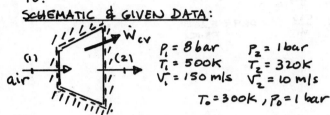

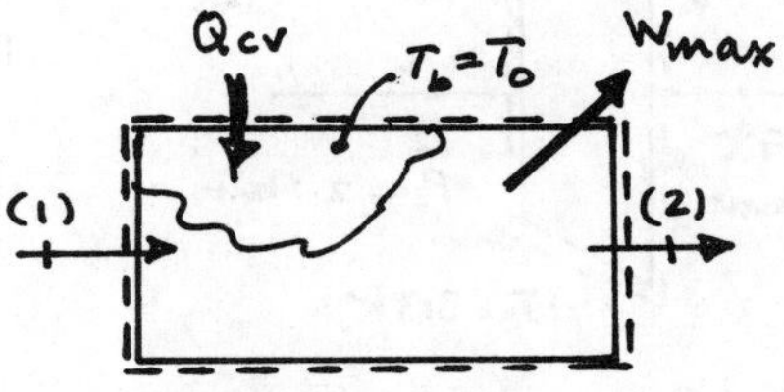

ASSUMPTIONS: (1) The control volumes shown are at steady state. (2) In part (a), $\dot{Q}_{cv} = 0$. In part (b), heat transfer occurs only at $T_b = T_0$. (3) Potential energy effects can be neglected. (4) The air is modeled as an ideal gas. (5) For the environment, $T_0 = 300K$, $p_0 = 1atm$.

ANALYSIS: The mass and energy rate balances reduce with assumptions (1)-(3) to give

$$0 = \cancel{\dot{Q}_{cv}}^{0} - \dot{W}_{cv} + \dot{m}\left[(h_1 - h_2) + \frac{(V_1^2 - V_2^2)}{2} + g\cancel{(z_1 - z_2)}^{0}\right] \Rightarrow \frac{\dot{W}_{cv}}{\dot{m}} = h_1 - h_2 + \frac{(V_1^2 - V_2^2)}{2} \qquad (1)$$

With data from Table A-22

$$\frac{\dot{W}_{cv}}{\dot{m}} = (503.02 - 320.29)\frac{kJ}{kg} + \left(\frac{150^2 - 10^2}{2}\right)\frac{m^2}{s^2}\left|\frac{1\,N}{1\,kg\cdot m/s^2}\right|\left|\frac{1\,kJ}{10^3\,N\cdot m}\right| = 193.93\,\frac{kJ}{kg} \longleftarrow (\dot{W}_{cv}/\dot{m})$$

The exergy destruction rate is found using $\dot{E}_d = T_0\dot{\sigma}_{cv}$ and an entropy balance.

$$0 = \cancel{\sum_j \left(\frac{\dot{Q}}{T}\right)_j}^{0} + \dot{m}(s_1 - s_2) + \dot{\sigma}_{cv} \Rightarrow \dot{E}_d/\dot{m} = T_0(s_2 - s_1) \qquad (2)$$

With ideal gas relations and data from Table A-22

$$\frac{\dot{E}_d}{\dot{m}} = T_0\left[s^\circ(T_2) - s^\circ(T_1) - R\ln p_2/p_1\right]$$

$$= (300K)\left[(1.76690 - 2.21952) - \left(\frac{8.314\,kJ}{28.97\,kg\cdot K}\right)\ln\left(\frac{1}{8}\right)\right] = 43.25\,kJ/kg \longleftarrow \dot{E}_d/\dot{m}$$

(b) In this case, the exergy rate balance reduces at steady state to give

$$0 = \cancel{\left[1 - \frac{T_0}{T_0}\right]}^{0}\dot{Q}_{cv} - \dot{W}_{cv} + \dot{m}(e_{f1} - e_{f2}) - \dot{E}_d$$

The work is maximum when there are no internal irreversibilities: $\dot{E}_d = 0$. Thus

$$\left(\frac{\dot{W}_{cv}}{\dot{m}}\right)_{max} = e_{f1} - e_{f2} = (h_1 - h_2) - T_0(s_1 - s_2) + \frac{(V_1^2 - V_2^2)}{2} + g\cancel{(z_1 - z_2)}^{0}$$

$$= (h_1 - h_2) - T_0\left[s^\circ(T_2) - s^\circ(T_1) - R\ln p_2/p_1\right] + \frac{(V_1^2 - V_2^2)}{2} \qquad (3)$$

$$= (503.02 - 320.29) - (300)\left[(2.21952 - 1.76690) - \frac{8.314}{28.97}\ln\left(\frac{8}{1}\right)\right] + \left(\frac{150^2 - 10^2}{2}\right)\left|\frac{1}{10^3}\right|$$

$$= 237.18\,kJ/kg \longleftarrow (\dot{W}_{cv}/\dot{m})_{max}$$

PROBLEM 7.61

KNOWN: Steady-state operating data are provided for an air compressor.

FIND: Determine the rate of exergy destruction and the rate of exergy transfer accompanying heat, each per unit mass of air flowing. Express each as a percentage of the compressor work input.

SCHEMATIC & GIVEN DATA:

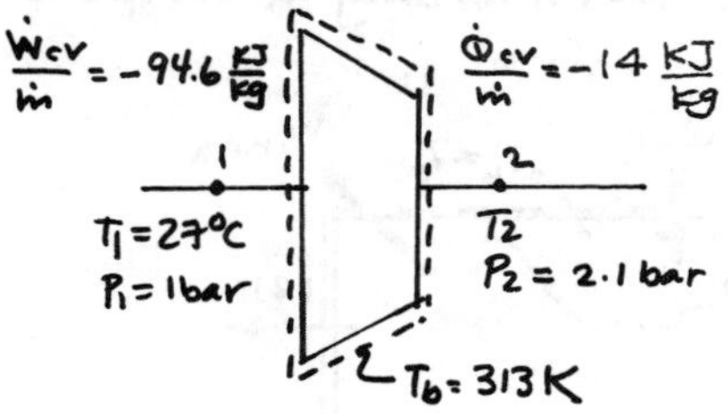

Using energy and entropy balances, the solution of Problem 6.116 gives

$T_2 = 350K$, $\dfrac{\dot{\sigma}_{cv}}{\dot{m}} = 0.0698 \dfrac{kJ}{kg \cdot K}$

ASSUMPTIONS: 1. The control volume shown in the schematic is at steady state. 2. Heat transfer occurs only at T_b. 3. Air is modeled as an ideal gas. 4. The effects of motion and gravity can be ignored. 5. $T_0 = 293K$ (20°C), $P_0 = 1$ atm.

ANALYSIS: The exergy destruction can be found by reducing an exergy balance. Alternatively,

$$\frac{\dot{E}_d}{\dot{m}} = T_0\left(\frac{\dot{\sigma}_{cv}}{\dot{m}}\right) = (293K)\left(0.0698 \frac{kJ}{kg \cdot K}\right) = 20.45 \frac{kJ}{kg} \quad \longleftarrow$$

With assumption 2

$$\begin{bmatrix}\text{exergy transfer accompanying} \\ \text{heat transfer per unit of} \\ \text{mass flowing}\end{bmatrix} = \left[1 - \frac{T_0}{T_b}\right]\left(\frac{\dot{Q}_{cv}}{\dot{m}}\right)$$

$$= \left[1 - \frac{293}{313}\right]\left(-14 \frac{kJ}{kg}\right)$$

$$= -0.89 \frac{kJ}{kg}$$

Expressed as percentages of the work input:

exergy destruction: $\left(\dfrac{20.45}{94.6}\right)(100) = 21.6\%$

exergy transfer: $\left(\dfrac{0.89}{94.6}\right)(100) = 0.9\%$ $\longleftarrow$

PROBLEM 7.62

KNOWN: Steady-state operating data are provided for a water-jacketed air compressor.

FIND: Perform an exergy accounting of the power input

SCHEMATIC:

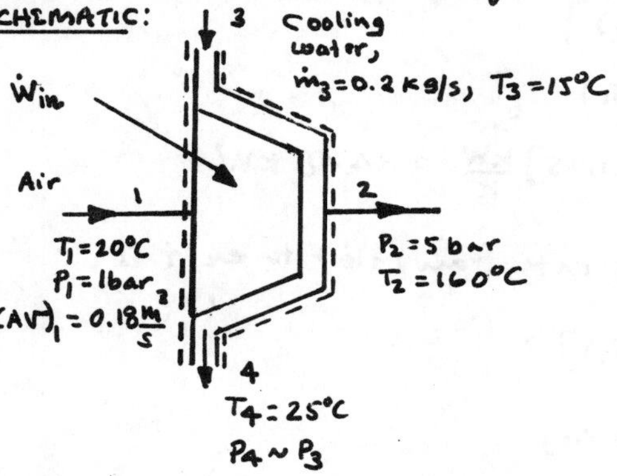

ASSUMPTIONS:

1. The control volume shown in the schematic operates at steady state with negligible kinetic and potential energy effects and $\dot{Q}_{cv} = 0$.
2. Air is modeled as an ideal gas.
3. Water is modeled as incompressible with $c = 4.19$ kJ/kg·K (Table A-19).
4. $T_0 = 293$ K (20°C), $p_0 = 1$ atm.

ANALYSIS: The exergy entering the control volume with the power input has the following disposition: (i) the exergy of the air stream is increased, (ii) the exergy of the water stream is increased, and (iii) exergy is destroyed by irreversibilities within the control volume. These quantities are now evaluated:

At steady state $\dot{m}_1 = \dot{m}_2$, $\dot{m}_3 = \dot{m}_4$. The energy rate balance reduces to give

$$\dot{W}_{cv} = \dot{m}_1 (h_1 - h_2) + \dot{m}_3 (h_3 - h_4)$$

With the ideal gas model equation of state

$$\dot{m}_1 = \frac{(AV)_1}{v_1} = \frac{(AV)_1}{(RT_1/p_1)} = \frac{(10^5 N/m^2)(0.18 m^3/s)}{\left(\frac{8314}{28.97} \frac{N \cdot m}{kg \cdot K}\right)(293 K)} = 0.214 \text{ kg/s}$$

Then, with h_1, h_2 from Table A-22 and Eq. 3.20b ($p_4 \sim p_3$)

$$\dot{W}_{cv} = \left(0.214 \frac{kg}{s}\right)[293.2 - 434.48] \frac{kJ}{kg} + \left(0.2 \frac{kg}{s}\right)\left(4.19 \frac{kJ}{kg \cdot K}\right)(-10 K)$$
$$= [(-30.23) + (-8.38)] \frac{kJ}{s} \left| \frac{1 kW}{1 kJ/s} \right| = -38.61 \text{ kW}$$

The rate of exergy destruction can be determined using an exergy rate balance or $\dot{E}_d = T_0 \dot{\sigma}$ and an entropy rate balance. At steady state the entropy rate balance reduces to give

$$0 = \sum_j \cancel{\dot{Q}_j}^{0} / T_j + \dot{m}_1 (s_1 - s_2) + \dot{m}_3 (s_3 - s_4) + \dot{\sigma}_{cv}$$

Or with $s°$ data from Table A-22 and Eq. 6.24

$$\dot{\sigma}_{cv} = \dot{m}_1 [s°(T_2) - s°(T_1) - R \ln p_2/p_1] + \dot{m}_3 \, c \ln \frac{T_4}{T_3}$$
$$= \left(0.214 \frac{kg}{s}\right)\left[2.07234 - 1.6783 - \frac{8.314}{28.97} \ln \frac{5}{1}\right] \frac{kJ}{kg \cdot K} + (0.2)(4.19) \ln \frac{298}{288}$$
$$= [(-0.0145) + (0.0286)] \frac{kJ/s}{K} \left| \frac{1 kW}{1 kJ/s} \right| = 0.0141 \frac{kW}{K}$$

$$\therefore \dot{E}_d = T_0 \dot{\sigma}_{cv} = (293 K)\left(0.0141 \frac{kW}{K}\right) = 4.13 \text{ kW}$$

For the air, the increase in flow exergy rate from inlet to exit is

$$\dot{E}_{f2} - \dot{E}_{f1} = \dot{m}_1(e_{f2} - e_{f1}) = \dot{m}_1[(h_2 - h_1) - T_0(s_2 - s_1)]$$

$$= \dot{m}_1(h_2 - h_1) - T_0\,\dot{m}(s_2 - s_1)$$

$$= 30.23\text{ kW} - (293\text{K})[-0.0145]\frac{\text{kW}}{\text{K}} = 34.48\text{ kW}$$

For the water, the increase in flow exergy rate from inlet to exit is

$$\dot{E}_{f4} - \dot{E}_{f3} = \dot{m}_3(e_{f4} - e_{f3}) = \dot{m}_3[(h_4 - h_3) - T_0(s_4 - s_3)]$$

$$= \dot{m}_3(h_4 - h_3) - T_0\,\dot{m}_3(s_4 - s_3)$$

$$= 8.38\text{ kW} - (293\text{ K})(0.0286)\frac{\text{kW}}{\text{K}} \sim 0$$

Exergy Accounting:

- Exergy input: power input — 38.61 kW
- Disposition:
  - Increase in exergy of the air stream — 34.48 kW (89.3%)
  - Increase in exergy of the water stream — ~0 kW (~0%)
  - Exergy destruction — 4.13 kW (10.7%)

Total: 38.61 kW

PROBLEM 7.63

KNOWN: Steady state operating data are provided for a steam turbine.

FIND: Plot versus the isentropic turbine efficiency ranging from 0 to 100%, the turbine exit temperature, power developed, and the rate of exergy destruction within the turbine.

SCHEMATIC & GIVEN DATA:

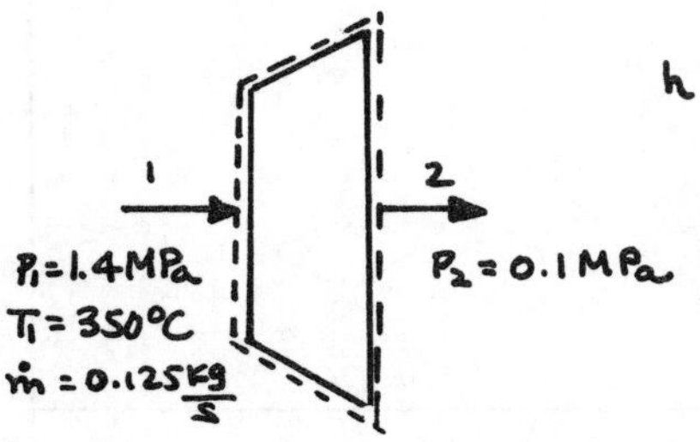

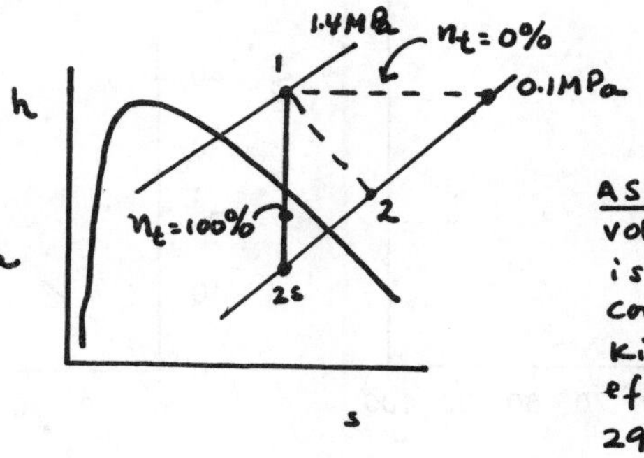

ASSUMPTIONS: 1. The control volume shown in the schematic is at steady state. 2. For the control volume, $\dot{Q}_{cv} = 0$ and kinetic and potential energy effects are negligible. 3. $T_0 = 293$ K (20°C), $p_0 = 1$ atm.

ANALYSIS: At steady state, mass and energy rate balances reduce to give $\dot{W}_{cv} = \dot{m}(h_1 - h_2)$. Introducing the isentropic turbine efficiency

$$\eta_t = \frac{h_1 - h_2}{h_1 - h_{2s}} \quad (1) \Rightarrow \dot{W}_{cv} = \dot{m}\,\eta_t (h_1 - h_{2s}) \quad (2)$$

Note that Eq. (1) gives h_2, which together with p_2 fixes T_2. Mass and entropy rate balances reduce to give

$$\dot{E}_d = \dot{m} T_0 (s_2 - s_1) \quad (3)$$

Sample Calculation: $\eta_t = 80\%$. From Table A-4, $h_1 = 3149.5 \frac{kJ}{kg}$, $s_1 = 7.1406$ kJ/kg·K (double interpolation). Then, with Eq. (1) and h_{2s} determined with $s_{2s} = s_1$

$$x_{2s} = \frac{s_{2s} - s_f}{s_g - s_f} = \frac{7.1406 - 1.3026}{7.3594 - 1.3026} = 0.964, \quad h_{2s} = 417.6 + (.964)(2258) = 2594.3 \frac{kJ}{kg}$$

$$h_2 = h_1 - \eta_t (h_1 - h_{2s}) = 3149.5 - .8(3149.5 - 2594.3) = 2705.3 \frac{kJ}{kg} \Rightarrow T_2 = 114.4°C,$$

$s_2 = 7.4373$ kJ/kg·K. So, with Eqs. (2), (3)

$$\dot{W}_{cv} = \left(0.125\frac{kg}{s}\right)(0.8)(3149.5 - 2594.3)\frac{kJ}{kg}\left|\frac{1\,kW}{1\,\frac{kJ}{s}}\right| = 55.52\,kW, \quad \dot{E}_d = (0.125)(293)(7.4373 - 7.1406) = 10.87\,kW$$

IT Code

```
p1 = 14  // bar
T1 = 350  // °C
p2 = 1  // bar
To = 293  // K
mdot = 0.125  // kg/s
eta = 0.8

h1 = h_PT("Water/Steam", p1, T1)
s1 = s_PT("Water/Steam", p1, T1)
h2s = h_Ps("Water/Steam", p2, s1)
h2 = h1 - eta * (h1 - h2s)
s2 = s_Ph("Water/Steam", p2, h2)
T2 = T_Ph("Water/Steam", p2, h2)
Wdotcv = mdot * (h1 - h2)
Edotcv = mdot * To * (s2 - s1)
```

IT Results for $\eta_t = 80\%$

$h_1 = 3149$ kJ/kg
$s_1 = 7.135$ kJ/kg·K
$h_{2s} = 2592$ kJ/kg
$h_2 = 2703$ kJ/kg
$s_2 = 7.433$ kJ/kg·K
$T_2 = 113.6$ °C
$\dot{W}_{cv} = 55.73$ kW
$\dot{E}_d = 10.9$ kW

PROBLEM 7.63 (Cont'd.)

PLOTS:

(1)

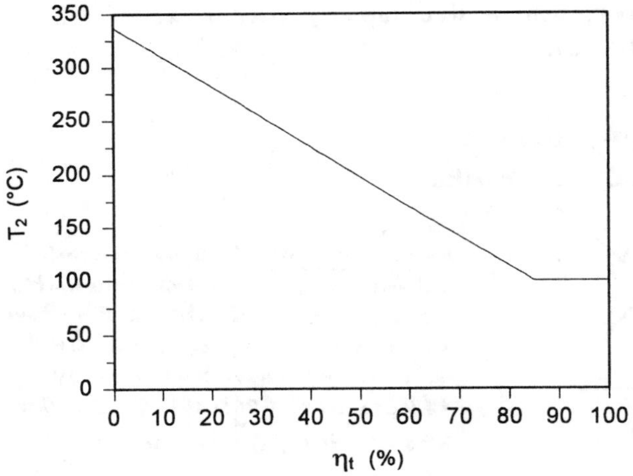

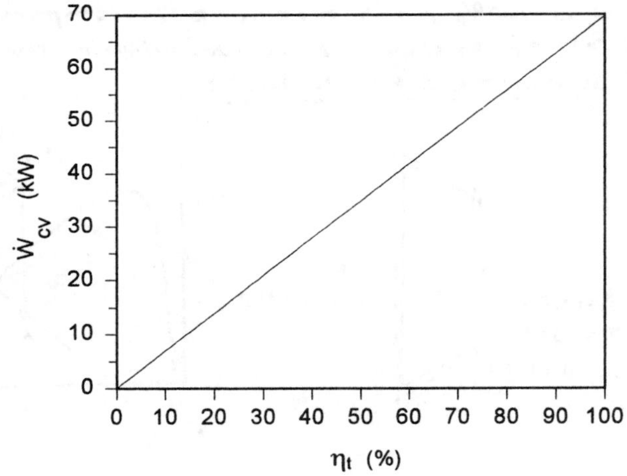

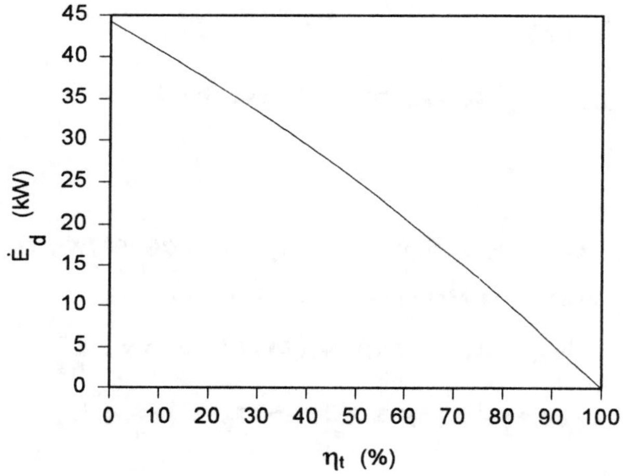

Discussion:

1. The case of $\eta_t = 100\%$ corresponds to maximum power and zero exergy destruction. The case of $\eta_t = 0$ corresponds to zero power and maximum exergy destruction.
2. In the case of $\eta_t = 0$, there is no work, and the turbine acts as a throttling process, with $h_2 = h_1$.

---

1. The flat portion of the plot corresponds to $x_2 \leq 1.0$. At these states the temperature corresponds to the saturation temperature at $p = 0.1$ MPa.

PROBLEM 7.64

KNOWN: Steady-state operating data are provided for a steam turbine.

FIND: (a),(b) If the isentropic efficiency is 80% and exergy is valued at 8 cents per kW·h, determine the value of the power produced and the cost of the exergy destroyed. (c) Plot the quantities of (a),(b) versus isentropic efficiency ranging from 80 to 100%

SCHEMATIC & GIVEN DATA:

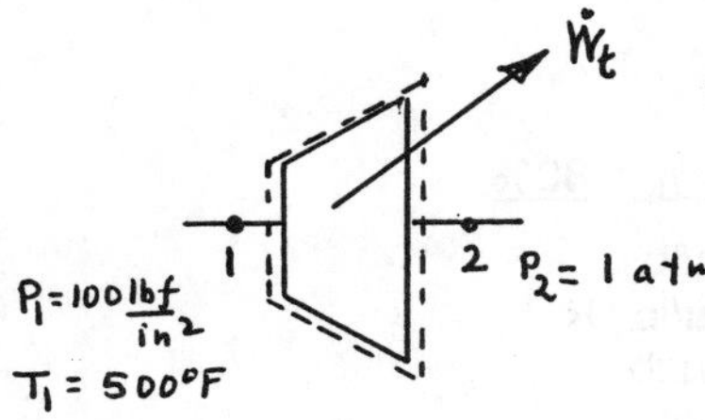

$\dot{m} = 3\times10^5$ lb/h

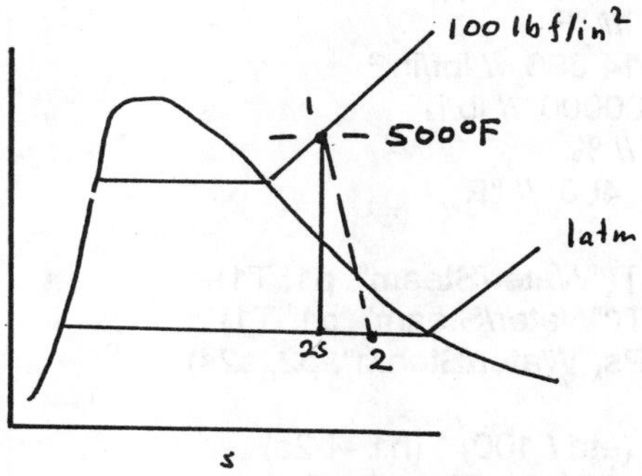

ASSUMPTIONS: (1) The control volume above is at steady state. (2) For the control volume $\dot{Q}_{cv} = 0$ and kinetic/potential energy effects can be ignored. (3) For the environment $T_0 = 530$°R, $p_0 = 1$ atm. (4) Exergy is valued at 8 cents per kW·h.

ANALYSIS: Reducing mass and energy rate balances and using the definition of isentropic turbine efficiency (Eq. 6.48)

$$\dot{W}_t = \dot{m}(h_1 - h_2) = \eta_t \dot{m}(h_1 - h_{2s}) \qquad (1)$$

Using $\dot{E}_d = T_0 \dot{\sigma}_{cv}$ and $\dot{\sigma}_{cv}$ from the mass and entropy rate balances

$$\dot{E}_d = T_0 \dot{m}(s_2 - s_1) \qquad (2)$$

(a) $\eta_t = 80\%$. From Table A-4E, $h_1 = 1279.1$ Btu/lb, $s_1 = 1.7085$ Btu/lb·°R. State 2s is fixed by $P_2 = 1$atm and x_{2s}:

$$x_{2s} = \frac{1.7085 - 0.3121}{1.4446} = 0.967$$

Then

$$h_{2s} = h_f + x_{2s}(h_g - h_f) = 180.15 + 0.967(970.4) = 1118.5 \text{ Btu/lb}$$

Substituting into Eq. (1)

$$\dot{W}_t = (0.8)\left(3\times10^5 \frac{\text{lb}}{\text{h}}\right)(1279.1 - 1118.5)\frac{\text{Btu}}{\text{lb}} = 385.4\times10^5 \frac{\text{Btu}}{\text{h}}$$

The value is

$$\dot{\$} = \left(385.4\times10^5 \frac{\text{Btu}}{\text{h}}\right)\left|\frac{\text{kW·h}}{3413 \text{ Btu}}\right|\left(\frac{0.08 \$}{\text{kW·h}}\right) = 903.4 \frac{\$}{\text{h}} \longleftarrow$$

(b) Continuing the discussion of the case $\eta_t = 80\%$, to fix state 2 use Eq. (1) to write $h_2 = h_1 - (\dot{W}_t/\dot{m}) = 1279.1 - (385.4\times10^5/3\times10^5) = 1150.6$ Btu/lb. Then, at $P_2 = 1$atm, state 2 is closely a saturated vapor for which $s_2 = 1.7567$ Btu/lb·°R. Substituting into Eq. (2)

$$\dot{E}_d = (530°\text{R})\left(3\times10^5 \frac{\text{Btu}}{\text{h}}\right)(1.7567 - 1.7085)\frac{\text{Btu}}{\text{lb·°R}} = 76.6\times10^5 \frac{\text{Btu}}{\text{h}}$$

The cost rate is

$$\dot{\$} = \left(76.6 \frac{\text{Btu}}{\text{h}}\right)\left|\frac{1 \text{ kW·h}}{3413 \text{ Btu}}\right|\left(\frac{0.08 \$}{\text{kW·h}}\right) = 179.5 \frac{\$}{\text{h}} \longleftarrow$$

PROBLEM 7.64 (Cond'd.)

The data required for the plot are generated using IT, as follows:

IT Code

```
p1 = 100  // lbf/in.²
T1 = 500  // °F
p2  = 1 * 14.696  // lbf/in.²
mdot = 300000  // lb/h
eta = 80  // %
To = 70 + 460  // °R

h1 = h_PT("Water/Steam", p1, T1)
s1 = s_PT("Water/Steam", p1, T1)
h2s = h_Ps("Water/Steam", p2, s2s)
s2s = s1
h2 = h1 - (eta / 100) * (h1 - h2s)
s2 = s_Ph("Water/Steam", p2, h2)

Wdot = mdot *  (h1 - h2)
CostW = Wdot * (0.08 / 3413)
Edot = mdot * To * (s2 - s1)
CostE = Edot * (0.08 / 3413)
```

IT Results for $\eta_t = 80\%$

$h_1 = 1279$ Btu/lb
$s_1 = 1.708$ Btu/lb·°R
$h_{2s} = 1118$ Btu/lb
$h_2 = 1150$ Btu/lb
$s_2 = 1.756$ Btu/lb
$\dot{W}_{cv} = 3.864 \times 10^7$ Btu/h
$\dot{\$}_{power} = 905.7$ \$/h
$\dot{E}_d = 7.623 \times 10^6$ Btu/h
$\dot{\$}_{exergy} = 178.7$ \$/h

PLOT:

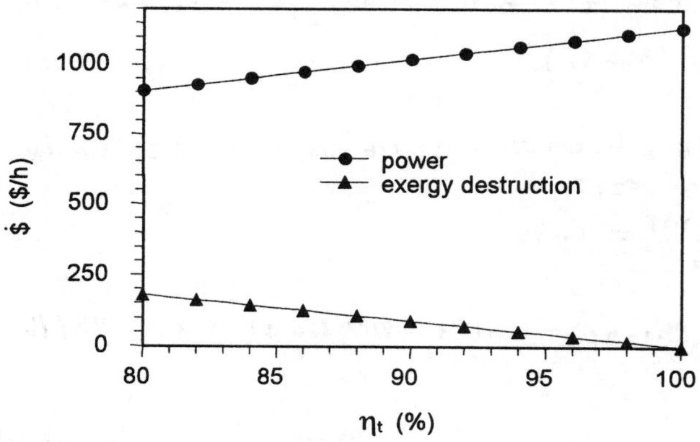

At higher values of η_t, there is less exergy destruction and greater power output, as expected.

PROBLEM 7.65

KNOWN: Operating data are provided for a water-jacketed air compressor at steady state.

FIND: Determine rate of exergy destruction, in kW, and express it as a percentage of the power input.

SCHEMATIC & GIVEN DATA:

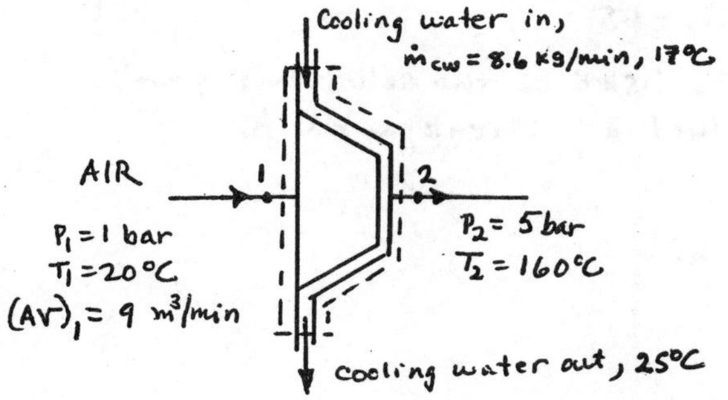

From solution to problem 6.118

$\dot{W}_{cv} = -30.01$ kW

$\dot{\sigma}_{cv} = 4.23 \times 10^{-3}$ kW/K

ASSUMPTIONS: (1) The control volume shown operates at steady state. (2) For the control volume $\dot{Q}_{cv} = 0$ and kinetic/potential energy effects are negligible. (3) Air is modeled as an ideal gas. (4) The cooling water is modeled as incompressible with constant specific heat c and a negligible change in pressure. (5) For the environment, $T_0 = 20°C$, $P_0 = 1$ bar.

ANALYSIS: The rate of exergy destruction is conveniently calculated in this case as

$$\dot{E}_d = T_0 \dot{\sigma} = (293\text{ K})\left(4.23 \times 10^{-3} \frac{\text{kW}}{\text{K}}\right) = 1.24 \text{ kW} \quad \leftarrow$$

When expressed as a percentage of the power input

$$\left(\frac{\dot{E}_d}{-\dot{W}_{cv}}\right) = \left(\frac{1.24 \text{ kW}}{30.01 \text{ kW}}\right)(100) = 4\% \quad \leftarrow$$

PROBLEM 7.66

KNOWN: Steady-state operating data are provided for water-jacketed compressors — Problems 7.62, 7.65.

FIND: Determine for each case, the hourly costs of the power input and the rate of exergy destruction.

SCHEMATIC & GIVEN DATA:

See solutions to Problems 7.62, 7.65

ASSUMPTIONS: 1. Refer to assumptions listed in the solutions for Problems 7.62, 7.65. 2. Exergy is valued at 8 cents per kW·h.

ANALYSIS:

(a) From the solution to Problem 7.62

$$-\dot{W}_{cv} = 38.61 \text{ kW}$$

$$\dot{E}_d = 4.13 \text{ kW}$$

$$\Rightarrow \quad \dot{\$} = (38.61 \text{ kW})\left(0.08 \frac{\$}{\text{kW}\cdot\text{h}}\right) = \$3.09/\text{h} \quad \leftarrow \text{power}$$

$$\dot{\$} = (4.13 \text{ kW})\left(0.08 \frac{\$}{\text{kW}\cdot\text{h}}\right) = \$0.33/\text{h} \quad \leftarrow \text{exergy destruction}$$

(b) From the solution to Problem 7.65

$$-\dot{W}_{cv} = 30.01 \text{ kW}$$

$$\dot{E}_d = 1.24 \text{ kW}$$

$$\Rightarrow \quad \dot{\$} = (30.01 \text{ kW})\left(0.08 \frac{\$}{\text{kW}\cdot\text{h}}\right) = \$2.40/\text{h} \quad \leftarrow \text{power}$$

$$\dot{\$} = (1.24 \text{ kW})\left(0.08 \frac{\$}{\text{kW}\cdot\text{h}}\right) = \$0.10/\text{h} \quad \leftarrow \text{exergy destruction}$$

PROBLEM 7.67

KNOWN: Steady-state operating data are provided for a throttling valve and turbine in series.

FIND: Plot the rates of exergy destruction for the valve and the turbine, each versus the quality of the steam at the turbine exit.

SCHEMATIC & GIVEN DATA:

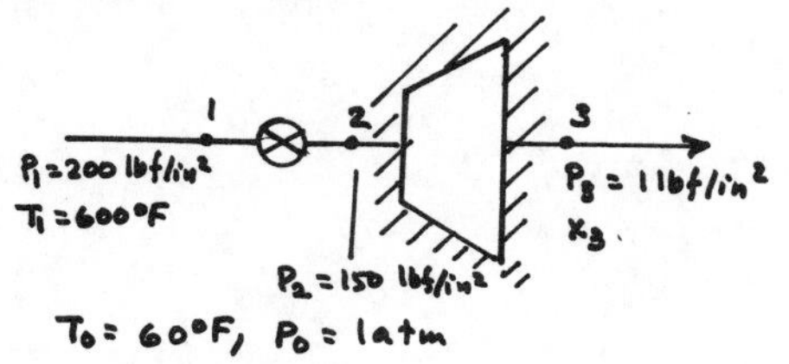

$T_0 = 60$°F, $p_0 = 1$ atm

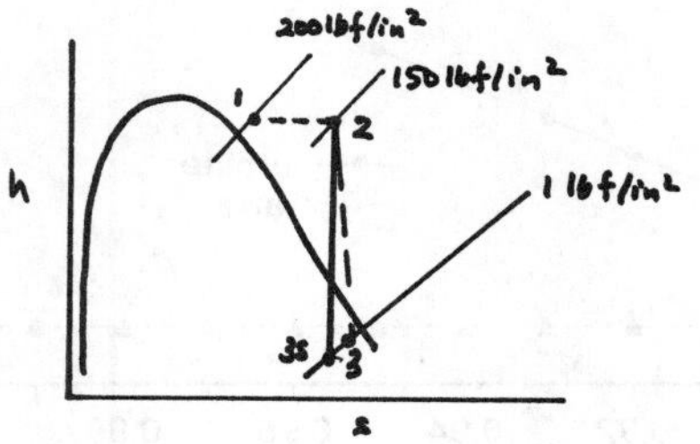

ASSUMPTIONS: (1) The turbine is insulated and at steady state. $\dot{Q}_{cv} = 0$. 2. The expansion across the valve is a throttling process. (3) For the environment, $T_0 = 60$ °F, $p_0 = 1$ atm.

ANALYSIS: For a 1-inlet, 1-exit control volume at steady state for which there is no heat transfer, an entropy rate balance reduces to give

$$0 = \sum \frac{\dot{Q}_j}{T_j}^{\,0} + \dot{m}(s_1 - s_2) + \dot{\sigma}_{cv} \Rightarrow \dot{\sigma}_{cv} = \dot{m}(s_2 - s_1)$$

Then, with $\dot{E}_d = T_0 \dot{\sigma}_{cv}$, we get

$$\text{Valve: } \frac{\dot{E}_d}{\dot{m}} = T_0(s_2 - s_1) \quad (1) \qquad \text{Turbine: } \frac{\dot{E}_d}{\dot{m}} = T_0(s_3 - s_2) \quad (2)$$

Sample calculation: $x_3 = 0.98$. From Table A-4E, $h_1 = 1322.1$ Btu/lb, $s_1 = 1.6767$ Btu/lb·°R. With assumption 2, $h_2 \sim h_1$. Interpolating in Table A-4E at 150 lbf/in² with $h_2 = 1322.1$ Btu/lb, $s_2 = 1.7078$ Btu/lb·°R. With data from Table A-3E, $s_3 = s_f + x_3 s_{fg}$ or $s_3 = 0.1327 + 0.98(1.8453) = 1.9411$ Btu/lb·°R. Inserting values into Eqs. (1), (2)

$$(\dot{E}_d/\dot{m})_{valve} = 520\,°R(1.7078 - 1.6767)\frac{\text{Btu}}{\text{lb}\cdot°R} = 16.17\ \frac{\text{Btu}}{\text{lb}}$$ ← valve

$$(\dot{E}_d/\dot{m})_{turbine} = 520\,°R(1.9411 - 1.7078)\frac{\text{Btu}}{\text{lb}\cdot°R} = 121.32\ \frac{\text{Btu}}{\text{lb}}$$

IT Code

```
p1 = 200  // lbf/in.²
T1 = 600  // °F
p2 = 150  // lbf/in.²
p3 = 1  //  lbf/in.²
To = 60 + 460  // °R
x3 = 0.98

h1 = h_PT("Water/Steam", p1, T1)
s1 = s_PT("Water/Steam", p1, T1)
h2 = h_Ps("Water/Steam", p2, s2)
h2 = h1
s3 = ssat_Px("Water/Steam", p3, x3)
Edvalve= To * (s2 - s1)
Edturb= To * (s3 - s2)
```

IT Results for $x_3 = 0.98$

h1= 1322 Btu/lb
s1 = 1.676 Btu/lb·°R
s2 = 1.707 Btu/lb·°R
s3 = 1.941 Btu/lb·°R
$(\dot{E}_d/\dot{m})_{valve}$ = 121.4 Btu/lb
$(\dot{E}_d/\dot{m})_{turbine}$ = 16.04 Btu/lb

PROBLEM 7.67 (Cont'd.)

PLOT:

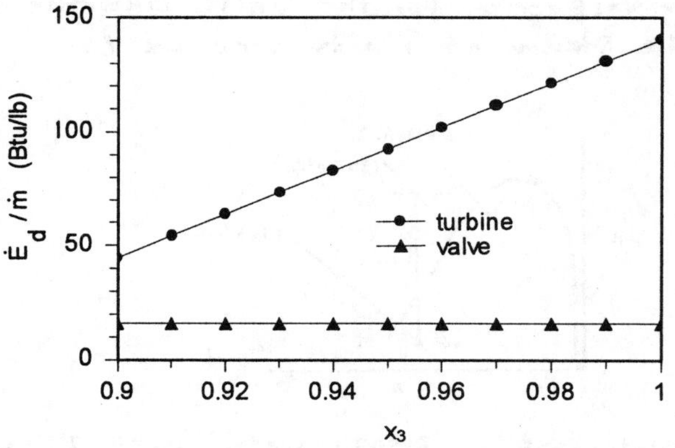

- Note that the exergy destruction in the valve is independent of x_3.
- The exergy destruction in the turbine increases as x_3 increases, corresponding to a decrease in isentropic turbine efficiency, as expected.

PROBLEM 7.68

KNOWN: A device for developing power using "waste heat" is described in Problem 6.102

FIND: If the device develops a net power output of 6 kW, determine (a) the rate exergy enters accompanying heat transfer, (b) the net rate exergy is carried in by the steam, (c) the rate of exergy destruction within the device.

SCHEMATIC & GIVEN DATA

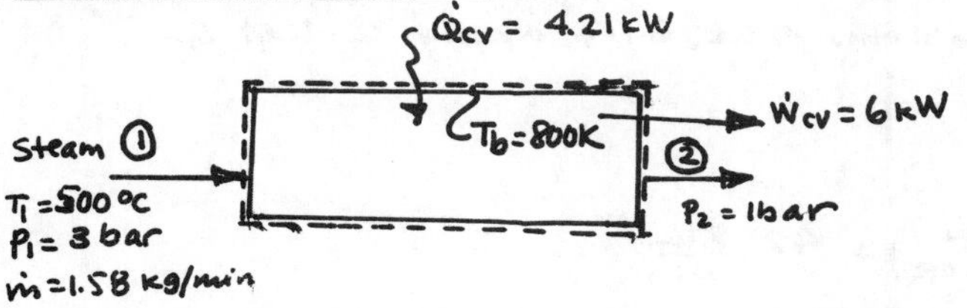

ASSUMPTIONS: 1. The control volume shown in the accompanying figure is at steady state. 2. Kinetic and potential energy effects can be ignored. 3. Heat transfer occurs only at T_b. 4. $T_0 = 293$ K, $p_0 = 1$ bar.

ANALYSIS: (a) The rate exergy enters accompanying heat transfer is obtained with Eq. 7.33:

$$\dot{E}_q = \left[1 - \frac{T_0}{T_b}\right]\dot{Q}_{cv} = \left[1 - \frac{293}{800}\right](4.21\,\text{kW}) = 2.67\,\text{kW} \longleftarrow$$

(b) To find $(\dot{E}_{f1} - \dot{E}_{f2})$ requires state 2 to be fixed. Begin with a mass and energy rate balance to obtain with $h_1 = 3486$ kJ/kg from Table A-4

$$0 = \dot{Q}_{cv} - \dot{W}_{cv} + \dot{m}[h_1 - h_2] \Rightarrow h_2 = h_1 + \frac{\dot{Q}_{cv} - \dot{W}_{cv}}{\dot{m}} = 3486\,\frac{\text{kJ}}{\text{kg}} + \frac{(4.21 - 6)\,\text{kJ/s}}{(1.58/60)\,\text{kg/s}} = 3418\,\frac{\text{kJ}}{\text{kg}}.$$

Then, from Table A-4, with $p_2 = 1$ bar and $h_2 = 3418$ kJ/kg, $s_2 = 8.7398$ kJ/kg·K. Also, Table A-4 gives $s_1 = 8.3251$ kJ/kg·K. With Eq. 7.36

$$(\dot{E}_{f1} - \dot{E}_{f2}) = \dot{m}\left[(h_1 - h_2) - T_0(s_1 - s_2)\right] = \left(\frac{1.58}{60}\,\frac{\text{kg}}{\text{s}}\right)\left[(3486 - 3418) - 293(8.3251 - 8.7398)\right]$$

$$= 4.99\,\text{kW} \longleftarrow$$

① (c) The rate of exergy destruction can be obtained from Eq. 7.32b

$$0 = \dot{E}_q - \dot{W}_{cv} + (\dot{E}_{f1} - \dot{E}_{f2}) - \dot{E}_d$$

② $$\Rightarrow \dot{E}_d = \dot{E}_q - \dot{W}_{cv} + (\dot{E}_f - \dot{E}_{f2}) = (2.67 - 6 + 4.99)\,\text{kW} = 1.66\,\text{kW} \longleftarrow$$

---

1. Alternatively, $\dot{E}_d = T_0\dot{\sigma}$, where

$$\dot{\sigma} = \left(-\frac{\dot{Q}_{cv}}{T_b}\right) + \dot{m}(s_2 - s_1) = -\frac{4.21\,\text{kW}}{800\,\text{K}} + \left(\frac{1.58}{60}\,\frac{\text{kg}}{\text{s}}\right)(8.7398 - 8.3251)\frac{\text{kJ}}{\text{kg}\cdot\text{K}}\left|\frac{1\,\text{kW}}{1\,\text{kJ/s}}\right|$$

$$\Rightarrow \dot{\sigma} = -0.00526 + 0.01092 = 0.00566\,\text{kW/K}$$

$$\Rightarrow \dot{E}_d = T_0\dot{\sigma} = (293\,\text{K})(0.00566\,\text{kW/K}) = 1.66\,\text{kW}$$

2. Exergy accounting

- Exergy supplied to the device
  - via heat transfer: 2.67 kW
  - via steam flow*: 4.99 kW
  - } 7.66 kW
- Disposition of exergy supplied
  - power developed: 6.00 kW (78.3%)
  - exergy destroyed: 1.66 kW (21.7%)

* The steam provides nearly 2/3 of the exergy supplied to the device.

PROBLEM 7.69

KNOWN: Steady-state operating data are provided for a duct system in Problem 4.68.

FIND: Determine the rate of exergy destruction.

SCHEMATIC & GIVEN DATA:

- See figure in solution of Problem 6.117
- From solution: $\dot{\sigma}_{cv} = 0.084 \frac{Btu}{min \cdot °R}$

ASSUMPTIONS: 1. Assumptions from Problems 4.68, 6.117 apply. 2. Let $T_0 = 500°R$, $p_0 = 1 bar$.

ANALYSIS: Using $\dot{E}_d = T_0 \dot{\sigma}_{cv}$, we get

$$\dot{E}_d = 500°R \left(0.084 \frac{Btu}{min \cdot °R}\right) = 42 \frac{Btu}{min}$$

PROBLEM 7.70

KNOWN: Steady-state operating data are provided for a vortex tube.

FIND: Evaluate the rate of exergy destruction per unit mass of air entering.

SCHEMATIC & GIVEN DATA: See Figure E 6.7

ASSUMPTIONS: 1. The assumptions of Example 6.7 apply. 2. $T_0 = 530°R$, $p_0 = 1 atm$.

ANALYSIS: Using $\dot{E}_d = T_0 \dot{\sigma}_{cv}$ and $\dot{\sigma}_{cv}/\dot{m}_1 = 0.1086 \, Btu/lb \cdot °R$ from the solution to Example 6.7, we get

$$\frac{\dot{E}_d}{\dot{m}_1} = T_0 \frac{\dot{\sigma}_{cv}}{\dot{m}} = (530 °R)(0.1086) \frac{Btu}{lb \cdot °R}$$

$$= 57.6 \frac{Btu}{lb}$$

The inventor claims the device operates without work or heat transfer. However, the exergy destroyed within the vortex tube can be traced to a device, or devices, located upstream of the vortex tube that would require some combination of work, heat transfer, and fuel input.

PROBLEM 7.71

KNOWN: Water and air flow on opposite sides of a counterflow heat exchanger operating at steady state. Operating data are given.

FIND: Determine (a) the change in flow exergy rate of each stream, and (b) the rate of exer y destruction.

SCHEMATIC & GIVEN DATA:

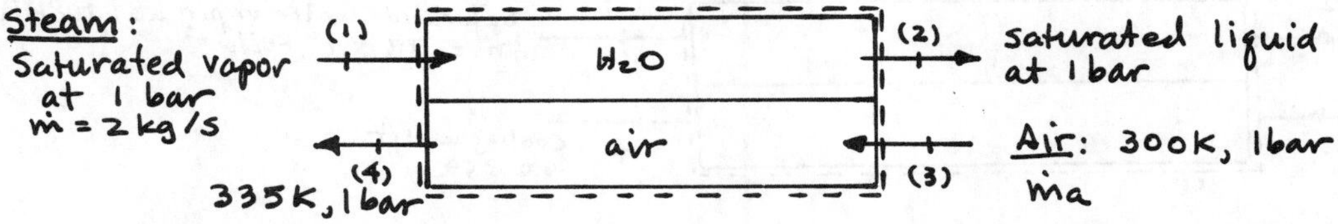

ASSUMPTIONS: (1) The control volume shown is at steady state. (2) For the control volume, $\dot{W}_{cv} = \dot{Q}_{cv} = 0$, and kinetic and potential effects are negligible. (3) The air is modeled as an ideal gas. (4) For the environment, $T_0 = 300\,K$, $p_0 = 1\,bar$.

ANALYSIS: First, we use mass and energy rate balances to determine $\dot{m}_a$. That is $\dot{m}_1 = \dot{m}_2 \equiv \dot{m}$ and $\dot{m}_3 = \dot{m}_4 \equiv \dot{m}_a$. Thus, with assumptions 1 and 2

$$0 = \cancel{\dot{Q}_{cv}}^{0} - \cancel{\dot{W}_{cv}}^{0} + \dot{m}(h_1 - h_2) + \dot{m}_a(h_3 - h_4) \qquad (1)$$

or, solving for $\dot{m}_a$ and inserting enthalpy data from Tables A-3 and A-22

$$\dot{m}_a = \left(\frac{h_1 - h_2}{h_4 - h_3}\right)\dot{m} = \frac{(2258\ kJ/kg)}{(335.38 - 300.19)\,kJ/kg}(2\ kg/s) = 128.3\ kg/s$$

(a) The change in flow exergy rate for the water stream is

$$\dot{E}_{f_2} - \dot{E}_{f_1} = \dot{m}(e_2 - e_1) = \dot{m}\left[(h_2 - h_1) - T_0(s_2 - s_1)\right]$$

$$= \left(2\,\frac{kg}{s}\right)\left[(-2258\ kJ/kg) - (300\,K)(1.3026 - 7.3594)\,\frac{kJ}{kg\cdot K}\right]\left|\frac{1\,kW}{1\,kJ/s}\right|$$

$$= -881.9\ kW \longleftarrow (\Delta\dot{E}_f)_{water}$$

For the air stream, using data from Table A-22

$$\dot{E}_{f4} - \dot{E}_{f3} = \dot{m}_a\left[(h_4 - h_3) - T_0(s_4 - s_3)\right] = \dot{m}_a\left[(h(T_4) - h(T_3)) - T_0\left(s^\circ(T_4) - s^\circ(T_3) - R\,\cancel{\ln p_4/p_3}^{0}\right)\right]$$

(1) $$= (128.3)\left[(335.38 - 300.19) - (300)(1.8129 - 1.70203)\right] = 247.5\ kW \longleftarrow (\Delta\dot{E}_f)_{air}$$

(b) At steady state, the exergy rate balance reduces to

$$0 = \sum_j\left[1 - \frac{T_0}{T_j}\right]\cancel{\dot{Q}_j}^{0} - \left[\cancel{\dot{W}_{cv}}^{0} - \cancel{p_0\frac{dV}{dt}}^{0}\right] + \dot{m}(e_{f_1} - e_{f_2}) + \dot{m}_a(e_{f_3} - e_{f_4}) - \dot{E}_d$$

Thus

$$\dot{E}_d = -\left[\dot{E}_{f_2} - \dot{E}_{f_1}\right] - \left[\dot{E}_{f4} - \dot{E}_{f3}\right] = -[-881.9] - [247.5] = 634.4\ kW \longleftarrow \dot{E}_d$$

---

1. The decrease in energy of the water stream exactly equals the increase in energy of the air stream, for energy is conserved — see Eq. (1). However, beause exergy is destroyed, the decrease in exergy of the water stream differs significantly from the increase of exergy of the air stream.

PROBLEM 7.72

KNOWN: Operating data are provided for a condenser at steady state.

FIND: For the cooling water stream determine the net rate, in MW, that (a) energy exits the plant, (b) exergy exits the plant. Discuss

SCHEMATIC & GIVEN DATA:

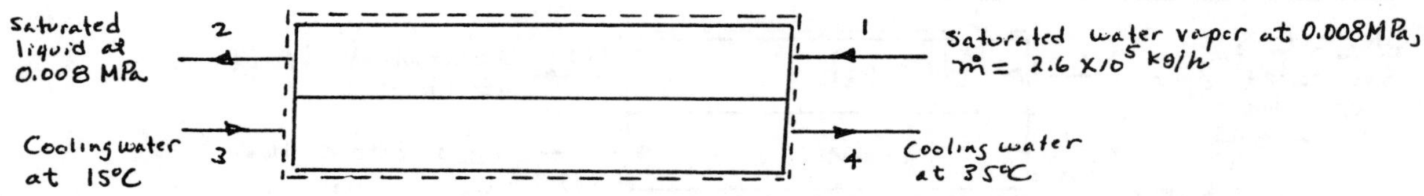

ASSUMPTIONS: (1) The control volume shown is at steady state. (2) For the control volume, $\dot{Q}_{cv} = \dot{W}_{cv} = 0$ and kinetic/potential energy effects are negligible. (3). For the cooling water there is a negligible change in pressure from inlet to exit. (4) For the environment, $T_0 = 20°C$, $p_0 = 0.1$ MPa.

ANALYSIS: The mass flow rate of the cooling water is needed as a preliminary. Thus, reducing mass and energy rate balances with indicated assumptions

$$\dot{m}_{cw} = \frac{\dot{m}(h_1 - h_2)}{h_4 - h_3}$$

Using $h \approx h_f(T)$ for the cooling water and steam table data

$$\dot{m}_{cw} = \frac{(2.6 \times 10^5 \text{ kg/h})(2403.1 \text{ kJ/kg})}{(146.7 - 63) \text{ kJ/kg}} = 7.46 \times 10^6 \text{ kg/h}$$

(a) The net rate energy is carried from the power plant by the cooling water is given by

$$\dot{m}_{cw}[h_4 - h_3] = \left(7.46 \times 10^6 \frac{\text{kg}}{\text{h}}\right)(146.7 - 63)\frac{\text{kJ}}{\text{kg}} = 6.24 \times 10^8 \frac{\text{kJ}}{\text{h}}$$

$$= \left(6.24 \times 10^8 \frac{\text{kJ}}{\text{h}}\right)\left(\frac{1 \text{ h}}{3600 \text{ s}}\right)\left(\frac{1 \text{ MW}}{10^3 \text{ kJ/s}}\right) = 173.3 \text{ MW} \longleftarrow \text{(a)}$$

(b) The net rate exergy is carried from the power plant by the cooling water is given by

$$\dot{E}_{f4} - \dot{E}_{f3} = \dot{m}_{cw}[h_4 - h_3 - T_0(s_4 - s_3)] = 6.24 \times 10^8 \frac{\text{kJ}}{\text{h}} - \left[\left(7.46 \times 10^6 \frac{\text{kg}}{\text{h}}\right)(293 \text{ K})(0.5053 - 0.2245)\frac{\text{kJ}}{\text{kg} \cdot \text{K}}\right]$$

$$= \left(0.1 \times 10^8 \frac{\text{kJ}}{\text{h}}\right)\left(\frac{\text{h}}{3600 \text{ s}}\right)\left(\frac{1 \text{ MW}}{10^3 \text{ kJ/s}}\right) = 2.8 \text{ MW} \longleftarrow \text{(b)}$$

where for the cooling water $s \approx s_f(T)$.

On an energy basis the quantity of energy exiting with the cooling water is significant. But when viewed in terms of exergy the potential for use of the effluent is seen to be limited: about 3% of the power developed. Many studies have shown that there are few practical uses for copious amounts of such low temperature condenser cooling water.

PROBLEM 7.73

KNOWN: Operating data are provider for a counterflow heat exchanger at steady state. One stream is R-134a and the other is air.

FIND: (a) For the R-134a, determine the rate of heat transfer, and (b) For each of the two streams, evaluate the change in flow exergy rate and compare.

SCHEMATIC & GIVEN DATA:

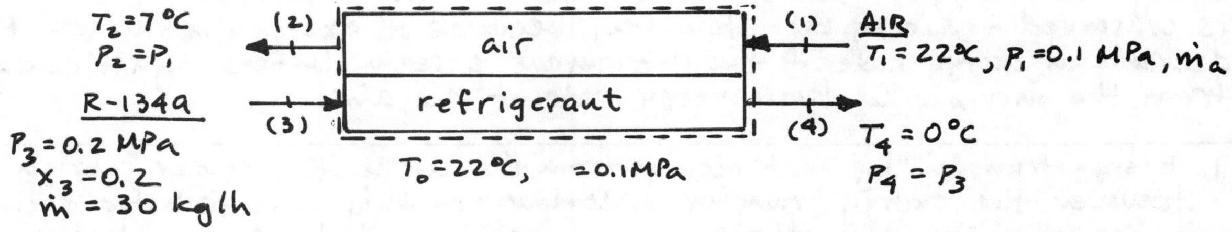

ASSUMPTIONS: (1) The control volume shown is at steady state, with $\dot{Q}_{cv} = 0$, $\dot{W}_{cv} = 0$, and negligible effects of motion and gravity. (2) For a control volume enclosing only the refrigerant stream, the foregoing applies, except $\dot{Q}_{cv} \neq 0$. (3) For each stream, pressure change is negligible. (4) The air is modeled as an ideal gas. (5) For the environment, $T_0 = 22°C = 295 K$, $P_0 = 0.1$ MPa.

(a) For a control volume enclosing only the refrigerant stream, the mass and energy rate balances reduce to give

$$0 = \dot{Q} - \dot{W}^{0} + \dot{m}(h_3 - h_4) \Rightarrow \dot{Q} = \dot{m}(h_4 - h_3) \quad (1)$$

From Table A-11 at 2 bar, $x_3 = 0.2$; $h_3 = h_{f3} + x_3 h_{fg3} = 36.84 + (.2)\,204.46 = 77.732$ kJ/kg
From Table A-12; $h_4 = 250.10$ kJ/kg. Thus

$$\dot{Q} = (30 \tfrac{kg}{h})(250.10 - 77.732)\,kJ/kg = 5171 \text{ kJ/h} \longleftarrow \dot{Q}$$

(b) The change in flow exergy rate for the refrigerant stream is, with data from Tables A-11 and A-12

$$(\Delta \dot{E}_f)_{ref} = \dot{m}[(h_4 - h_3) - T_0(s_4 - s_3)] = (30)[(250.1 - 77.732) - (295)(.9582 - .30354)]$$

① $$= -622.7 \text{ kJ/h} \longleftarrow (\Delta \dot{E}_f)_{ref}$$

To determine the change in flow exergy rate for the air, we first evaluate $\dot{m}_a$ using mass and energy rate balances for the overall control volume.

$$0 = \dot{Q}_{cv}^{0} - \dot{W}_{cv}^{0} + \dot{m}_a[h_1 - h_2] + \dot{m}[h_3 - h_4] \quad (1)$$

That is

$$\dot{m}_a = \dot{m}\left[\frac{h_4 - h_3}{h_1 - h_2}\right] = (30 \tfrac{kg}{h})\left[\frac{250.1 - 77.732}{295.17 - 280.13}\right]\frac{kJ}{kg} = 343.8 \text{ kg/h}$$

where he specific enthalpies for air are from Table A-22.

Now, the change in exergy rate of the air is

$$(\Delta \dot{E}_f)_{air} = \dot{m}_a(e_{f2} - e_{f1}) = \dot{m}[(h_2 - h_1) - T_0(s_2 - s_0)]$$

$$= \dot{m}_a[(h_2 - h_1) - T_0(s^\circ(T_2) - s^\circ(T_1) - R \ln p_2/p_1^{\,0})]$$

PROBLEM 7.73 (Cont'd.)

With $s^\circ(T)$ data from Table A-22

$$(\Delta \dot{E}_f)_{air} = (343.8)\left[(280.13 - 295.17) - (293)(1.63279 - 1.68515)\right]$$

$$= 139.7 \text{ kJ/h} \longleftarrow (\Delta \dot{E}_f)_{air}$$

Discussion. The change in energy of the air stream exactly equals the change in energy of the refrigerant stream in magnitude, for energy is conserved – see Eq. (1). However, because of exergy destruction, the decrease in exergy rate of the refrigerant stream differs significantly from the increase in flow exergy rate of the air.

---

1. Energy transfer by heat occurs from the air to the colder refrigerant. However, the exergy transfer accompanying this heat transfer occurs in the opposite direction; from the refrigerant to the air, because it takes place at temperatures below T_0. (See Sec. 7.3.2 for further discussion.) Even though its energy decreases, the flow exergy of the air increases, and conversely for the refrigerant.

PROBLEM 7.74

FIND: Determine the rate of exergy destruction for
(a) the computer of Example 4.8, when air exits at 32°C.
(b) the computer of Problem 4.70, ignoring the pressure change of the air.
(c) the water-jacketed electronics housing of Problem 4.71, when water exits at 24°C.

ANALYSIS: In each case, use $\dot{E}_d = T_0 \dot{\sigma}_{cv}$ where $T_0 = 293$ K and $\dot{\sigma}_{cv}$ is obtained from the counterpart Chap. 6 problem:

(a) From Problem 6.105, $\dot{\sigma}_{cv} = 0.328 \frac{W}{K}$

$$\therefore \dot{E}_d = (293 K)\left(0.328 \frac{W}{K}\right)\left|\frac{1 kW}{10^3 W}\right| = 0.096 \text{ kW} \leftarrow$$

(b) From Problem 6.106, $\dot{\sigma}_{cv} = 4.15 \times 10^{-4} \frac{kW}{K}$

$$\therefore \dot{E}_d = (293 K)(4.15 \times 10^{-4}) \frac{kW}{K} = 0.122 \text{ kW} \leftarrow$$

(c) From Problem 6.107, $\dot{\sigma}_{cv} = 8.5 \times 10^{-3} \frac{kW}{K}$

$$\therefore \dot{E}_d = (293 K)\left(8.5 \times 10^{-3} \frac{kW}{K}\right) = 2.491 \text{ kW} \leftarrow$$

PROBLEM 7.75

KNOWN: Steady-state operating data are provided for an electronics-laden cylinder.

FIND: Determine the rate of exergy destruction.

ANALYSIS: Using $T_0 = 293$ K and $\dot{\sigma}_{cv}$ from the solution to Problem 6.108

$$\dot{E}_d = T_0 \dot{\sigma}_{cv} = (293 K)\left(0.65 \frac{W}{K}\right)\left|\frac{1 kW}{10^3 W}\right| = 0.19 \text{ kW} \leftarrow$$

PROBLEM 7.76

KNOWN: Helium gas expands through an insulated nozzle at steady state.

FIND: Determine the exit velocity, the isentropic nozzle efficiency, and the exergy destruction per unit mass of gas flowing.

SCHEMATIC & GIVEN DATA:

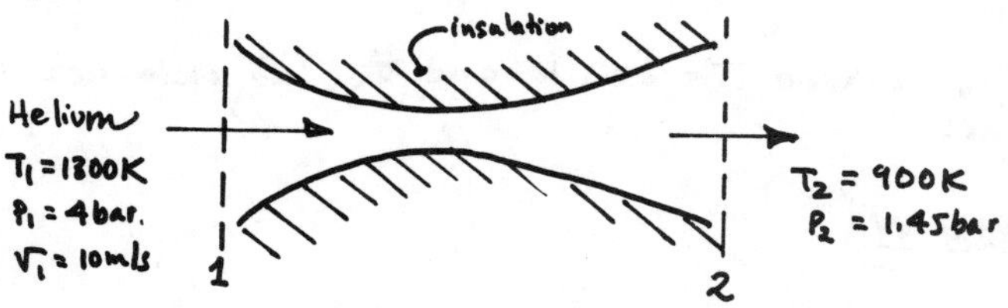

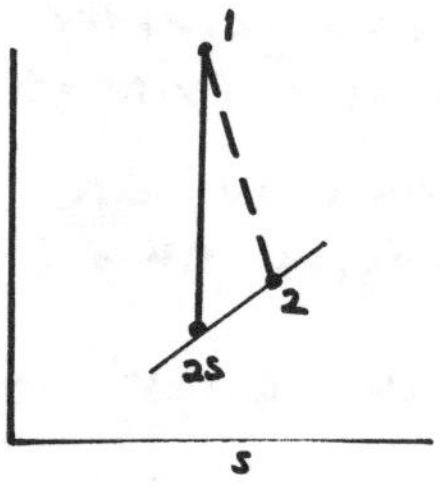

ASSUMPTIONS: (1) The nozzle is insulated and at steady state. (2) Potential energy effects can be ignored. (3) Argon is modeled as an ideal gas. (4) For the environment $T_0 = 20°C$, $P_0 = 1atm$.

ANALYSIS: Mass and energy balances reduce at steady state with assumption 2 to give

$$0 = \dot{Q}_{cv} - \dot{W}_{cv} + \dot{m}\left(h_1 - h_2 + \frac{(V_1^2 - V_2^2)}{2} + g(z_1 - z_2)\right)$$

Thus

$$V_2 = \sqrt{V_1^2 + 2(h_1 - h_2)} \qquad (1)$$

From Table A-21, for Helium $c_p = 5/2\,R$. Thus, $h_1 - h_2 = c_p(T_1 - T_2)$ and

$$V_2 = \sqrt{\left(10\frac{m}{s}\right)^2 + 2\left[(2.5)\frac{8314}{4.003}\frac{N\cdot m}{kg\cdot K}\right](1300K - 900K)\left|\frac{1\,kg\cdot m/s^2}{1N}\right|} = 2038\frac{m}{s} \quad \longleftarrow V_2$$

To determine the isentropic nozzle efficiency requires the temperature at state 2s. With Eq. 6.45 and $k = 1.667$

$$T_{2s} = T_1\left[\frac{P_2}{P_1}\right]^{\frac{k-1}{k}} = 1300\left[\frac{1.45}{4}\right]^{.667/1.667} = 866.2\,K$$

Then, with Eq. (1)

$$V_{2s} = \sqrt{V_1^2 + 2(h_1 - h_{2s})} = \sqrt{V_1^2 + 2c_p(T_1 - T_{2s})}$$

$$= \sqrt{(10)^2 + 2\left[\frac{(2.5)(8314)}{4.003}\right](1300 - 866.2)} = 2122\,m/s$$

The isentropic nozzle efficiency is

$$\eta_{nozzle} = \frac{V_2^2/2}{V_{2s}^2/2} = \left(\frac{2038}{2122}\right)^2 = 0.922 \quad (92.2\%) \quad \longleftarrow \eta_{nozzle}$$

The exergy destruction is given by $(\dot{E}_d/\dot{m}) = T_0(\dot{\sigma}_{cv}/\dot{m})$, where $\dot{\sigma}_{cv}/\dot{m}$ is the entropy production obtained from an entropy balance. Accordingly, as there is no heat transfer, we get using Eq. 6.23

$$\frac{\dot{E}_d}{\dot{m}} = T_0(s_2 - s_1) = T_0\left[c_p \ln\frac{T_2}{T_1} - R\ln\frac{P_2}{P_1}\right] = T_0 R\left[2.5\ln\frac{T_2}{T_1} - \ln\frac{P_2}{P_1}\right]$$

$$= (293K)\left(\frac{8.314}{4.003}\frac{kJ}{kg\cdot K}\right)\left[2.5\ln\frac{900}{1300} - \ln\frac{1.45}{4}\right]$$

$$= 58.1\frac{kJ}{kg} \quad \longleftarrow$$

PROBLEM 7.77

KNOWN: Oxygen enters a well-insulated nozzle operating at steady state at a given state and expands to a specified pressure. The isentropic nozzle efficiency is known.

FIND: Determine the exergy destruction rate, per kg of oxygen flowing.

SCHEMATIC & GIVEN DATA:

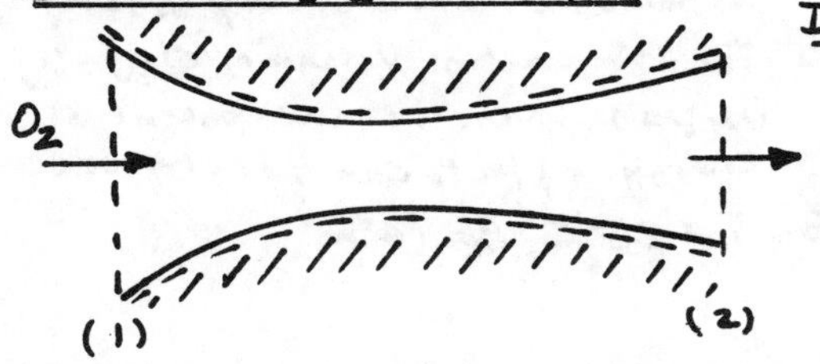

Inlet: $p_1 = 5$ bar
$T_1 = 600$ K
$V_1 = 20$ m/s

Exit: $p_2 = 1.5$ bar
$\eta_N = 0.92$

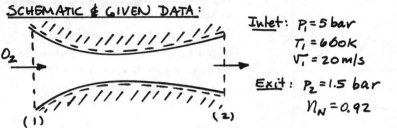

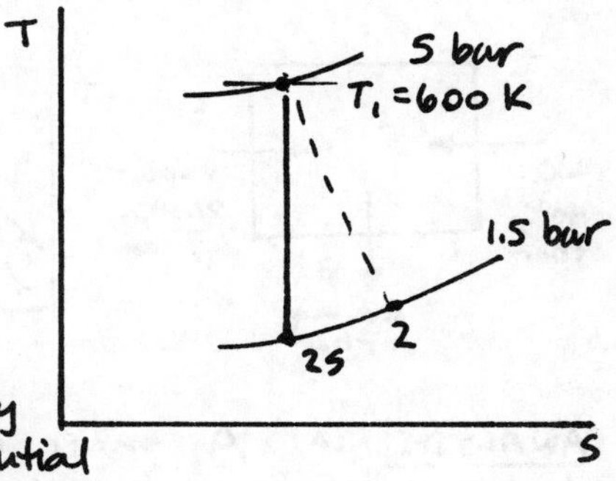

ASSUMPTIONS: (1) The control volume shown is at steady state. (2) For the control volume, $\dot{Q}_{cv} = \dot{W}_{cv} = 0$ and potential energy effects are negligible. (3) The oxygen behaves as an ideal gas. (4) For the environment, $T_0 = 20°C = 293$ K, $p_0 = 1$ atm.

ANALYSIS: To determine the exergy destruction rate, we use mass and entropy rate balances and $\dot{E}_d = T_0 \dot{\sigma}_{cv}$ to get

$$0 = \sum_j \left(\frac{\dot{Q}}{T}\right)_j^{\,0} + \dot{m}(s_1 - s_2) + \dot{\sigma}_{cv} \Rightarrow \dot{E}_d/\dot{m} = T_0(s_2 - s_1) \qquad (1)$$

To fix state 2, consider mass and energy balances which reduce as follows:

$$0 = \dot{Q}_{cv}^{\,0} - \dot{W}_{cv}^{\,0} + \dot{m}\left[\frac{(\bar{h}_1 - \bar{h}_2)}{M} + \left(\frac{V_1^2 - V_2^2}{2}\right) + g(z_1 - z_2)^{0}\right]$$

or

$$\bar{h}_2 = \bar{h}_1 + \left(\frac{V_1^2 - V_2^2}{2}\right) M \qquad (2)$$

To get V_2, we use $\eta_N = (V_2^2/2)/(V_2^2/2)_s$. For the isentropic process from 1 to 2s

$$0 = \bar{s}^\circ(T_{2s}) - \bar{s}^\circ(T_1) - R \ln p_2/p_1$$

or

$$\bar{s}^\circ(T_{2s}) = \bar{s}^\circ(T_1) + \bar{R} \ln p_2/p_1 = 229.430 + 8.314 \ln\left(\frac{1.5}{5}\right) = 219.42 \text{ kJ/kmol·K}$$

where the $\bar{s}^\circ$ values are from Table A-23. Interpolating in the table with $\bar{s}^\circ(T_{2s})$ gives:

$$T_{2s} = 481.5 \text{ K}, \quad h_{2s} = 14197 \text{ kJ/kmol}$$

Now

$$V_{2s} = \sqrt{\frac{2(\bar{h}_1 - \bar{h}_{2s})}{M} + V_1^2} = \sqrt{\frac{2(19870 - 14197)}{32} \frac{\text{kJ}}{\text{kg}} \left|\frac{10^3 \text{ N·m}}{1 \text{ kJ}}\right| \left|\frac{1 \text{ N}}{1 \text{ kg·m/s}^2}\right| + 20^2 \frac{\text{m}^2}{\text{s}^2}}$$

$$= 595.8 \text{ m/s}$$

Thus $V_2 = \sqrt{\eta_N}\, V_{2s} = 571.5$ m/s, and from (2)

$$\bar{h}_2 = 19870 \frac{\text{kJ}}{\text{kmol}} + \left(\frac{20^2 - 571.5^2}{2}\right) \frac{\text{m}^2}{\text{s}^2} \left|\frac{1 \text{ kJ}}{10^3 \text{ N·m}}\right| \left|\frac{1 \text{ N}}{1 \text{ kg·m/s}^2}\right| \left(\frac{32 \text{ kg}}{1 \text{ kmol}}\right) = 14651 \frac{\text{kJ}}{\text{kmol}}$$

From Table A-23; $\bar{s}^\circ(T_2) = 220.35$ kJ/kmol. Finally, from (1)

$$\dot{E}_d/\dot{m} = T_0 \left\{\bar{s}^\circ(T_2) - \bar{s}^\circ(T_1) - \bar{R} \ln p_2/p_1\right\} \frac{1}{M}$$

$$= (293 \text{ K})\left\{(220.35 - 229.430) - 8.314 \ln(1.5/5)\right\} \frac{\text{kJ}}{\text{kmol K}} \left(\frac{1 \text{ kmol}}{32 \text{ kg}}\right)$$

$$= 8.514 \text{ kJ/kg} \quad \longleftarrow \quad \dot{E}_d/\dot{m}$$

PROBLEM 7.78

KNOWN: Steady-state operating data are provided for an open feedwater heater.

FIND: Determine the ratio of the incoming mass flow rates, $\dot{m}_1/\dot{m}_2$, and the exergy destruction per unit mass exiting.

SCHEMATIC & GIVEN DATA:

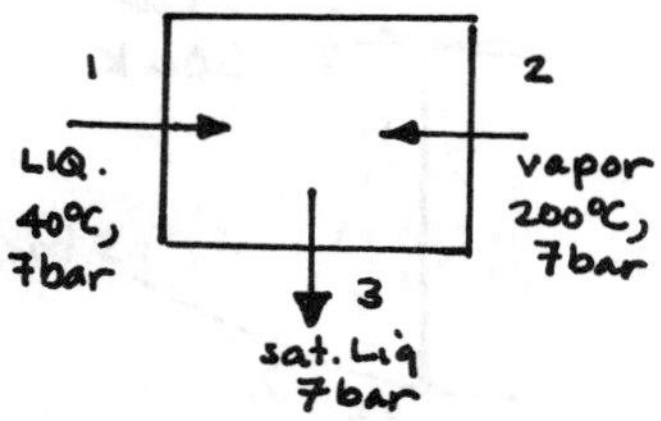

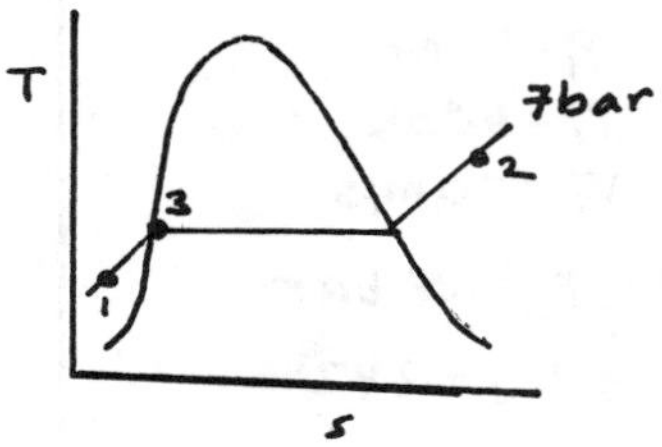

ASSUMPTIONS:

1. The control volume shown in the schematic is at steady state.
2. For the control volume, $\dot{Q}_{cv}=0$, $\dot{W}_{cv}=0$. Kinetic and potential energy effects can be ignored.
3. $T_0 = 25°C$, $P_0 = 1 atm$

ANALYSIS: (a) A mass rate balance reads, $\dot{m}_3 = \dot{m}_1 + \dot{m}_2$. An energy rate balance reduces with assumptions 2 to give

$$0 = \dot{m}_1 h_1 + \dot{m}_2 h_2 - \dot{m}_3 h_3 \Rightarrow 0 = \dot{m}_1 h_1 + \dot{m}_2 h_2 - (\dot{m}_1 + \dot{m}_2) h_3$$

$$\Rightarrow \frac{\dot{m}_1}{\dot{m}_2} = \frac{h_2 - h_3}{h_3 - h_1} = \frac{2844.8 - 697.22}{697.22 - 167.57} = 4.05 \quad \leftarrow$$

where data is from Tables A-2, 3, 4. At state 1, $h_1 \approx h_f(T_1)$.

(b) The exergy destruction can be found using $(\dot{E}_d/\dot{m}_3) = T_0(\dot{\sigma}_{cv}/\dot{m}_3)$, where $\dot{\sigma}_{cv}/\dot{m}_3$ is obtained from an entropy balance:

$$0 = \sum \frac{\dot{Q}_j}{T_j}^{\,0} + \dot{m}_1 s_1 + \dot{m}_2 s_2 - \dot{m}_3 s_3 + \dot{\sigma}_{cv}$$

$$\Rightarrow \frac{\dot{\sigma}_{cv}}{\dot{m}_3} = s_3 - \frac{\dot{m}_1}{\dot{m}_3} s_1 - \frac{\dot{m}_2}{\dot{m}_3} s_3$$

From part (a) $\dot{m}_1 = 4.05\,\dot{m}_2$, $\dot{m}_3 = \dot{m}_1 + \dot{m}_2 = 5.05\,\dot{m}_2$. Thus, $\dot{m}_2/\dot{m}_3 = 0.198$, $\dot{m}_1/\dot{m}_3 = 0.802$. With data from Tables A-2, 3, 4, At state 1, $s_1 \approx s_f(T_1)$

$$\frac{\dot{\sigma}_{cv}}{\dot{m}_3} = [1.9922 - (0.802)(0.5725) - (0.198)(6.8865)]\ \frac{kJ}{kg\cdot K} = 0.1696\ \frac{kJ}{kg\cdot K}$$

$$\Rightarrow \dot{E}_d/\dot{m}_3 = T_0(\dot{\sigma}_{cv}/\dot{m}_3) = (298K)(0.1696\ kJ/kg\cdot K) = 50.5\ kJ/kg \quad \leftarrow$$

PROBLEM 7.79

KNOWN: Steady-state operating data are provided in Problem 6.121 for an open feedwater heater.

FIND: Determine the cost of the exergy destroyed.

SCHEMATIC & GIVEN DATA: See solution to Problem 6.121

ASSUMPTIONS 1. The exiting mass flow rate is 1 kg/s. 2. 8000 hours of operation annually. 3. Exergy is valued 8 cents per kW·h. 4. $T_0 = 293\ K$ (20°C), $P_0 = 1\ atm$.

ANALYSIS: From the solution to Problem 6.121(a), $(\dot{\sigma}_{cv}/\dot{m}_3) = 0.218\ kJ/kg\cdot K$. Then, with $\dot{E}_d = T_0\dot{\sigma}_{cv}$, the cost rate is

$$\dot{\$} = [(293K)(1\ kg/s)(0.218\ kJ/kg\cdot K)]\left|\frac{1\ kW}{1\ kJ/s}\right|\left[\frac{8000\ h}{year}\right]\left[\frac{\$0.08}{kW\cdot h}\right] = \$40,879/year$$

PROBLEM 7.80

KNOWN: A proposal is made for providing steam at 2 MPa, 400°C from a source at 3 MPa, 700°C.

FIND: (a) Determine the total rate of exergy destruction that would result from the implementation of the proposal. (b) Determine the annual cost of the exergy destruction found in part (a). Discuss.

SCHEMATIC & GIVEN DATA:

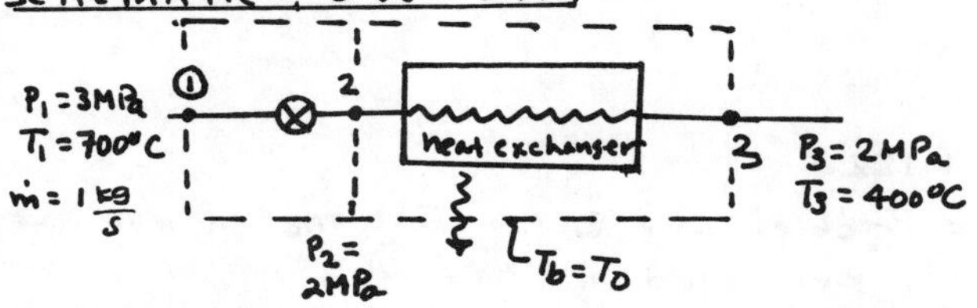

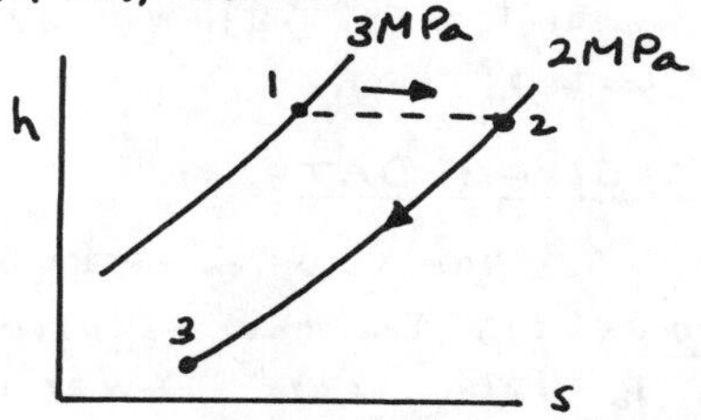

ASSUMPTIONS: 1. The control volumes shown in the schematic are at steady state. 2. The expansion across the valve is a throttling process. 3. For the control volume enclosing the heat exchanger, heat transfer takes place at $T_b = T_0$ only. 4. The effects of motion and gravity can be ignored. 5. Exergy is evaluated at 8 cents per kW·h, and there are 8000 h of operation annually. 6. $T_0 = 293K$, $P_0 = 0.1MPa$

ANALYSIS: There are two important sources of exergy destruction: Exergy is destroyed in the expansion across the valve. Exergy is also destroyed in the spontaneous heat transfer that takes place between the heat exchanger and the ambient at a lower temperature. These exergy destructions can be evaluated individually by study of each control volume shown in the schematic. The total exergy destruction is the sum of the individual contributions.

The total exergy destruction also can be found by considering a single control volume enclosing both components. Since there is no heat transfer associated with the valve, and heat transfer for the portion of the control volume enclosing the heat exchanger occurs at $T_b = T_0$, an exergy rate balance for the overall control volume takes the form

$$0 = \left[1 - \frac{T_0}{T_b}\right]^0 \dot{Q} - \dot{W}_{cv}^{\,0} + \dot{m}\,[e_{f1} - e_{f3}] - \dot{E}_d \Rightarrow \dot{E}_d = \dot{m}\,[e_{f1} - e_{f3}]$$

or

$$\dot{E}_d = \dot{m}\left[(h_1 - h_3) - T_0(s_1 - s_3)\right]$$

With data from Table A-4

$$\dot{E}_d = (1\,kg/s)\left[(3911.7 - 3247.6)\frac{kJ}{kg} - (293\,K)[7.7571 - 7.1271]\frac{kJ}{kg\cdot K}\right] = 479.51\,\frac{kJ}{s} \longleftarrow$$

(b) The cost rate is

$$\dot{\$} = (479.51\,kW)\left(8000\frac{h}{year}\right)\left(\frac{\$0.08}{kW\cdot h}\right) = \$306,886/year \longleftarrow$$

Discussion: The result of part (a) shows that considerable exergy is destroyed in the proposed method for achieving the lower-pressure, lower-temperature steam condition. This amount of exergy destruction is not necessary, however. For example, a power-recovery turbine could replace the valve and achieve a lower pressure and temperature while generating power, and rather than cooling to the ambient, process heat could be developed for some purpose while the main steam flow is cooled. The cost rate determined in part (b) provides an indicator of cost savings that might result from reduced exergy destruction. Of course there would be a tradeoff with the costs to acquire and operate the turbine, process heat exchanger, etc. that would be specified for more exergy-thrifty approaches. In the end, the decision would be governed by economic considerations.

PROBLEM 7.81

KNOWN: Steam expands through a turbine and then passes through a counter flow heat exchanger in which the other stream is air. Steady state operating data are provided.

FIND: Determine the exergy destruction rates for the turbine and heat exchanger, in Btu/s. Evaluating exergy at 8¢/kW·h, determine the hourly cost of each.

SCHEMATIC & GIVEN DATA:

See the solution to Problem 6.122

ASSUMPTIONS: (1) See the solution to Problem 6.122. (2) For the environment $T_0 = 500°R$, $p_0 = 1\,atm$. (3) Exergy is evaluated at 8¢/kW·h.

ANALYSIS: From the solution to Problem 6.122

turbine: $\dot{\sigma}_{cv} = 0.459 \frac{Btu/s}{°R}$

heat exchanger: $\dot{\sigma}_{cv} = 0.591 \frac{Btu/s}{°R}$

Accordingly, with $\dot{E}_d = T_0 \dot{\sigma}_{cv}$

① turbine: $\dot{E}_d = 500°R\left(0.459 \frac{Btu/s}{°R}\right) = 229.5 \frac{Btu}{s}$ ← turbine

heat exchanger: $\dot{E}_d = 500°R\left(0.591 \frac{Btu/s}{°R}\right) = 295.5 \frac{Btu}{s}$ ← HX

The corresponding cost rates are

turbine: $\dot{\$} = \left(229.5 \frac{Btu}{s}\right)\left|\frac{3600 s}{h}\right|\left|\frac{kW\cdot h}{3413\,Btu}\right|\left(0.08 \frac{\$}{kW\cdot h}\right) = 19.37 \frac{\$}{h}$ ← turbine

heat exchanger: $\dot{\$} = (295.5)|3600|\left|\frac{1}{3413}\right|(0.08) = 24.94 \frac{\$}{h}$ ← HX

---

1. The power developed by the turbine is 2852 hp. The exergy destruction rates expressed in terms of the horsepower for comparison are, respectively

turbine: $\dot{E}_d = \left(229.5 \frac{Btu}{s}\right)\left|\frac{3600 s}{h}\right|\left|\frac{1\,hp}{2545\,Btu/h}\right| = 325\,hp$

HX: $\dot{E}_d = (295.5)|3600|\left|\frac{1}{2545}\right| = 418\,hp$

PROBLEM 7.82

KNOWN: Operating data are provided for a gas turbine power plant at steady state in Problem 6.159(b).

FIND: For the compressor and turbine, determine the rates of exergy destruction per kg of air flowing. Express each as a percentage of the net work developed.

SCHEMATIC & GIVEN DATA:

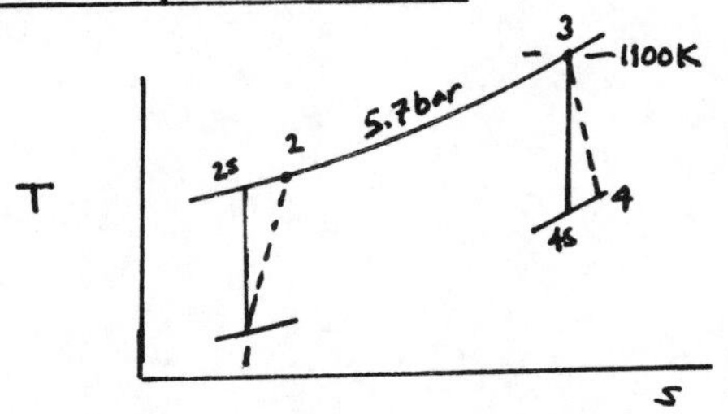

$p_1 = p_4 = 0.95$ bar
$p_2 = p_3 = 5.7$ bar
$T_1 = 22°C$, $T_3 = 1100K$
$\eta_c = 82\%$, $\eta_t = 85\%$

ASSUMPTIONS: (1) The gas turbine power plant operates at steady state. (2) The compressor and turbine operate adiabatically. (3) Kinetic and potential energy effects are negligible. (4) Air is modeled as an ideal gas. (4) For the environment, $T_0 = 22°C$, $p_0 = 0.95$ bar.

ANALYSIS: For the compressor and turbine the exergy destruction rates are conveniently determined using $\dot{E}_d/\dot{m} = T_0\,\dot{\sigma}_{cv}/\dot{m}$, where $\dot{\sigma}_{cv}$ is the rate of entropy production from an entropy balance. Thus

compressor: $\frac{\dot{E}_d}{\dot{m}} = T_0(s_2 - s_1)$, turbine: $\frac{\dot{E}_d}{\dot{m}} = T_0(s_4 - s_3)$

From the solution to Problem 6.159(b), $h_1 = 295.17$ kJ/kg, $h_3 = 1161.07$ kJ/kg, $h_{2s} = 493.03$ kJ/kg, $h_{4s} = 706.51$ kJ/kg, and

$$h_3 - h_4 = 386.38 \Rightarrow h_4 = 1161.07 - 386.38 = 774.69 \text{ kJ/kg}$$
$$h_2 - h_1 = 241.29 \Rightarrow h_2 = 295.17 + 241.29 = 536.46 \text{ kJ/kg}$$

Also, from the solution to Problem 6.159(b), $\dot{W}_{net}/\dot{m} = 145.09$ kJ/kg.

For the compressor,

$$\frac{\dot{E}_d}{\dot{m}} = T_0\left(s_2^\circ - s_1^\circ - R\ln p_2/p_1\right) = 295\left[2.2843 - 1.68515 - \frac{8.314}{28.97}\ln\frac{5.7}{0.95}\right] = 25.06\ \frac{\text{kJ}}{\text{kg}} \leftarrow \text{compressor}$$

where s_1° and s_2° are obtained from Table A-22 at T_1 and h_2, respectively.

For the turbine,

$$\frac{\dot{E}_d}{\dot{m}} = T_0\left(s_4^\circ - s_3^\circ - R\ln p_4/p_3\right) = 295\left(2.6571 - 3.07732 - \frac{8.314}{28.97}\ln\frac{0.95}{5.7}\right) = 27.73\ \frac{\text{kJ}}{\text{kg}} \leftarrow \text{turbine}$$

where s_3° and s_4° are obtained from Table A-22 at T_3 and h_4, respectively.

Expressed as a percentage of the net work developed

compressor: $\left(\frac{25.06}{145.09}\right)(100) = 17.3\%$ ← %

turbine: $\left(\frac{27.73}{145.09}\right)(100) = 19.1\%$

PROBLEM 7.83

KNOWN: Steady-state operating data are provided for a valve, flash chamber, and turbine in series.

FIND: For each component, determine the rate of exergy destruction and express it as a percentage of the turbine power.

SCHEMATIC & GIVEN DATA:

- See solution to Problem 6.160
- From the solution to Problem 6.160

$\dot{W}_t = 139.3$ Btu/s

flash chamber: $\dot{\sigma}_{cv} = 0.0054 \ \frac{\text{Btu/s}}{°R}$

valve: $\dot{\sigma}_{cv} = 0.0176 \ \frac{\text{Btu/s}}{°R}$

turbine: $\dot{\sigma}_{cv} = 0.0266 \ \frac{\text{Btu/s}}{°R}$

ASSUMPTIONS: 1. See solution to Problem 6.160. 2. $T_0 = 500°R$, $p_0 = 1\text{atm}$.

ANALYSIS: Since $\dot{\sigma}_{cv}$ values are known, it is convenient to find the exergy destruction rates using $\dot{E}_d = T_0 \dot{\sigma}_{cv}$

– flash chamber: $\dot{E}_d = 500°R\left(0.0054 \frac{\text{Btu/s}}{°R}\right)\left|\frac{3600\text{s}}{\text{h}}\right| = 9720 \frac{\text{Btu}}{\text{h}}$

– valve: $\dot{E}_d = 500°R\left(0.0176 \frac{\text{Btu/s}}{°R}\right)\left|\frac{3600\text{s}}{\text{h}}\right| = 31{,}680 \frac{\text{Btu}}{\text{h}}$ ←

– turbine: $\dot{E}_d = 500°R\left(0.0266 \frac{\text{Btu/s}}{°R}\right)\left|\frac{3600\text{s}}{\text{h}}\right| = 47{,}880 \frac{\text{Btu}}{\text{h}}$

Expressing each as a percentage of the power output,

– flash chamber: $\% = \frac{9720 \text{ Btu/h}}{(139.3 \times 3600) \text{ Btu/h}}(100) = 1.94$

– valve: $\% = \frac{(31680)(100)}{[139.3 \times 3600]} = 6.32$ ←

– turbine: $\% = \frac{(47{,}880)(100)}{[139.3 \times 3600]} = 9.55$

PROBLEM 7.84

KNOWN: Steady-state operating data are provided for two turbine stages and an interconnecting heat exchanger.

FIND: Determine the rates of exergy destruction. Place in rank order.

SCHEMATIC & GIVEN DATA:

- See Figure P4.82
- From the solution to Problem 6.125

$$\dot{\sigma}_{t1} = 3.1936\ \text{kW/K},\quad \dot{\sigma}_{t2} = 2.8649\ \frac{\text{kW}}{\text{K}},\quad \dot{\sigma}_{HX} = 3.1482\ \frac{\text{kW}}{\text{K}}$$

ASSUMPTIONS: 1. See solutions to Problems 4.82, 6.125. 2. $T_0 = 300\text{K}$, $p_0 = 1\text{bar}$.

ANALYSIS: Since the entropy production rates are known, it is convenient to determine the rates of exergy destruction using $\dot{E}_d = T_0 \dot{\sigma}_{cv}$. Thus,

$$(\dot{E}_d)_{t1} = (300\text{K})(3.1936\text{kW/K}) = 958.1\text{kW},\quad (\dot{E}_d)_{t2} = (300)(2.8649) = 859.5\text{kW},$$

$$(\dot{E}_d)_{HX} = 944.5\text{kW}$$

. In order, turbine 1, heat exchanger, turbine 2.

PROB 7.85

KNOWN: Steady-state operating data are provided for a simple vapor power plant.

FIND: Determine the rates of exergy destruction for the turbine, condenser, and pump. Place in rank order.

SCHEMATIC & GIVEN DATA:

- See Figure P4.85
- From the solution to Problem 6.126

$$\dot{\sigma}_{turb} = 87.51\ \frac{\text{kW}}{\text{K}},\quad \dot{\sigma}_{pump} = 1.47\ \frac{\text{kW}}{\text{K}},\quad \dot{\sigma}_{COND} = 35.4\ \frac{\text{kW}}{\text{K}}$$

ASSUMPTIONS: 1. See solutions to Problems 4.85, 6.126. 2. $T_0 = 293$, $p_0 = 1\text{bar}$.

ANALYSIS: Since the entropy production rates are known, it is convenient to determine the rates of exergy destruction using $\dot{E}_d = T_0 \dot{\sigma}_{cv}$. Thus,

turbine: $\dot{E}_d = (293\text{K})\left(87.51\ \frac{\text{kW}}{\text{K}}\right) = 25{,}640\text{kW}$

① pump: $\dot{E}_d = (293)(1.47) = 431\text{kW}$

condenser: $\dot{E}_d = (293)(35.4) = 10{,}372\text{kW}$

In order, turbine, condenser, pump.

---

1. Using results from the solution to Problem 4.85, the net power developed is

$$\dot{W}_{net} = \dot{m}\left[(h_1 - h_2) - (h_4 - h_3)\right] = \left(109\ \frac{\text{kg}}{\text{s}}\right)\left[(3425.1 - 2336.7) - (188.9 - 173.9)\right]\frac{\text{kJ}}{\text{kg}}\left|\frac{1\text{kW}}{1\text{kJ/s}}\right|$$

$$= 117{,}000\text{kW}$$

which provides a measure of the significance of each of the three exergy destructions determined here. Expressed as percentages of the net power output

$$\text{– turbine}\ \frac{25{,}640}{117{,}000}(100) = 22\%,\quad \text{– pump}\ \frac{(431)(100)}{117{,}000} = 0.4\%,\quad \text{– Condenser}\ \frac{(10{,}372)(100)}{117{,}000} = 9\%$$

PROBLEM 7.86

KNOWN: Steady-state operating data are provided for a gas turbine power plant in Problem 6.103. The actual power output is known.

FIND: Determine (a) the rate of exergy transfer accompanying heat transfer, (b) the net rate exergy is carried out by the flowing air, (c) the total rate of exergy destruction.

SCHEMATIC & GIVEN DATA:

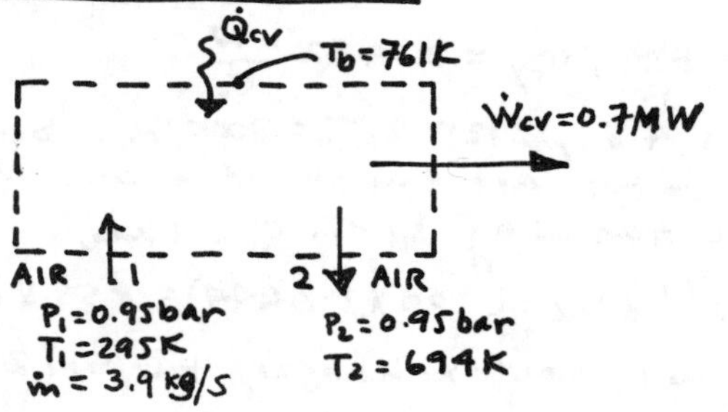

ASSUMPTIONS: 1. The control volume shown in the schematic is at steady state.
2. Heat transfer $\dot{Q}_{cv}$ takes place at temperature T_b only.
3. Air is modeled as an ideal gas.
4. $T_0 = 295\,K$, $P_0 = 0.95\,bar$. Effects of motion and gravity can be ignored.

ANALYSIS: (a) Mass and energy rate balances reduce to give

$0 = \dot{Q}_{cv} - \dot{W}_{cv} + \dot{m}(h_1 - h_2) \Rightarrow \dot{Q}_{cv} = \dot{W}_{cv} + \dot{m}(h_2 - h_1)$. Then with data from Table A-22

$$\dot{Q}_{cv} = 0.7\,MW + \left(3.9\,\frac{kg}{s}\right)\left[706.8 - 295.17\right]\frac{kJ}{kg}\left|\frac{1\,MW}{10^3\,kJ/s}\right| = 2.305\,MW$$

Using Eq. 7.33

$$\dot{E}_q = \left[1 - \frac{T_0}{T_b}\right]\dot{Q}_{cv} = \left[1 - \frac{295}{761}\right](2.305\,MW) = 1.411\,MW \quad \leftarrow$$

(b) With Eqs. 7.34, 7.36

$$\dot{E}_{f2} - \dot{E}_{f1} = \dot{m}\left[(h_2 - h_1) - T_0(s_2 - s_1)\right] = \dot{m}\left[(h_2 - h_1) - T_0\left(s^\circ(T_2) - s^\circ(T_1) - R\cancel{\ln\frac{p_2}{p_1}}^{\,0}\right)\right]$$

$$= \left(3.9\,\frac{kg}{s}\right)\left[(706.8 - 295.17) - 295\,(2.5635 - 1.68515)\right]\frac{kJ}{kg\cdot K}\left|\frac{1\,MW}{10^3\,kJ/s}\right|$$

$$= 0.595\,MW \quad \leftarrow$$

(c) The total rate of exergy destruction can be obtained by use of the exergy rate balance, Eq. 7.32b

$$\dot{E}_d = \dot{E}_q - \dot{W}_{cv} + (\dot{E}_{f1} - \dot{E}_{f2})$$

$$= (1.411 - 0.7 - 0.595)\,MW$$

① $$= 0.116\,MW \quad \leftarrow$$

---

1. Exergy balance sheet on a rate basis:

- Exergy rate in accompanying heat transfer: 1.411 MW
- Disposition of the exergy input
  - net power developed: 0.7 MW (49.6%)
  - net exergy carried out by air stream*: 0.595 MW (42.2%)
  - exergy destroyed: 0.116 MW (8.2%)

* Accordingly, the exergy carried out by the air stream is significant, and means should be considered for utilizing it cost effectively.

PROBLEM 7.87

KNOWN: Steady-state operating data are provided for a waste heat-recovery steam generator and turbine in Problem 4.83.

FIND: Develop a full exergy accounting of the net exergy carried in the oven exhaust.

SCHEMATIC & GIVEN DATA:

Data from the solution to Problem 4.83 are shown on the schematic.

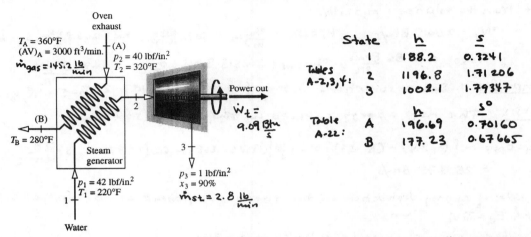

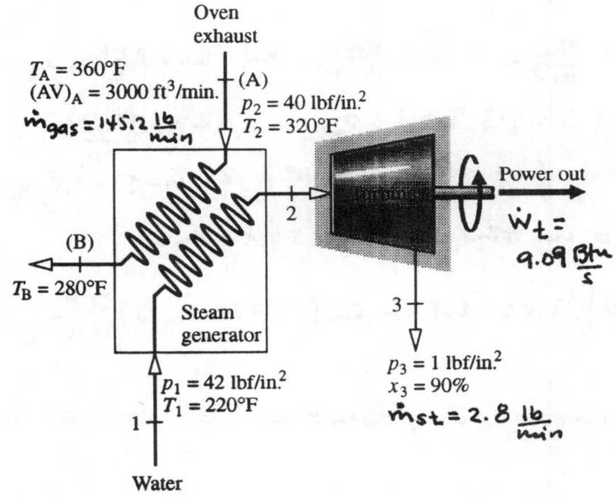

	State	h	s
Tables A-2,3,4:	1	188.2	0.3241
	2	1196.8	1.71206
	3	1002.1	1.79347
		h	s°
Table A-22:	A	196.69	0.70160
	B	177.23	0.67665

ASSUMPTIONS: 1. Control volumes enclosing the steam turbine and steam generator are at steady state. 2. $\dot{Q}_{cv} = 0$, and kinetic and potential energy effects can be ignored. 3. The oven gas is modeled as air as an ideal gas. There is no pressure change for the air. 4. T_0 = 540°R, p_0 = 1 atm.

ANALYSIS: The net rate exergy is carried in by the oven gas is

$$\dot{E}_{fA} - \dot{E}_{fB} = \dot{m}_{gas}\left[h_A - h_B - T_0\left(s_A^\circ - s_B^\circ - R \ln \frac{p_A}{p_B}\right)\right] \quad (\ln p_A/p_B = 0)$$

$$= \left(\frac{145.2}{60} \frac{lb}{s}\right)\left[196.69 - 177.23 - 540(0.70160 - 0.67665)\right] \frac{Btu}{lb} = 14.49 \frac{Btu}{s}$$

The net rate exergy is carried out by the water stream is

$$\dot{E}_{f3} - \dot{E}_{f1} = \dot{m}_{st}\left[h_3 - h_1 - T_0(s_3 - s_1)\right] = \left(\frac{2.8}{60} \frac{lb}{s}\right)\left[(1002.1 - 188.2) - 540(1.79347 - 0.3241)\right] \frac{Btu}{lb}$$

$$= 0.95 \frac{Btu}{s}$$

The rate of exergy destruction in the turbine is $(\dot{E}_d)_t = T_0(\dot{\sigma}_{cv})_t \Rightarrow$

$$(\dot{E}_d)_t = T_0 \dot{m}_{st}[s_3 - s_2] = (540\,°R)\left(\frac{2.8}{60} \frac{lb}{s}\right)\left[1.79347 - 1.71206\right] \frac{Btu}{lb \cdot °R} = 2.05\ Btu/s$$

The rate of the exergy destruction in the steam generator is $(\dot{E}_d)_{SG} = T_0(\dot{\sigma}_{cv})_{SG} \Rightarrow$

$$(\dot{E}_d)_{SG} = T_0\left[\dot{m}_{st}[s_2 - s_1] + \dot{m}_{gas}[s_B^\circ - s_A^\circ - R \ln p_B/p_A]\right] \quad (\ln p_B/p_A = 0)$$

$$= (540)\left[\left(\frac{2.8}{60}\right)(1.71206 - 0.3241) + \left(\frac{145.2}{60}\right)[0.67665 - 0.70160]\right] = 2.37\ Btu/s$$

Exergy Balance Sheet (rate basis):

- Net rate exergy is carried in by oven gas: 14.49 Btu/s
- Disposition of the exergy:

–	power developed	9.09 Btu/s	(62.7%)
–	net exergy carried out by water	0.95 Btu/s	(6.6%)
–	exergy destroyed – turbine	2.05 Btu/s	(14.1%)
–	exergy destroyed – steam generator	2.37 Btu/s	(16.4%)

PROBLEM 7.88

KNOWN: Steady-state operating data are provided in Problem 6.147 for a throttling valve in parallel with a steam turbine.

FIND: Determine a full exergy accounting of the net exergy carried in by the water stream.

SCHEMATIC & GIVEN DATA:

- See schematic in the solution to Problem 6.147.
- From the solution, $\dot{m}_1 = 25\,\text{lb/s}$

  $\dot{W}_t = 2045\,\text{Btu/s}$, $h_1 = 1350.6\,\frac{\text{Btu}}{\text{lb}}$, $s_1 = 1.5872\,\frac{\text{Btu}}{\text{lb}\cdot°\text{R}}$, $h_4 = 1268.8\,\frac{\text{Btu}}{\text{lb}}$, $s_4 = 1.6239\,\frac{\text{Btu}}{\text{lb}\cdot°\text{R}}$

  $\dot{\sigma}_{valve} = 0.6186\,\frac{\text{Btu/s}}{°\text{R}}$, $\dot{\sigma}_{turbine} = 0.2513\,\frac{\text{Btu/s}}{°\text{R}}$, $\dot{\sigma}_{mix} = 0.0476\,\frac{\text{Btu/s}}{°\text{R}}$

ASSUMPTIONS: 1. See assumptions listed for Problem 6.147. 2. $T_0 = 500°\text{R}$, $p_0 = 1\,\text{atm}$.

ANALYSIS: The net exergy carried in the water stream is

$$\dot{E}_{f1} - \dot{E}_{f4} = \dot{m}\left[(h_1 - h_4) - T_0(s_1 - s_4)\right] = 25\,\frac{\text{lb}}{\text{s}}\left[(1350.6 - 1268.8) - 500(1.5872 - 1.6239)\right]\frac{\text{Btu}}{\text{lb}\cdot°\text{R}}$$
$$= 2503.75\,\text{Btu/s}$$

The rates of exergy destruction in the three components can be evaluated using $\dot{E}_d = T_0\dot{\sigma}_{cv}$. Thus

valve: $(\dot{E}_d)_v = (500°\text{R})\left(0.6186\,\frac{\text{Btu/s}}{°\text{R}}\right) = 309.3\,\text{Btu/s}$

turbine: $(\dot{E}_d)_t = (500°\text{R})\left(0.2513\,\frac{\text{Btu/s}}{°\text{R}}\right) = 125.65\,\text{Btu/s}$

mixer: $(\dot{E}_d)_m = (500°\text{R})\left(0.0476\,\frac{\text{Btu/s}}{°\text{R}}\right) = 23.8\,\text{Btu/s}$

Exergy Balance Sheet (rate basis):

• net rate exergy is carried in by the water stream	2503.75 Btu/s	
• Disposition of the exergy carried in		
– power developed:	2045 Btu/s	(82 %)
– exergy destroyed:		
valve	309.3 Btu/s	(12 %)
turbine	125.65 Btu/s	(5 %)
mixer	23.8 Btu/s	(1 %)
	2503.75 Btu/s	

PROBLEM 7.89

KNOWN: Steady-state operating data are provided for a compressor and heat exchanger in series.

FIND: Develop a full exergy accounting of the compressor power input.

SCHEMATIC & GIVEN DATA

- See the schematic in the solution to Problem 6.124.
- From the solution

$$(-\dot{W}_{cv}) = 50.4\,kW, \quad \dot{m}_{air} = 0.5\,kg/s, \quad \dot{m}_{cw} = 0.403\,kg/s$$

$$\dot{\sigma}_{comp} = 0.0196\,\frac{kW}{K}, \quad \dot{\sigma}_{HX} = 0.0152\,\frac{kW}{K}$$

State	h	s°	State	h	s
1	300.19	1.70203	A	104.89	0.3674
3	350.49	1.85708	B	167.57	0.5725

ASSUMPTIONS: 1. See assumptions listed for Problem 6.124. 2. $T_0 = 300\,K$, $p_0 = 96\,kPa$.

ANALYSIS: The rates of exergy destruction can be found using $\dot{E}_d = T_0 \dot{\sigma}_{cv}$.

compressor: $(\dot{E}_d)_c = 300\,K\left(0.0196\,\frac{kW}{K}\right) = 5.88\,kW$

heat exchanger: $(\dot{E}_d)_{HX} = 300\,(0.0152) = 4.56\,kW$

The net rate exergy exits in the air stream

$$\dot{E}_{f3} - \dot{E}_{f1} = \dot{m}_{air}\left(h_3 - h_1 - T_0\left(s_3^\circ - s_1^\circ - R \ln p_3/p_1\right)\right)$$

$$= \left(0.5\,\frac{kg}{s}\right)\left[(350.49 - 300.19) - 300\left(1.85708 - 1.70203 - \frac{8.314}{28.97}\ln\frac{230}{96}\right)\right]\frac{kJ}{kg}\left|\frac{1\,kW}{1\,kJ/s}\right|$$

$$= 39.5\,kW$$

The net rate exergy exits in the water stream

$$\dot{E}_{fB} - \dot{E}_{fA} = \dot{m}_{cw}\left[(h_B - h_A) - T_0(s_B - s_A)\right] = \left(0.403\,\frac{kg}{s}\right)\left[(167.57 - 104.89) - 300(.5725 - 0.3674)\right]\frac{kJ}{kg}\left|\frac{1\,kW}{1\,kJ/s}\right|$$

$$= 0.46\,kW$$

Exergy Balance Sheet (rate basis)

• rate exergy is supplied by compressor power input	50.4 kW	
• disposition of the exergy		
– net exergy carried out by the air	39.5 kW	(78.4%)
– net exergy carried out by the water	0.46 kW	(0.9%)
– exergy destruction		
compressor	5.88 kW	(11.7%)
heat exchanger	4.56 kW	(9.0%)
	50.4 kW	

PROBLEM 7.90

KNOWN: Equation 7.39b

FIND: Plot ϵ versus T_u/T_o for $T_s/T_o = 8$ and selected values of η.

ANALYSIS:

Eq. 7.39b $$\epsilon = \eta\left[\frac{1 - T_o/T_u}{1 - T_o/T_s}\right] = \eta\left[\frac{1 - T_o/T_u}{1 - \frac{1}{8}}\right] = \frac{8}{7}\eta\left[1 - \frac{T_o}{T_u}\right]$$

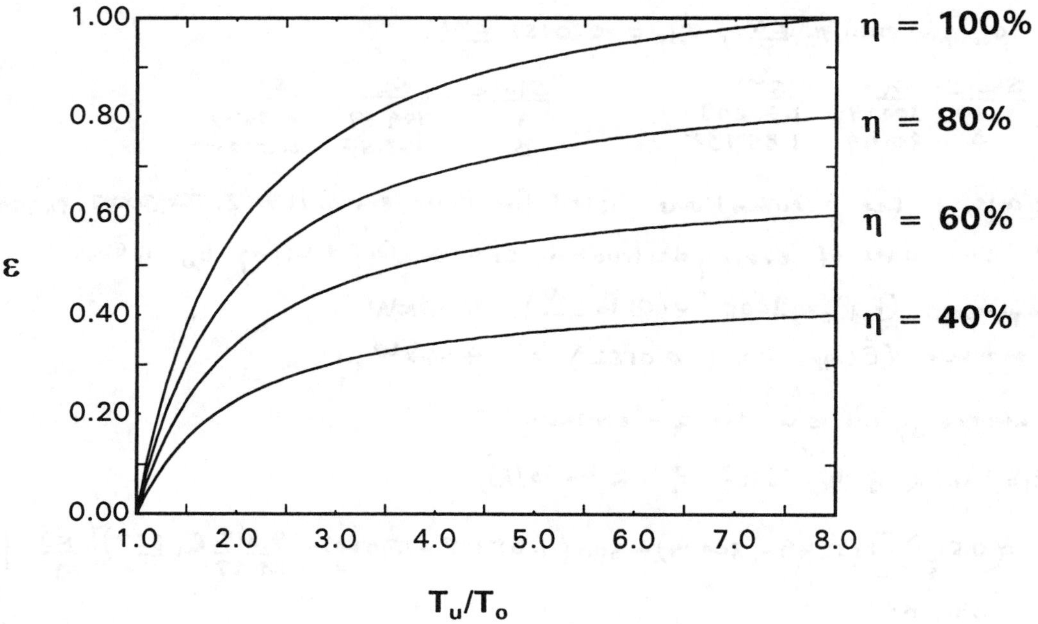

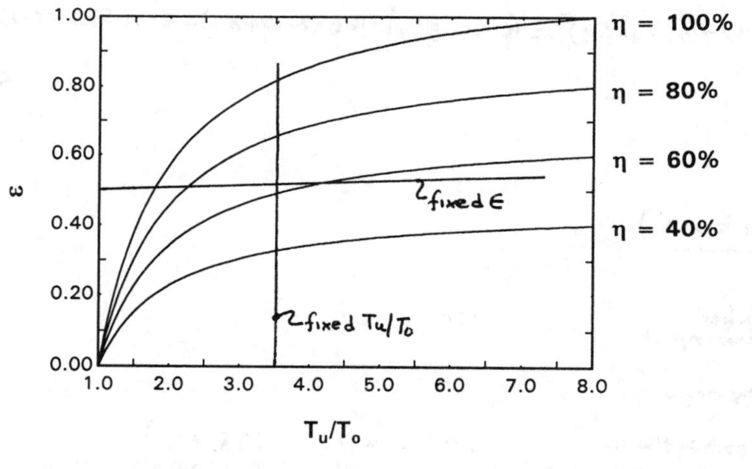

- when ϵ is fixed, as η increases, lower use temperatures are allowed.
- when T_u/T_o is fixed, as η increases, ϵ increases.

In sum, a high first-law efficiency, η, has practical benefits. So, both first-law and second-law considerations are important in proper energy resource utilization.

PROBLEM 7.91

KNOWN: Operating data are provided for an electrical water heater.

FIND: Devise and evaluate an exergetic efficiency.

SCHEMATIC & GIVEN DATA:

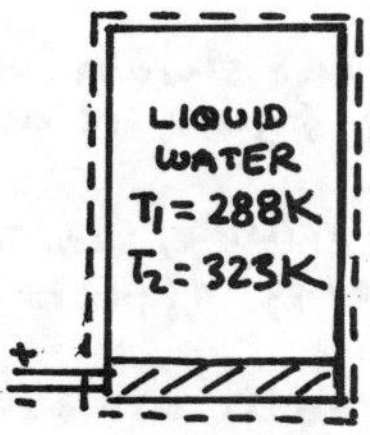

ASSUMPTIONS:

1. The system is shown in the accompanying schematic. $Q = 0$.
2. The water can be modeled as incompressible with constant specific heat c.
3. The states of the resistor and tank do not change.
4. $T_0 = T_1 = 288K$.

ANALYSIS: Applying a closed system exergy balance

$$\Delta E = \int_1^2 \left[1 - \frac{T_0}{T_b}\right]\delta Q^{\,0} - [W - p_0 \Delta V^{\,0}] - E_d$$

$$\Rightarrow \quad (-W) = \Delta E + E_d$$

That is, the exergy supplied to the system electrically, $(-W)$, is either stored, ΔE, or destroyed, E_d.

Thus, an exergetic efficiency in the form (exergy stored/exergy input) is

$$\epsilon = \frac{\Delta E}{(-W)} \qquad (1)$$

- The numerator can be evaluated using Eq. 7.10 and assumption 2 as

$$\Delta E = \Delta U + p_0 \Delta V^{\,0} - T_0 \Delta S = mc\,(T_2 - T_1) - T_0 mc \ln\frac{T_2}{T_1} \qquad (2)$$

- The work input can be found from an energy balance

$$\Delta U = Q^{\,0} - W \Rightarrow (-W) = mc\,(T_2 - T_1) \qquad (3)$$

Collecting Eqs 1,2,3

$$\epsilon = \frac{mc\,[(T_2 - T_1) - T_0 \ln(T_2/T_1)]}{mc\,[T_2 - T_1]} = 1 - \frac{T_0 \ln(T_2/T_1)}{(T_2 - T_1)}$$

Inserting $T_1 = T_0 = 288\,K$, $T_2 = 323K$

① $$\epsilon = 1 - \frac{288 \ln(323/288)}{(323 - 288)} = 0.056 \; (5.6\%)$$

---

1. This calculation shows that ordinary water heating makes poor use of the exergy input: electricity. The electricity supplied has a high potential for use, but only relatively low temperature water is obtained in exchange. The largest portion of the input is destroyed by irreversibilities within the water heater, leaving only 5.6% stored for later use.

PROBLEM 7.92

KNOWN: Steady-state operating data are provided for a hand-held hair dryer.

FIND: (a) Evaluate the power input. (b) Devise and evaluate an exergetic efficiency.

SCHEMATIC & GIVEN DATA:

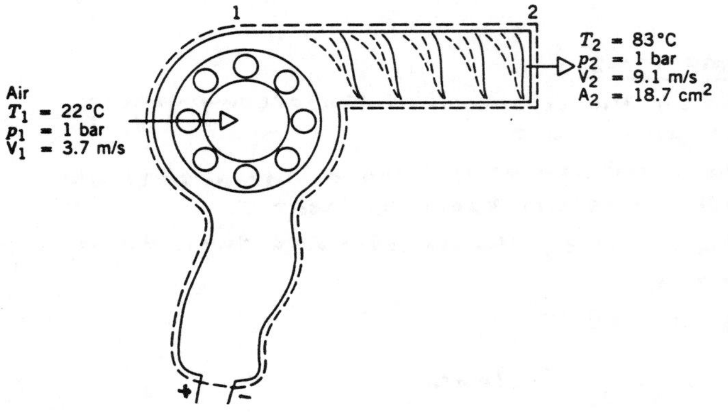

ASSUMPTIONS:

1. The control volume shown in the accompanying figure is at steady state.
2. For the control volume, $\dot{Q}_{cv} = 0$ and potential energy effects can be ignored.
3. Air is modeled as air as an ideal gas.
4. T_0 = 295K (22°C).

Analysis:

The power $\dot{W}_{cv}$ for the control volume in the accompanying figure can be evaluated using mass and energy rate balances. At steady state, the mass rate balance reduces to give $\dot{m}_1 = \dot{m}_2 = \dot{m}$. Using this relationship between the mass flow rates together with assumption 2, the energy rate balance reduces at steady state to give

$$\dot{W}_{cv} = \dot{m}\left(h_1 - h_2 + \frac{V_1^2 - V_2^2}{2}\right)$$

The mass flow rate $\dot{m}$ can be evaluated from the expression

$$\dot{m} = \frac{A_2 V_2}{v_2} = \frac{A_2 V_2}{RT_2/p_2}$$

$$\rightarrow \quad \dot{m} = \frac{\left[\dfrac{18.7\ \text{cm}^2}{(100\ \text{cm/m})^2}\right]\left(9.1\ \dfrac{\text{m}}{\text{s}}\right)\left(10^5\ \dfrac{\text{N}}{\text{m}^2}\right)}{\left(\dfrac{8314}{28.97}\ \dfrac{\text{N}\cdot\text{m}}{\text{kg}\cdot\text{K}}\right)(356\ \text{K})} = 0.0167\ \text{kg/s}$$

Then, with specific enthalpy values from Table A-22, h_1 = 295.2 kJ/kg and h_2 = 356.5 kJ/kg, the value of $\dot{W}_{cv}$ is obtained as follows:

$$\dot{W}_{cv} = 0.0167\ \frac{\text{kg}}{\text{s}}\left[295.2\ \frac{\text{kJ}}{\text{kg}} - 356.5\ \frac{\text{kJ}}{\text{kg}} + \frac{(3.7\ \text{m/s})^2 - (9.1\ \text{m/s})^2}{2}\left|\frac{1\ \text{N}}{1\ \text{kg}\cdot\text{m/s}^2}\right|\left|\frac{1\ \text{kJ}}{10^3\ \text{N}\cdot\text{m}}\right|\right]$$

$$= 0.0167\ \frac{\text{kg}}{\text{s}}\left(-61.3\ \frac{\text{kJ}}{\text{kg}} - 0.03\ \frac{\text{kJ}}{\text{kg}}\right)\left|\frac{1\ \text{kW}}{1\ \text{kJ/s}}\right| = -1.02\ \text{kW}$$

The minus sign reminds us that an electrical power *input* is required to operate the hair dryer. The change in kinetic energy between inlet and exit is seen to be negligible and is ignored in the remainder of the solution. Since $\dot{Q}_{cv} = 0$ and $\dot{m}_1 = \dot{m}_2$, the steady-state exergy rate balance reduces as follows:

$$0 = \sum_j \left[1 - \frac{T_0}{T_j}\right]\dot{Q}_j^{\,0} - \dot{W}_{cv} + \dot{m}(e_{f1} - e_{f2}) - \dot{E}_d \Rightarrow (-\dot{W}_{cv}) = \dot{m}(e_{f2} - e_{f1}) + \dot{E}_d$$

This result shows that the exergy input, $(-\dot{W}_{cv})$, either goes into increasing the exergy of the air stream, $\dot{m}(e_{f2} - e_{f1})$, or is destroyed, $\dot{E}_d$. Thus, an exergetic efficiency can be defined as

$$\varepsilon = \frac{\dot{m}(e_{f2} - e_{f1})}{(-\dot{W}_{cv})} = \frac{\dot{m}[(h_2 - h_1) - T_0(s_2 - s_1)]}{(-\dot{W}_{cv})} = \frac{\dot{m}[(h_2 - h_1) - T_0(s_2^\circ - s_1^\circ - R\ln p_2/p_1^{\,0})]}{(-\dot{W}_{cv})}$$

With data from Table A-22

① $$\varepsilon = \frac{(0.0167\ \text{kg/s})\left[(356.5 - 295.2)\frac{\text{kJ}}{\text{kg}} - 295\text{K}(1.874 - 1.685)\frac{\text{kJ}}{\text{kg}\cdot\text{K}}\right]}{(1.02\ \text{kW})}\left|\frac{1\ \text{kW}}{1\ \text{kJ/s}}\right| = 0.091\ (9.1\%)$$

---

1. In this device, a relatively high potential for use electrical input is used to obtain a stream of slightly warm air. The low value of the efficiency ε emphasizes that source and end use are not well matched.

PROBLEM 7.93

KNOWN: From an input of electricity, a furnace delivers energy by heat transfer at a rate of $\dot{Q}_u$ at a use temperature of T_u. The furnace operates at steady state.

FIND: (a) Devise an exergetic efficiency. (b) Plot this efficiency versus T_u ranging from 300 to 900K.

SCHEMATIC & GIVEN DATA:

ASSUMPTIONS: (1) The closed system is at steady state. (2) There are no significant energy transfers other than those shown. (3) For the environment, $T_0 = 20°C = 293K$.

ANALYSIS: (a) At steady state, the exergy balance with assumption 2 reads

$$0 = \dot{W}_e - \left[1 - \frac{T_0}{T_u}\right]\dot{Q}_u - \dot{E}_d$$

where the quantities $\dot{W}_e$ and $\dot{Q}_u$ have been taken as positive in the direction of the arrows. Rearranging

$$\dot{W}_e = \left[1 - \frac{T_0}{T_u}\right]\dot{Q}_u + \dot{E}_d$$

(input) (transfer out) (destroyed)

An exergetic efficiency in the form output/input is

$$\epsilon = \frac{\left[1 - \frac{T_0}{T_u}\right]\dot{Q}_u}{\dot{W}_e}$$

With $\dot{Q}_u = \dot{W}_e$ from an energy balance at steady state

$$\epsilon = \left[1 - \frac{T_0}{T_u}\right]$$ ← ϵ

(b) The following plot is generated using IT with $T_0 = 293$:

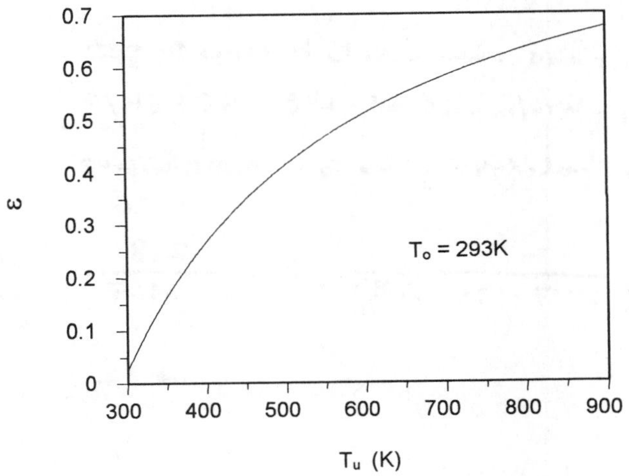

From the plot, we observe that as $T_u \to T_0$, $\epsilon \to 0$. Also, ϵ increases rapidly with T_u, since as T_u increases less exergy is destroyed, per unit of exergy input by electricity.

PROBLEM 7.94

KNOWN: Steady-state operating data are provided for a steam turbine.

FIND: Evaluate the isentropic turbine efficiency and the exergetic turbine efficiency.

SCHEMATIC & GIVEN DATA:

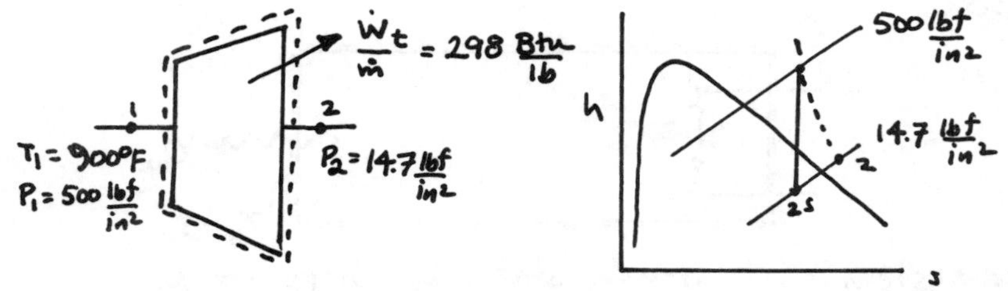

ASSUMPTIONS: 1. The control volume shown in the schematic is at steady state. 2. For the control volume, $\dot{Q}_{cv} = 0$, and kinetic and potential energy effects are negligible. 3. $T_0 = 520°R$, $p_0 = 14.7$ lbf/in².

ANALYSIS: Using Eq. 6.48, the isentropic turbine efficiency is

$$\eta_t = \frac{\dot{W}_{cv}/\dot{m}}{h_1 - h_{2s}}$$

From Table A-4E $h_1 = 1466.5$ Btu/lb, $s_1 = 1.6987$ Btu/lb·°R. With $s_{2s} = s_1$ at 14.7 lbf/in², Table A-3E gives

$$x_{2s} = \frac{s_{2s} - s_f}{s_g - s_f} = \frac{1.6987 - 0.3121}{1.4446} = 0.96$$

$$\Rightarrow h_{2s} = h_f + x_{2s} h_{fg} = 180.15 + 0.96(970.4) = 1111.7 \text{ Btu/lb}$$

Thus,

$$\eta_t = \frac{298 \text{ Btu/lb}}{[1466.5 - 1111.7] \text{ Btu/lb}} = \frac{298}{354.8} = 0.84 \quad (84\%) \quad \longleftarrow$$

Using Eq. 7.42, the exergetic turbine efficiency is

$$\epsilon = \frac{\dot{W}_{cv}/\dot{m}}{(h_1 - h_2) - T_0(s_1 - s_2)}$$

To fix the exit state, reduce the mass and energy rate balance to get

$$0 = \cancel{\dot{Q}_{cv}}^{0} - \dot{W}_{cv} + \dot{m}(h_1 - h_2) \Rightarrow h_2 = h_1 - \dot{W}_{cv}/\dot{m} = 1466.5 - 298 = 1168.5 \text{ Btu/lb}$$

Then, at 14.7 lbf/in², with h_2 known, Table A-4E gives $s_2 = 1.7828$ Btu/lb·°R.

Accordingly,

$$\epsilon = \frac{298 \text{ Btu/lb}}{[\underbrace{(1466.5 - 1168.5)}_{298} - 520(1.6987 - 1.7828)] \frac{\text{Btu}}{\text{lb}}} = \frac{298}{341.7}$$

$$= 0.872 \quad (87.2\%) \quad \longleftarrow$$

PROBLEM 7.95

KNOWN: Steady-state operating data are provided in Problem 7.67 for a valve and turbine in series.

FIND: For the turbine, plot the exergetic efficiency versus the steam quality at the turbine exit ranging from 90 to 100%.

SCHEMATIC & GIVEN DATA:

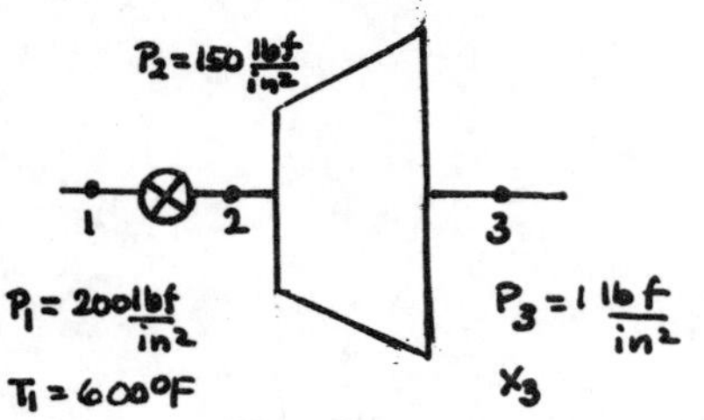

ASSUMPTIONS: 1. Control volumes enclosing the valve and turbine are at steady state.
2. For the control volumes, $\dot{Q}_{cv} = 0$, and the effects of motion and gravity can be ignored. The expansion across the valve is a throttling process.
3. $T_0 = 520°R$, $p_0 = 1 atm$.

ANALYSIS: The turbine exergetic efficiency is given by Eq. 7.42. With indicated assumptions, this can be expressed as

$$\epsilon = \frac{h_2 - h_3}{(h_2 - h_3) - T_0(s_2 - s_3)}$$

. Since $h_2 = h_1$ for the throttling process, we get

$$\epsilon = \frac{h_1 - h_3}{(h_1 - h_3) - T_0(s_2 - s_3)} \quad (1)$$

. In Eq. (1) h_1 is fixed by T_1, p_1, s_2 is fixed by $h_2 = h_1$, p_2, and h_3 and s_3 are fixed by p_3, x_3.

Sample Calculation: $x_3 = 0.98$. From Table A-4E, $h_1 = 1322.1$ Btu/lb. Interpolating in Table A-4E with $h_2 = h_1$ at 150 lbf/in.², $s_2 = 1.7078$ Btu/lb·°R. With data from Table A-3E at 1 lbf/in.², $h_3 = 69.74 + 0.98(1036) = 1085$ Btu/lb, $s_3 = 0.1327 + 0.98(1.8453) = 1.9411$ Btu/lb·°R. Inserting values, Eq. (1) gives

$$\epsilon = \frac{(1322.1 - 1085)}{(1322.1 - 1085) - 520(1.7078 - 1.9411)} = \frac{237.1}{358.4} = 0.662 \quad (66.2\%)$$

IT Code

```
p1 = 200  // lbf/in.²
T1 = 600  // °F
p2 = 150  // lbf/in.²
p3 = 1  // lbf/in.²
x3 = 0.98
To = 60 + 460  // °R

h1 = h_PT("Water/Steam", p1, T1)
s2 = s_Ph("Water/Steam", p2, h1)
s3 = ssat_Px("Water/Steam", p3, x3)
h3 = hsat_Px("Water/Steam", p3, x3)
eff = (h1 - h3) / ((h1 - h3) - To * (s2 - s3))
```

IT Results for $x_3 = 0.98$

eff	0.6613
h1	1322
h3	1085
s2	1.707
s3	1.941

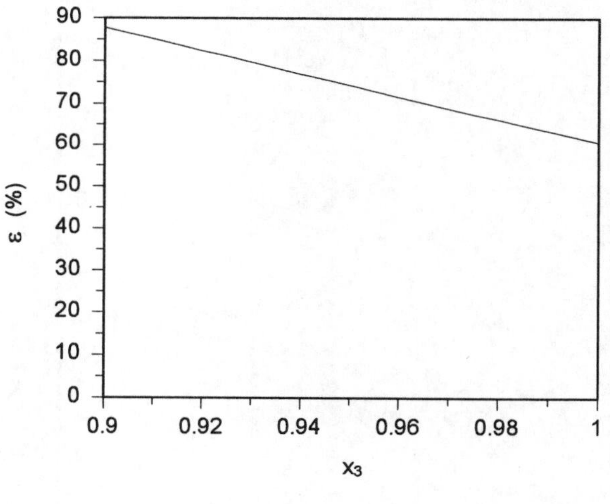

PROBLEM 7.96

KNOWN: Steady state operating data are provided for an air turbine.

FIND: Determine (a) the isentropic turbine efficiency, (b) the exergetic turbine efficiency.

SCHEMATIC & GIVEN DATA:

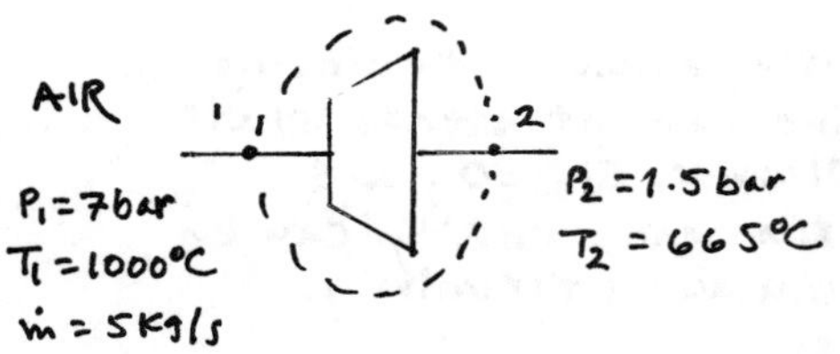

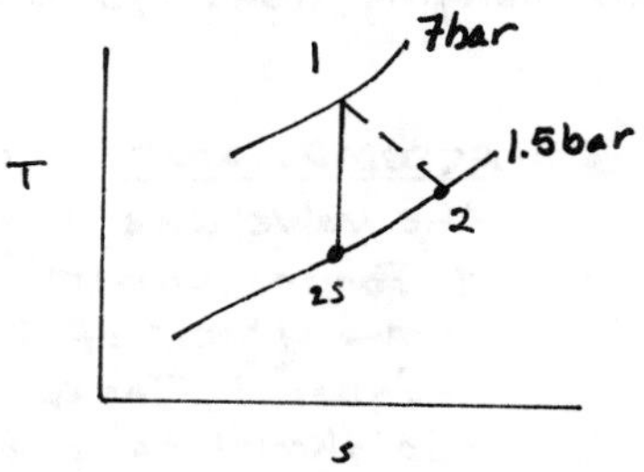

ASSUMPTIONS: (1) The control volume shown in the figure is at steady state. (2) For the control volume $\dot{Q}_{cv} = 0$, and kinetic/potential energy changes can be ignored. (3) Air is modeled as an ideal gas with $k = 1.35$. (4) The exergy reference environment is at $T_0 = 298$ K.

ANALYSIS: (a) The isentropic turbine efficiency is given by Eq. 6.48

$$\eta_t = \frac{h_1 - h_2}{h_1 - h_{2s}} = \frac{c_p(T_1 - T_2)}{c_p(T_1 - T_{2s})} = \frac{T_1 - T_2}{T_1 - T_{2s}}$$

With Eq. 6.45, $T_{2s} = T_1\left(\frac{P_{2s}}{P_1}\right)^{\frac{k-1}{k}} = (1273 K)\left(\frac{1.5}{7.0}\right)^{0.35/1.35} = 854 K$

$$\therefore \eta_t = \frac{1273 - 938}{1273 - 854} = 0.8 \quad (80\%) \longleftarrow \eta_t$$

(b) The exergetic turbine efficiency is given by Eq. 7.42, which becomes

$$\epsilon = \frac{h_1 - h_2}{(h_1 - h_2) - T_0(s_1 - s_2)} = \frac{c_p(T_1 - T_2)}{c_p(T_1 - T_2) - T_0\left[c_p \ln\frac{T_1}{T_2} - R \ln\frac{P_1}{P_2}\right]}$$

, Eq. 3.47a gives $R = c_p\left(\frac{k-1}{k}\right)$

$$\epsilon = \frac{T_1 - T_2}{T_1 - T_2 - T_0\left[\ln\frac{T_1}{T_2} - \left(\frac{k-1}{k}\right)\ln\frac{P_1}{P_2}\right]} = \frac{1273 - 938}{1273 - 938 - 298\left[\ln\frac{1273}{938} - \frac{0.35}{1.35}\ln\frac{7}{1.5}\right]} = 0.923 \ (92.3\%) \longleftarrow \epsilon$$

PROBLEM 7.97

KNOWN: An ideal gas with constant specific heat ratio k enters a turbine operating at steady state at T_1, P_1 and expands adiabatically to T_2, P_2.

FIND: Determine when the value of the exergetic turbine efficiency exceeds the value of the isentropic efficiency.

SCHEMATIC & GIVEN DATA:

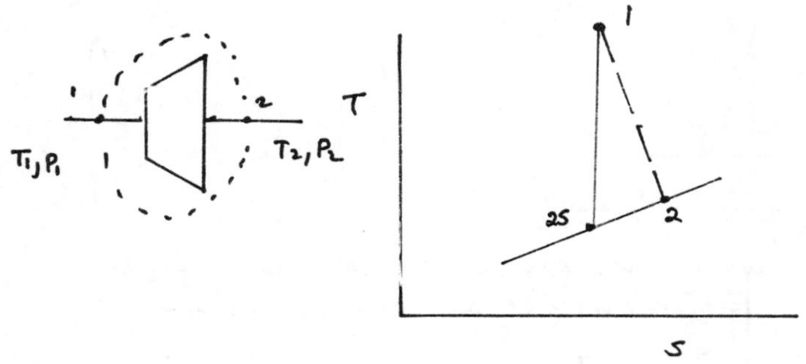

ASSUMPTIONS: (1) The control volume shown in the figure is at steady state. (2) For the control volume $\dot{Q}_{cv} = 0$, and the effects of kinetic/potential energy can be ignored. (3) The gas is modeled as an ideal gas with constant specific heat ratio k. (4) the exergy reference environment is at T_0.

ANALYSIS: The requirement is $\epsilon > \eta_t$. That is, with Eqs. 6.48 and 7.42

$$\frac{h_1 - h_2}{h_1 - h_2 - T_0(s_1 - s_2)} > \frac{h_1 - h_2}{h_1 - h_{2s}} \Rightarrow (h_1 - h_{2s}) > (h_1 - h_2) - T_0(s_1 - s_2)$$

$$\Rightarrow (h_2 - h_{2s}) > T_0(s_2 - \underbrace{s_1}_{= s_{2s}})$$

Thus, $\epsilon > \eta_t$ when

$$h_2 - h_{2s} > T_0(s_2 - s_{2s})$$

Since k is constant

$$c_p[T_2 - T_{2s}] > T_0\left[c_p \ln\frac{T_2}{T_{2s}} - \overset{\nearrow 0}{R \ln\frac{P_2}{P_{2s}}}\right]$$

$$\Rightarrow \quad (T_2 - T_{2s}) > T_0 \ln\frac{T_2}{T_{2s}}$$

or

$$\frac{T_2 - T_{2s}}{\ln(T_2/T_{2s})} > T_0 \qquad (1)$$

The left side is recognized as the logarithmic average of the temperature difference $T_2 - T_{2s}$.

Using the results of Problem 7.96 for illustration: $T_1 = 1273\,K$, $T_2 = 938\,K$, $T_{2s} = 854\,K$, $T_0 = 298$, Eq. (1) gives

$$\left[\frac{938 - 854}{\ln(938/854)}\right] = 895\,K > 298 \Rightarrow \epsilon > \eta_t$$

The solution to problem 7.96 gives $\epsilon = 92.3\%$, $\eta_t = 80\%$, which is in accord with the result elicited. [Note that the arithmetic average of T_2, T_{2s} is 896 K, only slightly different in this case from the logarithmic average.]

PROBLEM 7.98

KNOWN: Steady state operating data are provided for a pump.

FIND: Determine (a) exergy destruction per unit of mass flowing, (b) the exergetic pump efficiency.

SCHEMATIC & GIVEN DATA:

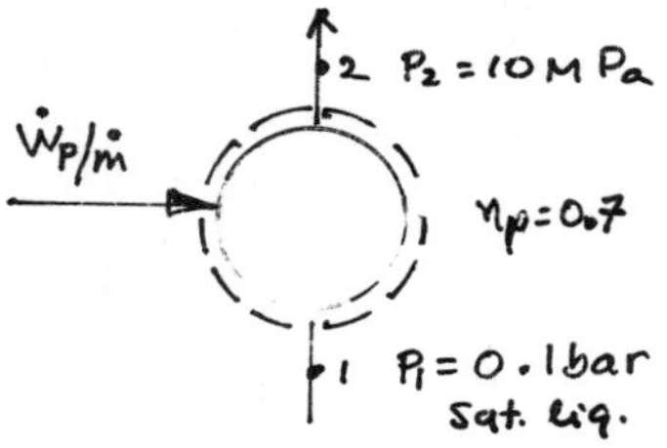

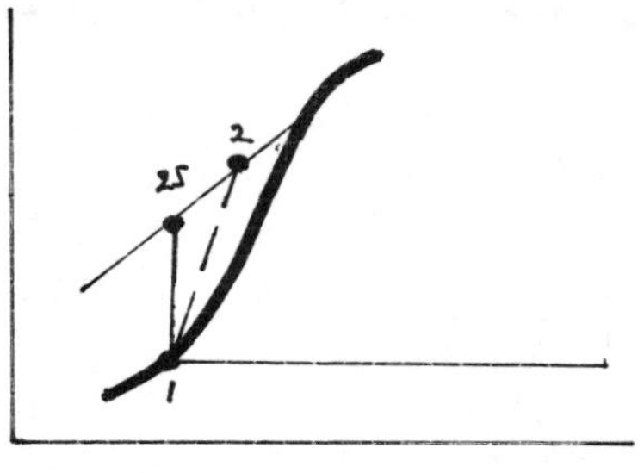

ASSUMPTIONS: (1) The control volume shown in the figure is at steady state. (2) For the control volume, $\dot{Q}_{cv} = 0$, and kinetic/potential energy effects are negligible. (3) $T_0 = 298$K.

ANALYSIS: (a) The exergy destruction per unit of mass flowing is conveniently found with

$$\frac{\dot{E}_d}{\dot{m}} = T_0 \frac{\dot{\sigma}_{cv}}{\dot{m}} \Rightarrow \frac{\dot{E}_d}{\dot{m}} = T_0(s_2 - s_1)$$

State 1 is fixed. State 2 is fixed by P_2 and h_2 obtained as follows using Eqs. 6.50, 6.53c together with an average value for the specific volume,

$$\eta_p = \frac{v_{ave}(P_2 - P_1)}{(-\dot{W}_{cv}/\dot{m})} = \frac{v_{ave}(P_2 - P_1)}{h_2 - h_1} \rightarrow h_2 = h_1 + \frac{v_{ave}(P_2 - P_1)}{\eta_p}$$

Here, v_{ave} is the average specific volume. Taking $v_{ave} \approx v_f(T_1)$, $h_1 = h_f(T_1)$, and data from Table A-3

$$h_2 = 191.83 \frac{kJ}{kg} + \frac{(1.0102/10^3 \ m^3/kg)[10 \times 10^6 - 10^4](N/m^2)}{0.7} \left| \frac{1 \ kJ}{10^3 N \cdot m} \right| = 206.2 \frac{kJ}{kg}$$

① Interpolating in Table A-5, $s_2 = 0.6582$ kJ/kg·K, $v_2 = 1.0072$ m³/kg. Also, $s_1 = s_f(T_1) = 0.6493$ kJ/kg·K. Then

$$\frac{\dot{E}_d}{\dot{m}} = 298K(0.6582 - 0.6493)\frac{kJ}{kg \cdot K} = 2.652 \ kJ/kg \quad \longleftarrow$$

(b) With Eq. 7.43

$$\epsilon = \frac{e_{f2} - e_{f1}}{(-\dot{W}_{cv}/\dot{m})} = \frac{(h_2 - h_1) - \overbrace{T_0(s_2 - s_1)}^{\dot{E}_d/\dot{m}}}{(h_2 - h_1)} = \frac{(206.2 - 191.83) - 2.652}{206.2 - 191.83}$$

$$= 0.818 \quad (81.8\%) \quad \longleftarrow$$

---

1. Comparing v_2 with $v_f(T_1)$, it is clear that specific volume does not vary appreciably in this case.

PROBLEM 7.99

KNOWN: Air enters an insulated turbine operating at steady state at a given state and with a known volumetric flow rate. The exit pressure and the isentropic turbine efficiency are specified.

FIND: Determine (a) the power developed and the exergy destruction rate. (b) the exergetic turbine efficiency.

SCHEMATIC & GIVEN DATA:

air
$p_1 = 4$ bar
$T_1 = 450$K
$(AV)_1 = 5\ m^3/s$

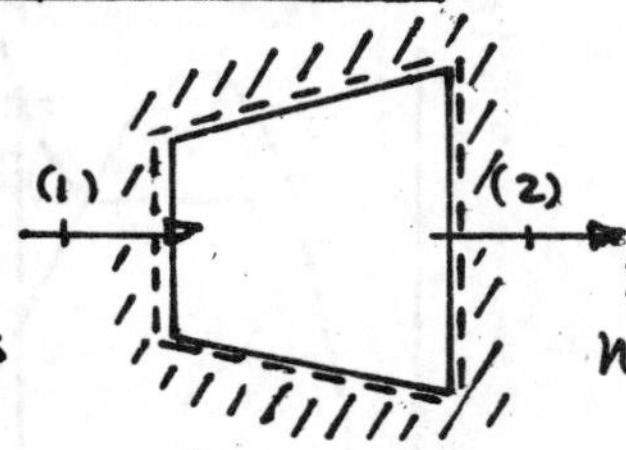

$p_2 = 1$ bar
$\eta_T = 84\%$

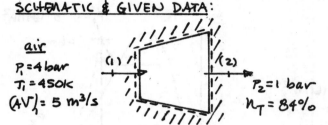

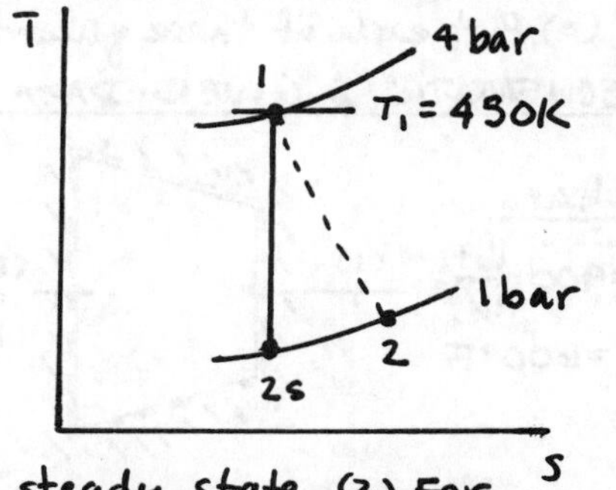

ASSUMPTIONS: (1) The control volume shown is at steady state. (2) For the control volume, $\dot{Q}_{cv} = 0$ and the effects of motion and gravity are neglected. (3) The air is modeled as an ideal gas. (4) For the environment, $T_0 = 20°C = 293$ K, $p_0 = 1$ bar.

ANALYSIS: (a) We begin by using the isentropic turbine efficiency to fix state 2. With data from Table A-22

$$p_r(T_{2s}) = \left(\frac{p_2}{p_1}\right) p_r(T_1) = \left(\frac{1}{4}\right) .775 = 1.44375 \Rightarrow \begin{cases} T_{2s} = 303.5\ K \\ h_{2s} = 303.71\ \frac{kJ}{kg} \end{cases}$$

Now

$$\eta_T = \frac{h_1 - h_2}{h_1 - h_{2s}} \Rightarrow h_2 = h_1 - \eta_T (h_1 - h_{2s}) = 451.8 - (.84)(451.8 - 303.71)$$
$$= 327.4\ kJ/kg$$

The mass flow rate of air is

$$\dot{m} = \frac{(AV)_1}{v_1} = \frac{p_1 (AV)_1}{RT_1} = \frac{(4\ bar)(5\ m^3/s)}{\left(\frac{8.314}{28.97}\ \frac{kJ}{kg \cdot K}\right)(450 K)} \left|\frac{10^5\ N/m^2}{1\ bar}\right| \left|\frac{1\ kJ}{10^3\ N\cdot m}\right|$$
$$= 15.49\ kg/s$$

From mass and energy rate balances: $0 = \cancel{\dot{Q}_{cv}}^{0} - \dot{W}_{cv} + \dot{m}\left[(h_1 - h_2) + \cancel{\left(\frac{V_1^2 - V_2^2}{2}\right)}^{0} + g\cancel{(z_1 - z_2)}^{0}\right]$

$$\dot{W}_{cv} = \dot{m}\left[h_1 - h_2\right] = \left(15.49\ \frac{kg}{s}\right)\left[451.8 - 327.4\right] \frac{kJ}{kg} \left|\frac{1\ kW}{1\ kJ/s}\right| = 1927\ kW \longleftarrow \dot{W}_{cv}$$

From mass and entropy balances: $0 = \cancel{\sum_j \left(\frac{\dot{Q}}{T}\right)_j}^{0} + \dot{m}(s_1 - s_2) + \dot{\sigma}_{cv} \Rightarrow \dot{\sigma}_{cv} = \dot{m}(s_2 - s_1)$

Thus

$$\dot{E}_d = T_0 \dot{\sigma}_{cv} = T_0 \dot{m}(s_2 - s_1) = T_0 \dot{m}\left[s°(T_2) - s°(T_1) - R \ln p_2/p_1\right]$$

Interpolating in Table A-22 with $h_2 = 327.4$; $s°(T_2) = 1.78886$.

$$\dot{E}_d = (293\ K)\left(15.49\ \frac{kg}{s}\right)\left[(1.78886 - 2.11161) - \frac{8.314}{28.97} \ln\left(\frac{1}{4}\right)\right] \frac{kJ}{kg \cdot K} \left|\frac{1\ kW}{1\ kJ/s}\right|$$
$$= 340.8\ kW \longleftarrow \dot{E}_d$$

(b) The exergetic turbine efficiency is

$$\epsilon = \frac{\dot{W}_{cv}}{\left[(h_1 - h_2) - T_0 (s_1 - s_2)\right]\dot{m}} = \frac{\dot{W}_{cv}}{\dot{W}_{cv} + \dot{E}_d} = \frac{1927}{1927 + 340.8} = 0.8497\ (85\%) \longleftarrow \epsilon$$

PROBLEM 7.100

KNOWN: Steam enters a well-insulated turbine operating at steady state at a specified state and exits at pressure p.

FIND: (a) For p = 50 lbf/in.², determine the exergy destruction rate per unit mass of steam flowing and the isentropic and exergetic turbine efficiencies.
(b) Plot each of these quantities versus p ranging from 1 to 50 lbf/in.²

SCHEMATIC & GIVEN DATA:

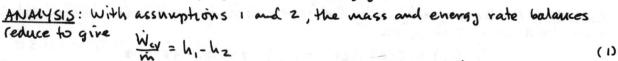

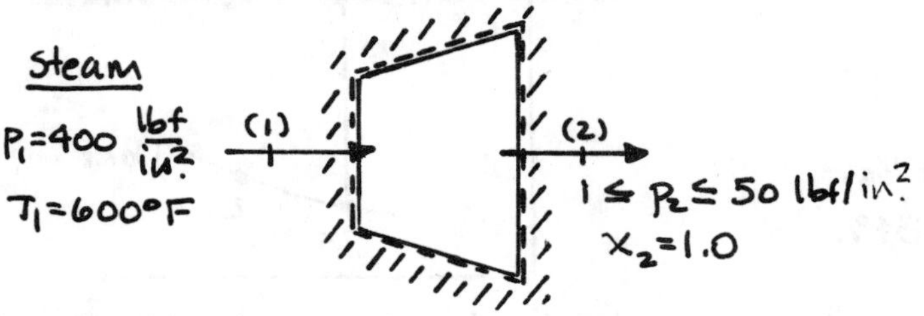

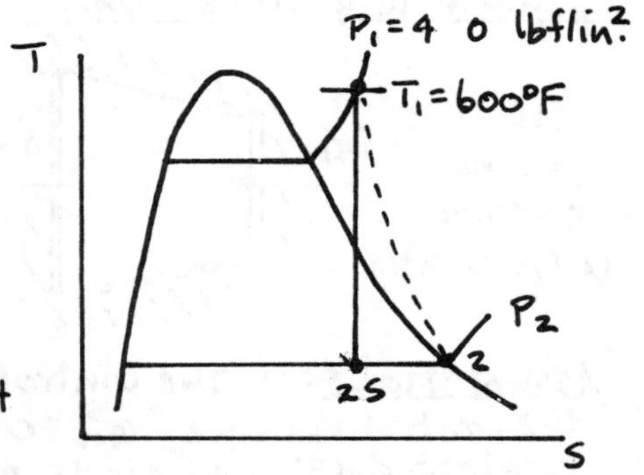

ASSUMPTIONS: (1) The control volume shown operates at steady state. (2) For the control volume, $\dot{Q}_{cv}=0$ and kinetic and potential energy effects are ignored.
(3) For the environment, $T_0 = 60°F = 520°R$, $p_0 = 1 atm$.

ANALYSIS: With assumptions 1 and 2, the mass and energy rate balances reduce to give

$$\frac{\dot{W}_{cv}}{\dot{m}} = h_1 - h_2 \qquad (1)$$

Further, mass and entropy rate balances reduce to give $\dot{\sigma}_{cv}/\dot{m} = s_2 - s_1$. The exergy destruction rate is $\dot{E}_d/\dot{m} = T_0(\dot{\sigma}_{cv}/\dot{m})$, or

$$\dot{E}_d/\dot{m} = T_0(s_2 - s_1) \qquad (2)$$

Furthermore, the isentropic and exergetic turbine efficiencies are, respectively

$$\eta_t = \frac{h_1 - h_2}{h_1 - h_{2s}} \qquad (3)$$

$$\epsilon = \frac{\dot{W}_{cv}/\dot{m}}{e_{f1} - e_{f2}} = \frac{h_1 - h_2}{(h_1 - h_2) - T_0(s_1 - s_2)} \qquad (4)$$

(a) From Table A-4E, $h_1 = 1306.6$ Btu/lb, $s_1 = 1.5892$ Btu/lb·°R. Further, from Table A-3E at 50 lbf/in.², $h_2 = h_g = 1174.4$ Btu/lb, $s_2 = s_g = 1.6589$ Btu/lb·°R. Accordingly

$$\dot{E}_d/\dot{m} = (520°R)(1.6589 - 1.5892)\frac{\text{Btu}}{\text{lb}\cdot°R} = 36.24 \text{ Btu/lb} \longleftarrow \dot{E}_d/\dot{m}$$

State 2s is fixed by p_2 and $s_{2s} = s_1$: $x_{2s} = (s_{2s} - s_f)/s_{fg} = (1.5892 - 0.4113)/1.2476 = 0.9441$. Therefore $h_{2s} = h_f + x_{2s} h_{fg} = 250.24 + (.9441)(924.2) = 1122.8$ Btu/lb. Thus

$$\eta_t = \frac{1306.6 - 1174.4}{1306.6 - 1122.8} = 0.719 \; (71.9\%) \longleftarrow \eta_t$$

Now, from Eq. (4)

$$\epsilon = \frac{(1306.6 - 1174.4)}{(1306.6 - 1174.4) - (520)(1.5892 - 1.6589)} = 0.785 \; (78.5\%) \longleftarrow \epsilon$$

PROBLEM 7.100 (Cont'd.)

(b) The data for the required plots are obtained using IT, as follows:

IT Code

```
p1 = 400  // lbf/in.²
T1 = 600  // °F
x2 = 1.0
p2 = 50  // lbf/in.²
To = 60 + 460  // °R

h1 = h_PT("Water/Steam", p1, T1)
s1 = s_PT("Water/Steam", p1, T1)
h2 = hsat_Px("Water/Steam", p2, x2)
s2 = ssat_Px("Water/Steam", p2, x2)
s2s = s_Ph("Water/Steam", p2, h2s)
s2s = s1

Ed = To * (s2 - s1)
eta = (h1 - h2) / (h1 - h2s)
eff = (h1 - h2) / ((h1 - h2) - To * (s1 -s2))
```

IT Results for p_2 = 50 lbf/in.²

h_1 = 1306 Btu/lb
h_2 = 1174 Btu/lb
h_{2s} = 1123 Btu/lb
s_1 = 1.589 Btu/lb·°R
s_2 = 1.659 Btu/lb·°R
s_{2s} = 1.589 Btu/lb·°R
$\dot{E}_d / \dot{m}$ = 36.26
η_t = 0.719 (71.9%)
ε = 0.7847 (78.47%)

PLOTS:

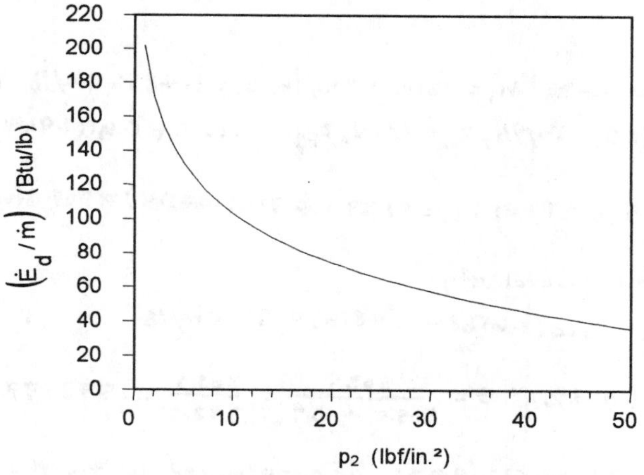

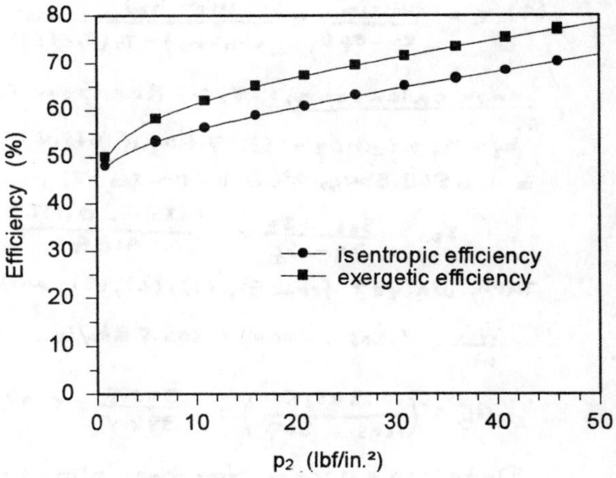

As expected, increased efficiency corresponds with decreased exergy destruction.

PROBLEM 7.101

KNOWN: Steady-state operating data are provided for a steam turbine.

FIND: Plot versus the steam quality at the turbine exit (a) the power developed and rate of exergy destruction, each per unit mass of steam flowing, (b) isentropic turbine efficiency, (c) exergetic turbine efficiency.

SCHEMATIC & GIVEN DATA:

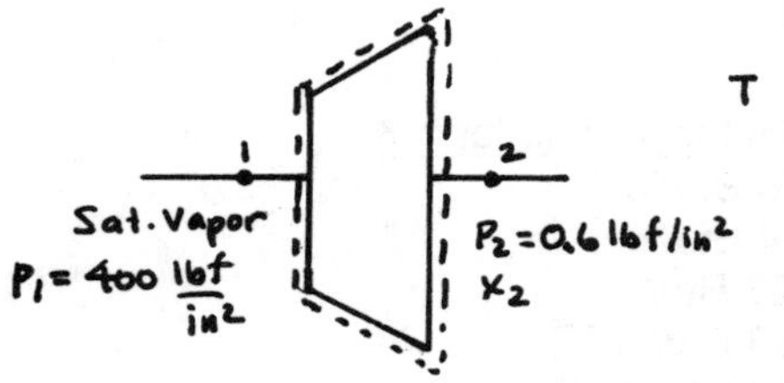

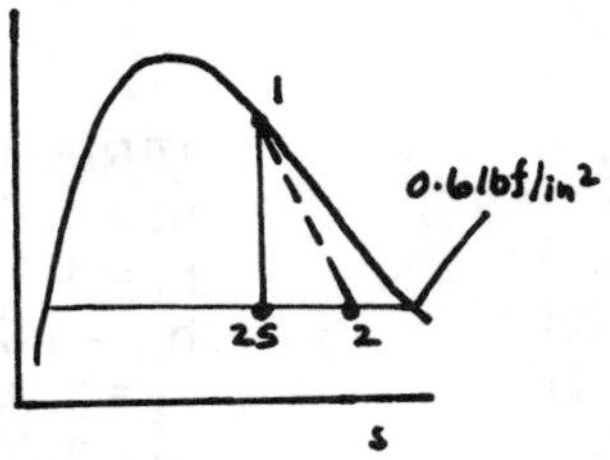

ASSUMPTIONS:
1. The control volume shown in the schematic is at steady state.
2. For the control volume, $\dot{Q}_{cv} = 0$, and kinetic and potential energy effects are negligible.
3. $T_0 = 520°R$, $p_0 = 1$ atm.

ANALYSIS: Reducing mass, energy, and entropy balances gives

$$\frac{\dot{W}_{cv}}{\dot{m}} = h_1 - h_2 \quad (1) \quad \text{and} \quad \frac{\dot{E}_d}{\dot{m}} = T_0 \frac{\dot{\sigma}_{cv}}{\dot{m}} = T_0(s_2 - s_1) \quad (2)$$

h_1 and s_1 are fixed by p_1, saturated vapor. h_2 and s_2 are fixed by p_2, x_2. Using Eq. 6.48,

$$(3) \quad \eta_t = \frac{(h_1 - h_2)}{(h_1 - h_{2s})}$$

where h_{2s} is fixed by p_2, $s_2 = s_1$. Then, with Eq. 7.42

$$(4) \quad \epsilon = \frac{\dot{W}_{cv}/\dot{m}}{e_{f1} - e_{f2}} = \frac{(h_1 - h_2)}{(h_1 - h_2) - T_0(s_1 - s_2)}.$$

Sample calculation: $x_2 = 0.81$. From Table A-3E $h_1 = 1205.5$ Btu/lb, $s_1 = 1.4856$ Btu/lb·°R,

$h_2 = h_f + x_2 h_{fg} = 53.27 + 0.81(1045.4) = 900$ Btu/lb, $s_2 = s_f + x_2 s_{fg} = 0.1029 + 0.81(1.9184)$
$= 1.6568$ Btu/lb·°R. With $s_{2s} = s_1$,

$$x_{2s} = \frac{s_{2s} - s_f}{s_g - s_f} = \frac{1.4856 - 0.1029}{1.9184} = 0.721 \Rightarrow h_{2s} = 53.27 + (0.721)(1045.4) = 807 \text{ Btu/lb}$$

Then, we get from Eq. (1), (2), (3), and (4), respectively

$$\frac{\dot{W}_{cv}}{\dot{m}} = (1205.5 - 900) = 305.5 \text{ Btu/lb}, \quad \frac{\dot{E}_d}{\dot{m}} = 520(1.6568 - 1.4856) = 89 \text{ Btu/lb}$$

$$\eta_t = \left(\frac{305.5}{1205.5 - 807}\right) = \frac{305.5}{398.5} = 0.767\ (76.7\%), \quad \epsilon = \frac{303.5}{303.5 + 89} = \frac{303.5}{392.5} = .773\ (77.3\%)$$

Data required for the plots on the next page are obtained with the following IT code:

IT Code

```
p1 = 400 // lbf/in.²
x1 = 1
p2 = 0.6  // lbf/in.²
x2 = 0.81
To = 60 + 460  // °R

h1 = hsat_Px("Water/Steam", p1, x1)
s1 = ssat_Px("Water/Steam", p1, x1)
h2 = hsat_Px("Water/Steam", p2, x2)
s2 = ssat_Px("Water/Steam", p2, x2)
s2s = s_Ph("Water/Steam", p2, h2s)
s2s = s1

Wcv = h1 - h2
Ed = To * (s2 - s1)
eta = Wcv / (h1 - h2s)
eff = Wcv / ((h1 - h2) - To * (s1 - s2))
```

IT Results for $x_2 = 0.81$

$h_1 = 1205$ Btu/lb
$h_2 = 899.8$ Btu/lb
$h_{2s} = 806.6$ Btu/lb
$s_1 = 1.485$ Btu/lb·°R
$s_2 = 1.656$ Btu/lb·°R
$\dot{W}_{cv}/\dot{m} = 305.5$ Btu/lb
$\dot{E}_d/\dot{m} = 88.97$ Btu/lb
$\varepsilon = 0.7744$ (77.44%)
$\eta_t = 0.7662$ (76.62%)

PROBLEM 7.101 (Cont'd.)

PLOTS:

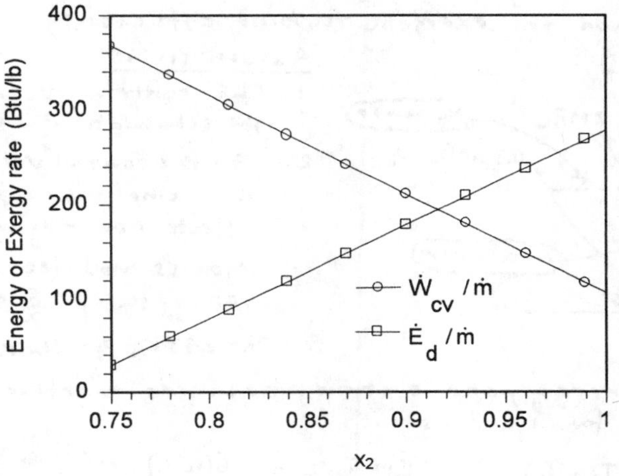

①

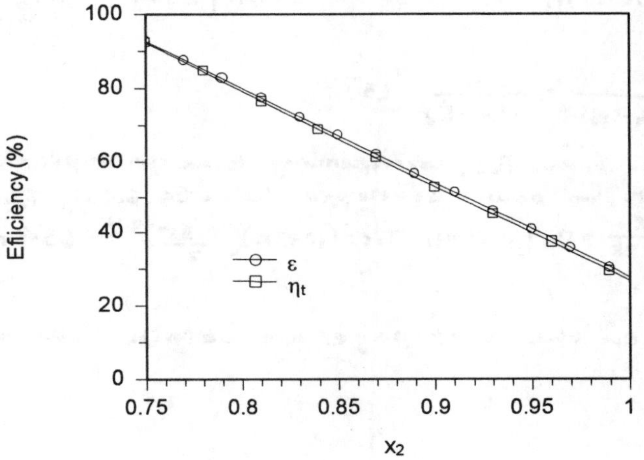

As expected, increased exergy destruction corresponds with decreased power output and efficiency.

1. In this case the numerical values of ϵ and η_t for each x_2 are nearly the same, but this is not observed generally.

PROBLEM 7.102

KNOWN: Steady-state operating data is provided for a turbine through which argon is expanding.

FIND: Plot versus the turbine exit temperature, the power developed, the rate of exergy destruction, the exergetic turbine efficiency.

SCHEMATIC & GIVEN DATA:

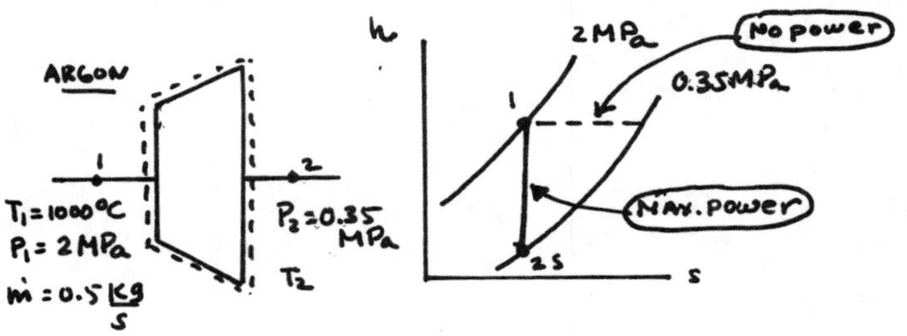

ASSUMPTIONS:

1. The control volume shown in the schematic is at steady state.
2. For the control volume, $\dot{Q}_{cv} = 0$, and kinetic and potential energy effects are negligible.
3. Argon is modeled as an ideal gas with $c_p = \frac{5}{2} R$ (Table A-21).
4. $T_0 = 293$ K, $p_0 = 1$ bar.

ANALYSIS: Reducing mass, energy, and entropy balances together with ideal gas model relations for c_p constant

$$\dot{W}_{cv} = \dot{m}(h_1 - h_2) = \dot{m} c_p (T_2 - T_1) = \dot{m}(2.5R)(T_2 - T_1) \quad (1)$$

$$\dot{E}_d = T_0 \dot{\sigma}_{cv} = \dot{m} T_0 (s_2 - s_1) = \dot{m} T_0 \left[c_p \ln \frac{T_2}{T_1} - R \ln \frac{P_2}{P_1}\right] = \dot{m} T_0 R \left[2.5 \ln \frac{T_2}{T_1} - \ln \frac{P_2}{P_1}\right] \quad (2)$$

$$\epsilon = \frac{\dot{W}_{cv}}{\dot{m}[e_{f1} - e_{f2}]} = \frac{\dot{W}_{cv}}{\dot{m}(h_1 - h_2 - T_0(s_1 - s_2))} = \frac{\dot{W}_{cv}}{\dot{W}_{cv} + \dot{E}_d} \quad (3)$$

Observe that T_2 can range from T_{2s}, corresponding to an isentropic expansion, to $T_2 = T_1$, corresponding to no power developed. With Eq. 6.45, $T_{2s} = T_1 (P_2/P_1)^{(k-1)/k}$. Using Eq. 3.47a, $\frac{k-1}{k} = \frac{R}{c_p} = \frac{R}{2.5R} = 0.4$. Thus, $T_{2s} = (1273 K)\left(\frac{0.35}{2}\right)^{0.4} = 634$ K. So, $634 K \le T_2 \le 1273 K$.

The data for the plots on the next page are obtained using IT as follows:

IT Code

```
p1 = 2 // MPa
T1 = 1000 + 273  // K
p2 = 0.35  // MPa
T2 = 1273  // K
To = 20 + 273  // K
mdot = 0.5  // kg/s
cp = 2.5 * R
R = 8.314 / 39.94  // kJ/kg·K

Wdot = mdot * cp * (T1 - T2)
delta_s = cp * ln(T2 / T1) - R * ln(p2 / p1)
Edot = To * mdot * delta_s
eff = Wdot / (Wdot + Edot)
```

IT Results

for T_2 = 634 K:	For T_2 = 1273 K:
$\dot{W}_{cv}$ = 166.3 kW	$\dot{W}_{cv}$ = 0
$\dot{E}_d \approx 0$	$\dot{E}_d$ = 53.15 kW
ε = 1 (100%)	ε = 0

PROBLEM 7.102 (Cont'd.)

PLOTS:

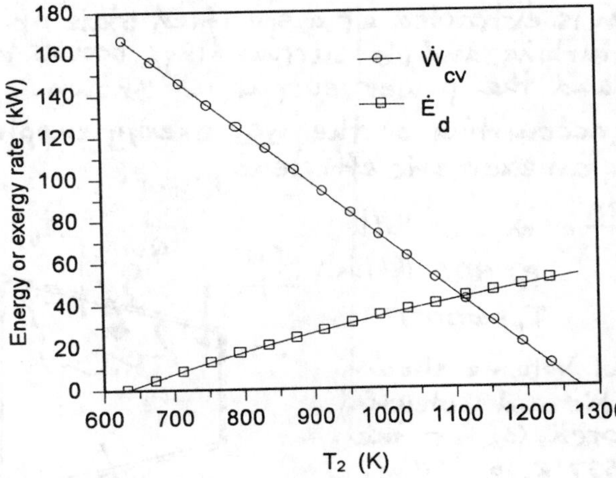

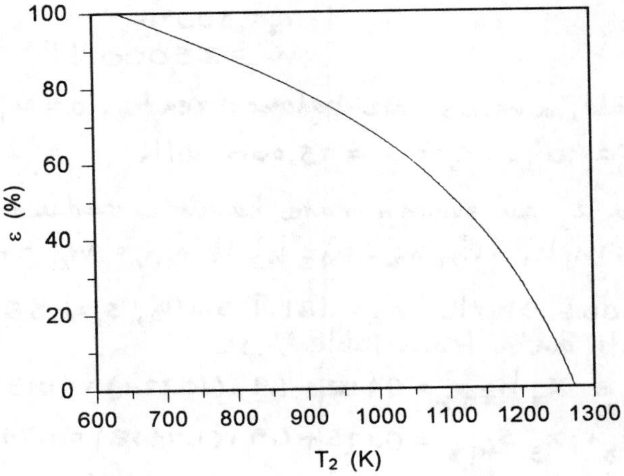

PROBLEM 7.103

KNOWN: Operating data are provided for a steam turbine operating at steady state from which steam is extracted at a specified state and flow rate. Heat transfer between the turbine and its surroundings occurs at a given average surface temperature, and the power output is known.

FIND: (a) Perform a full accounting of the net exergy supplied by the steam. (b) Devise and evaluate an exergetic efficiency.

SCHEMATIC & GIVEN DATA:

$\dot{m}_1 = 10^5$ lb/h
$p_1 = 400$ lbf/in.²
$T_1 = 600°F$

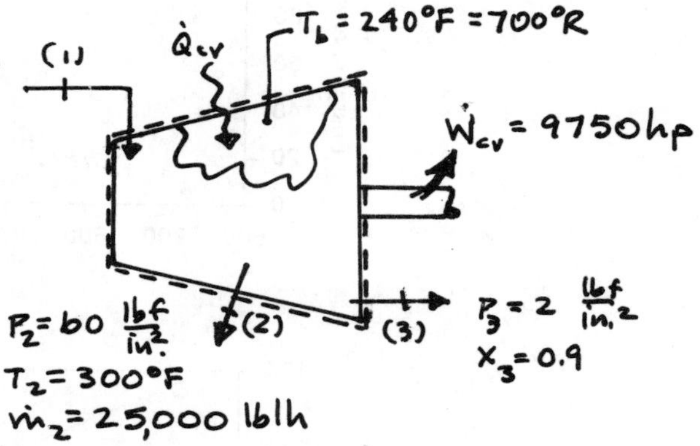

ASSUMPTIONS: (1) The control volume shown is at steady state. (2) Kinetic and potential energy effects can be ignored. (3) For the environment, $T_0 = 77°F = 537°R$, $p_0 = 1$ atm.

ANALYSIS: At steady state, a mass rate balance reads: $0 = \dot{m}_1 - \dot{m}_2 - \dot{m}_3$, or

$$\dot{m}_3 = \dot{m}_1 - \dot{m}_2 = 10^5 - 25,000 = 75,000 \text{ lb/h}$$

With assumptions 1 and 2, an energy rate balance reduced to

$$0 = \dot{Q}_{cv} - \dot{W}_{cv} + \dot{m}_1 h_1 - \dot{m}_2 h_2 - \dot{m}_3 h_3 \Rightarrow \dot{Q}_{cv} = \dot{W}_{cv} - \dot{m}_1 h_1 + \dot{m}_2 h_2 + \dot{m}_3 h_3$$

From Table A-4E, $h_1 = 1306.6$ Btu/lb, $h_2 = 1181.9$ Btu/lb, $s_1 = 1.5892$ Btu/lb·°R, and $s_2 = 1.6496$ Btu/lb·°R. With data from Table A-3E

$$h_3 = h_{f3} + x_3 h_{fg3} = 94.02 + (.9)(1022.1) = 1013.9 \text{ Btu/lb}$$

$$s_3 = s_{f3} + x_3 s_{fg3} = 0.175 + (.9)(1.7448) = 1.7453 \text{ Btu/lb·°R}$$

Inserting values into the expression for $\dot{Q}_{cv}$

$$\dot{Q}_{cv} = (9750 \text{ hp})\left|\frac{2545 \text{ Btu/h}}{1 \text{ hp}}\right| - (10^5 \tfrac{\text{lb}}{\text{h}})(1306.6)\tfrac{\text{Btu}}{\text{lb}} + (25000)(1181.9) + (75000)(1013.9)$$

$$= -2.5625 \times 10^5 \text{ Btu/h}$$

Also, $\dot{W}_{cv} = (9750)(2545) = 2.4814 \times 10^7$ Btu/h

(a) With these preliminary values determined, we turn to the exergy accounting. The net exergy supplied is

$$\begin{pmatrix}\text{net exergy}\\\text{supplied}\end{pmatrix} = \dot{E}_{f1} - \dot{E}_{f2} - \dot{E}_{f3} = \dot{m}_1 e_{f1} - \dot{m}_2 e_{f2} - \dot{m}_3 e_{f3}$$

$$= \dot{m}_1[(h_1 - \cancel{h_0}) - T_0(s_1 - \cancel{s_0})] - \dot{m}_2[(h_2 - \cancel{h_0}) - T_0(s_2 - \cancel{s_0})]$$
$$- \dot{m}_3[(h_3 - \cancel{h_0}) - T_0(s_3 - \cancel{s_0})]$$

$$= \dot{m}_1 h_1 - \dot{m}_2 h_2 - \dot{m}_3 h_3 - T_0(\dot{m}_1 s_1 - \dot{m}_2 s_2 - \dot{m}_3 s_3)$$

where the indicated terms cancel when the mass balance is applied.

$$\begin{pmatrix}\text{net exergy}\\ \text{supplied}\end{pmatrix} = (10^5)\left[(1)(1306.6) - (.25)(1181.9) - (.75)(1013.9) - (537)\{(1)(1.5892) - (.25)(1.6496) - (.75)(1.7453)\}\right]$$

$$= 3.217 \times 10^7 \text{ Btu/h}$$

The rate exergy is carried <u>out</u> with heat transfer is

$$\dot{E}_q = \left[1 - \frac{T_0}{T_b}\right]|\dot{Q}_{cv}| = \left[1 - \frac{537}{700}\right](2.5625\times10^5) = 5.97\times10^4 \text{ Btu/h}$$

The rate of exergy destruction is found from an exergy rate balance

$$0 = \left[1 - \frac{T_0}{T_b}\right]\dot{Q}_{cv} - \left[\dot{W}_{cv} - p_0\frac{dV}{dt}\right] + \dot{m}_1 e_{f_1} - \dot{m}_2 e_{f_2} - \dot{m}_3 e_{f_3} - \dot{E}_d$$

or

$$\dot{E}_d = (\dot{m}_1 e_{f_1} - \dot{m}_2 e_{f_2} - \dot{m}_3 e_{f_3}) - \dot{W}_{cv} - \dot{E}_q$$

$$= 3.217\times10^7 - 2.4814\times10^7 - 5.97\times10^4 = 7.296\times10^6 \text{ Btu/h}$$

<u>Exergy Accounting Summary</u>

net exergy supplied $= 3.217\times10^7 \frac{\text{Btu}}{\text{h}}$ (100%) ← exergy accounting

- power output $= 2.4814\times10^7$ (77.1%)
- transfer out with heat transfer $= 5.97\times10^4$ (0.2%)
- exergy destroyed $= 7.296\times10^6$ (22.7%)

3.217×10^7 (100%)

(b) An exergetic efficiency in the form (desired output / input) takes the form

$$\epsilon = \frac{\dot{W}_{cv}}{\begin{pmatrix}\text{net exergy}\\ \text{supplied}\end{pmatrix}} = \frac{2.4814\times10^7}{3.217\times10^7} = 0.771\ (77.1\%) \longleftarrow \epsilon$$

PROBLEM 7.104

KNOWN: Steady-state operating data are provided for an air compressor.

FIND: (a) Perform a full exergy accounting of the power input, (b) evaluate the exergetic efficiency.

SCHEMATIC & GIVEN DATA:

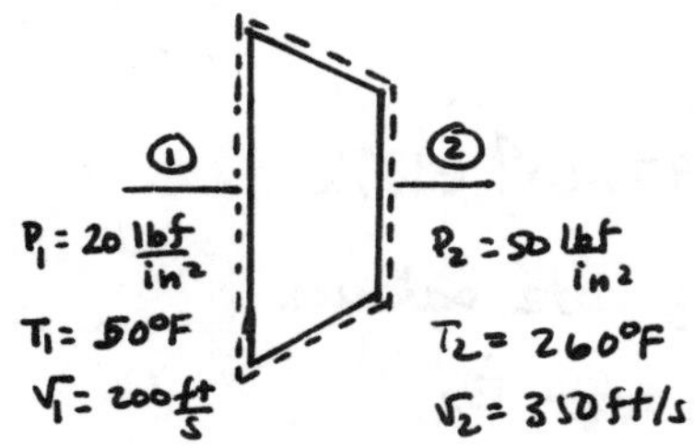

ASSUMPTIONS:

1. The control volume shown in the schematic is at steady state.
2. Neglect the effect of potential energy, $\dot{Q}_{cv} = 0$.
3. Air is modeled as an ideal gas.
4. $T_0 = 500°R$, $P_0 = 14.7\ \text{lbf/in}^2$.

ANALYSIS: (a) At steady state, mass and energy rate balances reduce to give

$$\frac{\dot{W}_{cv}}{\dot{m}} = h_1 - h_2 + \frac{V_1^2 - V_2^2}{2} = (121.875 - 172.39)\frac{\text{Btu}}{\text{lb}} + \frac{(200\,\text{ft/s})^2 - (350\,\text{ft/s})^2}{2}\left|\frac{1\ \text{lbf}}{32.2\,\text{ft/s}^2}\right|\left|\frac{1\ \text{Btu}}{778\ \text{ft·lbf}}\right|$$

$$= -52.16\ \text{Btu/lb}$$

Mass and entropy balances reduce to give $\dot{E}_d = T_0\dot{\sigma}_{cv} \Rightarrow \dot{E}_d/\dot{m} = T_0(s_2 - s_1)$.
or

$$\dot{E}_d/\dot{m} = T_0\left[s^\circ(T_2) - s^\circ(T_1) - R\ln\frac{P_2}{P_1}\right] = 500°R\left[0.67002 - 0.58703 - \frac{1.986}{28.97}\ln\frac{50}{20}\right] = 10.09\ \frac{\text{Btu}}{\text{lb}}$$

All h and s° data are from Table A-22E.

The increase in exergy of the air from 1 to 2 is given by Eq. 7.36

$$\frac{\dot{E}_{f2} - \dot{E}_{f1}}{\dot{m}} = h_2 - h_1 - T_0(s_2 - s_1) + (V_2^2 - V_1^2)/2 = \left(-\frac{\dot{W}_{cv}}{\dot{m}}\right) - \left(\frac{\dot{E}_d}{\dot{m}}\right)$$

$$= 52.16 - 10.09 = 42.07\ \text{Btu/lb}$$

Exergy Balance Sheet (unit mass basis)

- Exergy input — power   52.16 Btu/lb
- Disposition of the exergy
  - Increase in exergy of the air   42.07 Btu/lb (80.7%)
  - Exergy destroyed   10.09 Btu/lb (19.3%)

(b) With Eq. 7.43

$$\epsilon = \frac{e_{f2} - e_{f1}}{(-\dot{W}_{cv}/\dot{m})} = \frac{42.07}{52.16} = 0.807$$

This value can be seen by inspection of the exergy balance sheet of part (a).

PROBLEM 7.105

KNOWN: The states at the inlet and exit of an air compressor operating at steady state are known. The compressor power and the average surface temperature for heat transfer from the compressor to its surrounding are also specified.

FIND: (a) Perform a full exergy accounting of the power input. (2) Devise and evaluate an exergetic efficiency. (3) Determine the hourly costs of each exergy quantity.

SCHEMATIC & GIVEN DATA:

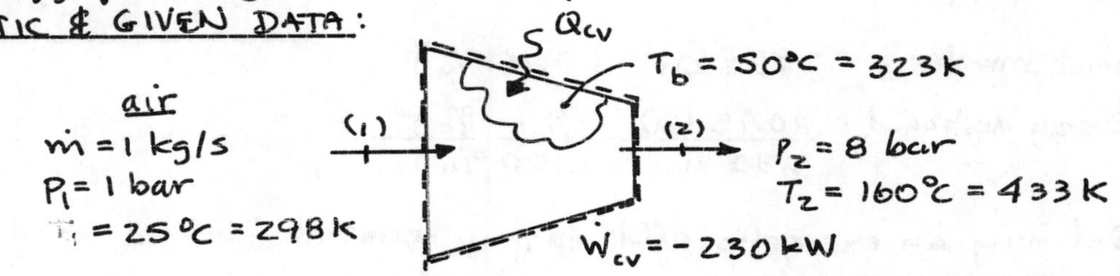

ASSUMPTIONS: (1) The control volume shown is at steady state. (2) For the control volume, kinetic and potential energy effects are negligible. (3) The air is modeled as an ideal gas. (4) For the environment, $T_0 = 25°C = 298\ K$, $p_0 = 1$ bar.

ANALYSIS: (a) First, we determine the heat transfer rate using mass and energy rate balances: $\dot m_1 = \dot m_2 \equiv \dot m$ and $0 = \dot Q_{cv} - \dot W_{cv} + \dot m(h_1 - h_2)$

$$\dot Q_{cv} = \dot W_{cv} + \dot m(h_2 - h_1) = -230\ kW + (1\ \tfrac{kg}{s})(434.48 - 298.18)\tfrac{kJ}{kg}\left|\tfrac{1\ kW}{1\ kJ/s}\right|$$

$$= -93.69\ kW$$

Exergy tranfer out with heat

$$\begin{pmatrix}\text{exergy out}\\ \text{with heat}\end{pmatrix} = \left(1 - \frac{T_0}{T_b}\right)(-\dot Q_{cv}) = \left(1 - \frac{298}{323}\right)(93.69) = \underline{7.252\ kW}$$

Net flow exergy increase of air

$$\dot E_{f2} - \dot E_{f1} = \dot m\left[(h_2 - h_1) - T_0(s_2 - s_1)\right]$$

$$= \dot m\left[(h_2 - h_1) - T_0\left(s^\circ(T_2) - s^\circ(T_1) - R \ln P_2/P_1\right)\right]$$

$$= (1\ \tfrac{kg}{s})\left[(434.48 - 298.18) - (298\ K)\left(2.07234 - 1.69528 - \frac{8.314}{28.97}\frac{kJ}{kg\cdot K}\ln\frac{8}{1}\right)\right] = \underline{201.8\ kW}$$

Exergy destroyed

Using an exergy rate balance

$$0 = \left[1 - \frac{T_0}{T_b}\right]\dot Q_{cv} - \dot W_{cv} + \dot m(e_{f1} - e_{f2}) - \dot E_d$$

$$\dot E_d = \left[1 - \frac{T_0}{T_b}\right]\dot Q_{cv} - \dot W_{cv} + \dot m(e_{f1} - e_{f2})$$

$$= (-7.252) - (-230) + (-201.8) = \underline{20.95\ kW}$$

PROBLEM 7.105 (Cont'd.)

Exergy Accounting Summary

Input

power = 230 kW (100%)

Output and Destruction

flow exergy = 201.8 kW (87.74%)

heat transfer = 7.252 kW (3.15%) ← exergy accounting

Exergy destroyed = 20.95 kW (9.11%)

230 kW (100%)

(b) Defining an exergetic efficiency in terms of (desired output/input)

$$\epsilon = \frac{\text{net flow exergy increase of air}}{\text{power input}} = \frac{201.8}{230} = 0.8774 \; (87.74\%) \leftarrow \epsilon$$

(c) If the exergy cost rate is \$0.08/kW·h

$$\begin{pmatrix}\text{hourly cost of}\\ \text{power input}\end{pmatrix} = (230\,\text{kW})(1\,\text{h})\left(0.08\frac{\$}{\text{kW}\cdot\text{h}}\right) = \$18.40/\text{h}$$

$$\begin{pmatrix}\text{hourly cost of}\\ \text{exergy loss due}\\ \text{to heat transfer}\end{pmatrix} = (7.252\,\text{kW})(1\,\text{h})\left(0.08\frac{\$}{\text{kW}\cdot\text{h}}\right) = \$0.58/\text{h}$$

$$\begin{pmatrix}\text{Cost of exergy}\\ \text{destruction}\end{pmatrix} = (20.95\,\text{kW})(1\,\text{h})\left(0.08\frac{\$}{\text{kW}\cdot\text{h}}\right) = \$1.66/\text{h}$$

← $\dot{\$}$

PROBLEM 7.106

KNOWN: Steady-state operating data is provided for a counterflow heat exchanger having air flowing on both sides.

FIND: Determine (a) the rate of exergy destruction, (b) the exergetic efficiency given by Eq. 7.45.

SCHEMATIC & GIVEN DATA

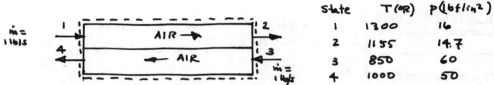

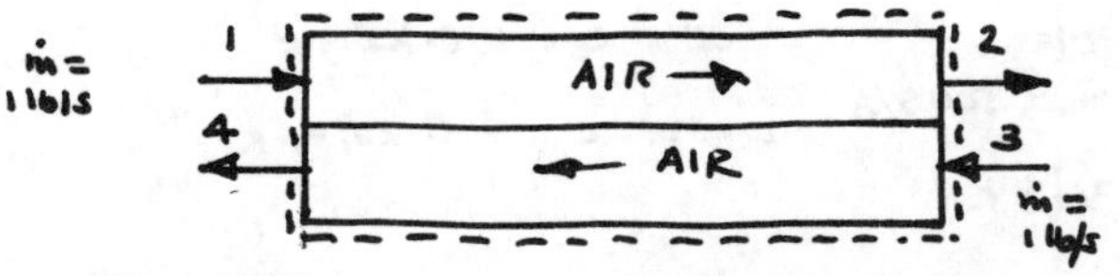

State	T (°R)	p (lbf/in²)
1	1300	16
2	1155	14.7
3	850	60
4	1000	50

ASSUMPTIONS: 1. The control volume shown in the schematic is at steady state. 2. For the control volume, $\dot{Q}_{cv} = \dot{W}_{cv} = 0$, and kinetic and potential energy effects are negligible. 3. Air is modeled as an ideal gas. 4. $T_0 = 520°R$, $p_0 = 14.7$ lbf/in².

ANALYSIS: Mass and energy rate balances reduce to give $0 = \cancel{\dot{Q}_{cv}} - \cancel{\dot{W}_{cv}} + \dot{m}(h_1 - h_2) + \dot{m}(h_3 - h_4)$. $\Rightarrow 0 = (h_1 - h_2) + (h_3 - h_4)$. With data from Table A-22E, $h_1 = 316.95$ Btu/lb, $h_2 = 279.88$ Btu/lb, $h_3 = 204.01$ Btu/lb, $h_4 = 240.98$ Btu/lb. These values check the energy balance, indicating that the given state data are consistent with the conservation of mass and energy principles.

Reducing an exergy rate balance at steady state

$$0 = \sum_j \left[1 - \frac{T_0}{T_j}\right]\dot{Q}_j^{\,0} - \dot{W}_{cv}^{\,0} + \dot{m}[e_{f1} - e_{f2}] + \dot{m}[e_{f3} - e_{f4}] - \dot{E}_d$$

$$\Rightarrow \frac{\dot{E}_d}{\dot{m}} = [h_1 - h_2 - T_0(s_1 - s_2)] + [h_3 - h_4 - T_0(s_3 - s_4)]$$

$$= \underbrace{[(h_1 - h_2) + (h_3 - h_4)]}_{=0} + T_0[(s_2 - s_1) + (s_4 - s_3)]$$

$$= T_0\left[\left(s°(T_2) - s°(T_1) - R \ln\frac{p_2}{p_1}\right) + \left(s°(T_4) - s°(T_3) - R \ln\frac{p_4}{p_3}\right)\right]$$

$$= 520°R\left[\left[0.78665 - 0.81678 - \frac{1.986}{28.97}\ln\frac{14.7}{16}\right]\frac{Btu}{lb \cdot °R} + \left[0.75042 - 0.71035 - \frac{1.986}{28.97}\ln\frac{50}{60}\right]\frac{Btu}{lb \cdot °R}\right]$$

$$= 520[-0.02432 + 0.05257] = 14.69 \text{ Btu/lb} \quad \Rightarrow \dot{E}_d = 14.69 \frac{Btu}{s}$$ ←

(b) The exergetic efficiency given by Eq. 7.45 is

① $$\epsilon = \frac{\dot{m}[e_{f4} - e_{f3}]}{\dot{m}[e_{f1} - e_{f2}]} = \frac{(h_4 - h_3) - T_0(s_4 - s_3)}{(h_1 - h_2) - T_0(s_1 - s_2)} = \frac{(240.98 - 204.01) - 520[0.05257]}{(316.95 - 279.88) - 520[0.02432]}$$

$$= \frac{9.634}{24.424} = 0.394 \quad (39.4\%)$$ ←

---

1. In accord with the discussion of Eq. 7.45, both streams are at temperatures above T_0.

PROBLEM 7.107

KNOWN: Steady state operating data are provided for a counterflow heat exchanger through which oil and water flow in separate streams.

FIND: Determine (a) the mass flow rate of the water, (b) the exergetic efficiency given by Eq. 7.45, (c) the hourly cost of exergy destruction.

SCHEMATIC & GIVEN DATA:

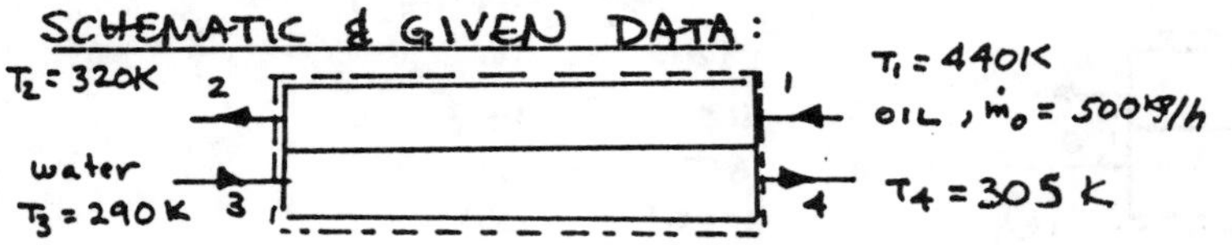

oil: $c_o = 2.0$ kJ/kg·K

water: $c_w = 4.0$ kJ/kg·K

ASSUMPTIONS: (1) The control volume shown in the accompanying figure is at steady state. (2) $\dot{Q}_{cv} = 0$ and the effects of kinetic and potential energy can be ignored. (3) The oil and water can each be regarded as incompressible with constant specific heats c_o and c_w, respectively, and negligible change in pressure in flowing through the heat exchanger. (4) For the environment $T_0 = 17°C$, $p_0 = 1$ atm. (5) exergy is valued at 8¢ per kW·h.

ANALYSIS: (a) At steady state, mass rate balances reduce to give $\dot{m}_1 = \dot{m}_2 \equiv \dot{m}_o$ and $\dot{m}_3 = \dot{m}_4 \equiv \dot{m}_w$. With assumptions 1 and 2, an energy balance gives

$$0 = \cancel{\dot{Q}_{cv}} - \cancel{\dot{W}_{cv}} + \dot{m}_o(h_1 - h_2) + \dot{m}_w(h_3 - h_4) \Rightarrow 0 = \dot{m}_o c_o (T_1 - T_2) + \dot{m}_w c_w (T_3 - T_4)$$

In writing the last expression, Eq. 3.20b was used together with assumption 3 to evaluate the enthalpy changes. Solving for $\dot{m}_w$

$$\dot{m}_w = \dot{m}_o \frac{c_o(T_1 - T_2)}{c_w(T_4 - T_3)} = 500 \frac{kg}{h} \frac{(2)(440-320)}{(4)(305-290)} = 2000 \text{ kg/h} \longleftarrow \quad \dot{m}_w$$

(b) The exergetic efficiency is, with Eq. 6.24

① $$\epsilon = \frac{\dot{m}_w(e_{f4} - e_{f3})}{\dot{m}_o(e_{f1} - e_{f2})} = \frac{\dot{m}_w c_w [(T_4 - T_3) - T_0 \ln(T_4/T_3)]}{\dot{m}_o c_o [(T_1 - T_2) - T_0 \ln(T_1/T_2)]}$$

$$= \frac{(2000)(4)[(305-290) - (290)\ln(305/290)]}{(500)(2)[(440-320) - (290)\ln(440/320)]} =$$

$$= \frac{3000.4 \text{ kJ/h}}{27650 \text{ kJ/h}} = 0.1085 \; (10.85\%) \longleftarrow \quad \epsilon$$

(c) The rate of exergy destruction is obtain using an exergy rate balance, as follows:

$$0 = \sum_j \cancel{\left[1 - \frac{T_0}{T_j}\right]\dot{Q}_j}^{\,0} - \cancel{\dot{W}_{cv}}^{\,0} + \dot{m}_o(e_{f1} - e_{f2}) + \dot{m}_w(e_{f3} - e_{f4}) - \dot{E}_d$$

or

$$\dot{E}_d = \dot{m}_o(e_{f1} - e_{f2}) - \dot{m}_w(e_{f4} - e_{f3})$$

With values from above, $\dot{E}_d = 27650 - 3000.4 = 24650$ kJ/h. Thus

$$\begin{pmatrix}\text{hourly}\\ \text{cost}\end{pmatrix} = \left(24650 \frac{kJ}{h}\right)\left|\frac{1 \text{ h}}{3600 \text{ s}}\right| \left|\frac{1 \text{ kW}}{1 \text{ kJ/s}}\right| \left(0.08 \frac{\$}{\text{kW·h}}\right) = \$0.55/h \longleftarrow \quad \$$$

---

1. In accord with the discussion of Eq. 7.45, both streams are at temperatures ≥ T_0.

PROBLEM 7.108

KNOWN: Steady-state operating data are provided for boiler tubes carrying water over which combustion gases flow.

FIND: Determine (a) the mass flow rate of the combustion gases, (b) the rate of exergy destruction, (c) the exergetic efficiency Eq. 7.45.

SCHEMATIC & GIVEN DATA:

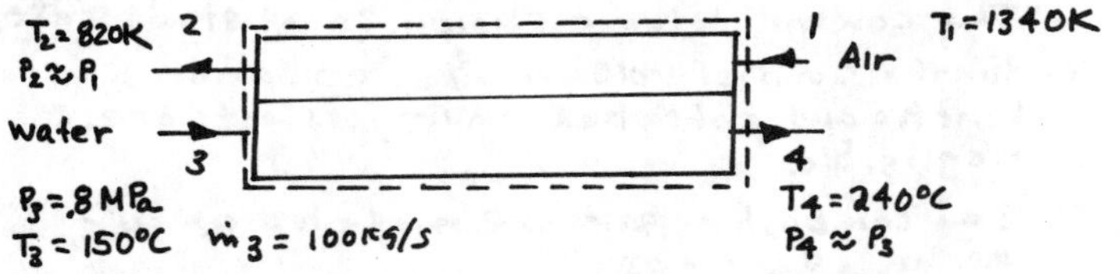

ASSUMPTIONS: (1) The control volume shown in the accompanying figure is at steady state with $\dot{Q}_{cv} = 0$ and negligible kinetic and potential energy effects. (2) The combustion gases are modeled as air as an ideal gas. (3) There is no significant change in pressure for either stream as it passes from inlet to exit. (4) For the environment, $T_0 = 298K$, $P_0 = 1 atm$.

ANALYSIS: At steady state mass rate balances give $\dot{m}_1 = \dot{m}_2$ and $\dot{m}_3 = \dot{m}_4$. An energy rate balance reduces with assumption 1 to give

$$0 = \cancel{\dot{Q}_{cv}}^{0} - \cancel{\dot{W}_{cv}}^{0} + \dot{m}_1(h_1 - h_2) + \dot{m}_3(h_3 - h_4) \Rightarrow \frac{\dot{m}_1}{\dot{m}_3} = \frac{h_4 - h_3}{h_1 - h_2}$$

With data from Tables A-2 and A-4, $h_4 = 2928.3 \, kJ/kg$ and $h_3 \approx h_f(T_3) = 632.2 \, kJ/kg$. Table A-22 gives $h_1 = 1443.6 \, kJ/kg$, $h_2 = 843.98 \, kJ/kg$.

$$\Rightarrow \dot{m}_1 = \left(100 \frac{kg}{s}\right)\left[\frac{2928.3 - 632.2}{1443.6 - 843.98}\right] = (100)\left[\frac{2296.1}{599.62}\right] = 382.93 \, kg/s \longleftarrow$$

The exergy rate balance reduces at steady state to give

$$\dot{E}_d = \dot{m}_1[e_{f1} - e_{f2}] + \dot{m}_3[e_{f3} - e_{f4}] = \dot{m}_1[(h_1 - h_2) - T_0(s_1 - s_2)] + \dot{m}_3[(h_3 - h_4) - T_0(s_3 - s_4)]$$

$$\dot{E}_d/\dot{m}_3 = \frac{\dot{m}_1}{\dot{m}_3}\left[h_1 - h_2 - T_0(s^\circ(T_1) - s^\circ(T_2) - R \cancel{\ln P_1/P_2}^{0}\right] + [(h_3 - h_4) - T_0(s_3 - s_4)]$$

With s° data from Table A-22, $s_3 \approx s_f(T_3)$ from Table A-2 and s_4 from Table A-4

$$\frac{\dot{E}_d}{\dot{m}_3} \quad 3.8293[599.62 - 298(3.30959 - 2.74504)] + [-2296.1 - 298(1.8418 - 7.0033)]$$

$$= 3.8293[431.38] + [-757.97] = 893.91 \frac{kJ}{kg} \Rightarrow \dot{E}_d = 8.939 \times 10^4 \frac{kJ}{s} \longleftarrow$$

(1) With Eq. 7.45 and previously calculated values

$$\epsilon = \frac{\dot{m}_3(e_{f4} - e_{f3})}{\dot{m}_1(e_{f1} - e_{f2})} = \frac{757.97}{3.8293(431.38)} = 0.459 \; (45.9\%) \longleftarrow$$

---

1. In accord with the discussion of Eq. 7.45, both streams are at temperatures above T_0.

PROBLEM 7.109

KNOWN: Steady-state operating data are provided for boiler tubes carrying water over which combustion gases flow.

FIND: Determine (a) the exit temperature of the combustion gases, (b) the rate of exergy destruction, (c) the exergetic efficiency given by Eq. 7.45.

SCHEMATIC & GIVEN DATA:

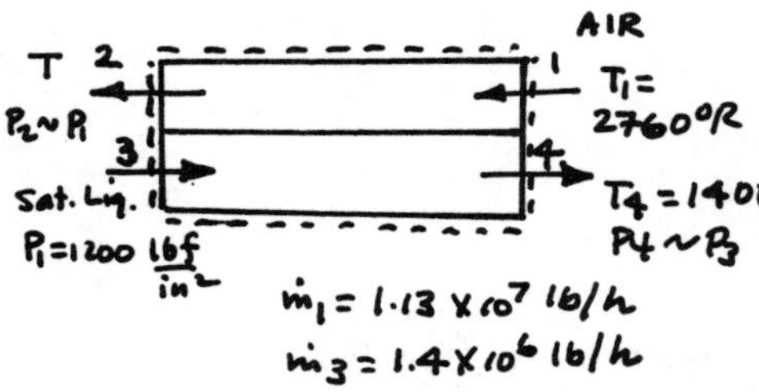

ASSUMPTIONS:
1. The control volume shown is at steady state.
2. For the control volume, $\dot{Q}_{cv} = 0$, and kinetic and potential energy effects are negligible.
3. The combustion gases are modeled as air as an ideal gas.
4. There is no significant change in pressure for either stream.
5. $T_0 = 530°R$, $P_0 = 1$ atm.

ANALYSIS: (a) Reducing mass and energy rate balances

$$0 = \dot{Q}_{cv}^{\,0} - \dot{W}_{cv}^{\,0} + \dot{m}_1(h_1 - h_2) + \dot{m}_3(h_3 - h_4) \Rightarrow h_2 = h_1 + \frac{\dot{m}_3}{\dot{m}_1}(h_3 - h_4)$$

with data from Table A-22E and Tables A-3 & 4E

$$h_2 = 720.73 + \left(\frac{1.4 \times 10^6}{1.13 \times 10^7}\right)(571.7 - 1730.7) = 577.14 \text{ Btu/lb}$$. Interpolating in Table A-22E,

$T_2 = 2259°R$. ←

(b) The rate of exergy destruction is obtained using an exergy rate balance, which reduces to give

$$\dot{E}_d = \dot{m}_1[e_{f1} - e_{f2}] + \dot{m}_3[e_{f3} - e_{f4}]$$
$$= \dot{m}_1[(h_1 - h_2) - T_0(s_1 - s_2)] + \dot{m}_3[(h_3 - h_4) - T_0(s_3 - s_4)]$$
$$= \dot{m}_1[(h_1 - h_2) - T_0(s_1^\circ - s_2^\circ - R \ln\frac{p_1}{p_2}^{\,0})] + \dot{m}_3[(h_3 - h_4) - T_0(s_3 - s_4)]$$

Inserting table data

$$\dot{E}_d = \left(1.13 \times 10^7 \frac{lb}{h}\right)\left[(720.73 - 577.14) - 530(1.02349 - 0.96609)\right]\frac{Btu}{lb} +$$
$$\left(1.4 \times 10^6 \frac{lb}{h}\right)\left[(571.7 - 1730.7) - 530(0.7712 - 1.7696)\right]\frac{Btu}{lb}$$
$$= (1.13 \times 10^7)[\underbrace{143.59 - 30.42}_{113.17}] + (1.4 \times 10^6)[\underbrace{-1159 - (-529.15)}_{-629.85}]$$
$$= 1.28 \times 10^9 + (-0.88 \times 10^9) = 4 \times 10^8 \frac{Btu}{h}$$

① (c) With Eq. 7.45 and previously calculated values

$$\epsilon = \frac{\dot{m}_3(e_{f4} - e_{f3})}{\dot{m}_1(e_{f1} - e_{f2})} = \frac{0.88 \times 10^9}{1.28 \times 10^9} = 0.688 \ (68.8\%).$$ ←

1. In accord with the discussion of Eq. 7.45, both streams are at temperatures above T_0.

PROBLEM 7.110

KNOWN: Steady state operating data are provided for a direct-contact heat exchanger.

FIND: Determine (a) the rate of destruction per unit mass exiting, (b) the exergetic efficiency given by Eq. 7.47.

SCHEMATIC & GIVEN DATA:

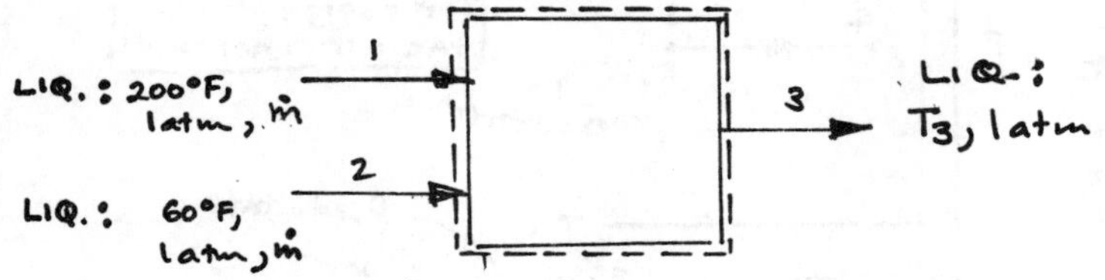

ASSUMPTIONS: (1) The control volume shown in the figure is at steady state. (2) For the control volume, $\dot{Q}_{cv}=0$, $\dot{W}_{cv}=0$, and kinetic/potential energy effects can be neglected. (3) For the exergy reference environment, $T_0 = 50°F$, $p_0 = 1atm$.

ANALYSIS: Mass and energy rate balances reduce to give

$$0 = \dot{Q}_{cv}^{\,0} - \dot{W}_{cv}^{\,0} + \dot{m}h_1 + \dot{m}h_2 - 2\dot{m}h_3 \Rightarrow h_3 = \frac{h_1+h_2}{2}.$$

Then, with saturated liquid data from Table A-2E, $h \approx h_f(T)$

$$h_3 = \frac{168.1 + 28.1}{2} = 98.1\ \frac{Btu}{lb.}$$

Interpolation in Table A-2E gives $T_3 \approx 130°F$

$s_3 \approx 0.1819$ Btu/lb·°R

The rate of exergy destruction is $\dot{E}_d = T_0\dot{\sigma}_{cv}$, where $\dot{\sigma}_{cv} = \dot{m}_3 s_3 - (\frac{\dot{m}_3}{2})s_1 - (\frac{\dot{m}_3}{2})s_2$.
giving with $s \approx s_f(T)$

$$\frac{\dot{E}_d}{\dot{m}_3} = T_0\left[s_3 - \left(\frac{s_1+s_2}{2}\right)\right] = 510°R\left[0.1819 - \left(\frac{0.05555+0.2940}{2}\right)\right]$$

$$= 3.63\ Btu/lb \longleftarrow$$

Applying Eq 7.47

$$\epsilon = \frac{\dot{m}\,[e_{f3}-e_{f2}]}{\dot{m}\,[e_{f1}-e_{f3}]} = \frac{(h_3-h_2)-T_0(s_3-s_2)}{(h_1-h_3)-T_0(s_1-s_3)} = \frac{(98.1-28.1)-510(0.1819-.05555)}{(168.1-98.1)-510(0.2940-0.1819)}$$

$$= \frac{5.56\ Btu/lb}{12.83\ Btu/lb} = 0.433\ (43.3\%) \longleftarrow \epsilon$$

PROBLEM 7.111

KNOWN: Steady-state operating data are provided for a counterflow heat exchanger.

FIND: (a) Sketch the variation of the temperature of each stream with with position. Locate T_0. (b) Determine the rate of exergy destruction. (c) Devise and evaluate an exergetic efficiency.

SCHEMATIC & GIVEN DATA:

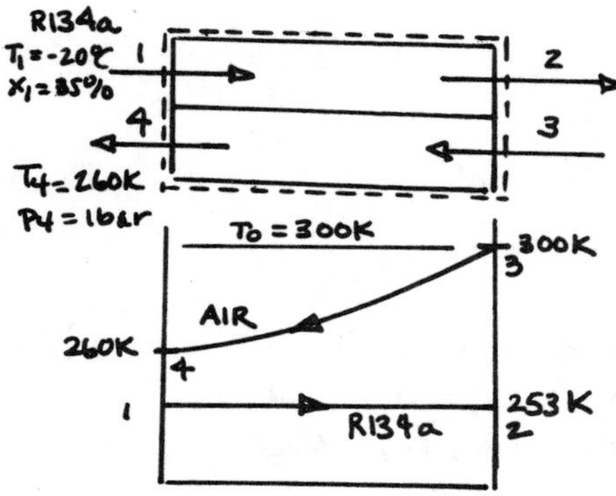

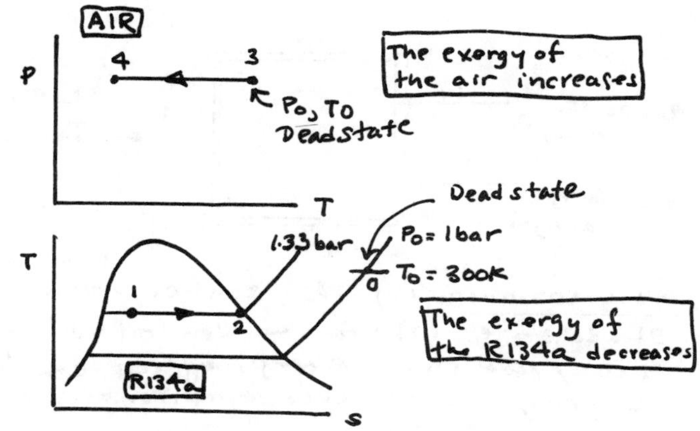

ASSUMPTIONS: 1. The control volume shown in the accompanying sketch is at steady state. 2. For the control volume, $\dot{Q}_{cv} = 0$, and kinetic and potential energy effects can be ignored. 3. Air is modeled as an ideal gas. 4. $T_0 = 300\,K$, $P_0 = 1\,bar$.

ANALYSIS: The refrigerant mass flow rate is required for parts (b), (c). Reducing mass and energy rate balances,

$$0 = \dot{Q}_{cv}^{\,0} - \dot{W}_{cv}^{\,0} + \dot{m}_1[h_1 - h_2] + \dot{m}_3[h_3 - h_4] \Rightarrow \text{with data from Tables A-10, A-22}$$

$$\dot{m}_1 = \frac{\dot{m}_3[h_3 - h_4]}{[h_2 - h_1]} = \frac{(4\,kg/s)[300.19 - 260.09]\,kJ/kg}{[235.31 - 98.13]\,kJ/kg} = 1.17\,kg/s$$

$$98.13 = [24.26 + 0.35(211.05)]$$

(b) The rate of exergy destruction can be determined by reducing an exergy rate balance:

$$0 = \sum\left[1 - \frac{T_0}{T_j}\right]\dot{Q}_j^{\,0} - \dot{W}_{cv}^{\,0} + \dot{m}_1[e_{f1} - e_{f2}] + \dot{m}_3[e_{f3} - e_{f4}] - \dot{E}_d$$

$$\Rightarrow \dot{E}_d = \dot{m}_1[(h_1 - h_2) - T_0(s_1 - s_2)] + \dot{m}_3[h_3 - h_4 - T_0(s_3 - s_4)]$$

$$= \dot{m}_1[(h_1 - h_2) - T_0(s_1 - s_2)] + \dot{m}_3[h_3 - h_4 - T_0(s_3^\circ - s_4^\circ - R\ln P_3/P_4^{\,0})]$$

$$0.3914 = 0.0996 + .35(0.9332 - 0.0996)$$

$$= 1.17[(98.13 - 235.31) - 300(0.3914 - 0.9332)] + 4[300.19 - 260.09 - 300(1.70203 - 1.55848)]$$

$$= [(+29.67) + (-11.86)]\left(\frac{kJ}{s}\right)\left|\frac{1\,kW}{1\,kJ/s}\right| = 17.8\,kW \leftarrow$$

(c) Since the air is brought from the dead state to a lower temperature, its exergy increases in the process. The state of the R134a is brought closer to the dead state, so its exergy decreases. This is confirmed by the calculations of part (b) that show

$$\dot{m}_3(e_{f4} - e_{f3}) = +11.86\,kW, \quad \dot{m}_1(e_{f2} - e_{f1}) = -29.67\,kW$$

Thus, the colder refrigerant provides the exergy that is either transferred to the air or destroyed by irreversibilities within the control volume.

(1) $$\Rightarrow \epsilon = \frac{\dot{m}_3(e_{f4} - e_{f3})}{\dot{m}_1(e_{f1} - e_{f2})} = \frac{11.86}{29.67} = 0.403 \quad (40.3\%)$$

---

1. Since the streams are each below T_0, this expression is formulated differently than Eq. 7.45, which regards the cold stream as <u>receiving</u> exergy <u>from</u> the hot stream. Here, the positions are reversed.

PROBLEM 7.112

KNOWN: Steady-state operating data are provided for two turbine stages and an interconnecting heat exchanger.

FIND: Determine the exergetic efficiencies of the two turbines and the heat exchanger.

SCHEMATIC & GIVEN DATA: See Fig. P4.82. Also, see Problem 7.84 solution for data.

ASSUMPTIONS: 1. See solutions to Probs. 4.82, 6.125, 7.84. 2 $T_0 = 300$ K, $p_0 = 1$ bar.

ANALYSIS: From the solution to Problem 7.84

$$(\dot{E}_d)_{t1} = 958.1\text{ kW},\quad (\dot{E}_d)_{t2} = 859.5\text{ kW},\quad (\dot{E}_d)_{HX} = 944.5\text{ kW}$$

with Eq 7.42, and invoking the assumptions for the turbines in this case

$$\epsilon_t = \frac{\dot{W}_t}{\dot{m}(e_{f1}-e_{f2})} = \frac{\dot{W}_t}{\dot{m}[h_1-h_2-T_0(s_1-s_2)]} = \frac{\dot{W}_t}{\dot{W}_t+\dot{E}_d}$$

Then, with $\dot{W}_{t1}$ and $\dot{W}_{t2}$ from the solution to Problem 4.82

$$\epsilon_{t1} = \frac{10{,}000\text{ kW}}{10{,}000\text{ kW}+958.1\text{ kW}}\ 0.913\ (91.3\%)\quad,\quad \epsilon_{t2} = \frac{10{,}570}{10{,}570+859.5} = .925\ (92.5\%)$$

With Eqs 7.44, 7.45

$$\epsilon_{HX} = \frac{\dot{m}_2(e_{f3}-e_{f2})}{\dot{m}_5(e_{f5}-e_{f6})} = \frac{\dot{m}_2(e_{f3}-e_{f2})}{\dot{m}_2(e_{f3}-e_{f2})+\dot{E}_d}$$

$$= \frac{\dot{m}_2[h_3-h_2-T_0(s_3-s_2)]}{\dot{m}_2[h_3-h_2-T_0(s_3-s_2)]+\dot{E}_d} \quad (1)$$

with data from the solutions to Problems 4.82 and 6.125

$$\dot{m}_2[h_3-h_2-T_0(s_3-s_2)] = (28.22)[1397.8-1161.07]-300(6.4265)$$
$$= 4752.6\text{ kW}$$

Then, Eq. (1) gives

$$\epsilon_{HX} = \frac{4752.6}{4752.6+944.5} = 0.834\quad (83.4\%)$$

PROBLEM 7.113

KNOWN: Steady-state operating data are provided for a steam generator and turbine.

FIND: Determine the exergetic efficiencies of the steam generator and turbine.

SCHEMATIC & GIVEN DATA: See solutions to Problems 4.83, 7.87.

ASSUMPTIONS: See assumptions listed for the solutions to Problems 4.83, 7.87.

ANALYSIS: With Eq. 7.42, and invoking the assumptions for the turbine in this case

$$\epsilon_t = \frac{\dot{W}_t}{\dot{m}(e_{f2}-e_{f3})} = \frac{\dot{W}_t}{\dot{m}[(h_2-h_3)-T_0(s_2-s_3)]} = \frac{\dot{W}_t}{\dot{W}_t + \dot{E}_d}$$

Inserting values from the solutions to Problems 4.83, 7.87

$$\epsilon_t = \frac{9.09 \text{ Btu/s}}{(9.09 + 2.05)\text{ Btu/s}} = 0.816 \quad (81.6\%)$$ ←

With Eqs. 7.44, 7.45, and data from the solution to Problem 7.87

$$\epsilon_{SG} = \frac{\dot{E}_{f2}-\dot{E}_{f1}}{\dot{E}_{fB}-\dot{E}_{fA}} = \frac{(\dot{E}_{fB}-\dot{E}_{fA})-\dot{E}_d}{(\dot{E}_{fB}-\dot{E}_{fA})} = \frac{14.49-2.37}{14.49} = 0.836 \quad (83.6\%)$$

PROBLEM 7.114

KNOWN: Steady-state operating data are provided for a heat pump system.

FIND: Determine the exergetic efficiencies of the compressor and condenser.

SCHEMATIC & GIVEN DATA: See Figs. E6.8, 14

ASSUMPTIONS: 1. See assumptions listed for Examples 6.8 and 6.14. 2. $T_0 = 273\text{ K}$, $p_0 = 1\text{ bar}$.

ANALYSIS: The exergetic efficiency of the compressor is obtained using Eq. 7.43 and data from the examples noted

① $$\epsilon = \frac{e_{f2}-e_{f1}}{(-\dot{W}_{cv}/\dot{m})} = \frac{h_2-h_1 - T_0(s_2-s_1)}{h_2-h_1} = 1 - \frac{T_0(s_2-s_1)}{(h_2-h_1)} = 1 - \frac{(273)(0.98225-0.9572)}{(294.17 - 249.75)}$$

$$= 0.846 \quad (84.6\%)$$ ←

For condenser, use Eq. 7.45 and data from the examples noted

$$\epsilon = \frac{\dot{m}_{air}[h_6-h_5 - T_0(s_6-s_5)]}{\dot{m}_{ref}[h_2-h_3-T_0(s_2-s_3)]} = \frac{0.5[30.15 - 273(0.098)]}{0.07[(294.17-79.05) - 273(0.98225-0.2936)]}$$

② $$= 0.894 \quad (89.4\%)$$ ←

---

1. Alternatively, with data from Example 7.9

$$\epsilon = \frac{(-\dot{W}_{cv}) - \dot{E}_d}{(-\dot{W}_{cv})} = \frac{(3.11\text{ kW}) - (0.478\text{ kW})}{3.11\text{ kW}} = 0.846$$

2. This calculation is sensitive to roundoff.

PROBLEM 7.115

KNOWN: Steady state operating data are provided for a compressor and heat exchanger in series.

FIND: Determine the exergetic efficiencies of the compressor and heat exchanger.

SCHEMATIC & GIVEN DATA: See the schematic from the solution to Problem 6.124. Also see Problem 7.89 solution for data.

ASSUMPTIONS: 1. See the assumptions listed for Problem 6.124.
2. $T_0 = 300K$, $P_0 = 96 kPa$.

ANALYSIS: With Eq. 7.43 and invoking the assumptions listed

$$\epsilon_{comp} = \frac{\dot{m}(e_{f2} - e_{f1})}{(-\dot{W}_{cv})} = \frac{\dot{m}[h_2 - h_1 - T_0(s_2 - s_1)]}{(-\dot{W}_{cv})} = \frac{(-\dot{W}_{cv}) - \dot{E}_d}{(-\dot{W}_{cv})}$$

With values from the solution to Problem 7.89

$$\epsilon_{comp} = \frac{50.4 - 5.88}{50.4} = 0.883 \ (88.3\%)$$ ←

With Eqs. 7.44, 7.45, and data from the solution to Problem 7.89

$$\epsilon_{HX} = \frac{\dot{E}_{fB} - \dot{E}_{fA}}{(\dot{E}_{fB} - \dot{E}_{fA}) + \dot{E}_d} = \frac{(0.46)}{(0.46) + 4.56} = 0.092 \ (9.2\%)$$ ←

PROBLEM 7.116

KNOWN: Steady-state operating data are provided for a steam generator and turbine.

FIND: Determine the exergetic efficiencies of the steam generator and turbine.

Schematic & Given Data: See figs. E4.10, E7.8.

① ASSUMPTIONS: 1. See assumptions for Examples 4.10, 7.8 2. $T_0 = 537°R$, $P_0 = 1atm$.

ANALYSIS: With Eqs. 7.41, 42, and data from Example 7.8

$$\epsilon_t = \frac{\dot{W}_{cv}}{\dot{E}_{f4} - \dot{E}_{f5}} = \frac{\dot{W}_{cv}}{\dot{W}_{cv} + \dot{E}_d} = \frac{49{,}610 \text{ Btu/min}}{(49{,}610 + 17{,}070) \text{ Btu/min}} = 0.744 \ (74.4\%)$$ ←

With Eqs. 7.44, 7.45 and data from Example 7.8

$$\epsilon_{SG} = \frac{\dot{E}_{f4} - \dot{E}_{f3}}{\dot{E}_{f1} - \dot{E}_{f2}} = \frac{(\dot{E}_{f1} - \dot{E}_{f2}) - \dot{E}_d}{(\dot{E}_{f1} - \dot{E}_{f2})} = \frac{100{,}300 - 22{,}110}{100{,}300} = 0.78 \ (78\%)$$ ←

---

1. In the first printing T_0 was given incorrectly as 540 °R.

PROBLEM 7.117

KNOWN: An expression is provided for the total cost rate $\dot{C}$ of a device as a function of pressure drop, $P_1 - P_2$.

FIND: (a) Sketch $\dot{C}$ versus $(P_1 - P_2)$. (b) At the point of minimum total cost, determine the percent contributions of the capital and operating cost rates.

ASSUMPTION: The total cost rate is given by

$$\dot{C} = \underbrace{c_1 (P_1 - P_2)^{-1/3}}_{\text{capital cost}} + \underbrace{c_2 (P_1 - P_2)}_{\text{operating cost}} \qquad (1)$$

where c_1 and c_2 are constants.

ANALYSIS: (a) Sketch of Eq. (1)

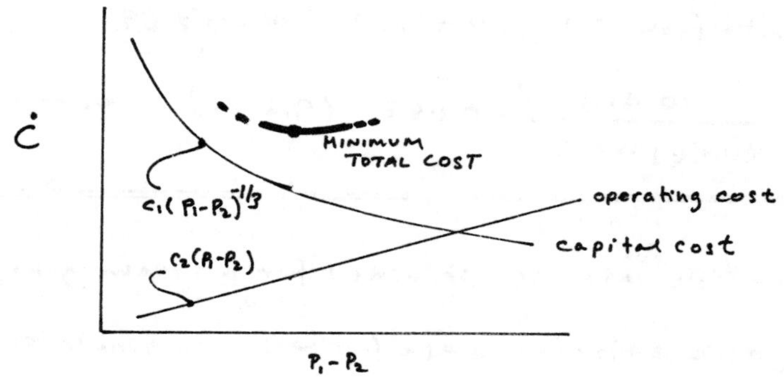

(b) $$\frac{\partial \dot{C}}{\partial (P_1 - P_2)} = -\frac{1}{3} c_1 (P_1 - P_2)^{-4/3} + c_2 = 0 \Rightarrow \frac{1}{3} c_1 (P_1 - P_2)^{-4/3} = c_2$$

or, on rearrangement, at the point of minimum total cost

$$\underbrace{c_1 (P_1 - P_2)^{-1/3}}_{\text{Cap. cost}} = \underbrace{3 c_2 (P_1 - P_2)}_{\text{op. cost}}$$

In other words, at this point the capital cost is three times the
① operating cost. Thus, at the point of minimum total cost, the percent contributions to the total cost rate are

% operating cost = 25%

% capital cost = 75%

---

1. For further discussion, see D. Steinmeyer, "Optimum Δp and ΔT in Heat Exchange," Hydrocarbon Processing, April, 1992, 53-56.

PROBLEM 7.118

KNOWN: Heat transfer and cost data are provided for an electric water heater.

FIND: (a) Determine the cost of the heat loss when the outer surface temperature is 535 °R. (b) Plot cost of heat loss versus outer surface temperature ranging from 535 to 570 °R.

SCHEMATIC & GIVEN DATA:

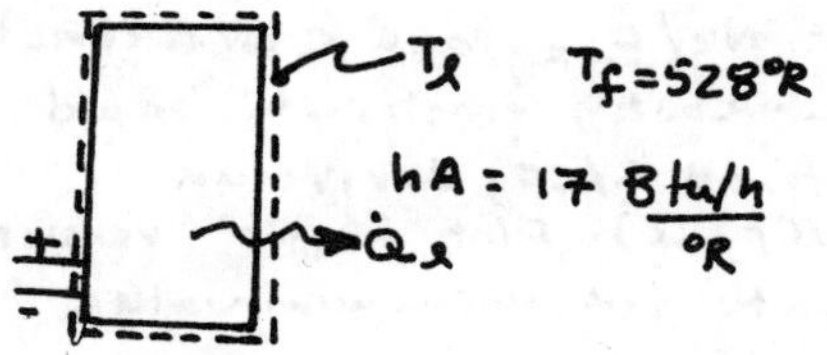

ASSUMPTIONS: 1. For the system shown in the accompanying schematic, heat transfer occurs only at T_l.
2. Electricity is evaluated at 8 cents/kW·h.
3. $T_0 = 528°R$.

ANALYSIS: Letting $\dot{Q}_l$ denote the rate of heat loss from the water heater,

$\dot{Q}_l = hA[T_l - T_f]\ \frac{Btu}{h}$. Following the discussion of Sec. 7.6.1 concerning costing of heat loss, and letting electricity play the role of the fuel, we get

$$\dot{\$}_l = c_e\left[1 - \frac{T_0}{T_l}\right]\dot{Q}_l = c_e\left[1 - \frac{T_0}{T_l}\right](hA)(T_l - T_f) \qquad (1)$$

(a) When $T_l = 535°R$, the annual cost $(24 \times 365 = 8760\,h)$ is

$$\dot{\$}_l = \left(\frac{\$.08}{kW\cdot h}\right)\left[1 - \frac{528}{535}\right]\left(17\ \frac{Btu}{h\cdot °R}\right)(7\ °R)\left|\frac{8760h}{year}\right|\left|\frac{1kW}{3413 Btu/h}\right| = \$0.32/year \leftarrow$$

(b) Plot

① 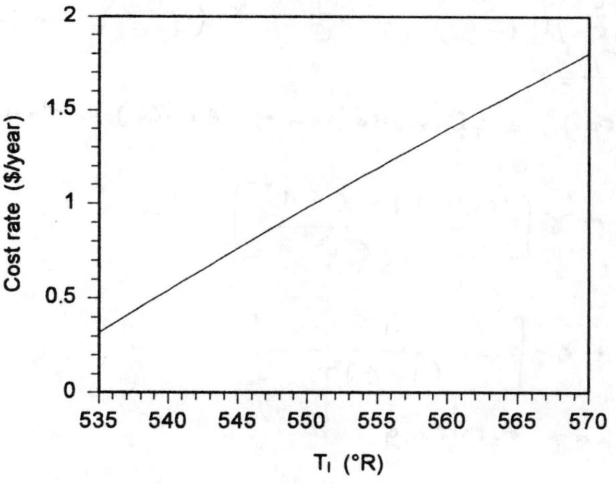

1. When considered over the several years life of such a water heater, the use of an insulating blanket to reduce the outer surface temperature is likely to be cost effective for the owner.

PROBLEM 7.119

KNOWN: Cost rate data are provided for a system operating at steady state. The system generates electricity at the rate $\dot{W}_e$. The total cost rate is

$$\dot{C} = \underbrace{c_F \dot{E}_{fF}}_{\text{fuel cost}} + \underbrace{c\left(\frac{\epsilon}{1-\epsilon}\right)\dot{W}_e}_{\text{owning and operating cost}} \qquad (1)$$

where c_F is the unit cost based on exergy, $\epsilon = \dot{W}_e/\dot{E}_{fF}$, and c is a constant.

FIND: (a) Derive an expression for c_e, the unit cost of electricity based on $\dot{W}_e$, in terms of ϵ, c_e/c_F, and c/c_F. (b) For fixed c/c_F, derive an expression for ϵ corresponding to minimum c_e/c_F. (c) Plot c_e/c_F versus ϵ for selected values of c/c_F. For each c/c_F, evaluate the minimum value for c_e/c_F and the corresponding ϵ.

SCHEMATIC & GIVEN DATA:

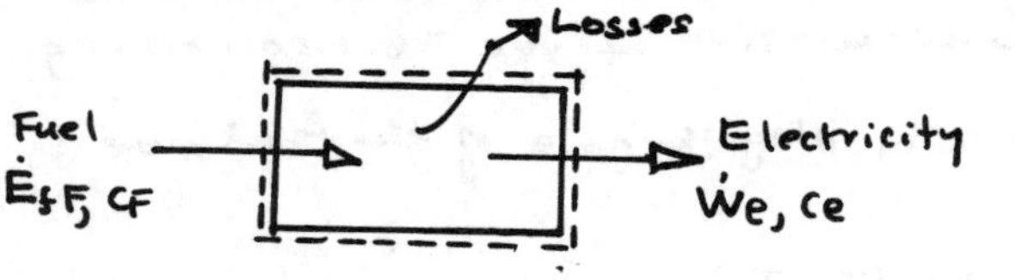

ASSUMPTIONS: 1. The system shown in the figure is at steady state. 2. In Eq. (1), c and c_F are constants. 3. Eq. (1) accounts for all significant contributors to the total cost rate.

ANALYSIS: (a) Beginning with Eq. (1), use $\dot{E}_{fF} = \dot{W}_e/\epsilon$ to obtain

$$\dot{C} = c_F\left[\frac{\dot{W}_e}{\epsilon}\right] + c\left[\frac{\epsilon}{1-\epsilon}\right]\dot{W}_e = \left(\frac{c_F}{\epsilon} + c\left[\frac{\epsilon}{1-\epsilon}\right]\right)\dot{W}_e$$

Then, with $\dot{C} = c_e \dot{W}_e$, the unit cost of electricity is

$$c_e = \frac{c_F}{\epsilon} + c\left(\frac{\epsilon}{1-\epsilon}\right)$$

or

$$\frac{c_e}{c_F} = \frac{1}{\epsilon} + \underbrace{\left(\frac{c}{c_F}\right)}_{\equiv \gamma}\left(\frac{\epsilon}{1-\epsilon}\right) = \frac{1}{\epsilon} + \gamma\left(\frac{\epsilon}{1-\epsilon}\right) \quad \longleftarrow \quad (a)$$

(b) For fixed $\gamma\ (= c/c_F)$, differentiation of Eq. (a) with respect to ϵ gives

$$\frac{\partial(c_e/c_F)}{\partial \epsilon} = -\frac{1}{\epsilon^2} + \gamma\left[\frac{1(1-\epsilon) - \epsilon(-1)}{(1-\epsilon)^2}\right]$$

$$= -\frac{1}{\epsilon^2} + \gamma\left[\frac{1}{(1-\epsilon)^2}\right]$$

Setting $\frac{\partial(c_e/c_F)}{\partial\epsilon} = 0$ and solving

$$\frac{1}{\epsilon^2} = \frac{\gamma}{(1-\epsilon)^2} \Rightarrow \left(\frac{1-\epsilon}{\epsilon}\right)^2 = \gamma \Rightarrow \frac{1-\epsilon}{\epsilon} = \sqrt{\gamma} \Rightarrow \epsilon = \frac{1}{1+\sqrt{\gamma}} \quad \longleftarrow \quad (b)$$

which is the value of ϵ corresponding to a minimum in c_e/c_F when γ is fixed. Using this in the result of part (a)

$$\left(\frac{c_e}{c_F}\right)_{MIN} = 1 + 2\sqrt{\gamma}$$

PROBLEM 7.119 (Cont'd.)

PLOT:

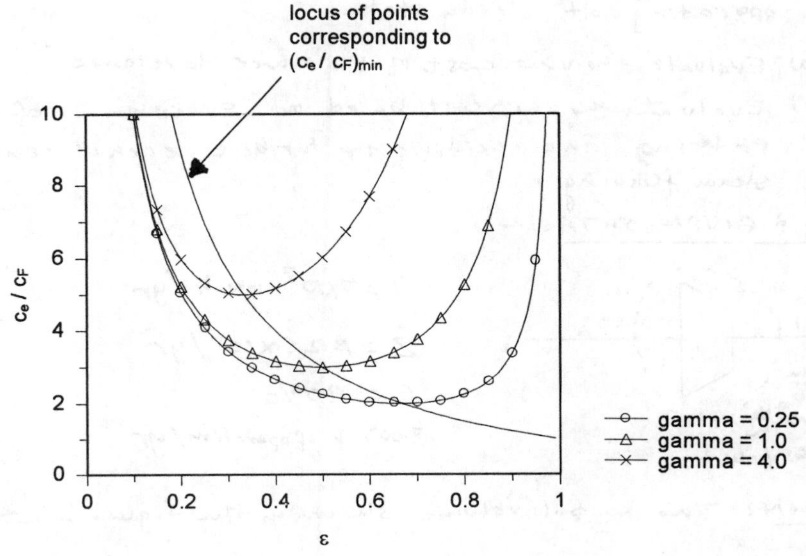

The following chart lists values of $(c_e / c_F)_{min}$ and the corresponding values of ϵ for $\gamma = 0.25$, 1, and 4:

γ	ε	$(c_e / c_F)_{min}$
.25	2/3	2
1	.5	3
4	1/3	5

PROBLEM 7.120

KNOWN: Operating and cost data are provided for a turbine operating at steady state.

FIND: (a) Evaluate the unit cost of the power developed.
(b) Evaluate the unit cost based on exergy of the steam entering and exiting the turbine, each in cents per lb of steam flowing.

SCHEMATIC & GIVEN DATA:

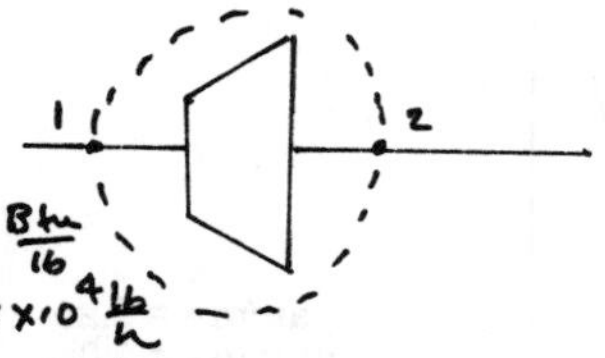

$e_{f1} = 559\ \frac{Btu}{lb}$

$\dot{m} = 12.55 \times 10^4\ \frac{lb}{h}$

$c_s = \$0.0165/kW\cdot h$

$\dot{W}_t = 7 \times 10^7\ kW\cdot h/yr$

$\dot{Z} = \$2.5 \times 10^5/yr$

$\epsilon = 90\%$

8000 h of operation/yr

ASSUMPTION: The control volume shown in the figure is at steady state.

ANALYSIS: (a) In this case, Eq. 7.52c is applicable:

$$c_e = \frac{c_s}{\epsilon} + \frac{\dot{Z}}{\dot{W}_e} = \left(\frac{0.0165\ \$/kW\cdot h}{0.9}\right) + \left(\frac{2.5\times10^5\ \$/yr}{7\times10^7\ kW\cdot h/yr}\right)$$

$$= (0.0183 + 0.0036)\ \$/kW\cdot h = 0.022\ \$/kW\cdot h = 2.2\ \frac{cents}{kW\cdot h} \leftarrow \text{(a)}$$

(b) By assumption leading to Eq. 7.52c, the same unit cost based on exergy applies to the steam at the inlet and at the exit: 1.65 $\frac{cents}{kW\cdot h}$. Thus the unit cost of steam expressed on a unit of mass basis is

$$\left(\begin{array}{c}\text{unit cost of steam,}\\ \text{per lb, entering the}\\ \text{turbine}\end{array}\right) = \left(1.65\ \frac{cents}{kW\cdot h}\right)\left[\left(559\ \frac{Btu}{lb}\right)\left|\frac{kW\cdot h}{3413\ Btu}\right|\right] = 0.27\ \frac{cents}{lb} \leftarrow$$

The same approach is used at the exit, but the value of e_{f2} is required. Using

$$\epsilon = \frac{\dot{W}_t}{\dot{m}(e_{f2}-e_{f1})} \Rightarrow e_{f2} = e_{f1} - \frac{\dot{W}_t}{\dot{m}\epsilon}$$

or

$$e_{f2} = 559\ \frac{Btu}{lb} - \frac{(7\times10^7\ kW\cdot h/yr)\,|3413\ Btu/kW\cdot h|}{\left(12.55\times10^4\ \frac{lb}{h}\right)\left(\frac{8000 h}{yr}\right)(0.9)} = 295\ \frac{Btu}{lb}$$

Then

$$\left(\begin{array}{c}\text{unit cost of steam,}\\ \text{per lb, exiting the}\\ \text{turbine}\end{array}\right) = \left(1.65\ \frac{cents}{kW\cdot h}\right)\left[\left(295\ \frac{Btu}{lb}\right)\left|\frac{kW\cdot h}{3413\ Btu}\right|\right] = 0.14\ \frac{cents}{lb} \leftarrow$$

PROBLEM 7.121

KNOWN: Operating and cost data are provided for the boiler of Fig. P7.121.

FIND: (a) Develop an expression for the unit cost, based on exergy of the steam exiting the boiler.

(b) Using the result of (a), determine the unit cost of the exiting steam in cents per kg of steam exiting the boiler.

SCHEMATIC & GIVEN DATA:

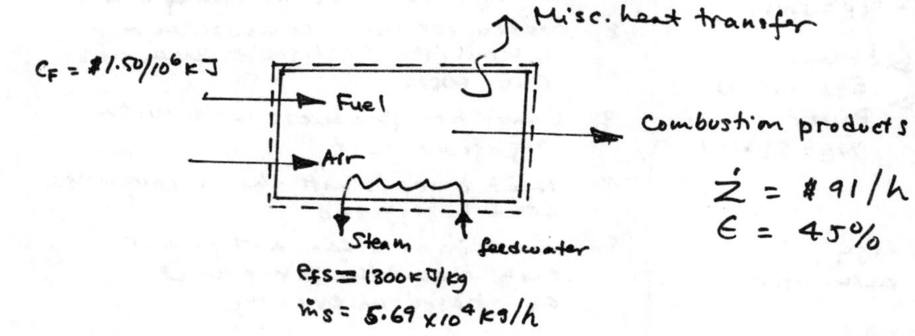

ASSUMPTIONS: (1) The control volume shown in the figure is at steady state. (2) The cost rates of the feedwater, combustion products, and heat transfer are ignored.

ANALYSIS: (a) With assumption (2), Equation 7.50b is applicable:

$$c_s = c_F\left(\frac{\dot{E}_{fF}}{\dot{E}_{fs}}\right) + \frac{\dot{Z}}{\dot{E}_{fs}}. \quad \text{With } \epsilon = \frac{\dot{E}_{fs}}{\dot{E}_{fF}}, \quad c_s = \frac{c_F}{\epsilon} + \frac{\dot{Z}}{\dot{E}_{fs}} \longleftarrow \text{(a)}$$

(b) Inserting values

$$c_s = \frac{(\$1.50/10^6\,\text{kJ})}{0.45} + \frac{91\,(\$/h)}{\left(5.69\times10^4\,\frac{\text{kg}}{\text{h}}\right)\left(1300\,\frac{\text{kJ}}{\text{kg}}\right)} = \left[\frac{(3.33+1.23)(\$/\text{kJ})}{10^6}\right]\left|\frac{3600\,\text{kJ}}{\text{kW}\cdot\text{h}}\right|$$

$$= 0.0164\ \$/\text{kW}\cdot\text{h}$$

This unit cost is on a per kW·h of exergy basis. It can be expressed on a per unit mass of steam basis as follows:

$$\left(\begin{array}{c}\text{unit cost of steam}\\ \text{per kg}\end{array}\right) = \left(0.0164\,\frac{\$}{\text{kW}\cdot\text{h}}\right)\left(1300\,\frac{\text{kJ}}{\text{kg}}\right)\left|\frac{\text{kW h}}{3600\,\text{kJ}}\right| = 0.0059\ \$/\text{kg}$$

① $$= 0.59\ \frac{\text{cents}}{\text{kg}} \longleftarrow \text{(b)}$$

---

1. When expressed on a per lb of steam basis, the result of part (b) is 0.27 cents/lb, which corresponds to the unit cost of the steam entering the turbine of Problem 7.120, part (a).

PROBLEM 7.122

KNOWN: Steady-state operating and cost data are provided for a cogeneration system.

FIND: (a) Determine the rate of exergy destruction. (b) Devise and evaluate an exergetic efficiency. (c) Evaluate the unit cost, and the cost rates of the power and steam produced, each on an exergy basis

SCHEMATIC & GIVEN DATA:

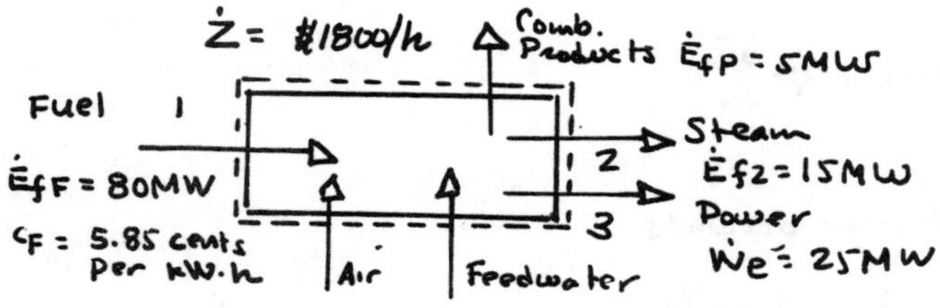

ASSUMPTIONS:

1. The control volume shown in the schematic is at steady state.
2. Feedwater and combustion air enter with negligible exergy and cost.
3. Combustion products exit with negligible cost.
4. Heat transfer with the surroundings can be ignored.
5. The exiting steam and power each have the same unit cost based on exergy.

ANALYSIS:

(a) At steady state, the exergy rate balance reduces with assumptions 2 and 4 to give

$$\dot{E}_d = \dot{E}_{fF} - \dot{E}_{fP} - \dot{E}_{f2} - \dot{W}_e$$

$$= 80 - 5 - 15 - 25 = 35\ \text{MW} \quad \leftarrow$$

(b) Regarding the exiting steam and power as the product of the system and the combustion products as a loss

$$\epsilon = \frac{\dot{E}_{f2} + \dot{W}_e}{\dot{E}_{fF}} = \frac{15 + 25}{80} = 0.5\ (50\%) \quad \leftarrow$$

(c) With assumptions 2-4, a cost rate balance reads

$$\dot{C}_2 + \dot{C}_3 = \dot{C}_1 + \dot{Z}$$

Then with assumption 5, $c\dot{E}_{f2} + c\dot{W}_e = c_F\dot{E}_{fF} + \dot{Z}$

Solving for the unit cost of the steam and the power

$$c = \frac{c_F\dot{E}_{fF} + \dot{Z}}{\dot{E}_{f2} + \dot{W}_e} = c_F\left[\frac{\dot{E}_{fF}}{\dot{E}_{f2} + \dot{W}_e}\right] + \frac{\dot{Z}}{[\dot{E}_{f2} + \dot{W}_e]}$$

$$= \frac{c_F}{\epsilon} + \frac{\dot{Z}}{[\dot{E}_{f2} + \dot{W}_e]}$$

$$= \left[\frac{5.85\ \text{¢/kW·h}}{0.5}\right] + \frac{\$1800/h}{[15+25]\ \text{MW}}\left|\frac{1\ \text{MW}}{10^3\ \text{kW}}\right|\left|\frac{10^2\ \text{cents}}{1\ \$}\right|$$

$$= [11.7 + 4.5]\ \frac{\text{cents}}{\text{kW·h}} = 16.2\ \frac{\text{cents}}{\text{kW·h}} \quad \leftarrow$$

The cost rates are

$$\dot{C}_2 = c\dot{E}_{f2} = \left(16.2\ \frac{\text{cents}}{\text{kW·h}}\right)(15\ \text{MW})\left|\frac{10^3\ \text{kW}}{1\ \text{MW}}\right|\left|\frac{1\ \$}{10^2\ \text{cents}}\right| = \$2430/h \quad \leftarrow$$

$$\dot{C}_3 = c\dot{W}_e = \left(16.2\ \frac{\text{cents}}{\text{kW·h}}\right)(25\ \text{MW})\left|\frac{10^3\ \text{kW}}{1\ \text{MW}}\right|\left|\frac{1\ \$}{10^2\ \text{cents}}\right| = \$4050/h \quad \leftarrow$$

PROBLEM 7.123

KNOWN: Steady-state operating and cost data are provided in Example 7.10 for a cogeneration system.

FIND: Taking an overall control volume and regarding the unit cost based on exergy as the same for the power and steam produced, evaluate the unit cost and compare with the values obtained in Example 7.10

SCHEMATIC & GIVEN DATA:

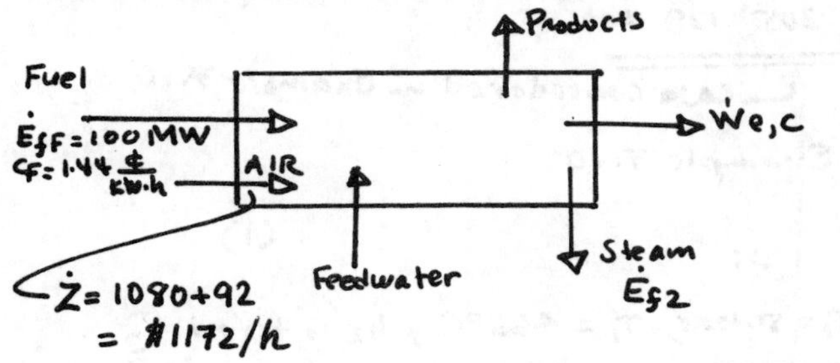

ASSUMPTIONS:
1. The control volume shown in the schematic is at steady state.
2. For the control volume, $\dot{Q}_{cv}=0$.
3. The feed water and combustion air enter with negligible exergy and cost.
4. The combustion products are discharged without cost.
5. The exiting steam and power each have the same unit cost based on exergy.

ANALYSIS: A cost rate balance reads

$$c\dot{W}_e + c\dot{E}_{f2} = c_F \dot{E}_{fF} + \dot{Z}$$

$$\Rightarrow \quad c = \frac{c_F \dot{E}_{fF} + \dot{Z}}{\dot{W}_e + \dot{E}_{f2}}$$

From the solution to Example 7.10, $\dot{W}_e = 12.75\text{ MW}$, $\dot{E}_{f2} = 20.67\text{ MW}$

$$c = \frac{(1.44¢/\text{kW·h})(100\text{ MW})}{(12.75+20.67)\text{MW}} + \frac{\$1172/\text{h}}{(12.75+20.67)\text{MW}}\left|\frac{1\text{ MW}}{10^3\text{ kW}}\right|\left|\frac{10^2 ¢}{\$}\right|$$

$$= (4.309 + 3.507) = 7.82\text{ cents/kW·h} \quad \longleftarrow$$

From the solution to Example 7.10, the unit cost for power is $8.81 cents/kW·h and for process steam 7.2 cents/kW·h. When the overall control volume is considered in the present case, the unit cost of the steam is higher because it now bears a part of the costs associated with the turbine in the analysis of Example 7.10. On the other hand, the unit cost of the power is lower because it is no longer required to bear the full burden of these costs. The approach of Example 7.10 is preferred because it takes into account important information related to the actual production processes.

PROBLEM 7.124

KNOWN: Steady-state operating and cost data are provided for the cogeneration system of Example 7.10.

FIND: Plot versus the pressure of the process steam (a) the power developed, (b) the unit costs of the power and process steam, per kW·h, (c) the unit cost of the process steam, per kg of process steam flowing.

SCHEMATIC & GIVEN DATA: See Figure E7.10.

$\dot{Z}_t = 7.2\dot{W}_e \frac{\$}{h}$, ($\dot{W}_e$ in MW)

p_2(bar)	40	30	20	9	5	2	1
T_2(°C)	436	398	349	262	205	128	sat.

(The 5 bar case is the case considered in Example 7.10.)

ASSUMPTIONS: See Example 7.10

ANALYSIS: Following the analysis of Example 7.10

$$\dot{W}_e = \dot{m}(h_1 - h_2)\left|\frac{1\,MW}{10^3\,kJ/s}\right| \qquad (1)$$

where $\dot{m} = 26.15$ kg/s, h_1 is fixed by $P_1 = 50$ bar, $T_1 = 466°C$, h_2 is fixed by P_2, T_2 from the table above.

The unit cost of the process steam is 7.2 cents/kW·h, as in Example 7.10. This value is determined by the boiler analysis together with the assumption underlying Eq. 7.52b. The unit cost of the power is determined by

$$c_e = c_2\left[\frac{\dot{E}_{f1} - \dot{E}_{f2}}{\dot{W}_e}\right] + \frac{\dot{Z}_t}{\dot{W}_e} \qquad (2)$$

where $c_2 = 7.2$ cents/kW·h, $\dot{W}_e$ is given by Eq.(1), $\dot{Z}_t = 7.2\dot{W}_e$, where $\dot{W}_e$ is in MW, and

$$\dot{E}_{f1} - \dot{E}_{f2} = \dot{m}[h_1 - h_2 - T_0(s_1 - s_2)]\left|\frac{1\,MW}{10^3\,kJ/s}\right| \qquad (3)$$

To determine the unit cost of the process steam on a unit mass basis

$$c_m = c_2\left[\frac{\dot{E}_{f2}}{\dot{m}}\right]\left|\frac{10^3\,kW}{1\,MW}\right|\left|\frac{1\,h}{3600\,s}\right| \qquad (4)$$

where $\dot{E}_{f2}$ is found using Eq.(3) and $\dot{E}_{f1} = 35$ MW.

The following IT code is used to generate data for each case in the above table. The data were imported to a spreadsheet program to obtain the plots on the next page.

IT Code

```
p1 = 50  // bar
T1 = 466  // °C
p2 = 40  // bar
T2 = 436  // °C
mdot = 26.15  // kg/s
To = 298  // K

h1 = h_PT("Water/Steam", p1, T1)
s1 = s_PT("Water/Steam", p1, T1)
h2 = h_PT("Water/Steam", p2, T2)
s2 = s_PT("Water/Steam", p2, T2)
s2s = s_Ph("Water/Steam", p2, h2s)
s2s = s1

Wdote = mdot * (h1 - h2) / 1000
Edotf1 - Edotf2 = mdot * ((h1 - h2) - To * (s1 - s2)) / 1000
Edotf1 = 35
ce = c2 * ((Edotf1 - Edotf2) / Wdote) + (Zdott / Wdote) / 1000
Zdott = 7.2 * Wdote
c2 = 7.2
cm = c2 * (Edotf2 / mdot) * (1000 / 3600)
```

PROBLEM 7.124 (Cont'd.)

PLOTS:

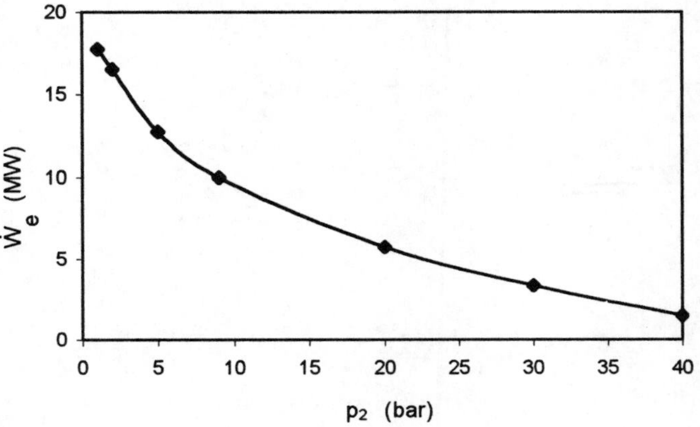

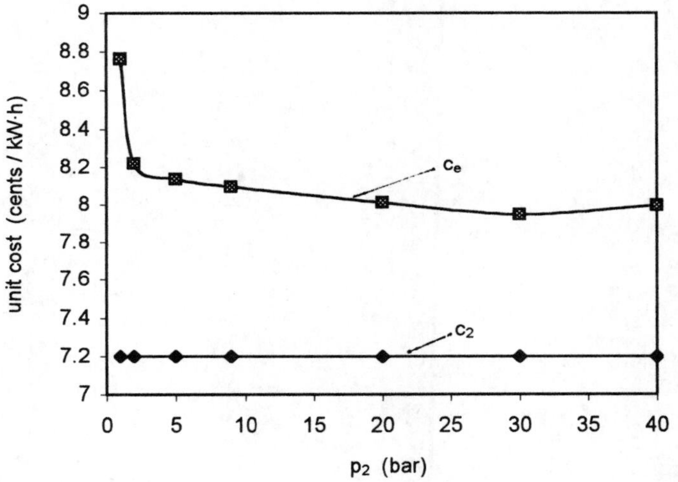

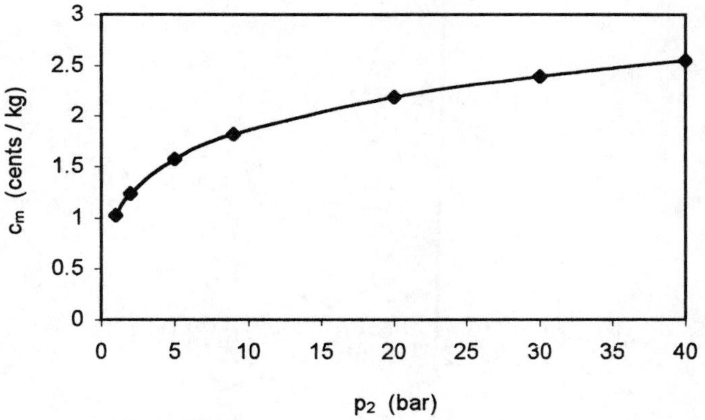